The Compendium of Statistical Tests

The Compendium of Statistical Tests

A Comprehensive Collection of Over 250 Statistical Tests

Maruthy Pannala, PhD

Pannala Series

2026

The Compendium of Statistical Tests: A Comprehensive Collection of over 250 Statistical Tests

First Edition

Published by
Pannala Series
Greensboro, North Carolina, USA
Email: pannala.series@gmail.com
Website: https://sites.google.com/view/pannala-series

Library of Congress Control Number: 2026914047

ISBN: 979-8-9961516-0-8 (Hardcover)

Cover design by the author.

Typeset in LaTeX using TeXmaker.
Printed and distributed by IngramSpark.

"What I cannot create, I do not understand."
— Richard Feynman

Preface

Statistical testing occupies a peculiar position in applied work. On the one hand, statistical tests are foundational tools—embedded deeply in modeling practice, diagnostics, validation, and scientific reporting. On the other hand, they are often invoked mechanically, selected by habit, software defaults, or partial recollection of formal training. The result is not a lack of tests, but a lack of clarity around their correct use, assumptions, interpretation, and limitations.

This book was written to address that gap.

The *Compendium of Statistical Tests* is a single-volume reference work that brings together over 250 statistical tests drawn from across the statistical literature and widely used software ecosystems. These tests are scattered across textbooks, journal articles, technical reports, and software documentation. Practitioners are often required to consult multiple sources—frequently under time pressure—to answer what should be straightforward questions: *Which test is appropriate here? What does it actually measure? What assumptions does it rely on? How should the result be interpreted?*

Each statistical test is presented in a standardized two-page format. The intent is deliberate. A practitioner should be able to open the book at any test and see, at a glance, all information required for correct and informed use. To that end, every test entry follows the same structure, capturing essential elements such as intended use, assumptions, formal test statistic, a plain-English explanation, warnings and best-practice guidance, hypotheses and decision rule, software references, worked example, common pitfalls, and original literature references. The tests are organized into twenty-five chapters, each preceded by a **Conceptual Primer**—a compact, self-contained background section introducing the key ideas, terminology, and mathematical framework needed for the tests that follow.

The selection of tests reflects two guiding principles: **depth** and **breadth**. For commonly encountered diagnostic goals—such as testing for normality, stationarity, or independence—multiple tests are included, each with distinct sensitivities, assumptions, and historical motivations. At the same time, the book spans a wide range of modeling domains: normality testing, general goodness-of-fit testing, classical parametric inference (means, variances, proportions, and rates), nonparametric and rank-based methods, categorical data analysis, correlation and association, regression diagnostics, generalized linear models, mixed models, multivariate analysis, time series analysis (diagnostics, stationarity, cointegration, and volatility), randomness and independence testing, survival and event-time analysis, spatial statistics, quality control and process monitoring, equivalence and non-inferiority testing, robust and resistant methods, permutation and bootstrap inference, outlier detection, and Bayesian inference.

This book is not intended to replace formal statistical training, nor does it attempt to provide full derivations or proofs. Instead, it is designed to serve as a desk-side reference for analysts, model developers, data scientists, econometricians, clinical researchers, quality engineers, environmental scientists, and researchers across disciplines who engage with statistical models and diagnostics as part of their professional work. The emphasis throughout is on correct application, conceptual understanding, and awareness of limitations.

Used as intended, this compendium should help practitioners move beyond rote application of statistical tests and toward more thoughtful, defensible, and transparent analytical decisions.

Acknowledgements

A work of this scope does not emerge in isolation. Many people, over many years, have shaped the thinking behind it.

My parents, Mr. Ramakrishna and Mrs. Indera, gave me the foundation that made everything else possible. The values they instilled—hard work, humility, and intellectual curiosity—have guided me far beyond the pages of this book.

My sister Sailaja has been a steady source of encouragement throughout my life, and this endeavor was no exception. I am grateful for her unwavering support.

My interest in statistics was seeded long before my doctoral work, by the faculty at the Indian Institute of Technology, Kanpur. Professors Debasis Kundu, Divakar Sharma, P. C. Joshi, and G. K. Shukla introduced me to the discipline and laid the statistical foundations I have built upon since. Professor U. B. Tiwari's rigorous instruction in measure theory and real analysis gave me the mathematical discipline underlying all of it. Professor Kundu deserves special mention: fresh from his own doctorate—earned under the supervision of the legendary C. R. Rao—and generous with his time, he encouraged me to pursue a PhD in the United States and helped make that path possible. I owe him more than a sentence can express.

My deepest professional debt belongs to my doctoral supervisor, Dr. K. Krishnamoorthy at the University of Louisiana at Lafayette. Statistics is, at its core, a mathematical discipline—but one largely driven by intuition. Before working with Dr. Krishnamoorthy, I approached the subject almost exclusively through a mathematical lens and, in doing so, missed much of its deeper meaning. His guidance opened a different way of seeing: that ideas come before mathematics, and that equations express intuitions already present. This shift was cemented through the simulation work at the heart of my doctoral dissertation, where I learned to reason about statistical behavior from first principles before writing down a formula. That instinct—to understand what a test is actually doing before asking how it is derived—runs through every page of this book, and is perhaps most visible in the *Plain English* section appearing in every test entry: an attempt to put the idea into words before, and independent of, the mathematics.

My first professional mentor, Dan Walker, helped me understand the business side of quantitative work—that numbers serve decisions, and that communicating them clearly to non-technical audiences is no less important than deriving them correctly. His patience with a young PhD entering the financial industry left a lasting impression.

Over a career spanning more than twenty-five years, I have had the privilege of working alongside many talented and thoughtful colleagues. It is through those interactions—explaining methods, fielding questions, and occasionally bridging the gap between statistical rigor and practical application—that I came to appreciate how much room there is to communicate statistical ideas more clearly across backgrounds and disciplines. That appreciation is, in no small part, what motivated this book.

I am also grateful to my childhood and lifelong friend Sesha Giri, whom I have known since fourth grade. His friendship and steadfast support through every critical juncture in my life have meant more than he probably knows.

Finally, and most importantly, I owe an immeasurable debt to my family. My wife Renukha's support and patience through the long hours this project demanded made it possible to complete. My daughters Soumya and Sneha, my nephew Teja, and my niece Akshara, provided the encouragement and energy that kept me going through the more difficult stretches of this work. My brother-in-law Ravi reviewed the worked examples throughout the book, and my sister-in-law Shobha carefully reviewed the references section—both offering corrections and observations that improved the final result. Their willingness to engage with the material so thoughtfully is something I deeply appreciate.

To all of you—and to Lucky, our Cockapoo, whose company through many late nights was a comfort—thank you.

A Note on Formulas, Variations, and Verification

Statistical knowledge accumulates gradually and sometimes inconsistently. For the majority of well-established tests, software implementations agree closely. But for a meaningful subset—particularly older or more specialized procedures—variations in formulas, finite-sample corrections, and computational approaches exist across both the literature and software. Where such variation exists, the most widely-adopted formulation has been selected. The goal throughout has been accuracy, consistency, and honest acknowledgment of complexity.

Readers should be aware of the following:

- **Approximations have limits.** Many p-value formulas and critical values rely on asymptotic approximations that may behave differently in finite samples. The worked examples and Common Pitfalls sections address these considerations directly.
- **The literature continues to evolve.** While I have endeavored to represent the state of knowledge as of 2026, newer refinements may have since emerged.

Verification. Every formula, worked example, and computational result has been verified through cross-referencing against original source papers and authoritative textbooks, and through independent calculation in Excel, R, Python, and SAS. Despite these efforts, occasional errors may remain. The primary value of this compendium lies in its explanation of test mechanics, assumptions, and proper interpretation—aspects that remain reliable guides regardless of minor formula variations. For critical applications, always verify results using established software and consult original sources when in doubt.

Reporting errors. If you identify any errors or inconsistencies, please contact the author at *pannala.series@gmail.com*. Confirmed errata will be maintained at *https://sites.google.com/view/pannala-series* and incorporated into future editions.

Statistical literature spans more than a century of scattered, sometimes conflicting contributions. This compendium is one attempt to gather that knowledge into a single accessible reference. It is offered in good faith—and with the hope that, with input from the community, it will improve over time.

— Maruthy Pannala
June 2026

Scope and Classification of Tests

This compendium covers **250+ statistical tests** organized into **25 chapters**, spanning classical inference, regression diagnostics, time series analysis, survival modeling, spatial statistics, and Bayesian procedures. Coverage reflects the diversity of settings in which statistical testing is employed in practice.

Test selection is inherently context-dependent: sample size, data structure, distributional assumptions, robustness requirements, and modeling objectives all influence which procedure is most appropriate. Related tests are therefore grouped together to facilitate comparison across competing approaches. Each chapter opens with a conceptual primer providing the mathematical background needed for the tests that follow; individual entries are presented in a standardized two-page reference format.

The classification scheme is a navigational aid rather than a rigid taxonomy. Many tests admit multiple valid interpretations depending on context, and several appear across more than one chapter. A test's placement should be understood as a primary lens, not a definitive category. Readers are encouraged to use both the table of contents and the Index when locating procedures across related modeling frameworks.

The twenty-five chapters are organized as follows:

1. **Normality Tests (NORM)**
   Tests for consistency with a normal distribution. Includes moment-based tests (skewness, kurtosis), EDF-based procedures, and regression/correlation methods applied to quantile plots, under both simple and composite hypothesis settings.

2. **Goodness-of-Fit Tests (GFIT)**
   Tests evaluating whether data conform to a fully or partially specified distribution. Includes EDF-based statistics (Anderson–Darling, Cramér–von Mises, Kolmogorov–Smirnov, Lilliefors), the chi-square goodness-of-fit test, Kuiper's test & Rao's spacing test for circular data, and Greenwood's test for uniformity of spacings.

3. **Means and Location Tests (MEANS)**
   Parametric inference for comparing population means. Covers one-sample, two-sample, and k-sample settings; pooled and Welch tests; one-way and two-way ANOVA; repeated measures; ANCOVA; and multiple comparison procedures (Tukey, Dunnett, Games–Howell, Scheffé).

4. **Variance and Dispersion Tests (VARI)**
   Tests for equality of variances and dispersion across groups. Ranges from the classical F-test through robust alternatives (Levene, Brown–Forsythe) to nonparametric procedures (Fligner–Killeen, Ansari–Bradley, Mood's, Siegel–Tukey).

5. **Tests for Proportions and Rates (PROP)**
   Inference for binomial proportions and Poisson rates. Covers one-sample proportion tests (z-test, exact binomial, Wilson score), two-sample tests (pooled z-test, Wald, Agresti–Caffo), the Farrington–Manning non-inferiority test, and one- and two-sample Poisson rate tests with exact and asymptotic procedures.

6. **Nonparametric and Rank-Based Tests (NPAR)**
   Distribution-free methods for location and trend inference. Covers two-sample (Mann–Whitney, Mood's median), paired (Wilcoxon signed-rank, sign test), k-sample (Kruskal–Wallis, Friedman), and ordered-alternative tests (Jonckheere–Terpstra, Page's trend test).

7. **Correlation and Association Tests (CORR)**
   Tests for the strength and direction of association across measurement scales. Includes Pearson, Spearman, Kendall's τ, distance correlation, Hoeffding's D, point-biserial, biserial, tetrachoric, polychoric, Cramér's V, partial correlation, and tools for comparing correlations (Fisher's z, Steiger's Z).

8. **Categorical Data and Contingency Tables (CTAB)**
   Tests for count data, proportions, and categorical outcomes. Covers independence and homogeneity (χ^2, likelihood ratio–G test), exact tests (Fisher's, Barnard's, Boschloo's), paired categorical data (McNemar, McNemar–Bowker, Stuart–Maxwell), repeated binary outcomes (Cochran's Q), stratified analyses (CMH, Breslow–Day, Woolf), trend tests (Cochran–Armitage), and inter-rater agreement (Cohen's κ, Weighted κ, Fleiss' κ, Krippendorff's α, Gwet's AC_1).

9. **Linear Regression Diagnostics (REGR)**
   Specification and diagnostic tests for the classical linear model. Covers heteroscedasticity (Breusch–Pagan, White, Goldfeld–Quandt, Glejser, Harvey–Godfrey), autocorrelation (Durbin–Watson, Breusch–Godfrey), functional form (RESET, Link, Box–Tidwell, Rainbow), structural stability (Chow), multicollinearity (VIF), and observation influence (leverage, Cook's distance, DFFITS, DFBETAS).

10. **Generalized Linear Models (GLMS)**
    Tests for fit, significance, and structural adequacy in GLMs (logistic, Poisson, Gamma, and related families). Covers goodness-of-fit (deviance, Pearson χ^2, Hosmer–Lemeshow, Le Cessie–van Houwelingen), the LRT/Wald/Score trinity, link function specification (Pregibon), overdispersion (Cameron–Trivedi, Dean's), excess zeros (zero-inflation score test, Vuong), and assumptions specific to multinomial logit (Hausman IIA, Small–Hsiao).

11. **Mixed Models (MIXD)**
    Inference in linear mixed-effects models. Covers likelihood-based tests for random effects (LRT, REML-LRT), score and Wald tests for variance components, the Breusch–Pagan LM test, fixed-effects tests (Wald, Kenward–Roger F), the intraclass correlation coefficient test, and the Hausman specification test.

12. **Multivariate Analysis (MULT)**
    Tests for mean vectors, covariance structure, and distributional form. Covers Hotelling's T^2 (one- and two-sample), MANOVA statistics (Wilks' Λ, Pillai's trace, Hotelling–Lawley trace, Roy's largest root), equality of covariance matrices (Box's M), sphericity (Bartlett, Mauchly, JNS), multivariate normality (Mardia's skewness and kurtosis, Henze–Zirkler), and Benjamini–Hochberg FDR for post-MANOVA multiplicity control.

13. **Time Series Diagnostics (TSDG)**
    Diagnostic tests for time series residuals and forecast evaluation. Covers portmanteau tests for residual autocorrelation (Box–Pierce, Ljung–Box) and the Diebold–Mariano test for comparing predictive accuracy.

14. **Stationarity and Unit Root Tests (STAT)**
    Tests for unit roots and non-stationarity in univariate and panel series. Covers classical tests (DF, ADF, Phillips–Perron), power-enhanced procedures (DF-GLS, Ng–Perron, Schmidt–Phillips), the KPSS stationarity test, structural break tests (Perron, Zivot–Andrews, Clemente–Montañés–Reyes, Lee–Strazicich), seasonal unit root tests (HEGY, Canova–Hansen), the KSS nonlinear unit root test, and panel unit root tests (LLC, IPS) with Pesaran's cross-sectional dependence test.

15. **Cointegration Tests (COINT)**
    Tests for long-run equilibrium among integrated series. Covers residual-based single-equation tests (Engle–Granger, Phillips–Ouliaris), the Johansen system procedures (trace and maximum eigenvalue), break-robust tests (Gregory–Hansen, Maki), alternative single-equation approaches (Shin Residual, Park's added variables), the ARDL bounds test for mixed integration orders, and panel cointegration tests (Pedroni, Westerlund).

16. **ARCH and Volatility Tests (ARCH)**
    Tests for conditional heteroskedasticity in time series residuals. Covers Engle's ARCH LM test, the McLeod–Li test, the Engle–Ng sign and size bias tests for asymmetric volatility, and the ICSS test for structural breaks in variance.

17. **Randomness and Independence Tests (RAND)**
Tests for patterns, trends, and dependence in ordered sequences. Covers runs-based tests (Wald–Wolfowitz, up-and-down runs), the turning points test, the difference-sign test, monotone trend tests (Mann–Kendall, Cox–Stuart), serial dependence tests (von Neumann ratio, Bartels rank test), Tsay's test for nonlinear dependence, and the BDS test for general nonlinear dependence.

18. **Survival and Event-Time Analysis (SURV)**
Tests for time-to-event data with censoring. Covers the log-rank family (Mantel–Cox, Gehan–Breslow, Tarone–Ware, Peto–Peto, Fleming–Harrington), Cox model inference (score test, LRT), proportional hazards diagnostics (Schoenfeld residuals, Grambsch–Therneau), overall adequacy (Cox–Snell residuals), and competing risks methods (Gray's test, Fine–Gray).

19. **Spatial Statistics (SPAT)**
Tests for spatial autocorrelation and model specification in areal data. Covers global measures (Moran's I, Geary's C, Getis–Ord General G), local statistics (Local Moran's I/LISA, Getis–Ord G_i^*), the Anselin LM diagnostic sequence (LM-Lag, LM-Error, Robust LM-Lag, Robust LM-Error), and the Join Count test for binary spatial data.

20. **Quality Control and Process Monitoring (QUAL)**
Statistical process control procedures for stability and capability monitoring. Covers Shewhart chart pattern rules (Western Electric, Nelson), attribute charts (p, np, c, u), memory-based charts (moving average, CUSUM, EWMA), change-point detection (Shiryaev–Roberts, Pettitt's test), and process capability indices (C_p, C_{pk}).

21. **Equivalence and Non-Inferiority Tests (EQVL)**
Tests for demonstrating practical equivalence or non-inferiority. Covers the TOST procedure for independent and paired samples (Schuirmann), non-inferiority and superiority-by-margin tests, equivalence testing for proportions, the Mann–Whitney equivalence test, and the Anderson–Hauck test.

22. **Robust Statistics (ROBU)**
Inferential procedures valid under outliers, heavy tails, and non-normality. Includes trimmed mean tests (Yuen's one-sample, two-sample, and ANOVA extensions), the Winsorised variance test, Welch–James test, Brunner–Munzel test, percentile bootstrap inference, Huber's M-estimator test, and the Theil–Sen slope estimator and test.

23. **Permutation and Randomization Tests (PERM)**
Exact and approximate inference from resampling. Covers permutation t-test and ANOVA, permutation test for correlation, Mantel matrix permutation test, randomization tests, bootstrap hypothesis tests, and bootstrap confidence interval tests.

24. **Outlier Detection Tests (OUTL)**
Formal tests for anomalous observations. Covers single-outlier parametric tests (Grubbs, Dixon's Q, Chauvenet's criterion), multiple-outlier procedures (Rosner's generalized ESD, Tietjen–Moore, Peirce's criterion), robust univariate detection (Iglewicz–Hoaglin modified z-score), and multivariate detection via Mahalanobis distance (classical and MCD-based).

25. **Bayesian Analysis (BAYES)**
Bayesian hypothesis testing and model comparison. Covers the Bayes Factor (BF_{10}) and the Savage–Dickey density ratio, Bayesian model comparison and averaging, Bayesian analogues of classical tests (t-test, ANOVA, correlation, regression, proportion), the HDI + ROPE framework for practical significance, and the posterior predictive p-value for model checking.

How to Use This Compendium

This book is designed as a working reference rather than a sequential read. Navigate directly to the relevant chapter based on your analytical task. Within each chapter, multiple tests addressing the same goal are presented side by side, making comparison straightforward.

Each chapter opens with a **Conceptual Primer**—a compact introduction to the key ideas, terminology, and mathematical framework needed for the tests that follow. Readers already familiar with the subject can proceed directly to the individual entries.

Anatomy of a Test Entry

Each test occupies a standardized two-page format. The elements are as follows:

Element	Content
Title & Subtitle	Name and primary purpose of the test.
Metadata Line	Classification code (e.g. NORM.1, REGR.7), Category, data type, null distribution.
When to Use	Scope, required data type, and inferential question answered. Guides selection and prevents misuse.
Assumptions	Explicitly stated conditions. Violations may invalidate the test or alter its null distribution.
Test Statistic	Formal mathematical definition in standard notation.
Plain English	Notation-free explanation of what the test measures and how departures from the null manifest in practice.
Warning	Practical cautions, finite-sample sensitivities, and modern alternatives where relevant.
Hypotheses & Decision Rule	Null and alternative hypotheses in both mathematical and/or verbal form, with the decision rule.
Software	Representative code in **Python**, **R**, and **SAS**. Code shown is illustrative; extended implementations are planned for *https://sites.google.com/view/pannala-series*.
Additional Notes	Supplementary technical material: derivations, asymptotic theory, connections to related tests. Skippable without loss of continuity.
Miscellanea	Historical background, etymology, and anecdotes.
References	Up to four or five key citations as a starting point for deeper study.
Worked Example	Complete numerical illustration from raw data to conclusion.
Common Pitfalls	Frequent errors in application or interpretation: assumption misuse, confusion with related tests, misreading of non-rejection.
Additional Material	Supplementary formulas or technical notes, included where warranted.

The uniform layout is intentional. After a few entries, readers will internalize the structure and extract what they need rapidly—whether the formula, the decision rule, the worked example, or the plain-English explanation. The *When to Use*, *Assumptions*, and *Warning* sections together constitute a decision-support framework embedded within every entry.

Contents

Notations and Abbreviations

The following notations and abbreviations are used consistently throughout this compendium. Where a symbol serves double duty (e.g., p as both p-value and dimension), the meaning is clear from context and noted below. log denotes the natural logarithm (base e) unless explicitly stated otherwise.

Greek Letters and Mathematical Symbols

α	Significance level (Type I error)
β	Type II error; regression coefficient
ϵ, η	Sphericity correction (Greenhouse–Geisser, Huynh–Feldt)
λ	Likelihood ratio statistic; Poisson rate
μ	Population mean
σ^2	Population variance
σ	Population standard deviation
τ	Kendall's rank correlation
χ^2_ν	Chi-square, ν degrees of freedom
$\Phi(\cdot)$	Standard normal CDF
$\phi(\cdot)$	Standard normal PDF
Σ	Population covariance matrix
θ	Generic parameter (or vector)
$\hat{\theta}$	MLE or point estimate of θ

Roman Symbols and Matrix Notation

F_{ν_1,ν_2}	F-distribution, ν_1 and ν_2 df
H_0	Null hypothesis
H_1	Alternative hypothesis
$\mathbf{I}_p$	$p \times p$ identity matrix
$\mathbf{S}$	Sample covariance matrix
$L(\theta)$	Likelihood function at θ
$\ell(\theta)$	Log-likelihood: $\log L(\theta)$
n	Sample size
$N(\mu, \sigma^2)$	Normal(μ, σ^2) distribution
p	p-value *or* dimension (context)
$\mathbb{R}^p$	p-dimensional Euclidean space
t_ν	t-distribution, ν df
$\mathrm{tr}(\cdot)$	Matrix trace
$z_\alpha, z_{1-\alpha}$	Standard normal critical values (tail direction is context-dependent)

Abbreviations

AIC	Akaike Information Criterion
ANOVA	Analysis of Variance
ANCOVA	Analysis of Covariance
ARL	Average Run Length
ARL_0	In-control Average Run Length
ARL_1	Out-of-control Average Run Length
BIC (SIC)	Bayesian (Schwarz) Information Criterion
CDF	Cumulative Distribution Function
CV	Critical Value
df	Degrees of freedom
DGP	Data Generating Process
FDR	False Discovery Rate
FTR	Failed to Reject H_0
FWER	Family-Wise Error Rate
GLM	Generalized Linear Model
GLMM	Generalized Linear Mixed Model
GLS	Generalized Least Squares
GOF	Goodness of Fit
HAC	Heteroscedasticity & Autocorrelation Consistent
i.i.d. (IID)	Independent and Identically Distributed
IIA	Independence of Irrelevant Alternatives
IV	Instrumental Variables
LM	Lagrange Multiplier (score test)
LOO	Leave-One-Out
LR	Likelihood Ratio
LRT	Likelihood Ratio Test
MANOVA	Multivariate Analysis of Variance
MIC	Maximal Information Coefficient
MLE	Maximum Likelihood Estimator
MR	Mean Reversion
MSE	Mean Squared Error
MVN	Multivariate Normal distribution
NB	Negative Binomial distribution
OLS	Ordinary Least Squares
PDF	Probability Density Function
PH	Proportional Hazards
PMF	Probability Mass Function
PSIS	Pareto Smoothed Importance Sampling
QC	Quality Control
SD	Standard Deviation
SE	Standard Error
SSCP	Sums of Squares and Cross-products
2SLS	Two-Stage Least Squares
ZIP	Zero-Inflated Poisson model
ZINB	Zero-Inflated Negative Binomial model

Distributions: N, χ^2, F, t denote Normal, Chi-square, F-, and Student's t-distributions.

Software: R packages as `pkg::fn()`; SAS procedures as `PROC NAME`.

Part 0

Foundations of Statistical Inference

Core principles for interpreting evidence, uncertainty, and decisions

Inference Foundations

INTRO - Foundations of Statistical Inference

Before delving into the tests themselves, this chapter builds the conceptual scaffolding that unifies every test in this compendium. Understanding these foundations transforms a collection of formulas into a coherent system for reasoning under uncertainty.

I. The Architecture of Inference: Signal, Noise, and Extraction

Statistical testing rests on a simple but profound idea: every empirical measurement blends **signal**—the underlying structure we seek—with **noise**, the random variation we must overcome to detect it.

1. All data are mixtures. Any observable Y can be expressed as

$$Y = f(X) + \varepsilon,$$

where $f(X)$ is the systematic component and ε the stochastic residue. Inference is difficult precisely when the signal-to-noise ratio is low: large noise can obscure meaningful differences, while high signal can make even small samples persuasive.

2. Models as filters. Statistical models act as filters separating structure from randomness. Estimation methods (OLS, MLE) extract the systematic component; standard errors quantify uncertainty due to noise. Hypothesis tests compare signal to noise through ratios such as

$$t = \frac{\text{signal}}{\text{noise}}.$$

A result becomes "statistically significant" only when the estimated signal rises clearly above background noise.

3. Design as a noise-reduction tool. Good experimental design improves inference not by changing the truth but by reducing uncertainty. Paired designs subtract subject-level variability; ANCOVA removes covariate effects. In both cases, the smaller noise in the denominator sharpens detection of the signal in the numerator.

4. When noise masquerades as signal. Overfitting occurs when a model becomes too flexible and begins to chase accidental patterns. Such models fit the current sample perfectly but generalize poorly. Regularization, cross-validation, parsimony, and diagnostic checking ensure that we learn the signal, not the static (noise).

Key insight: Every test in this compendium is a specific method for comparing signal to noise. The art of inference lies not merely in collecting more data, but in *shaping* data and models so that meaningful patterns emerge cleanly from background variability.

II. The Inferential Chain

Every test typically follows the same five-step logic:

1. **Model:** formulate a probability model for the data-generating process.
2. **Null hypothesis** H_0**:** specify a restricted version of the model (no effect, no difference, no association).
3. **Test statistic:** compute a scalar summary that measures deviation from H_0.
4. p**-value:** the probability, under H_0, of a statistic at least as extreme as the one observed.
5. **Decision:** reject H_0 if $p \le \alpha$; otherwise fail to reject it.

The decision is probabilistic—not a deterministic verdict about truth. Statistical significance reflects incompatibility with a model, not practical importance.

Direction of rejection: The rejection region (sometimes referred to as critical region) follows the direction of the alternative hypothesis. For right-tailed tests ($H_1 : \theta > \theta_0$), reject for large positive values ($Z > z_\alpha$); for left-tailed tests ($H_1 : \theta < \theta_0$), reject for large negative values ($Z < -z_\alpha$); and for two-sided tests ($H_1 : \theta \neq \theta_0$), reject for large deviations in either direction ($|Z| > z_{\alpha/2}$). In non-inferiority testing, the null asserts that the new treatment falls short by more than an acceptable margin δ: $H_0 : \theta \le \theta_0 - \delta$; reject if $Z > z_\alpha$, where Z is computed against the margin boundary $\theta_0 - \delta$, not the reference value θ_0 itself. For equivalence, two one-sided tests (TOST) are used: reject $H_{01} : \theta \le \theta_L$ using $Z_L = (\hat{\theta} - \theta_L)/\text{SE} > z_\alpha$, and separately reject $H_{02} : \theta \ge \theta_U$ using $Z_U = (\hat{\theta} - \theta_U)/\text{SE} < -z_\alpha$; conclude equivalence only if both nulls are rejected simultaneously.

III. Type I and Type II Errors

Every test balances two types of errors.

	H_0 True	H_0 False
Reject H_0	FP: Type I (α)	TP: Power ($1-\beta$)
Fail to reject H_0	TN: Correct ($1-\alpha$)	FN: Type II (β)

Here, TP-True Positive, FP-False Positive, TN-True Negative, and FN-False Negative.

- α: probability of a false positive (rejecting when H_0 is true).
- β: probability of a false negative (missing a true effect).
- Power $= 1 - \beta$: probability of correctly detecting a true effect.

Conservative vs. liberal tests: A test is *conservative* if the actual Type I error < α (fewer false positives, lower power); *liberal* if the actual Type I error > α (inflated false positives).

Key metrics from the confusion matrix:

$$\text{Recall} = \frac{TP}{TP + FN}, \quad \text{Precision} = \frac{TP}{TP + FP}.$$

Precision measures *how accurate positive predictions are*; **Recall** measures *how many positives are captured.*

The fundamental trade-off: for fixed n, lowering α (Type I) raises β (Type II). The only way to reduce both simultaneously is to increase the sample size or the effect size. Effect size (δ) quantifies the magnitude of departure from H_0 under H_1. Power depends on δ, n, α, and test directionality.

Warning: "Fail to reject" means *insufficient evidence*—not that H_0 is true. A study may simply be underpowered.

IV. Why Classical Tests Never Prove H_0

In classical hypothesis testing, we either *reject* H_0 or *fail to reject* it—never *accept* it. This is not pedantic language; it reflects a fundamental asymmetry in how statistical evidence works.

The direction of evidence. Tests are calibrated to answer one question:

> *If H_0 were true, how surprising would the observed data be?*

A small p-value indicates that the data are highly inconsistent with H_0, providing grounds to reject it. A large p-value indicates that the data are *not sufficiently incompatible* with H_0—it does not confirm H_0.

Absence of evidence is not evidence of absence *(Quote attributed to Carl Sagan)*. Failing to reject H_0 may reflect limited sample size, low power, high variability, or an insensitive test. The data may simply be uninformative.

The legal analogy. A defendant is found *guilty* or *not guilty*—never "innocent." A not-guilty verdict reflects insufficient evidence for conviction, not proof of innocence. Likewise, failing to reject H_0 reflects insufficient evidence *against* it, not proof *of* it.

What this means in practice:

- Non-rejection does not validate modeling assumptions.
- Large p-values do not imply equivalence.
- Power, effect size, and diagnostic plots remain essential regardless of the p-value.

Warning: "No evidence against" $H_0 \neq$ "evidence for" H_0. For methods that *can* establish equivalence, see the EQVL chapter.

V. The p-Value: Definition and Misuse

Definition:

$$p = P(T \geq t_{\text{obs}} \mid H_0).$$

What the p-value is:

- A measure of *compatibility* between the data and H_0.
- Sensitive to both effect size and n—a large n can make trivial effects produce small p-values.
- A continuous measure of evidence, not a binary gate: $p = 0.049$ and $p = 0.051$ carry essentially the same information.

What the p-value is not:

- $P(H_0 \mid \text{data})$—the probability H_0 is true (Bayesian concept; requires a prior; see § **VII**).
- The probability that results are "due to chance."
- A measure of effect size or practical importance.

ASA Statement (2016): $p < 0.05$ alone does not constitute scientific evidence. Conclusions require effect sizes, confidence intervals, replication, and subject-matter reasoning—not just a threshold crossing. (ASA: American Statistical Association; Wasserstein & Lazar (2016))

VI. The Symmetry of Evidence: p-Value Uniformity

Beneath the diversity of tests in this compendium lies a fundamental mathematical symmetry governed by the **Probability Integral Transform (PIT)**.

Under a correctly specified H_0, any *continuous* test statistic T maps via its CDF F_T to a uniform distribution:

$$p = 1 - F_T(T) \sim U(0, 1) \quad \text{under } H_0.$$

This means $P(p \leq \alpha \mid H_0) = \alpha$ for every $\alpha \in [0, 1]$—the calibration principle underlying every classical test.

Uniformity as a diagnostic tool. When running a battery of tests (e.g., in meta-analysis), the distribution of p-values is itself informative (e.g: see GFIT.7):

- **Clustering near 0:** genuine effects are present; the data contain signal beyond what is expected under the null.
- **Clustering near 1:** conservative procedures, overcorrection for multiplicity, or model misspecification producing systematically large p-values.
- **Non-uniform pattern:** calibration failure, violated assumptions, publication bias, or data snooping.

A single significant p-value may be a fluke; the *distribution* of p-values across independent tests provides a signature of statistical integrity. A histogram that departs substantially from flatness is often an early warning sign that assumptions have been violated or that systematic effects remain unmodeled.

This observation underlies many modern procedures, including false discovery rate (FDR) control and empirical Bayes methods, which model the distribution of p-values directly.

Technical caveats: exact uniformity requires (i) continuous test statistics, (ii) simple null hypotheses, and (iii) independent tests. Discrete tests (Fisher's exact, small-sample permutation) produce conservative, non-uniform p-values by construction. Composite nulls with estimated nuisance parameters may exhibit slight departures even when properly calibrated.

VII. The Bayesian Alternative: Quantifying Support

Classical inference asks $P(\text{data} \mid H_0)$. Bayesian inference asks what practitioners actually want: $P(H_0 \mid \text{data})$.

Bayes' theorem:

$$p(\theta \mid \mathbf{y}) \propto p(\mathbf{y} \mid \theta)\, p(\theta).$$

Prior beliefs $p(\theta)$ are updated by the likelihood $p(\mathbf{y}|\theta)$ to yield the posterior $p(\theta|\mathbf{y})$.

The Bayes Factor. The primary Bayesian tool for model comparison:

$$\mathrm{BF}_{10} = \frac{P(\mathbf{y} \mid H_1)}{P(\mathbf{y} \mid H_0)}.$$

Unlike the p-value, BF_{10} compares the likelihood of the data under *two competing models* simultaneously. $\mathrm{BF}_{01} = 10$ means the data are 10 times more likely under H_0 than H_1—providing direct, positive evidence *for* H_0, not merely a failure to reject it.

The legal analogy resolved. If the frequentist is a juror deciding if there is "reasonable doubt" to convict, the Bayesian is a detective weighing relative evidence for two suspects. In the Bayesian framework, "not guilty" can actually mean "innocent" if evidence for that hypothesis is sufficiently overwhelming.

Practical equivalence (ROPE). When a parameter is never exactly zero, Bayesians use the **Region of Practical Equivalence** (ROPE) to define a range of negligible values. If the entire 95% Highest Density Interval (HDI) falls within the ROPE, the null is accepted for practical purposes—something a non-significant p-value cannot provide.

Feature	Frequentist	Bayesian
Conclusion	Cannot accept H_0	Can support H_0
Measure	Surprise under H_0	H_0 vs. H_1 ratio
Ambiguity	Large p uninformative	Distinguishes absence of evidence from evidence of absence

VIII. Effect Size and Practical Significance

Statistical significance and practical importance are orthogonal. With large n, any non-zero effect—however tiny—will eventually be statistically significant. With small n, a large and important effect may not reach significance.

Always report an effect size:

Context	Measure	Small/Med/Large
Means	Cohen's d	0.2 / 0.5 / 0.8
ANOVA	η_p^2	0.01 / 0.06 / 0.14
Proportions	OR, RR, RD	context-dependent
Correlation	r	0.1 / 0.3 / 0.5
Regression	Cohen's f^2	0.02 / 0.15 / 0.35

Here, OR-odds ratio, RR-relative risk, RD-risk difference.

Effect sizes are also essential for sample size planning: determine n *before* data collection, based on the minimum effect of practical interest, not post-hoc.

IX Confidence Intervals as Inference

A $100(1-\alpha)\%$ CI is a random interval $[L, U]$ such that $P(L \le \theta \le U) = 1 - \alpha$ across repeated samples.

Correct interpretation: if this procedure were repeated many times, $1-\alpha$ of the resulting intervals would contain θ. This is a statement about the *procedure*, not the specific interval computed from the current data.

CIs provide more than p-values: direction, magnitude, and precision of the estimate. If the entire CI lies within a region of trivial effects, the finding is practically negligible even if $p < 0.05$.

CI–test duality: a two-sided test rejects $H_0 : \theta = \theta_0$ at level α if and only if θ_0 falls outside the $(1-\alpha)$ CI. The CI is the set of all null values not rejected.

Warning: a CI is not a Bayesian credible interval. They coincide numerically only under flat priors but carry fundamentally different interpretations.

X. Multiple Testing and the Forking Paths Problem

Multiple comparisons. When m tests are conducted simultaneously at level α, the probability of at least one false positive (family-wise error rate, FWER) is:

$$\mathrm{FWER} = 1 - (1-\alpha)^m.$$

For $m = 20$ and $\alpha = 0.05$: FWER $\approx 64\%$. Corrections—Bonferroni (α/m), Holm (step-down), Benjamini–Hochberg (FDR)—control different error rates.

The forking paths problem. A p-value is valid only when the analysis plan is fixed before seeing the data. Data-dependent choices—transformations, subsets, outlier removal, switching from one-sided to two-sided tests— constitute hidden multiplicity that inflates the true Type I error above the nominal level.

Two primary mechanisms:

1. **One-sided to two-sided (inflation):** switching after observing the direction of the result effectively doubles the rejection region, inflating α far above 0.05.

2. **Two-sided to one-sided (significance chasing):** if a two-sided test yields $p = 0.08$, switching to one-sided to report $p = 0.04$ uses the data to justify a more lenient test—a classic form of p-hacking.

The gold standard: the directional decision (one- vs. two-sided), the choice of test, and the analysis plan must be fixed *a priori*. If the direction of the effect is not known with certainty before the study begins, a two-sided test is the most robust and honest default.

Warning: post-hoc switching invalidates nominal significance levels. Integrity in analysis requires honoring the original hypothesis, regardless of the patterns that emerge in the data.

XI. Assumptions and Their Consequences

Every test in this compendium is valid under a set of assumptions. Violations do not necessarily invalidate the test, but they change its behavior.

Assumption	Consequence of violation
Independence	Biased SEs; inflated Type I error
Correct dist.	Non-standard null distribution
Equal variances	Inefficient; invalid SEs
No outliers	Inflated σ^2; masked effects
Correct model	Biased estimates

The appropriate response to an assumption violation depends on its severity:

- **Minor:** many tests are robust to mild violations (e.g. t-test under mild non-normality for large n).
- **Moderate:** use robust alternatives (Welch instead of pooled t; HC SEs instead of OLS SEs).
- **Severe:** switch to a fundamentally different method (non-parametric, bootstrap, or a correctly specified model).

Bottom Line

- Statistical tests assess compatibility with models—not truth.
- Evidence is asymmetric in classical inference: tests detect departures, not confirmations.
- p-values measure compatibility with H_0 under assumptions—not the probability H_0 is true, not effect size, not importance.
- Effect sizes and confidence intervals are always required alongside p-values.
- Under H_0, p-values are uniform on $[0, 1]$; deviations signal violated assumptions, data snooping, or unmodelled structure.
- Bayesian methods can quantify positive evidence for H_0; classical methods cannot.
- Pre-specification and transparency are not optional; they are the foundation of valid inference.
- Interpretation requires discipline beyond mechanical threshold crossing.

A Map of This Compendium

The tests in this book are organized by the type of inferential question they address. Every chapter primer provides the conceptual prerequisites for that chapter's tests.

Chapter	Core question/Context
NORM	Are the data normally distributed?
GFIT	Do the data fit a specified distribution?
MEANS	Do group means differ?
VARI	Do group variances differ?
PROP	Testing proportions and rates
NPAR	Rank-based alternatives to parametric tests
CORR	Are two variables associated?
CTAB	Are categorical variables independent?
REGR	Are regression assumptions satisfied?
GLMS	Does the GLM fit adequately?
MIXD	Mixed model: fixed and random effects
MULT	Multivariate mean and covariance tests
TS*	Time series: diagnostics (TSDG), stationarity (STAT), cointegration (COINT), volatility clustering (ARCH)
RAND	Randomness and independence
SURV	Time-to-event and censored data
SPAT	Spatial autocorrelation and regression
QUAL	Statistical process control
EQVL	Equivalence and non-inferiority
ROBU	Robust and resistant methods
PERM	Permutation and bootstrap inference
OUTL	Outlier detection
BAYES	Bayesian hypothesis testing and model comparison

Beneath their diverse formulas and assumptions, all statistical tests seek to quantify the tension between a model and the evidence.

PART I

Normality & Distributional Shape

Assessing distributional assumptions and shape characteristics

NORMALITY

I. NORM — A Conceptual Primer

1. Framing & Motivation

What Is Normality?

- Symmetric, bell-shaped distribution fully defined by (μ, σ^2).
- In practice an **approximation**, never exact.
- Distinguish: **data normality** (raw observations) vs. **residual normality** (model errors)—only the latter is required for valid regression and ANOVA inference.

Why It Matters

- Required for **exact finite-sample** t-, F-, and χ^2-inference.
- *Not* required for point estimation (OLS coefficients do not require normality; under the Gauss–Markov assumptions they remain BLUE).
- Large samples rely on the **CLT**; raw-data normality becomes progressively less critical.

Where Normality Is (and Is Not) Needed

Needed: Small-sample t-tests, ANOVA, regression inference, confidence intervals, prediction intervals.

Not needed: Point estimation, large-n inference, nonparametric methods, bootstrap-based inference.

2. Theoretical Foundation

The Normal Distribution

Density: $f(x) = \dfrac{1}{\sigma\sqrt{2\pi}} \exp\left\{-\dfrac{(x-\mu)^2}{2\sigma^2}\right\}$.

Key properties:

- Mean = Median = Mode = μ.
- Skewness $\gamma_1 = 0$; excess kurtosis $\gamma_2 = 0$.
- **Empirical rule:** 68% within $\mu \pm \sigma$; 95% within $\mu \pm 2\sigma$; 99.7% within $\mu \pm 3\sigma$.
- Tails decay as $e^{-x^2/2}$—much thinner than t or Cauchy.
- All odd central moments = 0; $E[(X-\mu)^4] = 3\sigma^4$ (benchmark for kurtosis).

Core Moment Formulas

$$\bar{x} = \frac{1}{n}\sum x_i, \; m_k = \frac{1}{n}\sum (x_i - \bar{x})^k, \; s^2 = \frac{1}{n-1}\sum (x_i - \bar{x})^2$$

$$\gamma_1 = \frac{m_3}{m_2^{3/2}}, \quad \gamma_2 = \frac{m_4}{m_2^2} - 3 \quad \text{(excess kurtosis)}$$

Useful Identities

- *Affine transform:* $aX + b \sim \mathcal{N}(a\mu + b, a^2\sigma^2)$.
- *Sum of independents:* $\sum X_i \sim \mathcal{N}(\sum \mu_i, \sum \sigma_i^2)$.
- *Sample mean:* $\bar{X} \sim \mathcal{N}(\mu, \sigma^2/n)$ exactly (or approx. via CLT).
- *Standardization:* $Z = (X-\mu)/\sigma \sim \mathcal{N}(0,1)$; $\Phi(-x) = 1 - \Phi(x)$. $\Phi(x)$ is the standard normal CDF.

Derived distributions (normality ⇒):

$$\textstyle\sum_{i=1}^k Z_i^2 \sim \chi_k^2, \quad \frac{(n-1)s^2}{\sigma^2} \sim \chi_{n-1}^2, \quad \frac{\bar{X}-\mu}{s/\sqrt{n}} \sim t_{n-1}.$$

These exact results underpin t-, F-, and χ^2-tests throughout the compendium.

CLT — Practitioner View

- The sampling distribution of $\bar{X}$ converges to $\mathcal{N}(\mu, \sigma^2/n)$ as $n \longrightarrow \infty$.
- Mild skew: $n \approx 30$ often sufficient.
- Heavy tails or outliers: larger n required.
- **CLT does not save you** in small samples or with extreme departures.

3. Structural Understanding of Normality

Characterizations of Normality

Three testable properties uniquely identify the normal distribution:

- **Moments:** $\gamma_1 = 0$ (symmetry) and excess kurtosis = 0 (mesokurtosis). *Caveat:* These conditions are necessary but not sufficient for normality; non-normal distributions can also have $\gamma_1 = 0$ and $\gamma_2 = 0$.
- **Quantile linearity:** order statistics $y_{(i)}$ are linear in expected normal scores $q_i = \Phi^{-1}((i - 0.375)/(n + 0.25))$ (Blom approximation).
- **CDF proximity:** $F_n(y)$ is uniformly close to $\Phi((y-\mu)/\sigma)$.

Skewness & Kurtosis — Shape Guide

- **Right skew** ($\gamma_1 > 0$): long right tail; mean > median; Q–Q bows below line at upper end.
- **Left skew** ($\gamma_1 < 0$): long left tail; mean < median; Q–Q bows above line at lower end.
- **Leptokurtosis** ($\gamma_2 > 0$): heavy tails, frequent extremes; Q–Q ends bend away (S-shape).
- **Platykurtosis** ($\gamma_2 < 0$): thin tails, few extremes; Q–Q ends bend toward the line.

4. Diagnostics

Q–Q Plot — Primary Diagnostic

Plot $y_{(i)}$ against q_i. A straight 45° line ⇒ consistent with normality.

Pattern	Interpretation
Points on line	Normality
S-shape (ends out)	Heavy tails (leptokurtic)
Reverse S (ends in)	Thin tails (platykurtic)
Bow above line	Left skew
Bow below line	Right skew
Isolated extremes	Outliers

Focus on *patterns*, not minor random scatter. The Q–Q plot reveals *where* and *how* the distribution departs from normality—information formal tests cannot convey.

5. Testing Framework

Simple vs. Composite Hypothesis

- **Simple H_0:** μ_0 and σ_0 known; standard K–S critical values apply.
- **Composite H_0:** μ and σ estimated from data (almost always the practical case); critical values must be adjusted. Using simple-hypothesis tables is **anti-conservative**. The Lilliefors correction is the canonical fix for K–S.

Taxonomy of Tests

Class	Based on	Examples
Regression	Q–Q linearity	Shapiro–Wilk, Shapiro–Francia
EDF	F_n vs. Φ distance	K–S, Lilliefors, Anderson–Darling, Cramér–von Mises
Moment	γ_1, γ_2	Jarque–Bera, D'Agostino–Pearson

Each class is sharpest against the departures it directly targets: moment tests detect skewness/kurtosis; EDF tests detect general distributional shifts; regression tests are sensitive to overall Q–Q non-linearity. **Shapiro–Wilk** has the best omnibus power across a wide range of alternatives and is the general-purpose default.

Power & Sample Size

- **Small n (< 30):** low power—tests most needed are least informative. Rely on Q–Q plot; use Shapiro–Wilk.
- **Large n (> 200):** very high power—trivial, practically irrelevant departures are flagged. CLT provides protection for downstream tests regardless.
- A non-significant result **does not confirm normality**; it only fails to reject it.

6. Consequences & Decision Layer

Impact of Violations

- Mild skew: usually acceptable, especially with balanced designs.
- Heavy tails: inflate Type I error; t-test is sensitive.
- Outliers: most damaging; distort both estimates and tests.
- Multimodality: signals model misspecification, not just non-normality.

When You Can Safely Ignore Normality

- Large sample size (CLT applies).
- No extreme outliers present.
- Mild departures only (modest skew, moderate kurtosis).
- Robust or bootstrapped standard errors are used.

7. Actionable Remedies

What To Do If Normality Fails

- **Transform:** $\log y$ (right skew), $\sqrt{y}$ (counts), Box–Cox λ (data-driven).
- **Nonparametric methods:** Wilcoxon, Kruskal–Wallis (see NPAR chapter).
- **Bootstrap inference:** no distributional assumption needed.
- **Robust methods:** trimmed means, M-estimators (see ROBU chapter).
- **Re-specify model:** non-normality often signals omitted variables or wrong functional form—the root cause, not a symptom.

Robust Measures (Normality-Free)

- Median, IQR = $Q_3 - Q_1$.
- MAD = $\text{median}(|x_i - \text{median}(\mathbf{x})|)$; multiply by 1.4826 for σ-consistent estimate under normality.
- Less sensitive to outliers and heavy tails than $\bar{x}$ and s.

8. Practitioner Guardrails & Closure

Common Pitfall

Testing normality of *raw predictors* in regression.
Fix: Always assess **residuals** $\hat{\varepsilon}_i$, not the original data.

Bottom Line

> Normality is a tool for inference, not a requirement to enforce. Most real data are non-normal—and that is often acceptable. Test *residuals*, not raw data. Combine Q–Q plots with formal tests. At large n, the CLT is your friend.

Shapiro-Wilk Test

Powerful Test for Assessing Normality

NORM.1 | **Category:** Normality | **Data:** Continuous | **Null:** Normal

When to Use

- Assess normality of a single continuous variable.
- Routine diagnostic prior to t-tests, ANOVA, and linear regression.
- Effective for sample sizes: $3 \leq n \leq 5000$.

Assumptions

- Random sample with i.i.d observations.
- Continuous data with few or no ties.
- Univariate measurement (one variable at a time).
- Complete, uncensored observations.

Test Statistic

$$W = \frac{\left(\sum_{i=1}^{n} a_i x_{(i)}\right)^2}{\sum_{i=1}^{n}(x_i - \bar{x})^2} \quad \Big| \quad \mathbf{a} = \frac{\mathbf{V}^{-1}\mathbf{m}}{\|\mathbf{V}^{-1}\mathbf{m}\|}, \quad m_i = \Phi^{-1}\left(\frac{i-a}{n+b}\right) \quad \Big| \quad 0 \text{ (non-normal)} < W \leq 1 \text{ (normal)}$$

Notation: $x_{(i)}$ are ordered sample values, $m_i = E[X_{(i)}]$, $V_{ij} = \text{Cov}(X_{(i)}, X_{(j)})$, Blom's approx: $a = 3/8$, $b = 1/4$

Plain English: The Shapiro-Wilk test measures how well ordered data align with what would be expected under normality. It quantifies the linearity of a normal Q-Q plot by comparing two variance estimates (R^2 like metric): one from regression of ordered values (numerator) on expected normal order statistics and another from sample variance (denominator).

⚠ Practitioner Warning

The test becomes highly sensitive in large samples ($n > 1000$); even trivial deviations from normality may yield significant results. Always complement with graphical diagnostics (Q-Q plots, histograms) and consider practical significance.

Hypotheses

- H_0: Data distribution is Normal
- H_1: Data distribution is **not** Normal

Decision Rule: Reject H_0 if $W < W_{\alpha,n}$, where $W_{\alpha,n}$ is the critical value for significance level α and sample size n.

Software References [Python, R, SAS]

`scipy.stats.shapiro` `shapiro.test` `PROC UNIVARIATE / NORMAL`

Additional Notes

The exact distribution of W under H_0 is complex and depends on n. Critical values for $\alpha = 0.01, 0.05, 0.10$ have been tabulated for $3 \leq n \leq 50$. For $n > 50$, Royston's (1992) approximations ($w \sim N(\mu, \sigma^2)$) are used. The test has excellent power against most alternatives to normality.

$4 \leq n \leq 11$

$$w = -\ln\left[\gamma - \ln(1-W)\right]; \ \gamma = -2.273 + 0.459\,n$$

$$\mu = 0.5440 - 0.39978\,n + 0.025054\,n^2 - 0.0006714\,n^3$$

$$\sigma = \exp\!\left(1.3822 - 0.77857\,n + 0.062767\,n^2 - 0.0020322\,n^3\right)$$

$12 \leq n \leq 2000$

$$w = \ln(1-W); \ x = \ln n$$

$$\mu = -1.5861 - 0.31082\,x - 0.083751\,x^2 + 0.0038915\,x^3$$

$$\sigma = \exp\!\left(-0.4803 - 0.082676\,x + 0.0030302\,x^2\right)$$

Prior normality tests (such as the χ^2 goodness-of-fit) were less powerful or required grouping data into bins, which lost information. Shapiro and Wilk's insight was to leverage the properties of order statistics to create a test that used the full dataset without binning. The test quickly became the gold standard due to its superior power. However, computing W by hand was tedious—the coefficients a_i required inverting covariance matrices of order statistics. Early adoption was slow until Royston (1982) provided efficient algorithms and extended the method to larger sample sizes. *Note:* A censored-data version of the test is available for assessing normality in partially observed samples.

Miscellanea: The original paper is titled *"An Analysis of Variance Test for Normality"*—a subtle misnomer, as the test is not an ANOVA in the usual sense. Interestingly, Wilk also co-invented (with R. Gnanadesikan, 1968) the Q–Q plot, so the test and its most widely used diagnostic share a common origin. *Mnemonic:* "W for Wilk, W for Well-behaved"—if $W \approx 1$, data are behaving *normally*.

Reference(s): Shapiro & Wilk (1965); Royston (1982, 1992)

Worked Example

Problem Statement Eight recovery times (days) are tested for normality before applying a parametric procedure:

Sorted Data and Sample Moments Sorted: $x_{(1)} \le \cdots \le x_{(8)}$:

$$3,\ 5,\ 6,\ 7,\ 9,\ 12,\ 18,\ 35 \qquad \bar{x} = 11.875, \quad SS = \sum_{i=1}^{8}(x_i - \bar{x})^2 = \mathbf{764.88}$$

H_0: the data are drawn from a normal distribution; H_1: the data are not normally distributed.

1. Shapiro-Wilk Coefficients & Numerator b

For $n = 8$, the tabulated coefficients a_i are applied to symmetric order-statistic differences $d_i = x_{(n+1-i)} - x_{(i)}$:

i	a_i	$x_{(9-i)}$	$x_{(i)}$	d_i
1	0.6052	35	3	32
2	0.3164	18	5	13
3	0.1743	12	6	6
4	0.0561	9	7	2

$$b = \sum_{i=1}^{4} a_i\, d_i = \mathbf{24.638}$$

2. W Statistic

$$W = \frac{b^2}{SS} = \frac{(24.638)^2}{764.875} = \frac{607.011}{764.875} = \boxed{0.7936}$$

$W \in (0, 1]$; a value near 1 is consistent with normality. Here $W = 0.794$ indicates substantial departure.

3. Results and Decision

Statistic	Value	CV (5%)	Decision
W	0.7936	0.818	Reject H_0

$p = 0.024$; $W = 0.794 < W_{\text{crit}}(n{=}8, 5\%) = 0.818$. For reference: $W_{\text{crit}}(n{=}8, 10\%) = 0.851$.

4. Comparison of SS contributions

x_i	3	5	6	7	9	12	18	35	$\sum(x_i - \bar{x})^2$
$(x_i - \bar{x})^2$	78.77	47.27	34.52	23.77	8.27	0.02	37.52	534.77	764.88

5. Interpretation

- Normality is rejected at 5%. The value $x = 35$ lies 2.21 standard deviations above the mean and inflates SS from ≈ 230 (without it) to 764.9, while the numerator b grows only moderately because $a_1 = 0.605$ is bounded. The resulting $W = 0.794$ signals right-skewed departure (sample skewness = 1.50).

Practitioner Insight: The Shapiro–Wilk statistic assesses how closely ordered data match expected normal order statistics. The coefficients a_i reflect normal spacings, so $W = b^2/SS$ is essentially the squared correlation in a normal probability plot, with $W = 1$ under perfect normality. It is sensitive to skewness and heavy tails, and is generally the most powerful normality test for $n \le 50$. For larger samples, Royston's extension provides analytic p-values. Rejection indicates non-normality but not its form, so graphical tools (e.g. Q–Q plots) are needed to diagnose departures.

Common Pitfalls

- **Non-rejection as confirmation.** Failing to reject H_0 indicates insufficient evidence *against* normality, not proof that the data are normal. Always complement the test with Q–Q plots and histograms.
- **Large-sample over-sensitivity.** With large n, even trivial deviations from normality can produce highly significant results. Assess whether departures are practically meaningful for the intended analysis.
- **Very small samples.** For very small samples ($n < 10$), the test has limited power and may fail to detect substantial non-normality. Results should be interpreted cautiously.
- **Tied observations.** Shapiro–Wilk assumes continuous data. Extensive ties—often due to rounding or discrete measurement—can distort the test statistic and invalidate inference.
- **Inherently non-normal data.** Counts, proportions, and bounded variables are not normally distributed. Normality testing is inappropriate; use models suited to the data type instead.

Shapiro-Francia Test

Simplified Approximation to Shapiro-Wilk for Normality

NORM.2 | **Category:** Normality | **Data:** Continuous | **Null:** Normal

When to Use

- Designed as a large-sample approximation to Shapiro–Wilk.
- Particularly suitable for moderate to large sample sizes.
- Useful when computational simplicity is desired.

Assumptions

- Random sample with i.i.d observations.
- Continuous data with few or no ties.
- Complete, uncensored observations.

Test Statistic

$$W' = \frac{\left(\sum_{i=1}^{n} b_i x_{(i)}\right)^2}{\sum_{i=1}^{n}(x_i - \bar{x})^2} \quad \Big| \quad \mathbf{b} = \frac{\mathbf{m}}{\|\mathbf{m}\|} = \frac{m_i}{\sum_{i=1}^{n} m_i^2}, \quad m_i = \Phi^{-1}\left(\frac{i-a}{n+b}\right) \quad \Big| \quad 0 \text{ (non-normal) } < W' \leq 1 \text{ (normal)}$$

Notation: $x_{(i)}$ are order statistics, $m_i = E[X_{(i)}]$, Blom's approx: $a = 3/8$, $b = 1/4$, Φ^{-1} is inv. standard normal CDF

Plain English: The Shapiro-Francia test measures how strongly the ordered data correlate with expected values from a normal distribution. It simplifies the Shapiro-Wilk calculation by using expected normal order statistics directly rather than computing inverse covariance matrices, making it more practical for larger samples.

⚠ Practitioner Warning

The Shapiro–Francia test is designed primarily to detect departures from normality through tail behaviour and overall distributional shape. While it performs well for large samples and approximately symmetric alternatives, it may be less sensitive than the Shapiro–Wilk test to certain subtle deviations from normality. As with all normality tests, statistical significance should be interpreted alongside graphical diagnostics such as Q–Q plots, since large samples can detect negligible departures and small samples may lack power to detect meaningful ones.

Hypotheses

- H_0: Data distribution is Normal
- H_1: Data distribution is **not** Normal

Decision Rule: Reject H_0 if $W' < W'_{\alpha,n}$, where $W'_{\alpha,n}$ is the critical value for significance level α and sample size n.

Software References [Python, R, SAS]

Python	R	SAS
`Not Native (Manual)`	`nortest::sf.test`	`Not Native (Manual)`

Additional Notes

The Shapiro–Francia statistic can be interpreted geometrically as the squared Pearson correlation coefficient between the ordered observations and their expected normal scores:

$$W' = \frac{\left(\sum_{i=1}^{n}(x_{(i)} - \bar{x})(m_i - \bar{m})\right)^2}{\sum_{i=1}^{n}(x_{(i)} - \bar{x})^2 \sum_{i=1}^{n}(m_i - \bar{m})^2}, \qquad \text{where } \bar{m} = \frac{1}{n}\sum_{i=1}^{n} m_i.$$

Francia's insight was that expected normal order statistics m_i could serve as reasonable approximations to the Shapiro-Wilk's optimal weights (a_i). While this sacrificed some statistical power, it made the test calculable with just a table of normal quantiles and a hand calculator. Royston (1993) showed that, under normality and for moderate to large sample sizes,

$$A = \ln(1 - W') \approx \mathcal{N}(\mu_n, \sigma_n^2), \quad \text{where } \mu_n = -1.2725 + 1.0521\,[\ln(\ln n) - \ln n], \quad \sigma_n = 1.0308 - 0.26758\,[\ln(\ln n) + \frac{2}{\ln n}].$$

Miscellanea: Francia introduced this test in 1972 as a "poor man's Shapiro–Wilk"—a computationally simpler alternative at a time when computing resources were limited, just seven years after the original Shapiro–Wilk test. Ironically, modern software now makes both tests equally easy to run. The method is also closely related to earlier work by de Wet and Venter, who studied correlation-type statistics based on normal order statistics; all these approaches compare ordered data to their expected normal counterparts, reflecting a broader regression-based view of normality testing.

Reference(s): Shapiro & Francia (1972); Royston (1983, 1993)

Worked Example

Problem Statement Ten ($n = 10$) hospital length-of-stay values (days) are tested for normality using the Shapiro-Francia test. Unlike Shapiro-Wilk, which derives weights from the full covariance matrix of normal order statistics, Shapiro-Francia uses expected normal scores directly. H_0: the data are drawn from a normal distribution; H_1: they are not.

1. Normal Scores via Blom Approximation

For order statistic $x_{(i)}$, the expected standard-normal quantile is:

$$m_i = \Phi^{-1}(p_i), \quad \text{where } p_i = \frac{i - 0.375}{n + 0.25}$$

i	p_i	m_i	$x_{(i)}$	$m_i x_{(i)}$
1	0.0610	−1.5466	5	−7.733
2	0.1585	−1.0005	7	−7.003
3	0.2561	−0.6554	8	−5.243
4	0.3537	−0.3755	9	−3.379
5	0.4512	−0.1226	10	−1.226
6	0.5488	+0.1226	12	+1.471
7	0.6463	+0.3755	15	+5.632
8	0.7439	+0.6554	20	+13.109
9	0.8415	+1.0005	28	+28.014
10	0.9390	+1.5466	52	+80.425
			$\sum m_i x_{(i)}$	104.065

Note: $\bar{m} = 0$ exactly by the symmetry of the Blom formula, so $S_{mx} = \sum m_i x_{(i)} - n\bar{m}\bar{x} = \sum m_i x_{(i)}$.

2. Sample Moments and Score Variance

$$\bar{x} = 16.6, \qquad SS_{xx} = \sum (x_i - \bar{x})^2 = \mathbf{1820.4}$$

$$SS_{mm} = \sum m_i^2 = \mathbf{7.957} \quad (\bar{m} = 0)$$

3. Q-Q Correlation and W'

$$r = \frac{S_{mx}}{\sqrt{SS_{mm} \cdot SS_{xx}}} = \frac{104.065}{\sqrt{7.957 \times 1820.4}} = \mathbf{0.8646}$$

$$W' = r^2 = (0.8646)^2 = \boxed{0.7476}$$

W' is the R^2 between the order statistics and their expected normal positions — equivalently, the R^2 of the Q-Q plot regression line through the origin.

4. Results and Decision

Statistic	Value	CV (1%)	Decision
W'	0.7476	0.793	Reject H_0

α	$W'_{\text{crit}}(n{=}10)$	Decision
10%	0.869	Reject
5%	0.842	Reject

5. Interpretation

- $W' = 0.748$ falls below all three critical values. The dominant signal is the outlying value $x = 52$ ($m_{10}x_{(10)} = 80.4$, accounting for 77% of S_{mx}), which inflates SS_{xx} far more than it inflates S_{mx}, pulling r below 1.

Practitioner Insight: W' is the squared correlation between ordered data $x_{(i)}$ and Blom normal scores m_i, so $W' = 1$ under perfect normality and decreases with curvature (skewness, heavy tails, outliers) in the Q–Q plot. Using m_i simplifies computation relative to Shapiro–Wilk, though it is slightly less powerful for small $n \le 11$. For $n > 50$, both tests perform similarly and W' is preferred for simplicity; p-values are obtained via the Royston approximation based on $\log(1 - W')$.

Common Pitfalls

- **Small-sample limitations.** Shapiro–Francia performs relatively poor for $n < 20$; use Shapiro–Wilk in this range due to markedly higher power.
- **Statistic mismatch.** The W' statistic is not interchangeable with Shapiro–Wilk's W; they have distinct null distributions and critical values.
- **Simplicity ≠ superiority.** Shapiro–Francia was created for computational ease; modern software makes Shapiro–Wilk just as easy and generally more powerful.
- **Overreliance on the test.** As with all normality tests, important structure may appear only in Q–Q plots or histograms; visual checks remain essential.
- **Inappropriate data types.** The test assumes continuous, unbounded data; discrete, rounded, or bounded variables (counts, proportions, Likert scales) violate this assumption and require alternative methods.

Anderson–Darling Test

Tail-Sensitive Test for Normality

NORM.3 | **Category:** Normality | **Data:** Continuous | **Null:** Normal

When to Use

- Test normality with emphasis on detecting deviations in distribution tails.
- Suitable for a wide range of sample sizes ($n \geq 5$).
- More powerful than Kolmogorov-Smirnov for detecting tail departures.

Assumptions

- Random sample with i.i.d observations.
- Continuous data with few or no ties.
- Complete, uncensored observations.
- When parameters are estimated, use modified version of the test (Anderson-Darling with estimated parameters).

Test Statistic

$$A_n^2 = -n - \frac{1}{n}\sum_{i=1}^{n}\left[(2i-1)\ln(\Phi(z_{(i)})) + (2n+1-2i)\ln(1-\Phi(z_{(i)}))\right] \quad \Big| \quad \text{(Stephens)}: A^{*2} = A_n^2\left(1 + \frac{0.75}{n} + \frac{2.25}{n^2}\right)$$

Notation: $z_{(i)}$ are the *standardized* order statistics, $\Phi(.)$ is the standard normal CDF, A_n^{2*} corrects for finite sample.

Plain English: It measures the weighted discrepancy between the empirical distribution and a normal distribution, giving extra weight to observations in the tails. This makes it particularly sensitive to departures that occur in the extremes of the distribution—exactly where violations often matter most.

⚠ Practitioner Warning

When using sample mean and variance to define the normal distribution, standard critical values are invalid. Use the modified Anderson-Darling test with adjusted critical values or directly obtain p-values from software.

Hypotheses

- H_0**:** Data distribution is Normal
- H_1**:** Data distribution is **not** Normal

Decision Rule: Reject H_0 if $A^{*2} > A_\alpha^{*2}$ (critical value) or p-value $< \alpha$. $A_{0.01}^{*2} = 1.035$, $A_{0.05}^{*2} = 0.752$, $A_{0.10}^{*2} = 0.631$

Software References [Python, R, SAS]

Python	R	SAS
`scipy.stats.anderson`	`nortest::ad.test`	`PROC UNIVARIATE / NORMAL`

Additional Notes

Anderson and Darling (1952) introduced the test as a modification of the Cramér-von Mises statistic, incorporating a weighting function that produces consistent tests against all alternatives. The AndersonDarling statistic (definition) for testing whether F_n (empirical CDF) comes from a continuous distribution F is:

$$A_n^2 = n\int_{-\infty}^{\infty}\frac{[F_n(x) - F(x)]^2}{F(x)[1-F(x)]}\,dF(x).$$

The weighting function $w(x) = [F(x)(1-F(x))]^{-1}$ gives more weight to observations in the tails, where $F(x)$ is close to 0 or 1. The test works for any general continuous distribution F. Subsequent work by Stephens (1974) provided critical values and modifications for parameters estimated from data, making the test practical for routine application.

Unlike the Kolmogorov-Smirnov test, which focuses on the maximum discrepancy between empirical and theoretical distributions, the Anderson-Darling test integrates squared discrepancies across the entire distribution, weighted to emphasize tail behavior. This weighting makes it one of the most powerful tests for normality, especially against alternatives with non-normal tail behavior.

Miscellanea: The test remained difficult to use in practice until 1974, when Michael A. Stephens derived modified critical values that made it distribution-free, transforming it from a theoretical curiosity into the widely-used test found in statistical software today. Stephens' modification was so successful that most modern implementations labeled "Anderson-Darling test" actually use his 1974 version. Unlike Shapiro-Wilk, Anderson-Darling is a general goodness of fit test (not just *normal*).

Reference(s): Anderson & Darling (1952); Stephens (1986); Marsaglia & Marsaglia (2004)

Worked Example

Problem Statement Eight service times (minutes) are tested for normality. The Anderson-Darling test is chosen over Kolmogorov-Smirnov because it weights deviations in the tails of the distribution more heavily than the center. H_0: data follow a normal distribution; H_1: they do not.

1. Standardize the Order Statistics $\bar{x} = 18.75,\ s = 8.972,\ z_{(i)} = (x_{(i)} - \bar{x})/s$.

i	$x_{(i)}$	$z_{(i)}$	$\Phi(z_{(i)})$	$1-\Phi_{n+1-i}$	$2i-1$	$\log\Phi(z_{(i)})$	$\log(1-\Phi_{n+1-i})$
1	10	−0.9752	0.1647	.0120	1	−1.804	−4.423
2	13	−0.6409	0.2608	.3586	3	−1.344	−1.026
3	14	−0.5294	0.2983	.4889	5	−1.210	−0.716
4	16	−0.3065	0.3796	.5773	7	−0.969	−0.549
5	17	−0.1950	0.4227	.6204	9	−0.861	−0.477
6	19	+0.0279	0.5111	.7017	11	−0.671	−0.354
7	22	+0.3622	0.6414	.7392	13	−0.444	−0.302
8	39	+2.2570	0.9880	.8353	15	−0.012	−0.180

2. AD Statistic (Alternative Formula)

$$A_n^2 = -n - \frac{1}{n}\sum_{i=1}^{n}(2i-1)\left[\log\Phi(z_{(i)}) + \log(1-\Phi(z_{(n+1-i)}))\right] = -8 - \tfrac{1}{8}(-69.496) = \mathbf{0.687}$$

3. Modified Statistic and Decision
Adjusting for estimated $\bar{x}$ and s:

$$A^{*2} = A_n^2\left(1 + \frac{0.75}{n} + \frac{2.25}{n^2}\right) = 0.687 \times 1.129 = \boxed{0.776}$$

Statistic	Value	CV (5%)	Decision
A^{*2}	0.776	0.752	Reject H_0

$p \approx 0.044$. Reject H_0 at 5%.

Practitioner Insight: The Anderson–Darling statistic weights observations by $(2i-1)$, placing greatest emphasis on the tails. Large contributions arise when tail probabilities are extreme, with asymmetry between Φ_i and $1-\Phi_{n+1-i}$ revealing skewness. Unlike Cramér–von Mises, which weights all points equally, AD is more sensitive to tail deviations. The adjustment $A^{*2} = A_n^2(1+0.75/n+2.25/n^2)$ corrects for estimating μ and σ. For large samples, even minor deviations can trigger rejection, so graphical checks (e.g. Q–Q plots) are essential; here, a single extreme value drives the result.

Common Pitfalls

- **Unadjusted critical values.** When μ and σ are estimated from the data, the raw A_n^2 statistic is invalid. Always use the adjusted statistic A^{*2} or software that applies the correction automatically.
- **Ties and discretization.** The Anderson–Darling test assumes continuous data. Extensive ties—often caused by rounding or discrete scales—can distort the statistic and invalidate inference.
- ***p*-value only interpretation.** The test indicates whether normality fails, not how. Always inspect Q–Q plots to identify whether deviations occur in the tails, center, or via skewness.
- **Large-sample over-sensitivity.** With large n, even negligible departures from normality can be statistically significant. Assess whether the deviation is practically meaningful.
- **Assumed dominance.** Anderson–Darling is especially sensitive to tail departures but is not uniformly most powerful. Tests such as Shapiro–Wilk may outperform it for other alternatives.

p-value Formulas

For $A^{*2} \geq 0.600$, $p = \exp(1.2937 - 5.709A^{*2} + 0.0186(A^{*2})^2)$; for $0.340 < A^{*2} < 0.600$, $p = \exp(0.9177 - 4.279A^{*2} - 1.38(A^{*2})^2)$; for $0.200 < A^{*2} \leq 0.340$, $p = 1 - \exp(-8.318 + 42.796A^{*2} - 59.938(A^{*2})^2)$; and for $A^{*2} \leq 0.200$, $p = 1 - \exp(-13.436 + 101.14A^{*2} - 223.73(A^{*2})^2)$.

Cramér–von Mises Test

Integrated Squared Distance Goodness-of-Fit Test

NORM.4 | **Category:** Normality | **Data:** Continuous | **Null:** Normal

When to Use

- Test normality with balanced sensitivity across the entire distribution.
- Preferred when no specific region (center vs. tails) is of primary concern.
- More powerful than Kolmogorov-Smirnov, simpler than Anderson-Darling.

Assumptions

- Random sample with i.i.d observations.
- Continuous data with few or no ties.
- Complete, uncensored observations.
- When parameters are estimated, use modified critical values.

Test Statistic

$$W_n^2 = \frac{1}{12n} + \sum_{i=1}^{n} \left[\frac{2i-1}{2n} - \Phi\left(\frac{x_{(i)} - \hat{\mu}}{\hat{\sigma}} \right) \right]^2 \qquad \text{(Stephens)}: W^{*2} = W_n^2 \left(1 + \frac{0.5}{n} \right)$$

This form is specific to testing normality, with (μ, σ^2) estimated from the data. Cramér–von Mises critical values are sample-size–free under a fully specified null (say, known μ_0, σ_0^2), but depend on it when (μ, σ^2) are estimated.

Notation: $x_{(i)}$ denotes the ith ordered observation, $\hat{\mu}$ and $\hat{\sigma}$ are the estimated sample mean and standard deviation, and $\Phi(\cdot)$ is the standard normal CDF. Note: Shapiro-Wilk uses W while Cramér–von Mises uses W^2 (for squared distance)

Plain English: The Cramér-von Mises test measures the integrated squared difference between the empirical and theoretical distributions, treating all regions equally. Unlike Anderson-Darling which emphasizes tails, this test is balanced—equally sensitive to departures anywhere in the distribution.

⚠ Practitioner Warning

By treating all regions equally, Cramér-von Mises sacrifices some power against specific alternatives (e.g., heavy tails) compared to targeted tests. Choose based on whether you care about all departures equally or have specific concerns about tails or center.

Hypotheses

- H_0: Data distribution is Normal
- H_1: Data distribution is **not** Normal

Decision Rule: Reject H_0 if $W^2 > W^2_{\alpha,n}$ (critical value) or p-value $< \alpha$. $W^2_{0.01} = 0.179$, $W^2_{0.05} = 0.126$, $W^2_{0.10} = 0.104$.

Software References [Python, R, SAS]

```
scipy.stats.cramervonmises        nortest::cvm.test        PROC UNIVARIATE / NORMAL
```

Additional Notes

The Cramér–von Mises test, independently introduced by Cramér (1928) and von Mises (1931), measures goodness of fit by integrating squared differences between the empirical and theoretical distribution functions (generic version),

$$W_n^2 = n \int_{-\infty}^{\infty} \left[F_n(x) - F(x) \right]^2 \, dF(x),$$

providing a smooth, omnibus alternative to supremum-based tests such as Kolmogorov–Smirnov. Under H_0, W_n^2 converges to a weighted sum of independent χ_1^2 variables, with critical values obtained from the asymptotic distribution or finite-sample tables.

Miscellanea: The Cramér–von Mises test is a classic example of scientific synchronicity. Harald Cramér (1928) and Richard von Mises (1931) independently derived essentially the same goodness-of-fit statistic while working in different countries and fields; their contributions were later unified under a single test. *Mnemonic:* CvM measures the *cumulative* (integrated) gap between distributions.

Reference(s): Cramér (1928); von Mises (1931); Anderson & Darling (1952)

Worked Example

Problem Statement Eight reaction times (milliseconds) are tested for normality using the Cramér-von Mises test. Unlike Anderson-Darling (which up-weights the tails) and Kolmogorov-Smirnov (which uses only the largest pointwise gap), CvM accumulates squared discrepancies between the empirical and fitted CDFs uniformly across all order statistics. H_0: data follow a normal distribution; H_1: they do not.

1. Sample Moments and Standardization $\bar{x} = 12.5, \quad s = 11.723, \quad z_{(i)} = (x_{(i)} - \bar{x})/s$

2. CvM Statistic

For each order statistic, compute the gap between the fitted normal CDF $F_i = \Phi(z_{(i)})$ and the empirical midpoint $p_i = (2i-1)/2n = (2i-1)/16$:

i	$x_{(i)}$	$z_{(i)}$	F_i	p_i	$(p_i - F_i)^2$
1	4	−0.7251	0.2342	1/16	0.02948
2	5	−0.6398	0.2612	3/16	0.00543
3	7	−0.4692	0.3195	5/16	0.00005
4	8	−0.3839	0.3505	7/16	0.00756
5	9	−0.2986	0.3826	9/16	0.03235
6	11	−0.1280	0.4491	11/16	0.05684
7	16	+0.2986	0.6174	13/16	0.03808
8	40	+2.3458	0.9905	15/16	0.00281
				$\sum(p_i - F_i)^2$	0.17260

$$W_n^2 = \frac{1}{12n} + \sum_{i=1}^{n}(p_i - F_i)^2 = \frac{1}{96} + 0.17260 = \mathbf{0.1830}$$

3. Modified Statistic

Adjusting for estimated $\bar{x}$ and s:

$$W^{*2} = W_n^2\left(1 + \frac{0.5}{n}\right) = \boxed{0.1945}$$

4. Results and Decision

Statistic	Value	CV (1%)	Decision
W^{*2}	0.1945	0.179	Reject H_0

Full critical value table (Stephens 1974):

α	CV	W^{*2}	Decision
10%	0.104	0.1945	Reject
5%	0.126	0.1945	Reject
1%	0.179	0.1945	Reject

$p < 0.01$. Normality is rejected at all standard levels.

5. Interpretation

- The dominant terms are $i = 6$ ($x = 11$, gap $= -0.238$), $i = 7$ ($x = 16$, gap $= -0.195$), and $i = 5$ ($x = 9$, gap $= -0.180$). The extreme value $x = 40$ contributes only 0.003 — the body of the distribution, not the tail, drives the rejection.

Practitioner Insight: The CvM statistic W_n^2 approximates $n\int[F_n(x) - F(x)]^2\,dF(x)$ by averaging squared gaps $(p_i - F_i)^2$ at the empirical CDF midpoints, giving equal weight to all observations. Unlike Anderson–Darling, which emphasizes tails, CvM is most sensitive to deviations in the body of the distribution; here, mid-range observations dominate while the extreme value contributes little. The adjustment $W^{*2} = W_n^2(1 + 0.5/n)$ corrects for estimating μ and σ, and Stephens' critical values (valid for $n \geq 5$) apply via this normalization.

Common Pitfalls

- **Using incorrect critical values when parameters are estimated.** Standard tables assume a fully specified null, so when parameters (e.g. μ, σ) are estimated, modified CVs are required; using unadjusted values inflates Type I error.
- **Expecting the test to be uniformly most powerful against all alternatives.** The equal-weighting nature of W^2 provides balanced sensitivity across the distribution, but it is often less powerful than specialized tests such as Anderson–Darling for tail deviations or Shapiro–Wilk for overall departures.
- **Applying the test to data with many ties or discrete structure.** The test assumes continuous data, and ties from rounding or discreteness can distort the null distribution, leading to conservative or anti-conservative results.
- **Relying solely on the test statistic without graphical diagnostics.** The W^2 statistic summarizes overall fit but does not indicate the nature of deviations, so Q–Q plots, histograms, and density estimates should also be examined.
- **Interpreting non-rejection as proof of normality.** Failing to reject H_0 indicates only that the data are consistent with the model, and limited power—especially in small samples—means meaningful departures may go undetected.

Kolmogorov–Smirnov Test (Lilliefors Test)

Distribution-Free Test Based on Maximum Deviation

NORM.5 | **Category:** Normality | **Data:** Continuous | **Null:** Distribution-free

When to Use

- Compare an empirical distribution to a fully specified theoretical distribution (not just normal distribution).
- Simple, distribution-free test requiring minimal assumptions.
- Suitable for any sample size, though power is moderate.
- Applicable beyond normality testing.

Assumptions

- Observations are independent and identically distributed.
- Data are continuous (test is conservative with ties).
- Null distribution must be **fully specified** (no parameter estimation).
- For normality with estimated parameters, use Lilliefors modification.

Test Statistic

$$D = \sup_x |F_n(x) - F_0(x)| = \max\left\{\max_i \left|\frac{i}{n} - F_0(x_{(i)})\right|, \max_i \left|F_0(x_{(i)}) - \frac{i-1}{n}\right|\right\}$$

Notation: $F_n(x)$ is empirical CDF, $F_0(x)$ is fully specified null CDF ($\Phi(x)$ for normal), supremum is maximum

Plain English: The Kolmogorov-Smirnov test measures the largest vertical gap between the empirical distribution and the hypothesized distribution. It asks: "What's the worst mismatch anywhere?" Unlike tests using squared distances, K-S focuses on the single point of greatest disagreement.

⚠ Practitioner Warning

The standard K-S test **cannot** be used when parameters are estimated from data. Doing so severely inflates Type I error. For normality testing with estimated μ and σ, use the Lilliefors correction instead, which adjusts the null distribution appropriately.

Hypotheses

- H_0: The data follow the specified distribution.
- H_1: The data do not follow the specified distribution.

Decision Rule: Reject H_0 if $D > D_{\alpha,n}$ (critical value) or p-value $< \alpha$. For large n, $D_{\alpha,n} \approx c_\alpha/\sqrt{n}$ where $c_{0.01} \approx 1.63$, $c_{0.05} \approx 1.36$, $c_{0.10} \approx 1.22$.

Software References [Python, R, SAS]

```
statsmodels.stats.
diagnostic.lilliefors
```

```
nortest::lillie.test
```

```
PROC UNIVARIATE / NORMAL
```

Additional Notes

Introduced by Kolmogorov (1933) and refined by Smirnov (1948), the Kolmogorov–Smirnov (K–S) test provides a distribution-free procedure for comparing an empirical distribution to a fully specified theoretical distribution, requiring minimal assumptions and making it broadly applicable. Finite-sample accuracy can be improved using Miller's scaling correction (μ and σ^2 known),

$$D_n^{(M)} = \left(\sqrt{n} + 0.12 + \frac{0.11}{\sqrt{n}}\right) D_n, \qquad p = P\left(\sup_{0\le t\le 1} |B(t)| \ge \sqrt{n}\, D_n\right) = 2\sum_{k=1}^{\infty}(-1)^{k-1}\exp\left(-2k^2(\sqrt{n}\, D_n)^2\right),$$

where p denotes the p-value computed from the asymptotic Kolmogorov distribution of the Brownian bridge $B(t)$. The scaling reduces the liberal behavior of the unadjusted test in small and moderate samples.

Miscellanea: Andrey Kolmogorov—who established the axiomatic foundations of probability theory (often called the Newton of probability theory) in 1933—derived the asymptotic distribution of the K–S test that same year as the supremum of a **Brownian bridge** (a continuous stochastic process constrained to start and end at zero). This limiting distribution is now known as the **Kolmogorov distribution**. Despite the joint name, they never co-authored a paper on the test. *Mnemonic:* "K–S: King of Supremum".

Reference(s): Kolmogorov (1933); Smirnov (1939, 1948); Miller (1951); Lilliefors (1967); Abdi & Molin (2007)

Kolmogorov–Smirnov Test (Lilliefors Test) NORM.5

Worked Example

Problem Statement Eight pollution concentration measurements (μg/m^3) are tested for normality. When μ and σ are estimated from the data, the standard KS critical values are too conservative; the Lilliefors (1967) correction must be used. H_0: data follow a normal distribution; H_1: they do not.

1. Sample Moments and Standardization $\bar{x} = 12.0, \quad s = 12.773, \quad z_{(i)} = (x_{(i)} - \bar{x})/s$

2. KS Statistic: Both One-Sided Gaps

At each order statistic $x_{(i)}$, the empirical CDF jumps from $(i-1)/n$ to i/n. Two gaps are computed:

$$D_i^+ = \frac{i}{n} - F_i, \quad D_i^- = F_i - \frac{i-1}{n}, \text{ where } F_i = \Phi(z_{(i)}).$$

i	$x_{(i)}$	$z_{(i)}$	F_i	i/n	D_i^+	D_i^-
1	3	−0.705	0.2405	0.125	−0.116	**0.241**
2	4	−0.626	0.2655	0.250	−0.016	0.141
3	6	−0.470	0.3193	0.375	0.056	0.069
4	7	−0.392	0.3477	0.500	0.152	−0.027
5	8	−0.313	0.3771	0.625	0.248	−0.123
6	10	−0.157	0.4378	0.750	**0.312**	−0.187
7	16	+0.313	0.6229	0.875	0.252	−0.127
8	42	+2.349	0.9906	1.000	0.009	0.116

3. Test Statistic

$$D^+ = \max_i D_i^+ = 0.3122 \quad \text{at } i = 6,\ x = 10$$

$$D^- = \max_i D_i^- = 0.2405 \quad \text{at } i = 1,\ x = 3$$

$$D = \max(D^+, D^-) = \mathbf{0.3122}$$

At $i = 6$: the empirical CDF has risen to $6/8 = 0.750$ but the fitted normal CDF is only $\Phi(-0.157) = 0.438$ — a gap of 0.312. This signals the distribution's bulk lies lower than a normal would predict, consistent with right skew.

4. Results and Decision

Statistic	Value	CV (5%)	Decision
D	0.3122	0.288	Reject H_0

Lilliefors CVs (Abdi et. al. 2007) for $n = 8$:

α	Lilliefors CV	Standard KS CV	Decision
10%	0.265		Reject
5%	0.288	0.454	Reject
1%	0.333		FTR

$p \approx 0.021$. Rejects at 5% but not 1%.

5. Interpretation

- $D = 0.312$ is driven by the gap at $x = 10$ (i=6), not by the extreme value $x = 42$. The right skew compresses the data's bulk into the lower range, causing the empirical CDF to outpace the fitted normal in the middle.

Practitioner Insight: The KS statistic D measures the largest vertical gap between F_n and the fitted CDF, so only the single worst deviation drives the test. This makes it less powerful than Cramér–von Mises and Anderson–Darling, which use all observations. When μ and σ are estimated, the Lilliefors correction is required: standard KS critical values are too large and tend to under-reject, whereas Lilliefors uses smaller thresholds and is more appropriate. In skewed data, the maximum gap often occurs in the middle rather than at extremes due to shifts in the fitted distribution.

Common Pitfalls

- **Using the K–S test for normality with estimated parameters.** Standard K–S critical values assume a fully specified distribution, so estimating μ and σ makes the test overly conservative; the Lilliefors correction must be used instead.
- **Assuming the test is among the most powerful goodness-of-fit procedures.** Because the statistic depends only on the maximum deviation, it discards information about the rest of the distribution and is typically less powerful than tests such as Shapiro–Wilk or Anderson–Darling.
- **Applying the test to discrete data or samples with many ties.** The K–S test assumes a continuous distribution, and ties distort the empirical distribution function, often making the test conservative and invalid under standard theory.
- **Interpreting the statistic as a measure of overall fit.** The statistic D captures only the largest pointwise discrepancy, so it may miss widespread moderate deviations or be driven by a single outlier; graphical diagnostics are essential.
- **Relying on asymptotic critical values in small samples.** The approximation $D_{\alpha,n} \approx c_\alpha/\sqrt{n}$ is inaccurate for small n, so exact critical values or software-based calculations should be used for reliable inference.

Jarque-Bera Test

Moment-Based Omnibus Test for Normality

NORM.6 | **Category:** Normality | **Data:** Continuous | **Null:** Chi-square

When to Use

- Test normality in large samples ($n > 50$, preferably $n > 100$).
- Quick computational check when only moments (skewness and kurtosis) are of concern.

Assumptions

- Observations are independent and identically distributed.
- Data have finite variance and fourth moment.
- Asymptotic approximation requires adequately large sample.

Test Statistic

$$JB = \frac{n}{6}\left[S^2 + \frac{(K-3)^2}{4}\right] \sim \chi^2_2 \text{ approximately under } H_0$$

where $S = m_3/m_2^{3/2}$ and $K = m_4/m_2^2$, with $m_k = (1/n)\sum(x_i - \bar{x})^k$.

Notation: S is sample skewness, K is sample kurtosis, normal has skewness = 0, kurtosis = 3.

Plain English: The Jarque–Bera test evaluates normality by measuring departures from the normal distribution's defining shape characteristics—symmetry and tail thickness (kurtosis). It combines information from sample skewness and kurtosis to assess whether the observed distribution is consistent with what would be expected under normality.

⚠ Practitioner Warning

The Jarque–Bera test relies on asymptotic approximations and assesses normality using only skewness and kurtosis, making it insensitive to other distributional features such as multimodality or localized deviations. As a result, it can exhibit poor power in small samples and may fail to detect important departures from normality. In modern practice, Shapiro–Wilk or Anderson–Darling tests are often preferred for diagnostic purposes, while Jarque–Bera remains useful as a simple, moment-based summary check.

Hypotheses

- H_0: Data distribution is Normal.
- H_1: Data distribution is **not** Normal.

Decision Rule: Reject H_0 if $JB > \chi^2_{2,\alpha}$ or p-value $< \alpha$.

$\chi^2_{2,0.01} = 9.210$, $\chi^2_{2,0.05} = 5.991$, $\chi^2_{2,0.10} = 4.605$.

Software References [Python, R, SAS]

```
scipy.stats.jarque_bera          tseries::jarque.bera.test          PROC AUTOREG (NORMAL)
```

Additional Notes

The Jarque–Bera test is a moment-based normality test that decomposes departures from normality into contributions from skewness and excess kurtosis, making its diagnostic interpretation straightforward. The reference χ^2_2 distribution is asymptotic and may be inaccurate in small samples, motivating finite-sample corrections (e.g., Urzúa's adjustment - given below) or simulation-based calibration. Because the test depends only on low-order moments, different non-normal distributions can yield similar Jarque–Bera statistics.

$$\mathrm{JB}_U = \frac{n}{6}\left(\frac{S^2}{\mathrm{Var}(S)} + \frac{(K-3)^2}{\mathrm{Var}(K)}\right), \qquad \mathrm{Var}(S) = \frac{6(n-2)}{(n+1)(n+3)}, \qquad \mathrm{Var}(K) = \frac{24n(n-2)(n-3)}{(n+1)^2(n+3)(n+5)}.$$

Miscellanea: The Jarque-Bera test exemplifies how statistical tests evolve through cumulative contributions. While Jarque and Bera (1980) are credited with the specific test statistic—derived as a Lagrange multiplier test in the regression context—the fundamental idea of combining sample skewness and kurtosis to test normality was explored by D'Agostino (1970-1973) and formalized graphically by Bowman and Shenton (1975) in their omnibus test contours.

Reference(s): Jarque & Bera (1980, 1987); Urzúa (1996); Bowman & Shenton (1975)

Jarque-Bera Test

NORM.6

Worked Example

Problem Statement Ten daily return percentages are tested for normality using the Jarque-Bera test, which combines sample skewness and excess kurtosis into a single moment-based statistic. Unlike the order-statistic tests (SW, SF, AD, CvM, KS), JB tests specific moment departures from normality rather than the overall shape of the empirical distribution. H_0: data are normally distributed ($S = 0$, $K = 3$); H_1: they are not.

1. Central Moments (divisor $n = 10$)

$\bar{x} = 13.0$; deviations $d_i = x_i - 13$:

x_i	d_i	d_i^2	d_i^3	d_i^4
3	−10	100	−1 000	10 000
5	−8	64	−512	4 096
6	−7	49	−343	2 401
7	−6	36	−216	1 296
9	−4	16	−64	256
10	−3	9	−27	81
12	−1	1	−1	1
15	+2	4	+8	16
21	+8	64	+512	4 096
42	+29	841	+24 389	707 281
$\sum$		1 184	22 746	729 524

$$m_2 = \frac{1184}{10} = 118.4, \quad m_3 = \frac{22746}{10} = 2274.6$$

$$m_4 = \frac{729524}{10} = 72952.4$$

2. Skewness and Kurtosis

$$S = \frac{m_3}{m_2^{3/2}} = 1.766, \qquad K = \frac{m_4}{m_2^2} = 5.204$$

$$K - 3 = 2.204 \quad \text{(excess kurtosis)}$$

3. JB Statistic Under H_0: $JB \xrightarrow{d} \chi^2(2)$.

$$JB = \frac{n}{6}S^2 + \frac{n}{6}\,\frac{(K-3)^2}{4} = 7.219$$

4. Results and Decision

Statistic	Value	CV (5%)	Decision
JB	7.219	5.991	Reject H_0

$p = 0.027$; CV at 1% = 9.210. Rejects at 5% but not 1%.

5. Contribution Breakdown

Component	Value	% of JB
Skewness	5.195	72%
Kurtosis	2.024	28%
Total JB	7.219	100%

Practitioner Insight: The Jarque–Bera test jointly assesses $S = 0$ (symmetry) and $K = 3$ (normal tails), with $JB \sim \chi^2(2)$ asymptotically from standardized skewness and kurtosis terms (the $1/4$ weight reflects the larger variance of K). Here skewness dominates, driven by a single extreme value that inflates both m_3 and m_4, producing positive skewness and excess kurtosis. While easy to compute and interpretable, JB is unreliable for small n (anti-conservative for $n < 50$); for larger samples it performs well, and for small samples a bootstrap p-value is preferable.

Common Pitfalls

- **Using the Jarque–Bera test with small samples.** The chi-square approximation is unreliable for small n, leading to low power and frequent non-rejection even for clearly non-normal data; alternative tests such as Shapiro–Wilk are preferred in small samples.
- **Assuming the test detects all forms of non-normality.** Because JB is based only on skewness and kurtosis, it can miss distributions that differ in shape but share the same third and fourth moments, so graphical diagnostics remain essential.
- **Over-interpreting the contribution of skewness and kurtosis components.** Although the statistic can be decomposed into skewness and kurtosis terms, these sample moments are correlated, so the decomposition provides only suggestive—not definitive—evidence about the source of deviation.
- **Ignoring sensitivity to outliers.** The reliance on higher-order moments makes the test highly sensitive to extreme values, where even a single outlier can dominate the statistic and drive rejection.

D'Agostino–Pearson K^2 Test

Omnibus Test Using Transformed Skewness and Kurtosis

NORM.7 | **Category:** Normality | **Data:** Continuous | **Null:** Chi-square

When to Use

- Test normality using refined skewness and kurtosis measures.
- Preferred over Jarque-Bera when better small-sample performance is needed.
- Suitable for moderate to large samples ($n \geq 20$).

Assumptions

- Observations are independent and identically distributed.
- Data have finite variance and fourth moment.
- Sample size should be at least $n \geq 20$ for adequate approximation.

Test Statistic

$$K^2 = Z_1^2 + Z_2^2$$

where transformations normalize skewness (b_1) and kurtosis (b_2) to approximate standard normal under H_0. Under H_0, $K^2 \sim \chi_2^2$ approximately. Specific transformations account for sample size and bias.

Notation: Z_1 is transformed skewness, Z_2 is transformed kurtosis, both approximately $N(0, 1)$

Plain English: D'Agostino–Pearson K^2 test improves upon Jarque-Bera by using more sophisticated transformations of sample skewness and kurtosis. These transformations better approximate normality under H_0, particularly for moderate sample sizes, yielding more accurate p-values and better power than the simpler Jarque-Bera approach.

⚠ Practitioner Warning

The D'Agostino–Pearson K^2 test relies on asymptotic normal-score transformations of skewness and kurtosis and may exhibit reduced power in small samples. Because it captures departures from normality only through low-order moments, it can miss localized or multimodal deviations. In modern practice, Shapiro–Wilk or Anderson–Darling tests are often preferred for general diagnostic use, while K^2 remains useful as an interpretable, moment-based summary test.

Hypotheses

- H_0: Data distribution is Normal.
- H_1: Data distribution is **not** Normal.

Decision Rule: Reject H_0 if $K^2 > \chi^2_{2,\alpha}$ or if p-value $< \alpha$.
$\chi^2_{2,0.01} = 9.210$; $\chi^2_{2,0.05} = 5.991$; $\chi^2_{2,0.10} = 4.605$.

Software References [Python, R, SAS]

`scipy.stats.normaltest`	`moments::agostino.test`	`Not Native (Manual)`

Additional Notes

The D'Agostino–Pearson K^2 test applies normal-score transformations to skewness and kurtosis, improving finite-sample behavior relative to the Jarque–Bera test while retaining a χ_2^2 reference distribution. Because $K^2 = Z_1^2 + Z_2^2$, practitioners often examine Z_1 and Z_2 separately to identify whether departures from normality are driven primarily by asymmetry or tail behavior. The χ_2^2 reference distribution is asymptotic; for very small samples, simulation-based calibration may provide more accurate size control.

Miscellanea: The 1990 paper that established the K^2 test as a standard tool was co-authored by Ralph D'Agostino, Albert Belanger, and Ralph D'Agostino Jr.—making it a rare example of father-son collaboration in statistical methodology. The senior D'Agostino had developed the component tests decades earlier (1970-1973), while the younger D'Agostino, trained in biostatistics at Harvard, contributed modern computational validation and power comparisons.

Note: Normalization transformations are quite messy and take up space. They are provided in the Technical Notes of this chapter.

Reference(s): D'Agostino (1970, 1971); D'Agostino & Pearson (1973); D'Agostino et al. (1990)

D'Agostino–Pearson K^2 Test

NORM.7

Worked Example

Problem Statement Ten patient biomarker values are tested for normality using the D'Agostino-Pearson K^2 test. Unlike the Jarque-Bera test which uses raw skewness and excess kurtosis directly, K^2 first transforms each moment statistic into an approximately standard-normal score, then combines them. This makes the $\chi^2(2)$ approximation substantially more accurate for small n. H_0: data are normally distributed; H_1: they are not.

$$6,\ 7,\ 8,\ 9,\ 10,\ 11,\ 12,\ 16,\ 23,\ 38$$

1. Moments (divisor n)

$\bar{x} = 14.0$; deviations $d_i = x_i - 14$; $m_2 = \sum d_i^2/n = 86.4$

$$m_3 = \frac{1}{n}\sum d_i^3 = 1326.6, \quad m_4 = \frac{1}{n}\sum d_i^4 = 34712.4$$

$$b_1 = \frac{m_3}{m_2^{3/2}} = 1.652, \quad b_2 = \frac{m_4}{m_2^2} = 4.650$$

2. Z_1: Skewness Transform (D'Agostino 1970)

$$Y = b_1\sqrt{\frac{(n+1)(n+3)}{6(n-2)}} = 2.851$$

$$\beta_2(b_1) = \frac{3(n^2+27n-70)(n+1)(n+3)}{(n-2)(n+5)(n+7)(n+9)} = 3.320$$

$$W^2 = -1 + \sqrt{2\beta_2 - 2} = 1.154, \quad \delta = \frac{1}{\sqrt{\ln\sqrt{W^2}}} = 3.734$$

$$\alpha = \sqrt{\frac{2}{W^2-1}} = 3.601, \ Z_1 = \delta \cdot \operatorname{arcsinh}\left(\frac{Y}{\alpha}\right) = 2.712$$

3. Z_2: Kurtosis Transform (Anscombe-Glynn 1983)

$$\mathrm{E}[b_2] = \frac{3(n-1)}{n+1} = 2.455, \quad \sqrt{\mathrm{Var}[b_2]} = 0.755$$

$$x = \frac{b_2 - \mathrm{E}[b_2]}{\sqrt{\mathrm{Var}[b_2]}} = 2.909$$

$$\beta_1(b_2) = 1.396, \quad A = 24.217$$

$$Z_2 = \sqrt{\frac{9A}{2}}\left[1 - \frac{2}{9A} - \left(\frac{1-2/A}{1+x\sqrt{2/(A-4)}}\right)^{1/3}\right] = 2.175$$

4. Results

$$K^2 = Z_1^2 + Z_2^2 = 7.354 + 4.731 = 12.084$$

Stat	Value	CV (1%)	Decision
K^2	12.084	9.210	Reject H_0

$p = 0.002$. Rejects at 1%.

Practitioner Insight: The K^2 test improves on JB by applying normalizing transforms to skewness and kurtosis, yielding more accurate inference in small samples. Here JB ($p = 0.058$) fails to reject, while K^2 ($p = 0.002$) strongly rejects, because JB's $\chi^2(2)$ approximation is overly conservative when n is small. The D'Agostino and Anscombe–Glynn transforms map b_1 and b_2 to approximately $N(0,1)$ variables, producing a better-calibrated test. As with JB, K^2 jointly detects skewness and kurtosis; here skewness dominates the departure from normality.

Common Pitfalls

- **Confusing D'Agostino's K^2 test with Jarque–Bera or chi-square goodness-of-fit tests.** D'Agostino's K^2, Jarque–Bera, and chi-square goodness-of-fit tests are often conflated. K^2 uses transformed skewness and kurtosis statistics for improved approximation, whereas chi-square tests rely on binning and address a different hypothesis.
- **Using the test with very small samples.** The normalizing transformations rely on asymptotic approximations. For small n, performance deteriorates and more reliable inference typically requires moderate to large samples.
- **Assuming the test detects all forms of non-normality.** The test depends only on skewness and kurtosis. Distributions with correct moments but incorrect shape may not be detected, so graphical diagnostics remain essential.
- **Ignoring which component drives rejection.** A significant K^2 statistic can arise from skewness, kurtosis, or both. Inspecting the individual transformed components helps interpret the source of deviation.
- **Attempting manual computation of the transformations.** The transformations involve multiple steps and approximations. In practice, statistical software should be used to ensure correct and stable implementation.

I. NORM — Raw Data & Charts

The raw data and corresponding Q–Q plots for all seven NORM tests are presented below. Each dataset is used in the worked example associated with its respective test.

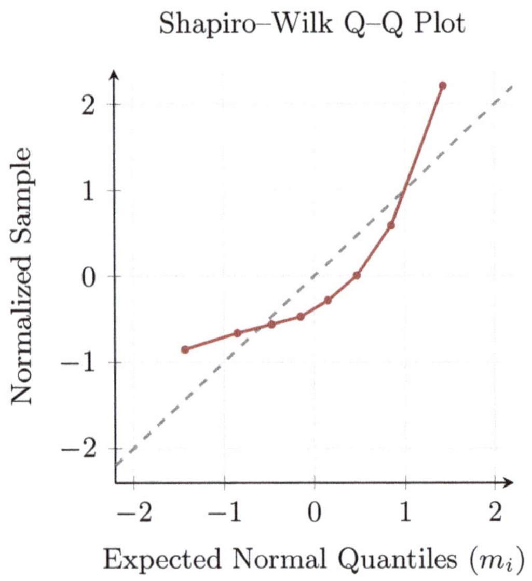

$X = \{3, 5, 6, 7, 9, 12, 18, 35\}$
$W = 0.7936$, Reject H_0

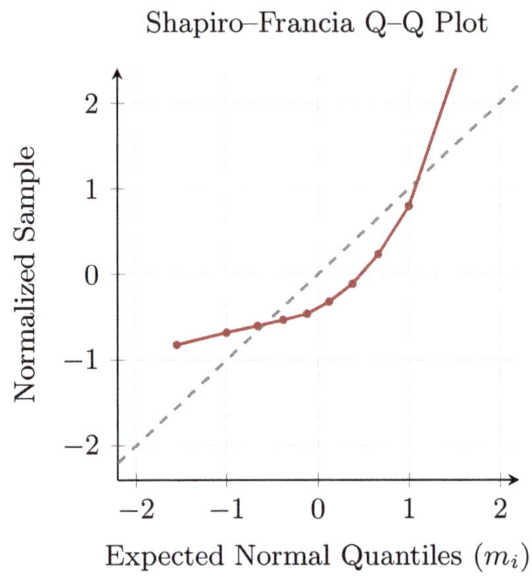

$X = \{5, 7, 8, 9, 10, 12, 15, 20, 28, 52\}$
$W' = 0.7476$, Reject H_0

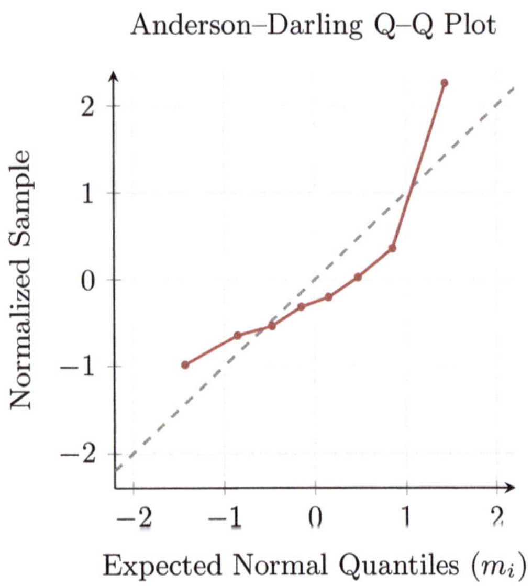

$X = \{10, 13, 14, 16, 17, 19, 22, 39\}$
$A^{*2} = 0.776$, Reject H_0

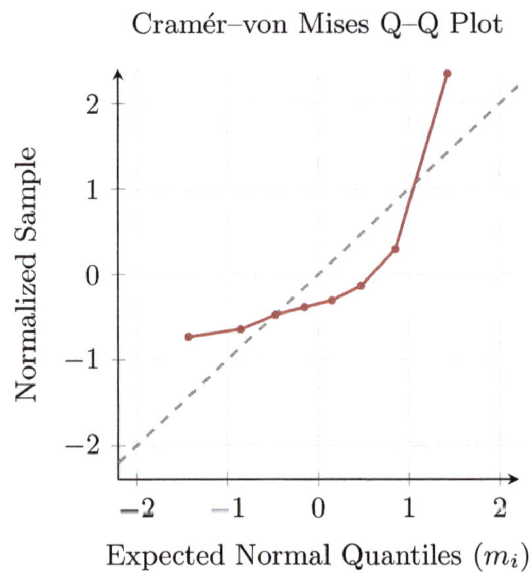

$X = \{4, 5, 7, 8, 9, 11, 16, 40\}$
$W^{*2} = 0.1945$, Reject H_0

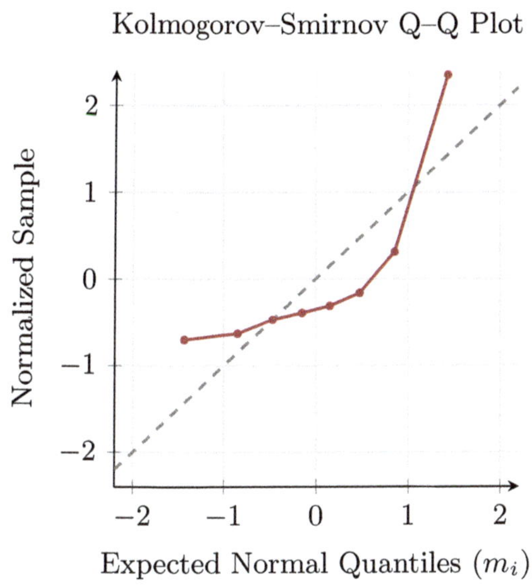

$X = \{3, 4, 6, 7, 8, 10, 16, 42\}$
$D = 0.3122$, Reject H_0

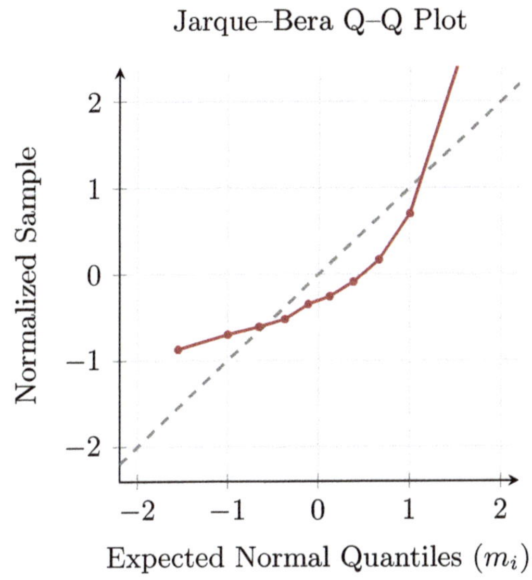

$X = \{3, 5, 6, 7, 9, 10, 12, 15, 21, 42\}$
$JB = 7.219$, Reject H_0

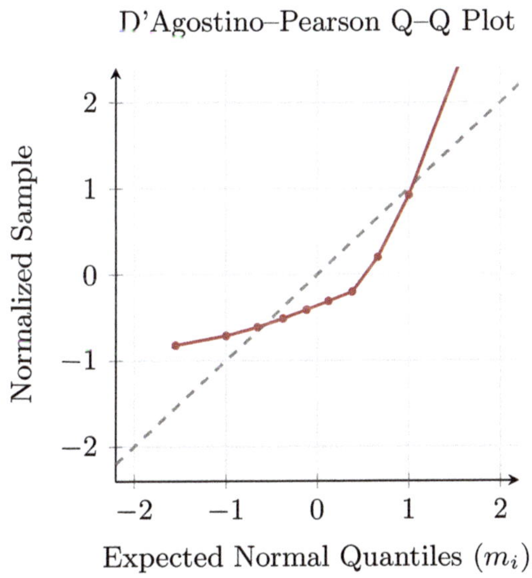

$X = \{6, 7, 8, 9, 10, 11, 12, 16, 23, 38\}$
$K^2 = 12.084$, Reject H_0

I. NORM — Technical Notes

D'Agostino's K^2 Test Formulation

The D'Agostino's K^2 omnibus test assesses normality by combining transformations of sample skewness (S) and kurtosis (K) into a single χ^2 statistic.

Step 1: Compute Sample Moments

Given a sample of size n, compute the k-th central moments $m_k = \frac{1}{n}\sum(x_i - \bar{x})^k$:

$$\text{Skewness: } b_1 = \frac{m_3}{m_2^{3/2}}, \quad \text{Kurtosis: } b_2 = \frac{m_4}{m_2^2}$$

Step 2: Transform Skewness to $Z_1 \sim N(0,1)$

$$Y = b_1\sqrt{\frac{(n+1)(n+3)}{6(n-2)}}$$

$$\beta_2(b_1) = \frac{3(n^2+27n-70)(n+1)(n+3)}{(n-2)(n+5)(n+7)(n+9)}$$

$$W^2 = -1 + \sqrt{2\beta_2(b_1) - 1}$$

$$\delta = \frac{1}{\sqrt{\ln W}}, \quad \alpha = \sqrt{\frac{2}{W^2 - 1}}$$

$$Z_1 = \delta \cdot \operatorname{arcsinh}\left(\frac{Y}{\alpha}\right), \text{ or}$$

$$Z_1 = \delta \cdot \ln\left(\frac{Y}{\alpha} + \sqrt{\left(\frac{Y}{\alpha}\right)^2 + 1}\right)$$

Step 3: Transform Kurtosis to $Z_2 \sim N(0,1)$

$$E[b_2] = \frac{3(n-1)}{n+1}$$

$$Var[b_2] = \frac{24n(n-2)(n-3)}{(n+1)^2(n+3)(n+5)}$$

$$X = \frac{b_2 - E[b_2]}{\sqrt{Var[b_2]}}$$

$$\sqrt{\beta_1(b_2)} = \frac{6(n^2-5n+2)}{(n+7)(n+9)}\sqrt{\frac{6(n+3)(n+5)}{n(n-2)(n-3)}}$$

$$A = 6 + \frac{8}{\sqrt{\beta_1(b_2)}}\left(\frac{2}{\sqrt{\beta_1(b_2)}} + \sqrt{1 + \frac{4}{\beta_1(b_2)}}\right)$$

$$Z_2 = \sqrt{\frac{9A}{2}}\left[1 - \frac{2}{9A} - \left(\frac{1 - 2/A}{1 + X\sqrt{2/(A-4)}}\right)^{1/3}\right]$$

Step 4: The Omnibus Statistic

$$K^2 = Z_1^2 + Z_2^2 \quad \sim \chi^2(2)$$

Part II

Model Fit & Adequacy Assessment

Evaluating agreement between observed data and assumed models

General Goodness-of-Fit

II. GFIT — A Conceptual Primer

1. Conceptual Foundation

What Is Goodness-of-Fit?

- Tests whether observed data are consistent with a **specified probability distribution** $F_0(x)$.
- Compares the **empirical distribution** of the sample against the **theoretical model**.
- Central question: *Is this distributional model plausible for these data?*
- Distinct from normality testing (NORM chapter): GFIT applies to *any* target distribution—exponential, Poisson, uniform, gamma, user-specified, etc.

The Null and Alternative Hypotheses

$$H_0 : F(x) = F_0(x)\ \forall x \qquad \text{(data follow } F_0\text{)}$$

- H_1 is typically **omnibus** (any departure): differences in location, scale, shape, tail weight, skewness, or multimodality.
- Unlike most tests in this compendium, GFIT tests are **non-directional**—they detect *that* the model fails, not *how* it fails.
- Tests such as Anderson–Darling remain omnibus but place greater emphasis on tail discrepancies than KS or CvM.

2. Distribution Comparison Framework

The Empirical CDF

The **empirical CDF** (ECDF) is the foundation of most GFIT tests:

$$F_n(x) = \frac{1}{n}\sum_{i=1}^{n} \mathbf{1}(x_i \le x).$$

By the Glivenko–Cantelli theorem, $\sup_x |F_n(x) - F(x)| \xrightarrow{\text{a.s.}} 0$, so F_n is a consistent estimator of the true CDF F. Different GFIT tests measure different aspects of the discrepancy between F_n and F_0.

The Fundamental Principle

Every GFIT procedure measures a discrepancy between the empirical distribution F_n and the hypothesized distribution F_0. The tests differ primarily in how that discrepancy is quantified: maximum distance (KS), integrated distance (CvM), tail-weighted distance (AD), spacing irregularity, or deviations in grouped frequencies (χ^2).

3. Hypothesis Structure

Simple vs. Composite Hypothesis

- **Simple H_0:** all parameters of F_0 are *fully specified* (e.g. $\text{Exp}(\lambda = 2)$). Standard critical values apply.
- **Composite H_0:** one or more parameters are *estimated from the data* (e.g. testing for exponentiality without specifying λ). Because parameters are estimated from the same data being tested, the discrepancy between F_n and F_0 is artificially reduced. Thus, ECDF is biased toward F_0, critical values shift substantially, and using simple-hypothesis tables is **anti-conservative**.
- Corrections exist for common distributions: Lilliefors (normal, exponential), modified AD and CvM tables. When in doubt, use a **parametric bootstrap** to obtain valid critical values.

4. Taxonomy of Methods

Classes of GFIT Tests

EDF-Based (Continuous Data)

- **Kolmogorov–Smirnov (KS):** maximum vertical distance between F_n and F_0.
- **Anderson–Darling (AD):** KS weighted by $[F_0(1 - F_0)]^{-1}$; emphasises tails.
- **Cramér–von Mises (CvM):** integrated squared distance; balanced sensitivity.

Frequency-Based (Discrete or Binned)

- **Pearson χ^2:** compares observed vs. expected cell counts; requires adequate expected counts (≥ 5).
- **Likelihood ratio (G-test):** $G^2 = 2\sum O_i \log(O_i/E_i)$; additive across subtables; often preferred for sparse tables because of its likelihood-based interpretation. Covered in CTAB.

Specialized

- **Probability integral transform (PIT):** if $X \sim F_0$, then $U = F_0(X) \sim \text{Uniform}(0, 1)$; reduces any GFIT problem to uniformity testing.
- **Spacing tests:** based on gaps $D_i = U_{(i)} - U_{(i-1)}$ between uniform order statistics; sensitive to clustering.

5. Core Mechanics

Core Test Statistics

KS: $D = \sup_x |F_n(x) - F_0(x)|$

CvM: $W^2 = \int (F_n - F_0)^2\, dF_0$

AD: $A^2 = \int \frac{(F_n - F_0)^2}{F_0(1 - F_0)}\, dF_0$

χ^2: $X^2 = \sum \frac{(O_i - E_i)^2}{E_i}$

G-test: $G^2 = 2\sum O_i \log(O_i/E_i)$ (covered in CTAB chapter)

A convenient computational form for EDF statistics uses order statistics $x_{(1)} \le \cdots \le x_{(n)}$:

$$D = \max_i \left\{ \left| \frac{i}{n} - F_0(x_{(i)}) \right|, \left| F_0(x_{(i)}) - \frac{i-1}{n} \right| \right\}.$$

6. Visual Diagnostics

The P-P Plot

The **probability–probability (P–P) plot** plots $F_0(x_{(i)})$ against i/n (or $(i-0.5)/n$). Deviations from the 45° line reveal where F_0 over- or under-estimates the observed frequencies.

- P–P plots are sensitive to **center** of the distribution.
- Q–Q plots are sensitive to **tails**.
- Use both: Q–Q for tail fit, P–P for bulk fit.

7. Performance & Power

Sensitivity and Power Comparison

Test	Sensitive to	Best used when
KS	Center	Quick general check
CvM	Overall	Balanced alternative
AD	Tails	Tail behavior critical
χ^2	Any (binned)	Discrete/categorical F_0

AD often exhibits higher power than KS and CvM, particularly when departures occur in the tails; it is the **default recommendation** when tail fit matters. χ^2 is the only option for purely discrete F_0.

Role of Sample Size

- **Small** n (< 30): low power; even substantial departures may go undetected. Visual diagnostics are essential.
- **Large** n (> 200): high power; trivial, practically irrelevant deviations are flagged as significant.
- **Key principle:** statistical significance ≠ practical significance. Ask whether the deviation matters for the downstream analysis, not merely whether it is detectable.

8. Chi-Square Specific Practicalities

Degrees of Freedom — χ^2 Test

- $df = k - 1$ (simple H_0; k cells).
- $df = k - 1 - m$ (composite H_0; m parameters estimated from grouped data).
- If parameters are estimated from *ungrouped* raw data (e.g. via MLE before binning), the χ^2 approximation is not exact; df is between $k - 1 - m$ and $k - 1$. Use simulation or bootstrap for reliable p-values.

Binning Strategy for χ^2

- **Expected count rule:** ensure all expected counts $E_i \ge 5$ (Cochran's guideline). Merge adjacent bins if needed; for small samples, $E_i \ge 3$ may be tolerated with caution.
- **Equal-probability bins (recommended):** choose bin boundaries so that $E_i = n/k$. This improves power and avoids sensitivity to arbitrary cutpoints.
- **Choosing number of bins k:** use a moderate number of bins (typically $k \approx 5$–15), balancing resolution with the $E_i \ge 5$ constraint.
- **Heuristic rules (use with caution):** $k \approx \sqrt{n}$ (square-root rule) or $k \propto n^{2/5}$ (asymptotic MISE-based rules). These ignore distributional shape and are not preferred in practice.
- **Key principle:** binning is part of the test design. Poor binning reduces power and can distort conclusions.

9. Decision & Action Layer

What To Do After Rejection

- **Diagnose first:** use Q–Q and P–P plots to identify whether the deviation is in the tails, center, skewness, or kurtosis.
- **Try alternative families:** e.g. if exponential fails, try gamma or Weibull; if normal fails, try t or log-normal.
- **Transform:** log, $\sqrt{\cdot}$, Box–Cox to induce approximate conformity with a simpler target family.
- **Use robust or nonparametric methods** downstream that do not require a specific distributional form.
- **Re-examine data quality:** outliers, measurement error, and data contamination are frequent causes of rejection.

10. Practitioner Guardrails & Closure

Common Pitfalls

- Using standard KS critical values when parameters are estimated (anti-conservative).
- Poor or arbitrary binning in χ^2 (results depend on bin boundaries).
- Over-relying on p-values; ignoring effect size.
- Treating rejection as *model failure*—a minor, inconsequential departure may cause rejection at large n.
- Ignoring visual diagnostics (Q–Q, P–P, histogram overlay).
- Using χ^2 for continuous data when EDF tests have higher power.

Bottom Line

> GFIT tests assess how well a model matches data—not whether the model is "true." All models are approximations; the goal is *adequacy for purpose*, not perfection. Combine formal tests with visual diagnostics, assess whether detected deviations are practically consequential, and correct for parameter estimation whenever the null distribution is not fully pre-specified.

Anderson–Darling Test

Goodness-of-Fit for Specified Distributions

GFIT.1 | **Category:** Goodness-of-Fit | **Data:** Continuous | **Null:** Varies by distribution

When to Use

- Test fit to any fully specified continuous distribution.
- Emphasize tail deviations from the null distribution.
- Alternative to Kolmogorov-Smirnov with higher power for tail departures.
- Suitable for sample sizes $n \geq 5$; works well across wide range.

Assumptions

- Observations are independent and identically distributed.
- Data are continuous (ties reduce test accuracy).
- Null distribution $F_0(x)$ is **fully specified** (all parameters known).
- For distributions with estimated parameters, use modified critical values.

Test Statistic

$$A^2 = -n - \sum_{i=1}^{n} \frac{2i-1}{n} \left[\ln F_0(X_{(i)}) + \ln(1 - F_0(X_{(n+1-i)})) \right]$$

Tail-weighting via $[F_0(1 - F_0)]^{-1}$ makes test sensitive to extreme values.

Notation: $F_n(x)$ is empirical CDF, $F_0(x)$ is specified CDF, weighting emphasizes tails

Plain English: The Anderson-Darling test measures weighted discrepancy between observed and hypothesized distributions, giving extra weight to observations in the tails. This version can test fit to any continuous distribution—not just normality. The tail emphasis makes it particularly valuable for distributions where extreme values matter.

⚠ Practitioner Warning

When distribution parameters are estimated from data, standard critical values are invalid. Use distribution-specific adjustments or modified critical values. Each distribution family (exponential, Weibull, gamma, etc.) has unique adjustments derived through simulation. While the Anderson–Darling test is powerful, it is highly sensitive to large sample sizes: even minor, practically insignificant deviations can yield p-values below conventional thresholds. For this reason, it is best used in conjunction with Q–Q plots rather than relying on the p-value alone.

Hypotheses

- H_0: Data follow the specified distribution $F_0(x)$.
- H_1: Data do **not** follow the specified distribution $F_0(x)$.

Decision Rule: Reject H_0 if $A^2 > A^2_{\alpha,n}$ (critical value specific to F_0) or p-value $< \alpha$.

Software References [Python, R, SAS]

Python	R	SAS
`scipy.stats.anderson`	`goftest::ad.test`	`PROC UNIVARIATE / NORMAL`

Additional Notes

Critical values are distribution-specific because the null distribution of A^2 changes with F_0. For common distributions (exponential, Weibull, logistic, extreme value), critical values have been extensively tabulated. The test extends naturally to censored data, making it valuable in survival analysis and reliability engineering. Finite sample corrections for a few different distributions are given below:

$$A^{2*} = A^2(1 + 0.75/n + 2.25/n^2)_{\text{Normal}} \; ; \; A^2(1 + 0.6/n)_{\text{Exponential}} \; ; \; A^2(1 + 0.2/\sqrt{n})_{\text{Weibull}} \; ; \; A^2(1 + 0.25/n)_{\text{Logistic}}.$$

Miscellanea: Although introduced in 1952, the Anderson–Darling test was not widely used for decades because its critical values depend on the specific null distribution (normal, exponential, etc.), unlike the Kolmogorov–Smirnov test with its universal tables. The test became practically accessible after Stephens (1974) tabulated and standardized these distribution-specific critical values, which is why many practitioners view the 1970s as its effective adoption period. The method was later extended to the k-sample setting (Scholz and Stephens, 1987), allowing comparison of multiple samples from an unknown distribution. Its creators, T.W. Anderson and D.A. Darling, were leading figures in 20th-century statistics, with Anderson's work on multivariate analysis remaining foundational to this day.

Reference(s): Anderson & Darling (1952); Stephens (1974, 1986)

Anderson–Darling Test
GFIT.1

Worked Example

Problem Statement A service system is claimed to have exponentially distributed service times with *known* rate $\lambda = 0.5$ (mean = 2 min). Eight observed service times (minutes) are recorded. The Anderson-Darling test (1952) assesses fit to the *fully specified* Exp(0.5) distribution: all parameters are given, none estimated from data.
H_0: $F(x) = 1 - e^{-0.5x}$; H_1: F is not Exp(0.5).

i	$x_{(i)}$	$F(x_{(i)})$	$1 - F(x_{(n+1-i)})$	$2i - 1$	$\log(F(x_{(i)}))$	$\log(1 - F(x_{(n+1-i)}))$
1	0.3	0.1393	0.0004	1	−1.971	−7.80
2	0.8	0.3297	0.0101	3	−1.110	−4.60
3	1.4	0.5034	0.0550	5	−0.686	−2.90
4	2.2	0.6671	0.1738	7	−0.405	−1.75
5	3.5	0.8262	0.3329	9	−0.191	−1.10
6	5.8	0.9450	0.4966	11	−0.057	−0.70
7	9.2	0.9899	0.6703	13	−0.010	−0.40
8	15.6	0.9996	0.8607	15	0.000	−0.15

1. AD Formula

$$A^2 = -n - \frac{1}{n}\sum_{i=1}^{n}(2i-1)\left[\ln F(x_{(i)}) + \ln(1-F(x_{(n+1-i)}))\right]$$

For Exp(λ): $F(x) = 1 - e^{-\lambda x}$, so $1 - F(x) = e^{-\lambda x}$ and

$$\ln(1 - F(x)) = -\lambda x$$

The second log-term is therefore exact: $-0.5\,x_{(n+1-i)}$.

2. AD Statistic and Decision

$$A^2 = -8 - (-10.930) = 2.930$$

Statistic	Value	CV (5%)	Decision
A^2	2.930	2.492	Reject H_0

$p \approx 0.031$; CV at 1%= 3.880 (FTR at 1%).
CVs: D'Agostino & Stephens (1986, pg. 105), fully specified.

Practitioner Insight: The general-fit Anderson–Darling test differs from the normality version in three key ways. First, all parameters must be *fully specified* in advance; estimating them (e.g. $\hat{\lambda} = 1/\bar{x}$) leads to a modified statistic with smaller values and different critical thresholds. Second, by the probability integral transform, $U_i = F(x_{(i)}) \sim \text{Uniform}(0, 1)$ under H_0, so the test reduces to assessing non-uniformity of the transformed data, regardless of the hypothesized continuous distribution. Third, critical values are much larger in the fully specified case (e.g. 2.492 vs. 1.341 at 5%) because the model has no flexibility to adapt to the data. In this example, the large sample mean and extreme observation $x_{(8)} = 15.6$ (with $F(x) = 0.9996$) drive the result, illustrating the test's strong tail sensitivity relative to the Kolmogorov–Smirnov test.

Common Pitfalls

- **Wrong critical values.** Anderson–Darling critical values are distribution-specific; values for normal, exponential, or Weibull models are not interchangeable.
- **Unadjusted parameter estimation.** When parameters are estimated from data, standard critical values are invalid; use distribution-specific adjusted versions or software corrections.
- **Ignoring tail sensitivity.** Rejection typically reflects tail mismatches rather than central fit. Always inspect Q–Q plots with emphasis on extremes.
- **Discrete data misuse.** The test assumes continuity; applying it to discrete data leads to distorted inference. Use discrete goodness-of-fit methods instead.
- **Overgeneralized conclusions.** The test evaluates fit to a single specified distribution; non-rejection does not imply optimality among alternatives.

Cramér-von Mises Test

Balanced EDF Test for Any Specified Distribution

GFIT.2 | **Category:** Goodness-of-Fit | **Data:** Continuous | **Null:** Varies by distribution

When to Use

- Test fit to any fully specified continuous distribution with balanced sensitivity.
- Alternative to Anderson-Darling when uniform weighting is desired.
- Suitable for sample sizes $n \geq 5$; reliable across wide range.

Assumptions

- Observations are independent and identically distributed.
- Data are continuous (ties reduce validity).
- Null distribution $F_0(x)$ is **fully specified** (all parameters known).
- For distributions with estimated parameters, use modified critical values.

Test Statistic

$$W^2 = n \int_{-\infty}^{\infty} [F_n(x) - F_0(x)]^2 \, dF_0(x) = \frac{1}{12n} + \sum_{i=1}^{n} \left[\frac{2i-1}{2n} - F_0(x_{(i)}) \right]^2$$

Notation: $F_n(x)$ is empirical CDF, $F_0(x)$ is specified CDF, equal weights across distribution

Plain English: The Cramér-von Mises test measures the integrated squared distance between observed and hypothesized distributions, treating all regions equally. Unlike Anderson-Darling which emphasizes tails, this test is balanced—equally sensitive to departures anywhere in the distribution. It extends naturally to any continuous distribution family.

⚠ Practitioner Warning

Critical values depend on the distribution being tested. The same W^2 value has different significance for exponential vs. normal vs. Weibull distributions. Always use appropriate critical values for the hypothesized F_0. When parameters are estimated, further adjustments are required. Finite sample corrections (Stephens (1974)) are needed.

Hypotheses

- H_0: Data follow the specified distribution $F_0(x)$.
- H_1: Data do **not** follow the specified distribution $F_0(x)$.

Decision Rule: Reject H_0 if $W^2 > W^2_{\alpha,n}$ (critical value for distribution F_0) or p-value $< \alpha$. Consult distribution-specific tables or use software for accurate critical values.

Software References [Python, R, SAS]

Python	R	SAS
`scipy.stats.cramervonmises`	`goftest::cvm.test`	`PROC UNIVARIATE / NORMAL`

Additional Notes

The Cramér–von Mises test uses an L^2 (squared) distance, measuring integrated deviation between F_n and F with equal weighting across the distribution. Under a fully specified continuous F, the probability integral transform (PIT) yields a distribution-free test based on Uniform$(0,1)$; however, when parameters are estimated (e.g. in normality testing), modified critical values are required. Stephens (1974) provided the practical critical values and approximations that enabled routine use of the test in such settings. For a fully specified F_0, asymptotic critical values are 0.347 ($\alpha = 0.10$), 0.461 ($\alpha = 0.05$), 0.581 ($\alpha = 0.025$), and 0.743 ($\alpha = 0.01$); for finite samples, use $W^{*2} = (W^2 - 0.4/n + 0.6/n^2)(1.0 + 1.0/n)$ with these thresholds.

The Cramér–von Mises test is closely related to Watson's U^2 test for circular data, which modifies the statistic to remove dependence on the choice of origin, making it invariant to rotation—a key requirement for directional and periodic data.

Miscellanea: The Cramér–von Mises test has independent origins: Harald Cramér (1928) and Richard von Mises (early 1930s) arrived at essentially the same integrated squared-distance idea from different directions. Its practical development, however, is largely due to Nikolai Smirnov, who derived the sampling distributions and produced usable tables, establishing the test alongside Kolmogorov–Smirnov methods. In recognition of this, some refer to the broader class as the Cramér–von Mises–Smirnov family.

Reference(s): Cramér (1928); von Mises (1931); Anderson (1962); Stephens (1974)

Worked Example

Problem Statement A random-number generator is claimed to produce values uniformly distributed on (0, 10). Six output values are recorded. H_0: $F(x) = x/10$ on (0, 10); H_1: F is not Uniform(0, 10).

i	$x_{(i)}$	$F(x_{(i)})$	$(2i-1)/12$	d_i	d_i^2
1	0.5	0.05	0.083	0.033	0.001
2	1.0	0.10	0.250	0.150	0.023
3	1.5	0.15	0.417	0.267	0.071
4	2.0	0.20	0.583	0.383	0.147
5	2.5	0.25	0.750	0.500	0.250
6	9.5	0.95	0.917	−0.033	0.001

1. CvM Statistic

With $F(x) = x/10$ and $n = 6$:

$$\frac{2i-1}{2n} = \frac{2i-1}{12}, \quad F(x_{(i)}) = \frac{x_{(i)}}{10}, \quad d_i = \frac{2i-1}{2n} - F(x_{(i)})$$

$$W^2 = \frac{1}{12n} + \sum_{i=1}^{n}\left[\frac{2i-1}{2n} - F(x_{(i)})\right]^2 = 0.507$$

2. Decision

Statistic	**Value**	**CV (5%)**	**Decision**
W^2	0.507	0.461	Reject H_0
W^{*2}	0.533	0.461	Reject H_0

$p \approx 0.036$; CV at 2.5%= 0.581 (FTR at 2.5%).
Critical values: Stephens (1974), fully specified.

Practitioner Insight: The CvM statistic $W^2 = 1/(12n) + n^{-1}\sum((2i-1)/(2n) - U_{(i)})^2$, with $U_{(i)} = F(x_{(i)})$, measures the average squared deviation of probability-transformed data from their expected uniform positions. In this example, most U_i cluster in the lower tail, so the cumulative deviations dominate W^2, while the single large observation contributes little. Unlike the Anderson–Darling test, CvM weights all deviations equally, yet both reach the same conclusion here because the lack of fit is spread across the bulk rather than concentrated in the tails. As with AD, fully specified models yield larger critical values since the model cannot adapt to the data.

Common Pitfalls

- **Wrong critical values.** Critical values depend on the distribution being tested; the null distribution of W^2 changes with F_0. Always use values calibrated for the specific family.
- **No adjustment for estimated parameters.** Estimating parameters reduces W^2, making standard critical values invalid. Use parameter-adjusted critical values (family-specific) when fitting θ from data.
- **Assuming universal power.** Because W^2 weights all deviations equally, it is balanced but not optimal. It may miss tail issues (AD) or moment deviations (SW). It is general-purpose, not most powerful.
- **Using with tied or discrete data.** The test assumes continuity; ties make it conservative. For inherently discrete data, use chi-square or discrete-specific GOF tests.
- **Interpreting non-rejection as confirmation.** Failing to reject only means compatibility with F_0. Several distributions may fit; complement with AIC/BIC and graphical diagnostics.

Asymptotic Critical Values

For large n and fully specified F_0: The limiting distribution can be expressed as

$$W^2 \xrightarrow{d} \int_0^1 [W(t)]^2\,dt \stackrel{d}{=} \sum_{k=1}^{\infty} \frac{Z_k^2}{k^2\pi^2}, \quad \text{where } W(t) \text{ is a Brownian bridge on } [0,1] \text{ and } Z_k \sim N(0,1) \text{ are independent.}$$

A Brownian bridge is a stochastic process obtained by conditioning Brownian motion to start and end at zero over a fixed interval (e.g: [0,1]).

Kolmogorov-Smirnov Test (Two-Sample)

Distribution-Free Test for Comparing Two Samples

GFIT.3 | **Category:** Goodness-of-Fit | **Data:** Continuous | **Null:** Kolmogorov Distribution

When to Use

- Compare entire distributions of two populations.
- Detect location, spread, or shape differences.
- Distribution-free alternative to parametric tests.
- Continuous data from two independent samples.
- Compare distributions without assuming normality or any other specific forms.

Assumptions

- Both samples are independent and randomly selected.
- Observations within each sample are independent.
- Data are continuous (test is conservative with ties).
- No assumption about the underlying distribution.
- Suitable for any sample sizes; most powerful for moderate to large samples.

Test Statistic

$$D_{m,n} = \sup_x |F_m(x) - G_n(x)| = \max\left\{\max_i \left|\frac{i}{m} - G_n(x_i)\right|, \max_j \left|F_m(y_j) - \frac{j}{n}\right|\right\}$$

Notation: $F_m(x)$ and $G_n(x)$ are empirical CDFs from samples of sizes m and n.

Plain English: The two-sample Kolmogorov-Smirnov test compares two distributions by measuring the largest gap between their empirical cumulative distribution functions. If the samples come from the same distribution, their CDFs should be similar everywhere. A large gap anywhere indicates differing distributions, regardless of location, spread, or shape.

⚠ Practitioner Warning

K-S focuses on maximum deviation, which often occurs in the center of distributions. It can miss differences concentrated in the tails. For detecting tail differences, consider Anderson-Darling two-sample test or permutation tests. K-S is most powerful for location shifts and scale changes.

Hypotheses

- H_0: $F(x) = G(x)$ Both samples come from same distn.
- H_1: $F(x) \neq G(x)$ Samples come from different distn's.

Decision Rule: Reject H_0 if $D_{m,n} > D_{\alpha,m,n}$ or p-value $< \alpha$. For large samples, $D_\alpha \approx c_\alpha\sqrt{\frac{m+n}{mn}}$ where $c_{0.05} \approx 1.36$, $c_{0.01} \approx 1.63$.

Software References [Python, R, SAS]

Python	R	SAS
`scipy.stats.ks_2samp`	`ks.test`	`PROC NPAR1WAY / EDF`

Additional Notes

The test is distribution-free under the null hypothesis, with critical values depending only on the sample sizes m and n, not on the underlying distributions F or G. It is symmetric with respect to the two samples and invariant under strictly monotone transformations. The test is sensitive to any distributional difference, including shifts in location, scale, or shape, but summarizes all departures through a single global discrepancy, which can reduce power relative to parametric tests when their assumptions are satisfied.

The two-sample Kolmogorov–Smirnov test is distribution-free under H_0, meaning its null distribution does not depend on the underlying continuous CDF; however, this property is lost under H_1, making power analysis difficult. The statistic is invariant under all strictly increasing transformations, a rare robustness property that ensures identical results under re-scaling (e.g. log or Box–Cox transforms). Despite its generality, the test is most sensitive near the center of the distribution and relatively insensitive in the tails, motivating alternatives such as the Anderson–Darling and Cramér–von Mises tests.

Miscellanea: Historically, the method originated with Kolmogorov (1933) and was extended by Smirnov (1939), with early adoption limited by computational difficulty before modern methods made exact and approximate calculations routine.

Reference(s): Smirnov (1939); Hodges (1957).

Worked Example

Problem Statement Recovery times (days) are recorded for patients treated at two hospitals: Hospital A ($n_1 = 5$) and Hospital B ($n_2 = 6$). The two-sample KS test (Smirnov 1948) tests whether the two samples are drawn from the same continuous distribution, with no distributional form assumed. H_0: $F_1(x) = F_2(x)$ for all x; H_1: $F_1(x) \neq F_2(x)$ for some x.

A: 1, 2, 3, 4, 8 **B:** 5, 6, 7, 9, 10, 11

1. Empirical CDFs at Each Combined Value

Pool and sort all $N = 11$ observations. At each value x:

$$\hat{F}_1(x) = \frac{\#\{\text{A obs} \le x\}}{5}, \quad \hat{F}_2(x) = \frac{\#\{\text{B obs} \le x\}}{6}$$

x	Src	$\hat{F}_1(x)$	$\hat{F}_2(x)$	$\hat{F}_1 - \hat{F}_2$	$\lvert\hat{F}_1 - \hat{F}_2\rvert$
1	A	1/5	0/6	+1/5	1/5
2	A	2/5	0/6	+2/5	2/5
3	A	3/5	0/6	+3/5	3/5
4	A	4/5	0/6	+4/5	**4/5**
5	B	4/5	1/6	+19/30	19/30
6	B	4/5	2/6	+7/15	7/15
7	B	4/5	3/6	+3/10	3/10
8	A	5/5	3/6	+1/2	1/2
9	B	5/5	4/6	+1/3	1/3
10	B	5/5	5/6	+1/6	1/6
11	B	5/5	6/6	0	0

2. Test Statistic

$$D = \sup_x |\hat{F}_1(x) - \hat{F}_2(x)| = \frac{4}{5} = 0.800$$

achieved at $x = 4$, where all four smallest combined observations belong to Hospital A while Hospital B has contributed nothing ($\hat{F}_2(4) = 0$).

3. Exact Critical Value and Decision

Under H_0, all $\binom{11}{5} = 462$ ways of assigning the five A-labels to the eleven combined positions are equally likely. The exact tail probability is:

$$p = P(D \ge \tfrac{4}{5} \mid H_0) = \frac{22}{462} = \frac{1}{21} \approx 0.048$$

Statistic	Value	CV (5%)	Decision
D	0.800	0.800	Reject H_0

$p = 1/21 \approx 0.048$; D equals the exact 5% CV, so the test just rejects at the 5% level. Asymptotic CV: $1.358\sqrt{11/30} = 0.822$ (conservative for small n).

Practitioner Insight: The two-sample KS statistic $D = \sup_x |\hat{F}_1(x) - \hat{F}_2(x)|$ is the maximum vertical gap between the empirical CDFs, evaluated at the observed data points. Here, the gap of 4/5 at $x = 4$ arises because all early observations from sample A precede those of sample B, after which the gap narrows as B accumulates. The test is distribution-free under H_0, with the null distribution determined by permutations of sample labels, yielding the exact p-value. Unlike the one-sample case, the two-sample distribution depends on (n_1, n_2), and asymptotic critical values can be conservative in small samples. Compared with AD and CvM, KS measures maximum pointwise deviation and is less sensitive to tail differences.

Common Pitfalls

- **Misusing K–S for location shifts.** When normality is plausible and only a mean shift is expected, a t-test is more powerful. Use K–S when shape or scale differences are possible or robustness is needed.
- **Weak tail sensitivity.** K–S often detects central rather than tail deviations. For tail-focused questions, prefer Anderson–Darling or tail-oriented permutation tests.
- **Misinterpreting rejection.** K–S is omnibus: rejection indicates some distributional difference, not specifically a mean shift. Use diagnostic plots to identify what differs.
- **Ignoring ties.** Ties make the test conservative. With rounded or discrete data, consider exact or permutation-based alternatives.
- **Small-sample approximations.** Asymptotic p-values are unreliable for small samples. Use exact or simulation-based methods when $\min(m, n)$ is small.

Lilliefors Test for Exponential Distribution

Modified K-S Test for Exponentiality with Estimated Rate

GFIT.4 | **Category:** Goodness-of-Fit | **Data:** Continuous | **Null:** Tables

When to Use

- Test whether data follow an exponential distribution with unknown rate parameter.
- Correct alternative to standard K-S when rate λ is estimated from data.
- Useful in reliability testing for constant failure rate assumption.

Assumptions

- Observations are independent and identically distributed.
- Data are continuous and positive (times, lifetimes, durations).
- Rate parameter λ is unknown and must be estimated.
- Suitable for sample sizes $n \geq 4$; better performance for $n \geq 20$.

Test Statistic

$$D = \sup_x |F_n(x) - \hat{F}_0(x)| = \max\left\{\max_i \left|\frac{i}{n} - (1 - e^{-x_{(i)}/\bar{x}})\right|, \max_i \left|(1 - e^{-x_{(i)}/\bar{x}}) - \frac{i-1}{n}\right|\right\}$$

Notation: $F_n(x)$ is empirical CDF, $\hat{F}_0(x)$ is exponential CDF with estimated rate $\hat{\lambda} = 1/\bar{x}$.

Plain English: The Lilliefors test for exponentiality checks whether data follow an exponential distribution when you don't know the rate parameter in advance. By estimating λ from the sample mean, the distribution is forced to fit the data better than if λ were fixed. This requires adjusted critical values that account for the parameter estimation, just like Lilliefors's correction for testing normality.

⚠ Practitioner Warning

The test has limited power for $n < 20$, often failing to detect departures from exponentiality. Power improves substantially with $n \geq 50$. For very small samples, consider graphical methods (exponential Q-Q plots) or combine with domain knowledge about the data-generating process. Critical values differ from standard K-S and are generally obtained via simulation for estimated λ.

Hypotheses

- H_0: Data $\sim$ Exponential(λ), λ, unspecified
- H_1: Data do **not** follow an exponential distribution.

Decision Rule: Reject H_0 if $D > D_{\alpha,n}$ (Lilliefors critical value for exponential) or if p-value $< \alpha$.

Software References [Python, R, SAS]

```
statsmodels.stats.
diagnostics.lilliefors
```

```
nortest::lillie.test
```

```
PROC UNIVARIATE / NORMAL
```

Additional Notes

The test uses $\hat{\lambda} = 1/\bar{x}$ (method of moments estimator equals MLE for exponential). Critical values are about 15–25% smaller than standard K-S values for the same sample size. The exponential distribution is memoryless: $P(X > s + t|X > s) = P(X > t)$, making it appropriate for constant hazard rate processes.

Miscellanea: Hubert Lilliefors introduced the exponential version of his test in a brief 1969 note (3-page JASA article), extending his earlier normality work to reliability and life-testing where exponential waiting times are fundamental. Motivated by the practical flaw that the Kolmogorov–Smirnov test assumes known parameters, he showed how to adjust it when the rate λ is estimated, making the method usable in real data. Stephens (1974) later provided a simple normalization $T^* = (D - 0.2/n)(\sqrt{n} + 0.26 + 0.5/\sqrt{n})$, yielding approximate n-invariant critical values (about 1.094 at 5% and 1.298 at 1%), which made the test far easier to apply in practice. The method has since been used beyond engineering, including in seismology to assess whether earthquake inter-event times follow the exponential pattern implied by a Poisson process.

Reference(s): Lilliefors (1969); Dallal & Wilkinson (1986)

Worked Example

Problem Statement Eight machine inter-failure times (hours) are recorded. The Lilliefors test (1969) assesses whether the data follow *some* exponential distribution, estimating the rate λ by maximum likelihood ($\hat{\lambda} = 1/\bar{x}$) and then applying a KS-type statistic with critical values adjusted for the fact that λ is estimated rather than specified in advance. H_0: $F(x) = 1 - e^{-\lambda x}$ for some $\lambda > 0$; H_1: F is not exponential.

1. MLE and Fitted CDF

$$\bar{x} = \frac{24.0}{8} = 3.0, \quad \hat{\lambda} = \frac{1}{\bar{x}} = \frac{1}{3}, \quad \hat{F}(x) = 1 - e^{-x/3}$$

2. KS Differences at Each Order Statistic

i	$x_{(i)}$	$\hat{F}(x_{(i)})$	$i/8$	$(i-1)/8$	D_i^+	D_i^-
1	0.1	0.0328	0.125	0.000	0.092	0.033
2	0.2	0.0645	0.250	0.125	0.186	−0.061
3	0.4	0.1248	0.375	0.250	0.250	−0.125
4	0.5	0.1535	0.500	0.375	0.347	−0.222
5	0.8	0.2341	0.625	0.500	**0.391**	−0.266
6	1.5	0.3935	0.750	0.625	0.357	−0.232
7	2.5	0.5654	0.875	0.750	0.310	−0.185
8	18.0	0.9975	1.000	0.875	0.003	**0.123**

where $D_i^+ = i/n - \hat{F}(x_{(i)})$ and $D_i^- = \hat{F}(x_{(i)}) - (i-1)/n$.

3. Test Statistic and Stephens Modification

$$D^+ = \max_i D_i^+ = 0.391 \quad (\text{at } i = 5)$$

$$D^- = \max_i D_i^- = 0.123 \quad (\text{at } i = 8)$$

$$D = \max(D^+, D^-) = 0.391$$

Apply the Stephens (1974) modification to stabilize the critical values across sample sizes:

$$T^* = (D - \frac{0.2}{n})\left(\sqrt{n} + 0.26 + \frac{0.5}{\sqrt{n}}\right) = 1.195$$

4. Decision

Statistic	Value	CV (5%)	Decision
T^*	1.195	1.094	Reject H_0

$p < 0.05$; CV at 1%= 1.298 (FTR at 1%).
Critical values: Stephens (1974), exponential with estimated λ.

Practitioner Insight: Three features define the Lilliefors exponential test. First, the rate λ is *estimated* ($\hat{\lambda} = 1/\bar{x}$), so the fitted CDF adapts to the data and requires smaller critical values; the Stephens adjustment $T^* = (D - 0.2/n)(\sqrt{n} + 0.26 + 0.5/\sqrt{n})$ yields n-invariant asymptotic cutoffs (5%: 1.094; 1%: 1.298). Second, the largest deviation $D^+ = 0.391$ at $x_{(5)} = 0.8$ reflects that the sample accumulates mass much faster than the fitted exponential ($\hat{F}_n = 0.625$ vs. $\hat{F} = 0.234$), driven by the extreme value 18.0 inflating $\bar{x}$ and producing a distribution with heavier tails than exponential. Third, unlike the two-sample KS test, Lilliefors is a one-sample test against a parametric family (not another empirical CDF) and, while simpler than AD, is generally less sensitive to tail deviations since AD up-weights extremes.

Common Pitfalls

- **Wrong critical values with estimated λ.** Estimating $\lambda = 1/\bar{x}$ invalidates standard K–S critical values by shrinking D. Use Lilliefors critical values (or Stephens' adjustment) or the test becomes severely underpowered.
- **Invalid support.** The exponential distribution is defined only for $x > 0$. Zeros or negatives rule out exponentiality; consider shifted models or alternative families.
- **Low power for small n.** With $n < 20$, K–S-based tests have limited power. Always inspect exponential Q–Q plots and trust clear graphical departures over non-rejection.
- **Shifted or truncated data.** Standard tests assume support starting at zero. Known shifts must be removed; truncated samples require specialized methods.
- **Overlooking the hazard-rate assumption.** Exponentiality implies a constant hazard. Even if not rejected, assess whether this assumption is substantively reasonable for the application.

Kuiper's Test

Rotation-Invariant K-S Test for Circular and Linear Data

GFIT.5 | **Category:** Goodness-of-Fit | **Data:** Continuous | **Null:** Varies by distribution

When to Use

- Test fit to distributions on circular data.
- Alternative to Kolmogorov-Smirnov when rotation invariance is needed. Particularly effective for periodic data.
- Useful for linear data when sensitivity near distribution boundaries matters.

Assumptions

- Observations are independent and identically distributed.
- Data are continuous (or circular/periodic).
- For circular data, distribution is defined on a circle (angles, times on a cycle). Suitable for sample sizes $n \geq 5$.
- For linear data, null distribution $F_0(x)$ is fully specified.

Test Statistic

$$V = D^+ + D^- = \max_i \left(\frac{i}{n} - F_0(x_{(i)}) \right) + \max_i \left(F_0(x_{(i)}) - \frac{i-1}{n} \right)$$

Notation: D^+ is maximum upward deviation, D^- is maximum downward deviation of empirical from theoretical CDF.

Plain English: Kuiper's test is a modification of Kolmogorov-Smirnov that sums the maximum upward (D^+) and downward (D^-) deviations instead of taking their maximum. This makes the test rotation-invariant for circular data and equally sensitive to departures anywhere in the distribution, including at the boundaries—addressing a key weakness of K-S.

⚠ Practitioner Warning

Kuiper's test is less commonly implemented than K-S or Anderson-Darling. Specialized packages may be needed, particularly for circular data applications. For linear data, K-S or Cramér-von Mises may be more readily available with similar performance.

Hypotheses

- H_0: Data follow the specified distribution F_0 (circular data)
- H_1: Data do **not** follow the specified distribution F_0.

Decision Rule: Reject H_0 if $V > V_{\alpha,n}$ or if p-value $< \alpha$. For large n, $V_{0.05,n} \approx 1.747/\sqrt{n}$; $V_{0.01,n} \approx 2.001/\sqrt{n}$.

Software References [Python, R, SAS]

`astropy.stats.kuiper`	`circular::kuiper.test`	`Not Native (Manual)`

Additional Notes

Kuiper's test was introduced by Nicolaas Kuiper in 1960 while studying goodness-of-fit problems on the circle, particularly in astronomy. Its defining feature—rotation invariance—makes it ideal for circular data (e.g., angles, time-of-day), since the test statistic remains unchanged regardless of where the circle is "cut," unlike the Kolmogorov–Smirnov test. This property made it especially valuable in fields like directional statistics and phase analysis.

The test gained prominence in astronomy through the analysis of periodic phenomena, such as pulsars and variable stars, where data naturally wrap around a cycle. In these applications, Kuiper's statistic is preferred over KS because it is equally sensitive across the entire domain, including near the boundaries—an area where KS tends to lose power.

Mathematically, Kuiper's statistic $V = D^+ + D^-$ can be viewed as a symmetrized version of the KS statistic, combining the largest positive and negative deviations. This makes it sensitive not only to shifts but also to changes in spread and shape. Despite its elegance, the test is less commonly taught than KS, partly because its critical values depend on more complex approximations, which historically limited its adoption before modern computing.

Miscellanea: Nicolaas Kuiper was a Dutch mathematician of remarkable range who later served as Director of the IHÉS in France. Although statisticians know him for Kuiper's test for circular uniformity, his deeper legacy lies in topology—most notably Kuiper's theorem showing that the unitary group of a separable infinite-dimensional Hilbert space is contractible.

Reference(s): Kuiper (1960); Stephens (1965)

Worked Example

Problem Statement Nine bird-migration departure directions (degrees clockwise from North) are recorded at a monitoring station. Kuiper's test (1960) assesses fit to the *fully specified* Uniform(0, 360) distribution—the null hypothesis that birds depart in all directions equally. Unlike the KS test, Kuiper's statistic $V = D^+ + D^-$ is invariant to the choice of origin on the circle, making it the correct analog for circular data. H_0: $F(\theta) = \theta/360$; H_1: Directions are not uniformly distributed.

1. Probability Transform and Differences

Under H_0: $u_i = \theta_{(i)}/360$. With $n = 9$, $i/n = i/9$ and $(i-1)/n = (i-1)/9$.

i	$\theta_{(i)}$	u_i	$i/9$	D_i^+	D_i^-
1	10	1/36	1/9	1/12	**+1/36**
2	20	1/18	2/9	1/6	−1/18
3	30	1/12	3/9	1/4	−5/36
4	40	1/9	4/9	1/3	−2/9
5	50	5/36	5/9	5/12	−11/36
6	70	7/36	6/9	17/36	−13/36
7	110	11/36	7/9	17/36	−13/36
8	130	13/36	8/9	**19/36**	−5/12
9	290	29/36	9/9	7/36	−1/12

where $D_i^+ = i/9 - u_i$ and $D_i^- = u_i - (i-1)/9$.

2. Kuiper Statistic

$$D^+ = \max_i D_i^+ = 19/36 \quad (\text{at } i = 8,\ \theta = 130)$$

$$D^- = \max_i D_i^- = 1/36 \quad (\text{at } i = 1,\ \theta = 10)$$

All other $D_i^- < 0$: the empirical CDF runs ahead of the uniform CDF at every observation except $\theta = 10$.

$$V = D^+ + D^- = \frac{19}{36} + \frac{1}{36} \approx 0.5556$$

3. Stephens Modification

$$V^* = V\left(\sqrt{n} + 0.155 + \frac{0.24}{\sqrt{n}}\right) \approx 1.797$$

4. Decision

Statistic	Value	CV (5%)	Decision
V^*	1.797	1.747	Reject H_0

$p < 0.05$; CV at 1%= 2.001 (FTR at 1%).
Critical values: Stephens (1970), fully specified.

Practitioner Insight (Condensed): Kuiper's statistic $V = D^+ + D^-$ combines the maximum upward and downward deviations, making it *rotation-invariant*: unlike KS, its value does not depend on where the circle is "cut." Here $D^+ = 19/36$ at 130° reflects strong clustering (8 of 9 observations in a narrow arc vs. 36% expected), while the lone outlier at 290° contributes little ($D^- = 1/36$). The choice $n = 9$ yields convenient exact fractions and $V = 5/9$. The Stephens adjustment $V^* = V(\sqrt{n} + 0.155 + 0.24/\sqrt{n})$ stabilizes critical values across n. Compared with the Rayleigh test (which is more powerful against unimodal concentration alternatives), Kuiper's test is more broadly sensitive to any departure from uniformity, including axial symmetry, multimodality, and gap patterns.

Common Pitfalls

- **Linear vs. circular misuse.** Kuiper is designed for circular data and rotation-invariant tests. For linear data, K-S (or A–D) is usually more powerful and better supported.
- **Incorrect handling of wrap-around.** Sorting circular data requires respecting angular continuity (e.g., 359° is near 1°). Software handles this; manual calculations must explicitly account for wrap-around.
- **Using with tied or discrete angles.** Kuiper assumes continuity; heavy rounding (e.g., nearest 10°) makes the test conservative. If discreteness is substantive, use discrete-specific circular tests.
- **Over-interpreting D^+ and D^-.** These components are not interpretable individually; the test statistic is the sum $V = D^+ + D^-$. Avoid directional over-interpretation.
- **Relying on asymptotic values for small n.** The approximation $V_\alpha \approx c_\alpha/\sqrt{n}$ is poor for $n \leq 20$. Use exact or simulation-based critical values for small samples.

Rao's Spacing Test

Circular Uniformity Test Based on Arc Lengths

GFIT.6 | **Category:** Goodness-of-Fit | **Data:** Continuous | **Null:** Simulated or Tables

When to Use

- Test uniformity of data on a circle (directions, angles).
- Detect clustering or concentration in circular data.
- Alternative to Rayleigh or Kuiper tests with different power characteristics.

Assumptions

- Observations are independent angular measurements. Data are measured on a circle (angles from 0° to 360°).
- No assumption about specific alternative distribution.
- Suitable for sample sizes $n \geq 4$; particularly effective for detecting clusters.

Test Statistic

$$U = \frac{1}{2}\sum_{i=1}^{n} |T_i - \lambda|, \qquad \text{where } T_i = \theta_{(i+1)} - \theta_{(i)} \text{ with } \theta_{(n+1)} = \theta_{(1)} + 360.$$

Notation: T_i are arc lengths (spacings) between consecutive ordered angles, $\lambda = 360/n$ is expected uniform spacing.

Plain English: Rao's spacing test examines whether angles are uniformly distributed around the circle by checking if the gaps between consecutive points are roughly equal. If points cluster together, some gaps will be small and others large, yielding a large test statistic. The test is specifically designed for circular data and has excellent power for detecting clustering.

⚠ Practitioner Warning

Points must be ordered around the circle. Unlike linear data where ordering is natural, circular data can be ordered starting from any point. Software for circular statistics handles this automatically, but manual calculation requires careful attention to wrap-around from 360° back to 0°.

Hypotheses

- H_0: Data are uniformly distributed on the circle.
- H_1: Data are **not** uniformly distributed on the circle.

Decision Rule: Reject H_0 if $U > U_{crit}$ or if p-value $< \alpha$. Critical values from simulation or tabled values.

Software References [Python, R, SAS]

Python	R	SAS
`Not Native (Manual)`	`circular::rao.spacing.test`	`Not Native (Manual)`

Additional Notes

Rao's spacing test is based entirely on arc lengths between successive observations, making it inherently rotation-invariant and free of any parameter estimation; the null distribution depends only on n. The exact sampling distribution of U under uniformity is known but analytically intractable, so practical use relies on tabulated critical values or simulation. Rao (1976) provided tables for common levels ($p = 0.10, 0.05, 0.01$) for n up to 200, later extended by Russell & Levitin (1995) to $n = 1000$ and broader significance levels. Because the test is spacing-based rather than EDF-based, it is relatively robust to ties caused by rounding (e.g., integer-degree measurements), where KS-type methods may degrade. Modern implementations typically compute p-values via Monte Carlo simulation or refined approximations rather than relying solely on tables.

Miscellanea: Rao's spacing test emerged from the rich research environment of the Indian Statistical Institute (ISI) in the late 1960s, where J.S. Rao (his PhD supervisor was J. Sethuraman) developed the method in his dissertation under the broader influence of C.R. Rao's (Director of ISI) statistical legacy. The test was designed in part to overcome a key limitation of the popular Rayleigh test, which can miss multimodal patterns (e.g., data clustered in opposite directions). By focusing on spacings rather than vector sums, Rao's statistic detects bimodal and irregular clustering with ease. Geometrically, U measures the "departure from a perfect ring," quantifying how far observed points are from forming a regular polygon on the circle. Beyond its origins in directional data, the test has found modern applications in fields like chronobiology, where it is used to detect non-uniform activity patterns over a 24-hour cycle.

Reference(s): Rao, J. S. (1969, 1976); Russell & Levitin (1995)

Rao's Spacing Test GFIT.6

Worked Example

Problem Statement Eight wind-direction readings (degrees clockwise from North) are recorded at a coastal station. Rao's spacing test (1976) assesses uniformity of circular data by examining whether the arcs between consecutive observations are consistent with a uniform distribution. Under H_0 each of the $n = 8$ spacings has expected length $\lambda = 360°/8 = 45°$. H_0: Directions are uniformly distributed on $(0°, 360°)$; H_1: Directions are not uniformly distributed.

i	$\theta_{(i)}$	$\theta_{(i+1)}$	$T_i = \theta_{(i+1)} - \theta_{(i)}$	$\frac{1}{2}\lvert T_i - \lambda\rvert$
1	15	30	15	15
2	30	45	15	15
3	45	60	15	15
4	60	75	15	15
5	75	90	15	15
6	90	105	15	15
7	105	285	180	67.5
8	285	15 + 360	90	22.5
			Sum	180

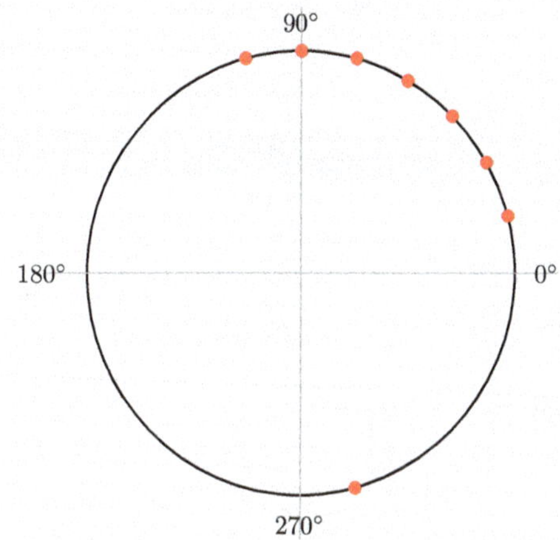

Circular data for Rao's Spacing Test

1. Compute Spacings T_i
Arrange the $n = 8$ sorted angles around the circle. The i-th spacing is the arc from $\theta_{(i)}$ to $\theta_{(i+1)}$; the final spacing wraps around: $T_n = 360° - \theta_{(n)} + \theta_{(1)}$.

2. Test Statistic

$$U = \frac{1}{2}\sum_{i=1}^{n} |T_i - \lambda| = 180°$$

3. Decision

Statistic	Value	CV (5%)	Decision
U	180°	175.7°	Reject H_0

$p < 0.05$; CV at 1%≈ 198° (FTR at 1%).
Critical values from Monte Carlo (Rao 1976; 10^6 reps).

Practitioner Insight: Rao's statistic $U = \frac{1}{2}\sum |T_i - \lambda|$ measures the mean absolute deviation of arc lengths from $\lambda = 360°/n$, with large U indicating clustering (many short spacings with compensating long gaps) and small U indicating regularity. Here, six tight 15° spacings and one large 180° gap reflect strong clustering. The identity $\sum |T_i - \lambda| = 360°$ follows from $\sum T_i = 360°$, which forces positive and negative deviations to balance. Unlike Kuiper's test, Rao's test penalizes each spacing, making it especially sensitive to clustering and also able to detect overly regular patterns. It is rotation-invariant and requires no parameter estimation, so its null distribution depends only on n.

Common Pitfalls

- **Wrap-around omission.** Failing to include the circular wrap-around spacing $T_n = (\theta_{(1)} + 360°) - \theta_{(n)}$ treats angular data as linear and invalidates the test. Always verify that $\sum_i T_i = 360°$ as a basic consistency check.
- **Applying the test to non-circular data.** Rao's spacing test is designed for angular or directional data. Mapping linear measurements into $[0, 360)$ does not make them circular; for linear data, use standard goodness-of-fit tests instead.
- **Over-interpreting the type of departure.** The test is omnibus: rejection indicates non-uniformity but does not identify whether the pattern is clustering, bimodality, or another structure. Inspect spacings and use circular plots (e.g., rose diagrams) to interpret the nature of the deviation.
- **Using chi-square approximations.** Rao's spacing statistic does not follow a chi-square distribution, even asymptotically. Always use tabulated or Monte-Carlo critical values specific to the sample size n.
- **Expecting power against smooth unimodal alternatives.** Rao's test is most powerful for clustering and multimodal patterns. For smooth unimodal concentration (e.g., von Mises alternatives), the Rayleigh test typically has higher power.

Greenwood Test for Uniformity

Spacing-Based Test for Uniformity on the Unit Interval

GFIT.7 | **Category:** Goodness-of-Fit Tests | **Data:** Continuous | **Null:** Simulated or Asymp. Normal

When to Use

- Test whether data follow a Uniform(0, 1) distribution.
- Assess p-value uniformity in simulation or meta-analysis.
- Evaluate randomness of pseudo-random number generators.
- Check uniform occurrence of rescaled event times on $[0, 1]$.

Assumptions

- Independent, continuous observations on $[0, 1]$.
- Under H_0, data follow Uniform(0, 1).
- Sample size $n \geq 3$ (power improves with n).
- No ties (ties suggest discreteness or model violation).

Test Statistic

$$G = \sum_{i=1}^{n+1} (x_{(i)} - x_{(i-1)})^2, \quad x_{(0)} = 0,\ x_{(n+1)} = 1, \quad \Big| \quad E[G] = \frac{2}{n+2}, \quad \text{Var}(G) = \frac{4n}{(n+2)^2(n+3)(n+4)}.$$

Notation: $x_{(1)} \leq \cdots \leq x_{(n)}$ are order statistics, d_i are spacings between consecutive $x_{(i)}$, n is sample size (spacings=(n+1))

Plain English: The Greenwood statistic assesses uniformity by examining how evenly observations partition the unit interval. Under uniform randomness, spacings are roughly $1/(n+1)$. Clustering creates uneven spacings and inflates G, while unusually regular spacing produces a small G. Either pattern signals deviation from uniformity.

⚠ Practitioner Warning

The test is two-sided: both large and small values of G can be significant, with critical regions in both tails. The null distribution depends on n and is typically obtained via tables, simulation, or asymptotic approximation. Data must be transformed to $[0, 1]$ before application (p-values already satisfy this). Compared to EDF tests such as Kolmogorov–Smirnov and especially Anderson–Darling (which emphasizes tail discrepancies), Greenwood is less sensitive to tail deviations but may be more powerful for multimodal or spacing-driven departures.

Hypotheses

- H_0**:** The observations are uniformly distributed on $[0, 1]$.
- H_1**:** The observations are either clustered or overly regular.

Decision Rule: Reject H_0 if $G < G_{\alpha/2,n}$ or $G > G_{1-\alpha/2,n}$ where critical values are obtained using tables or simulation.

Software References [Python, R, SAS]

Python	R	SAS
`Not Native (Manual)`	`Not Native (Manual)`	`Not Native (Manual)`

Additional Notes

The statistic arose from the idea that event times generated by a Poisson process, when properly scaled, should appear uniform on $[0, 1]$. Greenwood (1946) quantified deviation from this structure using spacing variability. Under H_0, $E[G] = 2/(n+2)$ and $\text{Var}(G) = 4n/[(n+2)^2(n+3)(n+4)]$ (corrected by Moran). Moran (1947) showed that a standardized version of G is approximately normal for large n, though exact or simulated critical values are preferable in small samples.

Spacing-based tests differ geometrically from EDF tests: EDF methods measure vertical CDF discrepancies, while Greenwood measures horizontal spacing variability. This yields complementary power characteristics and links naturally to Rao's spacing test for circular data. In modern applications, Greenwood is frequently used to assess p-value uniformity in meta-analysis and large-scale testing.

Miscellanea: Major Greenwood, a pioneer of British medical statistics at the London School of Hygiene and Tropical Medicine, developed the spacing statistic while studying whether disease outbreaks clustered beyond chance expectation. He emphasized that visual inspection is unreliable for judging randomness, advocating formal statistical diagnostics instead. Originally computed by hand for small datasets, the statistic later found unexpected uses in cryptography and random number testing—demonstrating how a tool devised for epidemiology became a general instrument for detecting hidden structure in supposedly random sequences.

Reference(s): Greenwood (1946); Moran (1947); Pyke (1965); Stephens (1981); D'Agostino & Stephens (1986)

Greenwood Test for Uniformity GFIT.7

Worked Example

Problem Statement A quality control engineer is investigating whether defects occur uniformly along a continuous production line. The production line operates for 24 hours, and 10 defects were detected during this period. To assess whether these defects are uniformly distributed in time (which might suggest random occurrence) or show clustering (which might indicate systematic problems during certain periods), the engineer records the time of each defect occurrence.

Data: The sorted times (in hours from the start of the production period) at which the 10 defects occurred are:

$$U_i = (2.3, 3.1, 5.7, 8.4, 11.2, 14.6, 16.1, 18.9, 20.5, 22.8), \text{ for } i = 1, \ldots, 10$$

Define $U_{(0)} = 0$, $U_{11} = 1$. The following plot shows the scaled points on the unit interval.

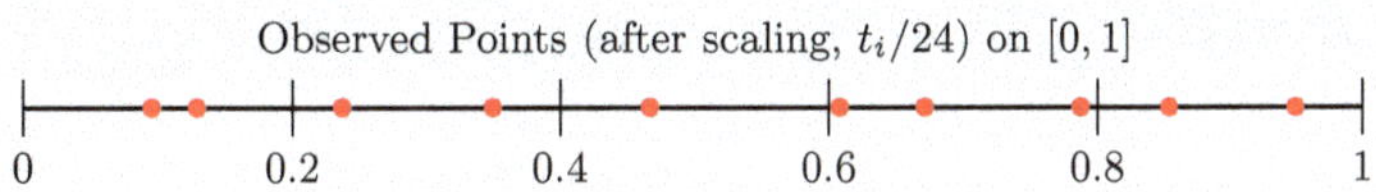

1. Greenwood statistic

$$G = \frac{1}{24^2}\sum_{i=1}^{n+1}(U_{(i)} - U_{(i-1)})^2 = \frac{1}{24^2}(2.3^2 + 0.8^2 + 2.6^2 + 2.7^2 + 2.8^2 + 3.4^2 + 1.5^2 + 2.8^2 + 1.6^2 + 2.3^2 + 1.2^2) = 0.1020$$

2. Critical Values and Decision (Monte Carlo, $n = 10$) Using 800,000 simulated samples under H_0 : Unif(0, 1) (with endpoint spacings), the two-sided $\alpha = 0.05$ critical values are: $G_{0.025} = 0.1164$, $G_{0.975} = 0.2658$.

Observed $G_{\text{obs}} = 0.1020$, so $G_{\text{obs}} < G_{0.025}$ and we **reject** H_0. Monte Carlo tail areas: $\hat{p}_L = \Pr(G^* \le G_{\text{obs}}) \approx 0.00046$, $\hat{p}_U \approx 0.99954$, hence $\hat{p}_{\text{2-sided}} = 2\min(\hat{p}_L, \hat{p}_U) \approx 0.00091$.

Practitioner Insight: Defect times are *not* consistent with uniform randomness over the 24-hour period; the unusually small G indicates *excessive regularity* (defects are more evenly spaced than a random uniform process would produce). This pattern is more consistent with a periodic/structured mechanism (e.g., scheduled cycles, batching, shift/maintenance effects, or reporting/inspection artifacts) than with random occurrence. The engineer should investigate time-structured drivers rather than clustering at specific hours.

Common Pitfalls

- **Failing to transform data to [0, 1].** Greenwood tests uniformity. For other distributions, apply the probability integral transform first. Using raw data invalidates the test.
- **Ignoring dependence.** The test assumes independent observations. Serial correlation (e.g., time series or clustered events) distorts spacings and inflates Type I error.
- **Applying to discrete or tied data.** Ties violate the continuous uniform assumption and artificially shrink spacings, biasing G.
- **Using asymptotic normal approximation for small n.** The null distribution of G depends on n; simulation or exact tables are preferable for moderate sample sizes.
- **Misinterpreting direction of departure.** Both unusually small and unusually large values of G indicate non-uniformity. Reporting only one tail misses half the diagnostic signal.
- **Confusing mixture alternatives with miscalibration.** In p-value analysis, non-uniformity may reflect genuine signals rather than flawed testing procedures.
- **Overlooking boundary effects.** Spacings include the intervals $(0, U_{(1)})$ and $(U_{(n)}, 1)$. Omitting endpoints changes the statistic.
- **Expecting strong tail sensitivity.** Greenwood is spacing-based and may miss alternatives concentrated only in extreme tails compared to Anderson–Darling.

Chi-Square Goodness-of-Fit Test

Pearson's Classic Test for Categorical Data

GFIT.8 | **Category:** Goodness-of-Fit | **Data:** Categorical/Discrete | **Null:** Chi-square

When to Use

- Test whether observed frequencies match expected frequencies from a specified distribution.
- Compare categorical data to a theoretical distribution or model.
- Appropriate for discrete distributions (binomial, Poisson, etc.) or grouped continuous data.

Assumptions

- Observations are independent. Total sample size is fixed.
- Data are counts in mutually exclusive categories.
- Expected frequency in each cell is at least 5 (preferably ≥ 5 in all cells).
- For estimated parameters, degrees of freedom must be adjusted.

Test Statistic

$$\chi^2 = \sum_{i=1}^{k} \frac{(O_i - E_i)^2}{E_i}. \qquad \text{Under } H_0,\ \chi^2 \sim \chi^2_{k-1-p}.$$

Notation: O_i are observed counts, E_i are expected counts under H_0, k is number of bins, p is the number of parameters estimated

Plain English: The chi-square goodness-of-fit test compares observed counts to expected counts by measuring how much the two differ, scaled by what we'd expect from random sampling variation. Large discrepancies (relative to expected counts) suggest the data don't follow the hypothesized distribution. It's one of the oldest and most versatile tests in statistics.

⚠ Practitioner Warning

The chi-square approximation deteriorates when expected frequencies are small. Traditional rule: all $E_i \geq 5$. With smaller expected frequencies, the test becomes conservative (under-rejects). Combine sparse categories to meet this requirement, but avoid over-combining which loses information.

Hypotheses

- H_0: Data follow the specified distribution
- H_1: Data do **not** follow the specified distribution.

Decision Rule: Reject H_0 if $\chi^2 > \chi^2_{k-1-p,\alpha}$ or if p-value $< \alpha$. df $= k - 1 - p$ where k is bins, p is estimated parameters.

Software References [Python, R, SAS]

Python	R	SAS
`scipy.stats.chisquare`	`chisq.test`	`PROC FREQ / CHISQ`

Additional Notes

The test is an asymptotic procedure for discrete data, with null distribution depending on the number of categories and any estimated parameters. The common guideline $E_j \geq 5$ is a rule of thumb rather than a strict requirement; smaller expected counts may be acceptable or can be handled by merging categories, though very small samples favor exact methods (e.g., Fisher's exact test). The test is asymptotically equivalent to the likelihood-ratio (G) test for multinomial models and can be viewed as a quadratic form of standardized residuals, linking it to modern diagnostics such as deviance residuals in generalized linear models. For continuous data, binning reduces power relative to EDF-based tests (KS, AD, CvM). For ordinal data; tests that account for ordering (e.g., Cochran–Armitage trend test) are more appropriate.

Miscellanea: Karl Pearson introduced the chi-square test in 1900 in a landmark paper often regarded as the birth of modern hypothesis testing. Originally motivated by problems in genetics, the test later played a central role in the famous Mendel–Fisher controversy: in 1936, R.A. Fisher showed that Mendel's pea-plant data were "too good to be true," sparking debate that continues today. The test was also at the heart of a long-standing dispute between Pearson and Fisher, who corrected the degrees-of-freedom formula to account for estimated parameters—a refinement now standard. Despite its simplicity, the chi-square statistic has deep connections to the normal distribution: it arises as the sum of squared standard normal variables. Today, the test remains ubiquitous, from genetics (Hardy–Weinberg equilibrium) to machine learning feature selection, illustrating its enduring impact across scientific disciplines.

Reference(s): Pearson (1900); Cochran (1952)

Worked Example

Problem Statement A quality-control engineer records the number of surface defects on each of $n = 100$ circuit boards. The Poisson distribution is a standard model for defect counts, but the rate λ is unknown and must be estimated from the data. The chi-square goodness-of-fit test assesses whether the observed frequency distribution is consistent with *some* Poisson distribution. H_0: counts follow Poisson(λ) for some $\lambda > 0$; H_1: counts do not follow any Poisson distribution.

j	$P(X=j)$	O_j	E_j	$O_j - E_j$	$(O-E)^2/E$
0	e^{-2}	22	13.53	+8.47	5.297
1	$2e^{-2}$	16	27.07	−11.07	4.525
2	$2e^{-2}$	26	27.07	−1.07	0.042
3	$\frac{4}{3}e^{-2}$	18	18.04	−0.04	0.000
4	$\frac{2}{3}e^{-2}$	12	9.02	+2.98	0.983
5+	$1 - F(4)$	6	5.27	+0.73	0.103
Sum	1	100	100	0	10.950

1. Estimate λ by MLE

For Poisson data, $\hat{\lambda} = \bar{x}$. Using the last grouped cell at $j = 5$ (approximation for $j \geq 5$):

$$\hat{\lambda} = \frac{\sum j\, O_j}{n} = 2.0$$

2. Expected Counts

$$E_j = n \cdot P(X=j;\, \hat{\lambda} = 2) = 100 \cdot \frac{2^j e^{-2}}{j!}$$

with $E_{5+} = 100 \cdot P(X \geq 5)$. All $E_j \geq 5$; total = 100.00 ✓

3. Chi-Square Statistic

$$X^2 = \sum_j \frac{(O_j - E_j)^2}{E_j} = 10.950$$

4. Degrees of Freedom and Decision

One parameter (λ) is estimated: df = bins − 1 − p = 4

Statistic	Value	CV (5%)	Decision
X^2	10.950	9.488	Reject H_0

$p = 0.027$; CV at 1%= 13.277 (FTR at 1%).

Practitioner Insight: The chi-square GoF test is unique in directly handling *discrete* distributions (e.g., Poisson, binomial), unlike EDF-based tests which require continuity. Its degrees of freedom df $= k - 1 - p$ adjust for parameter estimation, lowering critical values (e.g., from 11.07 to 9.49 here) and affecting rejection decisions. The $E_j \geq 5$ rule is a guideline—small cells can be merged. In this dataset, excess zeros and deficits at $j = 1$ suggest deviation from Poisson, consistent with a zero-inflated process.

Common Pitfalls

- **Small expected counts.** The chi-square approximation fails when any $E_i < 5$. Combine categories or use exact/simulation methods for small cells.
- **Incorrect degrees of freedom.** When parameters are estimated, use df $= k - 1 - p$. Ignoring the p parameter penalty gives invalid p-values.
- **Improper use with continuous data.** Chi-square requires categorical bins; binning loses information and results depend on cutpoints. Prefer EDF tests for continuous data.
- **Misreading omnibus rejection.** A significant result signals overall misfit, not which categories fail. Inspect individual contributions $(O_i - E_i)^2/E_i$ to locate problems.
- **Ignoring dependence.** The test assumes independent observations. For paired, clustered, or repeated data, use McNemar's test, Cochran's Q, or models accounting for dependence.

Part III

Location Differences & Central Tendencies

Testing contrasts in means, medians, and central location parameters

Means and Locations

III. MEAN — A Conceptual Primer

1. Problem Framing

What Are We Testing?

- Whether the **central location** (mean, trimmed mean, or median) of one or more populations equals a specified value or differs across groups.
- Core question: *Are observed differences in sample means real or due to sampling chance?*

Hypotheses

One-sample: $H_0 : \mu = \mu_0$ vs. $H_1 : \mu \neq \mu_0$ (or >, < for directional tests).

Two-sample: $H_0 : \mu_1 = \mu_2$ vs. $H_1 : \mu_1 \neq \mu_2$.

k-sample: $H_0 : \mu_1 = \cdots = \mu_k$ vs. H_1: at least one μ_i differs.

2. Methods

The t-Family of Tests

All t-tests pivot on the same structure: standardize a mean difference by its estimated standard error and compare to a t-distribution.

One-sample t-test:

$$T = \frac{\bar{X} - \mu_0}{s/\sqrt{n}} \sim t_{n-1}$$

Paired t-test: define $d_i = x_{1i} - x_{2i}$; apply one-sample t to $\{d_i\}$:

$$T = \frac{\bar{d}}{s_d/\sqrt{n}} \sim t_{n-1}$$

Pairing removes between-subject variability; it **increases power** when observations are naturally matched.

Two-sample pooled t-test (equal variances):

$$T = \frac{\bar{X}_1 - \bar{X}_2}{s_p\sqrt{1/n_1 + 1/n_2}} \sim t_{n_1+n_2-2}, \quad s_p^2 = \frac{(n_1-1)s_1^2 + (n_2-1)s_2^2}{n_1+n_2-2}$$

Welch's t-test (unequal variances; **default recommendation**):

$$T = \frac{\bar{X}_1 - \bar{X}_2}{\sqrt{s_1^2/n_1 + s_2^2/n_2}} \sim t_\nu$$

where the Welch–Satterthwaite degrees of freedom are

$$\nu = \frac{(s_1^2/n_1 + s_2^2/n_2)^2}{\dfrac{(s_1^2/n_1)^2}{n_1-1} + \dfrac{(s_2^2/n_2)^2}{n_2-1}}.$$

Welch's test is **uniformly valid** whether or not variances are equal; prefer it unless equal variance is strongly justified by design or prior evidence.

One-Way ANOVA

Extends the two-sample comparison to $k \geq 2$ groups by decomposing total variability:

$$\underbrace{\mathrm{SS_{Tot}}}_{=\sum_{ij}(y_{ij}-\bar{y})^2} = \underbrace{\mathrm{SS_B}}_{k\text{ groups}} + \underbrace{\mathrm{SS_W}}_{\text{within groups}}$$

$$F = \frac{\mathrm{MS_B}}{\mathrm{MS_W}} = \frac{\mathrm{SS_B}/(k-1)}{\mathrm{SS_W}/(N-k)} \sim F_{k-1,N-k} \quad \text{under } H_0.$$

Large F indicates between-group signal exceeds within-group noise. ANOVA tells you *that* at least one mean differs; it does not identify *which* groups differ.

Two-way ANOVA adds a second grouping factor (and possibly an interaction), partitioning $\mathrm{SS_{Tot}}$ further into main effects A, B, and $A \times B$.

3. Decision Layer

Choosing the Right Test

Design	Recommended test
1 sample vs. μ_0	One-sample t
2 matched/paired groups	Paired t
2 independent groups, equal var	Pooled t
2 independent groups, unequal var	Welch's t (default)
$k \geq 2$ groups, 1 factor	One-way ANOVA
k groups, 2 factors	Two-way ANOVA

4. Assumptions

Assumptions

Assumption	Consequence if violated
Independence	Biased p-values; invalid inference
Normality (n<30)	Exact t/F distributions fail
Equal variances	Pooled t / ANOVA inflated Type I error

Robustness notes:

- t-tests are robust to mild non-normality, especially for $n \gtrsim 30$ (CLT); less so for heavy tails or outliers.
- ANOVA is robust to mild non-normality with balanced designs; *not* robust to unequal variances—use Welch ANOVA (Brown–Forsythe or Welch F) when variances differ.
- Independence violations are the most damaging; no correction from the t/F family will help—use mixed models (MIXD) or GEE for clustered/longitudinal data.

Testing for Equal Variances (Pre-Test Caution)

Levene's test and Bartlett's test are commonly used to check the equal-variance assumption before choosing between pooled and Welch. **However**, this pre-testing strategy inflates overall Type I error and loses power. The better practice: **default to Welch's t-test always** and avoid the pre-test entirely. See Moser & Stevens (1992).

5. Inference

Confidence Intervals

Always report confidence intervals alongside tests; they convey both direction and magnitude of the estimated difference.

One-sample: $\bar{X} \pm t_{\alpha/2, n-1} \cdot s/\sqrt{n}$.

Two-sample (Welch): $(\bar{X}_1 - \bar{X}_2) \pm t_{\alpha/2, \nu} \cdot \sqrt{s_1^2/n_1 + s_2^2/n_2}$.

A CI that excludes zero is equivalent to rejecting H_0 at level α, but the CI additionally reveals *how large* the difference might plausibly be.

Effect Size — Practical Importance

Statistical significance alone is insufficient; always report effect size alongside p-values.

Cohen's d (standardized mean difference):

$$d = \frac{\bar{X}_1 - \bar{X}_2}{s_p}, \qquad \text{small: 0.2, medium: 0.5, large: 0.8.}$$

η^2 and partial η^2 (ANOVA proportion of variance):

$$\eta^2 = \frac{SS_B}{SS_{Tot}}, \qquad \eta_p^2 = \frac{SS_{effect}}{SS_{effect} + SS_W}.$$

Guidelines: $\eta^2 \approx 0.01$ small, 0.06 medium, 0.14 large.

ω^2 (less biased than η^2, especially for small n):

$$\omega^2 = \frac{SS_B - (k-1)MS_W}{SS_{Tot} + MS_W}.$$

6. Extensions

Multiple Comparisons (Post-Hoc)

ANOVA rejection of H_0 triggers post-hoc comparisons to identify *which* pairs of means differ. Running unadjusted t-tests inflates the family-wise error rate (FWER).

Method	Use case
Tukey HSD	All pairwise; controls FWER exactly
Bonferroni	Any set of m tests; conservative
Holm	Step-down Bonferroni; less conservative
Dunnett	Compare all groups to one control
Scheffé	All contrasts; most conservative
FDR (BH)	Large families; controls false discovery rate

Rule of thumb: use Tukey for all pairwise comparisons; Dunnett when one group is a control; Bonferroni/Holm for a pre-specified small set of contrasts.

Power and Sample Size

Power of a t-test depends on: effect size d, n, α, and directionality (one- vs. two-sided).

$$n \approx \frac{2(z_{\alpha/2} + z_\beta)^2}{d^2} \qquad \text{(approximate, two-sample, equal } n\text{)}$$

- Doubling n increases power but does not double it; power gains are concave in n.
- For ANOVA: power depends on the noncentrality parameter $\lambda = n \sum_i (\mu_i - \bar{\mu})^2/\sigma^2$.
- Unbalanced designs reduce power relative to balanced designs of the same total N.

7. Practice

When Assumptions Fail

- **Non-normality / ordinal data:** Wilcoxon signed-rank (one-sample/paired), Mann–Whitney (two-sample), Kruskal–Wallis (k-sample)—see NPAR chapter.
- **Outliers / heavy tails:** trimmed-mean tests (Yuen's t), M-estimator tests—see ROBU chapter.
- **Unequal variances (k groups):** Welch ANOVA (Brown–Forsythe F^* or Welch F).
- **Clustered / longitudinal data:** mixed models (MIXD chapter) or GEE.
- **Distribution-free inference:** bootstrap t test (percentile or BCa confidence intervals).

Common Pitfalls

- Using pooled t-test without verifying equal variances (use Welch by default).
- Ignoring pairing structure (treating paired data as independent loses power).
- Running multiple t-tests instead of ANOVA (inflates Type I error).
- Conflating statistical significance with practical importance; always report d or η^2.
- Multiple comparisons without adjustment (inflated FWER).
- Interpreting a non-significant result as evidence of equality (absence of evidence ≠ evidence of absence).

Bottom Line

Mean tests evaluate differences in location, but practical significance depends on effect size, not p-values alone. Default to Welch's t-test for two groups. Always accompany ANOVA with effect sizes and post-hoc tests. When assumptions are doubtful, nonparametric or robust alternatives are available with minimal power cost.

Z-Test (One-Sample)

Test for Population Mean with Known Variance

MEAN.1 | **Category:** Mean Comparison | **Data:** Continuous | **Null:** Standard Normal

When to Use

- Test mean against a hypothesized population value.
- Population variance is known or well established.
- Large samples support normal approximations.
- Continuous measurements from stable processes.
- Industrial, scientific, or quality-control studies.

Assumptions

- Observations are mutually independent.
- Population variance is known and fixed.
- Data are continuous measurements of interest.
- Population is normal or sample size is large.
- Sample is randomly drawn from population.

Test Statistic

$$Z = \frac{\bar{x} - \mu_0}{\sigma/\sqrt{n}}.$$

Under H_0, $Z \sim N(0,1)$ exactly (normal population) or approximately (large n).

Notation: $\bar{x}$ is sample mean, μ_0 is hypothesized mean, σ is known population standard deviation

Plain English: The one-sample z-test compares the observed sample mean to the hypothesized mean and expresses the difference in units of standard error. If the sample mean is far enough from the hypothesized value relative to the expected sampling variability, the null hypothesis is rejected. Because the test relies on a known population variance and the normal distribution, it is most often encountered in textbooks and theoretical examples rather than in everyday applied work.

⚠ Practitioner Warning

The requirement that σ is known (not estimated) is unrealistic in most applications. In practice, we estimate s from data and use the t-test instead. Use Z-test only when: (1) historical data firmly establishes σ, (2) very large samples make $s \approx \sigma$, or (3) theoretical/teaching contexts.

Hypotheses

- H_0: Population mean equals the hypothesized value: $\mu = \mu_0$
- H_1: μ_0. (two-sided), or $\mu > \mu_0$ (one-sided), or $\mu < \mu_0$ (one-sided).

Decision Rule: Reject H_0 if $|Z| > z_{\alpha/2}$ (two-sided) or if p-value $< \alpha$. For one-sided: reject if $Z > z_\alpha$ (upper) or $Z < -z_\alpha$ (lower). Standard (right-tail) critical values: $z_{0.025} = 1.96$, $z_{0.05} = 1.645$.

Software References [Python, R, SAS]

`scipy.stats.norm`	`BSDA::z.test`	`PROC MEANS`

Additional Notes

The one-sample z-test provides the conceptual foundation for hypothesis testing, as the test statistic follows the standard normal distribution exactly under known variance. This simplicity makes it a natural starting point for teaching inference, before introducing the t-test and the role of degrees of freedom. In practice, the requirement of a known population variance is rarely satisfied, which limits direct application. For large samples, however, replacing σ with the sample standard deviation yields results close to the z-test, reinforcing its role as a useful approximation framework.

Miscellanea: The z-test predates the t-test and is rooted in the early development of the normal distribution as a universal reference, supported by the Central Limit Theorem. Its known-variance assumption was more realistic in early applications such as astronomy, metrology, and industrial quality control than in most modern settings. The test became a canonical example in the Neyman–Pearson framework, where it is uniformly most powerful for one-sided hypotheses about a normal mean when the variance is known. It is also one of the few settings where hypothesis tests achieve exact Type I error control under ideal assumptions. With sufficiently large sample sizes, even trivial deviations from a hypothesized mean can become statistically significant—an enduring lesson in the distinction between statistical and practical significance.

Reference(s): Fisher (1932); Neyman & Pearson (1933); Lehmann (1986)

Z-Test (One-Sample) — MEAN.1

Worked Example

Problem Statement A standardized cognitive assessment has been normed on a large population with known mean $\mu_0 = 100$ and known standard deviation $\sigma = 10$. A researcher administers the test to $n = 25$ participants enrolled in an enrichment programs and records a sample mean of $\bar{x} = 105$. Test whether programs participants score differently from the general population. $H_0: \mu = 100$; $H_1: \mu \neq 100$.

1. Standard Error of the Mean

Because σ is known, the standard error is exact:

$$\text{SE} = \frac{\sigma}{\sqrt{n}} = \frac{10}{\sqrt{25}} = 2.$$

2. Test Statistic

$$Z = \frac{\bar{x} - \mu_0}{\text{SE}} = \frac{105 - 100}{2} = 2.50.$$

Under H_0: $Z \sim N(0, 1)$ exactly.

3. p-value

$$p = 2P(Z > 2.50) = 2\big(1 - \Phi(2.50)\big) = 0.0124.$$

4. Critical Values and Decision

α	$z_{\alpha/2}$	$\lvert Z\rvert = 2.50$	Decision
5%	1.96	>	**Reject H_0**
1%	2.576	<	FTR

Reject H_0 at 5% level; insufficient evidence to reject at 1%.

5. Confidence Interval

A 95% confidence interval for μ is

$$\bar{x} \pm z_{0.025}\,\text{SE} = 105 \pm 1.96 \times 2 = (101.08,\ 108.92).$$

The null value $\mu_0 = 100$ lies outside this interval, confirming rejection at 5%.

6. Effect Size

The standardized effect (Cohen's d):

$$d = \frac{\bar{x} - \mu_0}{\sigma} = \frac{105 - 100}{10} = 0.50 \quad \text{(medium effect)}.$$

$\bar{x}$	105
$\text{SE} = \sigma/\sqrt{n}$	2
Z	2.50
p-value (two-sided)	0.0124
95% CI	(101.08, 108.92)
Cohen's d	0.50

Practitioner Insight: The one-sample Z-test for the mean applies only when the population standard deviation σ is truly known (e.g., from well-characterized instruments, stable processes, or regulatory standards) When σ is estimated, the t-test with $n - 1$ degrees of freedom is appropriate. The statistic $Z = (\bar{x} - \mu_0)/(\sigma/\sqrt{n})$ measures how many standard errors the sample mean lies from the null; here $Z = 2.50$ exactly since SE = 2. The result rejects at 5% ($p = 0.012$) but not at 1%, illustrating that significance depends on the chosen level. The 95% confidence interval (101.08, 108.92) is the inversion of the test and agrees with it. Cohen's $d = 0.5$ indicates a moderate effect. The Z-test is the limiting case of the t-test as degrees of freedom increase; for example, $t_{24,0.025} = 2.064$ versus $z_{0.025} = 1.960$, implying slightly wider intervals when σ is estimated.

Common Pitfalls

- **Unknown variance.** The Z-test requires a *known* population standard deviation σ. If variability is estimated from the sample, a t-test must be used. Treating s as known understates uncertainty and inflates Type I error.
- **Small non-normal samples.** With small n, the Z-test depends on normality rather than the CLT. Skewed or heavy-tailed data can invalidate results; check normality and consider transformations or nonparametric alternatives.
- **Outlier sensitivity.** The sample mean is highly sensitive to extreme values, which can dominate the test statistic. Inspect data and consider robust methods when outliers are present.
- **Post-hoc one-sided testing.** One- vs. two-sided tests must be specified *a priori*. Changing tail direction after seeing data invalidates inference.

Useful Critical Values from Normal Distribution

	Two-Sided Critical Values							One-Sided (Upper-Tail) Critical Values						
α	0.20	0.10	0.05	0.02	0.01	0.005	0.001	0.20	0.10	0.05	0.02	0.01	0.005	0.001
z	1.282	1.645	1.960	2.326	2.576	2.807	3.291	0.842	1.282	1.645	2.054	2.326	2.576	3.090

One-Sample t-Test

Test for Population Mean with Unknown Variance

MEAN.2 | **Category:** Mean Comparison | **Data:** Continuous | **Null:** t-distribution

When to Use

- Test if a sample mean differs from a hypothesized mean.
- Population variance unknown; estimated from sample data.
- Small to moderate samples where normality is plausible.
- Benchmarking against a standard or target value.
- Most real-world mean testing where σ is unknown.

Assumptions

- Observations are independent and identically distributed.
- Data are approximately normal (critical for small samples).
- Sample is randomly drawn from the population.
- No extreme outliers that distort mean and variance.
- Large samples rely on CLT for robustness to non-normality.

Test Statistic

$$t = \frac{\bar{x} - \mu_0}{s/\sqrt{n}}.$$

Under H_0, t follows a t-distribution with $df = n - 1$ degrees of freedom.

Notation: $\bar{x}$ is sample mean, μ_0 is hypothesized mean, s is sample standard deviation, n is sample size

Plain English: The one-sample t-test asks whether a sample mean is far enough from a hypothesized value to conclude the population mean differs from that value. Unlike the Z-test, we estimate variability from the data itself, using the sample standard deviation. This introduces additional uncertainty captured by the t-distribution, which has heavier tails than the normal distribution, especially for small samples.

⚠ Practitioner Warning

The t-test assumes approximate normality, which matters most for small samples ($n < 15$). Severe skewness, heavy tails, or outliers can inflate Type I error rates. Check normality with Q-Q plots or Shapiro-Wilk test. For non-normal data with small samples, consider Wilcoxon signed-rank test, transformations, or bootstrap methods.

Hypotheses

- H_0: Population mean equals the hypothesized value: $\mu = \mu_0$
- H_1: Two-sided: $\mu \neq \mu_0$; One-sided: $\mu > \mu_0$ or $\mu < \mu_0$.

Decision Rule: Reject H_0 if $|t| > t_{\alpha/2,n-1}$ (two-sided) or if p-value $< \alpha$. One-sided: $t > t_{\alpha,n-1}$ or $t < -t_{\alpha,n-1}$.

Software References [Python, R, SAS]

```
scipy.stats.ttest_1samp        t.test()        PROC TTEST
```

Additional Notes

When the data are exactly normal, the t-test is exact. The t-distribution accounts for additional uncertainty from estimating variance. For large samples, t_{n-1}approaches the standard normal, making t-test and Z-test nearly equivalent. For non-normal data, it remains asymptotically valid but may lose accuracy in small samples. Two-sided critical values for different df ($\nu = n - 1$) are given below:

ν	1	2	3	4	5	6	7	8	9	10	15	20	25	30	50	∞
$t_{0.05,\nu}$ (10%)	6.314	2.920	2.353	2.132	2.015	1.943	1.895	1.860	1.833	1.812	1.753	1.725	1.708	1.697	1.676	1.645
$t_{0.025,\nu}$ (5%)	12.706	4.303	3.182	2.776	2.571	2.447	2.365	2.306	2.262	2.228	2.131	2.086	2.060	2.042	2.009	1.960
$t_{0.005,\nu}$ (1%)	63.657	9.925	5.841	4.604	4.032	3.707	3.499	3.355	3.250	3.169	2.947	2.845	2.787	2.750	2.678	2.576

Miscellanea: William Sealy Gosset, working at Guinness Brewery, published under the pseudonym "Student" due to company policy restricting employees from publishing under their own names. His identity was not widely known for several years. Gosset developed the t-test to address small-sample problems in industrial settings, where batch sizes were often limited. Combining mathematical analysis with extensive empirical investigation, he showed that replacing an unknown variance introduces additional uncertainty, leading to the heavier-tailed t-distribution. The letter "t" does not stand for "test" but was simply used as a notation for the standardized statistic. R. A. Fisher later formalized this work, introducing degrees of freedom and embedding the t-test into modern statistical inference.

Reference(s): Student (1908); Gosset (1942)

One-Sample t-Test — MEAN.2

Worked Example

Problem Statement A consumer watchdog monitors a product whose label claims a net weight of $\mu_0 = 12$ g. Five ($n = 5$) randomly selected packages yield weights (g): $x = (12,\ 14,\ 15,\ 16,\ 18)$. The population standard deviation is unknown. Test whether the true mean weight differs from the labeled value. $H_0: \mu = 12$; $\quad H_1: \mu \neq 12$.

1. Sample Mean and Variance

$$\bar{x} = \frac{\sum x_i}{n} = 15, \qquad s^2 = \frac{\sum (x_i - \bar{x})^2}{n-1} = 5.$$

2. Standard Deviation and Standard Error

$$s = \sqrt{s^2} = \sqrt{5}, \qquad \mathrm{SE} = \frac{s}{\sqrt{n}} = 1.$$

3. Test Statistic

$$t = \frac{\bar{x} - \mu_0}{\mathrm{SE}} = \frac{15 - 12}{1} = 3, \qquad \nu = n - 1 = 4 \text{ df}.$$

Under H_0: $t \sim t(4)$.

4. p-value and Decision

$$p = 2\,P\big(t(4) > 3\big) = 0.0399.$$

α	$t_{\alpha/2,4}$	$\lvert t\rvert = 3$	Decision
5%	2.776	>	**Reject H_0**
1%	4.604	<	FTR

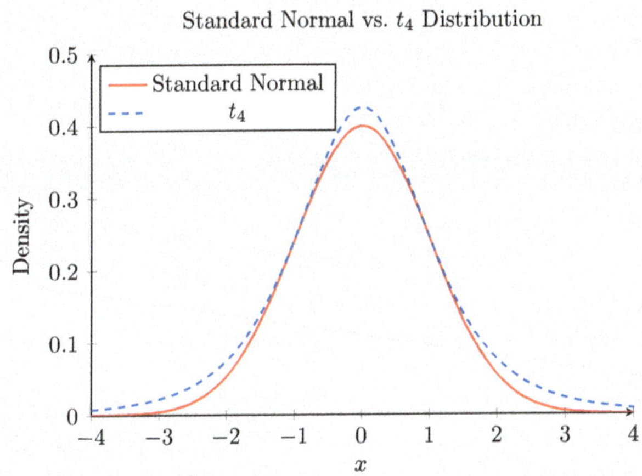

5. Confidence Interval

$$\bar{x} \pm t_{0.025,4} \cdot \mathrm{SE} = 15 \pm 2.776 \times 1 = (12.22,\ 17.78).$$

Since $\mu_0 = 12 < 12.22$, the null value lies outside the 95% CI, confirming rejection.

Practitioner Insight: The one-sample t-test differs from the Z-test by replacing known σ with the sample estimate s, introducing additional uncertainty captured by the $t(\nu)$ distribution with $\nu = n - 1$. Its heavier tails inflate critical values (e.g., 2.776 vs 1.96 at 5% for $\nu = 4$), yielding larger p-values for the same standardized deviation. Here $t = 3 > Z = 2.50$, yet $p = 0.040 > 0.012$, illustrating this effect. The data were constructed for clean arithmetic ($\bar{x} = 15$, $s^2 = 5$, SE $= 1$), giving symmetric deviations and SS= 20. Cohen's $d \approx 1.34$ indicates a large effect, but with $n = 5$ the evidence is only marginal at 5%, suggesting replication. As $\nu \to \infty$, $t(\nu) \to N(0,1)$, so the t-test converges to the Z-test.

Effect Size: Cohen's $d = (\mu - \mu_0)/\sigma$ is estimated by $\hat{d} = (\bar{x} - \mu_0)/s$, but is upward biased in small samples. Hedges' g corrects this via

$$g = \frac{\Gamma\left(\frac{n-1}{2}\right)}{\sqrt{\frac{n-1}{2}}\,\Gamma\left(\frac{n-2}{2}\right)}\,\hat{d} \approx \left(1 - \frac{3}{4n-5}\right)\hat{d}$$

Common Pitfalls

- **Ignoring dependence.** The test assumes independent observations. Paired, repeated, or clustered data require paired t-tests, mixed models, or cluster-robust methods.
- **Using small-sample non-normal data.** With small samples, strong skewness or outliers can distort the test. Check diagnostics or consider transformations, Wilcoxon, bootstrap, or robust methods.
- **Choosing the tail after seeing the data.** One- and two-sided alternatives should be specified *a priori*. Selecting a one-sided test post hoc inflates Type I error.
- **Reporting only p-values.** Statistical significance does not measure practical importance. Report effect sizes, confidence intervals, and descriptive summaries.
- **Using the t-test for variability.** The t-test evaluates the population mean, not variance. Use variance-specific procedures when spread is the quantity of interest.

Z-Test (Two-Sample)

Compare Two Population Means with Known Variances

MEAN.3 | **Category:** Mean Comparison | **Data:** Continuous | **Null:** Standard Normal

When to Use

- Compare means of two independent populations.
- Assess whether two population means differ.
- Large samples support normal approximations.
- Continuous measurements from stable processes.
- Industrial, scientific, or quality-control studies.

Assumptions

- Samples are mutually independent.
- Population variances are known and fixed.
- Data are continuous measurements of interest.
- Populations are normal or sample sizes are large.
- Samples are randomly drawn from populations.

Test Statistic

$$Z = \frac{(\bar{x}_1 - \bar{x}_2) - \Delta_0}{\text{SE}}, \text{ where SE} = \sqrt{\frac{\sigma_1^2}{n_1} + \frac{\sigma_2^2}{n_2}}. \quad \text{Under } H_0,\ Z \sim N(0,1) \text{ (exact if normal; asymptotic otherwise).}$$

Notation: $\bar{x}_1, \bar{x}_2$ are sample means; σ_1, σ_2 are known population SDs; n_1, n_2 are sample sizes

Plain English: The two-sample Z-test compares means from two groups by standardizing their difference using the known variability in each population. The denominator is the standard error of the difference, accounting for sample sizes and variances. In practice, knowing both σ_1 and σ_2 is unrealistic, making the two-sample t-test far more common.

⚠ Practitioner Warning

The requirement that both σ_1^2 and σ_2^2 are known is extremely rare outside quality control with extensive historical data. In nearly all applications, use the two-sample t-test (pooled or Welch's) which estimates variances from data. Using sample s_1, s_2 as if they were σ_1, σ_2 underestimates uncertainty and inflates Type I error.

Hypotheses

- H_0: $\mu_1 - \mu_2 = \Delta_0$ (often $\Delta_0 = 0$, testing equality).
- H_1: Two-sided: $\mu_1 - \mu_2 \neq \Delta_0$

Decision Rule: Reject H_0 if $|Z| > z_{\alpha/2}$ (2-sided) or p-value $< \alpha$. One-sided testing mirrors the one-sample Z-test.

Software References [Python, R, SAS]

Python	R	SAS
`scipy.stats.norm`	`BSDA::z.test`	`PROC TTEST`

Additional Notes

The two-sample Z-test extends one-sample logic to compare group means. The test statistic follows an exact standard normal distribution when populations are normal and variances are known, and it accommodates both equal and unequal variances. In practice, however, its use is limited because population variances are rarely known, making it largely pedagogical.

A common confusion is that while we subtract means, we add variances. For independent samples, uncertainty compounds rather than cancels, giving $\text{Var}(\bar{X}_1 - \bar{X}_2) = \sigma_1^2/n_1 + \sigma_2^2/n_2$. A general rule of thumb suggests using the Z test when $n_1, n_2 \geq 30$, but this threshold is heuristic; the adequacy of the normal approximation depends on tail behavior and underlying distribution, not a fixed cutoff.

When variances are unknown, the **pooled t-test** assumes equal variances and remains common in practice, while **Welch's t-test** relaxes this assumption and is generally preferred as the default. For non-normal or ordinal data, the **Mann–Whitney U test** provides a rank-based alternative without variance assumptions. In small samples or when assumptions are uncertain, **permutation tests** offer a fully nonparametric, resampling-based approach, while **bootstrap t-tests** extend this flexibility to more complex or non-standard settings.

Miscellanea: The two-sample Z-test was largely superseded in practice by Gosset's t-test and Fisher's developments, which accommodate unknown variances. The assumption of known population variances is rarely satisfied, making the method primarily pedagogical, though it remains useful in specialized settings such as quality control and calibration. Today, the Z-test is most often encountered as a conceptual benchmark, illustrating how inference simplifies under known variance before introducing more realistic t-based methods.

Reference(s): Fisher (1932); Snedecor & Cochran (1989)

Worked Example

Problem Statement A pharmaceutical manufacturer compares the active-ingredient content (mg per tablet) from two production lines. Long-running quality records give known process standard deviations $\sigma_1 = 6$ (line 1) and $\sigma_2 = 16$ (line 2). Independent random samples yield $n_1 = 25$ tablets from line 1 with $\bar{x}_1 = 524$ and $n_2 = 100$ tablets from line 2 with $\bar{x}_2 = 519$. Test whether the two lines produce the same mean content. $H_0: \mu_1 - \mu_2 = 0$; $H_1: \mu_1 - \mu_2 \neq 0$.

1. Standard Error of the Difference
With both σ's known:

$$\text{SE} = \sqrt{\frac{\sigma_1^2}{n_1} + \frac{\sigma_2^2}{n_2}} = \sqrt{\frac{36}{25} + \frac{256}{100}} = 2.$$

2. Test Statistic

$$Z = \frac{(\bar{x}_1 - \bar{x}_2) - 0}{\text{SE}} = \frac{524 - 519}{2} = 2.50.$$

Under H_0: $Z \sim N(0, 1)$ exactly.

3. p-value and Decision

$$p = 2P(Z > 2.50) = 2(0.0062) = 0.0124.$$

α	$z_{\alpha/2}$	$\lvert Z\rvert = 2.50$	Decision
5%	1.96	>	**Reject H_0**
1%	2.576	<	FTR

4. Confidence Interval
A 95% CI for $\mu_1 - \mu_2$:

$$(\bar{x}_1 - \bar{x}_2) \pm z_{0.025} \cdot \text{SE} = 5 \pm 1.96 \times 2 = (1.08,\ 8.92).$$

Zero lies outside: consistent with rejection at 5%.

5. Variance Contributions

	σ_i^2/n_i	Fraction	% of SE^2
Line 1	36/25	1.44	36%
Line 2	64/25	2.56	64%
Total	100/25	4.00	100%

Line 2 contributes 64% of the variance despite having $n_2 = 100$ (four times $n_1 = 25$): its larger $\sigma_2 = 16$ outweighs the larger sample size.

n_1, n_2	25, 100
σ_1, σ_2 (known)	6, 16
$\bar{x}_1, \bar{x}_2$	524, 519
$\text{SE} = \sqrt{\sigma_1^2/n_1 + \sigma_2^2/n_2}$	2
Z	2.50
p (two-sided)	0.0124
95% CI for $\mu_1 - \mu_2$	(1.08, 8.92)

Practitioner Insight: The two-sample Z-test requires both σ_1 and σ_2 to be known, a strong assumption rarely met outside well-characterized industrial settings; otherwise, the Welch t-test is preferred. The standard error $\text{SE} = \sqrt{\sigma_1^2/n_1 + \sigma_2^2/n_2}$ combines variance contributions from both groups. Despite $\sigma_2 \gg \sigma_1$, the larger n_2 tempers its impact, though it still dominates the SE. The test statistic $Z = 2.50$ matches the one-sample case, showing identical inference with only the SE formula differing. When $\sigma_1 = \sigma_2$ and $n_1 = n_2$, $\text{SE} = \sigma\sqrt{2/n}$, analogous to the one-sample case. The result ($p = 0.012$) is statistically significant at 5%, but practical significance depends on context.

Common Pitfalls

- **Using sample SDs in a Z-test.** Z-tests require *known* σ_1, σ_2 from external sources. If variances come from the data (s_1, s_2), use a two-sample t-test; treating them as known understates uncertainty.
- **Ignoring dependence.** The Z-test assumes independent groups. Paired or matched data require a paired t-test (paired Z-test only if the difference variance is known).
- **Equating large n with known variance.** Large samples improve precision but do not make variances "known." Use t-tests (Welch or pooled) unless true population variances are established historically.
- **Equal-variance confusion.** The Z-test does not assume $\sigma_1 = \sigma_2$, but pooled t-tests do. Do not carry the equal-variance assumption incorrectly across tests.
- **Statistical vs. practical significance.** With large samples, trivial mean differences yield tiny p-values. Always report effect sizes (Cohen's d) and confidence intervals to assess real-world impact.

Two-Sample t-Test (Independent)

Compare Means of Two Independent Groups ($\sigma_1^2 = \sigma_2^2$)

MEAN.4 | **Category:** Mean Comparison | **Data:** Continuous | **Null:** t-distribution

When to Use

- Compare means of two independent groups with unknown but approximately equal variances.
- Small to moderate samples where normality is reasonable.
- Balanced designs or similar sample sizes across groups.
- Classical setting where pooled variance improves efficiency.

Assumptions

- Two samples are independent.
- Population variances are equal ($\sigma_1^2 = \sigma_2^2$).
- Observations within each group are i.i.d.
- Data are approximately normal (critical for small samples).
- No extreme outliers affecting mean and variance estimates.

Test Statistic

$$t = \frac{(\bar{x}_1 - \bar{x}_2) - \Delta_0}{\text{SE}}, \quad \text{where SE} = s_p\sqrt{\frac{1}{n_1} + \frac{1}{n_2}}, \quad s_p^2 = \frac{(n_1 - 1)s_1^2 + (n_2 - 1)s_2^2}{n_1 + n_2 - 2}$$

Notation: $\bar{x}_1, \bar{x}_2$ are sample means; s_1, s_2 are sample SDs; n_1, n_2 are sample sizes, $df = n_1 + n_2 - 2$

Plain English: The two-sample t-test compares the means of two independent groups to determine whether they differ by more than would be expected from random sampling variation alone. It measures the observed difference between the sample means relative to the variability of the data and asks whether such a difference is plausible if the two population means are actually equal. The test can be performed using either a pooled variance estimate when the group variances are assumed equal or a separate-variance approach (Welch's t-test) when they are not. If the standardized difference is sufficiently large, the null hypothesis of equal population means is rejected.

⚠ Practitioner Warning

The pooled t-test assumes equal population variances ($\sigma_1^2 = \sigma_2^2$); Levene's test or variance ratio can check this assumption. When this assumption is violated, particularly in the presence of unequal sample sizes, the pooled variance estimate can distort the test's Type I error rate. The test also requires independent samples and is not appropriate for paired or repeated-measures data. When the equal-variance assumption is doubtful, Welch's t-test is generally a safer choice.

Hypotheses

- H_0: $\mu_1 - \mu_2 = \Delta_0$ (often $\Delta_0 = 0$, testing equality).
- H_1: Two-sided: $\mu_1 - \mu_2 \neq \Delta_0$

Decision Rule: Reject H_0 if $|t| > t_{\alpha/2,df}$ (2-sided) or p-value $< \alpha$. Use appropriate df for pooled or Welch.

Software References [Python, R, SAS]

Python	R	SAS
`scipy.stats.ttest_ind`	`t.test(var.equal=TRUE)`	`PROC TTEST`

Additional Notes

The pooled two-sample t-test combines information from both groups to estimate a common variance, s_p^2, and uses $n_1 + n_2 - 2$ degrees of freedom. When the equal-variance assumption holds, pooling stabilizes the variance estimate and can provide slightly greater power than methods that estimate the two variances separately. The test is fairly robust to moderate variance differences when the sample sizes are similar ($n_1 \approx n_2$). Under exact homoskedasticity, it is the uniformly most powerful unbiased test for comparing two normal means. Statistical significance depends on both the magnitude of the difference and the sample size.

Miscellanea: The pooled two-sample t-test builds directly on Gosset's original work, extending the small-sample logic from one sample to two groups under a shared-variance assumption. The idea of pooling variances reflects an early statistical insight: when populations share a common variability, combining information improves estimation efficiency—an idea that later influenced methods such as ANOVA. Historically, the equal-variance assumption was often taken for granted in controlled experiments, especially in early industrial and agricultural studies, where treatments were designed to have similar variability. Today, the pooled t-test is less commonly used than Welch's test, but it remains important as a conceptual bridge between simple mean comparisons and more general linear models.

Reference(s): Fisher (1932); Welch (1947)

Two-Sample t-Test (Independent)

MEAN.4

Worked Example

Problem Statement Researchers compare two teaching methods for mental arithmetic. Four students taught by method A and four by method B sit the same test (out of 30). Scores are: $A:\ 24, 24, 26, 26;\qquad B:\ 20, 22, 22, 24.$

The population standard deviation is unknown but assumed equal for both methods. Test whether the methods produce different mean scores. $H_0: \mu_A = \mu_B;\qquad H_1: \mu_A \neq \mu_B.$

1. Group Means and Deviations

$$\bar{x}_A = \frac{\sum x_{A,i}}{n_A} = 25, \quad \bar{x}_B = \frac{\sum x_{B,j}}{n_B} = 22.$$

Squared deviations:

$$\mathrm{SS}_A = \sum (x_{A,i} - \bar{x}_A)^2 = 4, \quad \mathrm{SS}_B = \sum (x_{B,j} - \bar{x}_B)^2 = 8.$$

2. Pooled Variance

$$s_p^2 = \frac{\mathrm{SS}_A + \mathrm{SS}_B}{n_A + n_B - 2} = 2 \quad \Rightarrow s_p = \sqrt{2}.$$

Equivalently, s_p^2 is the df-weighted average of the two group variances ($s_A^2 = 4/3$, $s_B^2 = 8/3$):

$$s_p^2 = \frac{3 \cdot \frac{4}{3} + 3 \cdot \frac{8}{3}}{6} = 2. \checkmark$$

3. Standard Error

$$\mathrm{SE} = s_p\sqrt{1/n_A + 1/n_B} = 1.$$

4. Test Statistic and Decision

$$t = \frac{\bar{x}_A - \bar{x}_B}{\mathrm{SE}} = 3, \quad \nu = n_A + n_B - 2 = 6.$$

$$p = 2\,P[t(6) > 3] = 0.024.$$

α	$t_{\alpha/2,6}$	$\lvert t\rvert = 3$	Decision
5%	2.447	>	**Reject H_0**
1%	3.707	<	FTR

5. Confidence Interval

A 95% CI for $\mu_A - \mu_B$:

$$(\bar{x}_A - \bar{x}_B) \pm t_{0.025,6} \cdot \mathrm{SE} = 3 \pm 2.447 \times 1 = (0.55,\ 5.45).$$

Zero lies outside: confirms rejection at 5%.

Practitioner Insight: The pooled two-sample t-test applies when the equal-variance assumption $\sigma_A = \sigma_B$ is reasonable, as with comparable groups from the same population. The pooled variance s_p^2 is a degrees-of-freedom–weighted average of group variances; with equal sample sizes it reduces to their simple average (here $s_p^2 = 2$). The construction yields SE = 1 and $t = 3$ exactly, enabling direct comparison with related tests. Equality of variances can be checked via an F-test or Levene's test; the observed variance ratio (2:1) is modest at small n but warrants scrutiny as n grows. If variances differ, the Welch t-test replaces s_p^2 and uses Satterthwaite degrees of freedom, typically giving a slightly larger p-value. Relative to the two-sample Z-test, replacing σ with s_p and $N(0,1)$ with t inflates critical values and p-values (e.g., 1.96 to 2.447; 0.012 to 0.024).

Common Pitfalls

- **Equal variances.** Pooled t-test assumes $\sigma_1^2 = \sigma_2^2$; violations inflate Type I error, especially with unequal n. Quick check: $s_{\max}^2/s_{\min}^2 > 2$. Welch's test is a safer default.
- **Independence.** Use paired t-test for matched or repeated measures; using the independent test on paired data loses power and misstates variability.
- **Small-sample normality.** For $n < 15$, non-normality and outliers can distort results. Check Q–Q plots; consider transformations, Mann–Whitney U, or bootstrapping. For larger samples, CLT provides robustness.
- **Multiple testing.** Running many pairwise t-tests inflates Type I error. Use ANOVA with corrected post-hoc comparisons (Tukey, Bonferroni) or adjust α.
- **Practical importance.** Large samples can make trivial differences "significant." Always report effect sizes, CIs, and context for magnitude.

Welch's t-Test

Compare Means of Two Independent Groups ($\sigma_1^2 \neq \sigma_2^2$)

MEAN.5 | **Category:** Mean Comparison | **Data:** Continuous | **Null:** t-distribution (approximate)

When to Use

- Compare means of two independent populations.
- Population variances may differ substantially.
- Unequal sample sizes are present across groups.
- Best when sample sizes are small and variances differ.
- Preferred default alternative to pooled t-test.

Assumptions

- The two samples are mutually independent.
- Data are continuous measurements of interest.
- Population distributions are approximately normal.
- Population variances may differ across groups.
- Samples are randomly drawn from populations.

Test Statistic

$$t = \frac{(\bar{x}_1 - \bar{x}_2) - \Delta_0}{\text{SE}}, \text{ where SE} = \sqrt{\frac{s_1^2}{n_1} + \frac{s_2^2}{n_2}}. \qquad \text{(Satterthwaite): } df = \frac{(s_1^2/n_1 + s_2^2/n_2)^2}{(s_1^2/n_1)^2/(n_1 - 1) + (s_2^2/n_2)^2/(n_2 - 1)}.$$

Notation: $\bar{x}_1, \bar{x}_2$ are sample means; s_1, s_2 are sample SDs; n_1, n_2 are sample sizes

Plain English: Welch's t-test compares two group means without assuming equal variances. Each group contributes its own variance estimate to the standard error, and degrees of freedom are adjusted based on how much each sample contributes to overall uncertainty. This makes the test robust across a wide range of variance ratios, performing well whether variances are equal or unequal. It's the modern default for two-sample comparisons.

⚠ Practitioner Warning

Welch's test assumes groups are independent. If observations are paired (before-after, matched subjects, repeated measures), use paired t-test instead. Treating paired data as independent loses substantial power and fails to account for within-subject correlation. Signs of pairing: same subjects twice, intentional matching, natural pairs.

Hypotheses

- H_0: $\mu_1 - \mu_2 = \Delta_0$ (often $\Delta_0 = 0$, testing equality).
- H_1: $\mu_1 - \mu_2 \neq \Delta_0$ (two-tailed)

Decision Rule: Reject H_0 if $|t| > t_{\alpha/2,df}$ (two-tailed) or if p-value $< \alpha$. Use Satterthwaite approximation for df.

Software References [Python, R, SAS]

Python	R	SAS
`scipy.stats.ttest_ind`	`t.test(var.equal=FALSE)`	`PROC TTEST`

Additional Notes

Welch's t-test replaces the pooled variance with separate variance estimates for each group, leading to the standard error SE $= \sqrt{s_1^2/n_1 + s_2^2/n_2}$. Because the two variance estimates contribute unequally, the test uses the Satterthwaite approximation for degrees of freedom, typically resulting in a non-integer value smaller than $n_1 + n_2 - 2$. This adjustment accounts for heteroskedasticity and yields more reliable inference when variances differ, especially with unequal sample sizes. Unlike the pooled t-test, Welch's method maintains appropriate Type I error control under variance inequality, which is why it is widely recommended as the default two-sample test. When sample sizes are equal and variances are similar, Welch's and pooled t-tests produce nearly identical results; the difference becomes important only when these conditions are violated. See Technical Notes for additional information.

Miscellanea: Welch's t-test was introduced by Bernard L. Welch in 1947 as a practical solution to the **Behrens–Fisher problem**—testing equality of means when variances are unequal. Earlier approaches struggled to provide an exact small-sample solution; Welch's insight was to approximate the distribution using an adjusted degrees-of-freedom formula (later popularized via Satterthwaite), enabling reliable inference without assuming equal variances. The test became widely adopted with the rise of computing, as its non-integer degrees of freedom are easily handled numerically. Today, it is commonly recommended as the default two-sample test in applied work.

Reference(s): Welch (1947); Satterthwaite (1946)

Worked Example

Problem Statement Reaction times (ms) are recorded for two groups of six participants each. Group 1 receives a standard stimulus; Group 2 receives a novel stimulus. Group 1 : 13, 14, 15, 15, 16, 17; Group 2 : 8, 11, 12, 12, 14, 15. Group 2 looks more variable. The population variances are unknown and not assumed equal. Test whether the groups have the same mean. $H_0: \mu_1 = \mu_2$; $H_1: \mu_1 \neq \mu_2$.

1. Group Summaries

$$\bar{x}_1 = \frac{\sum x_{1,i}}{n_1} = 15, \qquad \bar{x}_2 = \frac{\sum x_{2,i}}{n_2} = 12.$$

$$\mathrm{SS}_1 = \sum (x_{1,i} - \bar{x}_1)^2 = 10, \qquad s_1^2 = \frac{SS_1}{n_1 - 1} = 2,$$

$$\mathrm{SS}_2 = \sum (x_{2,j} - \bar{x}_2)^2 = 30, \qquad s_2^2 = \frac{SS_2}{n_2 - 1} = 6.$$

Group 2 variance is three times Group 1's.

2. Welch Standard Error and Test Statistic

$$\mathrm{SE}^2 = \frac{s_1^2}{n_1} + \frac{s_2^2}{n_2} = \frac{4}{3}, \qquad t_W = \frac{\bar{x}_1 - \bar{x}_2}{\sqrt{\mathrm{SE}^2}} \approx 2.598.$$

3. Satterthwaite Degrees of Freedom

Let $a = s_1^2/n_1 = 1/3$ and $b = s_2^2/n_2 = 1$.

$$\nu = \frac{(a+b)^2}{\dfrac{a^2}{n_1 - 1} + \dfrac{b^2}{n_2 - 1}} = 8.$$

4. p-value and Decision

$$p = 2P[t(8) > 2.598] = 0.032.$$

α	$t_{\alpha/2,8}$	$t_W \approx 2.60$	Decision
5%	2.306	>	**Reject H_0**
1%	3.355	<	FTR

5. Comparison with Pooled t-Test

	Welch	Pooled	Difference
s_p^2 or SE^2	4/3	4/3	identical
t	2.598	2.598	identical
df	$\nu = 8$	$n_1 + n_2 - 2 = 10$	Welch penalized
p-value	0.032	0.027	Welch larger
Decision (5%)	Reject	Reject	same

Note: Normality checks, confidence intervals, and effect sizes are omitted for brevity but should be evaluated in practice.

See Technical Notes at the end of this chapter for more information.

Practitioner Insight: The Welch t-test relaxes the equal-variance assumption by using separate variance terms s_1^2/n_1 and s_2^2/n_2. When $n_1 = n_2$, both Welch and pooled tests produce the same SE and the same t statistic ($t \approx 2.598$ for both tests). The *only* difference is the Satterthwaite df $\nu \leq n_1 + n_2 - 2$, which produces heavier t-distribution tails and a larger critical value, making Welch more conservative. When variances are equal ($s_1^2 = s_2^2$) the Satterthwaite formula returns $\nu = n_1 + n_2 - 2$ exactly, so Welch matches the pooled test with no penalty; the penalty grows with the variance ratio s_2^2/s_1^2, here equal to 3, reducing df from 10 to 8. The general rule is $\nu < n_1 + n_2 - 2$ whenever $s_1^2 \neq s_2^2$, and $\nu \to \min(n_1 - 1, n_2 - 1)$ in the extreme case $s_2^2/s_1^2 \to \infty$. Because Welch adapts to unequal variances without loss when they are equal, it is often recommended as the default, with the pooled test as a special case under verified homoscedasticity.

Common Pitfalls

- **Pre-testing variances.** Running Levene's or F-tests before choosing a method inflates Type I error and adds no benefit; modern practice is to *skip assumption tests* and use Welch's by default.
- **"Only for unequal variances."** Welch's is the recommended *general-purpose* two-sample test: it matches pooled t-test performance when variances are equal and is far superior when they differ.
- **Interpreting fractional *df*.** Welch–Satterthwaite degrees of freedom reflect an approximation to match heavier tails when one group dominates uncertainty. Do not round; software uses the fractional value directly.
- **Small-sample non-normality.** Welch's still requires approximate normality for $n < 15$. For skewed or heavy-tailed data, use transformations, Mann–Whitney U, permutation tests, or bootstrapping.
- **Independence violations.** Welch's test is for *independent* groups. Paired or matched samples require the paired t-test; using an independent test loses power and misstates variability.

Paired t-Test

Compare Means of Dependent Paired Observations

MEAN.6 | **Category:** Mean Comparison | **Data:** Continuous | **Null:** t-distribution

When to Use

- Repeated measurements obtained from the same subjects.
- Matched pairs formed using important subject traits.
- Assess average change between two related conditions.
- Before-and-after studies with naturally paired data.
- Each subject serves as his or her own control.

Assumptions

- Observations form naturally matched one-to-one pairs.
- Individual pairs are independent of one another.
- Paired differences are approximately normal in shape.
- Measurements are numeric with meaningful units.
- Extreme difference scores are not unduly influential.

Test Statistic

$$t = \frac{\bar{d} - \Delta_0}{\text{SE}}, \quad \text{where SE} = \frac{s_d}{\sqrt{n}}. \qquad \text{Under } H_0, t \sim t_{n-1}$$

Notation: $d_i = x_{1i} - x_{2i}$ is difference for pair i, $\bar{d}$ is mean difference, s_d is SD of differences, n is number of pairs.

Plain English: The paired t-test asks whether the average change within pairs is large enough to rule out random fluctuation. By focusing on differences, each subject serves as their own control, removing between-subject variability and sharpening the signal. The question becomes: are these within-pair changes systematic, or just noise?

⚠ Practitioner Warning

The paired t-test assumes each observation in one condition has exactly one matched observation in the other. Incorrect pairing (arbitrary matching, unequal sample sizes, independent samples) invalidates the test. If observations are truly independent, use independent samples t-test instead. Forcing pairing on independent data loses power and df.

Hypotheses

- H_0: $\mu_d = \Delta_0$ (typically $\Delta_0 = 0$ for no difference).
- H_1: Mean difference does not equal Δ_0 (two-sided).

Decision Rule: Reject H_0 if $|t| > t_{\alpha/2,n-1}$ (two-sided) or if p-value $< \alpha$. $df = n - 1$ (number of pairs minus one).

Software References [Python, R, SAS]

Python	R	SAS
`scipy.stats.ttest_rel`	`t.test(paired=TRUE)`	`PROC TTEST`

Additional Notes

Paired t-test reduces a two-measurement problem to a one-sample test on differences. Define $d_i = X_{i1} - X_{i2}$ and test whether the mean difference $\bar{d}$ equals zero using $t = \bar{d}/(s_d/\sqrt{n})$ with $n - 1$ degrees of freedom. By working with within-pair differences, the test removes between-subject variability, often leading to smaller variance and higher power than an independent-samples test. The gain depends on the correlation within pairs: stronger positive correlation yields greater variance reduction. All assumptions apply to the differences, not the raw measurements. In particular, approximate normality of d_i is important for small samples, and extreme outliers in the differences can distort inference. If pairing is weak or negative, the variance reduction disappears and the paired test offers no advantage over independent methods.

Miscellanea: Paired t-test is not a separate distributional result but a direct application of the one-sample t-test to within-pair differences. The idea of pairing traces back to early agricultural experiments. R. A. Fisher's work emphasized blocking and within-plot comparisons, recognizing that comparing treatments within the same unit could eliminate background variability (e.g., soil differences) and substantially increase power. Early applications in agriculture and medicine naturally produced paired data through before–after measurements and matched designs. The key insight is variance reduction: pairing removes shared variability, often improving efficiency without increasing sample size. This highlights a fundamental design principle—careful pairing or blocking can be as valuable as collecting more data. In intuitive terms, pairing acts like noise cancellation—subtracting shared variability to clarify the underlying signal.

Reference(s): Student (1908); Fisher (1932)

Worked Example

Problem Statement A clinician measures systolic blood pressure (mmHg) in five patients before and after a four-week drug trial. Each patient serves as their own control. Test whether the drug produces a non-zero mean reduction. $H_0: \mu_d = 0$; $H_1: \mu_d \neq 0$.

Patient	1	2	3	4	5
Before	140	138	136	132	129
After	139	136	133	128	124
d = Before − After	1	2	3	4	5

1. Summary of Differences

$$\bar{d} = \frac{\sum d_i}{n} = 3, \qquad SS_d = \sum (d_i - \bar{d})^2 = 10.$$

$$s_d^2 = \frac{SS_d}{n-1} = \frac{5}{2}, \quad SE^2 = \frac{s_d^2}{n} = \frac{1}{2}, \quad SE = \frac{1}{\sqrt{2}}.$$

2. Test Statistic

$$t = \frac{\bar{d}}{SE} = 3\sqrt{2} \approx 4.243, \qquad \nu = n - 1 = 4.$$

Under H_0: $t \sim t(4)$.

3. p-value and Decision

$$p = 2P[t(4) > 3\sqrt{2}] = 0.013.$$

α	$t_{\alpha/2,4}$	$t = 3\sqrt{2}$	Decision
5%	2.776	>	**Reject** H_0
1%	4.604	<	FTR

4. Confidence Interval

$$\bar{d} \pm t_{0.025,4} \cdot SE = 3 \pm \frac{2.776}{\sqrt{2}} = (1.04,\ 4.96).$$

Zero excluded: confirms rejection at 5%.

5. Why Pairing Matters: Unpaired Comparison

Applying the pooled two-sample t-test to the same data (ignoring the pairing):

$$s_p^2 = \frac{80 + 146}{8} = 28.25, \quad t_{\text{unpaired}} = 0.892 \text{ (df} = 8),$$

with $p = 0.398$. The unpaired test fails to reject. The paired test's p-value is 30 times smaller.

Correlation between Before and After is 0.9993 (very strong, positive), confirming that pairing meaningfully reduces variability.

Practitioner Insight: The paired t-test is simply a one-sample t-test applied to within-pair differences d_i, using $\bar{d}$ and s_d. Pairing removes between-subject variability from the error term; here baseline pressures vary widely, making the pooled unpaired variance much larger than $s_d^2 = 2.5$. The paired analysis therefore focuses on each patient's change, where $d_i = 1, 2, 3, 4, 5$ gives $\bar{d} = 3$, $SS_d = 10$, $SE = 1/\sqrt{2}$, and $t = 3\sqrt{2}$ exactly. The benefit of pairing depends on the before–after correlation: strong correlation yields major precision gains, while weak correlation offers little advantage. The 95% CI (1.04, 4.96) excludes zero, establishing a positive reduction, though its clinical size is modest. Treating the same data as independent groups would waste the design and give a nonsignificant result ($t \approx 0.89$, $p = 0.40$).

Common Pitfalls

- **Using the wrong design.** The paired t-test requires observations to be genuinely matched or repeatedly measured on the same unit. If the samples are independent, use the two-sample t-test instead.
- **Non-normal differences (small n).** The normality assumption applies to the *differences*, not the original measurements. With small samples ($n < 15$), check Q–Q plots; if strongly violated, consider Wilcoxon or resampling methods.
- **Misinterpreting the subtraction order.** Define and report the direction of the differences (e.g., After–Before). The sign of the mean difference and the interpretation of one-sided tests depend on this choice.
- **Using multiple paired tests.** Repeatedly applying paired t-tests across several time points inflates Type I error. Use repeated-measures ANOVA or adjust for multiple comparisons when appropriate.
- **Assuming pairing always helps.** Pairing improves precision only when within-pair correlation is positive. Weak or negative correlation can provide little benefit and may even reduce efficiency.

One-Way ANOVA F Test

Compare Means Across Three or More Independent Groups

MEAN.7 | **Category:** Mean Comparison | **Data:** Continuous | **Null:** F-distribution

When to Use

- Compare means across three or more independent groups.
- Test for any overall difference among group means.
- Control Type I error across multiple group comparisons.
- Extension of two-sample t-test to multiple groups.
- Designed or observational studies with categorical factors.

Assumptions

- Observations are independent within and across groups.
- Group populations are approximately normal for small n.
- Population variances are equal across groups ($\sigma_i^2 = \sigma^2$).
- Samples are randomly drawn or groups randomly assigned.
- No extreme outliers affecting group means or variances.

Test Statistic

$$F = \frac{\text{MSB (mean square between groups)}}{\text{MSW (mean square within groups)}} = \frac{\sum_{i=1}^{k} n_i(\bar{x}_i - \bar{x})^2/(k-1)}{\sum_{i=1}^{k}\sum_{j=1}^{n_i}(x_{ij} - \bar{x}_i)^2/(N-k)} \sim F_{k-1,N-k} \quad \text{(under } H_0\text{)}.$$

Notation: k is number of groups, n_i is size of group i, $N = \sum n_i$ is total sample size

Plain English: One-way ANOVA partitions total variability into two sources: variation between group means and variation within groups. If group means truly differ, between-group variability will be large relative to within-group variability, producing a large F-statistic. The test asks: "Is the variability among group means larger than we'd expect from random sampling alone?" It's an omnibus test—detecting *any* differences among means—but doesn't identify which specific groups differ.

⚠ Practitioner Warning

ANOVA tests only whether at least one group differs from others. A significant F-test doesn't identify which groups differ or by how much. When homoscedasticity is doubtful, Welch's ANOVA is preferred. Always follow significant ANOVA with post-hoc tests (e.g., Tukey HSD) or planned contrasts to determine specific group differences. Without follow-up, you know differences exist but not where.

Hypotheses

- H_0: All group means are equal: $\mu_1 = \mu_2 = \cdots = \mu_k$.
- H_1: At least one group mean differs from others.

Decision Rule: Reject H_0 if $F > F_{\alpha,k-1,N-k}$ or if p-value $< \alpha$.

Software References [Python, R, SAS]

`scipy.stats.f_oneway`	`aov()`	`PROC ANOVA / GLM`

Additional Notes

One-way ANOVA decomposes total variability as SST = SSB + SSW, where SSB captures between-group variation and SSW captures within-group noise. The F-statistic compares these components via $F = \text{MSB}/\text{MSW}$, with large values indicating that differences among group means are substantial relative to within-group variability. The test is omnibus: it detects whether at least one group mean differs but does not identify which groups differ. Significant results should therefore be followed by post-hoc comparisons (e.g., Tukey, Dunnett, or Games–Howell). ANOVA is reasonably robust to moderate departures from normality, especially in balanced designs with adequate sample sizes. However, unequal variances—particularly with unequal sample sizes—can inflate Type I error; Welch's ANOVA is then preferred.

Miscellanea: The term ANOVA ("analysis of variance") is paradoxical: it tests differences in means by comparing variances. R. A. Fisher introduced ANOVA in the 1920s at Rothamsted to analyze agricultural experiments with multiple treatments, formalizing both variance decomposition and the F-distribution. For two groups, ANOVA reduces exactly to the pooled two-sample t-test, with $F = t^2$. Fisher's framework provided a unified approach to testing multiple means without inflating Type I error and later became the foundation of the general linear model.

Reference(s): Fisher (1932); Scheffé (1959)

Worked Example

Problem Statement Do four teaching methods (Lecture, Discussion, Online, Hybrid) differ in mean test scores?

Data (0–100 scale). Method A: 72, 68, 75, 70, 73, 69, 71; Method B: 78, 82, 76, 80, 79, 81, 77, 83; Method C: 65, 68, 62, 67, 64, 66, 63; Method D: 74, 76, 75, 77, 73, 78, 76, 75.

1. Sample Summary and ANOVA

Total $N = 30$, $k = 4$ groups. Grand mean: $\bar{x} = 73.10$. Between-groups (effect): SSB $= \sum_i n_i(\bar{x}_i - \bar{x})^2 = 859.84$; $df = 4 - 1 = 3$. Within-groups (error): $SSW = \sum_i SS_i$, where $SS_i = \sum_j (x_{ij} - \bar{x}_i)^2$. Total SS: $SST = \sum_i \sum_j (x_{ij} - \bar{x})^2$. $SST = SSB + SSW$. $MS = SS/df$. F = ratio of mean sum of squares ($MS_{\text{Between}}/MS_{\text{Within}}$).

Method	n_i	$\bar{x}_i$	SS_i	df_i
A (Lecture)	7	71.14	34.86	6
B (Discussion)	8	79.50	42.00	7
C (Online)	7	65.00	28.00	6
D (Hybrid)	8	75.50	18.00	7
Total	30		122.86	26

Source	SS	df	MS	F	p
Between	859.84	3	286.61	60.66	< 0.0001
Within	122.86	26	4.73		
Total	982.70	29			

2. Decision and Conclusion

At $\alpha = 0.05$, $F_{0.05,3,26} \approx 2.98$. Since $F = 60.66 \gg 2.98$ ($p < 0.0001$), we *reject* H_0 and conclude that teaching method has a very strong effect on test scores.

3. Effect Size and Post-hoc Note

Effect size is $\eta^2 = 859.84/982.70 \approx 0.88$, implying an extremely large method effect. Tukey HSD (not shown) would likely detect significant differences for most pairs (B > D > A > C).

Common Pitfalls

- **Multiple t-tests.** Running $k(k-1)/2$ pairwise t-tests inflates Type I error (e.g., $k = 4$ gives 26% false-positive risk). Use ANOVA for the omnibus test, then apply corrected post-hoc comparisons (Tukey, Bonferroni).
- **Stopping after a significant F-test.** ANOVA only tells us that *at least one* mean differs. Follow with post-hoc tests, planned contrasts, or trend analyses to identify where differences occur.
- **Ignoring variance heterogeneity.** ANOVA assumes equal variances; violations (especially with unequal n) distort Type I error. Check via Levene's test, variance ratios, or plots. If unequal, use Welch's ANOVA or variance-stabilizing transformations.
- **Using ANOVA for two groups.** With two groups, ANOVA and the two-sample t-test are equivalent ($F = t^2$). Use the t-test unless the comparison is part of a broader factorial design.
- **Misinterpreting non-significance.** A non-significant F-test does not prove equality—it may reflect low power, small effects, or high within-group variability. Report effect sizes (e.g., η^2) and consider power analysis.

Comparison with Alternative Tests

Test	Best For	Key Feature	Robustness
One-Way ANOVA	3+ independent groups, equal variances	Tests equality of group means	Moderate to assumption violations
Welch's ANOVA	Unequal variances across groups	Mean comparison without homoskedasticity	High to variance inequality
Kruskal–Wallis	Ordinal or markedly non-normal data	Rank-based comparison of group locations	High to non-normality
Trimmed-Mean ANOVA	Heavy tails or outliers	Compares robust trimmed means	High to outliers
Permutation ANOVA	Small samples or doubtful assumptions	Uses randomization instead of theory	High under minimal assumptions

Welch's ANOVA

One-Way ANOVA Without Equal Variance Assumption

MEAN.8 | **Category:** Mean Comparison | **Data:** Continuous | **Null:** F-distribution (approx.)

When to Use

- Compare means across three or more groups with unequal variances. Protect against Type I error inflation
- Variance equality is doubtful or clearly violated.
- Unequal sample sizes with heteroskedasticity across groups.
- Robust alternative to classical ANOVA with unequal σ_i^2.

Assumptions

- Observations are independent within and across groups.
- Group populations are approximately normal for small n.
- Groups have adequate sample sizes for variance estimation.
- No extreme outliers affecting group means or variances.
- Samples are randomly drawn or groups randomly assigned.

Test Statistic

$$F_W = \frac{\sum_{i=1}^{k} w_i(\bar{x}_i - \bar{x}_w)^2/(k-1)}{1 + \frac{2(k-2)}{k^2-1}\sum_{i=1}^{k}\frac{(1-w_i/\sum w_j)^2}{n_i-1}} \sim F_{k-1,\nu} \text{ (under } H_0), \qquad \bar{x}_w = \frac{\sum w_i\bar{x}_i}{\sum w_i}, \qquad \nu = \frac{k^2-1}{3\sum_{i=1}^{k}\frac{(1-w_i/\sum w_j)^2}{n_i-1}}.$$

where $\bar{x}_w$ is weighted grand mean and ν is the adjusted degrees of freedom.

Notation: k is number of groups; n_i, $\bar{x}_i$, and s_i^2 are group sizes, means and variances; $w_i = n_i/s_i^2$ are weights.

Plain English: Welch's ANOVA compares group means while allowing each group to have its own variability. It gives more weight to precise (less variable) groups and downweights noisier ones, adjusting degrees of freedom accordingly. This yields a fairer comparison when variances differ, reducing distortion from high-variance groups, while producing results similar to classical ANOVA when variances are roughly equal.

⚠ Practitioner Warning

Welch's ANOVA is an omnibus test: it detects differences but not which groups differ. Significant results require follow-up comparisons using unequal-variance methods (e.g., Games–Howell); equal-variance procedures (e.g., Tukey HSD) can mislead. While robust to variance inequality, it does not guard against severe non-normality or influential outliers.

Hypotheses

- H_0: All group means are equal: $\mu_1 = \mu_2 = \cdots = \mu_k$.
- H_1: At least one group mean differs from others.

Decision Rule: Reject H_0 if $F_W > F_{\alpha,k-1,\nu}$ or if p-value $< \alpha$. numerator $df_1 = k - 1$, denominator $df_2 = \nu$.

Software References [Python, R, SAS]

`pingouin.welch_anova()`	`oneway.test`	`PROC GLM (WELCH option)`

Additional Notes

Welch's ANOVA modifies classical one-way ANOVA to handle unequal variances by weighting group means inversely to their variances, downweighting noisier groups. The resulting statistic compares weighted between-group to within-group variability, improving reliability under heteroskedasticity. Because variances differ, the test uses a Satterthwaite-type approximation for denominator degrees of freedom, typically yielding non-integer values and an approximate F-distribution. Welch's ANOVA provides better Type I error control when variances are unequal, especially with unequal sample sizes. When variances are similar and samples balanced, both methods give comparable results, making Welch's ANOVA a safe default with minimal efficiency loss. It is distinct from repeated-measures corrections (e.g., Greenhouse–Geisser), which address different variance structures.

Miscellanea: Welch's ANOVA is the multi-group extension of Welch's two-sample t-test, developed by Bernard L. Welch in response to the Behrens–Fisher problem of unequal variances. A recognizable feature is its fractional degrees of freedom. The method was designed to prevent high-variance groups from inflating the error term and masking differences among other groups—a known limitation of classical ANOVA. Although slower to gain adoption historically, Welch's ANOVA is now widely recommended when variance equality is uncertain, paralleling the role of Welch's t-test in two-sample analysis.

Reference(s): Welch (1951); Games & Howell (1976)

Worked Example

Problem Statement Do three exercise programs affect weight loss differently?

Dataset (Weight Loss in kg):

Low: 2.1, 3.2, 2.8, 3.5, 2.4, 3.0, 2.7, 3.1, 2.9
Mod: 5.2, 6.8, 7.1, 5.9, 6.3, 7.5, 6.0, 5.8
High: 8.2, 12.5, 9.7, 15.3, 10.8, 11.2, 13.6, 9.5, 14.1, 10.9

Number of groups: $k = 3$.

Note: The High Intensity group shows much greater variability, violating the assumption of homogeneity of variance. We must use Welch's ANOVA.

1. Summary Statistics

Group	n	Mean ($\bar{x}$)	Var (s^2)	$w_i = n/s^2$
Low	9	2.86	0.178	50.63
Mod	8	6.33	0.576	13.88
High	10	11.58	5.068	1.97
		$\bar{x}_w = 3.84$		$\sum w = 66.48$

2. The Welch F-Statistic

First, calculate the variance between groups (A) and the adjustment term (Λ):

$$A = \frac{1}{k-1}\sum w_i(\bar{x}_i - \bar{x}_w)^2 = \frac{1}{2}[50.63(2.86-3.84)^2 + 13.88(6.33-3.84)^2 + 1.97(11.58-3.84)^2] = \mathbf{126.48},$$

$$\Lambda = \frac{3\sum \frac{(1-w_i/\sum w)^2}{n_i-1}}{k^2-1} = \mathbf{0.075}, \qquad F_W = \frac{A}{1+\frac{2(k-2)}{3}\cdot\Lambda} = \mathbf{120.43},$$

$$\text{Num. } df: k-1 = \mathbf{2}, \qquad \text{Denom. } df: \nu = 1/\Lambda = \mathbf{13.26}, \qquad \text{Critical value: } F_{(2,13.26)} \approx \mathbf{3.79}.$$

3. Decision and Conclusion

Since $F_W > F_{(2,13.26)}$, we reject H_0. There is a statistically significant difference in weight loss across the three programs.

4. Comparison with Classical ANOVA

If we incorrectly applied the equal-variance ANOVA despite the severe heteroscedasticity (variance ratio $\approx 28{:}1$), we would obtain $F = 86.38$ with $df = (2, 24)$ and $p < 0.0001$ (critical value $F_{0.05,2,24} = 3.40$). Although the result remains significant, the inference would be unreliable: the classical ANOVA F is noticeably smaller than Welch's (86.38 vs. 120.43), Type I error would be inflated, confidence intervals would be misestimated, and the unequal precision across groups would be ignored. Welch's ANOVA provides valid inference under these conditions.

5. Post-hoc Testing (Games–Howell / Dunnett's T3)

When Welch's ANOVA indicates significant differences, follow-up comparisons should use procedures that allow unequal variances and unequal sample sizes. The **Games–Howell test** is generally preferred—more power with excellent Type I error control—while **Dunnett's T3** is slightly more conservative and useful when sample sizes are very small or variance ratios are extreme.

Common Pitfalls

- **Ignoring heteroscedasticity.** Standard ANOVA can severely inflate Type I error when variances differ, especially with unequal sample sizes. Welch's ANOVA maintains valid inference and should be the default for one-way comparisons.
- **Wrong post-hoc tests.** After Welch's ANOVA, equal-variance procedures (e.g., Tukey HSD) are inappropriate. Use Games–Howell or Dunnett's T3, which allow unequal variances and sample sizes.
- **Treating Welch's ANOVA as a special case.** Welch's ANOVA performs nearly as well as standard ANOVA under equal variances and far better when variances differ. There is little downside to routine use.
- **Pre-testing variances.** Preliminary tests (e.g., Levene's) create a two-stage procedure that inflates error rates. Modern practice recommends skipping assumption tests and using Welch's ANOVA directly.
- **Assuming full robustness.** Welch's ANOVA addresses unequal variances but still assumes independence and approximate normality. For severe non-normality, consider transformations, Kruskal–Wallis, or robust ANOVA methods.

Two-Way ANOVA

Compare Means Across Two Factors Simultaneously

MEAN.9 | **Category:** Mean Comparison | **Data:** Continuous | **Null:** F-distribution

When to Use

- Assess effects of two categorical factors jointly.
- Evaluate interaction between two experimental factors.
- Compare means across combinations of factor levels.
- Factorial designs improve efficiency and insight.
- Continuous outcomes measured under multiple conditions.

Assumptions

- Observations are independent within and across groups.
- Group populations are approximately normal for small n.
- Equal variances across all factor-level combinations.
- Data are balanced or reasonably balanced across cells.
- No extreme outliers affecting cell means or variances.

Test Statistic

$$F_A = \frac{\text{MS}_A}{\text{MS}_E} \sim F_{a-1,N-ab}, \quad F_B = \frac{\text{MS}_B}{\text{MS}_E} \sim F_{b-1,N-ab}, \quad F_{AB} = \frac{\text{MS}_{AB}}{\text{MS}_E} \sim F_{(a-1)(b-1),N-ab}$$

F_A tests main effect of Factor A, F_B tests main effect of Factor B, F_{AB} tests interaction.

Notation: a = levels of Factor A, b = levels of Factor B, n_{ij} = cell size, N = total sample, MS = mean square.

Plain English: Two-way ANOVA partitions total variability into four components: Factor A main effect, Factor B main effect, interaction effect (A×B), and error. Main effects test whether factor levels differ on average (collapsing across the other factor). Interaction tests whether the effect of one factor changes depending on levels of the other.

⚠ Practitioner Warning

A significant interaction means factors' effects are not additive—the effect of one factor depends on the other. When interaction is significant, main effects can be misleading and should be interpreted cautiously. Simple effects analysis (effect of one factor at each level of the other) is typically more informative than main effects when interaction is present.

Hypotheses

- H_0: Main Effect: $\alpha_i = 0, \beta_j = 0$. Interaction: $(\alpha\beta)_{ij} = 0$.
- H_1: At least one is nonzero in respective comparison.

Decision Rule: Reject H_0 if $F > F_{\alpha,df_1,df_2}$ or if $p < \alpha$. Test interaction first; if significant, interpret main effects cautiously.

Software References [Python, R, SAS]

`statsmodels.stats.anova.anova_lm` `aov()` `PROC ANOVA / PROC GLM`

Additional Notes

Two-way ANOVA can be viewed as a linear model with two categorical predictors and their interaction: $y_{ijk} = \mu + \alpha_i + \beta_j + (\alpha\beta)_{ij} + \epsilon_{ijk}$. The corresponding F-tests evaluate each term using mean squares relative to a common error term. Balanced designs (equal observations per cell) simplify computation, ensure orthogonality of effects (statistically independent), and make sums of squares uniquely defined. In unbalanced designs, results depend on the choice of Type I, II, or III sums of squares, requiring careful interpretation. Effect sizes such as partial η^2 are commonly reported for each factor and interaction. Diagnostic checks (residual plots, variance homogeneity) remain essential, as violations—especially unequal variances across cells—can affect inference.

Miscellanea: R. A. Fisher developed two-way ANOVA in the 1920s to analyze experiments with multiple factors (e.g., fertilizer and irrigation). His key contribution was formalizing *interaction*—recognizing that the effect of one factor may depend on another, moving science beyond "one-variable-at-a-time" experimentation. This insight transformed experimental design: interactions often carry more meaning than main effects, revealing how treatments work in combination. Fisher also championed factorial designs for efficiency, noting that each observation informs multiple effects. A practical constraint is that interaction requires replication: with only one observation per cell, it cannot be separated from random error. Beyond comparing means, two-way ANOVA also underpins variance component analysis in random-effects settings.

Reference(s): Fisher (1932, 1935)

Worked Example

Problem Statement Do teaching method and class size affect test scores, and is there an interaction between them?

Dataset (n = 5 replications per cell):

Lecture:
(S) 78, 82, 75, 80, 77
(M) 72, 75, 70, 73, 71
(L) 65, 68, 62, 66, 64

Discussion:
(S) 85, 88, 82, 86, 84
(M) 82, 85, 79, 83, 81
(L) 75, 78, 72, 76, 74

Design: 2 (Method) × 3 (Size) Factorial ANOVA (N = 30)

1. Cell & Marginal Means ($\bar{x}_{ij}$)

Method	Small	Med.	Large	**Marg.**
Lecture	78.4	72.2	65.0	71.87
Discussion	85.0	82.0	75.0	80.67
Marg.	81.7	77.1	70.0	$\bar{x}_G$ = **76.27**

$$SS_U = \sum_{i=1}^{2}\sum_{j=1}^{3}\sum_{k=1}^{5} x_{ijk}^2 = 175916$$

2. Correction Factor

$$C = N\bar{x}_G^2 = 30(76.27^2) = \mathbf{174{,}498.13}$$

3. Calculations of Sum of Squares (SS)

$$SS_A = 5\cdot 3\,(\bar{x}_{Lec}^2 + \bar{x}_{Disc}^2) - C = \mathbf{580.80}, \qquad SS_B = 5\cdot 2\,(\bar{x}_S^2 + \bar{x}_M^2 + \bar{x}_L^2) - C = \mathbf{694.87}.$$

$$SS_{AB} = 5\sum \bar{x}_{ij}^2 - C - SS_A - SS_B = \mathbf{18.20}, \qquad SS_E = \underbrace{(SS_U - C)}_{SS_{\text{Total}}} - SS_A - SS_B - SS_{AB} = \mathbf{124.00}.$$

4. ANOVA Summary Table

Source of Variation	SS	df	MS	F	F_{crit} (α = .05)	Conclusion
Teaching Method (A)	580.80	1	580.80	**112.41**	4.26	Significant
Class Size (B)	694.87	2	347.43	**67.25**	3.40	Significant
Interaction ($A \times B$)	18.20	2	9.10	**1.76**	3.40	Not significant
Error (Within-cell)	124.00	24	5.17			
Total	1417.87	29				

Common Pitfalls

- **Interaction overrides main effects.** A significant interaction means factors do not operate independently; main effects can be misleading. Examine *simple effects* and use interaction plots instead.
- **Using multiple one-way ANOVAs.** Separate one-way tests inflate Type I error, ignore interaction, and waste power. Two-way ANOVA tests both factors and their interaction in one coherent model.
- **Misunderstanding SS Types in unbalanced designs.** With unequal cell sizes, different SS types test different hypotheses. Type I (sequential and order-dependent), Type II (each main effect after controlling for the others but no interaction present), and Type III (each effect controlling for all others) differ. For balanced designs they coincide. Use Type II when interaction is absent, Type III when present.
- **Underpowered or uneven cell sizes.** Interaction needs more power than main effects; aim for 5–10 observations per cell, ideally balanced. Very small or unequal cells reduce power and complicate interpretation.
- **Ignoring variance homogeneity across all cells.** Two-way ANOVA requires similar variances across all *ab* cells. Check with Levene's test or variance ratios. If violated, consider transformations or nonparametric alternatives.

Variety of Effect Size Formulas

Eta-squared: $\eta_A^2 = SS_A/SS_{\text{total}}$; Partial eta-squared: $\eta_{p,A}^2 = SS_A/(SS_A + SS_E)$; Omega-squared: $\omega_A^2 = [SS_A + (a-1)MS_E)]/[SS_{\text{total}} + MS_E]$.

Repeated Measures ANOVA

Compare Means Across Repeated Measurements on Same Subjects

MEAN.10 | **Category:** Mean Comparison | **Data:** Continuous | **Null:** F-distribution

When to Use

- Compare means across repeated measurements on subjects.
- Assess change over time within the same subjects.
- Multiple conditions measured on each participant.
- Subject-to-subject variability should be controlled.
- Correlated observations arise from repeated measures.

Assumptions

- Observations from different subjects are independent.
- Residuals are approximately normally distributed.
- Response variable is measured on interval scale.
- Variances of pairwise differences are equal.
- Missing observations are minimal or absent.

Test Statistic

$$F = \frac{\text{MSB}}{\text{MSE}} = \frac{\sum_{j=1}^{k} n(\bar{x}_{\cdot j} - \bar{x})^2/(k-1)}{\text{SS}_{\text{error}}/(n-1)(k-1)} \sim F_{k-1,(n-1)(k-1)} \text{ (under } H_0\text{)}, \quad \text{where } \text{SS}_{\text{error}} = \sum_{i=1}^{n}\sum_{j=1}^{k}(x_{ij} - \bar{x}_{i\cdot} - \bar{x}_{\cdot j} + \bar{x}_{\cdot\cdot})^2$$

Notation: k = time points, n = subjects; MSB and MSE denote between-time and error mean squares.

Plain English: Repeated measures ANOVA extends the paired t-test to three or more time points by repeatedly measuring the same subjects. By removing between-subject variability from the error term, it increases power to detect time or treatment effects. Variability is partitioned into subject effects, time effects, and residual error. A key assumption is *sphericity*, requiring equal variances of all pairwise differences between time points.

⚠ Practitioner Warning

Sphericity is frequently violated, especially when measurements are unevenly spaced or variability changes over time. Violations inflate Type I error rates. Always test sphericity (e.g., Mauchly's test) and apply corrections when needed: Greenhouse–Geisser (conservative), Huynh–Feldt (less conservative), or use a multivariate approach that avoids the assumption entirely.

Hypotheses

- H_0: $\mu_1 = \mu_2 = \cdots = \mu_k$.
- H_1: At least one time point mean differs from others.

Decision Rule: Reject H_0 if $F > F_{\alpha,df_1,df_2}$ or if $p < \alpha$.

Software References [Python, R, SAS]

`statsmodels.stats.anova.AnovaRM` `car::Anova` `PROC GLM / REPEATED`

Additional Notes

Repeated measures ANOVA extends the paired t-test to three or more related measurements, modeling within-subject correlation by partitioning variability into between-subject and within-subject components. This removes subject-specific baseline effects and improves efficiency. The key additional assumption beyond one-way ANOVA is *sphericity*, which ensures valid F-tests for within-subject effects. When violated, corrections such as Greenhouse–Geisser or Huynh–Feldt adjust degrees of freedom to maintain appropriate Type I error control. Because measurements are taken on the same subjects, complete data are typically required, missing values complicate analysis and are better handled using mixed-effects models. More generally, repeated measures ANOVA is a special case of a linear mixed model, which offers greater flexibility for unequal variances, unbalanced designs, and complex correlation structures.

Miscellanea: The sphericity assumption emerged from early multivariate theory (e.g., Wilks, 1940s) but was not widely emphasized in repeated measures practice for some time. Mauchly (1940) proposed a formal test, though its routine use increased with the rise of computing and statistical software in the 1960s–70s. Recognition that sphericity violations inflate Type I error rates in within-subject tests led to the development of standard corrections: Greenhouse–Geisser (1959) and Huynh–Feldt (1976), which remain widely used today.

Reference(s): Mauchly (1940); Greenhouse & Geisser (1959)

Worked Example

Problem Statement Does a weight loss intervention produce sustained weight loss? Data ($n = 8, k = 4, N = nk$) below:

	Subj 1	Subj 2	Subj 3	Subj 4	Subj 5	Subj 6	Subj 7	Subj 8	$\bar{x}_{.j}$
Base	95	102	88	110	92	98	105	85	96.88
3m	88	94	82	102	85	91	97	79	89.75
6m	86	92	80	99	84	89	95	78	87.88
12m	87	94	82	101	85	90	96	79	89.25
$\bar{x}_{i.}$	89.00	95.50	83.00	103.00	86.50	92.00	98.25	80.25	$\bar{x}_G$ = **90.94**

$SS_T = N\sum\sum x_{ij}^2 - N\bar{x}_G^2$, $SS_{Time} = n\sum\bar{x}_{.j}^2 - N\bar{x}_G^2$, $SS_{Subj} = k\sum\bar{x}_{i.}^2 - N\bar{x}_G^2$, $SS_{Error} = SS_T - SS_{Time} - SS_{Subj}$.

1. ANOVA Table & Sphericity Correction

Source	SS	df	MS	F	p-value
Between Subjects	1686.10	7	240.87	–	–
Time (Within)	391.28	3	130.43	**326.91**	0.000
Error (Residual)	8.38	21	0.40	–	–

2. Assumption Check Pairwise difference variances range from -2.38 (6m - 12m) to 20.29 (base - 6m). Mauchly's test (W = 0.5038, p-value = 0.000) suggests sphericity violation. **Greenhouse-Geisser ($\hat{\epsilon} \approx 0.4617$)** [See Technical Notes] applied:

- Adjusted df : $df_1 = 3(0.4617) = \mathbf{1.385}$, $df_2 = 21(0.4617) = \mathbf{9.695}$
- Adjusted critical value: $F_{crit(1.385,9.695)} \approx 4.58$. Result: $F = 326.91$, $\mathbf{p < 0.0001}$.

3. Conclusion Evidence for weight change over time is significant both before and after sphericity correction ($p < 0.0001$). Weight decreases through 6 months, followed by slight regain at 12 months.

Common Pitfalls

- **Ignoring sphericity.** Violations inflate Type I error. Always run Mauchly's test; use Greenhouse–Geisser or Huynh–Feldt corrections when sphericity fails.
- **Using ANOVA with missing data.** Repeated measures ANOVA requires complete cases. With dropout or unbalanced data, use linear mixed models.
- **Treating repeated measures as independent.** One-way ANOVA ignores within-subject correlation and wastes power. Use within-subjects analysis.
- **Confusing repeated measures with blocking.** RM uses the same subjects across conditions; blocks use different subjects. Apply the correct framework.
- **Invalid post-hoc tests.** Between-subjects procedures (e.g., Tukey) may be inappropriate when sphericity fails. Use Bonferroni or mixed-model contrasts.

Comparison with Alternative Approaches

Approach	Best For	Missing Data	Assumptions
Repeated Measures ANOVA	Balanced design, complete data	Cannot handle	Sphericity + normality
Mixed Effects Model	Unbalanced, missing data, complex designs	Handles naturally	Approx. Normality, flex. correlation
MANOVA (Multivariate)	Few time points, no sphericity assumption	Requires complete data	Multivariate normality
Friedman Test	Non-normal, ordinal repeated measures	Cannot handle easily	Matched blocks; rank-based
GEE	Large samples, population-averaged effects	Handles MAR missingness	Working correlation structure

ANCOVA F Test

Compare Group Means While Controlling for Covariates

MEAN.11 | **Category:** Mean Comparison | **Data:** Continuous | **Null:** F-distribution

When to Use

- Compare group means while adjusting for covariates.
- Reduce unexplained variability using covariates.
- Baseline differences may confound treatment effects.
- Continuous outcomes influenced by known predictors.
- Improve precision of treatment-effect estimates.

Assumptions

- Observations are independent within and across groups.
- Group populations are approximately normally distributed.
- Covariate effects are linear across observations.
- Regression slopes are equal across all groups.
- Covariates are measured without substantial error.

Test Statistic

$$F = \frac{\text{MS}_{\text{groups (adj)}}}{\text{MS}_{\text{error (adj)}}} = \frac{\text{SS}_{\text{groups (adj)}}/(k-1)}{\text{SS}_{\text{error (adj)}}/(N-k-1)} \sim F_{k-1,N-k-1} \text{ (Under } H_0\text{)}$$

Notation: k groups, covariate X, outcome Y, adjusted means $\bar{y}_{i,\text{adj}}$ control for X

Plain English: ANCOVA blends regression and ANOVA by comparing group means after adjusting for one or more continuous covariates. It effectively levels the playing field, asking how the groups would compare if they had the same covariate values. By removing variation explained by the covariates, the test often increases precision and power, yielding adjusted means that better isolate the effect of interest. A key requirement is that the covariate–outcome relationship is similar across groups (parallel slopes).

⚠ Practitioner Warning

ANCOVA requires a common covariate–outcome slope across groups (homogenenous slopes). If slopes differ, adjusted means become invalid or undefined. Always test the covariate×group interaction first; if violated, use interaction models or report separate group-specific regressions instead of ANCOVA. The covariate should be related to the outcome but unaffected by treatment. Using post-treatment covariates can bias results and distort the interpretation of adjusted means.

Hypotheses

- H_0: $\mu_{1,\text{adj}} = \mu_{2,\text{adj}} = \cdots = \mu_{k,\text{adj}}$ (controlling for covariate).
- H_1: At least one adjusted mean differs from others.

Decision Rule: Reject H_0 if $F > F_{\alpha,k-1,N-k-1}$ or if $p < \alpha$. Follow with post-hoc tests on adjusted means if significant.

Software References [Python, R, SAS]

```
statsmodels.stats.anova.anova_lm    car::anova()    PROC GLM
```

Additional Notes

ANCOVA combines ANOVA and regression by modeling the outcome as a function of group indicators and continuous covariate(s), making it a special case of multiple regression. Group effects are assessed after adjusting for covariates, yielding *adjusted means* (least-squares means) that represent comparisons at a common covariate level. In simple settings, adjusted means can be expressed as $\bar{y}_{i,\text{adj}} = \bar{y}_i - b(\bar{x}_i - \bar{x})$, where b is the pooled regression slope. The covariate reduces residual variability, improving precision and statistical power, especially when it strongly predicts the outcome and differs across groups. This adjustment is most effective when covariates are measured reliably and are not influenced by the treatment. Interpretation relies on the assumption of parallel slopes across groups; when this holds, group differences correspond to vertical shifts in the outcome–covariate relationship, linking classical designs to modern regression frameworks.

Miscellanea: ANCOVA emerged in the mid-20th century as an extension of Fisher's ANOVA framework, incorporating regression ideas to adjust for continuous covariates while comparing group means. Early applications were prominent in agricultural and medical studies, where baseline differences needed adjustment without increasing sample size. ANCOVA highlighted an important design principle: controlling variability through modeling (covariates) can be as effective as collecting more data. Today, ANCOVA is recognized as a special case of the general linear model, bridging classical experimental design and modern regression analysis.

Reference(s): Fisher (1932); Cochran (1957)

ANCOVA F Test

MEAN.11

Worked Example

Problem Statement Three teaching methods (A, B, C) are compared on post-test scores (y), controlling for pre-test score (x) as a covariate. Test whether adjusted group means differ after removing the linear effect of x.

	(x_{ij}, y_{ij}) pairs				
A	(44,54)	(51,61)	(53,72)	(61,70)	(66,73)
B	(47,65)	(54,69)	(59,82)	(62,79)	(68,90)
C	(50,58)	(56,65)	(63,76)	(68,72)	(73,84)

Source	SS	df	MS	F
Groups (adj)	277.930	2	138.965	8.589
Covariate	898.018	1	898.018	55.501
Error	177.982	11	16.180	

1. Pooled Within-Group Slope

The common slope $\hat{\beta}$ is estimated from within-group deviations only (pooled across groups):

$$\hat{\beta} = \frac{\sum_i \sum_j (x_{ij} - \bar{x}_{i\cdot})(y_{ij} - \bar{y}_{i\cdot})}{\sum_i \sum_j (x_{ij} - \bar{x}_{i\cdot})^2} = \frac{894.0}{890.0} = 1.005$$

2. Sum of Squares Decomposition

ANOVA model (groups only): $y_{ij} = \mu_i + e_{ij}$

$$SS^{\text{anova}}_{\text{error}} = 1076.0 \quad (df = 12)$$

Regression model (covariate only): $y_{ij} = \mu + \beta x_{ij} + e_{ij}$

$$SS^{\text{reg}}_{\text{error}} = 455.912 \quad (df = 13)$$

Full model (groups + covariate): $y_{ij} = \mu_i + \beta(x_{ij} - \bar{x}_{\cdot\cdot}) + e_{ij}$

$$SS^{\text{full}}_{\text{error}} = 177.982 \quad (df = 11)$$

Groups-adjusted SS (reduction from adding groups):

$$SS_{\text{grp|cov}} = SS^{\text{reg}}_{\text{error}} - SS^{\text{full}}_{\text{error}} = 455.912 - 177.982 = 277.930$$

Covariate-adjusted SS (reduction from adding covariate):

$$SS_{\text{cov|grp}} = SS^{\text{anova}}_{\text{error}} - SS^{\text{full}}_{\text{error}} = 1076.0 - 177.982 = 898.018$$

3. ANCOVA F-Test

$$F = \frac{MS_{\text{grp|cov}}}{MS_{\text{error}}} = \frac{138.965}{16.180} \sim F(k-1, N-k-1) = F(2, 11)$$

Statistic	Value	CV (5%)	Decision
$F(2, 11)$	8.589	3.982	Reject H_0

4. Adjusted Group Means

At the grand mean $\bar{x}_{\cdot\cdot} = 58.33$: $\tilde{\mu}_i = \bar{y}_i - \hat{\beta}(\bar{x}_i - \bar{x}_{\cdot\cdot})$.

$$\tilde{\mu}_A = 69.35, \quad \tilde{\mu}_B = 77.33, \quad \tilde{\mu}_C = 67.32$$

5. Comparison with One-Way ANOVA

Test	F	df	p
ANOVA (no cov.)	1.691	(2, 12)	0.225
ANCOVA (with cov.)	8.589	(2, 11)	0.006

6. Interpretation

- After adjusting for pre-test score, method B produces significantly higher post-test scores than A or C ($\tilde{\mu}_B - \tilde{\mu}_C =$ 10.0 points). Without the covariate, group differences are obscured by pre-existing differences in ability.

Practitioner Insight: ANCOVA improves on one-way ANOVA by removing confounding and increasing power. Adjusted means account for baseline differences (e.g., pre-test scores), isolating treatment effects, while variance is reduced by partitioning SS_{within} into a covariate component and a smaller residual, shrinking the error term substantially. This gain depends on key assumptions: homogeneous slopes across groups, error-free covariate, and no treatment effect on the covariate. When x is strongly correlated with y within groups, ANCOVA can be far more powerful than ANOVA.

Common Pitfalls

- **Using covariates affected by treatment.** Covariates must be pre-treatment. If treatment influences the covariate, ANCOVA adjusts away real effects (Lord's paradox). Use only baseline covariates.
- **Causal overinterpretation in observational data.** ANCOVA controls only for *measured* covariates; unmeasured confounding remains. Randomized trials support adjustment; observational studies require caution.
- **Violating linearity.** Adjustment assumes a linear covariate–outcome relationship. Nonlinearity biases adjusted means. Check scatterplots; if curved, transform covariate or add polynomial terms.
- **Using noisy covariates.** Measurement error weakens adjustment and reduces power. Prefer reliable covariates or apply measurement-error correction methods.

Tukey's HSD Test

Post-Hoc Pairwise Comparisons After ANOVA

MEAN.12 | **Category:** Mean Comparison | **Data:** Continuous | **Null:** Studentized Range (q)

When to Use

- Follow-up analysis after a significant ANOVA result.
- Compare all pairwise differences among group means.
- Strong control of family-wise Type I error rate.
- Equal variances appear reasonable across groups.
- Balanced or nearly balanced group sizes occur.

Assumptions

- Observations are mutually independent.
- Group populations are approximately normally distributed.
- Variances are similar across treatment groups.
- Response variable is measured on interval scale.
- ANOVA model is correctly specified and fitted.

Test Statistic

$$|\bar{y}_i - \bar{y}_j| > \text{HSD} \quad \Big| \quad \text{(Balanced): HSD} = q_{\alpha,k,N-k}\sqrt{\frac{\text{MSE}}{n}} \quad \Big| \quad \text{(Unbalanced): HSD} = q_{\alpha,k,N-k}\sqrt{\frac{\text{MSE}}{2}\left(\frac{1}{n_i}+\frac{1}{n_j}\right)}.$$

Notation: $\bar{y}_i$, and $\bar{y}_j$ are group means, n is common sample size (balanced), n_i and n_j are individual sample sizes (unbalanced), $N = \sum n_i$, MSE is mean square error from ANOVA, q is from studentized range dist. $(k, N-k)$ df

Plain English: After ANOVA shows groups differ, Tukey's HSD identifies which specific pairs differ. It uses the studentized range distribution to set a threshold that controls the probability of any false positive across all comparisons at level α. The test is optimal when comparing all possible pairs and is more powerful than Bonferroni in this setting.

⚠ Practitioner Warning

Tukey's HSD assumes equal population variances and performs best with equal sample sizes. For unequal variances, use the Games–Howell test instead. For unequal group sizes, the Tukey–Kramer adjustment with harmonic mean (shown under "Unbalanced" in the Test Statistic box) extends Tukey's HSD while maintaining family-wise error control. Do not use HSD if ANOVA homogeneity assumptions are violated.

Hypotheses

- H_0: For each pair: $\mu_i = \mu_j$ (means are equal).
- H_1: For each pair: $\mu_i \neq \mu_j$ (means differ).

Decision Rule: Reject H_0 if $|\bar{y}_i - \bar{y}_j| > \text{HSD}$. With balanced designs, all $k(k-1)/2$ comparisons maintain family-wise α.

Software References [Python, R, SAS]

```
scipy.stats.tukey_hsd          TukeyHSD(aov_object)          PROC ANOVA (TUKEY option)
```

Additional Notes

Tukey's HSD performs all pairwise mean comparisons using the studentized range distribution, controlling the family-wise error rate (FWER) while providing simultaneous confidence intervals. The test is based on the statistic $q = (\bar{y}_{\max} - \bar{y}_{\min})/\sqrt{\text{MSE}/n}$, with the critical value depending on the number of groups and the ANOVA error variance (MSE). For unequal sample sizes, the Tukey–Kramer adjustment is used (often via the harmonic mean $n_h = k/(\sum 1/n_i)$), extending applicability beyond balanced designs. Results are typically reported as pairwise mean differences with simultaneous confidence intervals; intervals excluding zero indicate significant differences. Tukey's HSD is more powerful than Bonferroni when testing all pairwise comparisons, but less flexible for arbitrary contrasts. It performs best under homoscedasticity; when variances differ substantially, alternatives such as Games–Howell are preferred.

Miscellanea: John Tukey introduced the "honestly significant difference" (HSD) procedure as a method for controlling family-wise error in multiple comparisons. The name reflects his emphasis on providing intervals that maintain honest coverage across all pairwise differences. Tukey was known for coining memorable statistical terms, including "jackknife," "bit" (binary digit), and "exploratory data analysis." The HSD test became one of the most widely used post-hoc procedures following ANOVA due to its exact family-wise error control under equal variances and balanced designs.

Reference(s): Tukey (1949); Kramer (1956)

Tukey's HSD Test — MEAN.12

Worked Example

Problem Statement A quality engineer samples three components from each of three production machines and measures a critical dimension (mm): **A** = (7, 8, 9), **B** = (9, 10, 11), **C** = (11, 12, 13), giving $\bar{x}_A = 8$, $\bar{x}_B = 10$, $\bar{x}_C = 12$, $MSE = 1$, $\nu = 6$, $k = 3$, $n = 3$, $F = 12$, $p = 0.008$. The ANOVA rejected H_0 (all means are equal); apply Tukey's HSD to determine *which* pairs of machines differ significantly. $H_0^{(ij)}: \mu_i = \mu_j$ for each pair (i, j), controlling the family-wise error rate at $\alpha = 0.05$.

1. Standard Error for Pairwise Comparisons
With equal $n = 3$ and $MSE = 1$ from the ANOVA table:

$$\text{SE} = \sqrt{MSE/n} = 1/\sqrt{3}.$$

2. Critical Value and HSD
The studentized range critical value for $k = 3$ groups and $\nu = 6$:

$$q_{0.05,3,6} = 4.339.$$

The Honestly Significant Difference:

$$\text{HSD} = q_{0.05,3,6} \times \text{SE} \approx 2.505.$$

Flag pair (i, j) if $|\bar{x}_i - \bar{x}_j| > \text{HSD}$.

3. Pairwise Decisions

Pair	$\lvert\bar{x}_i - \bar{x}_j\rvert$	> HSD = 2.505?	Decision
A–B	2	No	FTR
A–C	4	**Yes**	**Reject**
B–C	2	No	FTR

Only machines A and C are significantly different ($p = 0.007$); machine B is not significantly different from either.

4. Simultaneous 95% Confidence Intervals

$\mu_A - \mu_B: \quad -2 \pm 2.505 \quad = (-4.50,\ 0.51).$ ✓0

$\mu_A - \mu_C: \quad -4 \pm 2.505 \quad = (-6.50,\ -1.49).$ excludes 0

$\mu_B - \mu_C: \quad -2 \pm 2.505 \quad = (-4.50,\ 0.51).$ ✓0

Practitioner Insight: Tukey's HSD controls the family-wise error rate at α for all $\binom{k}{2}$ pairwise comparisons (exactly when group sizes are equal) using the studentized range $q_{k,\nu}$ in place of the t-distribution. A single threshold $\text{HSD} = q_{\alpha,k,\nu}\sqrt{MSE/n}$ is computed and compared to each $|\bar{x}_i - \bar{x}_j|$, avoiding per-pair tests. The critical value $q_{0.05,3,6} = 4.339$ is larger than the two-sided $t_{0.025,6} = 2.447$ used in a single comparison, penalising for the $\binom{3}{2} = 3$ tests performed simultaneously; running three uncorrected t-tests at $\alpha = 0.05$ would inflate the family-wise error rate to approximately $1 - (0.95)^3 \approx 14\%$. Results may appear "non-transitive" (e.g., $A = B$, $B = C$, but $A \neq C$) due to limited power, not contradiction. Tukey is preferred when all pairwise comparisons are of interest; for targeted comparisons, Bonferroni or Dunnett may be more powerful. For unequal sizes, the Tukey–Kramer adjustment replaces n with the harmonic mean $\tilde{n} = k/\sum 1/n_i$.

Common Pitfalls

- **Using HSD without a significant ANOVA.** Post-hoc tests should follow a significant omnibus result. Running Tukey after a non-significant ANOVA is data-dredging and inflates Type I error.
- **Ignoring unequal sample sizes.** Standard Tukey HSD assumes balanced groups. For unequal n_i, use the Tukey–Kramer version; with strong imbalance, prefer Games–Howell.
- **Using HSD when only a few comparisons are needed.** Tukey protects all pairwise contrasts but loses power when interest is limited. Planned contrasts (Bonferroni or Dunnett) are more efficient in those cases.
- **Violating equal-variance assumptions.** Tukey requires homoscedasticity. Under heteroscedasticity, its error rates fail; Games–Howell or Dunnett's T3 should be used instead.
- **Misinterpreting family-wise error control.** Tukey controls the probability of *any* false positive across all comparisons, not α per comparison. Adjusted p-values or HSD thresholds reflect this.

Comparison with Alternative Tests

Test	Best For	Assumptions	Power
Tukey's HSD	All pairwise comparisons	Equal variances	High (optimal)
Games-Howell	Unequal variances and/or unequal n	Approx. normality	High under heteroscedasticity
Bonferroni	Few planned comparisons	Depends on base test	Lower for many tests
Scheffé	All possible contrasts (not just pairs)	Equal variances	Lowest (most conservative)
Dunnett	All groups vs. single control	Equal variances	Highest for control comparisons

Dunnett's Test

Compare Multiple Treatments to Single Control

MEAN.13 | **Category:** Mean Comparison | **Data:** Continuous | **Null:** Multivariate t (Dunnett)

When to Use

- Follow-up analysis after a significant ANOVA result.
- Compare treatment groups against a control group.
- Interest centers on a single reference treatment.
- Avoid unnecessary treatment-to-treatment comparisons.
- Greater power than all-pairs comparison procedures.

Assumptions

- Observations are mutually independent.
- Group populations are approximately normally distributed.
- Variances are similar across treatment groups.
- Response variable is measured on interval scale.
- A single control group is clearly identified.

Test Statistic

$$\text{MSD} = d_{\alpha,k,\nu}\sqrt{\text{MSE}\left(\frac{1}{n_i} + \frac{1}{n_c}\right)},$$ where k is number of treatment groups (not including control).

Notation: $\bar{y}_i$ is treatment mean, $\bar{y}_c$ is control mean, N is total sample size, $\nu = N - k - 1$ is error df, d is Dunnett CV.

Plain English: Unlike Tukey's HSD (which compares all pairs), Dunnett's test only compares each treatment mean directly to a common control while controlling the family-wise error rate (FWER) across all such comparisons. By ignoring treatment–treatment contrasts and exploiting the shared control group mean and variance, it achieves higher power than all-pairs procedures when only control comparisons matter. Dunnett's T3 handles unequal variances and is distinct from this test.

⚠ Practitioner Warning

The control must be defined by design or theory *before* seeing the data. Selecting the "best" or "worst" group post hoc invalidates error control. Dunnett's assumes one group has special status (placebo, standard treatment, baseline condition). If no group has special status, use an all-pairs method (e.g., Tukey).

Hypotheses

- H_0: For each i: $\mu_i = \mu_c$.
- H_1: For at least one i: $\mu_i \neq \mu_c$.

Decision Rule: Reject H_0 if $|\bar{y}_i - \bar{y}_c| > \text{MSD}$. Controls FWER at level α across all k contrasts.

Software References [Python, R, SAS]

`scipy.stats.dunnett` `DescTools::DunnettTest` `PROC GLM / DUNNETT`

Additional Notes

Dunnett's test performs multiple comparisons against a single control using a joint multivariate t-distribution, accounting for the correlation among treatment–control contrasts. This yields simultaneous confidence intervals and hypothesis tests that control the family-wise error rate. The test uses the pooled error variance (MSE) from the ANOVA and critical values that depend on the number of treatments and degrees of freedom. By restricting comparisons to control vs. treatments, Dunnett's test is typically more powerful than all-pairs procedures such as Tukey's HSD. Results are reported as differences between each treatment mean and the control, with simultaneous confidence intervals; intervals excluding zero indicate significant differences. One-sided Dunnett's test available when all treatments expected to be better (or worse) than control. Two-sided version tests for any difference. Dunnett's test is most appropriate under homoscedasticity and balanced or near-balanced designs; when variances differ substantially, alternative methods or adjusted procedures should be considered.

Miscellanea: Charles W. Dunnett introduced this test in 1955 while working at Lederle Laboratories in the pharmaceutical industry. He recognized that all-pairs procedures (e.g., Tukey's HSD) were inefficient when only treatment-vs-control comparisons were of interest, motivating a more powerful many-to-one approach. Dunnett's test became widely used in pharmaceutical and clinical trials, where treatments are routinely compared against a control or placebo. He later had a long academic career at McMaster University, contributing extensively to multiple comparisons and experimental design.

Reference(s): Dunnett (1955, 1964)

Worked Example

Problem Statement A factory evaluates two candidate replacement machines (B and C) against an established baseline machine (A, the control). Measurements from the same three-machine study:

$$\mathbf{A}=[7, 8, 9], \quad \mathbf{B}=[9, 10, 11], \quad \text{and } \mathbf{C}=[11, 12, 13]. \quad \bar{x}_A = 8, \quad \bar{x}_B = 10, \quad \text{and } \bar{x}_C = 12.$$

From the ANOVA: MSE = 1, $\nu = 6$. Test whether B or C differ from the control A, controlling the family-wise error rate at $\alpha = 0.05$.

1. Standard Error
With MSE = 1 and equal $n = 3$:

$$\text{SE} = \sqrt{\text{MSE} \cdot \frac{2}{n}} = \sqrt{\frac{2}{3}} = 0.8165.$$

2. Dunnett Statistics
For each treatment i vs control: $d_i = (\bar{x}_i - \bar{x}_{\text{ctrl}})/\text{SE}$.

$$d_B = \frac{10-8}{\sqrt{2/3}} \approx 2.449, \quad d_C = \frac{12-8}{\sqrt{2/3}} \approx 4.899.$$

Since the mean differences are 2 and $4 = 2\times 2$, the statistics satisfy $d_C = 2\,d_B$ exactly.

3. Critical Value and Decisions
Dunnett's two-sided critical value for $k-1 = 2$ treatments and $\nu = 6$:

$$d_{0.05,2,6} = 2.863, \text{ crit. diff} = (2.863)(0.8165) = 2.338.$$

Flag treatment i if $|d_i| > d_{\text{crit}}$.

Comparison	$\lvert d_i\rvert$	> 2.863?	Decision
B vs A	2.449	No	FTR ($p = 0.086$)
C vs A	4.899	**Yes**	**Reject** ($p = 0.005$)

Machine C differs significantly from the control; machine B does not.

4. Power Hierarchy for Control Comparisons

Method	Crit. diff	Comparisons
Dunnett	2.338	$k-1 = 2$ (vs control)
Tukey	2.505	$\binom{3}{2} = 3$ (all pairs)
Scheffé	2.619	all contrasts

Dunnett's smaller critical difference (2.338 vs Tukey's 2.505) makes it strictly more powerful than Tukey for detecting treatment-vs-control differences.

Practitioner Insight: Dunnett's test is tailored to experiments with one control and $k-1$ treatments, testing only control-vs-treatment differences rather than all $\binom{k}{2}$ pairs. By testing only $k-1$ comparisons instead of all $\binom{k}{2}$ pairwise combinations, Dunnett requires a smaller critical value to maintain the same family-wise error rate, making it strictly more powerful than Tukey for the control-vs-treatment question. The test statistic $d_i = (\bar{x}_i - \bar{x}_{\text{ctrl}})/(s_p\sqrt{2/n})$ uses the same pooled standard error as Tukey but is compared to the Dunnett distribution rather than the studentized range; the Dunnett distribution accounts for the fact that all d_i share the control mean $\bar{x}_{\text{ctrl}}$, inducing a correlation of 1/2 between any two statistics under equal sample sizes. Consequently, critical thresholds are lower (Dunnett (2.338) < Tukey (2.505) < Scheffé (2.619)), improving sensitivity for detecting treatment effects. In practice, Dunnett's test is used in drug trials (several doses vs placebo), agricultural experiments (several treatments vs untreated control), and quality engineering (candidate processes vs baseline); when there is no natural control group and all pairwise comparisons are of interest, Tukey's HSD is preferred.

Common Pitfalls

- **Post-hoc designation of a "control."** The control must be defined a priori. Choosing it after seeing group means invalidates error control; if no true control exists, use Tukey's HSD instead.
- **Using Dunnett's when treatment–treatment comparisons matter.** Dunnett only protects control–vs–treatment contrasts. If all pairwise comparisons are needed, switch to Tukey HSD or a closed-testing framework.
- **Ignoring unequal variances.** Dunnett's assumes homoscedasticity. With unequal variances, use Dunnett's T3 or another heteroscedastic method (e.g., Games–Howell) instead.
- **Mixing up one-sided and two-sided tests.** One-sided Dunnett is valid only with a pre-specified directional hypothesis. The decision must be made before seeing data to avoid inflating Type I error.
- **Running Dunnett's when ANOVA is non-significant.** As a post-hoc test, Dunnett typically follows a significant ANOVA. Exception: if control comparisons were pre-planned, Dunnett may serve as the primary analysis.

Games-Howell Test

Post-Hoc Pairwise Comparisons for Unequal Variances

MEAN.14 | **Category:** Mean Comparisons | **Data:** Continuous | **Null:** Studentized Range (approx.)

When to Use

- Follow-up analysis after a significant ANOVA result.
- Compare all pairwise differences among group means.
- Group variances appear unequal across treatments.
- Unequal sample sizes occur across comparison groups.
- Robust alternative to Tukey's HSD procedure.

Assumptions

- Observations are mutually independent.
- Group populations are approximately normally distributed.
- Variances may differ across treatment groups.
- Unequal sample sizes are permitted across groups.
- Response variable is measured on interval scale.

Test Statistic

$$q_{ij} = \frac{\bar{x}_i - \bar{x}_j}{SE_{ij}}, \quad SE_{ij} = \sqrt{s_i^2/n_i + s_j^2/n_j}, \quad df_{ij} = \frac{(s_i^2/n_i + s_j^2/n_j)^2}{(s_i^2/n_i)^2/(n_i - 1) + (s_j^2/n_j)^2/(n_j - 1)}.$$

Notation: q_{ij} is test statistic ($\sim$ studentized range dist.) for groups i, j, separate variances s_i^2, s_j^2, Welch-Satterthwaite df

Plain English: Games-Howell test conducts all pairwise comparisons between group means using separate variance estimates for each pair, avoiding the equal variance assumption. The test combines Welch's approach to unequal variances with Tukey's studentized range distribution for controlling family-wise error rate. Each pairwise comparison uses its own degrees of freedom calculated via Welch-Satterthwaite approximation.

⚠ Practitioner Warning

Commonly used post-hoc after ANOVA or Welch ANOVA, though an omnibus test is not strictly required. Slightly less powerful than Tukey HSD under equal variances and balanced designs, but often more powerful with unequal sample sizes. Performance declines with very small groups ($n_i < 10$) or extreme imbalance ($\max(n_i)/\min(n_i) > 4$). Requires studentized-range critical values with non-integer Welch–Satterthwaite degrees of freedom, so software is essential. Computationally heavier than pooled-variance methods and controls the family-wise error rate only approximately.

Hypotheses

- H_0: Group means equal for specific pair ($\mu_i = \mu_j$).
- H_1: Group means differ for specific pair ($\mu_i \neq \mu_j$).

Decision Rule: Reject H_0 if , for pair (i, j), $|q_{ij}| > q_{\alpha,k,df_{ij}}$ where q from modified studentized range.

Software References [Python, R, SAS]

`pingouin.pairwise_gameshowell` `rstatix::games_howell_test` `Not Native (Manual)`

Additional Notes

Games–Howell is designed for settings with unequal variances and unequal sample sizes, a situation closely related to the classical Behrens–Fisher problem of comparing means when variances are unknown and unequal. Because no exact finite-sample solution exists for the general Behrens–Fisher problem, the procedure relies on Welch–Satterthwaite approximations for the degrees of freedom. Simultaneous confidence intervals take the form $(\bar{x}_i - \bar{x}_j) \pm \frac{q_{\alpha,k,\nu_{ij}}}{\sqrt{2}} SE_{ij}$, where ν_{ij} is the Welch–Satterthwaite approximation for the comparison of groups i and j. Unlike Tukey's HSD, which assumes equal variances, Games–Howell remains valid under substantial heteroscedasticity. Compared with Dunnett's T3, it is typically less conservative. Power is similar to Tukey's method when variances are equal, but often superior when variances or sample sizes differ.

Miscellanea: Games and Howell (1976) introduced this method using Monte Carlo simulations to study post-hoc comparisons under unequal variances. The paper appeared in the *Journal of Educational Statistics*, now known as the *Journal of Educational and Behavioral Statistics*. Although statistically sound, the method saw broader adoption only after its inclusion in major statistical software packages (e.g., SPSS, SAS, Stata), highlighting how practical uptake depends on both methodology and computational accessibility.

Reference(s): Games & Howell (1976); Welch (1951)

Worked Example

Problem Statement An analyst conducts an ANOVA to compare plant growth across three fertilizers. A Levene's test rejects the null hypothesis of equal variances ($p < 0.05$), and sample sizes are unequal. The Games-Howell test is applied to perform pairwise comparisons while maintaining the family-wise error rate.

Data Summary

Group	n	Mean ($\bar{x}$)	Variance (s^2)
A (Traditional)	30	15.2	4.0
B (Experimental)	10	19.5	12.0
C (Organic)	20	16.8	7.0

1. Pairwise Comparison (A vs. B)
The standard error of the difference between Group A and B is:

$$SE_{A,B} = \sqrt{\frac{4.0}{30} + \frac{12.0}{10}} = 1.155$$

The adjusted degrees of freedom (Satterthwaite) is approximately 11.06. The q-statistic is:

$$q_{calc} = \frac{|\bar{x}_B - \bar{x}_A|}{SE_{A,B}} = \frac{|19.5 - 15.2|}{1.155} = 3.723$$

2. Critical Values and Results
We compare q_{calc} to the critical value from the Studentized Range distribution q for $k = 3$ groups and $df_{AB} = 11.06, df_{AC} = 33.09$, and $df_{BC} = 14.43$ at $\alpha = 0.05$.

Comp.	Diff	q-stat	Crit. q	Result
A vs B	4.3	3.723	3.817	Not Significant
A vs C	1.6	2.302	3.470	Not Significant
B vs C	2.7	2.160	3.689	Not Significant

3. Decision and Interpretation

- **Decision:** No pairwise differences are statistically significant ($p < 0.05$).
- **Interpretation:** While Fertilizer B (Experimental) shows the highest growth, it is not significantly better than Fertilizer A (Traditional) or Fertilizer C (Organic).

Practitioner Insight: The Games-Howell test is more conservative than Tukey HSD when variances are unequal. It prevents Type I errors (false positives) that often occur in ANOVA when high-variance groups (like Group B) are compared using a pooled variance estimate. Despite the large observed difference between Fertilizer A and B, the high uncertainty in Fertilizer B's performance prevents us from concluding they are truly different.

Common Pitfalls

- **Assuming an omnibus ANOVA is required.** Games–Howell controls the family–wise error rate directly; a significant ANOVA/Welch test is *not* a prerequisite. Omnibus tests are optional screens, not gatekeepers.
- **Using Games–Howell when Tukey HSD assumptions are met.** Under equal variances and balanced designs, Tukey HSD is more powerful. Games–Howell should be reserved for heteroscedastic or unbalanced groups.
- **Applying Games–Howell to heavily non-normal data.** The test is robust to unequal variances but still assumes approximate normality. Strong skew or outliers can inflate Type I error; consider transformations or nonparametric alternatives.
- **Ignoring small-sample conservatism.** With very small groups ($n_i < 10$), Games–Howell can be overly conservative, yielding nonsignificant results even for moderate effects. Report effect sizes and interpret $p > 0.05$ cautiously.
- **Using Games–Howell on dependent or repeated-measures data.** The method assumes independent groups. Repeated measures, matched pairs, or clustered designs require within-subject or mixed-model approaches—not Games–Howell.

Effect size: Unpooled Cohen's *d*

$$d = (|\bar{x}_i - \bar{x}_j|) \div \sqrt{\frac{s_i^2 + s_j^2}{2}}$$

Scheffé's Test

Conservative Post-Hoc Test for All Possible Contrasts

MEAN.15 | **Category:** Mean Comparisons | **Data:** Continuous | **Null:** F-distribution

When to Use

- Follow-up analysis after a significant ANOVA result.
- Compare all possible contrasts among group means.
- Unplanned comparisons selected after seeing data.
- Strong control of family-wise Type I error rate.
- Conservative inference preferred over power gains.

Assumptions

- Observations are mutually independent.
- Population distributions are approximately normal.
- Variances are similar across treatment groups.
- Response variable is measured on interval scale.
- Overall ANOVA model is correctly specified.

Test Statistic

$$F_C = \frac{(\sum_{i=1}^{k} c_i \bar{x}_i)^2}{SE_C} \sim F_{k-1,N-k} \text{ (under } H_0\text{)}. \quad \Big| \quad \sum_{i=1}^{k} c_i = 0 \text{ for contrast } C,\ SE_C = MS_E(k-1)\sum_{i=1}^{k}\frac{c_i^2}{n_i}.$$

Notation: F_C is F-statistic for contrast, c_i are contrast coefficients, MS_E is pooled error variance, N is total sample size

Plain English: Scheffé's test is a post-hoc procedure used after a significant ANOVA to determine which group comparisons are supported by the data. Unlike methods that focus only on pairwise differences, Scheffé's test can evaluate any contrast among group means, including averages of several groups versus others. A contrast is any linear combination with coefficients summing to zero. It adjusts the significance threshold to account for the many possible comparisons, ensuring that the overall probability of making one or more false discoveries remains controlled. This flexibility comes at the cost of reduced power, making the method conservative but especially useful when comparisons are chosen after examining the data.

⚠ Practitioner Warning

Highly conservative, with lower power than alternatives for pairwise comparisons. Not efficient when only pairwise tests are needed (prefer Tukey). Assumes homogeneity of variance and relies on pooled ANOVA error (MS_E); violations affect validity. Confidence intervals are wide. Typically used post-hoc; balanced designs improve efficiency.

Hypotheses

- H_0: $\sum_i c_i \mu_i = 0$ for specific contrast.
- H_1: Contrast differs from zero ($\sum_i c_i \mu_i \neq 0$).

Decision Rule: Reject H_0 if $F_C > F_{\alpha,k-1,N-k}$ where k is groups, N is total sample size.

Software References [Python, R, SAS]

Python	R	SAS
`Not Native (Manual)`	`DescTools::ScheffeTest`	`PROC GLM (LSMEANS SCHEFFE)`

Additional Notes

Scheffé's method provides simultaneous inference for all linear contrasts $C = \sum_{i=1}^{k} c_i \mu_i$ with $\sum c_i = 0$, including both pairwise and complex comparisons. Using the pooled ANOVA variance, the standard error is $SE_C = \sqrt{MS_E \sum c_i^2/n_i}$, and family-wise error is controlled through the critical value $(k-1)F_{\alpha,k-1,N-k}$, yielding simultaneous confidence intervals for every possible contrast. Geometrically, the procedure protects an entire confidence region for the vector of group means rather than individual comparisons. Any contrast—whether planned in advance or formulated after examining the data—corresponds to a direction through this region, which explains why unrestricted post-hoc exploration remains valid. Compared with Tukey's HSD (more powerful for pairwise comparisons) and Bonferroni (more efficient for a small number of planned comparisons), Scheffé's method is more conservative but uniquely supports data-driven and unplanned contrasts with valid family-wise error control. This flexibility is its primary advantage and the main reason it remains a classic post-hoc procedure.

Miscellanea: Scheffé (1953; 1959) developed a method for simultaneous inference on all linear contrasts in ANOVA, providing valid inference even when contrasts are chosen post hoc. Unlike procedures tailored to pairwise comparisons (e.g., Tukey), Scheffé's method remains valid for any contrast, though at the cost of reduced power. The method is closely tied to the general linear hypothesis framework and has been used in settings requiring flexible, complex comparisons, but is often unnecessarily conservative for simple pairwise tests.

Reference(s): Scheffé (1953, 1959); Maxwell et al. (2017)

Worked Example

Problem Statement A quality engineer samples three components from each of three production machines and measures a critical dimension (mm): $\mathbf{A} = (7, 8, 9)$, $\mathbf{B} = (9, 10, 11)$, $\mathbf{C} = (11, 12, 13)$, giving $\bar{x}_A = 8, \bar{x}_B = 10, \bar{x}_C = 12, \text{MS}_E = 1$, $\text{df}_E = 6, k = 3, n = 3, F = 12, p = 0.008$. The ANOVA rejected H_0 (all means are equal).

A process engineer asks two questions: (A) Do machines A and C differ? (B) Does machine A differ from the *combined average* of machines B and C? (B) is a *complex contrast* that Tukey's HSD cannot address. Apply Scheffé's test at $\alpha = 0.05$. Contrast: $L = \sum c_i \bar{x}_i$. $F_S = \frac{L^2}{\text{MS}_E \cdot \sum c_i^2/n_i}$.

1. Scheffé Critical Threshold
The Scheffé test uses the F-critical value from the original ANOVA:

$S = (k-1)\, F_{\alpha, k-1, \text{df}_E} = 2 \times F_{0.05, 2, 6} = 2 \times 5.143 = 10.286$.
Reject a contrast if its $F_S > S$.

2. Contrast 1: Machine A vs Machine C
Contrast coefficients $\mathbf{c} = (1, 0, -1)$, with $\sum c_i = 0$:

$$L_1 = -4, \qquad \sum \frac{c_i^2}{n_i} = \frac{2}{3}, \qquad F_{S1} = 24.$$

$F_{S1} = 24 > S = 10.286$: **reject** $H_0\colon \mu_A = \mu_C$.

3. Contrast 2: Machine A vs Average of B and C
Coefficients $\mathbf{c} = (1, -\frac{1}{2}, -\frac{1}{2})$, with $\sum c_i = 0$:

$$L_2 = -3, \qquad \sum \frac{c_i^2}{n_i} = \frac{1}{2}, \qquad F_{S2} = 18.$$

$F_{S2} = 18 > S = 10.286$: **reject** $H_0\colon \mu_A = \frac{1}{2}(\mu_B + \mu_C)$.

4. Summary

Contrast	L	$\sum c_i^2/n_i$	F_S	Decision
A vs B	-2	2/3	6	FTR
A vs C	-4	2/3	24	Reject
A vs $\frac{B+C}{2}$	-3	1/2	18	Reject

$S = (k-1)F_{0.05,2,6} = 10.286$

Machine A differs significantly from machine C, and also from the combined average of B and C; but A and B alone do not differ significantly.

5. Comparison with Tukey HSD
For a pairwise comparison ($\mathbf{c} = (1, 0, -1)$), the equivalent Scheffé mean-difference threshold is $\sqrt{S \cdot \text{MS}_E \cdot 2/n} = \sqrt{10.286 \cdot 2/3} \approx 2.619$, slightly larger than Tukey's HSD = 2.505; Scheffé is more conservative for pairwise tests, but uniquely extends to complex contrasts.

Practitioner Insight: Scheffé's test uniquely controls the family-wise error rate for *all* linear contrasts $L = \sum c_i \bar{x}_i$ with $\sum c_i = 0$, not just pairwise comparisons. While pairwise differences (coefficients ± 1) are common, contrasts can compare one group to averages or clusters of others, enabling more flexible, substantively meaningful questions. The test statistic $F_S = L^2/(\text{MS}_E \sum c_i^2/n_i)$ follows $F(1, \text{df}_E)$ for a fixed contrast, but the critical value is inflated by $(k-1)$ to account for all possible contrasts. This makes Scheffé more conservative than Tukey for pairwise comparisons (e.g., threshold 2.619 vs. 2.505 here), so Tukey is preferred when only pairwise tests are needed. The advantage is flexibility: Scheffé can detect patterns such as comparing one group to the average of others, which pairwise methods cannot address. Importantly, if any Scheffé contrast is significant, the overall ANOVA F-test must also be significant, since ANOVA corresponds to a particular contrast maximizing F_S.

Common Pitfalls

- **Using Scheffé for simple pairwise comparisons.** Scheffé protects all possible contrasts but is markedly less powerful than Tukey HSD or Bonferroni for pairwise tests. Use Scheffé only when testing complex or data-driven contrasts.
- **Applying Scheffé when ANOVA assumptions fail.** The method inherits ANOVA's requirements of normality and homoscedasticity; under variance heterogeneity, Type I error can inflate. Use Welch-based or heteroscedastic alternatives instead.
- **Treating post-hoc, data-driven contrasts as "planned."** Scheffé validly protects exploratory contrasts, but power is low. Pre-specified contrasts analyzed with focused tests (e.g., Bonferroni) are far more efficient.
- **Using incorrect critical values.** Scheffé relies on the F-distribution: critical value = $\sqrt{(k-1)F_{k-1, N-k, \alpha}}$, and not the studentized range q. Mixing distributions yields incorrect inferences.
- **Applying Scheffé to dependent or repeated-measures data.** The method assumes independent groups. Repeated-measures or clustered designs require within-subject or mixed-model contrasts—not Scheffé.

Welch vs. Pooled *t*-Test: Derivations and Key Cases

Both tests target $H_0\colon \mu_1 = \mu_2$. The difference lies entirely in how the standard error is estimated and how degrees of freedom are assigned.

The two standard errors. The pooled estimator assumes $\sigma_1^2 = \sigma_2^2 = \sigma^2$ and estimates the common variance by the weighted average:

$$s_p^2 = \frac{(n_1-1)s_1^2 + (n_2-1)s_2^2}{n_1+n_2-2},$$

giving

$$\boxed{\mathrm{SE}_p^2 = s_p^2(1/n_1 + 1/n_2)}$$

with exact $t_{n_1+n_2-2}$ null distribution under normality and equal variances.

The Welch estimator makes no variance assumption:

$$\boxed{\mathrm{SE}_w^2 = \frac{s_1^2}{n_1} + \frac{s_2^2}{n_2}}.$$

The Satterthwaite degrees of freedom approximate the distribution of SE_w^2 by a scaled χ^2, matching its first two moments:

$$\nu = \frac{\left(s_1^2/n_1 + s_2^2/n_2\right)^2}{\dfrac{(s_1^2/n_1)^2}{n_1-1} + \dfrac{(s_2^2/n_2)^2}{n_2-1}}.$$

By the Cauchy–Schwarz inequality, $\nu \le n_1 + n_2 - 2$, with equality iff $s_1^2/n_1 = s_2^2/n_2$.

Case 1: Equal variances ($s_1^2 = s_2^2 = s^2$).

$$\mathrm{SE}_w^2 = s^2\left(\tfrac{1}{n_1} + \tfrac{1}{n_2}\right) = \mathrm{SE}_p^2.$$

With $s_1^2 = s_2^2$, $s_p^2 = s^2$ exactly, so the test statistics are identical. For ν: each term in the denominator becomes $(s^2/n_i)^2/(n_i-1)$, and simplification gives $\nu = n_1+n_2-2$. *Conclusion:* Welch = Pooled exactly. No penalty for using Welch.

Case 2: Equal sample sizes ($n_1 = n_2 = n$).

$$\mathrm{SE}_w^2 = \frac{s_1^2+s_2^2}{n} = \frac{2s_p^2}{n} = s_p^2 \cdot \frac{2}{n} = \mathrm{SE}_p^2,$$

since $s_p^2 = (s_1^2+s_2^2)/2$ when $n_1 = n_2$. So $\mathrm{SE}_w = \mathrm{SE}_p$ and $t_w = t_p$ for *any* variance ratio when testing $H_0 : \mu_1 - \mu_2 = 0$. Only ν differs. With $n_1 = n_2 = n$:

$$\nu = \frac{(s_1^2+s_2^2)^2}{\dfrac{s_1^4+s_2^4}{n-1}} = \frac{(n-1)(s_1^2+s_2^2)^2}{s_1^4+s_2^4} \le 2(n-1).$$

The bound $2(n-1) = n_1+n_2-2$ is achieved iff $s_1^2 = s_2^2$. For unequal variances $s_1^4 + s_2^4 > (s_1^2+s_2^2)^2/2$, so $\nu < 2(n-1)$. *Conclusion:* Same t, smaller df under Welch $\Rightarrow$ larger p-value. Welch is conservative by the df penalty alone.

Case 3: Unequal variances and unequal n (general case). Now $\mathrm{SE}_w \ne \mathrm{SE}_p$ in general. Suppose $s_2^2 \gg s_1^2$ and $n_2 \ll n_1$ (large variance paired with small sample). Then s_p^2 is dominated by s_1^2 (the larger group) and underestimates the true SE for the noisier group. The pooled t is inflated and the test over-rejects. Welch correctly upweights the noisier sample in SE_w^2 and reduces ν accordingly. *Conclusion:* Pooled test can have severely distorted size; Welch maintains nominal level.

Case 4: Extreme variance imbalance ($s_2^2/s_1^2 \to \infty$). Let $r = s_2^2/s_1^2 \to \infty$ with n_1, n_2 fixed. The numerator of ν is dominated by $(s_2^2/n_2)^2$ and the denominator by $(s_2^2/n_2)^2/(n_2-1)$, giving:

$$\nu \longrightarrow n_2 - 1.$$

Similarly if $r \to 0$, $\nu \to n_1 - 1$. In general $\nu \to \min(n_1-1, n_2-1)$ as the variance ratio becomes extreme — inference is driven entirely by the noisier sample, correctly reflecting that the noisy group is the binding constraint on precision.

Summary. The pooled test is a special case of Welch, valid only when $\sigma_1^2 = \sigma_2^2$. Welch never loses when variances are equal (Case 1) and is strictly safer otherwise. The cost of using Welch unnecessarily is at most a few degrees of freedom; the cost of using pooled under heteroscedasticity can be severe Type I error inflation.

Effect Size Measures: Cohen's d, Hedges' g, and Glass' Δ

All three measures express the standardized mean difference between two groups. They differ in which standard deviation is used as the denominator — a choice that matters when sample sizes or variances are unequal.

Cohen's d.

$$d = \frac{\bar{x}_1 - \bar{x}_2}{s_p}, \qquad s_p = \sqrt{\frac{(n_1-1)s_1^2 + (n_2-1)s_2^2}{n_1+n_2-2}}.$$

Uses the pooled standard deviation s_p. Assumes equal population variances; biased upward in small samples because s_p underestimates σ. Cohen's (1988) benchmarks: $|d| = 0.2$ (small), 0.5 (medium), 0.8 (large) — treat as rough guides, not thresholds.

Hedges' g.

$$g = d \times J(m), \qquad J(m) \approx 1 - \frac{3}{4m-1},$$

where $m = n_1 + n_2 - 2$. $J(m) < 1$ is the small-sample correction factor that removes the upward bias in d. For $n_1 + n_2 \geq 20$, $g \approx d$; the correction is material only for very small samples. Hedges' g is preferred over Cohen's d in meta-analyses and wherever sample sizes are small.

Glass' Δ.

$$\Delta = \frac{\bar{x}_1 - \bar{x}_2}{s_{\text{control}}}.$$

Uses only the *control group* standard deviation as the denominator. Appropriate when the treatment is expected to change not just the mean but also the variance of the treated group — in that case pooling the two standard deviations would obscure the baseline variability. The choice of which group is "control" must be explicit and justified; the result is not symmetric.

Which to use.

Measure	Use when
Cohen's d	Large samples, equal variances assumed
Hedges' g	Small samples or meta-analysis
Glass' Δ	Unequal variances; clear control group

All three are reported with a confidence interval. For Cohen's d and Hedges' g the noncentral t-distribution provides exact CIs; the `effectsize` package in R computes all three with CIs directly.

ANOVA: Computational Recipes

Notation (all five procedures)

- k groups, n_i obs. in group i, $N = \sum n_i$ total
- y_{ij}: j-th obs. in group i; s_i^2: group sample variance
- $\bar{y}_{i\cdot}$: group mean; $\bar{y}_{\cdot\cdot}$: grand mean
- *Two-way only:* a levels of factor A, b levels of factor B, n obs. per cell; y_{ijk}: k-th obs. in cell (i, j); $\bar{y}_{ij\cdot}$: cell mean; $\bar{y}_{i\cdot\cdot}$, $\bar{y}_{\cdot j\cdot}$: marginal means for A and B; $\bar{y}_{\cdot\cdot\cdot}$: grand mean

1. One-Way ANOVA (Classical)

Step 1. Compute group and grand means:

$$\bar{y}_{i\cdot} = \frac{1}{n_i}\sum_j y_{ij}, \qquad \bar{y}_{\cdot\cdot} = \frac{1}{N}\sum_i\sum_j y_{ij}.$$

Step 2. Compute sums of squares:

$$\text{SS}_{\text{between}} = \sum_i n_i(\bar{y}_{i\cdot} - \bar{y}_{\cdot\cdot})^2$$

$$\text{SS}_{\text{within}} = \sum_i\sum_j (y_{ij} - \bar{y}_{i\cdot})^2$$

$$\text{SS}_{\text{total}} = \text{SS}_{\text{between}} + \text{SS}_{\text{within}}$$

Step 3. ANOVA table:

Source	SS	df	MS	F
Between	SS_B	$k-1$	MS_B	MS_B/MS_W
Within	SS_W	$N-k$	MS_W	
Total	SS_T	$N-1$		

Step 4. Reject H_0 if $F > F_{\alpha, k-1, N-k}$.

2. Welch's ANOVA *(unequal variances; replaces classical F)*

Step 1. Compute weights and weighted grand mean:

$$w_i = \frac{n_i}{s_i^2}, \qquad \tilde{y} = \frac{\sum_i w_i \bar{y}_{i\cdot}}{\sum_i w_i}.$$

Step 2. Compute Welch statistic:

$$W = \frac{\sum_i w_i(\bar{y}_{i\cdot} - \tilde{y})^2 / (k-1)}{1 + \frac{2(k-2)}{k^2-1}\sum_i \frac{(1 - w_i/\sum w_i)^2}{n_i - 1}}.$$

Step 3. Satterthwaite df:

$$\nu = \frac{k^2 - 1}{3\sum_i \frac{(1 - w_i/\sum w_i)^2}{n_i - 1}}$$

Step 4. Reject H_0 if $W > F_{\alpha, k-1, \nu}$.

3. Repeated Measures ANOVA *(n subjects, k conditions; balanced)*

Step 1. Compute means:

$$\bar{y}_{i\cdot} \text{ (condition)}, \quad \bar{y}_{\cdot j} \text{ (subject)}, \quad \bar{y}_{\cdot\cdot} \text{ (grand)}.$$

Step 2. Sums of squares:

$$\text{SS}_{\text{cond}} = n\sum_i (\bar{y}_{i\cdot} - \bar{y}_{\cdot\cdot})^2$$

$$\text{SS}_{\text{subj}} = k\sum_j (\bar{y}_{\cdot j} - \bar{y}_{\cdot\cdot})^2$$

$$\text{SS}_{\text{error}} = \text{SS}_{\text{total}} - \text{SS}_{\text{cond}} - \text{SS}_{\text{subj}}$$

Step 3. ANOVA table:

Source	SS	df	MS	F
Condition	SS_C	$k-1$	MS_C	MS_C/MS_E
Subject	SS_S	$n-1$	MS_S	
Error	SS_E	$(k-1)(n-1)$	MS_E	
Total	SS_T	$kn-1$		

Step 4. Apply Greenhouse-Geisser or Huynh-Feldt $\hat{\epsilon}$ correction to df if sphericity is violated: multiply both condition and error df by $\hat{\epsilon}$.

4. ANCOVA *(one covariate x; k groups)*

Step 1. Fit covariate: compute group and grand means of both y_{ij} and x_{ij}; estimate common within-group slope:

$$\hat{\beta} = \frac{SS_{xy,W}}{SS_{xx,W}},$$

where $SS_{xy,W} = \sum_i \sum_j (x_{ij} - \bar{x}_{i\cdot})(y_{ij} - \bar{y}_{i\cdot})$ and $SS_{xx,W} = \sum_i \sum_j (x_{ij} - \bar{x}_{i\cdot})^2$.

Step 2. Adjusted means:

$$\tilde{y}_{i\cdot} = \bar{y}_{i\cdot} - \hat{\beta}(\bar{x}_{i\cdot} - \bar{x}_{\cdot\cdot}).$$

Step 3. Adjusted sums of squares:

$$SS_{\text{adj,W}} = SS_{yy,W} - \hat{\beta}^2\, SS_{xx,W}$$
$$SS_{\text{adj,B}} = SS_{yy,T} - SS^2_{xy,T}/SS_{xx,T} - SS_{\text{adj,W}}$$

Step 4. ANOVA table:

Source	SS	df	MS	F
Covariate	—	1	MS_{cov}	$MS_{\text{cov}}/MS_{\text{adj,W}}$
Group (adj.)	$SS_{\text{adj,B}}$	$k-1$	$MS_{\text{adj,B}}$	$MS_{\text{adj,B}}/MS_{\text{adj,W}}$
Error (adj.)	$SS_{\text{adj,W}}$	$N-k-1$	$MS_{\text{adj,W}}$	

Step 5. Verify homogeneity of slopes before interpreting adjusted group means: test interaction group $\times\, x$; if significant, ANCOVA assumptions fail.

5. Two-Way ANOVA *(factor A: a levels; factor B: b levels; n obs. per cell; balanced)*

Step 1. Compute means:

$$\bar{y}_{ij\cdot}\ \text{(cell)},\ \bar{y}_{i\cdot\cdot}\ \text{(A)},\ \bar{y}_{\cdot j\cdot}\ \text{(B)},\ \bar{y}_{\cdot\cdot\cdot}\ \text{(grand)}.$$

Total observations: $N = abn$.

Step 2. Sums of squares:

$$SS_A = bn\sum_i(\bar{y}_{i\cdot\cdot} - \bar{y}_{\cdot\cdot\cdot})^2$$
$$SS_B = an\sum_j(\bar{y}_{\cdot j\cdot} - \bar{y}_{\cdot\cdot\cdot})^2$$
$$SS_{AB} = n\sum_i\sum_j(\bar{y}_{ij\cdot} - \bar{y}_{i\cdot\cdot} - \bar{y}_{\cdot j\cdot} + \bar{y}_{\cdot\cdot\cdot})^2$$
$$SS_E = \sum_i\sum_j\sum_k(y_{ijk} - \bar{y}_{ij\cdot})^2$$
$$SS_T = SS_A + SS_B + SS_{AB} + SS_E$$

Step 3. ANOVA table:

Source	SS	df	MS	F
A	SS_A	$a-1$	MS_A	MS_A/MS_E
B	SS_B	$b-1$	MS_B	MS_B/MS_E
$A\times B$	SS_{AB}	$(a-1)(b-1)$	MS_{AB}	MS_{AB}/MS_E
Error	SS_E	$ab(n-1)$	MS_E	
Total	SS_T	$N-1$		

Step 4. Each effect uses $F_{\alpha,\,\text{df}_{\text{effect}},\,ab(n-1)}$. Interpret main effects cautiously if the $A \times B$ interaction is significant — main effect means are then misleading without examining cell means directly.

Types of Sums of Squares in ANOVA

When cell frequencies are unequal (unbalanced designs), different SS types give different results. The choice matters.

- **Type I (Sequential).** Each effect is adjusted only for effects entered before it in the model. Order- dependent: SS for A ignores B; SS for B adjusts for A. *Use:* nested models, polynomial contrasts, or when a deliberate hierarchical entry order is theoretically justified. Avoid for factorial designs with unequal n.
- **Type II (Hierarchical).** Each main effect is adjusted for all other main effects but *not* for interactions. Assumes no interaction is present. *Use:* additive models where interaction terms are absent or theoretically ruled out. Most powerful test of main effects when interactions are zero.
- **Type III (Partial).** Each effect is adjusted for *all* other effects in the model, including interactions. Order-independent. *Use:* standard default in most software (SAS, SPSS, R `car::Anova`). Appropriate for factorial designs with unequal cell sizes and whenever interactions are included.
- **Type IV.** Designed for designs with *empty cells* (some factor combinations completely missing). Constructs estimable functions that remain valid despite missing cells. *Use:* only when empty cells exist; Type III suffices otherwise.

Practical rule: Balanced designs — all types agree. Unbalanced, no empty cells — use Type III. Empty cells — use Type IV. Additive model only — Type II is most powerful.

Sphericity Corrections in Repeated Measures ANOVA

When sphericity fails, the F-test is positively biased (over-rejects). All corrections multiply the numerator and denominator df by $\hat{\epsilon} \leq 1$, producing a more conservative test.

- **Greenhouse–Geisser (GG).** $\hat{\epsilon}_{GG}$ estimated from the sample covariance matrix; always $\geq 1/(k-1)$. Biased downward (conservative). *Use:* when $\hat{\epsilon}_{GG} < 0.75$; preferred when sphericity is substantially violated.
- **Huynh–Feldt (HF).** Bias-corrected version of GG; less conservative, higher power. Can exceed 1.0 in small samples — cap at 1. *Use:* when $\hat{\epsilon}_{GG} \geq 0.75$; mild violations.
- **Lower Bound.** Sets $\epsilon = 1/(k-1)$, the most extreme possible violation. Maximally conservative. *Use:* quick check — if F is significant even here, conclusion is robust regardless of sphericity.
- **Multivariate approach (MANOVA).** Makes no sphericity assumption; treats the $k-1$ contrasts as a multivariate outcome. *Use:* $n > k$; when sphericity is badly violated and n is adequate. Lower power than corrected univariate F when sphericity nearly holds.

Practical rule: Report GG by default. If GG and HF disagree materially, report both. Use the lower bound as a sensitivity check. Switch to MANOVA if $\hat{\epsilon}_{GG} < 0.5$ and n is sufficient.

Behind the Scenes. In the Repeated Measures ANOVA worked example (MEANS.10), the Greenhouse–Geisser correction factor $\hat{\varepsilon} = 0.4617$ was reported without derivation. The following supplemental calculation illustrates the intermediate steps, from the covariance matrix to the adjusted degrees of freedom, for readers interested in the mechanics of the correction.

Greenhouse–Geisser Correction. When sphericity is violated, the Greenhouse–Geisser correction adjusts the ANOVA degrees of freedom by an estimated factor

$$\hat{\varepsilon} = \frac{\{\mathrm{tr}(S_c)\}^2}{(k-1)\,\mathrm{tr}(S_c^2)}, \qquad S_c = \left(I - \frac{1}{k}J\right) S \left(I - \frac{1}{k}J\right).$$

For the repeated measurements ($k = 4$), the sample covariance matrix and the centering matrix ($P = I - \frac{1}{k}J$) arc

$$S = \begin{pmatrix} 73.268 & 66.536 & 62.268 & 63.607 \\ 66.536 & 60.500 & 56.536 & 57.786 \\ 62.268 & 56.536 & 52.982 & 54.036 \\ 63.607 & 57.786 & 54.036 & 55.357 \end{pmatrix},$$

$$P = \begin{pmatrix} 0.75 & -0.25 & -0.25 & -0.25 \\ -0.25 & 0.75 & -0.25 & -0.25 \\ -0.25 & -0.25 & 0.75 & -0.25 \\ -0.25 & -0.25 & -0.25 & 0.75 \end{pmatrix}.$$

Centering yields

$$S_c = \begin{pmatrix} 0.656 & 0.004 & -0.380 & -0.281 \\ 0.004 & 0.049 & -0.031 & -0.022 \\ -0.380 & -0.031 & 0.299 & 0.112 \\ -0.281 & -0.022 & 0.112 & 0.192 \end{pmatrix}.$$

and

$$\mathrm{tr}(S_c) = 1.1964, \qquad \mathrm{tr}(S_c^2) = 1.0335.$$

Using $k - 1 = 3$,

$$\hat{\varepsilon} = \frac{(1.1964)^2}{3(1.0335)} = \frac{1.4314}{3.1005} = 0.4617.$$

The original ANOVA degrees of freedom are

$$df_1 = 3, \qquad df_2 = 21.$$

Applying the correction,

$$df_1^* = 0.4617(3) = 1.385, \qquad df_2^* = 0.4617(21) = 9.696.$$

Thus the observed statistic $F = 326.91$ is evaluated against $F_{1.385,\,9.696}$ rather than $F_{3,\,21}$.

Part IV

Variability, Dispersion & Scale

Evaluating equality of variances and differences in spread

Variances and Scale

IV. VARI — A Conceptual Primer

1. Problem Framing

What Are We Testing?

- Whether the **spread** (variance, standard deviation, or a robust scale measure) is equal across two or more populations, or equals a specified value.
- Core question: *Is variability homogeneous across groups?*
- Distinct from location tests (MEANS): the focus is the **dispersion parameter** σ^2, not μ.

Key Terminology

- **Homoscedasticity:** $\sigma_1^2 = \sigma_2^2 = \cdots = \sigma_k^2$ (equal variances across groups).
- **Heteroscedasticity:** at least one σ_i^2 differs.
- Homoscedasticity is a key assumption of the pooled t-test, one-way ANOVA, and many regression procedures; violations produce incorrect standard errors and inflated Type I error.

Caution: Strictly speaking, variance and scale are not synonymous. This book may occasionally use the terms interchangeably, but scale is the more general concept.

Hypotheses

One-sample: $H_0 : \sigma^2 = \sigma_0^2$ vs. $H_1 : \sigma^2 \neq \sigma_0^2$ (or >, <).

Two-sample: $H_0 : \sigma_1^2 = \sigma_2^2$ vs. $H_1 : \sigma_1^2 \neq \sigma_2^2$.

k-sample: $H_0 : \sigma_1^2 = \cdots = \sigma_k^2$ vs. H_1: at least one variance differs.

2. Theory

Foundational Distributional Results

Under normality, these exact results underpin parametric variance tests:

$$\frac{(n-1)s^2}{\sigma^2} \sim \chi^2_{n-1}$$

$$\frac{s_1^2/\sigma_1^2}{s_2^2/\sigma_2^2} \sim F_{n_1-1,\,n_2-1}$$

Under $H_0 : \sigma_1^2 = \sigma_2^2$, the ratio simplifies to $F = s_1^2/s_2^2 \sim F_{n_1-1,\,n_2-1}$.

These results are **exact only under normality** and break down badly under departures—the key weakness of the F-test and Bartlett.

3. Methods

Core Test Statistics

One-sample χ^2 test:

$$\chi^2 = \frac{(n-1)s^2}{\sigma_0^2} \sim \chi^2_{n-1} \quad \text{under } H_0\text{, normality assumed.}$$

Two-sample F-test:

$$F = \frac{s_1^2}{s_2^2} \sim F_{n_1-1,\,n_2-1} \quad \text{under } H_0\text{, normality assumed.}$$

Place the larger variance in the numerator to use the upper-tail critical value; the two-sided p-value is then doubled.

Bartlett's test (k groups, normal data):

$$B = \frac{(N-k)\log s_p^2 - \sum_i (n_i - 1)\log s_i^2}{c} \xrightarrow{d} \chi^2_{k-1},$$

where $s_p^2 = \sum_i (n_i - 1)s_i^2/(N-k)$ is the pooled variance and $c = 1 + [(\sum_i 1/(n_i - 1)) - 1/(N-k)]/(3(k-1))$ is a continuity correction. Most powerful under normality but **severely inflated** by non-normality.

Levene's test (k groups, robust):

1. Compute deviations from group mean: $z_{ij} = |x_{ij} - \bar{x}_j|$.
2. Apply one-way ANOVA to $\{z_{ij}\}$.

$$W = \frac{(N-k)}{(k-1)} \cdot \frac{\sum_i n_i(\bar{z}_i - \bar{z})^2}{\sum_i \sum_j (z_{ij} - \bar{z}_i)^2} \xrightarrow{d} F_{k-1,\,N-k}.$$

Brown–Forsythe test: identical to Levene but uses **group medians** instead of means in the deviation step: $z_{ij} = |x_{ij} - \tilde{x}_j|$. More robust to skewness and outliers than Levene; the **recommended default** for non-normal or skewed data.

Fligner–Killeen test: rank-based; replaces raw deviations with their rank scores within the combined sample. Fully nonparametric; most resistant to outliers and heavy tails. Asymptotically χ^2_{k-1}.

Choosing the Right Test

Situation	Recommended test
1 sample vs. σ_0^2, normal	χ^2 test
2 groups, normal data	F-test
k groups, confirmed normal	Bartlett
k groups, mildly non-normal	Levene
k groups, skewed / outliers	Brown–Forsythe
k groups, heavy tails / robust	Fligner–Killeen

Practical default: Brown–Forsythe for most applied work. Reserve Bartlett and the F-test for situations where normality has been carefully verified.

4. Fragility

Sensitivity to Non-Normality

Test	Normality required?	Robust?
χ^2 / F-test	Strictly	No
Bartlett	Strictly	No
Levene (mean)	Approximately	Moderate
Brown–Forsythe	Not required	Yes
Fligner–Killeen	Not required	Yes

The F-test and Bartlett are among the **least robust** procedures in the compendium: even mild leptokurtosis causes severe over-rejection of H_0. Never use them without first confirming normality via Shapiro–Wilk or Anderson–Darling.

The Pre-Test Problem

A common workflow: test equal variances (or more broadly *scales*) first, then decide between pooled t and Welch. This **two-stage strategy inflates overall Type I error** and loses power because the variance test itself has uncertainty. The better practice: **default to Welch's t-test** for two-group mean comparisons and bypass the variance pre-test entirely (see MEANS primer).

5. Interpretation

Effect Size for Variance Ratios

The **variance ratio** σ_1^2/σ_2^2 is the natural effect size for two-group comparisons. For k groups, the **coefficient of variation** (CV = σ/μ) and the ratio of the largest to smallest group variance ($s^2_{\max}/s^2_{\min}$) are useful descriptive summaries. A ratio of 4:1 or greater ($s_{\max}/s_{\min} \geq 2$) is a common rule of thumb for practically meaningful heteroscedasticity in ANOVA.

Sample Size Effects

- **Small n:** variance tests have low power; even ratios of 4:1 may be missed. Do not interpret non-rejection as confirmation of equal variances.
- **Large n:** trivial, practically inconsequential differences are detected. Focus on the estimated ratio s_1^2/s_2^2, not just the p-value.
- Variance estimates are **inherently noisier** than mean estimates: $\mathrm{Var}(s^2) = 2\sigma^4/(n-1)$ under normality, so power grows slowly with n.

6. Action

Variance Stabilizing Transformations

When heteroscedasticity is structural (variance depends on the mean), a transformation can restore homoscedasticity before applying mean tests:

Relationship	Transformation
$\sigma \propto \mu$	$\log y$
$\sigma \propto \mu^{1/2}$	$\sqrt{y}$ (counts)
$\sigma \propto \mu^2$	$1/y$
y is proportion	$\arcsin \sqrt{y}$
General	Box–Cox λ

Transformations change the scale of the analysis and affect interpretation; document and justify any transformation applied.

What To Do When Variances Differ

- **Two groups:** use Welch's t-test (automatically accounts for unequal variances).
- **Multiple groups:** use Welch one-way ANOVA (Brown–Forsythe or Welch F); Games–Howell post-hoc for pairwise comparisons.
- **Regression:** use HC (heteroskedasticity-consistent) standard errors (HC3 recommended); see REGR primer.
- **Transform:** apply a variance-stabilizing transformation (see table above).
- **Robust / nonparametric methods:** if transformation is undesirable, see ROBU and NPAR chapters.

7. Practice

Common Pitfalls

- Using the F-test or Bartlett with non-normal data (false rejections are common).
- Using a variance pre-test to choose between pooled t and Welch (use Welch by default instead).
- Ignoring outliers—a single outlier can inflate s^2 dramatically and dominate the test result.
- Treating a non-significant variance test as proof of homoscedasticity (especially at small n).
- Neglecting to report the variance ratio alongside the p-value.

Bottom Line

> Variance tests detect differences in spread, but robustness to non-normality matters far more than parametric optimality. Default to Brown–Forsythe for k groups and Levene for mild departures. Avoid the F-test and Bartlett unless normality is firmly established. Report the variance ratio alongside p-values, and remember that Welch's t-test renders the variance pre-test unnecessary for mean comparisons.

F-Test for Equality of Two Variances

Test Whether Two Population Variances Are Equal

VARI.1 | **Category:** Variance Comparison | **Data:** Continuous | **Null:** F-distribution

When to Use

- Compare variances of two independent populations.
- Assess homogeneity of variance assumptions.
- Preliminary check before pooled *t*-tests.
- Normal-theory inference on population variances.
- Quality-control and process variability studies.

Assumptions

- Both populations are **normally distributed** (critical).
- Samples are mutually independent (not paired).
- Observations within samples are independent.
- Data are measured on interval or ratio scale.
- Data obtained through random sampling from populations.

Test Statistic

$$F = \frac{s_1^2}{s_2^2} \sim F_{n_1-1,n_2-1} \quad \text{(under } H_0\text{)}$$

Notation: s_1^2 and s_2^2 are sample variances, $s_1^2 > s_2^2$ (larger in numerator, convention), n_1, n_2 are sample sizes

Plain English: The F-test compares the ratio of two sample variances. If population variances are equal, sample variances should be similar (ratio near 1). Large ratios (much greater or less than 1) suggest population variances differ. The test is simple and widely used but extremely sensitive to non-normality—even slight departures from normality severely inflate Type I error rates. This makes it unreliable in practice except with very carefully validated normal data.

⚠ Practitioner Warning

The F-test for variances is **notoriously** non-robust to violations of normality—far more sensitive than t-tests or ANOVA are to their normality assumptions. Even mild departures from normality (slight skewness, modest kurtosis) cause actual Type I error rates of 15–20% when nominal rate is 5%. Modern consensus: avoid this test unless normality is carefully verified. Use Levene's test or Brown-Forsythe test instead (robust alternatives).

Hypotheses

- H_0: Population variances are equal: $\sigma_1^2 = \sigma_2^2$.
- H_1: $\sigma_1^2 \neq \sigma_2^2$ (two-sided) or $\sigma_1^2 > \sigma_2^2$ (one-sided).

Decision Rule: Reject H_0 if $F > F_{\alpha/2,n_1-1,n_2-1}$ (right tail). For one-sided, reject if $F > F_{\alpha,n_1-1,n_2-1}$ (right tail).

Software References [Python, R, SAS]

Python	R	SAS
`scipy.stats.f`	`var.test()`	`PROC TTEST`

Additional Notes

The F-test is a likelihood ratio test for variance equality under normality. Historically important but problematic in practice. Complications: (1) extreme sensitivity to non-normality, (2) using it as preliminary test before t-test creates two-stage procedure with inflated overall error, (3) power depends on sample sizes and true variance ratio. Modern practice: use robust alternatives (Levene, Brown-Forsythe) or proceed directly with Welch's t-test (doesn't assume equal variances).

Paired Variance Test (Pitman–Morgan): For paired data, the independent F-test is invalid. Instead,

$$\sigma_X^2 = \sigma_Y^2 \iff \text{Corr}(X+Y,\ X-Y) = 0.$$

Thus, equality of variances can be tested via the correlation between $(X+Y)$ and $(X-Y)$ (Pitman–Morgan test).

Miscellanea: The F-distribution is named for Ronald Fisher, though he initially called it "z." George Snedecor, Fisher's American colleague, renamed it "F" in Fisher's honor when tabulating critical values in the 1930s. Fisher was reportedly uncomfortable with the eponymous naming but the convention stuck. The F-test for variances predates ANOVA but became less prominent as ANOVA's importance grew. Despite its name (*Analysis of Variance*), ANOVA is designed to test differences in group means rather than variances.

Reference(s): Fisher (1924); Snedecor (1934)

Worked Example

Problem Statement Two calibration instruments measure the same reference standard. Instrument 1 ($n_1 = 8$ readings) and Instrument 2 ($n_2 = 10$ readings) are compared to determine whether their measurement precisions differ. H_0: $\sigma_1^2 = \sigma_2^2$; H_1: $\sigma_1^2 \neq \sigma_2^2$. Always, start with normality checks in practice.

Inst. 1: 46, 47, 50, 51, 55, 56, 59, 60 **Inst. 2:** 47, 47, 48, 48, 49, 51, 52, 52, 53, 53

1. Sample Means & Variances

$\bar{x}_1 = 53$; $s_1^2 = 28$; $\bar{x}_2 = 50$; $s_2^2 = 6$.

Since $s_1^2 > s_2^2$, place Instrument 1 in the numerator.

2. F-Statistic

$$F = \frac{s_1^2}{s_2^2} = 4.667 \sim F(n_1 - 1, n_2 - 1) = F(7, 9)$$

under H_0.

Note: $F_{0.025}(9,7) = 4.823$.

3. Critical Values and Decision ($F(7,9)$)

Test	Stat.	CV (5%)	Decision
Two-sided ($\sigma_1^2 \neq \sigma_2^2$)	4.667	4.197	Reject
One-sided ($\sigma_1^2 > \sigma_2^2$)	4.667	3.293	Reject

$p = 0.036$ (two-sided); $p = 0.018$ (one-sided).

4. 95% Confidence Interval for σ_1^2/σ_2^2

$$\left[\frac{F}{F_{0.025}(7,9)},\ F \cdot F_{0.025}(9,7)\right] = [1.11,\ 22.51]$$

5. Interpretation

- $F = 4.667 > 4.197$: the variance ratio is significantly greater than 1 at 5%. Instrument 1 is substantially less precise ($s_1^2 = 28$ vs $s_2^2 = 6$, ratio 4.7:1). The wide CI [1.11, 22.51] reflects the high variability of variance estimates at small n.

Practitioner Insight: The F-test for equality of variances is the simplest member of the variance-comparison family and the only one with an exact F distribution under normality. The convention of placing the *larger* sample variance in the numerator so $F \geq 1$ always — avoids the need to consult a lower tail critical value: only the upper critical value $F_{\alpha/2}(n_1 - 1, n_2 - 1)$ is needed for the two-sided test, with the factor of $\alpha/2$ (not α) reflecting that the pre-test choice to put the larger variance in the numerator already conditions on one tail.

The confidence interval uses asymmetric bounds $F/F_{\alpha/2}(n_1 - 1, n_2 - 1)$ and $F \times F_{\alpha/2}(n_2 - 1, n_1 - 1)$, with the *swapped* degrees of freedom $F_{\alpha/2}(n_2-1, n_1-1)$ for the upper bound; this follows from $F_{\alpha/2}(n_2-1, n_1-1) = 1/F_{1-\alpha/2}(n_1-1, n_2-1)$.

The test's main limitation is severe sensitivity to non-normality: even mild departures inflate the Type I error rate substantially, making Levene's test or Brown–Forsythe's test preferable when normality is in doubt. The wide CI here [1.11, 22.51] is typical for $n_1 = 8$ and $n_2 = 10$: reliable variance estimation requires much larger samples than mean estimation, with roughly $n > 30$ per group for a useful precision.

Common Pitfalls

- **Normality not checked.** The F-test is extremely sensitive to non-normality; even mild skewness inflates Type I error. Always assess normality (Shapiro–Wilk, Q–Q plots). If doubtful, use Levene or Brown–Forsythe tests.
- **Using F-test as a preliminary test.** Pre-testing variances before t-tests or ANOVA creates a two-stage procedure that inflates overall Type I error (*"Garden of Forking Paths"* paradox - Gelman & Loken (2013)). Avoid preliminary variance testing; use Welch's t-test or check diagnostics after ANOVA.
- **Interpreting non-significance as equality.** A non-significant F-test does not establish equal variances; power is often low with small samples. Report uncertainty and prefer variance-robust methods.
- **Insufficient sample size.** Small samples provide poor power for detecting variance differences. Plan adequate n (often ≥ 30 per group) or rely on robust alternatives.
- **Incorrect one-sided testing.** One-sided F-tests require pre-specified direction and correct numerator placement. Use one-sided tests only when direction is justified *a priori*; otherwise use two-sided tests.

Bartlett's Test

Test Equality of Variances Across Multiple Groups

VARI.2 | **Category:** Variance Comparison | **Data:** Continuous | **Null:** χ^2 distribution

When to Use

- Compare variances across two or more groups.
- Assess homogeneity of variance assumptions.
- Preliminary check before pooled t-tests.
- Normal-theory inference on population variances.
- When parametric alternative to Levene is needed.

Assumptions

- All populations are **normally distributed** (critical).
- Samples are mutually independent (not paired).
- Observations within samples are independent.
- Data are measured on interval or ratio scale.
- Data obtained through random sampling from populations.

Test Statistic

$$M = \frac{(N-k)\ln(s_p^2) - \sum_{i=1}^{k}(n_i - 1)\ln(s_i^2)}{1 + \frac{1}{3(k-1)}\left[\sum_{i=1}^{k}\frac{1}{n_i-1} - \frac{1}{N-k}\right]}, \text{ where } s_p^2 = \frac{\sum(n_i-1)s_i^2}{(N-k)}. \qquad M \approx \chi^2_{k-1} \text{ under } H_0$$

Notation: k groups, n_i is sample size of group i, $N = \sum n_i$, s_i^2 is sample variance, s_p^2 is pooled variance

Plain English: Bartlett's test extends the two-sample F-test to multiple groups, testing whether all group variances are equal. It uses a log-likelihood ratio that compares each group's variance to a pooled variance estimate. If variances are equal, the (log) sample variances should be similar; large deviations indicate that at least one group differs. The test is most powerful under normality but highly sensitive to departures from normality, often producing false positives when data are skewed or heavy-tailed.

⚠ Practitioner Warning

Bartlett's test shares the F-test's fatal flaw: extreme sensitivity to non-normality. Slight departures (modest skewness or kurtosis) cause actual Type I error rates of 15–20% when nominal rate is 5%. This sensitivity often exceeds violations it's meant to detect. Modern statisticians strongly discourage using Bartlett's test unless normality is rigorously verified. Use Levene's or Brown-Forsythe test instead—equally powerful under normality but robust to violations.

Hypotheses

- H_0: $\sigma_1^2 = \sigma_2^2 = \cdots = \sigma_k^2$.
- H_1: At least one σ_i^2 differs from others.

Decision Rule: Reject H_0 if $M > \chi^2_{\alpha,k-1}$ or if p-value $< \alpha$.
Degrees of freedom = $k - 1$ (number of groups minus one).

Software References [Python, R, SAS]

`scipy.stats.bartlett`	`bartlett.test()`	`PROC GLM (HOVTEST=BARTLETT)`

Additional Notes

Bartlett's test is the likelihood ratio test for homogeneity of variances under normality assumption. The chi-square approximation is accurate for moderate-to-large samples when normality holds. For small samples ($n_i < 5$), use exact permutation version. The test has maximum power when data are truly normal but becomes anti-conservative (liberal or inflated Type I error) under non-normality—often more severely than the violations it's testing for.

Miscellanea: Maurice S. Bartlett (1937) made a fundamental contribution to likelihood-based inference through what is now called *Bartlett's correction*. In likelihood ratio tests, the statistic $-2\log\Lambda$ is asymptotically χ^2, but in small samples this approximation is often biased upward, leading to overly frequent rejections. Bartlett showed that multiplying the statistic by a simple correction factor (typically of the form $(1 - c/n)$ dramatically improves the χ^2 approximation, yielding more accurate Type I error control. This idea—*adjust the test statistic, not the distribution*—became a cornerstone of higher-order asymptotic theory. Remarkably, while Bartlett's own variance test is highly sensitive to non-normality, his correction to likelihood ratio tests is designed precisely to fix small-sample distortions.

Reference(s): Bartlett (1937); Box (1953)

Worked Example

Problem Statement A pharmaceutical company tests consistency of tablet weights from four production lines. Are weight variances equal across lines (indicating similar process control)?

Dataset: Tablet weights (mg) from random samples of each line, sample statistics, and Shapiro-Wilk (W) statistic:

Line	Sample Weights (mg)	n_i	$\bar{x}_i$ (mg)	s_i^2 (mg^2)	W(p-value)
1	498, 502, 501, 499, 500, 503, 497, 501, 500, 502	10	500.3	3.57	0.9645 (0.8353)
2	496, 501, 504, 499, 502, 500, 498, 503, 501, 497	10	500.1	6.77	0.9752 (0.9347)
3	493, 502, 505, 491, 508, 495, 506, 494, 503, 493	10	499.0	40.89	0.8792 (0.1279)
4	499, 501, 500, 502, 498, 501, 500, 499, 502, 500	10	500.2	1.73	0.9419 (0.5745)

Notes: All lines ($k = 4$) show no evidence against normality

1. Pooled Variance & Scale Factor

$$s_p^2 = \frac{\sum(n_i - 1)s_i^2}{N - k} = 13.24 \text{ (pooled var.)}$$

$$C = 1 + \frac{\sum 1/(n_i - 1) - 1/(N - k)}{3(k - 1)} = \mathbf{1.0463}$$

2. Bartlett's Statistic

$$\text{Num} = (N - k)\ln(s_p^2) - \sum(n_i - 1)\ln(s_i^2) = \mathbf{25.99}$$

$$M = \frac{\text{Num}}{C} = \mathbf{24.84}$$

3. Critical Value and Decision At the 5% significance level with 3 degrees of freedom, the critical chi-square value is 7.815 (p-value < 0.0001). Since $T = 25.99 > 7.815$ (Critical Value), we **strongly reject** H_0.

Practitioner Insight: Bartlett's statistic compares the log of the pooled variance to the weighted average of log variances: if $s_i^2 = s_p^2$ for all i, then $M = 0$. Heterogeneity inflates M because log is concave (Jensen's inequality), with larger spread in s_i^2 producing larger values. Here $s_3^2/s_4^2 = 23.7$, and thus yielding strong rejection ($p < 0.0001$). The correction factor C adjusts the $\chi^2(k-1)$ approximation for small samples; its form simplifies when group sizes are equal. Unlike the exact two-sample F-test, Bartlett extends to $k \geq 2$ but is also highly sensitive to non-normality, often inflating Type I error. When normality is doubtful, Levene's test is preferred. Rejection indicates violation of homoscedasticity; use Welch's ANOVA, variance-stabilizing transformation, or a nonparametric alternative.

Common Pitfalls

- **Ignoring normality requirement.** Bartlett's test is *highly* sensitive to non-normality; even mild skewness or outliers inflate Type I error well above 5%. Always check normality (Shapiro–Wilk, Q–Q plots). If any doubt, use Levene's or Brown–Forsythe instead.
- **Using Bartlett as a preliminary filter before ANOVA.** A two-stage workflow (Bartlett → ANOVA) inflates overall Type I error and adds little value, as ANOVA is fairly robust to mild heterogeneity. Modern practice: run ANOVA, inspect residuals, and switch to Welch or transformations only if needed.
- **Interpreting non-significant results as equality of variances.** A large p-value does not confirm equal variances. With small samples, power is low, and non-normality invalidates the test. Report the result cautiously and consider Welch's ANOVA as a safer default.
- **Using Bartlett with unbalanced designs.** Unequal n_i degrade the chi-square approximation and reduce power; small groups contribute unstable variance estimates. Levene's and Brown–Forsythe handle unequal sample sizes more reliably.
- **Over-interpreting statistical significance.** With large samples, Bartlett detects trivial variance differences. Always examine variance ratios: < 2 usually negligible, 2–3 borderline, > 3 serious. Combine statistical significance with practical considerations.

Hartley's F_{max} Test

Test Equality of Variances Using Maximum Variance Ratio

VARI.3 | **Category:** Variance Comparison | **Data:** Continuous | **Null:** F_{max} distribution

When to Use

- Test equality of variances across multiple groups.
- Preliminary check before one-way ANOVA.
- Balanced designs with equal sample sizes.
- Identify the largest variance discrepancy.
- Normal-theory inference on group variances.

Assumptions

- All populations are **normally distributed** (critical).
- Random samples are mutually independent (not paired).
- Observations within samples are independent.
- Data are measured on interval or ratio scale.
- **Equal sample sizes** across all groups (or very nearly equal).

Test Statistic

$$F_{\max} = \frac{s^2_{\max}}{s^2_{\min}}$$

Notation: $\{s^2_{\max}, s^2_{\min}\}$ are largest & smalled sample variances, k is number of groups, n is common sample size

Plain English: Hartley's test uses the simplest possible approach: compute all sample variances, find the largest and smallest, take their ratio. If all population variances are equal, this ratio should be modest. Large ratios suggest at least one population variance differs. The test ignores intermediate variances entirely, using only the two extremes. This simplicity is both strength (easy to compute) and weakness (less powerful than tests using all information).

⚠ Practitioner Warning

Hartley's test has two critical requirements: (1) **Equal sample sizes** across groups—with unequal n_i, critical values are invalid and test fails. (2) **Exact normality**—like F-test and Bartlett's, extremely sensitive to non-normality. Even mild departures inflate Type I error to 15–20%. Hartley's F_{max} distribution differs from the standard F; dedicated critical value tables must be used. Modern practice: use Levene's test instead (handles unequal n, robust to non-normality, uses all data not just extremes).

Hypotheses

- H_0: $\sigma_1^2 = \sigma_2^2 = \cdots = \sigma_k^2$.
- H_1: At least one σ_i^2 differs from others.

Decision Rule: Reject H_0 if $F_{\max} > F_{\max,\alpha}(k, n-1)$ from Hartley's tables. CVs increase with k and decrease with n.

Software References [Python, R, SAS]

Python	R	SAS
`Not native (Manual)`	`SuppDists::maxFratio`	`Not native (Manual)`

Additional Notes

Hartley's test was designed for quick hand calculation when detailed tables weren't available. Its simplicity (using only two variances) made it popular in pre-computer era but wastes information by ignoring intermediate variances. Only valid with equal n_i; even slight imbalance invalidates tables. Like Bartlett's, extremely sensitive to non-normality. Modern consensus: historical interest only. Use Levene's test in practice—it handles unequal n, uses all data, and is robust. Nonetheless, it can serve as a quick diagnostic tool in some contexts.

Despite its flaws, the F-max (and Bartlett's) test helped statisticians recognize that testing variance homogeneity needed a more robust approach.

Miscellanea: The name "Hartley" occasionally causes confusion. H. Leon Hartley (who proposed the $F_{\max}$ test) is a different person from H. O. Hartley (Herman Otto Hartley), a prominent statistician known for contributions such as the Hartley estimator and work on ANOVA and F-distribution tables. Because both were active in related areas, the $F_{\max}$ test is sometimes mistakenly attributed to H. O. Hartley (the more "famous" one).

Reference(s): Hartley (1950); David (1952)

Worked Example

Problem Statement Four laboratories each make $n = 7$ replicate measurements of a reference material. Hartley's $F_{\max}$ test (1950) checks whether the measurement variances are equal across laboratories. The test requires equal group sizes and uses only the largest and smallest sample variances. H_0: $\sigma_1^2 = \sigma_2^2 = \sigma_3^2 = \sigma_4^2$; H_1: at least one σ_i^2 differs.

Lab	Measurements ($n = 7, \bar{x}_i = 50$)							s_i^2
1	40	46	48	50	52	54	60	**40**
2	43	47	49	50	52	53	56	18
3	45	48	49	50	52	52	54	9
4	47	48	49	51	51	52	52	**4**

1. Identify Extremes

$$s_{\max}^2 = 40 \text{ (Lab 1)}, \qquad s_{\min}^2 = 4 \text{ (Lab 4)}$$

Labs 2 and 3 play no further role in the test.

2. $F_{\max}$ Statistic

$$F_{\max} = \frac{s_{\max}^2}{s_{\min}^2} = \frac{40}{4} = 10$$

3. Critical Values and Decision
Critical values are read from the Hartley table indexed by $k = 4$ (groups) and $\nu = 6$ (df per group):

α	$F_{\max,\text{crit}}$	$F_{\max}$	Decision
0.05	7.80	10.0	Reject H_0
0.01	14.5	10.0	Failed to Reject H_0

Reject H_0 at 5% but not 1%.

Practitioner Insight: $F_{\max}$ is the simplest omnibus variance test: compute k sample variances, take the ratio of the largest to the smallest, and compare to a table. This simplicity comes at a cost. First, it ignores all intermediate variances (e.g., $s_2^2 = 18$, $s_3^2 = 9$ contribute nothing). Second, it requires *equal* sample sizes; unequal n_i invalidate the Hartley table, unlike Bartlett's test. Third, it is sensitive to non-normality. Compared with Bartlett on the same data: Bartlett gives $M = 7.53$, $p = 0.057$ (FTR at 5%), while $F_{\max} = 10$ rejects. This highlights that $F_{\max}$ is more sensitive when heterogeneity is concentrated in extremes, but less so when spread across groups. The Hartley table depends on k and common $\nu = n - 1$; with unequal sizes, a conservative choice is $\nu = \min(n_i) - 1$.

Common Pitfalls

- **Unequal sample sizes.** Hartley's $F_{\max}$ is valid *only* for equal n_i. Any imbalance invalidates tabulated critical values. If n_i differ, use Levene's or Brown–Forsythe instead.
- **Ignoring normality.** $F_{\max}$ is highly sensitive to non-normality; even mild skewness or outliers inflate Type I error. Always check normality (Shapiro–Wilk, Q–Q plots). If uncertain, default to Levene's test.
- **Assuming focus on extremes increases power.** Using only $s_{\max}^2$ and $s_{\min}^2$ discards information from intermediate groups. Levene/Brown–Forsythe generally have better power across realistic variance patterns.
- **Using Hartley's test in modern practice.** Designed for hand-calculated, balanced 1950s experiments. Modern standard is Levene's test: robust to non-normality, valid for any n_i, and uses all sample variances. Hartley's is mainly of historical interest.
- **Confusing $F_{\max}$ with two-sample F-test.** When $k = 2$, formulas coincide but distributions differ. For two groups, use the standard F-test (or better: Welch's t-test for means), not Hartley's critical values.
- **Using standard F-tables instead of $F_{\max}$ critical values.** Hartley's statistic follows the $F_{\max}$ distribution, not the ordinary F-distribution. Using critical values in a standard F-table, the thresholds will be far too low—greatly inflating Type I error and producing false claims of unequal variances.

Effect Size

The $F_{\max}$ ratio itself serves as an effect-size gauge:

values < 2 indicate mild heterogeneity, 2–3 moderate, 3–5 substantial, and > 5 severe variance heterogeneity.

Levene's Test

Robust Test for Equality of Scales Across Groups

VARI.4 | **Category:** Variance Comparison | **Data:** Continuous | **Null:** F-distribution

When to Use

- Test equality of scales (spreads) across groups.
- Robust alternative to Bartlett's and Hartley's tests.
- When normality cannot be assumed or verified.
- Preliminary check before ANOVA procedures.
- Works with any sample size pattern (equal or unequal).

Assumptions

- Samples are mutually independent (not paired).
- Observations within samples are independent.
- Data are measured on interval or ratio scale.
- Data obtained through random sampling from populations.
- Group centers (means) adequately summarize locations.

Test Statistic

$$F_L = \frac{(N-k)\sum_{i=1}^{k} n_i(\bar{z}_i - \bar{z})^2}{(k-1)\sum_{i=1}^{k}\sum_{j=1}^{n_i}(z_{ij} - \bar{z}_i)^2} \sim F_{k-1,N-k} \text{ approximately, under } H_0.$$

Notation: k groups, n_i sample sizes, $N = \sum n_i$, $z_{ij} = |y_{ij} - \bar{y}_i|$ are absolute deviations from group means, and $\bar{z}$ is grand mean of absolute deviations.

Plain English: Levene's test transforms the variance problem into a mean problem. Instead of directly comparing variances (which requires normality), it compares the average absolute deviations from group means. Groups with larger variances have larger average deviations. This transformation creates approximately normal data even when original data are not, making the test robust. Essentially, Levene's asks: "Do groups differ in how spread out their values are?" using ANOVA on spread measures.

⚠ Practitioner Warning

Levene's test is more robust than the classical F test but still depends on the choice of group center. The median-based version (Brown–Forsythe) is commonly preferred when data are skewed or heavy-tailed.

Hypotheses

- H_0: All k groups have the same scale or dispersion.
- H_1: At least one group differs in scale or dispersion.

Decision Rule: Reject H_0 if $F_L > F_{\alpha,k-1,N-k}$ or p-value $< \alpha$. Robust to non-normality, handles unequal n_i, uses all data.

Software References [Python, R, SAS]

```
scipy.stats.levene        car::leveneTest()        PROC GLM
```

Additional Notes

Levene's test is ANOVA on absolute deviations $z_{ij} = |y_{ij} - \bar{y}_i|$. Absolute deviations retain information about spread while improving robustness. For normal data, $E[|Y - \mu|] \approx 0.798\sigma$, and more generally they capture variability even under non-normality. Since $z_{ij} = |y_{ij} - \text{center}|$ are typically more symmetric, ANOVA on deviations is more robust than direct variance tests.

Three common versions: (1) Original Levene's: deviations from means (standard), (2) Brown-Forsythe: deviations from medians (more robust to skewness), (3) Deviations from trimmed means (compromise). The F-approximation is accurate for moderate samples ($n_i \geq 5$). For very small samples, exact permutation version available. Levene's generally preferred over Bartlett's and Hartley's due to robustness with minimal power loss under normality.

Miscellanea: Howard Levene (1914–2003) introduced what is now called Levene's test in a 1960 festschrift (collection of essays) honoring Harold Hotelling, rather than in a journal article. The method spread widely through citations to this book chapter and became a standard tool in applied statistics despite its informal publication route. Levene, a statistician and geneticist associated with Columbia University, made influential contributions to ANOVA methodology and robust variance testing. His simple variance test ultimately became one of the most widely used procedures for assessing homogeneity of variance.

Reference(s): Levene (1960); Brown & Forsythe (1974)

Levene's Test

VARI.4

Worked Example

Problem Statement Three manufacturing processes yield components measured at $k = 3$ sites with unequal group sizes ($n_1 = 6, n_2 = 6, n_3 = 4, N = 16$). H_0: $\sigma_1^2 = \sigma_2^2 = \sigma_3^2$; H_1: at least one σ_i^2 differs.

Setting	Observations	$\bar{x}_i$	s_i^2	$z_{ij} = \lvert x_{ij} - \bar{x}_i \rvert$	$\bar{z}_i$
G1	14, 16, 18, 22, 24, 26	20	22.40	6, 4, 2, 2, 4, 6	4
G2	27, 28, 29, 31, 32, 33	30	5.60	3, 2, 1, 1, 2, 3,	2
G3	24, 24, 26, 26	25	1.33	1, 1, 1, 1	1

1. Transform: Absolute Deviations $z_{ij} = |x_{ij} - \bar{x}_i|$

Grand mean of all z_{ij}:

$$\bar{z} = 2.5$$

2. Between-Group SS

$$SSB = \sum_i n_i(\bar{z}_i - \bar{z})^2 = 24$$

3. Within-Group SS

$$SSW = \sum_i \sum_j (z_{ij} - \bar{z}_i)^2 = 20$$

4. Levene F-Statistic

$$MSB = \frac{SSB}{k-1} = 12, \quad MSW = \frac{SSW}{N-k} = \frac{20}{13}$$

$$F_L = \frac{MSB}{MSW} = 7.8 \sim F(2, 13)$$

5. Results and Decision

Statistic	Value	CV (1%)	Decision
F_L	7.800	6.701	Reject H_0

$p = 0.006$; CV at 5%= 3.806 also exceeded. Variances are heterogeneous at both levels.

Practitioner Insight: Levene's test replaces each observation by its absolute deviation from the group center and applies one-way ANOVA to these deviations. Since $\mathrm{E}[|x_{ij} - \mu_i|]$ is proportional to σ_i (e.g., $\mathrm{E}[z_{ij}] = \sigma_i\sqrt{2/\pi}$ under normality), differences in variance translate into differences in group means of the deviations, which ANOVA detects.

Because the test operates on absolute deviations rather than squared deviations, it is substantially less sensitive to outliers and non-normality than Bartlett's log-likelihood test.

Here $SSW_3 = 0$ because all four Group 3 deviations are identical ($z_{3j} = 1$), making Group 3 contribute nothing to within-group variability; this is not a data problem but a reflection of the tight, symmetric structure of that group.

The Brown–Forsythe variant replaces the mean with the median, improving robustness for skewed data; the two coincide for symmetric distributions. Unlike $F_{\max}$, Levene's test naturally handles unequal group sizes, and unlike Bartlett, it remains reliable under non-normality.

Common Pitfalls

- **Using Levene as a screening test.** Preliminary testing before ANOVA inflates Type I error and complicates inference. Prefer Welch's ANOVA directly or check residuals after the fact.
- **Switching between Levene and Brown–Forsythe post hoc.** Choosing the version after seeing data is p-hacking. Pre-specify; Brown–Forsythe (medians) is a robust default.
- **Treating non-significant results as proof of equality.** Levene has low power with small samples; $p > 0.05$ does *not* confirm equal variances. Examine variance ratios and consider robust methods.
- **Using Levene with very discrete or tiny samples.** Discrete outcomes and $n < 5$ degrade the F-approximation. Prefer permutation Levene or median-based approaches in such settings.
- **Confusing statistical with practical significance.** Large samples detect trivial variance differences. Interpret Levene alongside variance or SD ratios to judge practical importance.

Brown-Forsythe Test

Median-Based Robust Test for Equality of Scales

VARI.5 | **Category:** Variance Comparison | **Data:** Continuous | **Null:** F-distribution

When to Use

- Compare scales across three or more groups.
- Normality doubtful but variance testing needed.
- Median-based alternative to Levene's test.
- Outliers may distort classical variance tests.
- Verify equal variances before ANOVA procedures.

Assumptions

- Samples are mutually independent (not paired).
- Observations within samples are independent.
- Data are measured on interval or ratio scale.
- Data obtained through random sampling from populations.
- Group centers (medians) adequately summarize locations.

Test Statistic

$$F_{\mathrm{BF}} = \frac{(N-k)\sum_{i=1}^{k} n_i(\bar{z}_i - \bar{z})^2}{(k-1)\sum_{i=1}^{k}\sum_{j=1}^{n_i}(z_{ij} - \bar{z}_i)^2} \sim F_{k-1,N-k} \quad \text{(approx., under } H_0\text{)}.$$

Notation: k groups, n_i sample sizes, $z_{ij} = |y_{ij} - \tilde{y}_i|$ are absolute deviations from group medians $\tilde{y}_i$, $N = \sum n_i$.

Plain English: The Brown–Forsythe test is a robust modification of Levene's test that measures variability using absolute deviations from group medians rather than means. Because medians are resistant to outliers and skewness, the test provides reliable variance comparisons for non-normal data. For symmetric distributions, it performs similarly to Levene's test; for skewed or heavy-tailed data, it is clearly superior and is often recommended as the default variance test.

⚠ Practitioner Warning

Brown-Forsythe test is slightly more conservative (lower power) than standard Levene's for symmetric distributions. The power loss is small (5–10%) and justified by robustness gains. For perfectly normal data, Levene's with means has marginally higher power. For any non-normal data (the more common case), Brown-Forsythe equals or exceeds Levene's power while maintaining better Type I error control.

Hypotheses

- H_0: All k groups have the same scale or dispersion.
- H_1: At least one group differs in scale or dispersion.

Decision Rule: Reject H_0 if $F_{\mathrm{BF}} > F_{\alpha,k-1,N-k}$ or $p < \alpha$. Robust to skewness, outliers, and unequal sample sizes.

Software References [Python, R, SAS]

```
scipy.stats.
levene(center="median")
```

```
car::leveneTest
(center="median")
```

```
PROC GLM (HOVTEST=BF)
```

Additional Notes

Brown–Forsythe is algebraically identical to Levene's test except for the choice of center. Using medians prevents extreme observations from distorting variability estimates, making the resulting F approximation reliable even for small samples and non-normal distributions. Brown–Forsythe is robust, maintaining nominal Type I error across diverse distributions (normal, lognormal, exponential, t). It is closely related to the median absolute deviation (MAD), MAD = median($|y_i - \tilde{y}|$): while MAD takes the median of absolute deviations, Brown–Forsythe uses their mean, with both resisting outliers. Since variance ratios are sensitive to extremes, a better effect size is RMAD = $\mathrm{MAD}_{\max}/\mathrm{MAD}_{\min}$, capturing typical spread. As a guide: RMAD < 2 (mild), 2–3 (moderate), > 3 (substantial).

Miscellanea: The Brown–Forsythe test was introduced in a 1974 JASA paper by Morton B. Brown and Alan B. Forsythe. Their work was motivated by practical problems in industrial and chemical research, where real-world data often showed skewness or heavy tails that made classical variance tests unreliable. Through extensive simulations, they demonstrated that replacing Levene's mean-based deviations with median-based deviations dramatically improved robustness, especially for skewed distributions. As a result, the Brown–Forsythe variant gradually became the preferred default form of Levene's test. It is often compared to the Fligner–Killeen test (1976), another highly robust method based on ranked absolute deviations.

Reference(s): Levene (1960); Brown & Forsythe (1974)

Worked Example

Problem Statement Three production lines are monitored with $n_i = 7$ measurements each ($N = 21$). The data are mildly right-skewed (mean > median within each group), making the Brown-Forsythe test (1974) preferable. H_0: $\sigma_1^2 = \sigma_2^2 = \sigma_3^2$; H_1: at least one σ_i^2 differs.

Line	Observations	Median, $\tilde{x}_i$	Mean, $\bar{x}_i$	s_i^2	$z_{ij} = \lvert x_{ij} - \tilde{x}_i \rvert$	Mean, $\bar{z}_i$
1	49, 49, 49, 50, 51, 52	50	50.14	1.60	0, 1, 1, 1, 1, 1, 2	1
2	98, 98, 98, 100, 102, 104	100	100.29	6.40	0, 2, 2, 2, 2, 2, 4	2
3	146, 146, 146, 150, 154, 154, 158	150	150.57	23.62	0, 4, 4, 4, 4, 4, 8	4

1. Transform: $z_{ij} = |x_{ij} - \tilde{x}_i|$
Each median is an observed value (odd n). Deviations:
Grand mean of all z_{ij}: $\bar{z} = \frac{7}{3}$

2. Between-Group SS

$$SSB = \sum_i n_i(\bar{z}_i - \bar{z})^2 = \frac{98}{3}, \quad MSB = \frac{SSB}{k-1} = \frac{49}{3}$$

3. Within-Group SS

$$SSW = \sum_i \sum_j (z_{ij} - \bar{z}_i)^2 = 42, \quad MSW = \frac{SSW}{N-k} = \frac{7}{3}$$

4. Brown-Forsythe F-Statistic

$$F = \frac{MSB}{MSW} = 7.0 \sim F(2, 18)$$

5. Results and Decision

Statistic	Value	CV (1%)	Decision
F	7.000	6.013	Reject H_0

$p = 0.006$; CV at 5%= 3.555 also exceeded. Rejects at both levels.

Practitioner Insight: Brown–Forsythe replaces the group mean with the median when computing absolute deviations. For symmetric data, results match Levene; for skewed data, differences emerge. Here each group has mean slightly above its median (right skew: mean > median by 0.14, 0.29, 0.57 respectively). Both tests reject (Levene $F = 9.29$, $p = 0.002$; BF $F = 7.00$, $p = 0.006$), with Levene slightly more sensitive under mild skew. The key advantage of BF appears with heavier skew or outliers: the median resists distortion, preserving large deviations for extreme values, whereas the mean can be pulled toward outliers and mask variability. In practice, the median-based BF test is widely recommended (e.g., NIST) as the default for variance homogeneity, with mean-based Levene mainly retained for comparison.

Common Pitfalls

- **Testing "typical" spread, not full variance.** Brown–Forsythe uses deviations from the *median*, so it captures typical variability and downweights outliers. Large variance differences driven by a few extremes may not be detected. If extremes matter substantively, supplement with outlier diagnostics.
- **Using Brown–Forsythe when Levene's is more powerful.** For symmetric, normal-like data with no outliers, mean-based Levene's is slightly more powerful. Use Brown–Forsythe when skewness or outliers are present; otherwise either is fine. If uncertain, Brown–Forsythe is the safer default.
- **Interpreting non-significance as evidence of equal variances.** Like all heteroscedasticity tests, Brown–Forsythe has low power with small samples ($n < 10$ per group). Non-significant results often reflect insufficient data, not equal variances. Report variance ratios and use heterogeneity-robust methods (e.g., Welch's ANOVA for means testing).
- **Applying Brown–Forsythe to dependent data.** The test assumes independent groups. Do not use for paired samples or repeated measures; variance structure is more complex. For dependent data, use mixed models or methods designed for correlated observations.
- **Using Brown–Forsythe as a preliminary gatekeeper.** Pre-testing inflates Type I error and complicates inference. Better practice: run the main analysis with Welch's ANOVA or other robust methods, and use Brown–Forsythe diagnostically rather than as a decision rule.

Mood's Dispersion Test

Early Rank-Based Test for Equality of Scale (Two Samples)

VARI.6 | **Category:** Variance Comparison | **Data:** Continuous / Ordinal | **Null:** Perm. / Normal (asymp.)

When to Use

- Compare variances between two groups.
- Distribution-free variance comparison procedure.
- Outliers may distort classical variance tests.
- Strong departures from normality are present.
- Verify equal variances before ANOVA procedures.

Assumptions

- Both Samples are mutually independent (not paired).
- Observations within samples are independent.
- Data are measured on interval or ratio scale.
- Data obtained through random sampling from populations.
- Group distributions have similar shapes.

Test Statistic

$$T = \sum_{i=1}^{m}(R_i - M)^2, \quad M = \frac{m+n+1}{2}, \quad E[T] = \frac{m(N^2-1)}{12}, \quad Var(T) = \frac{mn(N+1)(N^2-4)}{180}$$

where standardized T is approximately normal for large samples.

Notation: m, n sample sizes, $N = m + n$, R_i ranks in pooled sample, M median rank position.

Plain English: Mood's test measures dispersion using squared deviations of ranks from the median rank. After pooling and ranking both samples, ranks far from the center contribute large squared values. A sample with greater spread produces more extreme ranks, inflating the test statistic. Squaring emphasizes tail behavior, making the test sensitive to differences in scale rather than location.

⚠ Practitioner Warning

Variance is a specific measure of dispersion, whereas scale refers more generally to distributional spread. The test is therefore a test of scale rather than variance per se. Mood's test is statistically valid but inefficient compared to later methods. Ansari–Bradley provides simpler scoring for two groups, and Fligner–Killeen extends scale testing to multiple groups with better power. In modern applications, Mood's test is rarely recommended outside instructional or historical contexts.

Hypotheses

- H_0: Both populations have equal scale (dispersion).
- H_1: Populations have different scales (dispersions).

Decision Rule: Reject H_0 if standardized T is too large or too small. Use normal approximation for $m, n \gtrsim 20$; exact permutation distribution otherwise.

Software References [Python, R, SAS]

Python	R	SAS
`scipy.stats.mood`	`mood.test()`	`PROC NPAR1WAY (MOOD option)`

Additional Notes

Mood (1954) introduced one of the first distribution-free tests for scale. The squared-rank scoring $(R - M)^2$ strongly weights tail observations, directly linking the statistic to dispersion. Its asymptotic relative efficiency is about 0.76 under normality—comparable to other early rank-based scale tests but inferior to modern robust methods. *Asymptotic relative efficiency* (ARE) compares two tests by how many observations ("relative efficiency") each requires to achieve the same power as sample size grows ("asymptotic"); an ARE of 0.8 means the first test needs 25% more data, while an ARE of 1.2 means it needs 20% less.

Miscellanea: Alexander McFarlane Mood introduced his two-sample scale test in 1954 while studying the asymptotic efficiency of nonparametric procedures. His work provided one of the earliest formal treatments of rank-based tests for dispersion, establishing a conceptual framework that later researchers expanded. Ansari–Bradley (1960) developed a symmetric-score alternative in the same spirit, and Fligner–Killeen (1976) extended the rank-based scale-testing philosophy to multiple samples using normal-score transformations. Mood's early insights helped shape the development of robust statistics for decades.

Reference(s): Mood (1954); Ansari & Bradley (1960)

Worked Example

Problem Statement A manufacturer is testing two different machines ($n_1 = 5, n_2 = 5$) that cut steel rods. Both machines produce rods with the same median length, but the manager wants to know if Machine B is "less stable" (has higher variance) than Machine A. While the medians are identical, Mood's Dispersion Test is used to determine if the variances differ. We test $H_0 : \sigma_1^2 = \sigma_2^2$ against $H_a : \sigma_1^2 \neq \sigma_2^2$.

Data (in length units)

Machine A	10.1 (2)	10.2 (3.5)	10.3 (5.5)	10.4 (7.5)	10.5 (9)
Machine B	9.8 (1)	10.2 (3.5)	10.3 (5.5)	10.4 (7.5)	10.8 (10)

Ranks in red inside the parenthesis.

1. Calculation of the Test Statistic The Mood statistic (M) is the sum of squared deviations of the ranks of the first group from the average rank $\bar{R} = (N+1)/2$. Given $N = 10$, $\bar{R} = 5.5$:

$$T = \sum_{i=1}^{n_1} (R_i - 5.5)^2 = 32.5 \tag{1}$$

2. Standardized Statistic (Z) For larger samples, M is standardized. The expected value $E[M]$ and variance $Var(M)$ are:

$$E[T] = \frac{n_1(N^2-1)}{12} = 41.25,$$

$$Var(T) = \frac{n_1 n_2 (N+1)(N^2-4)}{180} = 146.67$$

The Z-score is:

$$Z = \frac{T - E[T]}{\sqrt{Var(T)}} = -0.722$$

3. Critical Values and Results

Metric	Value
Mood Statistic (T)	32.5
Standardized Z	-0.722
Critical $Z_{0.05}$ (Two-tailed)	1.96

4. Decision and Interpretation

- **Decision:** Since $|-0.722| < 1.96$, we **Fail to Reject** H_0.
- **Interpretation:** There is no statistically significant difference in the dispersion (variance) of the two machines.

Practitioner Insight: Mood's test is very robust against outliers because it uses ranks. However, it loses significant power if the medians of the two groups are not actually equal. If medians differ, the data should be centered by subtracting the respective group medians before ranking.

Common Pitfalls

- **Using Mood when better modern tests exist.** Mood is historically important but low-efficiency; Ansari–Bradley and Fligner–Killeen dominate in most settings. Reserve Mood for replication or theoretical comparison, not routine analysis.
- **Ignoring unequal medians.** Mood assumes equal medians—location differences contaminate scale inference. If medians differ, shift data first or use Fligner–Killeen; otherwise results are invalid.
- **Misreading the test statistic.** Mood uses squared rank deviations: large T implies greater dispersion in the first sample. Direction is not intuitive; specify one-sided alternatives carefully.
- **Relying on normal approximation with small samples.** Asymptotic normality requires $m, n \geq 20$; smaller samples need exact/permutation or bootstrap p-values. Normal approximation can misstate Type I error heavily.
- **Forgetting rank methods hide effect magnitude.** Mood detects scale differences but provides no magnitude estimate. Always pair with descriptive dispersion (IQR, MAD, SD ratios) and graphical comparisons.

Ansari–Bradley Test

Rank-Based Test for Equality of Scale (Two Samples)

VARI.7 | **Category:** Variance Comparison | **Data:** Continuous / Ordinal | **Null:** Perm. / Normal (asymp.)

When to Use

- Compare dispersions of two independent groups.
- Distribution-free alternative to the F test.
- Variance comparison under non-normality.
- Outliers may distort classical variance tests.
- Compare scale parameters of skewed data.

Assumptions

- **Critical:** Both populations have equal medians (or are median-adjusted).
- Random samples are mutually independent.
- Observations within samples are independent.
- Data are at least ordinal-scale measurements.

Test Statistic

$$Z = \frac{C - \mu_C}{\sigma_C}, \qquad C = \sum_{i=1}^{m} a(R_i), \qquad a(R) = \min(R, m + n + 1 - R)$$

where C is exact for small samples and approximately normal for large samples. Formulas are provided on the next page.
Notation: m, n sample sizes, R_i ranks in pooled sample, $a(R)$ are symmetric rank scores emphasizing central vs. extreme observations, μ_C and σ_C are expected mean and standard deviation of C

Plain English: The Ansari–Bradley test compares dispersion using ranks rather than raw values. After pooling and ranking both samples, symmetric scores assign low weight to extreme ranks and high weight to central ranks. A sample with greater spread contributes disproportionately to extreme ranks, shifting the score distribution. The test is fully distribution-free but valid only when the two samples share the same median.

⚠ Practitioner Warning

Variance is a specific measure of dispersion, whereas scale refers more generally to distributional spread. The test is therefore a test of scale rather than variance per se. The test *requires equal medians*. If medians differ, scale and location effects are confounded and results are invalid. Either verify median equality or shift one sample to equalize medians before testing. If median differences are present or uncertain, prefer Fligner–Killeen, which accommodates location and scale simultaneously.

Hypotheses

- H_0: Both populations have equal scale (dispersion).
- H_1: Populations have different scales (dispersions).

Decision Rule: Reject H_0 if C is too small or too large. Use exact tables for small samples; normal approximation for $m, n \gtrsim 25$.

Software References [Python, R, SAS]

`scipy.stats.ansari`	`ansari.test()`	`PROC NPAR1WAY (AB option)`

Additional Notes

Ansari–Bradley (1960) is one of several rank-based scale tests, alongside Mood (1954) and Klotz (1962). Its simple symmetric scoring makes it intuitive and easy to compute, with good power against scale alternatives when distributions have similar shapes. The test applies only to two samples; for $k \geq 3$ groups, use Fligner–Killeen.

Miscellanea: Ansari and Bradley introduced their two-sample rank test for scale in 1960 while both were at Florida State University. Their method was one of the earliest nonparametric alternatives to the classical F-test, using symmetric rank scores to detect differences in dispersion without assuming normality. Sixteen years later, Fligner and Killeen developed a multi-sample extension of the same rank-based philosophy, creating a fully distribution-free test built on ranked absolute deviations. Together, the Ansari–Bradley and Fligner–Killeen tests form an important historical bridge between early rank-based procedures and the modern era of robust scale testing.

Reference(s): Ansari & Bradley (1960); Mood (1954)

Ansari–Bradley Test

VARI.7

Worked Example

Problem Statement Do two customer–service teams differ in the *dispersion* of call–handling times?

Data. Team A ($m = 8$, median=5.05): 4.8, 5.2, 4.9, 5.5, 5.0, 4.7, 5.3, 5.1. Team B ($n = 7$, median=5.7): 4.4, 6.2, 5.8, 4.6, 6.0, 5.7, 4.5. As medians are close, Ansari-Bradley test can be performed.

1. Pooled data and center–outward assigned ranks.

Team:	B	B	B	A	A	A	A	A	A	A	A	B	B	B	B
Time:	4.4	4.5	4.6	4.7	4.8	4.9	5	5.1	5.2	5.3	5.5	5.7	5.8	6	6.2
R:	1	2	3	4	5	6	7	8	9	10	11	12	13	14	15
$m+n+1-R$:	15	14	13	12	11	10	9	8	7	6	5	4	3	2	1
$a(R)$:	1	2	3	4	5	6	7	8	7	6	5	4	3	2	1

Team A's ranks (original order): $5, 7, 6, 5, 7, 4, 6, 8$.

2. Ansari-Bradley statistic.

$$C = \sum \text{ranks}_A = 48, \quad \text{using exact tables, } p\text{-value} = 0.0006.$$

3. Decision and Conclusion. Reject H_0 as $p < 0.05$. There is strong evidence of unequal dispersion. Team B shows substantially greater variability in call times than Team A.

Practitioner Insight: The AB statistic, C, rewards observations near the center of the combined distribution with *high* scores and penalises extreme observations with *low* scores. A group with wide spread will contribute many extreme values, accumulating low C, suggesting high variance. The critical assumption is *equal medians*: if the two distributions differ in location as well as scale, the rank pattern conflates the two differences and the test is no longer a pure scale test. In practice, center each sample on its median before applying AB when medians are unequal. The AB test is asymptotically equivalent to the Siegel-Tukey test (which uses a different score assignment but tests the same hypothesis) and both are special cases of a broader class of rank-based scale tests.

Common Pitfalls

- **Requires equal medians.** Ansari–Bradley assumes identical medians; if medians differ, scale and location effects confound. Always check median equality (refer to NPAR). If medians differ, AB is inappropriate.
- **Lower power than alternatives.** AB is fully distribution-free but typically less powerful than Levene or Brown–Forsythe for continuous data. Reserve AB for ordinal data, severe non-normality, or when a distribution-free scale test is specifically required.
- **Two-sample only.** AB does not extend to $k > 2$ groups. Do *not* run multiple pairwise AB tests. For multi-group scale comparisons, use Fligner–Killeen or Levene/Brown–Forsythe instead.
- **Sensitive to ties.** Heavy ties degrade power and normal-approximation accuracy. For discrete or rounded data, use a permutation version of AB or switch to Fligner–Killeen.
- **Non-significant ≠ equal scales.** With small samples, AB has low power. Failure to reject H_0 may reflect insufficient data. Report scale ratios (e.g., IQR or variance ratios) and emphasize uncertainty rather than claiming equality.

Large Sample Approximation (Normal Distribution)

For larger samples ($m, n > 10$), the Ansari–Bradley statistic C approaches a normal distribution. A finite-sample continuity correction ± 0.5 may be applied.

Mean (μ_C) **Variance** (σ_C^2) **Z-score**

$$\mu_C = \begin{cases} \dfrac{m(N+2)}{4}, & N \text{ even,} \\[2mm] \dfrac{m(N+1)^2}{4N}, & N \text{ odd,} \end{cases} \qquad \sigma_C^2 = \begin{cases} \dfrac{mn(N-2)(N+2)}{48(N-1)}, & N \text{ even,} \\[2mm] \dfrac{mn(N+1)(3+N^2)}{48N^2}, & N \text{ odd,} \end{cases} \qquad Z = \frac{C - \mu_C}{\sigma_C}$$

Siegel-Tukey Test

Nonparametric Rank Test for Equality of Scale (Two Samples)

VARI.8 | **Category:** Variance Comparison | **Data:** Continuous / Ordinal | **Null:** Perm. / Normal (asymp.)

When to Use

- Compare dispersions of two independent groups.
- Rank-based alternative to the standard F test.
- Variance comparison under non-normality conditions.
- Outliers may distort classical variance tests meaningfully.
- Use a distribution-free scale/dispersion parameter test.

Assumptions

- Samples are mutually independent.
- Observations within samples are independent.
- Data are at least ordinal-scale measurements.
- Group distributions have similar shapes.
- Distributions differ only in scale (same shape and median).

Test Statistic

$$U = \sum_{i=1}^{n_1} R_i - \frac{n_1(n_1+1)}{2}, \quad E[U] = \frac{n_1 n_2}{2}, \quad \text{Var}[U] = \frac{n_1 n_2(n_1+n_2+1)}{12}. \qquad \text{Under } H_0,\ z = \frac{U - E[U]}{\sqrt{\text{Var}[U]}} \approx N(0,1).$$

Notation: R_i are Siegel-Tukey ranks (extremes ranked lowest), U is sum of ranks for group 1

Plain English: The Siegel-Tukey test ranks observations by their distance from the center of the combined sample, with extreme values (far from center) receiving lower ranks than central values. Values are ranked alternately from the extremes inward: smallest gets rank 1, largest gets rank 2, second-largest gets rank 3, second-smallest gets rank 4, and so on. The test then applies the Mann-Whitney U test to these specially assigned ranks. If one group has greater variability (dispersion), its observations will tend to be more extreme, receiving lower Siegel-Tukey ranks.

⚠ Practitioner Warning

Variance is a specific measure of dispersion, whereas scale refers more generally to distributional spread. The test is therefore a test of scale rather than variance per se. Assumes equal location (medians); differences can distort results, so centering may be needed. Less powerful than the F-test under normality but more robust otherwise. Ties reduce power, and small samples ($n_1, n_2 < 20$) require exact tests. Does not reliably indicate direction of variability difference; interpretation depends on ranking convention and can be distorted by location differences—use descriptive measures (e.g., sample variances). Alternative ranking schemes (Ansari-Bradley) may perform better under certain conditions.

Hypotheses

- H_0: Both populations have equal scale (dispersion).
- H_1: Populations have different scales (dispersions).

Decision Rule: Reject H_0 if $|z| > z_{\alpha/2}$ or p-value $< \alpha$ for two-sided test.

Software References [Python, R, SAS]

Python	R	SAS
`Not Native (Manual)`	`DescTools::SiegelTukeyTest`	`PROC NPAR1WAY (SIEGELTUKEY)`

Additional Notes

Procedure: combine samples, sort, assign alternating extreme ranks, and sum ranks by group. Some implementations reverse rank direction, but inference is unchanged. The test is computationally identical to Mann–Whitney applied to transformed ranks. Ansari-Bradley ranks symmetrically from extremes (1,2,2,1 pattern), Siegel-Tukey alternates (1,2,3,4,...); both have similar power; choice often based on software availability. Location adjustment: if medians differ substantially, subtract group medians before ranking or use preliminary location test. Exact test: for small samples, compute exact permutation distribution of rank sum rather than normal approximation.

Miscellanea: Siegel and Tukey (1960) introduced a nonparametric test for comparing dispersion by ranking observations based on their distance from the center. This extremeness-based ranking transforms a scale problem into a rank-based framework, extending nonparametric methods beyond location tests. Although less widely used than location-based procedures, it remains a useful distribution-free alternative when variance assumptions are questionable.

Reference(s): Siegel & Tukey (1960); Conover et al. (1981); Gibbons & Chakraborti (2011)

Worked Example

Problem Statement Same example as Ansari-Bradley's. Do two customer–service teams differ in the *dispersion* of call–handling times?

Data. Team A ($m = 8$, median=5.05): 4.8, 5.2, 4.9, 5.5, 5.0, 4.7, 5.3, 5.1. Team B ($n = 7$, median=5.7): 4.4, 6.2, 5.8, 4.6, 6.0, 5.7, 4.5. As medians are close, the Siegel-Tukey test can be performed.

1. Pooled data and alternating extreme-outward assigned ranks.
The array is ordered from smallest to largest. Ranks are assigned in a strict alternating pattern to the extremes: Rank 1 to the smallest, Ranks 2 and 3 to the two largest, Ranks 4 and 5 to the next two smallest, and so on.

Team:	B	B	B	A	A	A	A	A	A	A	A	B	B	B	B
Time:	4.4	4.5	4.6	4.7	4.8	4.9	5.0	5.1	5.2	5.3	5.5	5.7	5.8	6.0	6.2
$a(R)$ (S-T):	1	4	5	8	9	12	13	15	14	11	10	7	6	3	2

Team A's Siegel-Tukey ranks (original sample order): 9, 14, 12, 10, 13, 8, 11, 15. The rank sum $W = \sum \text{ranks}_A = 92$.

2. Siegel-Tukey statistic.
The Siegel-Tukey statistic, U, and its expected mean & variance under H_0 are:

$$U = W - \frac{m(m+1)}{2} = 56, \qquad E[U] = \frac{mn}{2} = 28, \qquad \text{Var}(U) = \frac{mn(m+n+1)}{12} = 74.667.$$

Using the continuity-corrected normal approximation:

$$Z = \frac{U - E[U] - 0.5}{\sqrt{\text{Var}(U)}} = 3.182 \implies p\text{-value} = 0.0015.$$

3. Decision and Conclusion. Reject H_0 as $p < 0.05$. There is highly significant statistical evidence of unequal dispersion. Team B shows substantially greater variability in call times than Team A.

Practitioner Insight: The Siegel-Tukey test maps dispersion testing directly into a standard Wilcoxon rank-sum framework. It achieves this by assigning *low* ranks to extreme values on both tails (1 to the minimum, 2 and 3 to the maximums) and *high* ranks to central values. Consequently, a group with high dispersion will systematically accumulate extreme positions, yielding low ranks and a significantly deflated rank-sum W.

Like the Ansari-Bradley test, it is critically sensitive to the *equal medians* assumption. If the locations differ, the alternating ranking scheme conflates location shifts with scale shifts. When medians are unequal, always center the samples by subtracting their respective medians ($\tilde{x}_i$) before assigning ranks.

Note: Test could also be performed directly using W. $E[W] = \frac{m(m+n+1)}{2}$ and $\text{Var}(W) = \text{Var}(U)$.

Common Pitfalls

- **Confounding scale with location differences.** Siegel–Tukey assumes equal medians; if groups differ in location, the test mixes location and scale effects. Always check median equality first or center groups before applying the test.
- **Using incorrect ranking scheme.** The test requires its special "extremes-inward" ranking (smallest, largest, 2nd largest, 2nd smallest, . . .). Using standard ascending ranks invalidates the statistic. Verify ties are handled with proper midranks.
- **Relying on asymptotic p-values with small samples.** Normal approximation is poor when $n < 20$ per group. Use exact/permutation versions for small samples to avoid inflated or deflated Type I error.
- **Applying to dependent or paired samples.** Siegel–Tukey requires independent groups. For paired data, use Pitman–Morgan or tests on absolute differences; rank-based two-sample tests are invalid under dependence.
- **Expecting it to diagnose specific distributional features.** Siegel–Tukey detects general scale differences, not outliers, skewness, or tail behavior specifically. Always supplement with descriptive dispersion measures and graphical checks.

Fligner-Killeen Test

Distribution-Free Test for Equality of Scale (k Samples)

VARI.9 | **Category:** Variance Comparison | **Data:** Continuous / Ordinal | **Null:** χ^2 distribution

When to Use

- Test equality of dispersions across two or more groups.
- Completely distribution-free (nonparametric) alternative.
- Data are severely non-normal, ordinal, or ranked.
- Want test that makes minimal distributional assumptions.

Assumptions

- Samples are independent across groups.
- Observations within each group are independent.
- Random sampling from each population.
- **No distributional assumptions** (fully nonparametric).

Test Statistic

$$FK = \frac{12\sum_{i=1}^{k} n_i(\bar{a}_i - \bar{a})^2}{N(N+1)} \sim \chi^2_{k-1} \text{ approximately under } H_0.$$

With ties : $$FK = \frac{\sum_{i=1}^{k} n_i\,(\bar{a}_i - \bar{a})^2}{\frac{1}{N-1}\sum_{i=1}^{k}\sum_{j=1}^{n_i}(a_{ij} - \bar{a})^2}$$

Notation: k groups, R_{ij} are ranks of $|y_{ij} - \tilde{y}_i|$, $a(R_{ij}) = \Phi^{-1}([1 + R_{ij}/(N+1)]/2)$ are half-normal quantile scores, $\bar{a}_i$ is mean score in group i, $\bar{a}$ is overall mean score, $N = \sum n_i$, FK is the standard ANOVA F statistic formula

Plain English: Fligner-Killeen test uses a clever two-step approach: (1) Compute absolute deviations from group medians (like Brown-Forsythe), (2) Rank these deviations; convert ranks to half-normal quantile scores; and use ANOVA on the quantile scores (like Levene's idea). By working with ranks, the test is completely free of distributional assumptions—valid whether data are normal, exponential, discrete, or anything else. The test is optimal among rank-based variance tests. Trade-off: lower power than Levene's/Brown-Forsythe for continuous data, but valid for ordinal data where those tests aren't.

⚠ Practitioner Warning

Although commonly described as a test of equal variances, Fligner-Killeen test is more accurately viewed as a test of equal scales, since it is based on ranked absolute deviations rather than sample variances. Fligner-Killeen test is distribution-free, making no assumptions about data shape. This generality costs power: with continuous normal data, Fligner-Killeen has about 15–20% less power than Brown-Forsythe or Levene's. Use Fligner-Killeen when data are ordinal, severely non-normal, or distributional assumptions truly unknown. For continuous data with mild non-normality, Brown-Forsythe is better choice (nearly as robust with higher power).

Hypotheses

- H_0: All k groups have the same scale or dispersion.
- H_1: At least one group differs in scale or dispersion.

Decision Rule: Reject H_0 if $FK > \chi^2_{\alpha,k-1}$ or if p-value $< \alpha$. Test valid for any distribution.

Software References [Python, R, SAS]

Python	R	SAS
`scipy.stats.fligner`	`fligner.test()`	Not Native (Manual)

Additional Notes

Fligner-Killeen (1976) is based on Mood's two-sample rank test (1954) extended to k samples. The test uses van der Waerden scores (half-normal quantile scores) by default: $a(R_{ij}) = \Phi^{-1}([1 + R_{ij}/(N+1)]/2)$ where Φ^{-1} is inverse standard normal CDF. Alternative scoring: simple ranks or squared ranks. van der Waerden scores are optimal for location testing under normality (different variant: $a(R_{ij}) = \Phi^{-1}(R_{ij}/(N+1))$), and Fligner-Killeen showed they're also efficient for scale testing

Miscellanea: Michael A. Fligner and George E. Killeen developed their variance test in 1976 while both were at Florida State University. Emerging during the 1970s "robustness revolution," the method paralleled Brown–Forsythe (1974) but went further by using ranks of absolute deviations, yielding a broadly distribution-free test. Whereas Brown–Forsythe replaced means with medians, Fligner–Killeen adopted a fully rank-based approach to maximize robustness. Both tests exemplify the era's push toward practical, nonparametric tools for variance comparison.

Reference(s): Fligner & Killeen (1976); Conover et al. (1981)

Worked Example

Problem Statement A quality-control engineer measures the tensile strength (in MPa) of metal coupons produced by three ($k = 3$) different furnaces. We test $H_0 : \sigma_1^2 = \sigma_2^2 = \sigma_3^2$ against H_a : At least one variance is different.

Data and Calculations (Use average rank for ties)

Furnace	X_{ij} MPa	$\tilde{X}_i$	Z_{ij}	Rank R_{ij}	a_{ij}
A	512	518	6	8.0	0.6745
A	530	518	12	11.0	1.0100
A	505	518	13	12.0	1.1503
A	525	518	7	9.0	0.7764
A	518	518	0	2.0	0.1573
B	498	506	8	10.0	0.8871
B	547	506	41	14.0	1.5341
B	489	506	17	13.0	1.3180
B	560	506	54	15.0	1.8627
B	506	506	0	2.0	0.1573
C	521	520	1	4.5	0.3601
C	519	520	1	4.5	0.3601
C	523	520	3	6.5	0.5334
C	517	520	3	6.5	0.5334
C	520	520	0	2.0	0.1573

Here, $\tilde{X}_i$ is the median for group i, $Z_{ij} = |X_{ij} - \tilde{X}_i|$. Group means are: $\bar{a}_1 = 0.7537$, $\bar{a}_2 = 1.1518$, and $\bar{a}_3 = 0.3889$. The grand mean is $\bar{a} = 0.7648$.

1. Test Statistic and Degrees of Freedom The Fligner-Killeen statistic (FK) is an LM-type statistic that follows a Chi-square distribution. In this example:

$$FK = \frac{\sum_{i=1}^{k} n_i (\bar{a}_i - \bar{a})^2}{\frac{1}{N-1}\sum_{i=1}^{k}\sum_{j=1}^{n_i}(a_{ij} - \bar{a})^2} = \frac{1.4563}{0.2769}$$

$$= \mathbf{5.259}, \qquad df = k - 1 = 3 - 1 = 2.$$

2. Critical Value and Decision Under H_0 with $k - 1 = 2$ degrees of freedom:

$$\chi^2_{0.95,2} = 5.991, \qquad \chi^2_{0.90,2} = 4.605.$$

The p-value is $P(\chi^2_2 > 5.259) \approx 0.0721$.

3. Decision and Interpretation

- **Decision:** Since $5.259 < 5.991$, we **fail to Reject** H_0 at the 5% significance level.
- **Interpretation:** However, at 10% level, there is enough evidence that the variances of the three furnaces are not equal.

Common Pitfalls

- **Using FK when more powerful tests suffice.** Fligner–Killeen is highly robust but less powerful than Brown–Forsythe under normal or mildly non-normal data. Prefer FK mainly for ordinal data, severe non-normality, or mixed discrete–continuous outcomes.
- **Interpreting FK as a "variance" test.** FK tests equality of *scale*, not variance per se. With skewed or heavy-tailed data, variance and scale diverge; report results as "scale differences," not strictly "variance differences."
- **Ignoring ties with discrete or coarse data.** Heavy ties weaken the chi-square approximation and can distort p-values. For discrete data with few unique values, use permutation versions or supplement with graphical dispersion summaries.
- **Applying FK to dependent or paired data.** FK assumes independent groups; it is invalid for paired, repeated-measures, clustered, or longitudinal designs. Use mixed models, GEE, or variance–structure modeling instead.
- **Over-interpreting non-significance in small samples.** FK has limited power with $n < 10$ per group; $p > 0.05$ often reflects low sensitivity, not equal scales. Report descriptive dispersion (MAD, IQR) and emphasize power limitations.

Normal-score transformation (Fligner–Killeen scores)

Convert each rank to a half-normal quantile score:

$$a_{ij} = \Phi^{-1}\left(\frac{1 + R_{ij}/(N+1)}{2}\right), \text{ where } \Phi^{-1} \text{ is the standard normal quantile function.}$$

Part V

Binary Outcomes & Event Rates

Testing differences in proportions, risks, and binomial parameters

Proportions and Rates

V. PROP — A Conceptual Primer

What Are Proportions and Rates?

Proportion p: the probability of success in a single Bernoulli trial; estimated by $\hat{p} = x/n$ where x successes are observed in n independent trials. The binomial distribution governs the count x.

Rate λ: the expected number of events per unit of *exposure* (time, area, population). The Poisson distribution governs counts $y \sim \text{Poisson}(\lambda t)$ over exposure t.

Key distinction:

- Proportions are bounded in $[0, 1]$; the denominator n is fixed and known.
- Rates are non-negative and unbounded; the exposure t may vary across observations.
- Do not confuse: a disease prevalence (proportion) and an incidence rate (events per person-year) are different estimands requiring different tests.

The Binomial Model

If x successes are observed in n independent trials with success probability p:

$$P(X = x) = \binom{n}{x} p^x (1-p)^{n-x},$$

$$E[X] = np, \qquad \text{Var}(X) = np(1-p).$$

The MLE is $\hat{p} = x/n$. By the CLT: $\hat{p} \xrightarrow{d} \mathcal{N}(p, p(1-p)/n)$, which underpins normal approximation tests. The approximation is reliable when $np \geq 5$ *and* $n(1-p) \geq 5$ (sometimes referred to as *np heuristic*); stricter conditions ($np(1-p) \geq 10$) are preferable for two-sided tests.

I. One-Sample Proportion Tests

Wald Z-test: $H_0 : p = p_0$ vs. $H_1 : p \neq p_0$:

$$Z = \frac{\hat{p} - p_0}{\sqrt{\hat{p}(1-\hat{p})/n}} \mathrel{\dot\sim} \mathcal{N}(0, 1).$$

Uses the *estimated* SE $\sqrt{\hat{p}(1-\hat{p})/n}$ (not the null SE $\sqrt{p_0(1-p_0)/n}$). Simple but **unreliable for small** n **or extreme** p_0 (near 0 or 1); the normal approximation to the binomial is poor in these regions.

Exact binomial test: Computes the exact p-value from the binomial distribution without any normal approximation:

$$p\text{-value} = 2 \min\Big[\sum_{k=0}^{x} \binom{n}{k} p_0^k (1-p_0)^{n-k},$$
$$\sum_{k=x}^{n} \binom{n}{k} p_0^k (1-p_0)^{n-k} \Big].$$

Exact and valid for any n or p_0. **Conservative** (actual size $\leq \alpha$) due to the discreteness of the binomial; the mid-p correction $p_{\text{mid}} = p\text{-value} - \frac{1}{2}P(X = x | p_0)$ reduces conservatism.

Score (Wilson) test: Uses the SE evaluated under H_0 rather than at $\hat{p}$, stabilizing the denominator when $\hat{p}$ is near 0 or 1:

$$Z_S = \frac{\hat{p} - p_0}{\sqrt{p_0(1-p_0)/n}} \mathrel{\dot\sim} \mathcal{N}(0, 1).$$

The Wilson CI is obtained by inverting this test — solving for the values of p_0 not rejected — and is

$$\frac{\hat{p} + z_{\alpha/2}^2/(2n) \pm z_{\alpha/2}\sqrt{\hat{p}(1-\hat{p})/n + z_{\alpha/2}^2/(4n^2)}}{1 + z_{\alpha/2}^2/n},$$

which has substantially better coverage than the Wald CI for small n or extreme p. **The Score test and Wilson CI are the recommended defaults** for one-sample proportion inference.

II. Two-Sample Proportion Tests

Two-sample Z-test (pooled): $H_0 : p_1 = p_2$ using pooled estimate $\hat{p} = (x_1 + x_2)/(n_1 + n_2)$:

$$Z = \frac{\hat{p}_1 - \hat{p}_2}{\sqrt{\hat{p}(1-\hat{p})(1/n_1 + 1/n_2)}} \mathrel{\dot\sim} \mathcal{N}(0, 1).$$

The pooled SE is appropriate under $H_0 : p_1 = p_2$; more efficient than the unpooled Wald version when H_0 is true.

Wald test for two proportions (unpooled): Uses separate SEs:

$$\text{SE} = \sqrt{\hat{p}_1(1-\hat{p}_1)/n_1 + \hat{p}_2(1-\hat{p}_2)/n_2}.$$

More appropriate when the test is framed as a CI for $p_1 - p_2$ rather than a test of $H_0 : p_1 = p_2$; still unreliable for small n.

Agresti–Caffo test: Adds $\tilde{n}_j = n_j + 2$ and $\tilde{x}_j = x_j + 1$ pseudo-observations to each group, then applies the unpooled Wald statistic to the adjusted proportions $\tilde{p}_j = (x_j + 1)/(n_j + 2)$:

$$Z_{\text{AC}} = \frac{\tilde{p}_1 - \tilde{p}_2}{\sqrt{\tilde{p}_1(1-\tilde{p}_1)/\tilde{n}_1 + \tilde{p}_2(1-\tilde{p}_2)/\tilde{n}_2}}.$$

Dramatically improves coverage for small samples and proportions near 0 or 1; computationally trivial. **Recommended over the standard Wald test** in practice.

Farrington–Manning test: Designed for **non-inferiority and equivalence** testing for two proportions (see EQVL chapter). Tests $H_0 : p_1 - p_2 \leq -\Delta$ using a score-type statistic that constrains the SE to be evaluated at the boundary of the null ($p_1 - p_2 = -\Delta$) rather than at $\hat{p}_1 - \hat{p}_2$. Provides better Type I error control than the Wald-based margin test for proportions.

III. Confidence Intervals for Proportions

The choice of CI method is as important as the test.

Method	Properties
Wald (normal)	$\hat{p} \pm z_{\alpha/2}\sqrt{\hat{p}(1-\hat{p})/n}$; simple; poor coverage for small n or extreme p.
Wilson (score)	Better coverage; recommended default; never extends outside $[0, 1]$.
Clopper–Pearson	Exact (conservative); based on beta distribution; guarantees $\geq 1 - \alpha$ coverage.
Agresti–Coull	Add 2 successes and 2 failures; close to Wilson; easy to compute.

For p near 0 or 1 (rare events), Clopper–Pearson or Wilson are strongly preferred. The Wald interval can include values outside $[0, 1]$ and has systematically poor coverage for p near 0 or 1.

IV. Poisson Rate Tests

One-sample Poisson rate test: $Y \sim \text{Poisson}(\lambda t)$ where t is the known exposure. Test $H_0 : \lambda = \lambda_0$:

Exact test: p-value from the Poisson CDF: $P(Y \geq y_{\text{obs}}|\lambda_0 t)$ for one-sided; exact and valid for any y.

Normal approximation:

$$Z = \frac{y - \lambda_0 t}{\sqrt{\lambda_0 t}} \stackrel{\cdot}{\sim} \mathcal{N}(0, 1),$$

reliable when $\lambda_0 t \geq 10$.

Two-sample Poisson rate test: Compare λ_1 and λ_2 from two independent groups with observed counts y_1, y_2 and exposures t_1, t_2. Test $H_0 : \lambda_1 = \lambda_2$ (i.e. $\lambda_1/\lambda_2 = 1$):

Conditional exact test: condition on the total $y_1 + y_2 = m$; the count $y_1|m$ follows $\text{Binomial}(m, t_1/(t_1 + t_2))$ under H_0. This is exact and is the preferred approach for small counts.

Wald test:

$$Z = \frac{\hat{\lambda}_1 - \hat{\lambda}_2}{\sqrt{\hat{\lambda}_1/t_1 + \hat{\lambda}_2/t_2}} \stackrel{\cdot}{\sim} \mathcal{N}(0, 1),$$

where $\hat{\lambda}_j = y_j/t_j$.

Poisson overdispersion: The Poisson model assumes $\text{Var}(Y) = E[Y] = \lambda t$. When the observed variance exceeds λt (overdispersion), standard Poisson tests are anti-conservative; use negative binomial regression or quasi-Poisson (see GLMS chapter).

Approximation Conditions

Test	Condition for reliable approx.
One-sample Z / Wald	$np_0 \geq 5$ and $n(1 - p_0) \geq 5$.
Two-sample pooled Z	$n_1\hat{p} \geq 5$, $n_1(1 - \hat{p}) \geq 5$, and same for n_2.
Poisson normal approx.	$\lambda t \geq 10$ (one-sample); $y_1 + y_2 \geq 20$ (two-sample).
Exact tests	Always valid; use when conditions fail.

Effect Sizes for Proportions

p-values alone are insufficient; report an effect measure:

- **Risk difference (RD):** $\hat{p}_1 - \hat{p}_2$; directly interpretable; NNT $= 1/|\text{RD}|$. (NNT = Number needed to treat).
- **Relative risk (RR):** $\hat{p}_1/\hat{p}_2$; valid for prospective designs; log-normal CI.
- **Odds ratio (OR):** $[\hat{p}_1/(1-\hat{p}_1)]/[\hat{p}_2/(1-\hat{p}_2)]$; symmetric; approximates RR when p is small.
- **Cohen's h:** $h = 2 \arcsin \sqrt{\hat{p}_1} - 2 \arcsin \sqrt{\hat{p}_2}$; effect size on the arcsine-transformed scale; small $|h| = 0.2$, medium 0.5, large 0.8.

Common Pitfalls

- **Wald CI for small n or extreme p:** notoriously poor coverage; use Wilson or Clopper–Pearson.
- **Unpooled SE for equality testing:** the pooled SE is more efficient under $H_0 : p_1 = p_2$; use pooled for tests, unpooled for CIs on $p_1 - p_2$.
- **Poisson with overdispersion:** Poisson tests are anti-conservative when $\text{Var}(Y) > E[Y]$; check residuals.
- **Confusing proportion and rate:** prevalence (p) and incidence rate (λ) require different models even when numerically similar.
- **No continuity correction for small n:** Yates' correction is sometimes applied to the Z-statistic but is generally over-conservative; prefer the exact binomial.
- **Ignoring exposure heterogeneity:** two-sample Poisson tests assume equal precision of $\hat{\lambda}_j$; if exposures $t_1 \ll t_2$, the rate estimates have very different variances.

Bottom Line

> Proportion and rate tests rely on binomial and Poisson models whose normal approximations can fail in small samples or with extreme parameters. Use Wilson (score) tests and CIs as the default for proportions; Agresti–Caffo for two-sample comparisons; exact binomial or Poisson tests when approximation conditions are not met. Always report effect sizes (RD, RR, OR, or Cohen's h) alongside p-values.

One-Sample Proportion Test (Wald)

Test a Population Proportion Against a Hypothesized Value

PROP.1 | **Category:** Proportions | **Data:** Binary | **Null:** Normal (approximate)

When to Use

- Test if a population proportion equals a hypothesized value.
- Large-sample setting where normal approximation is valid.
- Common in survey analysis, quality control, and clinical trials with binary outcomes.

Assumptions

- Independent random sample of binary outcomes.
- Constant probability p for each trial (Bernoulli trials).
- Fixed number of trials n (chosen *a priori*).
- Sample size sufficient for normal approximation: $n\hat{p} \geq 5$ and $n(1-\hat{p}) \geq 5$.

Test Statistic

$$Z = \frac{\hat{p} - p_0}{\text{SE}}, \quad \text{where SE} = \sqrt{\frac{\hat{p}(1-\hat{p})}{n}}. \qquad \text{With continuity correction: } Z = \frac{|\hat{p} - p_0| - \frac{1}{2n}}{\text{SE}}.$$

Notation: x is number of successes, n is sample size, $\hat{p} = x/n$ is sample proportion, p_0 is hypothesized proportion

Plain English: The one-sample Wald proportion test evaluates whether an observed sample proportion differs significantly from a claimed population proportion. The SE is estimated from the data using $\hat{p}$ rather than the null value p_0. The test statistic measures how many standard errors the sample proportion deviates from p_0; under the null hypothesis and with adequate sample size, it follows a standard normal distribution.

⚠ Practitioner Warning

Continuity (Yates) correction (adding ± 0.5 to numerator: $\hat{p} = (x \pm 0.5)/n$) improves accuracy for small samples but is often omitted with large samples. Add or subtract $0.5/n$ in the direction that moves $\hat{p}$ toward p_0. When normal approximation fails ($n\hat{p} < 5$ or $n(1-\hat{p}) < 5$), use exact binomial test instead. The Wald test can be anti-conservative near $p = 0$ or 1; the Score (Wilson) test is preferred in those settings.

Hypotheses

- H_0: $p = p_0$
- H_1: $p \neq p_0$ (two-sided), or $p > p_0$, or $p < p_0$ (one-sided).

Decision Rule: Reject H_0 if $|Z| > z_{\alpha/2}$ (two-sided) or p-value $< \alpha$. For one-sided: reject if $Z > z_\alpha$ (upper tail) or $Z < -z_\alpha$ (lower tail).

Software References [Python, R, SAS]

```
statsmodels::proportions_ztest     prop.test(correct=FALSE)     PROC FREQ (binomial)
```

Additional Notes

The Wald interval $\hat{p} \pm z_{\alpha/2}\sqrt{\hat{p}(1-\hat{p})/n}$ is the natural CI companion to this test, using $\hat{p}$ in the SE throughout. It can perform poorly near boundary values; the Wilson score interval and Agresti-Coull interval are more accurate alternatives in those settings and are generally preferred in practice. For very small samples or extreme proportions, use exact methods. The classic "$np > 5$" rule for the normal approximation traces to mid-20th-century authors such as Snedecor, who offered simple classroom heuristics before exact methods were widely accessible.

Miscellanea: The one-sample proportion Z-test was not originally introduced as a standalone method, but emerged from early work on the normal approximation to the binomial. De Moivre (1730s) first showed that binomial counts could be approximated by a normal curve, and Laplace later refined this result—making the Z-test a direct descendant of early central limit theorem ideas. Karl Pearson's 1900 χ^2 goodness-of-fit test provided a general framework for categorical data, and the proportion Z-test is algebraically equivalent to the one-degree-of-freedom case ($Z^2 = \chi^2_1$) under the normal approximation. More than a century later, debate continues about the Yates continuity correction—some sources always include it, others never do, and many recommend it only for small samples or borderline cases. Simple mnemonic: Wald = "Trust the data" ($\hat{p}$ in SE); Score = "Trust the null" (p_0 in SE).

Reference(s): Pearson (1900); Yates (1934); Agresti & Coull (1998)

One-Sample Proportion Test (Wald) PROP.1

Worked Example

Problem Statement A pharmaceutical company claims an 85% cure rate for a new antibiotic. In a trial of $n = 200$ patients, is the observed cure rate different from $p_0 = 0.85$? Two-sided test at $\alpha = 0.05$.

Data Summary

- $n = 200$; $x = 158$; $\hat{p} = 158/200 = 0.79$

1. Assumptions

- Independent patients: ✓
- Binary outcome: ✓
- $n\hat{p} = 158 \geq 5$, $n(1-\hat{p}) = 42 \geq 5$: ✓
- Normal approximation valid: ✓

2. Standard Error (uses $\hat{p}$)

$$SE = \sqrt{\frac{\hat{p}(1-\hat{p})}{n}} = \sqrt{\frac{0.79 \times 0.21}{200}} = 0.0288.$$

3. Z-statistic and p-value

$$Z = \frac{0.79 - 0.85}{0.0288} = -2.08.$$

$$p = 2\Phi(-|Z|) = 2\Phi(-2.08) = 2(0.0186) = 0.0372.$$

4. 95% CI (uses $\hat{p}$)

$$\hat{p} \pm z_{\alpha/2}\sqrt{\hat{p}(1-\hat{p})/n}.$$

$$0.79 \pm 1.96(0.0288) = [0.734,\ 0.846].$$

5. Effect Size (Cohen's h)

$$h = |2[\arcsin(\sqrt{0.79}) - \arcsin(\sqrt{0.85})]| = 0.157.$$

Small difference.

6. Continuity Correction

$$Z_c = \frac{(\hat{p} + 0.5/n) - p_0}{SE} = \frac{-0.0575}{0.0288} = -2.00.$$

$$p_c = 2\Phi(-2.00) = 0.046.$$

Still significant at $\alpha = 0.05$.

7. Decision and Interpretation. Because $|Z| = 2.08 > 1.96$ (equivalently, $p = 0.0372 < 0.05$), we reject H_0: the cure rate differs from the claimed 85%. The sample rate is 79%, a 6-percentage-point shortfall that is statistically significant and likely clinically meaningful (not as effective as claimed; requires subject matter expertise).

Common Pitfalls

- **Normal approximation misuse.** Using the Wald Z-test when $n\hat{p}$ or $n(1-\hat{p})$ is too small yields distorted p-values. Check adequacy; use exact binomial methods when conditions fail.
- **Wrong standard error.** The Wald test uses the estimated SE $\sqrt{\hat{p}(1-\hat{p})/n}$, not the null-based SE $\sqrt{p_0(1-p_0)/n}$. Using p_0 in the SE is the Score test a different procedure with better small sample behavior near boundaries.
- **Post-hoc one-sided testing.** Deciding direction after seeing the data invalidates Type I error control. Specify one- vs. two-sided tests *before* analysis.
- **Statistical vs. practical significance.** With large samples, even tiny differences can be "significant." Evaluate substantive relevance, not p-value alone.
- **Anti-conservative behavior near boundaries.** The Wald test and its CI perform poorly when $\hat{p}$ is near 0 or 1. Prefer the Score (Wilson) test in those settings and use Wilson, Agresti–Coull, or Clopper–Pearson intervals over Wald's.

Effect Size Interpretation (Cohen's $h = |2[\arcsin(\sqrt{\hat{p}}) - \arcsin(\sqrt{p_0})]|$)

|h|	Interpretation	Example
< 0.20	Small	50% vs. 55%
0.20-0.50	Small to medium	50% vs. 60%
0.50-0.80	Medium to large	50% vs. 70%
> 0.80	Large	50% vs. 80%

Exact Binomial Test

Exact Test for Population Proportion Without Approximations

PROP.2 | **Category:** Proportions | **Data:** Binary | **Null:** Binomial (exact)

When to Use

- Small sample sizes ($np_0 < 5$ or $n(1-p_0) < 5$) where normal approximation invalid. Exact inference required.
- Extreme proportions (near 0 or 1) regardless of n.

Assumptions

- Independent random sample of binary outcomes.
- Constant probability p for each trial (Bernoulli trials).
- Fixed number of trials n (chosen *a priori*).

Test Statistic

$$P(X \le x) = \sum_{k=0}^{x} \binom{n}{k} p_0^k (1-p_0)^{n-k}, \qquad P(X \ge x) = \sum_{k=x}^{n} \binom{n}{k} p_0^k (1-p_0)^{n-k}.$$

Notation: x is observed successes, n is sample size, p_0 is hypothesized proportion

Plain English: The exact binomial test calculates the probability of observing the data (or more extreme results) directly from the binomial distribution, without normal approximation. For one-sided tests, the p-value is simply the tail probability beyond the observed outcome. For two-sided tests, multiple definitions exist for determining "equally extreme" outcomes.

⚠ Practitioner Warning

Two-sided p-value calculation is not unique. Common approaches: (1) double the smaller one-sided p-value (may exceed 1), (2) sum probabilities ≤ observed probability, (3) minimum likelihood method. Different software packages may use different conventions, potentially yielding different p-values for the same data. Always verify which method is used.

Hypotheses

- H_0: $p = p_0$
- H_1: $p \ne p_0$ (two-sided), or $p > p_0$, or $p < p_0$ (one-sided).

Decision Rule: Reject H_0 if $p\text{-value} < \alpha$. One-sided p-value is straight forward. Two-sided p-value depends on convention.

Software References [Python, R, SAS]

Python	R	SAS
`scipy.stats.binomtest`	`binom.test`	`PROC FREQ (EXACT BINOMIAL)`

Additional Notes

Because the binomial distribution is discrete, p-values change in jumps rather than smoothly, which can affect inference near significance thresholds. Exact binomial tests therefore control Type I error at or below nominal levels but can be conservative. The mid-p adjustment reduces this conservatism by subtracting half the probability of the observed outcome:

$$p_{\text{mid}} = P(X < x) + \tfrac{1}{2}P(X = x) \text{ (one-sided).}$$

Exact confidence intervals (Clopper–Pearson) invert the test: the 95% CI contains all p_0 not rejected at $\alpha = 0.05$. These intervals guarantee at least nominal coverage but are conservative (actual coverage often exceeds 95%).

For discrete tests, two-sided p-values are not uniquely defined; different conventions (e.g., "minimum likelihood" summing probabilities $\le P(X = x)$) can yield different results. Computational complexity is negligible with modern software, even for $n > 1000$.

Miscellanea: The exact binomial test predates most modern statistical methods. Its foundations trace back to Jacob Bernoulli's *Ars Conjectandi* (1713), [*The Art of Conjecturing*], and de Moivre's 18th-century work, making it far older than Pearson's χ^2 test (1900).

In practice, however, exact calculations were tedious—often requiring printed tables such as the 1949 *Tables of the Binomial Probability Distribution*—so normal approximations (Z-tests) became dominant. With modern computing, exact methods regained prominence, especially in clinical trials and regulatory settings where strict control of Type I error is critical.

Despite the similar name, the exact binomial test (unconditional) is distinct from Fisher's exact test (conditional), which is based on the hypergeometric distribution.

Reference(s): Clopper & Pearson (1934); Hirji (2006)

Exact Binomial Test — PROP.2

Worked Example

Problem Statement A company claims a 70% response rate for a new drug. In a study of $n = 15$ patients, only $x = 7$ respond. Is the true response rate different from 70%? Two-sided test at $\alpha = 0.05$.

1. Why an exact test?
Since $n(1 - p_0) = 4.5 < 5$, the normal approximation is unreliable.

2. Model and probabilities
Under H_0, $X \sim \text{Bin.}(n = 15, p_0 = 0.70)$.
$SE_0 = \sqrt{p_0(1 - p_0)/n} \approx 0.118$.

3. Binomial probabilities under H_0

x	$P(X = x)$	$P(X \le x)$	x	$P(X = x)$	$P(X \le x)$
0	0.0000	0.0000	8	0.0811	0.1311
1	0.0000	0.0000	9	0.1472	0.2784
2	0.0000	0.0000	10	0.2061	0.4845
3	0.0001	0.0001	11	0.2186	0.7031
4	0.0006	0.0007	12	0.1700	0.8732
5	0.0030	0.0037	13	0.0916	0.9647
6	0.0116	0.0152	14	0.0305	0.9953
7	**0.0348**	0.0500	15	0.0047	1.0000

4. Two-sided extremeness
Observed outcome's probability: $P(X = 7) = 0.0348$.
Two-sided exact test: include outcomes within either tail

$$P(X = x) \le P(X = 7) \implies (X \le 7) \cup (X \ge 14).$$

5. Exact p-value.

$$p = P(X \le 7) + P(X \ge 14) = 0.0852.$$

6. Decision and interpretation
Since $p \approx 0.085 > 0.05$, we **fail to reject** H_0. The sample response rate is $\hat{p} = 7/15 \approx 46.7\%$, which is 23.3 percentage points below the claimed 70%.

With $n = 15$, this large apparent difference is not statistically significant and is compatible with random variation under $p = 0.70$.

7. Contrast with (incorrect) Z-test
If we incorrectly used the one-sample Z-test with normal approximation:

$$Z = \frac{0.467 - 0.70}{0.118} \approx -1.97,$$

giving $p_{\text{approx}} \approx 0.049$.

The Z-test would (incorrectly) reject H_0 at the 5% level, whereas the exact binomial test does not. This illustrates why the exact test is preferred when normal-approximation conditions fail.

8. Exact confidence interval (Clopper–Pearson)
A 95% exact Clopper–Pearson interval for p (based on $x = 7, n = 15$) is approximately

$$[0.213, \ 0.734],$$

which is wide due to the small sample and includes $p_0 = 0.70$, consistent with the test result.

Notes: With $p_0 = 0.7$, $P(X = 7) = \binom{15}{7}(0.7)^7(0.3)^8 \approx 0.0348$. In the exact two-sided test, outcomes are deemed "extreme" if they are at least as unlikely as the observed, i.e., $P(X = x) \le 0.0348$. From the table, this includes both tails: $X \le 7$ and $X \ge 14$.

Common Pitfalls

- **Conservatism of the exact test.** Exact binomial tests control Type I error but are often overly conservative due to discreteness, reducing power. Mid-p modifications provide better alignment with nominal α.
- **Unnecessary use for large samples.** For large n with adequate np_0 and $n(1 - p_0)$, exact tests add computation without benefit. Normal or Wilson approximations are accurate and far more efficient.
- **Mismatch between exact and Z-test p-values.** Exact tests yield discrete, typically larger p-values; Z-tests produce continuous approximations that may be anti-conservative for small samples. The two are not directly comparable.
- **Omitting the observed value from tails.** Two-sided exact tests must include the observed count when summing tail probabilities. Excluding it understates p and gives incorrect inference.
- **Overly conservative Clopper–Pearson intervals.** Exact CIs over-cover and are often much wider than needed. Wilson, Agresti–Coull, or Jeffreys intervals offer tighter estimates with reliable coverage.

Clopper–Pearson interval: Exact $(1 - \alpha)$ confidence interval for p based on inverting the Exact Binomial test

For $X = x$ successes out of n: $p_L = \text{BetaInv}(\frac{\alpha}{2}; x, \ n - x + 1)$, $\quad p_U = \text{BetaInv}(1 - \frac{\alpha}{2}; x + 1, \ n - x)$, where $\text{BetaInv}(q; a, b)$ is the qth quantile of a $\text{Beta}(a, b)$ distribution. For $x = 0$ or $x = n$, the formulas reduce to one-sided Beta limits.

Score Test (Wilson) for Proportions

Improved Test Using Null-Based Standard Error

PROP.3 | **Category:** Proportions | **Data:** Binary | **Null:** Normal (approximate)

When to Use

- Reliable confidence intervals are needed (prefer over Wald).
- Moderate to large samples with any p (including extremes).
- Recommended default for one-sample proportion inference.
- When consistency between testing and interval estimation is desired.

Assumptions

- Independent random sample of binary outcomes.
- Constant probability p for each trial (Bernoulli trials).
- Fixed number of trials n (chosen *a priori*).
- Sample size sufficient for normal approximation: $np_0 \geq 5$ and $n(1-p_0) \geq 5$.

Test Statistic

$$Z = \frac{\hat{p} - p_0}{\text{SE}}, \quad \text{where SE} = \sqrt{\frac{p_0(1-p_0)}{n}}, \quad \text{and } \hat{p} = \frac{x}{n} \text{ is sample proportion.}$$

Notation: x is number of successes, n is sample size, p_0 is hypothesized proportion

Plain English: The Score (Wilson) test evaluates whether an observed proportion differs from p_0 using the SE evaluated under the null — $\sqrt{p_0(1-p_0)/n}$ — rather than the data-estimated SE used by the Wald test. The Wilson confidence interval is obtained by inverting this test: finding all values of p_0 that the test would not reject. This produces an interval centered on an adjusted estimate $\tilde{p}$ rather than $\hat{p}$, with substantially better coverage for small samples or extreme proportions.

⚠ Practitioner Warning

Wilson intervals are intentionally asymmetric because they center on the adjusted estimate $\tilde{p} = (x + z^2_{\alpha/2}/2)/(n + z^2_{\alpha/2})$, which pulls extreme proportions toward 0.5 and ensures valid bounds. This asymmetry is crucial for proper coverage when $\hat{p}$ is near 0 or 1, where Wald intervals can produce impossible limits. Although Wilson intervals have far better performance, many software packages default to the Wald method; users must request the score-based interval explicitly. The "plus-four" Agresti–Coull version is a simplified approximation to the full Wilson score interval.

Hypotheses

- H_0: $p = p_0$
- H_1: $p \neq p_0$ (two-sided).

Decision Rule: Reject H_0 if $|Z| > z_{\alpha/2}$ (two-sided) or p-value $< \alpha$.

Software References [Python, R, SAS]

Python	R	SAS
`statsmodels::proportions_ztest`	`prop.test`	`PROC FREQ / BINOMIAL(CL=WILSON)`

Additional Notes

The Wilson interval is obtained by inverting the score-test inequality $(|p-\hat{p}|)/\sqrt{p(1-p)/n} \leq z_{\alpha/2}$, yielding a quadratic in p whose roots form the interval. The solution recenters the estimate at $\tilde{p} = (x + z^2/2)/(n + z^2)$, effectively adding $z^2/2$ pseudo-successes and z^2 total pseudo-observations. The Agresti–Coull interval is a simple approximation: for 95% confidence, $z_{\alpha/2} \approx 2$ so $z^2 \approx 4$, giving the "plus-four" adjustment $\tilde{p} = (x+2)/(n+4)$, followed by a Wald-style interval. The Wilson interval avoids the pathological behavior of the Wald interval, which can produce negative or degenerate bounds for small samples. For example, with $x = 0$ or 1 in $n = 10$, Wald can yield invalid lower limits, whereas Wilson remains within $[0, 1]$ and provides reliable coverage with much less conservatism than exact (Clopper–Pearson) intervals.

Miscellanea: Edwin Bidwell Wilson introduced this interval in 1927, but it remained largely overlooked while the inferior Wald method dominated practice. Interest revived only after Agresti and Coull's 1998 paper, *"Approximate is better than 'exact' for interval estimation of binomial proportions,"* and the influential 2001 article *"Interval Estimation for a Binomial Proportion."* This resurgence prompted the quip that Wilson's method was "statistically significant but practically ignored" for seven decades.

Reference(s): Wilson (1927); Agresti & Coull (1998); Brown et al. (2001)

Worked Example

Problem Statement In a quality-control study, a product must meet a conformance rate of at least 95%. A sample of $n = 50$ units shows $x = 46$ conforming (92%). Construct the 95% Wald and Wilson confidence intervals for comparison.

Data Summary

- $n = 50; x = 46; \hat{p} = 46/50 = 0.92$

Target specification: $p \geq 0.95$.

1. Wald (traditional) Interval
Standard error:

$$SE_{\text{Wald}} = \sqrt{\frac{\hat{p}(1-\hat{p})}{n}} = \sqrt{\frac{0.92 \times 0.08}{50}} = 0.0384.$$

95% interval:

$$\hat{p} \pm 1.96\, SE_{\text{Wald}} = 0.92 \pm 1.96(0.0384) = 0.92 \pm 0.075.$$

Wald CI: $[0.845,\ 0.995]$.

Key limitations (Wald).
Centered at $\hat{p}$; symmetric; can behave poorly when proportions are near 0 or 1. Useful baseline but known to under-cover.

2. Wilson (score) Interval
Wilson interval inverts the score test. Adjusted center:

$$\tilde{p} = \frac{x + z^2/2}{n + z^2} = \frac{46 + 1.96^2/2}{50 + 1.96^2} = \frac{47.92}{53.84} = 0.890.$$

Adjusted standard error:

$$SE_{\text{Wilson}} = \frac{\sqrt{\frac{\hat{p}(1-\hat{p})}{n} + \frac{z^2}{4n^2}}}{1 + \frac{z^2}{n}} = \frac{\sqrt{0.0384^2 + \frac{1.96^2}{4 \cdot 50^2}}}{1 + \frac{1.96^2}{50}} = 0.0400.$$

95% interval:

$$\tilde{p} \pm 1.96\, SE_{\text{Wilson}} = 0.890 \pm 1.96(0.0400) = 0.890 \pm 0.0784.$$

Wilson CI: $[0.812,\ 0.968]$.

3. Comparison

Method	Center	Lower	Upper
Wald	0.920	0.845	0.995
Wilson	0.890	0.812	0.968

4. Interpretation
Wilson interval is asymmetric and centers at $\tilde{p}$ (pulled toward 0.5). Its bounds stay within $[0, 1]$ and have more reliable coverage.

Wald is tighter and symmetric; Wilson shifts left and avoids overly optimistic upper bounds. Here, neither interval lies fully above the 95% requirement, so conformance cannot be claimed. Wilson gives the more reliable assessment.

Common Pitfalls

- **Asymmetry is intentional.** Wilson intervals are centered on the adjusted estimate $\tilde{p}$ rather than $\hat{p}$, so lack of symmetry around $\hat{p}$ is expected and correct.
- **Wald intervals under-cover.** Despite simplicity, Wald intervals often miss the nominal level—especially with small samples or extreme proportions. Wilson (or Agresti–Coull) provides far more reliable coverage.
- **"Exact" is not always better.** Clopper–Pearson intervals over-cover and are frequently too wide. Wilson intervals achieve near-nominal coverage and are appropriate for most routine applications.
- **Agresti–Coull is a practical approximation.** The plus-4 method is a simplified score-based interval and usually matches Wilson to two decimals, offering an easy hand-calculation alternative.
- **No continuity correction needed.** Wilson already handles discreteness well; adding a correction offers little benefit and is typically unnecessary.

Continuity-corrected Wilson's Formula

$$p_L = \frac{2(x-0.5) + z^2 - \sqrt{\left[2(x-0.5) + z^2\right]^2 - 4(n+z^2)\,\frac{(x-0.5)^2}{n}}}{2(n+z^2)}, \qquad p_U = \frac{2(x+0.5) + z^2 + \sqrt{\left[2(x+0.5) + z^2\right]^2 - 4(n+z^2)\,\frac{(x+0.5)^2}{n}}}{2(n+z^2)}.$$

Two-Sample Proportion Test (Pooled Wald Z-test)

Compare Proportions Between Two Independent Groups

PROP.4 | **Category:** Proportions | **Data:** Binary | **Null:** Normal (approximate)

When to Use

- Compare success rates between two independent groups (treatment vs. control, group A vs. B).
- Alternative to Fisher's exact test when normal approximation is valid.

Assumptions

- Independent binary outcomes within and between groups.
- Constant success probabilities p_1 and p_2 within each group.
- Fixed sample sizes n_1 and n_2 (chosen *a priori*).
- Sample sizes sufficient for normal approximation: $n_1\hat{p} \geq 5$, $n_1(1-\hat{p}) \geq 5$ where $\hat{p}$ is pooled; and similarly for n_2.

Test Statistic

$$Z = \frac{(\hat{p}_1 - \hat{p}_2) - \Delta_0}{\text{SE}}, \qquad \text{where SE} = \sqrt{\hat{p}(1-\hat{p})\left(\frac{1}{n_1} + \frac{1}{n_2}\right)}, \qquad \hat{p} = \frac{x_1 + x_2}{n_1 + n_2} \text{ (pooled).}$$

Notation: $\hat{p}_1, \hat{p}_2$ are sample proportions, x_1, x_2 are successes, n_1, n_2 are sample sizes

Plain English: The two-sample pooled Wald Z-test compares success rates between independent groups by standardizing their difference. Under the null hypothesis of equal proportions ($\Delta_0 = 0$), observations are pooled to estimate the common proportion, which provides the standard error. The test statistic follows an approximate normal distribution when sample sizes are adequate, enabling calculation of p-values and confidence intervals for the difference.

⚠ Practitioner Warning

Using the pooled standard error outside $H_0: p_1 = p_2$ leads to incorrect inference; it should not be used for estimation or non-inferiority/equivalence testing. Normal approximation can be unreliable with small samples or rare events; when expected counts are low, use Fisher's exact test or score-based methods instead. Imbalanced sample sizes further degrade approximation accuracy. Statistical significance does not imply practical importance—large samples can detect trivial differences. Always interpret effect sizes alongside p-values.

Hypotheses

- H_0: $p_1 - p_2 = \Delta_0$ (typically $\Delta_0 = 0$, testing equality).
- H_1: $p_1 - p_2 \neq \Delta_0$ (two-sided).

Decision Rule: Reject H_0 if $|Z| > z_{\alpha/2}$ (two-sided) or p-value $< \alpha$. One-sided alternatives allowed.

Software References [Python, R, SAS]

`statsmodels::proportions_ztest` `prop.test(correct=FALSE)` `PROC FREQ (RISKDIFF option)`

Additional Notes

The two-sample pooled Wald Z-test is algebraically equivalent to Pearson's χ^2 test for 2×2 tables: $Z^2 = \chi^2_1$ when testing $H_0: p_1 = p_2$. For confidence intervals on $(p_1 - p_2)$, the unpooled Wald interval uses $(\hat{p}_1 - \hat{p}_2) \pm z_{\alpha/2}\sqrt{\hat{p}_1(1-\hat{p}_1)/n_1 + \hat{p}_2(1-\hat{p}_2)/n_2}$ but performs poorly near boundary proportions or with small samples. Agresti and Caffo (2000) proposed a simple improvement: add one pseudo-success and one pseudo-failure to each group ($\tilde{p}_i = (x_i + 1)/(n_i + 2)$), then apply the standard unpooled Wald interval using $\tilde{p}_i$ in place of $\hat{p}_i$. This "plus-two" adjustment substantially improves coverage with minimal extra effort. Newcombe's hybrid interval combines separate Wilson intervals for p_1 and p_2, providing better coverage still, particularly for small samples, extreme proportions, or unequal group sizes.

Miscellanea: The 1954 Salk polio vaccine trial—one of the largest controlled medical experiments of its time—compared disease rates between vaccinated and control groups using large-sample proportion methods. With over 400,000 children in the randomized trial (and about 1.8 million participants overall), it demonstrated how large studies can detect differences with high statistical precision. The trial also highlighted an enduring principle: statistical significance must be interpreted alongside clinical importance.

Reference(s): Pearson (1900); Agresti & Caffo (2000); Newcombe (1998)

Two-Sample Proportion Test (Pooled Wald Z-test) PROP.4

Worked Example

Problem Statement In a clinical trial comparing two pain medications, Drug A reports relief in 82/120 patients and Drug B in 65/110 patients. Do relief rates differ? H_0: $p_1 = p_2$ vs. H_1: $p_1 \neq p_2$. Two-sided test at $\alpha = 0.05$.

Data summary

- Drug A: $n_1 = 120, x_1 = 82, \hat{p}_1 = 0.683$.
- Drug B: $n_2 = 110, x_2 = 65, \hat{p}_2 = 0.591$.
- Observed difference: $\hat{p}_1 - \hat{p}_2 = 0.092$.

1. Pooled proportion

$$\hat{p} = \frac{x_1 + x_2}{n_1 + n_2} = \frac{147}{230} = 0.639.$$

2. Assumptions
Independent samples; binary outcome; $n_i\hat{p} \geq 5$ and $n_i(1-\hat{p}) \geq 5$ for both groups. **All conditions satisfied.**

3. Standard error (pooled)

$$SE = \sqrt{\hat{p}(1-\hat{p})\left(\frac{1}{n_1} + \frac{1}{n_2}\right)} = 0.0634.$$

4. Z-statistic

$$Z = \frac{\hat{p}_1 - \hat{p}_2}{SE} = \frac{0.092}{0.0634} = 1.45.$$

5. p-value

$$p = 2P(Z > 1.45) = 2(0.0735) = 0.147.$$

6. Confidence interval (unpooled)

$$SE_{\text{unpooled}} = \sqrt{\frac{\hat{p}_1(1-\hat{p}_1)}{n_1} + \frac{\hat{p}_2(1-\hat{p}_2)}{n_2}} = 0.0632.$$

$$CI = 0.092 \pm 1.96(0.0632) = 0.092 \pm 0.124.$$

$$95\% \text{ CI: } [-0.032,\ 0.216].$$

Includes 0 → consistent with test result.

7. Continuity correction
Harmonic mean:

$$n_h = \frac{n_1 n_2}{n_1 + n_2} = 57.4. \quad \left[\text{idea} : \frac{1}{2n_h} = \frac{1}{2n_1} + \frac{1}{2n_2}\right]$$

Corrected statistic:

$$Z_c = \frac{|\hat{p}_1 - \hat{p}_2| - 1/(2n_h)}{SE} = \frac{0.0833}{0.0634} = 1.31,$$

$$p_c = 0.190 \quad \text{(still not significant).}$$

8. Chi-square equivalence

$$\chi^2 = Z^2 = (1.45)^2 = 2.10, \quad df = 1,$$

$$p = 0.147 \text{ (identical).}$$

9. Decision Because $|Z| = 1.45 < 1.96$ and the p-value of 0.147 exceeds 0.05, we fail to reject H_0; there is no statistically detectable difference in relief rates.

10. Interpretation and Conclusion Drug A's relief rate exceeds Drug B's by 9.2 percentage points, but this difference is not statistically significant. The 95% confidence interval includes zero, and both the Z-test and chi-square test indicate that the observed gap is compatible with sampling variation; evidence is insufficient to conclude the drugs differ in effectiveness.

Common Pitfalls

- **Pooled vs. unpooled.** Use pooled SE only for tests (only when $H_0 : p_1 = p_2$); CIs require unpooled SE.
- **Count requirements.** All four cell counts (success/failure in each group) must be ≥ 5; otherwise use Fisher's exact test.
- **Non-significant $\not\Rightarrow$ no difference.** Lack of significance may reflect low power—inspect the CI.
- **Multiple comparisons.** Pairwise tests inflate Type I error; apply corrections or use an omnibus test.
- **Paired data misuse.** Paired outcomes need McNemar's test—not an independent-samples Z-test.

Newcombe's hybrid (Method 10) forms the CI for $\Delta = p_1 - p_2$ using Wilson limits (L_1, U_1) and (L_2, U_2)

$$L = (\hat{p}_1 - \hat{p}_2) - \sqrt{(\hat{p}_1 - L_1)^2 + (U_2 - \hat{p}_2)^2}, \quad U = (\hat{p}_1 - \hat{p}_2) + \sqrt{(U_1 - \hat{p}_1)^2 + (\hat{p}_2 - L_2)^2}.$$

Simple to compute once Wilson limits are available and provides highly accurate coverage.

Wald Test for Proportions (General k-Sample)

General Matrix Framework for Testing Proportion Hypotheses

PROP.5 | **Category:** Proportions | **Data:** Binary | **Null:** Normal (approximate)

When to Use

- Compare proportions across $k \geq 1$ independent groups (subsumes one- and two-sample cases).
- Large-sample settings where normal approximation holds.
- Testing linear hypotheses on proportions via contrast matrices. Foundation for Wald tests in GLMs.
- Model-based alternative to Pearson's χ^2 for structured hypotheses.

Assumptions

- Independent binary outcomes within and between groups.
- Constant success probability p_i within each group.
- Fixed sample sizes n_i for each group (chosen *a priori*).
- Sample sizes sufficient for normal approximation in all groups: $n_i\hat{p}_i \geq 5$ and $n_i(1-\hat{p}_i) \geq 5$.
- No extreme boundary estimates ($\hat{p}_i = 0$ or 1) that make Wald SEs unstable.

Test Statistic

$$\text{Single group: } Z = \frac{\hat{p} - p_0}{SE(\hat{p})}, \quad SE(\hat{p}) = \sqrt{\frac{\hat{p}(1-\hat{p})}{n}}. \qquad \text{General: } W = \mathbf{c}^\top \hat{\boldsymbol{p}} \left[\mathbf{c}^\top \widehat{\mathbf{V}}\, \mathbf{c}\right]^{-1} \mathbf{c}^\top \hat{\boldsymbol{p}} \sim \chi^2_r.$$

Notation: $\hat{\boldsymbol{p}} = (\hat{p}_1, \ldots, \hat{p}_k)^\top$ are MLEs, $\widehat{\mathbf{V}} = \mathrm{diag}(\hat{p}_i(1-\hat{p}_i)/n_i)$ is estimated covariance, $\mathbf{c}$ is a contrast vector (or matrix), r is the rank of the contrast.

Plain English: The general Wald test evaluates hypotheses about one or more proportions by comparing maximum likelihood estimates to hypothesized values, standardized by the estimated covariance matrix. The single-group case reduces to a familiar Z-statistic; the k-group case uses a quadratic form in the contrast vector that follows a χ^2 distribution. The estimated SE uses $\hat{p}$ throughout — not the null value — distinguishing it from the Score test.

⚠ Practitioner Warning

The Wald test exhibits erratic behavior near boundary values (proportions near 0 or 1). Coverage of Wald confidence intervals can fall well below nominal levels, sometimes as low as 70% when claiming 95%. For critical applications, prefer score test or exact methods. The problem worsens with smaller samples and extreme proportions—precisely when conservatism matters most.

Hypotheses

- H_0: $\mathbf{c}^\top \boldsymbol{p} = \mathbf{0}$ (or $p = p_0$ for the single-group case).
- H_1: $\mathbf{c}^\top \boldsymbol{p} \neq \mathbf{0}$ (two-sided); directional alternatives for $k = 1$.

Decision Rule: Reject H_0 if $|Z| > z_{\alpha/2}$ (single group, two-sided) or $W > \chi^2_{r,\alpha}$ (general), or p-value $< \alpha$.

Software References [Python, R, SAS]

Python	R	SAS
`statsmodels...proportions_ztest`	Not Native (Manual)	PROC FREQ (RISKDIFF option)

Additional Notes

The general multiparameter Wald statistic is $W = (\mathbf{C}\hat{\boldsymbol{\theta}} - \boldsymbol{d})^\top [\mathbf{C}\widehat{\mathrm{Cov}}(\hat{\boldsymbol{\theta}})\mathbf{C}^\top]^{-1}(\mathbf{C}\hat{\boldsymbol{\theta}} - \boldsymbol{d}) \sim \chi^2_r$ where $\mathbf{C}$ is an $r \times k$ contrast matrix and $\boldsymbol{d}$ is the hypothesized value vector. The two-sample pooled Z-test (PROP.04) and one-sample Wald test (PROP.01) are special cases with $k = 2$ and $k = 1$ respectively. Wald intervals can be stabilized via a logit transformation: $\mathrm{logit}(\hat{p}) \pm z_{\alpha/2}\sqrt{\frac{1}{n\hat{p}(1-\hat{p})}}$, then back-transformed to the probability scale.

Miscellanea: Abraham Wald, a brilliant mathematician who escaped Austria in 1938, made fundamental contributions to decision theory and sequential analysis during World War II. His work on bomber armor placement—analyzing where bullet holes appeared on returning planes and deducing where armor was needed (where holes were absent)—became legendary. The simple Wald statistic used for proportions is just one instance of the broader asymptotic testing framework that bears his name.

Reference(s): Wald (1943); Agresti (2013)

Wald Test for Proportions (General k-Sample) PROP.5

Worked Example

Problem Statement Three production lines report defect counts. Are their defect rates equal? Line 1: 15/200, Line 2: 28/250, Line 3: 12/180.

Data Summary

$$\hat{p}_1 = \frac{15}{200} = 0.075, \quad \hat{p}_2 = \frac{28}{250} = 0.112, \quad \hat{p}_3 = \frac{12}{180} = 0.067.$$

Combined proportion:

$$\hat{p} = \frac{55}{630} = 0.087 \quad E_i = n_i \cdot p.$$

O_1=15; O_2=28; O_3=12 and E_1=17.5; E_2=21.8; E_3=15.7.

1. Hypotheses

$$H_0 : p_1 = p_2 = p_3, \qquad H_1 : \text{not all equal.}$$

Two independent constraints → degrees of freedom $k = 2$.

2. Estimated Proportions and Variances
Independent samples → diagonal covariance:

$$\hat{\Sigma} = \text{diag}\left(\frac{\hat{p}_1(1-\hat{p}_1)}{n_1}, \frac{\hat{p}_2(1-\hat{p}_2)}{n_2}, \frac{\hat{p}_3(1-\hat{p}_3)}{n_3}\right).$$

Numerically:

$$\hat{\Sigma} = \begin{pmatrix} 0.000347 & 0 & 0 \\ 0 & 0.000398 & 0 \\ 0 & 0 & 0.000346 \end{pmatrix}.$$

3. Wald Test Construction
Use contrasts to test equality:

$$C = \begin{pmatrix} 1 & -1 & 0 \\ 0 & 1 & -1 \end{pmatrix}, \qquad C\hat{\mathbf{p}} = \begin{pmatrix} -0.037 \\ 0.045 \end{pmatrix}.$$

Variance of contrasts:

$$V = C\hat{\Sigma}C^T = \begin{pmatrix} 0.000745 & -0.000398 \\ -0.000398 & 0.000745 \end{pmatrix}.$$

Wald statistic:

$$W = (C\hat{\mathbf{p}})^T V^{-1} (C\hat{\mathbf{p}}) = 3.07.$$

4. Decision

$$W = 3.07, \quad \chi^2_{2,0.05} = 5.99, \quad p = 0.216.$$

Fail to reject H_0: No evidence that defect rates differ across lines.

5. Comparison with Pearson Chi-square
2×3 table yields

$$\chi^2 = \sum_{i=1}^{3} \frac{(O_i - E_i)^2}{E_i} = 2.97, \quad p = 0.226,$$

nearly identical to the Wald result (as expected asymptotically).

Interpretation This example shows that multi-group proportion tests reduce to testing contrasts of estimated proportions. The covariance structure is diagonal due to independence, making the Wald statistic a weighted sum of squared differences. In practice, this produces results nearly identical to Pearson's chi-square test, highlighting that both approaches are asymptotically equivalent; the Wald formulation is mainly useful when extending to regression or more complex models.

Common Pitfalls

- **Non-invariance.** Wald tests depend on parameterization; results can differ on proportion vs. log-odds scales. Score and likelihood-ratio tests are invariant.
- **Poor performance with small n or extreme p.** Wald relies on large-sample normality and fails when counts are low or proportions near 0/1. Use score or exact tests in such settings.
- **Multiple unadjusted pairwise tests.** Testing all pairwise differences inflates Type I error. Use an omnibus Wald test first, then adjusted pairwise comparisons if needed.
- **Ignoring clustering.** Standard Wald tests assume independence; clustered data require cluster-robust SEs or hierarchical models.
- **Non-significance misinterpreted as equality.** Failing to reject H_0 does not prove equal proportions—may reflect low power. Only equivalence tests can support claims of similarity.

Farrington-Manning Test

Score Test for Non-Inferiority and Equivalence of Proportions

PROP.6 | **Category:** Proportions | **Data:** Binary | **Null:** Normal (approximate)

When to Use

- Test non-inferiority or equivalence of two proportions with a specified margin.
- Standard method in regulatory trials for binary endpoints.
- Preferred over simple difference tests when proving new treatment is "not worse."

Assumptions

- Observations are independent binary outcomes within and between groups.
- Random sampling or random assignment to groups.
- Sample sizes adequate for normal approximation.
- Non-inferiority margin δ is pre-specified based on clinical relevance.

Test Statistic

$$Z = \frac{(\hat{p}_1 - \hat{p}_2) - \delta}{\sqrt{\frac{\tilde{p}_1(1-\tilde{p}_1)}{n_1} + \frac{\tilde{p}_2(1-\tilde{p}_2)}{n_2}}}, \quad \text{where } \tilde{p}_1, \tilde{p}_2 \text{ are constrained by: } \tilde{p}_1 - \tilde{p}_2 = \delta.$$

Notation: δ is non-inferiority margin (negative) or equivalence bound, $\tilde{p}_i$ are constrained MLEs, n_1, n_2 are sample sizes

Plain English: The Farrington-Manning test evaluates whether the difference between two proportions falls within a pre-specified margin by computing a score statistic under the boundary of the null hypothesis. Unlike pooled tests that assume equality, this test finds the maximum likelihood estimates constrained to the margin ($p_1 - p_2 = \delta$), providing proper variance estimation for non-inferiority testing. This ensures correct Type I error control when testing against a margin rather than zero.

⚠ Practitioner Warning

The non-inferiority margin δ must be justified a priori based on clinical or practical significance—it cannot be chosen post-hoc to achieve desired results. Regulatory agencies scrutinize margin selection carefully. Using equality-based pooled variance (like standard two-sample Z-test) for non-inferiority testing is incorrect and anti-conservative. Always use constrained variance estimation. For equivalence testing, perform two one-sided tests (TOST) at margins $-\delta$ and $+\delta$.

Hypotheses

- H_0: $p_1 - p_2 \leq \delta$ (inferior by more than margin).
- H_1: $p_1 - p_2 > \delta$ (non-inferior, within acceptable margin).

Decision Rule: Reject H_0 if $Z > z_\alpha$ (one-sided test). Non-inferiority typically uses negative margin: $\delta < 0$.

Software References [Python, R, SAS]

Python	R	SAS
`Not Native (Manual)`	`DescrTab2::farrington.manning`	`PROC FREQ RISKDIFF(METHOD=FM)`

Additional Notes

Constrained MLEs $\tilde{p}_1, \tilde{p}_2$ are obtained by maximizing the binomial likelihood subject to the null boundary $\tilde{p}_1 - \tilde{p}_2 = \delta$. Unlike the equality case ($\delta = 0$), the total fitted number of successes is not generally preserved ($n_1\tilde{p}_1 + n_2\tilde{p}_2 \neq x_1 + x_2$) when $\delta \neq 0$; the estimates are found from the constrained score equations, typically by numerical solution. The test maintains appropriate Type I error by estimating variance under the null margin rather than under equality or the unconstrained sample proportions. For equivalence testing with symmetric margins, perform two one-sided tests: one at the lower margin and one at the upper margin; both must reject to claim equivalence.

Miscellanea: The Farrington–Manning test (1990) arose from vaccine trials where standard proportion tests failed under non-inferiority designs. Its key idea—estimating variance under the null constraint $p_1 - p_2 = \delta$ rather than equality—revealed a flaw in Wald methods, which can falsely declare non-inferiority due to using the wrong variance. By enforcing the null and using restricted MLEs, the FM test evaluates the least favorable scenario, making it more reliable and appropriately conservative. Though rarely named explicitly, it underlies many modern non-inferiority analyses in clinical trials. It remains the "judge and jury" for the FDA when deciding if a new, cheaper drug is "close enough" to the market leader.

Reference(s): Farrington & Manning (1990); Miettinen & Nurminen (1985)

Worked Example

Problem Statement New antibiotic vs. standard therapy for pneumonia. Standard cure rate: 85%. Non-inferiority margin: $\delta = -0.10$ (retain $\geq 75\%$ acceptable).

Data Summary

- New drug: $n_1 = 250, x_1 = 198$ ($\hat{p}_1 = 0.792$).
- Standard: $n_2 = 250, x_2 = 215$ ($\hat{p}_2 = 0.860$).
- Observed difference: $\hat{p}_1 - \hat{p}_2 = -0.068$.

1. Hypotheses (one-sided, $\alpha = 0.025$)

$$H_0 : p_1 - p_2 \leq -0.10, \quad H_1 : p_1 - p_2 > -0.10.$$

2. Constrained MLEs under H_0

Constraint:

$$\tilde{p}_1 - \tilde{p}_2 = -0.10.$$

Using a Lagrange multiplier λ, the score equations are

$$\frac{x_1 - n_1\tilde{p}_1}{\tilde{p}_1(1-\tilde{p}_1)} + \lambda = 0, \quad \frac{x_2 - n_2\tilde{p}_2}{\tilde{p}_2(1-\tilde{p}_2)} - \lambda = 0,$$

$$\tilde{p}_1 - \tilde{p}_2 + 0.10 = 0.$$

Solving this system numerically gives

$$\tilde{p}_1 \approx 0.772, \qquad \tilde{p}_2 \approx 0.872,$$

which satisfy $\tilde{p}_1 - \tilde{p}_2 = -0.10$.

3. Farrington–Manning Standard Error

$$SE_{FM} = \sqrt{\frac{\tilde{p}_1(1-\tilde{p}_1)}{n_1} + \frac{\tilde{p}_2(1-\tilde{p}_2)}{n_2}}.$$

$$SE_{FM} = \sqrt{\frac{0.770 \times 0.230}{250} + \frac{0.870 \times 0.130}{250}} = 0.0339.$$

4. Test Statistic

$$Z_{FM} = \frac{(\hat{p}_1 - \hat{p}_2) - \delta}{SE_{FM}} = \frac{-0.068 - (-0.10)}{0.0339} = 0.944.$$

5. Decision

One-sided upper-tail p-value:

$$p = P(Z > 0.944) \approx 0.173, \quad z_{0.025} = 1.96.$$

Since $Z_{FM} < 1.96$ and $p > 0.025$, we **fail to reject** H_0.

6. Interpretation

The observed cure rate for the new drug is 6.8 percentage points lower than standard, within the 10% margin numerically, but uncertainty is too large to rule out a loss exceeding 10%. Non-inferiority is *not* demonstrated.

Note: The numerator uses observed proportions $(\hat{p}_1, \hat{p}_2)$, while the standard error is computed using constrained MLEs $(\tilde{p}_1, \tilde{p}_2)$ under $H_0 : p_1 - p_2 = \delta$.

Common Pitfalls

- **Post-hoc margins.** Non-inferiority margins must be pre-specified and clinically justified; choosing δ after seeing results is data-driven and inflates Type I error.
- **Confusing NI with TOST.** TOST tests two bounds for equivalence, whereas non-inferiority tests only $p_1 - p_2 > -\delta$. They are not interchangeable; NI uses a one-sided test at α, not two tests at α each.
- **Using a non-significant difference test as NI evidence.** Failing to reject $p_1 = p_2$ does *not* show non-inferiority. NI requires rejecting $p_1 - p_2 \leq -\delta$ with a proper NI test.
- **Using Wald variance instead of FM constrained variance.** Farrington–Manning requires variance computed at the constrained MLEs; Wald's unconstrained variance is anti-conservative and can inflate false positives.
- **Ignoring margin when planning sample size.** Power depends on the gap between expected effect and the NI margin. Narrow margins require large n; using superiority-style calculations leads to underpowered NI trials.

Derivation of score equations (Farrington–Manning)

Under H_0, $p_1 - p_2 = \delta$. The binomial log-likelihood is: $\ell(p_1, p_2) = x_1 \log p_1 + (n_1 - x_1)\log(1-p_1) + x_2 \log p_2 + (n_2 - x_2)\log(1-p_2)$.

Impose the constraint via a Lagrange multiplier λ: $\mathcal{L}(p_1, p_2, \lambda) = \ell(p_1, p_2) + \lambda(p_1 - p_2 - \delta)$. To get score equations, set derivatives to zero:

$$\frac{\partial \mathcal{L}}{\partial p_1} = \frac{x_1 - n_1 p_1}{p_1(1-p_1)} + \lambda = 0, \quad \frac{\partial \mathcal{L}}{\partial p_2} = \frac{x_2 - n_2 p_2}{p_2(1-p_2)} - \lambda = 0, \quad \frac{\partial \mathcal{L}}{\partial \lambda} = p_1 - p_2 - \delta = 0.$$

Solving this system numerically yields the constrained MLEs $(\tilde{p}_1, \tilde{p}_2)$ used in the Farrington–Manning standard error.

One-Sample Poisson Rate Test

Test Event Rate Against Hypothesized Value

PROP.7 | **Category:** Rates | **Data:** Count | **Null:** Poisson / Normal (asymp.)

When to Use

- Test whether an observed event rate equals a specified value.
- Data are counts aggregated over a known exposure (time, area, or population).
- Events are infrequent relative to the exposure (rate interpretation is meaningful).
- Situations where inference is based on rates rather than proportions or continuous outcomes.

Assumptions

- Events occur independently over the observation period.
- The underlying event rate λ is constant over the exposure.
- Counts follow a Poisson distribution: $\text{Var}(X) = E[X]$.
- The exposure (time/area/population) is fixed, known, and measured without error.
- Observations are not subject to censoring, truncation, or systematic undercounting.

Test Statistic

$$Z = \frac{x - \mu_0}{\sqrt{\mu_0}}, \qquad P(X \le x) = \sum_{k=0}^{x} e^{-\mu_0}\mu_0^k/k! \text{ lower tail exact test.}$$

Notation: x is observed count, λ_0 is hypothesized rate, t is exposure time/area, $\mu_0 = \lambda_0 t$ is expected count under H_0

Plain English: The one-sample Poisson rate test evaluates whether observed counts are consistent with a specified rate per unit time or space. For moderate to large expected counts, especially $\mu_0 \ge 10$, the normal approximation is usually reasonable. For small counts, exact probabilities from the Poisson distribution provide accurate p-values. The test is fundamental in settings where events are discrete, random, and relatively rare.

⚠ Practitioner Warning

The constant-rate assumption is critical but often violated in practice. Overdispersion (variance exceeding the mean) indicates Poisson model inadequacy; consider negative binomial or quasi-Poisson models instead. Time-varying rates, clustering, dependence between events, or changing exposure intensity can distort p-values. With only one aggregate count, dispersion cannot be diagnosed directly; use subject-matter knowledge or replicated counts where available.

Hypotheses

- H_0: $\lambda = \lambda_0$ (rate equals hypothesized value).
- H_1: $\lambda \neq \lambda_0$ (two-tailed), or $\lambda > \lambda_0$, or $\lambda < \lambda_0$ (one-tailed).

Decision Rule: Reject H_0 if $|Z| > z_{\alpha/2}$ (asymptotic) or exact p-value $< \alpha$. If $\mu_0 < 10$, use Poisson. Otherwise, Normal.

Software References [Python, R, SAS]

```
statsmodels::test_poisson        poisson.test        PROC GENMOD (DIST=POISSON)
```

Additional Notes

Confidence interval for rate λ: exact intervals based on inverting the Poisson test, or asymptotic $\hat{\lambda} + z_{\alpha/2}\sqrt{\hat{\lambda}/t}$ where $\hat{\lambda} = x/t$. The exact interval uses relationship to chi-square distribution: $[\chi^2_{2x,\alpha/2}/(2t), \chi^2_{2(x+1),1-\alpha/2}/(2t)]$. For a single observed count, the test can be viewed as comparing observed to expected events under a Poisson model; multi-cell extensions lead to Poisson goodness-of-fit tests. With replicated counts, a variance-to-mean ratio substantially above 1 suggests *overdispersion*; below 1 suggests *underdispersion*.

Miscellanea: Siméon Denis Poisson derived the distribution in 1837 while studying jury verdicts and wrongful convictions in the French legal system, asking whether conviction rates varied by region or remained constant. Ironically, the distribution bearing his name found its greatest applications far from courtrooms—in radioactive decay (Rutherford, 1910), telephone calls (Erlang, 1909), and eventually web server requests. A key property behind its usefulness is the *law of rare events*: it arises as a limit of the binomial distribution when trials are many and success probability is small ($n \to \infty$ and $p \to 0$ with $np \to \mu$). While the Poisson distribution itself is not *memoryless*, the waiting times between events in a Poisson process follow the exponential distribution and that is *memoryless*.

Reference(s): Poisson (1837); Johnson et al. (2005)

One-Sample Poisson Rate Test

PROP.7

Worked Example

Research Question. Hospital monitors central-line infections. Historical rate: 3.5 per 1000 line-days. After new protocol: 12 infections in 5000 line-days. Test if rate decreased at $\alpha = 0.05$.

Data Summary
Historical rate: $\lambda_0 = 0.0035$ per line-day
Exposure: $t = 5000$
Expected under H_0: $\mu_0 = \lambda_0 t = 17.5$
Observed: $X = 12$
Observed rate: $\hat{\lambda} = 12/5000 = 0.0024$

1. Hypotheses

$$H_0 : \lambda = \lambda_0, \qquad H_1 : \lambda < \lambda_0.$$

2. Approximation Check

$\mu_0 = 17.5 \geq 10 \Rightarrow$ normal approximation valid.

3. Test Statistic and p-value

$$Z = \frac{X - \mu_0}{\sqrt{\mu_0}} = -1.31, \quad p = \Phi(-1.31) = 0.095.$$

4. Decision
Critical value: $z_{0.05} = -1.645$.

$$Z = -1.31 > -1.645 \Rightarrow \textbf{Fail to reject } H_0.$$

5. Conclusion Evidence insufficient to conclude the infection rate has decreased.

6. Interpretation
Observed: 12 vs. 17.5 expected (31% fewer). Reduction not statistically significant; may reflect chance.

7. Exact Poisson (comparison)

$$p = P(X \leq 12 \mid \mu_0 = 17.5) = \sum_{k=0}^{12} \frac{e^{-17.5} \cdot 17.5^k}{k!}$$

Exact (software): $p = 0.1117$
Close to normal approx; both non-significant.

The exact 95% Garwood CI for the rate λ is $[0.00124, 0.00419]$ per line-day, corresponding to $[1.24, 4.19]$ infections per 1000 line-days. As the historical rate of 3.5 lies within this interval, the observed data are consistent with the non-significant test result. With continuity correction ($X = 12 + 0.5$), $p_c = \Phi(-1.20) = 0.116$, closer to the exact p-value.

Common Pitfalls

- **Confusing rates with proportions.** Poisson models event *rates* over exposure (time/area); binomial models *proportions* out of fixed trials. Clarify if units can experience multiple events.
- **Using Poisson under overdispersion.** Event clustering inflates variance beyond the Poisson mean; standard tests become liberal. Use quasi-Poisson, negative binomial, or mixed-effects alternatives.
- **Applying normal Z-approximation prematurely.** The normal test requires $\mu_0 \geq 10$. Continuity correction (adding 0.5) can improve the approximation. Small expected counts need exact Poisson probabilities for valid p-values.
- **Comparing raw counts instead of rates.** Counts are meaningless without exposure adjustment. Always test X against $\lambda_0 t$, not against historical counts with different t.
- **Ignoring zero inflation.** Excess zeros signal a mixture process rather than a single Poisson rate. Zero-inflated Poisson or hurdle models give better fit and inference.

Rate vs. Proportion

Feature	Proportion (Binomial)	Rate (Poisson)
Data type	x of n (successes/trials)	x in t (events/exposure)
Bounded	Yes ($0 \leq x \leq n$)	No ($x \geq 0$, unbounded)
Multiple events	No (one per unit)	Yes (multiple per unit)

Garwood (Exact) Poisson Confidence Interval. For a one-sample Poisson rate test with observed count X over exposure t, the exact $(1 - \alpha)$ interval for the mean $\mu = \lambda t$ is obtained by inverting the Poisson test:

$$\left[\mu_L = \tfrac{1}{2}\chi^2_{2X,\,\alpha/2}, \quad \mu_U = \tfrac{1}{2}\chi^2_{2(X+1),\,1-\alpha/2}\right] \iff [\lambda_L = \mu_L/t, \quad \lambda_U = \mu_U/t].$$

Well-known property: For $X \sim \text{Poisson}(\mu)$, $P(X \leq x) = P\left(\chi^2_{2(x+1)} > 2\mu\right)$, $\quad P(X \geq x) = P\left(\chi^2_{2x} < 2\mu\right)$.

Two-Sample Poisson Rate Test

Compare Event Rates Between Two Independent Groups

PROP.8 | **Category:** Rates | **Data:** Count | **Null:** Poisson

When to Use

- Compare event rates between two independent groups with different exposures.
- Events occur randomly over time or space (Poisson setting).
- Applicable with equal or unequal observation periods.

Assumptions

- Events follow independent Poisson processes in each group.
- Event rates are constant within each group.
- Groups are independent (no pairing or matching).
- Exposure times t_1 and t_2 are fixed and known.

Test Statistic

$$Z = \frac{\hat{\lambda}_1 - \hat{\lambda}_2}{\text{SE}}, \text{ where SE} = \sqrt{\frac{\hat{\lambda}_1}{t_1} + \frac{\hat{\lambda}_2}{t_2}}, \quad \hat{\lambda}_i = \frac{x_i}{t_i}. \qquad \text{Under } H_0,\ X_1 \mid (X_1 + X_2) \sim \text{Binomial}(x_1 + x_2,\ p).$$

Notation: x_1, x_2 are observed counts, t_1, t_2 are exposure times, λ_1, λ_2 are true rates, and $\hat{\lambda}_1, \hat{\lambda}_2$ are estimated rates

Plain English: The two-sample Poisson rate test compares event intensities rather than raw counts, asking whether two groups generate events at the same underlying rate after accounting for exposure. By standardizing counts by time or space, the test isolates differences in process behavior rather than differences in observation opportunity. Inference is based on large-sample normal approximation, while exact conditional methods leverage the Poisson–binomial connection to provide valid results when counts are small.

⚠ Practitioner Warning

Differences in exposure can obscure interpretation: higher counts do not imply higher rates without proper normalization. Always report counts alongside exposure. Overdispersion (variance exceeding the mean) is common and leads to anti-conservative inference; consider negative binomial or quasi-Poisson models when present. Zero counts can destabilize log-rate ratios and standard errors, requiring small adjustments or exact methods. Time-varying rates, clustering, or dependence between events violate the Poisson assumption and can invalidate results.

Hypotheses

- H_0: $\lambda_1 = \lambda_2$ (rates are equal in both groups).
- H_1: $\lambda_1 \neq \lambda_2$ (two-sided), or $\lambda_1 > \lambda_2$, or $\lambda_1 < \lambda_2$.

Decision Rule: Reject H_0 if $|Z| > z_{\alpha/2}$ (asymptotic) or exact p-value $< \alpha$.

Software References [Python, R, SAS]

Python	R	SAS
`scipy.stats.poisson_means_test`	`poisson.test`	`PROC GENMOD`

Additional Notes

Rate ratio (RR) confidence interval (A) is derived via the delta method (large-sample approximation). For rate differences, use (B).

$$\text{(A): } \exp\left[\log(\hat{\lambda}_1/\hat{\lambda}_2) \pm z_{\alpha/2}\sqrt{\frac{1}{x_1} + \frac{1}{x_2}}\right] \qquad \text{(B): } (\hat{\lambda}_1 - \hat{\lambda}_2) \pm z_{\alpha/2}\sqrt{\frac{x_1}{t_1^2} + \frac{x_2}{t_2^2}}$$

The exact conditional test, conditioning on total count $x_1 + x_2$, removes the nuisance parameter and yields a binomial test with $p = t_1/(t_1 + t_2)$. When counts are small (especially $x_i = 0$), add a small adjustment (e.g., 0.5) before computing log ratios. Poisson regression (log link with offset) generalizes these methods to multiple groups and covariates.

Miscellanea: The exact conditional test for comparing Poisson rates became widely used in epidemiology in the 1960s–1970s for analyzing person-time data in cohort studies. Conditioning on the total count exploits a key Poisson property: given $X_1 + X_2$, the counts split binomially according to exposure proportions. Before modern computing, exact p-values were often impractical to obtain and approximations were routinely used; advances in computing made exact methods standard in applied work.

Reference(s): Przyborowski & Wilenski (1940); Gu et al. (2008)

Worked Example

Research Question. Did the defect rate decrease after the process improvement? Before: $X_1 = 28$ defects in $t_1 = 2000$ units. After: $X_2 = 15$ defects in $t_2 = 2500$ units. One–sided test ($\lambda_1 > \lambda_2$), $\alpha = 0.05$.

Data Summary

$$\hat{\lambda}_1 = \frac{28}{2000} = 0.014, \qquad \hat{\lambda}_2 = \frac{15}{2500} = 0.006$$

1. Rate Ratio

$$\widehat{RR} = \frac{0.014}{0.006} = 2.33.$$

2. Hypotheses

$$H_0 : \lambda_1 = \lambda_2, \qquad H_1 : \lambda_1 > \lambda_2.$$

3. Standard Error

$$SE = \sqrt{\frac{X_1}{t_1^2} + \frac{X_2}{t_2^2}} = \sqrt{\frac{28}{2000^2} + \frac{15}{2500^2}} = 0.00307.$$

4. Test Statistic

$$Z = \frac{\hat{\lambda}_1 - \hat{\lambda}_2}{SE} = \frac{0.008}{0.00307} = 2.61.$$

5. p-value (one-sided):

$$p = 1 - \Phi(2.61) = 0.0045.$$

7. Rate Ratio CI

$$\log(\widehat{RR}) = \log(2.33) = 0.847$$

$$SE_{\log(RR)} = \sqrt{\frac{1}{X_1} + \frac{1}{X_2}} = \sqrt{\frac{1}{28} + \frac{1}{15}} = 0.320.$$

$$0.846 \pm 1.96(0.320) = [0.220,\ 1.474].$$

Exponentiate:

$$RR \in [e^{0.219},\ e^{1.473}] = [1.25,\ 4.37].$$

8. Interpretation

Before rate is 1.25–4.37× higher than after; CI entirely > 1.

9. Exact Conditional Test

Total events: $T = 43$. Under H_0, conditional distribution:

$$X_1 \mid (X_1 + X_2) \sim \text{Binomial}\left(T,\ p = \frac{t_1}{t_1 + t_2} = 0.444\right).$$

$$p = P(X_1 \geq 28) = 0.0050$$

(consistent with Z-test).

6. Decision $Z > 1.645$ and $p < 0.05 \Rightarrow$ **Reject** H_0.

Conclusion Defect rate significantly reduced: Before = 14/1000; After = 6/1000 (57% reduction). Both asymptotic and exact tests show strong evidence of improvement.

Common Pitfalls

- **Ignoring exposure differences.** Rates require scaling by exposure (X/t); comparing raw counts can completely reverse conclusions when group exposures differ.
- **Using two-sample tests for paired data.** Before/after counts on the same units are *paired*; two-sample Poisson tests assume independence. Use paired or mixed-model approaches instead.
- **Assuming Poisson despite overdispersion.** When variance exceeds the mean (clustering, heterogeneity), standard Poisson tests become liberal. Use quasi-Poisson or negative binomial alternatives.
- **Relying solely on rate ratios.** Rate ratios show relative change; absolute rate differences determine practical impact. A "doubling" may be clinically trivial or substantial depending on baseline rates.
- **Overlooking differential ascertainment.** Unequal surveillance or reporting intensity biases rate comparisons. Valid inference requires comparable detection methods across groups.

Confidence Interval

For two independent Poisson counts X_1 and X_2 observed over exposures t_1 and t_2, the rate ratio is $\widehat{RR} = (X_1/t_1)/(X_2/t_2)$. Using the delta method with $\text{Var}(X_i) = \lambda_i t_i \approx X_i$, we obtain $\text{Var}(\widehat{RR}) \approx \widehat{RR}^2\left(\frac{1}{X_1} + \frac{1}{X_2}\right)$, so $SE(\widehat{RR}) = \widehat{RR}\sqrt{\frac{1}{X_1} + \frac{1}{X_2}}$. Because $\widehat{RR}$ is skewed, inference is performed on the log scale, where $SE\{\log(\widehat{RR})\} = \sqrt{\frac{1}{X_1} + \frac{1}{X_2}}$, yielding $\log(\widehat{RR}) \pm z_{\alpha/2}\sqrt{\frac{1}{X_1} + \frac{1}{X_2}}$, which is exponentiated to obtain the confidence interval for the rate ratio.

$$\text{Delta method: } g(\hat{\theta}) \approx g(\theta) + g'(\theta)(\hat{\theta} - \theta) \quad \Rightarrow \quad \text{Var}(g(\hat{\theta})) \approx [g'(\theta)]^2\,\text{Var}(\hat{\theta}) \qquad (\text{assumes } E(\hat{\theta}) = \theta)$$

Part VI

Rank-Based & Distribution-Free Inference

Testing differences without parametric assumptions

Nonparametric Statistics

VI. NPAR — A Conceptual Primer

What Are Nonparametric Methods?

Nonparametric (distribution-free) methods make no assumption about the functional form of the population distribution. Inference is based on the **ranks**, **signs**, or **order statistics** of observations rather than their raw values, making conclusions robust to outliers, heavy tails, and skewness.

When to use:

- Data are clearly non-normal or heavily skewed.
- Outliers are present and cannot be justified as errors.
- Data are **ordinal** (ranks, Likert scales).
- Sample sizes are small and normality cannot be verified.
- A distribution-free guarantee is required regardless of sample size.

The efficiency trade-off: Under ideal parametric conditions (e.g. true Gaussian data), nonparametric tests lose some power relative to their parametric counterparts. The **asymptotic relative efficiency** (ARE) quantifies this cost: the Wilcoxon signed-rank test has ARE $= 3/\pi \approx 0.955$ vs. the paired t-test under normality, and ARE ≥ 0.864 for *any* continuous symmetric distribution. The cost of robustness is thus very small.

The Rank Transformation

Replacing observations $y_1, \ldots, y_n$ with their ranks $R_1, \ldots, R_n$ (position in the ordered sample) is the central operation:

- **Tied observations** receive **midranks** (average of the ranks they would occupy); ties reduce power and require continuity corrections.
- Under H_0 of exchangeability (random group assignment), all $n!$ rank permutations are equally likely—the foundation for exact inference.
- For large n, most rank statistics satisfy a CLT: $(T - \mu_T)/\sigma_T \xrightarrow{d} \mathcal{N}(0, 1)$, enabling z-based approximations with a continuity correction $\pm\frac{1}{2}$.

What Are We Testing?

Nonparametric tests differ in their precise H_0:

Null	Interpretation
Identical distributions	Strongest; any difference (location, scale, shape) leads to rejection.
Location shift only	Groups have the same distribution up to a shift Δ; requires similar shapes.
$P(X > Y) = \frac{1}{2}$	Probability statement; does not require identical shapes; equivalent to stochastic equality.

Rejecting H_0 means *stochastic dominance*: one group tends to produce larger values. It does **not** specifically imply a median difference unless group shapes are assumed identical.

Two-Sample Tests: Independent Groups

Mann–Whitney U test (Wilcoxon rank-sum): Rank all $n_1 + n_2$ observations jointly; let W be the sum of ranks in group 1. The U statistic counts concordant pairs:

$$U = W - \frac{n_1(n_1 + 1)}{2}, \quad U \in [0, n_1 n_2].$$

$U/n_1 n_2$ estimates $P(X_1 > X_2)$—a natural effect size directly interpretable as a probability. Under H_0: $E[W] = n_1(n_1 + n_2 + 1)/2$. Asymptotically $\mathcal{N}(0, 1)$ after standardization; exact tables for small n.

Mood's median test: Compute the grand median $\tilde{m}$; form a $2 \times k$ contingency table of counts above/below $\tilde{m}$ for each group; apply Pearson χ^2. Requires only dichotomization—very robust but **wastes information**: it uses only whether $y_i > \tilde{m}$ or not, discarding rank magnitude. Power is substantially lower than Mann–Whitney; use primarily when extreme robustness is needed or for teaching purposes.

Two-Sample Tests: Paired / One-Sample

Wilcoxon signed-rank test: For differences $d_i = y_{1i} - y_{2i}$, rank the $|d_i|$ (excluding zeros), then form $W^+ = \sum_{d_i > 0} R_i^+$. Uses both **sign and magnitude**—more powerful than the sign test when the difference distribution is symmetric.

$$E[W^+] = \frac{n(n+1)}{4}, \quad \mathrm{Var}(W^+) = \frac{n(n+1)(2n+1)}{24} \text{ (no ties)}.$$

Assumes the distribution of d_i is **symmetric about its median**; tests H_0 : median$(d_i) = 0$.

Sign test: Counts $S^+ = \#\{d_i > 0\}$; under H_0, $S^+ \sim \text{Binomial}(n, \frac{1}{2})$. Uses **direction only**—no symmetry assumption required; valid for any continuous distribution. Very robust but **low power**: it discards all magnitude information. Prefer the signed-rank test unless symmetry is clearly untenable.

Choosing between them:

Test	Uses	ARE
Sign	Direction only	0.637
Signed-rank	Direction + magnitude	0.955
Paired t	Direction + magnitude + normality	1.000

k-Sample Tests: Independent Groups

Kruskal–Wallis test: Rank all $N = \sum n_i$ observations jointly; the test statistic is:

$$H = \frac{12}{N(N+1)} \sum_{i=1}^{k} \frac{R_i^2}{n_i} - 3(N+1) \xrightarrow{d} \chi^2_{k-1}.$$

Nonparametric analog of one-way ANOVA. H_0: all k groups have the same distribution. A **tie correction** divides H by $1 - \sum_t (t^3 - t)/(N^3 - N)$ where t is each tied-group size. Post-hoc pairwise Mann–Whitney tests with Bonferroni or Dunn's correction identify which groups differ.

k-Sample Tests: Repeated Measures / Blocks

Friedman test: Within each of b blocks, rank the k treatments 1 to k; the test statistic is:

$$Q = \frac{12}{bk(k+1)} \sum_{j=1}^{k} R_j^2 - 3b(k+1) \xrightarrow{d} \chi^2_{k-1}.$$

Nonparametric analog of two-way ANOVA (randomized complete block design). H_0: all k treatments have the same distribution within each block. More powerful than Kruskal–Wallis when blocking accounts for substantial between-subject variability. Post-hoc: pairwise Wilcoxon signed-rank tests with correction.

Ordered-Alternative Tests

Standard k-sample tests (Kruskal–Wallis, Friedman) test against an **omnibus** alternative: any difference in distributions. When the groups have a natural order and a **monotone trend** is hypothesized ($\mu_1 \leq \mu_2 \leq \cdots \leq \mu_k$, at least one strict), more powerful **trend tests** exploit this structure.

Jonckheere–Terpstra (JT) test: For independent ordered groups, counts concordant pairs across all group comparisons:

$$J = \sum_{i<j} U_{ij},$$

where U_{ij} is the Mann–Whitney U statistic comparing groups i and j. Asymptotically normal under H_0. **Substantially more powerful** than Kruskal–Wallis when a monotone ordering is plausible; use Kruskal–Wallis when no ordering is assumed.

Page's trend test: The repeated-measures (blocked) analog of JT; uses within-block ranks from the Friedman framework:

$$L = \sum_{j=1}^{k} j R_j,$$

where R_j is the rank sum for treatment j and j indexes the hypothesized order. Exact critical values for small b and k; asymptotically normal. **More powerful than Friedman** when a specific ordering of treatments is predicted in advance.

When to use ordered-alternative tests:

- Groups correspond to ordered doses, time points, or severity levels.
- A directional trend is hypothesized *before* data collection.
- Using JT/Page when no order is assumed inflates Type I error; always commit to the ordering *a priori*.

Effect Sizes for Nonparametric Tests

p-values alone are insufficient; always report an effect size:

Test	Effect size
Mann–Whitney	$\hat{P}(X_1 > X_2) = U/n_1n_2$; or rank-biserial $r = 1 - 2U/n_1n_2$
Signed-rank	Matched-pairs rank biserial $r = W^+/[n(n+1)/2] \cdot 2 - 1$
Kruskal–Wallis	$\eta^2_H = (H - k + 1)/(n - k)$; or $\epsilon^2 = H/(n^2 - 1)/(n + 1)$
Friedman	Kendall's $W = Q/[b^2(k^2 - 1)/k] \in [0, 1]$

Common Pitfalls

- **Median interpretation when shapes differ:** Mann–Whitney tests stochastic dominance, not median equality, when group distributions have different shapes.
- **Ignoring pairing structure:** applying Mann–Whitney to paired data (use signed-rank); applying Kruskal–Wallis to blocked data (use Friedman).
- **Sign test over signed-rank:** prefer signed-rank unless symmetry is clearly violated; the power loss from the sign test is substantial.
- **Omnibus when order is known:** using Kruskal–Wallis when JT is appropriate discards power; pre-specify the ordering.
- **Ties not corrected:** ignoring tie corrections distorts p-values, especially when ties are frequent.
- **No post-hoc after k-sample test:** Kruskal–Wallis or Friedman rejection identifies *that* groups differ; pairwise follow-up with multiplicity correction identifies *which*.

Bottom Line

> Nonparametric methods trade distributional assumptions for robustness—by focusing on order rather than magnitude, they provide reliable inference under minimal conditions. The efficiency cost relative to parametric methods is small (ARE ≈ 0.95 for the Wilcoxon tests) and is far outweighed by the protection against assumption violations. Match the test to the design: use signed-rank (not sign) for pairs, Friedman (not Kruskal–Wallis) for blocks, and JT/Page (not omnibus tests) when a monotone ordering is predicted.

Sign Test

Distribution-Free Test for Median Location

NPAR.1 | **Category:** Median Comparison | **Data:** Continuous or Ordinal | **Null:** Exact Bin. / Approx. Normal

When to Use

- Test whether a population median equals a target value.
- Paired data where only direction of change matters.
- Distribution-free alternative to one-sample tests.
- Small samples with doubtful distributional assumptions.
- Outliers may invalidate traditional mean-based tests.

Assumptions

- Observations or matched pairs are mutually independent.
- Data are measured on at least an ordinal scale.
- Positive and negative differences are identifiable.
- Tied observations are removed before analysis.
- Signs arise from a common population median value.

Test Statistic

$$S = \#\{i : d_i > 0\} \sim \text{Binomial}(n', 0.5) \quad \text{under } H_0$$

Notation: S = number of +ve signs (+ve d_i's), n' = number of nonzero observations ($d_i \neq 0$)

Plain English: The Sign Test ignores magnitudes and uses only direction. It counts how often observations fall above versus below a hypothesized median (or how often paired differences are +ve versus -ve). Zero differences (*non-informative*) are excluded in the sign test—they carry no directional information, unlike ties in rank-based tests. If the median is correct, +ve and -ve signs should occur equally often. The test is extremely robust but sacrifices power by discarding magnitude information.

⚠ Practitioner Warning

The Sign Test is conservative and has low power compared to rank-based alternatives. Because it ignores magnitudes, large effects and small effects are treated equally. When symmetry is plausible, the Wilcoxon Signed-Rank Test is usually preferred. The test can fail to detect real effects when sample sizes are small or when differences are consistently in one direction but modest in size. A large number of zero differences further reduces effective sample size, weakening inference. Interpret non-significant results cautiously, especially in low-power settings.

Hypotheses

- H_0: Population median difference is zero.
- H_1: Population median difference is not zero.

Decision Rule: Reject H_0 if S is too small or too large. Use exact binomial probabilities or normal approx. for large n.

Software References [Python, R, SAS]

Python	R	SAS
`scipy.stats.binomtest`	`DescTools::signTest()`	`PROC UNIVARIATE`

Additional Notes

The Sign test is implemented as an exact binomial test on the number of positive differences: if X is the count of positive signs out of n' nonzero differences, then $X \sim \text{Binomial}(n', 0.5)$ under H_0. Two-sided p-values are computed by doubling the smaller tail probability. For moderate to large samples, a normal approximation may be used with $\mu = n'/2$ and $\sigma^2 = n'/4$, typically with a continuity correction. Exact binomial inference is preferred when n' is small. The test can also be framed as a one-sample median test or a paired test via differences, linking it to simple contingency-style reasoning. Its minimal assumptions make it broadly applicable, though at the cost of efficiency relative to tests that use rank or magnitude information. Its asymptotic relative efficiency is about 0.637 relative to the paired t-test under normality, reflecting the loss of power from discarding magnitude information and thus substantially larger samples are needed to achieve comparable power.

Miscellanea: Arbuthnot (1710) analyzed London christening records (1629–1710) and observed that male births exceeded female births in all 82 years. Treating each year as a Bernoulli trial with equal probability, he computed the chance of such an outcome as $(1/2)^{82}$, astronomically small. This argument is widely regarded as one of the earliest examples of hypothesis testing and is equivalent to a sign (binomial) test. Modern applications include patient-preference studies ("which treatment do you prefer?") where only direction, not magnitude, is observed.

Reference(s): Arbuthnot (1710); Conover (1999)

Worked Example

Problem Statement Thirteen volunteers record their nightly sleep duration (hours) before and after a two-week sleep hygiene program. Because the distribution of differences cannot be assumed symmetric, apply the Sign test to decide whether the program changes median sleep duration at $\alpha = 0.05$. H_0: median difference = 0; H_1: median difference $\neq 0$.

Pair	1	2	3	4	5	6	7	8	9	10	11	12	13
Before	6.0	5.5	7.0	6.5	5.0	6.0	7.5	5.5	6.0	7.0	5.0	6.5	6.0
After	7.5	6.5	8.0	6.5	6.5	7.0	7.0	7.0	7.5	8.5	4.5	8.0	7.0
Sign of d_i	+	+	+	0	+	+	−	+	+	+	−	+	+

1. Assign Signs
Set $d_i = \text{After}_i - \text{Before}_i$. Pair 4 has $d_4 = 0$ and is dropped; the effective sample size is $n' = 12$.

2. Count Signs

$$S^+ = 10, \qquad S^- = 2, \qquad n' = 12.$$

Under H_0 each sign is equally likely, so $S^+ \sim \text{Bin}(12, \frac{1}{2})$.

3. Exact p-Value
The two-sided p-value uses the tail at least as extreme as $S^+ = 10$:

$$p = 2\,P(X \geq 10), \quad X \sim \text{Bin}\left(12, \tfrac{1}{2}\right).$$

$$P(X \geq 10) = \frac{\binom{12}{10} + \binom{12}{11} + \binom{12}{12}}{2^{12}} = \frac{79}{4096}, \; p = \frac{158}{4096}.$$

4. Normal Approximation
For reference, the large-sample statistic with continuity correction is

$$z = \frac{S^+ - \frac{1}{2} - \frac{n'}{2}}{\sqrt{n'/4}} = \frac{10 - 0.5 - 6}{\sqrt{3}} = 2.021.$$

5. Decision

α	Critical value	Decision
5%	$p < 0.05$	**Reject** ($p = 0.039$)
1%	$p < 0.01$	FTR ($p = 0.039$)

Reject H_0 at the 5% level: evidence that the program increases median sleep duration.

Practitioner Insight: The Sign test is the oldest and most assumption-free nonparametric paired test, requiring only independent pairs and continuously distributed differences. It discards magnitude, using only direction, and is therefore less powerful than the Wilcoxon signed-rank test when differences are symmetric about their median. In this example, Wilcoxon gives $z = 1.988$, $p = 0.047$, while the Sign test gives $p = 0.039$—an unusual reversal because the two negative differences ($d = -0.5$) are small and receive low ranks, so the signed-rank test weights them lightly ($W^- = 8$), approximating the Sign test's simple counts. When negative differences are large, the signed-rank test responds more strongly and typically yields smaller p-values. The Sign test's advantage is robustness to skewed or heavy-tailed differences and applicability when only direction is meaningful (e.g., ordinal ratings). Exact binomial p-values are preferred; the normal approximation with continuity correction ($p = 0.043$ here) is adequate for $n' \geq 25$. Zero differences are dropped, reducing effective n'; many ties may indicate an insensitive measurement scale.

Common Pitfalls

- **Using it when a signed-rank test is valid.** The sign test discards magnitude information and is much less powerful than the Wilcoxon signed-rank test under symmetric differences. Prefer Wilcoxon unless differences are clearly asymmetric or only directional information is reliable.
- **Improper handling of zero differences.** Zero differences offer no directional information and should be omitted, reducing the effective sample size. Counting zeros inconsistently or inflating sample size biases results, especially when many ties at zero occur.
- **Misusing one-sided tests.** One-sided sign tests require a strong, pre-specified directional hypothesis. Using them post hoc or without justification inflates Type I error. Default to two-sided unless theory or study design dictates direction.
- **Overlooking independence of paired differences.** The test assumes differences across pairs are independent. Correlation from repeated measures, clustering, or temporal patterns violates this assumption; such settings require clustered or permutation-based approaches.
- **Reporting results without effect size.** A p-value alone omits magnitude. Report the proportion of positive signs with a binomial confidence interval or an appropriate effect size to convey practical importance.

Wilcoxon Signed-Rank Test

Nonparametric Test for Paired or One-Sample Location

NPAR.2 | **Category:** Median Comparison | **Data:** Continuous or Ordinal | **Null:** Exact or Normal (approx.)

When to Use

- Compare paired measurements before and after treatment.
- Test median change using signed differences and ranks.
- Distribution-free alternative to the paired t-test.
- Small samples with doubtful normality assumptions.
- Magnitudes and directions of changes both matter.

Assumptions

- Observations or matched pairs are mutually independent.
- Data are measured on at least an ordinal scale.
- Differences are drawn from a **symmetric distribution**.
- Positive and negative differences are identifiable.
- Tied or zero differences are handled appropriately.

Test Statistic

$$z = \frac{W^+ - E[W^+]}{\sqrt{\text{Var}[W^+]}} \approx N(0,1) \text{ (under } H_0). \qquad W^+ = \sum_{i=1}^{n^+} R_i, \quad E[W^+] = \frac{n(n+1)}{4}, \quad \text{Var}[W^+] = \frac{n(n+1)(2n+1)}{24}$$

Notation: $d_i = x_i - y_i$ are paired differences, R_i are ranks of $|d_i|$, n^+ is number of +ve differences

Plain English: The Wilcoxon signed-rank test compares paired observations by ranking the magnitudes of their differences and assessing whether positive and negative differences balance around zero. Unlike the sign test, it uses both direction and magnitude, giving it substantially higher power when symmetry holds.

⚠ Practitioner Warning

The Wilcoxon signed-rank test assumes paired differences are **symmetrically distributed**. This is essential because it uses both signs and magnitudes (via ranks); when symmetry is violated (e.g., strong skewness), results can be misleading. Unlike the sign test, which requires no symmetry but ignores magnitude, the signed-rank test gains power at the cost of this assumption. It targets a median (location) shift only under symmetry and does not test means. Always inspect the difference distribution. If skewed, consider transforming the data (e.g., log or square root), using the sign test, or applying permutation or bootstrap methods. Avoid using the signed-rank test blindly on asymmetric differences.

Hypotheses

- H_0: Median difference equals zero.
- H_1: Median difference differs from zero.

Decision Rule: Reject H_0 if $|z| > z_{\alpha/2}$ or p-value $< \alpha$. Use exact (permutation-based) tables for small n.

Software References [Python, R, SAS]

Python	R	SAS
`scipy.stats.wilcoxon`	`wilcox.test()`	`PROC UNIVARIATE`

Additional Notes

The Wilcoxon signed-rank test is the paired-data analog of the Mann–Whitney test, analyzing differences D_i by ranking their absolute values $|D_i|$ and attaching original signs. The statistic W is the sum of signed ranks (often reported via the smaller of +ve and -ve rank sums). The null hypothesis is that the differences come from a symmetric distribution centered at zero. For small samples, the exact distribution of W is used; for larger samples, a normal approximation along with tie and zero-difference adjustments (and optional continuity correction). Ties are handled by assigning average ranks to equal absolute differences, and zero differences are typically omitted. The test assumes symmetry of the difference distribution; under this condition, it targets a median (location) shift. It is more powerful than the sign test because it uses magnitude information via ranks, while remaining more robust than the paired t-test under non-normality (ARE ≈ 0.955 vs. 0.637 for the sign test under normality).

Miscellanea: Frank Wilcoxon (1945) introduced rank-based methods for comparing two treatments in both unpaired and paired settings within the same paper. The unpaired procedure is now known as the **rank-sum test** (equiv. to Mann-Whitney), and the paired-difference procedure is the **signed-rank test**. These methods provided one of the earliest systematic nonparametric alternatives to the t-test.

Reference(s): Wilcoxon (1945); Lehmann (1975)

Worked Example

Problem Statement Ten patients complete a physiotherapy program for chronic lower-back pain. Pain is scored on a 100-point visual analog scale (higher = more pain) before and after the program. The differences are unlikely to be normally distributed. Apply the Wilcoxon Signed-Rank test to decide at $\alpha = 0.05$ (two-sided) whether the program reduces pain (H_0: median difference = 0; H_1: median difference $\neq 0$). Set $d_i = \text{Before}_i - \text{After}_i$; +ve values $\Rightarrow$ pain reduction.

1. Compute Signed Differences

Patient	Before	After	d_i	$\lvert d_i \rvert$	Rank
1	45	44	+1	1	+1
2	52	50	+2	2	+2
3	38	41	−3	3	−3
4	61	57	+4	4	+4
5	44	49	−5	5	−5
6	57	51	+6	6	+6
7	50	43	+7	7	+7
8	63	55	+8	8	+8
9	42	33	+9	9	+9
10	55	45	+10	10	+10

$\lvert d_i \rvert$: distinct, no ties. No d_i's are zero; all 10 pairs used.

2. Rank-Sum Statistics

$$W^+ = 47, \quad W^- = 8, \quad W^+ + W^- = 55 = \frac{n(n+1)}{2} \ \checkmark$$

3. Normal Approximation ($n = 10$)

$$E[W^+] = \frac{n(n+1)}{4} = 27.5$$

$$\text{Var}[W^+] = \frac{n(n+1)(2n+1)}{24} = 96.25$$

$$z = \frac{W^+ - E[W^+]}{\sqrt{\text{Var}[W^+]}} = \frac{47 - 27.5}{\sqrt{96.25}} = 1.988$$

4. Decision

$|z| = 1.988 > z_{0.025} = 1.960$, so $p = 0.047$.
Reject H_0 at $\alpha = 0.05$
The program produces a statistically significant reduction in pain scores.

Practitioner Insight: The Wilcoxon Signed-Rank test occupies the middle ground between the sign test and the paired t-test. The sign test uses only the direction of each difference, discarding magnitude entirely; the paired t-test uses the full numerical values but requires normally distributed differences. The Signed-Rank test uses the *relative magnitude* of differences through their ranks, making it more powerful than the sign test while remaining valid when the normality assumption is untenable. Because each difference contributes a rank rather than its raw value, a single extreme outlier shifts only the largest rank, not the entire statistic — a meaningful gain in robustness over the t-test. The test does assume that the differences are drawn from a continuous, symmetric distribution; it is not appropriate when the difference distribution is markedly skewed, in which case the sign test is the safer alternative. With $n = 10$ pairs and no ties the exact p-value ($p = 0.049$, from enumeration of all $2^{10} = 1024$ sign assignments) is available from tables; the normal approximation ($p = 0.047$) agrees closely and is standard for $n \geq 10$. A continuity correction of $\frac{1}{2}$ can be applied in the numerator of z to improve accuracy for small n. For $k > 2$ related samples the natural extension is the Friedman test.

Common Pitfalls

- **Using it for independent samples.** The signed-rank test requires *paired* observations. It is invalid for independent groups; use the Mann–Whitney (rank-sum) test instead. Always confirm the study design before selecting the test.
- **Ignoring asymmetry in the difference distribution.** The test assumes differences are *symmetric* about zero. Strong skewness inflates Type I error or reduces power. Inspect difference distributions and, if asymmetric, consider transformations, a sign test, or a permutation test on median differences.
- **Mishandling zeros and ties.** Zero differences provide no directional information and should be omitted. Ties in absolute differences should receive average ranks. With many ties (common in discrete data), power declines and exact methods are preferable.
- **Misinterpreting the null hypothesis.** The test evaluates symmetry of differences around zero—not solely median = 0. Asymmetric differences with median zero can still lead to rejection. Use a sign test for a pure median test without symmetry requirements.
- **Using normal approximation with very small n.** Asymptotic p-values are unreliable when the number of nonzero pairs is small. For $n < 10$, rely on exact tables or software's exact option rather than the normal approximation.

Mann–Whitney U Test

Distribution-Free Test for Two-Sample Location Differences

NPAR.3 | **Category:** Median Comparison | **Data:** Continuous or Ordinal | **Null:** Exact or Asymp. Normal

When to Use

- Compare two independent groups using ranks.
- Non-normal data where two-sample t-test assumptions fail.
- Ordinal or continuous data suitable for ranking.
- Presence of outliers affecting mean-based comparisons.
- Nonparametric alternative to two-sample t-test.

Assumptions

- Observations are independent within and across groups.
- Data are at least ordinal (rankable).
- Group distributions have similar shapes (location shift).
- Few or no ties, or tie correction applied.
- Samples are randomly drawn from populations.

Test Statistic

$$Z = \frac{U - \mu_U}{\sigma_U}, \quad \text{where } U = \min(U_1, U_2),\ U_i = n_1 n_2 + \frac{n_i(n_i+1)}{2} - R_i, i = 1, 2; \quad \mu_U = \frac{n_1 n_2}{2},\ \sigma_U^2 = \frac{n_1 n_2 (N+1)}{12}.$$

Notation: n_1, n_2 sample sizes, $N = n_1 + n_2$; R_1, R_2 rank sums, σ_U^2 (tie-adjusted if needed)

Plain English: The Mann–Whitney U test compares two independent groups by ranking all observations together and examining whether one group tends to receive higher ranks than the other. Instead of comparing means, it tests whether values from one population tend to be larger than values from the other, making it robust to skewness and outliers.

⚠ Practitioner Warning

Mann-Whitney tests whether one group tends to have larger values than the other (stochastic dominance), which equals median difference only if distributions have identical shapes. With different shapes (one skewed, one symmetric), significant result indicates some distributional difference but not necessarily median shift. For pure median test with different shapes, use Mood's median test. Interpret Mann-Whitney as location test only when shape assumption reasonable.

Hypotheses

- H_0: The two populations are stochastically equal.
- H_1: One population is stochastically larger.

Decision Rule: Reject H_0 if Reject H_0 if U (or $|Z|$) is extreme or $p < \alpha$.

Software References [Python, R, SAS]

Python	R	SAS
`scipy.stats.mannwhitneyu`	`wilcox.test()`	`PROC NPAR1WAY`

Additional Notes

The statistics satisfy $U_1 + U_2 = n_1 n_2$; U can be interpreted as the number of pairwise wins (plus half-ties) of one group over the other. With no ties, the exact (permutation) distribution of U is used for small samples. For larger samples, a normal approximation is used with variance $\sigma_U^2 = n_1 n_2 (N+1)/12$; a multiplicative correction for ties: $C = [1 - (\sum t_j^3 - t_j)/(N^3 - N)]$, where t_j are tie group sizes; a continuity correction may also be applied. The test has high efficiency: its asymptotic relative efficiency is about $3/\pi \approx 95\%$ of the two-sample t-test under normality, while remaining valid under severe non-normality. It can also be interpreted as testing whether $P(X > Y)$ differs from 0.5, linking directly to the AUC (area under the curve) in ROC (receiver operating characteristics) analysis.

Miscellanea: Mann Whitney U test (1947) and Wilcoxon rank-sum test (1945) were developed independently but are mathematically equivalent—they compute same p-value from different but algebraically related statistics: U (Mann-Whitney) vs. W (Wilcoxon). Relationship: $U = W - n_1(n_1+1)/2$ where W is sum of ranks. The statistic U has a natural interpretation: it counts how often observations from one group exceed those from the other, linking directly to $P(X > Y)$ and the area under the ROC curve (AUC). Early motivation came from problems where normality was doubtful, leading to one of the first widely used nonparametric alternatives to the two-sample t-test. The test is also known as the Wilcoxon–Mann–Whitney test, reflecting its dual origins and widespread adoption across disciplines. Despite its reputation as a "median test," it more generally detects differences in distributions; the median interpretation holds only when group shapes are similar.

Reference(s): Wilcoxon (1945); Mann & Whitney (1947)

Worked Example

Problem Statement Recovery times (days) are compared between a treatment group ($n_1 = 6$) and a control group ($n_2 = 6$). The Mann-Whitney U test is used as a non-parametric alternative to the two-sample t-test, requiring no normality assumption. H_0: the two populations are stochastically equal, $P(X > Y) = \frac{1}{2}$; H_1: $P(X > Y) \neq \frac{1}{2}$.

Treatment (T): 18, 22, 25, 28, 32, 36 **Control (C):** 7, 10, 12, 15, 19, 24

1. Combined Ranking Pool all $N = 12$ observations and rank from 1 to 12:

Rank	1	2	3	4	5	6	7	8	9	10	11	12
Value	7	10	12	15	18	19	22	24	25	28	32	36
Group	C	C	C	C	**T**	C	**T**	C	**T**	**T**	**T**	**T**

2. Rank Sums

$$R_1 = 5 + 7 + 9 + 10 + 11 + 12 = 54$$

$$R_2 = 1 + 2 + 3 + 4 + 6 + 8 = 24$$

$$R_1 + R_2 = 78 = \frac{N(N+1)}{2} = \frac{12 \times 13}{2} \checkmark$$

3. U Statistics

$$U_1 = n_1 n_2 + \frac{n_1(n_1 + 1)}{2} - R_1 = 36 + 21 - 54 = 3$$

$$U_2 = n_1 n_2 + \frac{n_2(n_2 + 1)}{2} - R_2 = 36 + 21 - 24 = 33$$

$$U_1 + U_2 = 36 = n_1 n_2 \checkmark \qquad U = \min(U_1, U_2) = 3$$

$U_1 = 3$: Treatment beats Control in only 3 of 36 pairwise comparisons, signaling that Control values tend to be lower.

4. Normal Approximation

$$\mu_U = \frac{n_1 n_2}{2} = 18, \quad \sigma_U = \sqrt{\frac{n_1 n_2 (N + 1)}{12}} = 6.245$$

$$z = \frac{U - \mu_U}{\sigma_U} = -2.402$$

$$p \text{ (two-sided)} = 2\Phi(-2.402) = 0.016$$

5. Results and Decision

Method	Value	CV (5%)	Decision
Exact (U vs table)	3	5	Reject H_0
Normal approx ($\lvert z \rvert$)	2.402	1.960	Reject H_0

Exact $p = 0.015$; $U = 3 \leq U_{crit} = 5$ for $n_1 = n_2 = 6$, $\alpha = 0.05$ two-sided.

Practitioner Insight: The U statistic counts pairwise wins. Under H_0, $\mu_U = 18$, so such extremes ($U_1 = 3$) occur with probability ≈ 0.015. The identity $U = W - n_1(n_1 + 1)/2$ shows equivalence with the Wilcoxon rank-sum test. For small samples ($n = 6$), exact critical values ($U_{crit} = 5$) are preferred over the normal approximation; a continuity correction gives slightly more conservative p-values. Tie adjustments modify σ_U, but are unnecessary here.

Common Pitfalls

- **Treating it as a median test.** Mann–Whitney tests *stochastic dominance*, not equality of medians unless the two distributions have the same shape. With differing spreads or skewness, significance reflects a general distributional shift, not necessarily a median difference.
- **Using it for paired data.** The test assumes two *independent* samples. It is invalid for before–after measurements, matched pairs, or repeated measures. Use the Wilcoxon *signed-rank* test instead.
- **Ignoring extensive ties.** Large numbers of ties reduce power and distort the normal approximation. When data are highly discrete (e.g., Likert scales), prefer exact methods or permutation tests and interpret results cautiously.
- **Relying on normal approximation with small samples.** Asymptotic p-values can be inaccurate when sample sizes are small. Use exact U-distribution calculations or software's exact option when either group is small.
- **Reporting significance without effect size.** A p-value alone is incomplete. Always include an effect size such as probability of superiority ($f = U_1/(n_1 n_2)$) or rank-biserial correlation ($r = 2f - 1$) for meaningful interpretation.

Mood's Median Test

Median-Based Nonparametric Test for Two or More Groups

NPAR.4 | **Category:** Location Comparison | **Data:** Continuous or Ordinal | **Null:** χ^2 (approx.) or Exact

When to Use

- Compare medians across two or more independent groups.
- Strongly non-normal data with heavy tails or skewness.
- Robust option when rank-based tests are shape-sensitive.
- Data with many outliers affecting rank-based comparisons.
- Simple, distribution-free test focused on medians.

Assumptions

- Observations are independent within and across groups.
- Data are at least ordinal (comparisons to median possible).
- Common median under null hypothesis across groups.
- Sufficient sample size for chi-square approximation.
- Few ties at the median or handled appropriately.

Test Statistic

$$\chi^2 = \sum_{i=1}^{2}\sum_{j=1}^{k}\frac{(O_{ij} - E_{ij})^2}{E_{ij}}, \quad i = 1(\text{above pooled median}),\ i = 2(\text{at or below pooled median})$$

Notation: k groups, O_{ij} observed counts, E_{ij} expected counts under equal medians, $\tilde{x}$ pooled median.

Plain English: Mood's median test compares groups by asking a simple question: do the groups differ in how many observations fall above or below a common median? Each value is reduced to a yes/no indicator—above the median or not. Instead of ranks or means, it reduces data to a contingency table around the median. This makes the test extremely robust to outliers and skewness, but also less sensitive to distributional differences beyond the median itself.

⚠ Practitioner Warning

Mood's median test is intentionally conservative and often has low power. It ignores magnitude information and responds only to median shifts. For most two-sample problems, rank-based tests (Mann–Whitney, Wilcoxon) are more efficient. Use Mood's test only when the median itself is the scientific estimand of interest.

Hypotheses

- H_0: All groups have the same population median.
- H_1: At least one group has a different median.

Decision Rule: Reject H_0 if $\chi^2 > \chi^2_{\alpha,k-1}$ or $p < \alpha$.

Software References [Python, R, SAS]

```
scipy.stats.median_test          mood.test()          PROC NPAR1WAY
```

Additional Notes

Mood's median test uses elegantly simple approach: (1) Find overall median of all observations combined, (2) For each group, count how many observations fall above vs. below this common median, (3) If groups have equal medians, each should have roughly 50% above and 50% below, (4) Test uses chi-square test on resulting 2×k table. Unlike Mann-Whitney, this tests medians specifically regardless of distribution shapes. Lower power than Mann-Whitney but more robust when shape assumption violated. Because it discards rank and distance information, its asymptotic efficiency is low relative to rank-based location tests. However, its logic is transparent, and it remains useful when extreme robustness or median-specific inference is required.

Miscellanea: Mood's median test was introduced by Alexander Mood (1950) and further developed with Brown (1951) as one of the earliest nonparametric methods focused explicitly on medians. Unlike rank-based tests (e.g., Mann–Whitney or Kruskal–Wallis), it reduces data to a binary classification relative to the grand median, making it one of the simplest distribution-free procedures. This simplicity comes at a cost: by discarding rank information, the test is typically less powerful than rank-based alternatives but gains robustness to extreme outliers. For two groups, the test reduces to a 2×2 contingency table, linking it directly to classical methods such as Pearson's χ^2 or Fisher's exact test. Despite its lower power, the test remains useful in settings where a clear, median-focused and highly robust comparison is desired.

Reference(s): Mood (1954)

Worked Example

Problem Statement Twelve athletes from three training programs (four per program) compete in a fitness test scored 1–12 (no ties). Scores are ordinal; normality is not assumed. Test whether the three programs have the same median at $\alpha = 0.05$. $H_0: M_1 = M_2 = M_3$; H_1: at least one median differs.

Program	Scores (x)			
1	1	2	3	4
2	5	6	7	8
3	9	10	11	12

1. Grand Median
The grand median of the pooled sample:

$$M = (6 + 7)/2 = 6.5.$$

2. Count Above and At/Below M
For each program count a_i ($x > 6.5$) and b_i ($x \leq 6.5$):

	Prog 1	Prog 2	Prog 3	Total
Above M	0	2	4	6
At/below M	4	2	0	6
Total	4	4	4	12

Program 1 lies entirely below M; Program 3 lies entirely above.

3. Expected Cell Counts

$$E(a_i) = n_i \cdot \frac{A}{N} = 4 \cdot \frac{6}{12} = 2, \qquad E(b_i) = 2.$$

All six expected counts equal 2.

4. Test Statistic

$$\chi^2 = \sum_{i=1}^{3} \left[\frac{(a_i - 2)^2}{2} + \frac{(b_i - 2)^2}{2} \right] = 8.$$

5. Decision
Under H_0: $\chi^2 \sim \chi^2(k-1) = \chi^2(2)$. $p = 0.018$.

α	$\chi^2_{\alpha,2}$	Decision
5%	5.991	**Reject** (8 > 5.991)
1%	9.210	FTR (8 < 9.210)

Practitioner Insight: Mood's median test reduces each observation to a binary indicator—above or at/below the grand median—and applies a Pearson χ^2 test to the resulting $2 \times k$ table. It directly tests equality of medians, whereas Kruskal–Wallis tests stochastic dominance and typically has higher power. The loss of power arises from discarding rank information: values far from and near the median contribute equally. The advantage is robustness to outliers—extreme values affect only whether they fall above or below the median, not their magnitude, making the test less sensitive to distortions that can impact rank-based methods. For $k = 2$, the test reduces to a 2×2 table (often analyzed via Fisher's exact test). The χ^2 approximation requires adequate expected counts; when sample sizes are small, exact methods are preferable.

Common Pitfalls

- **Using it when more powerful rank tests apply.** Mood's median test loses substantial power compared with Mann–Whitney or Kruskal–Wallis under simple location shifts. Use it when medians specifically matter or when shape differences make rank tests questionable.
- **Ambiguity from ties at the pooled median.** Observations exactly on the pooled median must be handled (omit, split, or reassign). With many such ties—common in discrete data—the test's power deteriorates sharply; document the tie-handling rule.
- **Reduced power with unbalanced designs.** Severely unequal sample sizes distort the pooled median and diminish sensitivity to group differences. Be cautious with small–large group comparisons and consider rank-based or permutation alternatives.
- **Violating χ^2 expected count assumptions.** Each cell in the 2×k table should have expected count ≥ 5 for the chi-square approximation to hold. With small samples or many groups, switch to Fisher's exact or permutation methods.
- **Unadjusted pairwise follow-ups.** After a significant overall test, running multiple pairwise median tests without correction inflates Type I error. Use multiplicity adjustments or rely on appropriate descriptive median comparisons.

Kruskal–Wallis Test

Rank-Based Nonparametric Alternative to One-Way ANOVA

NPAR.5 | **Category:** Median Comparison | **Data:** Continous or Ordinal | **Null:** Asymptotic chi-square

When to Use

- Compare medians across three or more independent groups.
- Non-normal data where ANOVA assumptions are violated.
- Ordinal data or continuous data analyzed via ranks.
- Presence of outliers affecting mean-based comparisons.
- Nonparametric alternative to one-way ANOVA.

Assumptions

- Observations are independent within and across groups.
- Data are measured on at least ordinal (rankable).
- Group distributions have similar shapes (location shift).
- Few or no ties, or tie correction applied.
- Samples are randomly drawn from populations.

Test Statistic

$$H = \frac{12}{N(N+1)} \sum_{i=1}^{k} \frac{R_i^2}{n_i} - 3(N+1) \approx \chi^2_{k-1} \text{ (under } H_0\text{)}$$

Notation: k groups, n_i group sizes, $N = \sum n_i$, R_i rank sum of group i

Plain English: Kruskal–Wallis extends the Mann–Whitney U test to more than two groups. All observations are pooled and ranked (1 = smallest, N = largest), then group rank sums are compared. If groups have same central tendency (median), each should accumulate similar rank sums (accounting for sample size). Groups with higher values get higher ranks, producing larger rank sums. The test detects whether at least one group tends to have larger or smaller values.

⚠ Practitioner Warning

Kruskal–Wallis tests for any difference among groups but does not identify which groups differ. Significant results require post-hoc pairwise tests (e.g., Dunn's test) with multiple-comparison correction. With different shapes (one skewed, one symmetric; different spreads), significant result indicates some distributional difference but not necessarily location shift. For pure median test with different shapes, use Mood's median test extended to k groups

Hypotheses

- H_0**:** All groups have same location (assuming similar shape).
- H_1**:** At least one group differs in location.

Decision Rule: Reject H_0 if $H > \chi^2_{\alpha,k-1}$ or $p < \alpha$.

Software References [Python, R, SAS]

Python	R	SAS
`scipy.stats.kruskal`	`kruskal.test()`	`PROC NPAR1WAY`

Additional Notes

Kruskal–Wallis is a rank-based, distribution-free test under H_0, robust to outliers, and reduces to the Mann–Whitney U test when $k = 2$. It replaces observations with pooled ranks and compares group rank sums via (**A**) which is approximately χ^2_{k-1} for moderate to large samples (exact permutation methods are available for small n). Under similar-shaped distributions, H tests equality of locations (often interpreted as medians); with differing shapes, it detects broader distributional differences. Ties are handled by assigning average ranks and applying a correction (**B**), where t_j are tie group sizes. For example, with $N = 10$ and tie groups of sizes 2 and 3, $\sum(t_j^3 - t_j) = 30$, $N^3 - N = 990$, $C = 32/33$, $H_{\text{adj}} = H/(32/33)$. Kruskal–Wallis is an omnibus test; significant results require post-hoc procedures (e.g., Dunn, Conover–Iman, or Dwass–Steel–Critchlow–Fligner) with multiplicity adjustment. It is less powerful than ANOVA when normality holds but more robust under non-normality and outliers.

$$\textbf{A}: H = \frac{12}{N(N+1)} \sum_{i=1}^{k} \frac{R_i^2}{n_i} - 3(N+1), \qquad \textbf{B}: H_{\text{adj}} = \frac{H}{C}, \text{ where } C = 1 - \frac{\sum(t_j^3 - t_j)}{N^3 - N}.$$

Miscellanea: William Kruskal and W. Allen Wallis introduced the test in 1952 as a generalization of rank-based two-sample methods. *The test assesses stochastic dominance—whether samples tend to yield larger values—rather than strictly comparing medians.* It quickly became the standard nonparametric alternative to one-way ANOVA and remains one of the most widely taught rank tests.

Reference(s): Kruskal & Wallis (1952); Conover (1999)

Worked Example

Problem Statement Twelve customers rate their satisfaction with three service formats on a 12-point scale, four customers per format. The scores happen to be all distinct and span 1 through 12. hence, the observed scores coincide with their ranks, and no separate ranking step is required. Because the scale is ordinal and normality is not assumed, apply the Kruskal–Wallis test. H_0: all three formats have the same distribution; H_1: at least one format differs.

Format	Scores				Ranks				Rank Sums
1	1	4	5	6	1	4	5	6	$R_1 = 16$
2	2	3	7	10	2	3	7	10	$R_2 = 22$
3	8	9	11	12	8	9	11	12	$R_3 = 40$

1. Assign Ranks

No need for this step here.

2. Rank Sums

Check: $R_1 + R_2 + R_3 = 78 = N(N+1)/2$. ✓

3. Kruskal–Wallis Statistic

$$H = \frac{12}{N(N+1)} \sum_{i=1}^{k} \frac{R_i^2}{n_i} - 3(N+1).$$

$$H = \frac{12}{12(12+1)} \left(\frac{16^2}{4} + \frac{22^2}{4} + \frac{40^2}{4} \right) - 3(12+1) = 6$$

Note: No tie correction is needed here, but it should be applied whenever ties are present.

4. Decision

Under H_0: $H \sim \chi^2(k-1) = \chi^2(2)$ approximately.

α	$\chi^2_{\alpha,2}$	Decision
5%	5.991	**Reject** ($H = 6 > 5.991$)
1%	9.210	FTR ($H = 6 < 9.210$)

$p = 0.050$. Reject H_0 at the 5% level: evidence that at least one format has a different distribution.

Practitioner Insight: The Kruskal–Wallis test is the nonparametric analog of one-way ANOVA, replacing observations with joint ranks and avoiding distributional assumptions. The statistic H measures whether group rank sums R_i deviate from their expected values $n_i(N+1)/2$ under random allocation, effectively capturing between-group variation in ranks. The formula $H = \frac{12}{N(N+1)} \sum R_i^2/n_i - 3(N+1)$ corresponds to scaled between-group variance of ranks and is approximated by a χ^2 distribution; the factor $12/[N(N+1)]$ normalises by the variance of a uniform discrete distribution on $\{1, \ldots, N\}$. Compared to ANOVA, it sacrifices power when normality holds (since magnitude information is discarded) but gains robustness under non-normality. For $k = 2$, it reduces to the Wilcoxon rank-sum (Mann–Whitney) test. Ties are handled by assigning average ranks and applying a correction factor in the denominator of H; here all 12 scores are distinct so no correction is needed. After a significant result, post-hoc comparisons are typically performed using Dunn's test with Bonferroni correction.

Common Pitfalls

- **Applying it to dependent or matched data.** Kruskal–Wallis requires *independent* samples. It is invalid for repeated measures, matched groups, or clustered designs. Use the Friedman test or correlation-aware methods for dependent data.
- **Running unadjusted pairwise tests after a significant result.** A significant omnibus test does not identify which groups differ. Unadjusted pairwise Mann–Whitney tests inflate Type I error. Use Dunn's test, Bonferroni-adjusted comparisons, or FDR-controlled procedures.
- **Interpreting it strictly as a median test.** The test detects any distributional differences—location, spread, or shape. Only when groups share similar shapes does it function as a location (median) test. Shape differences complicate interpretation.
- **Ignoring extensive ties in rank data.** Heavy ties reduce power and require tie corrections to the test statistic. With highly discrete data, consider exact permutation methods and interpret results cautiously due to limited rank resolution.
- **Using chi-square approximation with very small groups.** The chi-square approximation may be inaccurate when group sizes are very small (e.g., $n_i < 5$). For small samples, rely on exact permutation distributions or software's exact p-values.

Friedman Test

Nonparametric Test for Differences Across Related Samples

NPAR.6 | **Category:** Median Comparison | **Data:** Continuous or Ordinal | **Null:** χ^2 approx.

When to Use

- Compare three or more related treatments or conditions.
- Repeated-measures or block designs with non-normal data.
- Ordinal or continuous responses suitable for ranking.
- Within-subject designs with unknown or unequal variances.
- Nonparametric alternative to repeated-measures ANOVA.

Assumptions

- Blocks (subjects) are independent of each other.
- Observations within each block are mutually related.
- Responses are at least ordinal (rankable within blocks).
- Treatment effects are consistent across blocks.
- Few or no ties within blocks, or tie correction applied.

Test Statistic

$$Q = \frac{12}{nk(k+1)} \sum_{j=1}^{k} R_j^2 - 3n(k+1) \approx \chi^2_{k-1} \text{ (under } H_0\text{)}.$$

Notation: n subjects, k treatments, R_j sum of ranks for treatment j within subjects.

Plain English: Friedman's test compares multiple treatments measured on the same subjects by ranking responses *within each subject* and assessing whether rank sums differ systematically across treatments. By removing between-subject variability, the test focuses solely on treatment effects. It is the nonparametric counterpart to repeated-measures ANOVA when normality or sphericity assumptions are questionable.

⚠ Practitioner Warning

Friedman's test is an omnibus procedure: it detects overall differences among treatments but does not identify which treatments differ. Post-hoc comparisons (e.g., Wilcoxon signed-rank with Bonferroni, Nemenyi, or Conover) are required after rejection. The test assumes complete blocks: each subject must be observed under all k treatments. Missing data within blocks invalidate the standard procedure as within-block ranking requires all treatments to be present. In practice, one may omit incomplete blocks (with loss of power), use extensions for incomplete blocks, or adopt mixed-effects models that handle missingness more flexibly. Ties within blocks reduce power and require appropriate correction.

Hypotheses

- H_0: All treatments have identical effects.
- H_1: At least one treatment differs from others.

Decision Rule: Reject H_0 if $Q > \chi^2_{\alpha,k-1}$ or $p < \alpha$. Exact tables available for $n < 10$ or $k < 5$.

Software References [Python, R, SAS]

Python	R	SAS
`scipy.stats.friedmanchisquare`	`friedman.test()`	`PROC FREQ / CMH2`

Additional Notes

The Friedman test ranks responses within each block (subject) and sums ranks by treatment. The test is based on relative ordering within blocks, making it robust to non-normality and scale differences across subjects. A tie correction is applied when equal values occur within a block, adjusting the denominator of χ^2_F. When $k = 2$, the Friedman test reduces to the sign test. Compared to repeated-measures ANOVA, the Friedman test sacrifices power by using ranks instead of magnitudes but provides valid inference when distributional assumptions (e.g., normality, sphericity) are violated. Effect size measures such as Kendall's $W = Q/[n(k-1)]$ quantify agreement among subjects and are recommended for reporting practical significance. Its asymptotic relative efficiency (ARE) is slightly below repeated-measures ANOVA under normality (typically 0.85–0.95), but it can outperform it under non-normal or heavy-tailed distributions. For small k ($k \leq 4$) with highly variable block ranges, Quade's test may offer better power than Friedman's test and is worth considering as an alternative.

Miscellanea: Milton Friedman (1937) [economist, statistician, and later a Nobel Laureate (1976)] introduced this test as a rank-based method for repeated-measures (blocked) data. It serves as a nonparametric analog of randomized-block (repeated-measures) ANOVA and is widely used in longitudinal and crossover studies.

Reference(s): Friedman (1937); Conover (1999)

Worked Example

Problem Statement Six athletes each complete three training regimes (A, B, C) in randomized order. A blinded assessor records a performance score (higher = better) after each regime. Because the scores are unlikely to be normally distributed, apply the Friedman test to decide whether the three regimes produce the same median performance at $\alpha = 0.05$.

$$H_0: \text{all three regimes have the same median;} \quad H_1: \text{at least one median differs.}$$

1. Rank Within Each Block

Replace each row with ranks 1, 2, 3 assigned to the three regime scores in increasing order.

	Raw Scores			Ranks		
Block	A	B	C	A	B	C
1	55	68	79	1	2	3
2	48	61	74	1	2	3
3	62	70	83	1	2	3
4	57	75	66	1	3	2
5	71	64	80	2	1	3
6	66	59	77	2	1	3
R_j				8	11	17

Check: $R_A + R_B + R_C = 36 = \frac{nk(k+1)}{2}$. ✓

2. Friedman Statistic

With $n = 6$ blocks and $k = 3$ treatments:

$$\chi_F^2 = \frac{12}{nk(k+1)} \sum_{j=1}^{k} R_j^2 - 3n(k+1).$$

$$\chi_F^2 = \frac{12}{6 \cdot 3 \cdot 4}(8^2 + 11^2 + 17^2) - 3 \cdot 6 \cdot 4 = 7$$

3. Decision

Under H_0: $\chi_F^2 \sim \chi^2(k-1) = \chi^2(2)$ approximately.

α	$\chi^2_{\alpha,2}$	Decision
5%	5.991	**Reject** (7 > 5.991)
1%	9.210	FTR (7 < 9.210)

$p = 0.030$. Reject H_0 at the 5% level: evidence that at least one training regime produces a different median performance.

Regime C records the largest rank sum ($R_C = 17$), suggesting it yields the highest performance; Regime A records the smallest ($R_A = 8$). Effect size: Kendall's $W = \chi_F^2/[n(k-1)] = 7/12 \approx 0.583$ (moderately strong agreement).

Kendall's $W \in [0, 1]$: 0 (none), 0.1 (weak), 0.3 (moderate), 0.5 (moderately strong), 0.7 (strong), 0.9+ (very strong).

Practitioner Insight: The Friedman test (Friedman 1937) is the nonparametric analog of one-way repeated measures ANOVA, removing block effects by ranking *within* blocks so inter-subject differences do not inflate residual variation. The statistic χ_F^2 assesses whether column rank sums R_j deviate from their expected value $n(k+1)/2 = 12$, with scaling $12/[nk(k+1)]$ reflecting the variance of uniform ranks. The data were chosen so that $12/[nk(k+1)] = 1/6$ and $\sum R_j^2 = 474$ yield $\chi_F^2 = 7$, giving $p = 0.030$, between the 5% and 1% thresholds. Compared to one-way repeated measures ANOVA on the same data ($F(2, 10) = 9.29$, $p = 0.005$), the parametric test is more powerful because it uses magnitudes; Friedman sacrifices power by ranking but gains robustness when normality fails. Ties within blocks are handled by average ranks with a correction factor; none occur here. For significant results, post-hoc comparisons use Wilcoxon signed-rank with Bonferroni adjustment or the Nemenyi test. For $k = 2$, the Friedman test reduces to the sign test.

Common Pitfalls

- **Using it for independent groups.** Friedman requires repeated or matched measurements on the same subjects. It is invalid for independent samples; use Kruskal–Wallis in those designs.
- **Including incomplete blocks.** The test assumes every subject is measured under all treatments. Missing observations prevent proper within-block ranking. Omit incomplete blocks or use methods that explicitly handle missingness (e.g., mixed models or imputation).
- **…terpreting it as evidence for a specific trend.** Friedman is an omnibus test indicating that at least one treatment … It does not imply monotonic or ordered patterns. Use Page's test or planned contrasts for structured hypotheses.
- **…djusted pairwise comparisons.** Pairwise Wilcoxon signed-rank tests without multiplicity correction … Use Nemenyi, Bonferroni-adjusted comparisons, Conover procedures, or FDR-controlled methods …
- … Kendall's W measures the degree of agreement among blocks—not variance explained. … rankings across subjects; low W indicates heterogeneous responses.

Jonckheere–Terpstra Test

Nonparametric Test for Ordered Alternatives

NPAR.7 | **Category:** Trend Detection | **Data:** Continuous or Ordinal | **Null:** Exact or asymptotic normal

When to Use

- Test ordered differences across $k \geq 2$ independent groups.
- Detect monotone trends across ordered treatments or doses.
- More powerful than Kruskal–Wallis for ordered trends.
- Use when data is continuous or ordinal but non-normal.
- Detect dose–response or graded treatment effects.

Assumptions

- Observations are independent within and across groups.
- Groups have a pre-specified natural (monotonic) ordering.
- Data are measured on at least ordinal scale (rankable).
- Group distributions have similar shapes (location shift).
- Few or no ties, or tie correction applied.

Test Statistic

$$Z = \frac{J - E(J)}{\sqrt{\mathrm{Var}(J)}}, \qquad \text{where } J = \sum_{i<j} \sum_{x \in G_i} \sum_{y \in G_j} I(x < y), \quad E[J] = \frac{N^2 - \sum n_i^2}{4}, \; \mathrm{Var}[J] = \frac{N^2(2N+3) - \sum n_i^2(2n_i + 3)}{72}$$

Notation: G_i denotes group i in ordered sequence, k groups, J counts concordant pairs, $I(\cdot)$ is indicator function

Plain English: The Jonckheere–Terpstra test checks whether group values tend to increase (or decrease) in a prespecified order. Instead of asking whether groups differ in any way, it asks a sharper question: *do observations from later groups tend to be larger than those from earlier groups?* By focusing only on order-consistent pairwise comparisons, the test gains power over omnibus alternatives when a monotone trend is expected.

⚠ Practitioner Warning

The Jonckheere–Terpstra test is inherently **one-sided**, targeting an ordered alternative $\theta_1 \leq \theta_2 \leq \cdots \leq \theta_k$ with at least one strict inequality. The group ordering must be specified *a priori* based on theory (e.g., dose, time); choosing the order after inspecting the data invalidates the test. If no natural ordering exists, use the Kruskal–Wallis test instead. The power advantage comes from focusing on this directional hypothesis rather than an omnibus alternative. As with other rank-based tests, substantial differences in distributional shape (not just location) can complicate interpretation.

Hypotheses

- H_0: No monotonic trend across groups.
- H_1: Monotone trend exists across groups.

Decision Rule: Reject H_0 if Z is extreme in the direction of the specified order.

Software References [Python, R, SAS]

```
scipy.stats.jonckheere_terpstra    clinfun::jonckheere.test()    PROC NPAR1WAY (JT option)
```

Additional Notes

The Jonckheere–Terpstra statistic J is computed by summing Mann–Whitney counts U_{ij} over all ordered group pairs ($i < j$), where each U_{ij} counts the number of times observations in group j exceed those in group i (with $\frac{1}{2}$ added for ties). Thus, J measures the overall tendency for higher-ordered groups to dominate lower ones. For moderate to large samples, J is standardized to a normal Z using its mean and variance under H_0; exact permutation distributions are available for small samples. A tie correction is applied to the variance when ties are present. Normal approximation is generally appropriate when all groups contain at least five observations ($\min(n_i) \geq 5$ and $N = \sum n_i \geq 20$). The test is more powerful than Kruskal–Wallis when the ordered alternative holds, but both tests coincide for $k = 2$, reducing to the Mann–Whitney test. Because it is based on pairwise comparisons across groups, J can also be interpreted as an aggregate measure of stochastic dominance along the specified order.

Miscellanea: Terpstra (1952) and Jonckheere (1954) independently developed this test for detecting ordered alternatives across multiple groups, leading to its joint name. By focusing on monotone trends rather than arbitrary differences, it is particularly useful in dose–response and other structured experimental designs.

Reference(s): Terpstra (1952); Jonckheere (1954)

Worked Example

Problem Statement Twelve patients are randomly assigned to one of three physical therapy intensities—one, two, or three sessions per week—four patients per group. Pain reduction scores after four weeks (higher = more improvement) are recorded below. The distribution of scores cannot be assumed normal. Apply the Jonckheere–Terpstra test to decide at $\alpha = 0.05$ whether pain reduction increases with therapy intensity. $H_0: \theta_1 = \theta_2 = \theta_3$; $H_1: \theta_1 \le \theta_2 \le \theta_3$ (at least one strict).

Data by Group	Pain reduction scores			
Low (1 session)	6	10	14	18
Medium (2 sessions)	12	16	20	24
High (3 sessions)	19	23	27	31

1. Pairwise Mann–Whitney Counts
For each ordered pair of groups (i, j) with $i < j$, count $U_{ij} = \#\{(x, y) : x \in G_i,\ y \in G_j,\ x < y\}$.

U_{12} (Low vs. Medium):

	12	16	20	24
6	<	<	<	<
10	<	<	<	<
14	>	<	<	<
18	>	>	<	<

$U_{12} = 13$ (3 pairs not less, $16 - 3 = 13$).

U_{13} (Low vs. High):
$\max(G_1) = 18 < 19 = \min(G_3)$, so every Low score falls below every High score: $U_{13} = 16$.

U_{23} (Medium vs. High):
13 (3 pairs not less, $16 - 3 = 13$).

2. Test Statistic

$$J = U_{12} + U_{13} + U_{23} = 13 + 16 + 13 = 42.$$

3. Normal Approximation
With $n_1 = n_2 = n_3 = 4$ and $N = 12$:

$$E[J] = \frac{144 - 48}{4} = 24, \quad \mathrm{Var}[J] = \frac{3888 - 528}{72} = \frac{140}{3}.$$

$$z = \frac{J - E[J]}{\sqrt{\mathrm{Var}[J]}} = \frac{42 - 24}{\sqrt{140/3}} = 2.635.$$

4. Decision
The test is one-sided (upper tail).

α	z_α	Decision
5%	1.645	**Reject** (2.635 > 1.645)
1%	2.326	**Reject** (2.635 > 2.326)

$p = 0.004$. Reject H_0: pain reduction increases significantly with therapy intensity.

Practitioner Insight: The Jonckheere–Terpstra (JT) test targets ordered alternatives $\theta_1 \le \cdots \le \theta_k$, exploiting known group ordering (e.g., dose, time) ignored by Kruskal–Wallis (KW). The statistic J sums Mann–Whitney (MW) counts U_{ij} over all ordered pairs; large J indicates consistent dominance of higher-indexed groups. In the example, KW gives $H = 9.65$, $p = 0.008$ (2-sided, $df = 2$), while JT gives $p = 0.004$ (1-sided), reflecting greater power from focusing on the ordered tail rather than all orderings. This advantage increases with k; for $k = 2$, both reduce to the MW test. The normal approx. is used for illustration; exact enumeration is available for small samples. The data were constructed with overlapping adjacent groups ($U_{12} = U_{23} = 13$) and complete separation between Groups 1 and 3 ($U_{13} = 16$), yielding $J = 42$, well above the 1% CV. Ties are handled by adding $\frac{1}{2}$ per tied pair to U_{ij} and adjusting $\mathrm{Var}[J]$; none occur here.

Common Pitfalls

- **Determining the group ordering after examining the data.** The JT test requires a pre-specified ordering of groups based on theory or study design. Choosing the order from observed sample statistics inflates Type I error.
- **Using the test when the underlying trend may be non-monotonic.** The procedure is designed to detect monotonic increases or decreases across ordered groups. Non-monotonic relationships are better examined using more general rank-based or regression approaches.
- **Forgetting that the hypothesis is directional.** The Jonckheere–Terpstra test evaluates a specified increasing or decreasing trend. The direction of the alternative hypothesis should therefore be determined before analysis.
- **Applying the test to matched or repeated-measures data.** The method assumes independent samples across the ordered groups. Within-subject designs require alternative procedures designed for dependent observations.
- **Interpreting a significant result as evidence of causation or linearity.** A significant test indicates monotonic stochastic ordering among the groups. It does not establish causal relationships or imply a strictly linear effect.

Page's Trend Test

Ordered Alternative Test for Repeated Measures

NPAR.8 | **Category:** Trend Detection | **Data:** Continuous or Ordinal | **Null:** Exact or Asymp. Normal

When to Use

- Test ordered trend across related treatments or conditions.
- Repeated-measures or block designs with known ordering.
- Prior expectation of monotonic increase or decrease.
- More powerful than Friedman for ordered alternatives.
- Detect dose–response or time trend within subjects.

Assumptions

- Blocks (subjects) are independent of each other.
- Complete data: each block contains all treatments.
- Treatments have a pre-specified natural ordering.
- Responses are at least ordinal (rankable within blocks).
- Few or no ties within blocks, or tie correction applied.

Test Statistic

$$z = \frac{L - E[L]}{\sqrt{\text{Var}[L]}}, \text{ where } L = \sum_{j=1}^{k} j\, R_j, \quad E[L] = \frac{nk(k+1)^2}{4}, \quad \text{Var}[L] = \frac{nk^2(k+1)^2(k-1)}{144}.$$

Notation: k = number of ordered conditions, n = number of blocks, R_j = rank sum for condition j across blocks.

Plain English: Page's test detects whether responses tend to increase or decrease systematically across ordered conditions. Instead of asking whether all conditions differ (as Friedman does), it asks a sharper question: *Do the responses follow the expected order?* By weighting rank sums according to the hypothesized order, the test gains power when a monotonic trend truly exists.

⚠ Practitioner Warning

The ordering of conditions must be specified *before* data analysis. Using Page's test to discover a trend post hoc inflates Type I error. If no natural ordering exists, Friedman's test is more appropriate. Page's test detects monotonic trends only—it may miss non-monotone patterns. Violations of the assumed order (e.g., reversals in some blocks) can substantially reduce power and lead to non-significant results even when differences exist. Interpret non-significant results cautiously, especially when the ordering assumption is uncertain.

Hypotheses

- H_0: No ordered trend across conditions.
- H_1: Responses follow the specified monotonic order.

Decision Rule: Reject H_0 if L is sufficiently large. Use exact tables for small n, k or normal approx. for larger samples.

Software References [Python, R, SAS]

Python	R	SAS
`scipy.stats.page_trend_test`	`clinfun::page.test()`	`Not Native (Manual)`

Additional Notes

Page's statistic L is computed by weighting Friedman rank sums R_j by the hypothesized treatment order $j = 1, \ldots, k$, i.e., $L = \sum_{j=1}^{k} j\, R_j$. Large values of L indicate that higher-ordered conditions consistently receive higher within-block ranks. For moderate to large samples, L is standardized as $Z = (L - E(L))/\sqrt{\text{Var}(L)} \approx N(0,1)$, with mean and variance computed under H_0; exact critical values (Page, 1963) are available for small n and $k \leq 8$. Ties within blocks are handled using mid-ranks with a variance correction analogous to the Friedman adjustment. Page's test can be viewed as a directed (ordered) extension of Friedman's test: it concentrates power on the specified trend and can be substantially more powerful when the ordering is correct, but it does not test for general differences among conditions. For $k = 2$, it reduces to the sign test.

Miscellanea: Page's trend test (1963), due to E. B. Page, is unrelated to E. S. Page (1954), known for the CUSUM procedure, despite the shared surname. The test can be viewed as a natural extension of Friedman's test: instead of treating treatments equally, it incorporates prior ordering through simple linear weights, concentrating power on monotonic trends. Despite its simplicity, the test is widely used in areas such as sensory evaluation, clinical dose–response studies, and machine learning model comparisons across datasets.

Reference(s): Page (1963); Hollander et al. (2014)

Worked Example

Problem Statement Five patients each receive three rehabilitation intensities—Low (L), Medium (M), and High (H)—in randomized order, with a washout period between each. A mobility improvement index (higher = better) is recorded after each course. The scores are not assumed to be normally distributed. Apply Page's trend test to decide at $\alpha = 0.05$ whether median improvement increases monotonically across the three intensity levels.

$$H_0: \theta_L = \theta_M = \theta_H; \qquad H_1: \theta_L \le \theta_M \le \theta_H \text{ (at least one strict).}$$

1. Rank Within Each Block

Assign ranks 1, 2, 3 to the three scores within each patient, smallest to largest.

	Raw Scores			Ranks		
Block	L	M	H	L	M	H
1	12	18	25	1	2	3
2	10	16	22	1	2	3
3	15	20	28	1	2	3
4	19	14	24	2	1	3
5	23	17	21	3	1	2
R_j				8	8	14

Check: $R_L + R_M + R_H = 30 = nk(k+1)/2$. ✓

2. Page Statistic

Multiply each R_j by its hypothesized position $j = 1, 2, 3$:

$$L = \sum_{j=1}^{k} j\,R_j = 1 \times 8 + 2 \times 8 + 3 \times 14 = 66.$$

3. Normal Approximation

With $n = 5$ blocks and $k = 3$ treatments:

$$E[L] = \frac{nk(k+1)^2}{4} = \frac{5 \times 3 \times 16}{4} = 60.$$

$$\text{Var}[L] = \frac{nk^2(k+1)^2(k-1)}{144} = \frac{5 \times 9 \times 16 \times 2}{144} = 10.$$

$$z = \frac{L - E[L]}{\sqrt{\text{Var}[L]}} = \frac{66 - 60}{\sqrt{10}} = 1.897.$$

4. Decision

The test is one-sided (upper tail).

α	z_α	Decision
5%	1.645	**Reject** (1.897 > 1.645)
1%	2.326	FTR (1.897 < 2.326)

$p = 0.029$ (Exact $p = 0.0394$). Reject H_0 at the 5% level: evidence that median mobility improvement increases monotonically with rehabilitation intensity.

Practitioner Insight: Page's test is to the Friedman test what Jonckheere–Terpstra is to Kruskal–Wallis: it exploits a known ordering of k treatments to concentrate power on an ordered alternative, but is appropriate only when an *a priori* ordering is specified. The statistic L is the Friedman rank sums R_j weighted by treatment positions $j = 1, \ldots, k$; large L indicates higher-indexed treatments achieve higher within-block ranks. In the example, Friedman gives $\chi^2_F = 4.80$, $p = 0.091$ ($df = 2$, 2-sided), failing to reject at 5%, while Page's test gives $z = 1.897$ (1-sided) and $p = 0.029$ under the normal approximation, illustrating its power advantage. The design is instructive: Patients 1–3 follow the expected order (Low < Medium < High), Patient 4 reverses Low and Medium, and Patient 5 switches the Low and High; these deviations reduce L from its maximum of 70 by 4, leaving enough signal to reject at 5% but not 1%. For small n or k, exact critical values are preferred; the normal approximation improves as the number of blocks increases. Ties are handled by mid-ranks with an adjustment to the variance of L; none occur here. For $k = 2$, Page's test reduces to the sign test.

Common Pitfalls

- **Choosing post hoc ordering.** Page's test requires a *pre-specified* ordering of conditions. Ordering based on observed medians or ranks inflates Type I error. If no defensible a priori ordering exists, use the Friedman test instead.
- **Applying it to independent groups.** The test assumes repeated or matched measurements on the same subjects. For ordered but independent groups, the Jonckheere–Terpstra test is appropriate. Using Page's test with independent samples violates its core assumptions.
- **Ignoring the one-sided nature of the test.** Page's test targets a specific directional trend (increasing or decreasing). Direction must be declared before analysis; testing both directions or switching post hoc invalidates inference.
- **Using it when the true pattern is non-monotonic.** The test is designed for monotonic trends only. Biphasic, plateau, or inverted-U patterns may go undetected. Use Friedman or planned contrasts with non-monotonic alternatives.
- **Including incomplete blocks.** Page's test requires complete blocks where all subjects have data for all conditions. Missing values prevent proper ranking and require omitting incomplete blocks or using alternative modeling approaches.

VI. NPAR — Quade's Test

Quade's test addresses the same problem as the Friedman test and is a direct weighted refinement of it: the only structural difference is that Quade weights each block's contribution by the rank of its observed range, whereas Friedman treats all blocks equally. The author chose not to include it as a main-chapter entry for three reasons: it is rarely encountered outside specialised nonparametric textbooks; presenting it alongside Friedman would risk obscuring rather than clarifying that close relationship; and software support is uneven — R provides `quade.test()` *in base* `stats`*, but SAS, SPSS, and Python have no native implementation. The entry below walks through a complete example using the same data as the Friedman worked example.*

What Quade's Test Does Differently

The Friedman test assigns equal weight to every block: ranks from a tightly clustered block (range = 5) count the same as those from a widely spread block (range = 40). Quade (1979) argued that a larger spread signals a block in which treatment differences are more reliably expressed, and should carry more influence. He achieves this by multiplying each block's centred rank scores by the *rank* of that block's range across all blocks, producing weighted scores W_{ij}. The statistic follows an exact F distribution under H_0 rather than the asymptotic χ^2 approximation used by Friedman, offering better size control for small n.

Worked Example *Data* (same as Friedman — six athletes, three regimes, higher score is better):

Step 1: Rank within each block.
Replace each row with ranks 1, 2, 3 assigned to the three regime scores in increasing order.

	Raw Scores			Ranks		
Block	A	B	C	A	B	C
1	55	68	79	1	2	3
2	48	61	74	1	2	3
3	62	70	83	1	2	3
4	57	75	66	1	3	2
5	71	64	80	2	1	3
6	66	59	77	2	1	3
R_j				8	11	17

Step 3: Weighted scores $W_{ij} = S_i(r_{ij} - 2)$.
$(k + 1)/2 = 2$ is the mean rank for $k = 3$; centering makes each block's row sum to zero, then S_i scales it:

Blk	S_i	W_{iA}	W_{iB}	W_{iC}
1	5.0	−5.0	0	+5.0
2	6.0	−6.0	0	+6.0
3	4.0	−4.0	0	+4.0
4	2.5	−2.5	+2.5	0
5	1.0	0	−1.0	+1.0
6	2.5	0	−2.5	+2.5
$W_{\cdot j}$		−17.5	−1.0	+18.5

Check: $-17.5 - 1.0 + 18.5 = 0$. ✓

Step 2: Compute block ranges and rank them.
$Q_i = \max_j y_{ij} - \min_j y_{ij}$. Rank the $n = 6$ ranges smallest to largest; ties get average ranks. Call these ranks S_i:

Block	Min	Max	Q_i	S_i
1	55	79	24	5.0
2	48	74	26	6.0
3	62	83	21	4.0
4	57	75	18	2.5 (tied)
5	64	80	16	1.0
6	59	77	18	2.5 (tied)

Blocks 4 and 6 share range 18, so each receives $S_i = 2.5$.

Step 4: Quade F-statistic.

$$\text{SS}_{\text{tot}} = \textstyle\sum_{i,j} W_{ij}^2 = 181.00, \quad \text{SS}_{\text{trt}} = \frac{1}{n}\sum_j W_{\cdot j}^2 = 108.25$$

$$\text{SS}_{\text{err}} = 181.00 - 108.25 = 72.75, \quad F_Q = \frac{(n-1)\,\text{SS}_{\text{trt}}}{\text{SS}_{\text{err}}} = 7.44$$

Under H_0: $F_Q \sim F(k-1,\ (n-1)(k-1)) = F(2, 10)$.

α	$F_{0.05,\,2,10}$	$F_{0.01,\,2,10}$	Decision
5%	4.103	—	**Reject** (7.44 > 4.103)
1%	—	7.559	FTR (7.44 < 7.559)

$p = 0.011$. Reject H_0 at the 5% level.

Comparison with Friedman:

Test	Statistic	p-value	Distribution
Friedman	$\chi^2_F = 7.00$	0.030	$\chi^2(2)$ (approx.)
Quade	$F_Q = 7.44$	0.011	$F(2, 10)$ (approx.)

Both reject at 5%. Quade (1979) established that W is strictly distribution-free and has an asymptotic $\chi^2(k-1)$ null distribution; the $F(k-1,\ (n-1)(k-1))$ approximation used here follows Conover's textbook formulation and is implemented in R's `quade.test()`. It tends to give better agreement with the permutation distribution than the raw χ^2 approximation for small n, but it is an approximation in the same sense as Friedman's χ^2 — neither is exact. Quade's smaller p reflects that the widest-spread blocks ($S_i = 5, 6$) show the clearest A < B < C ordering, so upweighting them strengthens the signal. Regime C records the largest weighted column sum and Regime A the most negative, confirming the same ordering as Friedman.

Software. R: `quade.test()` in base `stats`; supply a matrix with blocks as rows. SAS, SPSS, Python: no native implementation. *Reference:* Quade (1979), *JASA* **74**(367), 680–683.

VI. NPAR — Corrections for Ties Corrections

When tied observations occur, midranks are assigned and a correction is applied to the variance or test statistic.
Notation: N = total sample size; n_i = group i size; g = tied groups; t_j = size of j-th tied group; b = blocks; k = treatments.

Test	Corrects	Formula and notes
Sign Test	N/A	Tied pairs (zero differences) are discarded; the reduced sample n' is used directly in the binomial or normal approximation.
Wilcoxon Signed-Rank	$\mathrm{Var}(T^+)$	Let g be the number of tied groups among the non-zero differences, with t_j the size of the j-th group. Zero differences are excluded before ranking: $\mathrm{Var}(T^+) = \dfrac{n(n+1)(2n+1)}{24} - \dfrac{1}{48}\sum_{j=1}^{g}(t_j^3 - t_j)$ where n = number of non-zero differences.
Mann-Whitney	$\mathrm{Var}(U)$	Let g be the number of tied groups in the combined sample $N = n_1 + n_2$: $\mathrm{Var}(U) = \dfrac{n_1 n_2}{12}\left[N + 1 - \sum_{j=1}^{g}\dfrac{t_j^3 - t_j}{N(N-1)}\right]$
Mood's Median	N/A	Chi-square based on a $k \times 2$ table; unaffected by ties. Ties at the grand median are excluded or split equally.
Kruskal-Wallis	Statistic H	The uncorrected H is divided by the correction factor $C = 1 - \dfrac{\sum_{j=1}^{g}(t_j^3 - t_j)}{N^3 - N}$ giving $H_c = H/C \geq H$. Refer H_c to χ^2_{k-1}.
Friedman	Statistic χ^2_r	The uncorrected χ^2_r is divided by the correction factor $C = 1 - \dfrac{\sum_{i=1}^{b}\sum_{j} t_{ij}(t_{ij}^2 - 1)}{bk(k^2-1)}$ giving $\chi^2_{r,c} = \chi^2_r / C \geq \chi^2_r$, where t_{ij} is the size of the j-th tied group within block i. Refer $\chi^2_{r,c}$ to χ^2_{k-1}.
Jonckheere- Terpstra	$\mathrm{Var}(J)$	Tied cross-group pairs are counted as $\frac{1}{2}$. The mean is unchanged: $E(J) = \frac{1}{2}\sum_{i<j} n_i n_j$. The tie-corrected variance is $\mathrm{Var}(J) = \dfrac{N(N-1)(2N+5) - \sum_i n_i(n_i-1)(2n_i+5) - \sum_j t_j(t_j-1)(2t_j+5)}{72} + \dfrac{\left[\sum_i n_i(n_i-1)(n_i-2)\right]\left[\sum_j t_j(t_j-1)(t_j-2)\right]}{36N(N-1)(N-2)} + \dfrac{\left[\sum_i n_i(n_i-1)\right]\left[\sum_j t_j(t_j-1)\right]}{8N(N-1)}$ where $N = \sum_i n_i$, and t_j is the size of tie group j in the pooled response values.
Page's Trend	$\mathrm{Var}(L)$ (asymptotic)	For the asymptotic approximation where $L = \sum_j jR_j$: $\mathrm{Var}(L) = \dfrac{bk^2(k^2-1)(k+1)}{144}\cdot\left(1 - \dfrac{\sum_{i=1}^{b}\sum_j t_{ij}(t_{ij}^2-1)}{bk(k^2-1)}\right)$ For the exact distribution, use midranks with a permutation argument.

References: Hollander, Wolfe & Chicken (2014), *Nonparametric Statistical Methods*, 3rd ed., Wiley. Conover (1999), *Practical Nonparametric Statistics*, 3rd ed., Wiley. Lehmann (2006), *Nonparametrics: Statistical Methods Based on Ranks*, Springer.

Part VII

Association, Dependence & Linear Relations

Measuring strength and direction of relationships between variables

Correlation and Association

VII. CORR — A Conceptual Primer

What Are We Measuring?

- The **strength** and **direction** of statistical association between two (or more) variables.
- Core question: *Do variables move together, and how strongly?*
- Notation: ρ (population correlation), r (sample estimate).
- **Critical distinction:** correlation $\neq$ causation. A strong association may be spurious, confounded, or driven by a third variable.

Three Levels of Association

- **Linear**: Y changes at a constant rate per unit of X; measured by Pearson's r.
- **Monotonic**: Y consistently increases (or decreases) with X, but not necessarily at a constant rate; measured by Spearman's ρ_s and Kendall's τ.
- **General dependence**: any functional or statistical relationship, including non-monotone and multivariate patterns; measured by distance correlation $\mathcal{R}$ and Hoeffding's D.

Pearson's r — Linear Association

$$r = \frac{\sum_i (x_i - \bar{x})(y_i - \bar{y})}{\sqrt{\sum_i (x_i - \bar{x})^2 \sum_i (y_i - \bar{y})^2}} = \frac{S_{xy}}{\sqrt{S_{xx} S_{yy}}}.$$

- $r \in [-1, 1]$; $r = \pm 1$ implies a perfect linear relationship.
- r^2 (**coefficient of determination**) is the proportion of variance in Y linearly explained by X.
- **Assumptions:** bivariate normality (for exact inference), linearity, no severe outliers (a single outlier can shift r dramatically).
- Independence $\Rightarrow r = 0$, but $r = 0$ does *not* imply independence (nonlinear dependence can exist with $r = 0$).

Rank-Based Measures

Spearman's ρ_s: Pearson's r applied to the **ranks** of x and y. When no ties:

$$\rho_s = 1 - \frac{6 \sum_i d_i^2}{n(n^2 - 1)}, \quad d_i = R(x_i) - R(y_i).$$

Kendall's τ: based on **concordant** (C) and **discordant** (D) pairs:

$$\tau_b = \frac{C - D}{\sqrt{(C + D + T_x)(C + D + T_y)}},$$

where T_x, T_y are within-variable tie counts. The probability interpretation $\tau = P(\text{concordant}) - P(\text{discordant})$ makes τ more intuitive than ρ_s; it converges more slowly to normality.

Both ρ_s and τ are more robust to outliers and non-normality; prefer them over Pearson when linearity is uncertain.

Specialized Correlation Measures

Measure	Variable types & key idea
Point-biserial r_{pb}	Continuous + natural binary (0/1); equals Pearson r on the binary variable; related to t-test: $t = r_{pb}\sqrt{(n-2)/(1-r_{pb}^2)}$.
Biserial r_b	Continuous + artificially dichotomized continuous; corrects r_{pb} upward assuming latent normality; $r_b = r_{pb} \cdot \sqrt{pq}/\phi(z_c)$.
Tetrachoric r_t	Binary × binary; assumes latent bivariate normal; estimates the *underlying* ρ before dichotomisation.
Polychoric r_{pc}	Ordinal × ordinal; generalizes tetrachoric to ordered categories; assumes latent bivariate normal with threshold model.
Cramér's V	Nominal × nominal; $V = \sqrt{X^2/[n(\min(r,c)-1)]} \in [0,1]$; generalizes ϕ to $r \times c$ tables.

General Dependence Measures

Distance correlation $\mathcal{R}(X, Y)$:

- $\mathcal{R} \in [0, 1]$; $\mathcal{R} = 0 \Leftrightarrow$ complete independence (unlike Pearson).
- Detects *any* dependence structure—linear, nonlinear, non-monotone.
- Computationally based on pairwise Euclidean distances; generalizes naturally to multivariate X and Y.

Hoeffding's D:

- Tests H_0: $X \perp Y$ vs. any form of dependence.
- Based on the difference between the joint and product-marginal CDFs; consistent against all alternatives with continuous distributions.
- Scaled to $[-0.5, 1]$; values near 1 indicate strong dependence; near 0 indicates independence.

Partial Correlation

Measures the linear association between X and Y **after removing the linear effect of a third variable** (or set of variables) $\mathbf{Z}$:

$$r_{xy \cdot z} = \frac{r_{xy} - r_{xz} r_{yz}}{\sqrt{(1 - r_{xz}^2)(1 - r_{yz}^2)}}.$$

Generalizes to multiple control variables via the inverse of the correlation matrix: $\mathbf{R}^{-1}$. The partial correlation between X_i and X_j controlling for all others is

$$r_{ij \cdot \text{rest}} = \frac{-R^{ij}}{\sqrt{R^{ii} R^{jj}}},$$

where R^{ij} are elements of $\mathbf{R}^{-1}$. Essential for disentangling direct from spurious associations in observational data.

Hypothesis Testing and Inference

Pearson t-test:

$$T = \frac{r\sqrt{n-2}}{\sqrt{1-r^2}} \sim t_{n-2} \quad (H_0 : \rho = 0).$$

Fisher's z-transformation: stabilizes the variance of r for any ρ:

$$z = \tfrac{1}{2}\ln\left(\frac{1+r}{1-r}\right) = \tanh^{-1}(r), \quad z \mathrel{\dot\sim} \mathcal{N}\left(\tfrac{1}{2}\ln\frac{1+\rho}{1-\rho}, \frac{1}{n-3}\right).$$

Use Fisher's z for: (i) constructing CIs for ρ, (ii) testing $H_0 : \rho = \rho_0 \neq 0$, and (iii) comparing two independent correlations.

Comparing two independent correlations (Fisher):

$$Z = \frac{z_1 - z_2}{\sqrt{1/(n_1-3) + 1/(n_2-3)}} \mathrel{\dot\sim} \mathcal{N}(0,1).$$

Steiger's Z-test (dependent correlations): tests whether two correlations sharing a common variable (e.g. ρ_{xy} vs. ρ_{xz} from the same sample) are equal. Requires the correlation between y and z (r_{yz}) and uses a corrected variance formula accounting for the dependency. Cannot be replaced by independent-samples Fisher Z.

Choosing the Right Measure (See Technical Notes)

Variable types / situation	Recommended measure
Continuous, linear, normal	Pearson r
Continuous, non-linear/outliers	Spearman ρ_s
Ordinal, probability interp.	Kendall τ_b
Any nonlinear dependence	Distance corr. $\mathcal{R}$
Omnibus independence test	Hoeffding's D
Continuous + natural binary	Point-biserial r_{pb}
Continuous + dich. continuous	Biserial r_b
Binary × binary	Tetrachoric r_t
Ordinal × ordinal	Polychoric r_{pc}
Nominal × nominal	Cramér's V
Controlling for confounders	Partial correlation
Compare 2 independent r's	Fisher's Z
Compare 2 dependent r's	Steiger's Z

General Interpretation Guidelines/Heuristics

$\lvert r\rvert$	Strength	Field note
< 0.10	Negligible	
0.10–0.29	Weak	Social science
0.30–0.49	Moderate	common range
0.50–0.69	Strong	
≥ 0.70	Very strong	Rare in observational data

Context matters: $r = 0.30$ is noteworthy in psychology but negligible in engineering. Always interpret alongside r^2 and a scatterplot.

Assumptions Summary

Method	Key assumptions
Pearson	Linearity; bivariate normality for exact t; no severe outliers
Spearman	Monotonic relationship; continuous or ordinal
Kendall	Monotonic; handles ties well
Point-biserial	Normality within binary groups
Tetrachoric	Latent bivariate normal; two thresholds
Polychoric	Latent bivariate normal; ordered thresholds
Distance corr.	Finite second moments; any dimension
Hoeffding	Continuous joint distribution; no ties

Common Pitfalls

- **Nonlinearity:** Pearson ≈ 0 does not mean no relationship—always plot first.
- **Outliers:** a single outlier can change r from 0 to ± 0.8; check with Spearman.
- **Range restriction:** a truncated sample underestimates $|\rho|$; correction formulas exist.
- **Ecological correlation:** correlations at the group level do not imply individual-level associations (Simpson's paradox, aggregation bias).
- **Scale mismatch:** applying Pearson to ordinal or binary data without the appropriate correction.
- **Comparing dependent correlations** with Fisher Z (requires Steiger's Z instead).
- **Ignoring confounders:** marginal correlation may vanish or reverse after controlling for $\mathbf{Z}$.

Visual Diagnostics

- **Scatterplot:** primary tool; reveals linearity, curvature, outliers, clusters, and heteroscedasticity.
- **Correlation matrix heatmap:** for multiple variables; quickly identifies clusters of associated pairs.
- **Partial regression (added-variable) plot:** visualizes the marginal relationship after controlling for other predictors.

Bottom Line

Correlation measures association, not causation. The choice of method depends on variable types, linearity, robustness needs, and whether confounders must be controlled. Default to **Spearman** when linearity is uncertain; use Pearson only when linearity is credible and outliers are absent. Always accompany any correlation with a scatterplot and a confidence interval, not just a p-value.

Pearson's Correlation Test

Linear Association Between Two Continuous Variables

CORR.1 | **Category:** Association | **Data:** Continuous | **Null:** t-distribution (exact under normality)

When to Use

- Measure strength of linear association between variables.
- Quantify direction and magnitude of linear relation.
- Continuous measurements on paired observations.
- Assess relationships before regression modeling.
- Test whether population correlation differs from zero.

Assumptions

- Paired observations are mutually independent.
- Relationship between variables is approximately linear.
- Variables are measured on interval or ratio scales.
- No influential outliers distort the association.
- Bivariate normality holds for exact inference.

Test Statistic

$$t = r\sqrt{\frac{n-2}{1-r^2}} \sim t_{n-2} \quad \text{under } H_0. \qquad r = \frac{\sum_{i=1}^{n}(x_i - \bar{x})(y_i - \bar{y})}{\sqrt{\sum_{i=1}^{n}(x_i - \bar{x})^2}\sqrt{\sum_{i=1}^{n}(y_i - \bar{y})^2}}$$

Notation: r sample correlation, n sample size, (x_i, y_i) paired data, $\bar{x}, \bar{y}$ sample means.

Plain English: Pearson's correlation quantifies how strongly two variables move together in a straight-line relationship. Values near $+1$ indicate strong positive association, near -1 strong negative association, and near 0 little linear relationship. The test evaluates whether the observed correlation could plausibly arise if the true population correlation were zero. Testing Pearson's correlation is equivalent to testing the slope in simple linear regression. Estimating the intercept and slope uses two degrees of freedom, leaving $n-2$ for variability. Hence, the correlation t-statistic follows a t_{n-2} distribution.

⚠ Practitioner Warning

Pearson's correlation measures only *linear* association. Strong nonlinear relationships can produce $r \approx 0$. The test is sensitive to outliers, which can dominate both the magnitude and sign of the correlation. Inference relies on approximate normality; when violated, consider rank-based alternatives such as Spearman or Kendall correlation. Correlation does not imply causation, and apparent relationships may arise from confounding variables or hidden subgroups. Restriction of range can substantially reduce the observed correlation, even when a strong association exists in the population.

Hypotheses

- H_0: $\rho = 0$ (no linear association)
- H_1: $\rho \neq 0$ (linear association exists)

Decision Rule: Reject H_0 if $|t| > t_{\alpha/2, n-2}$ or p-value $< \alpha$. Report r, confidence interval, and sample size.

Software References [Python, R, SAS]

```
scipy.stats.pearsonr          cor.test(method="pearson")          PROC CORR
```

Additional Notes

Pearson's r is invariant to location and scale changes but not to nonlinear transformations. The squared correlation, r^2, equals the coefficient of determination in simple linear regression and represents the proportion of variance explained. Confidence intervals and tests for the population correlation ρ typically use Fisher's z-transformation, which stabilizes the variance of r. For large samples, inference is generally robust to mild nonnormality. Correlation is symmetric: exchanging X and Y leaves r unchanged. Thus, Pearson's correlation measures the strength of linear association but does not imply directionality.

Miscellanea: Karl Pearson formalized the product–moment correlation in 1896 while extending Francis Galton's ideas on heredity and regression. Galton had discovered the "regression toward mediocrity" phenomenon but lacked the mathematical machinery; Pearson supplied it, turning Galton's empirical insight into a general statistical framework. Their collaboration later led to the founding of *Biometrika* (1901), the first journal devoted to statistical methodology. Pearson repeatedly warned that correlation measures association, not causation—a warning reinforced decades later by **Anscombe's quartet** (1973), where four datasets share identical correlations (and other summary statistics) yet exhibit strikingly different patterns when plotted.

Reference(s): Pearson (1896); Fisher (1921)

Pearson's Correlation Test

CORR.1

Worked Example

Problem Statement A researcher records eight paired observations of study hours (x) and exam score (y) for eight students. Test at $\alpha = 0.05$ whether study hours and exam score are linearly correlated. $H_0: \rho = 0$; $H_1: \rho \neq 0$.
$S_{xx} = \sum(x_i - \bar{x})^2$, $S_{yy} = \sum(y_i - \bar{y})^2$, $S_{xy} = \sum(x_i - \bar{x})(y_i - \bar{y})$.

Student	1	2	3	4	5	6	7	8	**Sum**
x_i	2	3	4	4	6	6	7	8	40
y_i	2	4	3	4	7	8	6	6	40

1. Corrected Sums of Squares and Cross-products

$$S_{xx} = 30, \quad S_{yy} = 30, \quad S_{xy} = 24.$$

Note: $S_{xx} = S_{yy} = 30$ because both variables have equal spread around their common mean $\bar{x} = \bar{y} = 5$.

2. Pearson Correlation Coefficient

$$r = \frac{S_{xy}}{\sqrt{S_{xx}\, S_{yy}}} = \frac{24}{\sqrt{30 \times 30}} = 0.8.$$

3. Test Statistic

Under $H_0: \rho = 0$, the statistic

$$t = \frac{r\sqrt{n-2}}{\sqrt{1-r^2}} = \frac{4\sqrt{6}}{3} \approx 3.266.$$

4. Decision

Under H_0: $t \sim t(n-2) = t(6)$.

α	$t_{(6),\,\alpha/2}$	Decision
5%	2.447	**Reject** (3.266 > 2.447)
1%	3.707	FTR (3.266 < 3.707)

$p = 0.017$. There is significant evidence of a positive linear association between study hours and exam score at the 5% level.

5. Coefficient of Determination Coefficient of determination is $r^2 = 0.64$. Study hours explain 64% of the variation in exam scores in this sample.

Practitioner Insight: Pearson's correlation coefficient r measures the strength and direction of a *linear* relationship between two continuous variables. Values of ± 1 occur only when all points lie exactly on a straight line; $r = 0$ indicates no linear association but does not rule out nonlinear dependence. Under bivariate normality, the statistic $r\sqrt{n-2}/\sqrt{1-r^2}$ follows an exact $t(n-2)$ distribution, providing a test of $H_0: \rho = 0$. Confidence intervals for ρ are usually obtained via Fisher's z-transformation, which stabilizes the variance of r. Pearson's r is sensitive to outliers and range restriction; for ordinal data or influential observations, Spearman's ρ or Kendall's τ are often preferable.

Common Pitfalls

- **Inferring causation from correlation.** Pearson's r measures association only. Significant correlation is compatible with multiple causal structures or confounding. Avoid causal language unless supported by design (temporal ordering, randomization) or additional evidence.
- **Applying it to non-linear relationships.** Pearson's r captures linear association. Strong non-linear or monotonic patterns can yield $r \approx 0$. Always inspect a scatterplot; use Spearman correlation, transformations, or non-linear models when relationships deviate from linearity.
- **Allowing outliers to distort results.** Pearson's r is highly sensitive to extreme points, which can create or mask correlations. Examine outliers carefully and consider robust alternatives such as Spearman or Kendall correlations when influential points are present.
- **Equating statistical significance with practical importance.** Large samples can produce tiny but statistically significant correlations; small samples can miss meaningful ones. Report and interpret both r and p-values, emphasizing effect magnitude and context rather than significance alone.
- **Ignoring the bivariate normality assumption.** The correlation test assumes joint normality of (X, Y). Severe non-normality can distort p-values and intervals. Use Spearman correlation, permutation tests, transformations, or bootstrap methods when normality is doubtful.

Correlation as a Cosine

Pearson's correlation has a geometric interpretation. After centering the data, r equals the cosine of the angle between the two data vectors: $r = \cos(\theta)$. Thus $r = 1, 0,$ and -1 correspond to angles of $0°$, $90°$, and $180°$, respectively.

Spearman's Rank Correlation Test

Nonparametric Measure of Monotonic Association

CORR.2 | **Category:** Association | **Data:** Ordinal or Continuous | **Null:** Exact (Perm.) / t or Normal

When to Use

- Assess strength of monotonic association between variables.
- Ordinal data or ranked measurements are available.
- Nonlinear but monotonic relationships are expected.
- Outliers may distort Pearson correlation estimates.
- Test whether rank correlation differs from zero.

Assumptions

- Paired observations are mutually independent.
- Relationship is monotonic over the observed range.
- Variables are measured on at least ordinal scale.
- Ranks are meaningful representations of ordering.
- Ties are absent or handled using corrections.

Test Statistic

$$r_s = 1 - \frac{6\sum_{i=1}^{n} d_i^2}{n(n^2-1)} \quad \text{(no ties)}. \quad \text{For large } n, t = \frac{r_s\sqrt{n-2}}{\sqrt{1-r_s^2}} \sim t_{n-2} \quad \text{or} \quad r_s \sim N(0, \frac{1}{n-1}).$$

Notation: n sample size, $R(X_i), R(Y_i)$ ranks, $d_i = R(X_i) - R(Y_i)$.

Plain English: Spearman's rank correlation measures how well two variables move together in a monotonic way. Instead of using raw values, it replaces data with ranks and computes correlation on those ranks. This makes the test robust to outliers and non-normality, and sensitive to any consistent increasing or decreasing relationship—even when the relationship is nonlinear.

⚠ Practitioner Warning

Spearman's test detects monotonic association, not linearity. A strong nonlinear monotonic relationship can produce a high r_s, while a non-monotonic relationship (e.g., U-shaped) may yield $r_s \approx 0$ despite clear dependence. With ties, r_s is computed as the Pearson correlation of rank variables. Large numbers of ties reduce power and require adjusted variance formulas.

Hypotheses

- H_0: $\rho_s = 0$ (no monotonic association).
- H_1: $\rho_s \neq 0$ (monotonic association present).

Decision Rule: Reject H_0 if $|r_s|$ is large or $p < \alpha$. Exact p-values for small n; normal approximation for large samples.

Software References [Python, R, SAS]

```
scipy.stats.spearmanr        cor.test(method="spearman")        PROC CORR (SPEARMAN)
```

Additional Notes

Spearman's r_s (1904) predates modern nonparametric statistics and was developed for settings where Pearson correlation assumptions fail. It is exactly Pearson's r applied to ranked data and is invariant under strictly monotone transformations. Under bivariate normality and linearity, it is slightly less efficient than Pearson's r (ARE = $9/\pi^2 \approx 0.91$), meaning it requires about 9% more data for the same power. However, under many real-world distributions it performs as well or better: ARE $\approx$ 1.00 (Uniform), $\approx$ 1.05 (Logistic), and $>$ 1.20 for heavy-tailed distributions (Laplace, low-df t), while Pearson's r can fail entirely for extreme cases like the Cauchy. Ties are handled via average ranks (standard practice). For any perfect monotone relationship (not necessarily linear), $r_s = \pm 1$. It is closely related to Kendall's τ, but is more widely used due to computational simplicity. Exact p-values are tabulated only up to $n \leq 30$. Under H_0, Spearman's r_s follows an exact permutation distribution over all $n!$ rank arrangements, yielding a discrete distribution symmetric about 0. Used for small samples to obtain exact p-values.

Miscellanea: Charles Spearman introduced rank correlation in 1904 while studying intelligence tests, seeking a measure of association that did not rely on normality or precise measurement scales. His early "foot-rule" statistic, based on absolute rank differences, soon evolved into the familiar ρ. Spearman's key insight—that ordering alone can reveal dependence—was decades ahead of modern nonparametric statistics. His work not only laid the foundation for robust correlation methods but also influenced his broader theory of a general intelligence factor ("g"), and later inspired developments such as Kendall's τ.

Reference(s): Spearman (1904); Conover (1999)

Spearman's Rank Correlation Test

Worked Example

Problem Statement Nine employees are ranked by years of experience (x) and their annual performance score (y, out of 100) is recorded. Test at $\alpha = 0.05$ whether experience and performance are monotonically associated using Spearman's rank correlation. $H_0: \rho_s = 0$; $H_1: \rho_s \neq 0$.

1. Assigning Ranks
For n pairs with no ties, rank both variables and compute $d_i = R(x_i) - R(y_i)$, the difference in ranks for i-th pair.

x_i	y_i	$R(x_i)$	$R(y_i)$	d_i	d_i^2
1	55	1	3	−2	4
2	45	2	1	+1	1
3	50	3	2	+1	1
4	65	4	5	−1	1
5	70	5	6	−1	1
6	90	6	9	−3	9
7	60	7	4	+3	9
8	80	8	8	0	0
9	75	9	7	+2	4
			Sum	0	**30**

(a necessary check: $\sum d_i = 0$ always holds).

2. Computing r_s
With $n = 9$ and $\sum d_i^2 = 30$:

$$r_s = 1 - \frac{6\sum_{i=1}^{n} d_i^2}{n(n^2-1)} = \frac{3}{4}.$$

3. Test Statistic

$$t = \frac{r_s\sqrt{n-2}}{\sqrt{1-r_s^2}} = 3.$$

4. Decision
Under H_0: $t \sim t(n-2) = t(7)$.

α	$t_{(7),\alpha/2}$	Decision
5%	2.365	**Reject** (3 > 2.365)
1%	3.499	FTR (3 < 3.499)

$p = 0.020$. There is significant evidence of a positive monotonic association between experience and performance at the 5% level.

5. Comparison with Pearson
Pearson's r on the raw scores gives $r = 0.716$. The slightly higher $r_s = 0.75$ indicates that the monotonic trend is somewhat stronger than the *linear* trend — Employee 6's unusually high score of 90 is influential in Pearson's calculation but is absorbed into a single rank in Spearman's.

Practitioner Insight: Spearman's r_s measures the strength of a *monotonic* relationship by applying Pearson's correlation to the ranks rather than the raw data. The shortcut formula is valid only when there are no ties; otherwise mid-ranks and the full rank-Pearson formula should be used. Compared with Pearson's r, Spearman's r_s is more robust to outliers and non-normality because only rank order matters. Values of ± 1 indicate perfect increasing or decreasing monotone association, even when the relationship is nonlinear. A limitation is that r_s cannot detect non-monotone patterns such as U-shaped relationships; scatter plots should always be examined alongside the test. Kendall's τ provides a related rank-based measure with a more direct probabilistic interpretation.

Common Pitfalls

- **Assuming it detects all nonlinear relationships.** Spearman's r_s measures monotonic association only. U-shaped, cyclic, or threshold relationships can produce weak correlations despite strong dependence.
- **Ignoring ties.** Tied observations reduce information and should receive average ranks. With many ties, Kendall's τ often provides more reliable inference.
- **Using it when Pearson is more appropriate.** Spearman's r_s is less powerful than Pearson's r for linear relationships under normality. Prefer Pearson when variables are continuous, approximately normal, and linearly related.
- **Confusing association with agreement.** A high Spearman correlation indicates similar rank ordering, not identical measurements. Agreement studies require methods such as ICC, CCC, Bland–Altman, or Cohen's κ.
- **Focusing only on significance.** Large samples can make trivial correlations significant, while small samples may miss meaningful effects. Report the correlation magnitude, confidence interval, and sample size alongside the p-value.

Kendall's τ Test

Rank-Based Measure of Ordinal Association

CORR.3 | **Category:** Association | **Data:** Ordinal or Continuous | **Null:** Exact (Perm.) / t or Normal

When to Use

- Assess strength of monotonic association between variables.
- Ordinal data or ranked measurements are available.
- Small samples where rank-based inference is preferred.
- Many tied values occur in one or both variables.
- Test whether ordinal association differs from zero.

Assumptions

- Paired observations are mutually independent.
- Relationship is monotonic over the observed range.
- Variables are measured on at least ordinal scale.
- Ranks or pairwise orderings are meaningful.
- Ties are absent or handled using corrections.

Test Statistic

$$\tau_a = \frac{C-D}{\binom{n}{2}} = \frac{C-D}{C+D}. \qquad \text{Under } H_0,\ E[\tau_a] = 0, \ \text{ and } \operatorname{Var}(\tau_a) = \frac{2(2n+5)}{9n(n-1)}.$$

Notation: n sample size; concordant: $(x_i - x_j)(y_i - y_j) > 0$; discordant: $(x_i - x_j)(y_i - y_j) < 0$; tie: $(x_i - x_j)(y_i - y_j) = 0$; C, D are the number of concordant and discordant pairs.

Plain English: Kendall's τ quantifies how often pairs of observations move in the same direction versus opposite directions. A value of $\tau = 1$ indicates perfect agreement in ordering, $\tau = -1$ perfect disagreement, and $\tau = 0$ no association. Unlike Pearson's correlation, τ depends only on ranks and is robust to outliers, nonlinearity, and skewed distributions.

⚠ Practitioner Warning

Kendall's τ measures *monotonic association*, not linear strength. A strong nonlinear but monotonic relationship can yield high τ, while non-monotonic patterns may yield values near zero. For large datasets, Tau is more conservative (lower magnitude) than Spearman's ρ but provides better probabilistic interpretation.

Hypotheses

- H_0: $H_0 : \tau = 0$ (no association)
- H_1: $H_1 : \tau \neq 0$ (association present)

Decision Rule: Reject H_0 if $p < \alpha$. Exact test for small n; normal approximation for large n.

Software References [Python, R, SAS]

```
scipy.stats.kendalltau        cor.test(method="kendall")        PROC CORR (KENDALL)
```

Additional Notes

Kendall's τ (1938) developed as alternative to Spearman with better statistical properties. Three main variants: τ_a (original, no tie correction), τ_b (adjusts for ties in both variables, most common), τ_c (for rectangular tables). τ_b used for correlation; τ_c for contingency tables (sometimes called Stuart's τ). Asymptotic relative efficiency (ARE) = 0.98 relative to Pearson under bivariate normality—higher than Spearman! More robust to outliers than Spearman. Better asymptotic normality (smaller sample approximation valid earlier). Historically, Kendall's τ was considered computationally intensive because it requires counting concordant and discordant pairs. Modern $O(n \log n)$ algorithms and computing power have largely eliminated this practical limitation.

$$\tau_b = \frac{C-D}{\sqrt{(C+D+T_x)(C+D+T_y)}}, \text{ where } T_x, T_y \text{ are ties}; \qquad \tau_c = \frac{2(C-D)}{n^2 \cdot \frac{m-1}{m}} \text{ (rectangular contingency tables).}$$

Kendall's τ has a direct probabilistic interpretation: $\tau = P(\text{concordance}) - P(\text{discordance})$, a measure of ordinal association. It is invariant under monotone transformations and less sensitive to ties than Spearman's ρ. Often preferred in theoretical work and small-sample settings.

Miscellanea: Maurice Kendall (1907–1983) developed τ while working at British civil service, initially for agricultural statistics. Kendall knighted (1974) for contributions to statistics, becoming Sir Maurice Kendall. He served as president of Royal Statistical Society and International Statistical Institute. Kendall's books (*The Advanced Theory of Statistics* trilogy with Stuart) educated generations of statisticians worldwide.

Reference(s): Kendall (1938); Kendall & Gibbons (1990)

Worked Example

Problem Statement Nine employees are classified by years of experience (x) and annual performance score (y, out of 100). Test at $\alpha = 0.05$ whether experience and performance are monotonically associated using Kendall's τ. $H_0 : \tau = 0$; $H_1 : \tau \neq 0$.

Employee	1	2	3	4	5	6	7	8	9
x_i (years)	1	2	3	4	5	6	7	8	9
y_i (score)	50	40	45	56	62	73	85	79	67

1. Kendall's τ and Concordance
For each of $\binom{n}{2}$ pairs (i, j) with $i < j$, classify the pair as:

- **Concordant** (C): $(x_j - x_i)(y_j - y_i) > 0$.
- **Discordant** (D): $(x_j - x_i)(y_j - y_i) < 0$.

2. Identifying Discordant Pairs
Since x is already sorted, a pair (i, j) with $i < j$ is discordant whenever $y_j < y_i$. There are exactly **6 discordant pairs**:

Pair (i, j)	$R(y_i)$	$R(y_j)$	Type
$(1, 2)$	3	1	D
$(1, 3)$	3	2	D
$(6, 9)$	7	6	D
$(7, 8)$	9	8	D
$(7, 9)$	9	6	D
$(8, 9)$	8	6	D

3. Computing τ
Total pairs: $\binom{9}{2} = 36$. With no ties, $C + D = 36$, so:

$$C = 36 - 6 = 30, \quad D = 6, \quad \tau = \frac{C - D}{\binom{n}{2}} = \frac{2}{3}.$$

4. Test Statistic
For $n \geq 8$, the normal approximation is:

$$z = \frac{\tau}{\sqrt{\frac{2(2n+5)}{9n(n-1)}}} \approx 2.502.$$

5. Decision
Under H_0: $z \sim \mathrm{N}(0, 1)$.

α	$z_{\alpha/2}$	Decision
5%	1.960	**Reject** (2.502 > 1.960)
1%	2.576	FTR (2.502 < 2.576)

$p = 0.012$. There is significant evidence of a positive monotonic association between experience and performance at the 5% level.

6. Comparison with Spearman
Applied to the same nine employees, Spearman's $r_s = 0.833$. The smaller magnitude of $\tau = 2/3 \approx 0.667$ is typical: the two measures are not numerically comparable, but they convey the same substantive conclusion.

Practitioner Insight: Kendall's τ has a direct probabilistic interpretation: $\tau = P(\text{concordant}) - P(\text{discordant})$, the difference between the probabilities that two randomly chosen observations have the same versus opposite ordering on both variables. Here $\tau = 2/3$ indicates substantially more concordant than discordant pairs. Unlike Spearman's r_s, which measures correlation between ranks, τ is based entirely on pairwise orderings. When ties are present, τ_b adjusts the denominator to account for tied pairs; the formula used here is the special case with no ties. Kendall's τ is typically smaller in magnitude than Spearman's r_s for the same data and is often preferred for ordinal data because of its intuitive interpretation.

Common Pitfalls

- **Ignoring ties.** Tied observations require τ_b, which adjusts for tied pairs in the denominator. Using τ_a with substantial ties typically underestimates association.
- **Using the wrong tau variant.** τ_a assumes no ties, τ_b is the standard choice for correlation, and τ_c is intended for rectangular contingency tables. With no ties, all variants coincide.
- **Comparing magnitudes to Pearson or Spearman.** Kendall's τ is naturally smaller in magnitude than Pearson's r or Spearman's r_s (exactly $\tau = \frac{2}{\pi} \arcsin(r_s)$; roughly $\tau \approx \frac{2}{3} r_s$). This reflects a different scale, not weaker association.
- **Using it when Pearson is more appropriate.** For continuous, approximately normal data, Pearson's r is generally more powerful. Kendall's τ is most useful for ordinal, non-normal, or outlier-prone data.
- **Expecting higher power than Spearman.** Kendall's τ and Spearman's r_s have similar efficiency in many settings. Kendall's advantages are principled tie handling, small-sample performance, and a probabilistic interpretation.

Distance Correlation Test

General Test of Dependence Between Random Variables

CORR.4 | **Category:** Dependence | **Data:** Continuous or Mixed | **Null:** Permutation

When to Use

- Detect nonlinear dependencies between variables.
- Pearson's r may miss important associations.
- Suitable for univariate or multivariate variables.
- No parametric model for the relationship is known.
- Detect non-monotonic relationships in the data.

Assumptions

- Joint distributions have finite first moments.
- Observations are independent random samples.
- Distances between observations are meaningful.
- Variables may be univariate or multivariate.
- Sample size supports reliable permutation inference.

Test Statistic

$$\mathcal{R}^2(X,Y) = \frac{\mathcal{V}_n(X,Y)}{\sqrt{\mathcal{V}_n(X,X)\mathcal{V}_n(Y,Y)}}; \ \mathcal{R}(X,Y) \in [0,1]. \quad \mathcal{V}_n(X,Y) = \frac{1}{n^2}\sum_{i,j} A_{ij}B_{ij}$$

Notation: A_{ij}, B_{ij} are double-centered pairwise distance matrices for X and Y, $\mathcal{V}_n$ is sample distance covariance

Plain English: Distance correlation measures dependence by comparing how pairwise distances in X align with pairwise distances in Y. If large distances in one variable tend to coincide with large distances in the other (and similarly for small distances), the variables are dependent. Unlike Pearson or Spearman correlation, distance correlation is zero *if and only if* the variables are independent, making it a powerful omnibus dependence test.

⚠ Practitioner Warning

Distance correlation detects dependence but does not describe its form or direction. Significant results must be followed by visualization or modeling to understand the relationship. The statistic can be biased upward in small samples; permutation testing is preferred for inference. Computational cost is $O(n^2)$, which may be prohibitive for very large datasets.

Hypotheses

- H_0: X and Y are independent ($\mathcal{R}(X,Y) = 0$).
- H_1: X and Y are dependent ($\mathcal{R}(X,Y) > 0$).

Decision Rule: Reject H_0 if permutation p-value $< \alpha$. See example for t approximation.

Software References [Python, R, SAS]

Python	R	SAS
`dcor.distance_correlation`	`energy::dcor.test()`	Not Native (Manual)

Additional Notes

Distance correlation provides a general measure of dependence with the defining property that it equals zero if and only if the variables are independent (under finite second moments). Unlike classical correlations, it extends naturally to multivariate X and Y (dimension of X and Y need not match) and is closely connected to distance covariance and kernel-based measures such as HSIC (Hilbert–Schmidt Independence Criterion) through characteristic functions and reproducing kernel Hilbert spaces.

Like Pearson correlation, distance correlation is invariant to scale changes (not invariant to general nonlinear transformations, again like Pearson), but its strength lies in detecting *any* form of dependence—linear or nonlinear, monotonic or not. Classical correlations can be zero even under strong deterministic relationships (e.g., $Y = X^2$), whereas distance correlation vanishes only under true independence.

Distance covariance is equivalent to Brownian covariance—the covariance of two stochastic processes obtained via Brownian motion embeddings—linking a geometric construction to deep probability theory.

Miscellanea: Distance correlation was introduced in 2007 by Gábor Székely. Maria Rizzo, and Nail Bakirov as a dependence measure that retains the intuitive appeal of correlation while capturing nonlinear relationships. Its zero-implies-independence property quickly made it influential in machine learning, genomics, and feature screening. Earlier distance-based ideas appear in Bickel and Breiman (1983), but lacked the normalization and generality that made distance correlation practical and widely adopted.

Reference(s): Székely, Rizzo & Bakirov (2007); Székely & Rizzo (2009)

Worked Example

Problem Statement Four paired observations of study time (X, hours) and test score (Y, out of 5) are recorded for four students. Apply the distance correlation test at $\alpha = 0.05$ to determine whether X and Y are statistically dependent.

For four students, the observed study hours and test scores are $(1, 1)$, $(2, 2)$, $(3, 4)$, and $(4, 3)$, where the first entry denotes hours studied and the second denotes test score. $H_0: \mathcal{R}(X, Y) = 0$; $H_1: \mathcal{R}(X, Y) > 0$. $T_n \sim t(df), df = \frac{n(n-3)}{2} - 1$.

1. Pairwise Distance Matrices

Compute the $n \times n$ distance matrices: $a_{kl} = |X_k - X_l|$ and $b_{kl} = |Y_k - Y_l|$.

$$\mathbf{a} = \begin{pmatrix} 0 & 1 & 2 & 3 \\ 1 & 0 & 1 & 2 \\ 2 & 1 & 0 & 1 \\ 3 & 2 & 1 & 0 \end{pmatrix}, \quad \mathbf{b} = \begin{pmatrix} 0 & 1 & 3 & 2 \\ 1 & 0 & 2 & 1 \\ 3 & 2 & 0 & 1 \\ 2 & 1 & 1 & 0 \end{pmatrix}$$

2. Double-Centring

The doubly-centred matrix is:

$$A_{kl} = a_{kl} - \bar{a}_{k\cdot} - \bar{a}_{\cdot l} + \bar{a}_{\cdot\cdot},$$

where $\bar{a}_{k\cdot}$ is row k mean, $\bar{a}_{\cdot l}$ is column l mean, and $\bar{a}_{\cdot\cdot}$ is the grand mean. Similarly for B_{kl} from $\mathbf{b}$.

For $\mathbf{a}$: row means = $(3/2, 1, 1, 3/2)$, col means = $(3/2, 1, 1, 3/2)$, grand mean = $5/4$.

For $\mathbf{b}$: row means = $(3/2, 1, 3/2, 1)$, col means = $(3/2, 1, 3/2, 1)$, grand mean = $5/4$.

$$\mathbf{A} = \frac{1}{4}\begin{pmatrix} -7 & -1 & 3 & 5 \\ -1 & -3 & 1 & 3 \\ 3 & 1 & -3 & -1 \\ 5 & 3 & -1 & -7 \end{pmatrix} \quad \mathbf{B} = \frac{1}{4}\begin{pmatrix} -7 & -1 & 5 & 3 \\ -1 & -3 & 3 & 1 \\ 5 & 3 & -7 & -1 \\ 3 & 1 & -1 & -3 \end{pmatrix}$$

3. Distance Covariance and Variance

$$\mathcal{V}^2(X, Y) = \frac{11}{16}, \quad \mathcal{V}^2(X) = \mathcal{V}^2(Y) = \frac{13}{16}.$$

4. Distance Correlation

$$\mathcal{R}(X, Y) = \sqrt{\frac{\mathcal{V}^2(X, Y)}{\sqrt{\mathcal{V}^2(X)\mathcal{V}^2(Y)}}} = \sqrt{\frac{11}{13}} \approx 0.920.$$

5. Test Statistic

The t-approximation (Székely & Rizzo 2013):

$$T_n = \sqrt{\frac{n(n-3)}{2} - 1} \cdot \frac{\mathcal{R}}{\sqrt{1 - \mathcal{R}^2}} = \sqrt{11/2} = 2.345.$$

6. Decision

$df = \frac{4 \times 1}{2} - 1 = 1$.

α	$t_{(1),\alpha}$	Decision
5%	6.314	FTR (2.345 < 6.314)
1%	31.82	FTR (2.345 < 4.032)

$p_t \approx 0.128$. Therefore, we fail to reject H_0 at the 5% level.

The exact permutation p-value is $p_{\text{perm}} = 6/24 = 0.25$, yielding the same conclusion. (The literature is somewhat unclear on one- versus two-sided use of the t-approximation; the permutation test avoids this issue.)

Practitioner Insight: Distance correlation $\mathcal{R}$ is unique among common dependence measures in that $\mathcal{R} = 0$ if and only if the variables are independent. Unlike Pearson's r or Spearman's ρ, it detects arbitrary nonlinear associations and extends naturally to multivariate data. The double-centering step removes marginal location effects, leaving only joint dependence structure. Distance correlation requires no normality assumptions, but computing the distance matrices requires $O(n^2)$ operations; for small samples, permutation tests are generally preferred to asymptotic approximations.

Common Pitfalls

- **Interpreting it like Pearson correlation.** Distance correlation measures the strength of dependence, not its direction. Positive and negative relationships can produce the same value.
- **Ignoring the relationship's form.** A significant distance correlation confirms dependence but does not reveal its nature. Follow up with plots and appropriate models to characterize the relationship.
- **Treating dCor^2 as variance explained.** Unlike r^2, dCor^2 is not a proportion of variance explained. It measures dependence strength rather than a variance decomposition.
- **Relying on asymptotic p-values.** Small and moderate samples can yield inaccurate asymptotic significance levels. Permutation tests are generally preferred when feasible.
- **Using it when simpler methods suffice.** For clearly linear or monotonic relationships, Pearson or rank correlations are often more interpretable and powerful. Distance correlation is most useful when the form of dependence is unknown.

Hoeffding's D Test

Nonparametric Test for General Dependence

CORR.5 | **Category:** Dependence | **Data:** Continuous or Ordinal | **Null:** Permutaion / Asymptotic

When to Use

- Detect general dependence beyond linear relationships.
- Non-monotonic associations may exist in the data.
- Rank-based alternative to correlation coefficients.
- Independence is questioned but form is unknown.
- Complex nonlinear patterns may be overlooked otherwise.

Assumptions

- Paired observations are mutually independent.
- Observations are randomly sampled from population.
- Variables are measured on at least ordinal scale.
- Rankings of observations are meaningful and valid.
- Sample size supports asymptotic or exact inference.

Test Statistic

$$D = \frac{1}{n^2}\sum_{i=1}^{n}\sum_{j=1}^{n}\left[F_{XY}(X_i, Y_j) - F_X(X_i)F_Y(Y_j)\right]^2 = \frac{[A - 2(n-2)B + (n-2)(n-3)C]}{n(n-1)(n-2)(n-3)(n-4)}. \quad D_{\text{scaled}} = 30 \cdot D$$

Notation: F_{XY} joint empirical CDF, F_X, F_Y marginal empirical CDFs, n sample size, A, B, C are f(bivariate ranks).

Plain English: Hoeffding's test checks whether two variables are independent by comparing the observed joint distribution to what would be expected if the variables were unrelated. Unlike Pearson (detects linear), Spearman (detects monotonic), or Kendall (detects monotonic), it is sensitive to *all* types of dependence—including nonlinear(e.g: inverted-U, sinusoidal), non-monotone, and structured relationships. A value near zero indicates independence; larger values signal dependence.

⚠ Practitioner Warning

Hoeffding's test is computationally intensive and has lower power for simple linear relationships compared to Pearson or Spearman. It is best used as a general diagnostic tool rather than a replacement for targeted correlation measures. Ties reduce sensitivity; permutation-based inference is preferred for small samples.

Hypotheses

- H_0: X and Y are independent ($F_{XY}(x, y) = F_X(x)F_Y(y)$).
- H_1: X and Y are dependent ($F_{XY}(x, y) \neq F_X(x)F_Y(y)$).

Decision Rule: Reject H_0 if D is sufficiently large. Use asymptotic critical values or permutation-based p-values.

Software References [Python, R, SAS]

`Not Native (Manual)`	`Hmisc::hoeffd()`	`PROC CORR (HOEFFDING)`

Additional Notes

Hoeffding's D is scale-free and invariant under strictly monotone transformations and symmetric in X and Y. It summarizes all departures from independence through a single global measure. While powerful against complex dependence, it is less efficient than Pearson or Spearman when the true relationship is simple and monotone. Often used alongside distance correlation or HSIC.

Definition via Ranks: Let $(X_1, Y_1), \ldots, (X_n, Y_n)$ be bivariate sample. Define: R_i = rank of X_i among all X values; S_i = rank of Y_i among all Y values; Q_i = number of points (X_j, Y_j) in quadrant (X_i, Y_i) (both $X_j < X_i$ and $Y_j < Y_i$). A, B, and C in Hoeffding's D are:

$$A = \sum_{i=1}^{n}(R_i - 1)(R_i - 2)(S_i - 1)(S_i - 2), \quad B = \sum_{i=1}^{n}(R_i - 2)(S_i - 2)(Q_i - 1), \quad C = \sum_{i=1}^{n}(Q_i - 1)(Q_i - 2),$$

Properties: D_{scaled} ranges from -0.5 to 1, with $E[D] = 0$ (no ties) under independence. Positive values indicate dependence, while negative values can occur by chance and are not interpreted as meaningful negative association.

Miscellanea: Wassily Hoeffding introduced this test in 1948 while developing the foundations of nonparametric statistics. The test anticipated modern dependence measures by decades, but its computational burden limited adoption until the rise of modern computing. Hoeffding later became a central figure in probability theory and large-sample statistics.

Reference(s): Hoeffding (1948); Blum et al. (1961); https://github.com/Dicklesworthstone/hoeffdings_d_explainer

Worked Example

Problem Statement An agronomist tests for a non-monotonic relationship between Soil Moisture (X) and Germination (Y) for $n = 10$ observations. Standard correlation measures fail due to the U-shaped nature of the data. We test H_0: X and Y are independent against H_a: X and Y are dependent.

Data

i	X_i	Y_i	R_i	S_i	$Q1_i$	$Q2_i$	$Q3_i$	$Q4_i$	Q_i
1	10	98	1	10	1	0	0	0	1
2	20	80	2	7	1	0	0	0	1
3	30	60	3	5	1	0	0	0	1
4	40	40	4	3	1	0	0	0	1
5	50	20	5	1	1	0	0	0	1
6	60	30	6	2	2	0	0	0	2
7	70	50	7	4	4	0	0	0	4
8	80	70	8	6	6	0	0	0	6
9	90	85	9	8	8	0	0	0	8
10	100	90	10	9	9	0	0	0	9

1. Calculation of the D-Statistic The Hoeffding's D statistic (or D_{scaled}) is calculated using R_i, S_i, and Q_i:
$A = 7440$, $B = 892$, $C = 124$,
$D_{\text{scaled}} = 3360/30240 = 0.1111$

Note: Definitions of $Q_{1i}, \ldots, Q_{4i}$ are provided at the bottom of this page.

2. Results Summary

Metric	Value	P-value
Pearson (r)	0.0364	0.9204
Spearman (ρ)	0.0424	0.9074
Hoeffding's D_{scaled}	**0.111**	**0.0940**

Note: 0.0940 is permutation-based.

3. Decision and Interpretation

- **Decision:** At $\alpha = 0.05$, we **Fail to Reject** H_0. However, at $\alpha = 0.10$, the result is significant.
- **Sensitivity:** Pearson and Spearman tests completely fail to detect the association ($p > 0.90$) due to the non-monotonic (U-shaped) nature of the data.
- **Practitioner Insight:** Hoeffding's D identifies a much stronger signal of dependency than linear or monotonic tests. The results suggest a complex relationship exists, though the small sample size limits the power to reach standard significance thresholds.

Common Pitfalls

- **Using it without examining the scatterplot.** Hoeffding's D is an omnibus dependence test and does not reveal the form or direction of the relationship. A significant result must be followed by visualization and directional correlations to interpret how variables are related.
- **Expecting high power for monotonic relationships.** Hoeffding's test trades specificity for generality. For monotonic patterns, Spearman or Kendall are more powerful; for non-monotonic patterns, Hoeffding is superior. Choose the test that matches the expected relationship structure.
- **Interpreting the magnitude of D as if it were a correlation coefficient.** D (unscaled version) ranges roughly from $-1/60$ to $1/30$ and is not centered symmetrically around zero. Values near zero suggest independence; larger values indicate stronger dependence. Do not compare magnitudes directly to Pearson or Spearman.
- **Ignoring sample-size and computational constraints.** With small samples, asymptotic p-values are unreliable; use permutation tests. With very large samples, computation becomes intensive and weak dependencies may appear significant. Hoeffding performs best for moderate sample sizes.
- **Relying on Hoeffding alone for interpretation.** Because D detects any form of dependence, it should be complemented with Pearson, Spearman, and graphical analysis to characterize linear, monotonic, or non-monotonic components of the relationship.

Calculation of Q_i

$Q1_i = 1 + \sum_{j=1}^{n} \mathbf{1}(R_j < R_i \wedge S_j < S_i)$, $\quad Q2_i = \frac{1}{4}\left(\sum_{j=1}^{n} \mathbf{1}(R_j = R_i \wedge S_j = S_i) - 1\right)$, $\quad Q3_i = \frac{1}{2}\sum_{j=1}^{n} \mathbf{1}(R_j = R_i \wedge S_j < S_i)$,
$Q4_i = \frac{1}{2}\sum_{j=1}^{n} \mathbf{1}(R_j < R_i \wedge S_j = S_i)$, $\quad Q_i = Q1_i + Q2_i + Q3_i + Q4_i$.

Point–Biserial Correlation Test

Association Between Binary and Continuous Variables

CORR.6 | **Category:** Association | **Data:** Binary & Continuous | **Null:** t-distribution

When to Use

- Relate a binary variable to a continuous outcome.
- Quantify group differences using a correlation measure.
- Binary grouping variable has exactly two categories.
- Assess association without fitting a full regression.
- Interpret effect size alongside a two-sample t-test.

Assumptions

- Binary variable contains exactly two categories.
- Observations are mutually independent.
- Continuous outcome is approximately normal by group.
- Variances are similar across the two groups.
- Sample is randomly drawn from the population.

Test Statistic

$$t = r_{pb}\sqrt{\frac{n-2}{1-r_{pb}^2}} \sim t_{n-2} \text{ (under } H_0\text{)}, \qquad \text{where } r_{pb} = \frac{\bar{y}_1 - \bar{y}_0}{s_y}\sqrt{\frac{n_1 n_0}{n^2}}$$

Notation: $\bar{y}_1, \bar{y}_0$ are group means, n_1, n_0 group sizes, s_y pooled SD (biased version, divisor n), $n = n_0 + n_1$.

Plain English: Point–biserial correlation is a bridge between group comparisons and correlation analysis. It measures how strongly membership in one of two groups is associated with higher or lower values of a continuous outcome. For example, it can assess whether patients receiving a treatment tend to have different recovery scores than those receiving a placebo. Although it tests the same hypothesis as a two-sample t-test, the result is reported as a correlation coefficient, providing an intuitive measure of effect size.

⚠ Practitioner Warning

Point–biserial correlation assumes the binary variable reflects a genuine dichotomy. Artificially dichotomizing a continuous variable discards information, reduces power, and usually weakens the observed correlation. As the measure is equivalent to Pearson's correlation with a binary predictor, a few extreme observations can substantially affect its magnitude and significance. Inference relies on assumptions similar to the two-sample t-test, including approximate normality within groups and reasonably equal variances. Highly unequal group sizes can also make results less reliable.

Hypotheses

- H_0: $\rho_{pb} = 0$ (no association between group & outcome)
- H_1: $\rho_{pb} \neq 0$

Decision Rule: Reject H_0 if $|t| > t_{\alpha/2, n-2}$ or $p < \alpha$. Equivalent to rejecting the two-sample t-test at level α.

Software References [Python, R, SAS]

`scipy.stats.pointbiserialr` | `ltm::biserial.cor()` | `PROC CORR`

Additional Notes

Point-biserial correlation is special case of Pearson correlation when one variable is dichotomous (coded 0/1). Karl Pearson (1909) introduced term "point-biserial" to distinguish from biserial correlation (different formula for artificially dichotomized normal variable, now rarely used). Squared point-biserial (r_{pb}^2) equals eta-squared (η^2) from t-test—proportion of variance explained by group membership. Relationship to t-statistic: $t^2 = r_{pb}^2 \times (n-2)/(1-r_{pb}^2)$. Common in psychometrics (item analysis: item score vs. total test score), educational testing, biomedical research (disease presence vs. biomarker level).

Miscellanea: Terman and Merrill (1937) extensively used point-biserial correlation for item analysis in Stanford-Binet Intelligence Scale. Each test item (pass/fail) correlated with total score (continuous) to assess item quality. Items with low point-biserial correlations (below 0.15–0.20) removed as non-discriminating. This item-total correlation approach became standard in psychological test construction, quality control for thousands of tests. Point-biserial correlation's role in intelligence testing may have affected more lives than any other correlation application—determining what items appear on IQ tests taken by millions.

Reference(s): Pearson (1909); Guilford (1954); Cohen (1988)

Point–Biserial Correlation Test CORR.6

Worked Example

Problem Statement Ten students sit an exam. Five received no tutoring ($X = 0$) and five received tutoring ($X = 1$). The point-biserial correlation measures the association between the binary tutoring variable and the continuous exam score Y. H_0: $\rho_{pb} = 0$ (no association); H_1: $\rho_{pb} \neq 0$. Under H_0: $t \sim t(n-2) = t(8)$. Here, $n_0 = 5$, and $n_1 = 5$.

Student	1	2	3	4	5	6	7	8	9	10
X (tutored)	0	0	0	0	0	1	1	1	1	1
Y (score)	3	4	5	6	7	7	8	9	10	11

1. Group Means, Grand Mean, and Biased SD

$$\bar{Y}_0 = 5, \qquad \bar{Y}_1 = 9, \qquad \bar{Y} = \frac{\bar{Y}_0 n_0 + \bar{Y}_1 n_1}{n} = 7$$

$$s_Y = \sqrt{\frac{\sum(y_i - \bar{Y})^2}{n}} = \sqrt{6}$$

2. Point-Biserial Correlation

$$r_{pb} = \frac{\bar{Y}_1 - \bar{Y}_0}{s_Y}\sqrt{\frac{n_1 n_0}{n^2}} = \frac{9-5}{\sqrt{6}}\sqrt{\frac{25}{100}} = 0.8165$$

$r_{pb}^2 = 2/3$: tutoring explains 67% of the variance in exam scores.

3. Significance Test

$$t = \frac{r_{pb}\sqrt{n-2}}{\sqrt{1 - r_{pb}^2}} = 4$$

Statistic	Value	CV (5%)	Decision
r_{pb}	0.817	—	—
$t(8)$	4.000	2.306	Reject H_0

$p = 0.004$; CV at 1%= 3.355 also exceeded.

4. Equivalence to Two-Sample t-test

$$t = \frac{\bar{Y}_1 - \bar{Y}_0}{s_p\sqrt{1/n_1 + 1/n_0}} = \frac{4}{s_p\sqrt{2/5}} = 4 \checkmark$$

Practitioner Insight: The point-biserial correlation is simply Pearson's correlation computed between a binary (0/1) variable and a continuous variable. The formula $r_{pb} = (\bar{Y}_1 - \bar{Y}_0)/s_Y \cdot \sqrt{n_1 n_0/n^2}$ is a simplified form of Pearson's r for a dichotomous predictor. Its significance test is algebraically identical to the pooled two-sample t-test, so both procedures always yield the same t-statistic and p-value. Different conventions for s_Y appear in the literature, but they produce the same inferential result. The quantity r_{pb}^2 has the usual R^2 interpretation; here $r_{pb}^2 = 2/3$, meaning two-thirds of the variability in scores is explained by group membership. A limitation is that highly unbalanced group sizes can substantially reduce the maximum attainable $|r_{pb}|$, even when the groups are well separated.

Common Pitfalls

- **Dichotomizing a continuous variable.** Point-biserial correlation is intended for genuine binary variables. Artificial splits (e.g., median splits) discard information, reduce power, and can distort relationships.
- **Ignoring distributional assumptions.** The test inherits the assumptions of Pearson correlation and the two-sample t-test. Strong non-normality or unequal variances may require Welch, robust, or rank-based alternatives.
- **Confusing point-biserial and biserial correlation.** Point-biserial correlation applies to true dichotomies. Biserial correlation assumes an underlying continuous latent variable and is appropriate only in specialized settings.
- **Treating it as different from the two-sample t-test.** Point-biserial correlation and the pooled two-sample t-test assess the same hypothesis. The choice is primarily one of presentation: association versus mean differences.
- **Ignoring group-size imbalance.** Highly unequal group sizes reduce the maximum attainable $|r_{pb}|$. Interpret magnitudes cautiously and consider complementary effect sizes such as Cohen's d.

Note on two versions of r_{pb}

Two formulas appear in the literature, differing only in whether s_Y uses divisor n or $n-1$. The form with divisor n equals Pearson $r(X, Y)$ exactly, yields $r_{pb}^2 = \eta^2 = SS_{\text{between}}/SS_{\text{total}}$, and produces a t-statistic identical to the pooled two-sample t-test; it is used by all modern software. The alternative form (divisor $n-1$), found in older psychometrics texts such as Glass & Stanley (1970), is smaller by $\sqrt{(n-1)/n}$ and, crucially, yields a t-statistic that does not correspond to any standard test when substituted into $t = r_{pb}\sqrt{n-2}/\sqrt{1-r_{pb}^2}$. The Pearson-consistent form should always be used; discrepancies from older hand calculations can be corrected by multiplying by $\sqrt{n/(n-1)}$.

Biserial Correlation

Correlation Between Continuous Variable and Artificially Dichotomized Variable

CORR.7 | **Category:** Association | **Data:** Binary & Continuous | **Null:** Asymptotic Normal

When to Use

- Relate an artificial dichotomy to a continuous variable.
- Binary groups arise from a latent continuous trait.
- Dichotomization obscures underlying information content.
- Estimate latent correlation before thresholding occurred.
- Educational and psychometric classification settings.

Assumptions

- Dichotomy results from thresholding a normal variable.
- Underlying latent variable is normally distributed.
- Observations are mutually independent.
- Continuous variable is approximately normal in shape.
- Linear relationship between the two variables.

Test Statistic

$$r_b = \frac{\bar{Y}_1 - \bar{Y}_0}{s_Y} \cdot \frac{pq}{\phi(z_p)}.$$

Modern practice: Test through r_{pb} using t-test; use r_b for effect size

Notation: $\bar{Y}_1, \bar{Y}_0$ are means of continuous variable Y for groups coded 1 and 0, s_Y is standard deviation of Y, $p = n_1/n$ and $q = 1 - p$ are proportions in each group, $\phi(z_p)$ is ordinate (height) of standard normal pdf at z_p where $\Phi(z_p) = p$

Plain English: Biserial correlation estimates what the correlation would be if a dichotomized variable were measured on its original continuous scale. For example, if a continuous depression score is split into high/low categories, biserial correlation approximates the relationship as if the full scale were used. It adjusts for information loss from dichotomization using a normal-based correction, making it larger than the corresponding point-biserial correlation.

⚠ Practitioner Warning

Assumes the underlying dichotomized variable is normally distributed; violations can cause serious bias. Estimates may exceed [-1,1] in small samples or when assumptions fail. Results are sensitive to the cutoff point, and imbalance (p far from 0.5) or small samples reduce reliability. Not appropriate for true dichotomies (use point-biserial), and should not replace the original continuous variable in regression or further analysis.

Hypotheses

- H_0: True correlation before dichotomization: $\rho_b = 0$.
- H_1: Correlation exists between the variables: $\rho_b \neq 0$.

Decision Rule: Reject H_0 if $|t_{pb}| > t_{\alpha/2, n-2}$ or $p < \alpha$. Equivalent to rejecting the two-sample t-test at level α.

Software References [Python, R, SAS]

Python	R	SAS
`Not Native (Manual)`	`ltm::biserial.cor`	`Not Native (Manual)`

Additional Notes

Biserial and point-biserial correlations are related by $r_b = r_{pb}\sqrt{pq}/\phi(z_p)$, where $\phi(z_p)$ is the standard normal density at the cutoff separating proportions p and q. The correction factor $pq/\phi(z_p)$ therefore depends on the split point. As splits become more extreme, $\phi(z_p)$ decreases and the correction factor increases, leading to instability. Older psychometric texts often present large-sample Z-tests for the biserial correlation. In modern practice, significance is usually assessed through the equivalent point–biserial t-test because $H_0 : \rho_b = 0$ is equivalent to $H_0 : \rho_{pb} = 0$, and the latter has a simple and well-known sampling distribution. The biserial correlation is then reported as the effect-size estimate, although large-sample theory also provides an approximately normal sampling distribution for r_b. The measure has been widely used in classical test theory for item discrimination, though it is often confused with the point-biserial correlation in practice.

Miscellanea: The biserial correlation emerged in early psychometrics as a method for relating dichotomous test items (e.g., correct/incorrect) to an underlying continuous ability. Unlike the point-biserial correlation, it assumes the binary variable reflects a latent normally distributed trait and estimates the correlation that would be observed without dichotomization. Although largely supplanted by modern methods such as item response theory, it remains in use due to its simplicity and interpretability in educational testing.

Reference(s): Tate (1954, 1955); Kraemer (2006); Sheskin (2011)

Worked Example

Problem Statement A medical researcher administers a continuous pain score (Y, scale 1–10) to ten patients who are also classified as non-responders ($X = 0$, $n_0 = 6$) or responders ($X = 1$, $n_1 = 4$) to a new therapy. The binary variable is assumed to be an artificial dichotomisation of an underlying continuous latent response. Compute the biserial correlation and test $H_0\colon \rho_b = 0$ at $\alpha = 0.05$. Note: Significance is assessed using t-test for the point–biserial correlation; r_b is reported as an effect-size estimate.

X_i	0	0	0	0	0	0	1	1	1	1
Y_i	2	3	4	4	5	6	5	6	6	7

1. Group Means and Unbiased Standard Deviation S_Y

$$M_0 = 4, \quad M_1 = 6, \quad \bar{Y} = \frac{6 \times 4 + 4 \times 6}{10} = 4.8.$$

$$S_Y^2 = \frac{21.60}{9} = \frac{12}{5}, \qquad S_Y \approx 1.549.$$

2. Cut-Point and Normal Ordinate

With $p = n_1/n = 4/10 = 2/5$ and $q = 3/5$: the cut point c is the value at which a standard normal variable has probability q below it:

$$c = \Phi^{-1}(q) = \Phi^{-1}(0.6) = 0.2533.$$

$$z(c) = \varphi(0.2533) = \frac{1}{\sqrt{2\pi}} e^{-0.0321} = 0.3863.$$

3. Biserial Correlation

$$r_b = \frac{M_1 - M_0}{S_Y} \cdot \frac{pq}{z(c)} = \frac{2}{1.549} \cdot \frac{0.24}{0.3863} = \mathbf{0.802}.$$

4. Test via Point-Biserial

Since $H_0\colon \rho_b = 0$ is equivalent to $H_0\colon \rho_{pb} = 0$, significance is assessed via the t-statistic based on r_{pb}. Using the biased SD gives $r_{pb} = 2/3$ exactly:

$$t = \frac{r_{pb}\sqrt{n-2}}{\sqrt{1 - r_{pb}^2}} = \frac{\frac{2}{3}\sqrt{8}}{\sqrt{5/9}} \approx 2.530.$$

5. Decision

Under H_0: $t \sim t(n-2) = t(8)$.

α	$t_{(8),\,\alpha/2}$	Decision
5%	2.306	**Reject** (2.530 > 2.306)
1%	3.355	FTR (2.530 < 3.355)

$p = 0.035$. There is significant evidence of a biserial association between therapy response and pain score at the 5% level.

Practitioner Insight: The biserial correlation r_b is intended for *artificial* dichotomies, where a binary variable arises from an underlying continuous trait. By adjusting for information lost through dichotomization, r_b is always at least as large as the corresponding point-biserial correlation (0.802 > 0.667 here). The correction grows as the split becomes more extreme, and in small samples or highly unbalanced splits r_b can even exceed 1, a known artifact rather than an error. Inference is typically based on the point-biserial correlation because its null distribution is known. When the dichotomy is naturally binary rather than artificially created, the point-biserial correlation is the appropriate measure.

Common Pitfalls

- **Confusing biserial correlation with point-biserial correlation.** Biserial assumes the binary variable arises from artificially dichotomizing an underlying continuous normal variable, whereas point-biserial treats the binary variable as truly categorical. Use biserial only when the dichotomy represents a threshold applied to a latent continuous variable.
- **Ignoring the distributional assumptions of the biserial estimator.** The method assumes the underlying latent variable is approximately normally distributed. Strong skewness or heavy tails can bias the estimate and may even produce impossible values exceeding ± 1.
- **Using biserial correlation with extreme split proportions or small samples.** When the dichotomous variable has highly unbalanced groups (e.g., 90/10 or 95/5) or when sample sizes are small, the correction factor in the biserial formula becomes unstable. This leads to large sampling variability and unreliable estimates.
- **Applying biserial correlation when both variables are dichotomous.** Biserial correlation requires one continuous variable. If both variables are binary indicators derived from underlying continuous variables, tetrachoric correlation is the appropriate method.
- **Ignoring the effect of the chosen dichotomization threshold.** The magnitude of the biserial correlation depends on where the continuous variable is split. Different cutpoints can yield different estimates, so the threshold should be reported and ideally justified in advance.

Tetrachoric Correlation Test

Latent Correlation Between Two Dichotomized Variables

CORR.8 | **Category:** Correlation | **Data:** Binary & Binary (Latent Continuous) | **Null:** Asymptotic normal

When to Use

- Relate two artificial dichotomies from latent traits.
- Binary variables arise from underlying continuums.
- Dichotomization attenuates latent correlation strength.
- Estimate correlation before thresholding occurred.
- Psychometric and diagnostic classification studies.

Assumptions

- Both dichotomies arise from latent normal variables.
- Latent variables follow a bivariate normal model.
- Observations are mutually independent.
- Thresholds create the observed binary outcomes.
- Sample is randomly drawn from the population.

Test Statistic

Pearson (1900): $r_t = \cos\left(\frac{\pi}{1+\sqrt{\omega}}\right)$ | Digby (1983): $r_t = \cos\left(\frac{\pi}{1+\omega^{0.45}}\right)$ | Edwards (1984): $r_t = \frac{\omega^{\pi/4}-1}{\omega^{\pi/4}+1}$

Notation: a, b, c, d are cell frequencies in 2×2 table, $\omega = (ad)/(bc)$ is the odds ratio, $N = a+b+c+d$.

Plain English: Tetrachoric correlation estimates how strongly two underlying continuous traits are related when only their binary outcomes are observed. Unlike Pearson or Phi correlations, it explicitly models the dichotomization process and recovers the latent association that would exist before thresholding. Some variants add 0.5 to cell frequencies. [Pearson ≈ Edwards]

⚠ Practitioner Warning

Tetrachoric correlation relies heavily on the assumption of latent bivariate normality. Violations (skewness, multimodality, unequal thresholds) can lead to biased or unstable estimates. Sparse or unbalanced contingency tables may cause convergence failures or inflated estimates. Many SE variants are available. Software implementations differ.

Hypotheses

- H_0: $\rho_t = 0$ (no latent association)
- H_1: $\rho_t \neq 0$ (latent association exists)

Decision Rule: Reject H_0 if $|Z| > z_{\alpha/2}$ or $p < \alpha$. Inference is asymptotic and sensitive to model assumptions.

Software References [Python, R, SAS]

`semopy.polycorr`	`psych::tetrachoric`	`PROC FREQ (PLCORR)`

Additional Notes

Tetrachoric correlation reduces to Pearson correlation when variables are continuous and observed directly. It is distinct from Phi (ϕ), which measures association between observed binaries without correcting for dichotomization. Estimates can exceed Phi substantially when thresholds are extreme. Three large-sample SE formulas are given below (x and y are the estimated thresholds):

$$\text{(Pearson): } SE(\hat{\rho}) = \frac{1}{N^{1/2} z_x z_y}\left[\hat{p}_{1\cdot}\,\hat{p}_{0\cdot}\,\hat{p}_{\cdot 1}\,\hat{p}_{\cdot 0} + \hat{\rho}\, z_x z_y \left(\hat{\rho}\, z_x z_y - \beta_1 \hat{p}_{\cdot 1}\,\hat{p}_{\cdot 0} - \beta_2 \hat{p}_{1\cdot}\,\hat{p}_{0\cdot}\right)\right]^{1/2}, \text{ where } \beta_1 = \frac{z_x}{\hat{p}_{1\cdot}}, \beta_2 = \frac{z_y}{\hat{p}_{\cdot 1}}$$

$$\text{(Edwards): } SE(\hat{\rho}) = \frac{\pi\alpha}{2(1+\alpha)^2}\sqrt{\left(\frac{1}{a}+\frac{1}{b}+\frac{1}{c}+\frac{1}{d}\right)}, \text{ where } \alpha = \omega^{\pi/4}, \qquad \text{(Hamdan): } SE(\hat{\rho}) = \frac{\sqrt{\hat{p}_{1\cdot}\,\hat{p}_{0\cdot}\,\hat{p}_{\cdot 1}\,\hat{p}_{\cdot 0}}}{N^{1/2} z_x z_y}$$

Here $\hat{p}_{1\cdot} = (a+b)/N$, $\hat{p}_{0\cdot} = 1 - \hat{p}_{1\cdot}$ $\hat{p}_{\cdot 1} = (a+c)/N$, $\hat{p}_{\cdot 0} = 1 - \hat{p}_{\cdot 1}$, $x = \Phi^{-1}(\hat{p}_{1\cdot})$, $y = \Phi^{-1}(\hat{p}_{\cdot 1})$, $z_x = \phi(x)$, $z_y = \phi(y)$.

Miscellanea: Karl Pearson introduced tetrachoric correlation in 1900 while studying heredity and contingency tables, proposing it as a way to estimate the correlation between two latent continuous variables observed only through dichotomization. Assuming each binary variable arises from thresholding a normal trait, Pearson showed that the underlying correlation can be recovered by solving a 2×2 likelihood equation. This was one of the earliest explicit latent-variable models, predating modern structural equation modeling (SEM) and forming the basis for later developments such as polychoric correlations and latent-trait models.

Reference(s): Pearson (1900); Olsson (1979); Muthén (1984)

Worked Example

Problem Statement A psychometrician assesses 100 respondents on two attitude items, each scored Agree/Disagree. Both items are assumed to be artificial dichotomizations of underlying continuous latent traits ξ_1 and ξ_2. The observed 2×2 frequency table is below. $H_0: \rho_t = 0$; $H_1: \rho_t \neq 0$.

$a = 31$ (Disagree, Disagree), $b = 19$ (Disagree, Agree), $c = 19$ (Agree, Disagree), $d = 31$ (Agree, Agree), $n = 100$.

1. Model and Cut Points
Since both splits are exactly 50/50:

$$p_1 = p_2 = 0.50 \Rightarrow h = k = \Phi^{-1}(0.50) = 0.$$

The normal ordinate at both cut points is:

$$z(h) = z(k) = \varphi(0) = \frac{1}{\sqrt{2\pi}} \approx 0.3989.$$

2. Exact Solution via the Median-Split Formula
When both marginals are at the median ($h = k = 0$), the bivariate normal probability of the upper-right cell has the closed form:

$$P(\xi_1 > 0, \xi_2 > 0; \rho_t) = \frac{1}{4} + \frac{\arcsin(\rho_t)}{2\pi}.$$

Setting this equal to $d/n = 31/100$:

$$\boxed{\hat{\rho}_t = \sin(0.12\pi) = \sin(21.6^\circ) \approx 0.368.}$$

3. Comparison with ϕ Coefficient
The simpler Pearson ϕ coefficient is:

$$\phi = \frac{ad - bc}{\sqrt{(a+b)(c+d)(a+c)(b+d)}} = 0.240.$$

As expected, $\hat{\rho}_t = 0.368 > \phi = 0.240$: the tetrachoric correlation corrects upward for the information lost in dichotomizing.

4. Standard Error of $\hat{\rho}_t$
The large-sample standard error ($p_k = 0.5$, $q_k = 0.5$ and $z(h) = z(k) = 1/\sqrt{2\pi}$) is:

$$\text{SE}(\hat{\rho}_t) = \frac{\sqrt{p_1 q_1 \, p_2 q_2}}{z(h)\, z(k)\, \sqrt{n}} \approx 0.157. \text{ (Hamdan)}$$

5. Test Statistic
Under $H_0: \rho_t = 0$:

$$z = \frac{\hat{\rho}_t}{\text{SE}(\hat{\rho}_t)} \approx 2.344.$$

6. Decision
Under H_0: $z \sim \text{N}(0, 1)$.

α	$z_{\alpha/2}$	Decision
5%	1.960	**Reject** (2.344 > 1.960)
1%	2.576	FTR (2.344 < 2.576)

$p = 0.019$. The two attitude items are significantly correlated at the latent-trait level at the 5% level; the estimated correlation of the underlying continuous traits is $\hat{\rho}_t \approx 0.368$.

7. Continuity Check
The chi-square test of independence gives $\chi^2 = (ad - bc)^2 n/[(a+b)(c+d)(a+c)(b+d)] = 5.76$, $p = 0.016$, consistent with the tetrachoric result.

Practitioner Insight: The tetrachoric correlation ρ_t estimates the Pearson correlation between two latent continuous variables observed only through binary cutpoints. When both variables are split at their medians, ρ_t has a simple closed-form solution; otherwise it is obtained numerically from the bivariate normal distribution. Because it adjusts for information lost through dichotomization, ρ_t is typically larger than the corresponding ϕ coefficient (0.368 vs. 0.240 here). The method assumes an underlying bivariate normal distribution and can be biased when this assumption fails. Tetrachoric correlation is appropriate for artificially dichotomized variables, whereas ϕ is preferred for naturally binary outcomes.

Common Pitfalls

- **Using it for genuine dichotomies.** Tetrachoric correlation assumes both binary variables arise from dichotomized continuous latent variables. For naturally binary outcomes, use the ϕ coefficient instead.
- **Ignoring the latent normality assumption.** The method assumes an underlying bivariate normal distribution, which cannot be verified from a 2×2 table. Estimates can be sensitive to violations of this assumption.
- **Using sparse or extreme tables.** Small samples, zero cells, or highly unbalanced marginals can produce unstable estimates. Continuity corrections, bootstrap intervals, or larger samples may be needed.
- **Comparing studies with different cutpoints.** Tetrachoric correlations depend on the dichotomization thresholds. Different cutpoints can yield different estimates even when the underlying association is unchanged.
- **Dichotomizing available continuous data.** If the original continuous measurements are available, use Pearson's correlation directly. Dichotomization discards information and reduces statistical power.

Polychoric Correlation

Correlation Between Ordinal Variables via Underlying Bivariate Normal

CORR.9 | **Category:** Correlation & Association | **Data:** Ordinal/Ordinal | **Null:** Normal or Chi-square

When to Use

- Relate two ordinal variables with ordered categories.
- Categories arise from latent continuous variables.
- Pearson correlation understates ordinal association.
- Estimate latent correlation behind categorized data.
- Likert-scale items are analyzed jointly.

Assumptions

- Ordinal categories arise from latent normal variables.
- Latent variables follow a bivariate normal model.
- Observations are mutually independent.
- Thresholds generate the observed ordinal categories.
- Sample is randomly drawn from the population.

Test Statistic

$$\chi^2 = -2[\ell(\rho = 0) - \ell(\hat{\rho_{pc}})] \sim \chi_1^2 \quad \Big| \quad z = \frac{\hat{\rho}_{pc}}{\widehat{\text{SE}}}, \ \widehat{\text{SE}}^2(\hat{\rho}_{pc}) = -\frac{\partial^2 \ell}{\partial \rho^2}\bigg|_{\hat{\rho}} \quad \Big| \quad \hat{\rho}_{pc} = \arg\max_{\rho} \mathcal{L}(\rho, \boldsymbol{\tau}_1, \boldsymbol{\tau}_2 | \text{data}).$$

Notation: ρ is correlation of bivariate normal latent variables, τ_i are threshold vectors, and $\mathcal{L}$ is likelihood function

Plain English: Polychoric correlation estimates the relationship that would be observed if ordinal variables were measured on their underlying continuous scales rather than grouped into categories. For example, Likert responses are treated as continuous attitudes divided into categories by unseen thresholds. The method estimates both these thresholds and the correlation between the latent variables. Because categorization reduces information, polychoric correlations are typically larger in absolute value than Pearson correlations computed on ordinal codes.

⚠ Practitioner Warning

Polychoric correlation assumes bivariate normal latent variables; violations can lead to substantial bias. Empirical data often deviate from this assumption, and estimates may even fall outside [-1,1] or fail to converge with sparse tables. Standard errors are typically larger than Pearson's, especially with few categories. Tests of bivariate normality are often uninformative. When assumptions fail, polychoric correlation can overestimate association and bias downstream analyses (e.g., factor analysis, SEM).

Hypotheses

- H_0: Underlying latent variables are uncorrelated: $\rho = 0$.
- H_1: Underlying latent variables are correlated: $\rho \neq 0$.

Decision Rule: Reject H_0 if $\chi^2 > \chi^2_{\alpha,1}$ or p-value $< \alpha$. Equivalently, z (normal) can be used.

Software References [Python, R, SAS]

Python	R	SAS
`semopy.polycorr`	`psych::polychoric`	`PROC CORR (PLCORR)`

Additional Notes

Polychoric correlation extends tetrachoric ideas to multi-category ordinal data, with practical methods developed by Olsson (1979). A common approach estimates thresholds from marginal proportions, then the correlation conditional on these thresholds; full ML jointly estimates both but is computationally heavier. The resulting correlation matrix may be only positive semi-definite, requiring smoothing for use in factor analysis. In practice, polychoric correlations often yield higher estimates than Pearson correlations on ordinal codes due to attenuation from categorization. Alternative models (e.g., skew-normal) have been proposed but are not widely implemented.

Miscellanea: Karl Pearson's 1900 paper introducing tetrachoric correlation was motivated by Francis Galton's studies of heredity, where traits like height were often recorded only as "tall/short" due to measurement limitations. Pearson wanted to estimate the correlation between parental and offspring height from such crude data, leading him to develop the underlying bivariate normal framework. It was later extended to polychoric correlation for multi-category ordinal data. Early computation was demanding, limiting practical use until modern computing made estimation feasible. Widely used in structural equation modeling, polychoric correlation relies on an underlying bivariate normal assumption, making its performance sensitive to deviations from normality.

Reference(s): Pearson (1900, 1922); Olsson (1979); Holgado-Tello et al. (2010)

Worked Example

Problem Statement Thirty respondents rate two Likert items (1 = Disagree, 2 = Neutral, 3 = Agree). Both ratings are treated as artificial categorisations of continuous latent traits ξ_1 (item A) and ξ_2 (item B), assumed jointly bivariate normal. Test at $\alpha = 0.05$ whether the latent traits are correlated. $H_0 : \rho_{pc} = 0$; $H_1 : \rho_{pc} \neq 0$.

$$(n_{11}, n_{12}, n_{13}; n_{21}, n_{22}, n_{23}; n_{31}, n_{32}, n_{33}) = (5, 4, 1; 3, 3, 4; 2, 3, 5), \qquad n = 30.$$

1. Estimate Thresholds from Marginals
Each marginal converts cumulative proportions to standard normal quantiles. With equal thirds ($n = 30$, 10 per row and column):

$$P(X \le 1) = \tfrac{10}{30} = \tfrac{1}{3}, \quad P(X \le 2) = \tfrac{20}{30} = \tfrac{2}{3}.$$

$$\hat{\tau}_{A,1} = \Phi^{-1}(\tfrac{1}{3}) = -0.431, \quad \hat{\tau}_{A,2} = \Phi^{-1}(\tfrac{2}{3}) = +0.431.$$

By symmetry (column marginals also 10/10/10):

$$\hat{\tau}_{B,1} = -0.431, \quad \hat{\tau}_{B,2} = +0.431.$$

2. Cell Probabilities
The probability of cell (i, j) under BVN(ρ) is:

$$\pi_{ij}(\rho) = \int_{\tau_{A,i-1}}^{\tau_{A,i}} \int_{\tau_{B,j-1}}^{\tau_{B,j}} \phi_2(x, y; \rho)\, dy\, dx.$$

where $\phi_2(x, y; \rho)$ is the standard bivariate normal pdf.

3. MLE via Score Equation
The log-likelihood $\ell(\rho) = \sum_{ij} n_{ij} \ln \pi_{ij}(\rho)$ is maximised by solving the score equation $\partial\ell/\partial\rho = 0$:

$$\sum_{ij} \frac{n_{ij}}{\pi_{ij}(\rho)} \cdot \frac{\partial \pi_{ij}}{\partial \rho} = 0.$$

ρ	0.30	0.40	**0.4253**	0.43
Score	+2.90	+0.67	**0.00**	−0.13

4. Fitted Cell Probabilities and Counts
At $\hat{\rho}_{pc} = 0.4253$ the fitted expected counts $n\hat{\pi}_{ij}$ are:

	B=1	B=2	B=3
A=1: $30\hat{\pi}_{ij}$	5.1	3.2	1.7
A=2: $30\hat{\pi}_{ij}$	3.2	3.6	3.2
A=3: $30\hat{\pi}_{ij}$	1.7	3.2	5.1

The diagonal structure of the expected counts confirms a positive latent correlation: high agreement on A tends to accompany high agreement on B.

5. Standard Error and Test Statistic
The large-sample SE via the observed information:

$$\widehat{\text{SE}}(\hat{\rho}_{pc}) = \left(-\frac{\partial^2 \ell}{\partial \rho^2}\bigg|_{\hat{\rho}} \right)^{-1/2} = 0.1908, \quad z = \frac{\hat{\rho}_{pc}}{\widehat{\text{SE}}} = 2.229.$$

6. Decision
Under H_0: $z \sim \text{N}(0, 1)$.

α	$z_{\alpha/2}$	Decision
5%	1.960	**Reject** (2.229 > 1.960)
1%	2.576	FTR (2.229 < 2.576)

$p = 0.026$. There is significant evidence of a +ve latent correlation between two latent traits at the 5% level.

Practitioner Insight: The polychoric correlation estimates the latent Pearson correlation between two ordinal variables by treating categories as discretizations of an underlying bivariate normal distribution. It generalizes tetrachoric correlation to variables with more than two ordered categories. Estimation proceeds by converting marginal proportions into threshold values and then numerically maximizing the bivariate normal likelihood over ρ. Because it adjusts for categorization, the polychoric correlation is often larger than Pearson's r computed from category codes (0.43 vs. 0.30 here). Its main limitation is the latent bivariate normal assumption; substantial departures from normality can bias the estimate.

Common Pitfalls

- **Ignoring latent normality.** Polychoric correlation assumes the observed categories arise from an underlying bivariate normal. Strong departures from this assumption can bias estimates and occasionally produce implausible values.
- **Using it for the wrong data type.** Polychoric correlation is intended for ordinal variables representing discretized continuous traits. It should not be used for nominal categories or variables observed on a continuous scale.
- **Confusing it with other correlations.** Polychoric correlation estimates a latent continuous association from ordinal data. Pearson uses observed values, Spearman uses ranks, and tetrachoric is for binary variables.
- **Using sparse tables.** Small samples, empty cells, or highly unbalanced categories can cause convergence problems and unstable estimates. Adequate cell counts are important for reliable maximum-likelihood estimation.
- **Ignoring convergence diagnostics.** Polychoric correlations are often inputs to factor analysis and SEM. Always check convergence warnings and matrix validity before interpreting results.

Cramér's V

Measure of Association for Nominal Variables in Contingency Tables

CORR.10 | **Category:** Correlation & Association | **Data:** Nominal/Nominal | **Null:** Chi-square based

When to Use

- Quantify association strength in contingency tables.
- Effect-size measure after a chi-square test.
- Nominal variables contain multiple categories.
- Compare association strengths across different studies.
- Table dimensions exceed 2×2 setting (else, same as ϕ).

Assumptions

- Observations are mutually independent.
- Variables are measured on a nominal scale.
- Counts represent frequencies, not percentages.
- Sample is randomly drawn from the population.
- Expected cell counts are reasonably adequate.

Test Statistic

$$V = \sqrt{\frac{\chi^2}{n \cdot \min(r-1, c-1)}},$$

Notation: χ^2 is Pearson chi-square statistic from test of independence, r is rows, c is columns, n is sample size

Plain English: Cramér's V measures the strength of association between two categorical variables on a scale from 0 to 1. A value of 0 means no relationship, while 1 indicates a perfect association. It is derived from the chi-square test but adjusted for sample size and table dimensions, making it an interpretable effect size. Unlike raw chi-square values, Cramér's V allows meaningful comparisons across different studies.

⚠ Practitioner Warning

Cramér's V can overestimate association strength, especially in small samples or large tables; bias corrections exist but are not always implemented. It is symmetric, so it cannot capture directional relationships. For 2×2 tables, the phi (ϕ) coefficient may be preferred. When variables are ordinal, measures like Spearman's rho or Kendall's tau are more appropriate. The statistic reaches 1 only under strong structural conditions (e.g., square tables). Some software may report misleading ϕ values for larger tables, though Cramér's V remains valid.

Hypotheses

- H_0: Two nominal variables are independent (no association).
- H_1: Variables are associated.

Decision Rule: Reject H_0 if $\chi^2 > \chi^2_{crit}$ or p-value$\leq \alpha$.
Interpret V: small (0.10), medium (0.30), large (0.50) effect.

Software References [Python, R, SAS]

`scipy::association` `rstatix::cramer_v` `PROC FREQ (CHISQ)`

Additional Notes

Cramér's V generalizes Karl Pearson's (1900) phi coefficient ($\phi^2 = \chi^2/n$) to $r \times c$ tables by normalizing with $\min(r-1, c-1)$: $V = \sqrt{\phi^2/\min(r-1, c-1)}$, ensuring values lie in $[0, 1]$. It can be viewed as the mean square canonical correlation between the variables. Finite-sample bias arises from the upward bias in χ^2, which propagates through the square root transformation; a common correction adjusts ϕ^2 before taking the square root: $\tilde{V} = \sqrt{\max\{0, \phi^2 - [(r-1)(c-1)/(n-1)]\}/\min(r-1, c-1)}$. The corrected version reduces bias (especially for small n or large tables) but targets a slightly different population quantity. In practice, $V = 1$ requires near-deterministic mapping between categories, which is rarely attainable in non-square tables.

Miscellanea: Harald Cramér was a Swedish mathematician and statistician who made foundational contributions to probability theory and mathematical statistics. Cramér's V, associated with Harald Cramér, provides a standardized measure of association for categorical variables by adjusting the chi-square statistic for table dimensions. It improves on earlier measures such as Pearson's contingency coefficient, whose maximum depends on table size. Although widely used, Cramér's V is upward biased in small samples, and confusion with the phi coefficient—valid only for 2×2 tables—remains a common issue in practice.

Reference(s): Cramér (1946); Cohen (1988); Bergsma (2013)

Worked Example

Problem Statement Twenty-seven employees are cross-classified by education level (Low/Medium/High) and job satisfaction (Low/Medium/High). Test at $\alpha = 0.05$ whether education and satisfaction are associated, and quantify the strength of association using Cramér's V. H_0: Education and Satisfaction are independent.

	Satisf. Low	**Satisf. Med**	**Satisf. High**	
Educ. Low	6	2	1	9
Educ. Med	2	5	2	9
Educ. High	1	2	6	9
	9	9	9	27

1. Expected Frequencies
All row totals and column totals equal 9, so every expected cell count is:

$$E_{ij} = \frac{n_{i\cdot}\,n_{\cdot j}}{n} = \frac{9 \times 9}{27} = 3.$$

2. Chi-Square Contributions $(O - E)^2/E$

	Satisf. L	Satisf. M	Satisf. H
L	$(6{-}3)^2/3 = 3$	$(2{-}3)^2/3 = \frac{1}{3}$	$(1{-}3)^2/3 = \frac{4}{3}$
M	$(2{-}3)^2/3 = \frac{1}{3}$	$(5{-}3)^2/3 = \frac{4}{3}$	$(2{-}3)^2/3 = \frac{1}{3}$
H	$(1{-}3)^2/3 = \frac{4}{3}$	$(2{-}3)^2/3 = \frac{1}{3}$	$(6{-}3)^2/3 = 3$

3. Chi-Square Statistic

$$\chi^2 = 34/3 \approx 11.33.$$

4. Cramér's V
With $n = 27$, $r = c = 3$, $\min(r, c) - 1 = 2$:

$$V = \sqrt{\frac{34/3}{27 \times 2}} \approx 0.458.$$

5. Decision
Under H_0: $\chi^2 \sim \chi^2((r{-}1)(c{-}1)) = \chi^2(4)$.

α	$\chi^2_{(4)}$	Decision
5%	9.488	**Reject** (11.33 > 9.488)
1%	13.277	FTR (11.33 < 13.277)

$p = 0.023$. Education and job satisfaction are significantly associated at the 5% level. The effect size $V \approx 0.458$ indicates a moderate-to-strong association.

6. Interpretation of V
Cohen's (1988) benchmarks for V with $\min(r, c) - 1 = 2$: small ≈ 0.10, medium ≈ 0.30, large ≈ 0.50. $V = 0.458$ is close to the large threshold, consistent with the visible diagonal pattern: low-education employees concentrate in low-satisfaction, and high-education employees concentrate in high-satisfaction.

Practitioner Insight: Cramér's V is the standard effect-size measure for chi-square tests of independence in $r \times c$ tables. The scaling by $n(\min(r, c) - 1)$ ensures $V \in [0, 1]$ regardless of table size; for a 2×2 table, V reduces to $|\phi|$. Unlike the chi-square test, which only assesses significance, V quantifies the strength of association. A limitation is that the maximum attainable value depends on the marginal distributions, so comparisons across tables should be made cautiously. Because V ignores category ordering, measures such as Spearman's ρ or Kendall's τ may be more informative for ordinal data.

Common Pitfalls

- **Using V with sparse tables.** Cramér's V inherits the same large-sample assumptions as the chi-square test. When many expected counts are small, both the test and the resulting effect size can be unreliable.
- **Treating V as significance.** Cramér's V measures the strength of association, not its statistical significance. Always report the chi-square test alongside V.
- **Applying correlation benchmarks.** Although V ranges from 0 to 1, it is not a correlation coefficient. Effect-size interpretation depends on the table dimensions and degrees of freedom.
- **Confusing V with ϕ.** The two measures are identical only for 2×2 tables. For larger tables, Cramér's V provides the proper standardized effect size.
- **Assuming V shows direction.** Cramér's V is always nonnegative and provides no information about the pattern of association. Examine the contingency table or standardized residuals to identify which cells drive the relationship.

Partial Correlation Test

Association Between Two Variables Controlling for Others

CORR.11 | **Category:** Correlation | **Data:** Continuous | **Null:** t-distribution

When to Use

- Measure association after controlling covariates.
- Remove confounding effects from correlation analysis.
- Assess direct relationships between variables.
- Multiple predictors may influence both variables.
- Correlation remains after adjustment of interest.

Assumptions

- Observations are mutually independent.
- Relationships among variables are approximately linear.
- Variables are measured on interval or ratio scales.
- No influential outliers distort adjusted relations.
- Variables are jointly normally distributed (exact inference).

Test Statistic

$$t = r_{XY\cdot Z}\sqrt{\frac{n-k-2}{1-r_{XY\cdot Z}^2}} \sim t_{n-k-2} \quad \Big| \quad r_{XY\cdot Z} = \frac{r_{XY} - r_{XZ}\cdot r_{YZ}}{\sqrt{(1-r_{XZ}^2)(1-r_{YZ}^2)}}$$

Notation: r_{XY}, r_{XZ}, r_{YZ} are Pearson correlations, Z denotes control variables ($k = \dim Z$), n is sample size.

Plain English: Partial correlation measures the association between two variables after removing the linear effects of one or more control variables. It answers: *Are X and Y still related once Z is held constant?* By isolating variation unique to each variable, it helps distinguish direct relationships from associations induced by confounding. A large ordinary correlation may disappear after adjustment if both variables are related primarily through the control variable, whereas a substantial partial correlation suggests that the relationship persists beyond the shared influence of Z.

⚠ Practitioner Warning

Partial correlation removes only *linear* effects; nonlinear confounding, interactions, and omitted variables may still produce misleading results. Strong multicollinearity among the control variables inflates variance and can make estimates unstable. Measurement error in the control variables weakens adjustment and may introduce residual confounding. As the number of controls grows, degrees of freedom decrease and statistical power can deteriorate rapidly. A nonzero partial correlation should not be interpreted as evidence of causation without strong design assumptions.

Hypotheses

- H_0: $\rho_{XY\cdot Z} = 0$ (no association after control)
- H_1: $\rho_{XY\cdot Z} \neq 0$

Decision Rule: Reject H_0 if $|t| > t_{\alpha/2, n-k-2}$ or $p < \alpha$. Degrees of freedom decrease with each control variable.

Software References [Python, R, SAS]

Python	R	SAS
`pingouin.partial_corr`	`ppcor::pcor.test`	`PROC CORR (PARTIAL)`

Additional Notes

Partial correlation is equivalent to correlating regression residuals: regress X and Y separately on the control variables Z, then compute the Pearson correlation between the residuals. In matrix form, partial correlations can also be obtained from the inverse covariance (precision) matrix, making them useful in graphical modeling and network analysis. The coefficient remains bounded between -1 and 1 and may be interpreted similarly to an ordinary correlation. When the control variables explain much of the variation in X and Y, the partial correlation can differ substantially from the marginal correlation, occasionally changing sign (a suppression effect). For rank-based robustness, partial Spearman and partial Kendall correlations are commonly used.

Miscellanea: George Udny Yule (1871–1951) introduced partial correlation while studying spurious associations. His famous 1926 paper, "Why do we sometimes get nonsense-correlations between time-series?", showed how unrelated variables can appear strongly correlated because both are driven by a common trend. One example involved correlations between Church of England marriages and mortality rates, which were largely explained by the passage of time. Yule's work helped establish the idea that association does not imply causation and laid important foundations for path analysis, structural equation modeling, and modern causal inference.

Reference(s): Yule (1907); Fisher (1924); Anderson (2003)

Partial Correlation Test

CORR.11

Worked Example

Problem Statement For $n = 53$ respondents, three pairwise Pearson correlations are computed between income (X), health score (Y), and years of education (Z) (given below). Income and health are positively correlated, but education is also correlated with both. Compute the partial correlation of X and Y *controlling for* Z, and test whether it is significantly different from zero at $\alpha = 0.05$. $H_0: \rho_{XY \cdot Z} = 0$; $\quad H_1: \rho_{XY \cdot Z} \neq 0$.

$$r_{XY} = \tfrac{1}{2}, \qquad r_{XZ} = \tfrac{1}{2}, \qquad r_{YZ} = \tfrac{1}{2}.$$

1. Partial Correlation

$$r_{XY \cdot Z} = \frac{r_{XY} - r_{XZ}\, r_{YZ}}{\sqrt{(1 - r_{XZ}^2)(1 - r_{YZ}^2)}} = \frac{1}{3}.$$

The zero-order correlation $r_{XY} = 1/2$ has been reduced to $r_{XY \cdot Z} = 1/3$ after controlling for education — part of the apparent income–health association was actually due to education driving both.

2. Test Statistic

Under $H_0: \rho_{XY \cdot Z} = 0$:

$$t = \frac{r_{XY \cdot Z}\sqrt{n - k - 2}}{\sqrt{1 - r_{XY \cdot Z}^2}} = \frac{5}{2} = 2.50,$$

where $k = 1$ control variable, so $df = n - 3 = 50$.

3. Decision

Under H_0: $t \sim t(df)$.

α	$t_{(50),\, \alpha/2}$	Decision
5%	2.009	**Reject** (2.500 > 2.009)
1%	2.678	FTR (2.500 < 2.678)

$p = 0.016$. The partial correlation between income and health, controlling for education, is significantly positive at the 5% level.

4. Comparison of Correlations

Measure	Value	Interpretation
r_{XY}	1/2	Zero-order (confounded)
$r_{XY \cdot Z}$	1/3	Partial (education removed)

The reduction from 1/2 to 1/3 quantifies the confounding effect of education: roughly one-third of the apparent income–health association was spurious.

Practitioner Insight: Partial correlation $r_{XY \cdot Z}$ (Yule 1907) measures the linear association between X and Y after removing the linear effect of Z from both variables. Geometrically, it is the correlation between the residuals obtained by regressing X and Y separately on Z. If $r_{XY} = r_{XZ} r_{YZ}$, the observed association is fully explained by the common relationship with Z and the partial correlation is zero. Here $r_{XY} = 0.5 > r_{XZ} r_{YZ} = 0.25$, indicating that income and health retain an independent association after controlling for education. Partial correlation measures association, not causation, and may miss nonlinear or mediated relationships.

Common Pitfalls

- **Controlling for mediators instead of confounders.** Adjusting for mediators removes the indirect pathways you may wish to estimate, distorting the total effect. Control confounders (common causes), not mediators (intermediate variables). Use causal diagrams to distinguish them.
- **Conditioning on colliders and inducing spurious associations.** Controlling a collider ($X \rightarrow C \leftarrow Y$) opens a false path, creating artificial correlation (Berkson's paradox). Identify colliders with DAGs and avoid including them as control variables.
- **Treating significant partial correlation as causal evidence.** Partial correlation only adjusts for measured variables and cannot eliminate unmeasured confounding, reverse causation, or residual bias. Interpret as association remaining after adjustment—not as proof of causality.
- **Over-controlling and introducing multicollinearity.** Including many highly correlated controls leads to unstable estimates and inflated standard errors. Examine correlations among controls, use VIFs, combine variables when appropriate, and select covariates parsimoniously.
- **Ignoring loss of degrees of freedom when adding controls.** Partial correlations reduce degrees of freedom ($df = n - k - 2$), lowering power. Non-significance may reflect limited sample size rather than absence of effect. Report both zero-order and partial correlations and ensure adequate sample size.

Fisher's Z Transformation Test

Inference and Comparison of Correlation Coefficients

CORR.12 | **Category:** Association | **Data:** Continuous | **Null:** Normal (via transformation)

When to Use

- Test whether a population correlation differs from zero.
- Construct confidence intervals for correlations.
- Compare correlations from independent samples.
- Correlation inference requires variance stabilization.
- Pearson correlation is the primary measure of interest.

Assumptions

- Paired observations are mutually independent.
- Relationship between variables is approximately linear.
- Variables follow a bivariate normal distribution.
- Sample is randomly drawn from the population.
- Correlations are estimated using Pearson's method.

Test Statistic

$$Z = \tanh^{-1}(\rho) = \frac{1}{2}\ln\left(\frac{1+r}{1-r}\right). \qquad \text{For testing } H_0 : \rho = \rho_0\text{: } Z_{\text{test}} = \frac{Z - Z_0}{1/\sqrt{n-3}} \sim N(0,1).$$

Notation: r sample correlation, ρ population correlation, n sample size, Z_0 Fisher transform of ρ_0.

Plain English: Fisher's test transforms the sample correlation into a scale where its sampling distribution is approximately normal. This stabilizes variance and allows standard normal inference. It underlies confidence intervals for correlations and formal tests comparing correlations across independent samples.

⚠ Practitioner Warning

The normal approximation depends on approximate bivariate normality. With small samples or heavy tails, accuracy deteriorates. Fisher's test applies only to Pearson correlations; rank-based correlations require different methods.

Hypotheses

- H_0: $\rho = \rho_0$ (or $\rho_1 = \rho_2$ for two samples).
- H_1: $\rho \neq \rho_0$ (or $\rho_1 \neq \rho_2$).

Decision Rule: Reject H_0 if $|Z_{\text{test}}| > z_{\alpha/2}$.

Software References [Python, R, SAS]

Python	R	SAS
`Not Native (Manual)`	`psych::r.test()`	`PROC CORR (FISHER)`

Additional Notes

The variance of the Fisher-transformed correlation (independent of ρ) is $1/(n-3)$—not $1/n$ or $1/(n-1)$—reflecting the effective loss of degrees of freedom in estimating correlation and arising naturally from Fisher's asymptotic derivation. The transformation maps the bounded interval $(-1, 1)$ onto $(-\infty, \infty)$, revealing a hidden symmetry: equal distances in z correspond to equal "information distances." This is why inference is conducted in z-space: confidence intervals are formed using variance $1/(n-3)$ and then transformed back via $\rho = \tanh(z)$, while comparisons of two correlations use standard error $\sqrt{1/(n_1-3) + 1/(n_2-3)}$.

Meta-analysis of correlations: Raw correlations r cannot be averaged directly because their sampling variance depends on r. Instead, each r is transformed to Fisher's z, averaged using weights proportional to $(n-3)$, and then back-transformed to r. This z-based pooling is the standard approach in meta-analysis of correlations.

Miscellanea: Before Fisher (1915–1921), inference on correlation coefficients was notoriously difficult because the sampling distribution of r is highly skewed, especially near ± 1. Fisher's key insight was to transform r via $z = \frac{1}{2}\ln\left(\frac{1+r}{1-r}\right)$, producing a statistic that is approximately normal with variance $1/(n-3)$—essentially independent of the true correlation. This variance-stabilizing transformation turned a messy, nonlinear problem into a standard normal inference task.

More broadly, the method exemplifies Fisher's philosophy: when a statistic has a complicated distribution, transform it into one that behaves nicely. This idea influenced later developments such as Box–Cox transformations and generalized linear models. Remarkably, over a century later, Fisher's z-transformation remains the standard approach for inference on correlations in modern statistical software.

Reference(s): Fisher (1915); Hotelling (1953)

Worked Example

Problem Statement A researcher measures the Pearson correlation between job performance (X) and salary (Y) separately for experienced employees ($n_1 = 53$) and new employees ($n_2 = 53$). The observed correlations are $r_1 = 4/5$ and $r_2 = 3/5$.

Test at $\alpha = 0.05$ whether the two population correlations are equal. $H_0: \rho_1 = \rho_2$; $H_1: \rho_1 \neq \rho_2$.

1. Transform Both Correlations
For $r_1 = 4/5$ and $r_2 = 3/5$:

$$z_1 = \frac{1}{2}\ln\frac{1+r_1}{1-r_1} \approx 1.099, \quad z_2 = \frac{1}{2}\ln\frac{1+r_2}{1-r_2} \approx 0.693.$$

2. Standard Error
The standard error of $z_1 - z_2$ under H_0:

$$\text{SE} = \sqrt{\frac{1}{n_1-3} + \frac{1}{n_2-3}} = \frac{1}{5}.$$

3. Test Statistic
Under $H_0: \rho_1 = \rho_2$:

$$z = \frac{z_1 - z_2}{\text{SE}} \approx 2.027.$$

4. Decision
Under H_0: $z \sim \text{N}(0, 1)$.

α	$z_{\alpha/2}$	Decision
5%	1.960	**Reject** (2.027 > 1.960)
1%	2.576	FTR (2.027 < 2.576)

$p = 0.043$. The two correlations differ significantly at the 5% level: the performance–salary relationship is stronger among experienced employees.

5. Confidence Intervals for Each ρ
As a supplement, 95% CIs for the individual population correlations use $z_r \pm 1.96/\sqrt{n-3}$, then $r = \tanh(z)$:

$$\text{SE}_{\text{single}} = \frac{1}{\sqrt{50}} = \frac{1}{5\sqrt{2}}.$$

Group	$z_r \pm 1.96/\sqrt{50}$	95% CI for ρ
Experienced	1.099 ± 0.277	(0.676, 0.880)
New	0.693 ± 0.277	(0.394, 0.749)

The CIs overlap modestly, consistent with a borderline-significant two-sample test.

Practitioner Insight: Fisher's z-transformation converts Pearson's r into a statistic that is approximately normal with variance $1/(n-3)$, regardless of the underlying correlation. This stabilizes the variance and removes much of the skewness present when $|r|$ is large, making confidence intervals and comparisons of correlations straightforward. The transformation is the standard method for testing differences between independent correlations and for constructing confidence intervals for ρ. Its accuracy declines for very small samples or correlations close to ± 1, where bootstrap or permutation methods may be preferable.

Common Pitfalls

- **Using Fisher's transformation for testing $\rho = 0$.** For testing independence, the simple t-test for Pearson's r is appropriate and equivalent. Fisher's z-transform is intended for comparing correlations, testing non-zero null hypotheses, or constructing confidence intervals.
- **Failing to back-transform confidence intervals.** Confidence intervals computed on the z-scale must be converted back to the correlation scale using the inverse hyperbolic tangent. Reporting z-scale intervals is uninterpretable for most readers.
- **Treating dependent correlations as independent.** The Fisher test assumes independent samples. When correlations share variables or come from the same sample, use Steiger's, Williams', or Dunn–Clark tests to account for dependence between correlation estimates.
- **Assuming the transformation fixes non-normal data.** The transform stabilizes the sampling distribution of r, but it does not correct violations of bivariate normality or linearity. Use nonparametric or robust methods when data depart from assumptions underlying Pearson correlation.
- **Interpreting non-significant differences as equality.** A non-significant result indicates insufficient evidence of difference, not equality of correlations. Report confidence intervals for the difference and consider power limitations before drawing substantive conclusions.

Steiger's Z Test

Comparing Dependent Correlation Coefficients

CORR.13 | **Category:** Association | **Data:** Continuous | **Null:** Asymptotic normal

When to Use

- Compare two correlations from the same sample.
- Correlations share one variable in common.
- Determine which relationship is significantly stronger.
- Dependent correlations prevent Fisher z methods.
- Psychological and behavioral studies of associations.

Assumptions

- Observations are mutually independent.
- Variables follow a multivariate normal distribution.
- Correlations are estimated from the same sample.
- Correlation estimates use Pearson's correlation.
- Sample size supports asymptotic normal inference.

Test Statistic

$$Z = \frac{(Z_{12} - Z_{13})\sqrt{n-3}}{\sqrt{2-2s}} \quad \Big| \quad Z_{ij} = \tanh^{-1}(r_{ij}) \quad \Big| \quad \bar{r} = \tanh\left(\frac{Z_{12} + Z_{13}}{2}\right) \quad \Big| \quad s = \frac{r_{23}(1-2\bar{r}^2) - \frac{1}{2}\bar{r}^2(1-2\bar{r}^2 - r_{23}^2)}{(1-\bar{r}^2)^2}$$

Notation: r_{12}, r_{13}, r_{23} are pairwise correlations, n is sample size, s accounts for dependency structure.

Plain English: Steiger's test evaluates whether two correlations computed from the same sample differ significantly when they share a common variable. Unlike Fisher's Z test for independent correlations, Steiger's method explicitly accounts for the dependency induced by the shared variable. It is commonly used to answer questions such as whether a predictor is more strongly associated with one outcome than another, or whether a measure relates more closely to one construct than a competing construct. The test determines whether the observed difference is larger than would be expected from sampling variation alone.

⚠ Practitioner Warning

Steiger's test applies only to *dependent* correlations with one variable in common. Using it for independent samples or non-overlapping correlations yields incorrect inference. Accuracy declines with small samples or strong non-normality. Because the test depends on all three pairwise correlations, extreme values (e.g., near ±1) can make the variance estimate unstable. Statistical significance does not imply that the difference between correlations is practically important.

Hypotheses

- H_0: $\rho_{12} = \rho_{13}$
- H_1: $\rho_{12} \neq \rho_{13}$

Decision Rule: Reject H_0 if $|Z| > z_{\alpha/2}$. Reject H_0 if the standardized difference exceeds the normal critical value.

Software References [Python, R, SAS]

Python	R	SAS
`Not Native (Manual)`	`psych::r.test()`	`Not Native (Manual)`

Additional Notes

Steiger's Z test addresses a subtle but important problem: comparing two correlations that share a common variable (e.g., testing whether r_{12} differs from r_{13} in the same sample). The naive approach of treating these as independent and using Fisher's Z transformation yields incorrect results because it ignores the correlation r_{23} between the two compared variables. Steiger's contribution was deriving a test statistic that properly accounts for this dependency by incorporating all three pairwise correlations. The test assumes multivariate normality of (X_1, X_2, X_3) and relies on an asymptotic normal approximation that performs well in moderate samples. For very high correlations ($|r| > 0.95$) or small samples, bootstrap methods provide more accurate p-values.

Miscellanea: Steiger's 1980 *Psychological Bulletin* paper became one of the most-cited methodological articles in psychology despite addressing a seemingly narrow problem. Researchers often need to determine whether a variable correlates more strongly with one outcome than another, yet standard methods assume independent correlations. Although related tests were proposed earlier by Williams (1959) and Dunn and Clark (1969), Steiger's clear exposition and worked examples drove widespread adoption. The paper's influence extended beyond psychology into education, medicine, and organizational research, illustrating how a well-communicated solution to a practical problem can have broad impact.

Reference(s): Steiger (1980); Meng et al. (1992)

Worked Example

Problem Statement From a single sample of $n = 53$ employees, three pairwise correlations are computed between salary (X_1), job satisfaction (X_2), and work–life balance (X_3) (given below). Salary is more strongly correlated with satisfaction than with work–life balance. Test at $\alpha = 0.05$ whether this difference is statistically significant, i.e. $H_0: \rho_{12} = \rho_{13}$.

$$r_{12} = \tfrac{3}{4}, \quad r_{13} = \tfrac{1}{2}, \quad r_{23} = \tfrac{1}{2}.$$

1. Why Steiger's Test?
Both r_{12} and r_{13} share variable X_1 and come from the *same* sample, so they are dependent. Fisher's two-sample Z-test (preceding example) is invalid here. Steiger's Z_H statistic corrects for this dependence using a factor s built from the average transformed correlation.

2. Fisher z-Transforms

$$Z_{r_{12}} = \tfrac{1}{2}\ln\tfrac{1+3/4}{1-3/4} = \tfrac{1}{2}\ln 7 \approx 0.9730,$$

$$Z_{r_{13}} = \tfrac{1}{2}\ln\tfrac{1+1/2}{1-1/2} = \tfrac{1}{2}\ln 3 \approx 0.5493.$$

3. Average Transformed Correlation
The mean Fisher z-value and the back-transformed average correlation are:

$$\bar{Z} = \tfrac{1}{2}(0.9730 + 0.5493) = 0.7611,$$

$$\bar{r} = \tanh(\bar{Z}) = 0.6417, \quad \bar{r}^2 = 0.4118.$$

Note: $\bar{r} = \tanh(\bar{Z}) \neq (r_{12} + r_{13})/2$ because the averaging is done on the z-scale.

4. Correction Factor s

$$s = \frac{r_{23}(1 - 2\bar{r}^2) - \frac{1}{2}\bar{r}^2(1 - 2\bar{r}^2 - r_{23}^2)}{(1 - \bar{r}^2)^2} = 0.2987.$$

5. Steiger's Z_H Statistic
With $n - 3 = 50$:

$$Z_H = \frac{\sqrt{n-3}\,(Z_{r_{12}} - Z_{r_{13}})}{\sqrt{2 - 2s}} \approx 2.529.$$

6. Decision
Under H_0: $Z_H \sim \mathrm{N}(0, 1)$.

α	$z_{\alpha/2}$	Decision
5%	1.960	**Reject** (2.529 > 1.960)
1%	2.576	FTR (2.529 < 2.576)

$p = 0.011$. The salary–satisfaction correlation ($r_{12} = 3/4$) is significantly stronger than the salary–work-life-balance correlation ($r_{13} = 1/2$) at the 5% level.

7. Comparison with Fisher's Z Test

Test	Statistic	p
Fisher (independent)	$Z = 5\ln(3/2) = 2.027$	0.043
Steiger (dependent)	$Z_H = 2.529$	0.011

Treating the same-sample correlations as independent (Fisher) would understate significance. Steiger's correction for dependence via s yields the correct, *stronger* result.

Practitioner Insight: Steiger's Z_H test compares two correlations that share a common variable within the same sample. Unlike Fisher's independent-sample Z-test, it adjusts for the dependence between correlations through the correction factor s, which incorporates the correlation between the non-shared variables. As this dependence increases, the comparison becomes more precise and the test gains power. The method produces a standard normal Z statistic and is widely implemented in software and online calculators. A different formula is required when the two correlations come from the same sample but share no common variable.

Common Pitfalls

- **Using Fisher's Z for dependent correlations.** Steiger's test is required when correlations come from the same sample and share variables. Fisher's Z assumes independence and underestimates standard errors, inflating Type I error.
- **Ignoring the overlap structure.** The correlation between the non-shared variables directly affects the variance of the comparison. Always report and interpret this correlation when presenting results.
- **Treating non-significance as equality.** A non-significant result does not prove that the correlations are equal. Low power, small samples, or strong overlap can mask meaningful differences.
- **Using very small samples.** Steiger's test relies on large-sample approximations and multivariate normality. With small samples, bootstrap or permutation methods are often preferable.
- **Ignoring multiplicity.** Comparing several correlations within the same dataset inflates the familywise error rate. Apply appropriate multiplicity adjustments when conducting multiple tests.

VII. CORR — Technical Notes

Variable Types and the Choice of Correlation Coefficient

Correlation coefficients differ in the assumptions they make about the measurement scale and underlying nature of each variable. The key distinctions are as follows.

Natural binary (true dichotomy). A variable that is genuinely two-valued with no underlying continuum—e.g., alive/dead, pass/fail, male/female. No latent continuous variable is assumed.

Artificial dichotomy. A naturally continuous variable that has been dichotomised at a threshold—e.g., hypertensive/normotensive from blood pressure, or high/low income from a continuous income distribution. An underlying latent continuous variable Z is assumed to exist.

Latent variable (Z). An unobserved continuous variable assumed to underlie an observed categorical or ordinal measure. The biserial and tetrachoric coefficients estimate the correlation between two such latent variables from their observed dichotomised proxies.

Continuous variable. A variable measured on an interval or ratio scale with no dichotomisation—e.g., height, test score, temperature.

Ordinal variable. A variable with a meaningful rank order but no assumption of equal intervals—e.g., Likert scale responses, educational attainment levels.

Polychotomy. A categorical variable with more than two unordered categories. No latent continuum is assumed unless explicitly modelled (polychoric coefficient).

Key practical rule. Use the point-biserial when the binary variable is a true dichotomy. Use the biserial when it is an artificial dichotomy with a plausible latent continuous variable. Misapplying the two inflates or deflates the estimated correlation systematically.

Summary: Variable Types and Correlation Coefficients

Coefficient	Variable X	Variable Y	Latent Z?	Notes
Pearson r	Continuous	Continuous	No	Assumes bivariate normality
Spearman ρ	Ordinal/continuous	Ordinal/continuous	No	Rank-based; no distributional assumption
Kendall τ	Ordinal/continuous	Ordinal/continuous	No	Concordance-based; robust to outliers
Point-Biserial r_{pb}	Natural binary	Continuous	No	Special case of Pearson r
Biserial r_b	Artificial dichotomy	Continuous	Yes (X)	Estimates latent-continuous/continuous correlation
Tetrachoric r_t	Artificial dichotomy	Artificial dichotomy	Yes (X & Y)	Estimates latent-to-latent correlation
Polychoric ρ_p	Ordinal (polychot.)	Ordinal (polychot.)	Yes (X & Y)	Generalises tetrachoric to $>$ 2 categories
Phi ϕ	Natural binary	Natural binary	No	Special case of Pearson r for 2×2 tables
Cramér's V	Nominal ($\geq$ 2 cat.)	Nominal ($\geq$ 2 cat.)	No	Generalises ϕ to larger contingency tables

Part VIII

Counts, Proportions & Categorical Structure

Testing independence, association, and fit for categorical outcomes

Categorical Data and Contingency Tables

VIII. CTAB — A Conceptual Primer

What Are Contingency Tables?

A **contingency table** is an $r \times c$ cross-tabulation of counts n_{ij} for $i = 1, \ldots, r$ row categories and $j = 1, \ldots, c$ column categories, with grand total $n = \sum_{i,j} n_{ij}$, row marginals n_{i+}, and column marginals n_{+j}. The central questions are:

- **Association / Independence:** are the row and column classifications related?
- **Homogeneity:** are conditional distributions the same across groups?
- **Agreement:** do two raters / instruments classify subjects consistently?
- **Trend:** does the response probability change monotonically across ordered categories?

Probability Models and Sampling Schemes

- **Multinomial:** n fixed; cell counts jointly multinomial. Used for independence testing.
- **Product-multinomial:** row totals fixed by design (separate samples from r groups); tests **homogeneity** of conditional distributions.
- **Poisson:** cell counts independent Poisson (μ_{ij}); conditioning on n recovers multinomial.
- **Hypergeometric:** both margins fixed; basis for Fisher's exact test.

Independence and homogeneity share the same test statistics but differ in sampling model and interpretation: independence asks whether two variables are associated; homogeneity asks whether the distribution of one variable is the same across levels of another.

Expected Counts and the Independence Hypothesis

Under independence, $\pi_{ij} = \pi_{i+}\pi_{+j}$. The MLE of the expected count is:

$$\hat{m}_{ij} = \frac{n_{i+}\, n_{+j}}{n}.$$

Cochran's rule: all $\hat{m}_{ij} \geq 5$ is a sufficient (not necessary) condition for the χ^2 approximation. When violated, use exact tests, collapse categories, or apply power-divergence (Cressie-Read) alternatives.

The 2 × 2 Table: Core Quantities

The 2×2 table underlies many tests in this chapter (Fisher, Barnard, Boschloo, McNemar, Mantel-Haenszel) and introduces the key effect measures:

	$Y = 1$	$Y = 0$	
$X = 1$	a	b	$a + b$
$X = 0$	c	d	$c + d$
	$a + c$	$b + d$	n

Odds ratio: $\text{OR} = ad/bc$; the ratio of the odds of $Y = 1$ in group $X = 1$ vs. $X = 0$. OR = 1 implies independence. Symmetric: OR for X given Y equals OR for Y given X.

Relative risk (risk ratio): $\text{RR} = [a/(a+b)]/[c/(c+d)]$; ratio of proportions. More intuitive than OR but only valid under prospective (cohort) sampling. RR = 1 implies independence.

Risk difference: $\text{RD} = a/(a + b) - c/(c + d)$; the absolute difference in event proportions. Most clinically interpretable; $1/\text{RD}$ is the **number needed to treat** (NNT).

OR vs. RR: OR ≈ RR when the outcome is rare ($a \ll a+b$ and $c \ll c + d$). For common outcomes, OR overstates the RR in the direction away from null; always report which measure is used.

Confidence intervals: all three measures are approximately log-normal; CIs are constructed on the log scale and back-transformed. For OR: $\widehat{\log \text{OR}} \pm 1.96/\sqrt{1/a + 1/b + 1/c + 1/d}$.

I. Independence Tests

Pearson χ^2 test:

$$X^2 = \sum_{i,j} \frac{(n_{ij} - \hat{m}_{ij})^2}{\hat{m}_{ij}} \xrightarrow{d} \chi^2_{(r-1)(c-1)} \text{ under } H_0.$$

The most widely used test; requires adequate expected counts. The likelihood ratio (G-test) $G^2 = 2\sum n_{ij} \log(n_{ij}/\hat{m}_{ij})$ is asymptotically equivalent and decomposes additively over subtables—preferred when hierarchical decomposition is needed.

Fisher's exact test (2×2): Conditions on both marginal totals (hypergeometric model); enumerates all tables with the same margins; p-value is the sum of hypergeometric probabilities at least as extreme as observed. Exact and valid for any n; the standard recommendation when $\min(\hat{m}_{ij}) < 5$. **Two-sided p-value**: sum all tables with probability ≤ that of the observed table (not simply doubling the one-sided value).

Barnard's exact test (2×2): Unconditional exact test; does not condition on column margins (treats column totals as random under product-binomial sampling). More powerful than Fisher's under the product-binomial model because it does not discard information by conditioning. Computationally intensive; maximises the p-value over the nuisance parameter (success probability under H_0).

Boschloo's test (2×2): Unconditional test that uses Fisher's p-value as the test statistic but evaluates it unconditionally. **Uniformly more powerful than Fisher's exact test** while maintaining exact size control—the recommended default for 2×2 tables when an unconditional exact test is desired.

II. Measures of Association

A significant χ^2 tells you *that* an association exists, not *how strong* it is. X^2 grows with n and cannot be compared across tables of different sizes. Always report an effect measure alongside the p-value.

Measure	Formula and range
ϕ	$\sqrt{X^2/n}$; $[0, 1]$ for 2×2; equals Pearson r between two binary variables.
Cramér's V	$\sqrt{X^2/[n(\min(r, c) - 1)]}$; $[0, 1]$; generalizes ϕ to $r \times c$.
Contingency C	$\sqrt{X^2/(X^2 + n)}$; $[0, \sqrt{(k-1)/k}]$; upper bound < 1, so less comparable.
Odds ratio	ad/bc; $[0, \infty)$; symmetric; log-OR $\approx \mathcal{N}$ for CIs.
Relative risk	$[a/(a+b)]/[c/(c+d)]$; prospective data only.
Risk difference	$a/(a+b) - c/(c+d)$; $[-1, 1]$; most clinically direct; NNT $= 1/\lvert\text{RD}\rvert$.

Interpretation of Cramér's V:

- $V \approx 0.10$: weak association.
- $V \approx 0.30$: moderate association.
- $V \approx 0.50$: strong association.

Context matters: $V = 0.15$ may be large in some fields and negligible in others.

III. Sparse Tables and Small-Sample Corrections

Yates' continuity correction (2×2): Subtracts $\frac{1}{2}$ from $|n_{ij} - \hat{m}_{ij}|$ before squaring; reduces Type I error inflation but is generally **over-conservative**—it over-corrects, reducing power unnecessarily. Prefer Boschloo's or Fisher's exact test for small n.

Structural vs. sampling zeros:

- **Structural zero:** the cell is logically impossible (e.g. male pregnancy); exclude from analysis and reduce df accordingly.
- **Sampling zero:** the cell is possible but unobserved due to small n; add a small constant (e.g. +0.5) for log-OR estimation; use exact tests for inference.

Power-divergence statistics: Cressie–Read family unifies χ^2 and G^2:

$$I^\lambda = \frac{2}{\lambda(\lambda+1)} \sum_{i,j} n_{ij} \left[\left(\frac{n_{ij}}{\hat{m}_{ij}} \right)^\lambda - 1 \right],$$

with $\lambda = 1$ giving X^2 and $\lambda \to 0$ giving G^2. The **Freeman–Tukey statistic** ($\lambda = -\frac{1}{2}$) is more robust to sparse tables than X^2 or G^2.

Mid-p correction for Fisher's exact test: Replaces the exact p-value $P(T \geq t_{\text{obs}})$ with $\frac{1}{2}P(T = t_{\text{obs}}) + P(T > t_{\text{obs}})$; reduces the inherent conservatism of Fisher's test; not strictly exact but has better size properties in practice.

IV. Paired and Matched-Data Tests

McNemar's test (2×2 paired): For matched pairs with binary outcomes, only the **discordant** cells matter:

$$\chi^2_{\text{McN}} = \frac{(b-c)^2}{b+c} \sim \chi^2_1,$$

where b = (positive, negative) pairs and c = (negative, positive) pairs. Concordant cells (a, d) contribute no information about change. Tests H_0: marginal homogeneity ($\pi_{1+} = \pi_{+1}$). For small $b + c$, use the exact binomial p-value.

McNemar–Bowker test ($k \times k$ paired): Generalizes McNemar to $k > 2$ nominal categories; tests symmetry of the off-diagonal counts:

$$S = \sum_{i<j} \frac{(n_{ij} - n_{ji})^2}{n_{ij} + n_{ji}} \sim \chi^2_{k(k-1)/2}.$$

Stuart–Maxwell test ($k \times k$ paired): Tests **marginal homogeneity** (not full symmetry) for matched k-category tables; more directly answers whether the marginal distributions shifted between conditions. Preferred over McNemar–Bowker when the goal is to detect directional change rather than any asymmetry.

Cochran's Q test (binary, k related samples): Generalizes McNemar to $k \geq 2$ related binary measurements on the same subjects (repeated binary outcomes):

$$Q = \frac{(k-1)\left[k \sum_j C_j^2 - \left(\sum_j C_j\right)^2\right]}{k \sum_i R_i - \sum_i R_i^2} \sim \chi^2_{k-1},$$

where C_j is the column (treatment) total and R_i is the row (subject) total. Nonparametric analog of repeated-measures ANOVA for binary outcomes.

V. Log-Linear Models

Log-linear models treat cell counts as Poisson responses and model the **log expected count** as a linear combination of main effects and interactions:

$$\log \mu_{ij} = \lambda + \lambda_i^A + \lambda_j^B + \lambda_{ij}^{AB}.$$

Setting $\lambda_{ij}^{AB} = 0$ for all (i, j) imposes **independence**. Estimated via IRLS; inference by G^2 differences between nested models.

Why log-linear matters for this chapter:

- The χ^2 and G^2 independence tests are special cases (saturated vs. independence model).
- The Mantel–Haenszel and CMH tests arise from fitting log-linear models with a stratum factor.
- Multi-way tables ($A \times B \times C$) require log-linear models to separate partial (conditional) from marginal associations.

- Model selection: **hierarchical** models (all lower-order relatives included) are standard; AIC and BIC guide selection among nested models.

Three-way tables and confounding: In an $A \times B \times C$ table, A and B may appear associated marginally but be independent conditional on C (**Simpson's paradox** / confounding). CMH stratification addresses this directly; see Section VI.

VI. Stratified Analysis: Mantel–Haenszel Framework

When data are collected across K strata (e.g. clinical sites, age groups, study periods), analyzing each stratum separately loses power and ignores confounding. The Mantel–Haenszel framework pools information across strata while controlling for the stratification variable.

Cochran–Mantel–Haenszel (CMH) test: A family of tests for $r \times c$ tables stratified over K strata. The general association statistic pools deviations from stratum-specific independence:

$$\text{CMH} = \frac{(\sum_k (n_{11k} - \hat{m}_{11k}))^2}{\sum_k \hat{V}_k} \sim \chi^2_1,$$

where $\hat{m}_{11k}$ and $\hat{V}_k$ are the stratum-specific expected value and variance. Three variants are used depending on the scale of the response variable:

- **General Association** ($df = (r-1)(c-1)$): nominal response; any association.
- **Row Mean Scores** ($df = r - 1$): ordinal response; mean score difference.
- **Correlation / Linear-by-Linear** ($df = 1$): both variables ordinal; linear trend.

Mantel–Haenszel test (common odds ratio): Estimates and tests a single pooled odds ratio $\widehat{\text{OR}}_{\text{MH}}$ across K strata:

$$\widehat{\text{OR}}_{\text{MH}} = \frac{\sum_k a_k d_k / n_k}{\sum_k b_k c_k / n_k},$$

where a_k, b_k, c_k, d_k are the 2×2 cell counts in stratum k. Valid when the odds ratio is homogeneous across strata (check with Breslow–Day or Woolf first).

VII. Tests for Odds Ratio Homogeneity

Before pooling with Mantel–Haenszel, verify that the odds ratio is consistent across strata.

Breslow–Day test: Tests H_0: $\text{OR}_1 = \text{OR}_2 = \cdots = \text{OR}_K$ (common odds ratio across all strata):

$$X^2_{\text{BD}} = \sum_{k=1}^{K} \frac{(a_k - \hat{A}_k)^2}{\hat{V}_k} \sim \chi^2_{K-1},$$

where $\hat{A}_k$ is the expected count under the common OR. Significant result ⇒ effect modification; pooling is inappropriate; stratified reporting required.

Woolf's test: Earlier alternative; regresses $\log \widehat{\text{OR}}_k$ on an intercept with weights $1/\hat{V}(\log \widehat{\text{OR}}_k)$; the χ^2_{K-1} homogeneity statistic is the weighted sum of squared deviations from the pooled log-OR. Requires all strata to have non-zero cells (add 0.5 correction if necessary).

VIII. Trend Tests for Ordered Categories

When the column (or row) variable is **ordinal**, tests that exploit the ordering are more powerful than the omnibus χ^2.

χ^2 test for trend ($2 \times c$): Assigns integer scores s_j to the c ordered columns; the trend statistic $T^2 = (\sum_j s_j n_{1j} - n_{1+}\bar{s})^2 / [n_{1+} n_{2+} \sum_j s_j^2 p_{+j}/n - \bar{s}^2]$ has approximately χ^2_1 under H_0 of no linear trend. Uses only 1 df regardless of c; substantially more powerful than the $(r-1)(c-1)$-df omnibus test when a monotone trend is the alternative.

Cochran–Armitage test: Formally equivalent to the χ^2 trend test for $2 \times c$ tables; commonly described as a score test for a logistic regression of the binary response on the ordered exposure scores. The standard reference for dose-response and exposure-response analyzes in epidemiology. Exact version available for small samples.

IX. Agreement Measures

Agreement measures quantify how consistently multiple raters (or instruments) classify the same subjects. Simple percentage agreement $P_o = \sum_i n_{ii}/n$ is inflated by chance; all measures below correct for this in different ways and for different designs.

Cohen's κ (two raters, nominal):

$$\kappa = \frac{P_o - P_e}{1 - P_e}, \quad P_e = \sum_{i=1}^{k} \frac{n_{i+} n_{+i}}{n^2}.$$

$\kappa = 0$: agreement no better than chance; $\kappa = 1$: perfect agreement; $\kappa < 0$: systematic disagreement.

Landis–Koch benchmarks:

κ	**Strength of agreement**
< 0.00	Poor (less than chance)
0.00–0.20	Slight
0.21–0.40	Fair
0.41–0.60	Moderate
0.61–0.80	Substantial
0.81–1.00	Almost perfect

Weighted κ (two raters, ordinal): Assigns partial credit for near-misses using a weight matrix $w_{ij} \in [0, 1]$ where $w_{ii} = 1$ (perfect agreement) and $w_{ij} < 1$ penalizes disagreement proportionally to the distance $|i - j|$:

$$\kappa_w = \frac{\sum_{ij} w_{ij}(p_{ij} - p_{i+}p_{+j})}{\sum_{ij} w_{ij}(p_{i+}p_{+j} - p_{i+}p_{+j})}.$$

Linear weights $w^L_{ij} = 1 - |i - j|/(k-1)$ give equal penalty per step; **quadratic weights** $w^Q_{ij} = 1 - (i-j)^2/(k-1)^2$ penalize larger disagreements more and make κ_w equivalent to the intraclass correlation coefficient. Always specify the weight scheme used.

Fleiss' κ ($m \geq 2$ raters, nominal): Extends Cohen's κ to m raters classifying n subjects into k categories. Each subject is rated by all m raters; let n_{ij} be the number of raters assigning subject i to category j:

$$\kappa_F = \frac{\bar{P} - \bar{P}_e}{1 - \bar{P}_e},$$

where $\bar{P} = \frac{1}{n}\sum_{i=1}^{n} P_i$, $P_i = \frac{1}{m(m-1)}\sum_{j=1}^{k} n_{ij}(n_{ij} - 1)$, and $\bar{P}_e = \sum_{j=1}^{k} p_j^2$ with $p_j = \frac{1}{nm}\sum_{i=1}^{n} n_{ij}$. Use Fleiss' κ instead of Cohen's whenever $m > 2$; Cohen's κ is not defined for more than two raters.

Krippendorff's α (any number of raters, any scale): The most general agreement measure; handles nominal, ordinal, interval, and ratio data; accommodates missing ratings:

$$\alpha = 1 - \frac{D_o}{D_e},$$

where $D_o = \frac{1}{n^*}\sum_i \sum_{c<c'} d^2(c, c')\, n_{ic}\, n_{ic'}$ is observed disagreement and $D_e = \frac{1}{n^*(n^*-1)}\sum_{c<c'} d^2(c, c')\, n_c\, n_{c'}$ is expected disagreement; n^* is the total number of (non-missing) ratings and $d(c, c')$ is the metric difference between categories c and c' (e.g. $|c - c'|$ for ordinal, $(c - c')^2$ for interval). $\alpha = 1$ is perfect agreement; $\alpha = 0$ is chance; $\alpha \geq 0.80$ is the conventional reliability threshold (Krippendorff 2004).

Gwet's AC_1 (two raters, nominal): Proposed as a kappa alternative robust to the **kappa paradox**: κ is artificially depressed when one category is very prevalent (or very rare) even when raters agree consistently. AC_1 replaces κ's chance-agreement term P_e with a term that is less sensitive to marginal distributions:

$$AC_1 = \frac{P_o - P_e^w}{1 - P_e^w}, \quad P_e^w = \frac{1}{k(k-1)}\sum_{i=1}^{k} p_i(1 - p_i),$$

where $p_i = (n_{i+} + n_{+i})/(2n)$ and k is the number of categories. When category distributions are balanced, $AC_1 \approx \kappa$; when distributions are highly skewed, AC_1 better reflects rater performance.

Choosing among agreement measures:

Situation	Measure
2 raters, nominal	Cohen's κ
2 raters, ordinal	Weighted κ
$m > 2$ raters, nominal	Fleiss' κ
Any raters, any scale	Krippendorff's α
Prevalence imbalance	Gwet's AC_1

Limitations of all kappa-type measures: κ and AC_1 are sensitive to prevalence — high or low prevalence of one category can inflate P_e and deflate κ even when P_o is high (the **kappa paradox**). Always report P_o alongside any agreement measure and verify that category distributions are not severely skewed before interpreting κ in isolation.

Test Selection Guide

Design / goal	Recommended test
Independence, large n	χ^2 (Pearson or G^2)
Independence, small n	Fisher's exact
Unconditional exact (2×2)	Boschloo's test
Paired binary	McNemar
Paired k-category symmetry	McNemar–Bowker
Paired k-category marginal	Stuart–Maxwell
Repeated binary (k times)	Cochran's Q
Stratified, any assoc.	CMH general association
Stratified, common OR	Mantel–Haenszel
OR homogeneity across strata	Breslow–Day / Woolf
Ordered categories, trend	Cochran–Armitage
2 raters, nominal agreement	Cohen's κ
2 raters, ordinal agreement	Weighted κ
$m > 2$ raters, nominal	Fleiss' κ
Any raters, any scale	Krippendorff's α
Prevalence-robust agreement	Gwet's AC_1

Common Pitfalls

- **Association ≠ effect size:** χ^2 grows with n; always report Cramér's V or OR.
- **OR vs. RR confusion:** OR overstates RR for common outcomes; specify which is reported.
- **χ^2 with small expected counts:** use Fisher's, Boschloo's, or collapse cells when any $\hat{m}_{ij} < 5$.
- **Yates over-correction:** prefer exact tests to Yates' correction for 2×2 tables.
- **Two-sided Fisher p-value:** do not simply double the one-sided value; sum all tables with probability $\leq$ that of the observed.
- **Ignoring pairing:** applying χ^2 to paired data (use McNemar); applying McNemar to independent data (use χ^2).
- **Pooling before homogeneity check:** always run Breslow–Day or Woolf before interpreting the MH common odds ratio.
- **Conflating symmetry and marginal homogeneity:** McNemar–Bowker tests symmetry; Stuart–Maxwell tests marginal homogeneity— these are different hypotheses.
- **Ignoring ordinal structure:** omnibus χ^2 is less powerful than Cochran–Armitage when a trend is the plausible alternative.
- **Prevalence effect on κ (kappa paradox):** κ can be low even when P_o is high when one category dominates; report P_o alongside κ and consider AC_1 or Krippendorff's α when distributions are highly skewed.

- **Using Cohen's κ for $m > 2$ raters:** Cohen's κ is defined only for two raters; use Fleiss' κ or Krippendorff's α for three or more raters.
- **Comparing κ across studies:** do not compare κ values across studies with different prevalences or numbers of raters; the same underlying agreement can yield very different κ values.
- **Simpson's paradox:** marginal association can reverse direction within strata; always examine stratified tables.

Bottom Line

The correct test is determined by the *structure* of the data: sampling scheme, pairing, stratification, ordering, and inferential goal. Always complement tests with effect measures (OR, RR, Cramér's V); check expected counts before applying χ^2; test homogeneity before pooling across strata; exploit ordinal structure with trend tests; and be alert to Simpson's paradox in unstratified analyses. For agreement studies, match the measure to the design: Cohen's κ for two raters, Fleiss' κ for more than two, Krippendorff's α when scale type or missing data are concerns, and Gwet's AC_1 when category prevalence is highly skewed.

Chi-Square Test of Independence

Association Between Two Categorical Variables (Unpaired Data)

CTAB.1 | **Category:** Contingency Tables | **Data:** Categorical | **Null:** Chi-square

When to Use

- Test independence between two categorical variables.
- Analyze contingency tables of counts/frequencies.
- Compare observed vs expected category patterns.
- Survey, clinical, or demographic cross-tabulations.
- Large-sample categorical data analysis.

Assumptions

- Observations are independent.
- Data are frequency counts, not percentages.
- Categories are mutually exclusive/exhaustive.
- Expected frequencies typically ≥ 5 per cell (rule of thumb).
- Sampling distribution follows asymptotic χ^2.

Test Statistic

$$\chi^2 = \sum_{i=1}^{r}\sum_{j=1}^{c} \frac{(O_{ij} - E_{ij})^2}{E_{ij}}, \qquad E_{ij} = \frac{(\text{row}_i)(\text{col}_j)}{n}. \qquad \text{Under } H_0, \chi^2 \sim \chi^2_{(r-1)(c-1)}.$$

Notation: O_{ij} = observed count; E_{ij} = expected count under independence (H_0)

Plain English: The chi-square test of independence evaluates whether two categorical variables are related. It compares the observed cell counts in a contingency table to the counts expected if the variables were independent. Large discrepancies between observed and expected counts provide evidence of association.

⚠ Practitioner Warning

Expected counts should generally satisfy $E_{ij} \geq 5$ in most cells; sparse tables can invalidate the chi-square approximation. Use Fisher's exact test for small 2×2 tables or combine categories when scientifically reasonable. A significant result indicates association, not causation, direction, or practical importance. The test is sensitive to sample size: very large samples may detect trivial associations, while small samples may miss meaningful ones. Report effect sizes such as Cramér's V or odds ratios alongside the p-value.

Hypotheses

- H_0: Variables are independent (no association).
- H_1: Variables are dependent (association exists).

Decision Rule: Reject H_0 if $\chi^2 > \chi^2_{\alpha,(r-1)(c-1)}$ or $p < \alpha$.

Cutoffs: $\chi^2_{0.05,1} = 3.84$; $\chi^2_{0.05,2} = 5.99$; $\chi^2_{0.05,3} = 7.81$.

Software References [Python, R, SAS]

```
scipy.stats.chi2_contingency      chisq.test      PROC FREQ; / chisq;
```

Additional Notes

Effect-size measures complement significance testing: Cramér's V ranges from 0 to 1, with rough guidelines of 0.1, 0.3, and 0.5 for small, medium, and large associations. For 2×2 tables, $\phi = \sqrt{\chi^2/n}$ equals Cramér's V. Yates' continuity correction reduces Type I error in small 2×2 tables but can be overly conservative. Cochran's conditions recommend that no expected count be below 1 and that at least 80% of cells have $E_{ij} \geq 5$. For ordered categories, the Cochran–Armitage trend test is often more powerful than the ordinary chi-square test because it exploits the ordinal structure.

$$V = \sqrt{\chi^2/(n\min(r-1, c-1))}, \qquad \chi^2_{\text{Yates}} = \sum \frac{(|O_{ij} - E_{ij}| - 0.5)^2}{E_{ij}}.$$

Miscellanea: Karl Pearson introduced the χ^2 test in 1900, originally as a goodness-of-fit procedure for comparing observed and expected frequencies. Ronald Fisher later clarified the correct degrees-of-freedom adjustments, particularly when parameters are estimated from the data, helping establish the modern $(r-1)(c-1)$ formulation for contingency tables. The Pearson–Fisher rivalry became one of the most famous disputes in statistical history and shaped much of early 20th-century mathematical statistics.

Reference(s): Pearson (1900); Cochran (1952)

Chi-Square Test of Independence **CTAB.1**

Worked Example

Problem Statement A clinical trial randomizes $n = 120$ patients equally between two treatment groups (Group A and Group B, 60 each). Each patient's response is classified as Low, Medium, or High. Test whether response level is independent of treatment group at $\alpha = 0.05$. H_0: response level and treatment group are independent; H_1: they are associated.

	Observed Counts				Expected Counts		
	Low	Medium	High	Total	Low	Medium	High
Group A	9	25	26	60	15	20	25
Group B	21	15	24	60	15	20	25
Total	30	40	50	120			

1. Expected Cell Counts

Under H_0: $E_{ij} = (\text{row } i \text{ total}) \times (\text{col } j \text{ total})/n = c_j/2$.
All $E_{ij} \geq 5$: the large-sample approximation is valid.

2. Cell Contributions $(O_{ij} - E_{ij})^2/E_{ij}$

	Low	Medium	High
Group A	$\frac{(9-15)^2}{15} = \frac{12}{5}$	$\frac{(25-20)^2}{20} = \frac{5}{4}$	$\frac{(26-25)^2}{25} = \frac{1}{25}$
Group B	$\frac{(21-15)^2}{15} = \frac{12}{5}$	$\frac{(15-20)^2}{20} = \frac{5}{4}$	$\frac{(24-25)^2}{25} = \frac{1}{25}$

Because the row totals are equal, Group A and Group B contribute *identical* terms.

3. Test Statistic

$$X^2 = \sum_{i,j} \frac{(O_{ij} - E_{ij})^2}{E_{ij}} = 2\left(\frac{12}{5} + \frac{5}{4} + \frac{1}{25}\right) = 7.38.$$

4. Degrees of Freedom and Decision

$df = (r-1)(c-1) = 2$. Under H_0: $X^2 \sim \chi^2(2)$.

α	$\chi^2_{0.05}(2)$	Decision
5%	5.991	**Reject** (7.38 > 5.991)
1%	9.210	FTR (7.38 < 9.210)

$p = 0.025$. Response level and treatment group are **not** independent.

5. Direction of Association

	Low	Medium	High
Group A	$9 < 15$	$25 > 20$	$26 \approx 25$
Group B	$21 > 15$	$15 < 20$	$24 \approx 25$

Practitioner Insight: The Pearson chi-square test of independence evaluates whether the conditional distribution of one categorical variable changes across levels of another. Under H_0, expected counts are $E_{ij} = n_{i\cdot}n_{\cdot j}/n$, with deviations weighted by $1/E_{ij}$, so smaller expected counts contribute more per unit discrepancy. Here the equal row totals force identical expected counts across rows, making the test driven mainly by the Low (4.8) and Medium (2.5) columns, while High contributes little (0.08). The primary difference is that Group B has far more Low responses (35% versus 15%). With all expected counts at least 15, the χ^2 approximation is reliable. A significant result establishes association but not its strength; Cramér's $V = 0.248$ indicates a moderate relationship. The test is asymptotically equivalent to the likelihood-ratio test G^2; for ordered categories, trend tests or ordinal logistic regression may provide greater power.

Common Pitfalls

- **Small expected counts.** The χ^2 approximation becomes unreliable when expected counts are very small (e.g., any $E_{ij} < 1$ or many $E_{ij} < 5$). Combine sparse categories or use Fisher's exact test for 2×2 tables.
- **Interpreting significance as strength.** A very large sample can make tiny, practically unimportant associations highly significant. Use Cramér's V to assess effect size.
- **Causation fallacy.** A significant test shows association, not causality. Confounding or reverse causation may explain observed patterns.
- **Ignoring table structure.** Collapsing categories or re-binning after examining the data changes the hypothesis being tested and inflates Type I error. Define categories *before* analysis.
- **Ignoring ordinal structure.** When categories have a natural order (e.g., low/medium/high), the χ^2 test ignores ordering. Consider linear-by-linear association or ordinal logistic models for greater power.
- **Violating independence of observations.** The test assumes each observation is from a distinct individual. Clustered data (e.g., couples, classrooms) invalidate the test and require mixed or GEE methods.

G-Test of Independence

Likelihood-Ratio Test for Contingency Tables

CTAB.2 | **Category:** Contingency Tables | **Data:** Categorical | **Null:** χ^2 Approximation

When to Use

- Test independence between two categorical variables in an $r \times c$ contingency table.
- Alternative to Pearson's χ^2 test based on likelihood ratios.
- Preferred when likelihood-based inference is desired.
- Common in log-linear modeling and GLMs.

Assumptions

- Observations must be completely independent of each other.
- Data must consist of raw counts, not percents or rates.
- The underlying variables must be categorical or discrete.
- Sampling mechanism must follow a multinomial structure.
- Expected frequency in every single cell must be at least 5.

Test Statistic

$$G = 2\sum_{i=1}^{r}\sum_{j=1}^{c} O_{ij}\ln\left(\frac{O_{ij}}{E_{ij}}\right) \sim \chi^2_{(r-1)(c-1)} \text{ under } H_0 \quad \Big| \quad E_{ij} = \frac{R_i \cdot C_j}{N}.$$

Notation: O_{ij} = observed count, E_{ij} = expected count under independence, r= rows, c= columns, N = total sample size

Plain English: The G-test compares the observed cell counts with the counts expected under independence using a likelihood-ratio framework. Large values indicate that the observed table is unlikely to have arisen if the row and column variables were independent. It answers the same practical question as Pearson's χ^2 test but measures discrepancy using logarithms of observed-to-expected ratios rather than squared differences.

⚠ Practitioner Warning

The χ^2 approximation may be poor when expected counts are very small. Cells with $O_{ij} = 0$ contribute zero to the statistic because $\lim_{x\to 0^+} x\ln(x) = 0$. Although asymptotically equivalent to Pearson's χ^2 test, the two statistics can differ noticeably in small samples. For sparse tables, Fisher's exact test or Monte Carlo methods may be preferable.

Hypotheses

- H_0: The row and column variables are independent.
- H_1: The row and column variables are associated.

Decision Rule: Reject H_0 if $G > \chi^2_{\alpha,(r-1)(c-1)}$ or if p-value $< \alpha$.

Software References [Python, R, SAS]

`scipy.stats.power_divergence()` `DescTools::GTest()` | `PROC FREQ / CHISQ`

Additional Notes

The G-statistic is a special case of the likelihood ratio statistic $-2\ln\Lambda$ derived from Wilks' theorem (1938), and equals the deviance in a saturated log-linear model. Pearson's χ^2 is obtained by a second-order Taylor expansion of G around $O_{ij} = E_{ij}$, which explains why the two statistics converge asymptotically but differ in finite samples. A distinctive property of G is exact additivity: if a table is partitioned into independent subtables, the G-values sum exactly to the overall G. Pearson's χ^2 only approximately shares this property. This additivity makes G the natural fit statistic in log-linear modeling and hierarchical model comparisons. Williams' (1976) correction improves the χ^2 approximation by dividing G by a factor $q > 1$, reducing over-rejection in moderate samples.

$$G_W = \frac{G}{q}, \qquad q = 1 + \frac{(N\sum_{i=1}^{r} R_i^{-1} - 1)(N\sum_{j=1}^{c} C_j^{-1} - 1)}{6N(r-1)(c-1)} \quad \text{(for 2D tables)}$$

From an information-theoretic perspective, G can be explicitly mapped to relative entropy, where $G = 2N \cdot D_{\mathrm{KL}}(P_{XY} \parallel P_X \otimes P_Y)$, measuring the Kullback–Leibler divergence between the observed joint empirical distribution and the product of its margins.

Miscellanea: The statistic is traditionally denoted by G since it originated as a likelihood-ratio goodness-of-fit measure. Long before GLMs became common, the G-test was already being used as a likelihood-based alternative to Pearson's χ^2 statistic for contingency-table analysis. Note that, G cannot be computed when any $E_{ij} = 0$. Workaround for when $O_{ij} = 0$ is handled using calculus limits.

Reference(s): Wilks (1938); Bishop, Fienberg & Holland (1975); Williams' (1976); Agresti (2013)

G-Test of Independence

CTAB.2

Worked Example

Problem Statement In a clinical trial, 60 patients are randomly assigned to one of two vaccine formulations (A or B) and monitored for side effects classified as None, Mild, or Severe. Test at $\alpha = 0.05$ whether the distribution of side-effect severity is independent of vaccine type. H_0: vaccine type and side-effect severity are independent; H_1: they are not independent.

1. Observed Counts

		Side Effects			Total
		None	Mild	Severe	
Vaccine	A	12	8	4	24
	B	8	12	16	36
Total		20	20	20	60

2. Expected Counts
Row totals: $R_A = 24$, $R_B = 36$; column totals all equal $C_j = 20$; $N = 60$. Only cells with $O_{ij} \neq E_{ij}$ contribute.

$$E_{Aj} = \frac{24 \times 20}{60} = 8 \quad \forall j, \qquad E_{Bj} = \frac{36 \times 20}{60} = 12 \quad \forall j.$$

The equal column totals force all $E_{Aj} = 8$ and all $E_{Bj} = 12$, so Mild cells have $O = E$ and contribute zero to G.

Practitioner Insight: The G-test is the likelihood-ratio counterpart of Pearson's χ^2 test. Both share the same asymptotic $\chi^2((r-1)(c-1))$ null distribution, but Pearson's χ^2 uses squared standardized deviations $(O-E)^2/E$, whereas G uses the likelihood-ratio form $O\ln(O/E)$, equivalent to the Kullback–Leibler divergence. Here $G = 6.90$ slightly exceeds Pearson's $\chi^2 = 6.67$, reflecting the greater sensitivity of the log-likelihood to large deviations. The Williams (1976) correction reduces small-sample bias and brings G_{adj} close (identical here) to χ^2, illustrating their asymptotic equivalence. A practical advantage of G is additivity across nested tables, making it useful in log-linear modeling. As with Pearson's χ^2, expected counts should generally exceed 5; otherwise Fisher's exact test is preferable.

3. Computing G
Nonzero contributions ($O_{ij} \neq E_{ij}$):

Cell	O	E	O/E	$2O\ln(O/E)$
(A, None)	12	8	3/2	$24\ln(3/2)$
(A, Severe)	4	8	1/2	$8\ln(1/2)$
(B, None)	8	12	2/3	$16\ln(2/3)$
(B, Severe)	16	12	4/3	$32\ln(4/3)$

Collecting $\ln 2$ and $\ln 3$ terms:

$$G = 24(2\ln 2 - \ln 3) = 24\ln\tfrac{4}{3} \approx 6.904.$$

4. Williams Correction (optional)
The first-order continuity correction for G:

$$q = 1 + \frac{(N\sum_{i=1}^{r} R_i^{-1} - 1)(N\sum_{j=1}^{c} C_j^{-1} - 1)}{6N(r-1)(c-1)} = 1.0352,$$

$$G_{\text{adj}} = \frac{G}{q} = \frac{24\ln(4/3)}{1.0352} \approx 6.670.$$

5. Decision
Under H_0: $G \sim \chi^2(2)$; df$= (2-1)(3-1) = 2$.

α	$\chi^2_{(2)}$	Decision
5%	5.991	**Reject** (6.904 > 5.991)
1%	9.210	FTR (6.904 < 9.210)

$p = 0.032$. Side-effect severity and vaccine type are not independent at 5%. Vaccine B is associated with more severe side effects.

Note: For 2×2 tables, Pearson's χ^2 with Yates' correction often performs better in small samples.

Common Pitfalls

- **Using G with sparse expected counts.** The χ^2 approximation becomes unreliable when expected counts are small. Apply Williams' correction for moderate sparsity or use Fisher's exact test for small tables.
- **Ignoring cells with zero counts.** Empty cells can destabilize the likelihood-ratio statistic and are handled differently across software. Check software conventions and consider combining sparse categories when appropriate.
- **Treating G and Pearson's χ^2 as identical.** The two tests are asymptotically equivalent but can differ noticeably in small or sparse samples. For small 2×2 tables, Fisher's exact test is often preferable.
- **Stopping at a significant G statistic.** Rejection only indicates association somewhere in the table. Examine standardized residuals to identify the cells driving the result.
- **Misusing the additivity property.** G statistics add exactly only for orthogonal, independent partitions. Applying additivity to overlapping subtables produces invalid results.

Fisher's Exact Test

Exact Test for 2×2 Contingency Tables

CTAB.3 | **Category:** Contingency Tables | **Data:** Categorical | **Null:** Hypergeometric (exact)

When to Use

- Test association in 2×2 tables when sample size is small.
- Preferred when expected frequencies fall below 5 (chi-square invalid).
- Provides exact p-values without relying on asymptotic approximations (gold standard for small samples).

Assumptions

- Observations are independent.
- Data arranged in 2×2 contingency table with fixed marginal totals.
- Marginal totals are treated as fixed by sampling design.
- All four cells contain non-negative integer counts.

Test Statistic

$$p_F = \sum_{i \in \mathcal{T}} \frac{\binom{a+b}{i}\binom{c+d}{k-i}}{\binom{n}{k}}, \quad \text{where } a, b, c, d \text{ are cell counts and } k = a + c$$

Notation: n is the total sample size. $\mathcal{T}$ is the set of tables as or more extreme than the observed table (1-sided or 2-sided).

Plain English: Fisher's exact test calculates the probability of observing the data (or more extreme configurations) directly from the hypergeometric distribution, given fixed row and column totals. Unlike chi-square, which relies on asymptotic approximation, this test provides exact inference valid for any sample size. The test conditions on marginal totals, treating them as fixed rather than random, which has sparked theoretical debate but ensures computational tractability.

⚠ Practitioner Warning

Fisher's exact test conditions on fixed marginal totals; if margins are random by design, this assumption may be questionable. As the hypergeometric distribution is discrete, the test is often conservative, with actual Type I error below the nominal level and reduced power relative to asymptotic methods in moderate or large samples. Different software may report different two-sided p-values as multiple definitions of "more extreme" tables exist. Unconditional exact tests (e.g., Barnard's, Boschloo's) can be more powerful but are computationally more intensive. Exact significance does not imply a practically important association, so effect sizes such as odds ratios and confidence intervals should also be reported.

Hypotheses

- H_0: Variables are independent (no association).
- H_1: Variables are dependent (association exists).

Decision Rule: Reject H_0 if $p_F < \alpha$. One-sided: sum tail probabilities. Two-sided: sum probabilities $\leq P(\text{observed})$.

Software References [Python, R, SAS]

`scipy.stats.fisher_exact`	`fisher.test`	`PROC FREQ / FISHER`

Additional Notes

Historically, Fisher's exact test was computationally burdensome because it required repeated factorial and hypergeometric probability calculations. Efficient network algorithms developed in the 1980s made exact p-values feasible even for moderately large tables, and modern software computes 2×2 Fisher tests almost instantly. Extensions to $r \times c$ tables exist but can become computationally expensive for large sparse tables. One-sided Fisher tests are unambiguous, but several competing definitions of the two-sided p-value exist because the hypergeometric distribution is discrete and asymmetric; software packages may therefore report slightly different two-sided results. Mid-p adjustments are sometimes used to reduce the conservativeness of Fisher's exact test, improving power at the cost of losing strict exactness.

Miscellanea: Ronald Fisher popularized exact inference through the famous "Lady Tasting Tea" experiment described in *The Design of Experiments* (1935). A colleague claimed she could distinguish whether milk or tea had been poured first into a cup. Fisher presented eight cups (four prepared each way); she identified all correctly. Under random guessing, the exact probability of this outcome is $p = 1/\binom{8}{4} = 1/70 \approx 0.014$. The example became one of statistics' most enduring examples of exact hypothesis testing for small samples.

Reference(s): Fisher (1935); Agresti (1992)

Fisher's Exact Test CTAB.3

Worked Example

Problem Statement A pilot trial for a rare disease randomizes 13 patients: 6 to a new treatment (T) and 7 to placebo (P). Recovery within 4 weeks is binary. Test whether recovery rates differ between groups at $\alpha = 0.05$ (one-sided: treatment improves recovery). H_0: recovery rate is the same in both groups; H_1: recovery rate is higher in Treatment.

Treatment: 5 recovered, 1 not recovered ($n_T = 6$); Placebo: 1 recovered, 6 not recovered ($n_P = 7$).

Why Fisher, not χ^2? All expected counts are below 5; the large-sample χ^2 approximation is unreliable.

1. Hypergeometric Probability Formula
With fixed marginals ($R_1 = 6, R_2 = 7, C_1 = 6, C_2 = 7$), the probability of exactly a recoveries in the Treatment group is:

$$P(a) = \binom{6}{a}\binom{7}{6-a} \div \binom{13}{6}, \quad a = 0, 1, \ldots, 6.$$

2. Reference Distribution

a	$\binom{6}{a}\binom{7}{6-a}$	$P(a)$	
0	$1 \times 7 = 7$	7/1716	
1	$6 \times 21 = 126$	126/1716	
2	$15 \times 35 = 525$	525/1716	
3	$20 \times 35 = 700$	700/1716	
4	$15 \times 21 = 315$	315/1716	
5	$6 \times 7 = 42$	42/1716	← observed
6	$1 \times 1 = 1$	1/1716	
Sum	$\binom{13}{6} = 1716$	1	

3. Observed Probability

$$P(a = 5) = 42/1716.$$

4. One-Sided p-Value
Sum over all tables *at least as extreme* as observed (i.e. $a \geq 5$):
$$p = P(a \geq 5) = (42 + 1)/1716 \approx 0.025.$$

5. Decision

α	Critical region	Decision
5%	$p < 0.05$	**Reject** ($0.025 < 0.05$)
1%	$p < 0.01$	FTR ($0.025 > 0.01$)

$p \approx 0.025$. The treatment group shows significantly higher recovery.

6. Effect Size

$$\hat{p}_T = 5/6 \approx 83\%, \quad \hat{p}_P = 1/7 \approx 14\%, \quad \hat{p}_T - \hat{p}_P = 0.69.$$
$$\text{Odds ratio} = (5 \times 6)/(1 \times 1) = 30.$$

The odds of recovery in the treatment group are estimated to be 30 times those in placebo.

Practitioner Insight: Fisher's exact test conditions on both row and column totals being fixed, giving a hypergeometric null distribution for the cell counts. Here each table probability equals $\binom{6}{a}\binom{7}{6-a}/1716$, and summing probabilities for $a \geq 5$ gives the exact one-sided $p \approx 0.025$ with no asymptotic approximation. The test is especially valuable for sparse 2×2 tables: all expected counts here are below 5, making the χ^2 approximation unreliable. The directional one-sided test is appropriate because treatment benefit was specified *a priori*; without a directional hypothesis, the two-sided test sums all tables at least as extreme as the observed one. Fisher's test is conservative for very small samples because the discrete distribution cannot attain exactly level α; mid-p variants reduce conservatism at the cost of exactness. Extensions to larger tables use exact multinomial or Monte Carlo permutation methods.

Common Pitfalls

- **Misinterpreting "exact" as always preferable.** Fisher's test is exact but often overly conservative, especially with moderate cell counts. For larger samples, χ^2 tests or logistic regression may be more informative.
- **Using two-sided p-values uncritically.** Several valid definitions of two-sided Fisher p-values exist, and results may differ slightly across software. If a clear directional hypothesis exists, use the one-sided version.
- **Ignoring study design and independence.** Fisher's test assumes independent observations. Paired or matched data require McNemar's test; clustered data require mixed or GEE models.
- **Using Fisher's test with large n.** With large margins, computation becomes slow and very small p-values may reflect trivial effects rather than practical importance. For large n, use logistic regression or χ^2 tests to quantify effect size.
- **Confusing statistical significance with association strength.** Fisher's test gives only a p-value—no effect size. Always accompany results with an odds ratio and confidence interval to understand magnitude.
- **Conditioning on margins when margins are not fixed.** Fisher's test conditions on row/column totals; if margins arise naturally from sampling (not fixed by design), the interpretation differs from χ^2. Report this assumption explicitly.

Barnard's Exact Test

Unconditional Exact Test for 2×2 Tables

CTAB.4 | **Category:** Contingency Tables | **Data:** Categorical | **Null:** Exact Uncond. Binomial

When to Use

- Exact inference for 2×2 contingency tables.
- Small, sparse, or unbalanced categorical samples.
- Often more powerful alternative to Fisher's exact test.
- Unconditional exact analysis of binary outcomes.

Assumptions

- Observations are independent within and across groups.
- Data are binary frequency counts in a 2×2 table.
- Categories are mutually exclusive and exhaustive.
- Row totals are fixed; column totals are unconditioned.

Test Statistic

$$p_{\text{Barnard}} = \sup_{p\in[0,1]} \sum_{a=0}^{n_1}\sum_{c=0}^{n_2}\left[\binom{n_1}{a}p^a(1-p)^{n_1-a} \times \binom{n_2}{c}p^c(1-p)^{n_2-c}\right]\cdot \mathbb{I}(S(a,c) \geq S(a_{\text{obs}}, c_{\text{obs}}))$$

where $S(a,c)$ is the chosen discrepancy statistic (e.g., Z_{score}) used to rank table $([a, n_1 - a], [c, n_2 - c])$ extremeness.

Notation: p is the nuisance parameter (common success probability), n_1, n_2 are fixed row totals, and $\mathbb{I}(\cdot)$ is the indicator function for tables as extreme as the observed.

Plain English: Barnard's exact test evaluates whether two proportions differ in a 2×2 table without conditioning on observed column totals. Unlike Fisher's exact test, it treats only the row totals as fixed and maximizes over a nuisance parameter (the common success probability under H_0) to obtain a valid exact p-value. As it avoids conditioning on the column totals, Barnard's test is often more powerful than Fisher's test in small samples, though historically it was computationally intensive.

⚠ Practitioner Warning

Barnard's exact test applies only to 2×2 tables and requires numerical maximization over a nuisance parameter, so different implementations and test statistics (score, Wald, likelihood ratio) may yield slightly different exact p-values. Although typically more powerful than Fisher's exact test, the unconditional framework can be computationally demanding for large samples. Statistical significance should be interpreted alongside odds ratios and confidence intervals.

Hypotheses

- H_0: Proportions are equal in both rows ($p_1 = p_2$).
- H_1: Proportions differ between rows ($p_1 \neq p_2$).

Decision Rule: Reject H_0 if $p_{\text{Barnard}} < \alpha$. Unconditional tail probability is maximized over the nuisance parameter p.

Software References [Python, R, SAS]

```
scipy.stats.barnard_exact          Exact::barnard.test          PROC FREQ / EXACT BARNARD
```

Additional Notes

Common ordering statistics include the score statistic (Suissa–Shuster, 1985; generally recommended and often exhibits strong power), the Wald statistic (common modern implementation but unstable when proportions are near 0 or 1), and the likelihood-ratio statistic (whose ordering may occasionally be irregular). Barnard's original procedure ordered tables using the observed difference in sample proportions. Modern network and optimization algorithms make the test practical for substantially larger sample sizes than originally feasible. Boschloo's test is a closely related unconditional procedure that uses Fisher's exact p-value as the ordering statistic.

$$T_{\text{score}} = \frac{\hat{p}_1 - \hat{p}_2}{\sqrt{\hat{p}(1-\hat{p})\left(\frac{1}{n_1}+\frac{1}{n_2}\right)}}, \text{ where } \hat{p} \text{ is pooled} \quad \& \quad T_{\text{Wald}} = \frac{\hat{p}_1 - \hat{p}_2}{\sqrt{\frac{\hat{p}_1(1-\hat{p}_1)}{n_1} + \frac{\hat{p}_2(1-\hat{p}_2)}{n_2}}}.$$

Miscellanea: George Barnard introduced his unconditional exact test in 1945 as an alternative to Fisher's exact test, arguing that conditioning on column margins (which are rarely fixed by design, unlike row totals) unnecessarily reduced power. The resulting debate over "what to condition on" became one of the classic controversies in exact inference. Barnard's test is often more powerful than Fisher's, but Fisher's exact test remains far more widely used because of its simplicity, historical precedence, and wide acceptance.

Reference(s): Barnard (1945); Suissa & Shuster (1985); Mehta & Senchaudhuri (2003)

Worked Example

Problem Statement A randomized pilot trial assigns 7 patients to a new treatment (T) and 7 to placebo (P). Observed recoveries are: Treatment: 5 recovered, 2 not recovered ($n_T = 7$); Placebo: 1 recovered, 6 not recovered ($n_P = 7$).

Fisher's exact test gives $p = 154/3003 \approx 0.051$ (FTR at 5%). Apply Barnard's exact test, which does not condition on the column marginals, to determine whether the conclusion changes.

1. Test Statistic
Barnard's test uses the score (or Wald) statistic for two proportions, evaluated at the *observed* pooled proportion:

$$T_{\text{score}} \approx 2.160,$$

with $\hat{p}_1 = 5/7$, $\hat{p}_2 = 1/7$, $\hat{p} = (5+1)/14 = 3/7$.

2. The Nuisance-Parameter p-Value Function
Unlike Fisher's test, the null distribution of T depends on the unknown common success probability p (the nuisance parameter). For each $p \in (0,1)$, compute the exact binomial probability that T achieves or exceeds T_{obs}:

$$g(p) = \sum_{\substack{(x_1,x_2):\\ T(x_1,x_2)\ge T_{\text{obs}}}} \binom{7}{x_1} p^{x_1}(1-p)^{7-x_1} \cdot \binom{7}{x_2} p^{x_2}(1-p)^{7-x_2}.$$

The Barnard p-value is $\sup_{p\in(0,1)} g(p)$.
Because $n_1 = n_2 = 7$, the function $g(p)$ is symmetric about $p = \frac{1}{2}$; the supremum is achieved at $p^* = \frac{1}{2}$.

3. Computing $g(1/2)$
At $p = \frac{1}{2}$, each $\binom{7}{x}p^x(1-p)^{7-x} = \binom{7}{x}/128$. The joint probability is $\binom{7}{x_1}\binom{7}{x_2}/128^2 = \binom{7}{x_1}\binom{7}{x_2}/16384$.
Pairs (x_1, x_2) with $T(x_1, x_2) \ge 2.160$:

x_1	x_2	T	$\binom{7}{x_1}\binom{7}{x_2}$	
7	0	3.742	$1\times 1 = 1$	
7	1	3.240	$1\times 7 = 7$	
6	0	3.240	$7\times 1 = 7$	
7	2	2.789	$1\times 21 = 21$	
5	0	2.789	$21\times 1 = 21$	
6	1	2.673	$7\times 7 = 49$	
7	3	2.366	$1\times 35 = 35$	
4	0	2.366	$35\times 1 = 35$	
5	1	2.160	$21\times 7 = 147$	← obs
6	2	2.160	$7\times 21 = 147$	
Sum			470	

$$g\left(\tfrac{1}{2}\right) = 470/16384 \approx 0.0287.$$

4. Decision

Test	p-value	Decision ($\alpha = 5\%$)
Fisher's exact	≈ 0.051	FTR
Barnard's exact	≈ 0.029	**Reject**

Barnard's test detects the treatment effect where Fisher's does not.

Practitioner Insight: Barnard's exact unconditional test differs from Fisher's exact test in what it conditions on. Fisher conditions on both row and column totals, producing a hypergeometric null distribution, whereas Barnard conditions only on the row totals and treats the common success probability p as a nuisance parameter. By avoiding conditioning on the column totals, Barnard's test uses more information and is uniformly more powerful for testing $H_0: p_1 = p_2$ when only the group sizes are fixed by design. The p-value is obtained by maximizing a probability function over p; here symmetry forces the maximum at $p^* = 1/2$, giving the exact value 470/16384. Barnard's advantage is often greatest for small or unbalanced tables, while the two methods converge for larger samples. The long-standing debate concerns whether conditioning on both margins, as Fisher does, is scientifically justified when only group sizes are fixed experimentally. Boschloo's test combines both approaches by embedding Fisher's p-value within Barnard's unconditional framework, often yielding even greater power.

Common Pitfalls

- **Confusing Barnard's with Fisher's test.** Fisher conditions on *both* margins; Barnard conditions on only one. They are not interchangeable—Barnard is typically more powerful.
- **Applying Barnard's when both margins are fixed by design.** If row and column totals are *structurally* fixed (e.g., case–control studies), Fisher's is the correct model. Barnard assumes only one fixed margin.
- **Two-sided p-value ambiguity.** Multiple definitions (Neyman, Sterne, unconditional "probability ≤ observed") exist. Software implementations differ; analysts often mix interpretations.
- **Interpreting non-significance with tiny samples.** Exact p-values remain discrete and "jumpy." Non-significant results may reflect discreteness, not absence of effect.

Boschloo's Test

Unconditional Exact Test for 2×2 Contingency Tables

CTAB.5 | **Category:** Contingency Tables | **Data:** Binary | **Null:** Exact Uncond. Binomial

When to Use

- Exact test for 2×2 contingency tables.
- Small-sample categorical data analysis.
- More powerful alternative to Fisher's Exact test.
- Unconditional exact inference using Fisher ordering.

Assumptions

- Observations are independent within and across groups.
- Data are binary frequency counts in a 2×2 table.
- Categories are mutually exclusive and exhaustive.
- Row totals are fixed; column totals are unconditioned.

Test Statistic

$$p_{\text{Boschloo}} = \sup_{\pi \in [0,1]} \sum_{a=0}^{n_1} \sum_{c=0}^{n_2} \left[\binom{n_1}{a} \pi^a (1-\pi)^{n_1-a} \times \binom{n_2}{c} \pi^c (1-\pi)^{n_2-c} \right] \cdot \mathbb{I}(p_F(a,c) \leq p_F(a_{\text{obs}}, c_{\text{obs}}))$$

where $p_F(a,c)$ is the p-value from Fisher's Exact Test calculated for a given table configuration (a,c).

Notation: π is the common success probability under H_0, n_1, n_2 are fixed row totals, and the indicator function identifies all tables with a Fisher p-value as small or smaller than the observed.

Plain English: Boschloo's test improves upon Fisher's exact test by using an unconditional approach rather than conditioning on marginal totals. While Fisher conditions on both margins (treating them as fixed), Boschloo recognizes only row totals are fixed by design in typical two-sample comparisons. The test computes the p-value by maximizing over the nuisance parameter (common probability under null) rather than conditioning it away. This unconditional approach yields higher power while maintaining exact error control.

⚠ Practitioner Warning

More complex than Fisher's exact test to implement and explain. Gain in power over Fisher modest in many practical settings. Requires specialized software; not available in all statistical packages. Different implementations may use different optimization algorithms affecting numerical precision. For very small samples or balanced designs, advantage over Fisher minimal. Most powerful for unbalanced designs with moderate total sample size. Conservative relative to asymptotic tests but less so than Fisher.

Hypotheses

- H_0: Probabilities equal in both groups ($p_1 = p_2$).
- H_1: Probabilities differ between groups ($p_1 \neq p_2$).

Decision Rule: Reject H_0 if $p_B < \alpha$. Unconditional probability of Fisher-extreme tables is maximized over p.

Software References [Python, R, SAS]

`scipy.stats.boschloo_exact`	`Exact::exact.test()`	`Not Native (Manual)`

Additional Notes

Boschloo (1970) introduced an unconditional exact test to overcome Fisher's conservativeness by conditioning only on sample sizes and treating column totals as random. For each null proportion p, compute the binomial tail probability of observing a table at least as extreme as the observed one (using Fisher's ordering), then take the supremum over p to obtain the unconditional p-value. The method is typically more powerful than Fisher and often outperforms Barnard's test, achieving exact Type I error while avoiding unnecessary conservatism —particularly with unbalanced margins. Efficient numerical optimization now makes the test practical in modern software.

Miscellanea: Boschloo introduced his test in 1970 in Statistica Neerlandica, arguing that Fisher's conditioning made the classical exact test overly conservative. His simulations showed notably higher power without violating Type I error control—an idea initially met with skepticism due to computational complexity. Interest increased only decades later when modern computing made unconditional tests feasible. Note: Boschloo's and Barnard's tests differ mainly in their ordering rule: Barnard orders tables by a test statistic, whereas Boschloo orders them by Fisher's exact p-value.

Reference(s): Boschloo (1970); Suissa & Shuster (1985); Mehta & Hilton (1993)

Worked Example

Problem Statement Apply Boschloo's exact test to the same pilot-trial data used in the Barnard example ($n_1 = n_2 = 7$): Observed recoveries are: Treatment: 5 recovered, 2 not recovered ($n_1 = 7$); Placebo: 1 recovered, 6 not recovered ($n_2 = 7$).

Fisher's exact test gives $p_F = 154/3003 \approx 0.051$ (FTR at 5%). Boschloo's test uses p_F as the *test statistic* within Barnard's unconditional framework.

1. Boschloo's Procedure

a) For every possible table (x_1, x_2) with $n_1 = 7, n_2 = 7$, compute Fisher's one-sided p-value $p_F(x_1, x_2)$.

b) Rank tables by p_F: the *extreme set* $\mathcal{E}$ contains all tables with $p_F(x_1, x_2) \le p_F(\text{observed}) = 154/3003$.

c) The Boschloo p-value is

$$p_B = \sup_{p\in(0,1)} \sum_{(x_1,x_2)\in\mathcal{E}} \binom{7}{x_1} p^{x_1}(1-p)^{7-x_1} \cdot \binom{7}{x_2} p^{x_2}(1-p)^{7-x_2}.$$

2. The Extreme Set $\mathcal{E}$

Ten tables satisfy $p_F \le 154/3003 \approx 0.051$:

(x_1, x_2)	C_1	$p_F(x_1, x_2)$	$\binom{7}{x_1}\binom{7}{x_2}$
(7,0)		$1/3432 \approx 0.000$	$1 \times 1 = 1$
(6,0)		$7/3003 \approx 0.002$	$7 \times 1 = 7$
(7,1)		$7/3003 \approx 0.002$	$1 \times 7 = 7$
(5,0)		$21/2002 \approx 0.010$	$21 \times 1 = 21$
(7,2)		$21/2002 \approx 0.010$	$1 \times 21 = 21$
(6,1)		$50/3432 \approx 0.015$	$7 \times 7 = 49$
(4,0)		$35/1001 \approx 0.035$	$35 \times 1 = 35$
(7,3)		$35/1001 \approx 0.035$	$1 \times 35 = 35$
(5,1)	← obs	$154/3003 \approx 0.051$	$21 \times 7 = 147$
(6,2)		$154/3003 \approx 0.051$	$7 \times 21 = 147$
			Sum = 470

3. Evaluating at the Supremum $p^* = \frac{1}{2}$

Because $n_1 = n_2 = 7$, the function $g(p)$ is symmetric about $p = \frac{1}{2}$, so $p^* = \frac{1}{2}$ exactly. At $p = \frac{1}{2}$:

$$\binom{7}{x} p^x (1-p)^{7-x} = \binom{7}{x}/128,$$

and the joint probability is $\binom{7}{x_1}\binom{7}{x_2}/16384$.

$$p_B = g\left(\tfrac{1}{2}\right) = 470/16384 \approx 0.029.$$

4. Decision and Comparison

Test	p-value	Decision
Fisher	$154/3003 \approx 0.051$	FTR
Barnard	$470/16384 \approx 0.029$	Reject
Boschloo	$470/16384 \approx 0.029$	Reject

Fisher alone fails to reject; both Boschloo and Barnard detect the treatment effect. Boschloo and Barnard coincide here.

Practitioner Insight: Boschloo's exact test is an unconditional exact test that uses Fisher's hypergeometric p-value as the ranking statistic within Barnard's maximization framework. Instead of ranking tables by a score statistic (Barnard) or conditioning on both margins (Fisher), Boschloo ranks them by Fisher's conditional extremeness, which generally yields greater power. Consequently, Boschloo's p-value is always no larger than Fisher's and often smaller than Barnard's. In this example, Barnard's score-statistic ordering and Boschloo's Fisher-p ordering generate the same rejection region (Extreme Set $\mathcal{E}$), producing identical p-values. Equal group sizes make these orderings closely aligned, though they are not identical in general; for other tables, even with $n_1 = n_2$, Barnard and Boschloo can differ slightly. The supremum over the nuisance parameter p guarantees an exact size-α test. Boschloo's procedure is often recommended for exact 2×2 inference when only the row totals are fixed by design, as in most randomized trials.

Common Pitfalls

- **Using Boschloo for paired data.** Boschloo assumes *independent* samples; applying it to matched or repeated-measures 2×2 tables violates independence and inflates Type I error. Use McNemar's (or exact McNemar) for paired designs.
- **Applying Boschloo to large samples unnecessarily.** The test is computationally intensive and offers little benefit once expected counts exceed 5–10; χ^2 or standard proportion tests are simpler and yield nearly identical results for large n.
- **Treating non-significance as evidence of equivalence.** Boschloo tests $p_1 = p_2$ vs difference—not equivalence. A non-significant result cannot justify claims of equivalence; use exact TOST procedures for equivalence hypotheses.
- **Assuming Boschloo and Fisher are interchangeable.** Boschloo uses Fisher's exact p-value as the ordering statistic within an unconditional testing framework. Often more powerful than Fisher's exact test, but the two are not identical.
- **Ignoring alternative direction (one- vs two-sided).** One-sided and two-sided Boschloo tests produce different p-values; the direction must be specified *before* seeing data to avoid post-hoc bias.

McNemar's Test

Test for Marginal Homogeneity in 2×2 Tables

CTAB.6 | **Category:** Contingency Tables | **Data:** Categorical | **Null:** Asymp. χ^2 / Binomial

When to Use

- Compare paired binary responses or outcomes.
- Before-after studies with dichotomous outcomes.
- Matched-pair categorical data analysis.
- Test marginal homogeneity in 2×2 tables.
- Diagnostic method or rater comparison studies.

Assumptions

- Observations are paired or meaningfully matched.
- Pairs are independent of one another.
- Outcomes are binary/dichotomous.
- Discordant pairs drive the inference.
- Sufficient discordant counts for asymptotics.

Test Statistic

$$\chi^2 = \frac{(b-c)^2}{b+c} \approx \chi^2_1 \text{ (under } H_0\text{)}, \qquad \text{Exact version uses binomial: } b \sim \text{Bin}(b+c, 0.5).$$

Notation: b is count of (0,1) discordant pairs, c is count of (1,0) discordant pairs

Plain English: McNemar's test examines whether the proportion of successes changes between paired measurements by focusing exclusively on discordant pairs—those that changed status between measurements. Concordant pairs (both success or both failure) provide no information about change and are ignored. The test evaluates whether the two types of discordance (0→1 vs. 1→0) occur with equal frequency, with substantial imbalance indicating a systematic shift in proportions.

⚠ Practitioner Warning

McNemar's test requires paired binary data; independent samples require Pearson's χ^2 test instead. Only discordant pairs (b, c) contribute to the statistic. With few discordant pairs (commonly $b + c < 25$), the asymptotic χ^2 approximation may be unreliable; the exact binomial version or *mid-p* adjustment is preferred. Mid-p adjustment reduces conservativeness in exact tests by subtracting half the probability of the observed outcome from the exact p-value. Edwards' continuity correction, $\chi^2_{\text{corr}} = (|b-c|-1)^2/(b+c)$, gives a more conservative approximation. McNemar's test assesses marginal homogeneity, not agreement; Cohen's κ is preferable for agreement studies.

Hypotheses

- H_0: $P(0 \to 1) = P(1 \to 0)$ (Equal change)
- H_1: $P(0 \to 1) \neq P(1 \to 0)$ (Unequal change)

Decision Rule: Reject H_0 if $\chi^2 > \chi^2_{\alpha,1}$ or p-value $< \alpha$.

Software References [Python, R, SAS]

`statsmodels::mcnemar`	`mcnemar.test`	`PROC FREQ; / agree;`

Additional Notes

The asymptotic McNemar statistic is (**A**). Under H_0, conditioning on the total number of discordant pairs gives $b \mid (b+c) \sim \text{Bin}(b+c, 0.5)$, forming the basis of the exact McNemar test. McNemar's test is equivalent to the sign test for paired binary outcomes and also to the score test from a conditional logistic regression model for matched pairs. Effect sizes are commonly summarized by the matched-pair odds ratio b/c or by the paired difference in proportions $\hat{d} = (b-c)/n$, with approximate CI is (**B**). Extensions include Cochran's Q test for more than two related binary measurements and Stuart–Maxwell tests for marginal homogeneity in multinomial matched tables.

$$(\mathbf{A}): \chi^2 = \frac{(b-c)^2}{b+c} \sim \chi^2_1, \qquad (\mathbf{B}): \hat{d} \pm z_{\alpha/2}\sqrt{\frac{b+c}{n^2} - \frac{(b-c)^2}{n^3}}.$$

Miscellanea: Quinn McNemar introduced this test in 1947 to analyze paired categorical data, particularly before–after responses and matched-pair designs where ordinary independence tests are inappropriate. His key insight was that only the discordant pairs contribute evidence about change, while concordant pairs provide no information about directional differences. Published in *Psychometrika*, the test quickly became a standard method for paired binary data in medicine, psychology, survey research, and epidemiology.

Reference(s): McNemar (1947); Edwards (1948)

Worked Example

Problem Statement Two screening tests (A and B) are applied to the same 100 patients. Test whether the positive rates of the two methods differ at $\alpha = 0.05$. $H_0: p_A = p_B$ (marginal rates equal); $H_1: p_A \neq p_B$.

Observed totals: $a = 60$ (A+, B+), $b = 15$ (A+, B-), $c = 5$ (A-, B+), and $d = 20$ (A-, B-).

Marginal positive rates: $\hat{p}_A = (60 + 15)/100 = 0.75$, $\hat{p}_B = (60 + 5)/100 = 0.65$.

1. Identify the Discordant Pairs

Only cells b (A+, B−) and c (A−, B+) are informative; the concordant pairs ($a = 60$, $d = 20$) carry no information about the difference in marginal rates.

$$b + c = 20 \text{ (total discordant)}, \quad b - c = 10.$$

The difference in marginal positive rates is $\hat{p}_A - \hat{p}_B = (b - c)/n = 10/100 = 0.10$, driven entirely by the discordant cells.

2. McNemar's Test Statistic

Under H_0: $b \mid (b + c) \sim \text{Bin}(20, \frac{1}{2})$. The asymptotic statistic is:

$$X^2 = \frac{(b - c)^2}{b + c} = \frac{(15 - 5)^2}{20} = 5.$$

Equivalently, the signed z-statistic:

$$z = \frac{b - c}{\sqrt{b + c}} = \frac{10}{\sqrt{20}} \approx 2.236.$$

3. With Edwards' Continuity Correction (CC)

$$X^2_{cc} = \frac{(|b - c| - 1)^2}{b + c} = \frac{(10 - 1)^2}{20} = 4.05.$$

4. Decision

Under H_0: $X^2 \sim \chi^2(1)$.

Statistic	Value	p-value
X^2 (no CC)	5.000	0.025
X^2_{cc}	4.050	0.044
Exact binomial	—	0.041

α	$\chi^2_{0.05}(1)$	Decision
5%	3.841	**Reject** (5.000 > 3.841)
1%	6.635	FTR (5.000 < 6.635)

The two screening methods have significantly different positive rates ($p = 0.025$); Test A detects more positives than Test B.

5. Exact Binomial p-Value

Under H_0: $b \mid (b + c) \sim \text{Bin}(20, \frac{1}{2})$.

$$p_{\text{exact}} = 2 \sum_{k=15}^{20} \binom{20}{k} \left(\frac{1}{2}\right)^{20} = \frac{43400}{1048576} \approx 0.041.$$

All three p-values reject H_0 at 5%. The exact test ($p = 0.041$) is more conservative (less likely to reject H_0) than the asymptotic approximation ($p = 0.025$) as the binomial is discrete. Note: The term *conservative* has two meanings: for a specific dataset it refers to producing a larger p-value, while in repeated sampling (the test) it refers to a lower-than-nominal Type I error rate.

Practitioner Insight: McNemar's test exploits paired binary data to test whether the marginal probabilities of two methods are equal. Only the discordant pairs contribute evidence: here $b = 15$ pairs are positive on A but negative on B, versus $c = 5$ with the reverse pattern. Concordant pairs ($a = 60$, $d = 20$) affect *prevalence* but not the difference between methods. Under H_0, the conditional distribution of $b \mid (b + c)$ is $\text{Bin}(20, \frac{1}{2})$, and the asymptotic statistic $X^2 = (b - c)^2/(b + c)$ is a Wald approximation to this exact binomial test. Here $X^2 = 5$ gives $p = 0.025$, while the continuity-corrected and exact binomial versions give larger p-values (0.041) and are preferred when $b + c < 25$. The difference illustrates how the normal approximation can overstate significance with few discordant pairs. McNemar's test is equivalent to a sign test on the discordant pairs and extends naturally to Cochran's Q test for more than two matched binary measurements.

Common Pitfalls

- **Using McNemar on unpaired data.** Test requires paired observations; independent samples must use χ^2 or Fisher's exact test.
- **Relying on asymptotic test with few discordant pairs.** When $b + c < 10$, the χ^2 approximation is unreliable—use the exact binomial or mid-p version.
- **Confusing marginal change with agreement.** McNemar tests marginal homogeneity, not rater agreement; use Cohen's kappa for agreement.
- **Misinterpreting non-significant results.** Low discordant counts reduce power; $p > 0.05$ does not imply "no change."
- **Overusing continuity corrections.** Edwards correction is conservative; report exact or mid-p for small samples.

McNemar-Bowker Test

Test for Symmetry in Square Contingency Tables

CTAB.7 | **Category:** Contingency Tables | **Data:** Categorical | **Null:** Asymptotic Chi-square

When to Use

- Compare paired multinomial categorical responses.
- Test symmetry in square contingency tables.
- Before-after studies with multiple categories.
- Matched-pair ordinal/nominal data analysis.
- Agreement studies with repeated classifications.

Assumptions

- Observations are paired or meaningfully matched.
- Pairs are independent of one another.
- Categories are mutually exclusive/exhaustive.
- Contingency table is square ($k \times k$).
- Expected discordant counts are sufficiently large.

Test Statistic

$$\chi^2 = \sum_{i<j} \frac{(n_{ij} - n_{ji})^2}{n_{ij} + n_{ji}} \approx \chi^2_{k(k-1)/2} \text{ (under } H_0\text{)}, \quad \text{summed over all off-diagonal pairs.}$$

Notation: n_{ij} is count in row i, column j; (i, j) and (j, i) form symmetric pair of off-diagonal cells

Plain English: The McNemar-Bowker test examines whether marginal distributions are equal in paired categorical data with multiple categories. It compares frequencies of each type of discordance (category i at time 1, category j at time 2) with its reverse (category j at time 1, category i at time 2). The test sums contributions from all pairs of off-diagonal cells, generalizing McNemar's focus on the single pair of discordant cells in 2×2 tables.

⚠ Practitioner Warning

Requires a square ($k \times k$) matched table with the same categories for both measurements; otherwise the symmetry hypothesis is undefined. With many categories, the degrees of freedom $k(k-1)/2$ increase rapidly, reducing power. Sparse tables with many small or zero off-diagonal counts can invalidate the asymptotic chi-square approximation; collapsing categories or exact methods may be preferable. The McNemar–Bowker test assesses symmetry ($p_{ij} = p_{ji}$), whereas the Stuart–Maxwell test evaluates the weaker condition of marginal homogeneity. Symmetry implies marginal homogeneity, but not conversely. Diagonal cells represent agreement and do not contribute to the test statistic.

Hypotheses

- H_0: $P(i \to j) = P(j \to i)$ for all $i \neq j$.
- H_1: $P(i \to j) \neq P(j \to i)$ for some pairs.

Decision Rule: Reject H_0 if $\chi^2 > \chi^2_{\alpha, k(k-1)/2}$ or p-value < α.

Software References [Python, R, SAS]

`statsmodels.stats...SquareTable` `mcnemar.test()` `PROC FREQ; / agree;`

Additional Notes

When $k = 2$, the McNemar–Bowker test reduces exactly to McNemar's test with $df = 1$. The test evaluates symmetry of the off-diagonal cell probabilities: under H_0, $p_{ij} = p_{ji}$ for all $i \neq j$. Each component $(n_{ij} - n_{ji})^2/(n_{ij} + n_{ji})$ has the same form as the McNemar statistic for the corresponding pair of categories. The Stuart–Maxwell and Bhapkar tests instead evaluate marginal homogeneity. Symmetry is stronger than marginal homogeneity: symmetry requires each off diagonal pair satisfy $p_{ij} = p_{ji}$, while marginal homogeneity requires only equality of the row and column margins ($p_{i\cdot} = p_{\cdot i}$). Sparse off-diagonal counts can make the asymptotic χ^2 approximation unreliable, particularly when k is large. For ordinal categories, methods that explicitly incorporate the ordering may provide greater power.

Miscellanea: Albert Bowker, one of the founding fathers of the Department of Statistics at Stanford, introduced his 1948 test of symmetry as a generalization of McNemar's test from 2 × 2 tables to matched multinomial tables. The procedure became widely known as the McNemar–Bowker test (sometimes as Bowker's test of symmetry), reflecting its role as the natural extension of McNemar's paired binary method. Bowker later became one of the leading figures in American statistics, serving as Chancellor of the City University of New York and helping shape modern statistical education.

Reference(s): Bowker (1948); Maxwell (1970)

Worked Example

Problem Statement A pain-management intervention is evaluated by asking 100 patients to rate their condition as Poor, Fair, or Good both before and after treatment. Test whether the marginal distributions of ratings changed at $\alpha = 0.05$. H_0: $P(\text{Before} = i, \text{After} = j) = P(\text{Before} = j, \text{After} = i)\ \forall i \neq j$; H_1: asymmetry.

		After		
		Poor	Fair	Good
	Poor	30	12	9
Before	Fair	4	24	6
	Good	3	2	10

Marginal Distributions

Time	Poor	Fair	Good
Before	51	34	15
After	37	38	25
Change	−14	+4	+10

1. McNemar–Bowker Statistic

For a $k \times k$ paired table, the test statistic sums a McNemar term for every off-diagonal pair:

$$Q = \sum_{i<j} \frac{(n_{ij} - n_{ji})^2}{n_{ij} + n_{ji}}, \quad \sim \chi^2\left(\tfrac{k(k-1)}{2}\right).$$

For $k = 3$ there are $\binom{3}{2} = 3$ discordant pairs, giving $df = 3$.

2. Three Discordant Pairs

Pair	n_{ij}	n_{ji}	Q_{ij}
(1,2) Poor ↔ Fair	12	4	$(12-4)^2/16 = 4$
(1,3) Poor ↔ Good	9	3	$(9-3)^2/12 = 3$
(2,3) Fair ↔ Good	6	2	$(6-2)^2/8 = 2$
			Sum $Q = 9$

All three pairs show improvement exceeds deterioration. The (1,2) pair contributes most to Q (44%) because its discordance is largest ($n_{12} - n_{21} = 8$ with $n_{12} + n_{21} = 16$).

3. Marginal Distributions

The marginals shift toward Fair and Good after treatment. Note: $-14 + 4 + 10 = 0$ (marginal totals both sum to 100).

4. Decision

Under H_0: $Q \sim \chi^2(3)$.

α	$\chi^2_{0.05}(3)$	Decision
5%	7.815	**Reject** (9 > 7.815)
1%	11.345	FTR (9 < 11.345)

$p = P(\chi^2(3) > 9) = 0.029$. The marginal rating distribution changed significantly after treatment.

5. Connection to McNemar

For the (1,2) pair alone: McNemar $Q_{12} = (12-4)^2/(12+4) = 4$, $p = 0.046$. The McNemar–Bowker test pools the evidence from all three pairs, increasing power. Each pair contributes a McNemar term; $Q = Q_{12} + Q_{13} + Q_{23}$.

Practitioner Insight: The McNemar–Bowker test extends McNemar's test to square $k \times k$ tables by applying McNemar's statistic to each off-diagonal pair (i, j) and (j, i) and summing the results. The null hypothesis is symmetry, H_0: $P(i, j) = P(j,i)\quad (i \neq j)$, which, for paired categorical data, corresponds to equal marginal distributions across time points or methods. Here $Q = 9$ arises because the three pairwise contributions simplify exactly to $4+3+2$. With $df = k(k-1)/2 = 3$, the result is significant at 5% ($p = 0.029$). The test is omnibus: it detects any asymmetry pattern, not necessarily a monotone improvement or deterioration. Although both McNemar-Bowker and Stuart-Maxwell treat categories as nominal, the latter often has greater power because it tests marginal homogeneity rather than the stronger symmetry hypothesis of McNemar–Bowker. When some discordant-pair totals are small, the χ^2 approximation becomes unreliable and exact or permutation methods may be preferable.

Common Pitfalls

- **Using independent samples.** McNemar–Bowker requires paired or matched observations with the same categories measured twice. Independent samples require Pearson's chi-square test instead.
- **Confusing symmetry with agreement.** The test evaluates symmetry ($p_{ij} = p_{ji}$), not overall agreement or reliability. High disagreement can still produce symmetry; use Cohen's κ for agreement assessment.
- **Sparse off-diagonal counts.** With many categories or small samples, off-diagonal counts may be too sparse for the asymptotic χ^2 approximation. Exact or permutation approaches may be preferable.
- **Ignoring the asymmetry pattern.** A significant result shows asymmetry exists but not which transitions drive it. Inspect the paired counts n_{ij} versus n_{ji} directly.
- **Ignoring ordinal structure.** Bowker's test treats categories as nominal and ignores ordering. For ordinal categories, methods exploiting order may provide greater power.

Stuart-Maxwell Test

Multivariate Test for Marginal Homogeneity in Square Tables

CTAB.8 | **Category:** Contingency Tables | **Data:** Categorical | **Null:** Chi-square

When to Use

- Test marginal homogeneity in paired categories.
- Paired multinomial or repeated-measures data.
- Before-after studies with multiple categories.
- Compare matched categorical classifications.
- Alternative to McNemar for $k > 2$ categories.

Assumptions

- Observations are paired or meaningfully matched.
- Pairs are independent of one another.
- Categories are mutually exclusive/exhaustive.
- Contingency table is square ($k \times k$).
- Expected discordant counts are sufficiently large.

Test Statistic

$$Q_{SM} = \mathbf{d}'\mathbf{S}^{-1}\mathbf{d} \sim \chi^2_{k-1} \text{ (under } H_0\text{)}, \quad \text{where } d_i = n_{i+} - n_{+i} \text{ and } \mathbf{S} \text{ is covariance matrix.}$$

Notation: n_{i+} is row i total, n_{+i} is column i total, **d** is vector of differences, **S** is covariance.

Plain English: The Stuart–Maxwell test evaluates marginal homogeneity by treating differences between row and column totals as a multivariate normal vector and testing whether this vector equals zero. Unlike McNemar–Bowker, which sums separate χ^2 components, Stuart–Maxwell accounts for correlations among marginal differences via the covariance matrix. This multivariate approach yields a more powerful test with fewer degrees of freedom ($k-1$ vs. $k(k-1)/2$), especially when marginal shifts are correlated across categories.

⚠ Practitioner Warning

Test requires paired or matched multicategory data arranged in a square ($k \times k$) table; applying it to independent samples is invalid. The covariance matrix used in the statistic may become near-singular with sparse tables or many categories, making the asymptotic χ^2 approximation unstable. The test evaluates marginal homogeneity, not full symmetry or agreement, and a non-significant result does not imply strong agreement between classifications. Because ordinal structure is ignored, methods that explicitly use category ordering may provide greater power for ordered outcomes.

Hypotheses

- H_0: Row and column marginals are equal.
- H_1: Marginal distributions differ.

Decision Rule: Reject H_0 if $Q_{SM} > \chi^2_{\alpha,k-1}$ or p-value $< \alpha$.

Software References [Python, R, SAS]

`statsmodels.stats...SquareTable` `DescTools::StuartMaxwellTest()` `Not Native (Manual)`

Additional Notes

The Stuart–Maxwell test evaluates marginal homogeneity by testing whether the vector of row–column marginal differences equals zero. If **d** is the vector of marginal differences and **S** its estimated covariance matrix, the statistic is $Q_{SM} = \mathbf{d}^\top\mathbf{S}^{-1}\mathbf{d} \sim \chi^2_{k-1}$. The covariance matrix **S** is constructed from the paired cell counts, with diagonal elements $S_{ii} = n_{i+} + n_{+i} - 2n_{ii}$ and off-diagonal elements $S_{ij} = -(n_{ij} + n_{ji})$. Because **S** has rank $k-1$, a generalized inverse (e.g. Moore–Penrose pseudoinverse) is used in computation. When $k = 2$, the Stuart–Maxwell test reduces exactly to McNemar's test. Symmetry implies marginal homogeneity, but not conversely. Under strong asymmetry, McNemar–Bowker may detect departures missed by Stuart–Maxwell because it tests full symmetry rather than only marginal homogeneity. Bhapkar's test is a closely related alternative based on a different covariance estimator and may provide greater power in some settings. The test is invariant to relabeling of categories but not to collapsing categories, which can alter both the covariance structure and power.

Miscellanea: Alan Stuart introduced his marginal homogeneity test in 1955 as a multinomial extension of McNemar's test for paired categorical data. Maxwell later derived an equivalent formulation in 1970, leading to the modern name "Stuart–Maxwell test." The procedure became widely used in medical and psychiatric research for comparing paired classifications across multiple categories.

Reference(s): Stuart (1955); Maxwell (1970); Bishop et al. (1975)

Worked Example

Problem Statement A clinical psychologist rates the mood severity of 58 patients on a three-category scale (Mild, Moderate, Severe) both at intake and after eight weeks of therapy. Test whether the marginal distribution of mood severity shifted between intake and follow-up at $\alpha = 0.05$. $H_0: r_i = c_i \ \forall i$; $H_1: r_i \neq c_i$ for some i, where r_i and c_i are the i-th row and column marginal totals.

		Follow-up		
		Mild	Moderate	Severe
	Mild	8	9	2
Intake	Moderate	2	14	8
	Severe	0	3	12

Marginal Differences

	Mild	Mod.	Severe
Row total r_i	19	24	15
Col total c_i	10	26	22
$d_i = r_i - c_i$	+9	−2	−7

1. Marginal Differences
For a $k \times k$ paired table, define $d_i = r_i - c_i$ for $i = 1, \ldots, k-1$ (the k-th difference is redundant since $\sum_i d_i = 0$).
With $k = 3$, $\mathbf{d} = (d_1, d_2)^\top = (9, -2)^\top$

2. Variance–Covariance Matrix and Its Inverse
The $(k-1) \times (k-1)$ matrix $\mathbf{V}$ has entries

$$V_{ii} = r_i + c_i - 2n_{ii}, \quad V_{ij} = -(n_{ij} + n_{ji}),\ i \neq j.$$

$$\mathbf{V} = \begin{pmatrix} 13 & -11 \\ -11 & 22 \end{pmatrix}, \quad \mathbf{V}^{-1} = \frac{1}{165}\begin{pmatrix} 22 & 11 \\ 11 & 13 \end{pmatrix}.$$

3. Test Statistic
The Stuart-Maxwell statistic is the quadratic form

$$Q_{SM} = \mathbf{d}^\top \mathbf{V}^{-1} \mathbf{d} \sim \chi^2(k-1).$$

$$Q_{SM} = \frac{1}{165}(9, -2)^\top \begin{pmatrix} 22 & 11 \\ 11 & 13 \end{pmatrix} \begin{pmatrix} 9 \\ -2 \end{pmatrix} \approx 8.715.$$

4. Decision
Under H_0: $Q_{SM} \sim \chi^2_{(2)}$. $p = P(\chi^2_{(2)} > 8.715) = 0.013$.

α	$\chi^2_{(2)}$	Decision
5%	5.991	**Reject** (8.715 > 5.991)
1%	9.210	FTR (8.715 < 9.210)

The marginal distribution of mood severity shifted significantly between intake and follow-up.

5. Marginal Shift

Time	Mild	Mod.	Severe
Intake	19	24	15
Follow-up	10	26	22
Change	−9	+2	+7

The net flow is away from Mild at intake toward Severe at follow-up, suggesting worsening on average despite individual variation on the diagonal.

Practitioner Insight: The Stuart–Maxwell test generalizes McNemar's test to paired $k \times k$ tables by testing marginal homogeneity across more than two categories. Instead of pooling discordant pairs into a single contrast, it forms a $(k-1)$-vector of marginal differences $\mathbf{d}$ and weights it by the inverse covariance matrix, producing a $\chi^2(k-1)$ statistic. The covariance terms are negative because counts contributing to one marginal difference reduce another. A significant result identifies overall marginal change but not which categories differ; follow-up McNemar tests on collapsed 2×2 tables help localize the shifts. Unlike the McNemar–Bowker test, which evaluates symmetry, Stuart–Maxwell tests marginal homogeneity directly; the two coincide only for 2×2 tables. With sparse counts the χ^2 approximation may be unreliable, and for ordered categories the Bhapkar test is often more powerful.

Common Pitfalls

- **Using independent samples.** Stuart–Maxwell requires paired or matched observations measured twice on the same categories. Independent samples require Pearson's chi-square test instead.
- **Ignoring sparse-table problems.** Sparse off-diagonal counts or small samples can make the covariance matrix unstable and the asymptotic χ^2 approximation unreliable. Exact or permutation methods may be preferable.
- **Confusing marginal homogeneity with agreement.** The test evaluates equality of marginal distributions, not agreement between raters or methods. Poor agreement may still coexist with marginal homogeneity.
- **Ignoring ordinal structure.** Stuart–Maxwell treats categories as nominal. Methods exploiting ordering may provide greater power for ordinal outcomes.
- **Overinterpreting non-significance.** Failure to reject may reflect low power rather than true stability, especially with many categories or rare transitions.

Cochran's Q Test

Test for Differences Across Multiple Paired Binary Outcomes

CTAB.9 | **Category:** Contingency Tables | **Data:** Categorical | **Null:** Chi-square

When to Use

- Compare related binary responses across treatments.
- Repeated-measures binary outcome studies.
- Extension of McNemar's test to $k > 2$ groups.
- Matched or blocked dichotomous data analysis.
- Before-after studies with multiple conditions.

Assumptions

- Observations are matched or repeatedly measured.
- Blocks/subjects are independent of one another.
- Outcomes are binary/dichotomous (e.g., 0/1 coding).
- Treatments/conditions are mutually exclusive.
- Sufficient sample size for χ^2 approximation.

Test Statistic

$$Q = \frac{k(k-1)\sum_{j=1}^{k}(C_j - \bar{C})^2}{\sum_{i=1}^{n} R_i(k - R_i)} \approx \chi^2_{k-1} \text{ (under } H_0) \quad \Big| \quad Q = \frac{(k-1)[k\sum_{j=1}^{k} C_j^2 - (\sum_{j=1}^{k} C_j)^2]}{k\sum_{i=1}^{n} R_i - \sum_{i=1}^{n} R_i^2}$$

Notation: X_{ij} is response (0 or 1) for subject i at time j, $C_j = \sum_{i=1}^{n} X_{ij}$, $R_i = \sum_{j=1}^{k} X_{ij}$, k is number of measurements, n is subjects

Plain English: Cochran's Q test evaluates whether the proportion of successes varies across multiple paired measurements by comparing variation among measurement occasions to variation within subjects. The test examines whether column totals (successes per occasion) differ more than expected given the pattern of individual subject responses. It is the nonparametric analog of one-factor repeated measures ANOVA for binary outcomes.

⚠ Practitioner Warning

All subjects must have complete data across all k measurements—missing data requires special handling or deletion. Test assumes common correlation structure across measurement pairs, which may not hold with unevenly spaced time points. Denominator equals zero if all subjects have identical response patterns (all 0s or all 1s)—test undefined. For $k = 2$, Cochran's Q equals McNemar's chi-square statistic. Post-hoc pairwise comparisons require multiple testing adjustment.

Hypotheses

- H_0: $p_1 = p_2 = \cdots = p_k$.
- H_1: At least one proportion differs from others.

Decision Rule: Reject H_0 if $Q > \chi^2_{\alpha,k-1}$ or p-value $< \alpha$.

Software References [Python, R, SAS]

`statsmodels::cochrans_q` `DescTools::CochranQTest()` `PROC FREQ / agree`

Additional Notes

Cochran's Q test is the multi-sample extension of McNemar's test for matched binary outcomes and may also be viewed as the binary-data analog of the Friedman test (rank-based). When $k = 2$, the statistic reduces exactly to McNemar's test. Under H_0, the test statistic is asymptotically distributed as χ^2_{k-1}. A significant result indicates that at least one condition differs but does not identify which pairs differ; post-hoc pairwise McNemar tests with multiplicity adjustments (e.g., Bonferroni or Holm correction) are commonly used. For k conditions, the Bonferroni-adjusted level is $\alpha/[k(k-1)/2]$ per comparison. The test assumes the same subjects are observed under all conditions with binary outcomes recorded each time. Extensions based on generalized estimating equations (GEE) or repeated-measures logistic regression allow covariates and more complex dependence structures.

Miscellanea: William Cochran introduced Cochran's Q test in 1950 as an extension of McNemar's test for matched binary data with more than two related conditions. The procedure became widely used in medical, psychological, and agricultural studies involving repeated yes/no outcomes. Cochran was one of the most influential statisticians of the twentieth century, known especially for his work on sampling, experimental design, and categorical data analysis.

Reference(s): Cochran (1950); Sheskin (2011)

Cochran's Q Test — CTAB.9

Worked Example

Problem Statement Twelve patients each undergo three diagnostic screening tests (A, B, C) for the same condition, with a binary outcome (1=detected, 0=not detected). Table shown transposed for compactness; thus R_i (Patients) and C_j (Tests) appear as column and row totals, respectively. Determine at $\alpha = 0.05$ whether the detection rates differ across the three tests. H_0: $p_A = p_B = p_C$; H_1: at least one detection rate differs.

	1	2	3	4	5	6	7	8	9	10	11	12	C_j
Test A	1	1	1	1	1	1	1	1	0	0	0	0	8
Test B	1	1	1	1	1	0	0	0	1	0	0	0	6
Test C	1	1	0	0	0	0	0	0	0	0	0	0	2
R_i	3	3	2	2	2	1	1	1	1	0	0	0	16

R_i	$k - R_i$	$R_i(k - R_i)$	Count
3	0	0	2
2	1	2	3
1	2	2	4
0	3	0	3

1. Cochran's Q Statistic

For n subjects and k binary conditions, define $C_j = \sum_i x_{ij}$ (Test totals) and $R_i = \sum_j x_{ij}$ (Patient totals). The statistic:

$$Q = \frac{k(k-1)\sum_{j=1}^{k}(C_j - \bar{C})^2}{\sum_{i=1}^{n} R_i(k - R_i)} \sim \chi^2(k-1).$$

2. Column Totals, Mean and Numerator

From the table: $C_A = 8$, $C_B = 6$, $C_C = 2$, total = 16.

$$\bar{C} = 16/3, \qquad \sum_j (C_j - \bar{C})^2 = 56/3.$$

$$k(k-1)\sum_j (C_j - \bar{C})^2 = 6 \times 56/3 = 112.$$

3. Row Sums and Denominator

The unique row-sum values and their frequencies:

$$\sum_{i=1}^{12} R_i(k - R_i) = 2(0) + 3(2) + 4(2) + 3(0) = 14.$$

4. Test Statistic

$$Q = \frac{112}{14} = 8.$$

5. Decision

Under H_0: $Q \sim \chi^2(2)$. $p = P(\chi^2(2) > 8) = 0.018$.

α	$\chi^2_{(2)}$	Decision
5%	5.991	**Reject** (8 > 5.991)
1%	9.210	FTR (8 < 9.210)

The three tests differ significantly in detection rate.

6. Detection Rates

	Test A	Test B	Test C
Detections C_j	8	6	2
Rate C_j/n	0.667	0.500	0.167

Test A detects at four times the rate of Test C; the monotone decline across tests drives the significant result.

Practitioner Insight: Cochran's Q test extends McNemar's test to $k \geq 3$ matched binary conditions. The numerator measures variability among the condition totals C_j, while the denominator measures within-subject variability; subjects with identical responses across all conditions contribute nothing because they provide no information about condition differences. Here the signal comes entirely from the eight subjects with mixed responses. The $\chi^2(k-1)$ approximation requires adequate effective sample size and sufficient total successes. Cochran's Q also assumes exchangeability across conditions within subjects, making it most appropriate when treatments are simultaneous or counterbalanced. A significant omnibus result should be followed by pairwise McNemar tests with multiplicity correction to identify which conditions differ. For paired multinomial responses, the Stuart–Maxwell test is the corresponding generalization.

Common Pitfalls

- **Ignoring the matched design.** Cochran's Q requires the same subjects measured under all conditions. Applying it to independent groups or incomplete repeated measurements invalidates the test.
- **Using sparse or small samples.** The asymptotic χ^2 approximation may be unreliable when n is small or success/failure patterns are sparse. Exact or permutation approaches may be preferable.
- **Using non-binary outcomes.** Cochran's Q applies only to dichotomous responses. Multinomial, ordinal, or continuous outcomes require different repeated-measures methods.
- **Overinterpreting the omnibus test.** A significant result shows that at least one condition differs but does not identify which pairs differ. Post-hoc McNemar tests with multiplicity correction are typically required.
- **Ignoring weak variability or low power.** Subjects with identical responses across all conditions contribute little information, and small marginal differences may be difficult to detect.

Cochran–Mantel–Haenszel Test

Framework for Stratified Association Tests

CTAB.10 | **Category:** Contingency Tables | **Data:** Categorical | **Null:** Chi-square

When to Use

- Test association while controlling for confounding strata.
- Analyze stratified contingency tables across subgroups.
- Pool evidence of association across multiple strata.
- Assess ordinal or nominal relationships conditionally.
- Extend Mantel–Haenszel methods beyond 2x2 tables.

Assumptions

- Strata are mutually independent from one another.
- Observations within each stratum are independent.
- Marginal totals are treated as conditionally fixed.
- Sample sizes should be reasonably large in each stratum.
- The form of association is consistent across strata.

Test Statistic

$$Q_{CMH} = (\mathbf{O} - \mathbf{E})^\top \mathbf{V}^{-1} (\mathbf{O} - \mathbf{E}), \qquad Q_{CMH} \sim \chi^2_\nu.$$

Notation: $\mathbf{O}$ is pooled observed score vector, $\mathbf{E}$ is pooled expected score vector under conditional independence, $\mathbf{V}$ is covariance matrix of $(\mathbf{O} - \mathbf{E})$, and ν depends on the CMH variant used.

Plain English: The Cochran–Mantel–Haenszel (CMH) framework tests whether two categorical variables are associated after controlling for stratification variables such as study center, age group, or matched sets. Rather than pooling raw counts directly, CMH compares observed and expected patterns of association within each stratum and then combines these deviations across strata into a single chi-square statistic. Different scoring schemes produce different CMH variants for nominal, ordinal, or linear association structures.

⚠ Practitioner Warning

CMH methods assume that the underlying association pattern is reasonably consistent across strata; strong interaction or effect modification can make pooled inference misleading. Sparse tables or zero counts may produce unstable variance estimates and poor chi-square approximations. The choice of row/column scores affects ordinal CMH variants, so scores should reflect the scientific ordering of categories. CMH tests detect conditional association, not causation, and important stratum-specific effects may still be masked by pooled summaries.

Hypotheses

- H_0: Conditional independence across strata.
- H_1: Conditional association exists.

Decision Rule: Reject H_0 if Reject H_0 if $Q_{CMH} > \chi^2_{\alpha,\nu}$ or if p-value $< \alpha$.

Software References [Python, R, SAS]

`statsmodels::StratifiedTable`	`mantelhaen.test`	`PROC FREQ (CMH)`

Additional Notes

The CMH framework contains four related variants. **General Association** is used for nominal row and column categories with degrees of freedom $(R-1)(C-1)$. **Row Mean Scores** assumes ordinal rows with nominal columns and has $C-1$ degrees of freedom. **Column Mean Scores** assumes nominal rows with ordinal columns and has $R-1$ degrees of freedom. **Correlation** assigns scores to both rows and columns, tests for linear trend, and has 1 degree of freedom. For $2 \times 2 \times K$ tables, the CMH statistic reduces to the classical Mantel–Haenszel chi-square: $\chi^2_{MH} = [|\sum_k (a_k - E[a_k])| - 0.5]^2 / \sum_k \text{Var}(a_k)$. The CMH framework can also be viewed as a family of stratified score tests arising from log-linear or conditional logistic models. Additional worked examples are provided at the end of this chapter.

Miscellanea: The CMH framework combines contributions from William Cochran, Nathan Mantel, and William Haenszel during the 1950s and 1960s, originally motivated by epidemiologic studies requiring adjustment for confounding variables such as age or study center. The Mantel–Haenszel procedure became one of the earliest widely used methods for obtaining an "adjusted" association estimate before generalized linear models became commonplace. The CMH framework later expanded into a broader family of stratified score tests capable of handling nominal, ordinal, and trend-based association structures.

Reference(s): Cochran (1954); Mantel & Haenszel (1959); Mantel (1963); Agresti (2013)

Worked Example

Problem Statement A multi-center drug trial records response (Yes/No) to three dose levels (Low, Medium, High) across $K = 3$ clinical centers. Unlike the standard Mantel-Haenszel test (which pools 2×2 tables), the CMH test handles these 3×2 stratified tables directly. Due to space limitation, sample data is provided at the end of this chapter.

H_0: dose and response are conditionally independent within each center; H_1: a common dose-response association exists across centers. Note: δ_{rs} denotes the Kronecker delta, equal to 1 when $r = s$ and 0 otherwise.

1. Deviation Vector $\mathbf{d}_k$ per Stratum

For a 3×2 table, $(I-1)(J-1) = 2$, so $\mathbf{d}_k$ is 2×1: deviations for (Low,Yes) and (Med,Yes).

Expected counts: $E_k[i,j] = n_{ik} \cdot n_{kj}/n_k$.

	Ctr A	Ctr B	Ctr C
$E[\text{Low,Yes}]$	11.714	9.667	5.250
$E[\text{Med,Yes}]$	14.643	9.667	7.875
$d_{k,1} = O - E[\text{Low}]$	−3.714	−4.667	−2.250
$d_{k,2} = O - E[\text{Med}]$	+0.357	+0.333	+0.125

Summing across strata:

$$\mathbf{d}' = \sum_k \mathbf{d}'_k = \begin{pmatrix} -10.631 & +0.815 \end{pmatrix}$$

2. Variance Matrix V_k per Stratum

For $J = 2$ the (r,s) element simplifies to:

$$V_k[r,s] = \frac{n_k}{n_k - 1} \cdot \frac{n_{k,\text{Yes}} \cdot n_{k,\text{No}}}{n_k^2} \cdot n_{rk}\left(\delta_{rs} - \frac{n_{sk}}{n_k}\right)$$

	Ctr A	Ctr B	Ctr C
$V_k[1,1]$	3.517	3.386	1.918
$V_k[2,2]$	3.956	3.386	2.398
$V_k[1,2]$	−1.758	−1.693	−0.959

3. Pooled Variance and Its Inverse

$$V = \sum_k V_k = \begin{pmatrix} 8.821 & -4.411 \\ -4.411 & 9.740 \end{pmatrix}, \quad V^{-1} = \begin{pmatrix} 0.1465 & 0.0664 \\ 0.0664 & 0.1327 \end{pmatrix}$$

4. CMH Statistic

$$\mathbf{d}' V^{-1} = \begin{pmatrix} -1.504 & -0.597 \end{pmatrix}, \quad CMH = \mathbf{d}' V^{-1} \mathbf{d} = 15.5$$

Under H_0: $CMH \sim \chi^2((I-1)(J-1)) = \chi^2(2)$.

5. Results and Decision

Statistic	Value	CV (5%)	Decision
CMH	15.500	5.991	Reject H_0

$p = 0.0004$. The dose-response association is highly significant after controlling for center. The negative $d_1 = -10.631$ (Low dose consistently below expectation) and near-zero $d_2 = +0.815$ (Medium dose near expectation) reflect a monotone dose effect driven mainly by the Low vs High contrast.

Practitioner Insight: The CMH test generalizes the Mantel–Haenszel procedure from 2×2 to $I \times J$ stratified tables by replacing the scalar deviation $O - E$ and variance with a $(I-1)(J-1)$-vector $\mathbf{d}_k$ and covariance matrix V_k. For 2×2 tables, the quadratic form $\mathbf{d}'V^{-1}\mathbf{d}$ reduces exactly to the usual MH statistic. Here, with $I = 3$ and $J = 2$, the test has $df = 2$ and captures both the Low-vs-High contrast (driven mainly by $d_1 = -10.631$) and the Low-vs-Medium contrast ($d_2 = 0.815$). As table dimensions increase, df grows and power is spread across more contrasts. When categories are ordered, the CMH correlation statistic is often preferred because it concentrates power into a single monotone-trend direction; stratified analog of Cochran–Armitage.

Common Pitfalls

- **Pooling heterogeneous strata.** CMH assumes a common odds ratio across strata. When stratum-specific odds ratios differ substantially, the pooled estimate may be misleading; assess homogeneity before pooling.
- **Using sparse strata.** Small or zero cell counts can produce unstable odds-ratio estimates and poor asymptotic approximations. Exact conditional methods or regression models may be preferable.
- **Confusing conditional association with confounding.** CMH tests association after stratification but does not itself prove absence of confounding. Compare crude and adjusted odds ratios to assess confounding impact.
- **Over-stratifying the data.** Stratifying on many variables creates numerous sparse strata and rapidly reduces efficiency. Multivariable logistic regression is often preferable for multiple confounders.
- **Using correlated observations.** CMH assumes independent observations within and across strata. Repeated-measures or clustered data require methods such as GEE or mixed-effects models.

Breslow-Day Test

Test for Homogeneity of Odds Ratios Across Strata

CTAB.11 | **Category:** Contingency Tables | **Data:** Categorical | **Null:** Chi-square

When to Use

- Test if odds ratios are constant across multiple strata.
- Check homogeneity before using Mantel-Haenszel pooling.
- Detect interaction effects in stratified binary outcomes.
- Verify that a common effect size exists across subgroups.
- Analyze data consistency in multi-center clinical trials.

Assumptions

- Each stratum represents an independent 2x2 table setup.
- The observations within each stratum are independent.
- The data must be binary for both the treatment and result.
- Sample sizes are large; avoid using for very sparse data.
- The underlying common odds ratio is fixed under the null.

Test Statistic

$$\chi^2_{BD} = \sum_{k=1}^{K} \frac{(a_k - \tilde{a}_k)^2}{\text{Var}(\tilde{a}_k)},$$ where $\tilde{a}_k$ is MLE under common odds ratio (OR). Under $H_0, \chi^2_{BD} \approx \chi^2_{K-1}$.

Notation: a_k is observed count in cell(1,1) of stratum k, $\tilde{a}_k$ is expected under homogeneity, K is strata

Plain English: The Breslow-Day test (a variant of Pearson's χ^2 under H_0) evaluates whether the strength of association (measured by odds ratio) between exposure and outcome is consistent across strata. It compares observed cell counts in each stratum to those expected if all strata shared a common odds ratio, using the Mantel-Haenszel estimate. Significant heterogeneity indicates effect modification—the association varies by stratum, making pooled estimates potentially misleading.

⚠ Practitioner Warning

The Breslow–Day test may have limited power for detecting moderate heterogeneity, especially with few strata or small samples, so a non-significant result does not guarantee homogeneous odds ratios. Sparse or zero cell counts can produce unstable odds-ratio estimates and poor chi-square approximations; Tarone's modification often improves small-sample performance. If significant heterogeneity is detected, pooled Mantel–Haenszel estimates may be misleading and stratum-specific odds ratios or interaction models should be reported instead. The test evaluates homogeneity of odds ratios, not the presence of an overall association, which is typically assessed separately using the CMH test.

Hypotheses

- H_0: $OR_1 = OR_2 = \cdots = OR_K$ (odds ratios are all equal).
- H_1: At least one odds ratio differs.

Decision Rule: Reject H_0 if $\chi^2_{BD} > \chi^2_{\alpha,K-1}$ or p-value $< \alpha$.
Degrees of freedom: $K - 1$ where K is number of strata.

Software References [Python, R, SAS]

`statsmodels::StratifiedTable` `DescTools::BreslowDayTest` `PROC FREQ (CMH)`

Additional Notes

For each stratum, the expected count $m_k = \mathbb{E}[a_k \mid \theta]$ is obtained by solving **(A)**, where $\widehat{\theta}_{MH}$ is the common Mantel–Haenszel odds ratio. Remaining expected counts are $b^*_k = r_{1k} - m_k$, $c^*_k = c_{1k} - m_k$, $d^*_k = n_k - r_{1k} - c_{1k} + m_k$, with variance **(B)**. Tarone's modification **(C)** improves the chi-square approximation when the common odds ratio is estimated rather than known. Significant heterogeneity suggests reporting stratum-specific odds ratios or fitting interaction terms instead of using a pooled estimate.

$$\textbf{(A)}: \frac{m_k(n_k - r_{1k} - c_{1k} + m_k)}{(r_{1k} - m_k)(c_{1k} - m_k)} = \widehat{\theta}_{MH}, \qquad \textbf{(B)}: v_k = \left[\frac{1}{m_k} + \frac{1}{b^*_k} + \frac{1}{c^*_k} + \frac{1}{d^*_k}\right]^{-1}, \qquad \textbf{(C)}: \chi^2_T = \chi^2_{BD} - \frac{[\sum_k (a_k - \tilde{a}_k)]^2}{\sum_k \text{Var}(\tilde{a}_k)}$$

Miscellanea: Norman Breslow and N.E. Day introduced this test in 1980 in their influential work *Statistical Methods in Cancer Research*. The procedure was developed to address an important practical problem in stratified epidemiologic studies: researchers often pooled odds ratios across strata without formally checking whether the association strength was actually consistent. The Breslow–Day test became a standard tool for assessing homogeneity of odds ratios before applying pooled Mantel–Haenszel estimates.

Reference(s): Breslow & Day (1980); Tarone (1985)

Worked Example

Problem Statement Before accepting the Mantel-Haenszel pooled estimate $\widehat{OR}_{\text{MH}} = 2.749$ from the previous example, the Breslow-Day test checks whether the three center-specific odds ratios (2.25, 3.14, 2.79) are consistent with a single common value. H_0: $OR_A = OR_B = OR_C$ (homogeneity); H_1: at least one OR_k differs.

	Center A ($n = 40$)		**Center B** ($n = 60$)		**Center C** ($n = 50$)	
	Treat	Ctrl	Treat	Ctrl	Treat	Ctrl
Recovered	12	8	22	8	18	7
Not recovered	8	12	14	16	12	13

1. Expected Cell Counts Under Common OR

For each stratum, solve for A_k — the value of a_k that gives $OR_k = \widehat{OR}_{\text{MH}} = 2.749$ exactly while preserving the observed marginals. A_k satisfies:

$$\frac{A_k\,(n_k - R_{1k} - C_{1k} + A_k)}{(R_{1k} - A_k)(C_{1k} - A_k)} = \widehat{OR}_{\text{MH}}$$

Rearranging gives the quadratic:

$$(1-\widehat{OR})\,A_k^2 + \left[D_k + \widehat{OR}(R_{1k}+C_{1k})\right]A_k - \widehat{OR}\,R_{1k}C_{1k} = 0$$

where $D_k = n_k - R_{1k} - C_{1k}$.

Center A ($R_1 = 20, C_1 = 20, D = 0$):

$$-1.749\,A^2 + 109.96\,A - 1099.6 = 0, \qquad \Longrightarrow\; A_A = 12.475$$

Similarly: $A_B = 21.555$, $A_C = 17.963$.

2. Variances and BD Terms

$\text{Var}_k = \left[A_k^{-1} + B_k^{-1} + C_k^{-1} + D_k^{-1}\right]^{-1}$ where $B_k = R_{1k} - A_k$, $C_k = C_{1k} - A_k$, $D_k \leftarrow n_k - R_{1k} - C_{1k} + A_k$.

	A	**B**	**C**
a_k (observed)	12	22	18
A_k (expected)	12.475	21.555	17.963
$a_k - A_k$	−0.475	+0.445	+0.037
Var_k	2.347	3.352	2.793
$(a_k - A_k)^2/\text{Var}_k$	0.0963	0.0590	0.0005

3. Breslow-Day Statistic

$$BD = \textstyle\sum_{k=1}^{K}(a_k - A_k)^2/\text{Var}_k = 0.156$$

Under H_0: $BD \sim \chi^2(K-1) = \chi^2(2)$.

4. Results and Decision

Statistic	**Value**	**CV (5%)**	**Decision**
BD	0.156	5.991	FTR H_0

$p = 0.925$. The three ORs are homogeneous; $\widehat{OR}_{\text{MH}} = 2.749$ is a valid pooled estimate.

Practitioner Insight: The Breslow–Day test should precede Mantel–Haenszel pooling: if it rejects, the stratum-specific odds ratios are heterogeneous and a single pooled $\widehat{OR}_{\text{MH}}$ is misleading. The expected count A_k is obtained by enforcing both the observed marginals and the common odds-ratio constraint $OR = \widehat{OR}_{\text{MH}}$, analogous to expected counts in Pearson's χ^2 test but under homogeneity rather than independence. The statistic has $K-1$ degrees of freedom because one constraint is used to estimate the common odds ratio. Here the deviations $a_k - A_k$ are all small, showing close agreement with the common $OR = 2.749$. Large differences among stratum-specific odds ratios would instead produce a large Breslow–Day statistic. Woolf's test is a related homogeneity procedure based on variance-weighted log-odds ratios, though it is less stable with sparse tables and requires non-zero cells. When Breslow–Day rejects, stratum-specific reporting or interaction modeling is preferable.

Common Pitfalls

- **Treating non-significance as proof of homogeneity.** Breslow–Day often has limited power with few strata or small samples. Failure to reject does not guarantee equal odds ratios across strata.
- **Pooling despite heterogeneity.** A significant Breslow–Day test indicates effect modification across strata. Pooling with the CMH test after significant heterogeneity can obscure important subgroup differences.
- **Using sparse strata.** Small or zero cell counts can produce unstable odds-ratio estimates and poor chi-square approximations. Exact methods, collapsing strata, or logistic regression with interaction terms may be preferable.
- **Confusing heterogeneity with association.** Breslow–Day tests consistency of odds ratios across strata, not whether an overall association exists. Conditional association is typically assessed separately using the CMH test.

Woolf's Test

Test for Homogeneity of Odds Ratios Across Strata

CTAB.12 | **Category:** Contingency Tables | **Data:** Categorical | **Null:** Chi-square

When to Use

- Test homogeneity of odds ratios across multiple strata.
- Verify that a common odds ratio exists for a meta-analysis.
- Test for interactions between a treatment and a covariate.
- Evaluate if the effect size is consistent across subgroups.
- Check if pooling data into one odds ratio is appropriate.

Assumptions

- The data in each stratum are independent and binary.
- A large sample size is required within each specific cell.
- The log odds ratios follow a normal distribution roughly.
- Individual observations within strata are independent.
- Total counts are high enough to avoid zero cell bias.

Test Statistic

$$\chi^2_W = \sum_{k=1}^{K} w_k(\ln\hat{\theta}_k - \ln\hat{\theta}_{MH})^2, \qquad w_k = 1/\mathrm{Var}(\ln\hat{\theta}_k). \quad \text{Under } H_0, \chi^2_W \sim \chi^2_{K-1}.$$

Notation: $\hat{\theta}_k$ is estimated odds ratio (OR) in stratum k, K is number of strata, $\hat{\theta}_{MH}$ is Mantel-Haenszel common OR

Plain English: Woolf's test compares stratum-specific odds ratios to the pooled Mantel–Haenszel estimate, weighting by their precision. If odds ratios are homogeneous, they should vary randomly around the pooled value; systematic deviations indicate heterogeneity, suggesting pooling may be inappropriate. The test uses log odds ratios due to their better statistical properties.

⚠ Practitioner Warning

Woolf's test relies on the approximate normality of log odds ratios and may perform poorly with sparse tables, small cell counts, or zero cells, where the log odds ratio becomes unstable. Continuity corrections or small-sample adjustments are often required in such settings. The test is also known to be conservative with limited power for detecting true heterogeneity, especially when the number of strata is small or odds ratios are extreme. A non-significant result does not establish homogeneous effects across strata, and alternative procedures such as the Breslow–Day test are often preferred for formal assessment of odds-ratio homogeneity.

Hypotheses

- H_0: $\theta_1 = \theta_2 = \cdots = \theta_K$.
- H_1: Odds ratios differ across strata.

Decision Rule: Reject H_0 if $\chi^2_W > \chi^2_{\alpha,K-1}$ or p-value $< \alpha$. If rejected, stratified analysis or interaction modeling needed.

Software References [Python, R, SAS]

Python	R	SAS
`Not Native (Manual)`	`vcd::woolf_test`	`Not Native (Manual)`

Additional Notes

Woolf's test provides a framework for assessing the homogeneity of odds ratios (OR) across k independent 2×2 strata. For each stratum i, let the observed cell counts be a_i, b_i, c_i, d_i. The test operates on the log-odds ratio scale, $y_i = \ln(\widehat{OR}_i) = \ln(a_i d_i / b_i c_i)$, which more closely approximates a normal distribution than the raw OR. The estimated large-sample variance of y_i is: $v_i = \mathrm{Var}(y_i) \approx 1/a_i + 1/b_i + 1/c_i + 1/d_i$, allowing simple inverse-variance weighting. A significant result indicates "interaction" or "effect modification," suggesting the odds ratio is not constant across strata. While computationally efficient, the test is susceptible to bias when cell counts are small (sparse data), as v_i becomes undefined if any cell is zero. In such cases, the **Breslow–Day test** or **Tarone's adjustment** is technically superior. Woolf's approach is fundamentally a fixed-effects meta-analysis technique, assuming underlying effect size is identical across strata under H_0.

Miscellanea: Woolf (1955) introduced a method for testing the homogeneity of odds ratios across 2×2 strata. The method was motivated by early blood-group genetics studies examining the association between blood group A and gastric cancer, where researchers observed different odds ratios across cities and populations. Woolf's procedure helped distinguish genuine effect modification from ordinary sampling variation. Although later methods such as the Breslow–Day test became more widely used, Woolf's approach remains historically important for highlighting the distinction between the odds ratio itself and the transformed scale used for inference.

Reference(s): Woolf (1955); Breslow & Day (1980); Gavaghan et al. (1996)

Worked Example

Problem Statement A multi-center case-control study of occupational dust exposure and respiratory disease is conducted across three sites (S1, S2, S3). Each site contributes a 2 × 2 table of cases (disease) and controls, cross-classified by exposure status (exposed / unexposed). The site-specific odds ratios appear to differ markedly. Test at $\alpha = 0.05$ whether the odds ratios are homogeneous across sites. H_0: $\text{OR}_1 = \text{OR}_2 = \text{OR}_3$; H_1: at least one OR differs.

	Site S1			Site S2			Site S3		
	Exp	Unexp	Tot	Exp	Unexp	Tot	Exp	Unexp	Tot
Case	20	20	40	20	10	30	15	5	20
Control	15	15	30	10	20	30	5	15	20

1. Woolf's Statistic

For k strata, denote cell counts (a_i, b_i, c_i, d_i) in the standard 2 × 2 layout. Log-odds ratio and variance for stratum i:

$$\widehat{\ln \text{OR}_i} = \ln \frac{a_i d_i}{b_i c_i}, \qquad v_i = \frac{1}{a_i} + \frac{1}{b_i} + \frac{1}{c_i} + \frac{1}{d_i}.$$

Set weights $w_i = 1/v_i$ and form the pooled estimate

$$\widehat{\ln \text{OR}_W} = \frac{\sum_i w_i \widehat{\ln \text{OR}_i}}{\sum_i w_i}.$$

The test statistic is

$$Q_W = \sum_{i=1}^{k} w_i \left(\widehat{\ln \text{OR}_i} - \widehat{\ln \text{OR}_W}\right)^2 \sim \chi^2(k-1).$$

2. Stratum Odds Ratios and Weights

Statistic	Site 1	Site 2	Site 3
Odds Ratio	1.00	4.00	9.00
ln(OR)	0.000	1.386	2.197
Var[ln(OR)] , v_i	0.233	0.300	0.533
$w_i = v_i^{-1}$	4.286	3.333	1.875

3. Pooled Log-Odds Ratio

$$\sum_i w_i = 4.494, \qquad \sum_i w_i \widehat{\ln \text{OR}_i} = 8.741,$$

$$\widehat{\ln \text{OR}_W} = 0.9207.$$

4. Test Statistic

$$Q_W = 3.633 + 0.723 + 3.055 = 7.411.$$

5. Decision

Under H_0: $Q_W \sim \chi^2(2)$.

α	$\chi^2_{(2)}$	Decision
5%	5.991	**Reject** (7.411 > 5.991)
1%	9.210	FTR (7.411 < 9.210)

$p = P(\chi^2(2) > 7.411) = 0.025$. The odds ratios are not homogeneous across sites; a pooled Mantel–Haenszel estimate is not appropriate without further investigation.

Practitioner Insight: Woolf's test checks homogeneity of stratum-specific odds ratios before applying the Mantel–Haenszel procedure. In the example, odds ratio increases ninefold from S1 to S3, with larger sites receiving higher weight. S3 has the most extreme OR but the lowest weight.A significant result indicates effect modification, implying that a single pooled odds ratio is misleading and stratum-specific estimates should be reported separately. Zero cell counts cause infinite variances and effectively remove strata from the analysis. The Breslow–Day test is a related alternative that performs better with sparse data. Because both tests have low power with few strata, a non-significant result does not rule out meaningful heterogeneity, so confidence intervals should also be inspected.

Common Pitfalls

- **Using sparse strata.** Woolf's test relies on large-sample normal approximations for log odds ratios. Small or zero cell counts can make variance estimates unstable and invalidate the χ^2 approximation.
- **Treating non-significance as proof of homogeneity.** Failure to reject does not guarantee identical odds ratios across strata. With few strata or moderate heterogeneity, the test may have limited power to detect effect modification.
- **Ignoring stratum-specific patterns.** A significant result only indicates heterogeneity somewhere among the strata. Inspect the individual odds ratios to determine whether the differences reflect trends, subgroup effects, or outlier strata.
- **Using ordered strata as nominal groups.** Woolf's test ignores ordering information in variables such as age or dose. Trend tests or regression models may provide more informative analyses.
- **Confusing with Cochran's Q.** Woolf's test evaluates heterogeneity of odds ratios within a stratified analysis, whereas Cochran's Q assesses heterogeneity across independent studies in meta-analysis.

Cochran-Armitage Test

Trend Test for Ordered Categorical Groups

CTAB.13 | **Category:** Contingency Tables | **Data:** Binary/Ordinal | **Null:** Normal / χ^2_1

When to Use

- Detect linear trends in proportions across ordered groups.
- Analyze dose-response relationships in clinical trials.
- Test for a monotonic trend in binary outcome frequencies.
- Study risk factor levels like age groups or exposure tiers.
- Evaluate if success rates increase with ordinal categories.

Assumptions

- Data consist of independent observations across all levels.
- The independent variable must have a natural rank order.
- The relationship between groups and outcome is linear.
- Binary outcomes are recorded for each independent subject.
- Large sample size ensures the test statistic is reliable.

Test Statistic

$$Z = \frac{\sum_{i=1}^{K} w_i(x_i - n_i\hat{p})}{\sqrt{\hat{p}(1-\hat{p})\sum_{i=1}^{K} n_i(w_i - \bar{w})^2}} \approx N(0,1) \text{ (under } H_0) \Leftrightarrow Z^2 \approx \chi^2_1. \qquad \hat{p} = \frac{\sum_{i=1}^{K} x_i}{\sum_{i=1}^{K} n_i}, \qquad \bar{w} = \frac{\sum_{i=1}^{K} n_i w_i}{\sum_{i=1}^{K} n_i}.$$

Notation: x_i = successes in group i, n_i = group size, w_i = assigned scores, $\hat{p}$ = overall proportion.

Plain English: The Cochran-Armitage test detects linear trends in proportions across ordered groups by assigning numeric scores to groups reflecting their ordering or dose levels. The test evaluates whether the proportion of successes increases or decreases linearly with these scores. Unlike the chi-square test which treats groups as unordered, Cochran-Armitage exploits the ordering to gain power for detecting monotonic trends. It measures correlation between group scores and outcomes, standardized by the variance.

Practitioner Warning

Requires meaningful group ordering specified a priori; post-hoc ordering invalidates results. Power depends on score choice—equally spaced scores are common, but actual doses are preferable. Detects only linear trends and may miss non-monotonic patterns. Less powerful than chi-square for unordered alternatives. One-sided tests are appropriate when direction is pre-specified. Small expected counts ($n_i p_i < 5$ or $n_i(1-p_i) < 5$) reduce validity; exact tests may be used. Does not estimate effect size—use logistic regression for that.

Hypotheses

- H_0: No linear trend in proportions across ordered groups.
- H_1: Linear trend exists.

Decision Rule: Reject H_0 if $Z^2 > \chi^2_{\alpha,1}$ (two-sided) or $|Z| > z_{\alpha/2}$ or p-value $< \alpha$.

Software References [Python, R, SAS]

`statsmodels::contingency_tables` `DescTools::CochranArmitageTest` `PROC FREQ (TREND)`

Additional Notes

Scores w_i may be equally spaced $(1, 2, \ldots, k)$ or reflect actual doses or midpoints; alternative schemes (e.g., ridit or optimal scores) are sometimes used. The test statistic is proportional to the covariance between scores and outcomes, and is equivalent to the score test for the slope in the linear probability model $p_i = \alpha + \beta s_i$. Mantel (1963) extended the method to stratified settings (CMH trend test). An exact version based on conditioning on margins is available for small samples. Logistic regression provides a complementary approach, estimating odds ratios per unit score with confidence intervals.

Miscellanea: Cochran (1954) and Armitage (1955) independently developed methods in the 1950s for testing trends in proportions across ordered groups, motivated by applications such as dose-response studies. By incorporating ordering information, the Cochran–Armitage test uses a 1-degree-of-freedom statistic, making it more powerful than the general chi-square test when a monotonic trend is present. It remains widely used in epidemiology, genetics, and toxicology for detecting dose-response relationships.

Reference(s): Cochran (1954); Armitage (1955); Agresti (2013)

Cochran-Armitage Test

CTAB.13

Worked Example

Problem Statement A dose-response study records the number of patients experiencing an adverse event across four dose groups with unequally spaced dose scores t_i. The Cochran-Armitage test (1954) detects a linear trend in the response proportion across the ordered groups. H_0: no linear trend in proportions; H_1: increasing trend. $X = \sum x_i$, $N = \sum n_i$.

Group	Placebo	Low	Medium	High
Dose score w_i	0	1	3	5
Sample size n_i	30	30	30	30
Events x_i	3	5	8	10
Proportion $\hat{p}_i$	0.100	0.167	0.267	0.333

1. Overall Proportion and Weighted Mean Score

$$\hat{p} = \frac{X}{N} = \frac{26}{120} = \frac{13}{60} \approx 0.217$$

Since all $n_i = 30$, the weighted mean score is simply

$$\bar{w} = \frac{0+1+3+5}{4} = \frac{9}{4} = 2.25$$

2. Numerator: Weighted Deviation Sum

Each $n_i\hat{p} = 30 \times 13/60 = 6.5$ exactly:

$$\mathrm{N} = \sum_{i=1}^{4} w_i(x_i - n_i\hat{p}) = 20.5$$

w_i	x_i	$x_i - 6.5$	$w_i(x_i - 6.5)$	$30(w_i - 2.25)^2$
0	3	−3.5	0.0	151.875
1	5	−1.5	−1.5	46.875
3	8	+1.5	+4.5	16.875
5	10	+3.5	+17.5	226.875

3. Denominator: Weighted Score Variance

$$S_{tt} = \sum_{i=1}^{4} n_i(w_i - \bar{w})^2 = 442.5, \mathrm{D} = \sqrt{\hat{p}(1-\hat{p})\, S_{tt}} = 8.666$$

4. Test Statistic and Decision

$$Z = \frac{\text{Num}}{\text{Denom}} = \frac{20.5}{8.666} = 2.366 \qquad T = Z^2 = 5.596$$

Form	Dist.	Stat.	CV (5%)	Decision
Z (one-sided)	$N(0,1)$	2.366	1.645	Reject
Z (two-sided)	$N(0,1)$	2.366	1.960	Reject
$T = Z^2$	$\chi^2(1)$	5.596	3.841	Reject

Practitioner Insight: The Cochran–Armitage test is a score test for trend in proportions, testing whether $\beta = 0$ in the linear probability model $p_i = \alpha + \beta t_i$. The Z form is preferred because it preserves the direction of the trend: here $Z = 2.366$ confirms an increasing dose-response, whereas the equivalent χ^2 statistic $T = Z^2$ loses the sign information. When a directional alternative is specified in advance, the one-sided test is more powerful than the two-sided version. With equally spaced scores $t_i = (1, 2, \ldots, k)$, the test coincides with Armitage's chi-square trend test. The choice of scores matters: using unequal scores such as $(0, 1, 3, 5)$ reflects actual dose spacing and gives greater weight to departures at higher doses. The Cochran–Mantel–Haenszel test can be viewed as a stratified extension of the Cochran–Armitage procedure.

Common Pitfalls

- **Applying the test when exposure levels are not truly ordinal.** Cochran–Armitage requires meaningful ordering of exposure categories. It is invalid for nominal groups or arbitrary orderings. Use standard chi-square tests when no intrinsic ordering exists.
- **Selecting exposure scores after seeing the data.** Scores must be chosen *a priori*. Searching for score assignments that maximize significance constitutes p-hacking. Use actual exposure values, midpoints, or pre-specified integer scores, and document the rationale.
- **Assuming monotonic trend when the true pattern is non-linear.** The test is powerful for monotonic trends but can miss U-shaped, J-shaped, or threshold effects. Inspect proportions across categories, assess deviations from linearity, and consider polynomial or spline models when non-linearity is plausible.
- **Using a two-sided test when direction is known a priori.** When biological or scientific theory predicts a clear direction, a one-sided Cochran–Armitage test is more powerful. Direction must be specified before analysis; switching post hoc invalidates inference.
- **Ignoring confounding in observational studies.** The test evaluates unadjusted trends and may yield misleading results when confounders are present. Use stratified analyses, logistic regression, or propensity-based methods to adjust for confounding before drawing causal conclusions.

Cohen's Kappa

Agreement Measure for Categorical Ratings Between Two Raters

CTAB.14 | **Category:** Agreement & Reliability | **Data:** Categorical | **Null:** Asymptotic Normal

When to Use

- Assess agreement between two raters on nominal scales.
- Evaluate diagnostic consistency for binary test results.
- Validate manual data labeling against a gold standard.
- Measure reliability for discrete classification tasks.
- Adjust raw agreement by accounting for chance success.

Assumptions

- The raters are independent and perform ratings blindly.
- The rating categories are mutually exclusive and fixed.
- Raters are non-unique; the same two judge every subject.
- Data consist of paired observations on the same subjects.
- Categories are nominal; use Weighted Kappa if ordinal.

Test Statistic

$$\hat{\kappa} = \frac{p_o - p_e}{1 - p_e}, \quad p_o = \frac{\sum_{i=1}^{k} n_{ii}}{n}, \quad p_e = \sum_{i=1}^{k} \frac{n_{i+}n_{+i}}{n^2}, \quad \mathrm{SE}(\hat{\kappa}) = \sqrt{\frac{p_o(1-p_o)}{n(1-p_e)^2}}. \quad \text{Under } H_0: z = \frac{\hat{\kappa}}{\mathrm{SE}(\hat{\kappa})} \approx N(0,1).$$

Notation: n_{ii} are diagonal counts, n_{i+}, n_{+i} are marginal totals, p_o is observed agreement, p_e is expected by chance

Plain English: Cohen's kappa measures agreement between two raters beyond what would occur by chance. It compares observed agreement (proportion of items both raters classified identically) to expected agreement (what would occur if raters assigned categories randomly but maintained their marginal distributions). Kappa of 1 indicates perfect agreement, 0 indicates agreement no better than chance, and negative values indicate systematic disagreement. The measure corrects for chance, providing more meaningful reliability assessment than raw percentage agreement.

⚠ Practitioner Warning

Sensitive to prevalence imbalance and marginal asymmetry (rater bias), producing the well-known "kappa paradox" in which high raw agreement may still yield a low κ. Assumes nominal categories; use weighted kappa for ordinal data. Does not identify which categories disagree—inspect the confusion matrix. Values depend on the number of categories, and the maximum may be below 1 with marginal heterogeneity. Asymptotic standard errors are unreliable for small samples; bootstrap methods are preferred. Measures reliability, not validity. Complex SE formula also exists.

Hypotheses

- H_0: No agreement beyond chance ($\kappa = 0$).
- H_1: Agreement is different from chance ($\kappa \neq 0$).

Decision Rule: Reject H_0 if $|z| > z_{\alpha/2}$ or p-value $< \alpha$.

Software References [Python, R, SAS]

`sklearn::cohen_kappa_score`	`irr::kappa2`	`PROC FREQ`

Additional Notes

Agreement is summarized in a $k \times k$ table with cell counts n_{ij}. Values of κ range from -1 to 1, where 1 indicates perfect agreement, 0 agreement no better than chance, and negative values agreement worse than chance. Inference typically uses asymptotic normality with Wald confidence intervals, though bootstrap methods are often preferred for small samples. Weighted κ extends the method to ordinal categories by assigning partial credit to near agreement; common choices are linear and quadratic weights. With quadratic weights, weighted κ is mathematically equivalent to certain forms of the intraclass correlation coefficient (ICC). Extensions include Fleiss' κ for multiple raters, PABAK for **p**revalence-**a**djusted **b**ias-**a**djusted **k**appa ($\kappa = 2p_o - 1$), and alternatives such as Krippendorff's α and Gwet's AC1.

Miscellanea: Jacob Cohen introduced kappa in 1960 to measure inter-rater agreement while adjusting for agreement expected by chance, addressing the tendency of raw agreement to overstate reliability when categories are imbalanced. The statistic became widely used in medicine, psychology, and social science research. Later interpretation guidelines by Landis and Koch (1977) popularized descriptive labels for agreement levels, though the cutoffs remain somewhat arbitrary.

Reference(s): Cohen (1960); Landis & Koch (1977); Fleiss et al. (2003)

Worked Example

Problem Statement Two radiologists independently classify 40 chest X-rays into one of three categories: Normal (N), Suspicious (S), or Abnormal (A). Their joint ratings are recorded in the table below. Compute Cohen's κ and test whether agreement exceeds chance at $\alpha = 0.05$. $H_0: \kappa = 0$ (agreement no better than chance); $H_1: \kappa > 0$.

		Rater 2		
		N	S	A
	N	14	2	0
Rater 1	S	2	12	2
	A	0	2	6

Marginal Totals

	N	S	A	Total
Rater 1 (row)	16	16	8	40
Rater 2 (col)	16	16	8	40

1. Observed and Expected Agreement
For a $k \times k$ agreement table with n subjects, let $p_{ij} = n_{ij}/n$. Define observed and expected agreement:

$$P_o = \sum_{i=1}^{k} p_{ii}, \qquad P_e = \sum_{i=1}^{k} p_{i+}\, p_{+i},$$

where p_{i+} and p_{+i} are the row and column marginal proportions.

$$P_o = (14 + 12 + 6)/40 = 4/5.$$

$$P_e = (16 \cdot 16 + 16 \cdot 16 + 8 \cdot 8)/40^2 = 9/25.$$

Note that both raters have identical marginal distributions (N:16, S:16, A:8), so P_e equals $\sum_i p_{i+}^2$.

2. Kappa Statistic

$$\kappa = \frac{P_o - P_e}{1 - P_e} = \frac{4/5 - 9/25}{1 - 9/25} = 0.6875.$$

3. Standard Error and Test Statistic
Under $H_0: \kappa = 0$:

$$\mathrm{SE}(\hat{\kappa}) = \frac{\sqrt{P_o(1 - P_o)}}{\sqrt{n}\,(1 - P_e)} = \sqrt{5/512} \approx 0.0988.$$

$$z = \frac{\kappa}{\mathrm{SE}} \approx 6.96.$$

4. Decision
Under H_0: $z \sim \mathrm{N}(0, 1)$ (one-sided).

α	$z_{(\alpha)}$	Decision
5%	1.645	**Reject** (6.96 > 1.645)
1%	2.326	**Reject** (6.96 > 2.326)

$p < 0.001$. Agreement is significantly better than chance. With $\kappa = 0.69$, agreement is in the *substantial* range.

Practitioner Insight: Cohen's κ adjusts observed agreement P_o for the agreement expected by chance under independent classification by the two raters. When the raters have similar marginal distributions, the chance agreement $P_e = \sum_i p_{i+}^2$ increases as one category dominates, reducing κ relative to P_o. In practice, the formal test of $H_0: \kappa = 0$ is often less important than the magnitude of κ, commonly interpreted using the Landis–Koch benchmarks from slight to almost perfect agreement. A major limitation is sensitivity to marginal distributions: identical agreement patterns can produce very different κ values under different prevalence structures, the so-called prevalence paradox. For ordered categories, weighted κ is preferred because it penalizes disagreements according to their distance from the diagonal. Unlike κ, the McNemar–Bowker test examines marginal symmetry rather than agreement strength.

Common Pitfalls

- **Prevalence and marginal imbalance effects.** Kappa can appear low even when raw agreement is high if one category is rare or the marginals are highly asymmetric (the "prevalence paradox"). Always report raw agreement and consider PABAK or Gwet's AC_1 when base rates are extreme.
- **Using unweighted kappa for ordinal ratings.** Standard kappa treats all disagreements equally. For ordered scales (severity, Likert), use weighted kappa (linear or quadratic) to account for degree of disagreement; otherwise agreement may be understated.
- **Applying Cohen's kappa to more than two raters.** Cohen's kappa applies only to two raters. For three or more raters use Fleiss' kappa, Krippendorff's alpha, or average pairwise kappas depending on design.
- **Relying on significance tests rather than magnitude.** With large samples, even trivial kappas (e.g., $\kappa \approx 0.15$) are "statistically significant." Emphasize magnitude and confidence intervals, not the p-value against $\kappa = 0$.
- **Misinterpreting kappa as validity.** Kappa measures inter-rater reliability, not accuracy. Two raters can agree perfectly yet both be systematically wrong. When possible, also assess validity against a gold standard.

Weighted Kappa

Chance-Corrected Agreement for Ordinal Ratings

CTAB.15 | **Category:** Agreement & Reliability | **Data:** Categorical | **Null:** Asymptotic Normal

When to Use

- Measure agreement between two ordinal raters.
- Penalize disagreements according to their severity.
- Analyze ordered categorical rating scales.
- Assess reliability beyond chance agreement alone.
- Compare clinical, survey, or grading consistency.

Assumptions

- Ratings are paired observations on the same subjects.
- Categories must possess a meaningful ordinal structure.
- The two raters operate independently of one another.
- The weighting scheme reflects disagreement severity.
- Marginal category frequencies are reasonably stable.

Test Statistic

$$\kappa_w = \frac{\sum_{i=1}^{C}\sum_{j=1}^{C} w_{ij}O_{ij} - \sum_{i=1}^{C}\sum_{j=1}^{C} w_{ij}E_{ij}}{1 - \sum_{i=1}^{C}\sum_{j=1}^{C} w_{ij}E_{ij}}.$$

Notation: O_{ij} is observed proportion in cell (i, j), E_{ij} is expected proportion under independence, w_{ij} is the weight assigned to agreement between categories i and j, and C is the number of ordered categories.

Plain English: Weighted Kappa extends Cohen's Kappa by accounting for the magnitude of disagreement between ordinal ratings. Exact agreement receives full credit, while partial disagreement is penalized according to a chosen weighting scheme. Small disagreements receive less penalty than large disagreements, making the procedure more suitable for ordered rating scales such as disease severity, educational grades, or Likert responses.

⚠ Practitioner Warning

Weighted Kappa depends strongly on the chosen weighting scheme; linear and quadratic weights can produce notably different values. Highly imbalanced marginal distributions may yield paradoxically low Kappa values despite high observed agreement. The statistic measures agreement, not accuracy, so two raters may agree consistently while both remain systematically biased. Interpretation thresholds such as "moderate" or "substantial" agreement should be treated only as rough guidelines rather than universal standards.

Hypotheses

- H_0: $\kappa_w = 0$ (Agreement equals chance expectation.)
- H_1: $\kappa_w > 0$ (Agreement exceeds chance expectation.)

Decision Rule: Reject H_0 if $z > z_\alpha$. Or, if the Lower Bound of the $(1 - \alpha)$ CI exceeds 0.

Software References [Python, R, SAS]

Python	R	SAS
`sklearn.metrics::cohen_kappa_score`	`irr::kappa2`	`PROC FREQ (AGREE)`

Additional Notes

Common weighting schemes include **linear** and **quadratic** weights. Linear weights penalize disagreements proportionally to category distance, whereas quadratic weights penalize larger disagreements more severely and often produce values closer to the intraclass correlation coefficient (ICC). Typical weights are:

$$w_{ij}^{(L)} = 1 - \frac{|i-j|}{C-1}, \qquad w_{ij}^{(Q)} = 1 - \frac{(i-j)^2}{(C-1)^2}.$$

Weighted κ reduces to ordinary Cohen's κ when only exact agreement receives weight 1 and all disagreements receive weight 0.

Miscellanea: Jacob Cohen introduced Kappa in 1960 as a chance-corrected measure of agreement between raters. Weighted versions were later developed to handle ordinal categories where some disagreements are more serious than others. Quadratic weighted Kappa became especially popular in medical research and educational assessment, where adjacent disagreements are often less consequential than extreme disagreements.

Reference(s): Cohen (1960); Cohen (1968); Landis & Koch (1977); Fleiss et al. (2003)

Weighted Kappa CTAB.15

Worked Example

Problem Statement Two radiologists independently grade the severity of lung lesions in 50 patients on a four-point ordered scale: 1 = None, 2 = Mild, 3 = Moderate, 4 = Severe. Their joint ratings are shown below. Compute quadratic-weighted κ_w and test whether agreement exceeds chance at $\alpha = 0.05$. $H_0\colon \kappa_w = 0$ (weighted agreement no better than chance); $H_1 : \kappa_w > 0$. Data is provided at the end of the chapter.

$$\mathbf{W} = \begin{pmatrix} 1 & \frac{8}{9} & \frac{5}{9} & 0 \\ \frac{8}{9} & 1 & \frac{8}{9} & \frac{5}{9} \\ \frac{5}{9} & \frac{8}{9} & 1 & \frac{8}{9} \\ 0 & \frac{5}{9} & \frac{8}{9} & 1 \end{pmatrix} \quad \mathbf{O} = \begin{pmatrix} 16 & 3 & 1 & 0 \\ 3 & 10 & 2 & 0 \\ 1 & 2 & 6 & 1 \\ 0 & 0 & 1 & 4 \end{pmatrix} \quad \mathbf{E} = \begin{pmatrix} 8 & 6 & 4 & 2 \\ 6 & \frac{9}{2} & 3 & \frac{3}{2} \\ 4 & 3 & 2 & 1 \\ 2 & \frac{3}{2} & 1 & \frac{1}{2} \end{pmatrix}$$

1. Quadratic Weights
For a k-category ordered scale, quadratic weights penalize disagreement proportionally to squared distance:

$$w_{ij} = 1 - [(i-j)/(k-1)]^2.$$

Constructed **W** matrix is shown above.

2. Weighted Observed Agreement

$$P_{o,w} = \frac{1}{n} \sum_{i,j} W_{ij} O_{ij} = 43/45.$$

3. Weighted Expected Agreement
The expected matrix **E** is constructed under the assumption of independent ratings using the marginal totals. Each entry is $e_{ij} = n\, p_{i+} p_{+j}$, where p_{i+} and p_{+j} are the row and column marginal proportions, respectively.

$$P_{e,w} = \frac{1}{n} \sum_{i,j} W_{ij} E_{ij} = 7/9.$$

4. Weighted Kappa

$$\hat{\kappa}_w = \frac{P_{o,w} - P_{e,w}}{1 - P_{e,w}} = \frac{43/45 - 7/9}{1 - 7/9} = \frac{4}{5}.$$

5. Standard Error and Test Statistic

$$\mathrm{SE}(\hat{\kappa}_w) = \frac{\sqrt{P_{o,w}(1 - P_{o,w})}}{\sqrt{n}\,(1 - P_{e,w})} = \frac{\sqrt{43}}{50} \approx 0.131.$$

6. Decision
Under H_0: $z \sim \mathrm{N}(0,1)$ (1-sided). $z = \hat{\kappa}_w / \mathrm{SE}(\hat{\kappa}_w) = 6.10$.

α	$z_{(\alpha)}$	Decision
5%	1.645	**Reject** (6.10 > 1.645)
1%	2.326	**Reject** (6.10 > 2.326)

$p < 0.001$. With $\kappa_w = 0.80$, weighted agreement is in the *almost perfect* range (Landis & Koch 1977).

Practitioner Insight: Weighted κ_w extends unweighted κ to ordered categories by assigning partial credit to near agreements: a one-grade disagreement is penalized less than a two-grade disagreement. Quadratic weights are common in clinical scales as they penalize larger disagreements more heavily, whereas linear weights treat disagreement costs as proportional to distance. Under quadratic weights, κ_w is numerically equivalent to the intraclass correlation coefficient under a one-way random-effects model. Here $\kappa_w = 0.80$ exceeds the unweighted κ as all disagreements are only one category apart and receive substantial partial credit. Since results depend on the weighting scheme, the choice of weights should be specified and justified *a priori*; reporting both linear and quadratic versions is often advisable. Unlike κ_w, the McNemar–Bowker and Stuart–Maxwell tests evaluate marginal symmetry rather than agreement strength.

For matrices $A, B \in \mathbb{R}^{m \times n}$, the Frobenius inner product is: $\langle A, B \rangle_F = \sum_{i,j} a_{ij} b_{ij} = \mathrm{tr}(A^\top B)$.

Common Pitfalls

- **Using unweighted Kappa for ordinal data.** Ordinary Kappa treats all disagreements equally. Weighted Kappa is usually more appropriate when category ordering carries practical meaning.
- **Choosing inappropriate weights.** Linear and quadratic weights can produce substantially different agreement values. The weighting scheme should reflect the real severity of disagreement.
- **Ignoring prevalence effects.** Highly imbalanced category frequencies may yield deceptively low Kappa values despite high observed agreement. Always examine the marginal distributions and raw agreement.
- **Interpreting agreement as accuracy.** Weighted Kappa measures consistency between raters, not correctness. Two raters may agree strongly while both remain systematically biased.
- **Using arbitrary interpretation cutoffs.** Labels such as "moderate" or "substantial" agreement are only rough guidelines. Interpretation should depend on the scientific context and consequences of disagreement.

Fleiss' Kappa

Chance-Corrected Agreement Among Multiple Raters

CTAB.16 | **Category:** Agreement & Reliability | **Data:** Categorical | **Null:** Asymptotic Normal

When to Use

- Measure agreement among multiple categorical raters.
- Assess inter-rater reliability beyond chance agreement.
- Analyze nominal classification consistency across judges.
- Evaluate diagnostic or coding agreement in studies.
- Summarize reliability for panel-based categorical ratings.

Assumptions

- Ratings are categorical and mutually exclusive.
- Each subject is rated independently of others.
- Raters evaluate subjects independently from one another.
- The same number of ratings is obtained per subject.
- Raters are viewed as interchangeable members of a pool.

Test Statistic

$$\kappa = \frac{\bar{P} - \bar{P}_e}{1 - \bar{P}_e}, \quad \text{where } \bar{P} = \frac{1}{n}\sum_{i=1}^{n} \frac{\sum_{j=1}^{k} n_{ij}(n_{ij}-1)}{m(m-1)}, \qquad \bar{P}_e = \sum_{j=1}^{k} p_j^2.$$

Notation: n is the number of subjects, m is the number of ratings per subject, k is the number of categories, n_{ij} is the number of raters assigning subject i to category j, and p_j is the overall proportion of ratings in category j.

Plain English: Fleiss' Kappa measures the extent to which multiple raters agree on categorical classifications beyond what would be expected by chance alone. The statistic compares the observed level of agreement among raters to the agreement expected if raters assigned categories randomly according to the overall category frequencies. Values near one indicate strong agreement, values near zero indicate chance-level agreement, and negative values suggest systematic disagreement.

⚠ Practitioner Warning

Fleiss' Kappa assumes that all raters are interchangeable and that each subject receives the same number of ratings. Highly imbalanced category frequencies can produce low Kappa values even when observed agreement appears high. The statistic measures agreement rather than accuracy, so raters may agree consistently while still being uniformly incorrect. Interpretation cutoffs such as "fair" or "substantial" agreement should be treated only as rough practical guidelines.

Hypotheses

- H_0: $\kappa = 0$ (Agreement equals chance expectation.)
- H_1: $\kappa > 0$ (Significant agreement beyond chance)

Decision Rule: Reject H_0 if $z > z_\alpha$. Or, if the Lower Bound of the $(1-\alpha)$ CI exceeds 0.

Software References [Python, R, SAS]

`statsmodels::fleiss_kappa` `irr::kappam.fleiss` `PROC FREQ (FLEISS)`

Additional Notes

Fleiss' Kappa generalizes Cohen's Kappa from two raters to multiple raters. Unlike Cohen's Kappa, raters need not rate all subjects identically, but each subject must receive the same number of ratings. The statistic is most appropriate for nominal categories; ordinal categories are often better analyzed using weighted agreement measures. A commonly cited interpretation scale classifies $\kappa < 0$ as *poor*, 0.00–0.20 as *slight*, 0.21–0.40 as *fair*, 0.41–0.60 as *moderate*, 0.61–0.80 as *substantial*, and 0.81–1.00 as *almost perfect* agreement. These interpretation ranges are heuristic guidelines rather than universally accepted standards.

Miscellanea: Joseph Fleiss introduced this extension of Kappa in 1971 to address agreement problems involving more than two raters. The method became widely used in psychology, medicine, education, and content analysis where multiple judges independently classify the same subjects into categorical outcomes. Fleiss' Kappa remains one of the most commonly reported inter-rater reliability measures for nominal data.

Reference(s): Fleiss (1971); Fleiss et al. (2003); Landis & Koch (1977)

Fleiss' Kappa

CTAB.16

Worked Example

Problem Statement Four clinicians independently classify 9 patient case notes into one of three diagnostic categories: Anxiety Disorder (A), Depressive Disorder (D), or Other (O). Each patient is reviewed by all four clinicians. The number of clinicians assigning each patient to each category is given below. Compute Fleiss' κ and test whether agreement exceeds chance at $\alpha = 0.05$. $H_0: \kappa = 0$ (agreement no better than chance); $H_1: \kappa > 0$. Data is provided at the end of the chapter.

$$p_j = \frac{1}{nm}\sum_{i=1}^{n} n_{ij}, \quad \bar{P}_e = \sum_{j=1}^{k} p_j^2, \quad P_i = \frac{1}{m(m-1)}\sum_{j=1}^{k} n_{ij}(n_{ij}-1), \quad \bar{P} = \frac{1}{n}\sum_{i=1}^{n} P_i, \quad \kappa = \frac{\bar{P}-\bar{P}_e}{1-\bar{P}_e}.$$

1. Fleiss' Kappa Formulas
For n subjects rated by m raters across k categories, let n_{ij} be the number of raters assigning subject i to category j. Formulas are given above.

2. Category Proportions
Column totals are all equal to 12, giving $nm = 9 \times 4 = 36$:

$$p_A = p_D = p_O = 12/36, \quad \bar{P}_e = 3 \times (12/36)^2 = 1/3.$$

3. Subject-Level Agreement P_i
For $m = 4$, the denominator is $m(m-1) = 12$.

Pattern	$\sum_j n_{ij}(n_{ij}-1)$	P_i
$(4,0,0)$: Patients 1–6	$4{\cdot}3 = 12$	1
$(2,2,0)$: Patient 7	$2{\cdot}1 + 2{\cdot}1 = 4$	$1/3$
$(0,2,2)$: Patient 8	$2{\cdot}1 + 2{\cdot}1 = 4$	$1/3$
$(2,0,2)$: Patient 9	$2{\cdot}1 + 2{\cdot}1 = 4$	$1/3$

4. Mean Observed Agreement

$$\bar{P} = (6 \times 1 + 3 \times \tfrac{1}{3})/9 = 7/9.$$

5. Kappa Statistic and Standard Error

$$\kappa = \frac{7/9 - 1/3}{1 - 1/3} = \frac{2}{3}, \quad \text{SE} = \sqrt{\frac{2\,\bar{P}_e(1-\bar{P}_e)}{n\,m(m-1)}} = \frac{1}{9\sqrt{3}}.$$

6. Test Statistic and Decision
Under H_0: $z \sim \mathrm{N}(0,1)$ (1-sided). $z = \kappa/\mathrm{SE} =\approx 10.39$.

α	$z_{(\alpha)}$	Decision
5%	1.645	**Reject** (10.39 > 1.645)
1%	2.326	**Reject** (10.39 > 2.326)

$p < 0.001$. With $\kappa = 2/3$, inter-rater agreement is in the *substantial* range (Landis & Koch 1977).

Practitioner Insight: Fleiss' κ (Fleiss 1971) generalizes Cohen's κ from two raters to $m \geq 2$ raters and does not require the same raters to evaluate every subject; each subject need only receive the same number of ratings. This makes it well suited to rotating-panel reliability studies. The subject-level agreement statistic P_i measures rating concentration: unanimous ratings give $P_i = 1$, while evenly split ratings approach chance agreement. Here patients 7–9 have 2–2 splits across categories, giving $P_i = 1/3 = \bar{P}_e$ and contributing no agreement beyond chance; the overall signal comes mainly from the six unanimous subjects. Equal marginal proportions ($p_j = 1/3$) maximize chance agreement and therefore make this a particularly difficult setting for obtaining a large κ. Unlike Cohen's κ, Fleiss' original formulation has no weighted version; for ordered categories, weighted ICC-type methods are typically preferred. Category-specific analyses can further identify which classes drive overall agreement.

Common Pitfalls

- **Using Fleiss' Kappa for ordinal categories.** Fleiss' Kappa treats all disagreements equally and ignores category ordering. Weighted agreement measures are usually preferable for ordinal scales.
- **Ignoring prevalence effects.** Highly imbalanced category frequencies can produce surprisingly low Kappa values despite high observed agreement. Always inspect marginal category distributions alongside Kappa.
- **Interpreting agreement as accuracy.** Fleiss' Kappa measures consistency among raters, not correctness. Multiple raters may agree strongly while still making the same systematic errors.
- **Using unequal numbers of ratings.** The classical Fleiss' Kappa formulation assumes the same number of ratings per subject. Unequal ratings require modified agreement procedures.
- **Over-relying on interpretation cutoffs.** Labels such as "moderate" or "substantial" agreement are only heuristic guidelines. Acceptable agreement levels depend on the scientific context and practical consequences.

Krippendorff's Alpha

Chance-Corrected Reliability for Multiple Raters

CTAB.17 | **Category:** Agreement & Reliability | **Data:** Categorical | **Null:** Perm. / Bootstrap

When to Use

- Measure reliability among multiple raters or coders.
- Handle nominal, ordinal, interval, or ratio scales.
- Analyze data with incomplete or missing ratings.
- Assess coding consistency in content-analysis studies.
- Evaluate agreement beyond chance across flexible designs.

Assumptions

- Ratings are assigned independently across raters.
- The distance function reflects disagreement severity.
- Subjects are sampled representatively from the population.
- Categories or measurements are consistently defined.
- Reliability is evaluated relative to chance disagreement.

Test Statistic

$$\alpha = 1 - \frac{D_o}{D_e}, \quad \text{where } D_o = \frac{\sum_c \sum_{c'} o_{cc'} \delta^2_{cc'}}{\sum_c \sum_{c'} o_{cc'}}, \qquad D_e = \frac{\sum_c \sum_{c'} e_{cc'} \delta^2_{cc'}}{\sum_c \sum_{c'} e_{cc'}}.$$

Notation: D_o is observed disagreement, D_e is expected disagreement under chance, $o_{cc'}$ and $e_{cc'}$ are observed and expected coincidence counts for categories c, c', and $\delta_{cc'}$ is the distance function between categories c and c'.

Plain English: Krippendorff's Alpha measures the reliability of ratings or coding assignments by comparing observed disagreement with the disagreement expected by chance. Unlike many traditional agreement measures, it accommodates multiple raters, incomplete data, and different measurement scales through flexible disagreement functions. Its coincidence-matrix framework evaluates all pairwise rating comparisons, making the method especially useful for complex and irregular datasets. Perfect agreement yields $\alpha = 1$, chance-level agreement gives $\alpha \approx 0$, and negative values indicate systematic disagreement.

⚠ Practitioner Warning

Krippendorff's Alpha depends heavily on the chosen distance metric, especially for ordinal or continuous data. Highly imbalanced category frequencies may produce unstable reliability estimates, particularly in small samples. The statistic measures reliability rather than validity, so raters may agree consistently while still applying incorrect coding rules. Interpretation thresholds should be treated cautiously and evaluated relative to the practical consequences of disagreement.

Hypotheses

- H_0: Reliability equals chance expectation.
- H_1: Reliability exceeds chance expectation.

Decision Rule: Reject H_0 if $\alpha > \alpha_{min}$ (typically 0.80 or 0.667). Use bootstrapped CIs (reject if excludes zero).

Software References [Python, R, SAS]

Python	R	SAS
`krippendorff::alpha`	`irr::kripp.alpha`	`Not Native (Manual)`

Additional Notes

Krippendorff's Alpha generalizes several classical agreement coefficients by supporting multiple raters, incomplete data, and flexible disagreement metrics. Different distance functions produce versions suitable for nominal, ordinal, interval, or ratio scales. For nominal data, disagreements are treated equally, whereas ordinal and interval versions penalize larger disagreements more heavily. Common interpretation guidelines classify $\alpha < 0.667$ as *low reliability*, 0.667–0.800 as permitting *tentative conclusions*, and $\alpha > 0.800$ as indicating *reliable agreement*. These thresholds are practical conventions rather than strict universal standards. Confidence intervals for Krippendorff's α are commonly obtained using bootstrap resampling, since analytic standard errors are difficult to derive for complex missing-data and multi-rater settings. Bootstrap percentile or BCa intervals are widely recommended in practice.

Miscellanea: Klaus Krippendorff developed Alpha in the late 1960s and early 1970s while studying reliability problems in communication and content-analysis research. The method became influential because it unified multiple agreement settings within a single framework and naturally handled missing data, which many earlier agreement coefficients could not accommodate. Krippendorff's Alpha is now widely used in computational linguistics, media analysis, psychology, and machine-learning annotation studies.

Reference(s): Krippendorff (1970, 2018); Hayes & Krippendorff (2007)

Worked Example

Problem Statement Four coders classify nine articles into one of three themes: Accuracy (A), Design (D), or Other (O). Some articles show perfect agreement, while others split evenly between two themes. Compute Krippendorff's nominal α.

	1	2	3	4	5	6	7	8	9	Total
A	4	4	0	0	0	0	2	0	2	12
D	0	0	4	4	0	0	2	2	0	12
O	0	0	0	0	4	4	0	2	2	12
m_u	4	4	4	4	4	4	4	4	4	36

1. Coincidence Matrix
Krippendorff's α is computed from ordered coder-pair coincidences. For unit u, each pair is weighted by $1/(m_u - 1)$. Since $m_u = 4$ here, each within-unit ordered pair receives weight $1/3$ (avoids double counting the rated-pair).

	A	D	O	Total
A	28/3	4/3	4/3	12
D	4/3	28/3	4/3	12
O	4/3	4/3	28/3	12
Total	12	12	12	36

The diagonal entries represent agreement; the off-diagonal entries represent disagreement.

2. Observed Disagreement
For nominal data, $\delta^2(c, c') = 0$ if $c = c'$ and $\delta^2(c, c') = 1$ otherwise.

$$D_o = \frac{\text{off-diagonal coincidences}}{\text{total coincidences}} = \frac{8}{36} = 0.2222.$$

3. Expected Disagreement
With category totals $n_A = n_D = n_O = 12$ and $N = 36$, the expected nominal disagreement is

$$D_e = 1 - \frac{\sum_c n_c(n_c - 1)}{N(N-1)}.$$

$$D_e = 1 - \frac{3(12)(11)}{36(35)} = 0.6857.$$

4. Krippendorff's Alpha

$$\alpha = 1 - \frac{D_o}{D_e} = 1 - \frac{0.2222}{0.6857} = 0.6759.$$

5. Interpretation
The estimated reliability is

$$\boxed{\alpha \approx 0.676}.$$

This indicates moderate-to-acceptable reliability, but it falls below the commonly recommended 0.80 threshold for strong agreement. Depending on the application, the data may support tentative conclusions but not highly confident coding reliability.

Practitioner Insight: Krippendorff's α measures disagreement rather than raw agreement: it asks whether observed disagreement is small relative to the disagreement expected by chance. The coincidence matrix is the computational core because it converts multi-rater data into weighted coder-pair counts, even when the ratings are ragged or incomplete. Unlike Cohen's or Fleiss' κ, the same framework accommodates nominal, ordinal, interval, and ratio data by changing the distance function $\delta^2(c, c')$.

Common Pitfalls

- **Using inappropriate distance functions.** Krippendorff's Alpha depends strongly on the chosen disagreement metric. Nominal, ordinal, and interval data require different distance structures.
- **Ignoring prevalence imbalance.** Highly skewed category frequencies can produce unstable or deceptively low reliability estimates. Always inspect category distributions alongside Alpha values.
- **Interpreting reliability as validity.** High Alpha indicates consistent coding, not necessarily correct coding. Multiple raters may agree strongly while applying flawed coding rules.
- **Using small or sparse samples.** Reliability estimates may become unstable when categories are rare or sample sizes are limited. Confidence intervals or bootstrap methods are often advisable.
- **Over-relying on interpretation thresholds.** Rules such as $\alpha > 0.80$ are practical conventions rather than universal standards. Acceptable reliability depends on the scientific context and consequences of disagreement.

Gwet's AC_1

Chance-Corrected Agreement Robust to Prevalence Effects

CTAB.18 | **Category:** Agreement & Reliability | **Data:** Categorical | **Null:** Asymptotic Normal

When to Use

- Measure inter-rater agreement beyond chance levels.
- Analyze nominal or ordinal categorical ratings.
- Reduce prevalence and marginal-distribution paradoxes.
- Assess reliability when Kappa appears unstable.
- Evaluate agreement in highly imbalanced categories.

Assumptions

- Ratings are assigned independently across raters.
- Categories are mutually exclusive and consistently defined.
- Subjects are independently sampled from the population.
- Raters evaluate the same underlying subjects or items.
- Chance agreement is estimated from marginal tendencies.

Test Statistic

$$AC_1 = \frac{P_o - P_e}{1 - P_e}, \quad \text{where } P_o = \frac{1}{N}\sum_{i=1}^{N} I(\text{agreement on subject } i), \qquad P_e = \frac{1}{C-1}\sum_{j=1}^{C} p_j(1-p_j).$$

Notation: P_o is observed agreement, P_e is expected agreement under Gwet's chance model, p_j is the marginal proportion for category j, C is the number of categories, and N is the number of subjects.

Plain English: Gwet's AC_1 is a chance-corrected agreement coefficient designed to overcome instability commonly observed with Kappa statistics when category frequencies are highly imbalanced. Rather than estimating chance agreement directly from observed marginal proportions alone, AC_1 uses an alternative chance model that tends to produce more stable and interpretable reliability estimates in practice. Values near one indicate strong agreement beyond chance, whereas values near zero indicate chance-level agreement.

⚠ Practitioner Warning

Although AC_1 is generally more stable than Kappa under prevalence imbalance, it remains sensitive to sparse categories and small samples. The statistic measures agreement rather than accuracy, so raters may agree consistently while still being systematically incorrect. Different software packages may implement slightly different variance estimators or weighting schemes, especially for ordinal extensions. Interpretation thresholds should be treated as practical guidelines rather than universal standards.

Hypotheses

- H_0: $AC_1 = 0$ (Agreement equals chance expectation).
- H_1: $AC_1 > 0$ (Significant agreement beyond chance).

Decision Rule: Reject H_0 if $z > z_\alpha$.Typically, $AC_1 > 0.80$ means excellent reliability. $z = AC_1/SE(AC_1)$.

Software References [Python, R, SAS]

`irrCAC::gwet.ac1.raw`	`irrCAC::gwet.ac1.dist`	`Not Native (Manual)`

Additional Notes

Gwet's AC_1 was developed to address the "Kappa paradox," where very high observed agreement may still produce unexpectedly low Kappa values when category prevalence is highly imbalanced. By modifying the estimation of expected agreement, AC_1 often provides more stable reliability estimates in practical settings. Weighted versions of AC_1 are available for ordinal data and operate similarly to weighted Kappa by assigning partial penalties to disagreements of different severity. In many applied studies, AC_1 and Kappa produce similar results when category frequencies are balanced, but AC_1 typically behaves more robustly under strong marginal imbalance.

Miscellanea: Kilem Gwet introduced the AC_1 statistic in the early 2000s while studying limitations of Cohen's Kappa in medical and survey reliability applications. The method became popular because it frequently avoided the "high agreement but low Kappa" paradox that frustrated practitioners interpreting inter-rater reliability. Today, AC_1 is widely used in healthcare, psychology, and annotation studies where prevalence imbalance is common.

Reference(s): Gwet (2002, 2008, 2014)

Gwet's AC_1 CTAB.18

Worked Example

Problem Statement Two reviewers classify 50 records as Positive (P) or Negative (N). The condition is rare, so most records are classified as Positive by both reviewers. Compute Gwet's AC_1 and test whether agreement exceeds chance at $\alpha = 0.05$. $H_0: AC_1 = 0$; $H_1: AC_1 > 0$.

		Reviewer 2 P	Reviewer 2 N	Total
Reviewer 1	P	45	3	48
	N	2	0	2
	Total	47	3	50

1. Observed Agreement

$$P_o = \frac{45 + 0}{50} = 0.900.$$

The reviewers agree on 45 of the 50 records.

2. Marginal Category Proportions
Gwet's AC_1 uses the overall marginal category proportions across both reviewers. There are 100 total ratings:

$$p_P = \frac{48 + 47}{100} = 0.95, \qquad p_N = \frac{2 + 3}{100} = 0.05.$$

3. Chance Agreement
For two nominal categories,

$$P_e = \sum_{j=1}^{C} \frac{p_j(1 - p_j)}{C - 1}.$$

Since $C = 2$,

$$P_e = 0.95(0.05) + 0.05(0.95) = 0.095.$$

4. Gwet's AC_1

$$AC_1 = \frac{P_o - P_e}{1 - P_e} = \frac{0.900 - 0.095}{1 - 0.095} = 0.8895.$$

Thus,

$$\boxed{AC_1 \approx 0.890}.$$

5. Test Statistic
Using a bootstrap or asymptotic standard error, suppose

$$\text{SE}(AC_1) \approx 0.053.$$

Then

$$z = \frac{AC_1}{\text{SE}(AC_1)} = \frac{0.8895}{0.053} \approx 16.78.$$

6. Decision
Under H_0, $z \sim N(0, 1)$ approximately.

$$z = 16.78 > 1.645.$$

Reject H_0 at $\alpha = 0.05$. The agreement is far greater than chance expectation.

Practitioner Insight: Gwet's AC_1 was designed to be less vulnerable than Cohen's κ to the prevalence paradox. In this example, observed agreement is very high ($P_o = 0.90$), but the marginal distributions are extremely imbalanced. Cohen's κ can behave poorly in such settings because its expected agreement is inflated by the dominant category. Gwet's AC_1 uses an alternative chance-agreement model, producing a more stable and practically interpretable reliability estimate.

Refer to https://agreestat.com/papers/onkrippendorffalpha_rev10052015.pdf for variance formulas.

Common Pitfalls

- **Assuming AC_1 eliminates all prevalence effects.** Although more stable than Kappa, AC_1 can still be influenced by extremely sparse or highly imbalanced categories. Always examine the marginal distributions.
- **Interpreting agreement as accuracy.** AC_1 measures consistency among raters, not correctness of the assigned categories. Strong agreement may still reflect systematic bias or shared errors.
- **Using nominal methods for ordinal data.** Ordinal categories often require weighted versions of AC_1 to reflect differing disagreement severities. Unweighted analyses may underestimate partial agreement.
- **Ignoring uncertainty in small samples.** Reliability estimates may become unstable when sample sizes are limited or categories are rare. Confidence intervals should be reported whenever possible.
- **Over-relying on interpretation cutoffs.** Labels such as "moderate" or "excellent" agreement are only heuristic guidelines. Practical interpretation depends on the scientific context and consequences of disagreement.

VIII. CTAB — Raw Data

Data used in selected worked examples are provided here to preserve the book's strict two-page-per-test format.

[CTAB.10] Cochran–Mantel–Haenszel Test

Problem Statement. A multi-center drug trial records response (Yes/No) to three dose levels (Low, Medium, High) across $K = 3$ clinical centers. Unlike the standard Mantel–Haenszel test (which pools 2×2 tables), the CMH test handles these 3×2 stratified tables directly. H_0: dose and response are conditionally independent within each center; H_1: a common dose–response association exists across centers.

	Center A ($n = 70$)			**Center B** ($n = 60$)			**Center C** ($n = 40$)		
Dose	Yes	No	n	Yes	No	n	Yes	No	n
Low	8	12	20	5	15	20	3	7	10
Med	15	10	25	10	10	20	8	7	15
High	18	7	25	14	6	20	10	5	15
Total	41	29	70	29	31	60	21	19	40

[CTAB.15] Weighted Kappa

Problem Statement. Two radiologists independently grade the severity of lung lesions in 50 patients on a four-point ordered scale: 1 = None, 2 = Mild, 3 = Moderate, 4 = Severe. Their joint ratings are shown below. Compute quadratic-weighted κ_w and test whether agreement exceeds chance at $\alpha = 0.05$: $H_0 : \kappa_w = 0$ vs. $H_1 : \kappa_w > 0$.

		Rater 2				
		None	Mild	Mod.	Sev.	Row total
Rater 1	None	16	3	1	0	20
	Mild	3	10	2	0	15
	Moderate	1	2	6	1	10
	Severe	0	0	1	4	5
Column total		20	15	10	5	50

[CTAB.16] Fleiss' Kappa

Problem Statement. Four clinicians independently classify 9 patient case notes into one of three diagnostic categories: Anxiety Disorder (A), Depressive Disorder (D), or Other (O). Each patient is reviewed by all four clinicians. The number of clinicians assigning each patient to each category is given below. Compute Fleiss' κ and test whether agreement exceeds chance at $\alpha = 0.05$: $H_0 : \kappa = 0$ vs. $H_1 : \kappa > 0$.

Patient	**A**	**D**	**O**	P_i
1	4	0	0	1
2	4	0	0	1
3	0	4	0	1
4	0	4	0	1
5	0	0	4	1
6	0	0	4	1
7	2	2	0	1/3
8	0	2	2	1/3
9	2	0	2	1/3
Column total	12	12	12	

VIII. CTAB — Worked Example: Mantel-Haenszel's Pooled Estimator

Problem Statement A binary outcome is compared between exposed and unexposed subjects across three study sites. Compute the Mantel–Haenszel pooled odds ratio and test whether there is an overall association after controlling for site. $H_0: OR_{MH} = 1$; $H_1: OR_{MH} \neq 1$.

	Site 1			Site 2			Site 3		
	Exp	Unexp	Tot	Exp	Unexp	Tot	Exp	Unexp	Tot
Case	20	20	40	20	10	30	15	5	20
Control	15	15	30	10	20	30	5	15	20

1. Stratum-Specific Odds Ratios
For each stratum,

$$OR_k = \frac{a_k d_k}{b_k c_k}.$$

Hence,

$$OR_1 = \frac{20(15)}{20(15)} = 1.00,$$

$$OR_2 = \frac{20(20)}{10(10)} = 4.00, \qquad OR_3 = \frac{15(15)}{5(5)} = 9.00.$$

The site-specific odds ratios suggest increasing association across sites.

2. Mantel–Haenszel Pooled OR
The Mantel–Haenszel estimator is

$$\widehat{OR}_{MH} = \frac{\sum_k a_k d_k / n_k}{\sum_k b_k c_k / n_k}.$$

The numerator is

$$\frac{20(15)}{70} + \frac{20(20)}{60} + \frac{15(15)}{40} = 16.577.$$

The denominator is

$$\frac{20(15)}{70} + \frac{10(10)}{60} + \frac{5(5)}{40} = 6.577.$$

Therefore,

$$\widehat{OR}_{MH} = \frac{16.577}{6.577} = 2.520.$$

3. CMH Test Statistic (MH is a special case)
For each stratum,

$$E(a_k) = \frac{r_{1k} c_{1k}}{n_k}, \qquad V(a_k) = \frac{r_{1k} r_{2k} c_{1k} c_{2k}}{n_k^2 (n_k - 1)}.$$

Here,

$$\sum_k \{a_k - E(a_k)\} = 10.000, \qquad \sum_k V(a_k) = 10.725.$$

Thus,

$$\chi^2_{MH} = \frac{[\sum_k \{a_k - E(a_k)\}]^2}{\sum_k V(a_k)} = \frac{10^2}{10.725} = 9.324.$$

4. Decision

Under H_0, $\chi^2_{MH} \sim \chi^2_1$. Since $9.324 > 3.841$, reject H_0 at $\alpha = 0.05$. There is evidence of an exposure–outcome association after controlling for site.

Practitioner Insight: The Mantel–Haenszel method pools evidence across stratified 2×2 tables while controlling for the stratification factor. The pooled estimate $\widehat{OR}_{MH} = 2.52$ suggests that the odds of being a case are about 2.5 times higher among exposed subjects after adjusting for site. However, because the stratum-specific odds ratios vary from 1.00 to 9.00, homogeneity should be checked using the Breslow–Day test before relying too heavily on the pooled estimate.

VIII. CTAB — Worked Example: CMH's Row Mean Scores Variant

Problem Statement Patients are classified into three ordered response categories: Low (L), Medium (M), and High (H). Treatment group is nominal with two levels: Drug and Control. Data are stratified by study site. Use the CMH Row Mean Scores test to determine whether the mean ordinal response differs between treatment groups after controlling for site. H_0: equal mean row scores across groups; H_1: mean row scores differ.

	Site 1		Site 2	
Response	Drug	Control	Drug	Control
L	18	10	20	12
M	12	14	10	16
H	5	11	4	10

1. Assign Row Scores
Use equally spaced ordinal scores:

$$s_L = 1, \qquad s_M = 2, \qquad s_H = 3.$$

For each stratum k, define

$$T_k = \sum_i s_i n_{i1k},$$

the total row-score contribution for the Drug column.

2. Site 1 Calculation
Site 1 row totals are

$$r = (28, 26, 16), \qquad c_{11} = c_{21} = 35, \qquad n_1 = 70.$$

The observed Drug score total is

$$T_1 = 1(18) + 2(12) + 3(5) = 57.$$

The overall site mean score is

$$\bar{s}_1 = \frac{1(28) + 2(26) + 3(16)}{70} = 1.8286.$$

Therefore,

$$E(T_1) = c_{11}\bar{s}_1 = 35(1.8286) = 64.0.$$

So,

$$T_1 - E(T_1) = 57 - 64 = \boxed{-7.0}.$$

The conditional variance is

$$V_1 = \frac{c_{11}c_{21}}{n_1(n_1-1)} \sum_i n_{i\cdot 1}(s_i - \bar{s}_1)^2$$

$$= \frac{35 \times 35}{70 \times 69}(41.943) = \boxed{10.638}.$$

3. Site 2 Calculation
Site 2 row totals are

$$r = (32, 26, 14), \qquad c_{12} = 34, \quad c_{22} = 38, \quad n_2 = 72.$$

The observed Drug score total is

$$T_2 = 1(20) + 2(10) + 3(4) = 52.$$

The overall site mean score is

$$\bar{s}_2 = \frac{1(32) + 2(26) + 3(14)}{72} = 1.7500.$$

Therefore,

$$E(T_2) = c_{12}\bar{s}_2 = 34(1.75) = 59.5.$$

Thus,

$$T_2 - E(T_2) = 52 - 59.5 = \boxed{-7.5}.$$

The conditional variance is

$$V_2 = \frac{c_{12}c_{22}}{n_2(n_2-1)} \sum_i n_{i\cdot 2}(s_i - \bar{s}_2)^2$$

$$= \frac{34 \times 38}{72 \times 71}(41.500) = \boxed{10.489}.$$

4. CMH Row Mean Scores Statistic
Using the conditional variance formula,

$$V_1 = 10.638, \qquad V_2 = 10.489.$$

Hence,

$$Q_{RMS} = \frac{[-7.0 - 7.5]^2}{10.638 + 10.489} = \frac{210.25}{21.127} = \boxed{9.95}.$$

5. Decision

Under H_0, the Row Mean Scores statistic follows approximately $Q_{RMS} \sim \chi^2_1$. Since $Q_{RMS} = 9.95 > 3.84$, reject H_0 at $\alpha = 0.05$. The treatment groups differ significantly in their ordinal response distributions after adjusting for site.

Practitioner Insight: The Row Mean Scores version of the CMH test is used when the row variable is ordinal and the column variable is nominal. Instead of treating the response categories as unrelated labels, it assigns numerical scores to the row levels and tests whether the average row score differs across column groups after stratification. In this example, the Drug group has consistently lower score totals than expected under conditional independence, indicating a shift toward lower response severity after controlling for site.

VIII. CTAB — Worked Example: CMH's Correlation Variant

Problem Statement Suppose disease severity and treatment response are both recorded on three ordered levels: Low (L), Medium (M), and High (H). The data are stratified by study site. Use the CMH Correlation statistic to test whether severity and response show a linear ordinal association after controlling for site. H_0: no linear ordinal association; H_1: linear ordinal association exists.

	Site 1 Response			Site 2 Response		
Severity	L	M	H	L	M	H
L	10	4	1	8	3	1
M	3	8	3	4	7	4
H	1	4	10	1	3	9

1. Assign Ordinal Scores

Use equally spaced row and column scores:

$$u_L = v_L = 1, \quad u_M = v_M = 2, \quad u_H = v_H = 3.$$

For stratum k, define the linear-by-linear score

$$T_k = \sum_i \sum_j u_i v_j n_{ijk}.$$

2. Site 1 Calculation

For Site 1,

$$T_1 = 194, \qquad n_1 = 44.$$

Row and column score totals & averages are:

$$\sum_i u_i n_{i\cdot 1} = 88, \qquad \sum_j v_j n_{\cdot j1} = 88.$$

$$\bar{u}_1 = \frac{\sum_i u_i n_{i\cdot 1}}{n_1} = 2, \quad \bar{v}_1 = \frac{\sum_j v_j n_{\cdot j1}}{n_1} = 2.$$

Hence the expected score under conditional independence is

$$E(T_1) = \frac{88(88)}{44} = 176.$$

Therefore,

$$T_1 - E(T_1) = 194 - 176 = \boxed{18}.$$

Variance Calculation (Site 1)

The conditional variance is

$$V_1 = \frac{\left[\sum_i n_{i\cdot 1}(u_i - \bar{u}_1)^2\right]\left[\sum_j n_{\cdot j1}(v_j - \bar{v}_1)^2\right]}{n_1 - 1}$$

$$= \frac{30 \times 28}{44 - 1} = \boxed{19.535}.$$

3. Site 2 Calculation

For Site 2,

$$T_2 = 179, \qquad n_2 = 40.$$

Row and column score totals & averages are:

$$\sum_i u_i n_{i\cdot 2} = 81, \qquad \sum_j v_j n_{\cdot j2} = 81.$$

$$\bar{u}_2 = \frac{\sum_i u_i n_{i\cdot 2}}{n_2} = 2.025, \quad \bar{v}_2 = \frac{\sum_j v_j n_{\cdot j2}}{n_2} = 2.025.$$

Hence the expected score under conditional independence is

$$E(T_2) = \frac{81(81)}{40} = 164.025.$$

Thus,

$$T_2 - E(T_2) = 179 - 164.025 = \boxed{14.975}.$$

Variance Calculation (Site 2)

The conditional variance is

$$V_2 = \frac{\left[\sum_i n_{i\cdot 2}(u_i - \bar{u}_1)^2\right]\left[\sum_j n_{\cdot j2}(v_j - \bar{v}_1)^2\right]}{n_2 - 1}$$

$$= \frac{24.975 \times 26.975}{40 - 1} = \boxed{17.274}.$$

4. CMH Correlation Statistic

$$Q_{\text{CORR}} = \frac{\left[\sum_k \{T_k - E(T_k)\}\right]^2}{\sum_k V_k}.$$

$$Q_{\text{CORR}} = \frac{(18 + 14.975)^2}{19.535 + 17.274} = \boxed{29.54}.$$

5. Decision

Under H_0, the CMH Correlation statistic follows approximately $Q_{\text{CORR}} \sim \chi^2_1$. Since $Q_{\text{CORR}} = 29.54 > 3.84$, reject H_0 at $\alpha = 0.05$. There is strong evidence of a positive ordinal association between severity and response after controlling for site.

Practitioner Insight: The CMH Correlation variant is used when both row and column variables are ordinal. It assigns numerical scores to the row and column categories and tests whether the cross-product score is systematically larger or smaller than expected under conditional independence. In this example, high severity tends to align with high response and low severity with low response in both strata, producing a large positive linear-by-linear association.

Part IX

Linear Model Assumptions & Influence

Evaluating fit, leverage, multicollinearity, and residual behavior

Regression Diagnostics

IX. REGR — A Conceptual Primer

The Linear Model and OLS

The classical linear regression model is

$$\mathbf{y} = \mathbf{X}\boldsymbol{\beta} + \boldsymbol{\varepsilon},$$

where $\mathbf{y}$ is $n \times 1$, $\mathbf{X}$ is $n \times p$ (including a column of ones), $\boldsymbol{\beta}$ is $p \times 1$, and $\boldsymbol{\varepsilon}$ is the error vector.

The OLS estimator minimises $\|\mathbf{y} - \mathbf{X}\boldsymbol{\beta}\|^2$:

$$\hat{\boldsymbol{\beta}} = (\mathbf{X}^\top\mathbf{X})^{-1}\mathbf{X}^\top\mathbf{y}, \quad \hat{\boldsymbol{\varepsilon}} = \mathbf{y} - \mathbf{X}\hat{\boldsymbol{\beta}} = \mathbf{M}\mathbf{y},$$

where $\mathbf{M} = \mathbf{I} - \mathbf{H}$ and $\mathbf{H} = \mathbf{X}(\mathbf{X}^\top\mathbf{X})^{-1}\mathbf{X}^\top$ is the **hat (projection) matrix**. The diagonal elements $h_{ii} = \mathbf{x}_i^\top(\mathbf{X}^\top\mathbf{X})^{-1}\mathbf{x}_i$ are the **leverages**. $\mathbf{M}$ is called the **residual maker** matrix.

Under the Gauss–Markov conditions (linearity, strict exogeneity, homoskedasticity, no perfect multicollinearity), OLS is BLUE. Adding normality of $\boldsymbol{\varepsilon}$ yields exact t- and F-distributions for inference.

Core Assumptions and Consequences of Violations

Assumption	Consequence of violation
Linearity	Biased $\hat{\boldsymbol{\beta}}$; wrong predictions
Exogeneity ($E[\boldsymbol{\varepsilon}\mid\mathbf{X}] = \mathbf{0}$)	Inconsistent OLS; IV/2SLS needed
Homoskedasticity	Inefficient OLS; SEs invalid
No autocorrelation	Inefficient OLS; SEs invalid
No multicollinearity	Inflated $\mathrm{Var}(\hat{\boldsymbol{\beta}})$; unstable
Normality (small n)	Exact t/F invalid

Diagnostic philosophy: each test targets a specific failure mode. No single test is sufficient; run the full suite and interpret jointly. **Plot first, test second**: graphical tools reveal the nature of violations; formal tests confirm and quantify them.

Residual Types

Raw residuals $\hat{e}_i = y_i - \hat{y}_i$ have $\mathrm{Var}(\hat{e}_i) = \sigma^2(1 - h_{ii})$ and are not identically distributed. Use standardized versions for diagnostics:

Type	Definition and use
Internally studentised	$r_i = \hat{e}_i/[\hat{\sigma}\sqrt{1 - h_{ii}}]$; approx. t_{n-p}
Externally studentised	$r_i^* = \hat{e}_i/[\hat{\sigma}_{(-i)}\sqrt{1 - h_{ii}}]$; exactly t_{n-p-1}; outlier tests
PRESS residuals	$e_{(-i)} = \hat{e}_i/(1 - h_{ii})$; leave-one-out prediction error without refitting

I. Graphical Diagnostics

Formal tests confirm what plots reveal. Always plot before testing; plots identify the *nature* of violations that p-values alone cannot convey.

Residuals vs. fitted ($\hat{e}_i$ vs. $\hat{y}_i$): The primary diagnostic plot. Under correct specification: random scatter around zero with constant spread.

- **Curvature / systematic pattern** ⇒ functional form misspecification (run RESET).
- **Fan or funnel shape** ⇒ heteroskedasticity (run BP/White).
- **Outlying points** ⇒ investigate with Cook's D and DFBETAS.

Scale-location plot ($\sqrt{|r_i|}$ vs. $\hat{y}_i$): Smoothed trend in residual magnitude; more sensitive to heteroskedasticity than the raw residual plot because the square root stabilises skewness. A rising trend confirms variance increasing with the fitted value.

Normal Q–Q plot of residuals: Plot r_i^* against t_{n-p-1} quantiles. Deviations from the 45° line signal non-normality: S-shape ⇒ heavy tails; one-sided bow ⇒ skewness; isolated points ⇒ outliers.

Partial regression (added-variable) plot for x_j: Plot residuals of y on $\mathbf{X}_{(-j)}$ against residuals of x_j on $\mathbf{X}_{(-j)}$. The slope equals $\hat{\beta}_j$ exactly; reveals the marginal contribution of x_j after controlling for all other predictors. Identifies influential observations specific to $\hat{\beta}_j$ and nonlinearity in that predictor.

Component-plus-residual (partial residual) plot: Plot $\hat{\beta}_j x_{ij} + \hat{e}_i$ against x_{ij}. Shows the functional form of the relationship more directly than the added-variable plot. A nonlinear pattern ⇒ apply Box–Tidwell or transform x_j.

Leverage-residual plot (h_{ii} vs. r_i^*): The definitive influence diagnostic. Four quadrants: safe (low leverage, small residual), outlier (low leverage, large residual), high leverage (high leverage, small residual), and **influential** (high leverage, large residual). Cook's D contours are often superimposed.

II. Heteroskedasticity Tests

What it is: $\mathrm{Var}(\varepsilon_i|\mathbf{x}_i) = \sigma_i^2 \neq \sigma^2$; variance depends on regressors or their functions. OLS $\hat{\boldsymbol{\beta}}$ remains unbiased but is no longer efficient and standard errors are incorrect.

Breusch–Pagan (BP) test: Regress $\hat{e}_i^2$ on $\mathbf{x}_i$; $\mathrm{LM} = nR^2 \sim \chi^2_{p-1}$ under H_0. Detects variance that is a linear function of the regressors. Sensitive to non-normality of ε.

White's test: Regress $\hat{e}_i^2$ on $\mathbf{x}_i$, $\mathbf{x}_i^2$, and all cross-products; $\mathrm{LM} = nR^2 \sim \chi^2_q$. More general than BP but loses power with many auxiliary regressors. A convenient special case regresses $\hat{e}_i^2$ on $\hat{y}_i$ and $\hat{y}_i^2$ (fewer df, often more power).

Goldfeld–Quandt (GQ) test: Sort by a suspected variance driver; omit c central observations; compare residual variances from the two end subsamples via $F_{(n-c)/2-p,\,(n-c)/2-p}$. Powerful when variance is monotone in one variable.

Glejser test: Regress $|\hat{e}_i|$ on candidate variables or functions (levels, square roots, inverses); significant coefficients indicate heteroskedasticity of the corresponding form. Asymptotically valid but has finite-sample size distortions.

Harvey–Godfrey test: Regress $\log \hat{e}_i^2$ on regressors; tests the multiplicative model $\sigma_i^2 = \exp(\mathbf{z}_i^\top \boldsymbol{\gamma})$. More appropriate when variance is multiplicative rather than additive.

III. Autocorrelation Tests

What it is: $\mathrm{Cov}(\varepsilon_i, \varepsilon_j) \neq 0$ for $i \neq j$; common in time-series regression. OLS remains unbiased but is inefficient; standard errors are incorrect.

Durbin–Watson (DW) test:

$$d = \frac{\sum_{t=2}^{n} (\hat{e}_t - \hat{e}_{t-1})^2}{\sum_{t=1}^{n} \hat{e}_t^2} \approx 2(1 - \hat{\rho}_1), \quad d \in [0, 4].$$

$d \approx 2$: no AR(1) autocorrelation; $d < d_L$: reject H_0. **Limitations:** AR(1) only; invalid with lagged dependent variables; inconclusive zone (d_L, d_U).

Breusch–Godfrey (BG) test: Regress $\hat{\boldsymbol{\varepsilon}}$ on $\mathbf{X}$ and m lagged residuals; $\mathrm{LM} = nR^2 \sim \chi_m^2$. **Preferred over DW:** tests AR(m)/MA(m); valid with lagged regressors; no inconclusive zone. Always use BG as the primary test; DW as a quick check only.

IV. Unified Robust Inference Framework

All robust standard errors share the same **sandwich structure**:

$$\widehat{\mathrm{Var}}(\hat{\boldsymbol{\beta}}) = (\mathbf{X}^\top \mathbf{X})^{-1} \hat{\boldsymbol{\Omega}} (\mathbf{X}^\top \mathbf{X})^{-1},$$

differing only in how the "meat" $\hat{\boldsymbol{\Omega}}$ is estimated:

Variant	Meat and use case
HC0	$\sum_i \hat{e}_i^2 \mathbf{x}_i \mathbf{x}_i^\top$; heteroskedasticity only; asymptotically valid.
HC3	$\sum_i [\hat{e}_i/(1 - h_{ii})]^2 \mathbf{x}_i \mathbf{x}_i^\top$; leverage-corrected; recommended for small n.
NW	$\hat{\boldsymbol{\Gamma}}_0 + \sum_j w_j (\hat{\boldsymbol{\Gamma}}_j + \hat{\boldsymbol{\Gamma}}_j^\top)$; heteroskedasticity *and* autocorrelation.
CR	$\sum_{g=1}^{G} \mathbf{X}_g^\top \hat{\mathbf{e}}_g \hat{\mathbf{e}}_g^\top \mathbf{X}_g$; within-cluster correlation.

Newey–West HAC estimator (NW):

$$\hat{\boldsymbol{\Omega}}_{\mathrm{NW}} = \hat{\boldsymbol{\Gamma}}_0 + \sum_{j=1}^{m} w_j \left(\hat{\boldsymbol{\Gamma}}_j + \hat{\boldsymbol{\Gamma}}_j^\top \right),$$

where $w_j = 1 - j/(m+1)$ (Bartlett kernel) and $\hat{\boldsymbol{\Gamma}}_j = T^{-1} \sum_{t=j+1}^{T} \hat{e}_t \hat{e}_{t-j} \mathbf{x}_t \mathbf{x}_{t-j}^\top$. Bandwidth: $m = \lfloor 4(T/100)^{2/9} \rfloor$ (Newey–West 1994) or Andrews (1991) automatic selection. Parzen and quadratic spectral kernels are asymptotically more efficient; QS is the optimal kernel under MSE criteria.

Cluster-robust (CR) SEs: When observations are independent *across* clusters but correlated *within* (students in schools, firms in industries):

$$\hat{\boldsymbol{\Omega}}_{\mathrm{CR}} = \frac{G(n-1)}{(G-1)(n-p)} \sum_{g=1}^{G} \mathbf{X}_g^\top \hat{\mathbf{e}}_g \hat{\mathbf{e}}_g^\top \mathbf{X}_g.$$

Reliable when $G \gtrsim 30$; for small G use wild cluster bootstrap or bias-corrected CR2/CR3 variants.

Decision guide:

- Cross-sectional, no clustering $\Rightarrow$ **HC3**.
- Time-series, serial correlation $\Rightarrow$ **Newey–West HAC**.
- Clustered / panel data $\Rightarrow$ **Cluster-robust**.
- Panel with serial *and* cross-sectional dependence $\Rightarrow$ two-way cluster-robust or Driscoll–Kraay.

V. Functional Form and Misspecification Tests

What it is: $E[y|\mathbf{x}]$ is not linear in $\mathbf{x}$, or relevant variables are omitted. OLS $\hat{\boldsymbol{\beta}}$ is biased and inconsistent.

Ramsey RESET test: Augment the regression with powers $\hat{y}_i^2, \hat{y}_i^3$ (and $\hat{y}_i^4$); test joint significance via F. Omnibus: detects misspecification but cannot identify its form.

Link test: Regress y on $\hat{y}$ and $\hat{y}^2$; under correct specification $\hat{y}^2$ should have zero coefficient. Used primarily in GLM diagnostics.

Box–Tidwell test: Augment with $x_j \log x_j$ terms to estimate the optimal power λ_j for each predictor and test whether $\lambda_j \neq 1$ (i.e. whether the linear form is appropriate). Requires $x_j > 0$.

Rainbow test: Compares residual variances from the central subset and full sample via F. Detects global nonlinearity without specifying its form.

VI. Transformations

Transformations serve two distinct purposes—do not conflate them:

- **Variance-stabilizing:** makes $\mathrm{Var}(\varepsilon_i)$ approximately constant.
- **Linearity-inducing:** makes $E[y|\mathbf{x}]$ approximately linear in transformed predictors.

Always re-run diagnostics after transforming.

Box–Cox transformation of y:

$$y^{(\lambda)} = \begin{cases} (y^\lambda - 1)/\lambda & \lambda \neq 0 \\ \log y & \lambda = 0 \end{cases}$$

λ is estimated by profile MLE. Common cases: $\lambda = 1$ (no transform), $\lambda = 0$ (log), $\lambda = \frac{1}{2}$ ($\sqrt{y}$), $\lambda = -1$ ($1/y$). **Requires** $y > 0$. Always report $\hat{\lambda}$ and its confidence interval.

Caution: OLS on $\log y$ minimises $\sum(\log y_i - \widehat{\log y_i})^2$, not $\sum(y_i - \hat{y}_i)^2$. Back-transformed predictions are medians, not means; apply the smearing correction $\exp(\hat{\sigma}^2/2)$ for unbiased mean predictions under log-normality.

Box–Tidwell transformation of x_j: Estimates the optimal power λ_j for each predictor via the Box–Tidwell test (see §V); requires $x_j > 0$.

Quick reference:

Pattern	Transform	Purpose
Right-skewed y	$\log y$	Both
Count data	$\sqrt{y}$	Var
Proportion $y \in (0, 1)$	$\arcsin\sqrt{y}$	Var
$\text{Var} \propto \mu^2$	$1/y$	Var
Nonlinear x_j (concave)	$\log x_j$	Lin
Nonlinear x_j (convex)	x_j^2	Lin
General y or x_j	Box–Cox/Tidwell	Both

VII. Structural Stability: Chow Test

Tests whether regression coefficients are the same across two subsamples separated at a known breakpoint k:

$$F = \frac{(\text{RSS}_{\text{pooled}} - \text{RSS}_1 - \text{RSS}_2)/p}{(\text{RSS}_1 + \text{RSS}_2)/(n - 2p)} \sim F_{p,n-2p} \text{ under } H_0.$$

Requires the breakpoint to be specified in advance. For an unknown break, use the **Quandt–Andrews test** (supremum of Chow F over candidate break dates with corrected critical values).

VIII. Model Selection Criteria

Diagnostic tests assess whether a fitted model is *adequate*; model selection criteria assess which of several candidate models is *best*, trading off goodness of fit against complexity. They are complementary: a model can pass all diagnostics yet still be improvable.

Akaike Information Criterion (AIC):

$$\text{AIC} = -2\ell(\hat{\boldsymbol{\beta}}, \hat{\sigma}^2) + 2p = n \log \hat{\sigma}^2 + 2p,$$

where ℓ is the maximized log-likelihood and p includes the intercept. AIC targets **predictive accuracy**: it is an asymptotically unbiased estimator of the expected Kullback–Leibler divergence between the fitted model and the truth. **Lower is better.** AIC does not penalize complexity heavily enough for consistent model selection; it tends to over-fit in small samples.

Corrected AIC (AICc):

$$\text{AICc} = \text{AIC} + \frac{2p(p+1)}{n - p - 1}.$$

Adds a finite-sample correction that becomes negligible as $n/p \to \infty$. **Preferred over AIC whenever** $n/p < 40$; defaults to AIC in large samples.

Bayesian Information Criterion (BIC):

$$\text{BIC} = -2\ell(\hat{\boldsymbol{\beta}}, \hat{\sigma}^2) + p \log n = n \log \hat{\sigma}^2 + p \log n.$$

Imposes a stronger penalty for complexity than AIC, scaling with $\log n$ rather than the constant 2. **Consistent**: selects the true model (if it is among the candidates) with probability $\to 1$ as $n \to \infty$. Tends to **under-fit** in small samples relative to AIC.

Adjusted R^2:

$$\bar{R}^2 = 1 - \frac{n-1}{n-p}(1 - R^2).$$

Penalizes for the number of predictors; can decrease when an irrelevant predictor is added (unlike R^2). **Not a formal information criterion**: it does not have a likelihood interpretation and should not be used as the sole selection tool.

Mallows' C_p:

$$C_p = \frac{\text{RSS}_p}{\hat{\sigma}^2_{\text{full}}} - n + 2p,$$

where RSS_p is the residual sum of squares for a model with p predictors and $\hat{\sigma}^2_{\text{full}}$ is the residual variance from the full model. Under the correct model, $E[C_p] \approx p$; models with $C_p \lesssim p$ are adequate. Useful for **subset selection** when a full model is available as a reference.

Practical summary:

Criterion	Goal	Best used when
AIC	Prediction	Large n; forecasting
AICc	Prediction	Small/moderate n ($n/p < 40$)
BIC	Consistency	Identifying true model
$\bar{R}^2$	Description	Quick comparison; not sole criterion
C_p	Subset sel.	Comparing subsets of full model

Key caution: all criteria assume the true model is among the candidates. If it is not, they select the best available approximation, not the truth. Never use information criteria as a substitute for substantive reasoning about which predictors belong in the model. Model selection after testing (e.g. using p-values to eliminate predictors and then reporting the final model as if pre-specified) inflates Type I error (**post-selection inference problem**).

IX. Multicollinearity: VIF

Near-linear dependence among columns of $\mathbf{X}$ inflates $(\mathbf{X}^\top\mathbf{X})^{-1}$ and hence $\text{Var}(\hat{\boldsymbol{\beta}})$.

$$\text{VIF}_j = \frac{1}{1 - R_j^2},$$

where R_j^2 is from regressing x_j on all other predictors.

- $\text{VIF}_j = 1$: no collinearity.
- $\text{VIF}_j > 5$: moderate concern.
- $\text{VIF}_j > 10$: severe; estimates unreliable.

Complementary diagnostics: condition number κ of $\mathbf{X}^\top\mathbf{X}$ ($\kappa > 30$ severe) and Belsley–Kuh–Welsch variance decomposition proportions. *Remedies:* ridge regression, LASSO, principal components regression, or dropping near-collinear predictors. Note: multicollinearity does not bias $\hat{\boldsymbol{\beta}}$; it only inflates uncertainty.

X. Outliers and Influence

An **outlier** has a large residual; a **high-leverage point** is extreme in predictor space; an **influential point** materially changes $\hat{\boldsymbol{\beta}}$ when deleted. These are distinct.

Leverage (hat values):

$$h_{ii} = \mathbf{x}_i^\top(\mathbf{X}^\top\mathbf{X})^{-1}\mathbf{x}_i, \quad h_{ii} \in [1/n, 1], \quad \bar{h} = p/n.$$

Rule of thumb: $h_{ii} > 2p/n$ flags high leverage. High leverage alone is not harmful; it becomes harmful when combined with a large residual.

Cook's distance:

$$D_i = \frac{r_i^2}{p} \cdot \frac{h_{ii}}{1 - h_{ii}}.$$

Measures the shift in *all* coefficients when observation i is deleted. The factored form shows D_i is the product of squared studentised residual (outlyingness in y-space) and leverage ratio (outlyingness in x-space). Rule of thumb: $D_i > 4/n$ or $D_i > 1$.

DFFITS:

$$\text{DFFITS}_i = r_i^* \sqrt{\frac{h_{ii}}{1 - h_{ii}}}.$$

standardized change in fitted value at i. Threshold: $|\text{DFFITS}_i| > 2\sqrt{p/n}$.

DFBETAS:

$$\text{DFBETAS}_{ij} = \frac{\hat{\beta}_j - \hat{\beta}_{j,(-i)}}{\hat{\sigma}_{(-i)}\sqrt{[(\mathbf{X}^\top\mathbf{X})^{-1}]_{jj}}}.$$

standardized change in $\hat{\beta}_j$ when i is deleted; identifies *which coefficients* an observation unduly influences. Threshold: $|\text{DFBETAS}_{ij}| > 2/\sqrt{n}$.

Influence vs. leverage vs. outlier:

Point type	Large $\lvert r_i \rvert$	Large h_{ii}	Influence?
Pure outlier	Yes	No	Possibly
Pure high leverage	No	Yes	Possibly
Influential	Yes	Yes	Yes
Good leverage	No	Yes	No

Never delete influential observations automatically. Investigate first: data entry error? A legitimately extreme but valid observation? Evidence of model misspecification?

Diagnostic Workflow

1. **Fit OLS;** compute residuals and leverages.
2. **Plot first:** residuals vs. fitted; scale-location; Q–Q; leverage-residual.
3. **Functional form:** if curvature seen, run RESET; follow up with Box–Tidwell; consider transformations (§VI).
4. **Heteroskedasticity:** if fan shape seen, run BP and White; apply HC3 or WLS as remedy.
5. **Autocorrelation:** run BG (primary) and DW (check); apply Newey–West HAC as remedy.
6. **Clustering:** if data are grouped, apply cluster-robust SEs; check $G \geq 30$.
7. **Multicollinearity:** VIF for each predictor; condition number of $\mathbf{X}^\top\mathbf{X}$.
8. **Influence:** Cook's D; DFFITS; DFBETAS; leverage-residual plot.
9. **Structural stability:** Chow test if time series or known subgroups; Quandt–Andrews if break date unknown.
10. **Correct and refit:** apply chosen remedy; re-run full diagnostic suite on corrected model.

Common Pitfalls

- **Testing before plotting:** running BP or RESET without first examining the residual plot misses the nature of the problem.
- **DW instead of BG:** DW tests AR(1) only and is invalid with lagged regressors.
- **Applying HC without testing:** HC SEs are consistent but less efficient than GLS when the variance structure is known; test first.
- **Confusing autocorrelation with heteroskedasticity:** both inflate residual patterns; BG/DW vs. BP/White separate the two.
- **Ignoring leverage in outlier detection:** moderate residual + extreme h_{ii} can still be highly influential.
- **Treating VIF > 10 as fatal:** collinearity inflates variance but does not bias $\hat{\boldsymbol{\beta}}$; harmless for prediction.
- **Over-relying on RESET:** RESET detects misspecification but cannot identify its form; follow up with Box–Tidwell or partial-residual plots.
- **Single threshold for Cook's D:** both $4/n$ and $D_i > 1$ are heuristics; examine the full distribution and investigate prominent spikes.
- **Smearing omission:** back-transforming $\log \hat{y}$ to $e^{\hat{y}}$ gives median predictions; multiply by $e^{\hat{\sigma}^2/2}$ for unbiased mean predictions.

Bottom Line

Regression diagnostics ensure that inference is valid. **Plot first, test second, then correct.** Each test targets a specific assumption failure. Heteroskedasticity and autocorrelation call for robust SEs (HC3 or Newey–West); clustering calls for cluster-robust SEs; functional form misspecification calls for transformation or model revision; influential observations require investigation, not automatic deletion. Re-diagnose after every correction.

Breusch-Pagan Test

Test for Heteroskedasticity in Linear Regression

REGR.1 | **Category:** Heteroskedasticity | **Data:** Continuous | **Null:** Chi-square

When to Use

- Detect heteroskedasticity related to model regressors.
- Assess whether error variance depends on predictors.
- Diagnose variance misspecification after OLS fitting.
- When a specific variance structure is hypothesized.
- Helps decide between OLS and WLS/GLS.

Assumptions

- Observations are independent.
- Linear mean model is correctly specified.
- Under H_0, error variance is constant.
- Errors have finite fourth moments.
- Regressors are not perfectly collinear.

Test Statistic

$$\text{(Orig.) } BP = \frac{ESS}{2\hat{\sigma}^4}; \text{ (Koenker) } LM = nR^2. \quad ESS \ \& \ R^2 \text{ from aux. reg.: } \hat{e}_i^2 = \gamma_0 + \sum_{j=1}^{q} \gamma_j x_{ij} + v_i,\ \hat{\sigma}^2 = \frac{\sum \hat{e}_t^2}{n}.$$

Notation: $\hat{e}_i$ are residuals from orig. reg., x's are predictors, q is number of x's, n is sample size. Under H_0, $BP, LM \approx \chi_q^2$

Plain English: The test checks whether regression error variance changes with predictor levels by regressing squared residuals on those predictors. If variance is constant (homoskedastic), the auxiliary regression should have little explanatory power. Significant explanatory power indicates heteroskedasticity. As a Lagrange Multiplier test, it is simple, quick, and widely used.

⚠ Practitioner Warning

Assumes normal errors; violations affect size/power. For non-normality, White's test is more robust, while Koenker's variant (nR^2) relaxes the normality assumption. The test detects linear forms of heteroskedasticity and may miss nonlinear or multiplicative patterns. Results are sensitive to mean model misspecification—omitted variables or incorrect functional form can induce spurious heteroskedasticity. With many predictors, df increase and power falls. A significant result indicates non-constant variance but not its form; follow up with residual plots and targeted modeling.

Hypotheses

- H_0: $\text{Var}(\epsilon_i) = \sigma^2$ for all i (homoskedasticity).
- H_1: $\text{Var}(\epsilon_i)$ depends on x_i (heteroskedasticity).

Decision Rule: Reject H_0 if $LM > \chi^2_{\alpha,q}$ or p-value $< \alpha$. Degrees of freedom = q (excludes intercept term).

Software References [Python, R, SAS]

Python	R	SAS
`statsmodels::het_breuschpagan`	`lmtest::bptest`	`PROC MODEL (option breusch)`

Additional Notes

Breusch–Pagan (1979): BP (**L**agrange **M**ultiplier or Score test; LM in short) is based on an auxiliary regression of $\hat{e}_i^2$ on the regressors and can be expressed as a scaled explained sum of squares (ESS) under normality. Koenker (1981): uses a studentized form $LM = nR^2$ from the same auxiliary regression, avoiding the normality assumption (See Technical Notes) and providing robustness to non-normal errors. A common one-degree-of-freedom variant regresses $\hat{e}_i^2$ on the fitted values $\hat{y}_i$. Remedies when heteroskedasticity is detected include heteroskedasticity-consistent (HC) standard errors, weighted least squares (WLS), or generalized least squares (GLS). Relationship to White test: Breusch–Pagan assumes a linear variance function in the regressors (x), whereas White's test allows a general quadratic form (in x).

Miscellanea: The Breusch–Pagan test was introduced by Trevor Breusch and Adrian Pagan (1979) as one of the earliest formal diagnostics for heteroskedasticity in regression models. The test is based on a Lagrange Multiplier (LM) framework, showing that heteroskedasticity can be detected using only residuals from the restricted (OLS) model without re-estimating a full alternative model. Koenker (1981) later proposed a studentized version (nR^2) that relaxes the normality assumption, making the test more robust and widely used in practice. The Breusch–Pagan idea influenced the development of more general tests such as White's test (1980), which allows arbitrary forms of heteroskedasticity via nonlinear functions of regressors. Despite its simplicity, the test highlights a key modeling insight: variance can carry systematic structure just like the mean, and ignoring it can lead to inefficient inference.

Reference(s): Breusch & Pagan (1979); Koenker (1981)

Breusch-Pagan Test

Worked Example

Problem Statement Twelve households are observed at monthly income levels $x = 1, 2, \ldots, 12$ (hundreds of pounds). Monthly consumption expenditure y (hundreds of pounds) is regressed on income. Apply the Breusch–Pagan test at $\alpha = 0.05$ to decide whether the error variance is constant or increases with income. ($\hat{\sigma}^2 = \sum \hat{e}_t^2/n = 0.1416$)
$H_0: \sigma^2$ constant; $\quad H_1: \sigma^2 = g(x'\gamma)$ for some $\gamma \neq \mathbf{0}$.

1. OLS Residuals

Fit $\hat{y} = \hat{\beta}_0 + \hat{\beta}_1 x$ by OLS to obtain $\hat{\beta}_0 = 2.024$, $\hat{\beta}_1 = 0.486$.

x	y	$\hat{y}$	$\hat{e}$	$\hat{u} = \hat{e}^2$
1	2.5	2.510	−0.010	0.000
2	3.0	2.996	0.004	0.000
3	3.3	3.482	−0.182	0.033
4	4.0	3.968	0.032	0.001
5	4.1	4.454	−0.354	0.125
6	5.3	4.940	0.360	0.129
7	5.4	5.426	−0.026	0.001
8	5.9	5.912	−0.012	0.000
9	6.6	6.398	0.202	0.041
10	7.3	6.884	0.416	0.173
11	7.9	7.370	0.530	0.281
12	6.9	7.856	−0.956	0.915

The squared residuals $\hat{e}_t^2$ are visibly larger at higher income levels, suggesting variance increases with x.

2. Auxiliary Regression

Regress $\hat{u}_t = \hat{e}_t^2$ on $(1, x)$ using OLS ($n = 12$ obs.):

$$\hat{u}_t = \hat{a}_0 + \hat{a}_1 x_t + v_t.$$

OLS yields $\hat{a}_0 = -0.160$, $\hat{a}_1 = 0.046$, with:

$$\text{ESS} = 0.307, \quad \text{TSS} = 0.740, \quad R^2 = \frac{0.307}{0.740} = 0.415.$$

3. Breusch–Pagan Statistic

The Koenker (1981) variant is:

$$LM = n \cdot R^2 = 12 \times 0.415 = 4.98.$$

4. Decision

Under H_0: $LM \sim \chi^2(q)$ where $q = 1$ is the number of variance regressors.

α	$\chi^2_\alpha(1)$	Decision
5%	3.841	**Reject** (4.98 > 3.841)
1%	6.635	FTR (4.98 < 6.635)

$p = 0.026$. Reject H_0 at the 5% level: evidence that the error variance increases with income.

Practitioner Insight: The Breusch–Pagan test detects heteroskedasticity by testing whether squared OLS residuals $\hat{e}_t^2$ depend on regressors. Under homoskedasticity, $\hat{e}_t^2$ is unpredictable from x_t, giving $R^2 \approx 0$; a large R^2 indicates a non-flat variance surface $\sigma_t^2 = g(x_t'\gamma)$. The original BP statistic is $\text{ESS}/(2\hat{\sigma}^4)$; here $BP = 7.66$, $p = 0.006$, stronger ($< p = 0.026$) than the Koenker version. Koenker's (1981) variant nR^2 is preferred in practice, as it is robust to non-normality, whereas the original BP requires Gaussian errors; $nR^2 \sim \chi^2(q)$ under finite fourth moments. White's test generalises BP by including squared regressors and cross-products, making it consistent against all forms of heteroskedasticity at the cost of more degrees of freedom; with one regressor x, it uses $(1, x, x^2)$ and $\chi^2(2)$ instead of $\chi^2(1)$. The degrees of freedom q equal the number of *variance regressors* (excluding intercept); with k original *mean regressors*, BP uses $\chi^2(k)$. A significant result calls for HC standard errors or feasible GLS based on the estimated variance function.

Common Pitfalls

- **Treating heteroskedasticity as fatal.** BP detects non-constant σ^2, but OLS coefficients remain unbiased. The issue is invalid standard errors. Use robust SEs or WLS; do *not* discard the model solely because BP is significant.
- **Using classical BP despite non-normal residuals.** BP's χ^2 approximation assumes normal errors. With skewed or heavy-tailed residuals, use the **Koenker** version (no normality assumption) or White's test.
- **Misattributing heteroskedasticity to variance issues rather than misspecification.** BP flags patterned residuals, but causes may be omitted variables, nonlinearity, or outliers. Always check residual plots and functional-form diagnostics before concluding true heteroskedasticity.
- **Using wrong degrees of freedom.** BP's $df = q$ (number of regressors excluding intercept). Using n or auxiliary-regression df $(n - q - 1)$ yields incorrect p-values. With intercept-only models ($q = 0$), BP is undefined.
- **Relying on BP without visual diagnostics.** BP may miss weak heteroskedasticity (small n) or detect trivial patterns (large n). Always inspect residuals vs. fitted and scale–location plots to understand pattern and source of σ^2 changes.

White's Test

General Test for Heteroskedasticity with Cross-Products

REGR.2 | **Category:** Heteroskedasticity | **Data:** Continuous | **Null:** Chi-square

When to Use

- Detect general heteroskedasticity without specifying its form. Prefer over Breusch–Pagan with unknown σ^2 form.
- Assess variance misspecification broadly after OLS fitting.
- When nonlinear variance patterns are suspected.

Assumptions

- Linear mean model (OLS) is correctly specified.
- Errors are independent with finite fourth moment.
- Under H_0, error variance is constant.
- Sufficient sample size for χ^2 approximation.

Test Statistic

$$\chi^2_W = nR^2, \quad \text{where } R^2 \text{ from auxiliary regression: } \hat{e}_i^2 = \alpha_0 + \sum \alpha_j x_{ij} + \sum\sum \alpha_{jk} x_{ij} x_{ik} + \sum \beta_j x_{ij}^2 + v_i.$$

Notation: $\hat{e}_i$ are OLS residuals, x's are predictors, q is number of terms in auxiliary regression (excluding intercept)

Plain English: White's test regresses squared residuals on all predictors, their squares, and all cross-products, capturing general forms of heteroskedasticity without assuming a specific functional relationship. The inclusion of squared terms and interactions allows detection of non-linear variance patterns that Breusch-Pagan might miss. As a bonus, the test also detects omitted variables and functional form misspecification, making it a comprehensive model diagnostic.

⚠ Practitioner Warning

With many predictors, number of terms in auxiliary regression grows (curse of dimensionality) rapidly ($q \approx k + k(k+1)/2$), consuming degrees of freedom and reducing power; multicollinearity can also cause numerical instability. A simplified White test uses only fitted values and their squares: $\hat{e}_i^2 = \alpha_0 + \alpha_1 \hat{y}_i + \alpha_2 \hat{y}_i^2 + v_i$. Rejection is non-specific: it may reflect heteroskedasticity, functional form errors, or both. The test is sensitive to mean model misspecification—omitted variables or incorrect functional form can induce spurious heteroskedasticity. In small samples, the χ^2 approximation can be oversized. A significant result indicates non-constant variance but not its structure; follow up with residual plots, alternative specifications, or use heteroskedasticity-consistent (HC) standard errors.

Hypotheses

- H_0: $\text{Var}(\epsilon_i) = \sigma^2 \, \forall i$; mean model is correctly specified.
- H_1: $\text{Var}(\epsilon_i) \neq \sigma^2 \, \forall i$ and/or mean model form is incorrect.

Decision Rule: Reject H_0 if $\chi^2_W > \chi^2_{\alpha,q}$ or p-value $< \alpha$.
$df = q$ is number of terms $(x_i, x_i^2, x_i x_j)$ in auxiliary regression.

Software References [Python, R, SAS]

```
statsmodels::het_white          skedastic::white_test          PROC REG (WHITE)
```

Additional Notes

White (1980) also developed heteroskedasticity-consistent (HC) standard errors, now called "White standard errors" or "robust standard errors." These correct inference without requiring homoskedasticity. HC0: $\widehat{\text{Var}}(\hat{\beta}) = (X'X)^{-1} X' \text{diag}(\hat{e}_i^2) X (X'X)^{-1}$. Improved versions: HC1 (degrees of freedom correction), HC2 (leverage adjustment), HC3 (more conservative, widely recommended), and HC4–HC7 (more aggressive adjustments for high-leverage points). The test statistic is typically computed as nR^2 from the auxiliary regression, following a Lagrange Multiplier (LM) framework, with degrees of freedom equal to the number of auxiliary regressors excluding the intercept. Breusch–Pagan can be viewed as a restricted (linear) version of White's test. Centering or scaling regressors before forming squares and cross-products can improve numerical stability in the auxiliary regression, especially when predictors differ widely in scale. A simplified version regressing on $\hat{y}_i$ and $\hat{y}_i^2$ is often used in practice, avoiding degrees-of-freedom explosion while retaining good power. A significant result motivates the use of robust standard errors or feasible GLS based on the estimated variance structure.

Miscellanea: Halbert White introduced both a general test for heteroskedasticity and heteroskedasticity-robust standard errors, allowing valid inference without specifying the variance structure. His approach became widely adopted in applied econometrics, with "White standard errors" now a standard tool in regression analysis.

Reference(s): White (1980); MacKinnon & White (1985)

Worked Example

Problem Statement The same twelve households from the Breusch–Pagan example are used here. Apply White's test at $\alpha = 0.05$ using the auxiliary regressors $(1, x, x^2)$. H_0: σ^2 constant; $H_1 : \sigma^2 = g(x, x^2)$ for some non-trivial g.

1. Auxiliary Regressor Matrix

Augment the original regressors with x^2.

x	x^2	y	$\hat{y}$	$\hat{e}$	$\hat{u} = \hat{e}^2$
1	1	2.5	2.510	−0.010	0.000
2	4	3.0	2.996	0.004	0.000
3	9	3.3	3.482	−0.182	0.033
4	16	4.0	3.968	0.032	0.001
5	25	4.1	4.454	−0.354	0.125
6	36	5.3	4.940	0.360	0.129
7	49	5.4	5.426	−0.026	0.001
8	64	5.9	5.912	−0.012	0.000
9	81	6.6	6.398	0.202	0.041
10	100	7.3	6.884	0.416	0.173
11	121	7.9	7.370	0.530	0.281
12	144	6.9	7.856	−0.956	0.915

2. Auxiliary Regression

Regress $\hat{u}_t = \hat{e}_t^2$ on $(1, x, x^2)$ by OLS ($n = 12$):

$$\hat{u}_t = \hat{a}_0 + \hat{a}_1 x_t + \hat{a}_2 x_t^2 + v_t.$$

OLS yields $\hat{a}_0 = 0.209$, $\hat{a}_1 = -0.112$, $\hat{a}_2 = 0.012$, with:

$$\text{ESS} = 0.504, \quad \text{TSS} = 0.740, \quad R^2 = \frac{0.504}{0.740} = 0.681.$$

3. White Statistic

$$LM = n \cdot R^2 = 12 \times 0.681 = 8.17.$$

4. Decision

Under H_0: $LM \sim \chi^2(q)$ where $q = 2$ is the number of non-constant auxiliary regressors (x and x^2).

α	$\chi^2_\alpha(2)$	Decision
5%	5.991	**Reject** (8.17 > 5.991)
1%	9.210	FTR (8.17 < 9.210)

$p = 0.017$. Reject H_0 at the 5% level: evidence of heteroskedasticity. $\hat{y} = 2.024 + 0.486 \cdot x$

The pattern in $\hat{e}_t^2$ — near zero for small x, large at $x = 11$ and $x = 12$ — is consistent with variance increasing in income.

Practitioner Insight: White's test is a general, omnibus test for heteroskedasticity requiring no assumed variance form. In the example, Breusch–Pagan with x alone gives $LM = 4.98$, $p = 0.026$ ($\chi^2(1)$), while White's test adds x^2, increasing R^2 from 0.415 to 0.681 and LM to 8.17, but using $\chi^2(2)$. The gain in R^2 (+0.266) indicates a nonlinear (quadratic) variance component missed by Breusch–Pagan. With k regressors, the auxiliary regression includes k levels, k squares, and $\binom{k}{2}$ cross-products, giving $q = 2k + \binom{k}{2}$ degrees of freedom ($k = 1 \Rightarrow q = 2$). A drawback is rapid growth in regressors (e.g. $k = 4 \Rightarrow q = 14$), reducing power; Breusch–Pagan with selected drivers may then be preferable. Auxiliary fitted values $\hat{u}_t$ can be negative (e.g. $t = 3$–6), since OLS imposes no non-negativity constraint; this does not affect $LM = nR^2$, which depends only on R^2. A significant result signals heteroskedasticity but not its form; here $\hat{a}_2 = 0.012$ suggests a quadratic variance, motivating WLS with weights $1/x^2$ or use of HC standard errors.

Common Pitfalls

- **Using full White's test with many predictors.** Full White requires $q \approx k + k(k+1)/2$ terms, which quickly inflates degrees of freedom and reduces power. With moderate k, consider: (a) a *reduced White* specification (e.g., squares only), or (b) Breusch–Pagan / Koenker when a simpler variance structure is plausible.
- **Interpreting rejection as pure heteroskedasticity.** White's test is an omnibus diagnostic: it primarily detects heteroskedasticity but may also reject under functional-form misspecification or omitted variables. Always examine residual plots and complement with specification tests (e.g., RESET) before concluding.
- **Assuming White is always more powerful than BP.** Breusch–Pagan has higher power when the variance depends linearly on regressors and errors are well-behaved. White is more robust to nonlinear variance patterns and distributional departures but can lose power due to additional degrees of freedom.
- **Applying White with insufficient sample size.** Reliability requires n large relative to the number of auxiliary regressors k. When k is large relative to n, the auxiliary regression becomes unstable and p-values unreliable. In such cases, prefer reduced forms or simpler tests.
- **Ignoring auxiliary regression structure.** The omnibus p-value masks the source of rejection. Inspect which terms (levels, squares, interactions) are significant to diagnose the variance pattern and guide remedies such as transformations, WLS, or model respecification.

Goldfeld-Quandt Test

Test for Heteroskedasticity via Variance Ratio of Subsamples

REGR.3 | **Category:** Heteroskedasticity | **Data:** Continuous | **Null:** F-distribution

When to Use

- Detect heteroskedasticity across ordered observations.
- Suspected variance increases or decreases with a predictor.
- Compare variance between low and high regimes.
- Simple alternative to Breusch–Pagan for monotonic variance patterns.

Assumptions

- Observations are independent.
- Linear mean model is correctly specified.
- Errors are normally distributed (critical for F distribution).
- Data can be meaningfully sorted by a key variable.
- Homoskedasticity holds under H_0.

Test Statistic

$$F = \frac{RSS_2/(n_2 - k)}{RSS_1/(n_1 - k)} \sim F_{n_2-k,n_1-k}, \text{ under } H_0. \qquad \text{Larger to smaller ratio, by default.}$$

Notation: RSS_i is residual sum of squares for subsample i, n_i is subsample size, k is number of parameters

Plain English: The Goldfeld–Quandt test orders observations by a variable suspected of driving heteroskedasticity, divides the data into low and high groups (often omitting middle observations), fits separate regressions (but with same regressors), and compares their residual variances using an F-test. The statistic is typically formed as the ratio of the larger to the smaller variance, so large values indicate heteroskedasticity. The test is intuitive and easy to implement but requires a plausible ordering variable and assumes normally distributed errors.

⚠ Practitioner Warning

Requires specifying ordering variable a priori—cannot test all predictors without multiple testing (e.g: Bonferroni correction) issues. Sensitive to normality assumption; use with caution if errors non-normal. Number of omitted central observations affects power: too few (low power), too many (loss of information). Typical recommendation: omit central 20-30% of observations. Test is one-sided by construction—assumes variance is monotonic with ordering variable. For decrease, reverse subsample ordering or use reciprocal F-ratio. Does not detect non-monotonic variance patterns.

Hypotheses

- H_0: $\sigma_1^2 = \sigma_2^2$ (constant variance, homoskedasticity).
- H_1: $\sigma_1^2 < \sigma_2^2$ (variance increases with ordering variable).

Decision Rule: Reject H_0 if $F > F_{\alpha,n_2-k,n_1-k}$ or p-value $< \alpha$.

Software References [Python, R, SAS]

Python	R	SAS
`statsmodels::het_goldfeldquandt`	`lmtest::gqtest`	`Not Native (Manual)`

Additional Notes

The test compares residual variances across two subsamples via an F-statistic, $F = \frac{RSS_2/(n_2-k)}{RSS_1/(n_1-k)}$, where k is the number of regressors (including intercept) and n_1, n_2 are subsample sizes. The ordering variable z is often chosen as a regressor suspected to influence variance or the fitted values $\hat{y}$, which serve as a proxy for the conditional mean. Omitting central observations increases separation between low- and high-variance regimes, improving power when variance changes monotonically, but reduces effective sample size. The choice of omitted fraction c is therefore a bias–variance trade-off; moderate trimming is typically effective in practice. The Harvey–Phillips "peak" variant orders observations by $|x - \bar{x}|$ to detect variance that increases away from the center rather than monotonically. A two-sided version can be implemented by comparing both F and its reciprocal to critical values, though the standard formulation targets monotonic increases in variance.

Miscellanea: Goldfeld and Quandt (1965) introduced this test in cross-sectional consumption models, where variance was suspected to increase with income. It can be viewed as a regression analog of the classical two-sample F-test for equality of variances, applied after sorting observations by a covariate. This perspective highlights its reliance on ordering rather than modeling the variance function.

Reference(s): Goldfeld & Quandt (1965); Harrison & McCabe (1979)

Worked Example

Problem Statement The same twelve households from the Breusch–Pagan example are used here. Apply the Goldfeld–Quandt test at $\alpha = 0.05$, dropping the $c = 2$ central observations ($x = 6, 7$) to form two equal subgroups. $H_0: \sigma_1^2 = \sigma_2^2$; $H_1: \sigma_1^2 < \sigma_2^2 (\sigma^2 \uparrow x)$. After dropping central observations, fit $\hat{y} = \hat{\beta}_0 + \hat{\beta}_1 x$ separately to each subgroup.

1. Form Two Subgroups

Group 1 ($x = 1, \ldots, 5$; $n_1 = 5$):
$\bar{x}_1 = 3$, $\bar{y}_1 = 3.38$, $S_{xx} = 10$, $S_{xy} = 4.20$.

$\hat{\beta}_1 = 4.20/10 = 0.42$, $\hat{\beta}_0 = 3.38 - 0.42 \times 3 = 2.12$.

x	y	$\hat{y}$	$\hat{e}$	$\hat{e}^2$
1	2.5	2.54	−0.04	0.0016
2	3.0	2.96	0.04	0.0016
3	3.3	3.38	−0.08	0.0064
4	4.0	3.80	0.20	0.0400
5	4.1	4.22	−0.12	0.0144
			RSS_1	0.0640

Group 2 ($x = 8, \ldots, 12$; $n_2 = 5$):
$\bar{x}_2 = 10$, $\bar{y}_2 = 6.92$, $S_{xx} = 10$, $S_{xy} = 3.30$.

$\hat{\beta}_1 = 3.30/10 = 0.33$, $\hat{\beta}_0 = 6.92 - 0.33 \times 10 = 3.62$.

x	y	$\hat{y}$	$\hat{e}$	$\hat{e}^2$
8	5.9	6.26	−0.36	0.1296
9	6.6	6.59	0.01	0.0001
10	7.3	6.92	0.38	0.1444
11	7.9	7.25	0.65	0.4225
12	6.9	7.58	−0.68	0.4624
			RSS_2	1.1590

2. Goldfeld–Quandt Statistic

With $df_1 = df_2 = n_i - k = 5 - 2 = 3$ equal, the F ratio simplifies:

$$F = RSS_2/RSS_1 = 1.159/0.064 = 18.11.$$

$p = 0.020$. Reject H_0 at the 5% level.

3. Decision

Under H_0: $F \sim F(df_2, df_1) = F(3, 3)$.

α	$F_\alpha(3,3)$	Decision
5%	9.277	**Reject** (18.11 > 9.277)
1%	29.457	FTR (18.11 < 29.457)

Practitioner Insight: The Goldfeld–Quandt test is the only heteroskedasticity test in this chapter that uses an F distribution rather than asymptotic χ^2, giving exact finite-sample validity under Gaussian errors since the ratio of independent χ^2 variates (scaled by their df) is exactly F. The cost is specificity: it has power only for variance differences between low-x and high-x groups, and misses U-shaped or non-monotone patterns. On this data, Breusch–Pagan gives $LM = 4.98$, $p = 0.026$; White gives $LM = 8.17$, $p = 0.017$; Goldfeld–Quandt gives $F = 18.11$, $p = 0.020$—all reject at 5%, indicating variance increases with income. The number of omitted central observations c is a tuning parameter: larger c sharpens group contrast (increasing power) but reduces $df_i = n_i - k$ and widens the F critical value; a common rule is $c \approx n/5$ (here $c = 2 \approx 12/5$). With equal subgroups and equal k, $F = RSS_2/RSS_1$. A 1-sided alternative $H_1: \sigma_1^2 < \sigma_2^2$ is used for increasing variance; a 2-sided test uses $\max(RSS_1, RSS_2)/\min(RSS_1, RSS_2)$ with doubled p-value.

Common Pitfalls

- **Arbitrarily choosing how many observations to omit.** Dropping middle observations is required, but the omission fraction affects power and degrees of freedom. Too much omission (e.g., 1/2) yields sharp contrast but low power; too little (e.g., 1/6) dilutes signal. Ensure each group is sufficiently large ($n_{group} \gtrsim 20$) and report the rule.
- **Using the test when no clear ordering variable exists.** Goldfeld–Quandt assumes a *single predictor* driving monotonic variance. When multiple predictors or complex patterns are present, power is low. Prefer Breusch–Pagan or White's test when the source is uncertain.
- **Treating non-significant results as evidence of homoskedasticity.** The test discards data and uses only two groups, making it low-powered—especially with small n or weak variance differences. Always supplement with BP/White tests and residual plots; non-significance often reflects low power, not true homogeneity.
- **Applying the F-test under non-normal residuals.** The Goldfeld–Quandt F-test assumes normality within each group. Skewed or heavy-tailed residuals distort Type I error. Check normality in both groups; if violated, use BP, White, transformations, or bootstrap inference.
- **Ignoring that the test is direction-specific.** Goldfeld–Quandt is typically one-sided (tests $\uparrow$ *or* $\downarrow \sigma^2$). Direction must be pre-specified; choosing it post hoc constitutes p-hacking. If unknown, use a two-sided F-ratio, noting reduced power.

Glejser Test

Test for Heteroskedasticity Using Absolute Residuals

REGR.4 | **Category:** Heteroskedasticity | **Data:** Continuous | **Null:** t-distribution

When to Use

- Detect heteroskedasticity via specific functional forms.
- Suspected variance depends on a predictor (e.g., $|x|$, $\sqrt{x}$).
- Explore candidate variance relationships in residuals.
- Complement Breusch–Pagan with other transformations.
- Model-driven investigation of variance structure.

Assumptions

- Observations are independent.
- Linear mean model is correctly specified.
- Errors have zero mean and finite variance.
- Absolute residuals relate linearly to chosen transformation.
- Correct functional form for variance proxy is specified.

Test Statistic

$$t = \frac{\hat{\beta}}{SE(\hat{\beta})} \sim t_{n-2} \text{ under } H_0, \qquad |\hat{e}_i| = \alpha + \beta f(z_i) + v_i.$$

Notation: $|\hat{e}_i|$ are absolute residuals, z_i is suspected heteroskedasticity variable, $f(\cdot)$ is functional form

Plain English: The Glejser test regresses absolute values of residuals on various functions of predictor variables to detect heteroskedasticity patterns. Unlike Breusch-Pagan which uses squared residuals and assumes linear variance relationship, Glejser uses absolute residuals and tests multiple functional forms (linear, quadratic, reciprocal, square root). Significant coefficients in any auxiliary regression indicate heteroskedasticity with that particular variance structure.

⚠ Practitioner Warning

Testing multiple functional forms without adjustment inflates Type I error—apply Bonferroni (use α/m where m is number of specifications) or similar correction. Using absolute residuals violates normality assumptions for test statistics, though asymptotically valid; squared residuals (Breusch–Pagan approach) are often preferred theoretically despite Glejser's empirical popularity. Results are sensitive to mean model misspecification: omitted variables or incorrect functional form can induce patterns in residual magnitudes unrelated to heteroskedasticity. The test is also sensitive to outliers, which can dominate absolute residual regressions and produce spurious significance. The test requires specifying candidate variance functions and may miss patterns not included in tested forms. With many predictors, specifications grow rapidly, and no consensus exists on a standard set of functional forms.

Hypotheses

- H_0: $\beta = 0$ for all tested specs (homoskedasticity).
- H_1: $\beta \neq 0$ for at least one spec (heteroskedasticity).

Decision Rule: Reject H_0 if $|t| > t_{\alpha/2,n-2}$ for any tested specification.

Software References [Python, R, SAS]

Python	R	SAS
`Not Native (Manual)`	`skedastic::glejser`	`Not Native (Manual)`

Additional Notes

Glejser regressions use $|e_i|$ as a proxy for σ_i, whereas tests based on e_i^2 target σ_i^2; thus, coefficient signs and functional form describe the shape of the standard deviation rather than the variance function. Significant terms indicate how dispersion scales with z, guiding construction of weights for WLS (e.g., if $|e_i| \propto z$, use weights $\propto 1/z^2$). Common specifications include z, z^2, $\sqrt{|z|}$, $1/z$ (for $z > 0$), and $\ln|z|$, with z chosen as a predictor, fitted value $\hat{y}_i$, or theoretically motivated driver of variance. In practice, Glejser is most useful when a specific variance pattern is hypothesized; for general detection, BP or White tests are typically preferred. The intercept and slope in Glejser regressions provide informal guidance on variance structure: a significant slope indicates dispersion varies with z, while the intercept reflects baseline variability. Together they suggest if variance is primarily z-driven or includes a constant component, guiding WLS specification.

Miscellanea: Glejser's use of $|e_i|$ contrasts with variance-based approaches that use squared residuals, reflecting an early attempt to balance sensitivity to heteroskedasticity with robustness to extreme observations—a theme that later became central in robust statistics.

Reference(s): Glejser (1969); Machado & Santos Silva (2000)

Worked Example

Problem Statement The same twelve households from the Breusch–Pagan example are used here. Apply the Glejser test at $\alpha = 0.05$ using the linear auxiliary function $f(x) = x$. H_0: σ constant; $H_1 : \sigma_t = \alpha_0 + \alpha_1 x_t$.

1. Absolute Residuals
The absolute residuals are noticeably larger at higher income levels, suggesting σ_t increases with x.

2. Auxiliary Regression
Regress $|\hat{e}_t|$ on $(1, x)$ by OLS ($n = 12$):

$$|\hat{e}_t| = \hat{\alpha}_0 + \hat{\alpha}_1 x_t + v_t.$$

OLS yields $\hat{\alpha}_0 = -0.113$, $\hat{\alpha}_1 = 0.057$, with RSS = 0.444.

3. Glejser t-Statistic

$$s = \sqrt{\frac{\text{RSS}}{n-2}} = 0.211,\ S_{xx} = \sum (x_t - \bar{x})^2 = 143,\ \bar{x} = 6.5.$$

$$\widehat{\text{SE}}(\hat{\alpha}_1) = \frac{s}{\sqrt{S_{xx}}} = 0.0176,\ t = \frac{\hat{\alpha}_1}{\widehat{\text{SE}}(\hat{\alpha}_1)} = 3.23.$$

4. Decision
Under H_0: $t \sim t(n-2) = t(10)$.

α	$t_{\alpha/2}(10)$	Decision
5%	2.228	**Reject** (3.23 > 2.228)
1%	3.169	**Reject** (3.23 > 3.169)

$p = 0.009$. Reject H_0 at both levels: evidence that σ_t increases with income.

$f(x)$	t	p
x	3.23	0.009
$\sqrt{x}$	2.83	0.018
$1/\sqrt{x}$	−2.01	0.072
$1/x$	−1.67	0.125

Practitioner Insight: The Glejser test regresses $|\hat{e}_t|$ rather than $\hat{e}_t^2$ on functions of regressors. Using absolute values improves robustness to outliers: large residuals inflate $\hat{e}_t^2$ quadratically but $|\hat{e}_t|$ only linearly, reducing distortion in R^2 and the test statistic. The trade-off is loss of the χ^2 LM framework of Breusch–Pagan and White; inference is based on a $t(n-2)$ test for $\hat{\alpha}_1$, though an approximate $LM = nR^2 = 12 \times 0.510 = 6.12$ ($p = 0.013$, $\chi^2(1)$) can be computed with weaker justification.
On this data: Breusch–Pagan $LM = 4.98$, $p = 0.026$; White $LM = 8.17$, $p = 0.017$; Goldfeld–Quandt $F = 18.11$, $p = 0.020$; Glejser $t = 3.23$, $p = 0.009$—the strongest rejection, as the signal at $t = 12$ ($|\hat{e}| = 0.956$) is less overwhelmed by near-zero residuals at $t = 1, 2, 4$ than under squaring.
The key choice is the auxiliary form $f(x)$; testing several is standard. Here linear and square-root forms reject (variance rising with income), while reciprocal forms do not (ruling out decreasing variance). As with White's auxiliary regression, OLS-fitted $|\hat{e}_t|$ can be negative (e.g. $t = 1$, fitted = −0.056), a harmless artifact. When multiple forms are tested, interpret the most significant cautiously due to data-snooping; here $f(x) = x$ was chosen *a priori* on subject-matter grounds (income-proportional heteroskedasticity).

Common Pitfalls

- **Multiple forms without correction (p-hacking).** Glejser proposed several functional forms, but testing all and reporting only the most significant inflates Type I error. If multiple forms are tried, use Bonferroni correction, pre-specify one form based on theory, or run a joint F-test. Never cherry-pick the lowest p-value.
- **Using Glejser when Breusch–Pagan is more powerful.** With approximately normal errors, BP (LM test) has higher power because squared residuals retain more information (efficient) than absolute residuals. Use Glejser mainly when outliers or heavy tails make absolute residuals more robust, or as a supplementary check.
- **Confusing detection with variance specification.** A significant Glejser relationship (e.g., $|e| \sim X$) only signals heteroskedasticity; it does *not* identify the correct variance model. Use residual plots and theory to choose WLS weights or variance-stabilizing transformations. Glejser suggests direction, not an exact form.
- **Allowing functional forms that produce negative predicted $|e|$.** Reciprocal or linear forms may yield negative fitted values for $|e|$—nonsensical by definition. Exclude any form generating negative predictions within the observed X range, or use strictly positive specifications such as $\exp(\alpha + \beta X)$.
- **Ignoring heteroskedasticity within the auxiliary regression.** The regression $|e_i| = \alpha + \beta X_i + u_i$ often has heteroskedastic u_i, making classical t-tests unreliable. Use heteroskedasticity-robust SEs (HC3) for $\hat{\beta}$ in the auxiliary regression to obtain valid inference.

Harvey-Godfrey Test

Multiplicative Heteroskedasticity Test with Log Transformation

REGR.5 | **Category:** Heteroskedasticity | **Data:** Continuous | **Null:** Chi-square

When to Use

- Detect multiplicative (log-linear) heteroskedasticity.
- Suspected variance follows $\sigma_t^2 = \exp(x_t'\gamma)$.
- When variance scales proportionally with predictors.
- Alternative to Breusch–Pagan for log-variance structures.
- Ensures strictly positive variance via log specification.

Assumptions

- Observations are independent.
- Linear mean model is correctly specified.
- Under H_0, error variance is constant.
- Log-variance is linear in regressors under H_1.
- Errors have finite variance and allow log transform.

Test Statistic

$$LM_{\text{HG}} = \frac{2}{\pi^2} ESS \approx \chi_k^2, \quad \text{(Koenker-style) } LM = nR^2 \approx \chi_k^2. \quad \ln(\hat{e}_i^2) = \alpha_0 + \sum_{j=1}^{k} \alpha_j z_{ij} + v_i.$$

Notation: $\hat{e}_i$ are OLS residuals, z's are variables affecting σ^2, k is number of regressors (excluding intercept)

Plain English: The Harvey–Godfrey test examines multiplicative heteroskedasticity by regressing the logarithm of squared residuals on variables suspected of influencing variance. The log transformation converts multiplicative variance structures into an additive (linear) form, making the test well-suited for log-linear variance patterns such as exponentially growing variance. In practice, the test is often implemented using the Koenker form nR^2 (robust form) from the auxiliary regression; the original formulation applies a scaling factor based on the variance of $\ln(\chi_1^2)$ [that is, $\text{Var}(\ln(\chi_1^2)) = \pi^2/2$].

⚠ Practitioner Warning

Log transformation requires strictly positive squared residuals—values of zero cause undefined logs (add a small constant if needed). Residuals close to zero generate large negative $\ln(\hat{e}_i^2)$ values, which can exert undue influence on the regression. The scaling factor $\pi^2/2 \approx 4.93$ arises under normality; deviations from normality can affect size and power, and robustness is less well-established than for Breusch–Pagan. The test is designed for multiplicative heteroskedasticity and may have low power against additive variance patterns. Some implementations use alternative scalings (e.g., 2 or $\pi^2/2$); check software documentation for the exact formulation.

Hypotheses

- H_0: $\text{Var}(\epsilon_i) = \sigma^2$ for all i (homoskedasticity).
- H_1: $\text{Var}(\epsilon_i) = \sigma^2 \exp(\alpha' z_i)$ with $\alpha \neq 0$ (multiplicative heteroskedasticity).

Decision Rule: Reject H_0 if $LM > \chi^2_{\alpha,k}$ or p-value $< \alpha$. Degrees of freedom equal number of variables k in auxiliary regression (excluding constant).

Software References [Python, R, SAS]

Python	R	SAS
`Not Native (Manual)`	`skedastic::harvey`	`Not Native (Manual)`

Additional Notes

In the auxiliary regression, $\ln(\hat{e}_i^2)$ serves as a proxy for $\ln(\sigma_i^2)$, so coefficients describe how *log-variance* scales with z. A positive coefficient implies σ_i^2 grows approximately exponentially with that regressor; the fitted model yields $\hat{\sigma}_i^2 = \exp(\hat{\alpha}_0 + \hat{\alpha}' z_i)$, which can be used to construct feasible GLS weights $w_i = 1/\hat{\sigma}_i^2$. The LM (nR^2) and coefficient-based (t/F) implementations are derived from the same auxiliary regression and are asymptotically equivalent, typically producing similar p-values in moderate to large samples. In practice, z is often chosen as the fitted values $\hat{y}_i$ or a small subset of theoretically motivated regressors to avoid overfitting the variance model; including all regressors is not required and may dilute interpretability.

Miscellanea: Andrew Harvey (1976) and Leslie Godfrey (1978) developed this approach in parallel while studying econometric models where variability scales with economic magnitude, such as consumption and income data. Their work reflects a broader shift in the 1970s toward explicitly modeling variance rather than treating heteroskedasticity solely as a nuisance.

Reference(s): Harvey (1976); Godfrey (1978)

Worked Example

Problem Statement Fifteen households are observed at monthly income levels $x = 1, \dots, 15$ (hundreds of pounds). Monthly consumption expenditure y (hundreds of pounds) is regressed on income. Apply the Harvey–Godfrey test at $\alpha = 0.05$ to test for multiplicative heteroskedasticity of the form $\sigma_t^2 = \sigma^2 \exp(\alpha_1 x_t)$. $H_0: \alpha_1 = 0$ (homoskedasticity); $H_1: \alpha_1 \neq 0$.

1. OLS Residuals and Log Squared Residuals
Fit $\hat{y} = \hat{\beta}_0 + \hat{\beta}_1 x$ by OLS, giving $\hat{\beta}_0 = 3.483$, $\hat{\beta}_1 = 0.390$.

t	x	y	$\hat{e}_t$	$\hat{e}_t^2$	$\ln(\hat{e}_t^2)$
1	1	3.6	−0.272	0.074	−2.600
2	2	4.4	0.138	0.019	−3.963
3	3	4.8	0.148	0.022	−3.818
4	4	4.4	−0.641	0.411	−0.888
5	5	6.0	0.569	0.324	−1.128
6	6	6.3	0.479	0.230	−1.471
7	7	6.6	0.390	0.152	−1.885
8	8	6.0	−0.600	0.360	−1.022
9	9	8.0	1.010	1.021	0.021
10	10	6.9	−0.479	0.230	−1.471
11	11	6.2	−1.569	2.462	0.901
12	12	9.1	0.941	0.886	−0.121
13	13	7.5	−1.048	1.099	0.094
14	14	9.2	0.262	0.069	−2.678
15	15	10.0	0.673	0.452	−0.794

2. Auxiliary Regression
Regress $\ln(\hat{e}_t^2)$ on $(1, x)$ by OLS ($n = 15$):

$$\ln(\hat{e}_t^2) = \hat{a}_0 + \hat{a}_1 x_t + v_t.$$

OLS yields $\hat{a}_0 = -2.845$, $\hat{a}_1 = 0.182$, with:

$$\text{ESS} = 9.283, \quad \text{TSS} = 27.761, \quad R^2 = \frac{9.283}{27.761} = 0.334.$$

3. Harvey–Godfrey Statistic (LM form)

$$LM = n \cdot R^2 = 15 \times 0.334 = 5.02.$$

For the equivalent t-form with $S_{xx} = \sum(x_t - \bar{x})^2 = 280$:

$$s = \sqrt{\frac{\text{RSS}}{n-2}} = 1.192, \quad \widehat{\text{SE}}(\hat{a}_1) = \frac{s}{\sqrt{S_{xx}}} = 0.0712.$$

$$t = \frac{\hat{a}_1}{\widehat{\text{SE}}(\hat{a}_1)} = 2.556.$$

4. Decision

Form	Stat	CV (5%)	Decision
$\chi^2(1)$	5.02	3.841	**Reject**
$t(13)$	2.556	2.160	**Reject**

$p = 0.025$. Reject H_0: evidence that σ_t^2 grows exponentially with income.

Practitioner Insight: The Harvey–Godfrey test targets multiplicative heteroskedasticity $\sigma_t^2 = \sigma^2 \exp(z_t'\gamma)$ via $\ln(\hat{e}_t^2) = \ln(\sigma_t^2) + \ln(\hat{e}_t^2/\sigma_t^2)$; regressing $\ln(\hat{e}_t^2)$ on z_t tests whether log-variance is linear in regressors. The exponential form ensures $\sigma_t^2 > 0$, unlike the linear BP model $\sigma_t^2 = z_t'\gamma$, and the log compresses extreme residuals, improving robustness to outliers. Under H_0, $v_t = \ln(\hat{e}_t^2/\sigma^2) \sim \ln(\chi_1^2)$ with mean -1.270 and variance $\pi^2/2 \approx 4.93$; the mean is absorbed by the intercept and does not affect inference on $\hat{a}_1$. Across tests on the same data: Breusch–Pagan $p = 0.092$, White $p = 0.146$, Goldfeld–Quandt $p = 0.081$, Glejser $p = 0.044$, Harvey–Godfrey $p = 0.025$—not all reject at 5%, while some indicating variance increases with income. The $LM = nR^2$ and coefficient-test forms are based on the same auxiliary regression and give similar conclusions here ($LM = 5.02$, $p = 0.025$; $t = 2.556$, $p \approx 0.024$). A limitation is that very small residuals produce large negative $\ln(\hat{e}_t^2)$, potentially distorting the regression; here the smallest residual is 0.138 so this concern is minor. A significant $\hat{a}_1 > 0$ implies $\hat{\sigma}_t^2 = \exp(\hat{a}_0)\exp(\hat{a}_1 x_t)$, providing weights for feasible GLS.

Common Pitfalls

- **Log transformation issues when residuals near zero.** Harvey–Godfrey uses $\log(e_i^2)$, undefined at $e_i = 0$ and extreme when $e_i \approx 0$. High-R^2 models or discrete outcomes often produce such residuals. Practical fixes include adding a small constant $\log(e_i^2 + c)$, trimming tiny residuals, or using Breusch–Pagan instead.
- **Assuming HG dominates Breusch–Pagan in all settings.** HG is most powerful for *multiplicative* heteroskedasticity, while BP is better for *additive* patterns. When the structure is unclear, report both or default to BP.
- **Overinterpreting auxiliary regression as exact variance function.** The model $\log(e^2) = \alpha_0 + \alpha_1 X + u$ is an approximation using noisy e^2. Significant coefficients indicate multiplicative heteroskedasticity, not the exact form of $\sigma^2(X)$. Use HG for detection; use plots, WLS, or formal models for specification.
- **Relying on HG when errors are non-normal.** HG relies on stronger normality due to the log transform. With skewed or heavy-tailed errors, Type I error and power may be distorted. If residuals are non-normal, use BP, or White.

Durbin-Watson Test

Test for First-Order Serial Correlation in Regression Residuals

REGR.6 | **Category:** Autocorrelation | **Data:** Continuous | **Null:** Tabled distribution

When to Use

- Detect first-order serial correlation in regression residuals.
- Time-series or ordered cross-sectional data after OLS.
- Suspected AR(1) error structure ($\rho \neq 0$).
- Quick diagnostic for positive or negative autocorrelation.
- Preliminary check before applying GLS or AR models.

Assumptions

- Linear regression model is correctly specified.
- Errors follow an AR(1) process under the alternative.
- Regressors are non-stochastic or strictly exogenous.
- No lagged dependent variable among regressors.
- Observations are ordered (typically time series).

Test Statistic

$$DW = \frac{\sum_{t=2}^{n}(\hat{e}_t - \hat{e}_{t-1})^2}{\sum_{t=1}^{n}\hat{e}_t^2} \approx 2(1-\hat{\rho}), \quad \text{where } \hat{\rho} = \frac{\sum_{t=2}^{n}\hat{e}_t\hat{e}_{t-1}}{\sum_{t=1}^{n}\hat{e}_t^2}. \qquad \text{Range: } 0 \leq DW \leq 4.$$

Notation: $\hat{e}_t$ are OLS residuals in time order, $\hat{\rho}$ is estimated first-order autocorrelation, n is sample size

Plain English: The Durbin-Watson test detects serial correlation by comparing consecutive residuals. The test statistic measures how much residuals change from one observation to the next relative to their overall variation. Values near 2 indicate no autocorrelation, values near 0 indicate strong positive autocorrelation (residuals persist in same direction), and values near 4 indicate negative autocorrelation (residuals oscillate). The test has an inconclusive region where results are ambiguous.

⚠ Practitioner Warning

Test has inconclusive region where no definitive decision regarding H_0 can be made—critical values d_L and d_U bound this region. Designed only for AR(1) errors; misses higher-order autocorrelation. Invalid with lagged dependent variables as regressors—use Durbin's h-test or Breusch-Godfrey instead. Assumes x's are non-stochastic and independent of the error process, making results less reliable with random predictors. Critical values depend on n, number of regressors, and presence of intercept—must consult tables. Most powerful at detecting AR(1), weak against MA or higher-order patterns.

Hypotheses

- H_0: $\rho = 0$ (no first-order autocorrelation in errors).
- H_1: $\rho > 0$ (positive autocorrelation present).

Decision Rule: Reject H_0 if $DW < d_L$ (+ve correlation); inconclusive if $d_L \leq DW \leq d_U$; fail to reject if $DW > d_U$.

Software References [Python, R, SAS]

Python	R	SAS
`statsmodels::durbin_watson`	`lmtest::dwtest`	`PROC REG (DW)`

Additional Notes

For testing negative autocorrelation, use $d^* = 4 - DW$ and compare d^* to the same (d_L, d_U) bounds; the tables are symmetric about 2. Exact finite-sample p-values depend on the design matrix X and are rarely tabulated; modern software may provide simulated or exact values, avoiding the inconclusive region. The statistic can be written as a quadratic form $\hat{\mathbf{e}}' A\hat{\mathbf{e}}/\hat{\mathbf{e}}'\hat{\mathbf{e}}$, where A encodes first differences, linking DW to general covariance-based diagnostics. When a lagged dependent variable (e.g: $y_t = \beta_0 + \beta_1 x_t + \beta_2 y_{t-1} + e_t$) is present, Durbin's h-test provides a valid alternative: $h = \hat{\rho}\sqrt{n/(1 - n\,\widehat{\text{Var}}(\hat{\beta}_{\text{lag}}))} \overset{a}{\sim} N(0,1)$. In practice, DW serves as a screening tool: if serial dependence is indicated, follow with ACF/PACF diagnostics and adopt robust inference or dynamic modeling as appropriate.

Miscellanea: Durbin and Watson (1950, 1951) derived their test by studying quadratic forms in regression residuals, yielding a statistic whose exact distribution depends on the design matrix X; this dependence led to the familiar lower and upper bounds tables. The statistic is closely related to the von Neumann ratio, both based on successive differences, while the Dickey–Fuller test addresses a deeper question of persistence by modeling the data-generating process directly. All three reflect serial dependence, but operate at different levels: descriptive (von Neumann), regression-based (DW), and model-based (DF).

Reference(s): Durbin & Watson (1950, 1951); Savin & White (1977)

Worked Example

Problem Statement Fifteen quarterly observations of a macroeconomic variable y_t are regressed on a linear time trend $t = 1, \ldots, 15$. OLS gives $\hat{\beta}_0 = 2.017$, $\hat{\beta}_1 = 0.583$. Apply the Durbin–Watson test at $\alpha = 0.05$ to test for positive first-order autocorrelation in the residuals. $H_0: \rho = 0; \quad H_1: \rho > 0$.

1. OLS Residuals

t	y_t	$\hat{e}_t$	$\hat{e}_t^2$	$(\hat{e}_t - \hat{e}_{t-1})^2$
1	3.2	0.600	0.3600	—
2	3.9	0.717	0.5142	0.0137
3	4.2	0.434	0.1886	0.0800
4	4.9	0.551	0.3040	0.0137
5	4.3	−0.631	0.3987	1.3990
6	4.6	−0.914	0.8359	0.0800
7	4.7	−1.397	1.9519	0.2331
8	6.2	−0.480	0.2304	0.8411
9	6.8	−0.463	0.2143	0.0003
10	7.7	−0.146	0.0212	0.1006
11	8.9	0.471	0.2222	0.3808
12	9.6	0.589	0.3464	0.0137
13	9.7	0.106	0.0112	0.2332
14	9.9	−0.277	0.0768	0.1465
15	11.6	0.840	0.7056	1.2479
Sum			6.381	4.784

The residuals show a clear pattern: positive in $t = 1$–4, negative in $t = 5$–10, then positive again—the hallmark of positive autocorrelation.

2. Durbin–Watson Statistic

$$d = \frac{\sum_{t=2}^{T}(\hat{e}_t - \hat{e}_{t-1})^2}{\sum_{t=1}^{T}\hat{e}_t^2} = \frac{4.784}{6.381} = 0.750.$$

The approximate first-order autocorrelation is:

$$\hat{\rho} \approx 1 - \frac{d}{2} = 1 - \frac{0.750}{2} = 0.625.$$

3. Decision

From the Durbin–Watson table at $\alpha = 0.05, T = 15, k = 1$:

$$d_L = 1.08, \qquad d_U = 1.36.$$

Region	Decision
$d < d_L = 1.08$	Reject H_0
$d_L \le d \le d_U$	Inconclusive
$d > d_U = 1.36$	FTR H_0

$d = 0.750 < d_L = 1.08$: **Reject** H_0. Evidence of significant positive first-order autocorrelation ($\hat{\rho} \approx 0.625$).

Practitioner Insight: The Durbin–Watson test uniquely uses tabulated bounds rather than a single critical value: the distribution of d depends on the regressor matrix X, so bounds d_L, d_U are provided for all X, creating an inconclusive region $[d_L, d_U]$. Here $[1.08, 1.36]$ and $d = 0.750 < d_L$ give clear rejection. The approximation $\hat{\rho} \approx 1 - d/2$ follows from $d = \sum(\hat{e}_t - \hat{e}_{t-1})^2 / \sum \hat{e}_t^2 = 2(1 - \hat{\rho}_1)$; the exact $\hat{\rho}_1 = \sum \hat{e}_t\hat{e}_{t-1} / \sum \hat{e}_t^2 = 0.542$ is different (below) the approximation 0.625 due to small sample ($T = 15$). The test assumes conditionally normal errors and non-stochastic regressors, and is invalid with lagged dependent variables (endogeneity biases d toward 2), for which Durbin's h or Breusch–Godfrey should be used. Negative autocorrelation is tested via $4 - d$ with the same bounds. For higher-order autocorrelation (AR(p), $p > 1$), the Breusch–Godfrey LM test—regressing $\hat{e}_t$ on X and lagged residuals—provides the natural extension.

Common Pitfalls

- **Using DW with lagged dependent variables.** DW is invalid when Y_{t-1} appears as a regressor: the statistic is biased toward 2 and can miss autocorrelation. Use Durbin's h-test for AR(1) with lagged Y, or Breusch–Godfrey LM tests (valid with lagged dependent variables and multiple lags).
- **Treating "inconclusive" as "no autocorrelation."** DW has an *indeterminate region* ($d_L < d < d_U$) where it gives no decision—not failure to reject. Proper response: use Breusch–Godfrey, exact DW tests (which use numerical inversion methods like the Imhof or Pan algorithms), or HAC standard errors as a conservative correction.
- **Assuming DW tests all forms of autocorrelation.** DW tests only AR(1) correlation and does not detect higher-order AR(p), MA(q), or seasonal patterns. For richer dependence: inspect the ACF, use Breusch–Godfrey with multiple lags, or Ljung–Box tests.
- **Ignoring detected autocorrelation and reporting classical OLS inference.** Positive autocorrelation leaves OLS coefficients unbiased but standard errors incorrect (often too small). Remedies: HAC/Newey–West standard errors, dynamic regression (ADL), or FGLS/Cochrane–Orcutt; significant DW should trigger correction.

Breusch-Godfrey Test

LM Test for Higher-Order Serial Correlation

REGR.7 | **Category:** Autocorrelation | **Data:** Continuous | **Null:** Chi-square

When to Use

- Detect higher-order serial correlation, AR(p), in residuals.
- Models with lagged dependent variables among regressors.
- General alternative to Durbin–Watson for autocorrelation.
- Time-series regressions requiring flexible error structures.
- Diagnose residual autocorrelation before GLS or ARMA.

Assumptions

- Observations are ordered (typically time series).
- Linear regression model is correctly specified.
- Errors are not serially correlated under H_0.
- Regressors are exogenous (no correlation with errors).
- Lag order p is correctly specified for testing.

Test Statistic

$$LM = (n-p)R^2 \sim \chi^2_p \text{ (under } H_0), \qquad \text{where } R^2 \text{ is from aux. regression: } \hat{e}_t = \alpha_0 + \sum_{j=1}^{k} \alpha_j x_{tj} + \sum_{i=1}^{p} \rho_i \hat{e}_{t-i} + v_t.$$

Notation: $\hat{e}_t$ are OLS residuals, x_{tj} are original regressors, p is order of autocorrelation tested, n is sample size

Plain English: The Breusch–Godfrey test regresses OLS residuals on the original regressors and their own lags up to order p, then tests whether the lagged residuals are jointly significant. Evidence of significance indicates serial correlation. The test generalizes Durbin–Watson by detecting autocorrelation of any specified order and remains valid even when the model includes lagged dependent variables or other non-strictly exogenous regressors.

⚠ Practitioner Warning

Choice of lag order p affects results—too small misses higher-order correlation, too large reduces power. Select p using data frequency, residual ACF/PACF, or information criteria rather than defaulting to $p = 1$. Lagging reduces the effective sample size by p. The test detects serial correlation but does not identify the underlying ARMA structure. An equivalent F-test form is: $F = \frac{(RSS_0 - RSS_1)/p}{RSS_1/(n^* - k - p)}$, where RSS_0 is from the restricted auxiliary regression (without lagged residuals), RSS_1 from the unrestricted auxiliary regression, and n^* is the effective sample size (p obs. lost due to lags $\hat{e}_{t-i}$).

Hypotheses

- H_0: $\rho_1 = \rho_2 = \cdots = \rho_p = 0$ (no serial correlation).
- H_1: At least one $\rho_i \neq 0$ (serial correlation present).

Decision Rule: Reject H_0 if $LM > \chi^2_{\alpha,p}$ or p-value $< \alpha$. Degrees of freedom equal lag order p tested.

Software References [Python, R, SAS]

Python	R	SAS
`statsmodels::acorr_breusch_godfrey`	`lmtest::bgtest`	`PROC AUTOREG (GOFFREY)`

Additional Notes

Relationship to Durbin–Watson: For $p = 1$ and exogenous regressors, BG and DW often yield similar conclusions, but are not equivalent. Also known as the LM test for serial correlation (Breusch 1978; Godfrey 1978), it evaluates whether lagged residuals add explanatory power without re-estimating the model under the alternative. In small samples, the LM statistic can over-reject; the F-form provides a better finite-sample approximation. A heteroskedasticity-robust version can be obtained using robust (e.g., White) standard errors in the auxiliary regression. The estimated lag coefficients ($\hat{\rho}_i$) provide a rough indication of the autocorrelation pattern, though they are not efficient AR parameter estimates. Under H_0, the auxiliary regression has low explanatory power ($R^2 \approx 0$); a large R^2 indicates predictable structure in residuals. Lag selection can be guided by residual ACF/PACF, sequential testing, or information criteria (AIC/BIC).

Miscellanea: Breusch and Godfrey independently developed this test as a Lagrange Multiplier (score) diagnostic for serial correlation, addressing limitations of the Durbin-Watson statistic such as handling lagged dependent variables and higher-order autocorrelation. As an LM test, it detects whether residuals contain predictable structure without re-estimating the model under the alternative. This idea extends beyond serial correlation—similar LM frameworks underlie tests for heteroskedasticity (Breusch–Pagan, White) and ARCH effects, making the Breusch-Godfrey test a canonical example of how a unified approach can diagnose diverse model violations.

Reference(s): Breusch (1978); Godfrey (1978)

Worked Example

Problem Statement The same fifteen quarterly observations from the Durbin–Watson example are used here. Apply the Breusch–Godfrey test with $p = 1$ lag at $\alpha = 0.05$ to test for first-order serial correlation. $H_0 : \rho_1 = 0$; $H_1 : \rho_1 \neq 0$.

1. Form the Auxiliary Regressors
Set $\hat{e}_0 = 0$. The auxiliary regression regresses $\hat{e}_t$ on the original regressors $(1, t)$ plus the lagged residual $\hat{e}_{t-1}$.

t	$\hat{e}_t$	$\hat{e}_{t-1}$	$\hat{v}_t$
1	0.600	0.000	0.671
2	0.717	0.600	0.393
3	0.434	0.717	0.021
4	0.551	0.434	0.302
5	−0.631	0.551	−0.970
6	−0.914	−0.631	−0.520
7	−1.397	−0.914	−0.839
8	−0.480	−1.397	0.368
9	−0.463	−0.480	−0.210
10	−0.146	−0.463	0.081
11	0.471	−0.146	0.483
12	0.589	0.471	0.194
13	0.106	0.589	−0.378
14	−0.277	0.106	−0.470
15	0.840	−0.277	0.874

2. Auxiliary Regression
OLS of $\hat{e}_t$ on $(1, t, \hat{e}_{t-1})$ yields $\hat{\rho} = 0.632$ for the lagged residual coefficient, with:

$$\text{RSS}_{\text{aux}} = 4.196, \quad \text{TSS}_{\text{aux}} = 6.382.$$

$$R^2 = 1 - \frac{4.196}{6.382} = 0.343.$$

3. Breusch–Godfrey Statistic

$$LM = T \cdot R^2 = 15 \times 0.343 = 5.14.$$

Equivalently, the t-test on $\hat{\rho}$ gives:

$$t = \frac{\hat{\rho}}{\widehat{\text{SE}}(\hat{\rho})} = \frac{0.632}{0.253} = 2.500.$$

4. Decision
Under H_0: $LM \sim \chi^2(p) = \chi^2(1)$.

Form	Stat	CV (5%)	Decision
$\chi^2(1)$	5.14	3.841	**Reject**
$t(12)$	2.500	2.179	**Reject**

$p = 0.023$. Reject H_0: evidence of first-order serial correlation ($\hat{\rho} = 0.632$).

Practitioner Insight: The Breusch–Godfrey test overcomes two key limitations of Durbin–Watson: it has no inconclusive region ($LM = TR^2 \sim \chi^2(p)$ gives a unique critical value), and it remains valid with lagged dependent variables since the auxiliary regression includes the full regressor matrix X (DW is biased toward 2 in that case, masking autocorrelation). On these data both agree: DW gives $d = 0.750 < d_L = 1.08$ and BG gives $LM = 5.14$, $p = 0.023$, both rejecting at 5%. BG is sharper due to the unique χ^2 critical value. A key procedural step is including the original regressors $(1, t)$ alongside $\hat{e}_{t-1}$; omitting them would confound serial correlation with mean misspecification. Initialization typically sets $\hat{e}_0 = 0$ (and $\hat{e}_{-1}, \ldots = 0$ for $p > 1$); alternatively, drop the first p observations, yielding $(T - p)R^2$ (here $14 \times 0.343 = 4.80$, $p = 0.029$), with the same conclusion. Extension to $p > 1$ is immediate by adding $\hat{e}_{t-2}, \ldots, \hat{e}_{t-p}$ and comparing TR^2 to $\chi^2(p)$, detecting AR(p) and MA(p) alternatives, whereas DW targets only AR(1). The t-test on $\hat{\rho}$ here gives $t = 2.500$, confirming the χ^2 result; for $p > 1$ only the joint LM statistic applies.

Common Pitfalls

- **Arbitrary lag order p.** Too few lags miss autocorrelation; too many waste degrees of freedom. Choose p using data frequency (annual: 1–2; quarterly: up to 4; monthly: up to 12), residual ACF/PACF, or sensitivity checks; avoid defaulting to $p = 1$ without justification.
- **Misspecified mean equation.** Significant BG may reflect omitted variables, incorrect functional form, or structural breaks—not true autocorrelation. Check specification first (RESET, residual plots, theory) before interpreting results.
- **No correction after detection.** BG detects but does not fix autocorrelation; OLS standard errors become invalid. Use HAC/Newey–West for inference or GLS/ARMA models for efficiency.
- **Ignoring stationarity.** Unit roots or omitted trends induce spurious autocorrelation, causing false rejection. Test for stationarity (ADF/PP/KPSS), difference or detrend as needed, and apply BG to stationary residuals.

Ramsey's RESET Test

Regression Specification Error Test for Functional Form

REGR.8 | **Category:** Functional Form | **Data:** Continuous | **Null:** F-distribution

When to Use

- Detect functional form misspecification in regressions.
- Check for omitted nonlinear transformations of predictors.
- Assess adequacy of linear specification after OLS fitting.
- General diagnostic when model form is uncertain.
- Complement residual plots and other specification tests.

Assumptions

- Observations are independent.
- Linear regression model is estimated via OLS.
- Errors have constant variance (for exact F inference).
- Errors are normally distributed (small-sample validity).
- Regressors are exogenous (uncorrelated with errors).

Test Statistic

$$F = \frac{(RSS_0 - RSS_1)/(m-1)}{RSS_1/(n-k-m)} \sim F_{m-1,n-k-m} \text{ (under } H_0). \qquad y = \beta_0 + \sum_{j=1}^{k} \beta_j x_j + \sum_{p=2}^{m} \gamma_p \hat{y}^p + \epsilon.$$

Notation: $\hat{y}$ are fitted values from original model, m is highest power tested (typically 3 or 4), RSS are residual sums

Plain English: Ramsey's RESET test checks whether the chosen regression form is adequate by adding powers of the fitted values (e.g., squared, cubed) to the model and testing whether they improve the fit. If the original specification is correct, these terms should have no explanatory power. A significant result indicates misspecification—such as omitted variables, incorrect functional form, or missing nonlinear relationships.

⚠ Practitioner Warning

RESET is an omnibus, non-specific diagnostic: rejection indicates misspecification but does not identify its source (e.g., omitted variables, incorrect functional form, missing interactions, or other specification errors). It may also reject due to heteroskedasticity or autocorrelation even when the mean function is correctly specified. Results depend on the choice of powers m; common practice uses low-order terms (e.g., $\hat{y}^2, \hat{y}^3$), as higher powers introduce multicollinearity and overfitting, reducing power. The test can have low sensitivity to certain types of misspecification. A significant result should prompt theory-driven refinement—adding variables, testing transformations, or considering nonlinear models—supported by residual diagnostics rather than relying on RESET alone.

Hypotheses

- H_0: Model is correctly specified ($\gamma_2 = \cdots = \gamma_m = 0$).
- H_1: Model is misspecified (functional form/omitted var.).

Decision Rule: Reject H_0 if $F > F_{\alpha,m-1,n-k-m}$ or p-value $< \alpha$.

Software References [Python, R, SAS]

Python	R	SAS
`statsmodels::linear_reset`	`lmtest::resettest`	`PROC AUTOREG (RESET)`

Additional Notes

RESET (**RE**gression **S**pecification **E**rror **T**est; Ramsey, 1969) augments the regression with powers of fitted values and tests their joint significance. The original formulation used $\hat{y}^2, \hat{y}^3, \hat{y}^4$, while modern practice typically uses low-order terms such as $\hat{y}^2$ (one-degree-of-freedom test) or $\hat{y}^2, \hat{y}^3$. Under H_0, these added terms should have no explanatory power; rejection signals omitted nonlinearities or interactions. Inference is usually based on an F test, with an equivalent large-sample LM form $nR^2 \sim \chi^2_{m-1}$. Thursby–Schmidt (1977) examined RESET variants using powers and cross-products of regressors, reflecting a truncated Taylor expansion that more directly captures nonlinear misspecification than fitted-value powers. When RESET rejects, follow-up should target specific structure: inspect residual plots, add quadratic or interaction terms, consider transformations (e.g., Box–Cox), or move to nonlinear or nonparametric models.

Miscellanea: Ramsey (1969) introduced RESET during a period when econometricians sought practical, broadly applicable diagnostics without committing to a specific alternative model. The test embodies a key econometric principle: when the true model is unknown, simple functional expansions can act as flexible proxies to uncover hidden structure.

Reference(s): Ramsey (1969); Thursby & Schmidt (1977)

Worked Example

Problem Statement Monthly revenue y_t (thousands of pounds) of a newly launched product is regressed on months since launch $t = 1, \dots, 12$ using a linear OLS model. The analyst suspects the product follows an inverted-U lifecycle (growth then decline) that a linear specification cannot capture. Apply Ramsey's RESET test with $q = 2$ added powers at $\alpha = 0.05$ to test for functional form misspecification. H_0: linear model correctly specified; H_1: functional form misspecification.

1. Restricted (Linear) OLS
Fit $\hat{y}_t = \hat{\beta}_0 + \hat{\beta}_1 t$, giving $\hat{\beta}_0 = 6.895$, $\hat{\beta}_1 = 0.747$.

t	y_t	$\hat{y}_t$	$\hat{e}_t$	$\hat{y}_t^2$
1	5.9	7.642	−1.742	58.40
2	7.6	8.389	−0.789	70.38
3	8.6	9.136	−0.536	83.47
4	9.2	9.883	−0.683	97.67
5	13.1	10.630	2.470	112.99
6	11.5	11.377	0.123	129.43
7	15.1	12.123	2.977	146.97
8	15.0	12.870	2.130	165.64
9	12.9	13.617	−0.717	185.43
10	14.0	14.364	−0.364	206.32
11	13.9	15.111	−1.211	228.34
12	14.2	15.858	−1.658	251.47

The residuals are negative for $t = 1$–4, positive for $t = 5$–8, and negative again for $t = 9$–12: the inverted-U curvature that a linear model cannot absorb.

2. Augmented Regression
Add $\hat{y}_t^2$ and $\hat{y}_t^3$ to the regressors and re-fit by OLS ($T = 12$, $k = 2$, $q = 2$):

$$y_t = \beta_0 + \beta_1 t + \gamma_2 \hat{y}_t^2 + \gamma_3 \hat{y}_t^3 + e_t.$$

OLS gives: Restricted (H_0) & Unrestricted (H_1)

$$\text{RSS}_R = \sum \hat{e}_t^2 = 28.786, \quad \text{RSS}_U = 11.446.$$

3. RESET F-Statistic

$$F = \frac{(\text{RSS}_R - \text{RSS}_U)/q}{\text{RSS}_U/(T-k-q)} = \frac{(28.786 - 11.446)/2}{11.446/8} = 6.06.$$

4. Decision
Under H_0: $F \sim F(q, T-k-q) = F(2, 8)$.

α	$F_\alpha(2,8)$	Decision
5%	4.459	**Reject** (6.06 > 4.459)
1%	8.649	FTR (6.06 < 8.649)

$p = 0.025$. Reject H_0: evidence of functional form misspecification. The linear model omits the curvature present in the data.

Practitioner Insight: Ramsey's RESET test detects functional form misspecification by testing whether powers of fitted values $\hat{y}_t^2, \hat{y}_t^3, \dots$ add explanatory power beyond X. Under H_0, $\hat{y}_t$ is linear in X, so $\hat{y}_t^j$ ($j \geq 2$) are nonlinear in X and uncorrelated with errors; correlation with residuals indicates misspecification. The choice of q is a tuning parameter: $q = 2$ (adding $\hat{y}^2, \hat{y}^3$) is standard and used here, but $q = 1$ (only $\hat{y}^2$) is common and gives $F(1, 9) = 13.58$, $p = 0.005$, stronger than $q = 2$ since one df captures the dominant concavity. RESET does not identify the source of misspecification. Here, knowing the true DGP (data generating process) is quadratic, adding t^2 yields $\hat{\beta}_2 < 0$ and removes significance. The added coefficients $\hat{\gamma}_2, \hat{\gamma}_3$ are not reported as they lack structural meaning. RESET complements heteroskedasticity and autocorrelation tests, which assume correct mean specification; RESET tests that assumption directly.

Common Pitfalls

- **Treating RESET as diagnosis.** RESET is an *omnibus* test: rejection signals misspecification but does not identify its source (functional form, omitted variables, heteroskedasticity, endogeneity, or breaks). Follow with residual plots, targeted tests, and theory-driven refinements.
- **Using too many powers.** High powers ($\hat{Y}^4, \hat{Y}^5, \dots$) induce multicollinearity and overfitting, especially in small samples. Standard practice uses $\hat{Y}^2$–$\hat{Y}^3$; higher orders rarely improve power.
- **Interpreting individual coefficients.** Coefficients on $\hat{Y}^2, \hat{Y}^3$ are not meaningful due to collinearity with $\hat{Y}$. RESET relies on the *joint* F-test; ignore individual t-statistics and signs.
- **Using RESET with small n.** Added powers quickly reduce degrees of freedom, yielding low power and unstable estimates. Prefer simpler versions (e.g., only $\hat{Y}^2$) or graphical and theory-based checks.
- **Ignoring scale with transformed Y.** RESET operates on the model's working scale (e.g., $\widehat{\log Y}$). Rejection may reflect misspecification within the transformed model or an inappropriate transformation; compare alternatives (AIC/BIC, Box–Cox).

Link Test

Specification Test for Single-Equation Models via Squared Prediction

REGR.9 | **Category:** Functional Form | **Data:** Continuous | **Null:** t-distribution

When to Use

- Check if the response variable needs a transformation.
- Quick post-estimation specification screen after any OLS.
- Detect nonlinearity in the relationship between y and $\hat{y}$.
- Applicable to any single-equation estimator, not only OLS.
- Complement to Ramsey RESET when a parsimonious single-df test is preferred.

Assumptions

- Observations are independent with exogenous regressors.
- Errors are homoscedastic and normally distributed for exact t-inference.
- No perfect collinearity between $\hat{y}$ and $\hat{y}^2$.
- Sample size is adequate for asymptotic inference ($n \gtrsim 30$ recommended).

Test Statistic

$$t = \frac{\hat{\gamma}_2}{SE(\hat{\gamma}_2)} \sim t_{n-3} \text{ under } H_0, \qquad y_i = \gamma_0 + \gamma_1 \hat{y}_i + \gamma_2 \hat{y}_i^2 + v_i.$$

Notation: $\hat{y}_i = \mathbf{x}_i^\top \hat{\beta}$ are OLS fitted values; $\hat{\gamma}_2$ tests for residual nonlinearity; under correct specification $\hat{\gamma}_1 \approx 1$ and $\hat{\gamma}_2 \approx 0$.

Plain English: The link test refits the original model using only the fitted values $\hat{y}$ and their square $\hat{y}^2$ as predictors, then tests whether $\hat{y}^2$ has any explanatory power. If the model is correctly specified, the fitted values alone capture all systematic variation in y, leaving nothing for their square to explain. A significant coefficient on $\hat{y}^2$ indicates that the model is missing structure — either the dependent variable needs a transformation, some predictors are nonlinearly related to y, or important variables have been omitted.

⚠ Practitioner Warning

Test is single-purpose diagnostic—only detects problems, does not identify solutions. A significant result could reflect a misspecified dependent variable (e.g. use $\log y$ instead of y), nonlinearity in a predictor, an omitted quadratic term, or a missing interaction. Under RESET with $q = 1$ power, the Link test and RESET are numerically identical when the model contains a single predictor; they diverge for multiple regressors since RESET retains $\mathbf{X}$ in the auxiliary regression while the Link test replaces $\mathbf{X}$ with $\hat{y}$ alone. High collinearity between $\hat{y}$ and $\hat{y}^2$ can inflate $SE(\hat{\gamma}_2)$ and reduce power.

Hypotheses

- H_0: $\gamma_2 = 0$: model correctly specified ($\hat{y}^2$ adds nothing).
- H_1: $\gamma_2 \neq 0$: model misspecified (link or functional form).

Decision Rule: Reject H_0 if $|t| > t_{\alpha/2,n-3}$ or p-value $< \alpha$. Two-sided; $df = n - 3$ (intercept, $\hat{y}$, $\hat{y}^2$).

Software References [Python, R, SAS]

Python	R	SAS
`Not Native (Manual)`	`Not Native (Manual)`	`Not Native (Manual)`

Additional Notes

Pregibon (1980) proposed a related but distinct procedure specifically for logistic regression — the *goodness-of-link test* — which augments the fitted GLM with $\hat{\eta}^2$ (the squared linear predictor on the link scale) rather than $\hat{y}^2$ (the squared predicted mean); this is the version in the GLMS chapter (GLMS.8). The REGR version here uses $\hat{y}$ directly on the response scale and produces a t_{n-3} statistic, making it universally applicable to any single-equation model. Under correct specification, the coefficient on $\hat{y}$ should equal 1: $\hat{\gamma}_1 \approx 1$ confirms the scale of prediction is right; $\hat{\gamma}_1 \neq 1$ alongside $\hat{\gamma}_2 \neq 0$ signals both scale and curvature problems simultaneously. This two-number diagnostic (the pair $(\hat{\gamma}_1, \hat{\gamma}_2)$) is more informative than the p-value alone.

Miscellanea: The link test traces to John Tukey's 1949 *Biometrics* paper "One Degree of Freedom for Non-Additivity," which carved a single degree of freedom from the error sum of squares in an unreplicated two-way ANOVA to detect interaction — an elegant trick at a time when computational resources made general interaction models prohibitively expensive. The core idea — that a squared fitted value should be uninformative if the model is right — was later generalized to regression by Pregibon.

Reference(s): Tukey (1949); Pregibon (1980, 1981)

Worked Example

Problem Statement Sixteen house sale prices y_i ($ in thousands) are regressed on floor area x_i (square feet) by OLS. The fitted values $\hat{y}_i$ are all positive. Apply the Link test at $\alpha = 0.05$ to assess whether the linear OLS specification is adequate, or whether y requires a transformation. $H_0: \gamma_2 = 0$ (OLS adequate); $H_1: \gamma_2 \neq 0$ (misspecified).

1. OLS Fit and Fitted Values

OLS: $y_i = \beta_0 + \beta_1 x_i$ gives $\hat{\beta}_0 = -110.36$, $\hat{\beta}_1 = 0.161$, $R^2 = 0.901$.

i	x_i	y_i	$\hat{y}_i$
1	800	44.7	18.2
2	900	56.1	34.3
3	1000	55.4	50.3
4	1100	72.5	66.4
5	1200	81.3	82.5
6	1300	87.2	98.5
7	1400	100.5	114.6
8	1500	115.4	130.7
9	1600	134.4	146.8
10	1700	138.3	162.8
11	1800	177.6	178.9
12	1900	163.7	195.0
13	2000	223.7	211.0
14	2100	193.2	227.1
15	2200	243.7	243.2
16	2400	347.8	275.3

The residuals increase with $\hat{y}$ — a sign that the relationship may be nonlinear (exponential growth).

2. Form $\hat{y}_i^2$ and Fit Auxiliary Regression

Regress y_i on $(1, \hat{y}_i, \hat{y}_i^2)$ ($n = 16$, $df = 13$):

$$y_i = \gamma_0 + \gamma_1 \hat{y}_i + \gamma_2 \hat{y}_i^2 + v_i.$$

OLS estimates:

$$\hat{\gamma}_0 = 53.36, \quad \hat{\gamma}_1 = -0.044, \quad \hat{\gamma}_2 = 0.00366.$$

Note $\hat{\gamma}_1 \approx 0 \neq 1$ and $\hat{\gamma}_2 \neq 0$ — both departures from correct specification.

3. t-Test on $\hat{\gamma}_2$

$$SE(\hat{\gamma}_2) = 0.000797, \qquad t = \frac{0.00366}{0.000797} = 4.60.$$

4. Decision

α	$t_{\alpha/2}(13)$	Decision
5%	2.160	**Reject**
1%	3.012	**Reject**

$p = 0.0005$. Strong evidence of misspecification.

5. Remedy

Refit with $\log(y_i)$ as the response. The log-linear model passes the link test: $\hat{\gamma}_2 \approx 0$, $\hat{\gamma}_1 \approx 1$.

Practitioner Insight: The Link Test assesses functional form by regressing y on $\hat{y}$ and $\hat{y}^2$ and testing whether the coefficient on $\hat{y}^2$ equals zero. Unlike RESET, it replaces the full regressor matrix X with $\hat{y}$, so the auxiliary regression always contains only three parameters regardless of the number of original predictors. It also provides a directional diagnostic: $\hat{\gamma}_2 < 0$ suggests concavity, whereas $\hat{\gamma}_2 > 0$ suggests convexity. RESET with $q = 1$ adds the same term $\hat{y}^2$ and therefore tests the identical null hypothesis. On these data, the Link Test gives $t(13) = 4.60$ ($p = 0.0005$), while RESET with $q = 1$ gives $F(1, 13) = 21.16$ ($p = 0.0005$). The two results agree exactly because $F = t^2$ when only one additional regressor is tested. RESET becomes more general when higher-order terms such as $\hat{y}^3$ and $\hat{y}^4$ are added. Under H_0, $\gamma_2 = 0$; the large positive estimate $\hat{\gamma}_2 = 0.00366$ provides strong evidence that the linear OLS model is misspecified. The formal Link Test focuses on $\hat{\gamma}_2$ alone.

Common Pitfalls

- **Conflating the REGR and GLMS versions.** The REGR version uses $\hat{y}$ (fitted means); the GLMS version uses $\hat{\eta}$ (linear predictor on the link scale). They test different things and should not be substituted.
- **Interpreting $\hat{\gamma}_1 \neq 1$ as the test statistic.** The formal test is on $\hat{\gamma}_2$ only. $\hat{\gamma}_1 \neq 1$ is an informative diagnostic but not part of the null hypothesis; do not use it as a separate test.
- **Assuming rejection identifies the remedy.** The test confirms misspecification but cannot distinguish a transformation of y, a nonlinear term in x, an omitted variable, or a structural break. Use residual plots and subject-matter knowledge to select the correction.
- **Stopping after a non-significant result.** A non-significant $\hat{\gamma}_2$ does not prove correct specification; the test has limited power against alternatives where the squared fitted value is not informative about the departure. Complement with residual plots, RESET, and Cook's distance.

Box-Tidwell Test

Test for Correct Power Transformation of Predictors

REGR.10 | **Category:** Functional Form | **Data:** Continuous | **Null:** t-distribution

When to Use

- Identify appropriate power transform of a continuous x.
- Detect nonlinearity in the effect of x on the response.
- When a monotonic but nonlinear relationship is suspected.
- Provide actionable alternative after RESET/Link rejection.
- Guide transform choice before fitting nonlinear models.

Assumptions

- Predictor x is strictly positive (or can be shifted to be +ve).
- Observations are independent. Linear model is correctly specified apart from transformation of x.
- Regressors are exogenous (uncorrelated with errors).
- Errors satisfy conditions for valid t-tests or large n.

Test Statistic

$$t = \frac{\hat{\lambda}_j - 1}{SE(\hat{\lambda}_j)}. \qquad y = \beta_0 + \sum_{j=1}^{k} \beta_j x_j^{\lambda_j} + \epsilon. \qquad \text{First order expansion: } x^{\lambda} \approx x + (\lambda - 1)x \ln x.$$

Notation: $x_j^{\lambda_j}$ is Box-Tidwell transformation, λ_j is power parameter ($\lambda_j = 1 \Rightarrow$ linear), $\hat{\lambda}_j$ is estimated λ_j

Plain English: The Box-Tidwell test determines whether continuous predictors should be transformed before entering the regression model. It estimates optimal power transformations for each predictor and tests whether the optimal power differs significantly from 1 (linear form). Powers less than 1 suggest concave relationships (log, square root), while powers greater than 1 suggest convex relationships (quadratic, cubic). The test helps identify non-linear predictor-response relationships.

⚠ Practitioner Warning

Requires strictly positive predictor values—negative or zero values invalidate power transformations. Computational complexity higher than other specification tests due to iterative nonlinear estimation. Convergence not guaranteed; may fail with poor starting values or multicollinearity. Testing multiple predictors simultaneously inflates Type I error—use sequential testing or Bonferroni correction. Estimated λ can be unstable with small samples or weak relationships. Preferable to have theoretical justification for transformations rather than purely data-driven selection.

Hypotheses

- H_0: $\lambda_j = 1$ for predictor j (linear form adequate).
- H_1: $\lambda_j \neq 1$ (predictor requires power transformation).

Decision Rule: Reject H_0 if $|t| > t_{\alpha/2,n-k-1}$ or p-value $< \alpha$. Separate test for each predictor.

Software References [Python, R, SAS]

Python	R	SAS
`Not Native (Manual)`	`car::boxTidwell`	`Not Native (Manual)`

Additional Notes

The test itself uses a simple but clever device: adding the term $x\ln(x)$ to the model; if it is significant, the assumed linear relationship is misspecified. Iterative procedure: (1) Start with $\lambda^{(0)} = 1$ (linear). (2) Compute $w_j = x_j \ln(x_j)$. (3) Regress y on $x_1, \ldots, x_k, w_1, \ldots, w_k$. (4) Update: $\lambda_j^{(1)} = 1 + \hat{\gamma}_j$ where $\hat{\gamma}_j$ is coefficient on w_j. (5) Iterate until convergence. Special values: $\lambda = 0$ corresponds to $\ln(x)$ (by continuity), $\lambda = 0.5$ is $\sqrt{x}$, $\lambda = -1$ is $1/x$. Relationship to Box-Cox: Box-Tidwell transforms predictors (x's); Box-Cox transforms response variable (y). Can combine both (Box Tidwell and Box Cox) for comprehensive transformation analysis (of both x's and y). Alternative: fractional polynomials provide more flexible functional forms without requiring positive predictors.

Miscellanea: George E. P. Box and Paul Tidwell (1962) developed a method for estimating power transformations of predictors within regression models, extending the broader use of transformation techniques in statistical modeling. Box's name also appears in the Box–Cox transformation and Box–Jenkins time series methods—a rare trifecta of foundational contributions. Despite its conceptual appeal, the Box–Tidwell approach is less widely used in practice, partly due to its computational complexity.

Reference(s): Box & Tidwell (1962); Box & Cox (1964)

Worked Example

Problem Statement Hourly wages y (pounds) for eleven workers with $x = 2, 3, \dots, 12$ years of experience are given below. An analyst fits a linear OLS model but suspects that the Mincer (1974) earnings equation calls for a power transformation of experience. Apply the Box–Tidwell test at $\alpha = 0.05$ to test whether $\lambda = 1$ (linear) is the correct power for x. $H_0: \lambda = 1$ (linear form correct); $H_1: \lambda \neq 1$.

1. Linear OLS and Augmenting Variable

Fit $\hat{y} = \hat{\beta}_0 + \hat{\beta}_1 x$, giving $\hat{\beta}_0 = 5.795$, $\hat{\beta}_1 = 0.626$.
Form the augmented variable $x^* = x \ln x$.

x	y	$\ln x$	$x^* = x \ln x$	$\hat{e}$
2	8.0	0.6931	1.386	0.955
3	8.2	1.0986	3.296	0.529
4	7.6	1.3863	5.545	−0.696
5	8.1	1.6094	8.047	−0.822
6	9.9	1.7918	10.751	0.353
7	10.1	1.9459	13.621	−0.073
8	10.5	2.0794	16.636	−0.298
9	10.1	2.1972	19.775	−1.324
10	12.3	2.3026	23.026	0.251
11	12.6	2.3979	26.377	−0.075
12	14.5	2.4849	29.819	1.200

2. Augmented Regression

Regress y on $(1, x, x^*)$ by OLS ($n = 11$, $df = 8$):

$$y = \beta_0 + \beta_1 x + \beta_2 x \ln x + v.$$

OLS yields $\hat{\beta}_0 = 9.815$, $\hat{\beta}_1 = -1.389$, $\hat{\beta}_2 = 0.701$, with:

$$\text{RSS}_U = 3.131, \quad s^2 = \frac{3.131}{8} = 0.391, \quad s = 0.626.$$

3. Box–Tidwell t-Statistic

$$\widehat{\text{SE}}(\hat{\beta}_2) = 0.267, \quad t = \frac{\hat{\beta}_2}{\widehat{\text{SE}}(\hat{\beta}_2)} = 2.626.$$

4. Estimated Power

$$\hat{\lambda} = 1 + \frac{\hat{\beta}_2}{\hat{\beta}_1} = 1 + \frac{0.701}{-1.389} = 0.496.$$

5. Decision

Under H_0: $t \sim t(n-3) = t(8)$.

α	$t_{\alpha/2}(8)$	Decision
5%	2.306	**Reject** (2.626 > 2.306)
1%	3.355	FTR (2.626 < 3.355)

$p = 0.030$. Reject H_0: linear x is not the correct functional form. The estimated power $\hat{\lambda} = 0.496 \approx \frac{1}{2}$ suggests a square-root transformation of experience. Note: Only one iteration is shown; the BT procedure is iterated until convergence.

Practitioner Insight: The Box–Tidwell test not only detects misspecification—as RESET and the Link Test do—but identifies the appropriate power transformation. It relies on the first-order expansion $x^\lambda \approx x + (\lambda - 1)x \ln x$, giving $\beta_2 = (\lambda - 1)\beta_1$ and $\hat{\lambda} = 1 + \hat{\beta}_2/\hat{\beta}_1$. Here $\hat{\lambda} = 0.496 \approx \frac{1}{2}$, aligning with the Mincer (1974) result that wages scale with $\sqrt{\text{experience}}$. The test requires $x > 0$; otherwise a shift $x + c$ is used. The sign reversal from $\hat{\beta}_1 = 0.626$ (restricted) to $\hat{\beta}_1 = -1.389$ (augmented) arises from near-collinearity between x and $x \ln x$ (correlation 0.998) and has no interpretation, though it does not affect the t-test on $\hat{\beta}_2$ or $\hat{\lambda}$. Here, RESET ($q = 2$) and the Link Test also reject ($p < 0.05$) but offer no remedy; BT suggests replacing x with $x^{0.5}$. Refitting $y = \alpha + \gamma\sqrt{x} + \varepsilon$ reduces residual variance and removes rejection by both RESET and Box–Tidwell. With multiple regressors, include one term $x_j \ln x_j$ per regressor and test each separately.

Common Pitfalls

- **Applying Box–Tidwell to predictors with zeros or negatives.** The test requires $X > 0$ because X^λ and $X \log X$ are undefined for nonpositive values. Shift predictors by a constant or use transformations that handle $X \leq 0$ (e.g., Yeo–Johnson). Never run Box–Tidwell on raw variables containing zeros.
- **Testing too many predictors simultaneously.** Joint estimation of multiple λ values is numerically unstable and often yields nonconvergent or inconsistent results. Test at most two, holding others fixed, or use a sequential procedure.
- **Over-interpreting precise $\hat{\lambda}$ estimates.** Small differences (e.g., $\hat{\lambda} = 0.47$ vs 0.50) are rarely meaningful. Use confidence intervals; if the CI includes a simple value (0, 0.5, 1), prefer the simpler, interpretable transformation.
- **Using Box–Tidwell for non-monotonic relationships.** Power transformations assume monotonicity; they cannot linearize U-shapes, inverted-U patterns, or threshold effects. Examine scatterplots first—polynomials, splines, or GAMs are required when the pattern is non-monotonic.
- **Including categorical predictors.** Box–Tidwell applies only to continuous predictors. Power transforms of 0/1 or ordinal codes are meaningless. Exclude categorical variables from the test and leave them untransformed in the model.

Rainbow Test

Test for Linearity Against Ordered Alternatives

REGR.11 | **Category:** Functional Form | **Data:** Continuous | **Null:** F-distribution

When to Use

- Detect functional form misspecification in linear regression.
- Suspected curvature or tail nonlinearity in predictor effects.
- Compare fit stability between central and peripheral data.
- Alternative to RESET without auxiliary regressions or transforms. Quick diagnostic for linearity of the mean.

Assumptions

- Linear regression model assumed under null hypothesis.
- Observations can be meaningfully ordered.
- Sample size adequate after splitting.
- Central subset adequately represents the linear structure.
- Errors are independent with constant variance.

Test Statistic

$$F = \frac{(\mathrm{RSS}_{\mathrm{FULL}} - \mathrm{RSS}_{\mathrm{CENTER}})/[(1-c)n]}{\mathrm{RSS}_{\mathrm{CENTER}}/(cn-k)} \sim F_{(1-c)n,cn-k} \text{ (under } H_0).$$

Notation: $\mathrm{RSS}_{\mathrm{FULL}}$ from all n obs., $\mathrm{RSS}_{\mathrm{CENTER}}$ from central cn fraction, (typically $c = 0.5$ or 0.6), k is number of predictors

Plain English: The rainbow test orders observations by a variable (often a predictor), fits a linear model to the central subset of observations, then tests whether this model adequately describes the outer observations. The name comes from the visual pattern: if you plot ordered residuals from the full model, non-linearity creates a rainbow-shaped arc. Poor fit in the extremes (tails) relative to center indicates non-linear relationships that the linear model misses.

⚠ Practitioner Warning

Requires specifying ordering variable and fraction of central observations to use—typical choice is central 50%. Results sensitive to these choices; different fractions may yield different conclusions. Power depends on whether non-linearity concentrated in tails versus center. Test designed for smooth non-linearity; may miss structural breaks or discontinuities. Low power with small samples or weak non-linearity. Assumes homoskedasticity across subsamples; heteroskedasticity can cause size distortions.

Hypotheses

- H_0: Linear model adequate.
- H_1: Linear model inadequate, particularly at extremes.

Decision Rule: Reject H_0 if $F > F_{\alpha,(1-c)n,cn-k}$ or p-value $< \alpha$.

Software References [Python, R, SAS]

Python	R	SAS
`statsmodels::linear_rainbow`	`lmtest::raintest`	`Not Native (Manual)`

Additional Notes

Unlike RESET or Link tests, the Rainbow test detects misspecification geometrically by comparing fit in central versus peripheral regions of the regressor space. Procedure: (1) order observations by predictor of interest (e.g., ascending x), (2) divide into center subset (middle 50% or 60%) and extremes (tails), (3) fit model on center, compute residual sum of squares $\mathrm{RSS}_{\mathrm{CENTER}}$, (4) fit model on all data, compute $\mathrm{RSS}_{\mathrm{FULL}}$, (5) test whether including extremes significantly worsens fit via an F-test. A significant result indicates lack of fit—the model describes the center adequately but performs poorly at the extremes, typical of functional-form misspecification. Fraction selection: 0.5 (middle half) is the usual default, but 0.4 or 0.6 may be tried for robustness. With multiple predictors, observation ordering is not unique; ordering by fitted values $\hat{y}$ is a common practical choice. The ordering variable is typically the predictor suspected of non-linearity or the fitted values for a more general check. Harvey–Collier provides a related recursive-residual approach.

Miscellanea: Jessica Utts (1982) developed the rainbow test as an intuitive diagnostic for regression misspecification. By comparing model fit on a central subset of the data to the full sample, the test detects nonlinearity and lack of fit. Its name reflects this "across-the-range" comparison, giving the method a memorable visual interpretation.

Reference(s): Utts (1982); Harvey & Collier (1977)

Worked Example

Problem Statement Average production cost y (pounds per unit) is recorded for sixteen output levels $x = 1, 2, \ldots, 16$ (hundreds of units). An analyst fits a linear OLS model to the full sample. Apply the Rainbow test with central fraction $h = 8$ at $\alpha = 0.05$ to test whether the linear specification is adequate. Data below is sorted by x.

x	1	2	3	4	5	6	7	8	9	10	11	12	13	14	15	16
y	18.1	13.4	13.9	12.0	9.1	7.7	12.0	9.0	8.8	10.6	12.9	10.8	11.8	15.1	14.6	16.6
	⟵ excluded ⟶				⟵ central subset ⟶								⟵ excluded ⟶			

H_0: linear model correctly specified; H_1: functional form misspecification.

1. Full-Sample OLS
Fit $\hat{y} = \hat{\beta}_0 + \hat{\beta}_1 x$ on all $n = 16$ observations. OLS gives $\hat{\beta}_0 = 11.800$, $\hat{\beta}_1 = 0.056$ (near-zero slope).

$$\text{RSS}_{\text{full}} = 128.428.$$

2. Central-Subset OLS
Drop the 4 lowest and 4 highest observations; retain the central $h = 8$ observations ($x = 5, \ldots, 12$). Refit OLS on the subset.

Central OLS gives $\hat{\beta}_0 = 6.723$, $\hat{\beta}_1 = 0.399$.

$$\text{RSS}_c = 15.169.$$

3. Rainbow Statistic
With $df_1 = n - h = 8$ and $df_2 = h - k = 6$:

$$F = \frac{(\text{RSS}_{\text{full}} - \text{RSS}_c)/df_1}{\text{RSS}_c/df_2} = 5.60.$$

4. Decision
Under H_0: $F \sim F(df_1, df_2) = F(8, 6)$.

α	$F_\alpha(8,6)$	Decision
5%	4.147	**Reject** (5.60 > 4.147)
1%	8.102	FTR (5.60 < 8.102)

$p = 0.025$. Reject H_0: the linear model fits the full sample substantially worse than the central subset, indicating a nonlinear (U-shaped) average cost curve.

Practitioner Insight: The Rainbow test exploits the idea that a correctly specified linear model should fit the full sample no worse per residual degree of freedom than a central subset: if the relationship is truly linear, the tails should not be systematically harder to fit than the center. When curvature is present, however, the tails diverge from the fitted line and RSS_{full} inflates relative to RSS_c. Here the full OLS slope is nearly flat ($\hat{\beta}_1 = 0.056$) because the high values at both extremes ($x = 1$–4 and $x = 13$–16) offset the lower central values, whereas the central subset recovers a stronger positive slope ($\hat{\beta}_1 = 0.399$) with much smaller residual variation ($\text{RSS}_c = 15.2$ vs $\text{RSS}_{\text{full}} = 128.4$). The subset size h acts as a tuning parameter: larger h retains more observations but reduces numerator degrees of freedom $(n - h)$ and sensitivity to tail nonlinearity; $h = n/2$ is the usual default. Unlike RESET, the Link Test, or Box–Tidwell, the Rainbow test requires no auxiliary regression or transformation, relying only on two OLS fits and their RSS values. It is also unique in focusing on the geometry of the regressor space (central vs peripheral observations) rather than residual structure. With multiple regressors, ordering becomes ambiguous; Utts (1982) recommended sorting by fitted values $\hat{y}$ rather than any single x.

Common Pitfalls

- **Forgetting to specify the ordering variable.** The test result depends on the ordering; always state which predictor was used. If multiple orderings are plausible, pre-specify or adjust for multiple testing.
- **Using the default center fraction without checking adequacy.** The choice of c affects power and stability. Use $c = 0.5$ for balanced samples, larger c (0.6–0.7) for small n, and avoid tuning c post-hoc.
- **Applying Rainbow to time-series data.** Rainbow assumes independent observations; autocorrelation or trending series invalidate the F-test. For time-series, use Chow, CUSUM, or other stability tests instead.
- **Interpreting rejection as a specific functional-form error.** Rainbow is omnibus—rejection may reflect nonlinearity, heteroskedasticity, or outliers at extremes. Follow up with residual plots and targeted diagnostics before modifying the model.
- **Using Rainbow with very small samples.** Splitting data into "center" and "extremes" yields unstable estimates when n is small. Prefer RESET or graphical checks unless $n \gtrsim 50$.

Chow Test

Test for Structural Break at Known Point

REGR.12 | **Category:** Structural Stability | **Data:** Continuous | **Null:** F-distribution

When to Use

- Detect structural breaks at a known breakpoint.
- Compare regression relationships across two subperiods.
- Test parameter stability before and after an event or policy change.
- Assess if separate regressions fit better than a pooled model.
- Time-series or grouped data with suspected regime shifts.

Assumptions

- Breakpoint is specified *a priori*.
- Linear regression model is correctly specified in each regime.
- Errors are independent within and across subsamples.
- Error variance is equal across subsamples.
- Each subsample has sufficient observations relative to k.

Test Statistic

$$F = \frac{(RSS_p - RSS_1 - RSS_2)/df_1}{(RSS_1 + RSS_2)/df_2}, \quad df_1 = k,\ df_2 = n_1 + n_2 - 2k. \qquad \text{Under } H_0, F \sim F_{df_1, df_2}.$$

Notation: RSS_p is pooled residual sum of squares, RSS_i for subsample i, k is number of parameters per model

Plain English: The Chow test compares the fit of a single pooled regression to the fit of separate regressions for each subsample. If coefficients are truly equal across groups, pooling should not substantially worsen fit. The test statistic measures the improvement in fit from allowing separate coefficients, standardized by degrees of freedom. Large F-values indicate that separate models fit significantly better than a pooled model, suggesting structural differences between subsamples.

⚠ Practitioner Warning

Break point must be specified exogenously based on theory or external events—data-driven break selection invalidates test. If break timing uncertain, use Quandt likelihood ratio test or similar methods. Sensitive to model misspecification: omitted variables or incorrect functional form can mimic structural breaks. Assumes equal error variances across subsamples; if heteroskedastic, use robust version or Wald test with heteroskedasticity-consistent covariances. Test requires sufficient observations in each subsample; very unbalanced splits reduce power. Rejection indicates some coefficients differ but does not identify which ones—examine individual coefficient differences post-hoc. Multiple testing across many candidate break points inflates Type I error if not adjusted.

Hypotheses

- H_0: $\beta_1 = \beta_2$ (no structural break).
- H_1: $\beta_1 \neq \beta_2$ (structural break present).

Decision Rule: Reject H_0 if $F > F_{\alpha, df_1, df_2}$ or p-value $< \alpha$. Here, $df_1 = k, df_2 = n_1 + n_2 - 2k$.

Software References [Python, R, SAS]

```
statsmodels::breaks_cusumolsresid   strucchange::sctest(type="Chow")   PROC REG (CHOW)
```

Additional Notes

The pooled regression fits all observations together ignoring group membership, yielding RSS_p. Separate regressions fit each subsample independently, yielding RSS_1 and RSS_2. The numerator $(RSS_p - RSS_1 - RSS_2)$ measures improvement from allowing different coefficients. Chow (1960) proposed this F-test framework for testing coefficient stability; it is a special case of the general linear hypothesis. Under normality and homoskedasticity, the test is exact; otherwise, inference relies on asymptotic approximations. For testing only subset of coefficients (e.g., slopes but not intercept), modified Chow test uses restricted and unrestricted models with partial equality constraints. The test can be reformulated as testing interaction terms: include dummy variable for group and its interactions with all predictors, then jointly test interactions equal zero. Predictive Chow test variant uses one subsample to estimate, other to validate when samples very unbalanced.

Miscellanea: Gregory Chow (1960) introduced a test for structural breaks in regression models, providing a formal method to assess whether coefficients differ across subsamples. The test became widely used in economics for analyzing policy changes and regime shifts.

Reference(s): Chow (1960); Quandt (1960)

Worked Example

Problem Statement Twenty quarterly observations of a firm's output y_t are regressed on a linear time trend $t = 1, \ldots, 20$. A policy change is suspected to have altered the trend after quarter 10. Apply the Chow test at $\alpha = 0.05$ to test for a structural break at $t = 10/11$. $H_0: \beta_0^{(1)} = \beta_0^{(2)},\ \beta_1^{(1)} = \beta_1^{(2)}$; H_1: at least one coefficient differs across subsamples.

t	1	2	3	4	5	6	7	8	9	10
y_t	4.6	3.3	2.7	5.0	5.9	5.1	5.3	6.2	8.1	8.9

t	11	12	13	14	15	16	17	18	19	20
y_t	13.6	11.6	12.9	13.4	13.6	13.9	17.0	17.6	19.3	16.4

1. Three OLS Regressions
Fit $\hat{y}_t = \hat{\beta}_0 + \hat{\beta}_1 t$ on the full sample, on Sub-sample 1 (t = 1–10), and on Sub-sample 2 (t = 11–20) separately.

Sample	$\hat{\beta}_0$	$\hat{\beta}_1$	RSS
Full ($n = 20$)	1.176	0.861	38.029
Sub-1 ($n_1 = 10$)	2.513	0.545	8.418
Sub-2 ($n_2 = 10$)	4.249	0.689	15.646

The output visible in the data: Sub-1 has a shallower trend ($\hat{\beta}_1 = 0.545$) with a lower base ($\hat{\beta}_0 = 2.513$), while Sub-2 has a steeper trend ($\hat{\beta}_1 = 0.689$) and higher intercept ($\hat{\beta}_0 = 4.249$)—consistent with a post-policy upward shift.

2. Restricted and Unrestricted RSS

$$\text{RSS}_U = \text{RSS}_1 + \text{RSS}_2, \qquad \text{RSS}_R = \text{RSS}_{\text{full}}.$$

3. Chow F-Statistic
With $k = 2$ parameters, $n_1 = n_2 = 10$:

$$F = \frac{(\text{RSS}_R - \text{RSS}_U)/k}{\text{RSS}_U/(n_1 + n_2 - 2k)} = 4.642.$$

4. Decision
Under H_0: $F \sim F(k,\ n_1 + n_2 - 2k) = F(2, 16)$.

α	$F_\alpha(2, 16)$	Decision
5%	3.634	**Reject** (4.642 > 3.634)
1%	6.226	FTR (4.642 < 6.226)

$p = 0.026$. Reject H_0 at the 5% level: evidence of a structural break after quarter 10.

Practitioner Insight: The Chow test compares a restricted model with common coefficients across subsamples (RSS_R) against an unrestricted model with separate regressions ($\text{RSS}_U = \text{RSS}_1 + \text{RSS}_2$). The F statistic measures the improvement in fit from allowing coefficients to differ, scaled by the number of restrictions k and the pooled unrestricted residual variance. The test requires $n_i > k$ in each subsample and assumes homoskedastic errors across groups; unequal variances invalidate the standard F distribution, in which case robust Wald-type procedures are preferred. Here the main break is in the intercept (+1.74), with a smaller slope increase (+0.14), though the Chow test identifies only whether coefficients change jointly, not which ones drive the break. After rejection, individual changes can be examined using a dummy-variable regression with a break indicator and interaction terms. The break date (t = 10/11 here) is assumed known; when unknown, the test must be repeated over candidate break points using a supremum (sup F) approach (Andrews 1993), since searching over breaks invalidates the standard F distribution.

Common Pitfalls

- **Ignoring heteroskedasticity.** The Chow test assumes equal error variance across subsamples, but structural breaks often coincide with variance changes. Check heteroskedasticity (e.g., Levene, Goldfeld–Quandt); if present, use robust Wald tests, WLS, or bootstrap methods.
- **Searching for breakpoints.** Chow requires a *pre-specified* breakpoint. Testing many candidate dates inflates Type I error and invalidates the standard F distribution. For unknown breaks, use Andrews' sup-F, Bai–Perron, or CUSUM-type methods.
- **Using small subsamples.** Each regime must contain enough observations relative to the number of parameters. Small subsamples produce unstable estimates, low power, and unreliable F statistics; balanced splits are preferable.
- **Ignoring autocorrelation.** Serial correlation violates Chow assumptions and distorts the F statistic. Diagnose autocorrelation first and use HAC-robust Wald tests, GLS, dynamic models, or block-bootstrap methods when needed.

VIF Test

Variance Inflation Factor for Multicollinearity Detection

REGR.13 | **Category:** Multicollinearity | **Data:** Continuous | **Null:** Rule of thumb

When to Use

- Diagnose multicollinearity among regression predictors.
- Assess inflation of standard errors due to collinearity.
- Identify unstable or poorly identified regression coefficients.
- Use as a routine check before interpreting regression results.

Assumptions

- Multiple predictor variables present in the model.
- No exact multicollinearity among regressors.
- Predictors are measured without severe error.
- Predictors are continuous or properly coded categorical.

Test Statistic

$$VIF_j = \frac{1}{1 - R_j^2}. \quad VIF_j = 1 \text{indicates no collinearity}, VIF_j > 10 \text{ suggests problematic collinearity.}$$

Notation: R_j^2 is coefficient of determination from auxiliary regression of x_j on other predictors

Plain English: The Variance Inflation Factor measures how much the variance of a regression coefficient is inflated due to collinearity with other predictors. For each predictor, VIF is calculated by regressing that predictor on all others and examining how well it can be predicted. High R^2 (predictor highly predictable from others) yields high VIF, indicating redundancy. VIF quantifies the multiplicative increase in coefficient variance compared to what it would be if predictors were orthogonal.

⚠ Practitioner Warning

VIF is a diagnostic measure, not a formal hypothesis test—no p-values or distributional theory. Common threshold of VIF > 10 is rule of thumb, not statistical criterion; some use VIF > 5 as cutoff. High VIF does not invalidate regression for prediction—only affects interpretation of individual coefficients. Multicollinearity is a data problem, not a model assumption violation. In real (especially observational) data, some degree of multicollinearity is inevitable. Removing variables solely based on VIF can introduce omitted variable bias. Always consider theoretical importance alongside VIF.

Hypotheses

- H_0: VIF near 1 (minimal collinearity).
- H_1: VIF substantially greater than 1 (collinearity present).

Decision Rule: Reject H_0 if VIF > 10 indicates serious collinearity; VIF > 5 suggests moderate concern.

Software References [Python, R, SAS]

`statsmodels::..._inflation_factor` `car::vif` `PROC REG (VIF)`

Additional Notes

VIF values are always positive and range from 1 (no multicollinearity) upward, with larger values indicating greater inflation of coefficient variance due to linear dependence among predictors. Common thresholds: VIF= 2 implies $R_j^2 = 0.50$, VIF= 3 implies $R_j^2 \approx 0.67$, VIF= 5 implies $R_j^2 = 0.80$, and VIF= 10 implies $R_j^2 = 0.90$. Tolerance is the reciprocal of VIF: Tolerance = $1/VIF_j = 1 - R_j^2$, with values near 0 indicating severe collinearity. The square root of VIF quantifies standard error inflation: $SE(\hat{\beta}_j) = \sqrt{VIF_j} \times SE(\hat{\beta}_j)_{\text{orthogonal}}$, so, for example, VIF= 4 doubles the standard error. The condition number (ratio of largest to smallest eigenvalue of $X'X$) provides a global diagnostic; values > 30 suggest potential collinearity. However, unlike VIF, it does not identify which specific predictors are responsible for the problem. Remedies for high VIF include collecting more data, removing redundant predictors, combining correlated predictors (e.g., principal components or factor scores), or accepting larger standard errors if all predictors are theoretically essential.

Miscellanea: The variance inflation factor emerged from regression theory in the 1960s-70s, with contributions from multiple researchers rather than single inventor. The concept formalized practitioners' intuition that "too-similar" predictors caused problems. Marquardt (1970) popularized VIF alongside ridge regression as solution to collinearity. The threshold VIF > 10 became conventional through textbooks in the 1980s, corresponding to $R^2 > 0.9$ for a predictor regressed on others—seemingly arbitrary but widely adopted.

Reference(s): Marquardt (1970); Belsley et al. (1980)

VIF Test

REGR.13

Worked Example

Problem Statement Annual household spending y (thousands of pounds) is regressed on three regressors: income x_1 (£k), wealth x_2 (£k), and age x_3 (years) for $n = 20$ households. Assess whether multicollinearity is present using VIFs.

x_1	20	24	27	30	33	36	40	43	46	50
x_2	47.8	45.8	54.2	61.9	62.4	72.0	80.0	77.9	96.7	102.8
x_3	51	43	57	23	57	46	35	24	65	22

x_1	22	26	29	34	37	41	44	47	52	55
x_2	41.1	51.2	60.3	66.8	72.9	75.3	90.6	94.6	105.3	102.9
x_3	55	54	38	43	62	39	64	45	26	57

1. Correlation Matrix
Inspect pairwise correlations among the regressors.

	x_1	x_2	x_3
x_1	1.000	0.982	−0.186
x_2	0.982	1.000	−0.174
x_3	−0.186	−0.174	1.000

$r(x_1, x_2) = 0.982$ signals severe collinearity between income and wealth; both are nearly uncorrelated with age.

3. VIF Summary

Regressor	R_j^2	VIF	Tolerance
Income (x_1)	0.964	28.0	0.036
Wealth (x_2)	0.964	27.9	0.036
Age (x_3)	0.036	1.0	0.964

2. Auxiliary Regressions
For each x_j, regress x_j on all other regressors and record R_j^2:

$$\mathrm{VIF}_j = \frac{1}{1 - R_j^2}.$$

4. Decision
Using the standard threshold VIF > 10:

Regressor	VIF	Decision
Income (x_1)	28.0	**Severe collinearity**
Wealth (x_2)	27.9	**Severe collinearity**
Age (x_3)	1.0	No collinearity

Income and wealth are severely collinear (VIF $\approx 28 \gg 10$); age is unaffected (VIF ≈ 1, Tolerance ≈ 1). The condition number of the standardized regressor matrix is $\kappa = \sqrt{40.88/0.36} = 10.6$, consistent with moderate-to-severe collinearity.

Practitioner Insight: The VIF measures how much the variance of $\hat{\beta}_j$ is inflated relative to the ideal case where x_j is orthogonal to the other regressors: $\mathrm{Var}(\hat{\beta}_j) = \sigma^2\,\mathrm{VIF}_j/(nS_{jj})$. With $\mathrm{VIF}_1 = 28$, the standard error of $\hat{\beta}_1$ is inflated by $\sqrt{28} \approx 5.3$, severely reducing precision. The collinearity here is structural: wealth is approximately twice income ($r = 0.982$), so the regressors contain nearly identical information and their separate effects are poorly identified—a design problem rather than a sample-size issue. Remedies include dropping one regressor, combining variables into an index or principal component, using ridge regression, or collecting data that breaks the near-linear dependence. VIF is a diagnostic, not a formal hypothesis test: thresholds such as 5 or 10 are rules of thumb, not critical values. The condition number κ is a complementary diagnostic: $\kappa < 10$ is mild, $10 \le \kappa < 30$ moderate, and $\kappa \ge 30$ severe (Belsley, Kuh & Welsch 1980); here $\kappa = 10.6$ is consistent with moderate-to-severe collinearity.

Common Pitfalls

- **Treating VIF > 10 as universal.** The cutoff is only a rule of thumb; acceptable levels depend on sample size, purpose, and required precision. Report actual VIFs and treat values around 3–5 as signals for investigation.
- **Dropping the highest-VIF variable mechanically.** VIF detects collinearity, not which predictor should be removed. Variable selection should be theory-driven; alternatives include principal components, or ridge regression.
- **Ignoring exact linear dependence first.** Perfect multicollinearity makes $R_j^2 = 1$ and VIF undefined. Remove structural redundancies (e.g., dummy-variable trap, derived variables) before computing VIF.
- **Assuming low VIF rules out collinearity.** VIF can miss subset or near-dependencies masked by unrelated predictors. Complement VIF with condition numbers, eigenvalues, or variance-decomposition diagnostics.
- **Expecting centering to fix collinearity.** Centering or scaling usually leaves VIF unchanged because predictor correlations remain the same. Only interaction or polynomial terms may substantially benefit.

Leverage (Hat Values) Test

Measure of Unusual Predictor Values

REGR.14 | **Category:** Influence Diagnostics | **Data:** Continuous | **Null:** Rule of thumb

When to Use

- Identify observations with unusual predictor values.
- Detect points capable of strongly influencing fitted slopes.
- Diagnose extrapolation risk in regression models.
- Complement Cook's distance and residual-based diagnostics.

Assumptions

- Linear regression model is estimated via OLS.
- Predictor matrix X is correctly specified.
- No exact multicollinearity among regressors.
- Observations are independent.
- Interpretation focuses on predictor-space extremeness.

Test Statistic

$h_{ii} = \mathbf{x}_i'(\mathbf{X}'\mathbf{X})^{-1}\mathbf{x}_i$. **X** includes a column of 1's for the intercept, Hat matrix: $\mathbf{H} = \mathbf{X}(\mathbf{X}'\mathbf{X})^{-1}\mathbf{X}'$.

Notation: $\mathbf{x}_i$ is a vector of predictors for i-th obs., h_{ii} is i-th diagonal element of **H**

Plain English: Leverage measures how far an observation's predictor values are from the center of the predictor distribution, quantifying its potential to influence the regression line. The hat value h_{ii} represents the weight given to observation i in determining its own fitted value: $\hat{y}_i = h_{ii}y_i + \sum_{j\neq i} h_{ij}y_j$. High leverage observations have unusual predictor combinations that give them the potential (but not guarantee) to strongly influence coefficient estimates. Leverage depends only on **X** values, independent of response y.

⚠ Practitioner Warning

Leverage is a diagnostic measure, not a formal hypothesis test—thresholds are guidelines, not statistical criteria. Common cutoffs $2k/n$ or $3k/n$ are conventions; examine distribution of leverage values rather than relying solely on cutoffs. High leverage alone does not mean observation is problematic—only indicates potential influence. An observation with high leverage but small residual may fit the model well and not be influential. Conversely, low leverage with large residual has limited influence. Leverage must be combined with residual information (as in Cook's distance or DFFITS) to assess actual influence.

Hypotheses

- H_0: h_{ii} near k/n (average leverage).
- H_1: h_{ii} substantially greater than k/n (high leverage).

Decision Rule: Reject H_0 if $h_{ii} > 2k/n$ suggests moderately high leverage; $h_{ii} > 3k/n$ suggests very high leverage.

Software References [Python, R, SAS]

`statsmodels::hat_matrix_diag` `stats::hatvalues` `PROC REG (INFLUENCE)`

Additional Notes

Properties of leverage: $0 \leq h_{ii} \leq 1$ for all observations, with $\sum_{i=1}^{n} h_{ii} = k$ where k is the number of parameters including intercept. Average leverage is therefore k/n. Leverage plays central role in influence diagnostics: variance of residual is $\text{Var}(\hat{u}_i) = \sigma^2(1 - h_{ii})$, and variance of fitted value is $\text{Var}(\hat{y}_i) = \sigma^2 h_{ii}$. High leverage points have small residual variance but large fitted value variance. Diagnostic plots: leverage vs. studentized residuals identifies four quadrants (low/high leverage × small/large residual), with high leverage and large residual being most concerning combination.

Miscellanea: The term "leverage" reflects a mechanical analogy: observations far from the center of the predictor space exert greater pull on the fitted line. In linear models, leverage is given by the diagonal elements of the projection ("hat") matrix **H**, which maps **y** to fitted values via $\hat{\mathbf{y}} = \mathbf{H}\mathbf{y}$. The term "hat matrix" is often attributed to John Tukey. Tukey also coined influential terms such as "bit," "software," and "exploratory data analysis," reflecting his broad impact on modern statistics and computing. While the projection framework was well established earlier, the systematic use of h_{ii} as leverage measures—and their combination with residuals to assess influence—was developed and widely popularized in the 1970s, notably by Hoaglin and Welsch (1978).

Reference(s): Hoaglin & Welsch (1978); Belsley et al. (1980)

Leverage (Hat Values) Test — REGR.14

Worked Example

Problem Statement Annual spending y ($k) is regressed on annual income x ($k) for twelve households. Eleven households earn between $4k and $16k; one earns $28k. Compute the leverage (hat) values, identify any high-leverage observations, and assess whether high leverage implies high influence. (Note: $\sum h_{ii} = k$, trace of hat matrix $\mathbf{H}$).

1. Key Quantities
OLS gives $\hat{\beta}_0 = 1.268$, $\hat{\beta}_1 = 0.577$, with $s = 0.738$.

$$\bar{x} = \frac{\sum x_i}{n} = 11.5, \quad S_{xx} = \sum x_i^2 - n\bar{x}^2 = 429.$$

2. Leverage Formula
For simple regression ($k = 2$):

$$h_{ii} = \frac{1}{n} + \frac{(x_i - \bar{x})^2}{S_{xx}} = \frac{1}{12} + \frac{(x_i - 11.5)^2}{429}.$$

i	x_i	y_i	$x_i - \bar{x}$	$(x_i - \bar{x})^2$	h_{ii}
1	4	4.8	−7.5	56.25	0.214
2	6	4.2	−5.5	30.25	0.154
3	7	5.2	−4.5	20.25	0.131
4	8	6.1	−3.5	12.25	0.112
5	9	5.8	−2.5	6.25	0.098
6	10	7.0	−1.5	2.25	0.089
7	11	7.6	−0.5	0.25	0.084
8	12	6.8	0.5	0.25	0.084
9	13	9.6	1.5	2.25	0.089
10	14	9.9	2.5	6.25	0.098
11	16	10.1	4.5	20.25	0.131
12	28	17.7	16.5	272.25	**0.718**
Sum					2.000

Check: $\sum h_{ii} = 2 = k$. ✓

3. Observation 12 in Detail

$$h_{12,12} = \frac{1}{12} + \frac{(28 - 11.5)^2}{429} = 0.718.$$

4. Decision
Average leverage: $\bar{h} = k/n = 2/12 = 0.167$. Flag threshold: $2k/n = 0.333$.

Obs	h_{ii}	Status
1 (x=4)	0.214	Below threshold
12 (x=28)	**0.718**	**High leverage**

5. Leverage vs Influence
For obs 12: $\hat{e}_{12} = 17.7 - 17.415 = 0.285$ (small). The internally studentised residual is

$$r_{12} = \frac{\hat{e}_{12}}{s\sqrt{1 - h_{12,12}}} = \frac{0.285}{0.738 \times \sqrt{0.282}} = 0.73.$$

High leverage but small standardized residual: obs 12 lies close to the fitted line and exerts little undue pull on the slope.

Practitioner Insight: Leverage measures how far an observation's regressor values lie from the centroid of the regressor space; it depends only on X, not on y. In simple regression, h_{ii} increases with $(x_i - \bar{x})^2$, so extreme x values have the highest leverage; observation 12 ($x = 28$) lies far from $\bar{x} = 11.5$, producing the largest h_{ii}. High leverage alone does not imply influence: points close to the fitted surface have little impact despite large h_{ii}, whereas moderate-leverage observations with large residuals may be influential. Here obs 12 has modest residual $r_{12} = 0.73$, while obs 8 ($h = 0.084$) has the largest studentised residual ($r_8 = -1.97$). Cook's distance combines leverage and residual size: $D_i = (\hat{e}_i^2/ks^2)\, h_{ii}/(1 - h_{ii})^2$; for obs 12, $D_{12} = 0.671 > 4/n = 0.333$, indicating meaningful influence despite the small residual. In multiple regression, leverage generalises to $h_{ii} = \mathbf{x}_i'(X'X)^{-1}\mathbf{x}_i$, with rules such as $2k/n$ or $3k/n$ commonly used to flag high leverage points.

Common Pitfalls

- **Leverage ≠ Influence.** High leverage only means unusual X-values; actual influence requires both high leverage *and* a large residual. Always pair h_i with Cook's D or DFFITS before drawing conclusions.
- **Using mechanical cutoffs (e.g., $h_i > 2k/n$).** These thresholds are rough heuristics and behave poorly for small n, large n, or many predictors. Compare h_i to the overall leverage distribution and the average k/n rather than using fixed rules.
- **Misinterpreting leverage as Y-outlier detection.** Leverage identifies outliers in *predictor space* only. Y-direction anomalies require studentized residuals; influence diagnostics must combine both.
- **Ignoring multivariate structure.** Typical values on each predictor can still yield high leverage if the *combination* of predictors is unusual. Severe collinearity also inflates h_i. Check VIF/condition number before interpreting leverage.
- **Deleting high-leverage points without investigation.** Extreme X-values may be valid and scientifically essential (rare subgroups, true extremes). Removing them without checking accuracy, context, or influence can bias the regression.

Cook's Distance Test

Measure of Influential Observations in Regression

REGR.15 | **Category:** Influence Diagnostics | **Data:** Continuous | **Null:** Rule of thumb

When to Use

- Identify influential observations in regression analysis.
- Assess combined impact of leverage and residual size.
- Detect observations that substantially alter fitted slopes.
- Diagnose model sensitivity to individual data points.
- Complement leverage, DFFITS, and studentized residuals.

Assumptions

- Linear regression model is estimated via OLS.
- Predictor matrix X is correctly specified.
- Observations are independent.
- No exact multicollinearity among regressors.
- Influence measures are interpreted diagnostically.

Test Statistic

$$D_i = \frac{(\hat{\mathbf{y}} - \hat{\mathbf{y}}_{(-i)})'(\hat{\mathbf{y}} - \hat{\mathbf{y}}_{(-i)})}{k \cdot MSE} = \frac{r_i^2}{k}\left(\frac{h_{ii}}{1-h_{ii}}\right). \qquad \mathbf{H} = \mathbf{X}(\mathbf{X}'\mathbf{X})^{-1}\mathbf{X}' \text{ (Hat matrix).}$$

Notation: $\hat{\mathbf{y}}_{(-i)}$ are fitted values with observation i deleted, r_i is standardized residual, h_{ii} is leverage (from Hat matrix)

Plain English: Cook's distance measures how much all fitted values change when a single observation is removed from the analysis. It combines information about how unusual an observation's predictor values are (leverage) with how unusual its response value is (residual). High value indicates an observation is both unusual in predictor space and poorly fit by the model, giving it substantial influence over coefficient estimates. It helps identify observations that need scrutiny or special handling.

⚠ Practitioner Warning

Cook's distance is a diagnostic, not a formal hypothesis test—no p-values or distributional reference. Threshold $D_i > 1$ is rough guideline; some use $D_i > 4/n$ or $D_i > 4/(n-k-1)$ for conservativeness. High value does not automatically mean observation should be removed—may represent legitimate extreme case or data quality issue requiring investigation. Influential observations are not necessarily errors; they may contain important information about relationship. Always examine influential points in context: check data entry, consider substantive meaning, assess impact on conclusions.

Hypotheses

- H_0: D_i near 0 (minimal influence on fitted model).
- H_1: D_i substantially greater than 0 (influences fitted model).

Decision Rule: Reject H_0 if $D_i > 1$ indicates substantial influence; $D_i > 4/n$ alternative threshold.

Software References [Python, R, SAS]

Python	R	SAS
`statsmodels::cooks_distance`	`cooks.distance`	`PROC REG (INFLUENCE)`

Additional Notes

Cook's distance combines leverage and residual magnitude through the formula **(A)** where leverage h_{ii} is the i-th diagonal element of hat matrix $\mathbf{H} = \mathbf{X}(\mathbf{X}'\mathbf{X})^{-1}\mathbf{X}'$. High leverage alone (unusual x values) or large residual alone (unusual y value) does not guarantee high influence; both are needed. DFBETAS measure influence on individual coefficients rather than overall fit: **(B)** with threshold $2/\sqrt{n}$. DFFITS measures influence on fitted value: **(C)** with threshold $2\sqrt{k/n}$. These measures complement Cook's distance for comprehensive influence diagnostics. Here, $e_i = \hat{y}_i - \hat{y}_{i(-i)}$ is raw residual, r_i is internally Studentized residual, and r_i^* is externally Studentized residual.

$$r_i = \frac{\hat{e}_i}{s\sqrt{1-h_{ii}}}, \quad r_i^* = \frac{\hat{e}_i}{s_{(-i)}\sqrt{1-h_{ii}}}, \quad \textbf{(A)}\ D_i = \frac{r_i^2}{k}\cdot\frac{h_{ii}}{1-h_{ii}}, \quad \textbf{(B)}\ DFBETAS_{j,i} = \frac{\hat{\beta}_j - \hat{\beta}_{j(-i)}}{SE(\hat{\beta}_j)_{(-i)}}, \quad \textbf{(C)}\ DFFITS_i = r_i^*\sqrt{\frac{h_{ii}}{1-h_{ii}}}.$$

Miscellanea: R. Dennis Cook (1977) introduced a diagnostic for identifying influential observations by combining leverage and residual information into a single measure, now known as Cook's distance. Widely used in regression analysis, the measure highlights observations that substantially affect fitted values. Common thresholds such as $D_i > 1$ are heuristic and should be interpreted with caution.

Reference(s): Cook (1977, 1979); Belsley et al. (1980)

Worked Example

Problem Statement The same twelve-household income–spending dataset from the Leverage example is used here. OLS gives $\hat{\beta}_0 = 1.268$, $\hat{\beta}_1 = 0.577$, $s^2 = 5.447/10 = 0.545$. Compute Cook's distance for each observation, identify influential points, and contrast with the leverage results.

1. Cook's Distance for All Observations

With $k = 2$ and $k\,s^2 = 2 \times 0.545 = 1.089$:

i	$\hat{e}_i$	h_{ii}	$\frac{\hat{e}_i^2}{ks^2}$	D_i
1	1.225	0.214	1.378	**0.479**
2	−0.528	0.154	0.256	0.055
3	−0.105	0.131	0.010	0.002
4	0.218	0.112	0.044	0.006
5	−0.658	0.098	0.398	0.048
6	−0.035	0.089	0.001	0.000
7	−0.012	0.084	0.000	0.000
8	−1.388	0.084	1.769	0.177
9	0.835	0.089	0.640	0.068
10	0.558	0.098	0.286	0.034
11	−0.395	0.131	0.143	0.025
12	0.285	0.718	0.074	**0.671**

2. Detailed Calculation for Flagged Observations

Obs 1 (x=4, residual-driven):

$$D_1 = \frac{(1.225)^2}{1.089} \cdot \frac{0.214}{(0.786)^2} = 0.479.$$

Obs 12 (x=28, leverage-driven):

$$D_{12} = \frac{(0.285)^2}{1.089} \cdot \frac{0.718}{(0.282)^2} = 0.671.$$

3. Decision

Threshold: $4/n = 4/12 = 0.333$.

Obs	D_i	h_{ii}	Driver
1 (x=4)	**0.479**	0.214	Large residual
12 (x=28)	**0.671**	0.718	High leverage
8 (x=12)	0.177	0.084	(not flagged)

Obs 1 and obs 12 exceed $4/n$. Neither exceeds 1.

4. Leave-One-Out Verification

Drop obs 12 and refit: $\hat{\beta}_0 = 1.630$, $\hat{\beta}_1 = 0.538$. Change: $\Delta\hat{\beta}_0 = -0.362$, $\Delta\hat{\beta}_1 = +0.039$. These are economically meaningful shifts confirming influence.

Practitioner Insight: Cook's distance combines the two components of influence—leverage and residual size—into a single diagnostic. The formula $D_i = (\hat{e}_i^2/ks^2)\, h_{ii}/(1 - h_{ii})^2$ shows this explicitly: the residual term measures lack of fit, while the leverage factor $h_{ii}/(1 - h_{ii})^2$ grows rapidly as $h_{ii} \to 1$. The observations here illustrate all cases: obs 1 ($h = 0.21$, $|\hat{e}| = 1.23$) has moderate leverage and a large residual, giving influential $D = 0.48$; obs 8 ($h = 0.08$, $|\hat{e}| = 1.39$) has the largest residual but low leverage, so $D = 0.18$ does not flag it; obs 12 ($h = 0.72$, $|\hat{e}| = 0.29$) has extreme leverage and a small residual, yet $D = 0.67$ because the leverage factor amplifies the residual ninefold. This illustrates why leverage and Cook's distance are complementary diagnostics. The common threshold $4/n$ is a rule of thumb (Bollen & Jackman 1985), while the stricter rule $D_i > 1$ flags severe influence; neither observation exceeds 1 here, indicating moderate but not extreme influence. Removing obs 12 changes the slope by $\Delta\hat{\beta}_1 = +0.039$ (6.7%), suggesting the high-income household slightly suppresses the estimated marginal propensity to consume. If influential points are data errors they should be corrected; if genuine but extreme, robust regression methods such as MM-estimation (MM: Multiple M-estimations) or least-trimmed squares reduce their impact.

Common Pitfalls

- **Deleting influential points automatically.** A large Cook's D signals influence, not error. Influential observations may be valid extremes or substantively important cases; investigate and report sensitivity analyses before removal.
- **Using Cook's D as an outlier test.** Cook's distance measures influence on fitted values, not outlyingness itself. Large residuals or high leverage alone do not necessarily imply large influence.
- **Ignoring leverage and residuals separately.** The same Cook's D can arise from very different leverage–residual combinations. Examine studentized residuals, leverage, and Cook's D together to understand why a point is influential.
- **Missing masked influence.** Multiple influential points can mask one another, appearing modest individually while jointly driving the fit. Sequential deletion, cluster inspection, or robust regression can help detect masking.
- **Using only the cutoff $D_i > 1$.** The classical threshold is often too conservative in large samples. Scaled rules such as $D_i > 4/n$ or $4/(n - k - 1)$ are usually more informative.

DFFITS Test

Measure of Influence on Fitted Values

REGR.16 | **Category:** Influence Diagnostics | **Data:** Continuous | **Null:** Rule of thumb

When to Use

- Identify observations strongly affecting fitted values.
- Detect influential points using leverage and residual size.
- Assess prediction sensitivity to individual observations.
- Diagnose observations distorting local model fit.
- Complement Cook's distance and leverage diagnostics.

Assumptions

- Linear regression model is estimated via OLS.
- Predictor matrix X is correctly specified.
- Observations are independent.
- No exact multicollinearity among regressors.
- Interpretation focuses on diagnostic influence.

Test Statistic

$$\text{DFFITS}_i = \frac{\hat{y}_i - \hat{y}_{i(-i)}}{SE(\hat{y}_i)} = r_i^* \sqrt{\frac{h_{ii}}{1-h_{ii}}}, \qquad r_i^* = \frac{\hat{e}_i}{s_{(-i)}\sqrt{1-h_{ii}}}, \qquad s_{(-i)}^2 = \frac{(n-k)s^2 - \hat{e}_i^2/(1-h_{ii})}{n-k-1} \text{ (error variance).}$$

Notation: $(-i)$ means without obs. i; $\hat{y}_i, \hat{y}_{i(-i)}$ are fitted values, h_{ii} is leverage, k is number of x's, n is sample size

Plain English: DFFITS measures how many standard errors the fitted value changes when an observation is deleted from the regression. It standardizes the change in fitted value by the standard error of prediction with that observation removed, making the measure scale-free. Unlike Cook's distance which measures overall influence on all fitted values, DFFITS focuses on influence on a single observation's own prediction. Large absolute DFFITS indicates the observation substantially affects its own fitted value.

⚠ Practitioner Warning

DFFITS is a diagnostic, not a formal hypothesis test—no distributional theory or p-values. Threshold $2\sqrt{k/n}$ is guideline for smaller datasets; alternative threshold $2\sqrt{k/(n-k)}$ used for larger samples. Sign of DFFITS matters: positive indicates observation pulls fitted value up, negative pulls it down. High DFFITS does not mandate removal—investigate whether observation is data error, legitimate extreme case, or theoretically important outlier. DFFITS more sensitive to individual observations than Cook's distance; both should be examined together for comprehensive influence.

Hypotheses

- H_0: $DFFITS_i$ near 0 (minimal influence).
- H_1: $|DFFITS_i|$ substantially > 0 (strong influence).

Decision Rule: Reject H_0 if $|DFFITS_i| > 2\sqrt{k/n}$. For large samples, use $|DFFITS_i| > 2\sqrt{k/(n-k)}$.

Software References [Python, R, SAS]

Python	R	SAS
`statsmodels::dffits`	`stats::dffits`	`PROC REG (INFLUENCE)`

Additional Notes

The relationship between DFFITS and Cook's distance follows from their formulations: both combine leverage and residual information but weight them differently. DFFITS equals the studentized residual multiplied by $\sqrt{h_{ii}/(1-h_{ii})}$, emphasizing leverage's role in prediction. For comparison, Cook's distance uses $(r_i^2/k) \times (h_{ii}/(1-h_{ii}))$, spreading influence across all k parameters. DFBETAS provides similar influence measure but for individual regression coefficients: $DFBETAS_{j,i} = (\hat{\beta}_j - \hat{\beta}_{j(-i)})/SE(\hat{\beta}_j)_{(-i)}$ with threshold $|DFBETAS_{j,i}| > 2/\sqrt{n}$. Studentized residuals $r_i^* = \hat{e}_i/(s_{(-i)}\sqrt{1-h_{ii}})$ account for both leverage and variance changes when observation deleted. Together, Cook's distance, DFFITS, and DFBETAS provide comprehensive influence diagnostics covering overall fit, predictions, and individual coefficients.

Miscellanea: DFFITS was introduced by Welsch and Kuh in their 1977 NBER technical report on linear regression diagnostics, which also presented related influence measures such as DFBETAS and COVRATIO. The name DFFITS reflects its definition as the "difference in fits" for an observation when it is included versus omitted from the model. The measure was later consolidated and widely disseminated through Belsley, Kuh, and Welsch's 1980 book *Regression Diagnostics*, which established DFFITS as a standard influence diagnostic.

Reference(s): Welsch & Kuh (1977); Belsley et al. (1980)

Worked Example

Problem Statement Twelve-household income–spending data from the Leverage example is used. OLS gives $\hat{\beta}_0 = 1.268$, $\hat{\beta}_1 = 0.577$, $s^2 = 0.545$. Compute DFFITS for each observation and identify influential points.

1. DFFITS for All Observations

With $n = 12$, $k = 2$, threshold $2\sqrt{k/n} = 2\sqrt{1/6} = 0.816$:

i	$\hat{e}_i$	h_{ii}	$s_{(-i)}$	r_i^*	DFFITS
1	1.225	0.214	0.627	2.205	**1.152**
2	−0.528	0.154	0.754	−0.762	−0.325
3	−0.105	0.131	0.777	−0.145	−0.056
4	0.218	0.112	0.774	0.299	0.106
5	−0.658	0.098	0.743	−0.933	−0.307
6	−0.035	0.089	0.778	−0.047	−0.015
7	−0.012	0.084	0.778	−0.016	−0.005
8	−1.388	0.084	0.609	−2.380	−0.720
9	0.835	0.089	0.721	1.213	0.378
10	0.558	0.098	0.753	0.781	0.257
11	−0.395	0.131	0.765	−0.554	−0.215
12	0.285	0.718	0.757	0.708	**1.129**

3. Decision

Obs	DFFITS	$\lvert\cdot\rvert$ vs 0.816	Driver
1 (x=4)	1.152	**Exceeds**	Large r_1^*
12 (x=28)	1.129	**Exceeds**	High h_{12}
8 (x=12)	−0.720	Below	(not flagged)

Leave-one-out: dropping obs 1 shifts $\hat{y}_1$ from 3.575 to 3.240, a change of 0.334.

2. Detailed Calculation for Flagged Observations

Obs 1 (x=4, residual-driven):

$$s^2_{(-1)} = \frac{10 \times 0.545 - 1.501/0.786}{9} = 0.393.$$

$$r_1^* = \frac{1.225}{0.627 \times 0.886} = 2.205.$$

$$\text{DFFITS}_1 = 2.205 \times \sqrt{\frac{0.214}{0.786}} = 1.152.$$

Obs 12 (x=28, leverage-driven):

$$s^2_{(-12)} = \frac{10 \times 0.545 - 0.081/0.282}{9} = 0.573.$$

$$r_{12}^* = \frac{0.285}{0.757 \times 0.531} = 0.708.$$

$$\text{DFFITS}_{12} = 0.708 \times \sqrt{\frac{0.718}{0.282}} = 1.129.$$

Practitioner Insight: DFFITS measures how much the *fitted value at observation i* changes when that observation is deleted, scaled by the leave-one-out standard error and by $\sqrt{h_{ii}}$ to account for leverage. Unlike Cook's distance, which measures the impact on *all* fitted values, DFFITS focuses on the local change in $\hat{y}_i$. On these data both diagnostics flag obs 1 and obs 12, while leaving obs 8 unflagged despite its large residual. For obs 12, the large leverage ($h_{12} = 0.718$) amplifies the otherwise modest external residual ($r_{12}^* = 0.708$) to $\text{DFFITS}_{12} = 1.129 > 0.816$.

The threshold $2\sqrt{k/n}$ scales with model complexity: as the number of regressors increases, observations have more opportunity to influence the fit and the cutoff becomes easier to exceed. When DFFITS and Cook's D disagree, both should be examined because an observation may influence its own fitted value differently from the overall regression surface.

Common Pitfalls

- **Using DFFITS alone.** DFFITS measures influence on an observation's *own* fitted value only; it does not capture influence on all fitted values (Cook's D) or on individual coefficients (DFBETAS). Use all three diagnostics together.
- **Equating large DFFITS with poor fit.** A large value reflects prediction *instability*, not necessarily poor prediction. High-leverage points can have small residuals yet large DFFITS as deleting them substantially changes their fitted value.
- **Applying the $2\sqrt{k/n}$ cutoff rigidly.** The rule can be too lenient in small samples and too strict in large ones. For small n, values near 1 are often informative; for large n, visual or percentile-based thresholds may work better.
- **Ignoring the sign.** The sign shows the direction of influence: positive values pull fitted values upward, negative values downward. Ignoring sign loses useful diagnostic insight.
- **Confusing DFFITS with studentized residuals.** Studentized residuals detect response outliers, whereas DFFITS measures influence by combining residual size with leverage. A point may be an outlier without being influential, or influential without being an outlier.

DFBETAS

Influence on Individual Regression Coefficients from Deleted Observations

REGR.17 | **Category:** Influence Diagnostics | **Data:** Continuous | **Null:** Rule of thumb

When to Use

- Identify observations strongly affecting individual slopes.
- Assess coefficient stability under case deletion.
- Diagnose localized influence masked by global measures.
- Diagnose which β's are most affected by individual points.
- Complement Cook's distance and leverage diagnostics.

Assumptions

- Linear regression model is estimated via OLS.
- Predictor matrix X is correctly specified.
- Observations are independent.
- No exact multicollinearity among regressors.
- Interpretation focuses on diagnostic influence.

Test Statistic

$$\text{DFBETAS}_{j,i} = \frac{\hat{\beta}_j - \hat{\beta}_{j(-i)}}{s_{(-i)}\sqrt{c_{jj}}}, \qquad c_{jj} = [(X'X)^{-1}]_{jj}, \text{ which denotes the } j\text{-th diagonal element.}$$

Notation: $\hat{\beta}_j$ is slope with all data, $\hat{\beta}_{j(-i)}$ & $s^2_{(-i)}$ are slope & error variance without obs. i

Plain English: DFBETAS measures how much each slope changes when a single observation is removed from the analysis, standardized by the coefficient's standard error. For each observation and each coefficient, DFBETAS quantifies whether that observation has unusual leverage on that specific parameter. Large DFBETAS values indicate the observation is pulling the coefficient estimate toward or away from where it would otherwise be. Unlike DFFITS which gives a single influence measure per observation, DFBETAS provides a vector of values showing which coefficients are most affected by each observation.

⚠ Practitioner Warning

DFBETAS yields p values per observation (one per slope), making interpretation more complex than scalar diagnostics like Cook's D. Cutoffs are heuristic—context and sample size matter. Large values indicate influence, not necessarily error. Removing influential points can shift influence to others. Assumes correct model specification; misspecification may cause misleading flags. Computation can be intensive, though efficient updating formulas avoid refitting.

Hypotheses

- H_0: No formal hypothesis test—diagnostic measure only.
- H_1: $|\text{DFBETAS}_{i,j}| > 2/\sqrt{n}$ warrant investigation.

Decision Rule: Reject H_0 if $|\text{DFBETAS}_{i,j}| > 2/\sqrt{n}$.

Software References [Python, R, SAS]

`statsmodels::dfbetas` `stats::dfbetas` `PROC REG (INFLUENCE DFBETAS)`

Additional Notes

DFBETAS measure the influence of individual observations on specific regression coefficients, extending influence diagnostics to parameter-level sensitivity. They quantify the standardized change in $\hat{\beta}_j$ when observation i is removed, allowing comparisons across coefficients. Efficient updating formulas avoid refitting, using leverage h_{ii} and residual information. The common cutoff $2/\sqrt{n}$ is a heuristic based on large-sample behavior; for small samples, a threshold of 1 is sometimes used. Unlike Cook's distance, DFBETAS reveal which coefficients are affected an observation may strongly influence one parameter but not others. Influence depends on both leverage and residual size: high-leverage points can affect coefficients even with small residuals, while low-leverage outliers may have limited impact. This granularity makes DFBETAS especially useful for diagnosing parameter-specific sensitivity.

Miscellanea: Belsley, Kuh, and Welsch (1980) introduced DFBETAS to assess the influence of individual observations on specific regression coefficients. By standardizing the change in each coefficient when an observation is removed, DFBETAS complement global measures such as Cook's distance.

Reference(s): Belsley, Kuh & Welsch (1980); Cook (1977); Chatterjee & Hadi (1988)

Worked Example

Problem Statement Twelve-household income–spending data from the Leverage example is used. OLS gives $\hat{\beta}_0 = 1.268$, $\hat{\beta}_1 = 0.577$. Compute DFBETAS for β_0 and β_1 to identify which observations significantly shift them when deleted.

1. Preliminary Quantities

$$X'X = \begin{pmatrix} 12 & 138 \\ 138 & 2016 \end{pmatrix}, \quad \sum x_i^2 = 2016, \; S_{xx} = 429, \; n = 12.$$

$$\sqrt{c_{00}} = \sqrt{\frac{\sum x_i^2}{n\,S_{xx}}} = 0.6258, \quad \sqrt{c_{11}} = \sqrt{\frac{1}{S_{xx}}} = 0.0483.$$

2. DFBETAS for All Observations

Threshold: $2/\sqrt{n} = 2/\sqrt{12} = 0.577$.

i	x_i	$s_{(-i)}$	$\text{DFBETAS}_{0,i}$	$\text{DFBETAS}_{1,i}$
1	4	0.627	**1.131**	**−0.901**
2	6	0.754	−0.305	0.220
3	7	0.777	−0.051	0.034
4	8	0.774	0.090	−0.054
5	9	0.743	−0.236	0.119
6	10	0.778	−0.010	0.004
7	11	0.778	−0.003	0.000
8	12	0.609	−0.278	−0.060
9	13	0.721	0.088	0.092
10	14	0.753	0.021	0.099
11	16	0.765	0.035	−0.129
12	28	0.757	**−0.764**	**1.062**

3. Detailed Calculation for Flagged Observations

Obs 1 (x=4): drop and refit.

$\hat{\beta}_{0(-1)} = 0.825$, $\hat{\beta}_{1(-1)} = 0.604$; $s_{(-1)} = 0.627$.

$$\text{DFBETAS}_{0,1} = \frac{1.268 - 0.825}{0.627 \times 0.6258} = 1.131.$$

$$\text{DFBETAS}_{1,1} = \frac{0.577 - 0.604}{0.627 \times 0.0483} = -0.901.$$

Obs 12 (x=28): drop and refit.

$\hat{\beta}_{0(-12)} = 1.630$, $\hat{\beta}_{1(-12)} = 0.538$; $s_{(-12)} = 0.757$.

$$\text{DFBETAS}_{0,12} = \frac{1.268 - 1.630}{0.757 \times 0.6258} = -0.764.$$

$$\text{DFBETAS}_{1,12} = \frac{0.577 - 0.538}{0.757 \times 0.0483} = 1.062.$$

4. Decision

Obs	DFBETAS_0	DFBETAS_1	Interpretation
1	1.13**	−0.90**	$\uparrow \hat{\beta}_0, \downarrow \hat{\beta}_1$
12	−0.76**	1.06**	$\downarrow \hat{\beta}_0, \uparrow \hat{\beta}_1$

**Exceeds threshold $|\cdot| > 2/\sqrt{12} = 0.577$.

Practitioner Insight: DFBETAS extends influence analysis from scalar diagnostics (Cook's D, DFFITS) to the coefficient level, producing a separate standardized measure for each coefficient $\hat{\beta}_j$. Here obs 1 and obs 12 both appear influential on earlier diagnostics, but DFBETAS reveals that they affect the regression line in opposite directions. Obs 1 ($x = 4$, large positive residual) increases the intercept ($\text{DFBETAS}_0 = +1.13$) and flattens the slope ($\text{DFBETAS}_1 = -0.90$), whereas obs 12 ($x = 28$, extreme leverage) anchors the right end of the line, so removing it lowers the slope and raises the intercept. Their opposing effects partially cancel in the full-sample fit. Unlike Cook's D or DFFITS, which produce one value per observation, DFBETAS yields one value for each (observation, coefficient) pair, making it especially useful in multiple regression for identifying which coefficients are most distorted. The common threshold $2/\sqrt{n}$ parallels the DFFITS rule $2\sqrt{k/n}$; DFFITS summarizes overall influence, whereas DFBETAS evaluates influence coefficient-by-coefficient.

Common Pitfalls

- **Using a fixed cutoff such as |DFBETAS| > 1.** Appropriate thresholds depend on sample size. A common guideline is $|\text{DFBETAS}_{i,j}| > 2/\sqrt{n}$, which adjusts for the natural variability of coefficient estimates.
- **Checking only one coefficient.** DFBETAS are computed for every β. An observation may strongly affect one predictor yet have little impact on others, so all coefficients—including interactions or polynomial terms—should be examined.
- **Deleting observations automatically.** Large DFBETAS values indicate influential observations, not necessarily bad data. Verify the observation and perform sensitivity analysis before considering removal.
- **Ignoring leverage and residuals.** Large DFBETAS values usually arise from high leverage, large residuals, or both. Interpret them alongside leverage measures and studentized residuals.

Breusch-Pagan Test: Two Variants

Setup: the shared auxiliary regression

Both tests share identical mechanics. Fit the main regression $y = X\beta + \varepsilon$, obtain OLS residuals $\hat{\varepsilon}$, and run the **auxiliary regression**

$$\hat{\varepsilon}_i^2 \;=\; \gamma_0 + z_i'\gamma + v_i,$$

where z_i are p regressors suspected of driving heteroskedasticity. The difference lies entirely in how the explained sum of squares (ESS) from this regression is scaled before comparing to a χ^2 distribution.

Breusch–Pagan (1979): the LM statistic

Breusch and Pagan (1979) derived their statistic as the exact **Lagrange multiplier (LM) test** from the Gaussian log-likelihood. Parameterise heteroskedasticity as $\sigma_i^2 = \sigma^2 h(z_i'\gamma)$ with $h(\cdot)$ smooth and $h(0) = 1$. The score of the log-likelihood with respect to γ, evaluated at $H_0\colon \gamma = 0$, is proportional to $\sum_i (e_i^2/\hat{\sigma}^2 - 1)\, z_i$. Forming the standard LM quadratic yields

$$\mathrm{LM}_{\mathrm{BP}} = \frac{\mathrm{ESS}_{\mathrm{aux}}/2}{\hat{\sigma}^4}, \qquad \hat{\sigma}^2 = \tfrac{1}{n}\sum \hat{\varepsilon}_i^2.$$

Under H_0 with **normality**, $\mathrm{LM}_{\mathrm{BP}} \sim \chi^2(p)$ asymptotically.

Why does normality matter? The denominator $\hat{\sigma}^4$ estimates $\frac{1}{2}\mathrm{Var}(\varepsilon_i^2)$ consistently *only* when errors are normal, because normality implies $\mathrm{Var}(\varepsilon_i^2) = 2\sigma^4$. For a general distribution with fourth central moment μ_4, $\mathrm{Var}(\varepsilon_i^2) = \mu_4 - \sigma^4$, which equals $2\sigma^4$ only when the excess kurtosis is zero. If $\mu_4 \neq 3\sigma^4$, the scaling $\hat{\sigma}^4$ is wrong, the null distribution of $\mathrm{LM}_{\mathrm{BP}}$ is no longer $\chi^2(p)$, and size distortions follow.

Koenker (1981): studentizing the test

Koenker (1981) observed that the BP statistic can be **studentized**—its denominator replaced by a direct empirical estimate of $\mathrm{Var}(\hat{\varepsilon}_i^2)$—to remove the normality requirement. The resulting statistic is simply

$$\mathrm{LM}_K = n\,R^2_{\mathrm{aux}},$$

where R^2_{aux} is from the same auxiliary regression. To see why this works, write

$$n\,R^2_{\mathrm{aux}} = \frac{\mathrm{ESS}_{\mathrm{aux}}}{\frac{1}{n}\sum_i(\hat{\varepsilon}_i^2 - \hat{\sigma}^2)^2}.$$

The denominator is the sample variance of the squared residuals — a direct, assumption-free estimate of $\mathrm{Var}(\varepsilon_i^2)$. Under H_0 with only finite fourth moments (no normality assumed), $\mathrm{LM}_K \xrightarrow{d} \chi^2(p)$ by standard central-limit arguments applied to the auxiliary score.

How the two statistics relate

Under normality, $\frac{1}{n}\sum(\hat{\varepsilon}_i^2 - \hat{\sigma}^2)^2 \approx 2\hat{\sigma}^4$, so the denominators coincide and

$$\mathrm{LM}_K \;\approx\; \mathrm{LM}_{\mathrm{BP}} \quad \text{(under normality).}$$

Away from normality the two diverge. $\mathrm{LM}_{\mathrm{BP}}$ over-rejects when the error distribution is leptokurtic ($\mu_4 > 3\sigma^4$) because $\hat{\sigma}^4$ underestimates the true variance of $\hat{\varepsilon}_i^2$, inflating the statistic. LM_K self-corrects by construction: the R^2 form absorbs whatever fourth-moment structure the residuals exhibit. Both statistics are $\chi^2(p)$ under H_0 with p the number of regressors in z_i.

Connection to the LM principle

The LM (score) principle states: restrict to the null, compute the gradient of the log-likelihood, and measure how far the gradient is from zero. A large gradient signals the null is locally inadequate.

$\mathrm{LM}_{\mathrm{BP}}$ is a *literal* LM statistic: it is the score inner product $s(\theta_0)'\,\mathcal{I}(\theta_0)^{-1}s(\theta_0)$ evaluated at the restricted (homoskedastic) MLE, with the Fisher information $\mathcal{I}$ computed under normality. Normality is baked into $\mathcal{I}$ through the $2\sigma^4$ term.

LM_K is a *robust* or *distribution-free* score statistic. Instead of $\mathcal{I}$ from the normal model, it uses the outer-product-of-scores (OPG) estimator of the information, which consistently estimates $\mathcal{I}$ under any distribution with the correct second moments. This is the same substitution that yields heteroskedasticity-robust standard errors in regression: replace the model-based sandwich filling with its empirical analogue. The asymptotic $\chi^2(p)$ result follows from the Slutsky theorem and a CLT, with no distributional shape assumption beyond finite fourth moments.

References: Breusch, T.S. & Pagan, A.R. (1979). A simple test for heteroscedasticity and random coefficient variation. *Econometrica*, **47**(5), 1287–1294. Koenker, R. (1981). A note on studentizing a test for heteroscedasticity. *Journal of Econometrics*, **17**(1), 107–112.

Part X

Inference for Non-Normal Response Models

Testing effects and diagnostics across logistic, Poisson, and related models

X. GLMS — A Conceptual Primer

What Is a GLM?

A **generalized linear model** (GLM) extends classical linear regression to response variables whose conditional distribution belongs to the **exponential family**. Three structural components define every GLM:

1. **Random component:** $Y_i \sim$ exponential family with conditional mean $\mu_i = E[Y_i|\mathbf{x}_i]$ and variance $\text{Var}(Y_i) = \phi V(\mu_i)$, where $V(\cdot)$ is the **variance function** and ϕ is the **dispersion parameter**.
2. **Systematic component:** $\eta_i = \mathbf{x}_i^\top \boldsymbol{\beta}$ (the linear predictor).
3. **Link function:** $g(\mu_i) = \eta_i$; a smooth monotone function connecting the mean to the linear predictor.

The **canonical link** sets $g(\mu_i) = \theta_i$ (the natural parameter), producing the simplest score equations and convenient sufficient statistics. Non-canonical links are valid and often more interpretable.

Common GLM Families

Family	Response	Canon. link	$V(\mu)$
Gaussian	Continuous	Identity	1
Binomial	Binary/prop	Logit	$\mu(1-\mu)$
Poisson	Count	Log	μ
Neg. Bin.	Overdispersed count	Log	$\mu + \mu^2/k$
Gamma	Positive	Inverse	μ^2
Inv. Gauss	Positive	$1/\mu^2$	μ^3

For Poisson and binomial, $\phi = 1$ is fixed by theory. When observed variance exceeds $\phi V(\mu)$, the model is **overdispersed**—a critical diagnostic concern.

Estimation: IRLS

GLM parameters are estimated by **maximum likelihood** via **iteratively reweighted least squares** (IRLS). At each iteration, a weighted least-squares problem is solved for working response $z_i = \hat{\eta}_i + (y_i - \hat{\mu}_i)g'(\hat{\mu}_i)$ with weights $w_i = [V(\hat{\mu}_i)\{g'(\hat{\mu}_i)\}^2]^{-1}$. The MLE $\hat{\boldsymbol{\beta}}$ is consistent and asymptotically normal under regularity conditions.

Deviance and Residuals

Deviance:

$$D = 2\big[\ell(\mathbf{y};\mathbf{y}) - \ell(\hat{\boldsymbol{\mu}};\mathbf{y})\big] = 2\sum_i d_i,$$

where $\ell(\mathbf{y};\mathbf{y})$ is the **saturated model** log-likelihood (one parameter per observation) and d_i is the ith deviance contribution. For Gaussian models, $D = \text{RSS}/\sigma^2$. Under the fitted model, $D \xrightarrow{d} \chi^2_{n-p}$ asymptotically—but **only when the grouping is coarse** (few distinct covariate patterns); for continuous covariates and ungrouped binary data, the χ^2 approximation fails.

Pearson χ^2 statistic:

$$X^2 = \sum_i \frac{(y_i - \hat{\mu}_i)^2}{V(\hat{\mu}_i)},$$

an alternative fit measure; $X^2/\phi \approx \chi^2_{n-p}$ asymptotically. The ratio $\hat{\phi} = X^2/(n-p)$ is the most common **empirical dispersion estimate**.

Residual types:

- **Pearson:** $r_i^P = (y_i - \hat{\mu}_i)/\sqrt{V(\hat{\mu}_i)}$.
- **Deviance:** $r_i^D = \text{sign}(y_i - \hat{\mu}_i)\sqrt{d_i}$; preferred for diagnostic plots.
- **Quantile (randomized):** PIT-based; approximately $\mathcal{N}(0,1)$ regardless of family; most powerful for detecting distributional misspecification.

I. Goodness-of-Fit Tests

Deviance GOF test: Tests H_0 that the fitted model is adequate against the saturated model. For grouped Poisson or grouped binomial data, $D \overset{a}{\sim} \chi^2_{n-p}$ under H_0. For logistic regression with individual binary observations, this chi-square approximation is generally unreliable and the deviance should not be interpreted as a formal goodness-of-fit test.

Pearson χ^2 GOF test: $X^2 \sim \chi^2_{n-p}$ under the same conditions as deviance. Both D and X^2 are **invalid GOF statistics for ungrouped binary logistic regression**—one of the most common errors in applied work.

Hosmer–Lemeshow test (logistic regression): Groups observations into g (typically 10) deciles of predicted probability; compares observed and expected event counts within each group via χ^2_{g-2}. Specifically designed for ungrouped logistic data where D and X^2 fail. **Limitations:** sensitive to the number of groups g; low power against smooth alternatives; does not test individual covariate effects.

Le Cessie–van Houwelingen test: Omnibus global test for logistic regression based on a smoothed sum of residuals; does not require grouping. More powerful than Hosmer–Lemeshow against smooth departures; detects missing nonlinear terms and interactions.

II. Tests of Parameter Significance

Three tests of $H_0 : \boldsymbol{\beta}_2 = \mathbf{0}$ (with $\boldsymbol{\beta}_1$ unrestricted) are available, all asymptotically equivalent under H_0 but differing in finite-sample performance:

Likelihood Ratio Test (LRT):

$$\Lambda = -2[\ell(M_0) - \ell(M_1)] \xrightarrow{d} \chi^2_q,$$

where q is the number of restrictions. Requires fitting *both* the null and full models. **Most reliable in small samples**; invariant to parametrization; preferred whenever computationally feasible.

Wald Test:

$$W = \hat{\beta}_2^\top [\widehat{\text{Var}}(\hat{\beta}_2)]^{-1} \hat{\beta}_2 \xrightarrow{d} \chi^2_q.$$

Requires only the full model; fast but **unreliable near parameter space boundaries** (e.g. large $|\hat{\beta}|$ in logistic regression causes Hauck–Donner effect: Wald p-values increase spuriously as the true effect grows large). Always confirm borderline Wald results with LRT.

Score (Rao) Test: Evaluates the gradient of ℓ at the **null model** $\hat{\beta}_0$; requires only the restricted model. Computationally attractive for large p or penalized models; asymptotically equivalent to LRT but can differ in small samples. Basis of Lagrange Multiplier tests.

Practical hierarchy: LRT $\succ$ Score $\succ$ Wald in reliability; use Wald for routine output, LRT for confirmation of key hypotheses.

Coefficient Interpretation by Family

On the **link scale**, $\hat{\beta}_j$ is the change in $g(\mu_i)$ per unit increase in x_j, holding other predictors fixed. Back-transforming via g^{-1} gives the effect on the **response scale**:

Family	Link	Interpretation of $e^{\hat{\beta}_j}$
Binomial	Logit	**Odds ratio**: multiplicative change in odds of $Y = 1$ per unit x_j.
Binomial	Probit	Change in $\Phi^{-1}(P)$; no simple multiplicative interpretation on probability scale.
Poisson	Log	**Incidence rate ratio** (IRR): multiplicative change in event rate per unit x_j.
Gamma	Log	Multiplicative change in mean; analogous to IRR.
Gaussian	Identity	Additive change in mean; same as OLS.

For a unit change in x_j: $e^{\hat{\beta}_j}$ multiplies the odds (logit), rate (log-Poisson), or mean (log-Gamma). A 1-unit change gives a $(e^{\hat{\beta}_j} - 1) \times 100\%$ change in the multiplicative quantity. For standardized predictors, multiply $\hat{\beta}_j$ by the SD of x_j before exponentiating to get a per-SD effect.

CIs on the response scale: exponentiate the CI on the link scale, not the SE. $\exp(\hat{\beta}_j \pm 1.96\,\widehat{\text{se}}(\hat{\beta}_j))$ gives a CI for the OR or IRR directly.

Marginal Effects and AMEs

In nonlinear GLMs, $\hat{\beta}_j$ is **not** the marginal effect of x_j on $E[Y|\mathbf{x}]$—it is the effect on the link scale $g(\mu)$. The marginal effect on the response scale is:

$$\frac{\partial \mu_i}{\partial x_{ij}} = \frac{\hat{\beta}_j}{g'(\hat{\mu}_i)} = \hat{\beta}_j \cdot [g^{-1}]'(\hat{\eta}_i),$$

which depends on the covariate values of observation i.

Average Marginal Effect (AME):

$$\widehat{\text{AME}}_j = \frac{1}{n}\sum_{i=1}^{n} \hat{\beta}_j \cdot [g^{-1}]'(\hat{\eta}_i).$$

Averages the marginal effect over the observed covariate distribution; the most defensible summary for heterogeneous samples.

Marginal Effect at the Mean (MEM): Evaluates $[g^{-1}]'$ at $\bar{\mathbf{x}}$:

$$\widehat{\text{MEM}}_j = \hat{\beta}_j \cdot [g^{-1}]'(\bar{\mathbf{x}}^\top \hat{\boldsymbol{\beta}}).$$

Computationally simpler but can be misleading when the average covariate vector is not representative (e.g. binary predictors).

For logistic regression (canonical link):

$$\frac{\partial P(Y = 1|\mathbf{x})}{\partial x_j} = \hat{\beta}_j \hat{\mu}_i (1 - \hat{\mu}_i),$$

so the marginal effect is largest near $\hat{\mu}_i = 0.5$ and shrinks toward zero as $\hat{\mu}_i \to 0$ or 1. **AME is preferred** over MEM and over reporting raw log-odds coefficients when communicating with non-technical audiences.

Pseudo-R^2 Measures

Unlike OLS, GLMs have no unique R^2. Several pseudo-R^2 measures exist; none is directly comparable to OLS R^2 or to each other across different families.

Measure	Formula and notes
McFadden's ρ^2	$1 - \ell(M)/\ell(M_0)$; ranges $[0, 1)$; $\rho^2 \in [0.2, 0.4]$ considered good fit for logistic; analogous to proportion of log-likelihood explained.
Adjusted McFadden	$1 - [\ell(M) - p]/\ell(M_0)$; penalises for the number of parameters p; preferred for model comparison.
Cox–Snell R^2	$1 - (L_0/L_M)^{2/n}$; never reaches 1 for discrete outcomes; upper bound < 1.
Nagelkerke R^2	Rescales Cox–Snell by its maximum; ranges $[0, 1]$; most widely reported but lacks clean interpretation.
Tjur's D	$\bar{\hat{\mu}}_1 - \bar{\hat{\mu}}_0$ (mean predicted probability in $Y = 1$ minus $Y = 0$ group); logistic only; $[0, 1]$; directly interpretable as discrimination.

Key caution: pseudo-R^2 measures are useful for comparing models fitted to the *same data and family*; they should never be compared across different families or datasets. For logistic regression, **the AUC (c-statistic)** and **Brier score** are better-calibrated measures of predictive performance.

Separation in Logistic Regression

Complete separation occurs when a linear combination of predictors perfectly predicts the outcome: $\exists\,\mathbf{c}$ such that $\mathbf{c}^\top \mathbf{x}_i > 0$ whenever $y_i = 1$ and < 0 whenever $y_i = 0$. The MLE does not exist: $\hat{\boldsymbol{\beta}} \to \infty$ and standard errors diverge.

Quasi-complete separation occurs when perfect separation holds for a subset of observations; MLE still diverges for the offending coefficient.

Symptoms:

- Implausibly large coefficients ($|\hat{\beta}_j| > 10$).
- Very large SEs or SEs much larger than $|\hat{\beta}_j|$.
- Convergence warnings from the fitting algorithm.
- Hauck–Donner effect: Wald p-values for the separated predictor paradoxically *increase* as the true effect grows.

Remedies:

- **Firth's penalized likelihood:** adds a Jeffreys prior penalty $-\frac{1}{2}\log|\mathcal{I}(\boldsymbol{\beta})|$ to $\ell(\boldsymbol{\beta})$; produces finite, consistent estimates; the standard remedy.
- **Exact logistic regression:** conditions on sufficient statistics; computationally intensive for large p.
- **Bayesian logistic regression:** weakly informative priors (e.g. Cauchy(0,2.5) on coefficients) regularize toward finite values.
- **Data-side:** merge sparse categories, collect more data, or drop the perfectly separating predictor if substantively justified.

III. Link Function Diagnostic

Pregibon link test: Tests whether the chosen link function $g(\cdot)$ is adequate by augmenting the linear predictor with $\hat{\eta}_i^2$ and testing its significance. Under the correct link, $\hat{\eta}_i^2$ should have zero coefficient. A significant result indicates link misspecification: the conditional mean is not a monotone transformation of a linear predictor on the current scale. Does not identify the correct alternative link; use AIC comparison of candidate link functions (logit, probit, cloglog, cauchit) to select.

IV. Overdispersion Tests

Overdispersion occurs when $\mathrm{Var}(Y_i) > \phi V(\mu_i)$—the observed variance exceeds what the distribution implies. For Poisson ($\phi = 1$, $V(\mu) = \mu$), this means $\mathrm{Var}(Y_i) > \mu_i$. Consequences: standard errors are **underestimated**; t- and χ^2-statistics are inflated; Type I error increases.

Cameron–Trivedi test: Tests $H_0 : \mathrm{Var}(Y_i) = \mu_i$ (Poisson) against $H_1 : \mathrm{Var}(Y_i) = \mu_i + \alpha h(\mu_i)$ for a specified function h. Common choices: $h(\mu_i) = \mu_i$ (NB1: variance linear in mean) and $h(\mu_i) = \mu_i^2$ (NB2: variance quadratic; standard negative binomial). The test regresses $(y_i - \hat{\mu}_i)^2 - y_i$ on $h(\hat{\mu}_i)$; significance implies overdispersion of the specified form.

Dean's test: Score test version of the Cameron–Trivedi approach; derived from the score equations of the negative binomial model evaluated at $\alpha = 0$; does not require fitting the NB model. Slightly more powerful than Cameron–Trivedi in some settings.

Remedies for overdispersion:

- **Quasi-likelihood:** estimate $\hat{\phi} = X^2/(n-p)$; multiply standard errors by $\sqrt{\hat{\phi}}$; F-tests replace χ^2.
- **Negative binomial regression:** models $\mathrm{Var}(Y_i) = \mu_i + \mu_i^2/k$; preferred when overdispersion is severe.
- **Robust (sandwich) SEs:** consistent regardless of distributional misspecification.

V. Excess Zeros and Model Comparison

Count data often contain more zeros than any single Poisson or NB distribution can accommodate—a phenomenon driven by a separate structural process generating zeros (e.g. non-participation).

Zero-inflated models: Mix a point mass at zero with a count distribution:

$$P(Y_i = 0) = \pi_i + (1 - \pi_i) f(0; \mu_i),$$

where π_i is the zero-inflation probability modeled via logistic regression. **ZIP** (zero-inflated Poisson) and **ZINB** (zero-inflated NB) are the standard variants. **Hurdle models** alternatively use separate models for the zero/non-zero boundary and the positive counts.

Zero-inflation test: Score test for $H_0 : \pi_i = 0$ (no excess zeros); based on the score function of the ZIP model evaluated under Poisson; significant result implies zero inflation.

Vuong test: Compares two **non-nested** models (e.g. Poisson vs. ZIP, or NB vs. ZINB) using the log-likelihood ratio per observation:

$$V = \frac{\sqrt{n}\,\bar{m}}{\hat{s}_m} \mathrel{\dot\sim} \mathcal{N}(0, 1),$$

where $m_i = \log[f_1(y_i)/f_2(y_i)]$ and $\bar{m}$, $\hat{s}_m$ are its sample mean and SD. $V > 1.96$ favours model 1; $V < -1.96$ favours model 2. **Limitation:** the Vuong test has incorrect size for comparing nested models and when one model nests the other as a boundary case; use LRT for nested comparisons.

VI. Tests for IIA (Multinomial Logit)

The multinomial logit model assumes **Independence of Irrelevant Alternatives** (IIA): the ratio of choice probabilities for any two alternatives is unaffected by the presence or attributes of other alternatives. Violations lead to biased coefficients.

Hausman–McFadden test: Compare estimates from the full model to those from a restricted model that drops one alternative:

$$H = (\hat{\beta}_R - \hat{\beta}_F)^\top (\hat{\mathbf{V}}_R - \hat{\mathbf{V}}_F)^{-1} (\hat{\beta}_R - \hat{\beta}_F) \mathrel{\dot\sim} \chi^2_k.$$

If IIA holds, both estimators are consistent and H should be small. Reject IIA if H is large. **Practical problem:** the matrix $\hat{\mathbf{V}}_R - \hat{\mathbf{V}}_F$ is frequently not positive definite in small samples, making H undefined; use Small–Hsiao as a more stable alternative.

Small–Hsiao test: Splits the sample randomly into two halves; estimates the restricted model on one half and compares to full model estimates on the other; avoids the matrix indefiniteness problem of the Hausman test. Asymptotically χ^2_k under IIA.

Alternatives to IIA: nested logit, mixed logit, probit (naturally IIA-free), or multinomial probit when IIA is rejected.

Diagnostic Workflow

1. **Fit the GLM;** check convergence and coefficient signs for plausibility.
2. **Goodness of fit:** deviance / Pearson χ^2 for grouped data; Hosmer–Lemeshow or Le Cessie–van Houwelingen for binary ungrouped data.
3. **Parameter significance:** LRT for key hypotheses; Wald for routine output.
4. **Link function:** Pregibon test; compare AIC across candidate links.
5. **Dispersion:** Cameron–Trivedi or Dean's test; if $\hat{\phi} > 1$, apply quasi-likelihood or NB.
6. **Zero inflation** (count models): score test; if significant, fit ZIP or ZINB and compare with Vuong.
7. **IIA** (multinomial logit): Hausman; use Small–Hsiao if matrix is indefinite.
8. **Residual plots:** deviance or quantile residuals vs. fitted values; partial residual plots for each predictor.

Common Pitfalls

- **Deviance GOF for binary logistic:** $D \not\sim \chi^2_{n-p}$ for ungrouped data; use Hosmer–Lemeshow or Le Cessie–van Houwelingen instead.
- **Wald test near boundaries:** Hauck–Donner effect causes Wald p-values to increase as effect size grows; always confirm with LRT.
- **Ignoring overdispersion:** deflated SEs inflate t-statistics; run Cameron–Trivedi or Dean's test routinely for count data.
- **Vuong for nested models:** incorrect; use LRT when one model nests the other.
- **Hosmer–Lemeshow sensitivity:** results change with g; always try $g = 10$ and $g = 8$; supplement with Le Cessie–van Houwelingen.
- **Ignoring zero inflation:** Poisson overdispersion may be partly or entirely driven by excess zeros; test separately before switching to NB.
- **IIA assumed without testing:** always test IIA in multinomial logit; violations are common when alternatives are similar.

Bottom Line

> GLMs extend regression to non-Gaussian outcomes, but correct inference requires validating each layer of the model separately: fit (deviance, Hosmer–Lemeshow), significance (LRT preferred over Wald), link function (Pregibon), dispersion (Cameron–Trivedi / Dean's), excess zeros (score test / Vuong), and IIA (Hausman / Small–Hsiao). A passing p-value at one layer does not guarantee adequacy at another.

Deviance Goodness-of-Fit Test

Test for Model Adequacy in Generalized Linear Models

GLMS.1 | **Category:** Generalized Linear Models | **Data:** Various | **Null:** Chi-square

When to Use

- Assess overall fit of a generalized linear model.
- Compare observed outcomes against model predictions.
- Detect lack of fit beyond random variation alone.
- Verify whether model assumptions appear reasonable.

Assumptions

- Observations are mutually independent.
- Response follows the specified GLM distribution.
- Mean structure and link function are correctly specified.
- Sample size supports chi-square approximations.

Test Statistic

$$D = 2\sum_{i=1}^{n}\left[y_i \log\left(\frac{y_i}{\hat{\mu}_i}\right) - (y_i - \hat{\mu}_i)\right], \quad \text{for Poisson; form varies by family.} \qquad \text{Under } H_0,\ D \approx \chi^2_{n-p}.$$

Notation: y_i are observed responses, $\hat{\mu}_i$ are fitted values, D is scaled deviance, $n - p$ are degrees of freedom

Plain English: The deviance measures the discrepancy between the fitted model and a saturated model (one parameter per observation, perfect fit). It is twice the difference in log-likelihoods between these models. Small deviance indicates good fit; large deviance suggests the model does not adequately describe the data. For normally distributed responses, deviance equals residual sum of squares. The deviance generalizes residual sums of squares to any exponential family distribution.

⚠ Practitioner Warning

Deviance goodness-of-fit test requires non-saturated models and adequate replication—problematic with continuous predictors or unique covariate patterns. For ungrouped Bernoulli data, the approximation is often poor and the test is usually not recommended; use Hosmer-Lemeshow instead. As a rule, lack-of-fit tests require replication at covariate patterns to separate random variation from systematic model error. Overdispersion inflates deviance and can cause spurious rejection. For Poisson and grouped binomial models, $\hat{\phi} = D/(n-p)$ should be near 1; large values suggest extra-Poisson/binomial variation or model misspecification. $|D|$ assesses overall fit; ΔD compare nested models.

Hypotheses

- H_0: The fitted GLM is adequate.
- H_1: The fitted GLM is not adequate.

Decision Rule: Reject H_0 if $D > \chi^2_{\alpha,n-p}$ or p-value $< \alpha$.
Nested models: reject if $\Delta D = D_0 - D_1 > \chi^2_{\alpha,p_1-p_0}$, $p_1 > p_0$.

Software References [Python, R, SAS]

```
statsmodels::GLM.deviance        summary(glm.object)$deviance        PROC GENMOD
```

Additional Notes

Deviance formulas vary by distributional family: (A) grouped binomial data with m_i trials, (B) gamma models, (C) Normal distribution yields the familiar residual sum of squares scaled by variance. Pearson chi-square statistic $X^2 = \sum(y_i - \hat{\mu}_i)^2/V(\hat{\mu}_i)$ provides alternative goodness-of-fit measure, also asymptotically χ^2_{n-p}. Deviance and Pearson statistics typically similar; large discrepancies suggest model problems. AIC uses deviance: AIC $= D + 2p$, penalizing model complexity. Deviance residuals $r_i^D = \text{sign}(y_i - \hat{\mu}_i)\sqrt{d_i}$ where d_i is contribution to total deviance, useful for residual diagnostics. General formula: $D = 2\{\ell(\text{saturated}) - \ell(\text{fitted})\}$, where ℓ is log-likelihood.

$$(A): 2\sum_{i=1}^{n}[y_i \log(y_i/\mu_i) + (m_i - y_i)\log((m_i - y_i)/(m_i - \hat{\mu}_i))], \quad (B): 2\sum_{i=1}^{n}[\log(\hat{\mu}_i/y_i) + (y_i - \hat{\mu}_i)/\hat{\mu}_i], \quad (C): \sum(y_i - \hat{\mu}_i)^2/\sigma^2$$

Miscellanea: The concept of deviance was formalized by John Nelder and Robert Wedderburn in their landmark 1972 paper introducing generalized linear models (GLMs). They developed a unified framework extending linear regression to non-normal responses, with deviance serving as a generalization of residual sums of squares via likelihood comparisons to a saturated model. Nelder was a strong critic of ad hoc data transformations, arguing that modeling should respect the natural distribution of the response. The GLM framework revolutionized applied statistics, with deviance becoming the standard measure of fit across exponential family models.

Reference(s): Nelder & Wedderburn (1972); McCullagh & Nelder (1989)

Worked Example

Problem Statement Disease-surveillance records for five regions show the number of new cases y_i and population n_i (thousands) during a reference period. A Poisson GLM with a log offset is fitted under H_0 that the disease rate is the same in every region: $\log \mu_i = \log n_i + \beta_0$ (intercept only, one parameter). The deviance tests whether this constant-rate model fits adequately against the saturated alternative. H_0: constant rate (λ) model adequate; H_1: rates differ across regions.

1. MLE and Fitted Values
The MLE of λ is $\hat{\lambda} = \sum y_i / \sum n_i = 40/100 = 0.40$ cases per 000, giving $\hat{\mu}_i = n_i \hat{\lambda}$.
$\sum \hat{\mu}_i = \sum y_i = 40$ ✓ (MLE constraint).

2. Deviance Formula
Because $\sum (y_i - \hat{\mu}_i) = 0$, the Poisson D simplifies:

$$D = 2\sum_i \left[y_i \log(y_i/\hat{\mu}_i) - (y_i - \hat{\mu}_i)\right] = 2\sum_i y_i \log(y_i/\hat{\mu}_i).$$

3. Term-by-Term Computation

i	n_i	y_i	$\hat{\mu}_i$	$\log(y_i/\hat{\mu}_i)$	$2y_i \log(y_i/\hat{\mu}_i)$
1	10	2	4	$-\ln 2$	$-4\ln 2$
2	20	16	8	$+\ln 2$	$+32\ln 2$
3	15	6	6	0	0
4	25	10	10	0	0
5	30	6	12	$-\ln 2$	$-12\ln 2$
Σ	100	40	40		$+16\ln 2$

The log ratios reduce to $\pm \ln 2$ or 0, so

$$D = (-4 + 32 - 12)\ln 2 = 16\ln 2 \approx 11.090.$$

4. Degrees of Freedom and Decision
$k = 5$, $p = 1$ parameter ($\hat{\lambda}$) estimated: $\text{df} = k - 1 = 4$.

Statistic	Value	CV (5%)	Decision
D	$16\ln 2 \approx 11.090$	9.488	Reject H_0

$p = 0.026$; CV at 1%= 13.277 (FTR at 1%).

Practitioner Insight: The deviance $D = 2(\ell_{\text{sat}} - \ell_{\text{fit}})$ measures how much log-likelihood is lost by imposing the model constraints. This parallels the chi-square GoF test in the GFIT chapter, but applies to fitted GLMs rather than raw data: the role of the fully specified null CDF is taken by the fitted model, and the role of bin counts is taken by y_i. Three features of this example are noteworthy. First, Region 2 drives the rejection: its 16 observed cases against an expected 8 contributes $32\ln 2 \approx 22.2$ to D, while Regions 3 and 4 contribute nothing (perfect fit). Second, the Pearson statistic $X^2 = \sum (y_i - \hat{\mu}_i)^2/\hat{\mu}_i = 1 + 8 + 0 + 0 + 3 = 12$ is close to $D = 16\ln 2 \approx 11.09$; for large counts the two statistics agree, but the deviance is the natural likelihood-ratio measure within the GLM framework. Third, the offset $\log n_i$ is *not* a covariate with an estimated coefficient: its coefficient is fixed at 1. Failing to include the offset would confound rate differences with population-size differences, inflating D spuriously.

Common Pitfalls

- **Using deviance goodness-of-fit for ungrouped binary logistic data.** For 0/1 responses with one trial per observation, the saturated model fits perfectly and the deviance test is not meaningful. Use calibration plots, Hosmer–Lemeshow, and residual diagnostics instead.
- **Treating significant deviance as major model failure.** With large n, even trivial departures produce significant p-values. Inspect the dispersion ratio D/df and practical effect size; values near 1–1.3 often reflect acceptable noise rather than substantive misspecification.
- **Expecting the deviance test to reveal the source of misfit.** The test is omnibus—rejects for wrong link, wrong variance function, omitted variables, interactions, or outliers but cannot distinguish among them. Always follow with deviance/Pearson residual plots, influence diagnostics, and alternative link/family checks.
- **Confusing absolute GOF deviance with deviance differences for LRTs.** Goodness-of-fit uses $df = n - p$ and evaluates absolute adequacy; nested model comparisons use ΔD with $df = \Delta p$. Mixing these leads to incorrect interpretations of p-values.
- **Ignoring overdispersion signaled by large D/df.** When D/df substantially exceeds 1, standard errors are too small and inference becomes anti-conservative. Address with quasi-likelihood scaling, negative binomial or beta-binomial families, or random-effects structures.

Pearson Chi-Square Goodness-of-Fit Test

GLM Goodness-of-Fit Test Using Squared Residuals

GLMS.2 | **Category:** Generalized Linear Models | **Data:** Various | **Null:** Chi-square

When to Use

- Assess overall fit of a generalized linear model.
- Compare observed outcomes against model predictions.
- Detect systematic departures from fitted values.
- Verify whether model assumptions appear reasonable.

Assumptions

- Observations are mutually independent.
- Response follows the specified GLM distribution.
- Mean structure and link function are correctly specified.
- Sample size supports chi-square approximations.

Test Statistic

$$X^2 = \sum_{i=1}^{n} \frac{(y_i - \hat{\mu}_i)^2}{V(\hat{\mu}_i)}, \quad \text{where } V(\hat{\mu}_i) \text{ is variance function.} \qquad \text{Under } H_0,\ X^2 \approx \chi^2_{n-p}.$$

Notation: y_i are observed responses, $\hat{\mu}_i$ are fitted means, $V(\cdot)$ is variance function, $n - p$ are degrees of freedom

Plain English: The Pearson chi-square statistic measures goodness-of-fit by summing squared standardized residuals, where standardization uses the variance function appropriate for the GLM family. For Poisson, variance equals mean (μ); for binomial, variance is $n\mu(1-\mu)$; for gamma, variance is proportional to squared mean. X^2 generalizes the classical Pearson χ^2 to any exponential family distribution. Like deviance, it tests whether observed data could plausibly arise from the fitted model.

⚠ Practitioner Warning

Test requires adequate expected frequencies—problematic with sparse data or continuous predictors creating unique covariate patterns. For ungrouped binary data (Bernoulli trials), test is generally not reliable (poor χ^2 approx.); use grouped data or alternative diagnostics like Hosmer-Lemeshow, or le Cessie–van Houwelingen. Overdispersion causes X^2 to exceed its expected value $(n-p)$; ratio $X^2/(n-p)$ estimates dispersion parameter (ϕ). If dispersion $\hat{\phi} > 1.5$ (rule of thumb), use quasi-likelihood adjustments to inflate standard errors. Pearson and deviance statistics usually give similar conclusions; substantial disagreement suggests model problems beyond simple lack of fit. A large X^2 relative to D often signals overdispersion, whereas a large D with modest X^2 suggests mean-structure misspecification.

Hypotheses

- H_0: The fitted GLM is adequate.
- H_1: The fitted GLM is not adequate.

Decision Rule: Reject H_0 if $X^2 > \chi^2_{\alpha,n-p}$ or p-value $< \alpha$. $\hat{\phi} = X^2/(n-p)$ should be near 1. $\hat{\phi} > 1.5 \Rightarrow$ overdispersion.

Software References [Python, R, SAS]

`statsmodels::pearson_chi2` `summary(glm.object)$pearson.chi2` `PROC GENMOD`

Additional Notes

Variance functions for common GLM families: Poisson $V(\mu) = \mu$, binomial $V(\mu) = \mu(1 - \mu/m)$ for m trials, gamma $V(\mu) = \mu^2$, inverse Gaussian $V(\mu) = \mu^3$, normal $V(\mu) = \sigma^2$ (constant). Pearson residuals are $r_i^P = (y_i - \hat{\mu}_i)/\sqrt{V(\hat{\mu}_i)}$, with the statistic being $X^2 = \sum(r_i^P)^2$. Pearson residuals are approximately mean-zero with variance near 1 under correct specification, though leverage and model structure can distort this. Standardized Pearson residuals $r_i^{SP} = r_i^P/\sqrt{1 - h_{ii}}$ account for leverage, where h_{ii} are hat matrix diagonal elements. For overdispersed data, quasi-likelihood estimates use $\hat{\phi} = X^2/(n-p)$ to inflate standard errors: $\widehat{SE}_{\text{quasi}}(\hat{\beta}) = \sqrt{\hat{\phi}} \cdot \widehat{SE}_{\text{ML}}(\hat{\beta})$. The Pearson statistic and deviance are asymptotically equivalent, both converging to χ^2_{n-p} under correct model specification.

Miscellanea: Karl Pearson (1900) introduced the chi-square goodness-of-fit test for categorical data, sparking his famous feud with Ronald Fisher over degrees of freedom. Within the GLM framework of Nelder and Wedderburn (1972), the statistic extends naturally as a sum of squared standardized residuals. McCullagh and Nelder (1983) established Pearson and deviance statistics as complementary measures of model fit. X^2 is a method-of-moments analog of the likelihood-based deviance, linking classical and modern inference.

Reference(s): Pearson (1900); McCullagh & Nelder (1989)

Worked Example

Problem Statement A toxicologist records the number of animals y_i (out of $n_i = 10$) showing a specific biomarker response at each of five dose levels. A logistic regression model is fitted: $\text{logit}(\pi_i) = \beta_0 + \beta_1 x_i$, giving MLE $\hat{\beta}_0 = -0.607$, $\hat{\beta}_1 = 0.202$. H_0: logistic model fits; H_1: model inadequate.

1. Fitted Probabilities and Expected Counts
From the MLE:
$\hat{\pi}_i = \{1 + \exp(-\hat{\beta}_0 - \hat{\beta}_1 x_i)\}^{-1}$, $\hat{\mu}_i = n_i\hat{\pi}_i = 10\hat{\pi}_i$.
$\sum \hat{\mu}_i = \sum y_i = 25$ ✓ (MLE score constraint).

2. Pearson Statistic

$$X^2 = \sum_{i=1}^{k} \frac{(y_i - \hat{\mu}_i)^2}{\hat{V}_i} = 9.56$$

Each term uses the *binomial* variance $\hat{V}_i = n_i\hat{\pi}_i(1-\hat{\pi}_i)$, not simply $\hat{\mu}_i$ as in the Poisson case.

3. Term-by-Term Computation

i	x	y_i	$\hat{\pi}_i$	$\hat{\mu}_i$	$\hat{V}_i$	$y_i-\hat{\mu}_i$	$(y_i-\hat{\mu}_i)^2/\hat{V}_i$
1	1	2	0.400	4.00	2.40	−2.00	1.67
2	2	8	0.450	4.50	2.47	+3.50	4.96
3	3	3	0.500	5.00	2.50	−2.00	1.60
4	4	7	0.550	5.50	2.47	+1.50	0.91
5	5	5	0.600	6.00	2.40	−1.00	0.42
						$X^2 =$	9.56

(Using exact MLE values: $X^2 = 9.551$.)

4. Degrees of Freedom and Decision
$k = 5$ groups, $p = 2$ parameters $(\hat{\beta}_0, \hat{\beta}_1)$ estimated: df $= k - p = 3$. $p = 0.023$; CV at 1%= 11.345 (FTR at 1%).

Statistic	Value	CV (5%)	Decision
X^2	9.551	7.815	Reject H_0

Practitioner Insight: The Pearson statistic X^2 in a GLM plays the same role as the deviance $D = 2(\ell_{\text{sat}} - \ell_{\text{fit}})$: both compare the fitted model to the saturated model and both follow $\chi^2(k-p)$ asymptotically under H_0. The difference is in the discrepancy measure — Pearson uses squared Pearson residuals $r_i = (y_i - \hat{\mu}_i)/\sqrt{\hat{V}_i}$, while the deviance uses a likelihood-ratio measure. For large counts the two are numerically close (here $X^2 = 9.551$, $D = 9.995$); for sparse data they can diverge, and the deviance is generally preferred within the GLM framework. The binomial variance function $\hat{V}_i = m_i\hat{\pi}_i(1-\hat{\pi}_i)$ in the denominator, which is quadratic in $\hat{\pi}_i$ and reaches its maximum at $\hat{\pi} = 1/2$, so a residual at a middle probability is *more* heavily divided than the same-sized residual at an extreme probability. The rejection here reflects the alternating low–high pattern ($y = 2, 8, 3, 7, 5$): the logistic model can only produce a monotone fitted curve ($\hat{\mu} = 4.0, 4.5, 5.0, 5.5, 6.0$), so it cannot track dose 2's spike or dose 3's dip, and these systematic deviations accumulate to exceed the critical value. A natural follow-up is to add a quadratic term or a batch indicator to the linear predictor.

Common Pitfalls

- **Using Pearson X^2 with continuous predictors without grouping.** With unique covariate patterns and no replication, expected counts ≈ 1 and the χ^2 approximation fails. Group predictors or rely on deviance and residual diagnostics instead of a formal Pearson test.
- **Interpreting $X^2/df > 1$ as automatic lack of fit.** Inflation can arise from overdispersion, clustering, or outliers—not only misspecification. Inspect Pearson residuals, examine variance structure, and separate mean-model issues from heterogeneity.
- **Relying on Pearson p-values in small samples or sparse data.** The χ^2 approximation is poor when expected counts are small (common in binary GLMs), giving unreliable p-values. Treat X^2/df as an informal dispersion check and use exact or resampling approaches when n is small.
- **Comparing Pearson statistics across different models or datasets.** X^2 is an absolute fit measure; it grows mechanically with n and decreases with added parameters. Compare *ratios* (X^2/df), use ΔX^2 only for nested models, or rely on AIC/BIC and cross-validation.
- **Assuming Pearson test checks full distributional fit.** Pearson X^2 assesses only the mean structure; it does not test variance assumptions, the link, or the underlying family. Complement with residual plots, overdispersion checks, link tests, and alternative distributions (e.g., NB vs. Poisson).

Hosmer-Lemeshow Test

Grouped Calibration Test for Logistic Regression Models

GLMS.3 | **Category:** Generalized Linear Models | **Data:** Binary | **Null:** Chi-square

When to Use

- Assess calibration of a fitted logistic regression.
- Compare observed and predicted event frequencies.
- Binary outcomes are modeled using logistic GLMs.
- Detect lack of fit across risk-score groupings.

Assumptions

- Observations are mutually independent.
- Response variable is binary in nature.
- Logistic model and link function are appropriate.
- Grouping of predicted risks is meaningful.

Test Statistic

$$C = \sum_{j=1}^{g} \left[\frac{(O_{1j} - E_{1j})^2}{E_{1j}} + \frac{(O_{0j} - E_{0j})^2}{E_{0j}} \right], \quad \text{where } E_{1j} = \sum_{i \in G_j} \hat{\pi}_i, \; E_{0j} = \sum_{i \in G_j} (1 - \hat{\pi}_i). \quad \text{Under } H_0, \hat{C} \approx \chi^2_{g-2}.$$

Notation: O_{yj}, E_{yj} are observed & expected counts in group j for outcome $y \in \{0, 1\}$, $\hat{\pi}_{ij}$ are fitted probabilities, divided into g groups.

Plain English: The Hosmer–Lemeshow test groups observations by fitted probability and compares observed and expected event counts within each group. If the model is well calibrated, predicted risks should agree with observed event proportions across these strata. The test is most sensitive to systematic over- or under-prediction in different parts of the risk range. Unlike deviance-based goodness-of-fit tests, it can be applied to individual binary data by creating groups from fitted values.

⚠ Practitioner Warning

Results are sensitive to the number and construction of groups; $g = 10$ (deciles) is common, but no universally optimal choice exists. Too few groups reduce sensitivity; too many can create sparse cells and weaken the χ^2 approximation. Expected counts should be reasonably large in each cell. The test may miss some forms of misspecification (e.g. omitted interactions or nonlinear effects), and in very large samples it may reject for minor departures of little practical importance. As it depends on arbitrary grouping, calibration plots and complementary diagnostics should be examined alongside.

Hypotheses

- H_0: The fitted logistic model is adequately calibrated.
- H_1: The fitted logistic model is not adequately calibrated.

Decision Rule: Reject H_0 if $C > \chi^2_{\alpha,g-2}$ or p-value $< \alpha$. Degrees of freedom: $g - 2$ where g is the number of groups.

Software References [Python, R, SAS]

```
statsmodels::HosmerLemeshowTest()  ResourceSelection::hoslem.test    PROC LOGISTIC (LACKFIT)
```

Additional Notes

Observations are first ordered by fitted probability and divided into g groups, often deciles. Within each group, observed and expected event/non-event counts are compared to form the HL statistic. Calibration plots (observed versus predicted by group) are a useful companion diagnostic, with perfect calibration lying on the 45° line. Because HL depends on grouping choice, alternatives such as the le Cessie–van Houwelingen test avoid this arbitrariness by using residual smoothing rather than grouped counts. HL is sensitive to grouping but insensitive to within-group structure: different models can yield identical HL statistics despite materially different calibration behavior. The Hosmer–Lemeshow test assesses calibration, not discrimination: a model can pass HL yet have poor predictive separation (e.g. low AUC), or fail HL while still ranking risks correctly. Always evaluate discrimination (e.g. ROC/AUC) alongside calibration.

Miscellanea: David Hosmer and Stanley Lemeshow (1980) developed a goodness-of-fit test for logistic regression based on grouping observations by predicted probabilities. They popularized this grouped goodness-of-fit test as a practical way to assess logistic-regression calibration when individual Bernoulli (ungrouped) data make classical deviance tests unreliable. The method compares observed and expected counts within groups and remains widely used despite sensitivity to the choice of grouping.

Reference(s): Hosmer & Lemeshow (1980); Hosmer et al. (1997)

Worked Example

Problem Statement A logistic regression model for 30-day hospital readmission is fitted to $n = 50$ patients using several clinical predictors. The Hosmer–Lemeshow test assesses calibration: whether predicted probabilities match observed event rates across the spectrum of risk. Patients are ranked by $\hat{\pi}_i$ and divided into $g = 5$ equal quintile groups ($n_j = 10$ each); the test statistic measures discrepancy between observed events O_j and expected events $E_j = \sum_{i \in j} \hat{\pi}_i$ within each group. H_0: model is well calibrated; H_1: systematic miscalibration.

1. Expected Events and Variance

Within each group: $E_j = n_j \bar{\pi}_j$, $V_j = n_j \bar{\pi}_j (1 - \bar{\pi}_j)$.

Quintile group j	1	2	3	4	5	Total
Mean fitted prob. $\bar{\pi}_j$	0.1	0.3	0.5	0.7	0.9	
Patients n_j	10	10	10	10	10	50
Observed events O_j	3	4	5	6	7	25
Expected events E_j	1	3	5	7	9	25
Variance V_j	0.9	2.1	2.5	2.1	0.9	

$\sum O_j = \sum E_j = 25$ ✓ (overall calibration holds).

The deviations $+2, +1, 0, -1, -2$ form a perfect arithmetic sequence: the model *overestimates* risk in the two high-probability groups and *underestimates* it in the two low-probability groups.

2. HL Statistic

$$C = \sum_{j=1}^{g} \frac{(O_j - E_j)^2}{V_j} = \frac{620}{63} \approx 9.841$$

3. Degrees of Freedom and Decision

df $= g - 2 = 5 - 2 = 3$.

Statistic	Value	CV (5%)	Decision
C	9.841	7.815	Reject H_0

$p = 0.020$; CV at 1%= 11.345 (FTR at 1%).

Practitioner Insight: The Pearson X^2 and deviance tests implicitly rely on replication at covariate patterns for their χ^2 approximation to be meaningful. Such replication arises naturally in grouped binomial data but not in individual-level binary data ($m_i = 1$). In this case, goodness-of-fit statistics are computed from single Bernoulli observations, and the χ^2 approximation is often poor. The Hosmer–Lemeshow test sidesteps this by *constructing* groups post hoc from ranked fitted probabilities, aggregating n observations into g groups with adequate expected counts. Here the pattern of deviations reveals a slope miscalibration: the fitted probabilities are too extreme (spread too wide), a common symptom of overfitting or failure to shrink coefficients. The identity $\sum O_j = \sum E_j = 25$ confirms intercept calibration (the model gets the overall event rate right) while the non-zero group deviations signal slope miscalibration. Two important limitations apply. First, the $\chi^2(g-2)$ distribution is an approximation whose accuracy depends on g and the E_j values; with $g = 10$ (the most common default) groups at the extremes often have small E_j, degrading the approximation. Second, the test has low power for certain miscalibration patterns and is sensitive to the choice of g and cut-points, so a non-significant result should not be taken as strong evidence of good calibration.

Common Pitfalls

- **Treating Hosmer–Lemeshow as the primary or only goodness-of-fit check.** Because it groups observations, the test loses information and often has low power, missing important misspecifications in functional form or omitted variables.
- **Interpreting a non-significant result as confirmation that the model is correct.** A large p-value merely indicates lack of evidence against calibration; it does not guarantee adequate fit or absence of systematic miscalibration.
- **Applying the test with small samples or sparse expected counts.** The chi-square approximation breaks down when many groups have fewer than five expected events or non-events, making the test unstable and unreliable.
- **Varying the number of groups to obtain a favorable p-value.** The test is highly sensitive to the chosen number of groups, and trying multiple g values risks p-hacking unless all results are reported transparently.
- **Using Hosmer–Lemeshow to evaluate discrimination.** The test assesses calibration only and should always be complemented with discrimination metrics such as ROC curves and AUC.

Le Cessie-van Houwelingen Test

Unweighted Sum of Squares Test for Logistic Regression

GLMS.4 | **Category:** Generalized Linear Models | **Data:** Binary | **Null:** Normal (asymptotic)

When to Use

- Assess overall fit of a logistic regression model.
- Avoid arbitrary grouping used by HL calibration tests.
- Detect subtle model misspecification and lack of fit.
- Continuous predicted probabilities are available.
- Calibration assessment using smoothed residual methods.

Assumptions

- Observations are mutually independent.
- Binary response variable follows logistic GLM form.
- Link function and predictor structure are appropriate.
- Predicted probabilities are estimated consistently.
- Sample size supports reliable asymptotic inference.

Test Statistic

$$Z = \frac{T - E(T)}{\sqrt{\mathrm{Var}(T)}}, \text{ where } T = \mathbf{r}^\top K \mathbf{r}. \qquad \text{Under } H_0,\ Z \approx N(0,1).$$

Notation: $r_i = y_i - \hat{\pi}_i$ are raw residuals, $\mathbf{r} = (r_1, \ldots, r_n)^\top$ is the residual vector, $\hat{\pi}_i$ are fitted probabilities, and K (kernel matrix, typically Guassian) is smoothing matrix, $\mathrm{E}[T \mid H_0] = \mathrm{tr}(K\,\mathrm{Cov}(\mathbf{r}))$, $\mathrm{Var}[T \mid H_0] = 2\,\mathrm{tr}[(K\,\mathrm{Cov}(\mathbf{r}))^2]$

Plain English: The le Cessie–van Houwelingen test assesses whether nearby observations in fitted-probability space show residual patterns that are too systematic to be explained by chance. Unlike Hosmer–Lemeshow, it does not require arbitrary grouping: instead, it smooths residual information across observations. Large positive standardized values indicate residual structure inconsistent with a correctly specified logistic model.

⚠ Practitioner Warning

The test is asymptotic and is most reliable in moderate to large samples. Its behavior depends on the smoothing construction used, so implementations may differ slightly. Like other omnibus goodness-of-fit procedures, it may miss certain localized or highly specific forms of misspecification. A non-significant result does not prove that the model is correct. Should be used alongside residual plots, calibration checks, and functional-form diagnostics.

Hypotheses

- H_0: The fitted logistic model is adequate.
- H_1: The fitted logistic model shows systematic lack of fit.

Decision Rule: Reject H_0 if $Z > Z_\alpha$ (one-tailed) or p-value $< \alpha$.

Software References [Python, R, SAS]

Python	R	SAS
`statsmodels::gof_..._discrete`	`rms::residuals.lrm(type="gof")`	`Not Native (Manual)`

Additional Notes

The test was proposed by le Cessie and van Houwelingen (1991) and further developed with Copas and Hosmer (1997) as alternative to grouped tests. While it is built from residuals $r_i = y_i - \hat{\pi}_i$, the LvH statistic is not simply the unweighted sum $\sum(y_i - \hat{\pi}_i)^2$, but a smoothed quadratic form that incorporates dependence between observations. The expectation and variance of the LvH statistic are derived using the full residual covariance matrix $\mathrm{Cov}(\mathbf{r}) = V - VX(X^\top VX)^{-1}X^\top V$ with $V = \mathrm{diag}(\hat{\pi}_i(1-\hat{\pi}_i))$, reflecting the fact that $\hat{\pi}_i$ are estimated. Borrowing strength across nearby observations in fitted-probability space, allows it to detect smooth calibration drift that grouped tests may miss. The test is closely related to a score test for adding smooth functions of the predictors, and a significant result indicates systematic miscalibration across the predictor space. Copas-Hosmer variant adds covariate-based terms to capture specific departures. The test can be extended to multinomial logistic regression by summing over all outcome categories. Calibration interpretation: significant result suggests predicted probabilities systematically misaligned with observed proportions across predictor space.

Miscellanea: Saskia le Cessie and Hans van Houwelingen (1991) developed this test as an ungrouped alternative to Hosmer–Lemeshow for logistic regression. By replacing arbitrary risk-group formation with residual smoothing, it offers a more stable omnibus assessment of fit.

Reference(s): le Cessie & van Houwelingen (1991, 1995); Copas & Hosmer (1989)

Worked Example

Problem Statement A logistic regression model for disease presence ($y_i \in \{0, 1\}$) is fitted to $n = 8$ patients using a single continuous predictor (risk score $x_i = 1, \ldots, 8$). H_0: logistic model fits; H_1: systematic residual structure.

1. Logistic Fit and Residuals

MLE: $\hat{\beta}_0 = -1.376, \hat{\beta}_1 = 0.306, \hat{\pi}_i = \{1 + e^{-(\hat{\beta}_0+\hat{\beta}_1 x_i)}\}^{-1}$, bandwidth $h = \text{sd}(\hat{\pi}) = 0.175$ (sd of $\hat{\pi}$ in col. 4 below).

i	x_i	y_i	$\hat{\pi}_i$	r_i	v_i
1	1	0	0.255	−0.255	0.190
2	2	1	0.318	+0.682	0.217
3	3	0	0.387	−0.387	0.237
4	4	0	0.462	−0.462	0.249
5	5	1	0.538	+0.462	0.249
6	6	1	0.613	+0.387	0.237
7	7	0	0.682	−0.682	0.217
8	8	1	0.745	+0.255	0.190

2. Kernel Matrix–Gaussian

$K_{ij} = \exp\!\left(-(\hat{\pi}_i - \hat{\pi}_j)^2/(2h^2)\right)$ for $i \neq j$, else $K_{ii} = 0$. Select entries:

	$j = 1$	2	3	4	5	6
$i = 1$	—	0.937	0.753	0.497	0.271	0.124
$i = 2$	0.937	—	0.925	0.713	0.454	0.242
$i = 3$	0.753	0.925	—	0.912	0.690	0.435

(Matrix is symmetric; by symmetry of $\hat{\pi}$.)

3. Test Statistic

$$T = \mathbf{r}^\top K \mathbf{r} = \sum_{i \neq j} K_{ij}\, r_i\, r_j = -1.778$$

Select entries with $|2K_{ij} r_i r_j| > 0.40$:

(i, j)	K_{ij}	r_i	r_j	$2K_{ij} r_i r_j$
(2, 3)	0.925	+0.682	−0.387	−0.488
(6, 7)	0.925	+0.387	−0.682	−0.488
(2, 4)	0.713	+0.682	−0.462	−0.449
(5, 7)	0.713	+0.462	−0.682	−0.449

4. Standardization and Decision

$\text{E}[T \mid H_0] = \text{tr}(K\,\text{Cov}(\mathbf{r})) = -1.085$ (uses the information-matrix correction for estimated parameters).
$\text{Var}[T \mid H_0] = 2\,\text{tr}[(K\,\text{Cov}(\mathbf{r}))^2] = 0.464$.

$$z = \frac{T - \text{E}[T]}{\sqrt{\text{Var}[T]}} = \frac{-1.778 - (-1.085)}{\sqrt{0.464}} = -1.018$$

Statistic	Value	Threshold (5%)	Decision
$|z|$	1.011	1.960	FTR H_0

$p = 0.312$. No evidence of systematic misfit at $n = 8$.

$\text{Cov}(\mathbf{r}) = V - VX(X^\top VX)^{-1}X^\top V, \quad V = \text{diag}\{\hat{\pi}_i(1 - \hat{\pi}_i)\}$.

Practitioner Insight: The LvH test occupies a different niche from both the Pearson X^2 and Hosmer–Lemeshow tests. The Pearson X^2 relies on replication at covariate patterns for its χ^2 approximation, while HL imposes post-hoc grouping of fitted probabilities. LvH avoids grouping entirely: each pair (i, j) contributes through the kernel weight K_{ij}, which is largest for nearby fitted probabilities and small for distant ones. The centered statistic is interpretable: $T > E[T]$ suggests local clustering of residual signs, consistent with smooth lack of fit, whereas $T < E[T]$ suggests alternating residual signs, more consistent with random variation. The information-matrix correction is essential because fitted probabilities and residuals are estimated from the same data; ignoring it can lead to incorrect calibration of the null distribution. With $n = 8$, the test has very low power, so failure to reject should be read only as absence of clear evidence of misfit, not as strong confirmation.

Common Pitfalls

- **Relying on asymptotic z–approximation in small samples.** The le Cessie–van Houwelingen statistic assumes large-sample normality; with $n < 150$–200 the p–value can be distorted. Use a parametric bootstrap for $n < 200$.
- **Misinterpreting the smoothing weights as residual-based.** The test uses leverage-based weights that reflect predictor–space density, not response errors. High-leverage observations dominate the statistic, so misfit in sparse regions may be amplified while dense regions contribute less.
- **Expecting the test to detect all forms of misspecification.** It is powerful for broad calibration drift and nonlinearity but may miss subgroup-specific issues, rare-pattern misfit, or wrong link choice. Always pair with calibration plots, residual diagnostics, and link / functional-form checks.
- **Over-interpreting the magnitude of the z statistic.** Unlike χ^2-based tests, the standardized z score has little intuitive scale; significance matters, not the absolute value. Interpretation should rely on p–values and calibration plots rather than the raw z magnitude.

Likelihood Ratio Test (LRT)

Compare Nested Generalized Linear Models

GLMS.5 | **Category:** Generalized Linear Models | **Data:** Various | **Null:** Chi-square

When to Use

- Compare nested generalized linear models formally.
- Assess contribution of predictors added to a model.
- Compare full and reduced models using likelihoods.

Assumptions

- Candidate models are properly nested within others.
- Observations are mutually independent.
- Link function and mean structure are appropriate.

Test Statistic

$$LR = -2\log\left(\frac{L_0}{L_1}\right) = -2(\ell_0 - \ell_1) = D_0 - D_1, \quad \text{where } D \text{ is deviance.} \qquad \text{Under } H_0,\ LR \approx \chi^2_{p_1-p_0}$$

Notation: L_0, L_1 are likelihoods, ℓ_0, ℓ_1 are log-likelihoods, D_0, D_1 are deviances, $p_0 < p_1$ are parameter counts

Plain English: The likelihood ratio test compares how well two nested models fit the same data by examining the ratio of their maximized likelihoods. If the simpler model is sufficient, its likelihood should not be much smaller than that of the more complex model. The test statistic is minus twice the log of this ratio, which for nested GLMs equals the difference in deviances. Large values indicate that the added parameters improve fit enough to reject the simpler model.

⚠ Practitioner Warning

Models must be truly nested: same response, same observations, same likelihood basis, and one model obtained by constraining parameters of the other. Using different subsets of data or changing the effective response structure invalidates the test. For non-nested models, use criteria such as AIC/BIC or predictive methods instead. The chi-square approximation may be poor in small samples or when parameters lie near the boundary of the parameter space. In overdispersed settings fit with quasi-likelihood, the usual LRT is not valid because a true likelihood is unavailable; use an F-type or other dispersion-adjusted procedure instead.

Hypotheses

- H_0: Simpler model is adequate.
- H_1: Complex model provides a better fit.

Decision Rule: Reject H_0 if $LR > \chi^2_{\alpha,p_1-p_0}$ or p-value $< \alpha$. df equal difference in number of parameters between models.

Software References [Python, R, SAS]

`statsmodels::compare_lr_test`   `anova(glm0, glm1, test="LRT")`   `PROC GENMOD`

Additional Notes

The likelihood ratio test is one of three classical large-sample tests alongside Wald and score (Rao's or Lagrange multiplier) tests. All three are asymptotically equivalent under H_0 but can differ substantially in finite samples. LRT generally preferred for its invariance to parameterization and better finite-sample properties. LRT measures relative fit between two models, not absolute adequacy: a significant result only indicates that the larger model fits better, not that it fits well in an absolute sense. Relationship to AIC: $\text{AIC} = -2\ell + 2p$, so $\text{AIC}_0 - \text{AIC}_1 = LR + 2(p_0 - p_1)$. Models with $\Delta\text{AIC} < 2$ (a rule of thumb) have similar support. Sequential testing: test hierarchy of nested models from simplest to most complex, stopping when test becomes non-significant. Bonett-Price adjustment for overdispersion: use $LR/\hat{\phi}$ with $F_{p_1-p_0,n-p_1}$ distribution where $\hat{\phi} = D_1/(n-p_1)$. For normal linear models, LRT reduces to F-test:

$$F = \frac{(RSS_0 - RSS_1)/(p_1 - p_0)}{RSS_1/(n-p_1)}.$$

Miscellanea: The likelihood ratio test builds on the hypothesis testing framework of Neyman and Pearson (1928) and the likelihood-based ideas of Fisher. Wilks (1938) established its asymptotic chi-square distribution for composite hypotheses, making it a central tool in maximum likelihood inference. Geometrically, the LRT compares the height of the log-likelihood at two points on the parameter surface, measuring how far the constrained model lies from the unconstrained optimum.

Reference(s): Neyman & Pearson (1928); Wilks (1938)

Likelihood Ratio Test (LRT) GLMS.5

Worked Example

Problem Statement Counts of bacterial colonies y_i are recorded at five increasing antibiotic concentrations $x_i \in \{0, 1, 2, 3, 4\}$ (multiples of the minimum inhibitory concentration). A Poisson GLM with log link is fitted. The LRT compares the null model M_0: $\log \mu = \beta_0$ (concentration has no effect) against the alternative M_1: $\log \mu = \beta_0 + \beta_1 x$ (linear log-dose response), testing whether adding concentration as a predictor significantly improves fit. H_0: $\beta_1 = 0$; H_1: $\beta_1 \neq 0$.

1. Null Model M_0

Under M_0 the MLE is $\hat{\mu}_i = \bar{y} = 25/5 = 5$ for all i. The deviance of M_0 against the saturated model is

$$D_0 = 2 \sum_i y_i \log\left(\frac{y_i}{\bar{y}}\right)$$

since $\sum(y_i - \bar{y}) = 0$ eliminates the linear term.

$$D_0 = 2 \times 2.750 = 5.501$$

i	x_i	y_i	$y_i \ln(y_i/\bar{y})$	$\hat{\mu}_i$	$y_i \ln(y_i/\hat{\mu}_i)$
1	0	2	−1.833	2.291	−0.272
2	1	3	−1.532	3.204	−0.197
3	2	5	0.000	4.480	0.549
4	3	7	2.355	6.265	0.777
5	4	8	3.760	8.761	−0.726
Σ			2.750		0.131

2. Full Model M_1

MLE (from score equations): $\hat{\beta}_0 = 0.829$, $\hat{\beta}_1 = 0.3353$.
$\sum \hat{\mu}_i = \sum y_i = 25$ ✓ (score constraint).

$$D_1 = 2 \times 0.131 = 0.263$$

For Λ, $p = 0.022$; CV at 1%= 6.635 (FTR at 1%).

3. LRT and Decision

$\Lambda = D_0 - D_1 = 5.501 - 0.263 = 5.238 \sim \chi^2(1)$

Statistic	Value	CV (5%)	Decision
Λ	5.240	3.841	Reject H_0

Practitioner Insight: The LRT $\Lambda = 2(\ell_1 - \ell_0) = D_0 - D_1$ is the deviance *difference* between two nested models: $M_0 \subset M_1$ with $\dim(M_1) - \dim(M_0) = q$ extra parameters. Under H_0, $\Lambda \sim \chi^2(q)$ asymptotically. This distinguishes the LRT from the deviance GoF test earlier in this chapter, where D_1 itself (not a difference) is compared to $\chi^2(n-p)$ to test adequacy of a single model against the saturated alternative. Here $D_0 = 5.501$ is compared to $\chi^2(4)$ when testing adequacy of M_0 against the saturated model: $\text{CV}_{5\%} = 9.488$, so M_0 is *not* rejected by that GoF test, yet the LRT rejects $\beta_1 = 0$. The two tests answer different questions. The LRT uses only one degree of freedom (the extra parameter β_1) rather than four, giving it more power against the specific one-dimensional alternative $\beta_1 \neq 0$. The residual deviance $D_1 = 0.263$ on $n - p = 3$ degrees of freedom confirms that M_1 fits well. The Wald test of $\beta_1 = 0$ ($z = \hat{\beta}_1/\text{se}(\hat{\beta}_1)$, compared to $N(0,1)$) and the score test are asymptotically equivalent to the LRT but use only the null or the full fit respectively; the LRT uses both and is generally preferred for small-to-moderate samples.

Common Pitfalls

- **Using LRT on non-nested models.** The test requires true nesting—one model must be a restricted version of the other. Comparing models with different predictors, links, or distributions voids the χ^2 reference distribution; use AIC/BIC, cross-validation, or Vuong tests instead.
- **Applying LRT to quasi-likelihood fits.** Quasi-Poisson and quasi-binomial models lack a true likelihood, so deviance differences cannot be interpreted as likelihood ratios. Use F-tests (scaled deviance), Wald tests, or switch to full-likelihood alternatives such as NB or beta-binomial.
- **Comparing models fit to different datasets.** Likelihoods are comparable only when both models use the exact same observations, response scale, and weights. Dropped cases, transformations, or differing weights make LRT invalid—even if software still reports a statistic.
- **Sequentially testing many nested models without adjustment.** Repeated LRTs (stepwise selection, testing many interactions or polynomials) inflate Type I error. Pre-specify comparisons, apply multiplicity corrections, or rely on penalized criteria (AIC/BIC) instead of multiple hypothesis testing.
- **Treating a non-significant LRT as proof the reduced model is "correct."** Failure to reject often reflects low power or small effect sizes, not true model equivalence. Inspect parameter estimates, consider practical effect sizes, and evaluate AIC/BIC rather than interpreting non-significant LRTs as confirmatory.

Wald Test

Test Parameter Constraints via Asymptotic Normality

GLMS.6 | **Category:** Generalized Linear Models | **Data:** Various | **Null:** Chi-square

When to Use

- Test significance of individual model coefficients.
- Assess predictor effects within a fitted GLM.
- Evaluate linear hypotheses on model parameters.
- Quick inference without refitting nested models.
- Large-sample coefficient testing is appropriate.

Assumptions

- Observations are mutually independent.
- Response distribution belongs to the GLM family.
- Link function and mean structure are appropriate.
- Parameter estimates are approximately normal.
- Sample size supports asymptotic inference theory.

Test Statistic

$$W = \frac{(\hat{\beta} - \beta_0)^2}{\widehat{Var}(\hat{\beta})}, \quad \text{or } W = (\hat{\boldsymbol{\beta}} - \boldsymbol{\beta}_0)'\widehat{Cov}(\hat{\boldsymbol{\beta}})^{-1}(\hat{\boldsymbol{\beta}} - \boldsymbol{\beta}_0) \text{ for multiple parameters.}$$

Under H_0, $W \sim \chi^2_1$ (single parameter) or χ^2_q (q parameters) asymptotically.

Notation: $\hat{\beta}$ is MLE, β_0 is null value (often 0), $\widehat{Var}$ is estimated variance from information matrix

Plain English: The Wald test evaluates whether a parameter differs from its hypothesized value by comparing the estimate to its standard error. It uses the asymptotic normality of maximum likelihood estimates, requiring only the fitted model without refitting under the null. The test squares the standardized difference between estimate and null value, yielding a chi-square statistic. For single parameters, the square root gives a z-test. The test is computationally convenient but can behave poorly in some settings.

⚠ Practitioner Warning

The Wald test is not invariant to parameterization: testing $\beta = 0$ versus $\exp(\beta) = 1$ can yield different conclusions, unlike the LRT. It performs poorly when the true parameter is near the boundary (e.g., variance near zero) or when the estimate lies far from the null, where the quadratic approximation to the likelihood is inaccurate. It can have inflated Type I error in small samples or sparse data (e.g., separation in logistic regression). For joint tests of multiple parameters, it can be less reliable than the likelihood ratio test. Wald confidence intervals may include implausible values (e.g., probabilities outside $[0, 1]$); profile likelihood intervals are often preferred.

Hypotheses

- H_0**:** $\beta = \beta_0$ or $\mathbf{C}\boldsymbol{\beta} = \mathbf{c}$.
- H_1**:** $\beta \neq \beta_0$ or constraints are violated.

Decision Rule: Reject H_0 if $W > \chi^2_{\alpha,q}$ or p-value $< \alpha$. $df = q$ equals number of constraints tested.

Software References [Python, R, SAS]

Python	R	SAS
`statsmodels.genmod.GLM.wald_test`	`summary(glm.object)`	`PROC GENMOD`

Additional Notes

For linear constraints $\mathbf{C}\boldsymbol{\beta} = \mathbf{c}$,

$$W = (\mathbf{C}\hat{\boldsymbol{\beta}} - \mathbf{c})'[\mathbf{C}\widehat{Cov}(\hat{\boldsymbol{\beta}})\mathbf{C}']^{-1}(\mathbf{C}\hat{\boldsymbol{\beta}} - \mathbf{c}).$$

Wald confidence intervals take the form $\hat{\beta} \pm z_{\alpha/2}SE(\hat{\beta})$. The Wald, likelihood ratio, and score tests are asymptotically equivalent but can differ in finite samples. The Hauck–Donner effect (logistic regression) causes the Wald test to lose power as the estimated effect becomes very large.

Miscellanea: Abraham Wald (1940s) developed the Wald test as part of his work on statistical decision theory. The test became widely used due to its computational simplicity, requiring only estimates from the fitted model. Wald's broader contributions, including wartime statistical analysis and decision theory, had a lasting impact on modern statistics.

Reference(s): Wald (1943); Hauck & Donner (1977)

Worked Example

Problem Statement Data from the Deviance GoF example is extended. A binary indicator z_i for Region 2 (city center) is added, giving model M_1: $\log \mu_i = \log n_i + \beta_0 + \beta_1 z_i$. H_0: $\beta_1 = 0$ (common rate, λ); H_1: $\beta_1 \neq 0$ (Region 2 differs).

1. MLE and Fitted Values

$\hat{\lambda}_a = \sum_{i:z_i=a} y_i / \sum_{i:z_i=a} n_i$; $\hat{\lambda}_0 = 24/80$, $\hat{\lambda}_1 = 16/20$,

i	n_i	y_i	z_i	$\hat{\mu}_i^{(1)} = n_i \hat{\lambda}_{z_i}$
1	10	2	0	3
2	20	16	1	16
3	15	6	0	4.5
4	25	10	0	7.5
5	30	6	0	9
$\sum$	100	40		40

$$\hat{\beta}_1 = \log(\hat{\lambda}_1/\hat{\lambda}_0) \approx 0.981$$

2. Observed Fisher Information

For Poisson log-linear: $I(\hat{\boldsymbol{\beta}}) = X^\top \operatorname{diag}(\hat{\boldsymbol{\mu}}^{(1)})\, X$.

$$I = \begin{pmatrix} \sum_i \hat{\mu}_i & \sum_i \hat{\mu}_i z_i \\ \sum_i \hat{\mu}_i z_i & \sum_i \hat{\mu}_i z_i^2 \end{pmatrix} = \begin{pmatrix} 40 & 16 \\ 16 & 16 \end{pmatrix}$$

3. Variance of $\hat{\beta}_1$

$$\det(I) = 40 \times 16 - 16^2 = 384$$

By the 2×2 inverse formula:

$$\widehat{\operatorname{Var}}(\hat{\beta}_1) = [I^{-1}]_{11} = \frac{I_{00}}{\det(I)} = \frac{40}{384} = \frac{5}{48}$$

$$\widehat{\operatorname{SE}}(\hat{\beta}_1) = \sqrt{5/48} \approx 0.3227$$

4. Wald Statistic and Decision

$$z = \frac{\hat{\beta}_1}{\widehat{\operatorname{SE}}(\hat{\beta}_1)} = 3.039, \quad W = z^2 = 9.235$$

Statistic	Value	CV (5%)	Decision
W	9.235	3.841	Reject H_0

$p = 0.002$; CV at 1%= 6.635 (also rejects at 1%).

$\lambda_{z=0} = e^{\beta_0}$ and $\lambda_{z=1} = e^{\beta_0+\beta_1} \implies \beta_1 = \log(\lambda_{z=1}/\lambda_{z=0})$.

Practitioner Insight: The Wald, LRT, and score tests are asymptotically equivalent — all converge to the same $\chi^2(1)$ null distribution — but they can differ in finite samples because they assess departure from H_0 at different points on the log-likelihood. The LRT compares the log-likelihood heights at $\hat{\beta}_1$ and at $\beta_1 = 0$; the Wald test uses the curvature at $\hat{\beta}_1$; the score test uses the slope at $\beta_1 = 0$. Here the Wald statistic ($W = 9.235$) and the deviance GoF statistic for M_0 ($D_{M_0} \approx 11.09$, df= 4) are not directly comparable: the Wald test targets a specific parameter ($\beta_1 = 0$), while the GoF test assesses the adequacy of the entire model. Both, however, provide evidence against the common-rate assumption. The practical advantage of the Wald test is computational: it requires only the fitted model M_1 and its information matrix, whereas the LRT additionally requires fitting M_0. One limitation is that the Wald statistic depends on parametrization: testing $\beta_1 = 0$ is equivalent to testing $e^{\beta_1} = 1$, but the corresponding Wald statistics differ, whereas the LRT is invariant.

Common Pitfalls

- **Using Wald tests in quasi-likelihood models without adjusting for dispersion.** Standard Wald z–statistics assume unit variance, but quasi-Poisson and quasi-binomial models inflate SEs by $\sqrt{\phi}$ and require t or F distributions; ignoring this produces anti-conservative p-values.
- **Relying on Wald tests when logistic regression exhibits (quasi-)complete separation.** With separation, $\hat{\beta}$ and its SE diverge and the Wald ratio becomes unreliable (Hauck–Donner effect), often yielding insignificant results even for strong effects; use penalized likelihood or LRT instead.
- **Testing joint hypotheses by examining coefficients one at a time.** Separate Wald z-tests ignore covariance among parameters and can miss or falsely detect joint significance; use the multivariate Wald χ^2 test based on the full covariance matrix.
- **Interpreting symmetric Wald intervals for highly non-linear parameters.** Odds ratios, risk ratios, and other exponentiated effects yield skewed distributions, and Wald intervals can give meaningless or impossible bounds; construct CIs on the link scale or use profile likelihood / bootstrap.
- **Comparing Wald results across different parameterizations.** Wald tests are not invariant to non-linear transformations, so testing $\beta = 0$ differs from testing $\exp(\beta) = 1$; use the link-scale parameterization for inference or prefer likelihood-ratio tests when invariance matters.

Score Test (Rao's)

Test Parameter Constraints Without Fitting Full Model

GLMS.7 | **Category:** Generalized Linear Models | **Data:** Various | **Null:** Chi-square

When to Use

- Assess predictor effects without fitting full models.
- Test additions to a reduced GLM specification.
- Computational efficiency is important in modeling.
- Variable screening during model-building stages.
- Large-sample inference under null-model fitting.

Assumptions

- Observations are mutually independent.
- Response distribution belongs to the GLM family.
- Link function and mean structure are appropriate.
- Reduced model under the null is correctly specified.
- Sample size supports asymptotic inference theory.

Test Statistic

$$S = \mathbf{U}(\tilde{\boldsymbol{\beta}})'\mathcal{I}(\tilde{\boldsymbol{\beta}})^{-1}\mathbf{U}(\tilde{\boldsymbol{\beta}}), \quad \text{where } \mathbf{U}(\boldsymbol{\beta}) = \frac{\partial \ell}{\partial \boldsymbol{\beta}}. \qquad \text{Under } H_0,\ S \approx \chi^2_q \text{ and } \tilde{\boldsymbol{\beta}} \text{ is MLE.}$$

Notation: $\mathbf{U}$ is score (gradient of log-likelihood), $\mathcal{I}$ is Fisher information, q is parameters tested

Plain English: The score test examines whether the gradient (score) of the log-likelihood is near zero at the restricted parameter estimates. If the null hypothesis is correct, the restricted estimates maximize the likelihood, making the score approximately zero. Large scores indicate the likelihood could be substantially increased by relaxing restrictions, suggesting the null hypothesis is false. The test requires fitting only the restricted model, making it computationally efficient when the alternative model is complex.

⚠ Practitioner Warning

Requires computing score function and information matrix at restricted estimates—more complex implementation than Wald test. Generally most conservative of three classical tests (Wald, LRT, score), with lowest power but best Type I error control. Less familiar to applied researchers than Wald or LRT despite theoretical advantages. Software support varies; not always readily available. Power advantage disappears as sample size increases—all three tests asymptotically equivalent. For testing variance components at boundary (e.g., $\sigma^2 = 0$), score test has better properties than Wald or LRT.

Hypotheses

- H_0: $\boldsymbol{\beta}_R = \boldsymbol{\beta}_0$ (restrictions hold, typically $\boldsymbol{\beta}_0 = \mathbf{0}$).
- H_1: At least one restriction violated (additional parameters non-zero).

Decision Rule: Reject H_0 if $S > \chi^2_{\alpha,q}$ or p-value $< \alpha$. Degrees of freedom q equals number of tested constraints.

Software References [Python, R, SAS]

```
statsmodels.genmod.GLM.score_test (varies by package)        Not Native (Manual)
```

Additional Notes

The score function (gradient) is $\mathbf{U}(\boldsymbol{\beta}) = \partial\ell/\partial\boldsymbol{\beta}$ where ℓ is log-likelihood. At the MLE, $\mathbf{U}(\hat{\boldsymbol{\beta}}) = \mathbf{0}$ by definition. Fisher information matrix is $\mathcal{I}(\boldsymbol{\beta}) = -E[\partial^2\ell/\partial\boldsymbol{\beta}\partial\boldsymbol{\beta}']$ or its observed version $\mathcal{J}(\boldsymbol{\beta}) = -\partial^2\ell/\partial\boldsymbol{\beta}\partial\boldsymbol{\beta}'$. For testing subset of q parameters: partition $\boldsymbol{\beta} = (\boldsymbol{\beta}_1, \boldsymbol{\beta}_2)$ where $\boldsymbol{\beta}_2$ are tested parameters. Under $H_0 : \boldsymbol{\beta}_2 = \mathbf{0}$, fit restricted model with only $\boldsymbol{\beta}_1$, evaluate score for $\boldsymbol{\beta}_2$ at $(\tilde{\boldsymbol{\beta}}_1, \mathbf{0})$. The score test is most sensitive to small (local) departures from H_0, as it evaluates the slope of the log-likelihood at the null rather than at the fitted alternative. The score test is also known as the Lagrange multiplier (LM) test or efficient score test. Many classical diagnostics, such as Breusch–Pagan and Breusch–Godfrey, are score-type tests.

Miscellanea: C. Radhakrishna Rao (1948) developed the score test, building on Fisher's likelihood framework. The test is locally most powerful under standard conditions and is equivalent to the Lagrange multiplier (LM) test, using only information from the null model. Its main appeal is conceptual elegance: instead of fitting the full alternative, it asks whether the null fit shows strong local pressure to move in the direction of a more complex model. It remains widely used due to its strong theoretical properties and computational simplicity.

Reference(s): Rao (1948); Silvey (1959)

Worked Example

Problem Statement Dataset from the deviance GoF and Wald examples is used. Unlike the Wald test, which requires fitting M_1 (Region 2 differs), the score test needs only M_0 — the model assumed true under H_0. H_0: $\beta_1 = 0$ (M_0: common rate); H_1: $\beta_1 \neq 0$ (Region 2 differs).

1. M_0 Fitted Values (Restricted MLE)
$\hat{\lambda} = 40/100 = 2/5$; $\hat{\mu}_i^{(0)} = n_i \times 2/5$:

i	n_i	y_i	z_i	$\hat{\mu}_i^{(0)}$	$y_i - \hat{\mu}_i^{(0)}$
1	10	2	0	4	−2
2	20	16	1	8	+8
3	15	6	0	6	0
4	25	10	0	10	0
5	30	6	0	12	−6
Σ	100	40		40	0

2. Score Vector at M_0
The score for β_0 vanishes at the MLE (first-order condition for $\hat{\beta}_0$):

$$U_0 = \sum_i (y_i - \hat{\mu}_i^{(0)}) = 0 \checkmark$$

The score for β_1 measures how much the log-likelihood would increase by allowing a Region-2 rate:

$$U_1 = \sum_i z_i(y_i - \hat{\mu}_i^{(0)}) = 1 \times (16 - 8) = 8$$

3. Fisher Information Under M_0
$I_0 = X^\top \operatorname{diag}(\hat{\mu}^{(0)})\, X$:

$$I_0 = \begin{pmatrix} \sum_i \hat{\mu}_i^{(0)} & \sum_i z_i \hat{\mu}_i^{(0)} \\ \sum_i z_i \hat{\mu}_i^{(0)} & \sum_i z_i^2 \hat{\mu}_i^{(0)} \end{pmatrix} = \begin{pmatrix} 40 & 8 \\ 8 & 8 \end{pmatrix}$$

4. Score Statistic and Decision
The effective information for β_1 (Schur complement):

$$I_0[1,1] - \frac{I_0[0,1]^2}{I_0[0,0]} = \frac{32}{5}, \qquad S = \frac{U_1^2}{32/5} = 10$$

Statistic	Value	CV (5%)	Decision
T	10	3.841	Reject H_0

$p = 0.002$; CV at 1%= 6.635 (also rejects at 1%).

Practitioner Insight: The score test closes the trinity of large-sample tests. Score $S = 10$, Wald $W = 9.235$, LRT LR = 8.372, all rejecting H_0 at the 1% level. The three tests are asymptotically equivalent but measure departure from H_0 at different points on the log-likelihood surface. The score test evaluates the *slope* at $\beta_1 = 0$ (M_0): a large positive $U_1 = 8$ means the log-likelihood is steeply rising at the restricted MLE, indicating that relaxing the constraint would substantially improve fit. The Wald test evaluates the *curvature* at $\hat{\beta}_1$ (M_1); the LRT compares the *heights* at both points. The score test's key practical advantage is that it requires only M_0: if the restricted model is cheap to fit but the extended model is not (e.g. high-dimensional M_1, convergence difficulties, or constrained optimization), the score test can assess whether the extension is warranted without ever fitting it. The ordering $S > W >$ LR observed here is data-specific; although the three tests are asymptotically equivalent, no universal finite-sample ordering is guaranteed.

Common Pitfalls

- **Implementing the score test with incorrect derivatives or information matrix.** Score tests require the correct score vector and full Fisher information matrix evaluated at the reduced-model MLE; using wrong derivatives or ignoring off-diagonal information terms leads to invalid test statistics.
- **Using the score test as a final inference tool rather than a screening device.** The score test is most powerful when the null is true but loses power when the alternative is meaningful; it should primarily be used for quick screening before confirming with a likelihood ratio test.
- **Assuming the score, Wald, and likelihood ratio tests give similar results in finite samples.** These tests are asymptotically equivalent, but with moderate or small samples they often disagree substantially due to non-quadratic likelihoods or weak information.
- **Applying the score test to a badly misspecified reduced model.** Because the score test evaluates the gradient at the reduced-model estimate, omitted-variable bias or misspecification in the base model distorts the score and inflates Type I error.
- **Confusing score tests with score residuals or diagnostic "scores."** Score residuals flag observation-level departures, whereas the score test is a global hypothesis test; the two are unrelated and should not be interpreted as substitutes.

Link Function Test (Pregibon)

Diagnostic Test for Link and Functional-Form Adequacy in GLMs

GLMS.8 | **Category:** Generalized Linear Models | **Data:** Various | **Null:** Normal or χ^2

When to Use

- Detect specification errors in a fitted GLM model.
- Assess whether important predictors were omitted.
- Verify adequacy of the chosen link function form.
- Logistic regression model diagnostics are needed.
- Quick screening for functional-form misspecification.

Assumptions

- Observations are mutually independent.
- Response distribution belongs to the GLM family.
- Fitted values summarize the model information well.
- Link function and mean structure are appropriate.
- Sample size supports asymptotic inference theory.

Test Statistic

$$Z = \frac{\hat{\gamma}_2}{SE(\hat{\gamma}_2)}. \quad \text{Auxiliary model: } g(\mu_i) = \gamma_0 + \gamma_1\hat{\eta}_i + \gamma_2\hat{\eta}_i^2.$$

Notation: $g(\cdot)$ is link function, $\hat{\eta}_i = \mathbf{x}_i^\top\hat{\boldsymbol{\beta}}$ is linear predictor, γ_2 indicates link misspecification

Plain English: Pregibon's link test augments the fitted GLM by adding the square of the linear predictor as an additional covariate, then tests whether this squared term has significant explanatory power. If the link function is correctly specified, the linear predictor alone should capture the relationship, leaving no systematic pattern for its square to explain. A significant squared term suggests that the model is missing curvature, which may reflect an inappropriate link, omitted nonlinear terms, or broader specification error.

⚠ Practitioner Warning

Test detects link misspecification but does not identify the correct alternative link. Rejection requires examining residuals and trying alternative links (probit vs. logit, log vs. square root). Test conflates link misspecification with other model problems (omitted variables, non-linearity in predictors). Low power to distinguish between similar links (logit vs. probit often indistinguishable). For binary data, link choice matters less than for count or continuous responses. Test assumes distributional family correct—wrong family can masquerade as wrong link.

Hypotheses

- H_0: $\gamma_2 = 0$ (chosen link / linear predictor appears adequate).
- H_1: $\delta \neq 0$ (link or functional-form misspecified).

Decision Rule: Reject H_0 if $|Z| > z_{\alpha/2,n-p-1}$ or p-value $< \alpha$. Equivalently, reject if the Wald statistic $W = Z^2$ exceeds $\chi^2_{\alpha,1}$.

Software References [Python, R, SAS]

Python	R	SAS
`Not Native (Manual)`	`Not Native (Manual)`	`Not Native (Manual)`

Additional Notes

Because $\hat{\eta}_i$ is estimated, inference is approximate; the test is asymptotic and may be unreliable in small samples. Common link functions by family: binomial (logit $g(\mu) = \log(\mu/(1-\mu))$, probit $\Phi^{-1}(\mu)$, cloglog $\log(-\log(1-\mu))$, Cauchit $g(\mu) = \tan(\pi(\mu - \frac{1}{2}))$), Poisson (log $\log(\mu)$, identity μ, square root $\sqrt{\mu}$), gamma (inverse $1/\mu$, log $\log(\mu)$), normal (identity μ). Canonical links (logit for binomial, log for Poisson) have desirable theoretical properties but may not fit best empirically. Box-Cox family of power transformations provides flexible link alternatives: $g(\mu) = (\mu^\lambda - 1)/\lambda$ for $\lambda \neq 0$, $\log(\mu)$ for $\lambda = 0$. Stukel (1988) proposed two-parameter link allowing separate flexibility in tails for binary data. After rejection, plausible remedies include adding nonlinear predictor terms, trying alternative links, or reconsidering the response family. Link test is special case of general RESET-type specification tests adapted to GLM context.

Miscellanea: David Pregibon (1980) developed diagnostic methods for generalized linear models, including the link test for detecting model misspecification. Its appeal lies in turning a broad question about model adequacy into a single added-term test. Pregibon's work also introduced influential diagnostic tools for GLMs, extending concepts such as leverage and influence beyond ordinary linear regression. Many modern logistic-regression diagnostic procedures trace their origins to this pioneering work.

Reference(s): Pregibon (1980); Stukel (1988)

Worked Example

Problem Statement In a dose–response experiment, $k = 5$ dose groups (coded $x = 0, 1, 2, 3, 4$) each contain $m = 20$ animals; y_i is the number showing a biomarker response. A logistic regression model M_0: $\text{logit}(\pi_i) = \beta_0 + \beta_1 x_i$ is fitted. The Pregibon (1980) link test assesses whether the linear predictor $\hat{\eta}_i = \hat{\beta}_0 + \hat{\beta}_1 x_i$ is sufficient. H_0: coefficient of $\hat{\eta}^2$ equals zero (link and linear predictor are correctly specified); H_1: coefficient of $\hat{\eta}^2$ differs from zero.

1. Original Logistic Fit (M_0)
MLE: $\hat{\beta}_0 = -0.701$, $\hat{\beta}_1 = 0.206$.

i	x_i	y_i	$\hat{\eta}_i$	$\hat{\pi}_i$	$\hat{\mu}_i = 20\hat{\pi}_i$	$\hat{\eta}_i^2$
1	0	4	−0.701	0.332	6.63	0.491
2	1	9	−0.494	0.379	7.58	0.244
3	2	11	−0.288	0.428	8.57	0.083
4	3	11	−0.081	0.480	9.59	0.007
5	4	8	+0.125	0.531	10.62	0.016

The fitted model is strictly increasing ($\hat{\beta}_1 > 0$) and cannot reproduce the observed decline from $x = 2$ to $x = 4$. Log-likelihood: $\ell_0 = -67.305$.

2. Augmented Model M_1
M_1 adds $\hat{\eta}_i^2$ as a second predictor:

$$\text{logit}(\pi_i) = \gamma_0 + \gamma_1 \hat{\eta}_i + \gamma_2 \hat{\eta}_i^2$$

with $\hat{\eta}_i$ fixed at M_0 values.

3. Augmented Fit and Wald Test
MLE of M_1: $\hat{\gamma}_0 = 0.026$, $\hat{\gamma}_1 = -2.687$, $\hat{\gamma}_2 = -6.634$. Log-likelihood: $\ell_1 = -64.776$. Fitted counts under M_1: $\{4.1, 8.7, 11.2, 11.0, 8.0\}$, tracking the observed rise-then-fall closely.

Wald test for $H_0: \gamma_2 = 0$:

$$z = \frac{\hat{\gamma}_2}{\widehat{\text{SE}}(\hat{\gamma}_2)} = \frac{-6.634}{3.021} = -2.196$$

Statistic	Value	CV (5%)	Decision
$\lvert z \rvert$	2.196	1.960	Reject H_0

$p = 0.028$; FTR at 1% (CV= 2.576).

4. LRT Confirmation

$$\text{LR} = 2(\ell_1 - \ell_0) = 2(-64.776 - (-67.305)) = 5.059$$

$\chi^2(1)$, $p = 0.025$. Consistent with Wald.

Practitioner Insight: The link test is distinctive in that it constructs a diagnostic predictor from the fitted model rather than comparing pre-specified nested models. By regressing $g(\mu)$ on $\hat{\eta}$ and $\hat{\eta}^2$, it tests both the correctness of the link function and the completeness of the linear predictor. Under correct specification, $\hat{\gamma}_1 \approx 1$ and $\hat{\gamma}_2 \approx 0$. Here $\hat{\gamma}_1 = -2.687$ and $\hat{\gamma}_2 = -6.634$, indicating that the original linear predictor is inadequate and that the model is capturing a non-monotone (inverted-U) pattern through the quadratic term. The rejection therefore reflects structural misspecification: a simple logistic model cannot capture a response that peaks at intermediate doses. Remedies include adding a quadratic term, using a non-monotone model, or investigating omitted variables. Unlike HL and LvH tests, which assess calibration and residual structure, the link test directly targets misspecification of the predictor's functional form.

Common Pitfalls

- **Mis-specifying the augmented model.** The link test must regress the outcome on $\hat{\eta}$ and $(\hat{\eta})^2$ *only*. Including the original predictors reintroduces perfect multicollinearity with $\hat{\eta} = X\hat{\beta}$, invalidates the test, and obscures link-function issues.
- **Treating a non-significant test as proof the link is correct.** Failure to reject merely means insufficient evidence of misspecification. Low power, small samples, or offsetting errors can mask real link problems—use the test as a screening tool, not definitive validation.
- **Using the link test as a full model diagnosis.** The test flags nonlinearity in $\hat{\eta}$ but cannot distinguish wrong link from wrong functional form, omitted variables, interactions, or outliers. Always pair it with residual plots, Box–Tidwell checks, GOF tests, and influence diagnostics.
- **Expecting the test to indicate which alternative link to use.** A significant $(\hat{\eta})^2$ term signals misspecification but provides no guidance on the appropriate replacement link. Compare candidate links (logit, probit, cloglog, cauchit), use AIC/BIC, and incorporate theory and domain knowledge.
- **Relying on the test in small samples or sparse binary data.** The asymptotic Wald test becomes unstable with $n < 100$, few events, or near separation. In such settings, supplement with cross-validation, AIC/BIC comparisons, or permutation-based variants, and interpret p-values conservatively.

Overdispersion Test (Cameron-Trivedi)

Test for Excess Variance in Count Models

GLMS.9 | **Category:** Generalized Linear Models | **Data:** Count | **Null:** Normal (asymptotic)

When to Use

- Assess overdispersion in count-data GLM models.
- Poisson variance assumption appears unrealistic.
- Residual variation exceeds model expectations.
- Decide whether negative binomial models are needed.
- Count outcomes exhibit unexplained extra variation.

Assumptions

- Observations are mutually independent.
- Mean structure and link function are appropriate.
- Response variable follows a count-data process.
- Baseline Poisson model is correctly specified.
- Sample size supports asymptotic inference theory.

Test Statistic

$$Z = \frac{\hat{\alpha}}{SE(\hat{\alpha})}, \quad \text{from auxiliary regression: } \frac{(y_i - \hat{\mu}_i)^2 - y_i}{\hat{\mu}_i} = \alpha\hat{\mu}_i + \epsilon_i.$$

Notation: y_i are counts, $\hat{\mu}_i$ are fitted means, α is overdispersion parameter

Plain English: The Cameron-Trivedi test examines whether the variance of count data exceeds what the Poisson model predicts by regressing a transformation of squared residuals on fitted values. Under true Poisson distribution, variance equals mean, making the regression slope zero. Positive slope indicates variance increases faster than mean, signaling overdispersion. The test provides formal evidence for switching from Poisson to negative binomial or quasi-Poisson models that accommodate extra variation.

⚠ Practitioner Warning

Test is one-sided (upper tail only) since focus is on overdispersion, not underdispersion. Low power with small samples or when overdispersion is modest. The auxiliary regression may have heteroskedastic errors, affecting standard error estimates; robust standard errors recommended. Test detects overdispersion but does not identify source (unobserved heterogeneity, omitted variables, clustering). If overdispersion detected, consider negative binomial model, quasi-Poisson with estimated dispersion, or mixed models with random effects. Zero-inflation can masquerade as overdispersion—examine zero counts separately.

Hypotheses

- H_0: $\alpha = 0$ (equidispersion).
- H_1: $\alpha > 0$ (overdispersion, variance exceeds mean).

Decision Rule: Reject H_0 if $Z > z_\alpha$ (one-tailed upper) or p-value $< \alpha$.

Software References [Python, R, SAS]

Python	R	SAS
`statsmodels::test_poisson_dispersion`	`AER::dispersiontest`	`Not Native (Manual)`

Additional Notes

The auxiliary regression uses transformed squared residuals $(y_i - \hat{\mu}_i)^2/\hat{\mu}_i - 1$ as response, regressing on $\hat{\mu}_i$ or functions thereof. Alternative formulations test $\text{Var}(y_i) = \hat{\mu}_i + \alpha\hat{\mu}_i^2$ (NB2 model) versus $\text{Var}(y_i) = \hat{\mu}_i + \alpha\hat{\mu}_i$ (NB1 model). A significant $\hat{\alpha} > 0$ is consistent with a negative binomial variance structure, but may also reflect unobserved heterogeneity or omitted predictors; adding covariates can sometimes eliminate apparent overdispersion. Dispersion parameter estimate from Pearson chi-square: $\hat{\phi} = X^2/(n-p)$ where $X^2 = \sum(y_i - \hat{\mu}_i)^2/\hat{\mu}_i$. Values $\hat{\phi} > 1$ suggest overdispersion; $\hat{\phi} > 1.5$ indicates serious concern. Dean's test provides alternative based on score statistic. For binomial data, similar test examines $\text{Var}(y_i/m_i) = p(1-p)/m_i + \alpha$ where m_i are trial counts. Negative binomial model parameterizes overdispersion: $\text{Var}(Y) = \mu + \mu^2/\theta$ where θ is dispersion parameter (smaller θ means more overdispersion).

Miscellanea: Colin Cameron and Pravin Trivedi (1990) developed a regression-based test for overdispersion in Poisson models, providing a simple and widely used diagnostic for count data. The approach uses an auxiliary regression to detect departures from the Poisson variance assumption.

Reference(s): Cameron & Trivedi (1990); Dean (1992)

Worked Example

Problem Statement An insurer records the number of claims y_i filed by each of $n = 6$ policyholders in a calendar year. An intercept-only Poisson GLM is fitted ($\hat{\mu}_i = \bar{y} = 4$ for all i). The Cameron–Trivedi (1990) test assesses whether the Poisson assumption $\text{Var}(Y_i) = \mu_i$ holds, against the alternative $\text{Var}(Y_i) = \mu_i + \alpha\mu_i$ (NB1-type overdispersion). H_0: $\alpha = 0$ (equidispersion); H_1: $\alpha > 0$ (overdispersion).

1. Auxiliary Variable z_i

The test is based on the auxiliary quantity $z_i = (y_i - \hat{\mu}_i)^2 - y_i$, which equals zero in expectation under the Poisson (H_0), and has positive expectation $\alpha\mu_i$ under the NB1 alternative.

i	y_i	$(y_i - \hat{\mu}_i)^2$	z_i	z_i^2
1	0	16	+16	256
2	0	16	+16	256
3	0	16	+16	256
4	6	4	−2	4
5	8	16	+8	64
6	10	36	+26	676
Σ	24		+80	1512

2. Regression Weight w_i

For the alternative $\text{Var} = \mu + \alpha\mu$, the weight is $w_i = g(\hat{\mu}_i)/\hat{\mu}_i$ where $g(\mu) = \mu$, so $w_i = 1$ for all i. Equivalently, the test regresses z_i on $\hat{\mu}_i$ (no intercept); since $\hat{\mu}_i = 4$ is constant, this reduces to testing whether $\bar{z} \neq 0$.

3. Slope Estimate and Robust SE

OLS estimate of α (no intercept):

$$\hat{\alpha} = \frac{\sum \hat{\mu}_i z_i}{\sum \hat{\mu}_i^2} = \frac{\bar{\mu}\sum z_i}{n\bar{\mu}^2} = \frac{\sum z_i}{n\bar{\mu}} = \frac{80}{6 \times 4} = \frac{10}{3}$$

Heteroscedasticity-robust (HC) standard error:

$$\widehat{\text{SE}}(\hat{\alpha}) = \frac{\sqrt{\sum \hat{\mu}_i^2 z_i^2}}{\sum \hat{\mu}_i^2} = \frac{\sqrt{\bar{\mu}^2 \sum z_i^2}}{n\bar{\mu}^2} = \frac{\sqrt{\sum z_i^2}}{n\bar{\mu}} \approx 1.620$$

4. Test Statistic and Decision

$$Z = \frac{\hat{\alpha}}{\widehat{\text{SE}}(\hat{\alpha})} = \frac{\sum z_i}{\sqrt{\sum z_i^2}} = \frac{80}{\sqrt{1512}} = \frac{40}{3\sqrt{42}} \approx 2.057$$

Statistic	Value	CV (5%)	Decision
Z	2.057	1.645	Reject H_0

$p = 0.020$ (one-sided); also rejects two-sided at 5% ($p = 0.040$, CV= 1.960).

Practitioner Insight: The test complements deviance and Pearson GoF tests by targeting the *variance function* rather than the mean structure. A model can fit the mean well yet exhibit overdispersion, which the CT test detects via the auxiliary variable $z_i = (y_i - \hat{\mu}_i)^2 - y_i$, for which $E[z_i] = 0$ under Poisson and $E[z_i] = \alpha\mu_i > 0$ under Negative Binomial (NB1). Here the large average of z_i reflects substantial variability driven by a bimodal pattern in the data. Robust (HC) standard errors are essential because z_i is inherently heteroscedastic. The simplified statistic $Z = \sum z_i / \sqrt{\sum z_i^2}$ applies only to intercept-only models; with covariates, the full regression-based formulation must be used. Alternative variance forms such as $g(\mu) = \mu^2$ (NB2) place greater weight on larger means and are preferred when overdispersion increases rapidly with μ.

Common Pitfalls

- **Confusing overdispersion with zero-inflation.** Cameron–Trivedi tests only $\text{Var}(Y) > \text{E}(Y)$; it does *not* test for excess zeros. Always check zero-inflation separately (ZIP/ZINB score tests, Vuong, observed vs expected zeros).
- **Mistaking misspecification for overdispersion.** Omitted predictors, wrong link, or missing nonlinear terms often inflate residual variance and trigger a false "overdispersion" signal. Verify Poisson specification first using residual plots and RESET/Link-type diagnostics.
- **Using CT with small samples or dependence.** The auxiliary regression is asymptotic; with $n < 100$ or clustered/panel data, the Z-statistic becomes unstable and overly liberal. Use bootstrap p-values, cluster-robust SEs, or random-effects Poisson/NB when dependence is present.
- **Assuming significance implies NB2 form.** CT detects *some* overdispersion but does not identify the correct variance function. Fit NB1 and NB2 and compare via AIC/BIC or mean–variance plots before choosing a model.
- **Applying CT outside standard Poisson settings.** CT assumes a conventional Poisson likelihood; it is invalid for ZIP/ZINB, hurdle, or zero-truncated counts, and for non-integer outcomes or rates without offsets. Use model-specific tests (ZIP→ZINB LR, Vuong, truncated Poisson vs truncated NB) in those settings.

Dean's Test

Score Test for Overdispersion in Poisson Regression

GLMS.10 | **Category:** Generalized Linear Models | **Data:** Count | **Null:** Chi-square

When to Use

- Assess overdispersion in fitted Poisson GLM models.
- Count-data variance appears larger than expected.
- Residual variability exceeds Poisson assumptions.
- Confirm findings from other overdispersion tests.
- Decide whether alternative count models are needed.

Assumptions

- Observations are mutually independent.
- Response variable follows a count-data process.
- Mean structure and link function are appropriate.
- Baseline Poisson model is correctly specified.
- Sample size supports asymptotic inference theory.

Test Statistic

$$D = \frac{\left[\sum_{i=1}^{n}(y_i - \hat{\mu}_i)^2 - \sum_{i=1}^{n} y_i\right]^2}{2\sum_{i=1}^{n} \hat{\mu}_i^2}, \quad \text{where } \hat{\mu}_i = \exp(\mathbf{x}_i'\hat{\boldsymbol{\beta}}). \qquad \text{Under } H_0,\ D \sim \chi_1^2.$$

Notation: y_i are observed counts, $\hat{\mu}_i$ are fitted means from Poisson model, numerator measures excess (Poisson) variation

Plain English: Dean's test evaluates overdispersion by comparing the observed sum of squared residuals to its expected value under the Poisson model. The numerator measures how far the observed squared residuals deviate from their expected value under equidispersion; the denominator scales this deviation by its theoretical variance. Large values indicate systematic excess variability beyond Poisson assumptions. As a score test, it requires only the fitted Poisson model without estimating alternative dispersion parameters.

⚠ Practitioner Warning

Test focuses on detecting overdispersion (upper tail); it is not designed to detect underdispersion. May have lower power than Cameron-Trivedi test in some settings but better theoretical properties as exact score test. Assumes Poisson mean structure correct—misspecified covariates can appear as overdispersion. The test detects presence of overdispersion but not its source (clustering, unobserved heterogeneity, missing covariates). If test rejects, examine whether negative binomial, zero-inflated Poisson, or hurdle models more appropriate. Very large samples make test overpowered—trivial overdispersion becomes significant.

Hypotheses

- H_0: Equidispersion holds (variance = mean, Poisson).
- H_1: Overdispersion present (variance > mean).

Decision Rule: Reject H_0 if $D > \chi^2_{\alpha,1}$ or p-value $< \alpha$. $df = 1$ always. For $\alpha = 0.05$, CV=3.84.

Software References [Python, R, SAS]

```
Not Native (Manual)          Not Native (Manual)          Not Native (Manual)
```

Additional Notes

Dean (1992) derived this as score (Lagrange multiplier) test for testing the dispersion parameter at the Poisson boundary (no overdispersion). The test statistic can be written as $D = S^2/(2V)$ where $S = \sum[(y_i - \hat{\mu}_i)^2 - y_i]$ is sum of adjusted squared residuals and $V = \sum \hat{\mu}_i^2$ is variance term. Thus, S measures deviation from the Poisson variance identity $E[(y_i - \mu_i)^2] = \mu_i$, and V provides the corresponding variance scaling under H_0. The test is closely related to the Pearson statistic, since $\sum r_i^2 = X^2$ with $r_i = (y_i - \hat{\mu}_i)/\sqrt{\hat{\mu}_i}$, but Dean's formulation explicitly centers the statistic to reflect expected variation under equidispersion. As a score test, it requires only the fitted Poisson model and avoids estimating an alternative (e.g. negative binomial) model. A significant result indicates variance exceeding the mean, motivating models with flexible variance functions such as quasi-Poisson or negative binomial. Breslow (1984) proposed similar score test for clustered binary data.

Miscellanea: C.B. Dean (1992) developed a score-based test for overdispersion in count data, derived from the negative binomial framework without requiring full model estimation. The test provides another alternative for assessing dispersion in Poisson-type models.

Reference(s): Dean (1992); Breslow (1984)

Worked Example

Problem Statement An insurer records the number of claims y_i filed by each of $n = 8$ policyholders in a calendar year. An intercept-only Poisson GLM is fitted, giving $\hat{\mu}_i = \bar{y} = 4$ for all i. Dean's (1992) score test assesses $H_0: \alpha = 0$ (equidispersion) against the NB1 alternative $\text{Var}(Y_i) = \mu_i + \alpha\mu_i$. H_0: $\alpha = 0$; H_1: $\alpha > 0$ (overdispersion).

1. Score Numerator

The score for α at the Poisson MLE is $S = \sum_i \mu_i[(y_i - \mu_i)^2/\mu_i - 1] = \sum_i[(y_i - \mu_i)^2 - \mu_i]$. With $\hat{\mu}_i = 4$ constant:

i	y_i	y_i−4	$(y_i-4)^2$	$(y_i-4)^2-4$
1	0	−4	16	12
2	0	−4	16	12
3	0	−4	16	12
4	6	+2	4	0
5	6	+2	4	0
6	6	+2	4	0
7	6	+2	4	0
8	8	+4	16	12
$\sum$	32	0	80	48

$$S = \sum_i (y_i - 4)^2 - n\bar{\mu} = 80 - 32 = 48$$

2. Theoretical Variance and Test Statistic

Under H_0, the asymptotic variance of S is $\text{Var}_0(S) = 2\sum_i \mu_i^2$. With $\hat{\mu}_i = 4$ constant:

$$\text{Var}_0(S) = 2\sum_i \hat{\mu}_i^2 = 2 \times n\bar{\mu}^2 = 2 \times 8 \times 16 = 256$$

Dean's score statistic:

$$P = \frac{S}{\sqrt{2\sum_i \hat{\mu}_i^2}} = \frac{48}{16} = 3$$

Statistic	Value	CV (5%)	Decision
P	3	1.645	Reject H_0

$p = 0.001$ (one-sided); also rejects two-sided at 1% ($p = 0.003$, CV= 2.576).

Practitioner Insight: Dean's test and the Cameron–Trivedi (CT) test target the same H_0 but use different variance standardizations. CT uses a robust (HC) standard error based on the sample, whereas Dean uses the theoretical Poisson variance under H_0. When the Poisson model is correctly specified, Dean's denominator is typically more efficient; CT trades some efficiency for robustness. In this example, Dean strongly rejects because the observed squared residuals are much larger than the Poisson variance identity allows. In regression settings with varying $\hat{\mu}_i$, both tests use observation-specific fitted means and no longer reduce to a simple ratio, though their asymptotic normal reference remains the same.

Common Pitfalls

- **Using Dean's test as a general model-fit test.** Dean's test targets *overdispersion* in Poisson regression, not overall adequacy of the mean structure. A significant result may reflect excess variance, but it does not identify whether the source is unobserved heterogeneity, clustering, zero inflation, or omitted predictors.
- **Assuming non-significant results prove equidispersion.** Dean's test can have limited power in small samples or when overdispersion is modest. Failure to reject should be read as lack of clear evidence of overdispersion, not confirmation that the Poisson model is correct.
- **Attributing overdispersion to stochastic variation without checking mean misspecification.** Significant Dean's test may arise from omitted covariates, wrong functional form, missing interactions, or incorrect exposure specification. Check the Poisson mean model first, then assess whether extra-Poisson variation remains.
- **Treating Dean's test as a substitute for alternative model fitting.** The test indicates whether overdispersion is present, but not which alternative model is best. After rejection, compare candidates such as quasi-Poisson, negative binomial, zero-inflated, hurdle, or random-effects count models as appropriate.
- **Interpreting Dean's statistic as a measure of effect size.** The statistic is a significance measure only. Quantify dispersion using measures such as $\hat{\phi} = X^2/(n-p)$ or a negative binomial dispersion parameter, and report uncertainty rather than relying on Dean's statistic alone.

Negative Binomial Alternatives

$$\text{NB1}: \text{Var}(Y_i) = \mu_i + \alpha\mu_i, \qquad \text{NB2}: \text{Var}(Y_i) = \mu_i + \alpha\mu_i^2.$$

Zero-Inflation Score Test

Score Test for Excess Zeros in Count Data Models

GLMS.11 | **Category:** Generalized Linear Models | **Data:** Count | **Null:** Asymptotic χ^2_1

When to Use

- Count data contain more zeros than expected by model.
- Poisson or NB models underpredict observed zeros.
- Structural zeros may exist beyond sampling variation.
- Evaluate need for a zero-inflated count model.

Assumptions

- Observations are mutually independent.
- Response variable follows a count-data process.
- Baseline count model is correctly specified.
- Sample size supports asymptotic inference theory.

Test Statistic

$$S = \frac{\left[\sum_{i=1}^{n} \frac{I(y_i=0)-p_{0i}}{p_{0i}}\right]^2}{\sum_{i=1}^{n} \frac{1-p_{0i}}{p_{0i}} - \sum_{i=1}^{n} \hat{\mu}_i}. \quad \text{Special case } (\mu_i = \mu)\text{: } S = \frac{(n_0 - n\hat{p}_0)^2}{n\hat{p}_0(1-\hat{p}_0) - n\bar{y}\,\hat{p}_0^2}. \qquad \text{Under } H_0,\ S \approx \chi^2_1.$$

Notation: $I(y_i = 0)$ is an indicator for zero outcomes, $\hat{\mu}_i$ are fitted means, $p_{0i} = \exp(-\hat{\mu}_i)$ are fitted zero probabilities, n is the sample size, $n_0 = \sum_{i=1}^{n} I(y_i = 0)$ is the observed number of zeros

Plain English: The zero-inflation test compares the number of observed zeros to the number expected under the fitted count model. Standard Poisson (P) or negative binomial (NB) models predict specific zero probabilities based on estimated means. If observed zeros substantially exceed predictions, this suggests a separate zero-generating process—some observations are "structural zeros" that could never be non-zero, not just sampling zeros from the count distribution. Significant excess zeros indicate need for zero-inflated (ZIP, ZINB) or hurdle-type models.

⚠ Practitioner Warning

Test detects excess zeros but does not prove a zero-inflation mechanism—other model misspecifications can also produce excess zeros (overdispersion, omitted variables, wrong distributional family). High correlation between overdispersion and zero-inflation complicates diagnosis. Hurdle models provide an alternative with a different interpretation. The test assumes the fitted model is otherwise correctly specified; poor mean structure can masquerade as excess zeros. With very small means ($\hat{\mu}_i < 0.1$), nearly all observations should be zero under Poisson, making the test uninformative.

Hypotheses

- H_0: Zero-inflation not indicated by the baseline count model.
- H_1: Zero-inflation suggested by the baseline count model.

Decision Rule: Reject H_0 if $S > \chi^2_{\alpha,1}$ or p-value $< \alpha$.

Software References [Python, R, SAS]

Python	R	SAS
`Not Native (Manual)`	`countreg::scoretest`	`PROC COUNTREG (ZIP);`

Additional Notes

For Poisson model, probability of zero is $P(Y_i = 0) = \exp(-\hat{\mu}_i)$. For negative binomial with dispersion θ: $P(Y_i = 0) = (1 + \hat{\mu}_i/\theta)^{-\theta}$. The NB version of the (above) score statistic uses these fitted zero probabilities in place of the Poisson p_{0i}, with the denominator adjusted accordingly. Zero-inflated models have mixture structure: $P(Y = 0) = \pi + (1-\pi)P_{\text{count}}(Y = 0)$ where π is probability of structural zero. Score test for zero-inflation: test $H_0 : \pi = 0$ in ZIP model without fitting full ZIP. The Vuong (1989) test compares non-nested models (e.g., Poisson vs. ZIP) via a normalized log-likelihood ratio. Wilson (2015) demonstrated that the Vuong test is unreliable specifically for zero-inflation testing and recommends the score test instead. Detailed explanation is provided at the end of this chapter. Hurdle models separate zero and positive count processes: $P(Y = 0) = \pi$, $P(Y = k \mid Y > 0) \propto P_{\text{count}}(Y = k)$ for $k > 0$.

Miscellanea: Diane Lambert (1992) introduced the zero-inflated Poisson (ZIP) model, providing a framework for separating structural zeros from sampling zeros. Van den Broek (1995) later developed a score test for formally assessing whether such a zero-inflated mechanism is needed.

Reference(s): Lambert (1992); Vuong (1989); van den Broek (1995); Wilson (2015)

Zero-Inflation Score Test

Worked Example

Problem Statement A quality engineer inspects $n = 10$ manufactured items and counts y_i, the number of surface defects on each. An intercept-only Poisson GLM is fitted, giving $\hat{\mu}_i = \bar{y} = 1$ defect per item on average. Seven of the ten items have zero defects — far more than a Poisson distribution with mean 1 predicts. H_0: Poisson model (no zero-inflation); H_1: zero-inflated Poisson ($\phi > 0$).

Item i	1	2	3	4	5	6	7	8	9	10	Total
Defects y_i	0	0	0	0	0	0	0	2	3	5	10
$\hat{\mu}_i$	1	1	1	1	1	1	1	1	1	1	10

1. Poisson MLE and Predicted Zero Rate

MLE under H_0: $\hat{\mu} = \bar{y} = 10/10 = 1$.

Predicted probability of zero under Poisson($\hat{\mu} = 1$):

$$\hat{p}_0 = e^{-\hat{\mu}} = e^{-1} \approx 0.3679$$

Expected number of zeros in $n = 10$ observations:

$$E_0 = n\,\hat{p}_0 = 10\,e^{-1} \approx 3.679$$

Observed zeros: $n_0 = 7$.

The observed zero rate (70%) is almost twice the Poisson prediction (36.8%), suggesting either overdispersion or genuine zero-inflation.

2. Score Statistic

The van den Broek score statistic tests whether the ZIP mixing weight ϕ departs from zero. Under H_0, the score for ϕ at $\phi = 0$ is:

$$U = n_0\,(e^{\hat{\mu}} - 1) - (n - n_0) = 7(e - 1) - 3 \approx 9.028$$

The van den Broek score statistic is:

$$S = \frac{(n_0 - n\hat{p}_0)^2}{n\hat{p}_0(1 - \hat{p}_0) - n\bar{y}\,\hat{p}_0^2} = 11.347 \sim \chi^2(1)$$

Statistic	Value	CV (5%)	Decision
S	11.347	3.841	Reject H_0

$p = 0.001$ (χ^2_1); also rejects at 0.1% (CV= 10.83, $p =$ 0.0008).

Practitioner Insight: The zero-inflation test targets excess zeros specifically, unlike Cameron–Trivedi or Dean tests, which assess overall overdispersion. A dataset may exhibit zero-inflation without general variance inflation, or vice versa. Here $n_0 = 7$ versus $E_0 = 3.68$ signals clear excess zeros, while the non-zero counts $(2, 3, 5)$ remain plausible under Poisson(1). The denominator $n\hat{p}_0(1 - \hat{p}_0) - n\bar{y}\,\hat{p}_0^2$ includes a correction for estimating $\hat{\mu}$, reducing the effective information; omitting it (using the binomial variance) understates the evidence. In regression settings, the test generalizes using $\sum_i (I(y_i = 0) - \hat{p}_{0i})/\hat{p}_{0i}$ with the corresponding adjusted denominator, retaining an asymptotic $\chi^2(1)$ null. A significant result motivates fitting a ZIP model and assessing structural zeros via covariates.

Common Pitfalls

- **Treating many zeros as automatic evidence of zero-inflation.** High zero counts may arise from low means, overdispersion, or misspecification. Always compare observed vs. expected zeros under Poisson/NB before invoking a zero-inflated alternative.
- **Skipping the overdispersion check.** Negative binomial overdispersion can often explain apparent excess zeros without a separate zero-inflation component. Check whether a more flexible baseline count model already accounts for the zero frequency before adding structural-zero machinery.
- **Using informal zero comparisons instead of formal tests.** Statements such as "observed 60% versus expected 40% zeros" are descriptive only. Use a formal score test or another appropriately derived model-comparison procedure rather than relying on raw percentage differences alone.
- **Assuming zero-inflation is constant across observations.** Zero-inflation probability may vary with covariates; failing to model $\pi(Z)$ leads to poor fit and misleading tests. Test constant vs. covariate-dependent inflation.
- **Using zero-inflation methods outside their intended setting.** Standard zero-inflation tests assume independent integer-valued count data. For semicontinuous outcomes with many zeros, use two-part models; for clustered, panel, or time-series counts, use methods that account for dependence.

Vuong Test

Test for Non-Nested Model Selection

GLMS.12 | **Category:** Generalized Linear Models | **Data:** Various | **Null:** Normal (asymptotic)

When to Use

- Choose among competing count-model specifications.
- Compare predictive adequacy using likelihood values.
- ZIP and NB models provide rival explanations.
- Several plausible models fit the data reasonably.
- Information criteria alone give unclear guidance.

Assumptions

- Observations are mutually independent.
- Candidate models are estimated on identical data.
- Models are non-nested and separately specified.
- Observation-level log-likelihoods are available.
- Sample size supports asymptotic normal inference.

Test Statistic

$$V = \frac{\sqrt{n}\bar{m}}{\hat{\sigma}_m}, \quad \text{where } \bar{m} = \frac{1}{n}\sum_{i=1}^{n} m_i, \quad m_i = \log\frac{L_{1i}}{L_{2i}}, \quad \hat{\sigma}_m^2 = \frac{1}{n}\sum_{i=1}^{n}(m_i - \bar{m})^2. \qquad \text{Under } H_0,\ V \approx N(0,1).$$

Notation: L_{ji} is likelihood of observation i under model j, m_i is log-likelihood ratio for observation i

Plain English: The test compares non-nested models by examining the average log-likelihood ratio across individual observations, standardized by its estimated standard deviation. If one model consistently fits better across observations, the test statistic becomes large in magnitude. Positive values favor the first model, negative values favor the second, and values near zero suggest models fit equally well. Unlike likelihood ratio tests for nested models, this test applies when neither model is a special case of the other.

⚠ Practitioner Warning

Test assumes models are non-nested and not observationally equivalent; if models are nearly indistinguishable, the test has low power. The test is inherently two-sided; the sign of V indicates which model is favored, but inference should be based on $|V|$. Wilson (2015) showed inflated Type I error when using Vuong for zero-inflation testing; score-based tests are preferred in that setting. The test does not inherently penalize model complexity; corrected versions (AIC- or BIC-adjusted Vuong) are often used. If models are nested or overlapping, standard likelihood ratio tests are more appropriate. Rejection identifies the *better approximation*, not a correct model—both candidates may still be misspecified.

Hypotheses

- H_0: H_0: Models fit equally well ($E(m_i) = 0$).
- H_1: H_1: One model fits better than the other.

Decision Rule: Reject H_0 if $|V| > Z_{\alpha/2}$ or p-value $< \alpha$. $V > 0$ favors model 1; $V < 0$ favors model 2.

Software References [Python, R, SAS]

`Not Native (Manual)`	`pscl::vuong`	`Not Native (Manual)`

Additional Notes

The test statistic can be written as $V = \frac{1}{\sqrt{n}}\sum_{i=1}^{n}\frac{m_i}{\hat{\sigma}_m}$ where individual contributions $m_i = \log(L_{1i}/L_{2i})$ are standardized by estimated standard deviation. Vuong (1989) proved the test is consistent for distinguishing between non-nested models under regularity conditions. The test requires computing observation-level likelihoods L_{ji} for each model, not just overall log-likelihoods. For comparing ZIP to Poisson: $L_{\text{ZIP},i} = \pi\mathbb{I}(y_i = 0) + (1-\pi)f_{\text{Poisson}}(y_i \mid \mu_i)$ versus $L_{\text{Poisson},i} = f_{\text{Poisson}}(y_i \mid \mu_i)$. Clarke (2007) test provides distribution-free alternative based on sign of m_i rather than magnitude. Calibrated Vuong test applies variance correction when models have different numbers of parameters. The test extends to other non-nested comparisons: logit vs. probit, different link functions, competing distributional families.

Miscellanea: Quang Vuong (1989) developed a likelihood-based test for comparing non-nested models, using a standardized average of log-likelihood ratios. The method is widely applied in model selection, though later work has highlighted limitations, particularly in testing zero-inflated models.

Reference(s): Vuong (1989); Clarke (2007); Wilson (2015)

Worked Example

Problem Statement Eight experimental plots are inspected for a soil pathogen; the count y_i of colonies per plate is recorded. Four plates show zero colonies and four show exactly three colonies each. For illustration, an intercept-only Poisson model f_1 and a Zero-Inflated Poisson (ZIP) model f_2 are fitted and compared using per-observation log-likelihood ratios $m_i = \log f_1(y_i \mid \hat{\mu}_1) - \log f_2(y_i \mid \hat{\phi}, \hat{\mu}_2)$. H_0: both models equally close to truth; H_1: one model is strictly closer.

Plot i	1	2	3	4	5	6	7	8	Total
Colonies y_i	0	0	0	0	3	3	3	3	12

1. Model MLEs

Poisson (f_1): $\hat{\mu}_1 = \bar{y} = 12/8 = 3/2$.

ZIP (f_2): $\hat{\phi} = 0.4684$, $\hat{\mu}_2 = 2.8214$. The ZIP MLE (intercept only) always satisfies $\hat{\phi} + (1-\hat{\phi})e^{-\hat{\mu}_2} = n_0/n = 1/2$ and $(1-\hat{\phi})\hat{\mu}_2 = \bar{y} = 3/2$.

2. Per-Observation Log-Likelihood Ratio m_i

For $y_i = 0$ (plots 1–4):

$$\log f_1(0) = -\hat{\mu}_1 = -\tfrac{3}{2}$$

At the ZIP MLE, $f_2(0) = n_0/n = \frac{1}{2}$, so:

$$\log f_2(0) = \log \tfrac{1}{2} = -\ln 2$$

$$m_0 = \log f_1(0) - \log f_2(0) = -\tfrac{3}{2} + \ln 2 \approx -0.807$$

For $y_i = 3$ (plots 5–8):

$$\log f_1(3) = 3\ln\tfrac{3}{2} - \tfrac{3}{2} - \ln 6 \approx -2.075$$

$$\log f_2(3) = \ln(1-\hat{\phi}) + 3\ln\hat{\mu}_2 - \hat{\mu}_2 - \ln 6 \approx -2.133$$

$$m_3 = -2.075 - (-2.133) \approx +0.058$$

3. Vuong Statistic

Since $n_0 = n_1 = 4$ (equal groups):

$$\bar{m} = \frac{m_0 + m_3}{2} = \frac{-0.807 + 0.058}{2} = -0.375$$

The sample standard deviation, using the equal-group simplification $s_m = \sqrt{8/7} \cdot |m_0 - m_3|/2 \approx 0.462$:

$$V = \frac{\sqrt{n}\,\bar{m}}{s_m} = \frac{\sqrt{8} \times (-0.375)}{0.462} = -2.292$$

Statistic	Value	CV (5%)	Decision
$\lvert V\rvert$	2.292	1.960	Reject H_0

$p = 0.022$ (two-sided); $V < 0$ favours f_2 (ZIP).
FTR at 1% (CV= 2.576).

Practitioner Insight: The Vuong test is designed for comparing *non-nested* models by asking which is closer to the true data-generating process in Kullback–Leibler divergence, without assuming either model is correct. The key quantity m_i is the per-observation log-likelihood ratio: $m_i > 0$ favors Poisson, while $m_i < 0$ favors ZIP. In this example, zero-count observations contribute negatively ($m_0 = \ln 2 - 3/2 < 0$), reflecting ZIP's advantage in modeling zeros, whereas the non-zero counts contribute slightly positively ($m_3 \approx 0.058$), favoring Poisson. The negative mean $\bar{m}$ shows that the zero-count advantage dominates overall. However, this example is only an illustration of the mechanics: with $n = 8$, the asymptotic normal approximation is weak, and Poisson is a boundary case of ZIP, so Vuong is not appropriate here as a formal zero-inflation test. Results should therefore be interpreted descriptively rather than inferentially.

Common Pitfalls

- **Using Vuong for nested or overlapping models.** Vuong is designed for *strictly non-nested* comparisons. If one model is a restricted version of the other, use a likelihood-ratio or score-based procedure instead. (The worked example above is included only to illustrate the mechanics of the statistic.)
- **Ignoring that Vuong has low reliability in small samples.** Vuong's statistic is asymptotic; with small n, p-values can be unstable and the normal approximation may be poor. In such settings, rely more heavily on AIC/BIC or other supporting diagnostics.
- **Interpreting non-significance as model equivalence.** A non-significant Vuong result means the models are *indistinguishable with current sample size*, not "equally good." Low power or both models being misspecified can produce the same outcome.
- **Comparing models on different samples or scales.** Vuong requires identical observations and the *same dependent variable*. Do not compare models fitted to different missing-data patterns or transformed responses without the appropriate likelihood adjustment.
- **Letting Vuong select overfitted models.** Vuong uses raw log-likelihoods and imposes no complexity penalty—complex models may "win" by overfitting. Always supplement Vuong with AIC/BIC or cross-validated predictive checks.

Hausman Test for IIA

Test Independence of Irrelevant Alternatives in Multinomial Models

GLMS.13 | **Category:** Generalized Linear Models | **Data:** Categorical | **Null:** Chi-square

When to Use

- Assess IIA validity in multinomial choice models.
- Verify stability of choices after option removal.
- Multinomial logit model assumptions are questioned.
- Alternatives may be close substitutes for others.
- Choice probabilities may depend on available options.

Assumptions

- Observations are mutually independent.
- Full and restricted models are consistently estimated.
- Models are fitted to the same underlying sample.
- Choice alternatives are clearly and uniquely defined.
- Sample size supports asymptotic chi-square inference.

Test Statistic

$$H = (\hat{\beta}_r - \hat{\beta}_f)'[\widehat{Var}(\hat{\beta}_r) - \widehat{Var}(\hat{\beta}_f)]^{-1}(\hat{\beta}_r - \hat{\beta}_f). \qquad \text{Under } H_0, H \approx \chi^2_k.$$

Notation: $\hat{\beta}_f$ from full choice set, $\hat{\beta}_r$ from restricted set, k is number of parameters compared

Plain English: The Hausman test for IIA compares parameter estimates from the full multinomial logit model (all alternatives included) to estimates from a restricted model (subset of alternatives removed). If IIA holds, removing alternatives should not systematically change coefficients for remaining alternatives—both models estimate the same underlying parameters. Substantial differences indicate IIA violation, suggesting that relative choice probabilities depend on the broader choice set, not only on the attributes of the remaining alternatives.

⚠ Practitioner Warning

Test requires estimating model twice: once with full choice set, once with restricted set. Choice of which alternatives to drop affects test results—no consensus on optimal strategy. Negative variance difference can occur (matrix not positive definite), invalidating test; suggests computational issues or severe IIA violation. Test has low power with small samples or when IIA violations are subtle. IIA violations common in practice (red bus/blue bus problem), but alternatives to multinomial logit (nested logit, mixed logit) computationally demanding. It is a specification test, not goodness-of-fit.

Hypotheses

- H_0: IIA holds (coefficients stable across choice sets).
- H_1: IIA violated (coefficients change with alternatives).

Decision Rule: Reject H_0 if $H > \chi^2_{\alpha,k}$ or p-value $< \alpha$. df = number of parameters compared.

Software References [Python, R, SAS]

Python	R	SAS
`Not Native (Manual)`	`mlogit::hmftest`	`Not Native (Manual)`

Additional Notes

Independence of irrelevant alternatives (IIA) means the ratio of choice probabilities for any two alternatives depends only on attributes of those alternatives: $P(i)/P(j) = \exp(\mathbf{x}_i'\beta - \mathbf{x}_j'\beta)$, independent of other alternatives in choice set. Classic violation: red bus/blue bus problem where adding blue bus as alternative to car/red bus splits bus riders but IIA predicts equal split among all three. Hausman (1978) proposed a general specification test comparing two estimators that are both consistent under the null, but only one is efficient. Under the null, their difference should be small. In the multinomial-logit IIA setting, full- and restricted-choice estimators should agree asymptotically if IIA holds. Small-Hsiao (1985) test provides alternative: randomly partition sample, estimate on subset, predict on complement, compare predicted to actual choices for dropped alternatives. Nested logit relaxes IIA by allowing correlation within nests (e.g., all bus types correlated). Mixed logit (random parameters) most flexible but computationally intensive.

Miscellanea: Jerry Hausman (1978) introduced a general specification test based on comparing estimators under different assumptions. The framework was later adapted to multinomial logit models to test the IIA assumption. The famous red bus/blue bus example, associated with McFadden, remains the canonical cautionary tale about taking IIA too literally in discrete-choice analysis.

Reference(s): Hausman (1978); Small & Hsiao (1985); McFadden (1974)

Worked Example

Problem Statement Ten travellers each choose between three transport modes — Car (A), Bus (B), and Train (C) — based on journey cost ($). A conditional logit model $P(j \mid \mathbf{c}_i) = \exp(-\beta c_{ij})/\sum_k \exp(-\beta c_{ik})$ is fitted with a single cost coefficient β. The Hausman–McFadden (1984) test checks the IIA assumption: under IIA, removing Train from the choice set should not change the estimated cost sensitivity for Car vs Bus. H_0: IIA holds (β is stable); H_1: IIA is violated (β shifts when Train is removed). Example data can be found at the end of this chapter.

1. Full Model: All Three Modes
Maximize the conditional log-likelihood over all $n = 10$ observations using $J = 3$ alternatives. MLE (numerical): $\hat\beta_F = 2.084$. Fisher information $\mathcal{I}_F = \sum_i \left[\overline{c_i^2} - (\bar c_i)^2\right] = 0.970$, where $\bar c_i$ and $\overline{c_i^2}$ are the fitted mean and mean-square cost under the full model. Hence:

$$\widehat{\mathrm{Var}}(\hat\beta_F) = \mathcal{I}_F^{-1} = \frac{1}{0.970} = 1.031$$

2. Restricted Model: Drop Train
Observations 2, 3, and 10 chose Train; they are excluded. The remaining $n_R = 7$ observations (choosing Car or Bus) are used to fit the binary conditional logit with $J = 2$ alternatives {Car, Bus}. MLE: $\hat\beta_R = 1.508$. Fisher information $\mathcal{I}_R = 0.920$:

$$\widehat{\mathrm{Var}}(\hat\beta_R) = \mathcal{I}_R^{-1} = \frac{1}{0.920} = 1.087$$

3. Hausman–McFadden Statistic
Under IIA, $\hat\beta_R$ and $\hat\beta_F$ estimate the same parameter; any difference is sampling error.

$$\hat d = \hat\beta_R - \hat\beta_F = 1.508 - 2.084 = -0.577$$

The variance of $\hat d$ under H_0:

$$\widehat{\mathrm{Var}}(\hat d) = \widehat{\mathrm{Var}}(\hat\beta_R) - \widehat{\mathrm{Var}}(\hat\beta_F) = 1.087 - 1.031 = 0.055$$

$$H = \frac{\hat d^2}{\widehat{\mathrm{Var}}(\hat d)} = \frac{(-0.577)^2}{0.055} = \frac{0.333}{0.055} = 5.994$$

Statistic	Value	CV (5%)	Decision
H	5.994	3.841	Reject H_0

$p = 0.014$; FTR at 1% (CV= 6.635).
The cost coefficient shifts from 2.084 to 1.508 when Train is removed — a relative change of 28%, inconsistent with IIA.

Practitioner Insight: The Hausman–McFadden test targets the *IIA assumption* underlying multinomial logit, not overall model fit. IIA requires relative choice probabilities (e.g. Car vs Bus) to remain unchanged when other alternatives are removed. Here, removing Train alters the Car–Bus tradeoff, indicating violation of IIA and suggesting that Bus and Train are closer substitutes. The test uses the Hausman form $H = \hat d^2/\widehat{\mathrm{Var}}(\hat d)$, where the variance difference reflects the fact that the restricted estimator is less efficient under the null than the full estimator. Under H_0, this difference estimates $\mathrm{Var}(\hat d)$; a negative value usually signals finite-sample or numerical instability rather than substantive support for IIA; many implementations then treat the test as undefined or set $H = 0$ by convention. Rejection motivates alternatives such as nested logit or mixed logit; if theory suggests that some options are closer substitutes (e.g. Bus and Train), a nested structure may be especially natural.

Common Pitfalls

- **Confusing the IIA Hausman test with the panel-data Hausman test.** They share a name but test different hypotheses; the IIA version checks stability of multinomial logit coefficients, not random vs. fixed effects. Always specify which test.
- **Dropping alternatives arbitrarily.** IIA tests depend heavily on which option is removed; arbitrary drops can hide or create violations. Drop theoretically similar or substitutable alternatives, not random ones.
- **Using the test with small or rare alternatives.** Sparse choices make variance estimates unstable and can yield negative Hausman statistics. The test needs reasonable counts per alternative; very rare categories undermine aymptotic validity.
- **Treating negative statistics as evidence for IIA.** A negative statistic signals numerical failure (non–positive-definite variance difference), not support for IIA. Treat such outcomes as invalid and rely on alternative tests (Small–Hsiao test, nested logit comparisons).
- **Interpreting non-significance as confirmation of IIA.** The test has limited power; failure to reject does not validate the IIA assumption. Check multiple dropped alternatives and corroborate with other IIA diagnostics.

Small-Hsiao Test

Test for Independence of Irrelevant Alternatives in Multinomial Logit

GLMS.14 | **Category:** Generalized Linear Models | **Data:** Categorical | **Null:** Chi-square (likelihood ratio)

When to Use

- Assess IIA validity in multinomial choice models.
- Multinomial logit assumptions appear questionable.
- Use as an alternative to Hausman-type IIA tests.

Assumptions

- Observations are mutually independent.
- Choice alternatives are clearly and uniquely defined.
- Random subsamples adequately represent the population.

Test Statistic

$$\chi^2_{SH} = -2[\ell_S - \ell_F]. \qquad \text{Under } H_0,\ \chi^2_{SH} \sim \chi^2_p \text{ where } p \text{ is number of parameters.}$$

Notation: ℓ_S is subsample log-likelihood with alternative omitted, ℓ_F is subsample log-likelihood with all alternatives

Plain English: The Small–Hsiao test randomly divides the sample into two parts. The first part is used to estimate the full multinomial logit model (all alternatives included), producing a set of parameter estimates. Those estimates are then used to construct a weighted log-likelihood on the second part. Separately, the restricted model—with one alternative removed—is re-estimated on the second part alone. The test compares these two log-likelihoods using a likelihood ratio statistic. If IIA holds, dropping an alternative should not alter substitution patterns among the remaining alternatives, so the two log-likelihoods should be similar. Results can vary across different random splits of the same data.

⚠ Practitioner Warning

Performance of the test depends critically on random data splitting, and results can vary across different random splits of the same dataset. The test has been shown in simulation studies to have unstable size and power properties, with actual Type I error rates sometimes differing substantially from nominal levels. Cheng and Long (2007) found that both Small-Hsiao and Hausman-McFadden tests performed poorly enough to conclude they are unsatisfactory for applied work. The test requires adequate sample size in both subsamples to estimate all parameters reliably—splitting small samples produces unstable estimates. Unlike Hausman-McFadden, Small-Hsiao does not produce negative chi-square values, but this computational advantage does not address fundamental statistical performance issues. When Small-Hsiao and Hausman-McFadden disagree, neither provides definitive evidence, and model specification should be reconsidered.

Hypotheses

- H_0: IIA holds: odds ratios unaffected by alternatives.
- H_1: IIA is violated: odds ratios affected by alternatives.

Decision Rule: Reject H_0 if $\chi^2_{SH} > \chi^2_{\alpha,p}$ or p-value $< \alpha$.

Software References [Python, R, SAS]

Python	R	SAS
`Not Native (Manual)`	`mlogit::mlogit`	`Not Native (Manual)`

Additional Notes

Small and Hsiao (1985) proposed a likelihood-based IIA test that uses data splitting to avoid the non-positive-definite variance issues of the Hausman–McFadden test. The statistic is asymptotically χ^2 under H_0. The test is performed separately for each alternative j, producing J test statistics. Simulation evidence (Cheng and Long, 2007) shows that both the Hausman–McFadden and Small–Hsiao tests can suffer from size distortions and low, unstable power against practically relevant IIA violations. Moreover, when IIA fails, restricted-model estimates may themselves be inconsistent, complicating interpretation. In practice, results should be interpreted cautiously and supplemented with structural reasoning about the choice process and, where appropriate, estimation of more flexible alternatives such as nested logit, mixed logit, or multinomial probit.

Miscellanea: Kenneth Small and Cheng Hsiao (1985) developed a test for the IIA property in multinomial logit models based on sample splitting. Motivated by limitations of existing approaches, the method provides an alternative diagnostic for IIA violations, though it is sensitive to the partitioning of the data. The red bus/blue bus example illustrates the practical importance of testing the IIA assumption.

Reference(s): Small & Hsiao (1985); Cheng & Long (2007); Fry & Harris (1996, 1998)

Worked Example

Problem Statement Twelve travellers chose a transport mode from three options — Car (A), Bus (B), Train (C) — based on journey cost. The sample is split at random into two equal halves: S_1 (travellers 1–6, full choice set) and S_2 (travellers 7–12, also full). The Small–Hsiao (1985) test checks IIA, H_0: IIA holds; H_1: IIA violated. See next page for data.

1. Full Model on S_1: Estimate $\hat{\beta}_1$
Fit $P(j \mid \mathbf{c}_i) = \exp(-\beta c_{ij})/\sum_k \exp(-\beta c_{ik})$ on all six S_1 observations using all three modes. Five travellers chose the cheapest option; traveller 4 chose Bus at cost 4 when Train (cost 3) was cheaper. This single 'irrational' choice prevents complete separation and yields a finite MLE:

$$\hat{\beta}_1 = 2.092$$

2. Restricted Model on S_{2R}: Estimate $\hat{\beta}_2$
S_{2R} contains the six travellers in S_2 who chose Car or Bus; Train is excluded. All six face identical costs (Car = 1, Bus = 2) and the sample splits evenly: 3 chose Car, 3 chose Bus. The binary conditional logit MLE satisfies $P(\text{Car}) = 3/6 = 1/2$, so:

$$\hat{\beta}_2 = 0 \quad \text{(exactly, by symmetry)}$$

3. Weighted Average and Log-Likelihoods on S_{2R}
Equal half-samples give equal weights (illustration):

$$\hat{\beta}_W = \tfrac{1}{2}\hat{\beta}_1 + \tfrac{1}{2}\hat{\beta}_2 = \tfrac{1}{2}(2.092) + \tfrac{1}{2}(0) = 1.046$$

Both $\hat{\beta}_1$ and $\hat{\beta}_W$ are now evaluated on S_{2R} using the Car/Bus log-likelihood $L_R(\beta) = 3\log P(\text{Car} \mid \beta) + 3\log P(\text{Bus} \mid \beta)$:

$$L_R(\hat{\beta}_1) = 3\log P_A(2.092) + 3\log P_B(2.092) = -6.973$$

$$L_R(\hat{\beta}_W) = 3\log P_A(1.046) + 3\log P_B(1.046) = -4.944$$

4. Test Statistic and Decision

$$\text{SH} = -2\left[L_R(\hat{\beta}_1) - L_R(\hat{\beta}_W)\right]$$

$$= -2(-6.973 - (-4.944)) = -2(-2.029) = 4.058$$

Statistic	Value	CV (5%)	Decision
SH	4.058	3.841	Reject H_0

$p = 0.044$ (χ^2_1); FTR at 1% (CV= 6.635).

Practitioner Insight: The Small–Hsiao test addresses the same IIA hypothesis as the Hausman–McFadden (HM) test but avoids its numerical instability by using *separate subsamples*. Estimating $\hat{\beta}_1$ on S_1 and $\hat{\beta}_2$ on S_{2R} ensures a well-defined variance, whereas HM can fail when $V_R - V_F$ is not positive semi-definite. The trade-off is reduced power due to sample splitting. The statistic compares how well $\hat{\beta}_1$ versus a pooled estimate $\hat{\beta}_W$ fits S_{2R}; under IIA, both estimates coincide and SH $\approx$ 0. Here the large discrepancy ($\hat{\beta}_1 = 2.092$ vs. $\hat{\beta}_2 = 0$) drives rejection, with $\hat{\beta}_2 = 0$ reflecting no cost signal in S_{2R}. A limitation is sensitivity to the random split, so results are often averaged over multiple splits or adjusted (e.g. McFadden–Train correction). In general, weighting formula is: $\hat{\beta}_W = \left(V_1^{-1} + V_2^{-1}\right)^{-1}\left(V_1^{-1}\hat{\beta}_1 + V_2^{-1}\hat{\beta}_2\right)$

Common Pitfalls

- **Confusing the Small–Hsiao test with the Hausman IIA test.** Both assess the Independence of Irrelevant Alternatives (IIA) assumption in multinomial logit models, but they rely on different ideas. The Hausman test compares parameter estimates across models, whereas the Small–Hsiao test evaluates prediction consistency using sample splitting.
- **Applying the test with small samples.** The Small–Hsiao procedure requires splitting the data into estimation and prediction subsamples. If the total sample is small, each subsample may contain too few observations to estimate stable multinomial logit parameters, leading to unreliable test results.
- **Ignoring sensitivity to the random sample split.** Because the method relies on randomly partitioning the data, different splits can yield different likelihood ratio statistics. Robust practice is to repeat the test with multiple random splits and check that conclusions remain consistent.
- **Interpreting a non-significant result as proof that IIA holds.** Failing to reject the null hypothesis only indicates insufficient evidence against the IIA assumption. With small samples or weak violations, the test may simply lack power to detect departures from IIA.
- **Applying the test when alternatives are unavailable for some individuals.** The Small–Hsiao test assumes that all individuals face the same choice set. When choice sets vary across observations (e.g., some alternatives are unavailable), the multinomial logit model and the resulting IIA test may be invalid.

X. GLMS — Raw Data

Data used in selected worked examples are provided here to preserve the book's strict two-page-per-test format.

[GLMS.13] Hausman Test for IIA

Problem Statement Ten travellers each choose between three transport modes — Car (A), Bus (B), and Train (C) — based on journey cost ($). A conditional logit model $P(j \mid \mathbf{c}_i) = \exp(-\beta c_{ij})/\sum_k \exp(-\beta c_{ik})$ is fitted with a single cost coefficient β. The Hausman–McFadden (1984) test checks the IIA assumption: under IIA, removing Train from the choice set should not change the estimated cost sensitivity for Car vs Bus. H_0: IIA holds (β is stable); H_1: IIA is violated (β shifts when Train is removed).

Traveller i	1	2	3	4	5	6	7	8	9	10
Cost_{Car}	1	4	5	4	4	1	3	2	2	4
Cost_{Bus}	3	5	2	3	5	2	3	3	1	5
$\text{Cost}_{\text{Train}}$	3	1	1	5	5	1	4	3	5	4
Choice	Car	Train	Train	Bus	Car	Car	Bus	Car	Car	Train

[GLMS.14] Small-Hsiao's Test for IIA

Problem Statement Twelve travellers chose a transport mode from three options — Car (A), Bus (B), Train (C) — based on journey cost. The sample is split at random into two equal halves: S_1 (travellers 1–6, full choice set) and S_2 (travellers 7–12, also full). The Small–Hsiao (1985) test checks IIA, H_0: IIA holds; H_1: IIA violated.

	Cost ($)			
Traveller	Car	Bus	Train	Choice
S_1: full model				
1	3	2	1	Train
2	3	2	1	Train
3	4	3	4	Bus
4	4	4	3	Bus*
5	3	2	4	Bus
6	3	4	4	Car

	Cost ($)			
Traveller	Car	Bus	Train	Choice
S_{2R}: restricted (Car/Bus only)				
7	1	2	–	Car
8	1	2	–	Car
9	1	2	–	Car
10	1	2	–	Bus
11	1	2	–	Bus
12	1	2	–	Bus

**Train (cost 3) was cheapest; Bus (cost 4) chosen.*

Wilks's Theorem & the χ^2 Approximation

Every likelihood ratio test in this compendium that produces an asymptotic χ^2 null distribution traces its justification to a single 1938 paper by Samuel Stanley Wilks. In "The Large-Sample Distribution of the Likelihood Ratio for Testing Composite Hypotheses" (*Annals of Mathematical Statistics*, 9(1):60–62), Wilks proved that under a set of regularity conditions—including correct model specification, parameter identifiability, and the null hypothesis holding in the interior of the parameter space—twice the log-likelihood ratio statistic,

$$\Lambda = -2\log\left(\frac{\sup_{\theta\in\Theta_0} L(\theta)}{\sup_{\theta\in\Theta} L(\theta)}\right) = 2\big[\ell(\hat{\theta}) - \ell(\hat{\theta}_0)\big],$$

converges in distribution to χ^2_ν as $n \to \infty$, where ν is the number of restrictions imposed by H_0—equivalently, the difference in the number of free parameters between the unrestricted and restricted models. The elegance of the result lies in its generality: the limiting distribution depends only on the dimension of the constraint, not on the specific functional form of the likelihood or the values of the true parameters.

The regularity conditions Wilks required are worth naming, because they are where the theorem can fail in practice. The likelihood must be twice continuously differentiable in a neighborhood of the true parameter, the Fisher information matrix must be positive definite (ruling out non-identifiable parameters), and—critically—the null hypothesis must not place the true parameter on the *boundary* of the parameter space. This last condition is violated, for example, when testing whether a variance component equals zero, whether a mixing proportion $\pi = 0$, or whether a regression coefficient is exactly at a constraint boundary. In such cases the χ^2_ν approximation does not hold and the correct null distribution is typically a mixture of chi-square distributions. The zero-inflation score test (GLMS.11) and the boundary issues with the Vuong test (GLMS.12) discussed next in this appendix are direct illustrations of this failure mode.

Wilks published this result at age 32, three years after joining the faculty at Princeton, where he would spend his entire career. The paper is three pages long. Despite its brevity, it became one of the most cited results in mathematical statistics, underpinning the likelihood ratio tests used across virtually every chapter of this compendium—from goodness-of-fit tests in generalized linear models to score tests in survival analysis. Wilks went on to make foundational contributions to multivariate analysis, including the lambda criterion (Λ_W) bearing his name in MANOVA, before his death in 1964. The 1938 theorem, however, remains his most consequential legacy: the quiet guarantee behind every p-value derived from $-2\log L$.

A Note on the Vuong Test for Zero-Inflation

The Vuong (1989) test is frequently cited in applied work as a diagnostic for zero-inflation—a way to decide whether a zero-inflated Poisson (ZIP) or zero-inflated negative binomial (ZINB) model is preferable to its standard counterpart. Despite its wide use, the test has a fundamental methodological flaw in this application, identified formally by Wilson (2015), and its use as a zero-inflation diagnostic should be avoided.

What the Vuong test was designed to do. Vuong (1989) proposed a model *comparison* test framed around the Kullback–Leibler (KL) divergence. Given two non-nested, separately estimated models M_1 and M_2, the test asks: which model is closer to the true data-generating process in the KL sense? The null hypothesis is that both models are *equidistant* from the truth. The test statistic is a normalized sum of individual log-likelihood ratios,

$$V = \frac{\frac{1}{\sqrt{n}}\sum_{i=1}^{n}\log\left(\frac{\hat{f}_1(y_i)}{\hat{f}_2(y_i)}\right)}{\hat{\omega}},$$

where $\hat{\omega}$ is the sample standard deviation of the individual log-likelihood differences, and $V \xrightarrow{d} N(0,1)$ under the null.

The boundary null problem. The trouble arises from repurposing the Vuong test to answer a different question: *Is a ZIP model necessary at all?* This amounts to testing the point null $H_0\colon \pi = 0$, where π is the structural-zero mixing probability in the ZIP model. At $\pi = 0$, the ZIP model collapses exactly to the Poisson model—the null parameter lies on the *boundary* of the ZIP parameter space, and the Poisson model is a *degenerate special case* of the ZIP, not a genuinely non-nested alternative.

This violates a core assumption of Vuong's framework. The asymptotic $N(0,1)$ distribution of V was derived under the condition that the two models are truly non-nested and separated. When one model is a boundary case of the other, the distribution of V under H_0 is no longer standard normal; it is a mixture distribution whose form depends on unknown nuisance parameters. The result is that the nominal Type I error rate is not achieved, and the test can be either anticonservative (over-rejecting when Poisson is correct) or have power that behaves erratically across the parameter space.

Practical consequence. In simulation studies, the Vuong test has been shown to produce spuriously significant results—favoring the ZIP model—even in data generated from a pure Poisson process. Researchers who use the Vuong test as a pre-test before reporting model results may systematically adopt the more complex ZIP or ZINB model when it is not warranted, inflating apparent model complexity without genuine inferential justification.

The preferred alternative. The van den Broek (1995) score test (GLMS.11) is designed specifically for the boundary null $H_0: \pi = 0$. It derives its test statistic from the score function of the ZIP log-likelihood evaluated at $\pi = 0$, using only the fitted baseline Poisson (or negative binomial) model—no ZIP estimation is required. Because the score test is derived from the correct likelihood geometry at the boundary, its χ^2_1 asymptotic null distribution is valid. For practitioners who need a formal test of zero-inflation, the score test is the appropriate tool.

For model *selection* between two separately fitted, genuinely non-nested models (e.g., ZIP vs. hurdle-Poisson), the Vuong test may be appropriate provided neither model is a boundary case of the other. In that narrower application, the test's conditions are satisfied. The problem is specifically and solely its misuse as a zero-inflation diagnostic.

References. Lambert (1992); van den Broek (1995); Vuong (1989); Wilson (2015).

A Note on Testing the IIA Assumption

Cheng and Long (2007) conducted a systematic simulation study of the two most widely used tests for the independence of irrelevant alternatives (IIA) assumption in multinomial logit models: the Hausman–McFadden (1984) test and the Small–Hsiao (1985) test. Their findings were unambiguous and discouraging. Both tests performed poorly under conditions where IIA violations are known to exist, and neither test reliably distinguished between models that satisfy IIA and those that do not. The Hausman–McFadden test produced non-positive-definite variance matrices in a substantial fraction of simulations—an algebraic failure that renders the test statistic undefined or negative, a well-known pathology that had been noted in applied work but not previously quantified at scale. The Small–Hsiao test, which was proposed partly to address the variance estimation problems in the Hausman approach, fared no better: its null distribution was highly sensitive to the random subsample partition used in estimation, producing results that varied materially across replications of the same dataset.

The deeper implication of Cheng and Long's findings is that passing either test provides little assurance that IIA holds, and failing either test is similarly uninformative about the severity or nature of the violation. Both tests have low and erratic power against the alternatives of practical interest. The authors conclude that applied researchers should not rely on formal IIA tests as gatekeepers for model selection, and instead recommend substantive reasoning about the choice set—whether alternatives are genuinely distinct in the perception of decision-makers—combined with estimation of alternative models (nested logit, mixed logit, multinomial probit) as a more reliable diagnostic strategy. The Hausman–McFadden and Small–Hsiao tests remain in the literature and in standard software, but their results should be interpreted with considerable caution.

X. GLMS — Likelihoods, MLEs, and Key Quantities

1. General GLM Framework

A GLM specifies a conditional distribution from the exponential family and a link function:

$$Y_i \sim \text{Exp. Family}, \qquad g(\mu_i) = \eta_i = x_i^\top \beta$$

- $\mu_i = E[Y_i]$; $g(\cdot)$: link function; η_i: linear predictor

Exponential family density/pmf:

$$f(y_i;\theta_i,\phi) = \exp\left(\frac{y_i\theta_i - b(\theta_i)}{a(\phi)} + c(y_i,\phi)\right)$$

$$E[Y_i] = b'(\theta_i), \qquad \text{Var}(Y_i) = b''(\theta_i)\,a(\phi)$$

2. Score, Fisher Information, and IRLS

Log-likelihood and score:

$$\ell(\beta) = \textstyle\sum_i \ell_i(\beta), \qquad U(\beta) = X^\top W(y-\mu) = 0$$

IRLS working weights and adjusted response:

$$W_i = \frac{1}{\text{Var}(Y_i)}\left(\frac{d\mu_i}{d\eta_i}\right)^2, \qquad z_i = \eta_i + (y_i - \mu_i)\frac{d\eta_i}{d\mu_i}$$

Fisher (expected) information:

$$\mathcal{I}(\beta) = X^\top W X, \qquad \text{Var}(\hat\beta) \approx \mathcal{I}(\hat\beta)^{-1}$$

IRLS (iterative reweighted least squares) update:

$$\beta^{(t+1)} = (X^\top W^{(t)} X)^{-1} X^\top W^{(t)} z^{(t)}$$

3. Normal (Identity Link)

$$Y_i \sim N(\mu_i,\sigma^2), \quad \mu_i = x_i^\top\beta$$

$$\ell(\beta) = -\tfrac{1}{2\sigma^2}\textstyle\sum(y_i-\mu_i)^2$$

$$\hat\beta = (X^\top X)^{-1}X^\top y, \qquad \text{Var}(\hat\beta) = \sigma^2(X^\top X)^{-1}$$

4. Binomial / Logistic Model

$$Y_i \sim \text{Bin}(m_i,\pi_i), \quad \log\frac{\pi_i}{1-\pi_i} = x_i^\top\beta$$

$$\ell(\beta) = \textstyle\sum\big[y_i\log\pi_i + (m_i - y_i)\log(1-\pi_i)\big]$$

$$\pi_i = \frac{e^{\eta_i}}{1+e^{\eta_i}}, \qquad W_i = m_i\pi_i(1-\pi_i)$$

5. Poisson Model

$$Y_i \sim \text{Poisson}(\mu_i), \quad \log\mu_i = x_i^\top\beta$$

$$\ell(\beta) = \textstyle\sum\big[y_i\log\mu_i - \mu_i - \log(y_i!)\big]$$

$$\mu_i = e^{\eta_i}, \qquad W_i = \mu_i$$

Offset for exposure n_i: $\log\mu_i = \log n_i + x_i^\top\beta$

6. Negative Binomial (NB2) - Quadratic Variance

$$\text{Var}(Y_i) = \mu_i + \alpha\mu_i^2 \quad (\alpha > 0\text{: overdispersion})$$

$$\ell = \textstyle\sum\Big[\log\Gamma(y_i+\alpha^{-1}) - \log\Gamma(\alpha^{-1}) - \log(y_i!) + \alpha^{-1}\log\frac{1}{1+\alpha\mu_i} + y_i\log\frac{\alpha\mu_i}{1+\alpha\mu_i}\Big]$$

As $\alpha \to 0$, NB2 $\to$ Poisson. α estimated jointly with β.

7. Gamma Model

$$Y_i \sim \text{Gamma}(\mu_i,\nu), \quad \log\mu_i = x_i^\top\beta$$

$$\ell(\beta) = \textstyle\sum\Big[\nu\log\frac{\nu y_i}{\mu_i} - \log y_i - \log\Gamma(\nu) - \frac{\nu y_i}{\mu_i}\Big]$$

$$\text{Var}(Y_i) = \mu_i^2/\nu, \qquad W_i = \nu \ \ (\nu = \text{shape})$$

Used for positive continuous outcomes with constant CV.

8. Multinomial Logit

$$P(Y_i = j) = \frac{e^{x_i^\top\beta_j}}{\sum_k e^{x_i^\top\beta_k}}, \qquad \ell = \textstyle\sum_i\sum_j I(y_i{=}j)\log P(Y_i{=}j)$$

Identifiability: set $\beta_J = 0$ for baseline category J.

9. Zero-Inflated Poisson (ZIP)

$$P(Y=0) = \phi + (1-\phi)e^{-\mu}, \qquad P(Y = y > 0) = (1-\phi)\,\frac{e^{-\mu}\mu^y}{y!}$$

Two-component likelihood: structural zeros (prob. ϕ) plus Poisson count process. Estimated via EM or direct maximization.

10. Deviance

For a fitted model vs. the saturated model:

$$D = 2\big[\ell(\hat\mu_{\text{sat}}) - \ell(\hat\mu)\big] = \textstyle\sum_i d_i$$

Approximate χ^2_{n-p} under correct model specification and regularity conditions.; used for model comparison and goodness of fit.

11. Canonical Links and Variance Functions

Family	Canon. Link	Var(Y)	Range
Normal	Identity	σ^2	$(-\infty,\infty)$
Binomial	Logit	$\mu(1-\mu)/m$	$[0,1]$
Poisson	Log	μ	$[0,\infty)$
Neg. Bin.	Log	$\mu + \alpha\mu^2$	$[0,\infty)$
Gamma	Reciprocal	μ^2/ν	$(0,\infty)$
Inv. Gauss	$1/\mu^2$	μ^3/λ	$(0,\infty)$

Non-canonical links (e.g., log for Binomial, identity for Poisson) are valid but may cause convergence issues.

Part XI

Multilevel & Random-Effects Modeling

Assessing variance components, clustering, and hierarchical structure

Mixed/Hierarchical Models

XI. MIXD — A Conceptual Primer

What Are Mixed Models?

- Combine **fixed effects $\boldsymbol{\beta}$** (population-level parameters of direct interest) and **random effects b** (unit-specific deviations treated as draws from a distribution).
- Natural framework for **clustered**, **hierarchical**, **longitudinal**, and **repeated-measures** data, where observations within the same group are correlated.
- The model form (suppressing group subscripts):

$$\mathbf{y} = \mathbf{X}\boldsymbol{\beta} + \mathbf{Z}\mathbf{b} + \boldsymbol{\varepsilon},$$

  with $\mathbf{b} \sim \mathcal{N}(\mathbf{0}, \mathbf{G})$, $\boldsymbol{\varepsilon} \sim \mathcal{N}(\mathbf{0}, \mathbf{R})$, and $\mathbf{b} \perp \boldsymbol{\varepsilon}$.
- Marginal covariance: $\mathbf{V} = \mathbf{Z}\mathbf{G}\mathbf{Z}^\top + \mathbf{R}$.

Fixed vs. Random Effects — The Key Distinction

	Fixed	Random
Levels	Complete set of interest	Sample from a population
Estimator	OLS/GLS (no shrinkage)	BLUP (shrunk $\to 0$)
Inference	t, F, Wald tests	LRT, score, Wald on **G**
Examples	Treatment, sex, time	Subjects, schools, clinics

BLUP (best linear unbiased predictor): $\hat{\mathbf{b}} = \mathbf{G}\mathbf{Z}^\top\mathbf{V}^{-1}(\mathbf{y} - \mathbf{X}\hat{\boldsymbol{\beta}})$. Shrinkage increases as $\mathbf{G} \to \mathbf{0}$ or $n_i \to 0$.

What Are We Testing?

Three distinct inferential targets, each requiring a different approach:

- **Fixed effects:** $H_0 : \mathbf{R}\boldsymbol{\beta} = \mathbf{r}$ (significance of predictors, contrasts).
- **Variance components:** $H_0 : \tau^2 = 0$ or $\mathbf{G} = \mathbf{0}$ (is a random effect needed at all?).
- **Model specification:** are random effects correlated with regressors (Hausman)?

ML vs. REML — The Critical Choice

- **ML** maximises the full marginal likelihood over $(\boldsymbol{\beta}, \mathbf{G}, \mathbf{R})$ simultaneously. Variance component estimates are **downward biased** because no correction is made for estimating $\boldsymbol{\beta}$.
- **REML** marginalises out $\boldsymbol{\beta}$ first, then maximises the residual (restricted) likelihood over $(\mathbf{G}, \mathbf{R})$. Produces **unbiased** variance component estimates; the default in most software.
- **Golden rule:**
  - Compare **random-effects structures** $\Rightarrow$ use REML (same fixed effects in both models).
  - Compare **fixed-effects structures** $\Rightarrow$ use ML (REML likelihoods with different fixed effects are *not* comparable).

The Marginal Likelihood

Integrating out the random effects **b**, the marginal log-likelihood under the LMM is

$$\ell(\boldsymbol{\beta}, \mathbf{G}, \mathbf{R}) = -\frac{1}{2}\Big[\log|\mathbf{V}| + (\mathbf{y} - \mathbf{X}\boldsymbol{\beta})^\top\mathbf{V}^{-1}(\mathbf{y} - \mathbf{X}\boldsymbol{\beta}) + n\log(2\pi)\Big]$$

The **REML log-likelihood** removes the dependence on $\boldsymbol{\beta}$ by working with $n - p$ error contrasts $\mathbf{K}^\top\mathbf{y}$ (where $\mathbf{K}^\top\mathbf{X} = \mathbf{0}$):

$$\ell_R(\mathbf{G}, \mathbf{R}) = -\frac{1}{2}\Big[\log|\mathbf{V}| + \log|\mathbf{X}^\top\mathbf{V}^{-1}\mathbf{X}| + \mathbf{y}^\top\mathbf{P}\mathbf{y} + (n-p)\log 2\pi\Big]$$

where $\mathbf{P} = \mathbf{V}^{-1} - \mathbf{V}^{-1}\mathbf{X}(\mathbf{X}^\top\mathbf{V}^{-1}\mathbf{X})^{-1}\mathbf{X}^\top\mathbf{V}^{-1}$ is the residual projection matrix.

Under $H_0 : \tau^2 = 0$ (no random effect), $\mathbf{G} = \mathbf{0}$ and $\mathbf{V}_0 = \mathbf{R} = \sigma^2\mathbf{I}$, so both likelihoods reduce to the OLS log-likelihood:

$$\ell_0 = -\frac{n}{2}\log\sigma^2 - \frac{1}{2\sigma^2}(\mathbf{y} - \mathbf{X}\hat{\boldsymbol{\beta}})^\top(\mathbf{y} - \mathbf{X}\hat{\boldsymbol{\beta}}).$$

The LRT statistic $\Lambda = -2(\ell_0 - \ell_1)$ measures how much the log-likelihood improves by adding the random effect; the Score test evaluates the gradient of ℓ (or ℓ_R) with respect to **G** at $\mathbf{G} = \mathbf{0}$ without fitting the full model.

The Boundary Problem

Testing $H_0 : \tau^2 = 0$ places the null on the **boundary** of the parameter space (variances cannot be negative). The standard χ^2_1 approximation for the LRT is **too conservative**:

$$-2\log\Lambda \xrightarrow{d} \tfrac{1}{2}\chi^2_0 + \tfrac{1}{2}\chi^2_1 \quad \text{under } H_0.$$

Practical fix: divide the nominal p-value from χ^2_1 by 2. For testing multiple variance components simultaneously, more complex mixture distributions apply; simulation or parametric bootstrap p-values are most reliable.

Tests for Random Effects / Variance Components

Likelihood Ratio Test (LRT): Fit the full model (with random effect) and the restricted model (without) by **ML**; compare $\Lambda = -2[\ell_0 - \ell_1] \stackrel{.}{\sim} \frac{1}{2}\chi^2_0 + \frac{1}{2}\chi^2_1$ (boundary correction applied).

Restricted LRT (REML-based): Same logic but both models fitted by REML. Valid *only* when the fixed-effects specification is identical in both models; produces less biased variance estimates than the ML version.

Breusch–Pagan LM Test: Score (Lagrange Multiplier) test for the presence of random effects in a panel or mixed model; does *not* require fitting the full random-effects model. Based on OLS residuals:

$$\text{LM} = \frac{n}{2(T-1)}\left[\frac{\sum_i(\sum_t \hat{e}_{it})^2}{\sum_i \sum_t \hat{e}_{it}^2} - 1\right]^2 \mathrel{\dot\sim} \chi_1^2.$$

Computationally attractive as a pre-screening test.

Score Test for Variance Components: Evaluates the gradient of the REML log-likelihood at $\mathbf{G} = \mathbf{0}$; asymptotically equivalent to the LRT at the null but requires only the null model. Computationally advantageous near the boundary since it requires only the null model; asymptotic behavior is similar to the LRT in large samples.

Wald Test for Variance Components: $W = \hat{\tau}^2/\widehat{\text{se}}(\hat{\tau}^2)$; compared to $\mathcal{N}(0,1)$ or χ_1^2. **Unreliable** near the boundary ($\tau^2 \approx 0$) because the normal approximation to $\hat{\tau}^2$ breaks down; prefer LRT or score test for variance components.

Tests for Fixed Effects

Wald Test:

$$W = (\mathbf{R}\hat{\boldsymbol\beta} - \mathbf{r})^\top[\mathbf{R}\widehat{\text{Var}}(\hat{\boldsymbol\beta})\mathbf{R}^\top]^{-1}(\mathbf{R}\hat{\boldsymbol\beta} - \mathbf{r}) \mathrel{\dot\sim} \chi_q^2 \text{ or } F_{q,\nu}.$$

Fast and simple but relies on the asymptotic normality of $\hat{\boldsymbol\beta}$ and accurate estimation of $\text{Var}(\hat{\boldsymbol\beta})$—both can fail in small samples.

F-Test with Kenward–Roger (KR) Approximation: Inflates $\widehat{\text{Var}}(\hat{\boldsymbol\beta})$ to account for uncertainty in estimated variance components, then approximates denominator degrees of freedom by matching moments of the F-distribution. **Preferred for small n or unbalanced designs**; computationally intensive but substantially better Type I error control than naive Wald.

Satterthwaite Approximation: A simpler df approximation for F-tests; faster than KR but less accurate when variance component estimation is uncertain. Default in some software (e.g. `lmerTest`).

Intraclass Correlation Coefficient (ICC)

Quantifies the proportion of total variance attributable to between-group differences:

$$\text{ICC} = \frac{\tau^2}{\tau^2 + \sigma^2} \in [0, 1].$$

- ICC ≈ 0: negligible clustering; OLS is adequate.
- ICC ≈ 1: near-perfect within-group homogeneity; group membership fully determines outcomes.
- **Design effect:** DEFF $= 1 + (n_c - 1)\cdot$ ICC where n_c is cluster size; inflates the effective sample size needed to achieve a given power.
- Testing H_0 : ICC $= 0$ is equivalent to testing $H_0 : \tau^2 = 0$ (boundary problem applies).

Hausman Specification Test

Tests whether the random-effects assumption—that group-level random effects $\mathbf{b}_i$ are **uncorrelated** with the regressors $\mathbf{X}$—holds:

$$H = (\hat{\boldsymbol\beta}_{\text{FE}} - \hat{\boldsymbol\beta}_{\text{RE}})^\top[\widehat{\text{Var}}(\hat{\boldsymbol\beta}_{\text{FE}}) - \widehat{\text{Var}}(\hat{\boldsymbol\beta}_{\text{RE}})]^{-1}(\hat{\boldsymbol\beta}_{\text{FE}} - \hat{\boldsymbol\beta}_{\text{RE}}) \mathrel{\dot\sim} \chi_k^2.$$

- **Fail to reject:** RE estimator is consistent and efficient; random effects model preferred.
- **Reject:** RE estimator is inconsistent; fixed effects model required (within-group estimator).
- A negative test matrix (making H undefined or negative) signals model misspecification; use robust Hausman variants.
- Primarily used in panel data (econometrics); less common in biostatistical mixed-model practice.

Decision Guide: Which Test to Use?

Question	Recommended test
Is any random effect needed?	Breusch–Pagan LM
Is $\tau^2 > 0$? (confirmatory)	LRT or REML-LRT
How large is clustering?	ICC test
Test fixed-effect coefficient	Wald (F or χ^2)
Fixed effects, small n	KR-adjusted F
Variance component $\neq 0$?	Score or LRT
RE vs FE model choice	Hausman

Common Pitfalls

- **Ignoring clustering:** OLS on clustered data underestimates standard errors; inflates Type I error.
- **ML for variance component LRT:** using ML instead of REML inflates Λ and over-rejects H_0.
- **REML for fixed-effects comparison:** REML likelihoods differ with different fixed effects; LRT is invalid.
- **Ignoring boundary:** using the raw χ_1^2 p-value for $H_0 : \tau^2 = 0$ is conservative by a factor of 2.
- **Wald test for variance components:** highly unreliable near $\tau^2 = 0$; always prefer LRT or score test.
- **KR vs. Satterthwaite:** Satterthwaite can be anti-conservative in very small or severely unbalanced designs; KR generally performs better in these settings, though neither is uniformly reliable with extreme imbalance.

Bottom Line

> Mixed model tests separate fixed-effects signal from random variability, but correct inference hinges on three things: choosing ML vs. REML appropriately, applying the boundary correction for variance component tests, and using KR-adjusted F-tests for fixed effects in small samples. The Hausman test is the gateway between random-effects and fixed-effects specifications in panel settings.

Likelihood Ratio Test for Random Effects

Test for Presence of Random Effects in Mixed Models

MIXD.1 | **Category:** Mixed Models | **Data:** Continuous (Clus. / Long.) | **Null:** Chi-square (mixture)

When to Use

- Test necessity of random effects in mixed models.
- Compare nested mixed-effects model structures.
- Assess subject or cluster-level variability.
- Longitudinal or hierarchical data are present.
- Select fixed-only versus mixed-effects models.

Assumptions

- Models being compared are properly nested.
- Observations are independent across clusters.
- Random effects follow normal distributions.
- Variance components are estimated reliably.
- Maximum-likelihood estimation is used.

Test Statistic

$$LR = -2(\ell_0 - \ell_1) = -2\log\left(\frac{L_0}{L_1}\right), \qquad \text{Under } H_0, LR \sim \frac{1}{2}\chi_0^2 + \frac{1}{2}\chi_1^2 \text{ (boundary problem).}$$

Notation: ℓ_0 from fixed effects model, ℓ_1 from mixed model (fixed + random effects)

Plain English: The likelihood ratio test for random effects compares a mixed model containing random effects to a simpler model with only fixed effects, using the ratio of their maximized likelihoods. If random effects are unnecessary, the mixed model likelihood should not substantially exceed the fixed effects likelihood.

⚠ Practitioner Warning

Standard χ^2 distribution does not apply because null hypothesis $\sigma_u^2 = 0$ is on boundary of parameter space (variance cannot be negative). Correct distribution is 50:50 mixture of χ_0^2 (point mass at zero) and χ_1^2, so critical values and p-values must be adjusted. Halve the p-value from standard χ_1^2 table, or use mixture critical values. REML (restricted maximum likelihood) preferred over ML for estimating variance components; use REML for both models when testing. Test assumes random effects normally distributed; violations can affect validity. With small number of groups ($J < 30$), asymptotic approximation poor.

Hypotheses

- H_0: $\sigma_u^2 = 0$ (no random effects).
- H_1: $\sigma_u^2 > 0$ (random effects present).

Decision Rule: Reject H_0 if $LR > c_\alpha$ where c_α from 50:50 mixture of χ_0^2 and χ_1^2. Equivalently: halve p-value from χ_1^2.

Software References [Python, R, SAS]

`statsmodels::MixedLM` | `lme4::anova` | `PROC MIXED (COVTEST)`

Additional Notes

Self and Liang (1987) derived the mixture distribution: $P(LR \le x) = 0.5P(\chi_0^2 \le x) + 0.5P(\chi_1^2 \le x)$ where χ_0^2 is point mass at zero. For testing multiple variance components simultaneously, mixture becomes more complex. Testing q variance components yields mixture of $\chi_0^2, \chi_1^2, \ldots, \chi_q^2$ with weights from multivariate normal orthant probabilities. Stram and Lee (1994) provided tables for testing multiple variance components. Alternative approach: parametric bootstrap simulating null distribution by repeatedly fitting models to data generated under H_0. Information criteria (AIC, BIC) provide alternative to formal testing, penalizing model complexity.

Miscellanea: The boundary problem in variance component testing plagued statisticians for decades. Early researchers naively applied standard χ^2 tests, yielding conservative results (too few rejections). Self and Liang's 1987 paper resolved the issue by deriving the mixture distribution, though the solution was foreshadowed by Chernoff (1954) in different context. Modern software now automatically applies corrections, but many analysts remain unaware of the underlying boundary issue.

Reference(s): Self & Liang (1987); Stram & Lee (1994)

Worked Example

Problem Statement A researcher measures test scores from $G = 5$ schools, $n = 4$ students per school ($N = 20$ total). The one-way random effects model $y_{ij} = \mu + u_i + e_{ij}$, with $u_i \sim N(0, \sigma_u^2)$ and $e_{ij} \sim N(0, \sigma_e^2)$, is compared to OLS ($\sigma_u^2 = 0$) via the likelihood ratio test. H_0: $\sigma_u^2 = 0$ (no school effect); H_1: $\sigma_u^2 > 0$.

i	Test Scores y_{ij}				$\bar{y}_{i\cdot}$	$\sum y_{ij}$	$\sum y_{ij}^2$	$n(\bar{y}_{i\cdot} - \bar{y})^2$	$\sum_j (y_{ij} - \bar{y})^2$	$(\sum_j (y_{ij} - \bar{y}))^2$	ℓ_i
1	8.2	10.5	9.1	10.3	9.525	38.1	366.39	6.452	9.939	25.806	−7.005
2	10.5	12.8	11.4	11.9	11.650	46.6	545.66	2.924	5.694	11.696	−6.163
3	8.6	10.2	9.5	11.3	9.900	39.6	395.94	3.204	7.104	12.816	−6.921
4	11.3	13.1	12.0	13.2	12.400	49.6	617.54	10.304	12.804	41.216	−6.783
5	9.2	11.5	10.4	10.9	10.500	42.0	443.86	0.348	3.208	1.392	−5.944
	Grand mean $\bar{y}$ = 10.795				Sum ⇒	215.9	2369.39	23.232			−32.816

1. Sum of Squares Decomposition

$$SS_T = \sum_{i,j} y_{ij}^2 - N\bar{y}^2 = 38.750, \quad SS_B = 23.232, \quad SS_W = SS_T - SS_B = 15.518.$$

2. OLS Log-Likelihood Under H_0, all $N = 20$ observations are i.i.d. with $\hat{\mu} = \bar{y}$:

$$\hat{\sigma}_e^2\big|_{H_0} = \frac{SS_T}{N} = 1.938, \quad \ell_{OLS} = -\frac{N}{2}\log(2\pi\hat{\sigma}_e^2|_{H_0}) - \frac{N}{2} = -34.993$$

3. Random Effects MLEs (H_1) Balanced one-way RE MLEs:

$$\hat{\sigma}_e^2 = \frac{SS_W}{N} = 0.776, \quad \hat{\sigma}_u^2 = \frac{SS_B/G - \hat{\sigma}_e^2}{n} = 0.968, \quad ICC = \frac{\hat{\sigma}_u^2}{(\hat{\sigma}_u^2 + \hat{\sigma}_e^2)} = 0.555, \quad \hat{\sigma}_e^2 + n\hat{\sigma}_u^2 = 4.646$$

4. RE Log-Likelihood Each school's marginal covariance: $\mathbf{V}_i = \hat{\sigma}_e^2 \mathbf{I}_4 + \hat{\sigma}_u^2 \mathbf{J}_4$, $\det(\mathbf{V}_i) = \hat{\sigma}_e^{2(n-1)}(\hat{\sigma}_e^2 + n\hat{\sigma}_u^2) = 2.170$.

$$\ell_i = -\frac{n}{2}\log(2\pi) - \frac{1}{2}\log(\det(\mathbf{V}_i)) - \frac{1}{2\hat{\sigma}_e^2}\left[\sum_j (y_{ij} - \bar{y})^2 - \frac{\hat{\sigma}_u^2}{\hat{\sigma}_e^2 + n\hat{\sigma}_u^2} \times (\sum_j (y_{ij} - \bar{y}))^2\right], \quad \ell_{RE} = \sum_i \ell_i = -32.816$$

5. LRT and Boundary Correction Because $H_0 : \sigma_u^2 = 0$ lies on the boundary ($\sigma_u^2 \geq 0$), the null distribution is the mixture $\frac{1}{2}\chi^2(0) + \frac{1}{2}\chi^2(1)$. CV = 2.706 (the $\chi^2(1)$ 10th-percentile (2α) upper tail). We **reject** H_0 at $\alpha = 0.05$.

$$LR = 2(\ell_{RE} - \ell_{OLS}) = 2(-32.816 - (-34.993)) = \boxed{4.354}, \quad p = \tfrac{1}{2}P(\chi^2(1) > 4.354) = \tfrac{1}{2}(0.037) = \boxed{0.018}$$

Practitioner Insight: The likelihood gain reflects how well the model separates variation into within- and between-group components: OLS uses $\hat{\sigma}_e^2 = SS_{\text{total}}/N$, while the RE model decomposes it into $\hat{\sigma}_e^2 = SS_W/N$ and $\hat{\sigma}_u^2$, capturing between-group heterogeneity. In practice, variance components are typically estimated via REML to adjust for fixed effects. However, LRTs comparing models with different *fixed* effects must use ML, whereas tests on *random* effects (as here) may use either REML or ML with the boundary correction.

Common Pitfalls

- **Comparing models fitted with REML instead of ML.** Likelihood ratio tests require comparable likelihoods. When testing the presence of random effects, both models must be fitted using **maximum likelihood (ML)** rather than REML.
- **Applying the test with too few clusters.** Power and validity depend primarily on the number of clusters (m), not the total number of observations (n). With few clusters ($m < 20$), the asymptotic reference distribution may be unreliable and results should be interpreted cautiously.
- **Using the LR test to compare fixed-effects and random-effects models.** Fixed-effects and random-effects estimators are not nested likelihood models. Comparisons between them should be performed using the **Hausman test** instead.
- **Changing fixed effects while testing random effects.** LR tests isolate the contribution of random effects only when both models have *identical fixed effects*. Differences in fixed predictors confound the interpretation of the LR statistic.

Restricted Likelihood Ratio Test (REML)

LRT for Variance Components Using REML Estimation

MIXD.2 | **Category:** Mixed Models | **Data:** Continuous (Clus. / Long.) | **Null:** Chi-square (mixture)

When to Use

- Test necessity of random effects in mixed models.
- Compare nested random-effects model structures.
- Assess subject or cluster-level variability.
- Longitudinal or hierarchical data are present.
- Compare variance components under REML fitting.

Assumptions

- Models being compared are properly nested.
- Fixed-effects structures are identical.
- Observations are independent across clusters.
- Random effects follow normal distributions.
- Restricted maximum likelihood is used.

Test Statistic

$$LR_{\text{REML}} = -2(\ell_{\text{REML},0} - \ell_{\text{REML},1}), \quad \text{where } \ell_{\text{REML}} \text{ is restricted log-likelihood.}$$

Notation: $\ell_{\text{REML},0}$ from restricted model, $\ell_{\text{REML},1}$ from full model, q is difference in variance parameters

Plain English: The REML-based LRT compares nested random effects structures using restricted likelihood, which removes the influence of fixed effects when estimating variance components. REML provides less biased variance estimates than maximum likelihood, especially with small sample sizes, by accounting for degrees of freedom consumed by fixed effects estimation. The test compares restricted likelihoods rather than full likelihoods, making it appropriate when models differ in random effects but share fixed effects.

⚠ Practitioner Warning

REML likelihoods cannot be compared across models with different fixed effects—use ML instead. Testing variance components at boundary (e.g., $\sigma_u^2 = 0$) still requires mixture distribution correction; halve p-value from standard χ^2 table. With small number of groups or highly unbalanced data, asymptotic approx. may be poor; consider parametric bootstrap. REML assumes random effects and errors are normally distributed; violations can affect test validity. When testing multiple variance components simultaneously, mixture distribution becomes complex; consult Stram-Lee tables or use simulation.

Hypotheses

- H_0: Simpler random effects structure sufficient.
- H_1: Additional variance components improve fit.

Decision Rule: Reject H_0 if $LR_{\text{REML}} > \chi^2_{\alpha,q}$ where q is difference in variance parameters.

Software References [Python, R, SAS]

`statsmodels::MixedLM` `lme4::lmer` `PROC MIXED (REML COVTEST)`

Additional Notes

REML maximizes likelihood of error contrasts (linear combinations orthogonal to fixed effects) rather than full likelihood. This effectively integrates out fixed effects, yielding unbiased variance component estimates regardless of fixed effects. REML log-likelihood differs from ML by term depending on fixed effects design matrix: $\ell_{\text{REML}} = \ell_{\text{ML}} - \frac{1}{2}\log|\mathbf{X}'\mathbf{V}^{-1}\mathbf{X}|$ where $\mathbf{X}$ is fixed effects design matrix and $\mathbf{V}$ is marginal covariance matrix. For balanced designs, REML estimates coincide with ANOVA-based estimates. Information criteria for REML: use $\text{AIC}_{\text{REML}} = -2\ell_{\text{REML}} + 2p$ where p counts variance parameters only, not fixed effects. Kenward-Roger approximation adjusts degrees of freedom for fixed effects tests with REML estimation.

Practitioner insight: REML adjusts for the loss of df from estimating fixed effects, producing less biased variance estimates than ML, especially in small samples. However, REML likelihoods cannot be used to compare models with different fixed effects—such comparisons require ML.

Miscellanea: Patterson and Thompson (1971) introduced REML, aiming to obtain less biased variance component estimates in the presence of fixed effects. The method gained widespread adoption in mixed model software despite computational complexity.

Reference(s): Patterson & Thompson (1971); Verbeke & Molenberghs (2000)

Worked Example

Problem Statement The same five-school dataset from LRT test ($G = 5$, $n = 4$, $N = 20$) is re-analyzed using REML log-likelihood. REML conditions on a set of linear contrasts that remove the fixed effects, yielding unbiased variance component estimates. H_0: $\sigma_u^2 = 0$; H_1: $\sigma_u^2 > 0$. Sum of Squares: $SS_W = 15.518$, $SS_B = 23.232$, $SS_T = 38.750$

1. ANOVA Quantities and REML Estimates

$$\hat{\sigma}_e^2 = MS_W = \frac{SS_W}{N-G} = 1.034, \quad MS_B = \frac{SS_B}{G-1} = 5.808$$

REML divides by unbiased degrees of freedom:

$$\hat{\sigma}_u^2 = \frac{MS_B - \hat{\sigma}_e^2}{n} = 1.193, \quad ICC = \frac{\hat{\sigma}_u^2}{(\hat{\sigma}_u^2 + \hat{\sigma}_e^2)} = 0.536$$

Under H_0 the REML estimator uses $N - 1 = 19$ degrees of freedom (one fixed effect partialled out):

$$\hat{\sigma}_e^2|_{H_0} = \frac{SS_T}{N-1} = 2.039$$

2. REML Log-Likelihood Formula

$$\ell_{\text{REML}} = \ell_{\text{ML}}(\hat{\mu}, \hat{\sigma}^2) \underbrace{- \tfrac{1}{2}\log|\mathbf{X}'V^{-1}\mathbf{X}|}_{\text{REML adj.}}$$

For an intercept-only model, $\mathbf{1}_N' V^{-1}\mathbf{1}_N = N/(\hat{\sigma}_e^2 + n\hat{\sigma}_u^2)$, so the REML adjustment is:

$$-\tfrac{1}{2}\log\frac{N}{\hat{\sigma}_e^2 + n\hat{\sigma}_u^2} = +\tfrac{1}{2}\log\left(\hat{\sigma}_e^2 + n\hat{\sigma}_u^2\right) - \tfrac{1}{2}\log N$$

	$\hat{\sigma}_e^2$	$\hat{\sigma}_u^2$	LR	p
ML	0.776	0.968	4.354	0.018
REML	1.034	1.193	5.995	0.007

3. School-Level ML Contributions (H_1)

With $\hat{\sigma}_e^2 = 1.034, \hat{\sigma}_u^2 = 1.193$: $\det(V_i) = 1.034^3 \times 5.808 = 6.430$, $\log\det = 1.861$.

School	$\sum(y_{ij} - \bar{y})^2$	$(\sum(y_{ij} - \bar{y}))^2$	$\mathbf{y}'V^{-1}\mathbf{y}$	ℓ_i
1	9.939	25.806	4.482	−6.847
2	5.694	11.696	3.181	−6.197
3	7.104	12.816	4.322	−6.767
4	12.804	41.216	4.191	−6.702
5	3.208	1.392	2.825	−6.019

$$\ell_{\text{ML}}(H_1) = \textstyle\sum_i \ell_i = -32.531$$

REML adj (H_1): $-\frac{1}{2}\log(20/5.808) = -0.618$

$$\ell_{\text{REML}}(H_1) = -32.531 + (-0.618) = -33.150$$

Under H_0 ($\hat{\sigma}_e^2 = 2.039$, $\sigma_u^2 = 0$): $\ell_{\text{ML}}(H_0) = -35.006$, REML adj $= -\frac{1}{2}\log(20/2.039) = -1.142$,

$$\ell_{\text{REML}}(H_0) = -35.006 + (-1.142) = -36.147$$

4. REML LRT and Decision

$$LR_{\text{REML}} = 2(\ell_{\text{REML}}(H_1) - \ell_{\text{REML}}(H_0)) = 2(2.998) = \boxed{5.995}$$

$$p = \tfrac{1}{2}P(\chi^2(1) > 5.995) = \tfrac{1}{2}(0.014) = \boxed{0.007}$$

The boundary-adjusted critical value 2.706. We therefore reject H_0, indicating significant between-group variability.

Practitioner Insight: REML and ML differ in both estimates and log-likelihood. REML uses $N - G$ and $G - 1$ (vs. N and G for ML), increasing variance estimates (here $\hat{\sigma}_e^2$: $0.776 \to 1.034$, $\hat{\sigma}_u^2$: $0.968 \to 1.193$). The REML adjustment penalizes H_0 more than H_1, yielding $LR_{\text{REML}} = 5.995 > LR_{\text{ML}} = 4.354$. The boundary correction applies equally to both. In practice, use REML for estimating variance components, but ML for comparing models with different fixed effects, since REML likelihoods are not comparable across differing fixed effects.

Common Pitfalls

- **Comparing models with different fixed effects using REML likelihood ratios.** REML likelihoods are not comparable when the fixed-effect structure differs. When testing or comparing fixed effects, models must be estimated using ML.
- **Using REML when the primary goal is fixed-effects inference.** REML is designed to estimate variance components with reduced bias, not to test fixed effects. For hypothesis tests on fixed effects, ML-based likelihood ratio tests or appropriate Wald tests should be used.
- **Applying REML in very small samples or with few clusters.** Although REML reduces bias relative to ML, reliable variance component estimation still requires adequate cluster sizes and numbers of clusters. With small J or sparse data, estimates can be highly unstable.
- **Ignoring convergence diagnostics after estimation.** REML relies on iterative optimization, and failure to converge indicates the estimates may not correspond to the true likelihood maximum. Convergence messages, gradients, and Hessian diagnostics should always be checked before interpreting results.

Breusch-Pagan LM Test (Random Effects)

Lagrange Multiplier Test for Panel Random Effects

MIXD.3 | **Category:** Mixed Models | **Data:** Continuous (Panel) | **Null:** Chi-square

When to Use

- Test for random effects in panel-data models.
- Distinguish pooled OLS from random-effects models.
- Repeated observations on the same units exist.
- Assess panel-level heterogeneity formally.
- Preliminary step before random-effects estimation.

Assumptions

- Panel observations are correctly identified.
- Errors are independent across panel units.
- Random effects are uncorrelated with regressors.
- Variance components are estimated reliably.
- Panel dimensions are adequate for asymptotics.

Test Statistic

$$LM = \frac{nT}{2(T-1)}\left[\frac{\sum_{i=1}^{n}\left(\sum_{t=1}^{T}\hat{u}_{it}\right)^2}{\sum_{i=1}^{n}\sum_{t=1}^{T}\hat{u}_{it}^2} - 1\right]^2, \quad \text{where } \hat{u}_{it} \text{ are pooled OLS residuals.} \quad \text{Under } H_0, LM \sim \chi_1^2.$$

Notation: n is individuals, T is time periods per individual, $\hat{u}_{it}$ are OLS residuals

Plain English: The Breusch-Pagan LM test detects random effects by examining whether residuals are correlated within individuals. It compares the sum of squared individual-specific residual means to total residual variation. If no random effects exist, within-individual residual means should be small relative to overall variation. Large individual-specific means indicate persistent individual differences (random effects) beyond what fixed predictors explain. The test requires only pooled OLS estimation, avoiding the complexity of fitting random effects models.

⚠ Practitioner Warning

Originally derived for balanced panels; extensions are needed for unbalanced data. Assumes random effects are uncorrelated with predictors; if violated, random effects estimates are inconsistent and fixed effects are preferred—use the Hausman test to assess this. The test has low power with small T or n. As a score test, it is effectively one-sided, detecting positive variance components. Assumes homoskedastic errors; violations can affect size and power. Requires panel structure ($T > 1$); not applicable to pure cross-sectional data.

Hypotheses

- H_0: No random effects (pooled OLS adequate, $\sigma_u^2 = 0$).
- H_1: Random effects present ($\sigma_u^2 > 0$, within correlation).

Decision Rule: Reject H_0 if $LM > \chi^2_{\alpha,1}$ or p-value $< \alpha$.
Critical value: $\chi^2_{0.05,1} = 3.84$. One degree of freedom always.

Software References [Python, R, SAS]

`statsmodels::het_breuschpagan` `plm::plmtest(type="bp")` `PROC PANEL (RANONE BP)`

Additional Notes

The test statistic can be written in terms of correlation structure. Define $\bar{u}_i = T^{-1}\sum_{t=1}^{T}\hat{u}_{it}$ as individual-specific mean residual. The numerator $\sum_{i=1}^{n}(\sum_{t=1}^{T}\hat{u}_{it})^2 = T^2\sum_{i=1}^{n}\bar{u}_i^2$ measures between-individual variation. Under random effects, this exceeds what would be expected from independent errors. Breusch and Pagan (1980) derived the test as score (Lagrange Multiplier) test for variance component σ_u^2 in one-way error components model: $y_{it} = \mathbf{x}_{it}'\beta + u_i + \epsilon_{it}$ where $u_i \sim N(0, \sigma_u^2)$ are individual effects. Honda (1985) and King-Wu (1997) proposed modifications for unbalanced panels. For two-way random effects (individual and time effects), test becomes χ_2^2. Typical workflow: (1) Breusch-Pagan LM test for random effects vs. pooled OLS, (2) if reject, Hausman test for random vs. fixed effects.

Miscellanea: Breusch and Pagan (1980) introduced a Lagrange Multiplier test for detecting random effects in panel data, providing a computationally simple alternative to fitting full random effects models. The test uses OLS residuals to assess whether unobserved heterogeneity across groups is significant, and it remains a standard diagnostic tool in panel data econometrics.

Reference(s): Breusch & Pagan (1980); Honda (1985)

Breusch-Pagan LM Test (Random Effects) MIXD.3

Worked Example

Problem Statement Five-school dataset from LRT test ($n = 5, T = 4, N = 20$) is tested for random school effects using the Breusch-Pagan Lagrange Multiplier test (1980). Student dimension is treated as Time dimension to mimic panel structure. Unlike the ML and REML LRTs, the BP LM test is a *score test* evaluated under H_0: it requires only OLS residuals and never fits the random effects model. H_0: $\sigma_u^2 = 0$; H_1: $\sigma_u^2 > 0$.

1. OLS Residuals Fit OLS (intercept only): $\hat{\mu} = \bar{y}_{..} = 10.795$. Residuals $\hat{e}_{ij} = y_{ij} - 10.795$ (See table below)

2. The Two Key Quantities

$$A = \sum_{i=1}^{n}\left(\sum_{j=1}^{T}\hat{e}_{ij}\right)^2 = 92.926$$

$$B = \sum_{i=1}^{n}\sum_{j=1}^{T}\hat{e}_{ij}^2 = SS_T = 38.750, \quad \frac{A}{B} = 2.398$$

Sch	$\hat{e}_{i1}$	$\hat{e}_{i2}$	$\hat{e}_{i3}$	$\hat{e}_{i4}$	$\sum_j \hat{e}_{ij}$	$(\sum_j \hat{e}_{ij})^2$
1	−2.595	−0.295	−1.695	−0.495	−5.080	25.806
2	−0.295	+2.005	+0.605	+1.105	+3.420	11.696
3	−2.195	−0.595	−1.295	+0.505	−3.580	12.816
4	+0.505	+2.305	+1.205	+2.405	+6.420	41.216
5	−1.595	+0.705	−0.395	+0.105	−1.180	1.392

3. LM Statistic Under H_0, $LM \xrightarrow{d} \chi^2(1)$ (no boundary correction).

$$LM = \frac{nT}{2(T-1)}\left(\frac{A}{B} - 1\right)^2 = \boxed{6.516}$$

4. Results and Decision

Test	Stat	CV (5%)	p	Null dist.
BP LM	6.516	3.841	0.011	$\chi^2(1)$
ML LRT	4.354	2.706	0.018	$\frac{1}{2}\chi^2(0) + \frac{1}{2}\chi^2(1)$
REML LRT	5.995	2.706	0.007	$\frac{1}{2}\chi^2(0) + \frac{1}{2}\chi^2(1)$

5. Interpretation

- All three tests reject H_0. The BP LM test confirms significant between-school variation without fitting the random effects model. $A/B = 2.398 \gg 1$ indicates that group sums of residuals are disproportionately large relative to the total sum of squares, the signature of positive intraclass correlation.

Practitioner Insight: The BP LM statistic has a clear interpretation: $A/T = \sum_i T\,\bar{e}_i^2$ represents between-group residual variation, while $B = \sum_{ij}\hat{e}_{ij}^2$ is total variation, so A/B measures their relative share. Under H_0, $A/B \approx 1$, and the scaling $N/(2(T-1))$ makes $(A/B - 1)^2 \sim \chi^2(1)$. Unlike the LRT, the LM test uses the standard $\chi^2(1)$ (no boundary correction) since it evaluates the score at $\sigma_u^2 = 0$. The larger LM statistic (6.516 vs. 4.354) reflects differing null distributions, not just magnitude. The Honda refinement uses $z = \sqrt{LM} \sim N(0,1)$, preferred when -ve ICC is impossible, yielding $p = 0.005$.

Common Pitfalls

- **Confusing the Breusch–Pagan LM random-effects test with the Breusch–Pagan heteroskedasticity test.** Despite sharing the same authors, the two tests address different problems. The panel-data BP-LM test evaluates whether random effects are present ($\sigma_u^2 = 0$), whereas the classical BP test evaluates heteroskedasticity in cross-sectional regression.
- **Interpreting non-rejection as proof that random effects are absent.** Failure to reject $H_0 : \sigma_u^2 = 0$ indicates insufficient statistical evidence, not proof that individual effects do not exist. Low power (especially with small T or weak variance components) may prevent detection of meaningful random effects.
- **Using the test with very small time dimension (T).** The LM statistic relies on separating within- and between-individual variation. When T is very small, variance components are estimated imprecisely and the test may have low power.
- **Concluding that a random-effects model is appropriate solely because BP-LM is significant.** The test only distinguishes pooled OLS from models with individual effects. If the LM test rejects pooled OLS, a Hausman test is still required to determine whether random effects or fixed effects are appropriate.
- **Applying the test after transformations that remove individual effects.** The BP-LM test must be applied to pooled OLS residuals in levels. If the data have already been first-differenced or within-transformed, individual effects have been removed and the test becomes meaningless.

Score Test for Variance Components

Lagrange Multiplier Test for Random Effects Variance

MIXD.4 | **Category:** Mixed Models | **Data:** Continuous (Clus. / Long.) | **Null:** Chi-square (modified)

When to Use

- Test variance components in mixed-effects models.
- Assess necessity of random effects efficiently.
- Longitudinal or hierarchical data are present.
- Fit reduced model only before testing.
- Alternative to LRT for boundary-value problems.

Assumptions

- Reduced model is correctly specified.
- Observations are independent across clusters.
- Random effects follow normal distributions.
- Variance components are estimated reliably.
- Null places variance components on boundaries.

Test Statistic

$$T = \frac{1}{2}\mathbf{S}'\mathcal{I}_{22}^{-1}\mathbf{S}, \quad \text{where } \mathbf{S} = \left.\frac{\partial \ell}{\partial \theta}\right|_{\theta=0}, \quad \mathcal{I}_{22} = -E\left[\frac{\partial^2 \ell}{\partial\theta\partial\theta'}\right] \quad \text{Under } H_0, T \sim \frac{1}{2}\chi_0^2 + \frac{1}{2}\chi_1^2 \text{ (one random effect).}$$

Notation: $\mathbf{S}$ is score vector for variance parameters, $\mathcal{I}_{22}$ is information matrix block, $\boldsymbol{\theta}$ are variance components

Plain English: The score test for variance components evaluates whether random effects are needed by examining the gradient (score) of the log-likelihood with respect to variance parameters, evaluated at the null values (zero variance). If the null hypothesis is correct, this gradient should be near zero. Large scores indicate the likelihood could be substantially increased by adding random effects. The test requires fitting only the fixed effects model, avoiding computational burden of mixed model estimation under the alternative.

⚠ Practitioner Warning

Test suffers from boundary problem since null hypothesis $\sigma_u^2 = 0$ is on parameter space boundary. Correct distribution is 50:50 mixture of χ_0^2 and χ_1^2, not standard χ_1^2; halve p-value from standard chi-square table. Testing multiple variance components simultaneously yields more complex mixtures; consult Stram-Lee tables. Score function requires computing derivatives of likelihood, more complex than simple residual-based tests. Test assumes null model correctly specifies fixed effects; misspecified fixed effects can masquerade as need for random effects. Generally less powerful than LRT but computationally attractive when alternative model difficult to fit.

Hypotheses

- H_0: $\sigma_u^2 = 0$ (no random effects).
- H_1: $\sigma_u^2 > 0$ (random effects present).

Decision Rule: Reject H_0 if $T > c_\alpha$ from 50:50 mixture of χ_0^2 and χ_1^2. Equiv: halve p-value from standard χ_1^2 table.

Software References [Python, R, SAS]

`statsmodels::MixedLM` `RLRsim::RLRTSim` `PROC GLIMMIX`

Additional Notes

For linear mixed model $\mathbf{y} = \mathbf{X}\boldsymbol{\beta} + \mathbf{Z}\mathbf{u} + \boldsymbol{\epsilon}$ where $u \sim N(\mathbf{0}, \sigma_u^2\mathbf{I})$, the score for σ_u^2 evaluated at $\sigma_u^2 = 0$ is $q = \frac{1}{2}\text{tr}(\mathbf{PZZ}'\mathbf{PZZ}') - \frac{1}{2\sigma^2}\hat{\boldsymbol{\epsilon}}'\mathbf{PZZ}'\mathbf{P}\hat{\boldsymbol{\epsilon}}$ where $\mathbf{P} = \mathbf{I} - \mathbf{X}(\mathbf{X}'\mathbf{X})^{-1}\mathbf{X}'$ is projection onto residual space and $\hat{\boldsymbol{\epsilon}}$ are fixed effects residuals. Silvapulle and Silvapulle (1995) provided simplified computational forms. For testing q variance components simultaneously, mixture distribution involves weights from multivariate normal orthant probabilities. Crainiceanu and Ruppert (2004) developed restricted likelihood ratio test (RLRT) combining score and LR test ideas. Bootstrap methods provide alternative to mixture distributions when analytical results unavailable.

Miscellanea: C. Radhakrishna Rao's (1948) score test framework underlies many tests for variance components, but boundary constraints (e.g., testing a variance equal to zero) complicate standard inference. Self and Liang (1987) showed that such tests follow mixture chi-square distributions rather than the usual χ_1^2, resolving a long-standing issue in likelihood-based inference. Score tests remain valuable in settings where fitting alternative models is computationally demanding.

Reference(s): Rao (1948); Self & Liang (1987); Silvapulle & Silvapulle (1995)

Score Test for Variance Components

MIXD.4

Worked Example

Problem Statement Five-school dataset from LRT test ($G = 5$, $n = 4$, $N = 20$) is tested for random school effects using the Score Test for Variance Components. Both the ML and REML versions are derived from OLS quantities alone — no random effects model is fitted. H_0: $\sigma_u^2 = 0$; H_1: $\sigma_u^2 > 0$.

1. OLS Residuals Fit OLS (intercept only): $\hat{\mu} = \bar{y}_{..} = 10.795$. Residuals $\hat{e}_{ij} = y_{ij} - 10.795$ (See Breusch-Pagan LM)

2. OLS Quantities Under H_0

$$A = \sum_{i=1}^{G}\left(\sum_{j=1}^{n} \hat{e}_{ij}\right)^2 = 92.926$$

$$B = \sum_{i=1}^{G}\sum_{j=1}^{n} \hat{e}_{ij}^2 = SS_T = 38.750,$$

$$\hat{\sigma}_e^2\big|_{H_0} = \frac{B}{N} = 1.938$$

2. ML Score and Information at H_0
At $\sigma_u^2 = 0$, $V = \hat{\sigma}_e^2 I_N$:

$$S_{\text{ML}}(0) = \frac{A - B}{2\hat{\sigma}_e^4} = \frac{92.928 - 38.750}{2(1.938)^2} = \frac{54.178}{7.508} = 7.217$$

$$\mathcal{I}_{\text{ML}}(0) = \frac{G\,n^2}{2\hat{\sigma}_e^4} = \frac{5 \times 16}{2(1.938)^2} = \frac{80}{7.508} = 10.656$$

$$T_{\text{ML}} = \frac{S_{\text{ML}}(0)^2}{\mathcal{I}_{\text{ML}}(0)} = \frac{(7.217)^2}{10.656} = \boxed{4.887}$$

Under H_0: $T \xrightarrow{d} \chi^2(1)$ for both versions.

3. REML Score and Information at H_0
The REML log-likelihood adds a correction for the fixed intercept. Its score gains an adjustment term:

$$S_{\text{REML}}(0) = S_{\text{ML}}(0) + \frac{n}{2\hat{\sigma}_e^2} = 8.249$$

The REML information replaces G by $G - 1$ (one fixed effect absorbed):

$$\mathcal{I}_{\text{REML}}(0) = \frac{(G-1)\,n^2}{2\hat{\sigma}_e^4} = 8.525,$$

$$T_{\text{REML}} = \frac{S_{\text{REML}}(0)^2}{\mathcal{I}_{\text{REML}}(0)} = \boxed{7.982}$$

4. Five-Test Comparison

Test	Stat	Null dist.	p	CV (5%)
ML Score	4.887	$\frac{1}{2}\chi^2$ mix	0.014	2.706
REML Score	7.982	$\frac{1}{2}\chi^2$ mix	0.002	2.706
ML LRT	4.354	$\frac{1}{2}\chi^2$ mix	0.018	2.706
REML LRT	5.995	$\frac{1}{2}\chi^2$ mix	0.007	2.706
BP LM	6.516	$\chi^2(1)$	0.011	3.841

Practitioner Insight: The score test evaluates the log-likelihood gradient at H_0: a large $S(0)$ indicates that increasing σ_u^2 would improve model fit. The ML form $S_{\text{ML}}(0) \propto (A - B)$ captures excess between-group variation, while the REML adjustment typically increases the statistic (often $T_{\text{REML}} > T_{\text{ML}}$) by accounting for fixed-effects estimation. Unlike likelihood ratio tests, the score test requires only the null (fixed-effects) model and avoids fitting the full mixed model. However, like LRTs, it is affected by boundary constraints, so inference is based on the mixture distribution $\frac{1}{2}\chi_0^2 + \frac{1}{2}\chi_1^2$. Compared to the Breusch–Pagan LM test, it uses a difference $(A - B)$ rather than a ratio $(A/B - 1)$ with different scaling.

Common Pitfalls

- **Using a standard χ_1^2 reference distribution.** Variance component hypotheses lie on the boundary of the parameter space under the null. The correct reference distribution is the mixture $\frac{1}{2}\chi_0^2 + \frac{1}{2}\chi_1^2$, not the ordinary χ_1^2.
- **Applying the test when the null model is misspecified.** The score test assumes that the fixed-effects structure is correctly specified under the null model. Omitted covariates can produce residual structure that mimics random effects.
- **Testing several variance components without accounting for boundary complications.** When multiple variance components are tested simultaneously, the null distribution becomes a mixture of several chi-square distributions. Sequential testing or specialized software is typically required for valid inference.
- **Confusing the score test with the likelihood ratio test.** The score test evaluates derivatives of the likelihood using only the null model fit. Likelihood ratio tests instead require estimation of both the null and alternative models.
- **Applying the test with very few groups or clusters.** The asymptotic mixture distribution assumes a sufficient number of independent groups. With very small group counts, simulation-based or exact procedures may prove more reliable.

Wald Test for Variance Components

Test Variance Parameters Using Asymptotic Normality

MIXD.5 | **Category:** Mixed Models | **Data:** Continuous (Clus. / Long.) | **Null:** Chi-square or Normal

When to Use

- Test variance components in mixed-effects models.
- Assess significance of random-effect variances.
- Longitudinal or hierarchical data are present.
- Full mixed-effects model has been fitted.
- Quick inference using parameter estimates.

Assumptions

- Full model is correctly specified.
- Observations are independent across clusters.
- Random effects follow normal distributions.
- Variance estimates are approximately normal.
- Variance components are not near boundaries ($\sigma^2 > 0$).

Test Statistic

$$W = \frac{(\hat{\sigma}_u^2 - \sigma_{u,0}^2)^2}{\widehat{Var}(\hat{\sigma}_u^2)}, \quad \text{or } W = (\hat{\boldsymbol{\theta}} - \boldsymbol{\theta}_0)'\widehat{Cov}(\hat{\boldsymbol{\theta}})^{-1}(\hat{\boldsymbol{\theta}} - \boldsymbol{\theta}_0). \qquad \text{Under } H_0, W \sim \chi_q^2 \text{ (q variances) asymp.}$$

Notation: $\hat{\sigma}_u^2$ is variance component estimate, $\sigma_{u,0}^2$ is null value, $\hat{\boldsymbol{\theta}}$ is the estimator of variance parameters

Plain English: The Wald test for variance components evaluates hypotheses about variance parameters by examining whether estimated variances differ significantly from hypothesized values relative to their standard errors. The test uses asymptotic normality of variance component estimates from maximum likelihood or REML, requiring only the fitted model without refitting under constraints. The approach provides convenient inference on variance structure after model fitting.

⚠ Practitioner Warning

Wald test performs very poorly when testing variance at boundary ($\sigma^2 = 0$)—use LRT or score test with mixture distribution instead. Not recommended for testing presence of random effects, only for comparing non-zero variance values. Standard errors of variance components can be poorly estimated with small samples, leading to unreliable test. The test is not invariant to reparameterization; testing $\sigma_u^2 = 0$ versus $\sigma_u = 0$ can yield different results. Generally less reliable than LRT for variance components due to skewed sampling distribution of variance estimates.

Hypotheses

- H_0: $\sigma_u^2 = \sigma_{u,0}^2$ (typically, $\sigma_{u,0}^2 > 0$).
- H_1: $\sigma_u^2 \neq \sigma_{u,0}^2$ (variance differs from hypothesized value).

Decision Rule: Reject H_0 if $W > \chi_{\alpha,q}^2$ or p-value $< \alpha$. Do not use for testing $\sigma^2 = 0$; use LRT with mixture dist.

Software References [Python, R, SAS]

`statsmodels::MixedLMResults` `lme4::lmerTest` `PROC MIXED (COVTEST WALD)`

Additional Notes

Variance of variance component estimate obtained from information matrix: $\widehat{Var}(\hat{\sigma}_u^2) = [\mathcal{I}^{-1}]_{\theta\theta}$ where $\mathcal{I}$ is expected or observed information matrix from likelihood (REML uses restricted likelihood). Asymptotic distribution of variance estimates often highly skewed, especially for small true variances, making Wald test unreliable. A log transformation with the delta method can improve normality. Testing equality of two variance components: $W = (\hat{\sigma}_1^2 - \hat{\sigma}_2^2)^2/\widehat{Var}(\hat{\sigma}_1^2 - \hat{\sigma}_2^2)$ where covariance between estimates must be included. Profile likelihood confidence intervals for variance components generally superior to Wald intervals, avoiding boundary and skewness issues. Bartlett correction can improve performance by adjusting for skewness but rarely implemented in most software.

Miscellanea: Wald's (1943) test assumes parameters lie in the interior of the parameter space, an assumption violated when testing variance components at zero. This invalidates the usual normal approximation at the boundary. While early software reported Wald tests for variances, modern practice discourages their use in this setting, favoring likelihood ratio or score-based methods. Wald tests remain appropriate for parameters away from the boundary.

Reference(s): Wald (1943); Verbeke & Molenberghs (2000)

Wald Test for Variance Components **MIXD.5**

Worked Example

Problem Statement The same five-school dataset from LRT test ($G = 5$, $n = 4$, $N = 20$) is tested for random school effects using the Wald test applied directly to the REML variance component estimate $\hat{\sigma}_u^2$. The Wald test divides the estimate by its standard error and compares to a reference distribution. H_0: $\sigma_u^2 = 0$; H_1: $\sigma_u^2 > 0$. (χ^2 mixture applied).

1. REML Estimates (ANOVA Mean Squares)

Source	SS	df	MS
Between	23.232	4	5.808
Within	15.518	15	1.034

$$\hat{\sigma}_e^2 = MS_W = 1.034$$

$$\hat{\sigma}_u^2 = \frac{MS_B - MS_W}{n} = \frac{5.808 - 1.034}{4} = 1.193$$

2. Variance of $\hat{\sigma}_u^2$

$\hat{\sigma}_u^2 = (MS_B - MS_W)/n$ is a linear combination of two independent mean squares:

$$\text{Var}(\hat{\sigma}_u^2) = \frac{1}{n^2}\left[\underbrace{\frac{2\,MS_B^2}{G-1}}_{\text{Var}(MS_B)} + \underbrace{\frac{2\,\hat{\sigma}_e^4}{N-G}}_{\text{Var}(MS_W)}\right] = 1.063$$

$$\widehat{SE}(\hat{\sigma}_u^2) = \sqrt{1.063} = 1.031$$

3. Wald Statistic

$$z = \frac{\hat{\sigma}_u^2}{\widehat{SE}(\hat{\sigma}_u^2)} = \frac{1.193^2}{1.063} = 1.157 \qquad (\sim N(0,1))$$

$$W = z^2 = (1.157)^2 = \boxed{1.340} \qquad (\sim \tfrac{1}{2}\chi^2(1))$$

4. Results and Decision

Test	Stat	p	CV	Dec
REML Score	7.982	0.002	2.706	Reject
REML LRT	5.995	0.007	2.706	Reject
BP LM	6.516	0.011	3.841	Reject
ML LRT	4.354	0.018	2.706	Reject
ML Score	4.887	0.014	2.706	Reject
Wald	**1.340**	**0.124**	2.706[†]	**FTR**

Two-sided $p = 0.247$; one-sided $p = 0.124$ (using $z = 1.157$, CV= 1.645). Neither rejects at 5%. 95% CI: [0.000, 3.214] (lower bound truncated at zero).

The SE is nearly as large as the estimate itself, signaling high uncertainty from only $G = 5$ groups.

Practitioner Insight: The Wald test for variance components is typically weak because $\hat{\sigma}_u^2$ is highly uncertain when G is small. Its SE, driven by $\text{Var}(MS_B) = 2MS_B^2/(G-1)$, remains large regardless of N, so adding observations per group offers limited improvement—more groups are needed. Unlike LRT and score tests, which leverage the ratio MS_B/MS_W, the Wald test relies on the less sensitive difference $MS_B - MS_W$. The resulting wide CI (e.g., [0, 3.21]) and non-rejection, despite other tests rejecting, reflect its known conservatism and poor Gaussian approximation of the likelihood curvature.

Common Pitfalls

- **Using the Wald test to test whether a variance component equals zero[†].** Variance parameters lie on the boundary of the parameter space under the null hypothesis. Although Wald tests are also affected by boundary issues, no simple mixture χ^2 correction restores valid inference. LRT or score tests with appropriate χ^2 mixture are generally preferred.
- **Applying the Wald test when the estimated variance is near the boundary.** The normal approximation underlying the Wald statistic performs poorly when variance estimates are very small. In such cases, likelihood-based or simulation-based inference often provides more reliable results.
- **Treating reported standard errors of variance components as fully reliable.** Standard errors are typically based on asymptotic approximations that may be unstable in small samples or unbalanced designs. Profile likelihood or bootstrap approaches can provide more dependable uncertainty estimates.
- **Testing multiple variance components without adjustment.** Conducting several Wald tests increases the probability of false discoveries. Family-wise error control or joint testing procedures should be considered when evaluating multiple components.
- **Using Wald tests in models with very few clusters or groups.** The chi-square approximation underlying the Wald statistic assumes a sufficiently large number of groups. With small cluster counts, simulation-based likelihood ratio methods often yield more reliable inference.

Wald Test for Fixed Effects

Test Fixed Effect Parameters in Mixed Models

MIXD.6 | **Category:** Mixed Models | **Data:** Continuous (Clus. / Long.) | **Null:** Chi-square or F-distribution

When to Use

- Test significance of fixed effects in mixed models.
- Assess predictors in longitudinal data analyses.
- Hierarchical or clustered data are present.
- Jointly test one or more fixed-effect parameters.
- Inference after fitting a mixed-effects model.

Assumptions

- Mixed-effects model is correctly specified.
- Observations are independent across clusters.
- Random effects follow normal distributions.
- Parameter estimates are approximately normal.
- Sample size is adequate for asymptotic theory.

Test Statistic

$$W = (\mathbf{C}\hat{\boldsymbol{\beta}} - \mathbf{c})'[\mathbf{C}\widehat{Cov}(\hat{\boldsymbol{\beta}})\mathbf{C}']^{-1}(\mathbf{C}\hat{\boldsymbol{\beta}} - \mathbf{c}). \qquad \text{Under } H_0, W \sim \chi^2_q \text{ asymp. or } W/q \sim F_{q,\nu} \text{with df correction.}$$

Notation: $\mathbf{C}\boldsymbol{\beta} = \mathbf{c}$ specifies linear constraints, q is number of constraints, $\hat{\boldsymbol{\beta}}$ are fixed effects estimates

Plain English: The Wald test for fixed effects evaluates hypotheses about regression coefficients in mixed models by examining whether estimated parameters differ significantly from hypothesized values relative to their standard errors. The test accounts for correlation structure induced by random effects when computing standard errors. For testing single parameters, the test reduces to squared t-statistic. For testing multiple parameters jointly, the quadratic form measures distance between estimates and null values, weighted by inverse covariance matrix.

⚠ Practitioner Warning

Degrees of freedom for F-distribution version contentious in mixed models—no consensus on best method. Options include denominator df from Satterthwaite approximation, Kenward-Roger adjustment, or containment method. Chi-square version assumes large samples; F-version with approximate df preferred for small samples or unbalanced designs. Test assumes variance components correctly specified; misspecified random effects can bias fixed effects inference. REML estimation introduces additional uncertainty in variance components not fully reflected in asymptotic standard errors. With small number of groups, asymptotic approximations poor; consider parametric bootstrap.

Hypotheses

- H_0: $\mathbf{C}\boldsymbol{\beta} = \mathbf{c}$ (linear constraints on fixed effects hold).
- H_1: $\mathbf{C}\boldsymbol{\beta} \neq \mathbf{c}$ (at least one constraint violated).

Decision Rule: Reject H_0 if $W > \chi^2_{\alpha,q}$ (large sample) or $W/q > F_{\alpha,q,\nu}$ (small sample with df adjustment).

Software References [Python, R, SAS]

`statsmodels::wald_test` `lme4::lmer` `PROC MIXED (CONTRAST)`

Additional Notes

For single parameter test $H_0 : \beta_j = 0$, Wald statistic is $W = \hat{\beta}_j^2/\widehat{Var}(\hat{\beta}_j)$ with $t = \hat{\beta}_j/SE(\hat{\beta}_j)$ being the t-version. Square of t-statistic equals F-statistic with numerator df = 1. General linear hypothesis testing uses contrast matrix $\mathbf{C}$ to specify constraints. Example: testing all slopes equal zero uses $\mathbf{C} = [\mathbf{0}\,\mathbf{I}_{p-1}]$ excluding intercept. Satterthwaite (1946) approximation for denominator df: $\nu \approx 2[\mathbf{c}'\widehat{Cov}(\hat{\boldsymbol{\beta}})\mathbf{c}]^2/\mathbf{c}'\widehat{Var}[\widehat{Cov}(\hat{\boldsymbol{\beta}})]\mathbf{c}$. Kenward-Roger (1997) method adjusts both variance estimates and df, generally recommended for small samples. Containment method uses df from smallest containing random effect, conservative and simple but often too conservative. Type I/II/III sums of squares from ANOVA extend to mixed models with varying interpretations and df.

Miscellanea: Wald (1943) developed a general framework for hypothesis testing, but its application to mixed models introduced challenges due to ambiguous degrees of freedom. Various approximations, including Satterthwaite and Kenward–Roger, were developed to address this issue, and different statistical software adopted different defaults. Despite advances, no single approach has universal acceptance, reflecting the inherent complexity of inference in mixed models.

Reference(s): Wald (1943); Kenward & Roger (1997); Satterthwaite (1946)

Worked Example

Problem Statement A student-level prior score z_{ij} is added as a fixed effect to the five-school random intercepts model: $y_{ij} = \beta_0 + \beta_1 z_{ij} + u_i + e_{ij}$, with $u_i \sim N(0, \sigma_u^2)$ and $e_{ij} \sim N(0, \sigma_e^2)$. Using variance components from the previous examples ($\hat{\sigma}_u^2 = 1.193$, $\hat{\sigma}_e^2 = 1.034$), the Wald test assesses whether z significantly predicts y. H_0: $\beta_1 = 0$; H_1: $\beta_1 \neq 0$.

$$X_i = \begin{pmatrix} 1 & z_{i1} \\ 1 & z_{i2} \\ 1 & z_{i3} \\ 1 & z_{i4} \end{pmatrix} \quad V_i^{-1} = \begin{pmatrix} 0.768 & -0.199 & -0.199 & -0.199 \\ -0.199 & 0.768 & -0.199 & -0.199 \\ -0.199 & -0.199 & 0.768 & -0.199 \\ -0.199 & -0.199 & -0.199 & 0.768 \end{pmatrix}$$

School (i)	Prior score z_{ij}				$\bar{z}_{i\cdot}$	$\bar{y}_{i\cdot}$
1	5.0	7.0	6.5	6.0	6.125	9.525
2	7.5	8.0	6.5	7.0	7.250	11.650
3	5.5	6.0	7.0	5.5	6.000	9.900
4	7.0	9.0	7.5	8.0	7.875	12.400
5	6.5	8.0	6.0	7.5	7.000	10.500

1. GLS Estimator via V^{-1}
The marginal covariance of each school's vector $\mathbf{y}_i$ is $V_i = \hat{\sigma}_e^2 I_4 + \hat{\sigma}_u^2 J_4$, with inverse (shown above):

$$V_i^{-1} = a\, I_4 - b\, J_4,\ a = \frac{1}{\hat{\sigma}_e^2} = 0.967,\ b = \frac{\hat{\sigma}_u^2}{\hat{\sigma}_e^2(\hat{\sigma}_e^2 + n\hat{\sigma}_u^2)} = 0.199$$

The GLS estimator stacks contributions from all groups:

$$\sum_i X_i' V_i^{-1} X_i = \begin{pmatrix} 3.444 & 23.588 \\ 23.588 & 172.589 \end{pmatrix},\ \sum_i X_i' V_i^{-1} \mathbf{y}_i = \begin{pmatrix} 37.172 \\ 263.204 \end{pmatrix}$$

2. GLS Estimates

$$\begin{pmatrix} \hat{\beta}_0 \\ \hat{\beta}_1 \end{pmatrix} = \Big(\sum_i X_i' V_i^{-1} X_i\Big)^{-1} \sum_i X_i' V_i^{-1} \mathbf{y}_i = \begin{pmatrix} 5.462 \\ 0.779 \end{pmatrix}$$

$$\widehat{\mathrm{Var}}(\hat{\beta}_1) = \Big(\sum_i X_i' V_i^{-1} X_i\Big)^{-1}_{[1,1]} = 0.0908, \widehat{SE}(\hat{\beta}_1) = 0.301$$

3. Wald Statistic Under H_0, $W \xrightarrow{d} \chi^2(1)$.

$$W = \left(\hat{\beta}_1 / \widehat{SE}(\hat{\beta}_1)\right)^2 = \boxed{6.675}$$

4. Results and Decision Since $W = 6.675 > 3.841$, so we reject H_0 at 5% ($p = 0.010$). This effect is significant after accounting for school-level random variation. Each unit increase in z is associated with a 0.779 increase in y.

Practitioner Insight: It retains the form $W = (\hat{\beta}/SE)^2$ but uses the GLS estimator, incorporating the covariance V from the random-effects structure. With $V^{-1} = aI - bJ$, the positive diagonal and negative off-diagonal terms reflect within-group correlation, down-weighting between-group information. The test uses $\chi^2(1)$ asymptotically; for small G, a t-reference (e.g., t_{14}) is more conservative. Variance components are treated as known (plug-in), so SEs are slightly understated—addressed by Kenward–Roger adjustments. See Five-school dataset from LRT test for rest of the data.

Common Pitfalls

- **Confusing the fixed-effects Wald test with a general Wald test for coefficient restrictions.** In panel models, the Wald test for fixed effects specifically tests whether all individual (or time) effects are jointly equal, implying pooled OLS is adequate. It's distinct from a general Wald test that evaluates arbitrary linear restrictions on regression slopes.
- **Using the test with very small time dimension.** The fixed-effects estimator relies on within-individual variation across time. When T is very small, slope estimates become imprecise and the Wald test has low power to detect meaningful individual heterogeneity.
- **Concluding pooled OLS is appropriate when the test fails to reject.** Failure to reject the null indicates insufficient evidence for fixed effects, not proof that individual heterogeneity is absent. Limited within variation, weak effects, or modest sample sizes can easily lead to low test power.
- **Ignoring heteroskedasticity or serial correlation when conducting the test.** The standard Wald test assumes homoskedastic and uncorrelated errors. If these assumptions fail, standard errors and the resulting test statistic may be biased unless robust or cluster-adjusted variance estimators are used.
- **Applying joint Wald tests in high-dimensional two-way fixed-effects models.** Models with both individual and time effects involve testing many dummy parameters simultaneously. When the number of restrictions becomes large, the joint test may become numerically unstable and difficult to interpret.

F-Test with Kenward-Roger Approximation

Small-Sample F-Test for Fixed Effects with DF Adjustment

MIXD.7 | **Category:** Mixed Models | **Data:** Continuous (Clus. / Long.) | **Null:** F-distribution

When to Use

- Test fixed effects in mixed-effects models.
- Small-sample longitudinal data analyses.
- Hierarchical or clustered data are present.
- Improve inference beyond standard Wald tests.
- Adjust degrees of freedom for finite samples.

Assumptions

- Mixed-effects model is correctly specified.
- Observations are independent across clusters.
- Random effects follow normal distributions.
- Covariance structure is correctly specified.
- Variance components are estimated reliably.

Test Statistic

$$F_{KR} = \frac{(\mathbf{C}\hat{\boldsymbol{\beta}} - \mathbf{c})'[\mathbf{C}\tilde{\boldsymbol{\Phi}}\mathbf{C}']^{-1}(\mathbf{C}\hat{\boldsymbol{\beta}} - \mathbf{c})}{q},$$ where $\tilde{\boldsymbol{\Phi}}$ is adjusted covariance. Under H_0, $F_{KR} \sim F_{q,\nu_{KR}}$.

Notation: $\tilde{\boldsymbol{\Phi}}$ is bias-corrected covariance, ν_{KR} are adjusted denominator df, q is numerator df

Plain English: The Kenward-Roger F-test improves upon standard Wald tests by adjusting both the covariance matrix of fixed effects estimates and the reference distribution's degrees of freedom. The adjustments account for uncertainty in variance component estimation and small-sample bias in covariance estimates. This produces more accurate Type I error rates and confidence interval coverage in small samples or unbalanced designs where standard methods are too liberal. The method provides proper calibration when sample size is limited relative to model complexity.

⚠ Practitioner Warning

Computationally intensive, especially with large datasets or complex random effects structures. Requires REML estimation; ML-based tests incompatible with Kenward-Roger adjustments. Method assumes normality of random effects and errors; robust to mild violations but severe non-normality affects performance. With very small samples ($n < 30$ total or very few groups), even Kenward-Roger may be inadequate; parametric bootstrap preferred. Not applicable when variance components constrained or at boundary. ν_{KR} can be non-integer, sometimes causing confusion in reporting.

Hypotheses

- H_0: $\mathbf{C}\boldsymbol{\beta} = \mathbf{c}$ (linear constraints on fixed effects hold).
- H_1: $\mathbf{C}\boldsymbol{\beta} \neq \mathbf{c}$ (at least one constraint violated).

Decision Rule: Reject H_0 if $F_{KR} > F_{\alpha,q,\nu_{KR}}$ or p-value $< \alpha$.
q =constraints tested; ν_{KR} can be non-integer.

Software References [Python, R, SAS]

Python	R	SAS
`Not Native (Manual)`	`pbkrtest::KRmodcomp`	`PROC MIXED (REML DDFM=KR)`

Additional Notes

The Kenward-Roger adjustment has two components: covariance inflation factor correcting bias in $\widehat{Cov}(\hat{\boldsymbol{\beta}})$ due to estimating variance components, and degrees of freedom approximation matching moments of quadratic form distribution. Adjusted covariance: $\tilde{\boldsymbol{\Phi}} = \boldsymbol{\Phi}(\mathbf{I} + \mathbf{A})$ where $\mathbf{A}$ is bias correction matrix derived from second derivatives of REML likelihood. Degrees of freedom: ν_{KR} from Satterthwaite-type approximation applied to adjusted distribution, often substantially smaller than naive df, especially with complex random effects. Computation requires derivatives of variance-covariance matrix with respect to variance parameters. For simple models (random intercept only), Kenward-Roger often similar to Satterthwaite; differences emerge with multiple variance components or complex structures. Schaalje et al. (2002) simulation study showed Kenward-Roger maintaining nominal Type I error rates where other methods failed.

Miscellanea: Michael Kenward and James Roger (1997) developed a method to improve small-sample inference in mixed models by adjusting the covariance matrix of fixed effects and the associated degrees of freedom. Building on earlier work such as Satterthwaite's approximation, their approach reduces the tendency of standard methods to produce overly liberal tests. It has been widely adopted in applied settings, particularly in clinical trials and repeated-measures analyses.

Reference(s): Kenward & Roger (1997); Verbeke & Molenberghs (2000); Schaalje et al. (2002)

F-Test with Kenward-Roger Approximation

MIXD.7

Worked Example

Problem Statement The Wald test for $\beta_1 = 0$ from the previous example used the asymptotic $\chi^2(1)$ reference distribution, treating variance components as known. The Kenward-Roger (1997) approximation corrects this by replacing the asymptotic reference with an $F(1, \nu_{\text{KR}})$ distribution, where ν_{KR} is a Satterthwaite-type denominator df that accounts for uncertainty in $\hat{\sigma}_u^2$ and $\hat{\sigma}_e^2$. Same model, data, and estimates as the Wald example.

1. Starting Point from Wald Test
From the previous example:

$$\hat{\beta}_1 = 0.779,\ \widehat{\text{Var}}(\hat{\beta}_1) = V_0 = 0.0908,\ F = \hat{\beta}_1^2/V_0 = \boxed{6.675}$$

This F statistic is held fixed. The KR procedure changes only the *reference distribution.*

2. Derivatives of V_0 w.r.t. Variance Components
$V_0 = \widehat{\text{Var}}(\hat{\beta}_1)$ is a function of $(\hat{\sigma}_u^2, \hat{\sigma}_e^2)$. Its partial derivatives (evaluated numerically at the REML estimates) are:

$$g_1 = \frac{\partial V_0}{\partial \hat{\sigma}_u^2} = 0.00970, \quad g_2 = \frac{\partial V_0}{\partial \hat{\sigma}_e^2} = 0.07658$$

$g_2 \gg g_1$: V_0 is much more sensitive to $\hat{\sigma}_e^2$ than to $\hat{\sigma}_u^2$, because within-group variation directly governs precision of the student-level predictor.

4. Delta-Method Variance of V_0
Propagate uncertainty in $(\hat{\sigma}_u^2, \hat{\sigma}_e^2)$ to uncertainty in V_0 via the delta method:

$$\widehat{\text{Var}}(V_0) = g_1^2\,\widehat{\text{Var}}(\hat{\sigma}_u^2) + g_2^2\,\widehat{\text{Var}}(\hat{\sigma}_e^2) + 2\,g_1 g_2\,\widehat{\text{Cov}} = 0.000936$$

5. Satterthwaite Denominator df

$$\nu_{\text{KR}} = 2\,V_0^2/\widehat{\text{Var}}(V_0) = \boxed{17.62}$$

6. Results and Decision

Test	Ref. dist.	p	CV (5%)
Naive Wald	$\chi^2(1)$	0.010	3.841
$t(N{-}G{-}1)$	$t(14)$	0.022	2.145
KR F-test	$F(1, 17.62)$	0.019	4.428

3. Uncertainty in Variance Components From the inverse of the exact REML Fisher information matrix $\mathcal{I}$:

$$\widehat{\text{Var}}(\hat{\sigma}_u^2) = 1.149, \qquad \widehat{\text{Var}}(\hat{\sigma}_e^2) = 0.153, \qquad \widehat{\text{Cov}}(\hat{\sigma}_u^2, \hat{\sigma}_e^2) = -0.046$$

Practitioner Insight: The Kenward–Roger correction addresses the bias in the Wald test when variance components are estimated, making $V_0 = \widehat{\text{Var}}(\hat{\beta}_1)$ random. Treating it as fixed understates variability and leads to anti-conservative inference. KR adjusts this via an effective df ν_{KR}, replacing the $\chi^2(1)$ reference with $F(1, \nu_{\text{KR}})$, increasing the critical value (e.g., 3.841 → 4.428) while leaving the statistic unchanged. Here $\nu_{\text{KR}} = 17.62 > 14$ because information comes from both within- and between-group variation; for purely group-level predictors, ν_{KR} would be much smaller ($\approx G - 2$). The full KR also inflates V_0; using only Satterthwaite df (as here) is slightly less conservative.

Common Pitfalls

- **Confusing Kenward–Roger with Satterthwaite degrees-of-freedom approximations.** Both adjust inference in mixed models, but Satterthwaite adjusts only the denominator degrees of freedom, whereas Kenward–Roger adjusts both the F-statistic and the degrees of freedom to correct small-sample bias. The two methods can yield noticeably different p-values in small or unbalanced samples.
- **Applying the Kenward–Roger approximation with extremely small samples.** KR improves small-sample inference but remains an approximation. With very small samples (e.g., $n < 20$), variance component estimates are unstable and the adjusted F distribution may still provide unreliable inference.
- **Using Kenward–Roger tests for model comparison instead of parameter testing.** KR F-tests are designed for testing fixed-effect coefficients within a specified model. They are not appropriate for comparing models with different fixed-effect structures, where likelihood ratio tests using ML estimation should be used instead.
- **Applying the approximation to models estimated with ML instead of REML.** The Kenward–Roger adjustment was derived for REML estimation of variance components. Applying it to ML-estimated models produces incorrect adjustments and invalid inference.
- **Using Kenward–Roger in inappropriate model settings.** The method is designed for linear mixed models with normally distributed errors. It does not apply to generalized linear mixed models (e.g., logistic or Poisson), where Wald or likelihood ratio tests should be used instead.

Intraclass Correlation Coefficient Test

Test and Estimate Within-Group Correlation

MIXD.8 | **Category:** Mixed Models | **Data:** Continuous (Clus. / Long.) | **Null:** F-distribution

When to Use

- Quantify similarity of observations within same group.
- Assess whether grouping structure induces meaningful correlation.
- Determine sample size requirements for clustered designs.
- Test reliability of measurements/agreement between raters.

Assumptions

- Observations nested within groups (subjects, raters).
- Random effects model appropriate: observations within groups correlated.
- Equal group sizes preferred; unequal sizes require weights.
- Normally distributed outcomes and random effects.

Test Statistic

$$\rho_I = \frac{\sigma_u^2}{\sigma_u^2 + \sigma_\epsilon^2}, \quad F = \frac{MS_B}{MS_W}, \quad \text{where } MS_B, MS_W \text{ from one-way ANOVA.} \quad \text{Under } H_0 : \rho_I = 0, F \sim F_{J-1,J(n-1)}.$$

Notation: σ_u^2 is between-group variance, σ_ϵ^2 is within-group variance, J is groups, n is observations per group

Plain English: The intraclass correlation coefficient (ICC) measures the proportion of total variance attributable to grouping structure, representing the expected correlation between any two observations from the same group. Values near 0 indicate observations within groups are no more similar than observations from different groups. Values near 1 indicate strong within-group clustering with most variation occurring between groups. The F-test evaluates whether ICC significantly exceeds zero, testing for meaningful group effects.

⚠ Practitioner Warning

ICC interpretation depends on context: reliability studies (higher is better), clustered trials (higher reduces effective n), multilevel models (variance partitioning). With unequal group sizes, several ICC estimators exist with different properties; no consensus on best choice. Negative ICC estimates can occur with small samples or when between-group variance small; typically set to zero. ICC alone does not indicate whether random or fixed effects model more appropriate—use Hausman test. Very small ICC (< 0.01) suggests clustering negligible, but formal test still recommended.

Hypotheses

- H_0: $\rho_I = 0$ (no within-group correlation).
- H_1: $\rho_I > 0$ (within-group correlation exists).

Decision Rule: Reject H_0 if $F > F_{\alpha,J-1,J(n-1)}$ or p-value $< \alpha$.

Software References [Python, R, SAS]

Python	R	SAS
`pingouin.intraclass_corr`	`ICC::ICCest`	`PROC MIXED (RANDOM INTERCEPT)`

Additional Notes

For unbalanced designs, use harmonic mean of group sizes or more complex estimators (Searle-Casella-McCulloch). Relationship to variance components: $\sigma_u^2 = (MS_B - MS_W)/n$ and $\sigma_\epsilon^2 = MS_W$ in balanced case. Design effect for clustered sampling: Deff $= 1 + (n-1)\rho_I$ quantifies effective sample size reduction; accounts for ICC when calculating power. Multiple forms of ICC exist: ICC(1) for single rater/measurement, ICC(2) for mean of multiple raters, ICC(3) for specific raters. Confidence intervals for ICC available via variance-stabilizing transformations or bootstrapping. Fleiss-Shrout (1979) taxonomy distinguishes ICC variants for different study designs.

Miscellanea: Fisher's early work on correlations among relatives (1918) laid the foundation for what later became intraclass correlation, used to quantify similarity within groups such as families. The concept was later adopted in psychometrics as a measure of reliability, leading to multiple formulations for different study designs. Shrout and Fleiss (1979) clarified these variants, establishing a widely used framework for ICC in reliability analysis.

Reference(s): Fisher (1918); Shrout & Fleiss (1979)

Intraclass Correlation Coefficient Test MIXD.8

Worked Example

Problem Statement Five-school dataset from LRT test ($G = 5, n = 4, N = 20$) is used to estimate the intraclass correlation coefficient (ICC) and test whether it is significantly greater than zero. The ICC measures the proportion of total score variance attributable to school membership. H_0: ICC= 0 ($\sigma_u^2 = 0$); H_1: ICC> 0.

1. One-Way ANOVA Decomposition

Source	SS	df	MS
Between	23.232	4	5.808
Within	15.518	15	1.034
Total	38.750	19	

2. Variance Components and ICC

By the method of moments (REML-equivalent):

$$\hat{\sigma}_e^2 = 1.034,\ \hat{\sigma}_u^2 = \frac{MS_B - MS_W}{n} = 1.193$$

$$F = \frac{MS_B}{MS_W} = \boxed{5.614} \sim F(G{-}1,\ N{-}G) = F(4, 15)$$

$$\widehat{\text{ICC}} = \frac{\hat{\sigma}_u^2}{\hat{\sigma}_u^2 + \hat{\sigma}_e^2} = \frac{F-1}{F+n-1} = 0.536$$

3. 95% Confidence Interval Shrout-Fleiss (1979) method:

$$F_{0.025}(4, 15) = 0.116, \quad F_{0.975}(4, 15) = 3.804$$

$$F_L = \frac{F}{F_{0.975}} = 1.476, \quad F_U = \frac{F}{F_{0.025}} = 48.60$$

$$\text{ICC}_L = \frac{F_L - 1}{F_L + n - 1} = 0.106, \ \text{ICC}_U = \frac{F_U - 1}{F_U + n - 1} = 0.923$$

$$\widehat{\text{ICC}} = \boxed{0.536}, \quad 95\%\ \text{CI: } [0.106,\ 0.923]$$

Test	Stat	Null dist.	p	CV (5%)
ICC F-test	5.614	$F(4, 15)$	0.006	3.056
REML LRT	5.995	$\frac{1}{2}\chi^2(0) + \frac{1}{2}\chi^2(1)$	0.007	2.706
BP LM test	6.516	$\chi^2(1)$	0.011	3.841
ML LRT	4.354	$\frac{1}{2}\chi^2(0) + \frac{1}{2}\chi^2(1)$	0.018	2.706

(Reject H_0 at $\alpha = 0.05$)

4. Interpretation

- School membership explains 53.6% of total score variance. Wide CI reflects uncertainty with only $G = 5$ groups.

Practitioner Insight: The ICC test is equivalent to the one-way ANOVA F-test but reframes the question as the proportion of variance between groups. The identity $\widehat{\text{ICC}} = (F-1)/(F+n-1)$ shows ICC is a monotone transform of F: $F > 1$ implies positive ICC, and $F = 1$ gives ICC= 0. Unlike LRT and BP LM tests, the ICC test uses the exact $F(G-1, N-G)$ distribution, avoiding asymptotics. The Shrout-Fleiss CI is obtained by inverting the F-pivot and mapping bounds back via $(F-1)/(F+n-1)$. Wide intervals (e.g. $[0.11, 0.92]$) are common with small G, reflecting that accurate ICC estimation requires many groups.

The true intra-class correlation (ICC) is non-negative since it is a ratio of variance components. Negative estimates can arise from sampling variability (e.g., when $MS_B < MS_W$) and are typically interpreted as zero, indicating no meaningful clustering.

Common Pitfalls

- **Using the wrong ICC formulation for the study design.** Different ICC definitions apply to different designs (e.g., ICC(1), ICC(2), ICC(3)). The correct choice depends on whether reliability refers to single measurements or averages and whether raters are random or fixed.
- **Treating ICC as equivalent to Pearson's r.** Pearson's r measures consistency of ranking, whereas ICC measures absolute agreement. Systematic differences between raters (mean or scale shifts) can leave r high but reduce ICC.
- **Using ICC with very small numbers of subjects or raters.** Reliable estimation of between- and within-subject variance requires adequate sample sizes. With very small n or few raters, ICC estimates become unstable and confidence intervals extremely wide.
- **Reporting only the ICC point estimate without confidence intervals.** An ICC estimate alone does not convey precision. Confidence intervals are essential because the same point estimate can correspond to either precise or highly uncertain reliability.
- **Applying ICC to inappropriate data types.** Standard ICC assumes continuous approximately normal measurements. For ordinal or categorical ratings, agreement measures such as weighted kappa or related statistics are more appropriate.

Hausman Specification Test

Test Random Effects vs. Fixed Effects Specification

MIXD.9 | **Category:** Mixed Models | **Data:** Continuous (Panel) | **Null:** Chi-square

When to Use

- Choose between fixed- and random-effects models.
- Assess correlation of effects with regressors.
- Panel-data regression with repeated observations.
- Compare consistent and efficient estimators.
- Evaluate random-effects exogeneity assumption.

Assumptions

- Fixed- and random-effects models are fitted.
- Fixed-effects estimator is consistent.
- Random effects are uncorrelated with regressors.
- Panel observations are correctly identified.
- Variance estimates are computed consistently.

Test Statistic

$$H = (\hat{\beta}_{FE} - \hat{\beta}_{RE})'[\widehat{Var}(\hat{\beta}_{FE}) - \widehat{Var}(\hat{\beta}_{RE})]^{-1}(\hat{\beta}_{FE} - \hat{\beta}_{RE}).$$

Under H_0, $H \approx \chi^2_k$ with k time-varying regressors.

Notation: $\hat{\beta}_{FE}$ from fixed effects, $\hat{\beta}_{RE}$ from random effects, k excludes time-invariant predictors

Plain English: The Hausman test compares coefficient estimates from fixed effects and random effects models. If random effects are uncorrelated with predictors (random effects assumption valid), both estimators are consistent and should yield similar coefficients. Substantial differences indicate correlation between random effects and predictors, violating the random effects assumption. The test measures whether efficiency gains from random effects come at cost of inconsistency, helping choose appropriate specification for panel data analysis.

⚠ Practitioner Warning

Variance difference matrix may be non-positive definite, invalidating the test and indicating computational or specification issues. Fixed effects excludes time-invariant variables. Rejection favors fixed effects, but failure to reject does not confirm random effects (possible low power, especially with small T). The Hausman test assesses specification, not the presence of random effects—use the Breusch–Pagan LM test first. Cluster-robust standard errors are recommended for both models.

Hypotheses

- H_0: REs uncorrelated with predictors ($Cov(u_i, \mathbf{x}_{it}) = 0$).
- H_1: Random effects (REs) correlated with predictors.

Decision Rule: Reject H_0 if $H > \chi^2_{\alpha,k}$ or p-value $< \alpha$. Degrees of freedom equal number of time-varying regressors.

Software References [Python, R, SAS]

`linearmodels::wu_hausman`	`plm::phtest`	`PROC PANEL`

Additional Notes

Hausman (1978) compares an efficient but potentially inconsistent estimator with a consistent alternative; under correct specification, their difference is asymptotically zero. In panel data, fixed effects uses within transformation to ensure consistency regardless of correlation between u_i and $\mathbf{x}_{it}$, while random effects assumes $Cov(u_i, \mathbf{x}_{it}) = 0$ for efficiency. Mundlak (1978) provides an equivalent approach by adding group means $\bar{\mathbf{x}}_i$ to the random effects model and testing their coefficients. If the variance difference matrix is not positive definite, auxiliary regression or bootstrap methods may be used. Hybrid (Mundlak–Chamberlain) models allow partial correlation while retaining random effects structure.

Miscellanea: Hausman (1978) developed a general specification test based on comparing an efficient estimator with a consistent but less efficient alternative. Originally motivated by problems in instrumental variables and simultaneous equations, the framework was later widely adopted in panel data analysis to choose between fixed and random effects models. The test remains a standard tool for detecting violations of key modeling assumptions.

Reference(s): Hausman (1978); Mundlak (1978)

Worked Example

Problem Statement Five-school dataset (LRT) with prior score z_{ij} is used to demonstrate Hausman's specification test for the mixed model $y_{ij} = \beta_0 + \beta_1 z_{ij} + u_i + e_{ij}$. Under H_0 (RE consistent), $u_i \perp z_{ij}$ and RE is efficient; under H_1, $\text{Cov}(u_i, z_{ij}) \neq 0$ and only FE is consistent. $G = 5$, $n = 4$, $N = 20$.

$$S_{zy,W} = \sum_{ij}(z_{ij} - \bar{z}_{i\cdot})(y_{ij} - \bar{y}_{i\cdot}) = 6.288, \quad S_{zz,W} = \sum_{ij}(z_{ij} - \bar{z}_{i\cdot})^2 = 9.625, \quad SS_{\text{FE}} = 11.410, \quad SS_{\text{BE}} = 0.514$$

1. Fixed Effects (Within) Estimator

Within-group (demeaned) OLS:

$$(y_{ij} - \bar{y}_{i\cdot}) = \beta_{\text{FE}}\,(z_{ij} - \bar{z}_{i\cdot}) + \varepsilon_{ij}$$

$$\hat{\beta}_{\text{FE}} = \frac{S_{zy,W}}{S_{zz,W}} = 0.653, \quad \hat{\sigma}_e^2 = \frac{SS_{\text{FE}}}{N - G - 1} = 0.815,$$

$$\widehat{\text{Var}}(\hat{\beta}_{\text{FE}}) = \frac{\hat{\sigma}_e^2}{S_{zz,W}} = 0.0847$$

2. Swamy–Arora Variance Components

Between regression on group means $\bar{y}_{i\cdot} = \alpha + \beta\,\bar{z}_{i\cdot} + \eta_i$:

$$\hat{\sigma}_{\text{BE}}^2 = \frac{SS_{\text{BE}}}{G - 2} = 0.171, \quad \hat{\sigma}_u^2 = \max(0, \hat{\sigma}_{\text{BE}}^2 - \frac{\hat{\sigma}_e^2}{n}) = 0$$

As $\theta = 1 - \sqrt{\hat{\sigma}_e^2/(\hat{\sigma}_e^2 + n \cdot \hat{\sigma}_u^2)} = 0$, the RE estimator collapses to **pooled OLS**.

3. Random Effects Estimator (Pooled OLS, $\theta = 0$)

$$y_{ij} = \alpha + \beta_{\text{RE}}\, z_{ij} + \varepsilon_{ij}$$

$$\hat{\beta}_{\text{RE}} = 1.063, \quad \widehat{\text{Var}}(\hat{\beta}_{\text{RE}}) = 0.0473$$

4. Classic Hausman Statistic

$$H = \frac{(\hat{\beta}_{\text{FE}} - \hat{\beta}_{\text{RE}})^2}{\widehat{\text{Var}}(\hat{\beta}_{\text{FE}}) - \widehat{\text{Var}}(\hat{\beta}_{\text{RE}})} = \boxed{4.498} \sim \chi^2(1)$$

5. Mundlak Device (Augmented Regression)

Add $\bar{z}_{i\cdot} - \bar{z}_{\cdot\cdot}$ as a regressor to pooled OLS:

$$y_{ij} = \alpha + \beta z_{ij} + \pi(\bar{z}_{i\cdot} - \bar{z}_{\cdot\cdot}) + \varepsilon_{ij}$$

$$\hat{\beta} = 0.653\ (= \hat{\beta}_{\text{FE}}), \quad \hat{\pi} = 0.808$$

$$\hat{s}^2 = 0.792\ (df = 17), \quad \widehat{SE}(\hat{\pi}) = 0.403$$

$$W_\pi = (\hat{\pi}/\widehat{SE}(\hat{\pi}))^2 = \boxed{4.022} \sim \chi^2(1)$$

6. Results and Decision

Version	Stat	CV (5%)	p	Decision
Hausman H	4.498	3.841	0.034	Reject
Mundlak W_π	4.022	3.841	0.045	Reject

Both methods agree: FE is preferred.

Practitioner Insight: The Swamy–Arora estimate $\hat{\sigma}_u^2 = -0.032$ is truncated to 0, implying $\theta = 0$ and reducing RE to pooled OLS ($\hat{\beta}_{\text{RE}} = 1.063$). This exceeds the FE estimate ($\hat{\beta}_{\text{FE}} = 0.653$), reflecting upward bias from between-school sorting (high-z schools also have higher outcomes). The large gap $\hat{\delta} = (\hat{\beta}_{\text{FE}} - \hat{\beta}_{\text{RE}}) - 0.410$ yields $H = 4.498$, rejecting RE at 5%. The Mundlak test confirms this: $\hat{\pi} = 0.808$ indicates significant between-group effects, a sign of endogeneity. Despite small $G = 5$, both tests agree in favor of FE.

Common Pitfalls

- **Interpreting a non-significant result as proof that random effects are correct.** Failure to reject H_0 only indicates insufficient evidence against the random-effects model. Because the test may have low power, theoretical considerations about exogeneity should also guide model choice.
- **Confusing negative or undefined statistics with evidence for the null.** Finite-sample variability can produce a non–positive-definite variance difference matrix, yielding negative test values. This reflects numerical instability rather than support for the random-effects specification.
- **Testing only a subset of slopes without support.** Hausman test is designed to compare estimators jointly across parameters. Testing only selected coefficients can miss broader specification problems unless motivated by clear theory.
- **Ignoring the possibility that both estimators are inconsistent.** The test assumes one estimator is consistent under both hypotheses. If both models are misspecified (e.g., omitted variables or measurement error), the test may fail to detect the problem.
- **Applying the test mechanically without examining sources of endogeneity.** The Hausman test detects correlation between regressors and unobserved effects. Understanding the substantive mechanism behind potential endogeneity is essential for interpreting results and selecting remedies.

Part XII

High-Dimensional & Joint Distribution Inference

Testing multivariate means, covariance structures, and dependence

Multivariate Analysis

XII. MULT — A Conceptual Primer

What Is Multivariate Analysis?

Multivariate analysis concerns data where each observation is a **vector** $\mathbf{x}_i \in \mathbb{R}^p$, with $p \geq 2$ variables measured simultaneously. The joint distribution of the p variables—not just their marginals—is the object of inference.

Why not just run p univariate tests?

- Ignores correlations among variables; a joint signal can be invisible in any single variable.
- Inflates Type I error: running p tests at $\alpha = 0.05$ gives a family-wise error rate far exceeding 5%.
- Multivariate tests are more powerful when effects are spread across correlated variables.

The Multivariate Normal Distribution

The **MVN** $\mathbf{X} \sim \mathcal{N}_p(\boldsymbol{\mu}, \boldsymbol{\Sigma})$ is the cornerstone of classical multivariate inference:

$$f(\mathbf{x}) = (2\pi)^{-p/2}|\boldsymbol{\Sigma}|^{-1/2}\exp\{-\tfrac{1}{2}(\mathbf{x}-\boldsymbol{\mu})^\top\boldsymbol{\Sigma}^{-1}(\mathbf{x}-\boldsymbol{\mu})\}.$$

Key properties: (i) all marginals and conditionals are normal; (ii) uncorrelatedness implies independence; (iii) any linear combination $\mathbf{a}^\top\mathbf{X}$ is univariate normal.

Mahalanobis distance:

$$D^2(\mathbf{x}) = (\mathbf{x}-\boldsymbol{\mu})^\top\boldsymbol{\Sigma}^{-1}(\mathbf{x}-\boldsymbol{\mu}) \sim \chi^2_p \quad \text{under MVN.}$$

D^2 is the covariance-adjusted squared distance from the mean; it accounts for variable scales and correlations. Under MVN, ordered $D^2_{(i)}$ should follow χ^2_p quantiles—the basis of the χ^2 Q–Q plot for multivariate normality.

Sample estimates:

$$\bar{\mathbf{x}} = \frac{1}{n}\sum_{i=1}^{n}\mathbf{x}_i, \quad \mathbf{S} = \frac{1}{n-1}\sum_{i=1}^{n}(\mathbf{x}_i-\bar{\mathbf{x}})(\mathbf{x}_i-\bar{\mathbf{x}})^\top.$$

$\mathbf{S}$ is positive semi-definite; positive definite when $n > p$. Tests requiring $\mathbf{S}^{-1}$ degrade rapidly when p is large relative to n.

I. Testing Mean Vectors: Hotelling's T^2

One-sample T^2: Tests $H_0 : \boldsymbol{\mu} = \boldsymbol{\mu}_0$:

$$T^2 = n(\bar{\mathbf{x}}-\boldsymbol{\mu}_0)^\top\mathbf{S}^{-1}(\bar{\mathbf{x}}-\boldsymbol{\mu}_0),$$

with exact distribution: $\frac{n-p}{p(n-1)}T^2 \sim F_{p,n-p}$ under H_0 and MVN. The F-conversion is exact in finite samples.

Two-sample T^2: Tests $H_0 : \boldsymbol{\mu}_1 = \boldsymbol{\mu}_2$ using the pooled covariance matrix $\mathbf{S}_p$:

$$T^2 = \frac{n_1 n_2}{n_1+n_2}(\bar{\mathbf{x}}_1-\bar{\mathbf{x}}_2)^\top\mathbf{S}_p^{-1}(\bar{\mathbf{x}}_1-\bar{\mathbf{x}}_2),$$

$$\mathbf{S}_p = \frac{(n_1-1)\mathbf{S}_1+(n_2-1)\mathbf{S}_2}{n_1+n_2-2}.$$

$\frac{n_1+n_2-p-1}{p(n_1+n_2-2)}T^2 \sim F_{p,n_1+n_2-p-1}$. Assumes equal covariance matrices $\boldsymbol{\Sigma}_1 = \boldsymbol{\Sigma}_2$ (check with Box's M). For unequal covariances, a multivariate Welch analog exists but is less standard.

Paired observations: Apply one-sample T^2 to the difference vectors $\mathbf{d}_i = \mathbf{x}_{1i} - \mathbf{x}_{2i}$; no equal-covariance assumption required.

II. MANOVA: Comparing k Mean Vectors

For $k \geq 2$ groups, the total SSCP matrix decomposes: $\mathbf{T} = \mathbf{H} + \mathbf{E}$, where $\mathbf{H}$ (hypothesis) captures between-group variation and $\mathbf{E}$ (error) captures within-group variation. Four test statistics summarize the eigenvalues $\lambda_1 \geq \cdots \geq \lambda_s > 0$ of $\mathbf{E}^{-1}\mathbf{H}$, where $s = \min(p, k-1)$:

Statistic	Formula and properties
Wilks' Λ	$\prod_{i=1}^{s} 1/(1+\lambda_i) = \lvert\mathbf{E}\rvert/\lvert\mathbf{H}+\mathbf{E}\rvert$; likelihood ratio test; most common; exact F for special cases.
Pillai's Trace	$\sum_i \lambda_i/(1+\lambda_i)$; most **robust** to MVN violations and unequal $\boldsymbol{\Sigma}$; preferred when robustness matters.
Hotelling–Lawley	$\sum_i \lambda_i = \mathrm{tr}(\mathbf{E}^{-1}\mathbf{H})$; efficient when all eigenvalues contribute.
Roy's Largest Root	$\lambda_1/(1+\lambda_1)$; most **powerful** when the true difference is one-dimensional (single dominant effect); most sensitive to MVN violations. Sometimes $\theta = \lambda_1/(1+\lambda_1)$ is used.

All four are equivalent under a single non-zero eigenvalue ($s = 1$; two groups, $k = 2$). For $s > 1$:

- **Use Pillai** when assumptions are uncertain.
- **Use Wilks'** Λ as the default likelihood-ratio reference.
- **Use Roy** only when a one-dimensional group structure is theoretically justified; its upper-bound F-approximation is conservative.

III. Covariance Structure Tests

Box's M test (equality of covariance matrices): Tests $H_0 : \boldsymbol{\Sigma}_1 = \cdots = \boldsymbol{\Sigma}_k$ across k groups. Based on the log-determinant of each group's covariance matrix relative to the pooled estimate:

$$M = (N-k)\log|\mathbf{S}_p| - \sum_i (n_i-1)\log|\mathbf{S}_i|,$$

with a $\chi^2_{p(p+1)(k-1)/2}$ (after correction, (1-C)) or F approximation. Violation of equal covariance is an assumption of the two-sample T^2 and MANOVA. **Warning:** Box's M is extremely sensitive to MVN departure; a significant result often reflects non-normality rather than genuinely unequal covariances. Use as a rough guide, not a definitive gatekeeper.

Bartlett's test of sphericity: Tests $H_0 : \mathbf{\Sigma} = \sigma^2\mathbf{I}$ (all variances equal, all covariances zero):

$$\chi^2 = -\left(n - 1 - \frac{2p+5}{6}\right)\log|\mathbf{R}| \mathrel{\dot\sim} \chi^2_{p(p-1)/2},$$

where $\mathbf{R}$ is the sample correlation matrix. Rejection confirms that variables are correlated—a prerequisite for PCA and factor analysis to be meaningful. **With large** n, Bartlett rejects even trivially small correlations; always inspect $\mathbf{R}$ directly.

Mauchly's test of sphericity (repeated measures): In repeated-measures designs, sphericity requires $\mathbf{\Sigma}_D = \sigma^2\mathbf{I}$ for the **difference score** covariance (equal variances for all pairwise differences). Violation inflates Type I error in repeated-measures ANOVA. Remedies: Greenhouse–Geisser ($\hat\varepsilon_{GG}$) or Huynh–Feldt ($\hat\varepsilon_{HF}$) corrections multiply the degrees of freedom by $\hat\varepsilon \in [1/(p-1), 1]$; MANOVA on the difference scores is an assumption-free alternative.

JNS (John–Nagao–Sugiura) test of sphericity: An omnibus test of $H_0 : \mathbf{\Sigma} = \sigma^2\mathbf{I}$ based on the statistic $T = \frac{p(n-1)}{2}\left[\frac{p\,\mathrm{tr}(\mathbf{S}^2)}{[\mathrm{tr}(\mathbf{S})]^2} - 1\right] \mathrel{\dot\sim} \chi^2_{p(p+1)/2-1}$. More powerful than Bartlett in many settings and less sensitive to the choice of normalization; used when both equal variances and zero covariances are of interest.

IV. Multivariate Normality Tests

All classical multivariate tests (Hotelling's T^2, MANOVA, Box's M) assume MVN. Univariate normality of each variable is necessary but not sufficient for MVN; the joint distribution must be tested.

χ^2 Q–Q plot: Plot ordered Mahalanobis distances $D^2_{(i)}$ against χ^2_p quantiles. Departures from linearity signal non-normality or outliers. Essential graphical complement to any formal test.

Mardia's tests: Based on multivariate extensions of skewness and kurtosis:

$$b_{1,p} = \frac{1}{n^2}\sum_{i,j}[(\mathbf{x}_i - \bar{\mathbf{x}})^\top\mathbf{S}^{-1}(\mathbf{x}_j - \bar{\mathbf{x}})]^3,$$

$$b_{2,p} = \frac{1}{n}\sum_{i}[(\mathbf{x}_i - \bar{\mathbf{x}})^\top\mathbf{S}^{-1}(\mathbf{x}_i - \bar{\mathbf{x}})]^2.$$

Under MVN: $nb_{1,p}/6 \sim \chi^2_{p(p+1)(p+2)/6}$ and $b_{2,p} \approx \mathcal{N}(p(p+2), 8p(p+2)/n)$. Mardia's skewness is sensitive to non-normal joint structure; kurtosis is sensitive to heavy tails and outliers. **Both** should be tested.

Henze–Zirkler test: Based on the L^2 distance between the empirical and MVN characteristic functions; an omnibus test with good power against a broad range of alternatives:

$$\mathrm{HZ} = \frac{1}{n}\sum_{i,j} e^{-\frac{\beta^2}{2}D^2_{ij}} - 2\left(1+\beta^2\right)^{-p/2}\sum_i e^{-\frac{\beta^2}{2(1+\beta^2)}D^2_i} + n\left(1+2\beta^2\right)^{-p/2},$$

where $\beta = 1/\sqrt{2}(n(2p+1)/4)^{1/(p+4)}$ and D^2_{ij} are Mahalanobis distances between pairs. Asymptotically normal; among the most powerful omnibus tests for MVN.

Practical guidance:

- Use the χ^2 Q–Q plot always.
- Run Mardia's tests (skewness and kurtosis jointly) as the primary formal test.
- Use Henze–Zirkler for confirmation; it has better power against many alternatives.
- At large n, tests may reject MVN for trivial departures; assess practical severity from the Q–Q plot.

V. Multiple Testing: Benjamini–Hochberg FDR

After MANOVA or multivariate normality testing, follow-up univariate tests on individual variables are often needed. Running p uncorrected tests at α inflates the family-wise error rate (FWER).

Bonferroni correction: reject H_{0j} if $p_j < \alpha/p$; controls FWER but is conservative (especially when tests are positively correlated, as they typically are in multivariate data).

Benjamini–Hochberg (BH) procedure: Controls the **false discovery rate** FDR $= E[\text{false rejections}/\text{total rejections}]$ at level q:

1. Order p-values: $p_{(1)} \le p_{(2)} \le \cdots \le p_{(p)}$.
2. Find the largest j such that $p_{(j)} \le jq/p$.
3. Reject all $H_{0(1)}, \ldots, H_{0(j)}$.

BH is **less conservative than Bonferroni**: it allows more rejections when many tests are truly significant, at the cost of allowing a controlled fraction of false discoveries. **Preferred in high-dimensional settings** (e.g. post-MANOVA follow-up, genomics). Controls FDR exactly under independence and under positive dependence (PRDS condition), which covers most multivariate data.

Assumptions and Robustness

Test	Sensitivity to violations
Hotelling's T^2	Robust to mild MVN deviation for large n (CLT); sensitive to outliers via $\mathbf{S}^{-1}$.
MANOVA (Wilks)	Moderate MVN sensitivity; invalid with severe non-normality or $p \approx n$.
MANOVA (Pillai)	Most robust to non-normality and unequal $\boldsymbol{\Sigma}$; the safe default.
Box's M	Extremely sensitive to non-normality; significant result often reflects non-MVN, not unequal $\boldsymbol{\Sigma}$.
Mauchly / JNS	Moderate sensitivity; supplement with ε corrections if violated.

The p relative to n problem: When p is large relative to n, $\mathbf{S}$ is ill-conditioned and $\mathbf{S}^{-1}$ is unreliable. As a rough guide: all classical tests degrade noticeably when $p > n/5$. High-dimensional alternatives include regularized T^2 (replacing $\mathbf{S}^{-1}$ with a regularized inverse), Srivastava's test, and Chen–Qin test which do not require $p < n$.

Diagnostic Workflow

1. **Inspect the data:** χ^2 Q–Q plot; scatterplot matrix; check for outliers.
2. **Test MVN:** Mardia's skewness and kurtosis; Henze–Zirkler for confirmation.
3. **Test covariance structure:** Bartlett (sphericity / correlations exist); Box's M (equal $\boldsymbol{\Sigma}$ across groups); Mauchly / JNS (sphericity in repeated measures).
4. **Test mean vector(s):** Hotelling T^2 (one or two groups); MANOVA (Pillai as default, Wilks for LR reference, Roy for one-dimensional alternatives).
5. **Follow-up:** if MANOVA is significant, run univariate tests with BH-FDR correction; inspect discriminant functions to understand the nature of the group separation.

VI. Principal Component Analysis (PCA)

PCA finds orthogonal linear combinations of the p variables—**principal components** (PCs)—that maximize successive variance. It is a dimension-reduction tool, not a hypothesis test, but it underlies Bartlett's sphericity test and precedes many multivariate analyses.

The kth PC is $z_k = \mathbf{e}_k^\top \mathbf{x}$, where $\mathbf{e}_k$ is the kth eigenvector of $\boldsymbol{\Sigma}$ (in practice $\mathbf{S}$), with associated eigenvalue λ_k:

$$\mathbf{S}\mathbf{e}_k = \lambda_k \mathbf{e}_k, \quad \lambda_1 \geq \lambda_2 \geq \cdots \geq \lambda_p \geq 0.$$

$\mathrm{Var}(z_k) = \lambda_k$; total variance $= \mathrm{tr}(\mathbf{S}) = \sum_k \lambda_k$. Proportion explained by the first m PCs: $\sum_{k=1}^{m} \lambda_k / \sum_k \lambda_k$.

Covariance vs. correlation matrix:

- Use $\mathbf{S}$ when variables are on the same scale; PCs depend on variable variances.
- Use $\mathbf{R}$ (standardize first) when scales differ; the more common choice in practice.

Number of components to retain:

- **Scree plot:** retain before the "elbow."
- **Kaiser rule:** $\lambda_k > 1$ ($\mathbf{R}$-based PCA); tends to retain too many.
- **Cumulative variance:** $\geq 80\%$ (or $\geq 70\%$ for exploratory work).
- **Bartlett–Lawley LRT** for formal tests on the number of components.

PCA is not a model—no distributional assumption and no hypothesis test on the components themselves.

VII. Linear Discriminant Analysis (LDA)

LDA finds linear combinations of the p variables that **maximally separate** k groups—the same objective as MANOVA, expressed as a classification rule.

Fisher's discriminant functions: The discriminant functions are the eigenvectors of $\mathbf{E}^{-1}\mathbf{H}$—the same matrix whose eigenvalues define the MANOVA test statistics. The first discriminant function $\mathbf{a}_1^\top \mathbf{x}$ maximizes the ratio of between- to within-group variance.

Linear classification rule (equal covariances): Assign $\mathbf{x}$ to group g minimizing:

$$d_g(\mathbf{x}) = (\mathbf{x} - \bar{\mathbf{x}}_g)^\top \mathbf{S}_p^{-1} (\mathbf{x} - \bar{\mathbf{x}}_g) - 2 \log \pi_g,$$

where π_g is the prior for group g. For equal priors this reduces to the nearest group centroid in Mahalanobis distance; the decision boundary is a **linear hyperplane**.

Quadratic Discriminant Analysis (QDA): When $\boldsymbol{\Sigma}_g$ differ (Box's M significant), use group-specific $\mathbf{S}_g$; the decision boundary becomes a **quadratic surface**. Requires more data per group ($p(p+1)/2$ free parameters per $\mathbf{S}_g$).

Assessing classification performance:

- **Resubstitution rate:** classify training data; optimistically biased.
- **Leave-one-out (LOO) error:** less biased.
- **Confusion matrix:** $k \times k$ table of true vs. predicted group; reveals which groups are confused.

Connection to MANOVA: a significant MANOVA establishes that groups are separable; LDA identifies *how* they separate and classifies future observations. Both are based on the eigenstructure of $\mathbf{E}^{-1}\mathbf{H}$.

VIII. Effect Sizes for Multivariate Tests

Statistical significance alone is insufficient; always report effect size alongside MANOVA p-values.

Measure	Formula and interpretation
η^2	$SS_{between}/SS_{total}$ per variable; upward biased; not adjusted for n.
Partial η^2	$SS_{effect}/(SS_{effect} + SS_{error})$; most widely reported MANOVA effect size.
ω^2	$(SS_B - (k-1)MS_W)/(SS_T + MS_W)$; less biased than η^2; preferred for small n.
$1 - \Lambda^{1/s}$	Multivariate analog of f^2; overall effect across all variables; $s = \min(p, k-1)$.

Guidelines (Cohen, per univariate follow-up): small $\eta_p^2 \approx 0.01$; medium ≈ 0.06; large ≈ 0.14. No universal multivariate benchmark exists.

IX. Post-MANOVA Follow-Up Strategy

A significant MANOVA omnibus test establishes that at least one variable differs across groups; it does not identify *which* variables or *which* groups. Structured follow-up is required.

Discriminant function analysis: Examine the canonical discriminant functions (eigenvectors of $\mathbf{E}^{-1}\mathbf{H}$) and their loadings to identify which variables contribute most to group separation.

Roy–Bargmann step-down tests: Test each variable in order of priority, adjusting for all higher-priority variables already in the model. The jth step-down test is the ANCOVA F-test for variable j with variables $1, \ldots, j-1$ as covariates. Maintains Type I error control without Bonferroni conservatism but requires a *pre-specified* variable ordering.

Univariate F-tests with BH correction: Run one-way ANOVA for each of the p variables; apply Benjamini–Hochberg at level q. Simpler than step-down but ignores correlation structure. **Never run uncorrected univariate tests** after a significant MANOVA.

Decision guide:

- *How* groups differ $\Rightarrow$ discriminant function loadings.
- *Which* variables drive the difference $\Rightarrow$ Roy–Bargmann (pre-specified order) or BH-corrected F.
- Classify future observations $\Rightarrow$ LDA or QDA.

X. Canonical Correlation Analysis (CCA)

CCA finds pairs of linear combinations $u = \mathbf{a}^\top \mathbf{x}$ and $v = \mathbf{b}^\top \mathbf{y}$ from two variable sets $\mathbf{x} \in \mathbb{R}^p$ and $\mathbf{y} \in \mathbb{R}^q$ that are maximally correlated:

$$(\mathbf{a}_1, \mathbf{b}_1) = \arg\max_{\mathbf{a},\mathbf{b}} \text{Cor}(\mathbf{a}^\top \mathbf{x}, \mathbf{b}^\top \mathbf{y}).$$

Subsequent pairs are orthogonal to all previous pairs. The canonical correlations $r_1 \geq \cdots \geq r_s > 0$ ($s = \min(p, q)$) are the square roots of the eigenvalues of $\mathbf{\Sigma}_{xx}^{-1}\mathbf{\Sigma}_{xy}\mathbf{\Sigma}_{yy}^{-1}\mathbf{\Sigma}_{yx}$.

Testing significance: $\Lambda_k = \prod_{i=k}^{s}(1 - r_i^2)$ (Wilks' Λ for remaining correlations) has a $\chi^2_{(p-k+1)(q-k+1)}$ approximation via Bartlett's factor. Testing $k = 1$ is equivalent to testing $\mathbf{\Sigma}_{xy} = \mathbf{0}$ (no association between variable sets).

CCA as the master framework:

- $q = 1$ (scalar $\mathbf{y}$) $\Rightarrow$ **multiple regression**.
- $\mathbf{y}$ = group indicators $\Rightarrow$ **LDA / MANOVA**.
- $\mathbf{x} = \mathbf{y} \Rightarrow$ **PCA**.

CCA unifies regression, PCA, LDA, and MANOVA within a single framework. Regularized CCA (ridge or sparse) is preferred when $n \ll p + q$.

Common Pitfalls

- **Repeated univariate tests instead of MANOVA:** ignores correlations and inflates FWER.
- **Box's M as a strict gatekeeper:** it is over-sensitive to non-normality; a significant result does not automatically invalidate MANOVA.
- **p close to n:** $\mathbf{S}^{-1}$ becomes unstable; all tests are unreliable; use regularized alternatives.
- **Roy's root without justification:** use only when a single dominant dimension of group separation is theoretically expected.
- **No follow-up after MANOVA:** a significant omnibus test tells you *that* groups differ; post-hoc univariate tests with multiplicity control identify *where*.
- **Ignoring sphericity in repeated measures:** apply Greenhouse–Geisser or Huynh–Feldt correction, or switch to MANOVA on difference vectors.
- **Covariance vs. correlation PCA:** using $\mathbf{S}$ when variable scales differ produces PCs dominated by high-variance variables; standardize to $\mathbf{R}$ when in doubt.
- **No multiplicity correction after MANOVA:** never run uncorrected univariate F-tests; apply BH or use Roy–Bargmann step-down.

Bottom Line

Multivariate analysis evaluates the joint behavior of p variables simultaneously, accounting for their correlation structure. The Mahalanobis distance is the foundational geometric object. Test MVN and covariance structure before testing means; use Pillai's trace as the robust default for MANOVA; apply Benjamini–Hochberg for follow-up univariate tests; use LDA to understand group separation; and use high-dimensional alternatives when p is large relative to n.

Hotelling's T² Test (One-Sample)

Multivariate Extension of One-Sample t-Test

MULT.1 | **Category:** Multivariate Analysis | **Data:** Continuous | **Null:** F-distribution

When to Use

- Test a multivariate mean vector against a target.
- Simultaneous inference on several variables.
- Multivariate alternative to one-sample t-test.
- Correlated response variables are analyzed.
- Quality-control or profile-comparison studies.

Assumptions

- Observations are mutually independent.
- Data follow a multivariate normal distribution.
- Variables are measured on interval scale.
- Covariance matrix is positive definite.
- Sample size exceeds the number of variables ($n > p$).

Test Statistic

$$T^2 = n(\bar{\mathbf{x}} - \mu_0)'\mathbf{S}^{-1}(\bar{\mathbf{x}} - \mu_0), \quad F = \frac{n-p}{p(n-1)}T^2. \qquad \text{Under } H_0, F \sim F_{p,n-p}.$$

Notation: $\bar{\mathbf{x}}$ is sample mean vector, μ_0 is hypothesized mean, $\mathbf{S}$ is sample covariance, n is sample size, p is variables

Plain English: Hotelling's T² test generalizes the one-sample t-test to multiple variables by measuring the squared Mahalanobis distance between the sample mean vector and hypothesized mean vector. The Mahalanobis distance accounts for correlations among variables and scales by the inverse covariance matrix, providing a multivariate measure of how far the sample mean is from the hypothesized value. Large values indicate the sample mean vector differs significantly from the null hypothesis in the multivariate space.

⚠ Practitioner Warning

Requires multivariate normality; violations affect test validity, especially with small samples. Sensitive to outliers which can inflate covariance matrix and distort distances. Sample size must exceed number of variables ($n > p$); test undefined otherwise. Power decreases as p increases relative to n. Assumes covariance structure homogeneous across observations. Rejection indicates mean vector differs but not which variables drive difference—examine individual variables post-hoc. Simultaneous confidence regions more complex than univariate intervals.

Hypotheses

- H_0: $\mu = \mu_0$
- H_1: $\mu \neq \mu_0$

Decision Rule: Reject H_0 if $F > F_{\alpha,p,n-p}$ or p-value $< \alpha$.
Numerator $df = p$ (variables), Denominator $df = n - p$.

Software References [Python, R, SAS]

`pingouin.hotelling` `Hotelling::hotelling.test` `PROC GLM (MANOVA)`

Additional Notes

Mahalanobis distance: $D^2 = (\mathbf{x} - \mu)'\mathbf{\Sigma}^{-1}(\mathbf{x} - \mu)$ measures distance accounting for covariance structure. Hotelling's $T^2 = nD^2$ with $\mathbf{\Sigma}$ replaced by sample covariance $\mathbf{S}$. Relationship to χ^2: as $n \to \infty$, $T^2 \to \chi^2_p$ asymptotically, but exact finite-sample distribution is F-based. Simultaneous confidence region: $\{(\bar{\mathbf{x}} - \mu)'\mathbf{S}^{-1}(\bar{\mathbf{x}} - \mu) \leq \frac{p(n-1)}{n(n-p)}F_{\alpha,p,n-p}\}$ forms ellipsoid in p-dimensional space. Individual variable tests: if T^2 significant, examine univariate t-tests or compute simultaneous confidence intervals using Bonferroni correction α/p. Connection to discriminant analysis: T^2 equivalent to testing separation between sample mean and null value in discriminant space. For $p = 1, T^2 = t^2$ exactly (squared one-sample t-statistic).

Miscellanea: Harold Hotelling (1931) introduced the T^2 statistic as a multivariate extension of the t-test, allowing joint inference on multiple correlated variables. A key contribution was establishing its exact finite-sample relationship to the F-distribution, enabling practical use beyond asymptotic approximations. The test highlights the importance of accounting for correlation in multivariate analysis.

Reference(s): Hotelling (1931); Anderson (2003)

Worked Example

Problem Statement A researcher evaluates whether the mean vector of Protein and Carbohydrate content in a sample ($n = 5$) differs significantly from the manufacturer's specification $\mu_0 = (20, 50)^\top$. $H_0 : \mu = \mu_0$.
Sample Data ($n = 5$): $(19, 48)$, $(17, 52)$, $(20, 51)$, $(18, 47)$, $(16, 47)$, where x_1 denotes protein and x_2 denotes carbohydrates (expressed as row vectors).

$$\bar{x} = \begin{pmatrix} 18.0 \\ 49.0 \end{pmatrix}, \qquad S = \frac{1}{n-1}\sum_{i=1}^{n}(x_i - \bar{x})(x_i - \bar{x})' = \begin{pmatrix} 2.5 & 1.0 \\ 1.0 & 5.0 \end{pmatrix}.$$

1. Calculation of the T^2 Statistic Given the sample mean vector $\bar{x} = [18, 49]^\top$ and the sample covariance matrix S, we calculate the multivariate distance:

$$T^2 = n(\bar{x} - \mu_0)^\top S^{-1}(\bar{x} - \mu_0)$$

Assuming $S^{-1} = \begin{pmatrix} 0.4314 & -0.0784 \\ -0.0784 & 0.1961 \end{pmatrix}$, the calculation yields:

$$T^2 = 5 \times (-2, -1)\begin{pmatrix} 0.4314 & -0.0784 \\ -0.0784 & 0.1961 \end{pmatrix}\begin{pmatrix} -2 \\ -1 \end{pmatrix} \approx 8.04$$

2. Conversion to F-statistic To find the p-value, T^2 is transformed into an F-distribution where p is the number of variables (2) and n is the sample size (5):

$$F = \frac{n-p}{p(n-1)}T^2 = \frac{5-2}{2(5-1)} \times 8.04 = \frac{3}{8} \times 8.04 = 3.01$$

The degrees of freedom are $df_1 = p = 2$, $df_2 = n - p = 3$.

3. Results

Metric	Value
Hotelling's T^2	8.04
Transformed $F(2, 3)$	3.01
Critical $F_{0.05,2,3}$	9.55
P-value	**0.192**

4. Decision and Interpretation

- **Decision:** Since $3.01 < 9.55$, we **Fail to Reject** H_0.
- **Interpretation:** The nutritional content of the energy bars does not significantly deviate from the specifications when Protein and Carbohydrates are considered simultaneously.
- **Practitioner Insight:** Hotelling's T^2 is sensitive to the correlation between variables. If Protein and Carbs were highly correlated, a small shift in both might be more significant than a large shift in just one. It provides a "joint confidence region" (an ellipse) rather than two separate "confidence intervals" (a rectangle).

Common Pitfalls

- **Running multiple univariate t-tests instead of Hotelling's T^2.** Separate tests inflate the overall type I error rate and ignore correlations among variables. Hotelling's T^2 provides a single multivariate test for the mean vector.
- **Applying the test when the number of variables approaches or exceeds the sample size.** The statistic requires inversion of the sample covariance matrix, which is unstable when n is close to p. Reliable inference generally requires substantially more observations than variables.
- **Assuming rejection identifies which variables differ from the hypothesized mean.** Hotelling's T^2 is an omnibus test of the entire mean vector. Follow-up procedures such as simultaneous confidence intervals are needed to identify specific components.
- **Ignoring the multivariate normality assumption.** The validity of the test depends on approximate multivariate normality of the observations. Departures from this assumption can affect the accuracy of the test statistic and its distribution.
- **Interpreting the magnitude of T^2 as an effect size.** Since $T^2 = n\,D^2$, it scales with sample size and covariance structure, making its magnitude difficult to interpret. The squared Mahalanobis distance D^2 provides a more interpretable, scale-free measure of multivariate effect.

Hotelling's T^2 Test (Two-Sample)

Compare Multivariate Mean Vectors Between Two Groups

MULT.2 | **Category:** Multivariate Analysis | **Data:** Continuous | **Null:** F-distribution

When to Use

- Compare multivariate mean vectors of two groups.
- Simultaneous inference on several variables.
- Multivariate alternative to two-sample t-test.
- Correlated response variables are analyzed.
- Profile-comparison or quality-control studies.

Assumptions

- Samples are mutually independent.
- Data follow multivariate normal distributions.
- Variables are measured on interval scale.
- Population covariance matrices are equal.
- Sample size exceeds the number of variables ($n_1, n_2 > p$).

Test Statistic

$$T^2 = \frac{n_1 n_2}{n_1 + n_2}(\bar{\mathbf{x}}_1 - \bar{\mathbf{x}}_2)'\mathbf{S}_p^{-1}(\bar{\mathbf{x}}_1 - \bar{\mathbf{x}}_2), \quad F = \frac{n_1 + n_2 - p - 1}{p(n_1 + n_2 - 2)}T^2. \quad \text{Under } H_0, F \sim F_{p, n_1+n_2-p-1}.$$

Notation: $\bar{\mathbf{x}}_i$ are group mean vectors, $\mathbf{S}_p$ is pooled covariance, n_i are sample sizes, p is variables

Plain English: The two-sample Hotelling's T^2 test extends the two-sample t-test to multiple correlated variables by computing the squared Mahalanobis distance between group mean vectors. The pooled covariance matrix accounts for within-group variability and correlations among variables. The test provides omnibus assessment of whether groups differ on all p variables jointly, automatically accounting for correlations that would inflate Type I error if variables tested separately.

⚠ Practitioner Warning

Assumes equal covariance matrices across groups; violation affects test validity. Use Box's M test to check homogeneity; if rejected, consider separate covariance versions or non-parametric alternatives. Requires multivariate normality; robust to mild violations with moderate samples but sensitive with small n or high p. Power decreases as p increases relative to total sample size. Rejection indicates groups differ somewhere but not which variables—requires post-hoc univariate tests or discriminant analysis. Unbalanced designs ($n_1 \neq n_2$) reduce power; balanced designs preferred. Very large p relative to n can make $\mathbf{S}_p$ near-singular.

Hypotheses

- H_0: $\boldsymbol{\mu}_1 = \boldsymbol{\mu}_2$
- H_1: $\boldsymbol{\mu}_1 \neq \boldsymbol{\mu}_2$

Decision Rule: Reject H_0 if $F > F_{\alpha, df_1, df_2}$ or p-value $< \alpha$
$df_1 = p$, and $df_2 = n_1 + n_2 - p - 1$

Software References [Python, R, SAS]

Python	R	SAS
`pingouin.hotelling`	`Hotelling::hotelling.test`	`PROC GLM (MANOVA)`

Additional Notes

Pooled covariance matrix: $\mathbf{S}_p = \frac{(n_1-1)\mathbf{S}_1+(n_2-1)\mathbf{S}_2}{n_1+n_2-2}$ where $\mathbf{S}_i$ are group covariance matrices. This assumes homogeneity; alternative uses separate covariances but loses exact F-distribution. Relationship to MANOVA: two-sample Hotelling's T^2 is special case of one-way MANOVA with two groups. For $p = 1$, $T^2 = t^2$ where t is two-sample t-statistic assuming equal variances. Discriminant analysis connection: the linear combination maximizing group separation has coefficients $\mathbf{a} \propto \mathbf{S}_p^{-1}(\bar{\mathbf{x}}_1 - \bar{\mathbf{x}}_2)$ (Fisher's linear discriminant function). Post hoc testing: if T^2 significant, examine univariate t-tests with Bonferroni correction (α/p) or compute simultaneous confidence intervals. Roy's largest root provides alternative multivariate test focusing on single linear combination. Unequal covariances: Yao (1965) and Johansen (1980) proposed approximate solutions not assuming $\boldsymbol{\Sigma}_1 = \boldsymbol{\Sigma}_2$.

Miscellanea: Harold Hotelling extended the T^2 framework to two-sample problems, providing a multivariate analog of the two-sample t-test. The method assumes equal covariance matrices across groups, yielding an exact F-distribution under this assumption. Although this requirement can be restrictive, the test remains widely used due to its simplicity and theoretical tractability.

Reference(s): Hotelling (1931); Anderson (2003)

Hotelling's T² Test (Two-Sample) MULT.2

Worked Example

Problem Statement A study compares the physiological outcomes of two exercise groups (Yoga: $n_1 = 12$, HITT: $n_2 = 15$) across two dependent variables: Flexibility (x_1) and Heart Rate Recovery (x_2). We test the null hypothesis $H_0 : \mu_1 = \mu_2$.

1. Calculation of the Pooled Statistic Given the difference in sample mean vectors $\mathbf{d} = (15, -20)^\top$ and the pooled covariance matrix S_{pl}, the T^2 statistic is:

$$T^2 = \left(\frac{n_1 n_2}{n_1 + n_2}\right) \mathbf{d}^\top S_{pl}^{-1} \mathbf{d},$$

$$\text{where } S_p = \frac{(n_1 - 1)S_1 + (n_2 - 1)S_2}{n_1 + n_2 - 2}$$

Assuming a pooled covariance inverse $S_p^{-1} = \begin{pmatrix} 0.05 & 0.01 \\ 0.01 & 0.04 \end{pmatrix}$:

$$T^2 = \left(\frac{180}{27}\right)(15, -20)\begin{pmatrix} 0.05 & 0.01 \\ 0.01 & 0.04 \end{pmatrix}\begin{pmatrix} 15 \\ -20 \end{pmatrix} \approx 141.67$$

2. Transformation to F-Distribution With $p = 2$ variables and $N = 27$ total subjects, the T^2 value is converted to an F-statistic:

$$F = \frac{n_1 + n_2 - p - 1}{p(n_1 + n_2 - 2)} T^2 \approx 68.00$$

Degrees of freedom: $df_1 = p = 2$, $df_2 = n_1 + n_2 - p - 1 = 24$.

3. Results

Metric	Value
Two-Sample Hotelling's T^2	141.67
Transformed $F(2, 24)$	68.00
Critical $F_{0.05,2,24}$	3.40
P-value	**< 0.001**

4. Decision and Interpretation

- **Decision:** Since 68.00 > 3.40, we **Reject** H_0.
- **Interpretation:** There is a statistically significant difference between the mean physiological profiles of the Yoga and HIIT groups.
- **Practitioner Insight:** The HIIT group likely performed better on heart rate recovery, while the Yoga group dominated flexibility. Hotelling's T^2 captures joint differences across variables. Even if individual variables show moderate differences, their combined multivariate effect—accounting for covariance—may be highly significant.

Common Pitfalls

- **Assuming equal covariance matrices without verification.** The classical two-sample Hotelling's T^2 test assumes the covariance matrices of the two populations are equal. Large differences between group covariance structures can invalidate the pooled estimate used in the test.
- **Applying the test when the total sample size is small relative to the number of variables.** The pooled covariance matrix must be invertible to compute the statistic. Reliable results generally require substantially more observations than variables.
- **Performing multiple follow-up univariate tests without adjustment.** After rejecting the global Hotelling's T^2 test, researchers often examine individual variables. Multiplicity adjustments are required to control the family-wise error rate.
- **Interpreting rejection as evidence that all variables differ between groups.** The Hotelling's T^2 test is an omnibus test comparing the two mean vectors. A significant result indicates that at least one component differs but does not identify which.
- **Using the two-sample test for paired or repeated-measures data.** The standard two-sample formulation assumes independent observations between groups. Paired or matched designs should instead be analyzed using a one-sample T^2 test on the difference vectors.

Wilks' Lambda Test

Likelihood Ratio Test for Multivariate Group Differences

MULT.3 | **Category:** Multivariate Analysis | **Data:** Continuous | **Null:** F-distribution (approximate)

When to Use

- Test multivariate mean differences among groups.
- MANOVA with multiple correlated responses.
- Assess overall group separation multivariately.
- Control Type I error across several outcomes.
- Profile-comparison and experimental studies.

Assumptions

- Observations are mutually independent.
- Responses follow multivariate normality.
- Covariance matrices are equal across groups.
- Variables are measured on interval scale.
- Sample size exceeds the number of variables ($n_i > p$).

Test Statistic

$$\Lambda = \frac{|\mathbf{W}|}{|\mathbf{W}+\mathbf{B}|} = \prod_{i=1}^{s} \frac{1}{1+\lambda_i}, \quad \text{where } s = \min(p, g-1).$$

Notation: $\mathbf{B}$ is between-groups SSCP, $\mathbf{W}$ is within-groups SSCP, λ_i are eigenvalues of $\mathbf{W}^{-1}\mathbf{B}$

Plain English: Wilks' Λ measures the proportion of total variance not explained by group differences, based on the ratio of determinants of within-groups to total sum of squares and cross-products (SSCP) matrices. Values near 1 indicate little evidence of multivariate group separation; values near 0 indicate strong separation. Small Λ indicates substantial group differences. The test is derived from likelihood ratio principles, making it the multivariate analog of the F-test in ANOVA.

⚠ Practitioner Warning

Exact F-distribution only for special cases ($p = 1, p = 2, g = 2$, or $g = 3$); otherwise uses Rao's F-approximation. Sensitive to violations of multivariate normality and homogeneity of covariance, especially with small samples. Determinant-based statistic can be numerically unstable with near-singular covariance matrices. Less robust than Pillai's trace to assumption violations but potentially more powerful under ideal conditions. Cannot identify which groups differ or which variables drive differences—requires post-hoc analysis. Performance degrades with very unequal sample sizes.

Hypotheses

- H_0: $\boldsymbol{\mu}_1 = \cdots = \boldsymbol{\mu}_g$.
- H_1: At least one $\boldsymbol{\mu}_i$ differs from others.

Decision Rule: Reject H_0 if $F > F_{\alpha, df_1, df_2}$ or p-value $< \alpha$.
Use $df_1 = p(g-1)$, Rao's approx. (below) for df_2.

Software References [Python, R, SAS]

`statsmodels.MANOVA` `stats::summary.manova` `PROC GLM (MANOVA)`

Additional Notes

Wilks' Λ is the likelihood-ratio statistic comparing the fit of the null model (equal group means) to the unrestricted multivariate model. It depends only on the latent roots of $\mathbf{W}^{-1}\mathbf{B}$, which also underlie other MANOVA criteria. Exact F distributions exist only in special low-dimensional cases (e.g., $p = 1$, $p = 2$, or $g = 2$, where the test reduces to Hotelling's T^2). A common large-sample alternative is Bartlett's chi-square approximation: $-[N-1-(p+g)/2]\ln(\Lambda) \sim \chi^2_{p(g-1)}$. Wilks' Λ, Pillai's trace, Hotelling–Lawley trace, and Roy's largest root are all functions of the same latent roots but differ in robustness and power under various alternatives. Discriminant analysis uses these same roots to construct canonical variates, and significant omnibus results are typically followed by targeted multivariate contrasts or pairwise comparisons. Rao's F-approximation refines inference in general settings:

$$F = \frac{1-\Lambda^{1/t}}{\Lambda^{1/t}} \cdot \frac{df_2}{df_1}, \quad \text{with } t = \sqrt{\frac{p^2(g-1)^2-4}{p^2+(g-1)^2-5}} \text{ when } p^2+(g-1)^2 > 5, \quad df_1 = p(g-1), \; df_2 = t\left[N-1-\frac{p+g}{2}\right] - \frac{df_1-2}{2}.$$

Miscellanea: Samuel S. Wilks, founding editor of the *Annals of Mathematical Statistics*, introduced the Λ statistic as a likelihood ratio test for multivariate normal models. Closely related to determinants and eigenvalues, it became a standard tool in MANOVA. Later work by Rao provided practical F-approximations, making the test widely usable despite early computational challenges.

Reference(s): Wilks (1932); Rao (1948); Anderson (2003); Johnson & Wichern (2015)

Worked Example

Problem Statement A researcher evaluates the effect of sleep duration (3 groups, $g = 3$) on two psychological outcomes ($p = 2$): Focus and Mood. The null hypothesis $H_0 : \mu_{Short} = \mu_{Normal} = \mu_{Long}$ is tested using Wilks' Λ. Each group had 10 participants ($n_i = 10$; total $N = 30$) in the study.

Formulas

$$\mathbf{W} = \sum_{i=1}^{3}\sum_{j=1}^{n_i}(\mathbf{x}_{ij} - \bar{\mathbf{x}}_i)(\mathbf{x}_{ij} - \bar{\mathbf{x}}_i)', \qquad \mathbf{B} = \sum_{i=1}^{3} n_i(\bar{\mathbf{x}}_i - \bar{\mathbf{x}})(\bar{\mathbf{x}}_i - \bar{\mathbf{x}})'$$

$$t = \sqrt{\frac{p^2(g-1)^2 - 4}{p^2 + (g-1)^2 - 5}} = 2, \qquad df_1 = p(g-1) = 4, \qquad df_2 = t\left[N - 1 - \frac{p+g}{2}\right] - \frac{df_1 - 2}{2} = 52,$$

1. Calculation of the Λ Statistic From the MANOVA output, the Error (**W**) and Hypothesis (**B**) matrices are obtained. Using the sample data, the determinants are:

$$|\mathbf{W}| = 450.0, \quad |\mathbf{W} + \mathbf{B}| = 1250.0$$

The Wilks' Lambda statistic is:

$$\Lambda = \frac{|\mathbf{W}|}{|\mathbf{W} + \mathbf{B}|} = \frac{450}{1250} = 0.36$$

This indicates that 36% of the total variance is unexplained by the sleep groups, while 64% is attributable to the differences between groups.

2. Significance Testing (Rao's F-Approximation) Because Λ is not intuitively distributed, it is converted to an F-statistic. For $g = 3$ and $p = 2$:

$$F = \left(\frac{1 - \sqrt{\Lambda}}{\sqrt{\Lambda}}\right)\left(\frac{df_2}{df_1}\right) \approx \left(\frac{1 - 0.6}{0.6}\right) \times 13 = 8.67$$

using Rao's degrees of freedom $df_1 = 4$ and $df_2 = 52$.

3. Results

Metric	Value
Wilks' Lambda (Λ)	0.360
Approximate $F(4, 52)$	8.67
Critical $F_{0.05}$	2.55
P-value	**< 0.001**

4. Decision and Interpretation

- **Decision:** Since $F = 8.67 > 2.55$, we **Reject** H_0.
- **Interpretation:** Sleep duration has a statistically significant effect on the combined vector of Focus and Mood.
- **Practitioner Insight:** Wilks' Λ is the exact likelihood ratio test statistic under multivariate normality, and has a direct interpretation as the proportion of total generalized variance *not* explained by group membership — the multivariate analog of $1 - R^2$. A value near 0 indicates strong group separation; a value near 1 indicates the groups are alike.

Common Pitfalls

- **Interpreting the magnitude of Wilks' Λ without context.** The statistic ranges from 0 to 1 but its interpretation depends on dimensionality and sample size. Statistical conclusions should be based on the corresponding F transformation and significance test.
- **Applying Wilks' Λ when MANOVA assumptions are clearly violated.** The statistic performs best when covariance matrices are homogeneous and sample sizes are balanced. Robust alternatives such as Pillai's trace are often preferable when these assumptions are doubtful.
- **Computing Λ with poorly conditioned covariance matrices.** Highly correlated variables or small samples can produce unstable determinants. Standardizing variables and using numerically stable software implementations helps avoid computational errors.
- **Overlooking relationships with simpler test statistics.** In special cases Wilks' Λ reduces to familiar tests such as Hotelling's T^2 or the univariate F test. Recognizing these connections can provide useful consistency checks during analysis.
- **Interpreting rejection as evidence that all variables differ between groups.** Wilks' Λ tests equality of the group mean vectors as a whole. Additional analyses are required to determine which variables or combinations drive the difference.

Pillai's Trace Test

Sum of Squared Canonical Correlations for MANOVA

MULT.4 | **Category:** Multivariate Analysis | **Data:** Continuous | **Null:** F-distribution (approximate)

When to Use

- Test multivariate mean differences among groups.
- MANOVA with multiple correlated responses.
- Assess overall group separation multivariately.
- Control Type I error across several outcomes.
- Profile-comparison and experimental studies.

Assumptions

- Observations are mutually independent.
- Responses follow multivariate normality.
- Covariance matrices are equal across groups.
- Variables are measured on interval scale.
- Sample size exceeds the number of variables ($n_i > p$).

Test Statistic

$$V = \mathrm{tr}[\mathbf{B}(\mathbf{W}+\mathbf{B})^{-1}] = \sum_{i=1}^{s} \frac{\lambda_i}{1+\lambda_i}, \quad \text{where } s = \min(p, g-1). \qquad \text{Range: } 0 \le V \le s.$$

Notation: **B** is between-groups SSCP, **W** is within-groups SSCP, λ_i are eigenvalues, N is total sample size

Plain English: Pillai's trace sums the proportion of variance explained by group differences across all discriminant dimensions, measuring the overall strength of multivariate association between groups and dependent variables. It equals the sum of squared canonical correlations between group membership and the outcome variables. Large values (approaching maximum s) indicate strong group separation. The statistic is more intuitive than Wilks' Λ—it directly measures explained variance rather than unexplained variance— the multivariate analog of summing the R-squared values across all canonical dimensions.

⚠ Practitioner Warning

Generally less powerful than Wilks' Λ when assumptions perfectly satisfied, but more robust to violations. Conservative (maintains Type I error control) under heterogeneity of covariance and non-normality. Can be more powerful than Wilks when group differences spread across multiple dimensions rather than concentrated in one. F-distribution is approximate, not exact even in special cases. With perfect discrimination ($V = s$), test statistic is undefined (division by zero issue in some formulations). Does not identify which variables or linear combinations drive group separation.

Hypotheses

- H_0: All group mean vectors equal ($\mu_1 = \cdots = \mu_g$).
- H_1: At least one group mean vector differs from others.

Decision Rule: Reject H_0 if $F > F_{\alpha, df_1, df_2}$ or p-value $< \alpha$.
Pillai-Samson: $df_1 = s(2m + s + 1), df_2 = s(2n + s + 1)$.

Software References [Python, R, SAS]

`statsmodels.MANOVA` `stats::summary.manova` `PROC GLM (MANOVA)`

Additional Notes

Pillai's trace can be expressed in terms of canonical correlations: $V = \sum_{i=1}^{s} R_i^2$, where $R_i^2 = \lambda_i/(1+\lambda_i)$ are squared canonical correlations between discriminant functions and group membership. These λ_i are the same latent roots of $\mathbf{W}^{-1}\mathbf{B}$ used by other MANOVA criteria, but combined differently—Wilks' Λ uses a product, while Pillai's trace uses a sum. Equivalently, $V = \mathrm{tr}\left[\mathbf{H}(\mathbf{H}+\mathbf{E})^{-1}\right]$, where **H** is the hypothesis SSCP and **E** is the error SSCP (alternative notation for **B** and **W**). Pillai's trace is generally the most robust MANOVA statistic under violations of multivariate normality and covariance homogeneity (Olson, 1976, 1979). In terms of power, it performs best when effects are distributed across multiple dimensions; Roy's largest root is most powerful for single dominant effects, with Wilks' Λ typically intermediate. Pillai–Samson (1959) F-approximation:

$$F = \frac{2n+s+1}{2m+s+1} \cdot \frac{V}{s-V}, \text{ with } s = \min(p, g-1), m = \frac{|p-(g-1)|-1}{2}, n = \frac{N-p-g-1}{2}, df_1 = s(2m+s+1), df_2 = s(2n+s+1).$$

Miscellanea: K.C.S. Pillai (1950s) introduced the trace statistic as an alternative to Wilks' Λ in MANOVA. Known for its robustness to violations of model assumptions, it gained wider recognition through later studies highlighting its favorable properties. Today, Pillai's trace is often recommended when robustness is a primary concern.

Reference(s): Pillai (1955); Olson (1976, 1979)

Worked Example

Problem Statement An agricultural research firm is testing two new bio-organic fertilizers to improve crop health. Forty-five ($N = 45$)identical tomato plants (*Solanum lycopersicum*) are randomly assigned to three groups ($g = 3$): **Control**, **Formula X** (high-nitrogen organic), and **Formula Y** (balanced micro-nutrient organic), with 15 plants per group. Plant health is measured by **Stem Height** (y_1, cm) and **Leaf Chlorophyll Content** (y_2, SPAD units) ($p = 2$). A **one-way MANOVA** is used to test whether the fertilizer treatment changes the mean response vector $[\mu_{height}, \mu_{chlorophyll}]$ across groups.

1. Statistical Matrices After analyzing the plant data, the eigenvalues of the matrix $\mathbf{W}^{-1}\mathbf{B}$ are calculated to be:

- $\lambda_1 = 0.425$ (First canonical dimension)
- $\lambda_2 = 0.085$ (Second canonical dimension)

2. Calculation of Pillai's Trace (V) The statistic is the sum of the proportion of explained variance across all dimensions:

$$V = \sum \frac{\lambda_i}{1 + \lambda_i} == 0.3765$$

3. F-Approximation and Significance To determine the p-value, we use the parameters $s = \min(p, g - 1) = 2$, $m = (|p-(g-1)|-1)/2 = -0.5$, and $n = (N-p-g-1)/2 = 19.5$.

$$F_{(4,84)} \approx \frac{(2n + s + 1)}{(2m + s + 1)} \cdot \frac{V}{(s - V)} = 4.88, \quad p = 0.0014$$

4. Results Summary

Metric	Value
Pillai's Trace (V)	0.3765
Approximate F-Value	4.88
Degrees of Freedom	(4, 84)
P-value	**0.0014**

5. Interpretation

- **Decision:** Since $p < 0.05$, we reject the null hypothesis. The fertilizer type significantly affects the combined plant health metrics (Height and Chlorophyll).

Practitioner Insight: Pillai's Trace is preferred here because it provides a conservative estimate. Even if the groups had slightly unequal variances (common in biological data), we can be confident in this significant result.

Common Pitfalls

- **Using Pillai's trace automatically without considering assumptions.** Although Pillai's statistic is relatively robust to assumption violations, it may sacrifice some power under ideal conditions. When assumptions clearly hold, other MANOVA statistics such as Wilks' Λ may be more efficient.
- **Interpreting the trace value without considering dimensionality.** The magnitude of Pillai's trace depends on the number of response variables and group contrasts. Effect size interpretation often benefits from standardizing the statistic relative to the number of dimensions.
- **Applying the test with extremely small samples relative to the number of variables.** Reliable estimation of covariance matrices requires sufficient observations relative to the dimensionality. Very small sample sizes can lead to unstable results regardless of the MANOVA statistic used.
- **Ignoring discrepancies between Pillai's trace and other MANOVA statistics.** Different statistics summarize the eigenvalue structure of the hypothesis and error matrices differently. Substantial disagreement among tests may signal assumption violations or unusual data structure.
- **Assuming Pillai's trace protects against all assumption violations.** The statistic is more robust than alternatives to certain violations such as covariance heterogeneity. However, issues such as dependence among observations or severe non-normality still require separate attention.

Sum of Squares and Cross-products (SSCP) - E = Error; W = Within; H = Hypothesis; B = Between

$$\mathbf{E} \text{ or } \mathbf{W} = \sum_{i=1}^{g}\sum_{j=1}^{n_i}(\mathbf{x}_{ij} - \bar{\mathbf{x}}_i)(\mathbf{x}_{ij} - \bar{\mathbf{x}}_i)', \qquad \mathbf{H} \text{ or } \mathbf{B} = \sum_{i=1}^{g} n_i(\bar{\mathbf{x}}_i - \bar{\mathbf{x}})(\bar{\mathbf{x}}_i - \bar{\mathbf{x}})'$$

Hotelling-Lawley Trace Test

Sum of Eigenvalues for Multivariate Group Differences

MULT.5 | **Category:** Multivariate Analysis | **Data:** Continuous | **Null:** F-distribution (approximate)

When to Use

- Test multivariate mean differences among groups.
- MANOVA with multiple correlated responses.
- Assess overall group separation multivariately.
- Control Type I error across several outcomes.
- Profile-comparison and experimental studies.

Assumptions

- Observations are mutually independent.
- Responses follow multivariate normality.
- Covariance matrices are equal across groups.
- Variables are measured on interval scale.
- Sample size exceeds the number of variables ($n_i > p$).

Test Statistic

$$U = \mathrm{tr}(\mathbf{B}\mathbf{W}^{-1}) = \mathrm{tr}(\mathbf{W}^{-1}\mathbf{B}) = \sum_{i=1}^{s} \lambda_i, \quad \text{where } s = \min(p, g-1). \qquad \text{McKeon's approx: } F = \frac{2(s \cdot n + 1)}{s^2(2m + s + 1)} \cdot U$$

Notation: $\mathbf{B}$ is between-groups SSCP, $\mathbf{W}$ is within-groups SSCP, λ_i are eigenvalues, U is trace

Plain English: The Hotelling-Lawley trace sums the eigenvalues of the matrix product $\mathbf{W}^{-1}\mathbf{B}$, which represent the ratios of between-group to within-group variance along each discriminant dimension. Unlike Wilks (product of ratios) or Pillai (sum of proportions), Hotelling-Lawley directly sums the variance ratios. Larger values indicate stronger group separation. The statistic generalizes Hotelling's T^2 from two groups to multiple groups, maintaining connection to classic multivariate theory.

⚠ Practitioner Warning

Less robust to violations than Pillai's trace but potentially more powerful (like Wilks' Λ) than Pillai under ideal conditions. Sensitivity to heterogeneity of covariance similar to Wilks' Λ. F-distribution (McKeon, 1974) is approximate; exact distribution only for special cases. Can be undefined when **W** is singular (more variables than observations in smallest group). Range unbounded above (unlike Pillai or Wilks), making magnitude interpretation less intuitive. Less commonly used than Wilks or Pillai in practice; no clear advantage over these alternatives in most situations.

Hypotheses

- H_0: All group mean vectors equal ($\mu_1 = \cdots = \mu_g$).
- H_1: At least one group mean vector differs from others.

Decision Rule: Reject H_0 if $F > F_{\alpha, df_1, df_2}$ or p-value $< \alpha$. McKeon's approx. $df_1 = s(2m + s + 1)$, $df_2 = 2(sn + 1)$.

Software References [Python, R, SAS]

`statsmodels.MANOVA` | `stats::summary.manova` | `PROC GLM (MANOVA)`

Additional Notes

Relationship among MANOVA statistics: for eigenvalues $\lambda_1, \ldots, \lambda_s$ of $\mathbf{W}^{-1}\mathbf{B}$, Wilks' $\Lambda = \prod(1+\lambda_i)^{-1}$, Pillai's trace $= \sum \lambda_i/(1+\lambda_i)$, and the Hotelling–Lawley trace $U = \sum \lambda_i$. Thus, all criteria depend on the same latent roots but combine them differently. The trace formulation $U = \mathrm{tr}(\mathbf{W}^{-1}\mathbf{B})$ sums the diagonal elements of the matrix product and reflects total multivariate separation across dimensions. For two groups ($g = 2$), U is directly related to Hotelling's T^2: $T^2 = (n_1 n_2)U/(n_1 + n_2)$ when $s = p$. In terms of power, the Hotelling–Lawley trace is generally intermediate between Pillai's trace (more conservative, robust) and Roy's largest root (more sensitive to dominant effects), and often behaves similarly to Wilks' Λ in balanced settings. However, it is less directly interpretable than Pillai (bounded, variance-explained view) or Wilks (likelihood-based), which contributes to its less frequent use in practice. Historically, Lawley (1938) developed the asymptotic distribution theory, and the statistic is named jointly with Hotelling to reflect their foundational contributions to multivariate inference. McKeon's (1974) F-approximation: $s = \min(p, g-1)$, $m = (|(g-1) - p| - 1)/2$, and $n = ((N-g) - p - 1)/2$.

Miscellanea: The Hotelling–Lawley trace is one of the standard MANOVA test statistics, related to Hotelling's T^2 and later developments in multivariate test theory by Lawley. Based on the sum of eigenvalues of a matrix product, it provides an alternative to Wilks' Λ and Pillai's trace. While less emphasized in practice, it remains part of the classical set of MANOVA criteria.

Reference(s): Lawley (1938); McKeon (1974); Pillai (1960)

Worked Example

Problem Statement A technology company is evaluating two experimental keyboard layouts (**Layout A** and **Layout B**) against a **Standard** keyboard ($g = 3$). Sixty users ($N = 60$) are randomly assigned to the three layouts (20 per group). Performance is measured using two variables ($p = 2$): **Typing Speed** (WPM) and **Error Rate** (%). Using **Hotelling–Lawley's Trace** in a one-way MANOVA framework, the objective is to test whether keyboard layout affects the joint performance vector $[\mu_{\text{speed}}, \mu_{\text{errors}}]$. $p = 2, g = 3, N = 60$.

1. Statistical Matrices After analyzing the keyboard performance data, the eigenvalues of the matrix $\mathbf{W}^{-1}\mathbf{B}$ are calculated to be:

- $\lambda_1 = 0.582$ (First canonical dimension)
- $\lambda_2 = 0.114$ (Second canonical dimension)

2. Calculation of Hotelling-Lawley's Trace (T) The Hotelling-Lawley Trace is the sum of the eigenvalues:

$$T = \sum \lambda_i = 0.582 + 0.114 = 0.696$$

3. F-Approximation and Significance To determine statistical significance, we convert T to an F-statistic. With $s = \min(p, g-1) = 2$, $n = ((N-g) - p - 1)/2 = 27$, and $m = (|(g-1) - p| - 1)/2 = -1/2$:

$$F_{(4,110)} \approx \frac{2(sn+1)}{s^2(2m+s+1)} \cdot T = 9.57, \quad p < 0.001$$

4. Results Summary

Metric	Value
Hotelling-Lawley Trace (T)	0.696
Approximate F-Value	9.57
Degrees of Freedom	(4, 110)
P-value	**< 0.001**

5. Interpretation

- **Decision:** Since $p < 0.001$, we reject the null hypothesis. The keyboard layout significantly influences the user's combined speed and accuracy profile.

Practitioner Insight: Hotelling-Lawley Trace is generally more sensitive than Pillai's Trace when the differences between groups are large and the assumptions of MANOVA (normality and equal covariance) are strictly met.

Common Pitfalls

- **Using the Hotelling–Lawley trace as a default MANOVA statistic.** The test performs best when assumptions such as multivariate normality and homogeneous covariance matrices are satisfied. When assumptions are uncertain, more robust alternatives such as Pillai's trace are generally preferred.
- **Interpreting the magnitude of the trace statistic directly.** The Hotelling–Lawley statistic has no fixed upper bound and depends on model dimensionality. Inference should rely on the associated F transformation and corresponding significance test.
- **Expecting identical conclusions across MANOVA statistics.** Hotelling–Lawley, Wilks' Λ, Pillai's trace, and Roy's largest root emphasize different aspects of the eigenvalue structure. Moderate differences among their results can occur even when the same hypothesis is tested.
- **Applying the statistic in strongly unbalanced designs.** Substantial differences in group sample sizes can amplify the impact of covariance heterogeneity. Robust MANOVA statistics are often preferable when group sizes differ markedly.
- **Misinterpreting the trace as a property of the raw hypothesis matrix.** The statistic is based on the eigenvalues of the matrix $\mathbf{W}^{-1}\mathbf{B}$. This formulation ensures the test reflects separation relative to the error covariance structure.

Sum of Squares and Cross-products (SSCP) - E = Error; W = Within; H = Hypothesis; B = Between

$$\mathbf{E} \text{ or } \mathbf{W} = \sum_{i=1}^{g}\sum_{j=1}^{n_i}(\mathbf{x}_{ij} - \bar{\mathbf{x}}_i)(\mathbf{x}_{ij} - \bar{\mathbf{x}}_i)', \qquad \mathbf{H} \text{ or } \mathbf{B} = \sum_{i=1}^{g} n_i(\bar{\mathbf{x}}_i - \bar{\mathbf{x}})(\bar{\mathbf{x}}_i - \bar{\mathbf{x}})'$$

Roy's Largest Root Test

Maximum Eigenvalue Test for Multivariate Differences

MULT.6 | **Category:** Multivariate Analysis | **Data:** Continuous | **Null:** Roy's distribution or F

When to Use

- Test multivariate mean differences among groups.
- MANOVA with multiple correlated responses.
- Assess overall group separation multivariately.
- Control Type I error across several outcomes.
- Profile-comparison and experimental studies.

Assumptions

- Observations are mutually independent.
- Responses follow multivariate normality.
- Covariance matrices are equal across groups.
- Variables are measured on interval scale.
- Sample size exceeds the number of variables ($n_i > p$).

Test Statistic

$\lambda_{\max} = \max(\lambda_1, \ldots, \lambda_s)$, largest eigenvalue of $\mathbf{W}^{-1}\mathbf{B}$. Upper bound: $F_{\text{ub}} = \dfrac{N-s-1}{s}\lambda_{\max} \approx F_{s,\,N-s-1}$.

Notation: $\mathbf{B}$ between-groups SSCP, $\mathbf{W}$ within-groups SSCP, $s = \min(p, g-1)$ [F_{ub} matches Stata version]

Plain English: Roy's largest root focuses on the single strongest dimension of group separation by using only the maximum eigenvalue of $\mathbf{W}^{-1}\mathbf{B}$. This eigenvalue represents the largest ratio of between-group to within-group variance achievable by any linear combination of variables. The test is most powerful when group differences are primarily unidimensional—when one discriminant function captures most separation. It ignores weaker dimensions that the three previous tests incorporate.

⚠ Practitioner Warning

Roy's test is the most aggressive of the four standard MANOVA criteria and is least robust to covariance heterogeneity and non-normality. Can be overly optimistic when multiple weak dimensions combine to create significant overall difference. Critical values difficult to obtain; Roy's (and Heck's) tables limited, approximations imprecise. Most powerful only when single dimension truly dominates; can have lower power than alternatives when effects diffuse across dimensions. Does not provide omnibus test of all group differences—misses orthogonal separation patterns. Generally not recommended unless strong theoretical reason to focus on single dimension.

Hypotheses

- H_0: $\mu_1 = \cdots = \mu_g$ (no separation on any dimension).
- H_1: Groups differ along at least one linear combination.

Decision Rule: Reject H_0 if $F_{\text{ub}} > F_{\alpha,s,N-s-1}$ using upper bound approximation.

Software References [Python, R, SAS]

Python	R	SAS
`statsmodels.MANOVA`	`stats::summary.manova`	`PROC GLM (MANOVA)`

Additional Notes

Roy's statistic is the largest eigenvalue $\lambda_{\max}$ of $\mathbf{W}^{-1}\mathbf{B}$, corresponding to the first canonical discriminant function—the linear combination maximizing the between-group to within-group variance ratio. The transformed quantity $\lambda_{\max}/(1+\lambda_{\max}) = R_1^2$ equals the largest squared canonical correlation and provides an interpretable measure of association for this dominant dimension. As noted by Kuhfeld (1986), some software report $\lambda_{\max}/(1+\lambda_{\max})$ instead of $\lambda_{\max}$, leading to confusion. Relationship to other MANOVA tests: Roy uses $\max(\lambda_i)$, Hotelling–Lawley uses $\sum\lambda_i$, Pillai uses $\sum\lambda_i/(1+\lambda_i)$, and Wilks uses $\prod(1+\lambda_i)^{-1}$; all are functions of the same latent roots. Roy's test follows from the union–intersection principle, corresponding to the maximum univariate F-statistic over all linear combinations of the dependent variables. In terms of power, Roy is most sensitive when one eigenvalue dominates ($\lambda_1 \gg \lambda_2, \ldots, \lambda_s$) and least effective when effects are distributed across dimensions. Roy (1953) derived the distribution theory; tabled critical values depend on $s = \min(p, g-1)$, $m = (|p-(g-1)|-1)/2$, and $n = (N-g-p-1)/2$. The F-statistic is an upper bound and the associated p-value a lower bound.

Miscellanea: S.N.Roy (1953) introduced the largest root test, based on the maximum eigenvalue of a multivariate test matrix and linked to union–intersection principles. The test is most powerful when a single dominant dimension drives group differences, but it is less balanced than alternatives such as Wilks' Λ or Pillai's trace. It remains one of the standard MANOVA test statistics.

Reference(s): Roy (1953); Heck (1960); Olson (1976); Kuhfeld (1986)

Worked Example

Problem Statement A sports scientist compares three training intensities (**Low, Moderate, High**; $g = 3$) to evaluate their effect on two performance outcomes ($p = 2$): **Vertical Jump Height** (y_1) and **Sprinting Speed** (y_2). Thirty athletes ($N = 30$) are randomly assigned to the three groups ($n = 10$ per group). Because the expected difference between groups is primarily in vertical jump performance, the researcher uses **Roy's Maximum Root** test within a one-way MANOVA to determine whether the mean performance vectors differ across training intensities.

1. Statistical Matrices After analyzing the athletic performance data, the eigenvalues of the matrix $\mathbf{W}^{-1}\mathbf{B}$ are calculated to be:

- $\lambda_1 = 0.845$ (First canonical dimension)
- $\lambda_2 = 0.012$ (Second canonical dimension)

2. Calculation of Roy's Statistic (θ) Using $\lambda_{\max} = 0.845$, the corresponding squared canonical correlation is:

$$\theta = \frac{\lambda_{max}}{1 + \lambda_{max}} = 0.458$$

3. F-Approximation and Significance To determine significance, we calculate the F-statistic using the parameter $s = \min(p, g - 1) = 2$:

$$F_{(2,27)} \approx \frac{(N - s - 1)}{s} \cdot \lambda_{max} = 11.408, \quad p < 0.001$$

4. Results Summary

Metric	Value
Roy's Largest Root (λ_{max})	0.845
Roy's θ	0.458
Approximate F-Value	11.408
Degrees of Freedom	(2, 27)
P-value	**< 0.001**

5. Interpretation

- **Decision:** Since $p < 0.001$, we reject the null hypothesis. Training intensity has a significant effect on athletic performance.

Note: Some texts and software (e.g., SPSS MANOVA) report Roy's statistic as $\theta_1 = \lambda_{\max}/(1+\lambda_{\max})$, a squared canonical correlation bounded in $[0, 1]$; this book follows Stata and Rencher in using $\lambda_{\max}$ directly.

Practitioner Insight: Roy's Largest Root is the most powerful test here because the treatment effect was "concentrated" ($\lambda_1 \ggg \lambda_2$). If the effect had been spread equally across both Jump Height and Sprinting Speed, Wilks' Λ or Pillai's Trace might have been more appropriate.

Common Pitfalls

- **Using Roy's test when multiple dimensions contribute to group differences.** Roy's statistic focuses on the largest eigenvalue of $\mathbf{W}^{-1}\mathbf{B}$. When several dimensions contribute comparably, statistics such as Wilks' Λ or Pillai's trace provide a more balanced test.
- **Interpreting Roy's significance when other MANOVA statistics do not agree.** Roy's test concentrates on the strongest single dimension and may reject when others do not. Such discrepancies should prompt examination of eigenvalues, canonical variates, and potential data irregularities.
- **Using Roy's test repeatedly in exploratory analyses.** Because the statistic selects the largest eigenvalue, it effectively performs an implicit search over dimensions. Repeated use across multiple hypotheses can substantially inflate the probability of false positives.
- **Assuming a larger Roy F statistic indicates stronger evidence.** Roy's statistic often produces larger F values because it focuses on the dominant dimension. This reflects a different testing perspective rather than stronger evidence of overall group differences.
- **Applying Roy's test when only two groups are compared.** With two groups there is only one non-zero eigenvalue in the MANOVA decomposition. In this situation Roy's statistic becomes equivalent to the other multivariate test statistics.

Sum of Squares and Cross-products (SSCP) - E = Error; W = Within; H = Hypothesis; B = Between

$$\mathbf{E} \text{ or } \mathbf{W} = \sum_{i=1}^{g}\sum_{j=1}^{n_i}(\mathbf{x}_{ij} - \bar{\mathbf{x}}_i)(\mathbf{x}_{ij} - \bar{\mathbf{x}}_i)', \qquad \mathbf{H} \text{ or } \mathbf{B} = \sum_{i=1}^{g} n_i(\bar{\mathbf{x}}_i - \bar{\mathbf{x}})(\bar{\mathbf{x}}_i - \bar{\mathbf{x}})'$$

Box's M Test

Test Equality of Covariance Matrices Across Groups

MULT.7 | **Category:** Multivariate Analysis | **Data:** Continuous | **Null:** Chi-square (approximate)

When to Use

- Test equality of covariance matrices across groups.
- Check assumptions before MANOVA procedures.
- Assess covariance homogeneity multivariately.
- Multiple correlated response variables exist.
- Preliminary diagnostic before discriminant analysis.

Assumptions

- Observations are mutually independent.
- Responses follow multivariate normality.
- Variables are measured on interval scale.
- Covariance matrices are estimated reliably.
- Sample sizes are adequate within groups.

Test Statistic

$$M = (N-g)\ln|\mathbf{S}_p| - \sum_{i=1}^{g}(n_i-1)\ln|\mathbf{S}_i|, \quad \chi^2 \approx (1-C)\,M. \qquad \text{Under } H_0, \chi^2 \approx \chi^2_{df},\ df = p(p+1)(g-1)/2.$$

Notation: $\mathbf{S}_p$ is pooled covariance, $\mathbf{S}_i$ are group covariances, C is Box's correction for df

Plain English: Box's M test compares the pooled covariance matrix (weighted avg. across groups) to individual group covariance matrices using a likelihood ratio approach. It measures the discrepancy between assuming a common covariance structure versus allowing each group its own structure. Large values indicate covariances differ, violating the homogeneity assumption required for standard multivariate procedures. It's very sensitive, often rejecting with even minor differences.

⚠ Practitioner Warning

Extremely sensitive to departures from multivariate normality—can reject due to non-normality rather than true covariance differences. Very liberal (over-rejects) with large samples, detecting trivial differences. Overly conservative (under-rejects) with small samples or when p large relative to n_i. Degrees of freedom df can be very large, making test unwieldy. Rejection common but not necessarily fatal—MANOVA reasonably robust to moderate heterogeneity, especially with balanced designs. Many statisticians recommend ignoring this test unless violation appears severe in exploratory analysis. Better to examine variance ratios directly than rely solely on Box's M.

Hypotheses

- H_0: $\mathbf{\Sigma}_1 = \mathbf{\Sigma}_2 = \cdots = \mathbf{\Sigma}_g$ (equal covariance matrices).
- H_1: At least one $\mathbf{\Sigma}_i$ is different from others.

Decision Rule: Reject H_0 if $\chi^2 > \chi^2_{\alpha,df}$ or p-value $< \alpha$. Use conservative α (e.g., 0.01) or treat as rough diagnostic.

Software References [Python, R, SAS]

`pingouin.box_m`	`biotools::boxM`	`PROC DISCRIM (POOL)`

Additional Notes

Box's correction factor: $C = \frac{2p^2+3p-1}{6(p+1)(g-1)}\left[\sum_{i=1}^{g}\frac{1}{n_i-1} - \frac{1}{N-g}\right]$ improves chi-square approximation. For large samples, $C \to 0$ and correction vanishes. Pooled covariance: $\mathbf{S}_p = \sum_{i=1}^{g}(n_i-1)\mathbf{S}_i/(N-g)$ weights group covariances by degrees of freedom. The M statistic is $(N-g)$ times log-likelihood ratio comparing pooled to separate covariances. Box (1949) developed test and correction factor for finite samples. Alternative: examine variance ratios $\max(\sigma_i^2)/\min(\sigma_i^2)$ for each variable; rule of thumb is ratios < 10 acceptable. For two groups, Bartlett's test alternative but equally sensitive. Modern view: use Box's M as rough screening; if wildly significant, examine covariances graphically; MANOVA often robust to moderate violations with balanced n_i. Unequal covariances more problematic with unbalanced designs. Remedies: use separate covariance methods, transform data, or use robust multivariate procedures.

Miscellanea: George Box (1949) introduced the M test for equality of covariance matrices as part of his work on multivariate methods. Box's M is Bartlett's test in higher dimensions—variance becomes covariance, scalar becomes matrix. Although widely used, the test is sensitive to departures from normality and large sample sizes, so results should be interpreted with caution. Reflecting his broader philosophy, famously remarked, "All models are wrong, but some are useful," underscoring the practical limits of statistical models.

Reference(s): Box (1949); Anderson (2003)

Worked Example

Problem Statement A researcher compares two different training programs (**Method A** and **Method B**) based on two outcome variables: **Task Speed** (X_1) and **Accuracy** (X_2). To use MANOVA, the researcher must verify that the covariance between Speed and Accuracy is the same for both training groups.

1. Hypotheses and Parameters We test the null hypothesis $H_0 : \Sigma_1 = \Sigma_2 = \cdots = \Sigma_g$ (Covariance matrices are equal across groups).

- **Number of Variables** (p)**:** 2 (Speed, Accuracy)
- **Number of Groups** (g)**:** 2 (Method A, Method B)
- **Sample Size per Group:** $n_1 = 50, n_2 = 50$

2. Determinant Calculation Let S_1 and S_2 be the sample covariance matrices. We calculate their determinants $|S_i|$:

$$S_1 = \begin{pmatrix} 15.2 & 2.1 \\ 2.1 & 5.4 \end{pmatrix}, \qquad |S_1| = 77.67$$

$$S_2 = \begin{pmatrix} 14.8 & 2.3 \\ 2.3 & 5.9 \end{pmatrix}, \qquad |S_2| = 82.03$$

3. Calculation of the M Statistic Given the pooled determinant $|S_p| \approx 79.91$:

$$S_p = \frac{(n_1 - 1)S_1) + (n_2 - 1)S_2}{(n_1 + n_2 - 2)}$$

$$M = (n_1 + n_2 - 2)\ln(|S_p|) - [(n_1 - 1)\ln(|S_1|) + (n_2 - 1)\ln(|S_2|)] = 0.110$$

4. Results and Decision

Metric	Value
Box's M Statistic	0.110
Box's Correction C	0.022
F-approximation / χ^2	0.108
Degrees of Freedom	3
P-value	**0.991**
Decision	**Fail to Reject H_0**

5. Interpretation

- **Decision:** Since $p = 0.991 > 0.05$, we fail to reject the null hypothesis.
- **Conclusion:** The assumption of homogeneity of covariance matrices is met.
- **Practitioner Insight:** Box's M is extremely sensitive to large samples. In many software packages (like SPSS), a p-value of 0.001 is often used as the threshold instead of 0.05 to avoid over-rejecting H_0 due to minor, non-influential differences.

Common Pitfalls

- **Treating Box's M as a strict gatekeeping test at $\alpha = 0.05$.** The test is highly sensitive and often rejects for minor covariance differences. Slight heterogeneity rarely invalidates MANOVA, particularly when robust statistics such as Pillai's trace are used.
- **Interpreting rejection without checking multivariate normality.** Box's M is strongly affected by departures from multivariate normality. Non-normal data can cause rejection even when covariance matrices are equal.
- **Focusing solely on the p-value rather than the magnitude of differences.** Statistical significance does not necessarily imply meaningful heterogeneity. Inspecting variance ratios and covariance matrices often provides a clearer assessment of practical impact.
- **Using the test with extremely small or very large samples.** Small samples provide unstable covariance estimates and reduce test reliability. Large samples can detect trivial differences that have little practical consequence.
- **Assuming Box's M evaluates all MANOVA assumptions.** The test examines only the equality of covariance matrices across groups. Other assumptions such as multivariate normality and independence must be assessed separately.

Box's Correction Factor (C) & Pooled Covariance Matrix ($\mathbf{S}_p$)

$$C = \frac{2p^2 + 3p - 1}{6(p+1)(g-1)}\left[\sum_{i=1}^{g}\frac{1}{n_i - 1} - \frac{1}{N-g}\right], \qquad \mathbf{S}_p = \frac{\sum_{i=1}^{g}(n_i - 1)\mathbf{S}_i}{N - g}$$

Bartlett's Test of Sphericity

Test Whether Correlation Matrix is Identity Matrix

MULT.8 | **Category:** Multivariate Analysis | **Data:** Continuous | **Null:** Chi-square

When to Use

- Assess suitability for factor-analysis methods.
- Test whether variables are sufficiently correlated.
- Preliminary diagnostic before PCA or EFA.
- Evaluate correlation matrix structure formally.
- Multiple continuous variables are analyzed.

Assumptions

- Observations are mutually independent.
- Variables are measured on interval scale.
- Data follow multivariate normality.
- Correlation matrix is estimated reliably.
- Sample size is adequate for factor analysis.

Test Statistic

$$\chi^2 = -[(n-1) - \frac{2p+5}{6}] \ln |\mathbf{R}|. \qquad \chi^2 \sim \chi^2_{p(p-1)/2} \text{ (under } H_0).$$

Notation: $|\mathbf{R}|$ is determinant of sample correlation matrix, n is sample size, p is number of variables

Plain English: Bartlett's test of sphericity evaluates whether the observed correlation matrix could plausibly be an identity matrix (all correlations zero, and, under multivariate normality, variables independent). The test uses the determinant of the correlation matrix—for an identity matrix, determinant equals 1; for correlated variables, determinant is less than 1. The logarithm of the determinant, scaled by sample size, provides a chi-square statistic. Rejection indicates correlations are present, justifying multivariate techniques like factor analysis that exploit correlation structure.

⚠ Practitioner Warning

Sensitive to departures from multivariate normality. Very sensitive to sample size—power increases with n, so even trivial correlations are often detected in moderate to large samples. As a result, the test almost always rejects for large n, and statistical significance may not imply practically meaningful correlations. Primarily used as a screening test before factor analysis; however, rejection does not guarantee that factor analysis will be successful. The test does not assess adequacy of the correlation structure for specific techniques (use the KMO measure for that). In practice, examine correlation magnitudes directly rather than relying solely on this test.

Hypotheses

- H_0: $\mathbf{R} = \mathbf{I}$ (variables uncorrelated).
- H_1: $\mathbf{R} \neq \mathbf{I}$ (variables are correlated).

Decision Rule: Reject H_0 if $\chi^2 > \chi^2_{\alpha,p(p-1)/2}$ or p-value $< \alpha$. df: $p(p-1)/2$ is number of unique correlations.

Software References [Python, R, SAS]

`Not Native (Manual)`	`psych::cortest.bartlett`	`PROC FACTOR`

Additional Notes

Determinant interpretation: $|\mathbf{R}|$ measures generalized variance. For identity matrix, $|\mathbf{I}| = 1$. As correlations increase, determinant decreases toward 0 (perfect multicollinearity yields $|\mathbf{R}| = 0$). The test statistic $-\ln |\mathbf{R}|$ increases as correlations strengthen. Bartlett (1950, 1951) developed correction factor $(n-1) - (2p+5)/6$ improving chi-square approximation. For exact identity matrix, $\ln |\mathbf{R}| = \ln(1) = 0$ yielding $\chi^2 = 0$. Large correlations make $|\mathbf{R}| \ll 1$, so $-\ln |\mathbf{R}| \gg 0$ yielding large χ^2. Relationship to factor analysis: rejection indicates correlations exist but not that common factors adequately explain them. Kaiser-Meyer-Olkin (KMO) measure of sampling adequacy provides better assessment of factor analysis suitability. Anti-image correlation matrix shows partial correlations; if these are small, factor analysis appropriate. Modern practice: report Bartlett's test alongside KMO and interpret both; with large n, prioritize KMO over Bartlett.

Miscellanea: Maurice Bartlett (early 1950s) introduced the test of sphericity to assess whether a correlation matrix differs from identity, providing a diagnostic for the suitability of factor analysis. When variables are uncorrelated, factor extraction is not meaningful. Although the test often rejects with large samples, it remains a standard preliminary check, commonly reported alongside measures such as KMO.

Reference(s): Bartlett (1950, 1951); Kaiser (1970)

Worked Example

Problem Statement A researcher develops a survey with three questions ($p = 3$) intended to measure *Brand Loyalty*. Before performing a Factor Analysis, the researcher must verify that the questions are sufficiently correlated with each other. We test the null hypothesis $H_0 : R = I$ (the correlation matrix is an identity matrix) against $H_a : R \neq I$. Sample size (n) is 100. Degrees of freedom (df): $p(p-1)/2 = 3$.

1. Determinant of the Correlation Matrix For the 3×3 matrix R, the determinant $|R|$ is calculated:

$$R = \begin{pmatrix} 1.00 & 0.70 & 0.65 \\ 0.70 & 1.00 & 0.80 \\ 0.65 & 0.80 & 1.00 \end{pmatrix}, \qquad |R| = 0.1755$$

2. Calculation of the χ^2 Statistic

$$\chi^2 = -\left[(n-1) - \frac{2(p)+5}{6}\right] \ln(|R|) \approx 169.07$$

3. Results and Decision

Metric	Value
Determinant $\|R\|$	0.1755
Bartlett's χ^2	169.07
Degrees of Freedom	3
Critical $\chi^2_{0.05,3}$	7.815
P-value	**< 0.0001**
Decision	**Reject** H_0

4. Interpretation

- **Decision:** Since 169.07 > 7.815, we reject the null hypothesis of sphericity.
- **Conclusion:** The variables are significantly correlated. The dataset is suitable for structure reduction techniques.
- **Practitioner Insight:** While Bartlett's test tells you if the matrix is "factorable," it is usually paired with the **KMO (Kaiser-Meyer-Olkin) measure**, which tells you *how well* the data is suited for factor analysis on a scale of 0 to 1.

Common Pitfalls

- **Treating Bartlett's rejection as sufficient justification for factor analysis.** The test only indicates that some correlations among variables are statistically different from zero. Additional diagnostics such as KMO and inspection of the correlation matrix are needed to assess factorability.
- **Concluding factor analysis is impossible when Bartlett's test is not significant.** Failure to reject may result from small sample size or moderate correlations that lack statistical power. Examining the correlation matrix and sample size is necessary before dismissing factor analysis.
- **Ignoring assumptions underlying Bartlett's test.** The test assumes approximate multivariate normality and adequate sample size. Severe non-normality or outliers can distort correlations and produce misleading results.
- **Focusing only on the overall test result.** Bartlett's statistic evaluates whether the entire correlation matrix differs from the identity matrix. Detailed inspection of individual correlations is needed to understand the underlying structure.
- **Applying the test directly to ordinal or categorical variables.** Bartlett's test is derived for continuous variables using Pearson correlations. Ordinal data often require polychoric correlations or alternative methods for assessing factor suitability.

Kaiser–Meyer–Olkin (KMO) measure of sampling adequacy

KMO assesses whether a dataset is suitable for **factor analysis**. It compares the magnitude of observed correlations with the magnitude of corresponding **partial correlations**. If variables share common latent factors, correlations should be relatively large while partial correlations should be small. The statistic is computed as

$$\text{KMO} = \frac{\sum_{i\neq j} r_{ij}^2}{\sum_{i\neq j} r_{ij}^2 + \sum_{i\neq j} p_{ij}^2},$$

where r_{ij} denotes the simple correlation and p_{ij} the partial correlation between variables i and j. The measure ranges from 0 to 1, with larger values indicating greater suitability for factor analysis. Kaiser suggested the following interpretation: values of **0.90–1.00** are *superb*, **0.80–0.89** *meritorious*, **0.70–0.79** *middling*, **0.60–0.69** *mediocre*, **0.50–0.59** *miserable*, and **below 0.50** *unacceptable*. Researchers typically report both the overall KMO statistic and variable-specific measures of sampling adequacy before performing factor analysis.

Mauchly's Test of Sphericity

Test Sphericity Assumption for Repeated Measures ANOVA

MULT.9 | **Category:** Multivariate Analysis | **Data:** Continuous | **Null:** Chi-square (approximate)

When to Use

- Test sphericity in repeated-measures ANOVA.
- Check assumptions before within-subject analyses.
- Multiple measurements on the same subjects.
- Determine need for GG/HF corrections.
- Longitudinal or crossover study designs.

Assumptions

- Observations across subjects are independent.
- Repeated measurements are continuous variables.
- Data follow multivariate normality.
- Covariance matrix is estimated reliably.
- Sample size is adequate for approximation.

Test Statistic

$$W = \frac{|\mathbf{A}|}{(\mathrm{tr}(\mathbf{A})/(p-1))^{p-1}}, \quad \chi^2 \approx -\left[(n-1) - \frac{2p^2-3p+3}{6(p-1)}\right]\ln(W). \qquad \text{Under } H_0, \chi^2 \sim \chi^2_{df}, \ df = \frac{p(p-1)}{2} - 1.$$

Notation: $\mathbf{A} = CSC'$ is the covariance of orthonormalized contrasts, p is the number of repeated-measures levels, and n is the number of subjects.

Plain English: Mauchly's test evaluates whether the covariance matrix of difference scores between repeated measures satisfies sphericity—the condition that all pairwise differences have equal variance. Sphericity is required for univariate repeated measures ANOVA F-tests to have correct Type I error rates. The test compares the observed covariance structure to the spherical pattern using a ratio of determinants. Violation of sphericity inflates Type I error, requiring corrections (Greenhouse-Geisser, Huynh-Feldt) to degrees of freedom.

⚠ Practitioner Warning

Extremely sensitive to departures from multivariate normality—can reject for non-normality rather than true sphericity violation. Liberal with large samples (over-rejects for trivial violations); conservative with small samples. Rejection very common in practice. When test rejects, use multivariate approach (MANOVA) or apply epsilon corrections to univariate F-test. For two levels, sphericity automatically satisfied (test not applicable). Does not indicate severity of violation; examine epsilon estimates ($\hat{\epsilon}_{GG}, \hat{\epsilon}_{HF}$) for magnitude. Many practitioners report GG- or HF-corrected results routinely.

Hypotheses

- H_0: Sphericity holds ($\mathrm{Var}(Y_i - Y_j) = \sigma^2$ for all $i \neq j$).
- H_1: Sphericity violated.

Decision Rule: Reject H_0 if $\chi^2 > \chi^2_{\alpha,df}$ or p-value $< \alpha$. If reject, use $\hat{\epsilon}_{GG} \cdot df$ or $\hat{\epsilon}_{HF} \cdot df$ in place of df.

Software References [Python, R, SAS]

```
pingouin.sphericity        stats::mauchly.test        PROC GLM (REPEATED)
```

Additional Notes

Sphericity condition: variances of all pairwise differences equal, $\mathrm{Var}(Y_i - Y_j) = \sigma^2$ for all $i \neq j$. Compound symmetry (equal variances *and* equal covariances) implies sphericity but is strictly stronger — a covariance matrix can satisfy sphericity without satisfying compound symmetry. Compound symmetry is therefore sufficient but not necessary for sphericity. Epsilon (ϵ) measures degree of sphericity: $\epsilon = 1$ indicates perfect sphericity, $\epsilon < 1$ indicates violation. Greenhouse-Geisser (GG) epsilon is always $\geq 1/(p-1)$; Huynh-Feldt (HF) epsilon is less conservative than GG. Corrections: multiply both degrees of freedom by $\hat{\epsilon}$. Alternative: use multivariate approach (no sphericity assumption) at cost of power and requires $n > p$. Lower bound correction: $\epsilon = 1/(p-1)$ is maximally conservative.

Miscellanea: John Mauchly introduced a test of sphericity used in repeated measures ANOVA to assess whether variance–covariance structures satisfy required assumptions. Later work by Box and others developed corrections for violations, making the test a standard diagnostic tool. Despite sensitivity to sample size, it remains widely reported in repeated measures analyses.

Reference(s): Mauchly (1940); Greenhouse & Geisser (1959); Huynh & Feldt (1976)

Worked Example

Problem Statement A sports science clinic compares three recovery protocols following high-intensity interval training (HIIT): **Passive Rest (PR)**, **Active Recovery (AR)**, and **Cold Water Immersion (CWI)**. Ten professional athletes ($n = 10$) complete all three conditions in a **one-way repeated measures design**. Before performing the repeated measures ANOVA, the **sphericity assumption** must be tested.

	1	2	3	4	5	6	7	8	9	10
PR	88	92	85	90	86	94	89	87	91	88
AR	84	89	82	88	83	91	85	84	87	86
CWI	79	82	78	80	77	85	81	76	83	80

1. Covariance structure and Orthogonal difference space. Applying the transformation $A = CSC'$ produces the covariance matrix of the orthogonal contrasts:

$$S = \begin{pmatrix} 7.778 & 7.667 & 7.000 \\ 7.667 & 8.100 & 6.678 \\ 7.000 & 6.678 & 7.656 \end{pmatrix}, \qquad C = \begin{pmatrix} \frac{1}{\sqrt{2}} & -\frac{1}{\sqrt{2}} & 0 \\ \frac{1}{\sqrt{6}} & \frac{1}{\sqrt{6}} & -\frac{2}{\sqrt{6}} \end{pmatrix}, \qquad A = CSC' = \begin{pmatrix} 0.272 & -0.279 \\ -0.279 & 1.187 \end{pmatrix}.$$

The matrix A contains the variances and covariances of the independent pairwise contrasts among the recovery conditions and forms the basis for computing Mauchly's W statistic.

2. Compute Mauchly's W. For $m = p - 1 = 2$,

$$W = \frac{|A|}{\left(\frac{\text{tr}(A)}{m}\right)^m} = \frac{0.245}{(1.459/2)^2} \approx 0.461.$$

Under perfect sphericity, the transformed covariance matrix is proportional to the identity, so W should be close to 1.

3. Convert W to a chi-square statistic.

$$\chi^2 \approx -(n-1)(1-f)\ln(W) \approx 6.20,$$

where $f = (2m^2 + m + 2)/(6m(n-1)) = 0.111$. The degrees of freedom are $df = m(m+1)/2 - 1 = 2$.. Thus, $p \approx 0.045$. Alternative χ^2 formula (uses m instead of p) but identical to the one on the previous page.

4. Decision and Interpretation. At the $\alpha = 0.05$ level, $p = 0.045 < 0.05$, so we reject the null hypothesis of sphericity. The variances of the pairwise condition differences are not consistent with sphericity. Therefore, the repeated measures ANOVA should be reported with a df correction such as **Greenhouse–Geisser** or **Huynh–Feldt**.

Common Pitfalls

- **Using Mauchly's test as a strict decision rule at $\alpha = 0.05$.** The test can be underpowered with small samples and overly sensitive with large samples. Epsilon estimates and corrected tests should be examined rather than relying solely on the p-value.
- **Misunderstanding what the sphericity assumption requires.** Sphericity concerns the equality of variances of all pairwise differences among repeated measures. It does not require equal variances or equal correlations across measurement occasions.
- **Applying the test when only two repeated measurements are available.** With two time points there is only one difference, so the sphericity condition is automatically satisfied. In such cases the test is unnecessary and provides no meaningful information.
- **Choosing among sphericity corrections arbitrarily.** When sphericity is violated, corrections such as Greenhouse–Geisser or Huynh–Feldt adjust the degrees of freedom. The choice should be guided by sample size and the severity of the violation rather than statistical convenience.
- **Ignoring multivariate alternatives when violations are severe.** Severe departures from sphericity may limit the reliability of univariate repeated-measures corrections. Multivariate approaches provide inference without requiring the sphericity assumption.

John–Nagao–Sugiura Test for Sphericity

Multivariate Test for Spherical Covariance Structure

MULT.10 | **Category:** Multivariate Tests | **Data:** Continuous (multivariate) | **Null:** χ^2 (asymptotic)

When to Use

- Test sphericity in repeated-measures designs.
- Alternative to Mauchly's sphericity test.
- Small-sample repeated-measures studies.
- Check assumptions before repeated-measures ANOVA.
- Determine need for GG/HF corrections.

Assumptions

- Observations across subjects are independent.
- Repeated measurements are continuous variables.
- Data follow multivariate normality.
- Covariance matrix is estimated reliably.
- Sample size is adequate for approximation.

Test Statistic

$$T = \frac{(n-1)p}{2}\left[\frac{p \cdot \mathrm{tr}(\mathbf{S}^2)}{[\mathrm{tr}(\mathbf{S})]^2} - 1\right] \qquad \text{Under } H_0, T \sim \chi^2_{df},\ df = \frac{p(p+1)}{2} - 1.$$

Notation: $\mathbf{S}$ is sample covariance matrix, $\mathrm{tr}(\mathbf{S}^2) = \sum_{i,j} s_{ij}^2$ is sum of all squared elements, $\mathrm{tr}(\mathbf{S}) = \sum_i s_{ii}$, n is sample size

Plain English: The JNS test examines whether all variables have the same variance and are mutually uncorrelated by comparing two measures of variability in the sample covariance matrix. The first measure, trace of the squared matrix, captures the total sum of all squared covariances and variances. The second measure, the squared trace divided by the number of variables, represents what this sum would be under perfect sphericity. Under sphericity, where all variables have the same variance and zero covariances, these two measures should match up in a predictable way. Large deviations indicate either unequal variances across variables, non-zero correlations between variables, or both. [$T \equiv$ (3.6) in Nagao (1973)]

⚠ Practitioner Warning

Requires multivariate normality; violations can distort results, especially in small samples. The test is sensitive to both variance heterogeneity and correlation, so rejection does not identify the source—inspect the covariance matrix or use separate diagnostics. The χ^2 approximation needs adequate sample size (rule of thumb: $n \geq 3p$); performance degrades when $p \gg n$, and the test is undefined if the covariance matrix is singular. Power is low for large p unless n is very large. Not equivalent to Mauchly's test, which addresses a different sphericity concept in repeated measures ANOVA.

Hypotheses

- H_0: $\mathbf{\Sigma} = \sigma^2\mathbf{I}_p$ (covariance matrix is spherical).
- H_1: $\mathbf{\Sigma} \neq \sigma^2\mathbf{I}_p$ (variances differ).

Decision Rule: Reject H_0 if $T > \chi^2_{\alpha,df}$ or p-value $< \alpha$.

Software References [Python, R, SAS]

`Not Native (Manual)`	`Not Native (Manual)`	`Not Native (Manual)`

Additional Notes

Geometrically, equal variances and zero covariances implies spherical (not elliptical) contours of the multivariate normal distribution. The test, developed by John (1971, 1972) with refinements by Nagao and Sugiura, is often jointly attributed to all three. Degrees of freedom df reflect the reduction from the unrestricted covariance matrix to a single variance parameter under sphericity. Bartlett and JNS both test *sphericity* ($\Sigma = \sigma^2 I$), combining equal variances and zero correlations; JNS differs in construction and is typically more stable in high-dimensional settings. The two are related but not identical, offering alternative approaches to assessing sphericity. In PCA, sphericity implies equal eigenvalues (no dominant directions), while rejection indicates meaningful structure. In practice, sphericity rarely holds, but testing it provides insight into covariance structure and can guide method choice (e.g., robustness of Pillai's trace in MANOVA).

Miscellanea: The John–Nagao–Sugiura test, developed in the early 1970s, provides an alternative approach to testing covariance structure based on functions of the sample covariance matrix such as trace and determinant. Although theoretically well-founded, it is less commonly used in practice than Bartlett's test and is not routinely implemented in standard statistical software.

Reference(s): John (1971, 1972); Nagao (1973); Sugiura (1973); Anderson (2003)

Worked Example

Problem Statement A hardware engineer is evaluating a new array of four sensors ($p = 4$) to assess whether the manufacturing process produces a **spherical covariance structure**, meaning equal variances and zero correlations among sensors. A sample of $n = 50$ units is collected. For the sample covariance matrix S, suppose $\mathrm{tr}(S) = 12.0$ and $\mathrm{tr}(S^2) = 38.5$.

The objective is to use the **John–Nagao–Sugiura (JNS) test** to determine whether the covariance matrix is proportional to the identity matrix: $H_0 : \Sigma = \sigma^2 I$ (Sphericity).

1. Calculation of the JNS Statistic (U) First, we find the squared ratio of the traces:

$$U = \frac{p \cdot tr(S^2)}{[tr(S)]^2} - 1 = 0.0694$$

2. Test Statistic and Significance The JNS statistic is transformed to a χ^2 distribution:

$$T_{JNS} = \frac{(n-1)p}{2} U = 6.801$$

Degrees of Freedom (df):

$$\frac{p(p+1)}{2} - 1 = 9.$$

3. Results Summary

Metric	Value
JNS U Statistic	0.0694
Calculated χ^2 (T_{JNS})	6.801
Critical $\chi^2_{0.05,9}$	16.919
P-value	**0.658**

4. Interpretation Since $6.801 < 16.919$, we fail to reject the null hypothesis. The covariance matrix is considered spherical. This implies that all sensors have nearly identical variances and that there is no significant correlation between them, confirming sensor uniformity.

Common Pitfalls

- **Confusing the JNS (John–Nagao–Sugiura) test with other sphericity tests such as Mauchly's or Bartlett's tests.** The JNS test evaluates whether the covariance matrix is proportional to the identity ($\Sigma = \sigma^2 I$). Bartlett's test of sphericity tests the same null using a likelihood-ratio framework, whereas Mauchly's test pertains to repeated-measures sphericity and is not equivalent.
- **Applying the JNS test with very small sample sizes relative to the number of variables.** Because the test relies on asymptotic χ^2 approximations based on covariance eigenvalues, small n relative to p leads to unstable estimates and unreliable Type I error rates.
- **Ignoring the multivariate normality assumption underlying the test.** Departures from multivariate normality—especially heavy tails or strong skewness—distort the sampling distribution of the covariance eigenvalues and invalidate the reference χ^2 approximation.
- **Interpreting non-rejection of the test as proof that variables are independent.** Failure to reject $\Sigma = \sigma^2 I$ simply indicates insufficient evidence against sphericity and may occur when correlations are weak or sample size is too small to detect departures.
- **Applying the test directly to variables measured on very different scales.** Large differences in measurement units inflate covariance magnitudes and can cause rejection due to scale effects rather than genuine correlation structure, making standardization or use of the correlation matrix preferable.

Derivation: Nagao (3.6) to Standard Form

From Nagao (1973): $T_2 = \frac{p^2 n}{2} \mathrm{tr}\{S(\mathrm{tr}\, S)^{-1} - p^{-1} I\}^2$.

Expanding: $\left[\frac{S}{\mathrm{tr}\, S} - \frac{I}{p}\right]^2 = \frac{S^2}{[\mathrm{tr}\, S]^2} - \frac{2S}{p \cdot \mathrm{tr}\, S} + \frac{I}{p^2}$.

Taking trace (using $\mathrm{tr}(I) = p$): $\frac{\mathrm{tr}(S^2)}{[\mathrm{tr}\, S]^2} - \frac{2}{p} + \frac{1}{p} = \frac{\mathrm{tr}(S^2)}{[\mathrm{tr}\, S]^2} - \frac{1}{p}$.

Substituting: $T_2 = \frac{p^2 n}{2} \left[\frac{\mathrm{tr}(S^2)}{[\mathrm{tr}\, S]^2} - \frac{1}{p}\right] = \frac{pn}{2} \left[\frac{p \cdot \mathrm{tr}(S^2)}{[\mathrm{tr}\, S]^2} - 1\right]$.

Since Nagao uses $n = N - 1$, in standard notation: $T = \frac{(n-1)p}{2} \left[\frac{p \cdot \mathrm{tr}(S^2)}{[\mathrm{tr}(S)]^2} - 1\right]$. □

Mardia's Test (Multivariate Normality)

Test for Multivariate Skewness and Kurtosis

MULT.11 | **Category:** Multivariate Analysis | **Data:** Continuous | **Null:** Chi-square and Normal

When to Use

- Assess multivariate normality of several variables.
- Validate assumptions before MANOVA procedures.
- Preliminary diagnostic before discriminant analysis.
- Check skewness and kurtosis jointly.
- Multiple continuous variables are analyzed.

Assumptions

- Observations are mutually independent.
- Variables are measured on interval scale.
- Covariance matrix is positive definite.
- Sample size is adequate for approximation.
- Data arise from a continuous distribution.

Test Statistic

$$\chi^2 = \frac{nb_{1,p}}{6} \sim \chi^2_{df};\ z = \frac{b_{2,p} - p(p+2)}{\sqrt{8p(p+2)/n}} \sim N(0,1).\quad d_{ij} = (\mathbf{x}_i - \bar{\mathbf{x}})'\mathbf{S}^{-1}(\mathbf{x}_j - \bar{\mathbf{x}}),\ b_{1,p} = \frac{1}{n^2}\sum_{i=1}^{n}\sum_{j=1}^{n} d_{ij}^3,\ b_{2,p} = \frac{1}{n}\sum_{i=1}^{n} d_{ii}^4.$$

Notation: $b_{1,p}, b_{2,p}$ are multivariate skewness & kurtosis, d_{ij} are Mahalanobis distances, $df = p(p+1)(p+2)/6$

Plain English: Mardia's test assesses multivariate normality by examining third and fourth moments of the distribution via skewness and kurtosis statistics. Multivariate skewness measures asymmetry in the joint distribution, while multivariate kurtosis measures tail heaviness relative to normal. Both statistics are based on Mahalanobis distances between observations. The tests detect departures from normality that affect multiple variables jointly, not just marginal normality of individual variables. Rejection indicates multivariate non-normality requiring transformation or robust methods.

⚠ Practitioner Warning

Skewness test has large df $[p(p+1)(p+2)/6$ grows rapidly with $p]$, reducing power. Kurtosis test more powerful for detecting heavy tails or outliers. Tests assume random sampling; violations of independence affect validity. With large n, tests can reject for trivial departures. Does not identify which variables or combinations drive non-normality. Marginal normality (univariate tests on each variable) neither necessary nor sufficient for multivariate normality. Graphical diagnostics (Q-Q plots, chi-square plots) recommended alongside formal tests. Moderate violations often acceptable for robust procedures.

Hypotheses

- H_0: Data follow multivariate normal distribution.
- H_1: Data depart from multivariate normality.

Decision Rule: Reject H_0 if skewness $\chi^2 > \chi^2_{\alpha,df}$ or kurtosis $|z| > Z_{\alpha/2}$ or p-values $< \alpha$. Can test each separately or jointly (reject if either significant).

Software References [Python, R, SAS]

`pingouin.multivariate_normality` `MVN::mvn(type="mardia")` `MultNorm macro`

Additional Notes

Multivariate skewness measures third moment: $b_{1,p} = E[\{(\mathbf{x}-\boldsymbol{\mu})'\boldsymbol{\Sigma}^{-1}(\mathbf{y}-\boldsymbol{\mu})\}^3]$ where $\mathbf{x}, \mathbf{y}$ are independent copies. Zero under multivariate normality. Multivariate kurtosis measures fourth moment: $b_{2,p} = E[\{(\mathbf{x}-\boldsymbol{\mu})'\boldsymbol{\Sigma}^{-1}(\mathbf{x}-\boldsymbol{\mu})\}^2]$, equals $p(p+2)$ for multivariate normal. Mahalanobis distance: $d_{ij}^2 = (\mathbf{x}_i - \bar{\mathbf{x}})'\mathbf{S}^{-1}(\mathbf{x}_j - \bar{\mathbf{x}})$ measures distance accounting for covariance structure. Under normality, d_{ii}^2 approximately χ^2_p. Chi-square Q-Q plot of d_{ii}^2 versus χ^2_p quantiles provides graphical normality check. Alternative tests: Henze-Zirkler (energy-based), Royston (extends Shapiro-Wilk), Doornik-Hansen (transforms to normality). Small sample corrections exist for Mardia's statistics. Mardia (1970, 1974) developed these measures systematically.

Miscellanea: Kanti Mardia (1970) introduced multivariate measures of skewness and kurtosis as extensions of their univariate counterparts, using Mahalanobis-based formulations. These statistics provide a widely used approach to testing multivariate normality, capturing different types of departures from the normal distribution.

Reference(s): Mardia (1970, 1974); Mardia et al. (1979)

Mardia's Test (Multivariate Normality) MULT.11

Worked Example

Problem Statement A researcher analyzes data from 80 participants ($n = 80$) measured on two personality traits: **Extraversion** (y_1) and **Openness** (y_2) ($p = 2$). Before applying MANOVA, the researcher must verify the assumption of **multivariate normality**. This is assessed using **Mardia's test**, which evaluates multivariate skewness and kurtosis of the joint distribution of (y_1, y_2). Sample skewness: $b_{1,2} = 0.45$, and kurtosis: $b_{2,2} = 8.15$

2. Multivariate Skewness Calculation The calculated skewness statistic for this dataset is $b_{1,2} = 0.45$. To test for significance, we use the χ^2 transformation:

$$\chi^2_{skew} = \frac{n}{6} b_{1,p} = 6.00$$

With degrees of freedom $df = p(p+1)(p+2)/6 = 4$:

$$p_{skew} = P(\chi^2_4 > 6.00) = 0.199$$

3. Multivariate Kurtosis Calculation The calculated kurtosis statistic is $b_{2,2} = 8.15$. In a multivariate normal distribution, the expected kurtosis is $p(p+2) = 2(4) = 8$. We use the standard normal z-test:

$$z_{kurt} = \frac{b_{2,p} - p(p+2)}{\sqrt{8p(p+2)/n}} = 0.168$$

$$p_{kurt} = 2 \times P(Z > 0.168) = 0.866$$

4. Results Summary

Component	Statistic	p-value	Result
Mardia's Skewness	0.450	0.199	Normal
Mardia's Kurtosis	8.150	0.866	Normal

5. Interpretation

- **Decision:** Since both p-values exceed 0.05, we conclude that there is no significant departure from multivariate normality. Failed to reject H_0.
- **Practitioner Insight:** Mardia's test is more rigorous than running individual Shapiro-Wilk tests on each variable, as it accounts for the correlation structure between variables. If the sample size was small ($n < 20$), a small-sample correction for the skewness χ^2 would be required.

Common Pitfalls

- **Assuming univariate normality implies multivariate normality.** Individual variables can appear normally distributed while the joint distribution is not. Multivariate normality must therefore be assessed using dedicated tests such as Mardia's.
- **Examining only one component of Mardia's test.** Mardia's procedure evaluates multivariate skewness and kurtosis separately. Both statistics should be reported because they diagnose different types of departures from normality.
- **Interpreting non-significant results with very small samples as confirmation of normality.** With limited observations, the test has low power to detect moderate departures from normality. Graphical diagnostics and additional checks are advisable when sample sizes are small.
- **Not investigating the source of a significant test result.** Rejection indicates that the joint distribution deviates from multivariate normality but does not explain why. Further inspection of marginal distributions, Mahalanobis distances, and scatterplots can help identify the cause.
- **Assuming that violation of multivariate normality invalidates all multivariate analyses.** Many multivariate methods remain reasonably robust to moderate departures from normality. When violations occur, robust statistics or resampling approaches may provide reliable inference.

Calculation Logic

Mardia's test for multivariate normality is based on pairwise Mahalanobis distances computed from the centered observations. Let $\mathbf{x}_i$ denote the i-th observation, $\bar{\mathbf{x}}$ the sample mean vector, and $\mathbf{S}$ the sample covariance matrix. First compute the Mahalanobis distance between observations i and j. These distances are then used to construct measures of multivariate skewness and kurtosis.

$$\bar{\mathbf{x}} = \frac{1}{n}\sum_{i=1}^{n}\mathbf{x}_i, \qquad \mathbf{S} = \frac{1}{n-1}\sum_{i=1}^{n}(\mathbf{x}_i - \bar{\mathbf{x}})(\mathbf{x}_i - \bar{\mathbf{x}})^\top, \qquad d_{ij} = (\mathbf{x}_i - \bar{\mathbf{x}})^T\mathbf{S}^{-1}(\mathbf{x}_j - \bar{\mathbf{x}}), \qquad b_{1,p} = \frac{1}{n^2}\sum_{i=1}^{n}\sum_{j=1}^{n} d_{ij}^3, \qquad b_{2,p} = \frac{1}{n}\sum_{i=1}^{n} d_{ii}^2.$$

Henze-Zirkler Test

Consistent Test for Multivariate Normality

MULT.12 | **Category:** Multivariate Analysis | **Data:** Continuous | **Null:** Lognormal (approximate)

When to Use

- Test multivariate normality (before multivariate procedures) with better power than Mardia's test in many settings.
- Consistent test detecting wide range of departures from normality.

Assumptions

- Random sample (i.i.d) from common multivariate distribution (no missing data and $n > p$).
- Sample size adequate for stable test ($n \geq 50$ recommended).
- Smoothing parameter β chosen appropriately.

Test Statistic

$$HZ = \frac{1}{n}\sum_{i=1}^{n}\sum_{j=1}^{n}\exp\left(-\frac{\beta^2}{2}d_{ij}^2\right) - 2(1+\beta^2)^{-p/2}\sum_{i=1}^{n}\exp\left(-\frac{\beta^2}{2(1+\beta^2)}d_i^2\right) + n(1+2\beta^2)^{-p/2}.$$

Notation: β is smoothing parameter, d_{ij}^2 and d_i^2 are squared Mahalanobis distances, n is sample size, p is the dimension

Plain English: The Henze–Zirkler test compares the empirical characteristic function of the data with that of a multivariate normal distribution using a weighted distance. A smoothing parameter controls sensitivity to different kinds of departure. The test is consistent against all fixed alternatives to multivariate normality, so with enough data it will eventually detect non-normality. Unlike Mardia's test, which targets specific moments, Henze–Zirkler is a more global check of overall distributional shape.

⚠ Practitioner Warning

Power depends on the smoothing parameter β, though most software uses a standard or data-dependent default. The null distribution is not simple; p-values are usually obtained from a lognormal approximation, and some implementations may use simulation or bootstrap methods. The test is more computationally intensive than moment-based procedures because it uses all pairwise distances. It is an omnibus test, so rejection does not identify whether the departure is driven by skewness, kurtosis, tail weight, or multimodality. As with other normality tests, large samples may detect trivial departures, so graphical diagnostics remain important.

Hypotheses

- H_0: Data arise from multivariate normal distribution.
- H_1: Data do not arise from multivariate normal.

Decision Rule: Reject H_0 if $HZ >$ CV from lognormal or p-value $< \alpha$. CVs depend on n, p, β.

Software References [Python, R, SAS]

Python	R	SAS
`pingouin.multivariate_hz`	`MVN::mvn(type="hz")`	Not Native (Manual)

Additional Notes

Characteristic function approach: test measures $\int |\phi_n(t) - \phi_0(t)|^2 w(t)dt$ where ϕ_n is empirical characteristic function, ϕ_0 is multivariate normal characteristic function, $w(t)$ is weight function. Henze-Zirkler use Gaussian weight with parameter β. Smoothing parameter: default $\beta = (1/\sqrt{2})[n(2p+1)/4]^{1/(p+4)}$ data-dependent choice balancing bias and variance. Mahalanobis distances: $d_{ij}^2 = (\mathbf{x}_i - \mathbf{x}_j)'\mathbf{S}^{-1}(\mathbf{x}_i - \mathbf{x}_j)$ or centered versions using $\bar{\mathbf{x}}$. Lognormal approximation: $\log(HZ)$ approximately normal with moments estimated from data. Henze and Zirkler (1990) proved consistency and derived asymptotic distribution. Power comparisons: generally more powerful than Mardia against heavy-tailed alternatives; similar or less powerful against skewed alternatives. Royston (1983) test extends Shapiro-Wilk to multivariate setting, providing another alternative.

Miscellanea: Norbert Henze and Bernhard Zirkler (1990) introduced a test for multivariate normality based on the characteristic function, providing an alternative to moment-based approaches such as Mardia's test. The method captures broader distributional features and performs well across a range of alternatives. Its adoption increased with advances in computing and software implementation.

Reference(s): Henze & Zirkler (1990); Henze (1994)

Worked Example

Problem Statement A geologist collects soil samples from 25 locations ($n = 25$) and measures the concentration of three minerals ($p = 3$): **Zinc** (y_1), **Magnesium** (y_2), and **Lead** (y_3). Before applying cluster analysis or MANOVA, the geologist must verify that the three-dimensional data follow a **multivariate normal distribution**. This assumption is assessed using the **Henze–Zirkler test** for multivariate normality. Smoothing Parameter (β): ≈ 1.213 (calculated based on n and p).

Key Inputs: The Henze–Zirkler statistic uses two averaged weighted distance terms. $\bar{D}_{ij} = 0.310$, and $\bar{D}_i = 0.800$

1. Calculation of the HZ Statistic The Henze-Zirkler statistic is calculated by comparing the weighted distance between the observed characteristic function and the theoretical normal function:

$$HZ = n\left[\bar{D}_{ij} - 2(1+\beta^2)^{-p/2}\bar{D}_i + (1+2\beta^2)^{-p/2}\right] = 0.650$$

where $\bar{D}_{ij}$ represents the mean of the exponentiated Mahalanobis distances between all pairs.

2. Log-Normal Transformation for Significance The HZ statistic is transformed to a z-score using the mean ($E[HZ]$) and variance ($Var[HZ]$) expected under normality:

$$\mu = E[HZ] = 0.596, \quad \sigma^2 = Var[HZ] = 0.0204$$

Using the corresponding lognormal approximation, $p = 0.314$

3. Results Summary

Metric	Value
HZ Statistic	0.650
Smoothing Parameter (β)	1.213
Mean of HZ under H_0	0.596
P-value	**0.314**

4. Interpretation

- **Decision:** Since $p = 0.314 > 0.05$, we fail to reject the null hypothesis of multivariate normality.
- **Conclusion:** The data are consistent with multivariate normality, supporting the use of parametric multivariate methods.

Practitioner Insight: The HZ test is highly recommended by statisticians because it has good power against a wide range of non-normal alternatives. If the HZ value is large and the p-value is small, it indicates the data "shape" is too far from the expected Gaussian ellipsoid. (See Technical Notes section at the end of this chapter for additional formulas.)

Common Pitfalls

- **Tuning the smoothing parameter β to obtain a desired result.** Different values of β emphasize different types of departures from multivariate normality. The parameter should be chosen using the standard default or pre-specified criteria rather than post hoc selection.
- **Comparing Henze–Zirkler and Mardia test results without understanding their differences.** The Henze–Zirkler test provides a single omnibus assessment of multivariate normality. Mardia's tests separately evaluate skewness and kurtosis, offering additional diagnostic insight.
- **Ignoring computational cost for very large samples.** The test requires computing pairwise distances among observations, which scales quadratically with sample size. Large datasets may require subsampling or alternative normality tests.
- **Assuming the test evaluates only marginal normality.** The Henze–Zirkler test assesses the full joint distribution rather than each variable separately. Variables may appear normal individually while still violating multivariate normality.
- **Applying the test with extremely small samples or high dimensional data.** Small sample sizes reduce the reliability of asymptotic approximations used in the test. High-dimensional settings can also weaken power and complicate covariance estimation.

Calculation Logic

The Henze–Zirkler (HZ) test evaluates multivariate normality using pairwise squared Mahalanobis distances between observations. A smoothing parameter β, determined by the sample size n and the number of variables p, controls the kernel weighting in the statistic. The HZ statistic aggregates exponential functions of these distances and, under the null hypothesis of multivariate normality, follows approximately a log-normal distribution.

$$d_{ij} = (x_i - \bar{x})^T S^{-1}(x_j - \bar{x}), \quad \beta = \frac{1}{\sqrt{2}}\left[\frac{n(2p+1)}{4}\right]^{1/(p+4)}, \quad \bar{D}_{ij} = \frac{1}{n^2}\sum_{i=1}^{n}\sum_{j=1}^{n}\exp\left(-\frac{\beta^2}{2}d_{ij}^2\right), \quad \bar{D}_i = \frac{1}{n}\sum_{i=1}^{n}\exp\left(-\frac{\beta^2}{2(1+\beta^2)}d_i^2\right).$$

Benjamini-Hochberg (FDR)

False Discovery Rate Control for Multiple Testing

MULT.13 | **Category:** Multivariate Analysis | **Data:** Any | **Null:** Procedure (stepwise)

When to Use

- Control false discovery rate (some false positives acceptable) when testing many hypotheses simultaneously.
- More powerful alternative to Bonferroni when testing hundreds or thousands of hypotheses.

Assumptions

- Multiple hypothesis tests conducted "simulataneously".
- P-values must be uniformly distributed under true nulls.
- Tests independent or exhibit positive dependence.
- Desired false discovery rate level q specified.

Test Statistic

$$\text{Order p-values: } p_{(1)} \leq p_{(2)} \leq \cdots \leq p_{(m)}, \quad \text{find } k = \max\{i : p_{(i)} \leq \frac{i}{m} q\} \text{ (BH inequality).}$$

Notation: $p_{(i)}$ are ordered p-values, m is number of tests, q (commonly 0.05 or 0.10) is target FDR level, k is cutoff index

Plain English: The Benjamini-Hochberg procedure controls the false discovery rate—the expected proportion of false positives among rejected hypotheses. Order all p-values from smallest to largest, then find the largest rank k where the p-value is at most k/m times the target FDR level q. Reject all hypotheses with p-values up to and including $p_{(k)}$. The threshold becomes less stringent for larger ranks, allowing more rejections while controlling the overall false discovery rate. Unlike family-wise error rate methods (Bonferroni), FDR accepts some false positives while controlling their expected proportion.

⚠ Practitioner Warning

Controls FDR (not FWER), so some false positives are expected. Valid under independence or positive dependence; under arbitrary dependence, FDR control may fail; the Benjamini–Yekutieli adjustment provides a conservative correction. The threshold depends on the total number of tests m; changing m post-hoc invalidates results. FDR is controlled in expectation across repeated studies, not estimated for a single analysis. Comparability of p-values is assumed; heterogeneous tests complicate interpretation. More liberal than Bonferroni, suitable when some false discoveries are acceptable. Does not reflect differing importance of hypotheses; weighted FDR methods can address this.

Hypotheses

- H_0**:** Individual null hypotheses H_{0i} for $i = 1, \ldots, m$.
- H_1**:** Reject a subset while controlling FDR at level q

Procedure: Find largest k satisfying the BH inequality; reject $H_{(1)}, \ldots, H_{(k)}$.

Software References [Python, R, SAS]

`statsmodels::multipletests`	`p.adjust(method="BH")`	`PROC MULTTEST (FDR)`

Additional Notes

The false discovery rate is defined as FDR $= E\left[\frac{V}{\max(R,1)}\right]$, the expected proportion of false rejections/discoveries (V) among all rejections (R). The BH procedure guarantees FDR $\leq q$ under independence or certain positive dependence conditions. Extensions include the Benjamini–Yekutieli adjustment for arbitrary dependence using $(i/m)q/c_m$ with $c_m \approx \log(m) + 0.577$, and adaptive methods that estimate the proportion of true nulls to improve power. Weighted BH allows prioritizing hypotheses via weights. Compared to Bonferroni, BH is far more powerful for large m, enabling discovery in high-dimensional settings. The q-value (Storey, 2002) provides the minimum FDR level at which a test becomes significant, analogous to a p-value. FDR methods are now standard in genomics, neuroimaging, and other large-scale inference problems. Note: $c_m = \sum_{k=1}^{m} \frac{1}{k} \sim \ln m + \gamma$, where $\gamma = 0.577$ is the Euler-Mascheroni constant.

Miscellanea: Yoav Benjamini and Yosef Hochberg (1995) introduced false discovery rate (FDR) control to address large-scale multiple testing problems where traditional methods were overly conservative. By controlling the expected proportion of false positives rather than eliminating them entirely, the approach enabled more powerful inference in fields such as genomics and neuroscience, and has become a standard tool in modern statistical practice.

Reference(s): Benjamini & Hochberg (1995); Benjamini & Yekutieli (2001); Storey (2002)

Worked Example

Problem Statement A pharmacologist tests $m = 10$ candidate gene targets for differential expression under a new compound, obtaining one p-value per gene from independent two-sample t-tests. Apply the Benjamini–Hochberg procedure at FDR level $q^* = 0.05$. The BH procedure controls the *false discovery rate*—the expected fraction of rejections that are false positives—rather than the familywise error rate controlled by Bonferroni or Holm.

1. Sort and Rank

Order the p-values from smallest to largest:

k	Gene	$p_{(k)}$	$k\,q^*/m$	$\leq$?	
1	G4	0.001	0.005	✓	
2	G2	0.003	0.010	✓	
3	G8	0.007	0.015	✓	
4	G10	0.009	0.020	✓	
5	G6	0.013	0.025	✓	
6	G9	0.028	0.030	✓	$\leftarrow k^*$
7	G3	0.038	0.035	×	
8	G7	0.130	0.040	×	
9	G1	0.260	0.045	×	
10	G5	0.510	0.050	×	

2. BH Critical Values

The BH threshold at rank k is $k\,q^*/m = 0.005k$. All thresholds are multiples of 0.005: 0.005, 0.010, ..., 0.050.

3. Find k^* and Reject

Find the *largest* k for which $p_{(k)} \leq kq^*/m$. Here, $k^* = 6$ and so reject $H_{(1)}, \ldots, H_{(6)}$: G4, G2, G8, G10, G6, G9.

4. Comparison with Bonferroni

Bonferroni threshold: $\alpha/m = 0.05/10 = 0.005$. Only G4 (0.001) and G2 (0.003) are rejected.

Method	Threshold	Rejections
Uncorrected	0.050	G4,G2,G8,G10,G6,G9,G3 (7)
BH ($q^* = 0.05$)	$0.005k$	G4,G2,G8,G10,G6,G9 (6)
Bonferroni	0.005	G4,G2 (2)

Gene G3 ($p = 0.038$) is rejected without correction but correctly retained by BH: $0.038 > 7 \times 0.005 = 0.035$.

Practitioner Insight: The BH procedure is a *step-up* rule: find the largest k^* such that $p_{(k)} \leq kq^*/m$ and reject all hypotheses $1, \ldots, k^*$. Unlike Bonferroni and Holm, which control the familywise error rate (FWER), BH controls the *false discovery rate* (FDR), making it more powerful when m is large (here 6 rejections vs. 2). Geometrically, BH compares sorted p-values to the line kq^*/m (slope q^*/m). For $m = 10$, $q^* = 0.05$, the thresholds are $0.005, 0.010, \ldots, 0.050$. The cutoff occurs at $k = 6$ ($0.028 \leq 0.030$) but fails at $k = 7$ ($0.038 > 0.035$). The guarantee holds under independence or positive dependence (PRDS); a conservative adjustment (Benjamini–Yekutieli) using $c_m = \sum 1/i$ extends validity to arbitrary dependence at some loss of power. **PDRS** stands for *Positive Regression Dependence on a Subset*. It means, when one test statistic becomes more "significant," it should not make other tests less likely to be significant. In other words, the dependence is positively aligned—signals tend to move together rather than in opposite directions.

Common Pitfalls

- **Confusing FDR with FWER.** BH controls the *expected proportion* of false positives among rejections, not the probability of making any false positive. Use FDR for large exploratory screens; use FWER (Bonferroni/Holm) when even one false positive is unacceptable.
- **Ignoring dependence structure.** Standard BH assumes independence or positive dependence (PRDS). For arbitrary/unknown dependence, use the Benjamini–Yekutieli (BY) correction. BH may become liberal under strong negative or complex correlations.
- **Applying BH to unsorted p-values.** BH requires sorting p-values in ascending order before applying thresholds $p_{(i)} \leq (i/m)\alpha$. Applying thresholds to unsorted p-values produces incorrect rejections.
- **Misinterpreting the BH rejection rule.** BH is a *sequential* stopping rule: find the largest i^* meeting the criterion, and reject the first i^* hypotheses only. Do *not* treat $(i^*/m)\alpha$ as a global cutoff.
- **Misinterpreting q-values as posterior probabilities.** A q-value is the smallest FDR level at which a test would be rejected. It is *not* the probability that a specific finding is a false positive.
- **Assuming non-significant results validate the null.** BH controls false discoveries among rejections, not false negatives among non-rejections. Non-significant hypotheses may still correspond to true effects (especially in low-power settings).

Epsilon Corrections for Sphericity Violations: Greenhouse–Geisser and Huynh–Feldt

When the sphericity assumption is violated in a repeated-measures ANOVA, the F-statistic is becomes liberal because the nominal degrees of freedom are too large, inflating Type I error. Greenhouse and Geisser (1959) proposed correcting both the numerator and denominator degrees of freedom by a factor $\hat{\epsilon} \in [1/(p-1), 1]$, where p is the number of repeated conditions. Their estimator is

$$\hat{\epsilon}_{GG} = \frac{p^2(\bar{s}_{..} - \bar{s}_{i.})^2}{(p-1)\left[\sum_i \sum_j s_{ij}^2 - 2p \sum_i \bar{s}_{i.}^2 + p^2 \bar{s}_{..}^2\right]},$$

where s_{ij} are entries of the sample covariance matrix of the repeated measures, $\bar{s}_{i.}$ are row means, and $\bar{s}_{..}$ is the grand mean. The value $\hat{\epsilon}_{GG} = 1$ indicates perfect sphericity; $\hat{\epsilon}_{GG} = 1/(p-1)$ is the most severe possible violation. Multiplying both degrees of freedom by $\hat{\epsilon}_{GG}$ yields a corrected F-test that maintains approximately the nominal size. Because $\hat{\epsilon}_{GG}$ is a biased downward estimator of the true ϵ, the correction is conservative — it over-corrects in large samples, reducing power unnecessarily.

Huynh and Feldt (1976) addressed this conservatism by proposing a less biased estimator,

$$\hat{\epsilon}_{HF} = \frac{n(p-1)\hat{\epsilon}_{GG} - 2}{(p-1)\left[n - 1 - (p-1)\hat{\epsilon}_{GG}\right]},$$

where n is the number of subjects. Because $\hat{\epsilon}_{HF}$ corrects upward for the negative bias in $\hat{\epsilon}_{GG}$, it is less conservative and yields higher power, but can exceed 1.0 in small samples, in which case it is conventionally capped at 1. Many practitioners prefer GG when $\hat{\epsilon}_{GG} < 0.75$ and HF when $\hat{\epsilon}_{GG} \geq 0.75$. When $\hat{\epsilon}_{GG}$ is close to $1/(p-1)$, the lower-bound correction — using $\epsilon = 1/(p-1)$ exactly — provides a maximally conservative benchmark.

Henze–Zirkler Test: Formulas and Computational Recipe

The Henze–Zirkler test (1990) assesses multivariate normality by measuring how closely the empirical characteristic function of the data matches the theoretical characteristic function of a multivariate normal distribution, using a weighted L^2 distance. The key formulas are collected below, followed by the step-by-step procedure.

Smoothing parameter.

$$\beta = \frac{1}{\sqrt{2}}\left[\frac{n(2p+1)}{4}\right]^{1/(p+4)}$$

Intermediate kernel averages.

$$\bar{D}_{ij} = \frac{1}{n^2}\sum_{i=1}^{n}\sum_{j=1}^{n} e^{-\frac{\beta^2}{2}D_{ij}}, \qquad \bar{D}_i = \frac{1}{n}\sum_{i=1}^{n} e^{-\frac{\beta^2}{2(1+\beta^2)}d_i}$$

Test statistic.

$$HZ = n\left[\bar{D}_{ij} - 2(1+\beta^2)^{-p/2}\,\bar{D}_i + (1+2\beta^2)^{-p/2}\right]$$

Asymptotic moments under H_0. Let $a = 1 + 2\beta^2$ and $c = 1 + 4\beta^2$. Then:

$$\mu_{HZ} = 1 - a^{-p/2}\left(1 + \frac{p\beta^2}{a} + \frac{p(p+2)\beta^4}{2a^2}\right)$$

$$\sigma_{HZ}^2 = 2\left[1 + c^{-p/2}\left(1 + \frac{2p\beta^2}{c} + \frac{2p(p+2)\beta^4}{c^2}\right) - 2(1+\beta^2)^{-p}\left(1 + \frac{p\beta^2}{1+\beta^2} + \frac{p(p+2)\beta^4}{2(1+\beta^2)^2}\right)\right]$$

Lognormal parameters and p-value.

$$\zeta^2 = \ln\left(1 + \frac{\sigma_{HZ}^2}{\mu_{HZ}^2}\right), \qquad \xi = \ln(\mu_{HZ}) - \tfrac{1}{2}\zeta^2$$

$$p\text{-value} = 1 - \Phi\left(\frac{\ln(HZ) - \xi}{\zeta}\right)$$

Computational Recipe. Given an $n \times p$ data matrix $\mathbf{X}$:

1. **Standardize.** Compute the sample mean vector $\bar{\mathbf{x}}$ and covariance matrix $\mathbf{S}$. Form $\mathbf{z}_i = \mathbf{S}^{-1/2}(\mathbf{x}_i - \bar{\mathbf{x}})$ so that the transformed data have identity covariance.
2. **Compute distances.** Form the $n \times n$ matrix of squared pairwise distances $D_{ij} = (\mathbf{z}_i - \mathbf{z}_j)^\top(\mathbf{z}_i - \mathbf{z}_j)$ and the vector of squared norms $d_i = \mathbf{z}_i^\top \mathbf{z}_i$.
3. **Select the bandwidth.** Compute β from the smoothing parameter formula above.
4. **Compute $\bar{D}_{ij}$.** Apply the Gaussian kernel to each of the n^2 pairwise distances D_{ij} and average as shown above.
5. **Compute $\bar{D}_i$.** Apply the bandwidth adjusted kernel to each squared norm d_i and average over n observations as shown above.
6. **Assemble HZ.** Combine $\bar{D}_{ij}$, $\bar{D}_i$, and the constant term using the test statistic formula above.
7. **Compute asymptotic moments.** Substitute β, p, and the shorthand $a = 1 + 2\beta^2$, $c = 1 + 4\beta^2$ into the formulas for μ_{HZ} and σ_{HZ}^2 above.

8. **Obtain the p-value.** Compute ζ^2 and ξ, then refer $(\ln(HZ) - \xi)/\zeta$ to the standard normal as shown above. Reject multivariate normality at level α if p-value $< \alpha$.

In practice, steps 1–8 are handled by `MVN::mvn()` in R (`mvnTest = "HZ"`) or `HZ.test()` in the `mvnormtest` package, so manual computation is rarely required. The test is consistent against all fixed alternatives to multivariate normality and performs well across a range of p and n, though power naturally decreases as p grows relative to n. A rough guideline is $n \geq 10p$; below this ratio the lognormal approximation to the null distribution becomes unreliable and permutation-based p-values are preferable.

Part XIII

Temporal Dependence & Model Adequacy

Checking autocorrelation patterns and structural dynamics

Time Series Diagnostics

XIII. TSDG — A Conceptual Primer

What Is a Time Series?

- A sequence of observations $\{y_t\}_{t=1}^{T}$ indexed by time, where the **ordering carries information**.
- Unlike cross-sectional data, observations are **dependent**: past values inform the present.
- The statistical challenge: classical inference assumes i.i.d. observations; time series violates this fundamentally.

Two Dimensions of a Time Series

A time series is a single realization of an underlying **stochastic process** $\{Y_t\}$.

Ensemble dimension:

- The hypothetical collection of *all possible realizations* the process could have produced under identical conditions.
- **Probability distributions are defined on this dimension**: $E[Y_t]$, $\text{Var}(Y_t)$, and $\text{Cov}(Y_t, Y_{t-k})$ are ensemble averages.
- We never observe the ensemble directly.

Temporal dimension:

- The single observed path $\{y_1, y_2, \ldots, y_T\}$ is all we have.
- All estimation must be done from **one realization**—there is no second dataset to fall back on.

The bridge — Ergodicity:

- Under **ergodicity**, time averages converge to ensemble averages as $T \to \infty$: $\bar{y}_T \xrightarrow{p} \mu = E[Y_t]$.
- This is what justifies estimating population moments from a single long realization. Without ergodicity (assumption), inference from one path is impossible. In most applications, ergodicity is assumed rather than proven.
- Sufficient condition: the autocovariance function is absolutely summable, $\sum_{k=-\infty}^{\infty} |\gamma(k)| < \infty$.

Autocorrelation: The Binding Glue

Autocorrelation is the mechanism that links the ensemble and temporal dimensions—it encodes the **memory** of the process and is the central quantity in time series diagnostics.

Autocovariance at lag k:

$$\gamma(k) = \text{Cov}(Y_t, Y_{t-k}) = E[(Y_t - \mu)(Y_{t-k} - \mu)].$$

Under stationarity, $\gamma(k)$ depends only on the lag k, not on t. Note $\gamma(0) = \text{Var}(Y_t)$. Intuitively speaking, stationarity $\Rightarrow$ memory is time-invariant.

Autocorrelation function (ACF):

$$\rho_k = \frac{\gamma(k)}{\gamma(0)}, \quad \rho_k \in [-1, 1], \quad \rho_0 = 1.$$

Sample ACF (estimated from data):

$$r_k = \frac{\sum_{t=k+1}^{T} (y_t - \bar{y})(y_{t-k} - \bar{y})}{\sum_{t=1}^{T} (y_t - \bar{y})^2}.$$

Under the null of no autocorrelation ($H_0 : \rho_k = 0$), $r_k \mathrel{\dot\sim} \mathcal{N}(0, 1/T)$ for large T, giving approximate 95% confidence bands at $\pm 1.96/\sqrt{T}$ on ACF plots.

Partial Autocorrelation Function (PACF)

The PACF at lag k measures the **direct** linear dependence between Y_t and Y_{t-k} after removing the influence of all intermediate lags $Y_{t-1}, \ldots, Y_{t-k+1}$:

$$\alpha(k) = \text{Corr}(Y_t - \hat{Y}_t^{(k)},\ Y_{t-k} - \hat{Y}_{t-k}^{(k)}),$$

where $\hat{Y}_t^{(k)}$ is the projection onto $\{Y_{t-1}, \ldots, Y_{t-k+1}\}$. Computed via the Yule–Walker equations or Levinson–Durbin recursion.

ACF vs. PACF diagnostic patterns:

Process	ACF	PACF
AR(p)	Decays (tails off)	Cuts off at lag p
MA(q)	Cuts off at lag q	Decays (tails off)
ARMA	Decays	Decays
WN	All ≈ 0	All ≈ 0

These patterns guide model identification in the Box–Jenkins framework. For diagnostic purposes, if **residual** ACF/PACF spikes exceed the $\pm 1.96/\sqrt{T}$ bands, the fitted model has not fully captured the dependence.

The Backshift Operator and ARMA Models

The **backshift (lag) operator** B is defined by $B^k y_t = y_{t-k}$, so $\Delta y_t = (1 - B)y_t = y_t - y_{t-1}$. It converts lag notation into compact polynomial algebra. Sometimes L (lag) is also used.

An **ARMA(p, q)** process combines autoregressive and moving-average components (AR: $\phi(B)$, MA: $\theta(B)$):

$$\phi(B)y_t \;=\; c + \theta(B)\varepsilon_t,$$

where $\phi(B) = 1 - \phi_1 B - \cdots - \phi_p B^p$ and $\theta(B) = 1 + \theta_1 B + \cdots + \theta_q B^q$, with $\varepsilon_t \sim \text{WN}(0, \sigma^2)$.

- **AR(p):** $\theta(B) = 1$; current value is a weighted sum of p past values plus noise. Stationary when all roots of $\phi(z) = 0$ lie *outside* the unit circle.
- **MA(q):** $\phi(B) = 1$; current value is a weighted sum of q past shocks; always stationary. Invertible when all roots of $\theta(z) = 0$ lie outside the unit circle.
- **ARMA(p, q):** parsimonious combination of both; ACF and PACF both decay geometrically rather than cutting off cleanly.

After fitting an ARMA model, the residuals $\hat{\varepsilon}_t = \phi(B)\hat{y}_t - \theta(B)\hat{\varepsilon}_{t-1}$ should behave as white noise—which is precisely what the portmanteau tests in this chapter verify.

White Noise (WN) — The Diagnostic Target

A process $\{\varepsilon_t\}$ is **white noise** if:

$$E[\varepsilon_t] = 0, \quad \mathrm{Var}(\varepsilon_t) = \sigma^2, \quad \mathrm{Cov}(\varepsilon_s, \varepsilon_t) = 0 \;\forall\, s \neq t.$$

White noise is uncorrelated but need not be independent. **Strong (i.i.d.) white noise** additionally requires independence and is a stronger condition.

Well-fitted time series models should produce residuals that are indistinguishable from white noise. Diagnostic tests in this chapter formally assess whether this holds.

The null hypothesis for portmanteau tests:

$$H_0 : \rho_1 = \rho_2 = \cdots = \rho_h = 0 \quad \text{(residuals are white noise).}$$

Portmanteau Tests

Portmanteau (omnibus) tests assess the **joint significance** of sample autocorrelations at lags 1 through h, rather than testing each lag individually (which inflates Type I error).

Box–Pierce statistic:

$$Q_{\mathrm{BP}} = T \sum_{k=1}^{h} r_k^2 \xrightarrow{d} \chi_h^2 \quad \text{under } H_0.$$

Ljung–Box statistic (preferred; better finite-sample approx):

$$Q_{\mathrm{LB}} = T(T+2) \sum_{k=1}^{h} \frac{r_k^2}{T-k} \xrightarrow{d} \chi_h^2 \quad \text{under } H_0.$$

Q_{LB} up-weights autocorrelations at longer lags (dividing by $T - k < T$), correcting the downward bias of r_k at large k in finite samples.

After ARMA fitting: if the model has $p + q$ estimated parameters, the degrees of freedom reduce to χ^2_{h-p-q} because $p + q$ autocorrelations are absorbed by the model. Using χ_h^2 without this correction is **anti-conservative**.

Choosing h:

- Common default: $h = \min(10, T/5)$ for non-seasonal data; $h = 2s$ for seasonal data with period s.
- Rule: $h \approx \ln(T)$ (Tsay).
- Too small: miss dependence at longer lags.
- Too large: power diluted across uninformative lags.
- Run at multiple values of h for robustness.

Diebold–Mariano Test

Compares the **predictive accuracy** of two competing forecast models, accounting for serial correlation in forecast errors—a correction that standard t-tests omit.

Let e_{1t} and e_{2t} be the forecast errors of models 1 and 2 at horizon h, and $L(\cdot)$ a loss function (typically squared error e^2 or absolute error $|e|$). Define the **loss differential**:

$$d_t = L(e_{1t}) - L(e_{2t}). \quad H_0 : E[d_t] = 0 \text{ (equal pred. accuracy).}$$

The DM statistic:

$$\mathrm{DM} = \bar{d} / \sqrt{\hat{V}(\bar{d})} \stackrel{\cdot}{\sim} \mathcal{N}(0, 1),$$

where $\hat{V}(\bar{d}) = T^{-1} \hat{f}_d(0)$ uses a HAC (long-run variance) estimator of the spectral density of $\{d_t\}$ at frequency zero to account for autocorrelation in the loss differential (which is inevitable when $h > 1$).

Key points:

- Does *not* require nested models; both can be misspecified.
- Loss function choice matters: squared error penalises large errors; absolute error is more robust.
- For multi-step ahead forecasts ($h > 1$), d_t is serially correlated by construction (MA($h-1$)); HAC estimation is essential, not optional.
- Harvey–Leybourne–Newbold (HLN) small-sample correction replaces $\mathcal{N}(0, 1)$ with t_{T-1}; adjusts finite-sample bias.

Interpreting Diagnostic Results

- **Reject H_0 (Ljung–Box):** residual autocorrelation remains; model has not captured all temporal dependence. Try higher AR/MA orders, seasonal components, or a structural break.
- **Fail to reject:** residuals are consistent with **WN**; model is adequate in terms of linear dependence (nonlinear dependence may still be present— check squared residuals separately).
- **Reject H_0 (DM):** the two models have significantly different pred. accuracy; the model with lower avg. loss is preferred.

Common Pitfalls

- Applying portmanteau tests to **raw data** rather than model **residuals** (the test has a different null distribution in each case).
- Using χ_h^2 after ARMA fitting instead of χ^2_{h-p-q} (anti-conservative).
- Choosing h too large (power loss) or too small (missed dependence); report results for several h.
- Ignoring seasonality: for monthly data, always check lags 12 and 24 explicitly.
- Using a standard t-test instead of DM when comparing forecasts (ignores serial correlation in d_t).
- Treating a non-significant Ljung–Box as proof of *independence* (it only tests linear dependence).

Bottom Line

Time series diagnostics assess whether a fitted model has adequately captured the temporal dependence in the data. Residual autocorrelation signals misspecification. Always apply Ljung–Box after model fitting; use Diebold–Mariano—not a t-test—when comparing forecasts. Complement formal tests with ACF/PACF plots.

Portmanteau Test

General Framework for Testing Multiple Autocorrelations

TSDG.1 | **Category:** Time Series Diagnostics | **Data:** Time Series | **Null:** Chi-square

When to Use

- Test overall autocorrelation in time-series or model residuals.
- Assess adequacy of fitted ARIMA or regression models.
- Detect remaining serial dependence across multiple lags.
- Validate white-noise assumption after model fitting.
- Useful when no specific autocorrelation pattern is assumed.

Assumptions

- Series is stationary or residuals from a well-specified model.
- Observations are sequential and equally spaced in time.
- Sample size is adequate for chi-square approximation.
- Lag order is appropriate relative to sample size (e.g., $n > 4h$).
- Autocorrelations are approximately independent under H_0.

Test Statistic

$$Q(h) = n \sum_{k=1}^{h} w_k \hat{\rho}_k^2, \qquad \text{where } w_k \text{ are lag-dependent weights, and } \hat{\rho}_k = \frac{\sum_{t=k+1}^{n}(x_t - \bar{x})(x_{t-k} - \bar{x})}{\sum_{t=1}^{n}(x_t - \bar{x})^2}.$$

Box-Pierce: $w_k = 1$. Ljung-Box: $w_k = (n+2)/(n-k)$. Under H_0, $Q \sim \chi_h^2$ asymptotically.

Notation: $\hat{\rho}_k$ is sample autocorrelation at lag k, h is maximum lag, w_k are weights

Plain English: Portmanteau tests constitute a family of diagnostic tools that aggregate information across multiple autocorrelation lags into a single test statistic. The term "portmanteau" (French for "suitcase") reflects the test's ability to carry multiple pieces of diagnostic information simultaneously. Different weighting schemes produce different members of the family, with Ljung-Box being the most commonly used variant due to superior small-sample properties.

⚠ Practitioner Warning

The portmanteau test is omnibus—powerful against many alternatives but may lack power against specific patterns. For known seasonal periods, testing specific seasonal lags may be more powerful. Results are sensitive to lag selection: automated selection ($h = \ln(n)$ or $h = \sqrt{n}$) may not suit all contexts. When comparing nested models, likelihood ratio tests may be more appropriate than portmanteau diagnostics.

Hypotheses

- H_0: $\rho_1 = \rho_2 = \cdots = \rho_h = 0$ (series is white noise).
- H_1: At least one $\rho_k \neq 0$ (some autocorrelation present).

Decision Rule: Reject H_0 if $Q(h) > \chi^2_{\alpha,h-p}$ or p-value $< \alpha$. Adjust degrees of freedom by p (number of fitted parameters) when testing residuals from estimated models.

Software References [Python, R, SAS]

```
statsmodels::acorr_ljungbox        Box.test                    (varies by implementation)
```

Additional Notes

Monti's test (1994) modifies the portmanteau framework for multivariate time series, testing cross-correlations in addition to autocorrelations. Hosking's (1980) test uses different weights optimized for residuals from fitted ARMA models. The asymptotic chi-square approximation deteriorates when $h/n > 0.25$—avoid testing too many lags relative to sample size. For seasonal data with period m, consider testing at multiples of m: lags $m, 2m, 3m, \ldots$. Modified versions exist for heteroskedastic series (ARCH effects) and long-memory processes.

Miscellanea: Although Box and Pierce introduced the Q statistic in 1970, the "portmanteau test" moniker was popularized in the 1978 Ljung–Box paper. George Box later confirmed the term was a whimsical allusion to Lewis Carroll's Through the Looking-Glass. Just as Humpty Dumpty describes a portmanteau word as having multiple meanings "packed into one," this test packs a range of residual autocorrelations into a single statistic to evaluate overall model fit. The statistical community warmly embraced this literary metaphor, and it remains the standard classification for "omnibus" goodness-of-fit tests in time series analysis.

Reference(s): Box & Pierce (1970); Ljung & Box (1978); Hosking (1980); Monti (1994)

Portmanteau Test TSDG.1

Worked Example

Problem Statement Quarterly earnings-growth deviations ($n = 12$ quarters) from a fitted model are tested for residual autocorrelation up to lag $m = 3$. The Box-Pierce portmanteau test (1970) is the original version of this family; it uses the simplest possible weighting — equal weight r_k^2 at every lag. H_0: $\rho_1 = \rho_2 = \rho_3 = 0$ (white noise); H_1: at least one $\rho_k \neq 0$.

$$\underbrace{3,\ 5,\ 5,\ 3,\ 0,\ -2,\ -4,\ -4,\ -3,\ -1,\ 1,\ -3}_{t=1,\ldots,12}$$

1. Sample Summary

$$\bar{x} = 0, \qquad SS = \sum_{t=1}^{12} x_t^2 = 124$$

$$\sum_{t=1}^{12} x_t x_{t-1} = 90, \quad \sum_{t=3}^{12} x_t x_{t-2} = 48, \quad \sum_{t=4}^{12} x_t x_{t-3} = 2$$

2. Sample Autocorrelations r_k

$$r_k = \frac{\sum_{t=k+1}^{n} x_t x_{t-k}}{SS}$$

$$r_1 = 0.7258, \quad r_2 = 0.3871, \quad r_3 = 0.0161$$

3. Box-Pierce Statistic Under H_0: $Q \xrightarrow{d} \chi^2(m) = \chi^2(3)$.

$$Q = n \sum_{k=1}^{m} r_k^2 = 8.123$$

4. Results and Decision

Statistic	Value	CV (5%)	Decision
Q	8.123	7.815	Reject H_0

$p = 0.044$. Residual autocorrelation is detected.

Contribution to Q: r_1^2 (77.82%), r_2^2 (22.14%), r_3^2 (0.04%).

5. Comparison with Ljung-Box

Test	Formula	Q	p
Box-Pierce	$n \sum r_k^2$	8.123	0.044
Ljung-Box	$n(n+2) \sum \frac{r_k^2}{n-k}$	10.568	0.014

Practitioner Insight: The Box–Pierce statistic $Q = n \sum r_k^2$ is a score test for MA(m) structure, treating each lag equally and assuming $r_k \approx N(0, 1/n)$ so that $nr_k^2 \sim \chi^2(1)$. Its simplicity leads to a downward bias for small n since each r_k is based on only $n - k$ observations, making the $\chi^2(m)$ approximation conservative. The Ljung–Box correction rescales terms by $1/(1 - k/n)$ to account for this loss, increasing Q and improving performance. As a portmanteau test, Q summarizes joint evidence across lags, with m typically chosen based on sample size or seasonality.

Common Pitfalls

- **Applying portmanteau tests to non-stationary data.** Portmanteau tests assume covariance stationarity of the series or residuals being analyzed. Trends, seasonality, or unit roots induce spurious autocorrelation, leading to systematic over-rejection unless the data are properly transformed first.
- **Not adjusting degrees of freedom for fitted models.** When applied to residuals from estimated models, the reference distribution depends on the number of fitted parameters. Failing to use df $= h - m$ (where m is the number of estimated coefficients) makes the test anti-conservative and inflates false rejections.
- **Selecting the lag horizon h through significance hunting.** Trying multiple values of h and reporting only favorable results introduces implicit multiple testing. The lag choice should be pre-specified based on theory, seasonality, or sample size considerations, with transparent reporting if multiple values are examined.
- **Treating non-rejection as evidence of model adequacy.** A non-significant result indicates only that no linear autocorrelation was detected up to lag h. It does not rule out nonlinear dependence, higher-lag effects, heteroskedasticity, or other forms of misspecification.
- **Interchanging Box–Pierce and Ljung–Box interpretations.** Both tests target the same null but differ in finite-sample behavior. Ljung–Box provides a better small-sample approximation and is generally preferred, while Box–Pierce can underperform unless the sample size is large.

Ljung-Box Test

Test for Autocorrelation in Time Series Residuals

TSDG.2 | **Category:** Time Series Diagnostics | **Data:** Time Series | **Null:** Chi-square

When to Use

- Test overall autocorrelation in time-series or model residuals.
- Assess adequacy of fitted ARIMA or regression models.
- Detect remaining serial dependence across multiple lags.
- Validate white-noise assumption after model fitting.
- Useful when no specific autocorrelation pattern is assumed.

Assumptions

- Series is stationary or residuals from a well-specified model.
- Observations are sequential and equally spaced in time.
- Sample size is adequate for chi-square approximation.
- Lag order is appropriate relative to sample size (e.g., $n > 4h$).
- Autocorrelations are approximately independent under H_0.

Test Statistic

$$Q_{LB}(h) = n(n+2)\sum_{k=1}^{h}\frac{\hat{\rho}_k^2}{n-k}, \qquad \text{where } \hat{\rho}_k = \frac{\sum_{t=k+1}^{n}(x_t-\bar{x})(x_{t-k}-\bar{x})}{\sum_{t=1}^{n}(x_t-\bar{x})^2}. \qquad \text{Under } H_0,\ Q_{LB} \sim \chi^2_{h-p}$$

Notation: $\hat{\rho}_k$ is sample autocorrelation at lag k, n is sample size, h is number of lags, p is model parameters.

Plain English: The Ljung-Box test examines whether a group of autocorrelations of a time series are significantly different from zero. It tests the joint hypothesis that all autocorrelations up to lag h are zero, providing an overall assessment of model adequacy. The test statistic weights squared autocorrelations by sample size, with finite-sample correction $(n+2)/(n-k)$ improving performance over the original Box-Pierce test.

⚠ Practitioner Warning

Choice of maximum lag h substantially affects results. Too few lags may miss autocorrelation; too many lags reduce power. Common guidelines: $h \approx \ln(n)$ for non-seasonal data, or $h = 2m$ where m is seasonal period. When testing residuals from fitted models, adjust degrees of freedom by subtracting number of estimated parameters p. Testing raw data versus residuals requires different interpretations.

Hypotheses

- H_0: $\rho_1 = \rho_2 = \cdots = \rho_h = 0$ (series is white noise).
- H_1: At least one $\rho_k \neq 0$ (some autocorrelation present).

Decision Rule: Reject H_0 if $Q_{LB} > \chi^2_{\alpha,h-p}$ or p-value $< \alpha$. Here, p is fitted model parameters (raw series: $p = 0$)

Software References [Python, R, SAS]

Python	R	SAS
`statsmodels::acorr_ljungbox`	`Box.test(type="Ljung-Box")`	`Not Native (Manual)`

Additional Notes

The original Box-Pierce test (1970) used $Q_{BP} = n\sum_{k=1}^{h}\hat{\rho}_k^2$ without the $(n+2)$ finite-sample correction. Ljung-Box modification (1978) provides better small-sample approximation to chi-square distribution. For ARIMA(p,d,q) model residuals, use $df = h - p - q$. The test has good power against slowly decaying autocorrelation but may miss high-frequency patterns. McLeod Li test extends Ljung-Box to test for ARCH effects by applying it to squared residuals. Seasonal autocorrelation requires testing at seasonal lags specifically. Monti's test (1994) is a Ljung–Box–style portmanteau test that uses partial autocorrelations instead of autocorrelations, providing a PACF-based check of residual independence in ARMA models.

Miscellanea: Ljung and Box (1978) refined the earlier Box–Pierce (1970) portmanteau test by introducing a small-sample correction that improves finite-sample performance and reduces the tendency of the original test to over-reject. The correction modifies the test statistic through a factor involving $(n+2)/(n-k)$, yielding more accurate Type I error control when sample size is limited. As a result, the Ljung–Box test became the standard portmanteau diagnostic in time-series analysis, though practical choices such as the number of lags remain context-dependent.

Reference(s): Box & Pierce (1970); Ljung & Box (1978); Monti (1994)

Worked Example

Problem Statement Monthly temperature anomalies (°C) over $n = 12$ periods are tested for autocorrelation up to lag $m = 3$. A rejection indicates that the series is not white noise and that past values carry predictive information about future values. H_0: $\rho_1 = \rho_2 = \rho_3 = 0$; H_1: at least one $\rho_k \neq 0$.

$$3,\ 4,\ 3,\ 1,\ -1,\ -3,\ -3,\ -2,\ -1,\ 0,\ 1,\ -2$$

1. Sample Summary

$$\bar{x} = 0, \qquad SS = \sum_{t=1}^{12} x_t^2 = 64$$

$$\sum_{t=1}^{12} x_t x_{t-1} = 44, \quad \sum_{t=3}^{12} x_t x_{t-2} = 18, \quad \sum_{t=4}^{12} x_t x_{t-3} = 8$$

2. Sample Autocorrelations r_k

$$r_k = \frac{\sum_{t=k+1}^{n} x_t x_{t-k}}{SS}$$

$$r_1 = 0.6875, \quad r_2 = 0.2813, \quad r_3 = -0.1250$$

3. Ljung-Box Statistic

$$Q = n(n+2)\sum_{k=1}^{m} \frac{r_k^2}{n-k} = 8.839$$

k	r_k	r_k^2	$n-k$	$r_k^2/(n-k)$
1	0.6875	0.4727	11	0.042969
2	0.2813	0.0791	10	0.007910
3	−0.1250	0.0156	9	0.001736
			Sum	0.052615

Contribution to Q: r_1^2 (81.67%), r_2^2 (15.03%), r_3^2 (3.30%).

Under H_0: $Q \xrightarrow{d} \chi^2(m) = \chi^2(3)$.

4. Results and Decision

Statistic	Value	CV (5%)	Decision
Q	8.839	7.815	Reject H_0

$p = 0.032$. Significant autocorrelation exists up to lag 3.

5. Comparison with Box-Pierce

Test	Statistic	p	Decision
Box-Pierce	6.809	0.078	FTR
Ljung-Box	8.839	0.032	Reject

Practitioner Insight: The Ljung–Box statistic improves Box–Pierce by replacing r_k^2 with $r_k^2/(n-k)$ and scaling by $n(n+2)$, upweighting higher lags and correcting small-sample bias. This can materially change conclusions (e.g., FTR under Box–Pierce vs. rejection under Ljung–Box). The pattern of autocorrelations (e.g., $r_1 > r_2 > |r_3|$) suggests underlying dynamics such as AR(1). Choice of m balances sensitivity and approximation quality, with rules like $m \approx \ln(n)$ or $\min(10, n/5)$. The test should be applied to residuals, not raw data, to avoid confounding with trend or seasonality.

Common Pitfalls

- **Not adjusting degrees of freedom for fitted models.** When applied to residuals, the reference distribution must account for estimated parameters via df $= h - m$. Using df $= h$ ignores parameter estimation and leads to anti-conservative inference with inflated rejection rates.
- **Arbitrary choice of lag horizon h.** The maximum lag determines the range of autocorrelation being tested and directly affects power. Too small misses longer dependencies, while too large dilutes signal, so h should be chosen based on sample size, frequency, or theoretical structure.
- **Applying the test to non-stationary or raw series.** The Ljung–Box test assumes stationarity of the series being tested. Applying it to trending or unit root data produces spurious rejections, so it should be used on properly transformed series or model residuals.
- **Interpreting non-rejection as model validation.** A non-significant result indicates only a lack of evidence for autocorrelation up to lag h. It does not confirm model adequacy or rule out other forms of misspecification such as nonlinearity or heteroskedasticity.
- **Using the test uncritically in small samples.** The chi-square approximation underlying the statistic is asymptotic and can be unreliable for small n. With limited data, both size and power may deviate from nominal levels, requiring cautious interpretation or alternative approaches.

Diebold-Mariano Test

Test for Predictive Accuracy Comparison Between Forecasting Models

TSDG.3 | **Category:** Time Series Diagnostics | **Data:** Time Series | **Null:** t-dist. / Normal (asymptotic)

When to Use

- Compare predictive accuracy of two competing forecasting models. Assess whether one model consistently outperforms another.
- Test whether mean forecast loss differs significantly.
- Compare models using loss functions (e.g., MSE, MAE).

Assumptions

- Loss differential series is stationary with finite variance.
- Forecast errors computed on the same evaluation sample.
- Observations are sequential and equally spaced in time.
- Serial correlation in loss differential is accounted for.
- Sample size is sufficient for normal approximation.

Test Statistic

$$DM = \frac{\bar{d}}{\sqrt{\hat{V}(\bar{d})/T}}, \quad \bar{d} = \frac{1}{T}\sum_{t=1}^{T} d_t, \quad d_t = L(e_{1t}) - L(e_{2t}).$$ Under H_0, $DM \approx N(0,1)$ or t_{T-1} for small samples.

Notation: d_t is loss differential, $L(\cdot)$ is loss function, e_{it} are forecast errors, $\hat{V}(\bar{d})$ is HAC variance estimator

Plain English: The Diebold–Mariano test compares the forecasting performance of two models by examining the series of loss differentials across time. For each period, compute both forecast errors, apply a chosen loss function (such as squared or absolute error, asymmetric LINEX loss, or any other measure), and take their difference. The test evaluates whether the average loss differential differs significantly from zero—negative values favor model 1, positive values favor model 2. The procedure naturally handles autocorrelation that arises when multi-step forecasts overlap.

⚠ Practitioner Warning

Conclusions depend on the specified loss function; different losses can favor different models. Loss differentials must be stationary for the test's asymptotic justification. Overlapping forecasts create serial correlation, requiring HAC variance estimation. Small samples reduce reliability, and modified versions improve finite-sample performance. The test cannot be used when models are nested (requires Clark–West or encompassing tests). Non-significance does not imply equal practical performance, and the test offers no insight into why one model outperforms another.

Hypotheses

- H_0: Models have equal predictive accuracy ($E[d_t] = 0$)
- H_1: Models differ in predictive accuracy ($E[d_t] \neq 0$)

Decision Rule: Reject H_0 if $|DM| > Z_{\alpha/2}$ (large samples) or $|DM| > t_{\alpha/2,T-1}$ (small samples), or p-value $< \alpha$.

Software References [Python, R, SAS]

Python	R	SAS
`Not Native (Manual)`	`forecast::dm.test`	`Not Native (Manual)`

Additional Notes

The Diebold–Mariano framework helped establish forecast comparison as a model-free exercise, applicable across domains from macroeconomic indicators to volatility and probability forecasts. It also motivated related tools such as forecast-encompassing tests, model-confidence-set procedures, and rolling-window analyses that track changes in relative accuracy over time. DM statistics are often reported alongside loss plots or stability diagnostics to assess whether forecast performance shifts across regimes or sample periods.

Miscellanea: Diebold and Mariano introduced their test in the early 1990s to formalize the comparison of forecast accuracy, moving beyond informal MSE comparisons. Their 1995 paper quickly became influential and is now standard in macroeconomic and financial forecasting. A key innovation is the use of a heteroskedasticity- and autocorrelation-consistent (HAC) variance estimator, allowing valid inference even when loss differentials are serially correlated, as in multi-step forecasts. The test is widely used in both academic research and policy analysis, where rigorous evaluation of competing forecasts is essential.

Reference(s): Diebold & Mariano (1995); Harvey et al. (1997); Clark & West (2007)

Worked Example

Problem Statement Two competing models produce one-step-ahead forecasts of a monthly economic indicator over $T = 10$ periods. Model 1 (complex autoregressive model) and Model 2 (simpler benchmark) are compared under squared-error loss. The Diebold-Mariano test assesses whether the mean loss differential is significantly different from zero. H_0: $E[d_t] = 0$ (equal predictive accuracy); H_1: $E[d_t] \neq 0$. $MSE_1 = 0.927$ vs $MSE_2 = 0.239$.

1. Loss Differential
Under squared-error loss:

$$d_t = L(e_t^{(1)}) - L(e_t^{(2)}) = (e_t^{(1)})^2 - (e_t^{(2)})^2$$

$d_t > 0$ means Model 1 had larger squared error at time t.
Note $d_t < 0$ at $t = 3, 6$ (Model 2 worse).

2. Mean and Variance of d_t

$$\bar{d} = \frac{\sum d_t}{T} = 0.688, \quad s_d = \sqrt{\frac{\sum(d_t - \bar{d})^2}{T-1}} = 0.855$$

$$\bar{d} = \overline{MSE_1} - \overline{MSE_2} = 0.927 - 0.239 = 0.688 \checkmark$$

3. DM Statistic ($h = 1$) Under H_0: $DM \xrightarrow{d} N(0,1)$.
For one-step-ahead forecasts ($h = 1$), errors are uncorrelated, so no HAC correction is needed:

$$DM = \frac{\bar{d}}{s_d/\sqrt{T}} = \frac{0.688}{0.855/\sqrt{10}} = \frac{0.688}{0.270} = 2.545$$

4. Harvey-Leybourne-Newbold Correction
For small T, the HLN-corrected statistic uses a t reference distribution ($t_{(T-1)}$):

$$DM^* = DM \times \sqrt{\frac{T + 1 - 2h + h(h-1)/T}{T}} = 2.415$$

5. Results and Decision

Version	Dist.	Stat.	CV(5%)	Dec.
DM	$N(0,1)$	2.545	1.960	Reject
DM*	$t(9)$	2.415	2.262	Reject

$p_{DM} = 0.011$; $p_{DM^*} = 0.039$. Model 2 is significantly more accurate at 5% under both versions.

6. Interpretation

- $\bar{d} = 0.688 > 0$: on average Model 1's $(e_t^{(1)})^2$ exceed Model 2's by 0.688 per period. Model 2 is preferred.

Practitioner Insight: The Diebold–Mariano test reduces forecast comparison to a one-sample t-test on loss differentials d_t, with $DM = \bar{d}/(s_d/\sqrt{T}) \sim N(0,1)$ under equal expected loss (H_0). It is flexible in allowing any loss function, and for multi-step forecasts ($h > 1$) accounts for serial correlation in d_t via a Newey–West HAC variance; for $h = 1$ this reduces to s_d^2/T. While broadly applicable, DM can be biased for nested models, where Clark–West is preferred.

Common Pitfalls

- **Using in-sample or contaminated evaluation data.** The DM framework assumes genuinely out-of-sample forecasts. Using in-sample fitted values or data involved in model selection introduces overfitting bias and invalidates inference.
- **Applying DM to nested model comparisons.** The standard DM test assumes non-nested models, where neither specification is a restricted version of the other. When models are nested, the null lies on the boundary and the DM statistic has invalid asymptotics; use Clark–West or related corrections instead.
- **Relying on asymptotic CVs in small samples.** The DM statistic is asymptotically normal in large out-of-sample samples P. For small P, size distortions can be severe, so Harvey et al. corrections or bootstrap inference are preferred.
- **Overgeneralizing significant results.** A significant DM test indicates superior average performance over the specific evaluation sample. It does not imply universal dominance across time periods, regimes, assets, or future data.
- **Confusing Diebold–Mariano with other forecast comparison frameworks.** The Diebold–Mariano (DM) test evaluates equal *unconditional* forecast accuracy via $H_0 : E[d_t] = 0$. It differs from Giacomini–White (conditional predictive ability), Model Confidence Set procedures (multiple-model selection), and simple MSE comparisons (descriptive only), so the testing objective must be clearly specified.

Newey–West (NW) Variance (with Serial Correlation)

$$\widehat{\sigma}^2_{NW} = \gamma_0 + 2\sum_{k=1}^{q} w_k \gamma_k, \quad \text{where } \gamma_k \text{ is the sample autocovariance at lag } k, \quad w_k = 1 - \frac{k}{q+1} \text{ (Bartlett kernel)},$$

The triangular weights downweight higher lags, ensuring positive semidefiniteness. This yields a consistent variance estimate even with heteroskedasticity and serial correlation.

XIII. TSDG — Raw Data & Charts

Data used in selected worked examples are provided here to preserve the book's strict two-page-per-test format.

[TSDG.3] Diebold-Mariano Test

Problem Statement Two competing models produce one-step-ahead forecasts of a monthly economic indicator over $T = 10$ periods. Model 1 (complex autoregressive model) and Model 2 (simpler benchmark) are compared under squared-error loss. The Diebold-Mariano test assesses whether the mean loss differential is significantly different from zero. H_0: $\mathrm{E}[d_t] = 0$ (equal predictive accuracy); H_1: $\mathrm{E}[d_t] \neq 0$. $\mathrm{MSE}_1 = 0.927$ vs $\mathrm{MSE}_2 = 0.239$.

t	y_t	$\hat{y}_t^{(1)}$	$\hat{y}_t^{(2)}$	$e_t^{(1)}$	$e_t^{(2)}$	$(e_t^{(1)})^2$	$(e_t^{(2)})^2$	d_t
1	3.5	4.6	3.9	−1.1	−0.4	1.21	0.16	+1.05
2	5.0	5.9	4.7	−0.9	+0.3	0.81	0.09	+0.72
3	2.0	2.4	2.8	−0.4	−0.8	0.16	0.64	−0.48
4	4.5	5.8	4.3	−1.3	+0.2	1.69	0.04	+1.65
5	3.0	3.4	3.4	−0.4	−0.4	0.16	0.16	0.00
6	5.5	5.9	4.7	−0.4	+0.8	0.16	0.64	−0.48
7	2.5	3.6	2.9	−1.1	−0.4	1.21	0.16	+1.05
8	4.0	4.9	3.5	−0.9	+0.5	0.81	0.25	+0.56
9	3.5	5.0	3.8	−1.5	−0.3	2.25	0.09	+2.16
10	5.0	5.9	4.6	−0.9	+0.4	0.81	0.16	+0.65
					Σ	**0.927**	**0.239**	**0.688**

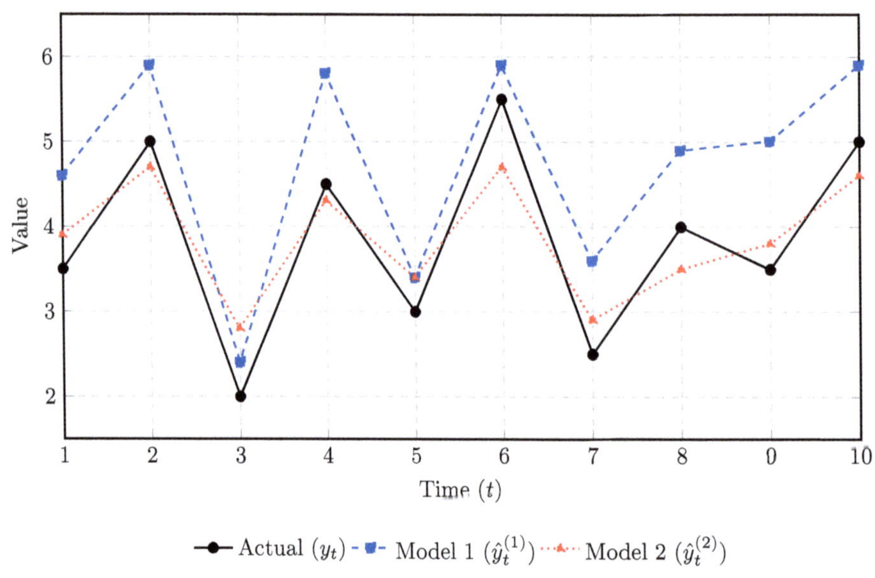

Part XIV

Trend, Persistence & Stochastic Structure

Determining stationarity, unit roots, and long-run equilibrium behavior

Stationarity and Unit Roots

XIV. STAT — A Conceptual Primer

What Is Stationarity?

A stochastic process $\{Y_t\}$ is **weakly (covariance) stationary** if its first two moments are time-invariant:

$$E[Y_t] = \mu, \quad \text{Var}(Y_t) = \sigma^2, \quad \text{Cov}(Y_t, Y_{t-k}) = \gamma(k),$$

where μ, σ^2, and $\gamma(k)$ are all independent of t. **Strict stationarity** requires the full joint distribution to be time-invariant—a stronger condition rarely invoked in practice.

Stationarity is the **regularity condition** that allows classical inference: it ensures ACF, spectral density, and autoregressive representations are well-defined, and that time averages converge to ensemble means (ergodicity). Without it, OLS standard errors are invalid, t-statistics diverge, and regressions among unrelated $I(1)$ series appear significant (**spurious regression**, Granger–Newbold 1974).

Types of Nonstationarity

Not all departures from stationarity are alike—each requires a different remedy and a different test.

Type	Characterization
Unit root (stochastic trend)	$Y_t = Y_{t-1} + \varepsilon_t$; variance grows as $t\sigma^2$; shocks are permanent; remedy: *difference*.
Deterministic trend	$E[Y_t] = \mu + \beta t$; variance is stable; remedy: *detrend* (OLS).
Structural break	Abrupt shift in mean, trend, or variance at time τ; mimics a unit root in standard tests.
Seasonal nonstationarity	Periodic nonstationarity at frequency $2\pi s/S$ for seasonal period S; must difference at lag S.
Nonlinear nonstationarity	Mean reversion is regime-dependent or threshold-based; missed by linear unit root tests.

Why Testing Is Difficult

- **One realisation only:** probability distributions are defined on the ensemble but we observe a single path. There is no repeated sampling; all inference rests on temporal dependence within that path.
- **Near-unit-root problem:** a stationary AR(1) with $\phi = 1 - c/T$ (local-to-unity) is virtually indistinguishable from a unit root in finite samples. Standard tests have power barely above size against such alternatives.
- **Structural breaks inflate non-rejection:** a stationary series with a mean shift looks like a unit root to ADF and PP tests—they are biased toward non-rejection in the presence of breaks.
- **Small samples:** most unit root tests have low power for $T < 100$; finite-sample critical values differ substantially from asymptotic ones.
- **Deterministic specification:** including or omitting a constant or trend in the test regression changes both the null distribution and the power profile.

Two Opposing Frameworks

Tests differ fundamentally in what they place under H_0:

Framework	H_0	Examples
Unit root tests	$I(1)$	ADF, PP, DF-GLS, Ng-Perron
Stationarity tests	$I(0)$	KPSS, Schmidt-Phillips

Key implication: failing to reject H_0 in a unit root test is *not* evidence of stationarity—it may simply reflect low power. Failing to reject H_0 in KPSS is not evidence of a unit root either. Only when ADF *fails* to reject *and* KPSS *also* fails to reject can one tentatively conclude stationarity. When they conflict, the data are non-informative and larger T or additional tests are needed.

The ADF Maintained Regression

The core of most unit root tests is the ADF regression:

$$\Delta y_t = \alpha + \beta t + \gamma y_{t-1} + \sum_{j=1}^{p} \phi_j \Delta y_{t-j} + \varepsilon_t,$$

where $H_0 : \gamma = 0$ (unit root) vs. $H_1 : \gamma < 0$ (stationarity). The lagged differences absorb serial correlation (whiten residuals); p is chosen by AIC, BIC, or sequential testing. Critical values are **non-standard** (left-skewed; Dickey–Fuller tables), not t_{n-2}.

Three deterministic specifications:

	Regressors	Use when
I	None	Zero-mean series
II	Constant α	Nonzero mean
III	Constant + trend βt	Trending series

Over-including deterministics reduces power; omitting them distorts size. Visual inspection guides specification.

Taxonomy of Tests in This Chapter

Classical unit root tests:

- **DF / ADF:** parametric correction for serial correlation via lagged differences; the standard baseline.
- **Phillips–Perron (PP):** nonparametric (HAC) correction for serial correlation; does not augment the regression. More size-distorted than ADF under MA errors.

- **DF-GLS:** GLS-detrends the series before applying ADF; substantially higher power against local alternatives than ADF. Recommended over ADF as the default.
- **Ng–Perron:** builds on DF-GLS with modified test statistics (MZ_α, MZ_t, MSB, MPT); correct size and high power even under MA roots near -1, where ADF and PP over-reject badly.
- **Schmidt–Phillips:** LM-based; jointly estimates the unit root parameter and deterministic components; invariant to nuisance parameters under H_0.

Stationarity test (reversed null):

- **KPSS:** based on the partial-sum process of OLS residuals; $\eta = T^{-2} \sum_t S_t^2 / \hat{\sigma}_\infty^2$ where $S_t = \sum_{s=1}^{t} \hat{e}_s$. Rejects stationarity for large η. Use alongside ADF/PP, never alone.

Structural break tests:

- **Perron (1989):** allows a *known* break date τ in the null and alternative; three models (crash, changing growth, both).
- **Zivot–Andrews (ZA):** endogenously estimates a *single* unknown break date as the date minimizing the ADF t-statistic; critical values adjusted for the search.
- **Clemente–Montañés–Reyes (CMR):** extends ZA to *two* unknown break dates; additive outlier (AO) or innovational outlier (IO) models.
- **Lee–Strazicich (LS):** LM-based with one or two endogenous breaks; unlike ZA, does not spuriously reject H_0 when breaks occur under stationarity.

Seasonal unit root tests:

- **HEGY:** tests for unit roots at zero frequency *and* all seasonal frequencies simultaneously in quarterly/monthly data.
- **Canova–Hansen (CH):** reverses the null (H_0: seasonal stationarity); complements HEGY the same way KPSS complements ADF.

Nonlinear unit root test:

- **Kapetanios–Shin–Snell (KSS):** tests for a unit root against a globally stationary ESTAR (**E**xponential **S**mooth-**T**ransition **A**uto**R**egressive) alternative; detects nonlinear mean reversion that linear tests miss.

Panel unit root tests:

- **Levin–Lin–Chu (LLC):** pools the ADF regression across N units; assumes a *common* δ under H_1. Requires cross-sectional independence.
- **Im–Pesaran–Shin (IPS):** averages individual ADF statistics; allows *heterogeneous* δ_i under H_1. More flexible than LLC; still requires cross-sectional independence.
- **Pesaran's CD test:** does not test for unit roots but for **cross-sectional dependence** (CSD) in panel data. If CSD is present, LLC and IPS are invalid; second-generation panel tests (CIPS, PANIC) must be used. Always run CD before panel unit root tests.

Why No Single Test Is Sufficient

Each test is designed against a specific alternative and has blind spots:

Test	Power against	Weak against
ADF/PP	Linear $I(0)$	Near unit root, breaks
DF-GLS	Local alt.	Breaks
Ng-Perron	MA errors	Breaks
KPSS	$I(1)$	Near unit root
ZA/CMR/LS	Breaks	Nonlinear dynamics
KSS	ESTAR	Structural breaks

Common Pitfalls

- **Wrong deterministic specification:** omitting a trend when the series is trending inflates rejection; including one unnecessarily reduces power.
- **Lag length in ADF:** too few lags leave serial correlation in residuals (size distortion); too many reduce power. Use AIC or BIC.
- **Ignoring structural breaks:** ADF and PP have near-zero power against stationary series with breaks; always plot the series first.
- **Relying on a single test:** use ADF/DF-GLS + KPSS as a minimum; add ZA or LS when breaks are visually apparent.
- **Panel tests without CD check:** LLC and IPS are invalid under cross-sectional dependence; always run Pesaran's CD test first.
- **Treating borderline p-values rigidly:** near $\alpha = 0.05$, power is low; small changes in lag or sample can flip conclusions.

Bottom Line

Stationarity is not binary—it is a spectrum of conditions, each violated in different ways. No single test detects all forms of nonstationarity. A robust strategy combines a unit root test (ADF or DF-GLS), a stationarity test (KPSS), and a break-robust test (ZA or LS) when breaks are plausible. In panels, always screen for cross-sectional dependence before applying unit root tests.

Dickey-Fuller (DF) Test

Test for Unit Root in Autoregressive Time Series

STAT.1 | **Category:** Stationarity | **Data:** Time Series | **Null:** Dickey-Fuller distribution

When to Use

- Determine if a time series has a stochastic unit root.
- Verify stationarity before fitting an ARMA or ARIMA.
- Assess if a series needs differencing for mean reversion.
- Test for unit roots in models with a drift or time trend.
- Check the order of integration for cointegration tests.

Assumptions

- The error terms are independent and identically distributed.
- The variance of the white noise process remains constant.
- The data is collected at equally spaced time intervals.
- The underlying process follows a first-order autoregression.
- Residuals lack serial correlation; otherwise use ADF test.

Test Statistic

$$\tau = \frac{\hat{\gamma}}{SE(\hat{\gamma})} = \frac{\hat{\rho} - 1}{SE(\hat{\rho})} \quad \Big| \quad \text{First Difference: } \Delta y_t = \delta + \gamma y_{t-1} + \epsilon_t \quad \Big| \quad \text{Levels: } y_t = \delta + \rho y_{t-1} + \epsilon_t$$

Notation: $\Delta y_t = y_t - y_{t-1}$ is first difference, γ is tested parameter ($\gamma = \rho - 1$), δ is drift term

Plain English: The Dickey–Fuller test determines whether a time series contains a unit root, making it nonstationary. A unit root implies shocks persist indefinitely rather than fading over time (memory is infinite — every shock leaves a permanent scar). The test examines whether the series tends to return toward a stable mean (stationary) or instead behaves like a persistent random walk. Regressions involving nonstationary series can produce misleading statistical inference and spurious relationships.

⚠ Practitioner Warning

The basic DF test assumes an AR(1) process with no serial correlation. If higher-order autocorrelation is present, the test becomes misspecified; use the Augmented Dickey–Fuller (ADF) test instead. Critical values depend on specification (no constant, constant only, or constant plus trend), so using the wrong version can produce incorrect conclusions. The test also has low power against near-unit-root alternatives ($\rho \approx 1$), causing highly persistent stationary series to be mistaken for nonstationary processes. Failure to reject H_0 does not prove a unit root; the test may simply lack power. Modern software typically uses MacKinnon's response surface approximations for finite-sample critical values and p-values.

Hypotheses

- H_0: $\gamma = 0$, equivalently $\rho = 1$ (unit root, non-stationary).
- H_1: $\gamma < 0$, equivalently $\rho < 1$ (stationary process).

Decision Rule: Reject H_0 if $\tau < t_\alpha$ (left-tailed test) or p-value $< \alpha$.

Software References [Python, R, SAS]

Python	R	SAS
`statsmodels::adfuller`	`urca::ur.df`	`PROC ARIMA (stationarity)`

Additional Notes

Three standard specifications are commonly used: (1) no constant or trend, $\Delta y_t = \gamma y_{t-1} + \epsilon_t$; (2) constant only, $\Delta y_t = \delta + \gamma y_{t-1} + \epsilon_t$; and (3) constant plus trend, $\Delta y_t = \delta + \beta t + \gamma y_{t-1} + \epsilon_t$. The test statistic follows the nonstandard Dickey–Fuller distribution under H_0, so special critical values are required. Typical (asymptotic) critical values (1%, 5%, 10%) are $(-2.56, -1.94, -1.62)$ with no constant/trend, $(-3.43, -2.86, -2.57)$ with constant only, and $(-3.96, -3.41, -3.12)$ with constant and trend. A unit-root process is often called integrated of order one, denoted $I(1)$; first differencing may produce stationarity. Structural breaks can bias the test toward non-rejection, making stationary series appear to contain unit roots. Nonstationary $I(1)$ series may still form stable long-run relationships through cointegration.

Miscellanea: David A. Dickey developed the unit-root test in his 1976 Iowa State dissertation under Wayne A. Fuller. A key contribution was showing that the test statistic follows a nonstandard limiting distribution rather than the usual Student's t. Their 1979 *JASA* paper became one of the foundational works in econometrics. An earlier conceptual precursor is von Neumann's ratio test (1940s), which detects serial dependence by comparing successive differences to overall variance. Although not designed for unit roots, it reflects a similar intuition: highly persistent series exhibit relatively small successive changes relative to their overall variability.

Reference(s): Dickey & Fuller (1979); MacKinnon (1996)

Dickey-Fuller (DF) Test **STAT.1**

Worked Example

Problem Statement An economist tests whether a monthly stock price index contains a unit root. The DF test is applied under all three deterministic specifications to illustrate how model choice affects the result. $T = 20$, no lag augmentation ($k = 0$). H_0: $\gamma = 0$ (unit root, $\rho = 1$); H_1: $\gamma < 0$ (stationary, $\rho < 1$).

$y = \{10.0, 10.1, 11.4, 11.6, 12.6, 12.5, 13.7, 13.8, 14.9, 15.1, 16.4, 16.4, 17.5, 17.6, 18.8, 19.0, 20.3, 20.3, 21.4, 21.5\}$

1. Three DF Model Specifications

The three DF models differ in their deterministic components:

- **M1**: $\Delta y_t = \gamma y_{t-1} + e_t$ (no drift, no trend)
- **M2**: $\Delta y_t = \alpha + \gamma y_{t-1} + e_t$ (with drift)
- **M3**: $\Delta y_t = \alpha + \beta t + \gamma y_{t-1} + e_t$ (drift with trend)

2. Model 2 OLS (Primary Specification)

With $n = T - 1 = 19$ observations of $(\Delta y_t, y_{t-1})$:

- $\sum y_{t-1} = 293.4$, $\sum y_{t-1}^2 = 4760.2$
- $\sum \Delta y_t = 11.5$, $\sum y_{t-1}\Delta y_t = 174.75$

OLS gives:

$$\text{slope: } \hat{\gamma} = -0.0124, \qquad \text{intercept: } \hat{\alpha} = 0.796$$

$$\text{error variance: } \hat{s}^2 = \frac{RSS}{n-2} = \frac{\sum \hat{e}_t^2}{n-2} = \frac{5.754}{17} = 0.339$$

$$SE(\hat{\gamma}) = \frac{\hat{s}}{\sqrt{\sum \tilde{y}_{t-1}^2}} = 0.0384, \qquad \tau = \frac{\hat{\gamma}}{SE(\hat{\gamma})} = -0.322$$

3. Results Across All Three Models

Critical values (τ, MacKinnon 1996, $T = 20$):

Model	τ	CV(5%)	CV(1%)	Decision
M1	+4.270	−1.96	−2.69	FTR H_0
M2	−0.322	−3.02	−3.81	FTR H_0
M3	−11.957	−3.66	−4.50	Reject H_0

4. Interpretation

- **M1** gives $\tau = +4.27$ (positive!) because this trending series cannot be modeled without a constant — M1 is misspecified.
- **M2** gives $\tau = -0.322 \gg -2.86$: we **fail to reject** H_0. The stock price index contains a unit root.
- **M3** rejects because including the trend regressor absorbs the series' drift, making $\hat{\gamma}$ spuriously negative. M3 is appropriate only when a deterministic trend exists under H_1 (trend-stationarity), not for a random walk with drift.
- **Correct model**: M2. Series is $I(1)$.

Practitioner Insight: The DF τ statistic does NOT follow a standard t-distribution under H_0 — using t-table critical values would severely over-reject the null. The Dickey-Fuller critical values are more negative (further from zero) because the OLS estimator of γ is biased toward zero under unit root. Model selection should be guided by economic reasoning before looking at the data: M2 is appropriate when a drift exists under H_0 (often seen in most economic series); M3 when a deterministic trend exists under H_1. The DF test assumes i.i.d. errors — if serial correlation is present, use the ADF test instead, which adds lagged Δy terms to whiten the residuals.

Common Pitfalls

- **Using t-critical values.** DF statistics are *not* t-distributed; using t-tables severely over-rejects. Always use DF critical values matched to the chosen deterministic terms (constant/trend).
- **Misreading the hypotheses.** H_0 is a unit root; failing to reject implies non-stationarity, not stationarity. Only strong evidence (very negative τ) supports stationarity.
- **Ignoring serial correlation.** Basic DF assumes i.i.d. errors. If residuals show autocorrelation, use the ADF test with lagged differences; otherwise size and power are distorted.
- **Wrong deterministic specification.** Including/excluding constant or trend incorrectly reduces power or biases results. Choose based on data behavior: none (mean-zero), constant (level stationary), constant+trend (trend stationary).
- **Relying on DF alone.** DF has low power. Supplement with PP, KPSS, plots, and theory to avoid false non-rejection of unit roots.

The Counterintuitive Result

DF test statistic (τ), despite looking like a t-ratio, follows a completely different distribution under the null. Under H_0, both numerator and denominator of τ involve non-stationary processes with different rates of convergence. Their ratio doesn't normalize to something normal—it converges to a ratio of Brownian motion functionals. A functional is a function whose **input is a function** and whose **output is a scalar** (a real or complex number).

Augmented Dickey-Fuller (ADF) Test

Unit Root Test with Lagged Differences for Serial Correlation

STAT.2 | **Category:** Stationarity | **Data:** Time Series | **Null:** Dickey-Fuller distribution

When to Use

- Test for a unit root when errors have serial correlation.
- Verify stationarity before fitting high-order ARMA models.
- Determine the integration order of complex time series.
- More reliable than basic DF for higher-order AR processes.
- Verify $I(1)$ preconditions for Johansen cointegration.

Assumptions

- The underlying process follows an AR(p) model structure.
- Lag order p correctly specified or reasonably approximated.
- The variance of the innovations remains constant over time.
- Data points are collected at fixed, equally spaced intervals.
- Series exhibits no structural breaks in intercept or trend.

Test Statistic

$$t = \frac{\hat{\gamma}}{SE(\hat{\gamma})} = \frac{\hat{\rho} - 1}{SE(\hat{\rho})} \quad \Big| \quad \Delta y_t = \delta + \beta t + \gamma y_{t-1} + \sum_{i=1}^{p} \phi_i \Delta y_{t-i} + \epsilon_t \quad \Big| \quad y_t = \delta + \beta t + \rho y_{t-1} + \sum_{i=1}^{p} \phi_i \Delta y_{t-i} + \epsilon_t$$

Notation: $\gamma = \rho - 1$ where ρ is AR(1) coefficient, p is number of augmenting lags (absorb serial correlation), δ, β optional

Plain English: The Augmented Dickey-Fuller test extends the basic Dickey-Fuller test by including lagged differences of the dependent variable to capture serial correlation. This "augmentation" ensures the error term is white noise, satisfying the DF test's assumptions even when the underlying process has complex dynamics. The test maintains the same null hypothesis (unit root) and non-standard distribution as the basic DF test, but gains validity for broader class of processes.

⚠ Practitioner Warning

Lag-length selection critically affects test performance. Too few lags leave residual autocorrelation, causing misspecification and distorted inference; too many lags reduce power and consume degrees of freedom. Common selection methods include AIC, BIC, and general-to-specific testing. Some software uses Schwert's rule, $p = \lfloor 12(T/100)^{1/4} \rfloor$, as an upper bound for lag search. Residuals should be approximately white noise after fitting. Trend specification (none, constant, or constant+trend) should reflect the deterministic structure of the series.

Hypotheses

- H_0: $\gamma = 0$ or $\rho = 1$ (unit root, non-stationary).
- H_1: $\gamma < 0$ or $\rho < 1$ (stationary around deterministic trend).

Decision Rule: Reject H_0 if $t < t_\alpha$ (left-tailed test) or p-value $< \alpha$. Use original Dickey-Fuller critical values.

Software References [Python, R, SAS]

`statsmodels::adfuller`	`tseries::adf.test`	`PROC ARIMA (stationarity)`

Additional Notes

Information criteria for lag selection include AIC $= -2\ln(L) + 2k$ and BIC $= -2\ln(L) + k\ln(n)$, where k is the number of estimated parameters. BIC typically selects fewer lags than AIC, favoring parsimony over power. A common sequential approach starts with a maximum lag order and repeatedly removes the highest-order insignificant lag. Modified AIC (MAIC) was proposed by Ng and Perron (2001) to improve finite-sample performance and often outperforms conventional criteria in simulations. The asymptotic critical values are unchanged by lag augmentation under suitable regularity conditions. Test power increases with sample size and as the process moves farther from the unit-root boundary. Structural breaks can bias the test toward non-rejection, motivating extensions such as the Zivot–Andrews and Perron tests. A unit-root process is often called integrated of order one, denoted $I(1)$; first differencing may produce stationarity. Regressions involving nonstationary series may produce spurious high R^2 values and misleading significance tests.

Miscellanea: Said and Dickey's 1984 paper formalized the Augmented Dickey–Fuller (ADF) test by showing that adding lagged differences to the regression can eliminate higher-order autocorrelation without changing the asymptotic Dickey–Fuller distribution under the null hypothesis. Although practitioners had already begun adding lagged differences after the original 1979 Dickey–Fuller paper, the ADF framework provided the rigorous theoretical foundation that made the procedure standard in econometrics and time-series analysis.

Reference(s): Said & Dickey (1984); Schwert (1989); Ng & Perron (2001)

Worked Example

Problem Statement An economist re-examines a monthly exchange rate index for a unit root. The underlying process has MA(1) errors, so the plain DF test ($k = 0$) leaves serially correlated residuals. The ADF test adds lagged Δy terms to whiten the residuals without changing the critical values. $T = 20$, constant-only specification, lag selected by AIC.

$y = \{10.0, 11.0, 11.3, 12.2, 12.7, 13.5, 13.8, 14.6, 15.0, 15.8, 16.8, 17.1, 18.0, 18.5, 19.3, 19.6, 20.4, 20.8, 21.6, 22.6\}$

1. Lag Selection via AIC

Run ADF for $k = 0, 1, 2$ and compare AIC and residual autocorrelation:

k	$\hat{\gamma}$	τ	AIC	$\hat{\gamma}_1^{res}$
0	+0.001	+0.068	−47.20	−0.686
1	+0.010	+0.711	**−55.20**	−0.030
2	+0.008	+0.456	−49.04	−0.002

At $k = 0$ the residual lag-1 autocorrelation $\hat{\gamma}_1^{res} = -0.686$ signals serial correlation — the DF regression is misspecified. AIC favors $k^* = 1$ (lowest AIC = −55.20) and the $\hat{\gamma}_1^{res}$ drops to −0.030, confirming whitening.

2. ADF Regression ($k = 1$)

$\Delta y_t = \alpha + \gamma y_{t-1} + c_1 \Delta y_{t-1} + e_t$, using $n = T - 1 - k = 18$ observations ($t = 3 \ldots 20$): $RSS = 0.601$,

$$\hat{\alpha} = 0.916,\ \hat{\gamma} = 0.010,\ \hat{c}_1 = -0.678,\ \hat{s}^2 = \frac{0.601}{15} = 0.040$$

$$SE(\hat{\gamma}) = 0.0143, \quad \tau_{ADF} = \hat{\gamma}/SE(\hat{\gamma}) = 0.711$$

3. Results and Decision

Critical values (τ, constant, MacKinnon 1996, $T = 20$):

Statistic	Value	CV (5%)	Decision
$\tau_{ADF}(k = 1)$	+0.711	−3.02	FTR H_0

For reference: CV at 10% = −2.65; CV at 1% = −3.81.

4. Comparison with DF ($k = 0$)

Test	τ	$\hat{\gamma}_1^{res}$	Valid?
DF ($k = 0$)	+0.068	−0.686	**No**
ADF ($k = 1$)	+0.711	−0.030	**Yes**

5. Interpretation

- **Decision:** Since $\tau_{ADF} = 0.711 \gg -3.02$, we **fail to reject** H_0. The exchange rate index contains a unit root. Adding Δy_{t-1} absorbs the MA structure without affecting the critical values.

Practitioner Insight: The ADF test adds k lags of Δy_t to the DF regression to eliminate residual autocorrelation — without changing the Dickey-Fuller critical values, which remain valid regardless of k. The coefficient $\hat{c}_1 = -0.678$ is not itself tested; it is a nuisance parameter whose sole purpose is to whiten the errors. Lag selection by AIC or BIC is preferred over the *ad hoc* rule $k = \lfloor (T-1)^{1/3} \rfloor$; both criteria selected $k = 1$ here. Note that $\tau_{ADF} = +0.711$ (positive) is far from the rejection region — the unit root null is not even close to being rejected. Contrast this with the PP test which corrects for serial correlation non-parametrically; both typically give similar conclusions when applied correctly to the same series.

Common Pitfalls

- **Underfitting lag order.** Too few lags leave residual autocorrelation, invalidating the ADF assumptions. Check residuals (e.g., Ljung–Box) and increase p until they are approximately white noise.
- **Overfitting lag order.** Too many lags drain degrees of freedom and sharply reduce power. Use AIC/BIC, sequential tests, or Schwert's rule to balance parsimony and adequacy.
- **Ignoring structural breaks.** Breaks in mean/trend can mimic unit roots and cause spurious non-rejection. When breaks are plausible, use break-robust tests (Perron; Zivot–Andrews) or model the break explicitly.
- **Misinterpreting non-rejection.** Failing to reject does *not* confirm a unit root. ADF has low power; non-rejection may reflect near-unit-root behavior, breaks, or short samples. Complement with KPSS and graphical analysis.
- **Using wrong deterministic terms.** Omitting a needed trend biases toward non-rejection; including an unnecessary trend reduces power. Always inspect plots and choose constant/trend specification consistent with data.

Why Lagged Differences Do Not Change the DF Distribution.

Under H_0, y_t behaves as a random walk, so y_{t-1} is non-stationary and grows at rate $\sqrt{T}$, whereas the lagged differences Δy_{t-i} are stationary and contribute only $O(T^{-1/2})$ information. In the ADF regression, the stationary regressors are asymptotically dominated by y_{t-1}, so the limiting distribution of the t-ratio for γ is identical to the original Dickey–Fuller distribution (Said–Dickey, 1984).

Phillips-Perron (PP) Test

Non-Parametric Unit Root Test Robust to Heteroskedasticity

STAT.3 | **Category:** Stationarity | **Data:** Time Series | **Null:** Dickey-Fuller distribution

When to Use

- Test for unit roots when errors have heteroscedasticity.
- Alternative to ADF that avoids parametric lag selection.
- Apply when serial correlation structure is fully unknown.
- Suitable for series with structural breaks in variance.
- Robust alternative to Dickey–Fuller tests.

Assumptions

- Observations are collected at fixed, equi-spaced intervals.
- Time series follows a unit-root process under H_0.
- Errors are weakly dependent and possibly heteroskedastic.
- The long-run variance (at frequency zero) is positive.
- The deterministic specification(constant/trend)is correct.

Test Statistic

$$\text{(I) } Z_\tau = \sqrt{\frac{\hat{\gamma}_0}{\hat{\lambda}^2}}\, t_\gamma - \frac{1}{2}\cdot\frac{\hat{\lambda}^2 - \hat{\gamma}_0}{\hat{\lambda}}\cdot\frac{n\hat{\sigma}}{s} \quad \Big| \quad \text{(II) } Z_\gamma = n\hat{\gamma} - \frac{1}{2}\cdot\frac{n^2\hat{\sigma}^2}{s^2}\left(\hat{\lambda}^2 - \hat{\gamma}_0\right) \quad \Big| \quad t_\gamma = \frac{\hat{\gamma}}{\hat{\sigma}},\ \hat{\sigma} = \text{OLS } SE(\hat{\gamma})$$

Notation: $RSS = \sum \hat{e}_t^2$; $s^2 = RSS/(n-k)$; $\hat{\gamma}_0 = n^{-1}\sum \hat{e}_t^2$ short-run variance; $\hat{\lambda}^2$ long-run (Newey–West) variance.
Models: (A) $\Delta y_t = \gamma y_{t-1} + e_t,\ k = 1$; (B) $\Delta y_t = \delta + \gamma y_{t-1} + e_t\ \ k = 2$; (C) $\Delta y_t = \delta + \beta t + \gamma y_{t-1} + e_t,\ k = 3$.

Plain English: The Phillips-Perron test modifies the Dickey-Fuller t-statistic using non-parametric correction for serial correlation and heteroskedasticity. Rather than adding lagged differences (ADF approach), PP estimates the long-run variance of residuals and adjusts the test statistic accordingly. This makes the test robust to general forms of autocorrelation and changing variance without explicitly modeling lag dynamics through augmented regressions.

⚠ Practitioner Warning

Performance depends critically on bandwidth selection for long-run variance estimation. Common choices: Newey-West automatic bandwidth, Andrews kernel-based selection, or Bartlett kernel with fixed bandwidth. Too small bandwidth: inadequate correction for autocorrelation. Too large bandwidth: loss of power. The test has lower power than ADF when errors follow low-order AR process, but outperforms ADF under MA errors or structural breaks.

Hypotheses

- H_0: $\gamma = 0$ equivalently $\rho = 1$ (unit root, non-stationary).
- H_1: $\gamma < 0$ equivalently $\rho < 1$ (stationary process).

Decision Rule: Reject H_0 if $Z_\tau < t_\alpha$ or p value $< \alpha$. Use same Dickey-Fuller critical values as DF and ADF tests.

Software References [Python, R, SAS]

```
statsmodels::pp_test          urca::ur.pp          PROC ARIMA (stationarity=(pp))
```

Additional Notes

The PP test uses the same null hypothesis and asymptotic critical values as the Dickey–Fuller test despite employing nonparametric corrections. Two commonly reported statistics are Z_τ (modified t statistic) and Z_ρ (modified normalized-bias statistic). Long-run variance estimation typically uses HAC estimators: $\hat{\lambda}^2 = \hat{\gamma}_0 + 2\sum_{j=1}^{q} w_j\hat{\gamma}_j$, where $\hat{\gamma}_j$ are sample autocovariances and w_j are kernel weights (e.g., Bartlett: $w_j = 1 - j/(q+1)$). Automatic bandwidth selection is often based on Newey–West rules. A unit-root process is called integrated of order one, denoted $I(1)$; first differencing may produce stationarity. Regressions involving nonstationary series may exhibit spurious high R^2 values and misleading significance tests. When errors are i.i.d., $\hat{\lambda}^2 = \hat{\gamma}_0$ and both corrections vanish, recovering the DF statistics.

Miscellanea: Peter Phillips and Pierre Perron introduced their test in 1988 as an alternative to the Augmented Dickey–Fuller procedure, motivated partly by the practical difficulty of lag-length selection in ADF testing. Rather than adding lagged differences, the Phillips–Perron test uses nonparametric corrections to account for serial correlation and heteroskedasticity in the errors while preserving the same asymptotic Dickey–Fuller distributions. The procedure became widely used in econometrics because it reduced reliance on explicit dynamic specification.

Reference(s): Phillips & Perron (1988); Newey & West (1987)

Worked Example

Problem Statement An economist tests whether a monthly inflation rate series is stationary. The Phillips-Perron test is preferred over ADF because it corrects for serial correlation non-parametrically, avoiding the need to choose a lag length. A constant-only model is used with $T = 20$, Newey-West (NW) bandwidth $\ell = 2$.

$$y = \{16.0, 15.2, 15.7, 15.0, 15.5, 14.8, 15.3, 14.7, 15.2, 14.6, 15.1, 14.6, 15.2, 14.7, 15.3, 14.8, 15.4, 14.8, 15.3, 14.7\}$$

1. Step 1 — OLS Regression

Regress Δy_t on $[1, \ y_{t-1}]$ using $n = T - 1 = 19$ obs.

$$\hat{\delta} = 19.779, \quad \hat{\gamma} = -1.313, \quad s^2 = \frac{RSS}{n-2} = 0.1004$$

$$\hat{\sigma} = SE(\hat{\gamma}) = \sqrt{\frac{s^2}{\sum \tilde{y}_{t-1}^2}} = 0.1941, \quad t_\gamma = \frac{\hat{\gamma}}{SE(\hat{\gamma})} = -6.766$$

t_γ is the standard Dickey-Fuller t-statistic, uncorrected for serial correlation.

2. Step 2 — Long-Run Variance ($\ell = 2$ NW)

From OLS residuals $\hat{e}_t$:

- $\hat{\gamma}_0 = 0.0898$ (short-run variance $= RSS/n$)
- $\hat{\gamma}_1 = 0.0099$, $\hat{\gamma}_2 = 0.0574$ (divide by n)

$$\hat{\lambda}^2 = \hat{\gamma}_0 + 2\left(\tfrac{2}{3}\right)\hat{\gamma}_1 + 2\left(\tfrac{1}{3}\right)\hat{\gamma}_2 = 0.1413$$

4. Results and Decision

Critical values (constant only, MacKinnon 1996):

Stat	Value	CV (5%)	Decision
Z_τ	-6.191	-3.58	Reject H_0
Z_γ	-28.07	—	Reject H_0

$\hat{e}_t = \Delta y_t - \hat{\delta} - \hat{\gamma}\, y_{t-1}$; $\quad \hat{\gamma}_j = \frac{1}{n}\sum_{t=j+2}^{20} \hat{e}_t\, \hat{e}_{t-j}$

Bartlett's weight: $w_j = 1 - (j/3)$; $w_1 = 2/3, w_2 = 1/3$

3. Step 3 — PP Corrections

$$Z_\tau = \sqrt{\frac{\hat{\gamma}_0}{\hat{\lambda}^2}}\, t_\gamma - \frac{1}{2} \cdot \frac{\hat{\lambda}^2 - \hat{\gamma}_0}{\hat{\lambda}} \cdot \frac{n\hat{\sigma}}{s} = -6.191$$

$$Z_\gamma = n\hat{\gamma} - \frac{1}{2} \cdot \frac{n^2\hat{\sigma}^2}{s^2}\left(\hat{\lambda}^2 - \hat{\gamma}_0\right) = -28.435$$

5. Interpretation

- **Decision:** Both PP statistics fall well below their 5% critical values (−3.580). The inflation series is stationary.
- **Practitioner Insight:** The inflation series strongly rejects the unit-root null: $Z_\tau = -6.191$ is well below the 5% critical value of −3.58. The raw Dickey–Fuller statistic $t_\gamma = -6.766$ is adjusted by 0.797 to account for positive serial correlation in the residuals, particularly the relatively large lag-2 autocovariance $\hat{\gamma}_2 = 0.057$ compared with $\hat{\gamma}_0 = 0.090$. The correction is modest because the series oscillates around the mean, so serial dependence is present but not dominant. The companion statistic $Z_\rho = -28.435$ also strongly rejects the null. The large negative estimate $\hat{\gamma} = -1.313$ reflects rapid mean reversion, with shocks largely corrected within one period.

OLS: $\Delta y_t = \delta + \gamma\, y_{t-1} + e_t$; $n = 19$, $RSS = 1.706$

$\sum \tilde{y}_{t-1}^2 = \sum(y_{t-1} - \bar{y}_{t-1})^2 = 2.665$, $\bar{y}_{t-1} = 15.116$

In unit-root regressions, ρ typically denotes the autoregressive coefficient in levels, whereas $\gamma = \rho - 1$ is the coefficient in the first-difference representation.

Common Pitfalls

- **Bandwidth choice for long-run variance.** PP relies on consistent long-run variance estimation; bandwidths that are too small undercorrect (invalid size) and those too large overcorrect (power loss). Use data-driven selectors (Newey–West, Andrews) and check sensitivity across reasonable bandwidths.
- **Assuming PP is uniformly superior to ADF.** PP avoids lag selection and is robust to heteroskedasticity, but ADF often performs better in small samples or when AR structure is well-modeled. In practice, apply both tests and compare conclusions.
- **Underrejection in small or MA-heavy samples.** PP can be markedly conservative with $n < 100$ or strong negative MA components, leading to low power. Non-rejection in such settings should be interpreted cautiously or complemented with size-adjusted variants (e.g., Perron–Ng).
- **Ignoring structural breaks.** Like most classical unit root tests, PP is invalid under structural breaks, which can mimic non-stationarity. When breaks are plausible, use break-robust tests (Perron, Zivot–Andrews) or models incorporating regime shifts.
- **Over-interpreting the correction formula.** The PP statistic applies nonparametric bias and variance adjustments to the DF regression. Users should rely on the final test statistic and critical values—software handles the complex corrections.

DF-GLS Test

Efficient Dickey-Fuller Test with GLS Detrending

STAT.4 | **Category:** Stationarity | **Data:** Time Series | **Null:** Tabled distribution

When to Use

- Test for unit roots when sample sizes are small to moderate.
- Apply when standard ADF or PP tests lack statistical power.
- Evaluate stationarity near the unity root boundary zone.
- Analyze series with unknown deterministic drift or trend.
- Preliminary check before cointegration analysis.

Assumptions

- Observations are recorded at fixed, uniform time intervals.
- Errors are weakly dependent and stationary.
- Error term variance remains constant across the horizon.
- GLS detrending model is correctly specified.
- Series features no structural breaks in trend parameters.

Test Statistic

$$t = \frac{\hat{\gamma}}{SE(\hat{\gamma})} \quad \Big| \quad \Delta y_t^d = \gamma y_{t-1}^d + \sum_{i=1}^{p} \phi_i \Delta y_{t-i}^d + \epsilon_t.$$

Notation: y_t^d is GLS-detrended series using $\bar{\alpha}$ (local-to-unity parameter), $\bar{\alpha} = 1 - 7/T$ (constant) or $1 - 13.5/T$ (trend).

Plain English: The DF-GLS test improves upon the ADF by using Generalized Least Squares to remove deterministic components (constant and/or trend) before testing for a unit root. This GLS detrending is more efficient than OLS detrending, especially when the series is close to having a unit root. The result is a test with substantially higher power against stationary alternatives while preserving reliable Type I error control.

⚠ Practitioner Warning

Although DF–GLS generally has higher power than ADF, performance still depends critically on lag selection. Too few lags leave residual autocorrelation, whereas too many reduce power and consume degrees of freedom. The GLS detrending step assumes the deterministic spec (constant or trend) is correctly chosen; misspecification can substantially distort inference. Critical values differ from standard ADF values, so using ADF critical values invalidates the test. Structural breaks may bias the procedure toward non-rejection, and residuals should be approximately white noise after fitting. The power advantage over ADF largely disappears when the true process contains no deterministic components.

Hypotheses

- H_0: $\gamma = 0$ equivalently $\rho = 1$ (unit root, non-stationary).
- H_1: $\gamma < 0$ equivalently $\rho < 1$ (stationary).

Decision Rule: Reject H_0 if $t < t_\alpha^{GLS}$ or p-value $< \alpha$. Use DF–GLS CVs, which differ from standard ADF values

Software References [Python, R, SAS]

`arch.unitroot::dfgls`	`urca::ur.ers`	`PROC ARIMA`

Additional Notes

The DF–GLS test first removes deterministic components using generalized least-squares (GLS) detrending and then applies an ADF-style regression to the transformed series. Quasi-differenced data are constructed as $\tilde{y}_t = y_t - \bar{\alpha} y_{t-1}$, where $\bar{\alpha} = 1 + \bar{c}/n$ is chosen to maximize local asymptotic power. Recommended values are $\bar{c} = -7$ for models with an intercept and $\bar{c} = -13.5$ for models with intercept and trend. The detrended series is then obtained from $y_t^d = y_t - z_t'\hat{\beta}$, where z_t contains deterministic regressors. The test retains the same null hypothesis as ADF but often achieves substantially higher power near the unit-root boundary. Lag selection typically follows AIC, BIC, or modified criteria such as MAIC. As with ADF, residuals should be approximately white noise after fitting.

Miscellanea: Graham Elliott, Thomas Rothenberg, and James Stock introduced the DF–GLS test in 1996 as a more powerful alternative to the Augmented Dickey–Fuller procedure, particularly when the autoregressive root is close to unity. The method combines generalized least-squares detrending with a Dickey–Fuller regression, substantially improving power in many near-unit-root settings. Although the test is sometimes called the "ERS test" after the authors' initials, the name "DF–GLS" became more common because it emphasizes the test's connection to the familiar Dickey–Fuller framework.

Reference(s): Elliott et al. (1996); Ng & Perron (2001)

Worked Example

Problem Statement An economist tests the same stock price index examined by ADF, now using the DF-GLS test which replaces OLS detrending with GLS quasi-differencing. The GLS transformation is more efficient, yielding a more powerful test at less negative critical values. Trend specification ($\bar{c} = -13.5$), $T = 20$, lag selected by AIC.

$y = \{10.0, 10.8, 11.3, 12.5, 13.0, 13.6, 14.8, 15.2, 16.5, 17.1, 17.8, 18.9, 19.4, 20.6, 21.1, 22.4, 22.9, 24.1, 24.6, 25.9\}$

1. Step 1 — GLS Quasi-Differencing

Set $\bar{\alpha} = 1 + \bar{c}/T = 1 + (-13.5)/20 = 0.325$. Form quasi-differenced series ($z_t = [1, t]'$):

$$\tilde{y}_t = y_t - \bar{\alpha}\, y_{t-1}, \quad \tilde{z}_t = z_t - \bar{\alpha}\, z_{t-1}$$

with $\tilde{y}_1 = y_1 = 10.0$, $\tilde{z}_1 = [1,\ 1]'$. Select values:

- $\tilde{y}_2 = 10.8 - 0.325 \times 10.0 = 7.550$
- $\tilde{z}_2 = [1 - 0.325,\ 2 - 0.325] = [0.675,\ 1.675]'$

2. Step 2 — GLS Estimates of Trend

OLS of $\tilde{y}$ on $\tilde{z}$ gives:

$$\hat{\mu}_{GLS} = 8.923, \qquad \hat{\beta}_{GLS} = 0.830$$

3. Step 3 — GLS-Detrended Series

$$y_t^d = y_t - \hat{\mu}_{GLS} - \hat{\beta}_{GLS} \cdot t$$

Selected values: $y_1^d = +0.247$, $y_8^d = -0.367$, $y_{20}^d = +0.368$. The detrended series oscillates tightly around zero (max $|y_t^d| = 0.37$), confirming the trend was removed.

4. Step 4 — ADF on y^d (Lag Selection)

Regress $\Delta y_t^d = \rho\, y_{t-1}^d + \sum c_j\, \Delta y_{t-j}^d + e_t$ (no intercept — already removed):

k	τ_{GLS}	AIC	$\hat{\rho}_1^{res}$
0	−5.675	−58.07	−0.037
1	−2.276	**−59.17**	−0.189

AIC selects $k^* = 1$. At $k = 1$:

$$\hat{\rho} = -0.795, \quad \hat{c}_1 = -0.482, \quad \hat{s}^2 = 0.034$$

$$SE(\hat{\rho}) = 0.349, \quad \tau_{GLS} = \frac{-0.795}{0.349} = -2.276$$

5. Results and Decision

Critical values (ERS 1996, trend spec, $T = 20$):

Stat	Value	CV (5%)	Decision
τ_{GLS} ($k = 1$)	−2.276	−3.030	FTR H_0

For reference: CV at 10% = −2.74; CV at 1% = −3.58.

6. Interpretation

- **Decision:** The series contains a unit root — consistent with the ADF conclusion.

Practitioner Insight: DF-GLS replaces OLS detrending with GLS quasi-differencing at $\bar{\alpha} = 1 + \bar{c}/T$, which is more efficient under the local-to-unity alternative. The payoff is a less negative 5% critical value (−3.03 vs. −2.86 for ADF) — but because the test is more powerful, $\tau_{GLS} = -2.28$ is much closer to its boundary than $\tau_{ADF} = +1.78$ is to its own. Had the series been weakly stationary, DF-GLS would likely have rejected while ADF did not. GLS detrending also reduces the optimal lag length: here $k = 1$ suffices even though the MA(1) errors required $k = 1$ in ADF; with some series GLS detrending reduces the required k to 0.

Note: In local-to-unity asymptotics, $\rho_n = 1 + c/n$ with $c < 0$, so the process remains stationary but approaches the unit root boundary as $n \to \infty$.

Common Pitfalls

- **Using standard DF/ADF critical values instead of DF–GLS values.** The DF–GLS statistic has a different reference distribution due to GLS detrending. Always use ERS critical values rather than standard DF tables.
- **Misinterpreting the local-to-unity parameter $\bar{\alpha}$.** The quantity $\bar{\alpha} = 1 - c/T$ is a tuning parameter used in the GLS detrending step, not an estimate of the autoregressive coefficient. The parameter actually tested for a unit root is the coefficient in the subsequent ADF regression.
- **Applying DF–GLS in very small samples.** The efficiency gains of DF–GLS over ADF appear mainly in moderate or large samples. With small samples, finite-sample distortions can reduce its reliability.
- **Ignoring the assumption of homoskedastic errors.** DF–GLS, like the ADF test, assumes constant innovation variance. When heteroskedasticity or volatility clustering is present, alternatives such as Phillips–Perron or bootstrap-based tests may be preferable.
- **Including inappropriate deterministic components.** The test can be run with a constant or with a constant plus trend. Including an unnecessary trend reduces power, while omitting a needed trend biases the test toward non-rejection.

Ng-Perron Test

Modified Unit Root Tests with Improved Size and Power

STAT.5 | **Category:** Stationarity | **Data:** Time Series | **Null:** Tabled distribution

When to Use

- Test for unit roots when data has severe serial correlation.
- Apply when standard tests suffer from size distortions.
- Evaluate series where the autoregressive root is near one.
- Efficient unit-root testing in small samples.
- Establish highly reliable baselines for cointegration.

Assumptions

- Observations are recorded at fixed, uniform time intervals.
- The series is detrended via a localized GLS framework.
- Structural innovation variance remains constant over time.
- Timeline contains no structural breaks in trend parameters.
- Lag selection is optimized using the modified AIC (MAIC).

Test Statistic

(1) $MZ_\alpha = \dfrac{L - s_{AR}^2}{2K}$ | (2) $MSB = \sqrt{\dfrac{K}{s_{AR}^2}}$ | (3) $MZ_t = MZ_\alpha \times MSB$ | (4) $MPT = \begin{cases} \frac{c^2K - cL}{s_{AR}^2} & \text{for constant} \\ \frac{c^2K + (1-c)L}{s_{AR}^2} & \text{for trend} \end{cases}$

Notation: $\tilde{y}_t$ = locally GLS-detrended series $(y_t - \hat{\beta}' z_t)$, T = sample size, c = local-to-unity tuning parameter (−7.0 for constant, −13.5 for trend), $K = \frac{1}{T^2}\sum_{t=2}^{T}\tilde{y}_{t-1}^2$, $L = \frac{1}{T}\tilde{y}_T^2$, and s_{AR}^2 = autoregressive estimate of the long-run variance

Plain English: The Ng–Perron tests improve upon earlier unit-root procedures by combining GLS detrending with a more reliable long-run variance estimator. This substantially reduces the size distortions that affect ADF and Phillips–Perron tests under strong persistence or moving-average errors. Four complementary statistics are reported, providing more stable inference in difficult near-unit-root settings.

⚠ Practitioner Warning

Four statistics are reported (MZ_α, MSB, MZ_t, and MPT), and their conclusions may occasionally differ in borderline cases. Critical values differ across statistics and depend on deterministic specification. Modified AIC (MAIC) is generally preferred for lag selection; standard AIC or BIC may perform poorly in persistent series. Structural breaks can bias the tests toward non-rejection, and the procedure is computationally more intensive than ADF.

Hypotheses

- H_0: Series has unit root (non-stationary).
- H_1: Series is stationary around deterministic trend.

Decision Rule: Reject H_0 if MZ_α or MZ_t < critical value (left-tailed), or MSB or MPT < critical value.

Software References [Python, R, SAS]

Python	R	SAS
`Not Native (Manual)`	`Not Native (Manual)`	`PROC AUTOREG (NP)`

Additional Notes

A key distinction is between the residual variance $\hat{\sigma}^2$ from the ADF-type regression and the autoregressive spectral density estimate $s_{AR}^2 = \hat{\sigma}^2/(1 - \sum_{i=1}^{k}\hat{\phi}_i)^2$, which estimates long-run variance using fitted AR coefficients $\hat{\phi}_i$. The MPT statistic is designed for strong local asymptotic power against near-unit-root alternatives. Standard lag selection criteria like AIC or BIC systematically select lags that are too short when negative MA errors are present. Ng and Perron introduced the Modified AIC: $\text{MAIC}(k) = \ln(\hat{\sigma}_k^2) + 2(\tau_T(k) + k)/(T - k_{\max})$ where the correction term $\tau_T(k)$ is dynamically driven by the squared OLS coefficient of the lagged level. This forces the selection of longer lag lengths when necessary, safely whitening the residuals. Compared with ADF and Phillips–Perron, Ng–Perron tests generally provide better size control and improved power in highly persistent series. Structural breaks may still bias the tests toward non-rejection.

Miscellanea: Serena Ng and Pierre Perron (2001) developed modified unit-root tests to address the severe finite-sample distortions observed in earlier procedures under strong serial correlation and negative moving-average errors. Their refinements greatly improved empirical reliability and are now widely regarded as among the strongest classical unit-root procedures in econometrics.

Reference(s): Ng & Perron (2001); Perron & Ng (1996)

Ng-Perron Test

Worked Example

Problem Statement An economist tests whether a monthly consumer price index (CPI) series for a small open economy contains a unit root. The series spans $T + 1 = 15$ observations. A finding of non-stationarity would indicate that shocks to CPI are permanent. The GLS-demeaned series (constant only) is used throughout, with no lag augmentation ($k = 0$).

$$y = \{100.0, 99.7, 100.3, 100.1, 100.6, 100.9, 100.8, 101.2, 100.9, 101.5, 101.7, 101.3, 101.8, 102.1, 101.9\}$$

1. GLS Demeaning

For the constant-only Ng–Perron test:

$$\bar{c} = -7, \qquad T = 14, \qquad \bar{\alpha} = 1 + \frac{\bar{c}}{T} = 0.5$$

The GLS-demeaned mean estimate is:

$$\hat{\mu}_{\text{GLS}} = \frac{\sum z_t^* y_t^*}{\sum (z_t^*)^2} = \frac{454.175}{4.5} = 100.928.$$

Thus: $\tilde{y}_t = y_t - \hat{\mu}_{\text{GLS}}$ giving:

$$\tilde{y} = \left\{ \begin{array}{ccccc} -0.928, & -1.228, & -0.628, & -0.828, & -0.328, \\ -0.028, & -0.128, & 0.272, & -0.028, & 0.572, \\ 0.772, & 0.372, & 0.872, & 1.172, & 0.972 \end{array} \right\}$$

2. DF Regression on GLS Series

With $k = 0$, estimate: $\Delta\tilde{y}_t = \gamma\tilde{y}_{t-1} + e_t, \ t = 2, \ldots, 15$

Number of usable observations: $n = T = 14$

OLS gives:

$$\hat{\gamma} = \frac{\sum \tilde{y}_{t-1}\Delta\tilde{y}_t}{\sum \tilde{y}_{t-1}^2} = \frac{-0.973}{6.844} = -0.1421$$

Residual sum of squares: $RSS = \sum \hat{e}_t^2 = 1.8917$

Since one slope parameter is estimated:

$$s_{AR}^2 = \frac{RSS}{T - k} = \frac{1.8917}{14} = 0.1351$$

3. Intermediate Quantities

Compute:

$$K = \frac{1}{n^2}\sum_{t=2}^{15} \tilde{y}_{t-1}^2 = \frac{6.8446}{14^2} = 0.03492, \quad \frac{\tilde{y}_T^2}{n} = \frac{0.972^2}{14} = 0.0675$$

4. Ng–Perron Statistics

$$MZ_\alpha = \frac{n^{-1}\tilde{y}_T^2 - s_{AR}^2}{2K} = -0.968, \quad MSB = \sqrt{\frac{K}{\hat{s}_{AR}^2}} = 0.508,$$

$$MZ_t = MZ_\alpha \cdot MSB = -0.492$$

$$MPT = \frac{\bar{c}^2 K - \bar{c}\, n^{-1}\tilde{y}_T^2}{\hat{s}_{AR}^2} = 16.16$$

5. Decision and Interpretation

Typical 5% critical values:

$$MZ_\alpha = -8.1, \ MSB = 0.233, \ MZ_t = -1.98, \ MPT = 3.17$$

Since:

$$-0.968 > -8.1, \ 0.508 > 0.233, \ -0.492 > -1.98, \ 16.16 > 3.17,$$

we fail to reject the unit-root null.

- **Conclusion:** The CPI series behaves as nonstationary in this small sample.
- **Practitioner Insight:** Ng–Perron tests improve upon Phillips–Perron by combining GLS detrending with modified unit-root statistics designed to reduce severe size distortions in small samples. The statistics MZ_α and MZ_t behave similarly to PP-type statistics, whereas MSB and MPT provide additional stabilization and power improvements. Use Ljung-Box on $\hat{e}_t$ to re-adjust k.

GLS demeaning with constant only

$$y_t^* = \begin{cases} y_1, & t = 1, \\ y_t - \bar{\alpha} y_{t-1}, & t = 2, \ldots, T, \end{cases} \quad z_t^* = \begin{cases} 1, & t = 1, \\ 1 - \bar{\alpha}, & t = 2, \ldots, T. \end{cases}$$

Common Pitfalls

- **Using standard AIC/BIC instead of modified criteria.** Ng–Perron tests rely on lag selection via *Modified AIC (MAIC)* or *Modified BIC (MBIC)*, designed for unit root testing. Using ordinary AIC/BIC can produce size distortions and unreliable inference.
- **Estimating the long-run variance inconsistently.** The spectral density at zero must be estimated using the same lag structure selected for the test regression. Using different lag orders or ad-hoc bandwidth choices invalidates the statistic.
- **Selective reporting of test statistics.** The Ng–Perron framework produces four related statistics (MZ_α, MZ_t, MSB, MP_T). Standard practice reports MZ_α and MZ_t as the primary statistics rather than choosing whichever rejects.
- **Assuming Ng–Perron always dominates DF–GLS.** Although Ng–Perron improves finite-sample size and power, its advantage diminishes in very large samples. DF–GLS may be adequate when sample sizes are large or implementation simplicity is preferred.
- **Ignoring structural breaks in the series.** Structural breaks can cause stationary series to appear nonstationary, biasing unit-root tests toward non-rejection. Break-adjusted procedures should be used when regime changes or level shifts are suspected.

Schmidt-Phillips Test

Unit Root Test Robust to Serial Correlation via Nonparametric Variance

STAT.6 | **Category:** Stationarity | **Data:** Time Series | **Null:** Tabled distribution

When to Use

- Test for unit roots under parameterization-free trends.
- Apply when the test statistic must be invariant to drift.
- Evaluate series where trend parameters may shift under H_0.
- Analyze data requiring robust power under linear trends.
- Establish invariant baseline metrics for cointegration.

Assumptions

- The data is detrended via a parameter-less OLS framework.
- Structural innovation variance remains constant over time.
- The regression errors are independent and white noise.
- Observations are recorded at fixed, uniform time intervals.
- The underlying path contains no structural trend breaks.

Test Statistic

$$\textbf{(1)}\ Z(\tilde{\rho}) = \frac{\tilde{\rho}}{\hat{\omega}^2} \quad \Big| \quad \textbf{(2)}\ Z(\tilde{\tau}) = \frac{\tilde{\tau}}{\hat{\omega}} \quad \Big| \quad \hat{\xi} = \frac{y_T - y_1}{T-1},\ \hat{\psi}_X = y_1 - \hat{\xi} \quad \Big| \quad S_t = y_t - \hat{\psi}_X - \hat{\xi}t, \quad t = 1, \ldots, T$$

Notation: $\hat{\xi}$ restricted MLE of drift; S_t LM residual series ($S_1 = S_T = 0$ by construction); $\tilde{\rho} = T\hat{\phi}$ and $\tilde{\tau} = \hat{\phi}/SE(\hat{\phi})$. Long-run variance ratio from DF levels regression residuals $\hat{e}_t$: $\hat{\omega}^2 = S^2/s^2$, where $s^2 = T^{-1}\sum \hat{e}_t^2$ and S^2 is the Newey–West long-run variance. **Test Regression:** $\Delta y_t = \alpha + \hat{\phi} S_{t-1} + e_t, \quad t = 2, \ldots, T$.

Plain English: The Schmidt–Phillips test examines whether a time series follows a permanent random-walk-type process (unit root) or eventually returns to a deterministic trend path. Unlike standard Dickey–Fuller methods, Schmidt and Phillips decompose the series into a fixed level, a fixed slope, and a stochastic component whose structure remains unchanged under both the null and alternative hypotheses. This approach reduces the power losses and instability that conventional unit-root tests can experience in strongly trending series. The test constructs cumulative deviations from the estimated deterministic path and checks whether these deviations help predict future changes. If large deviations tend to reverse over time, the series is trend-stationary and the unit root is rejected; otherwise shocks persist permanently.

⚠ Practitioner Warning

Critical values depend on the deterministic trend specification and differ from standard Dickey–Fuller distributions. Misspecifying the trend order (polynomial trend) can substantially distort size and power. Structural breaks may bias the test toward non-rejection, and finite-sample performance can deteriorate in short series or highly persistent processes.

Hypotheses

- H_0: $\phi = 0$ (unit root; non-stationary with drift).
- H_1: $\phi < 0$ (trend-stationary; reverts to deterministic trend).

Decision Rule: Reject H_0 if $Z(\tilde{\rho}) < C_{\tilde{\rho},\alpha}$ or $Z(\tilde{\tau}) < C_{\tilde{\tau},\alpha}$ (left-tailed test), or if the empirical p-value $< \alpha$.

Software References [Python, R, SAS]

Python	R	SAS
`Not Native (Manual)`	`urca::ur.sp()`	`Not Native (Manual)`

Additional Notes

The test constructs an LM-based unit-root statistic after estimating deterministic components under H_0. A key feature is that the detrending framework remains structurally invariant under both hypotheses, avoiding the parameterization inconsistencies that can affect standard Dickey–Fuller procedures. The detrended residual process S_t represents the level deviations from the restricted specification—behaving asymptotically as a partial sum process under the null—and the test is then based on whether S_{t-1} helps predict future changes in the series. Long-run variance estimation follows same kernel methods as PP: $\hat{\lambda}^2 = \hat{\gamma}_0 + 2\sum_{j=1}^{q} k(j/M)\hat{\gamma}_j$ where $k(\cdot)$ is kernel and M is bandwidth.

Miscellanea: Peter Schmidt and Peter C. B. Phillips (1992) introduced LM-based unit-root tests that explicitly allow higher-order polynomial deterministic trends while retaining nonstandard unit-root asymptotics. The procedure became particularly relevant for long-run macroeconomic and historical time series, where nonlinear deterministic growth patterns are common. Although theoretically influential, the test remains less widely used in practice than ADF or Phillips–Perron procedures.

Reference(s): Schmidt & Phillips (1992); Stock (1999)

Worked Example

Problem Statement An economist tests whether a quarterly real GDP index for an emerging economy contains a unit root. A unit root would imply that demand shocks have permanent effects on output levels. The series spans $T = 12$ quarters and is analyzed under the constant-plus-trend specification using a Newey-West bandwidth of $\ell = 1$.

$$y = \{100.0,\ 100.9,\ 101.7,\ 102.4,\ 102.6,\ 102.9,\ 103.3,\ 104.1,\ 104.8,\ 105.0,\ 105.3,\ 105.9\}$$

1. Restricted MLE Under H_0
Impose $\beta = 1$. The restricted MLE of the drift ξ is the mean first difference (SP eq. 5):

$$\hat{\xi} = \frac{y_T - y_1}{T-1} = \frac{105.9 - 100.0}{11} = 0.5\overline{36}.$$

The restricted level estimate (SP eq. 6):

$$\hat{\psi}_X = y_1 - \hat{\xi} = 100.0 - 0.5\overline{36} = 99.46\overline{36}.$$

2. Restricted Residuals S_t
$S_t = y_t - \hat{\psi}_X - \hat{\xi}\,t$ (SP eq. 7). By construction $S_1 = S_T = 0$.

t	S_t	t	S_t	t	S_t
1	0.000	5	+0.455	9	+0.509
2	+0.364	6	+0.218	10	+0.173
3	+0.627	7	+0.082	11	−0.064
4	+0.791	8	+0.345	12	0.000

3. Regression (SP eq. 10)
Regress Δy_t on an intercept and S_{t-1} for $t = 2, \ldots, T$ ($n = 11$ observations): $\Delta y_t = \hat{\alpha} + \hat{\phi}\, S_{t-1} + \hat{\varepsilon}_t$. OLS yields:

$$\hat{\alpha} = 0.690, \quad \hat{\phi} = -0.482, \quad \text{RSS} = 0.5202.$$

$$\hat{\sigma}^2 = RSS/(n-2) = 0.05780, \quad SE(\hat{\phi}) = 0.2851.$$

4. Raw SP Statistics (SP eqs. 11–12)

$$\tilde{\rho} = T\hat{\phi} = -5.785, \quad \tilde{\tau} = \hat{\phi}/SE(\hat{\phi}) = -1.691.$$

For i.i.d. errors, $\tilde{\rho}$ and $\tilde{\tau}$ are the final statistics. With possible serial correlation, correct using the variance ratio.

5. Variance Ratio Correction (SP Section IV)
Estimate $\hat{\omega}^2 = S^2/s^2$ from residuals of the DF levels regression $y_t = c + \delta t + \rho y_{t-1} + e_t$ (SP eqs. 19–20, 24–25):

$$s^2 = \hat{\gamma}_0 = 0.02628, \quad \hat{\gamma}_1 = 0.00410.$$

Bartlett weight for $\ell = 1$: $w_1 = 1 - 1/(1+1) = 0.5$:

$$S^2 = \hat{\gamma}_0 + 2(0.5)\hat{\gamma}_1 = 0.03038, \quad \hat{\omega}^2 = S^2/s^2 = 1.1560,$$

6. Corrected Statistics and Decision
Per SP eq. (26), **divide** by $\hat{\omega}^2$ and $\hat{\omega}$:

$$Z(\tilde{\rho}) = \tilde{\rho}/\hat{\omega}^2 = -5.005, \quad Z(\tilde{\tau}) = \tilde{\tau}/\hat{\omega} = -1.573.$$

Critical values (SP 1992, Table 1A, $p = 1, T = 25$):

Stat.	Value	CV(5%)	CV(1%)	Decision
$Z(\tilde{\rho})$	−5.005	−15.7	−20.4	FTR H_0
$Z(\tilde{\tau})$	−1.573	−3.18	−3.90	FTR H_0

Fail to reject H_0: the GDP series is consistent with a unit root at all conventional levels.

Practitioner Insight: The Schmidt–Phillips test differs fundamentally from ADF and PP tests in how deterministic components are removed. Rather than estimating the trend from a potentially spurious levels regression, it imposes $\beta = 1$ and estimates the drift from first differences, producing restricted residuals S_t with $S_1 = S_T = 0$ by construction. The raw statistics $\tilde{\rho}$ and $\tilde{\tau}$ come from regressing Δy_t on S_{t-1}; serial correlation is then corrected by dividing $\tilde{\rho}$ by $\hat{\omega}^2 = S^2/s^2$ and $\tilde{\tau}$ by $\hat{\omega}$. Since positive serial correlation inflates the long-run variance S^2, $\hat{\omega}^2 > 1$ makes the corrected statistics less negative than the raw values. Here $\hat{\omega}^2 = 1.156$, so the correction is modest. With only $T = 12$, power is low: $Z(\tilde{\rho}) = -5.0$ remains far above the 5% critical value −15.7, so stronger conclusions require a longer series. ($\hat{c} = 66.864$, $\hat{\delta} = 0.309$, $\hat{\rho} = 0.335$).

Common Pitfalls

- **Using Dickey–Fuller critical values.** Schmidt–Phillips statistics follow different asymptotic distributions than standard DF/ADF tests. Applying incorrect critical values leads to invalid inference.
- **Misspecifying deterministic trends.** The test explicitly models deterministic components under the null hypothesis. Omitting relevant trends or using the wrong polynomial order can substantially reduce power and distort size.
- **Ignoring bandwidth sensitivity.** Long-run variance estimation requires kernel methods and bandwidth selection, similar to Phillips–Perron tests. Poor bandwidth choices can materially affect finite-sample performance.
- **Assuming higher power in all settings.** Although theoretically attractive, the SP test does not uniformly dominate ADF or Phillips–Perron procedures. Relative performance depends on trend structure, persistence, and sample size.

KPSS Test

Test for Stationarity (Null) vs. Unit Root (Alternative)

STAT.7 | **Category:** Stationarity | **Data:** Time Series | **Null:** Tabled distribution

When to Use

- Test stationarity as H_0 (opposite of ADF, PP tests).
- Complement other unit root tests by reversing hypotheses.
- Identify long-memory or fractionally integrated processes.
- Verify stationarity before fitting standard ARMA models.
- Confirm if a time series requires first-differencing step.

Assumptions

- Observations are recorded at fixed, uniform time intervals.
- The series is decomposed into trend, walk, and error.
- The stationary error process exhibits weak dependence.
- The random walk component has zero variance under H_0.
- The underlying data contains no structural regime breaks.

Test Statistic

(1) $\eta_\mu = \dfrac{\sum_{t=1}^{T} S_t^2}{T^2 \hat{\lambda}^2(\ell)}$, **Level Residuals:** $e_t = y_t - \bar{y}$ | (2) $\eta_\tau = \dfrac{\sum_{t=1}^{T} S_t^2}{T^2 \hat{\lambda}^2(\ell)}$, **Trend Residuals:** $e_t = y_t - \hat{\alpha} - \hat{\beta} t$

Notation: $S_t = \sum_{i=1}^{t} e_i$ is running partial sum, T is the sample size, and $\bar{y}$ is the sample mean. $\hat{\alpha}$ and $\hat{\beta}$ are OLS coefficients from a linear time-trend regression. The denominator $\hat{\lambda}^2(\ell)$ is Newey–West long-run estimator

Plain English: The KPSS test reverses the typical unit root testing framework by making stationarity the null hypothesis and unit root the alternative. It measures how far the series deviates from a level or trend-stationary process by examining cumulative sums of residuals from deterministic regression. Large values indicate the presence of a unit root component, leading to rejection of stationarity. This complementary perspective helps avoid ambiguous conclusions when used with traditional tests.

⚠ Practitioner Warning

Interpreting KPSS jointly with ADF or Phillips–Perron tests can produce four common outcomes: both reject (possible structural breaks or near-unit-root behavior), both fail to reject (inconclusive), KPSS rejects while ADF/PP fail to reject (evidence consistent with a unit root), or ADF/PP reject while KPSS does not (evidence consistent with stationarity). Bandwidth selection for long-run variance estimation critically affects results; automatic methods (e.g., Newey–West) or sensitivity analysis are recommended. The test is also sensitive to the chosen deterministic specification.

Hypotheses

- H_0: Series is level-stationary (or trend-stationary).
- H_1: Series has unit root component (non-stationary).

Decision Rule: Reject H_0 if $\eta_\mu > C_{\mu,\alpha}$ (or $\eta_\tau > C_{\tau,\alpha}$) or p-value $< \alpha$.

Software References [Python, R, SAS]

`statsmodels::kpss` `tseries::kpss.test` `PROC ARIMA (KPSS)`

Additional Notes

Separate KPSS versions exist for level stationarity ($y_t = \mu + u_t$) and trend stationarity ($y_t = \mu + \beta t + u_t$). The statistic is based on the partial-sum process of residuals **(A)** after removing deterministic components. Long-run variance is typically estimated using the Newey–West HAC estimator **(B)** where $\hat{\gamma}_j$ is the sample autocovariance at lag j. Automatic lag selection often uses **(C)**. Under H_0, the statistic follows a nonstandard distribution derived from Brownian-motion functionals. Critical values: level-stationary (0.347, 0.463, 0.739 at 10%, 5%, 1%) or trend-stationary (0.119, 0.146, 0.216).

$$\textbf{(A)}: \; S_t = \sum_{i=1}^{t} \hat{e}_i, \qquad \textbf{(B)}: \; \hat{\lambda}^2(\ell) = \hat{\gamma}_0 + 2\sum_{j=1}^{\ell}\left(1 - \frac{j}{\ell+1}\right)\hat{\gamma}_j. \qquad \textbf{(C)}: \; \ell = \lfloor 4(T/100)^{2/9} \rfloor$$

Miscellanea: Denis Kwiatkowski, Peter Phillips, Peter Schmidt, and Yongcheol Shin (1992) introduced the KPSS test to complement traditional unit-root procedures by reversing the null hypothesis. Their insight was that failure to reject a unit root does not imply stationarity, motivating a direct test of stationarity itself.

Reference(s): Kwiatkowski et al. (1992); Hobijn et al. (2004)

Worked Example

Problem Statement A financial analyst tests whether a monthly stock price index is level-stationary. Unlike the ADF or PP tests, the KPSS test takes stationarity as H_0, making it a useful complement when unit root tests are inconclusive. The series spans $T = 14$ months and is tested under the level-stationarity specification ($\ell = 2$ Bartlett kernel).

$$y = \{100.0,\ 101.2,\ 102.8,\ 101.5,\ 103.4,\ 105.1,\ 106.3,\ 105.8,\ 107.2,\ 108.9,\ 108.1,\ 109.7,\ 111.2,\ 110.5\}$$

1. Parameters and Preliminary Quantities

Detrend by mean: $\hat{e}_t = y_t - \bar{y}$, with $\bar{y} = 105.836$.

t	$\hat{e}_t$	S_t	t	$\hat{e}_t$	S_t
1	−5.836	−5.836	8	−0.036	−20.586
2	−4.636	−10.471	9	+1.364	−19.221
3	−3.036	−13.507	10	+3.064	−16.157
4	−4.336	−17.843	11	+2.264	−13.893
5	−2.436	−20.279	12	+3.864	−10.029
6	−0.736	−21.014	13	+5.364	−4.664
7	+0.464	−20.550	14	+4.664	0.000

$$S_t = \sum_{t=1}^{T} \hat{e}_t, \qquad \sum_{t=1}^{T} S_t^2 = 3289.27$$

2. Long-Run Variance ($\ell = 2$ Bartlett NW)

- $\hat{\gamma}_0 = T^{-1} \sum \hat{e}_t^2 = 12.292$
- $\hat{\gamma}_1 = T^{-1} \sum \hat{e}_t \hat{e}_{t-1} = 9.418$
- $\hat{\gamma}_2 = T^{-1} \sum \hat{e}_t \hat{e}_{t-2} = 6.637$
- Bartlett weights: $w_1 = \frac{2}{3}$, $w_2 = \frac{1}{3}$

$$\hat{\lambda}^2(\ell) = \hat{\gamma}_0 + 2w_1\hat{\gamma}_1 + 2w_2\hat{\gamma}_2 = 29.274$$

3. KPSS Statistic

$$\eta_\mu = \frac{\sum_{t=1}^{T} S_t^2}{T^2\, \hat{\lambda}^2(\ell)} = \frac{3289.27}{196 \times 29.274} = 0.573$$

4. Results and Decision

Critical values at (level stationarity, KPSS 1992):

Statistic	Value	CV (5%)	Decision
η_μ	0.573	0.463	Reject H_0

For reference: CV at 10% = 0.347; CV at 1% = 0.739.

5. Interpretation

- **Decision:** Since $\eta_\mu = 0.573 > 0.463$, we **reject** H_0 of level stationarity at the 5% level (but not at 1%). The stock price index is non-stationary.
- **Practitioner Insight:** The KPSS test is the only test in this category where H_0 is stationarity — rejecting H_0 here is consistent with failing to reject H_0 in ADF/PP/SP/Ng-Perron tests on the same series. Note that $S_{14} \approx 0$ always, since $\sum_{t=1}^{T} \hat{e}_t = 0$ by construction of the mean-demeaned residuals. The large negative S_t values in the middle of the sample reflect the systematic upward drift of the series relative to its overall mean — precisely the signature of a random walk that KPSS detects. The bandwidth ℓ materially affects $\hat{\lambda}^2(\ell)$; increasing ℓ here would inflate the denominator and could push η_μ below the critical value, so sensitivity analysis across ℓ is advisable in practice.

Common Pitfalls

- **Misinterpreting the null hypothesis.** Unlike most unit root tests, KPSS takes *stationarity* as the null hypothesis and nonstationarity as the alternative. Rejecting the test therefore suggests a unit root, not stationarity.
- **Using KPSS as a stand-alone test.** KPSS is most informative when paired with unit-root tests such as ADF or Phillips–Perron. Joint interpretation helps distinguish cases of clear stationarity, clear unit roots, or inconclusive evidence.
- **Ignoring bandwidth selection in long-run variance estimation.** The KPSS statistic depends on a consistent estimate of the long-run variance using kernel methods. Bandwidth choice affects the test's size and power and should be selected carefully or reported with sensitivity checks.
- **Applying the test without considering structural breaks.** Structural breaks in level or trend can cause stationary series to appear nonstationary. Break-adjusted tests or segmented analysis should be considered when regime shifts are suspected.
- **Using incorrect deterministic specifications.** KPSS can be applied under either level stationarity (constant) or trend stationarity (constant plus trend). Misspecifying the deterministic components can distort inference.

Perron Test

Unit Root Test with Known Structural Break

STAT.8 | **Category:** Stationarity | **Data:** Time Series | **Null:** Tabled distribution

When to Use

- Test for unit roots when a structural break is suspected.
- Avoid spurious unit-root non-rejection from breaks.
- Evaluate series with known structural shifts in intercept.
- Analyze data with an exogenous change in trend slope.
- Avoid false non-stationarity findings due to regime shifts.

Assumptions

- Observations are recorded at fixed, uniform time intervals.
- Errors are weakly dependent and stationary.
- The error term variance remains constant across the horizon.
- Timing (T_B) of the structural break is known a priori.
- The break occurs instantly rather than over a transition.

Test Statistic

$$t_{\hat{\alpha}} = \frac{\hat{\alpha}}{SE(\hat{\alpha})} \quad \Big| \quad \Delta y_t = \mu + \beta t + \theta DU_t + \gamma DT_t + \delta DTB_t + \alpha y_{t-1} + \sum_{i=1}^{k} \phi_i \Delta y_{t-i} + \epsilon_t.$$

Notation: T_B is break point, $\alpha = \rho - 1$, $DU_t = 1$ if $t > T_B$ (level shift), $DT_t = t - T_B$ if $t > T_B$ (trend change), $DTB_t = 1$ if $t = T_B + 1$ (one-time shock)

Plain English: The Perron test extends the ADF framework by including dummy variables that capture structural breaks at a predetermined time point. Standard unit root tests have low power against trend-stationary alternatives with breaks, often incorrectly concluding a unit root exists. By explicitly modeling the break, this test distinguishes between true non-stationarity and apparent non-stationarity caused by regime changes. Three break models accommodate different types of structural changes.

⚠ Practitioner Warning

Break date T_B must be specified exogenously based on economic/historical knowledge, not by searching for minimum test statistic (data-dependent selection invalidates critical values). If break timing is uncertain, use Zivot-Andrews or similar tests that endogenize break selection. Critical values (CVs) differ substantially from standard ADF depending on break type and relative break location $\lambda = T_B/T$. Test has low power when break is small relative to noise.

Hypotheses

- H_0: $\alpha = 0$ equivalently $\rho = 1$ (unit root).
- H_1: $\alpha < 0$ (stationary around broken trend/level).

Decision Rule: Reject H_0 if $t_\alpha < t_{\alpha,\text{crit}}$ or p-value $< \alpha$. Critical values in Perron (1989) more negative than ADF.

Software References [Python, R, SAS]

Python	R	SAS
`Not Native (Manual)`	`urca::ur.pp (modified)`	`Not Native (Manual)`

Additional Notes

Perron (1989) proposed three principal structural break models: (A) the Crash model, allowing an abrupt intercept shift through DU_t; (B) the Changing Growth model, allowing a change in slope through DT_t; and (C) a hybrid model permitting simultaneous level and trend changes. Each specification can be formulated under either an Innovational Outlier (IO) framework, where the break propagates gradually through the error process, or an Additive Outlier (AO) framework, where the deterministic component shifts instantaneously at the break date. CVs depend strongly on the break fraction $\lambda = T_B/T$, with breaks near sample endpoints producing different asymptotic behavior than mid-sample breaks. Perron (1997) later extended the framework to allow unknown break dates selected endogenously from the data.

Miscellanea: Pierre Perron (1989) showed that structural breaks, such as the Great Depression and oil shocks, can lead standard unit root tests to incorrectly indicate nonstationarity. By incorporating exogenous break points, his work challenged the prevailing view that many macroeconomic series contain unit roots. Criticism of fixed break dates led to the development of tests with endogenous break selection, such as Zivot–Andrews.

Reference(s): Perron (1989, 1997); Vogelsang & Perron (1998)

Worked Example

Problem Statement An economist tests whether a monthly industrial production index is trend-stationary, allowing for a known structural break at period $T_b = 10$ (a policy change). The series spans $T = 20$ observations. The Perron **Model C** (crash + slope change) is used, with break fraction $\lambda = T_b/T = 0.50$ and no lag augmentation ($k = 0$).

$$y = \{2.6, 2.2, 3.4, 2.9, 3.9, 3.3, 4.4, 4.0, 5.2, 4.8, 2.5, 1.7, 2.6, 2.0, 3.0, 2.3, 3.1, 2.3, 3.2, 2.7\}$$

1. Dummy Variables (Model C)
Three dummies capture the break at $T_b = 10$:

- $DU_t = \mathbf{1}(t > T_b)$ (level shift)
- $DT_t = (t - T_b) \cdot \mathbf{1}(t > T_b)$ (slope change)
- $DTB_t = \mathbf{1}(t = T_b + 1)$ (impulse at break)

2. Step 1 — Detrending Regression
OLS of y_t on $[1, t, DU_t, DT_t, DTB_t]$:

$$\hat{\mu} = 2.080, \quad \hat{\beta} = 0.289, \quad \hat{\theta} = -3.076$$

$$\hat{\gamma} = -0.181, \quad \hat{\delta} = 0.497$$

Selected OLS residuals $\hat{e}_t$:

- $\hat{e}_1 = 0.231$, $\hat{e}_{10} = -0.171$
- $\hat{e}_{11} = 0.000$ (by construction at break)
- $\hat{e}_{20} = -0.278$

3. Step 2 — ADF on Residuals ($k = 0$)
Regress $\Delta\hat{e}_t = \alpha\,\hat{e}_{t-1} + v_t$:

$$\hat{\alpha} = \frac{\sum \hat{e}_{t-1}\Delta\hat{e}_t}{\sum \hat{e}_{t-1}^2} = \frac{-5.356}{2.847} = -1.881$$

$$\hat{s}^2 = 0.0367, \qquad SE(\hat{\alpha}) = \sqrt{\frac{\hat{s}^2}{\sum \hat{e}_{t-1}^2}} = 0.114$$

$$t_{\hat{\alpha}} = \frac{\hat{\alpha}}{SE(\hat{\alpha})} = \frac{-1.881}{0.114} = -16.57$$

4. Results and Decision
Critical values (Model C, $\lambda = 0.50$, Perron 1989):

Statistic	Value	CV (5%)	Decision
$t_{\hat{\alpha}}$	−16.57	−4.24	Reject H_0

For reference: CV at 10% = −3.96; CV at 1% = −4.90.

5. Interpretation

- **Statistical Decision:** Since $t_{\hat{\alpha}} = -16.57 \ll -4.24$, we **reject** H_0. The series is trend-stationary around a broken trend.
- **Practitioner Insight:** The Perron test requires the break date T_b to be specified *a priori* — using a break date chosen by inspecting the data inflates the rejection rate and invalidates the critical values (use Zivot-Andrews instead if the break is unknown). The three Model C coefficients carry direct interpretations: $\hat{\theta} = -3.076$ estimates the permanent level drop after the policy change, $\hat{\gamma} = -0.181$ the reduction in trend slope, and $\hat{\delta} = 0.497$ the one-period impulse at the break itself. Here $k = 0$ is used for transparency; in practice k is chosen by AIC/BIC, and a larger k will reduce $|t_{\hat{\alpha}}|$ toward more moderate values while leaving the critical values unchanged.

Common Pitfalls

- **Searching for the break date using the same data.** The Perron test assumes the break date T_B is known *a priori* from institutional or historical information. Selecting the break that maximizes the test statistic introduces severe pre-testing bias; use Zivot–Andrews or related procedures when the break is unknown.
- **Misspecifying the type of structural break.** Perron models allow different break structures: intercept (Model A), slope (Model B), or both (Model C). Including unnecessary components reduces power, while omitting relevant ones biases the test toward non-rejection.
- **Ignoring the effect of break timing.** The distribution of the test statistic depends on the break fraction $\lambda = T_B/T$. Breaks near the beginning or end of the sample provide less information and require different critical values.
- **Concluding stationarity without verifying the break effect.** Rejecting the unit root null implies trend-stationarity with a break only if the break parameters themselves are statistically meaningful. The significance of level or trend break coefficients should be examined before attributing structural change.
- **Applying the test when multiple breaks are present.** The Perron framework allows only a single structural break. When multiple regime shifts occur, tests designed for multiple breaks (e.g., Bai–Perron) are more appropriate.

Zivot-Andrews Test

Unit Root Test with Endogenous Single Structural Break

STAT.9 | **Category:** Stationarity | **Data:** Time Series | **Null:** Tabled distribution

When to Use

- Unit-root testing with unknown structural breaks.
- Time series with possible regime or trend shifts.
- Endogenously determine structural break dates.
- Avoid spurious unit-root non-rejection from breaks.
- Preliminary check before cointegration analysis.

Assumptions

- Observations are recorded at fixed, uniform time intervals.
- The error terms are serially correlated but stationary.
- The data features at most a single structural break point.
- The structural break date is chosen endogenously by data.
- The underlying break occurs instantly at the chosen date.

Test Statistic

$$t_{\min} = \min_{T_B} t_{\hat{\alpha}}(T_B),\ T_B \in [\tau_1 T, \tau_2 T] \quad \Big| \quad \Delta y_t = \mu + \beta t + \theta DU_t(T_B) + \gamma DT_t(T_B) + \alpha y_{t-1} + \sum_{i=1}^{k} \phi_i \Delta y_{t-i} + \epsilon_t.$$

Notation: T_B is the candidate break date, $\hat{T}_B$ is the endogenous break date estimate, $DU_t(T_B) = \mathbf{1}(t > T_B)$ captures an intercept shift, $DT_t(T_B) = (t - T_B)\mathbf{1}(t > T_B)$ captures a trend shift, $t_{\hat{\alpha}}(T_B) = \hat{\alpha}/\mathrm{SE}(\hat{\alpha})$ is the standard OLS t-ratio

Plain English: The Zivot–Andrews test extends Perron's structural-break framework by treating the break date as unknown rather than fixed in advance. The procedure searches over a trimmed range of possible break dates and selects the one producing the most negative unit-root test statistic. This endogenous break selection avoids the criticism that break dates may be chosen subjectively after inspecting the data. Because the break date is estimated from the sample itself, the resulting test statistic follows a different distribution and requires more negative critical values than standard ADF or Perron tests.

⚠ Practitioner Warning

Critical values are substantially more negative than standard ADF values because the procedure selects the most negative statistic across many candidate break dates. Using ADF or Perron CVs can therefore cause severe over-rejection. The test is computationally intensive since the regression must be re-estimated for each admissible break point. Zivot–Andrews assumes a single structural break; multiple-break settings require extensions such as Lumsdaine–Papell or Clemente–Montañés–Reyes tests. Wider trimming regions improve endpoint stability but may reduce power.

Hypotheses

- H_0: $\alpha = 0$ (unit root).
- H_1: $\alpha < 0$ (trend-stationary with one structural break).

Decision Rule: Reject H_0 if $t_{\min} < t^{ZA}_{\alpha,\text{crit}}$ or p-value $< \alpha$.
CVs ≈ −4.80 (A), −4.42 (B), −5.08 (C) at 5% (asymp.).

Software References [Python, R, SAS]

Python	R	SAS
`arch.unitroot.ZivotAndrews`	`urca::ur.za`	`Not Native (Manual)`

Additional Notes

Three standard specifications are commonly used: Model A allows a break in intercept only (DU_t), Model B allows a break in trend slope only (DT_t), and Model C allows simultaneous breaks in both intercept and trend. Break-date search is typically restricted to the trimmed range ($0.15T$, $0.85T$) to reduce endpoint instability. Lag selection may be based on information criteria (AIC/BIC) or fixed maximum-lag procedures for each candidate break date. Although the estimated break date $\hat{T}_B$ is often reported, it may be imprecisely estimated in finite samples. Finite- sample critical values from Perron–Vogelsang (1992) are generally more accurate than purely asymptotic approximations.

Miscellanea: Eric Zivot and Donald Andrews (1992) developed a unit root test with an endogenous structural break, motivated by concerns that earlier approaches, such as Perron's, relied on exogenously chosen break dates that could introduce bias. By allowing the data to determine the break point, their method reduced pre-test bias and became widely used in time series analysis.

Reference(s): Zivot & Andrews (1992); Perron & Vogelsang (1992)

Worked Example

Problem Statement An economist tests whether a monthly industrial output index is stationary around a trend with an *unknown* structural break. Unlike the Perron test, the break date is determined endogenously by searching across all candidate dates. **Model A** (intercept break only) is used with $T = 20$, one lag augmentation ($k = 1$), and a 15% trim on each end (candidates $T_B = 3 \ldots 18$).

$$y = \{5.1,\ 4.2,\ 6.2,\ 5.5,\ 7.3,\ 6.4,\ 8.0,\ 7.3,\ 9.3,\ 8.6,\ 10.1,\ 13.2,\ 15.2,\ 14.5,\ 16.3,\ 15.4,\ 17.0,\ 16.3,\ 18.3,\ 17.5\}$$

T_B	3	5	7	9	10	**11**	12	14	16	18
$\hat{\alpha}$	−0.518	−0.559	−0.416	−0.400	−0.536	**−1.017**	−1.963	−0.205	−0.455	−0.484
$t(\hat{\alpha})$	−1.637	−1.617	−1.168	−1.354	−2.135	**−6.753**	−3.954	−0.580	−1.529	−1.558

1. ZA Model A Regression (per candidate T_B)
For each T_B, the following ADF-type regression is estimated:

$$\Delta y_t = \mu + \beta t + \theta DU_t(T_B) + \alpha y_{t-1} + \phi_1 \Delta y_{t-1} + \varepsilon_t$$

where $DU_t(T_B) = \mathbf{1}(t > T_B)$. The t-statistic on $\hat{\alpha}$ is recorded at each T_B (see table above).

2. Optimal Break and Final Regression
The minimum $t(\hat{\alpha})$ is found at $T_B^* = 11$ ($\hat{\lambda}^* = 11/20 = 0.55$):

$$\hat{\mu} = 4.481,\ \hat{\beta} = 0.487,\ \hat{\theta} = 4.335,\ \hat{\alpha} = -1.017,\ \hat{\phi}_1 = -0.357$$

$$SE(\hat{\alpha}) = 0.151, \qquad t^* = \frac{-1.017}{0.151} = -6.753$$

3. Results and Decision
Critical values (Model A, ZA 1992 Table 2):

Statistic	Value	CV (5%)	Decision
$\inf_\lambda\ t(\hat{\alpha})$	−6.753	−4.800	Reject H_0

4. Interpretation

- **Decision:** Since $t^* = -6.753 < -4.800$, we **reject** H_0 of a unit root. The series is trend-stationary with a level break at $T_B^* = 11$.
- **Practitioner Insight:** The key distinction from the Perron test is that $T_B^* = 11$ is *discovered* from the data, not assumed in advance — the sharp drop in $t(\hat{\alpha})$ from −2.1 at $T_B = 10$ to −6.8 at $T_B = 11$ marks the structural break precisely. The critical values (−4.80 at 5%) are more negative than standard ADF values to account for the pre-testing bias from searching across all candidates; using standard ADF critical values here would severely over-reject. The $\hat{\theta} = 4.335$ estimates the permanent level shift at $T_B^* = 11$, directly interpretable as the magnitude of the structural change. Trim boundaries (15% each end) protect against spurious breaks detected in data-sparse regions near the sample edges.

For reference: CV at 10% = −4.58; CV at 1% = −5.34.

Common Pitfalls

- **Using standard DF or Perron critical values.** Because the break date is chosen endogenously, the test statistic has a more extreme distribution than standard unit root tests. Always use Zivot–Andrews critical values corresponding to the chosen break specification.
- **Over-interpreting the estimated break date.** The estimated break $\hat{T}_B$ is selected because it produces the most negative test statistic, not necessarily because it represents the true structural break. Inference should focus primarily on whether the unit root null is rejected.
- **Applying the test when multiple breaks exist.** The Zivot–Andrews procedure allows only a single structural break. When several regime shifts are present, alternative tests designed for multiple breaks should be used.
- **Searching for breaks at the sample endpoints.** Breaks very near the beginning or end of the sample leave too few observations in one regime. Standard practice trims the search region (e.g., 10–15% at each end) to maintain reliable inference.
- **Interpreting non-rejection as evidence of no structural break.** Failure to reject H_0 does not imply that no break exists. The series may contain breaks but still behave as a unit-root process or the test may lack sufficient power.

Clemente-Montañés-Reyes Test

Unit Root Test with Two Endogenous Structural Breaks

STAT.10 | **Category:** Stationarity | **Data:** Time Series | **Null:** Tabled distribution

When to Use

- Test for unit roots when double structural breaks occur.
- Apply when structural break dates are completely unknown.
- Evaluate series with two sudden shifts in intercept levels.
- Analyze data with two gradual, smooth shifts in the trend.
- Prevent false unit-root findings under double regimes.

Assumptions

- The series contains a maximum of two structural breaks.
- The dual break dates are endogenously chosen by data.
- The error innovations follow a stationary ARMA process.
- Observations are recorded at fixed, uniform time intervals.
- The regime shifts follow an additive or innovative path.

Test Statistic

$$t_{\min} = \min_{T_{b1}, T_{b2} \in \Lambda} t_{\hat{\alpha}}(T_{b1}, T_{b2}) \quad \Big| \quad \Delta y_t = \mu + \sum_{i=1}^{2} \theta_i DU_{it} + \sum_{i=1}^{2} \gamma_i DTB_{it} + \alpha y_{t-1} + \sum_{j=1}^{k} \phi_j \Delta y_{t-j} + \epsilon_t.$$

Notation: T_{b1}, T_{b2} are break points, $DU_{it} = 1$ if $t > T_{bi}$ (level shift), $DTB_{it} = 1$ if $t = T_{bi} + 1$ (one-time shock), $t_{\hat{\alpha}}(T_{b1}, T_{b2}) = \hat{\alpha}/\text{SE}(\hat{\alpha})$

Plain English: The Clemente-Montañés-Reyes test extends endogenous break detection to two breaks by searching over all possible break date combinations within a trimmed range. Two models are available: the Additive Outlier (AO) model assumes immediate level shifts, while the Innovational Outlier (IO) model allows gradual changes through the dynamics. The test selects break dates that provide strongest evidence against the unit root hypothesis.

⚠ Practitioner Warning

The test is computationally intensive because it requires a grid search over pairs of candidate break dates. Minimum spacing between breaks is typically imposed to avoid spurious detection of nearby breaks. Critical values are more negative than those of single-break procedures such as Zivot–Andrews due to the additional break-date search dimension. Separate critical values are required for the Additive Outlier (AO) and Innovational Outlier (IO) specifications. Power may deteriorate in short samples, when breaks are small, or when breaks occur near sample endpoints.

Hypotheses

- H_0: $\alpha = 0$ (nonstationary with structural mean break(s)).
- H_1: $\alpha < 0$ (stationary process around shifted means).

Decision Rule: Reject H_0 if $t_{\min} < C_\alpha$ or p-value $< \alpha$. Use CMR (1998) critical values.

Software References [Python, R, SAS]

Python	R	SAS
`Not Native (Manual)`	`Not Native (Manual)`	`Not Native (Manual)`

Additional Notes

The CMR procedure searches jointly over two break fractions $\lambda_1 = T_{b1}/T$ and $\lambda_2 = T_{b2}/T$, selecting the pair that minimizes the unit-root test statistic. Standard implementations trim the search region (commonly excluding the first and last 15% of the sample) and impose minimum spacing between breaks to avoid unstable or overlapping estimates. The Additive Outlier (AO) specification models abrupt level shifts through impulse-type break dummies, whereas the Innovational Outlier (IO) specification allows break effects to propagate gradually through the autoregressive dynamics. In practice, AO models are often used for sudden shocks or regime changes, while IO models are more appropriate for gradual adjustment processes. Lee–Strazicich (2003) later developed LM-based multiple-break tests designed to avoid spurious rejections under the null with structural breaks.

Miscellanea: Jesús Clemente, Antonio Montañés, and Marcelo Reyes (1998) extended the Perron framework to allow two structural breaks in the mean of a time series rather than just one. Their work was motivated by macroeconomic series that experienced multiple major shocks, such as oil crises, wars, or policy regime changes, where single-break tests often falsely indicated a unit root simply because one break was insufficient.

Reference(s): Clemente et al. (1998); Lee & Strazicich (2003)

Worked Example

Problem Statement An economist tests whether a quarterly interest rate series is stationary, allowing for two unknown instantaneous level shifts. The CMR **Additive Outlier (AO)** model is used. $T = 20$, $k = 0$; break candidates trimmed to $T_{b1}, T_{b2} \in \{3, \dots, 18\}$ with $T_{b2} > T_{b1} + 1$.

$$y = \{5.1,\ 4.8,\ 5.3,\ 5.0,\ 4.7,\ 5.2,\ 10.1,\ 9.8,\ 10.3,\ 10.0,\ 9.7,\ 10.2,\ 10.5,\ 7.1,\ 6.8,\ 7.3,\ 7.0,\ 6.7,\ 7.2,\ 7.5\}$$

1. Two-Step AO Procedure (per candidate pair)

Step 1 — OLS of y_t on deterministic components:

$$y_t = \mu + \theta_1 DU_{1t} + \theta_2 DU_{2t} + z_t$$

where $DU_{jt} = \mathbf{1}(t > T_{bj})$. Residuals $\hat{z}_t$ are retained.

Step 2 — ADF on $\hat{z}_t$ (no intercept/trend, $k = 0$):

$$\Delta\hat{z}_t = \alpha\,\hat{z}_{t-1} + \varepsilon_t$$

Record $t(\hat{\alpha})$ at each (T_{b1}, T_{b2}) pair.

Snapshot of Double-minimization grid $t(\hat{\alpha})$:

$T_{b1} \backslash T_{b2}$	11	12	**13**	14	15
5	−3.28	−3.91	−4.16	−4.31	−3.67
6	−2.76	−4.14	**−4.89**	−4.60	−2.96
7	−3.57	−4.28	−4.45	−4.22	−3.45
8	−2.57	−2.85	−2.57	−2.75	−2.48

Min. at $(T^*_{b1}, T^*_{b2}) = (6, 13)$, $\hat{\lambda}^*_1 = 0.30$, $\hat{\lambda}^*_2 = 0.65$.

2. Optimal-Break Regression

At $(T^*_{b1}, T^*_{b2}) = (6, 13)$, Step 1 OLS yields:

$$\hat{\mu} = 5.017, \quad \hat{\theta}_1 = +5.069, \quad \hat{\theta}_2 = -3.000.$$

The residuals $\hat{y}_t$ have zero mean by construction.

Step 2 ADF on $\hat{z}_t$ ($k = 0$, no intercept):

$$\hat{\alpha} = \frac{\sum \hat{z}_{t-1}\Delta\hat{z}_t}{\sum \hat{z}^2_{t-1}} = \frac{-1.2569}{1.0338} = -1.2158,$$

$$RSS = 1.1504, \quad s^2 = RSS/18 = 0.06391,$$

$$SE(\hat{\alpha}) = \sqrt{\frac{s^2}{\sum \hat{z}^2_{t-1}}} = 0.2486, \quad t^* = \frac{\hat{\alpha}}{SE(\hat{\alpha})} = -4.89$$

3. Results and Decision

Critical values (CMR 1998, Table 2, AO model, $T = 50$, $k = 0$):

Statistic	Value	CV (5%)	Decision
$\min t(\hat{\alpha})$	−4.89	−5.60	FTR H_0

For reference: CV at 10% = −5.28; CV at 1% = −6.32.

4. Interpretation

- **Decision:** The series is consistent with a unit root with two level shifts at $t = 6$, and 13.
- **Practitioner Insight:** The CMR model includes only an intercept and two step-dummy regressors in Step 1 — no time trend — because the test targets mean-shift stationarity rather than trend stationarity. Adding a trend term would misspecify the model and distort the residual-based statistics. The minimum statistic $t^* = -4.89$ at $(\hat{\lambda}_1, \hat{\lambda}_2) = (0.30, 0.65)$ is less extreme than comparable Zivot–Andrews values because searching over two break dates imposes a larger pre-testing penalty, making CMR critical values more negative (about −5.5 at 5%). The estimated shifts $\hat{\delta}_1 = +5.07$ and $\hat{\delta}_2 = -3.00$ indicate a sharp upward shift followed by a partial reversal, but with only $T = 20$ observations the evidence remains insufficient to reject the unit-root null.

Common Pitfalls

- **Confusing additive outlier (AO) and innovations outlier (IO) specifications.** The AO model assumes an abrupt level shift, whereas the IO model assumes the break propagates gradually through the innovation process. Choice should reflect the expected nature of the structural change.
- **Over-interpreting the estimated break dates.** Break dates are selected because they minimize the test statistic, not necessarily because they represent the true structural change. The primary inference concerns rejection of the unit root, not the exact timing of the breaks.
- **Applying the test when more than two breaks are present.** The CMR framework allows exactly two structural breaks. When several regime shifts occur, procedures designed for multiple breaks (e.g., Bai–Perron) are more appropriate.
- **Allowing candidate breaks too close together.** Breaks occurring too near each other create regimes with too few obs., reducing reliability. Standard practice imposes min. spacing between candidate break dates.
- **Using computational shortcuts without recognizing their limitations.** Sequential break searches are faster but may fail to identify the globally optimal pair of break dates. Exhaustive search provides more reliable inference.

Lee-Strazicich Test

LM-Based Unit Root Test with Endogenous Breaks

STAT.11 | **Category:** Stationarity | **Data:** Time Series | **Null:** Tabled distribution

When to Use

- Test for unit roots with breaks under the alternative.
- Avoid spurious unit-root non-rejection from breaks.
- Evaluate series with up to two unknown endogenous breaks.
- Analyze data with shifts in both the intercept and trend.
- Prevent size distortions common to Zivot-Andrews tests.

Assumptions

- Observations are recorded at fixed, uniform time intervals.
- The unit-root null hypothesis includes structural breaks.
- The dual break dates are endogenously chosen by data.
- The error terms are serially correlated but stationary.
- Break model (level, trend, or both) correctly specified.

Test Statistic

$$LM_\tau = \inf_\lambda \tilde{\tau}(\lambda),\ \lambda = \left(\frac{T_{b1}}{T}, \frac{T_{b2}}{T}\right), \quad \Big| \quad \Delta y_t = \tilde{\delta}'\Delta Z_t + \phi \tilde{S}_{t-1} + \sum_{i=1}^{k} c_i \Delta \tilde{S}_{t-i} + u_t \quad \Big| \quad \tilde{S}_t = y_t - \tilde{\psi}_x - Z_t\tilde{\delta}, \quad \tilde{\psi}_x = y_1 - Z_1\tilde{\delta}$$

Notation: For break dates $T_{b1} < T_{b2}$: $D_{jt} = \mathbf{1}(t \geq T_{bj} + 1)$ captures level shifts, $DT_{jt} = (t - T_{bj})\mathbf{1}(t \geq T_{bj} + 1)$ captures trend shifts, $Z_t = [1,\ t,\ D_{1t},\ D_{2t},\ DT_{1t},\ DT_{2t}]'$, and k is the lag augmentation order used to whiten residuals, $\tilde{\tau} = \hat{\phi}/SE(\hat{\phi})$. **Note:** The alternative LM_ρ statistic ($T\hat{\phi}$) is rarely reported in practice; most applications use LM_τ.

Plain English: The Lee-Strazicich test uses a Lagrange Multiplier (LM) framework rather than earlier ADF-type break tests, allowing structural breaks under both the null and alternative hypotheses. This prevents spurious rejections that occur in ADF-type tests when the null hypothesis itself contains a break. The test searches for break dates that minimize the unit root statistic (τ), providing valid inference regardless of whether breaks exist in the data-generating process.

⚠ Practitioner Warning

Computational burden can be substantial due to the grid search over one or two endogenous break dates. Critical values depend on the break specification and number of breaks; using incorrect tables invalidates inference. Although the test allows breaks under both H_0 and H_1, distinguishing a unit root with breaks from a stationary process with large shifts may still be difficult in finite samples. Power declines when breaks are small relative to innovation variance or when break dates occur near sample endpoints. Results can also be sensitive to trimming choices and lag selection.

Hypotheses

- H_0: $\phi = 0$ (unit root with structural break(s).
- H_1: $\phi < 0$ (trend-stationary around structural break(s).

Decision Rule: Reject H_0 if $LM_\tau < \tau_{\alpha,\text{crit}}$ or p-value $< \alpha$. Use Lee & Strazicich (2003) for critical values.

Software References [Python, R, SAS]

Python	R	SAS
`Not Native (Manual)`	`Not Native (Manual)`	`Not Native (Manual)`

Additional Notes

Two main models: Model A (one break in level): $Z_t = [1, t, D_{1t}]'$ where $D_{1t} = 1$ if $t > T_{B1}$. Model C (two breaks in level and trend): $Z_t = [1, t, D_{1t}, D_{2t}, DT_{1t}, DT_{2t}]'$. Additional models for breaks in level only or trend only. The LM framework tests $\phi = 0$ in data de-trended under the null, whereas ADF-type tests (Zivot-Andrews, CMR) impose unit root only under null but stationarity under alternative's break specification. This asymmetry causes spurious rejections when true process is unit root with break. Lee-Strazicich (2004) extended to two-break case with comprehensive Monte Carlo evidence.

Miscellanea: Junsoo Lee and Mark Strazicich developed this test after recognizing that many earlier endogenous-break unit-root tests could spuriously reject the null because structural breaks were excluded under H_0. Their LM framework incorporated breaks under both the null and alternative, greatly reducing this size-distortion problem. Although methodologically influential, the test's computational burden limited widespread adoption relative to simpler ADF-type break tests.

Reference(s): Lee & Strazicich (2003, 2004); Nunes et al. (1997)

Worked Example

Problem Statement An economist tests whether a quarterly productivity index is stationary, allowing for two unknown structural breaks. The Lee-Strazicich **Model AA** (two level breaks) is used. Unlike ZA/CMR, the LS null hypothesis is a *unit root with breaks*, so rejection cannot be attributed to breaks alone. $T = 20$, $k = 0$, trim $\{3, \ldots, 18\}$.

$y = \{5.8, 5.3, 6.8, 6.2, 7.5, 6.8, 8.2, 11.9, 13.1, 12.6, 14.0, 13.3, 14.6, 14.0, 13.0, 12.5, 13.6, 14.2, 13.7, 15.1\}$

1. LM Three-Step Procedure (per candidate pair)

Step 1 — Regress Δy_t on ΔZ_t to obtain $\tilde{\delta}$, where $Z_t = [1, t, D_{1t}, D_{2t}]$, $D_{jt} = \mathbf{1}(t \geq T_{bj} + 1)$.

Step 2 — Construct detrended series:

$$\tilde{S}_t = y_t - \tilde{\psi}_x - Z_t\tilde{\delta}, \quad \tilde{\psi}_x = y_1 - Z_1\tilde{\delta}$$

Note: $\tilde{S}_1 = \tilde{S}_T = 0$ by construction.

Step 3 — Regress Δy_t on $[\Delta Z_t, \tilde{S}_{t-1}]$ and test $H_0 : \phi = 0$.

Snapshot of Double-minimization grid $t(\hat{\phi})$:

$T_{b1} \backslash T_{b2}$	12	13	**14**	15	16
5	−1.27	−1.49	−1.53	−1.48	−1.25
6	−1.70	−2.21	−2.32	−2.16	−1.68
7	−1.94	−3.45	**−4.04**	−3.32	−2.07
8	−2.04	−2.18	−2.27	−2.13	−1.68
9	−1.69	−1.63	−1.65	−1.60	−1.35

Minimum at $(T_{b1}^*, T_{b2}^*) = (7, 14)$, $\hat{\lambda}_1^* = 0.35$, $\hat{\lambda}_2^* = 0.70$.

2. Optimal-Break Calculations

At $(T_{b1}^*, T_{b2}^*) = (7, 14)$, Step 1 yields:

$$\tilde{\delta} = [0, \ \ 0.388, \ \ 3.312, \ \ -1.388]$$

($\tilde{\delta}_3 = 3.312$: 1st level shift; $\tilde{\delta}_4 = -1.388$: 2nd level shift.)

Step 2 detrending: $\tilde{\psi}_x = 5.412$. Selected $\tilde{S}_t$:

- $\tilde{S}_1 = 0$, $\tilde{S}_{20} = 0$ (by construction)
- $\tilde{S}_7 = 0.071$, $\tilde{S}_{14} = -0.159$

Step 3 regression (Δy_t on $[\Delta Z_t, \tilde{S}_{t-1}]$):

$$\hat{\phi} = -1.076, \quad SE(\hat{\phi}) = 0.267, \quad t^* = \frac{-1.076}{0.267} = -4.039$$

3. Results and Decision

Critical values (LS 2003, Model AA, $\lambda_1 \approx 0.4$, $\lambda_2 \approx 0.6$):

Statistic	Value	CV (5%)	Decision
$\min t(\hat{\phi})$	−4.039	−3.842	Reject H_0

For reference: CV at 10% = −3.504; CV at 1% = −4.545.

4. Interpretation

- **Decision:** The series is level-stationary around two structural breaks at $t = 7$ and $t = 14$.
- **Practitioner Insight:** Rejection cannot be driven by breaks alone, eliminating the spurious-rejection problem of ZA/CMR in trending data. The $\tilde{S}_1 = \tilde{S}_T = 0$ property is a built-in arithmetic check on every regression in the grid. Critical values are tabulated by (λ_1, λ_2) in LS (2003) Table 2; use the row nearest to $(\hat{\lambda}_1^*, \hat{\lambda}_2^*)$ — interpolation is acceptable for applied work. The $\tilde{\delta}$ coefficients from Step 1 directly estimate the break magnitudes: +3.31 and −1.39.

Common Pitfalls

- **Treating Lee-Strazicich and Zivot-Andrews tests as interchangeable.** The two tests have different null hypotheses. Zivot–Andrews assumes a unit root *without* breaks, whereas Lee-Strazicich allows breaks under the null. As a result, Lee–Strazicich is generally provides better size control and avoids certain spurious rejections.
- **Misinterpreting the notion of "spurious rejection."** A Zivot–Andrews rejection may occur when the true process is a unit root with structural breaks, because the ZA null excludes breaks. The Lee-Strazicich test addresses this by incorporating breaks into the null specification.
- **Ignoring the LM framework underlying the test.** The Lee-Strazicich procedure is an LM test constructed from the restricted model rather than a Dickey–Fuller–type regression. Correct implementation requires estimating break parameters under the null and detrending the series accordingly.
- **Applying the test when more breaks exist than modeled.** Standard versions allow one or two structural breaks only. If additional regime shifts occur, the test becomes misspecified and other multi-break procedures should be used.
- **Assuming the test always has lower power than Zivot–Andrews.** While the test is designed to control size when breaks occur under the null, its power relative to ZA depends on the true data-generating process. The choice between tests should therefore depend on the hypothesized structure of the series.

HEGY Test

Seasonal Unit Root Test at Multiple Frequencies

STAT.12 | **Category:** Stationarity | **Data:** Time Series | **Null:** Tabled distribution

When to Use

- Test for unit roots at zero and seasonal frequencies.
- Apply when evaluating quarterly or monthly time series.
- Identify specific seasons driving non-stationary paths.
- Determine if seasonal differencing is truly required.
- Establish foundational baselines for seasonal cointegration.

Assumptions

- Observations are recorded at fixed, uniform time intervals.
- Structural innovation variance remains constant over time.
- Errors are weakly dependent and stationary.
- The data exhibits a fixed, regular seasonal periodicity.
- The underlying series contains no structural regime breaks.

Test Statistic

For Quarterly data: $t(\hat{\pi}_1) = \dfrac{\hat{\pi}_1}{\text{SE}(\hat{\pi}_1)}$ | $t(\hat{\pi}_2) = \dfrac{\hat{\pi}_2}{\text{SE}(\hat{\pi}_2)}$ | $F(\hat{\pi}_3, \hat{\pi}_4) = \dfrac{(\text{RSS}_R - \text{RSS}_U)/2}{\text{RSS}_U/(T-K)}$

$$\Delta_4 y_t = \mu_t + \pi_1 y_{1,t-1} + \pi_2 y_{2,t-1} + \pi_3 y_{3,t-1} + \pi_4 y_{4,t-1} + \textstyle\sum_{i=1}^{k} \phi_i \Delta_4 y_{t-i} + \epsilon_t$$

Notation: seasonally differenced series: $\Delta_4 y_t = y_t - y_{t-4}$; auxiliary series: $y_{1,t-1}$, $y_{2,t-1}$, $y_{3,t-1}$, $y_{4,t-1}$; seasonal dummies + time trend: $\mu_t = \sum_{j=1}^{4} \alpha_j D_{jt} + \beta t$; lag augmentation: $\sum \phi_i \Delta_4 y_{t-i}$ to whiten residuals

Plain English: HEGY extends standard unit root testing to seasonal data by examining unit roots at multiple frequencies: zero frequency (non-seasonal), annual, and harmonics. For quarterly data ($S = 4$), it tests four unit roots corresponding to frequencies 0, π, $\pi/2$, and $3\pi/2$. Each frequency corresponds to different seasonal patterns—the test determines which frequencies exhibit non-stationary behavior requiring differencing versus stationary patterns needing only seasonal dummies.

⚠ Practitioner Warning

Interpretation requires understanding frequency domain concepts. Multiple testing (one per frequency) without adjustment can inflate Type I error—joint tests recommended for overall seasonal unit root presence. Critical values differ across frequencies and depend on deterministic specification. For monthly data ($S = 12$), test involves 12 frequencies and becomes cumbersome. Insufficient lag augmentation leaves residual autocorrelation, causing substantial size distortions.

Hypotheses

- H_0: Unit roots exist at one or more seasonal frequencies.
- H_1: No unit roots at seasonal frequencies.

Decision Rule: Reject H_0 if $|t_{\pi_i}| > t^{(i)}_{\alpha,\text{crit}}$ (for π frequency) or $F_{\pi_i,\pi_{i+1}} > F_{\alpha,\text{crit}}$ (for harmonic pairs).

Software References [Python, R, SAS]

`statsmodels::HEGY` | `uroot::hegy.test` | `(PROC ARIMA (HEGY)`

Additional Notes

General HEGY representation: $\phi(B)(1-B^S)y_t = \mu_t + \sum_{i=1}^{S} \pi_i y^{(i)}_{S,t-1} + \epsilon_t$, where the auxiliary variables $y^{(i)}_{S,t}$ isolate distinct seasonal frequencies. For quarterly data ($S = 4$) auxiliary variables: $y^{(1)}_{4,t} = (1 + B + B^2 + B^3)y_t$ (Zero Frequency / Long-run Root), $y^{(2)}_{4,t} = -(1 - B + B^2 - B^3)y_t$ (Semi-annual Frequency / π Root), $y^{(3)}_{4,t} = -(1 - B^2)y_t$ (Quarterly Frequency / Real component), $y^{(4)}_{4,t} = (1 - B^2)By_t$ (Quarterly Frequency / Lagged component). Rejection at all frequencies implies stationarity; failure only at the zero-frequency root suggests ordinary differencing, whereas failure at seasonal frequencies indicates seasonal differencing may be required. Beaulieu–Miron (1993) extended the test to monthly data.

Miscellanea: Svend Hylleberg, Robert Engle, Clive Granger, and Byung Sam Yoo (1990) developed a test for seasonal unit roots that decomposes seasonal dynamics into distinct components. Their approach allows researchers to identify which seasonal patterns require differencing, avoiding the distortions that can arise from indiscriminate seasonal differencing. The HEGY test became a standard tool for analyzing quarterly macroeconomic and financial time series.

Reference(s): Hylleberg et al. (1990); Beaulieu & Miron (1993); Franses & Hobijn (1997)

Worked Example

Problem Statement An economist tests for seasonal unit roots in a quarterly retail sales index ($T = 24$, 6 years). The HEGY test examines unit roots simultaneously at three frequencies: zero (long-run), Nyquist (π), and annual ($\pi/2$). A constant and three seasonal dummies (D_{Q1}, D_{Q2}, D_{Q3}) are included; no lag augmentation ($k = 0$).
$y = \{14.9, 8.6, 13.0, 9.8, 14.9, 11.0, 15.0, 10.1, 18.3, 13.5, 16.5, 13.4, 19.9, 14.8, 18.8, 14.3, 21.9, 17.1, 21.2, 16.3, 24.8, 19.1, 22.7, 21.1\}$

1. HEGY Filter Variables

From y_t, 4 auxiliary series are computed for $t = 5\ldots24$:

$$\Delta_4 y_t = y_t - y_{t-4},$$

$$[1]:\ y_{1,t-1} = y_{t-1} + y_{t-2} + y_{t-3} + y_{t-4},$$

$$[2]:\ y_{2,t-1} = -y_{t-1} + y_{t-2} - y_{t-3} + y_{t-4},$$

$$[3]:\ y_{3,t-1} = -y_{t-1} + y_{t-3},$$

$$[4]:\ y_{4,t-1} = -y_{t-2} + y_{t-4}$$

2. HEGY Regression

OLS of $\Delta_4 y_t$ on $[1, y_{1,t-1}, y_{2,t-1}, y_{3,t-1}, y_{4,t-1}, D_{Q1}, D_{Q2}, D_{Q3}]$ using 20 obs. ($df = 12$, $\hat{s}^2 = 0.356$):

Param	Coef	SE	t
$\hat{\mu}$	−5.452	1.810	−3.01
$\hat{\pi}_1$	+0.019	0.012	+1.57
$\hat{\pi}_2$	−0.389	0.150	−2.60
$\hat{\pi}_3$	+0.485	0.233	+2.08
$\hat{\pi}_4$	−0.710	0.234	−3.04
$\hat{d}_1$	+10.464	3.343	+3.13
$\hat{d}_2$	+3.687	0.874	+4.22
$\hat{d}_3$	+8.593	2.607	+3.30

3. Test Statistics at Each Frequency

Zero frequency (t-test on $\hat{\pi}_1$, one-sided): $t(\hat{\pi}_1) = +1.571$

Nyquist frequency (t-test on $\hat{\pi}_2$, one-sided): $t(\hat{\pi}_2) = -2.596$

Annual frequency (F-test: $\pi_3 = \pi_4 = 0$ jointly):
The covariance matrix of OLS estimates is:

$$\widehat{\mathrm{Cov}}(\hat{\beta}) = \hat{\sigma}^2(X'X)^{-1}.$$

We only require the $(\hat{\pi}_3, \hat{\pi}_4)$ block:

$$R\widehat{\mathrm{Cov}}R' = \begin{pmatrix} 0.05437 & 0.03282 \\ 0.03282 & 0.05460 \end{pmatrix}.$$

$$F(\hat{\pi}_3, \hat{\pi}_4) = \frac{(\hat{\pi}_3, \hat{\pi}_4)[R\widehat{\mathrm{Cov}}R']^{-1}(\hat{\pi}_3, \hat{\pi}_4)'}{2} = 16.615$$

4. Results and Decision CVs (const+seasonal dummies):

Test	Stat	CV (5%)	Decision
$t(\hat{\pi}_1)$	+1.571	−2.81	FTR H_0
$t(\hat{\pi}_2)$	−2.596	−2.80	FTR H_0
$F(\hat{\pi}_3, \hat{\pi}_4)$	6.62	3.06	Reject H_0

5. Interpretation

- **Decision:** A unit root exists at the *zero frequency* and *Nyquist*, but unit root at the *annual seasonal* frequency is **rejected**. Both ordinary and biannual seasonal differencing may therefore be required.
- **Practitioner Insight:** The HEGY test decomposes $(1 - B^4) = (1 - B)(1 + B)(1 + B^2)$ into three components — t-tests for real roots and a joint F-test for complex roots. This mixed outcome (unit root in levels but not seasonal) is common in trended series: applying Δ_4 would over-difference and lose information, so Δy_t suffices. Critical values depend on the deterministic specification (constant, trend, seasonal dummies).

Common Pitfalls

- **Using standard DF or F critical values.** HEGY statistics ($t_1, t_2, F_{3:4}$, etc.) follow nonstandard distributions because of the seasonal transformations in the auxiliary regression. Inference therefore requires HEGY-specific critical values.
- **Confusing deterministic and stochastic seasonality.** Deterministic seasonality is captured by seasonal dummies, whereas stochastic seasonality implies seasonal unit roots and may require seasonal differencing. The HEGY procedure is specifically designed to distinguish these cases.
- **Ignoring joint seasonal-root tests.** The procedure is inherently sequential: joint tests first determine whether seasonal unit roots exist, after which individual statistics identify the affected frequencies. Individual t-statistics should not be interpreted in isolation.
- **Applying quarterly HEGY to monthly data.** The standard HEGY decomposition is derived for quarterly data ($S = 4$). Monthly data ($S = 12$) require modified transformations and different critical values.
- **Ignoring structural changes in seasonality.** The HEGY framework assumes stable seasonal behavior over time. Structural breaks or evolving seasonal patterns can bias the test and should be investigated separately.

Canova-Hansen Test

Test for Seasonal Stability in Deterministic Seasonality

STAT.13 | **Category:** Stationarity | **Data:** Time Series | **Null:** Tabled distribution

When to Use

- Test for stability of seasonal patterns over the timeline.
- Apply when the null hypothesis assumes stationary seasons.
- Identify if seasonal dummies change as a random walk.
- Evaluate seasonal stability as a check against HEGY tests.
- Analyze monthly or quarterly macro-financial time series.

Assumptions

- The data features a deterministic seasonal base structure.
- Structural innovation variance remains constant over time.
- Error terms are stationary and weakly dependent.
- Observations are recorded at fixed, uniform time intervals.
- The underlying data contains no structural regime breaks.

Test Statistic

$$L(\omega_j) = \frac{1}{T^2}\,\mathrm{tr}\big[\mathbf{F}'\mathbf{F}\hat{\Omega}^{-1}\big], \quad \mathbf{F}_t = \sum_{k=1}^{t}\mathbf{x}_k(\omega_j), \quad \hat{\Omega} = \hat{\Gamma}_0 + \sum_{h=1}^{\ell} w_h(\hat{\Gamma}_h + \hat{\Gamma}_h'), \quad \hat{\Gamma}_h = \frac{1}{T}\sum_{t=h+1}^{T}\mathbf{x}_t\mathbf{x}_{t-h}'$$

$$\mathbf{x}_t(\omega_j) = \big[\cos(\omega_j t),\ \sin(\omega_j t)\big]'\hat{e}_t, \quad 0 < \omega_j < \pi, \qquad x_t(\pi) = \cos(\pi t)\hat{e}_t = (-1)^t\hat{e}_t, \quad L_{\text{joint}} = \textstyle\sum_j L(\omega_j)$$

Notation: $\hat{e}_t$: OLS residuals from regression on trend and seasonal dummies; $\mathbf{F}_t$: cumulative partial sums; $\hat{\Omega}$: Newey–West long-run covariance matrix; $w_h = 1 - \frac{h}{\ell+1}$: Bartlett weights; ℓ: truncation lag / bandwidth

Plain English: The Canova–Hansen test examines whether seasonal patterns remain stable over time by testing the constancy of seasonal effects in a time series. It evaluates cumulative instability in seasonal components across distinct seasonal frequencies. Large test statistics indicate that seasonal behavior changes over the sample period, suggesting that fixed seasonal dummies may be inadequate. Test is flexible enough to capture both gradual evolution and structural changes in seasonality.

⚠ Practitioner Warning

Test assumes meaningful deterministic seasonality is present; if seasonal structure is weak or absent, the test provides limited diagnostic value. Multiple frequency-specific statistics are often reported, requiring careful interpretation of both joint and individual seasonal components. Results can be sensitive to bandwidth selection in long-run variance estimation. The test is designed to assess the stability of deterministic seasonality and should not be viewed as a substitute for HEGY-type tests of seasonal unit roots.

Hypotheses

- H_0: Seasonal pattern is stable (time-invariant).
- H_1: Seasonal pattern evolves over time (time-varying).

Decision Rule: Reject H_0 if $L > L_{\alpha,\text{crit}}$. CVs depend on seasonal frequency ω_j and number of frequencies tested.

Software References [Python, R, SAS]

Python	R	SAS
`Not Native (Manual)`	`uroot::ch.test`	`Not Native (Manual)`

Additional Notes

Two common parameterizations are used for seasonal components. Trigonometric seasonality employs harmonic terms such as $f_j(\tau) = \cos(2\pi j\tau)$ and $\sin(2\pi j\tau)$, whereas dummy-variable seasonality uses seasonal indicator variables D_{jt}. The joint statistic aggregates instability across seasonal frequencies, although individual frequency statistics may also be examined separately. The procedure is flexible with respect to the form of seasonal instability and can detect both gradual evolution and discrete breaks. For higher-frequency data (e.g., monthly series), multiple harmonic seasonal frequencies are tested. Rejection often motivates time-varying seasonal models such as state-space or regime-switching frameworks.

Miscellanea: Canova and Hansen (1995) developed a test for the stability of deterministic seasonal patterns, motivated by evidence that seasonality in economic and macroeconomic series often evolves over time. Their work emphasized that standard seasonal models can become misleading when seasonal effects are not constant across the sample period.

Reference(s): Canova & Hansen (1995); Franses (1996)

Worked Example

Problem Statement An economist tests whether the seasonal pattern of a quarterly retail sales index is stable over time (seasonally stationary). Unlike HEGY, the CH test takes seasonal stationarity as H_0 and detects evolving seasonal effects. $T = 24$ (6 years), $\ell = 2$ Bartlett kernel. Q1 amplitude grows markedly across years; Q2–Q4 remain stable.

$y = \{13.4, 7.9, 12.8, 7.9, 15.3, 9.0, 13.9, 9.2, 17.5, 10.3, 15.1, 10.2, 19.3, 11.4, 16.4, 11.6, 21.4, 12.5, 17.4, 12.7, 23.5, 13.9, 18.7, 13.8\}$

1. Step 1 — Detrending Regression

Regress y_t (no intercept) on $[t,\ D_{Q1},\ D_{Q2},\ D_{Q3},\ D_{Q4}]$:

$$\hat{\beta} = 0.348; \quad \hat{\delta}_1 = 14.572, \quad \hat{\delta}_2 = 6.657$$

$$\hat{\delta}_3 = 11.192, \qquad \hat{\delta}_4 = 6.028$$

Selected OLS residuals $\hat{e}_t$ (note $\sum \hat{e}_t = 0$):

- $\hat{e}_1 = -1.520$ (Q1, yr 1) $\hat{e}_5 = -1.012$ (Q1, yr 2)
- $\hat{e}_{17} = +0.912$ (Q1, yr 5) $\hat{e}_{21} = +1.620$ (Q1, yr 6)
- Q2–Q4 residuals remain small throughout

2. Step 2 — $L(\pi)$ at Nyquist Frequency

Define $x_t^{(\pi)} = \cos(\pi t) \cdot \hat{e}_t = (-1)^t \hat{e}_t$ and $F_t^{(\pi)} = \sum_{k=1}^{t} x_k^{(\pi)}$ (partial sums).

$$\hat{\gamma}_j = \frac{1}{T} \sum_{t=j+1}^{T} x_t x_{t-j}, \quad \hat{S}^2(\pi) = \hat{\gamma}_0 + \tfrac{4}{3}\hat{\gamma}_1 + \tfrac{2}{3}\hat{\gamma}_2 = 0.4068$$

$$L(\pi) = \frac{1}{T^2} \cdot \frac{\sum_{t=1}^{T} [F_t^{(\pi)}]^2}{\hat{S}^2(\pi)} = \frac{171.663}{576 \times 0.4068} = 0.733$$

3. Step 3 — $L(\pi/2)$ at Annual Frequency

Define

$$\mathbf{x}_t^{(\pi/2)} = [\cos(\tfrac{\pi}{2}t), \sin(\tfrac{\pi}{2}t)] \cdot \hat{e}_t, \quad \mathbf{F}_t = \sum_{k=1}^{t} \mathbf{x}_k^{(\pi/2)}.$$

Matrix NW estimator:

$$\hat{\Gamma}_j = \frac{1}{T} \sum_{t=j+1}^{T} \mathbf{x}_t \mathbf{x}_{t-j}', \quad \hat{\Omega} = \hat{\Gamma}_0 + \tfrac{2}{3}(\hat{\Gamma}_1 + \hat{\Gamma}_1') + \tfrac{1}{3}(\hat{\Gamma}_2 + \hat{\Gamma}_2')$$

Long-run covariance matrix (Newey-West, $\ell = 2$):

$$\hat{\Omega} = \begin{pmatrix} 0.0364 & 0.0323 \\ 0.0323 & 0.4200 \end{pmatrix}, \quad L(\pi/2) = \operatorname{tr}\left[\frac{1}{T^2} \mathbf{F}'\mathbf{F}\, \hat{\Omega}^{-1}\right] = 0.87$$

4. Results and Decision

Critical values (CH 1995, Table 1, quarterly):

Statistic	Value	CV (5%)	Decision
$L(\pi/2)$	0.870	0.749	Reject H_0
$L(\pi)$	0.733	0.470	Reject H_0
L_{joint}	1.603	1.010	Reject H_0

5. Interpretation

- **Decision:** Both frequency-specific statistics and the joint statistic (= $L(\pi/2) + L(\pi)$)exceed their 5% critical values. We **reject** H_0 of seasonal stationarity: the seasonal pattern is **unstable** and contains seasonal unit roots.
- **Practitioner Insight:** The CH test is the seasonal analog of the KPSS test. Rejection means the seasonal pattern drifts over time. The growing Q1 residuals ($\hat{e}_1 = -1.52$ rising to $\hat{e}_{21} = +1.62$) are the signature: the partial sums $F_t^{(\pi)}$ and $\mathbf{F}_t^{(\pi/2)}$ wander persistently rather than oscillating near zero. Note that $F_T^{(\pi)} = F_T^{(\pi/2)} = 0$ by construction (since $\sum \hat{e}_t = 0$). Used together with HEGY, CH completes a full picture: HEGY tests whether a seasonal unit root is present; CH tests whether the seasonal pattern is stable conditional on stationarity being imposed.

Common Pitfalls

- **Confusing seasonal instability with seasonal unit roots.** The Canova–Hansen test evaluates whether deterministic seasonal patterns remain stable over time, not whether seasonal unit roots are present. Seasonal unit roots are instead tested using procedures such as HEGY.
- **Failing to remove nonseasonal trends.** Trend components should be removed before testing seasonal stability. Otherwise, instability in the trend may be mistaken for changing seasonal behavior.
- **Interpreting rejection as evidence for seasonal differencing.** Rejection implies evolving seasonal patterns rather than stochastic seasonality. Such behavior is typically modeled using time-varying seasonal effects instead of seasonal differencing.
- **Ignoring frequency-specific diagnostics.** After rejecting the joint stability test, examining individual seasonal-frequency statistics helps identify which cycles are unstable and guides seasonal model specification.
- **Using the test without clear seasonality.** The CH procedure assesses the stability of existing seasonal structure. If meaningful seasonality is absent, the test provides limited diagnostic information.

Kapetanios-Shin-Snell Test

Unit Root Test Against Nonlinear STAR Alternative

STAT.14 | **Category:** Stationarity | **Data:** Time Series | **Null:** Tabled distribution

When to Use

- Test for unit roots against non-linear alternatives.
- Apply when the speed of mean reversion varies over time.
- Evaluate series exhibiting smooth exponential adjustments.
- Analyze real exchange rates under transaction cost bands.
- Identify stationarity hidden from standard linear tests.

Assumptions

- Data-generating path follows a non-linear framework.
- Structural mean reversion follows a smooth ESTAR path.
- Residual innovation variance remains stable over time.
- Observations are recorded at fixed, uniform time intervals.
- Underlying process is free of structural trend breaks.

Test Statistic

$$t_{\text{KSS}} = \frac{\hat{\delta}}{\text{SE}(\hat{\delta})} \quad \Big| \quad \Delta \tilde{y}_t = \delta \tilde{y}_{t-1}^3 + \sum_{i=1}^{k} \gamma_i \Delta \tilde{y}_{t-i} + \epsilon_t \quad \Big|$$

Notation: Δ is the first-difference operator, cubic term from Taylor expansion of ESTAR model, $\sum \gamma_i \Delta \tilde{y}_{t-i}$ is the parametric correction added to control for serial correlation if errors are not white noise.

Plain English: The KSS test detects nonlinear mean reversion, where the speed of adjustment depends on how far the series deviates from equilibrium. Standard linear unit-root tests assume constant adjustment speed and may miss processes that behave almost like a random walk near equilibrium but revert strongly when deviations become sufficiently large. KSS captures this behavior using a simple nonlinear transformation derived from Smooth Transition AutoRegressive (STAR) dynamics.

⚠ Practitioner Warning

Test is specifically designed for ESTAR-type (Exponentially Smooth Transition AutoRegressive) nonlinear adjustment and may have low power against other nonlinear processes such as threshold or Markov-switching models. Because H_0 lies on the boundary ($\theta = 0$), the test has a nonstandard asymptotic distribution requiring specially simulated critical values; using standard ADF/Dickey–Fuller tables invalidates inference and can substantially over-reject the null. The cubic transformation may also be sensitive to outliers and can perform poorly in short samples or weakly nonlinear settings.

Hypotheses

- H_0: $\delta = 0$ (Unit root with linear/no nonlinear ESTAR MR).
- H_1: $\delta < 0$ (globally stationary with nonlinear ESTAR MR).

Decision Rule: Reject H_0 if $t_{\text{KSS}} < t^{\text{KSS}}_{\alpha,\text{crit}}$ or p-value $< \alpha$. CVs are more negative with trend or demeaned data.

Software References [Python, R, SAS]

Python	R	SAS
`Not Native (Manual)`	`NonlinearTSA::KSS_Unit_Root`	`Not Native (Manual)`

Additional Notes

The cubic specification arises from first-order Taylor expansion of the ESTAR transition function: $(1 - \exp(-\theta y_{t-1}^2))y_{t-1} \approx \theta y_{t-1}^3$ for small θ. Three common specifications: (1) raw data **(A)**, (2) demeaned residuals **(B)**, and (3) detrended residuals **(C)**. Augmented versions include lagged differences **(D)** to remove serial correlation. ESTAR dynamics imply symmetric adjustment: positive and negative deviations revert at the same speed, unlike threshold models with asymmetric regimes. The auxiliary cubic regression is only a local approximation to the nonlinear ESTAR process, not the structural model itself. KSS showed substantial power gains over ADF tests when nonlinear mean reversion is present but weak near equilibrium. Sollis (2009) later extended the framework to asymmetric nonlinear adjustment.

$$\textbf{(A)} :: \Delta y_t = \delta y_{t-1}^3 + \epsilon_t \qquad \textbf{(B)} :: (y_{t-1} - \bar{y})^3 \qquad \textbf{(C)} :: \tilde{y}_t = y_t - \hat{\alpha} - \hat{\beta} t \qquad \textbf{(D)} :: \Delta y_t = \delta y_{t-1}^3 + \sum_{i=1}^{p} \phi_i \Delta y_{t-i} + \epsilon_t$$

Miscellanea: George Kapetanios, Yongcheol Shin, and Andy Snell (2003) developed their test to detect nonlinear mean reversion that standard linear unit-root tests such as ADF often miss. Inspired by smooth-transition autoregressive (STAR) models, the procedure allows weak deviations from equilibrium to behave almost like a unit root while large deviations trigger stronger mean-reverting dynamics.

Reference(s): Kapetanios et al. (2003); Sollis (2009)

Kapetanios-Shin-Snell Test

STAT.14

Worked Example

Problem Statement An economist tests whether a price index is a unit root process or globally stationary with nonlinear (ESTAR) mean reversion — where deviations from equilibrium trigger stronger reversion the further the series strays. The KSS test targets this alternative directly; standard ADF has no power against it. $T = 20$, demeaned specification, $k = 0$ (no augmentation).

$y = \{20.0, 14.7, 18.3, 14.8, 16.9, 12.7, 16.1, 13.2, 16.1, 13.8, 17.6, 13.5, 16.4, 13.4, 16.8, 14.2, 17.8, 13.6, 16.4, 13.4\}$

1. ESTAR Framework and Test Equation

The ESTAR model is:

$$\Delta y_t = \phi\, y_{t-1} \underbrace{\left[1 - e^{-\theta y_{t-1}^2}\right]}_{\text{transition}} + \varepsilon_t$$

Under H_0: $\theta = 0$ (unit root, no transition). A first-order Taylor expansion of the transition function around $\theta = 0$ gives:

$$\Delta \tilde{y}_t \approx \delta\, \tilde{y}_{t-1}^3 + \varepsilon_t, \quad H_0 : \delta = 0$$

where $\tilde{y}_t = y_t - \bar{y}$ is the demeaned series.

2. Step 1 — Demeaning & Key Sums

$$\bar{y} = 15.485, \qquad \tilde{y}_t = y_t - 15.485$$

Selected demeaned values:

$\tilde{y}_1 = +4.515$, $\tilde{y}_6 = -2.785$, $\tilde{y}_{11} = +2.115$, $\tilde{y}_{17} = +2.315$.

$$\sum_{t=2}^{T} \tilde{y}_{t-1}^3 \Delta \tilde{y}_t = -888.926, \quad \sum_{t=2}^{T} \tilde{y}_{t-1}^6 = 10{,}051.592$$

4. Step 3 — KSS Regression ($k = 0$)

OLS of $\Delta \tilde{y}_t$ on $\tilde{y}_{t-1}^3$ (no intercept):

$$\hat{\delta} = \frac{\sum \tilde{y}_{t-1}^3 \Delta \tilde{y}_t}{\sum \tilde{y}_{t-1}^6} = \frac{-888.926}{10{,}051.592} = -0.0884$$

$$\hat{s}^2 = \frac{\sum \hat{\varepsilon}_t^2}{n-1} = 8.046, \qquad SE(\hat{\delta}) = \frac{\hat{s}}{\sqrt{\sum \tilde{y}_{t-1}^6}} = 0.0283$$

$$t_{KSS} = \frac{\hat{\delta}}{SE(\hat{\delta})} = \frac{-0.0884}{0.0283} = -3.126$$

5. Results and Decision

Critical values (KSS 2003, Table 1, demeaned):

Statistic	Value	CV (5%)	Decision
t_{KSS}	−3.126	−2.93	Reject H_0

For reference: CV at 10% = −2.66; CV at 1% = −3.48.

6. Interpretation

- **Decision:** The price index exhibits globally stationary ESTAR-type nonlinear mean reversion.
- **Practitioner Insight:** The KSS test uses $\tilde{y}_{t-1}^3$ as the regressor — the cubic term naturally captures the ESTAR mechanism where reversion strengthens with the cube of the deviation. The negative $\hat{\delta} = -0.088$ confirms this: large deviations (large $|\tilde{y}|$) generate a strong negative pull back toward the mean. Standard ADF would not detect this pattern because it tests linearly against y_{t-1}, not y_{t-1}^3. The demeaned specification is the most commonly used; the detrended version should be used when a deterministic trend is present, which requires the more negative critical values (−3.40 at 5%). Lag augmentation ($k > 0$) is selected by AIC/BIC if residual autocorrelation is suspected; here $k = 0$ suffices.

Common Pitfalls

- **Interpreting rejection as evidence of linear stationarity.** Rejection of the KSS null indicates *nonlinear* mean reversion consistent with ESTAR dynamics, not linear stationarity. Linear ARMA models may miss the nonlinear adjustment mechanism implied by the test.
- **Using the KSS test as a stand-alone unit root test.** KSS is designed to complement conventional tests such as ADF. Comparing both results helps distinguish linear stationarity, nonlinear stationarity, and genuine unit-root behavior.
- **Ignoring structural breaks in the series.** Structural breaks can reduce the power of nonlinear unit root tests and lead to misleading conclusions. Break-adjusted procedures should be considered when regime shifts are suspected.
- **Confusing ESTAR alternatives with other nonlinear dynamics.** KSS test is specifically designed to detect ESTAR-type nonlinear mean reversion. Other nonlinear processes (TAR, LSTAR, etc.) require different tests.
- **Using DF critical values instead of KSS tables.** The cubic transformation in the auxiliary regression produces a different asymptotic distribution from DF tests. Correct inference requires CVs reported in KSS (2003).

Levin-Lin-Chu Test

Panel Unit Root Test with Homogeneous Alternative

STAT.15 | **Category:** Stationarity | **Data:** Panel Time Series | **Null:** Normal (asymptotic)

When to Use

- Test for unit roots in balanced panel data setups.
- Apply when both dimensions are moderate and $N/T \to 0$.
- Increase power by pooling info. across panel members.
- Assess common persistence across panel members.
- Preliminary check before panel cointegration modeling.

Assumptions

- Panel individuals share a homogenous AR parameter (δ).
- Individual effects and time trends can vary across units.
- Cross-sectional units are independent of one another.
- Observations are recorded at fixed, uniform time intervals.
- Individual series contain no structural regime breaks.

Test Statistic

$$t_\delta^* = \frac{t_\delta - N\widetilde{T}\,\bar{S}_N\,\hat{\sigma}^{-2}\,\mathrm{SE}(\hat{\delta})\,\mu^*_{m\widetilde{T}}}{\sigma^*_{m\widetilde{T}}} \xrightarrow{d} N(0,1) \quad \Big| \quad \tilde{e}_{it} = \delta\tilde{v}_{i,t-1} + \tilde{\epsilon}_{it}, \quad \hat{\delta} = \frac{\sum_{i=1}^N \sum_{t=k+2}^T \tilde{v}_{i,t-1}\tilde{e}_{it}}{\sum_{i=1}^N \sum_{t=k+2}^T \tilde{v}_{i,t-1}^2}, \quad t_\delta = \frac{\hat{\delta}}{\mathrm{SE}(\hat{\delta})},$$

$$\bar{S}_N = \frac{1}{N}\sum_{i=1}^N \hat{s}_i, \; \hat{s}_i = \frac{\hat{\sigma}_{yi}}{\hat{\sigma}_{\epsilon i}} \quad \Big| \quad \hat{\sigma}^2 = \frac{1}{N\widetilde{T}}\sum_{i=1}^N \sum_{t=k+2}^T (\tilde{e}_{it} - \hat{\delta}\tilde{v}_{i,t-1})^2 \quad \Big| \quad \mathrm{SE}(\hat{\delta}) = \frac{\hat{\sigma}}{\sqrt{\sum_{i=1}^N \sum_{t=k+2}^T \tilde{v}_{i,t-1}^2}}$$

Notation: $\widetilde{T} = T - k - 1$ is the usable time length per unit after k lags; $\tilde{e}_{it}$ and $\tilde{v}_{i,t-1}$ are unit-wise orthogonalized and standardized auxiliary residuals; $\bar{S}_N$ is the mean long-run to short-run standard-deviation ratio, $N \times T$ are panel data dimensions

Plain English: The Levin–Lin–Chu test pools information across panel units to test whether all series contain unit roots against the alternative that all are stationary with a common autoregressive parameter. By combining data from N units over T periods ($N/T \to 0$), LLC gains power relative to separate unit-root tests. The multi-step procedure uses unit-specific standardization and bias corrections to produce a standardized normal limiting distribution.

⚠ Practitioner Warning

The homogeneous alternative ($\delta_i = \delta \; \forall \, i$) is restrictive; heterogeneous dynamics can reduce power or distort size. Cross-sectional dependence (common shocks, spatial spillovers) invalidates the LLC asymptotics, requiring second-generation tests such as Pesaran's CIPS. LLC also assumes balanced panels and sufficiently large N and T with $N/T \to 0$.

Hypotheses

- H_0: $\delta_i = 1$ for all i (all units have unit roots).
- H_1: $\delta_i = \delta < 1$ for all i (stationary and common δ).

Decision Rule: Reject H_0 if $t_\delta^* < -z_\alpha$ (left-tailed) or p-value $< \alpha$.

Software References [Python, R, SAS]

`linearmodels::LevinLinChu` `plm::purtest(type="levinlin")` `PROC PANEL (LLC)`

Additional Notes

Three-step procedure: (1) For each unit, select k_i via ADF and obtain orthogonalized residuals $\hat{e}_{it}$ and $\hat{v}_{i,t-1}$ from auxiliary regressions; standardize to $\tilde{e}_{it}$ and $\tilde{v}_{i,t-1}$. (2) Estimate the long-run/short-run variance ratio $\hat{s}_i$ using a kernel-based long-run variance estimator to compute the panel average $\bar{S}_N$. (3) Pool the $N\widetilde{T}$ standardized observations to estimate $\hat{\delta}$, compute t_δ^*, and apply the LLC finite-sample adjustment. The adjustment constants $\mu^*_{m\widetilde{T}}$ and $\sigma^*_{m\widetilde{T}}$ depend on the deterministic specification m and effective sample size $\widetilde{T}$. Canonical specifications: no effects ($m = 1$), intercepts only ($m = 2$), and intercepts with trends ($m = 3$). Related panel unit-root procedures include Harris–Tzavalis (1999) and Breitung (2000).

Miscellanea: Andrew Levin, Chien-Fu Lin, and Chia-Shang James Chu developed one of the earliest panel unit root tests, published in 2002, based on a pooled framework that assumes a common unit root process across panels. While this homogeneity assumption is restrictive, the approach offers increased power and became widely used in empirical work.

Reference(s): Levin et al. (2002); Breitung (2000)

Worked Example

Problem Statement An economist tests whether consumer price indices across four regions ($N = 4$) share a common unit root. The LLC test imposes a *homogeneous* AR coefficient under H_1 and gains power by pooling across units. The panel spans $T = 10$ periods per region; a fixed-effects specification is used with $k_i = 0$ (no lag augmentation), giving $\widetilde{T} = T - k_i - 1 = 9$ usable observations per unit. Sample data is at the end of the chapter

1. Auxiliary Regressions
For each unit i, with $k_i = 0$ and individual-specific intercepts ($m = 2$), the LLC procedure runs two auxiliary regressions separately: regress Δy_{it} on a constant to get $\hat{e}_{it}$, and regress $y_{i,t-1}$ on a constant to get $\hat{v}_{i,t-1}$ (LLC eqs. 2–3). These reduce to within-demeaning:

$$\hat{e}_{it} = \Delta y_{it} - \overline{\Delta y}_i, \quad \hat{v}_{i,t-1} = y_{i,t-1} - \bar{y}_{i,\text{lag}}$$

The individual $\hat{\delta}_i$ and $\hat{\sigma}_i^2$ follow from regressing $\hat{e}_{it}$ on $\hat{v}_{i,t-1}$ (LLC eq. 4, denominator $\widetilde{T} = 9$): $\hat{\sigma}_i = RSS_i/\widetilde{T}$. Then normalize: $\tilde{e}_{it} = \hat{e}_{it}/\hat{\sigma}_i$, $\tilde{v}_{i,t-1} = \hat{v}_{i,t-1}/\hat{\sigma}_i$.

2. Long-Run Variance Ratios
With $k_i = 0$ (no serial correlation), the long-run variance of each unit reduces to the variance of demeaned first differences (LLC eq. 5, $m = 2$):

$$\hat{\sigma}_{yi}^2 = \frac{1}{\widetilde{T}}\sum_{t=2}^{T}(\Delta y_{it} - \overline{\Delta y}_i)^2, \quad \hat{s}_i = \hat{\sigma}_{yi}/\hat{\sigma}_i, \quad \bar{S}_N = \frac{1}{N}\sum_{i=1}^{N}\hat{s}_i$$

3. Step 3 — Pooled Regression and t_δ
Pool the $N\widetilde{T} = 36$ normalized observations $(\tilde{e}_{it}, \tilde{v}_{i,t-1})$ and run the regression (LLC eq. 7):

$$\hat{\delta} = \frac{\sum_i \sum_t \tilde{v}_{i,t-1}\tilde{e}_{it}}{\sum_i \sum_t \tilde{v}_{i,t-1}^2} = \frac{-57.35}{41.74} = -1.374$$

The pooled residual variance uses denominator $N\widetilde{T} = 36$ exactly as in LLC eq. (11):

$$\hat{\sigma}^2 = \frac{1}{N\widetilde{T}}\sum_i\sum_t(\tilde{e}_{it} - \hat{\delta}\,\tilde{v}_{i,t-1})^2 = \frac{36.36}{36} = 1.010$$

$$\text{SE}(\hat{\delta}) = \frac{\hat{\sigma}}{\sqrt{\sum_i \sum_t \tilde{v}_{i,t-1}^2}} = 0.156, \quad t_\delta = \frac{\hat{\delta}}{\text{SE}(\hat{\delta})} = -8.833$$

	$\sum \hat{v}\hat{e}$	$\sum \hat{v}^2$	$\hat{\delta}_i$	$\hat{\sigma}_i$	$\hat{\sigma}_{yi}$	$\hat{s}_i$
P1	−1.842	1.449	−1.272	0.379	0.636	1.677
P2	−1.579	1.042	−1.515	0.303	0.598	1.976
P3	−1.563	1.140	−1.371	0.325	0.587	1.804
P4	−1.882	1.429	−1.317	0.388	0.653	1.682
					$\bar{S}_N$	1.784

4. LLC Adjustment (eq. 12)
For $m = 2$ (individual effects), the Table 2 adjustment values at $\widetilde{T} = 25$ (the closest tabulated value; $\widetilde{T} = 9$ is below the table range) are $\mu^* = -0.554$, $\sigma^* = 0.919$:

$$\text{adj} = N\widetilde{T}\,\bar{S}_N\,\hat{\sigma}^{-2}\,\text{SE}(\hat{\delta})\,\mu^*_{m\widetilde{T}} = -5.481, \quad t^*_\delta = -3.647$$

5. Results and Decision

Statistic	Value	CV (5%)	Decision
t_δ (raw)	−8.833	—	—
t^*_δ (adjusted)	−3.647	−1.645	Reject H_0

Practitioner Insight: The adjustment corrects the downward bias in t_δ that arises under H_0 when unit-specific intercepts are included: even under H_0, the expected value of t_δ is substantially -ve. Here $\hat{s}_i > 1$ indicates that long-run variance exceeds innovation variance, reflecting serial dependence in the oscillating data, while the large $\bar{S}_N = 1.784$ further amplifies the correction. Both $t_\delta = -8.833$ and the adjusted statistic $t^* = -3.647$ fall well below -1.645, strongly rejecting the unit-root null and supporting stationarity with a common AR coefficient across panels. A key limitation of LLC is its homogeneous alternative: all units share the same δ. If some panels are stationary and others are not, the test loses power, making the heterogeneous IPS test a useful complement. With only $N = 4$ and $\widetilde{T} = 9$, the example is primarily illustrative only.

Common Pitfalls

- **Ignoring cross-sectional dependence.** The LLC test assumes cross-sectional independence or only weak dependence among units. Strong common shocks or correlations can substantially inflate rejection rates and invalidate inference.
- **Assuming rejection $\Longrightarrow$ universal stationarity.** LLC imposes a homogeneous alternative (common δ) across units. Rejection therefore supports panel-level stationarity under this restriction, not stationarity of every individual series.
- **Using LLC under heterogeneous dynamics.** The procedure assumes similar persistence behavior across panel units. When dynamics differ materially across series, tests such s IPS or Fisher-type procedures are often more appropriate.
- **Applying LLC to unbalanced panels.** The standard LLC framework is derived for balanced panels with equal time observations for each unit. Unbalanced panels may require adjustments or alternative panel unit-root procedures.

Im-Pesaran-Shin (IPS) Test

Panel Unit Root Test with Heterogeneous Alternative

STAT.16 | **Category:** Stationarity | **Data:** Panel Time Series | **Null:** Normal (asymptotic)

When to Use

- Panel-data unit-root testing with heterogeneity.
- Allow different persistence across panel members.
- Large-N, moderate-T panel time-series analysis.
- More flexible alternative to Levin-Lin-Chu.
- Preliminary check before panel cointegration tests.

Assumptions

- Observations are recorded at fixed, uniform time intervals.
- The panel individuals are cross-sectionally independent.
- The error terms can have heterogeneous variance profiles.
- Errors are weakly dependent and stationary.
- The data contains no unmodeled structural regime breaks.

Test Statistic

$$\bar{t} = \frac{1}{N}\sum_{i=1}^{N} t_{i,T_i}, \quad W = \frac{\sqrt{N}(\bar{t} - E[\bar{t}])}{\sqrt{\text{Var}[\bar{t}]}} \xrightarrow{d} N(0,1). \qquad \text{Here, } E[\bar{t}], \text{Var}[\bar{t}] \text{ are from simulations.}$$

Notation: N is number of units, T_i is time periods for unit i, t_{i,T_i} is ADF t-statistic for unit i, $\bar{t}$ is average ADF statistic

Plain English: The Im-Pesaran-Shin test averages individual ADF t-statistics across panel units and standardizes the mean to obtain asymptotic normality. Unlike LLC, IPS allows each unit to have different autoregressive parameters under the alternative, testing whether at least some units are stationary rather than requiring all units to share common dynamics. This heterogeneous alternative is more realistic for diverse panel datasets.

⚠ Practitioner Warning

Cross-sectional dependence (common shocks, spillovers, or spatial correlation) can produce serious size distortions and invalid inference. Cross-sectionally augmented procedures such as CADF or CIPS are often preferred when dependence is suspected. Standardization of the IPS statistic relies on tabulated finite-sample moments for specific (T_i, k_i) combinations, making unbalanced panels more difficult to handle. The asymptotic approx. performs best when $N \gg T$. Test power also depends on the proportion of stationary units and may be weak when only a small fraction of the panel is stationary.

Hypotheses

- H_0: $\gamma_i = 0$ for all i (all units have unit roots).
- H_1: $\gamma_i < 0$ for some i (some units are stationary).

Decision Rule: Reject H_0 if $W < -z_\alpha$ (left-tailed) or p-value $< \alpha$.

Software References [Python, R, SAS]

`linearmodels::ImPesaranShin` `plm::purtest(type="ips")` `PROC PANEL (IPS)`

Additional Notes

Procedure: (1) Run individual ADF regressions for each unit: $\Delta y_{it} = \alpha_i + \gamma_i y_{it-1} + \sum_{j=1}^{k_i} \phi_{ij}\Delta y_{it-j} + \epsilon_{it}$. (2) Extract t_{i,T_i} statistics testing $\gamma_i = 0$. (3) Compute the average $\bar{t}$ and standardize using tabulated finite-sample moments. Lag orders k_i can vary across units (select via AIC/BIC). The IPS test assumes a balanced panel because the standardized mean-group statistic relies on common T_i. To handle unbalanced panels, Maddala and Wu (1999) proposed Fisher-type tests that combine individual p-values rather than t-statistics, and Choi (2001) developed modified Fisher tests with improved small-sample performance. IPS is generally more powerful than LLC when individual heterogeneity is substantial, but for unbalanced or highly heterogeneous panels, Fisher-type tests are recommended.

Miscellanea: Kyung So Im, M. Hashem Pesaran, and Yongcheol Shin (2003) developed the IPS panel unit-root test as an extension of the LLC framework to accommodate heterogeneous autoregressive coefficients across units. Unlike LLC, which imposes a common persistence parameter, the IPS test allows each cross-sectional unit to have its own autoregressive root under the alternative hypothesis, enhancing flexibility and robustness in panels with diverse dynamics. This approach made IPS one of the most widely cited and practical panel unit-root tests in applied econometrics

Reference(s): Im et al. (2003); Maddala & Wu (1999)

Worked Example

Problem Statement An economist tests for unit roots in GDP indices across four countries ($N = 4, T = 10$). Unlike LLC, the IPS test allows a *heterogeneous* alternative: some countries may be stationary while others are not. A constant-only ADF regression is run separately for each panel ($k = 0$), then individual t-statistics are averaged and adjusted.

	Type	$t = 1$	$t = 2$	$t = 3$	$t = 4$	$t = 5$	$t = 6$	$t = 7$	$t = 8$	$t = 9$	$t = 10$
Panel 1	Stat.	10.0	9.6	10.4	10.0	10.6	9.9	10.6	10.0	10.6	10.1
Panel 2	Stat.	12.0	11.5	12.2	11.8	12.4	11.9	12.5	12.0	12.5	12.0
Panel 3	I(1)	8.0	8.1	9.0	9.0	9.7	9.7	10.5	10.6	11.3	11.3
Panel 4	I(1)	15.0	14.9	15.7	15.8	16.5	16.5	17.3	17.4	18.1	18.0

1. Step 1 — Individual ADF Regressions

For each panel i, estimate: $\Delta y_{it} = \alpha_i + \gamma_i y_{i,t-1} + \varepsilon_{it}$ and record $t_{\hat{\gamma}_i}$.

Panel	$\hat{\alpha}_i$	$\hat{\gamma}_i$	SE_i	t_i
1	15.440	−1.514	0.318	−4.758
2	17.049	−1.410	0.345	−4.092
3	1.185	−0.086	0.128	−0.672
4	1.534	−0.073	0.135	−0.543

Panels 1–2 show strong mean reversion ($t \approx -4$); Panels 3–4 show near-zero $\hat{\gamma}_i$ ($t \approx -0.6$), consistent with unit roots.

2. Step 2 — Average t-Stat $\bar{t}_{NT} = \frac{1}{N}\sum_{i=1}^{N} t_i = -2.516$

3. Step 3 — IPS Adjustment

Adjustment moments from IPS (2003) Table 3 ($T = 10$, $k = 0$, constant):

$$E[t_i \mid H_0] = -1.520, \qquad \mathrm{Var}[t_i \mid H_0] = 1.066$$

$$W_{\bar{t}} = \frac{\sqrt{N}\left(\bar{t}_{NT} - E[t_i \mid H_0]\right)}{\sqrt{\mathrm{Var}[t_i \mid H_0]}} = -1.930$$

Under H_0, $W_{\bar{t}} \xrightarrow{d} N(0,1)$.

4. Results and Decision

Statistic	Value	CV (5%)	Decision
$\bar{t}_{NT}$ (raw)	−2.516	—	—
$W_{\bar{t}}$ (adj.)	−1.930	−1.645	Reject H_0

5. Interpretation

- **Decision:** We **reject** H_0 that all panels contain a unit root. At least some of the panels are stationary.
- **Practitioner Insight:** The IPS test's key advantage over LLC is the heterogeneous alternative — here Panels 1–2 drive the rejection while Panels 3–4 individually would not reject. LLC's pooled regression would average these out and potentially lose power; IPS preserves the information in the cross-sectional distribution of t_i values. Note that $E[t_i \mid H_0] \approx -1.52$ (not zero) reflects the Dickey-Fuller small-sample bias — without this correction, $\bar{t}_{NT} = -2.52$ would already look significant relative to the standard normal, leading to spurious rejection. The IPS adjustment moments depend on both T and k for each unit; if units have different lag lengths k_i, use unit-specific moments.

Common Pitfalls

- **Interpreting rejection as proof that all units are stationary.** The IPS test evaluates average behavior across units ($E[\gamma_i] < 0$). Rejection implies that *some* series are stationary on average, not that every individual unit is stationary.
- **Using incorrect tabulated moments or critical values.** The IPS statistic depends on the model specification (no constant, constant, constant+trend), sample size T_i, and lag order. Using values corresponding to a different specification leads to size distortions.
- **Ignoring cross-sectional dependence.** Standard IPS assumes cross-sectional independence. Strong dependence across units (e.g., countries or firms exposed to common shocks) can inflate rejection rates and invalidate inference.
- **Misinterpreting differences between IPS and LLC results.** IPS allows heterogeneous autoregressive parameters across units, whereas LLC imposes a common persistence parameter. Different conclusions do not indicate contradiction—they reflect different alternative hypotheses.
- **Applying the test with highly uneven time dimensions.** IPS allows unbalanced panels, but extreme heterogeneity in T_i can reduce efficiency because each unit contributes equally to the average statistic.

Pesaran's CD Test

Test for Cross-Sectional Dependence in Panel Data

STAT.17 | **Category:** Stationarity | **Data:** Panel Time Series | **Null:** Standard Normal (asymptotic)

When to Use

- Test for cross-sectional dependence in panel data.
- Residual diagnostics for panel-data models.
- Financial or macroeconomic panel datasets.
- Check validity of panel independence assumptions.
- Preliminary check before panel unit-root tests.

Assumptions

- Data are recorded at fixed, uniform time intervals.
- Cross-sectional units are properly identified.
- Panel dimensions are sufficiently large.
- Errors are weakly stationary over time.
- Pairwise correlations are well defined.

Test Statistic

$$CD = \sqrt{\frac{2T}{N(N-1)}}\left(\sum_{i=1}^{N-1}\sum_{j=i+1}^{N}\hat{\rho}_{ij}\right), \quad \hat{\rho}_{ij} = \frac{\sum_{t=1}^{T} e_{it}e_{jt}}{\sqrt{\sum_{t=1}^{T} e_{it}^2}\sqrt{\sum_{t=1}^{T} e_{jt}^2}}. \qquad \text{Under } H_0,\ CD \approx N(0,1).$$

Notation: CD is cross-sectional dependence statistic, $\hat{\rho}_{ij}$ is correlation between units i and j, e_{it} are residuals

Plain English: Pesaran's CD test detects cross-sectional dependence by examining correlations among residuals across panel units. For each pair of units, the residual correlation is computed over time and then averaged across all pairs. Under cross-sectional independence, these correlations should average near zero. Significant positive or negative average correlation indicates dependence across units.

⚠ Practitioner Warning

Requires residuals from a fitted panel model (e.g., pooled OLS, fixed effects, or differenced regressions). Assumes average pairwise correlation is zero under H_0; heterogeneous dependence structures can affect power. Performs best in panels with moderate or large N; very small panels may exhibit size distortions. The test detects dependence but does not identify its source or structure (e.g., common factors, spatial spillovers, or network effects). Asymptotically robust to non-normality and heteroskedasticity. Unbalanced panels require modified formulas using pair-specific T_{ij}.

Hypotheses

- H_0: $E[\hat{\rho}_{ij}] = 0$ for all $i \neq j$.
- H_1: Cross-sectional dependence present.

Decision Rule: Reject H_0 if $|CD| > z_{\alpha/2}$ or p-value $< \alpha$.
+ve CD: +ve dependence common; -ve CD: -ve dependence.

Software References [Python, R, SAS]

Python	R	SAS
`Not Native (Manual)`	`plm::pcdtest`	`PROC PANEL (CDTEST)`

Additional Notes

Pesaran (2004) proposed the CD test as a simple alternative to earlier methods (e.g., Breusch–Pagan, Frees) that perform poorly in large-N panels. The test averages pairwise residual correlations $\hat{\rho}_{ij}$ and scales them so that $CD \to_d N(0,1)$ under cross-sectional independence. It remains valid for large N even when T is small or moderate (Pesaran, 2015). Alternative procedures include the Breusch–Pagan LM test (best when $T \gg N$) and Frees' rank-based statistic. A bias-corrected version (Pesaran, 2021) improves finite-sample performance. Cross-sectional dependence may arise from common shocks, spatial spillovers, or network effects. If dependence is detected, use models accounting for common factors (e.g., CCE, spatial, or factor models) and second-generation panel tests such as CIPS.

Miscellanea: Pesaran (2004) developed the CD test in response to the growing use of large panel datasets where cross-sectional independence was often unrealistic. Based on average pairwise residual correlations, the test is simple, scalable, and helped popularize panel methods that explicitly account for cross-sectional dependence.

Reference(s): Pesaran (2004, 2015, 2021)

Worked Example

Problem Statement An economist estimates a consumption-income relationship for $N = 4$ countries over $T = 8$ periods. A global demand shock (common factor) is suspected to create cross-sectional dependence in the residuals. Pesaran's CD test is used to detect this dependence before applying panel unit root or cointegration tests that assume independence.

	$t=1$	$t=2$	$t=3$	$t=4$	$t=5$	$t=6$	$t=7$	$t=8$
P1	(10, 13.7)	(11, 13.4)	(12, 15.3)	(13, 15.0)	(14, 16.7)	(15, 16.4)	(16, 18.5)	(17, 18.2)
P2	(8, 11.9)	(9, 11.5)	(10, 13.7)	(11, 13.5)	(12, 15.5)	(13, 15.2)	(14, 17.3)	(15, 17.0)
P3	(12, 15.6)	(13, 15.2)	(14, 17.0)	(15, 17.0)	(16, 18.5)	(17, 18.2)	(18, 20.2)	(19, 19.9)
P4	(9, 12.1)	(10, 11.5)	(11, 13.6)	(12, 13.1)	(13, 15.1)	(14, 14.8)	(15, 17.1)	(16, 16.5)

1. Step 1 — Individual OLS Residuals

Estimate $y_{it} = \alpha_i + \beta_i x_{it} + \hat{e}_{it}$ separately for each panel:

	$\hat{\alpha}_i$	$\hat{\beta}_i$	$\hat{e}_{it}$
P1	5.936	0.738	[+0.383, −0.655, +0.507, ..., −0.283]
P2	4.702	0.848	[+0.417, −0.831, +0.521, ..., −0.417]
P3	6.592	0.717	[+0.408, −0.708, +0.375, ..., −0.308]
P4	4.642	0.767	[+0.558, −0.808, +0.525, ..., −0.408]

Selected residuals $\hat{e}_{it}$ (alternating sign pattern reveals the common shock not captured by income).

2. Step 2 — Pairwise Correlations among Residuals

	(1,2)	(1,3)	(1,4)	(2,3)	(2,4)	(3,4)
Num.	2.504	2.153	2.803	2.357	3.047	2.603
Den.	2.541	2.180	2.828	2.388	3.099	2.658
$\hat{\rho}_{ij}$	0.985	0.988	0.991	0.987	0.983	0.980
Sum						5.914

Under H_0, $CD \xrightarrow{d} N(0,1)$ as $N, T \to \infty$.

3. Step 3 — CD Statistic

$$CD = \sqrt{\frac{2T}{N(N-1)}} \sum_{i=1}^{N-1} \sum_{j=i+1}^{N} \hat{\rho}_{ij} = 6.829$$

4. Interpretation

- **Decision:** Since $|CD| = 6.829 \gg 1.960$, we **reject** H_0 of cross-sectional independence. Strong positive dependence exists across all six (4 choose 2) country pairs ($\bar{\hat{\rho}} = 0.986$), consistent with a global common factor.
- **Practitioner Insight:** The CD test uses a *two-sided* critical region (± 1.960) because cross-sectional dependence can be positive (common global shocks) or negative (competition effects). The mean pairwise correlation $\bar{\hat{\rho}} = 0.986$ directly signals the practical severity of dependence values near 1 indicate that a single common factor drives most of the within-period variation. Failing to account for this would invalidate LLC and IPS tests (which assume cross-sectional independence); the Pesaran CIPS test is the appropriate unit root test to apply instead. Unlike the older Breusch-Pagan LM test, CD remains valid when $N > T$, making it the standard diagnostic for modern unbalanced macro panels.

Common Pitfalls

- **Small N undermines validity.** The CD test relies on large-N asymptotics; with $N < 10$ the null distribution may be poorly approximated by normality and size distortions can be substantial. Prefer $N \geq 20$ or use LM tests, bias-corrected CD, or permutation critical values when N is small.
- **Short T yields very noisy correlations.** Pairwise correlations estimated from very short panels ($T < 10$) have high sampling variability, producing both false positives and low power. Reliable inference generally requires $T \geq 15$.
- **Using CD on unbalanced panels.** Standard CD assumes a balanced panel; missing data cause unequal sample sizes for $\hat{\rho}_{ij}$ and invalidate asymptotics. Use modified CD tests for unbalanced panels or restrict to a balanced subsample.
- **Testing residuals before fixing model misspecification.** Cross-sectional dependence in residuals may reflect omitted time effects, common shocks, or wrong functional form—not true dependence. Include time FE or common-factor controls before applying the CD test.
- **Interpreting non-rejection as proof of independence.** Failure to reject often reflects low power (especially when dependence is weak or T is small). Treat non-significant results as inconclusive rather than evidence for independence; consider robust SE (e.g., Driscoll–Kraay).

XIV. STAT — Raw Data

Data used in selected worked examples are provided here to preserve the book's strict two-page-per-test format.

[STAT.15] Levin-Lin-Chu Test

Problem Statement An economist tests whether consumer price indices across four regions ($N = 4$) share a common unit root. The LLC test imposes a *homogeneous* AR coefficient under H_1 and gains power by pooling across units. The panel spans $T = 10$ periods per region; a fixed-effects specification is used with $k_i = 0$ (no lag augmentation), giving $\widetilde{T} = T - k_i - 1 = 9$ usable observations per unit.

	$t = 1$	$t = 2$	$t = 3$	$t = 4$	$t = 5$	$t = 6$	$t = 7$	$t = 8$	$t = 9$	$t = 10$
Panel 1	10.0	9.5	10.4	9.9	10.5	9.9	10.7	10.2	10.8	10.2
Panel 2	12.0	11.4	12.2	11.9	12.4	11.7	12.4	12.0	12.5	11.8
Panel 3	8.0	7.6	8.3	7.8	8.5	8.0	8.6	8.0	8.7	8.2
Panel 4	15.0	14.3	15.2	15.0	15.4	14.6	15.5	15.2	15.6	14.8

Three Variance Estimators: s^2, Newey–West, and the AR Spectral Density

Three distinct variance concepts appear throughout this chapter. They target different quantities and are appropriate in different settings.

1. Regression Variance s^2

The OLS residual variance:

$$s^2 = \frac{RSS}{n-k} = \frac{\sum_{t=1}^{n} \hat{u}_t^2}{n-k},$$

where k is the number of estimated parameters. s^2 estimates $\sigma^2 = \text{Var}(u_t)$, the variance of a *single* innovation, under the assumption that u_t is i.i.d. (or at least serially uncorrelated).

When to use: standard regression with i.i.d. or homoscedastic uncorrelated errors; forming the standard t-statistic $t = \hat{\beta}/\sqrt{s^2/\sum x_t^2}$.

Limitation: if u_t is serially correlated, s^2 consistently estimates σ^2 but the OLS t-statistic is no longer valid — its denominator ignores the covariance structure of the estimator.

2. Newey–West Long-Run Variance

When errors are serially correlated (and possibly heteroscedastic), the relevant quantity for inference is the **long-run variance** $\lambda^2 = \sum_{j=-\infty}^{\infty} \gamma_j$, where $\gamma_j = \text{Cov}(u_t, u_{t-j})$. The Newey–West (1987) kernel estimator is:

$$\hat{\lambda}^2_{\text{NW}} = \hat{\gamma}_0 + 2 \sum_{j=1}^{q} \underbrace{\left(1 - \frac{j}{q+1}\right)}_{\text{Bartlett weight}} \hat{\gamma}_j,$$

where $\hat{\gamma}_j = n^{-1} \sum_{t=j+1}^{n} \hat{u}_t \hat{u}_{t-j}$ and q is the bandwidth.

Note $\lambda^2 = \sigma^2$ when u_t is i.i.d. (all $\gamma_j = 0$ for $j \neq 0$). In general: positive serial correlation gives $\lambda^2 > \sigma^2$; negative gives $\lambda^2 < \sigma^2$.

When to use: HAC-robust inference in regression; Phillips–Perron unit root test; any setting where serial correlation is possible but you prefer not to model it parametrically. The Bartlett kernel guarantees $\hat{\lambda}^2_{\text{NW}} \geq 0$.

Bandwidth: default $q = \lfloor 4(n/100)^{2/9} \rfloor$ (Andrews automatic) or $q = \lfloor n^{1/3} \rfloor$.

Limitation: nonparametric; imprecise in small samples; unreliable when errors have a large negative MA component (causes over-rejection in unit root tests per Schwert 1989).

3. AR Spectral Density Estimator

A parametric alternative: fit an AR(k) to the residuals and recover the spectral density at frequency zero (Berk 1974):

$$\hat{\sigma}^2_{\text{AR}} = \frac{\hat{\sigma}^2_k}{\left(1 - \hat{\beta}(1)\right)^2}, \quad \hat{\beta}(1) = \textstyle\sum_{j=1}^{k} \hat{\beta}_j,$$

where $\hat{\sigma}^2_k = (T-k)^{-1} \sum_{t=k+1}^{T} \hat{e}^2_{tk}$ is the residual variance of the AR(k) and $\hat{\beta}_j$ are its coefficients (Ng–Perron 2001 convention: divisor $T-k$, counting usable differences). With $k = 0$: $\hat{\sigma}^2_{\text{AR}} = \hat{\sigma}^2_0 = T^{-1} \sum \hat{u}_t^2$.

When to use: DF-GLS and Ng–Perron M^{GLS} tests; any unit root test using GLS-detrended data. The AR spectral estimator is preferred over Newey–West in these settings because (i) it is invariant to the slope of the trend function under GLS detrending; and (ii) the MAIC lag selection explicitly accounts for the cost of underfitting the AR order.

Limitation: requires lag order selection k; misspecified k distorts size, especially under large negative MA errors — the central motivation for Ng–Perron's modified information criterion (MAIC).

Comparison Summary

	s^2	NW	AR spec.
Targets	σ^2	λ^2	λ^2
Approach	Parametric	Nonparametric	Parametric
Handles s.c.	No	Yes	Yes
Tuning	None	Bandwidth q	Lag k
Always ≥ 0	Yes	Yes	Yes
Use for	OLS t/F	PP test	DF-GLS, NP

When errors are i.i.d., all three coincide: $s^2 \approx \hat{\lambda}^2_{\text{NW}} \approx \hat{\sigma}^2_{\text{AR}}$. A large divergence among them is diagnostic evidence of serial correlation.

References. Newey & West (1987); Berk (1974); Andrews (1991); Ng & Perron (2001).

Phillips–Perron Test: Statistics and Supporting Formulas

Overview

The Phillips–Perron (PP) test (1988) modifies the Dickey–Fuller test statistics nonparametrically to account for serial correlation and heteroscedasticity in the error process, without requiring the explicit specification of lag terms. The test fits the same OLS regression as the DF test but applies a correction to the resulting statistics using a consistent estimate of the long-run variance.

Regression Model

Fit by OLS (deterministic components may be excluded or added):

$$y_t = \alpha + \rho\, y_{t-1} + \delta\, t + u_t,$$

where α (constant) and δt (trend) are included or excluded depending on the model specification. Let $\hat\rho$ denote the OLS estimate of ρ, $\hat\sigma$ the OLS standard error of $\hat\rho$, $\hat u_t$ the OLS residuals, k the number of regressors, and n the number of usable observations.

Step 1: Short-Run Variance Estimate

$$s_n^2 = \frac{1}{n-k}\sum_{t=1}^{n} \hat u_t^2.$$

This is the OLS residual mean square — the standard regression variance estimate using the degrees-of-freedom correction $n - k$.

Step 2: Autocovariance Estimates

For lag $j = 0, 1, \ldots, q$:

$$\hat\gamma_{j,n} = \frac{1}{n}\sum_{t=j+1}^{n} \hat u_t\, \hat u_{t-j}.$$

Note the denominator is n (not $n - j$) — this is a biased autocovariance estimator, consistent with the original paper.

Step 3: Long-Run Variance Estimate (Newey–West)

$$\hat\lambda_n^2 = \hat\gamma_{0,n} + 2\sum_{j=1}^{q} \underbrace{\left(1 - \frac{j}{q+1}\right)}_{\text{Bartlett weight}} \hat\gamma_{j,n},$$

where q is the truncation lag (bandwidth). A common automatic choice is $q = \lfloor 4(n/100)^{2/9} \rfloor$ (Newey–West automatic bandwidth). The Bartlett kernel ensures $\hat\lambda_n^2 > 0$.

The Two PP Test Statistics

Z_γ statistic (also written Z_α):

$$Z_\rho = n(\hat\rho - 1) - \frac{n^2}{2\,S_{xx}}\left(\hat\lambda_n^2 - \hat\gamma_{0,n}\right).$$

Z_τ statistic (also written Z_t):

$$Z_\tau = \sqrt{\frac{\hat\gamma_{0,n}}{\hat\lambda_n^2}}\, t_\gamma - \frac{\hat\lambda_n^2 - \hat\gamma_{0,n}}{2\hat\lambda_n} \cdot \frac{n}{\sqrt{S_{xx}}},$$

where $t_\gamma = \hat\gamma/\widehat{SE}(\hat\gamma)$ is the OLS t-ratio for $\hat\gamma = \hat\rho - 1$, and $S_{xx} = \sum_{t=1}^{n} \tilde y_{t-1}^2$ is the sum of squared demeaned lagged values (the OLS denominator for $\hat\rho$).

Interpretation of the Correction Terms

Each PP statistic has two parts:

- The **first term** is the standard DF statistic: $n(\hat\rho - 1)$ for Z_ρ; the scaled t-ratio $\sqrt{\hat\gamma_{0,n}/\hat\lambda_n^2}\, t_\gamma$ for Z_τ.
- The **second term** is the nonparametric correction. When errors are i.i.d., $\hat\lambda_n^2 = \hat\gamma_{0,n}$ and the correction vanishes, recovering the DF statistics exactly. When serial correlation is present, $\hat\lambda_n^2 > \hat\gamma_{0,n}$ and the correction shifts the statistic to account for the bias in the DF t-ratio.

The correction terms use S_{xx} rather than $\hat\sigma$ and s_n separately, because the OLS variance of $\hat\rho$ satisfies $\widehat{\text{Var}}(\hat\rho) = s_n^2/S_{xx}$, so $n^2\hat\sigma^2/s_n^2 = n^2/S_{xx}$ exactly. This substitution makes the formulas self-contained in terms of directly computable OLS quantities.

Notation Summary

Symbol	Definition
$\hat\rho, \hat\gamma$	OLS estimates; $\hat\gamma = \hat\rho - 1$
t_γ	OLS t-ratio $\hat\gamma/\widehat{SE}(\hat\gamma)$
$\hat u_t$	OLS residuals from the test regression
k	Number of regressors (incl. constant/trend)
n	Usable observations
s_n^2	Short-run variance: $RSS/(n-k)$
S_{xx}	$\sum_{t=1}^{n} \tilde y_{t-1}^2$, demeaned lag SS
$\hat\gamma_{j,n}$	Lag-j autocovariance: $(1/n)\sum \hat u_t \hat u_{t-j}$
$\hat\gamma_{0,n}$	Lag-0: $(1/n)\sum_{t=1}^{n} \hat u_t^2$
$\hat\lambda_n^2$	Long-run variance (Newey–West)
q	Bandwidth (Newey–West lags)
Z_ρ	PP statistic based on $n(\hat\rho - 1)$
Z_τ	PP statistic based on the t-ratio

Notes on Implementation

- The three model specifications (no constant; constant only; constant and trend) use the same correction formulas but different critical values from the DF/MacKinnon tables.
- p-values use MacKinnon's (1994) response surface; the same surface applies as for the ADF test.
- Davidson and MacKinnon (2004) document that PP can have severely distorted size in finite samples under MA errors — the ADF test is generally preferred for practical use.

References. Phillips & Perron (1988); Newey & West (1987); MacKinnon (1994); Davidson & MacKinnon (2004).

Part XV

Long-Run Equilibrium Between Time Series

Testing shared stochastic trends and cointegrating relationships

Cointegration

XV. COINT — A Conceptual Primer

What Is Cointegration?

Two or more $I(1)$ series are **cointegrated** if a linear combination of them is $I(0)$—stationary—even though each series individually drifts as a random walk. Formally, $(y_t, x_t) \sim \mathrm{CI}(1,1)$ if $\exists \beta$ such that

$$u_t = y_t - \beta x_t \sim I(0).$$

The vector $(1, -\beta)^\top$ is the **cointegrating vector**; u_t is the **equilibrium error**. The series share a **common stochastic trend**: their individual random-walk components cancel, leaving only transitory deviations.

Superconsistency: the OLS estimator $\hat{\beta}$ of the cointegrating vector converges at rate T (not $\sqrt{T}$), so estimation error in $\hat{\beta}$ is asymptotically negligible for inference on other parameters.

Why it matters:

- Avoids **spurious regression**: two unrelated $I(1)$ series have $R^2 \to 1$ and t-statistics diverging to $\pm\infty$; cointegration distinguishes genuine long-run relationships from statistical artifacts.
- Provides the foundation for **error-correction models (ECMs)**: the Granger Representation Theorem guarantees that cointegrated series have an ECM representation and vice versa.

The Error Correction Model (ECM)

By the Granger Representation Theorem, if (y_t, x_t) are CI(1,1), each variable has an ECM representation. For the bivariate case there are *two* equations, one for each variable:

$$\Delta y_t = \alpha_y u_{t-1} + \sum_{j=1}^{p-1} \Gamma_j^y \Delta y_{t-j} + \sum_{j=1}^{p-1} \Pi_j^y \Delta x_{t-j} + \varepsilon_{y,t},$$

$$\Delta x_t = \alpha_x u_{t-1} + \sum_{j=1}^{p-1} \Gamma_j^x \Delta y_{t-j} + \sum_{j=1}^{p-1} \Pi_j^x \Delta x_{t-j} + \varepsilon_{x,t},$$

where $u_{t-1} = y_{t-1} - \beta x_{t-1}$ is the lagged equilibrium error and α_y, α_x are the **adjustment (loading) coefficients** for y_t and x_t respectively, governing how quickly each variable reverts toward equilibrium after a shock. Cointegration requires that *at least one* of (α_y, α_x) be non zero; if both are zero, neither variable responds to deviations from equilibrium and no cointegrating relationship exists. In practice $\alpha_y < 0$ and $\alpha_x > 0$ are the expected signs: y_t falls and x_t rises when the system is above equilibrium ($u_{t-1} > 0$).

The multivariate generalization is the **VECM**:

$$\Delta \mathbf{Y}_t = \mathbf{\Pi} \mathbf{Y}_{t-1} + \sum_j \mathbf{\Gamma}_j \Delta \mathbf{Y}_{t-j} + \boldsymbol{\varepsilon}_t,$$

where $\mathrm{rank}(\mathbf{\Pi}) = r$ equals the number of cointegrating relations and $\mathbf{\Pi} = \boldsymbol{\alpha}\boldsymbol{\beta}^\top$.

What Are We Testing?

Framework	Hypothesis
Residual-based	H_0: no cointegration ($u_t \sim I(1)$)
System-based	H_0: $\mathrm{rank}(\mathbf{\Pi}) \le r$ (at most r cointegrating vectors)
ECM-based	H_0: $\alpha = 0$ (no error correction)
Null-reversal	H_0: cointegration ($u_t \sim I(0)$)

All tests require that each variable is confirmed $I(1)$ before applying cointegration tests. Including an $I(0)$ or $I(2)$ variable invalidates the test and its critical values.

Residual-Based Tests

Engle–Granger (EG) two-step:

1. Estimate the **cointegrating regression** by OLS: $\hat{u}_t = y_t - \hat{\alpha} - \hat{\beta} x_t$.
2. Apply an ADF-type test to $\{\hat{u}_t\}$.

Critical values are non-standard (must use EG or MacKinnon tables, not standard ADF tables) because $\hat{u}_t$ is a generated regressor. **Limitations:** restricted to a single cointegrating vector; sensitive to which variable is placed on the left-hand side (normalization); two-step propagates first-step estimation error.

Phillips–Ouliaris (PO): Applies a nonparametric (HAC) correction to the residuals rather than augmenting the regression; two statistics (Z_α and Z_t) are available. Advantage over EG: more robust to serial correlation in $\hat{u}_t$ without choosing a lag length.

Both EG and PO test H_0: *no* cointegration. Running them alongside the Shin test (which reverses the null) provides stronger triangulated evidence.

System-Based Tests: Johansen Framework

Johansen's reduced-rank regression is the **preferred multivariate method**: estimates $\boldsymbol{\alpha}$, $\boldsymbol{\beta}$, and $\mathbf{\Gamma}_j$ jointly from the VECM via canonical correlation between $\Delta \mathbf{Y}_t$ and $\mathbf{Y}_{t-1}$ after partialling out lagged differences.

Let $\hat{\lambda}_1 > \hat{\lambda}_2 > \cdots > \hat{\lambda}_K$ be the ordered squared canonical correlations.

Trace test: tests $H_0: r \le r_0$ vs. $H_1: r > r_0$:

$$\lambda_{\text{trace}}(r_0) = -T \sum_{i=r_0+1}^{K} \log(1 - \hat{\lambda}_i).$$

Maximum eigenvalue test: tests $H_0: r = r_0$ vs. $H_1: r = r_0 + 1$ (one additional cointegrating vector):

$$\lambda_{\max}(r_0, r_0+1) = -T \log(1 - \hat{\lambda}_{r_0+1}).$$

Critical values depend on $K - r_0$ and the specification of deterministics; five cases (Johansen 1995) range from no intercept to unrestricted trend. **Trace** has more power in general; **max eigenvalue** is sharper when the alternative is exactly one additional vector. Lag length p must be selected before testing (AIC/BIC on the unrestricted VAR in levels).

Deterministics in Cointegrated Systems

	Specification	Typical use
H_2^*	No intercept or trend	Rare
H_1^*	Restricted intercept (in CE only)	No drift in $\Delta\mathbf{Y}$
H_1	Unrestricted intercept	Linear trend in levels
H^*	Restricted linear trend (in CE)	Trend in CE
H	Unrestricted linear trend	Quadratic trend

CE = cointegrating equation ($u_t = y_t - \boldsymbol{\beta}' x_t$); "restricted" means the term enters only the CE, not the VAR equations freely. Misspecifying deterministics shifts critical values and can substantially distort inference on r.

Structural Break Cointegration Tests

Standard tests assume the cointegrating vector $\boldsymbol{\beta}$ is stable over time. Regime changes (policy shifts, financial crises) can cause $\boldsymbol{\beta}$ to shift, making a genuinely cointegrated relationship appear non-cointegrated.

Gregory–Hansen (GH): Extends EG to allow a *single unknown* structural break in the cointegrating vector, the intercept, or both. The test statistic is the *infimum* of the ADF statistic over all candidate break dates $\tau \in (0.15T, 0.85T)$; critical values adjusted for the break-date search. Three models: level shift (C), level shift with trend (C/T), and regime shift (C/S, allowing β to change).

Maki test: Generalizes GH to *multiple unknown break dates* (up to 5); selects breaks sequentially as dates minimizing the test statistic. Substantially more power than GH when more than one regime change has occurred.

Alternative Single-Equation Tests

Shin Residual Cointegration Test: A two-step procedure combining efficient estimation with null-reversal testing. **Step 1** uses Stock and Watson's (1993) Dynamic OLS (DOLS) estimator: the cointegrating regression is augmented with leads and lags of Δx_t to purge endogeneity and serial correlation, yielding an asymptotically efficient $\hat{\beta}$ with standard normal inference. **Step 2** applies Shin's (1994) statistic to the DOLS residuals—a KPSS-type test with H_0: *cointegration* (residuals are $I(0)$). Shin tabulates three variants keyed to the deterministics in Step 1: C (no intercept or trend), C_μ (intercept only), and C_τ (intercept and trend); critical values differ across variants and must be matched to the regression specification. Unlike EG and PO, small values support cointegration; large values reject it. Running Shin alongside PO provides triangulated evidence: if PO rejects no-cointegration *and* Shin fails to reject cointegration, the evidence is strong. The $C/C_\mu/C_\tau$ distinction is bookkeeping, not a conceptual one—all three share the same KPSS-type limiting distribution, differing only in which deterministics have been projected out.

Park's Added Variables Test: Augments the static cointegrating regression (same as CE) with superfluous deterministic functions of time—specifically polynomial terms t^j for integer powers $j \geq 2$—and tests their joint significance by a χ^2 Wald statistic. The logic is asymmetric: under cointegration the added variables are $I(0)$-equivalent in the regression and their coefficients converge to zero, so the statistic is $O_p(1)$ (the H test); under no cointegration the equilibrium error is $I(1)$, the polynomial terms do not vanish, and the statistic diverges. Park's test thus provides an independent regression-based check on cointegration without requiring residual extraction or kernel corrections.

ARDL Bounds Test: Proposed by Pesaran, Shin, and Smith (2001), the bounds test detects a levels relationship without requiring prior knowledge of the integration order of the regressors—the defining advantage over all other tests here. An ARDL(p, q) model is estimated by OLS and an F-statistic is computed for the joint significance of the lagged level terms. Inference uses two asymptotic critical value bounds: a lower bound (all regressors $I(0)$) and an upper bound (all $I(1)$). If F exceeds the upper bound, cointegration is confirmed regardless of integration orders; if F falls below the lower bound, it is rejected; if F falls between, the result is inconclusive. **Restriction:** no variable may be $I(2)$ or higher; confirm with unit root pre-tests. A companion t-statistic on the error-correction term provides an alternative bounds decision.

Panel Cointegration Tests

When T is limited but N (cross-sections) is large, panel tests pool information across units for higher power.

Pedroni test: Seven residual-based statistics generalizing EG to panels; allows heterogeneous cointegrating vectors and short-run dynamics across units. Four statistics pool within-dimension (panel statistics); three average between-dimension (group statistics). Robust to heterogeneity but assumes cross-sectional independence (CSD).

Westerlund test: ECM-based; tests $H_0 : \alpha_i = 0$ (no error correction) for each unit via the one-step levels ECM $\Delta y_{it} = \delta_i + \alpha_i y_{i,t-1} + \lambda_i x_{i,t-1} + \theta_i \Delta x_{it} + e_{it}$. Two panel statistics ($P_\alpha$, P_τ) pool α_i across units; two group-mean statistics (G_α, G_τ) average individual t-statistics. The normalizing quantity $\alpha_i(1) = 1 - \sum_{j=1}^{p_i} \alpha_{ij}$ equals 1 when no AR lags are included ($p_i = 0$), simplifying both α-statistics. **Advantage:** does not restrict the common factor structure; bootstrapped p-values correct for cross-sectional dependence without requiring independence.

Pre-test for panel: always run Pesaran's CD test (see Stationarity primer) before panel cointegration tests; cross-sectional dependence invalidates Pedroni and requires bootstrapped critical values for Westerlund.

Why Tests Differ and When to Use Each

Test	Best used when
Engle–Granger	Single equation, quick check
Phillips–Ouliaris	As above; avoids lag selection
Johansen	Multiple variables, full system
Gregory–Hansen	Single break suspected
Maki	Multiple breaks suspected
Shin (DOLS)	Efficient estimate + null-reversal check
Park	Regression-based confirmation
ARDL Bounds	Mixed $I(0)/I(1)$ regressors
Pedroni	Heterogeneous panel, no CSD
Westerlund	Panel with possible CSD

Common Pitfalls

- **Skipping integration pre-tests:** cointegration requires all variables to be $I(1)$; mixing $I(0)$ and $I(1)$ variables invalidates all tests except ARDL Bounds (which tolerates the mixture but not $I(2)$).
- **Wrong EG critical values:** using standard ADF tables instead of EG/MacKinnon tables for residuals is anti-conservative.
- **normalization sensitivity:** EG results depend on which variable is placed on the left; always try both and use Johansen for robustness.
- **Deterministic misspecification in Johansen:** wrong case shifts critical values; plot the data and choose the case deliberately.
- **Confusing Shin's null direction:** $C/C_\mu/C_\tau$ has H_0: cointegration—large values reject, not small. Applying EG/PO critical regions to Shin gives the wrong decision.
- **Wrong Shin variant:** use C_μ when the regression has an intercept only; use C_τ when it also includes a trend. Critical values differ materially between the two columns of Shin's Table 1.
- **Ignoring structural breaks:** standard tests under-reject H_0 when breaks are present; use GH or Maki when regime changes are plausible.
- **ARDL with $I(2)$ variables:** the bounds framework assumes no variable exceeds $I(1)$; confirm with unit root pre-tests.
- **Panel tests without CD check:** cross-sectional dependence inflates Pedroni; bootstrap Westerlund.

Bottom Line

> Cointegration formalizes long-run equilibrium among nonstationary variables, but conclusions are sensitive to normalization, lag length, deterministic specification, and structural stability. Confirm $I(1)$ status first—except for ARDL Bounds, which tolerates a mixture of $I(0)$ and $I(1)$ regressors. Use Engle–Granger for a quick single-equation check; Phillips–Ouliaris as a nonparametric complement; and run both alongside the Shin test to triangulate from opposing nulls. Use Johansen for a full multivariate system analysis. Apply Gregory–Hansen or Maki when regime breaks are plausible. In panels, screen for cross-sectional dependence before applying Pedroni or Westerlund.

Engle-Granger Two-Step Test

Test for Cointegration Between Time Series

COINT.1 | **Category:** Cointegration | **Data:** Time Series | **Null:** Tabled distribution

When to Use

- Test cointegration between nonstationary $I(1)$ series.
- Assess long-run equilibrium relationships.
- Distinguish spurious regression from true cointegration.
- Preliminary step before Error Correction modeling.
- Macroeconomic or financial time-series analysis.

Assumptions

- Observations are recorded at fixed, uniform time intervals.
- All series are integrated of order one, I(1).
- Long-run cointegrating relationship is linear in levels.
- Residuals from cointegrating regression are stationary.
- No endogeneity feedback affects the main regressors.

Test Statistic

$$\tau_{EG} = \frac{\hat{\gamma}}{SE(\gamma)}$$

Step 1: $y_t = \beta_0 + \beta_1 x_t + u_t$. Step 2: $\Delta\hat{u}_t = \gamma\hat{u}_{t-1} + \sum_{i=1}^{p} \phi_i \Delta\hat{u}_{t-i} + \epsilon_t$.

Notation: $\hat{u}_t$ are residuals from step 1, γ tests for unit root in residuals, special critical values apply

Plain English: The Engle–Granger test uses two steps: estimate the long-run relationship between nonstationary I(1) series, then apply the augmented Dickey–Fuller test to the residuals. If the residuals are stationary (mean-reverting), the series are cointegrated and share a stable long-run equilibrium relationship. Nonstationary residuals instead suggest a spurious regression with no genuine long-run connection between the series.

⚠ Practitioner Warning

CVs differ from standard ADF because residuals are estimated—use Engle–Granger or MacKinnon CVs. Test assumes a single cointegrating relationship and is not symmetric: reversing the dependent and independent variables can produce different results. For multiple I(1) variables, use Johansen. Pre-test for unit roots; applying to mixtures of I(0) and I(2) series invalidates inference. Lag selection in step 2 affects results. Power is low when alternatives are near a unit root.

Hypotheses

- H_0: $\gamma = 0$ (residuals contain a unit root; no cointegration).
- H_1: $\gamma < 0$ (stationary residuals; cointegrated variables).

Decision Rule: Reject H_0 if $t_\gamma < t_{\alpha,\text{EG}}$. Use Engle–Granger/MacKinnon critical values.

Software References [Python, R, SAS]

`statsmodels::coint` `urca::ca.jo` `PROC AUTOREG`

Additional Notes

MacKinnon (1991) developed response-surface approximations for Engle–Granger critical values, which depend on sample size, deterministic specification, and the number of regressors in the cointegrating equation. The framework extends naturally to multiple regressors: $y_t = \beta_0 + \beta_1 x_{1t} + \cdots + \beta_k x_{kt} + u_t$. Engle and Granger (1987) showed that if the series are cointegrated, the OLS estimator of the cointegrating vector is *super-consistent*, converging at rate T despite the regressors being nonstationary. Cointegration also implies an error correction representation: $\Delta y_t = \alpha(y_{t-1} - \beta' x_{t-1})$ + short-run dynamics, where $\alpha < 0$ measures the speed of adjustment toward long-run equilibrium. Although simple and widely used, the two-step Engle–Granger procedure is generally less efficient than system-based methods such as Johansen's maximum-likelihood approach. Phillips–Ouliaris (1990) later proposed alternative residual-based cointegration tests using Phillips–Perron-type statistics.

Miscellanea: Robert Engle and Clive Granger introduced cointegration theory in their influential 1987 *Econometrica* paper, a contribution that later earned them the 2003 Nobel Prize in Economics. The idea emerged from attempts to resolve the problem of "spurious regressions," where unrelated nonstationary series can exhibit misleadingly high correlations and significant t-statistics. Granger's key insight was that although individual series may be nonstationary, certain linear combinations can still be stationary, implying a stable long-run equilibrium relationship. The Engle–Granger framework fundamentally changed how economists model and interpret nonstationary time-series data.

Reference(s): Engle & Granger (1987); MacKinnon (1991)

Engle-Granger Two-Step Test

Worked Example

Problem Statement An economist tests whether a commodity price index (y_t) and its input cost index (x_t) are cointegrated. Both series must first be confirmed $I(1)$. $T = 20$, constant in Step 1, lag selected by AIC in Step 2.

	$t=1$	$t=2$	$t=3$	$t=4$	$t=5$	$t=6$	$t=7$	$t=8$	$t=9$	$t=10$
x_t	5.0	5.1	6.1	6.1	7.0	7.1	8.1	8.1	9.0	9.1
y_t	9.9	9.3	11.6	10.7	12.8	12.1	14.6	13.8	16.0	15.4
	$t=11$	$t=12$	$t=13$	$t=14$	$t=15$	$t=16$	$t=17$	$t=18$	$t=19$	$t=20$
x_t	10.1	10.1	11.0	11.1	12.1	12.1	13.0	13.1	14.1	14.1
y_t	17.5	16.8	18.8	18.2	20.5	19.8	22.0	21.2	23.4	22.6

0. Pre-Test: Confirm Both Series are $I(1)$
ADF ($k = 0$, constant) on each level series:

- x_t: $\tau = -0.384 > -2.86 \Rightarrow$ FTR, $I(1)$ ✓
- y_t: $\tau = -0.788 > -2.86 \Rightarrow$ FTR, $I(1)$ ✓

1. Step 1 — Cointegrating Regression (OLS)
Regress y_t on $[1, x_t]$:

$$\hat{\alpha} = 2.073, \quad \hat{\beta} = 1.491, \quad \hat{u}_t = y_t - \hat{\alpha} - \hat{\beta} x_t$$

Selected residuals: $\hat{u}_1 = +0.372, \hat{u}_2 = -0.378, \hat{u}_{11} = +0.367, \hat{u}_{20} = -0.497$.
The residuals oscillate around zero — a preliminary signal of stationarity (cointegration).

2. Step 2 — Lag Selection for ADF on $\hat{u}_t$
ADF on $\hat{u}_t$ **without constant** (already demeaned by OLS):

k	$\hat{\gamma}$	τ_{EG}	AIC	$\hat{\rho}_1^{res}$
0	−1.956	−23.75	−18.78	+0.586
1	−0.701	−1.690	−23.04	+0.461
2	−1.317	−3.810	**−31.18**	−0.029

AIC selects $k^* = 2$; $\hat{\rho}_1^{res}$ (residual autocorrelation) drops to −0.029 (white noise).

3. ADF Regression on $\hat{u}_t$ ($k = 2$)
$\Delta\hat{u}_t = \gamma\,\hat{u}_{t-1} + c_1\,\Delta\hat{u}_{t-1} + c_2\,\Delta\hat{u}_{t-2} + e_t, \; n = 17$:

$$\hat{\gamma} = -1.317, \quad \hat{c}_1 = 0.413, \quad \hat{c}_2 = 0.751$$

$$\hat{s}^2 = 0.0080, \; SE(\hat{\gamma}) = 0.346, \; \tau_{EG} = \frac{-1.317}{0.346} = -3.810$$

4. Interpretation
Critical values (MacKinnon 1991, $n = 1$ regressor, $T \approx 20$):

- **Decision:** Since $\tau_{EG} = -3.810 < -3.370$, we **reject** H_0 of no cointegration. The residuals $\hat{u}_t$ are $I(0)$: a stable long-run relationship $y_t \approx 2.07 + 1.49 x_t$ exists.
- **Practitioner Insight:** Two features distinguish EG Step 2 from a standard ADF test. First, no constant is included in the ADF regression because OLS in Step 1 already forces $\sum \hat{u}_t \approx 0$. Second, the critical values are *more negative* than standard ADF critical values (e.g. −3.37 vs −2.86 at 5%) because $\hat{u}_t$ are estimated residuals, not observed series — using ADF tables would lead to systematic over-rejection of H_0. The number of regressors in Step 1 also shifts the critical values: for two regressors ($n = 2$), the 5% CV tightens further to approximately −3.80. The main limitation of EG is that it assumes a unique cointegrating vector; the Johansen test should be used when more than two $I(1)$ series are involved.

Common Pitfalls

- **Using standard ADF critical values.** Residuals are generated regressors, so standard ADF tables are invalid and lead to severe over-rejection. Always use Engle–Granger/MacKinnon critical values appropriate to the number of variables and deterministic terms.
- **Testing when variables are not all I(1).** Engle–Granger requires all series to be I(1). Mixed integration orders invalidate inference. Pre-test each variable for unit roots before proceeding.
- **Ignoring normalization dependence.** Results depend on which variable is chosen as dependent. With two variables, test both normalizations; with multiple potential cointegrating vectors, prefer Johansen.
- **Overlooking structural breaks.** The test assumes a stable long-run relationship. Regime shifts reduce power and may mask true cointegration. Use break-adjusted tests when instability is suspected.
- **Mistaking high R^2 for cointegration.** Trending I(1) variables often produce high R^2 even when unrelated. Only stationarity of residuals (Step 2) establishes cointegration.

Phillips-Ouliaris Test

Non-Parametric Cointegration Test

COINT.2 | **Category:** Cointegration | **Data:** Time Series | **Null:** Tabled distribution

When to Use

- Test cointegration between nonstationary $I(1)$ series.
- Useful when error structure is unknown or complex.
- Preliminary step before Error Correction modeling.
- Avoid lag-selection sensitivity in ADF residual tests.

Assumptions

- All series are integrated of order one, I(1).
- Long-run cointegrating relationship is linear in levels.
- Long-run variance is consistently estimated.
- No endogeneity feedback affects the main regressors.

Test Statistic

$$\text{(I) } Z_\tau = \sqrt{\frac{\hat{\gamma}_0}{\hat{\lambda}^2}}\, t_\gamma - \frac{1}{2}\cdot\frac{\hat{\lambda}^2 - \hat{\gamma}_0}{\hat{\lambda}}\cdot\frac{n\,\hat{\sigma}}{s_n} \quad \Big| \quad \text{(II) } Z_\gamma = n\hat{\gamma} - \frac{n^2}{2}\cdot\frac{\hat{\sigma}^2}{s_n^2}\left(\hat{\lambda}^2 - \hat{\gamma}_0\right) \quad \Big| \quad t_\gamma = \frac{\hat{\gamma}}{\hat{\sigma}},\ \hat{\sigma} = \text{OLS } SE(\hat{\gamma})$$

Notation: $s_n^2 = RSS/(n-p)$ short-run variance; $\hat{\gamma}_0 = n^{-1}\sum \hat{\epsilon}_t^2$; $\hat{\lambda}^2$ long-run variance, n is eff. sample size, p is estimated parameters. COINTEGRATING REG.: $y_{1t} = \hat{\mu} + \beta t + \hat{\boldsymbol{\phi}}'\mathbf{x}_t + \hat{u}_t$; AUXILIARY RESIDUAL REG.: $\Delta\hat{u}_t = \gamma\hat{u}_{t-1} + \epsilon_t$. CONFIGURATIONS: (A) Intercept-free residuals; (B) Demeaned residuals ($\hat{\mu}$ included in Step 1); (C) Demeaned/detrended residuals ($\hat{\mu}$ and linear trend βt included in Step 1).

Plain English: The Phillips–Ouliaris test examines whether nonstationary $I(1)$ variables share a stable long-run equilibrium relationship by testing the stationarity of regression residuals. Like the Engle–Granger approach, the procedure first estimates the long-run relationship and then tests whether the residuals are mean-reverting. However, instead of adding lagged differences as in ADF-based residual tests, Phillips–Ouliaris uses nonparametric corrections for serial correlation and heteroskedasticity through long-run variance estimation. This avoids explicit lag selection while preserving the same cointegration framework.

⚠ Practitioner Warning

Critical values differ from standard ADF values because residuals are estimated rather than observed. Although the procedure avoids explicit lag selection, results remain sensitive to bandwidth and kernel choices in long-run variance estimation. The test assumes a single cointegrating relationship and is not symmetric (switching roles of y and x) with respect to normalization; may lose power under structural breaks or near-unit-root alternatives.

Hypotheses

- H_0: $\gamma = 0$ No cointegration (residuals non-stationary).
- H_1: $\gamma < 0$ Cointegration exists (residuals stationary).

Decision Rule: Reject H_0 if $Z_\tau < Z_{\tau,\alpha}$ or $Z_\gamma < Z_{\gamma,\alpha}$ (left-tailed tests). Use Phillips-Ouliaris critical values.

Software References [Python, R, SAS]

`statsmodels::coint`	`urca::ca.po`	`Not Native (Manual)`

Additional Notes

The long-run variance estimator is $\hat{\lambda}^2 = \hat{\gamma}_0 + 2\sum_{j=1}^{q} w_j\hat{\gamma}_j$, where $\hat{\gamma}_j$ are residual autocovariances and w_j are kernel weights (e.g., Bartlett: $w_j = 1 - j/(q+1)$, Quadratic Spectral and Parzen are alternatives). While Phillips–Ouliaris (1990) also introduced alternative multivariate variance ratio tests (P_u and P_z), the residual-based Z_τ and Z_α metrics remain the primary standards for practitioner-focused workflows. MacKinnon (1991) provided response surface critical values based on regressors and deterministic terms. Conceptually, the test parallels Phillips–Perron for unit roots: a residual-based test using nonparametric corrections rather than parametric lag augmentation.

Miscellanea: Peter Phillips and Sam Ouliaris (1990) extended Phillips–Perron methodology to residual-based cointegration testing as an alternative to the Engle–Granger approach. Their procedure replaced parametric lag augmentation with nonparametric long-run variance estimation, avoiding the need to specify lag length directly in the residual regression. Although this reduced sensitivity to lag specification, the method instead introduced bandwidth-selection issues through the long-run variance estimator.

Reference(s): Phillips & Ouliaris (1990); MacKinnon (1991)

Worked Example

Problem Statement An economist re-tests the same commodity price / input cost panel from the Engle-Granger example using the Phillips-Ouliaris test. Unlike EG which uses lag augmentation to correct for serial correlation, PO applies a non-parametric Newey-West correction — directly analogous to the PP unit root test. $T = 20$, $\ell = 2$ Bartlett kernel.

	$t=1$	$t=2$	$t=3$	$t=4$	$t=5$	$t=6$	$t=7$	$t=8$	$t=9$	$t=10$
x_t	5.0	5.1	6.1	6.1	7.0	7.1	8.1	8.1	9.0	9.1
y_t	9.9	9.3	11.6	10.7	12.8	12.1	14.6	13.8	16.0	15.4
	$t=11$	$t=12$	$t=13$	$t=14$	$t=15$	$t=16$	$t=17$	$t=18$	$t=19$	$t=20$
x_t	10.1	10.1	11.0	11.1	12.1	12.1	13.0	13.1	14.1	14.1
y_t	17.5	16.8	18.8	18.2	20.5	19.8	22.0	21.2	23.4	22.6

1. Step 1 — Cointegrating Regression (OLS)
Identical to EG Step 1:

$$\hat\alpha = 2.073,\ \hat\beta = 1.491,\ \hat u_t = y_t - 2.073 - 1.491\,x_t$$

2. Step 2 — OLS of $\Delta\hat u_t$ on $\hat u_{t-1}$
No constant, no lags ($k = 0$ — PP style), $n = T - 1$:

$$\hat\gamma = \frac{\sum \hat u_{t-1}\,\Delta\hat u_t}{\sum \hat u_{t-1}^2} = -1.956,\ RSS = 0.3727,\ s_n^2 = \frac{RSS}{19-1}$$

$$\hat s_n^2 = 0.0207,\quad \hat\sigma = SE(\hat\gamma) = 0.0825,\quad t_{\hat\gamma} = -23.702$$

3. Step 3 — NW Long-Run Variance ($\ell = 2$)
From residuals $\hat e_t = \Delta\hat u_t - \hat\gamma\,\hat u_{t-1}$:

- $\hat\gamma_h = \sum_{t=h+1}^{n} \hat e_t \hat e_{t-h}/19,\ h = 0, 1, 2$ (MLE-style)
- $\hat\gamma_0 = 0.0196, \hat\gamma_1 = 0.0108, \hat\gamma_2 = -0.00108$
- $\hat\lambda^2 = 0.0196 + \frac{4}{3}(0.0108) + \frac{2}{3}(-0.00108) = 0.0333$

4. Step 4 — PO Corrections
Let $\sum \tilde u_{t-1}^2 = 3.041$ (demeaned lag sum of squares).

$$Z_\tau = \sqrt{\frac{\hat\gamma_0}{\hat\lambda^2}}\,t_\gamma - \frac{1}{2}\cdot\frac{\hat\lambda^2 - \hat\gamma_0}{\hat\lambda}\cdot\frac{n\,\hat\sigma}{s_n} = -18.60$$

$$Z_\alpha = n\hat\gamma - \frac{n^2}{2}\cdot\frac{\hat\sigma^2}{s_n^2}\left(\hat\lambda^2 - \hat\gamma_0\right) = -37.97$$

5. Results and Decision
Critical values (PO 1990, $n = 1$ regressor, $T \approx 20$):

Stat	Value	CV (5%)	Decision
Z_τ	−18.60	−3.370	Reject H_0
Z_α	−37.97	−20.70	Reject H_0

Formula: $\hat\sigma = SE(\hat\gamma) = s_n/\sqrt{\sum \tilde u_{t-1}^2}$. s_n^2 is the usual unbiased variance estimator.

6. Interpretation

- **Decision:** Both PO statistics are far below their 5% critical values. The series share a stable long-run equilibrium.
- **Practitioner Insight:** The PO test is the cointegration analog of the PP unit root test: Z_τ and Z_α mirror the $Z(\hat t)$ and $Z(\hat\alpha)$ formulas exactly, with $\hat u_t$ replacing y_t. Like PP vs ADF, PO avoids selecting a lag length — but the NW bandwidth ℓ plays an equivalent role. Here $\hat\lambda^2 > \hat\gamma_0$ (positive residual autocorrelation) so the correction shifts both statistics further negative, strengthening the rejection. The PO critical values are more negative than standard PP critical values (e.g. −3.37 vs −2.86 at 5% for Z_τ) for the same reason as EG: estimating the cointegrating regression alters the null distribution of the residual-based statistics. Both EG and PO use the same cointegrating regression in Step 1 and test the same H_0; their results should be consistent, as they are here. Note: Phillips–Perron and Phillips–Ouliaris typically use the *biased* (MLE-style) Newey–West autocovariance estimator, dividing by T rather than $T - k$ when computing sample autocovariances.

Common Pitfalls

- **Using Phillips–Perron critical values.** PO statistics resemble PP, but critical values differ because residuals are generated regressors (from cointegrating regression). Using PP tables causes serious over-rejection. Always use Phillips–Ouliaris/MacKinnon critical values.
- **Poor bandwidth selection.** Bandwidth choice directly affects long-run variance estimates. Too small over-rejects; too large reduces power. Use automatic selection (e.g., Andrews) and report bandwidth; check robustness.
- **Applying PO in small samples.** Nonparametric variance estimation performs poorly for $T < 100$. In small samples, Engle–Granger may have better size and power properties.
- **Assuming PO and Engle–Granger must agree.** They test the same null with different corrections; disagreement often signals lag or bandwidth misspecification. Re-check specifications if results conflict.
- **Ignoring single-vector limitation.** Like EG, PO tests only one cointegrating vector. For multiple vectors, use Johansen.

Johansen Trace Test

Test for Number of Cointegrating Relationships

COINT.3 | **Category:** Cointegration | **Data:** Multivariate Time Series | **Null:** Tabled distribution

When to Use

- Determine number of cointegrating relationships.
- Multivariate cointegration analysis in VAR systems.
- Multiple nonstationary time-series variables.
- Macroeconomic or financial system modeling.
- Preliminary step before VECM estimation.

Assumptions

- Variables are integrated of order one.
- VAR lag structure is correctly specified.
- Cointegrating relationships are linear.
- Underlying structural VAR errors must be white noise.
- No higher-order deterministic trends exist over time.

Test Statistic

$$\lambda_{\text{trace}}(r) = -T \sum_{i=r+1}^{p} \ln(1 - \hat{\lambda}_i) \quad \Big| \quad \textbf{VECM: } \Delta \mathbf{y}_t = \mathbf{c} + \mathbf{\Pi} \mathbf{y}_{t-1} + \sum_{i=1}^{p-1} \mathbf{\Gamma}_i \Delta \mathbf{y}_{t-i} + \boldsymbol{\epsilon}_t \text{ (vector ECM)}$$

Notation: r is hypothesized cointegration rank, p is number of variables, $\hat{\lambda}_i$ are eigenvalues $(1 > \hat{\lambda}_1 > \cdots > \hat{\lambda}_p \geq 0)$

Plain English: The Johansen trace test determines how many linearly independent cointegrating relationships exist among a set of I(1) variables by testing the rank of the long-run coefficient matrix in a VECM. The test sums logs of eigenvalues from a reduced rank regression, with small eigenvalues (near 0) corresponding to unit roots and large eigenvalues (near 1) indicating stationarity. Testing proceeds sequentially: first test no cointegration ($r = 0$), then at most one ($r \leq 1$), continuing until failing to reject. The first non-rejection identifies the cointegration rank.

⚠ Practitioner Warning

Critical values depend on number of variables, deterministic specification (constant or trend inside/outside the cointegrating relation), and hypothesized rank. Sequential testing inflates overall Type I error; some recommend Bonferroni or other corrections. Lag length selection critical; information criteria (AIC, BIC) or sequential testing on VAR residuals. Small sample bias can lead to over-rejection; use Bartlett small-sample correction or adjust critical values. Test assumes all variables I(1); mixing I(0) and I(1) variables invalidates inference. Power decreases as number of variables increases relative to sample size.

Hypotheses

- H_0: Rank of cointegration matrix ($\mathbf{\Pi}$), $r \leq r_0$.
- H_1: Rank of cointegration matrix ($\mathbf{\Pi}$), $r > r_0$.

Decision Rule: Reject H_0 if $\lambda_{\text{trace}}(r) > CV_{\alpha,r}$ or p value $< \alpha$. Test sequentially from $r_0 = 0$ until first non-rejection.

Software References [Python, R, SAS]

`statsmodels::coint_johansen`	`urca::ca.jo(type="trace")`	`PROC VARMAX (COINT JOHANSEN)`

Additional Notes

The Johansen procedure estimates a VECM:

$$\Delta \mathbf{y}_t = \mathbf{\Pi} \mathbf{y}_{t-1} + \sum_{i=1}^{p-1} \mathbf{\Gamma}_i \Delta \mathbf{y}_{t-i} + \boldsymbol{\epsilon}_t,$$

where $\mathbf{\Pi} = \boldsymbol{\alpha}\boldsymbol{\beta}'$ with rank r. The matrix $\boldsymbol{\beta}$ contains cointegrating vectors and $\boldsymbol{\alpha}$ adjustment coefficients. Cointegration rank is determined via eigenvalues from a reduced-rank regression problem, leading to trace and maximum eigenvalue tests. Different deterministic specifications (intercepts/trends) must be chosen appropriately. Finite-sample adjustments (e.g., Bartlett correction) and critical values (Osterwald-Lenum, 1992) are used for inference.

Miscellanea: Søren Johansen developed a maximum-likelihood framework for cointegration in the late 1980s, formulating the problem as a reduced-rank regression solved through eigenvalue analysis. Further developed with Katarina Juselius, the approach provided a unified system-based method for detecting and estimating multiple cointegrating relationships and quickly became a standard tool in multivariate econometrics.

Reference(s): Johansen (1988, 1991); Osterwald-Lenum (1992)

Johansen Trace Test

COINT.3

Worked Example

Problem Statement An economist tests for cointegrating relationships among a bivariate system ($n = 2$): commodity price y_{1t} and input cost y_{2t}. Unlike EG and PO, the Johansen test handles multiple cointegrating vectors simultaneously via maximum likelihood on a VAR. A VAR(1) with restricted constant is estimated ($T = 20$, $p = 1 \Rightarrow$ no lagged differences in the VECM).

	$t=1$	$t=2$	$t=3$	$t=4$	$t=5$	$t=6$	$t=7$	$t=8$	$t=9$	$t=10$
y_{1t}	9.9	9.3	11.6	10.7	12.8	12.1	14.6	13.8	16.0	15.4
y_{2t}	5.0	5.1	6.1	6.1	7.0	7.1	8.1	8.1	9.0	9.1
	$t=11$	$t=12$	$t=13$	$t=14$	$t=15$	$t=16$	$t=17$	$t=18$	$t=19$	$t=20$
y_{1t}	17.5	16.8	18.8	18.2	20.5	19.8	22.0	21.2	23.4	22.6
y_{2t}	10.1	10.1	11.0	11.1	12.1	12.1	13.0	13.1	14.1	14.1

1. Reduced-Form Residuals ($p = 1, T_{\text{eff}} = 19$)
With $p = 1$: $R_{0t} = \Delta Y_t - \overline{\Delta Y}$ and $R_{1t} = Y_{t-1} - \bar{Y}_{t-1}$ (demeaned). Selected rows:

t	R_{01}	R_{02}	R_{11}	R_{12}
2	−1.268	−0.379	−6.121	−4.337
3	+1.632	+0.521	−6.721	−4.237
4	−1.568	−0.479	−4.421	−3.237

2. Cross-Product Matrices $S_{ij} = \frac{1}{T_{\text{eff}}} R_i' R_j$:

$$S_{00} = \begin{pmatrix} 2.156 & 0.665 \\ 0.665 & 0.207 \end{pmatrix}, \quad S_{11} = \begin{pmatrix} 17.380 & 11.424 \\ 11.424 & 7.578 \end{pmatrix}$$

$$S_{01} = \begin{pmatrix} -1.148 & -0.383 \\ -0.350 & -0.116 \end{pmatrix}, \quad T_{eff} = 19$$

3. Eigenvalue Problem
Solve $|\lambda S_{11} - S_{10} S_{00}^{-1} S_{01}| = 0$. The ordered eigenvalues of $S_{11}^{-1} S_{10} S_{00}^{-1} S_{01}$ are: $\hat{\lambda}_1 = 0.9695$, $\hat{\lambda}_2 = 0.0002$

4. Trace Statistics (Sequential Testing)

$$LR_{\text{trace}}(r) = -T_{\text{eff}} \sum_{i=r+1}^{n} \ln(1 - \hat{\lambda}_i)$$

H_0	LR_{trace}	CV (5%)	Decision
$r \le 0$	66.296	15.410	Reject H_0
$r \le 1$	0.003	3.760	FTR H_0

5. Interpretation

- **Decision:** Reject $r \le 0$, FTR $r \le 1 \Rightarrow$ exactly $r = 1$ cointegrating vector.
- **Cointegrating vector:** The eigenvector for $\hat{\lambda}_1$, normalized to unity on y_{1t}: $\hat{\beta} = [1, -1.493]'$, closely recovering the true $[1, -1.5]'$.

Practitioner Insight: The Johansen trace test is applied *sequentially*: test $r \le 0$ first, if rejected, test $r < 1$; stop at the first non-rejection — that value is $\hat{r}$. The trace statistic combines *all* eigenvalues beyond rank r via the sum of $-\ln(1 - \hat{\lambda}_i)$ terms; here $\hat{\lambda}_2 \approx 0$ makes $LR_{\text{trace}}(r = 1) \approx 0$, a clear signal of no second cointegrating vector. The CVs depend on $n - r$ (common stochastic trends under H_0) and the deterministic specification — always match the CV table to the constant/trend case used in estimation. Unlike EG and PO which require choosing a dependent variable, Johansen is symmetric across all n series and simultaneously identifies both $\hat{\beta}$ (cointegrating space) and $\hat{\alpha}$ (adjustment speeds).

Common Pitfalls

- **Using incorrect critical values for the deterministic specification.** Johansen critical values depend on the deterministic components included in the VAR (e.g., constants or trends). Using tables corresponding to a different specification can produce substantial size distortions.
- **Misinterpreting a rank of $k - 1$.** When $\hat{r} = k - 1$ (with k variables), the system still contains one common stochastic trend. This represents the maximum number of cointegrating relationships, not full stationarity in levels.
- **Applying the test in very small samples.** The trace statistic relies on asymptotic approximations that can perform poorly when the sample size is limited. Small samples may lead to over-rejection and spurious evidence of cointegration.
- **Ignoring sensitivity to VAR lag length.** The estimated cointegration rank depends on the lag structure of the underlying VAR. Lag order should be selected carefully and robustness across nearby lag choices should be examined.

Johansen Maximum Eigenvalue Test

Alternative Sequential Test for Cointegration Rank

COINT.4 | **Category:** Cointegration | **Data:** Multivariate Time Series | **Null:** Tabled distribution

When to Use

- Determine number of cointegrating relationships.
- Multivariate cointegration analysis in VAR systems.
- Multiple nonstationary time-series variables.
- Macroeconomic or financial system modeling.
- Preliminary step before VECM estimation.

Assumptions

- Variables are integrated of order one.
- VAR lag structure is correctly specified.
- Cointegrating relationships are linear.
- Underlying structural VAR errors must be white noise.
- No higher-order deterministic trends exist over time.

Test Statistic

$$\lambda_{\max}(r) = -T\ln(1-\hat{\lambda}_{r+1})$$

$$\textbf{VECM: } \Delta\mathbf{y}_t = \mathbf{c} + \mathbf{\Pi}\mathbf{y}_{t-1} + \sum_{i=1}^{p-1}\mathbf{\Gamma}_i\Delta\mathbf{y}_{t-i} + \boldsymbol{\epsilon}_t \quad \text{(vector ECM)}$$

Notation: r is hypothesized cointegration rank, $\hat{\lambda}_{r+1}$ is $(r+1)$-th largest eigenvalue, T is sample size

Plain English: The Johansen maximum eigenvalue test examines whether the cointegration rank increases from r to $r+1$ by focusing on the $(r+1)$-th eigenvalue of the system. Unlike the trace test, which cumulatively evaluates all remaining eigenvalues beyond rank r, the maximum eigenvalue test isolates a single eigenvalue at each step. This produces a sharper alternative hypothesis—that the rank is specifically $r+1$ rather than merely greater than r. Sequential testing proceeds similarly to the trace procedure until the null hypothesis is no longer rejected.

⚠ Practitioner Warning

Critical values differ from those of the trace test and depend on the number of variables, deterministic specification, and hypothesized cointegration rank. The trace and maximum eigenvalue procedures can produce different rank estimates in finite samples, so applied studies often report both. The maximum eigenvalue test may have stronger power against specific rank alternatives, whereas the trace statistic is often viewed as more stable across broader alternatives. Both procedures can over-reject in small samples, motivating Bartlett corrections or adjusted critical values. As with the trace test, careful lag selection and integration-order assessment are essential.

Hypotheses

- H_0: Rank of cointegration matrix $(\mathbf{\Pi})$, $r = r_0$.
- H_1: Rank of cointegration matrix $(\mathbf{\Pi})$, $r = r_0 + 1$.

Decision Rule: Reject H_0 if $\lambda_{\max}(r) > CV_{\alpha,r}$ or p-value $< \alpha$.
Test sequentially from $r_0 = 0$ until first non-rejection.

Software References [Python, R, SAS]

```
statsmodels::coint_johansen        urca::ca.jo(type="eigen")        PROC VARMAX (COINT JOHANSEN)
```

Additional Notes

The trace statistic is the sum of maximum eigenvalue statistics: $\lambda_{\text{trace}}(r) = \sum_{i=r+1}^{n}\lambda_{\max}(i-1)$. The maximum eigenvalue test targets H_1 : rank $= r+1$, while the trace test uses H_1 : rank $> r$, making the former more powerful for detecting a single additional vector and the latter more general. Finite-sample adjustments (e.g., Bartlett correction) and simulation evidence suggest the trace test is often more robust, while the maximum eigenvalue test can be sharper in specific cases. In practice, both are reported—agreement strengthens conclusions, while disagreement warrants further investigation. Bootstrap methods can improve inference when asymptotic assumptions are questionable.

Miscellanea: Søren Johansen introduced both the trace and maximum eigenvalue tests as part of his unified maximum-likelihood framework for cointegration. Whereas the trace statistic tests whether the cointegration rank is at most r, the maximum eigenvalue statistic focuses specifically on the increment from r to $r+1$ cointegrating relationships. Because the two tests can sometimes produce different rank estimates in finite samples, applied studies often report and compare both.

Reference(s): Johansen (1991); Osterwald-Lenum (1992); Cheung & Lai (1993)

Worked Example

Problem Statement The same bivariate system (y_{1t}: commodity price, y_{2t}: input cost, $T = 20$, $p = 1$) is now tested using the Johansen Maximum Eigenvalue statistic. While the Trace test asks "is the rank *at most* r?", the Max-Eigenvalue test asks "is the rank *exactly* r?" — a sharper, one-step-at-a-time alternative hypothesis. See Johansen's Trace test for data.

1. Steps 1–3: Identical to Trace (previous test entry)
The S_{ij} matrices and eigenvalues are computed exactly as in the Trace example ($p = 1$, $T_{\text{eff}} = 19$, restricted constant):

$$S_{00} = \begin{pmatrix} 2.156 & 0.665 \\ 0.665 & 0.207 \end{pmatrix}, \quad S_{11} = \begin{pmatrix} 17.380 & 11.424 \\ 11.424 & 7.578 \end{pmatrix}$$

$$S_{01} = \begin{pmatrix} -1.148 & -0.383 \\ -0.350 & -0.116 \end{pmatrix}$$

Ordered eigenvalues of $S_{11}^{-1} S_{10} S_{00}^{-1} S_{01}$:

$$\hat{\lambda}_1 = 0.9695, \qquad \hat{\lambda}_2 = 0.0002$$

2. Max-Eigenvalue Statistic
Unlike Trace, each statistic uses *only one* eigenvalue:

$$LR_{\max}(r) = -T_{\text{eff}} \ln(1 - \hat{\lambda}_{r+1})$$

$$LR_{\max}(r = 0) = -19 \ln(1 - 0.9695) = 66.293$$

$$LR_{\max}(r = 1) = -19 \ln(1 - 0.0002) = 0.003$$

3. Additive Relationship with Trace

$$LR_{\text{trace}}(r = 0) = LR_{\max}(r = 0) + LR_{\max}(r = 1)$$

$$66.296 = 66.293 + 0.003 \quad \checkmark$$

4. Max-Eigenvalue Statistics (Sequential Testing)

H_0	$LR_{\max}$	CV (5%)	Decision
$r = 0$ vs $r = 1$	66.293	14.070	Reject H_0
$r = 1$ vs $r = 2$	0.003	3.760	FTR H_0

5. Interpretation

- **Decision:** Reject $H_0: r = 0$ against $H_1: r = 1$; FTR $H_0: r = 1$ against $H_1: r = 2$. Conclusion: exactly $r = 1$ cointegrating vector — consistent with the Trace result.
- **Cointegrating vector** (normalized on y_{1t}): $\hat{\beta} = [1, \; -1.493]'$, implying a long-run equilibrium $y_{1t} \approx \text{const} + 1.493\, y_{2t}$.

6. Comparison with Trace

	Max-Eig		Trace	
H_0	Stat	CV(5%)	Stat	CV(5%)
$r = 0 \,/\, r \leq 0$	66.293	14.07	66.296	15.41
$r = 1 \,/\, r \leq 1$	0.003	3.76	0.003	3.76

Practitioner Insight: The Max-Eigenvalue and Trace tests share the same eigenvalues and cross-product matrices; they differ only in how the eigenvalues are aggregated and in their critical values. The final step ($r = 1$) is identical in both tests because $LR_{\max}(1) = -T_{\text{eff}} \ln(1 - \hat{\lambda}_2)$ and $LR_{\text{trace}}(1)$ contains only the same remaining eigenvalue. The additive identity $LR_{\text{trace}}(r) = \sum_{i=r+1}^{n} LR_{\max}^{(i)}$ provides a useful arithmetic check: $66.296 \approx 66.293 + 0.003$. Practically, the Max-Eigenvalue test asks whether adding one more cointegrating vector significantly improves the system rank, making it sharper but sometimes less stable in small samples. Trace is generally more robust when multiple cointegrating relationships may exist. When the two tests disagree, the discrepancy itself can be informative and often reflects sensitivity to lag order p, deterministic specification, or small-sample distortions.

Common Pitfalls

- **Using trace critical values for the max-eigen statistic.** The trace and maximum eigenvalue tests have different limiting distributions and critical values. Always use the correct tables corresponding to the max-eigen statistic.
- **Accepting a full-rank result without checking integration order.** If the estimated rank equals the number of variables, the system is stationary in levels. This contradicts the typical assumption that variables are $I(1)$ and should prompt re-examination of unit root results.
- **Ignoring the magnitude of the estimated eigenvalues.** The size of the eigenvalues provides insight into the strength of potential cointegrating relationships. Large drops in eigenvalues can help confirm the appropriate rank.
- **Treating sequential rank tests as independent.** The tests use the same set of estimated eigenvalues and must be applied sequentially. Testing ranks independently without the sequential procedure can distort inference.

Gregory-Hansen Test (Single Break)

Cointegration Test with Structural Break

COINT.5 | **Category:** Cointegration | **Data:** Time Series | **Null:** Tabled distribution

When to Use

- Test long-run cointegration under structural breaks.
- Evaluate regimes where equilibrium shifts unexpectedly.
- Detect stability drops without knowing exact break dates.
- Apply when standard Engle–Granger shows low test power.
- Avoid false rejection due to ignored breaks.

Assumptions

- All individual time-series components must be purely $I(1)$.
- A single structural break occurs during the sample path.
- The timing of the regime shift is strictly exogenous.
- Variables follow a single, left-hand normalization loop.
- Cointegrating relationship is piecewise linear.

Test Statistic

$$GH^* = \inf_{\tau \in [\tau_1, \tau_2]} ADF(\tau), \quad \text{where } \tau = T_B/T \text{ is break fraction.}$$

Notation: τ is break fraction, $ADF(\tau)$ is augmented Dickey-Fuller statistic for given break date, $[\tau_1, \tau_2]$ typical trimming

Plain English: The Gregory-Hansen test extends Engle-Granger cointegration testing by allowing the long-run relationship to shift at an unknown break point. It estimates cointegrating regressions with dummy variables for breaks at all candidate dates, testing residuals for stationarity at each potential break. The test statistic is the minimum ADF statistic across all break dates, with the minimizing date identifying the estimated break. This endogenous break selection requires more stringent critical values than standard cointegration tests to avoid spurious break detection.

⚠ Practitioner Warning

Critical values substantially more negative than Engle-Granger due to searching over break dates. Break date selection changes test distribution; using standard cointegration critical values leads to severe over-rejection. Three break models available: level shift (C), level shift with trend (C/T), regime shift (C/S) where slope coefficients change. Results depend on break model chosen; no consensus on selection. Trimming required to avoid breaks near endpoints; typical range $[0.15T, 0.85T]$. Power low when break is small or sample size modest. Multiple breaks not accommodated; see Hatemi-J for extensions.

Hypotheses

- H_0: No cointegration even allowing for one structural break.
- H_1: Cointegration exists with structural break at $\hat{\tau}$.

Decision Rule: Reject H_0 if $GH^* < GH^*_\alpha$ (left-tailed) or p-value $< \alpha$. Use Gregory-Hansen CVs.

Software References [Python, R, SAS]

Python	R	SAS
`Not Native (Manual)`	`Not Native (Manual)`	`Not Native (Manual)`

Additional Notes

Three break models: Model C (level shift): $y_t = \mu_1 + \mu_2 D_t(\tau) + \beta' x_t + u_t$ where $D_t(\tau) = 1$ if $t > T\tau$. Model C/T (level shift with trend): adds trend term $\beta_1 t$. Model C/S (regime shift): $y_t = \mu_1 + \mu_2 D_t(\tau) + \beta_1' x_t + \beta_2' x_t D_t(\tau) + u_t$ allowing all coefficients to change. Gregory-Hansen (1996) provided asymptotic CVs via simulation for each model type. Break date estimate: $\hat{\tau} = \arg\min_\tau ADF(\tau)$ identifies structural break timing. Confidence intervals for break date available but wide with typical sample sizes. Hatemi-J (2008) extended to multiple breaks; computational burden increases rapidly. Alternative statistics: Z_α and Z_t (Phillips-Ouliaris type) can also be minimized over break dates. The test nests Engle-Granger: if no break, minimization yields standard cointegration test (but with different CVs).

Miscellanea: Allan Gregory and Bruce Hansen (1996) extended residual-based cointegration testing to allow a structural break in the long-run relationship itself, addressing a major limitation of Engle–Granger and Phillips–Ouliaris procedures. Their work showed that ignoring regime shifts in the cointegrating equation can falsely suggest no cointegration even when a stable long-run relationship exists before and after the break.

Reference(s): Gregory & Hansen (1996); Hatemi-J (2008)

Worked Example

Problem Statement An economist suspects the long-run cointegrating relationship between a commodity price index (y_t) and its input cost index (x_t) shifted permanently at an *unknown* date — a regime change that would cause EG/PO to miss cointegration. The GH test searches over all candidate break dates and takes the most favorable (minimum) ADF statistic. Model C (level shift), $T = 20$, $k = 0$, trim 15%–85%. Note the visible level jump between $t = 10$ and $t = 11$.

	$t=1$	$t=2$	$t=3$	$t=4$	$t=5$	$t=6$	$t=7$	$t=8$	$t=9$	$t=10$
x_t	5.0	5.1	6.1	6.1	7.0	7.1	8.1	8.1	9.0	9.1
y_t	10.2	9.3	11.9	11.0	13.0	12.1	14.6	13.8	16.2	15.6
	$t=11$	$t=12$	$t=13$	$t=14$	$t=15$	$t=16$	$t=17$	$t=18$	$t=19$	$t=20$
x_t	10.1	10.1	11.0	11.1	12.1	12.1	13.0	13.1	14.1	14.1
y_t	20.7	19.7	22.1	21.4	23.8	22.5	24.8	24.4	26.8	25.7

0. Pre-Test: Both Series $I(1)$

ADF ($k = 0$, constant): $x_t: \tau = -0.384$; $y_t: \tau = -0.724$.
Both $> -2.86 \Rightarrow$ FTR, both $I(1)$.

1. Model C Regression at Each Candidate Break

For each $t_b \in \{4, \dots, 17\}$, define $D_t(t_b) = \mathbf{1}(t \geq t_b)$ and run $y_t = \mu_1 + \mu_2 D_t + \beta x_t + \hat{u}_t$, then apply ADF ($k = 0$, no constant) to $\hat{u}_t$. Select output shown below:

t_b	τ_b	$\hat{\mu}_1$	$\hat{\mu}_2$	$\hat{\beta}$	τ_{ADF}
9	0.45	+0.269	+1.076	1.782	−3.999
10	0.50	+0.818	+1.603	1.700	−6.293
11	**0.55**	**+2.203**	**+2.932**	**1.495**	**−16.834**
12	0.60	+0.777	+1.400	1.731	−5.454
13	0.65	+0.275	+0.864	1.813	−3.678

2. GH Statistic and Selected Break

The minimum ADF statistic across all candidate break dates gives: $\tau^* = \min_{t_b} \tau_{ADF}(t_b) = -16.834$at $t_b^* = 11$.

3. ADF Detail at $t_b^* = 11$

OLS: $\hat{\mu}_1 = 2.203$, $\hat{\mu}_2 = 2.932$, $\hat{\beta} = 1.495$.
Residuals oscillate tightly around zero; in particular: $\hat{u}_1 = +0.524$, $\hat{u}_{11} = +0.469$, $\hat{u}_{20} = -0.509$.
ADF ($k = 0$, no constant) on $\hat{u}_t$:

$$\hat{\rho} = -1.879, \quad \hat{s}^2 = 0.0565, \quad SE(\hat{\rho}) = 0.112$$

$$\tau_{ADF}(t_b = 11) = \frac{-1.879}{0.112} = -16.834$$

4. Results and Decision

Critical values (Gregory-Hansen 1996, Model C):

Statistic	Value	CV (5%)	Decision
τ^* ($t_b^* = 11$)	−16.834	−4.610	Reject H_0

5. Interpretation

- **Decision:** We **reject** H_0 of no cointegration. A stable long-run equilibrium exists, with a level shift at $t_b^* = 11$.

Practitioner Insight: The search table reveals the key pattern of the GH procedure: 12 of the 14 candidate break dates give $|\tau_{ADF}| < 4.17$ (FTR even at 10%), while evidence concentrates sharply at $t_b \in \{10, 11, 12\}$ on the true regime change. The GH CVs are considerably more negative than EG CVs (−4.61 vs −3.37 at 5%) to guard against data-snooping from the break-date search — applying EG tables to τ^* would spuriously reject H_0 at nearly every dataset with a trend shift.

Common Pitfalls

- **Using Engle–Granger critical values instead of Gregory–Hansen values.** GH critical values are substantially more negative because the test searches over all possible break dates. Using Engle–Granger tables leads to severe over-rejection; always use the GH critical values for the chosen model.
- **Failing to trim the sample when searching for break dates.** Break dates must be restricted to the interior of the sample (typically $[0.15T, 0.85T]$). Else, the asymptotic theory is invalid and estimation near endpoints is unreliable.
- **Treating the estimated break date $\hat{\tau}$ as exact.** The estimated break location identifies where cointegration evidence is strongest, but it is prone to sampling uncertainty. Interpret $\hat{\tau}$ as an approx. break period rather than a precise date.
- **Selecting the model (C, C/T, C/S) based on outcomes.** Model choice should be guided by theory or prior diagnostics, not by significance. Testing several specifications and reporting only the rejecting one leads to data-mining bias.
- **Applying the test when multiple structural breaks are present.** Gregory–Hansen allows at most one regime shift. If several breaks exist, the test may lose power or produce misleading results; multi-break methods should be considered.

Maki Test (Multiple Breaks)

Cointegration Test with Endogenous Multiple Structural Breaks

COINT.6 | **Category:** Cointegration | **Data:** Time Series | **Null:** Tabled distribution

When to Use

- Test for cointegration with multiple structural breaks.
- Model regimes with up to five unknown equilibrium shifts.
- Isolate exact break dates endogenously across a sample.
- Apply when Gregory–Hansen fails due to multiple switches.
- Avoid false rejection due to ignored breaks.

Assumptions

- All individual time-series components must be purely $I(1)$.
- Structural breaks occur strictly as distinct regime steps.
- The timing of each breakthrough date is highly exogenous.
- Variables follow a single, left-hand normalization loop.
- Cointegrating relationship is piecewise linear.

Test Statistic

$$MK_k = \inf_{\tau} ADF_k(\tau), \quad \text{where } \tau = (\tau_1, \ldots, \tau_k) \text{ are break fractions.}$$

Notation: k is maximum number of breaks allowed, τ are break fractions, ADF_k is statistic with k breaks

Plain English: The Maki test extends Gregory-Hansen's single-break framework to multiple breaks by searching over all valid combinations of up to k break dates. For each configuration of break dates satisfying minimum separation constraints, the test estimates a cointegrating regression with dummy variables for all breaks, then computes the ADF statistic on residuals. The global minimum ADF statistic across all configurations provides evidence for cointegration, with minimizing break dates identifying structural change timing. More stringent critical values account for extensive searching over break date combinations.

⚠ Practitioner Warning

Computationally intensive due to combinatorial explosion of break date possibilities. For $k = 5$ breaks with typical trimming, thousands of configurations must be evaluated. Critical values increasingly negative as k increases to control for data mining across break combinations. Test assumes number of breaks k is specified; no formal method for selecting k, though sequential testing proposed. Minimum separation between breaks (typically $0.15T$) necessary to prevent spurious detection of nearby breaks. Power decreases as k increases relative to sample size.

Hypotheses

- H_0: No cointegration even with up to k structural breaks.
- H_1: Cointegration exists with up to k structural breaks.

Decision Rule: Reject H_0 if $MK_k < MK_{k,\alpha}$ (left-tailed) or p-value $< \alpha$. Use Maki CVs depending on k and break model.

Software References [Python, R, SAS]

Python	R	SAS
`Not Native (Manual)`	`Not Native (Manual)`	`Not Native (Manual)`

Additional Notes

Maki's framework generalizes Gregory–Hansen cointegration testing by allowing multiple unknown structural breaks in the long-run relationship. Four common specifications are considered: Model 0 (level shifts only), Model 1 (level shifts with trend), Model 2 (regime shifts in slope coefficients), and Model 3 (regime shifts with both slopes and trend). For k breaks, the sample is partitioned into $k + 1$ regimes. A typical specification is $y_t = \mu_0 + \sum_{i=1}^{k} \mu_i D_i(\tau_i) + \beta' x_t + u_t$, with extensions allowing trend and slope changes. Maki (2012) tabulated critical values for up to five structural breaks using simulation methods. Estimated break dates may exhibit substantial uncertainty, especially in short samples or when break magnitudes are small. Hatemi-J (2008) proposed related tests with different break specifications.

Miscellanea: Maki (2012) extended structural-break cointegration testing to allow multiple unknown breaks in the long-run relationship, addressing a major limitation of Gregory–Hansen and other single-break procedures. The test became particularly useful for macroeconomic and financial series spanning long periods with repeated regime changes, crises, or policy shifts.

Reference(s): Maki (2012); Hatemi-J (2008)

Worked Example

Problem Statement An economist suspects the commodity price–input cost long-run relationship experienced *two* permanent level shifts — a first-quarter supply disruption at $t \approx 8$ and a policy change at $t \approx 14$. $T = 20$, $k = 0$.

	$t=1$	$t=2$	$t=3$	$t=4$	$t=5$	$t=6$	$t=7$	$t=8$	$t=9$	$t=10$
x_t	5.0	5.1	6.1	6.1	7.0	7.1	8.1	8.1	9.0	9.1
y_t	9.9	9.3	11.6	10.7	12.8	12.1	14.6	16.8	19.0	18.4
	$t=11$	$t=12$	$t=13$	$t=14$	$t=15$	$t=16$	$t=17$	$t=18$	$t=19$	$t=20$
x_t	10.1	10.1	11.0	11.1	12.1	12.1	13.0	13.1	14.1	14.1
y_t	20.5	19.8	21.8	23.2	25.5	24.8	27.0	26.2	28.4	27.6

0. Pre-Test: Both Series $I(1)$

ADF (lags= 0, const): $x_t : \tau = -0.384$; $y_t : \tau = -0.724$. Both FTR at 5% (−2.86) ⇒ both $I(1)$.

1. Two-Break Search (91 Combinations)

For each (t_{b1}, t_{b2}) with $4 \le t_{b1} < t_{b2} \le 17$, build $D_{1t} = \mathbf{1}(t \ge t_{b1})$, $D_{2t} = \mathbf{1}(t \ge t_{b2})$, regress $y_t = \mu_0 + \mu_1 D_{1t} + \mu_2 D_{2t} + \beta x_t + \hat{u}_t$, and apply ADF (lags= 0, no constant) to $\hat{u}_t$. Top results (ranked by τ_{ADF}):

(t_{b1}, t_{b2})	$\hat{\mu}_0$	$\hat{\mu}_1$	$\hat{\mu}_2$	$\hat{\beta}$	τ_{ADF}
(8, 14)	**1.539**	**2.747**	**1.614**	**1.578**	**−12.567**
(8, 15)	1.066	2.691	1.205	1.653	−5.458
(8, 13)	0.912	2.428	1.138	1.677	−5.169
(7, 14)	0.663	1.990	1.496	1.715	−4.025
(9, 14)	0.779	1.957	1.082	1.741	−3.964
⋮ (86 remaining combinations, all $\tau_{ADF} > -4.0$)					

2. Maki Statistic and Selected Breaks

$$\tau^* = \min_{(t_{b1}, t_{b2})} \tau_{ADF} = -12.567 \text{at } (t^*_{b1}, t^*_{b2}) = (8, 14)$$

3. ADF Detail at $(t^*_{b1}, t^*_{b2}) = (8, 14)$

OLS: $\hat{\mu}_0 = 1.539$, $\hat{\mu}_1 = 2.747$, $\hat{\mu}_2 = 1.614$, $\hat{\beta} = 1.578$. Residuals oscillate around zero; selected: $\hat{u}_1 = +0.470$, $\hat{u}_8 = -0.269$, $\hat{u}_{14} = -0.217$, $\hat{u}_{20} = -0.552$.
ADF (lags= 0, no constant) on $\hat{u}_t$:

$$\hat{\rho} = \frac{\sum \hat{u}_{t-1} \Delta \hat{u}_t}{\sum \hat{u}^2_{t-1}} = \frac{-5.075}{2.804} = -1.810$$

$$\hat{s}^2 = 0.0582, \quad SE(\hat{\rho}) = 0.144, \quad \tau^* = \frac{-1.810}{0.144} = -12.567$$

4. Results and Decision

Critical values (Maki 2012, Table 1, Model 0, $k = 2$):

Statistic	Value	CV (5%)	Decision
τ^* (8, 14)	−12.567	−4.893	Reject H_0

5. Interpretation

- **Decision:** We **reject** H_0. A stable cointegrating relationship holds across three regimes (pre-t_8, t_8–t_{13}, post-t_{14}).
- **Practitioner Insight:** With $k = 2$ breaks and 14 eligible dates, $\binom{14}{2} = 91$ combinations must be evaluated—manageable for small T but typically computed algorithmically. CVs become more negative as k increases (Model 0, 5%: −4.602 for $k = 1$, −4.893 for $k = 2$, −5.083 for $k = 3$), reflecting greater data-snooping. The Maki test nests GH ($k = 1$) and reduces to it when $k = 1$. Four variants parallel GH: Model 0 (level shifts), Model 1 (level + trend), Model 2 (level + slope), and Model 3 (level + slope + trend). The concentration around (8, {13, 14, 15}) rather than a single minimum indicates $t^*_{b1} = 8$ is robust, while $t^*_{b2} = 14$ shows local uncertainty.

Common Pitfalls

- **Using incorrect critical values.** Maki CVs differ substantially from EG and GH values because the test searches over multiple break combinations. Always use the Maki (2012) CVs corresponding to the chosen number of breaks.
- **Testing too many structural breaks for the available sample size.** Each additional break reduces effective observations within regimes. If too many breaks are allowed relative to T, estimation becomes unstable and power deteriorates.
- **Treating estimated break dates as precise.** Estimated break locations are subject to substantial sampling uncertainty. Interpret the break dates as approximate regime shifts rather than exact change points.
- **Reporting Maki results without comparison to simpler tests.** The Maki test extends EG and GH by allowing multiple breaks. Comparing results helps determine if cointegration depends on the presence of structural breaks.

Shin Residual Cointegration Test

KPSS Test Using Efficient Cointegrating Residuals

COINT.7 | **Category:** Cointegration | **Data:** Time Series | **Null:** Tabled distribution

When to Use

- Directly test the null hypothesis of cointegration.
- Confirm long-run equilibrium among I(1) series.
- Complement Engle–Granger or Phillips–Ouliaris tests.
- Use efficient first-stage cointegrating estimation.
- Detect stable equilibrium despite nonstationary levels.

Assumptions

- Variables are integrated of order one, I(1).
- Cointegrating relationship is linear in levels.
- No cointegration among regressors themselves.
- Residual long-run variance is consistently estimated.
- Efficient first-stage estimation is preferred.

Test Statistic

$$C_T = \frac{1}{T^2}\frac{\sum_{t=1}^{T} S_t^2}{\hat{\lambda}^2}$$

Step 1 : $y_t = \alpha + \beta' x_t + \hat{u}_t$

Optional DOLS: $y_t = \alpha + \beta' x_t + \sum_{j=-q}^{q} \delta_j \Delta x_{t-j} + u_t^*$

Notation: $\hat{u}_t$ are residuals from Step 1, $S_t = \sum_{j=1}^{t} \hat{u}_j$, $\hat{\lambda}^2$ is long-run (Newey-West) variance, T is effective sample size

Plain English: Shin's test directly tests the cointegration hypothesis. In step 1, an efficient long-run cointegrating regression is estimated to obtain the residuals. In step 2, the KPSS stationarity framework is applied to these residuals by forming cumulative partial sums. If the residuals are stationary and mean-reverting, the cumulative sums remain small, supporting cointegration. Large cumulative deviations instead suggest that the residuals contain a unit root and no stable long-run equilibrium exists.

⚠ Practitioner Warning

Critical values depend on the deterministic specification (no intercept, intercept only, or intercept plus trend). Shin emphasized efficient first-stage estimation because ordinary OLS residuals can introduce nuisance-parameter dependence under endogeneity and serial correlation. DOLS, Fully Modified OLS (FM-OLS), or related efficient estimators are therefore preferred. Poor HAC bandwidth selection can materially affect finite-sample results.

Hypotheses

- H_0: H_0: Residuals are stationary; variables are cointegrated.
- H_1: H_1: Residuals contain a unit root; no cointegration.

Decision Rule: Reject H_0 if $C_T > CV_\alpha$. Large statistics reject the null of cointegration. Use Shin (1994) for CVs.

Software References [Python, R, SAS]

`cointReg::cointRegD`	`urca`	`statsmodels`

Additional Notes

Shin (1994) extended the KPSS stationarity framework to cointegrating residuals, producing one of the few classical procedures that directly tests the null of cointegration rather than no cointegration. Consequently, small values of the statistic support cointegration, whereas large values reject it. The test supports three deterministic specifications: **(1)** no intercept or trend (test statistic - C), **(2)** intercept only (test statistic - C_μ), and **(3)** intercept with linear trend (test statistic - C_τ). The limiting distribution depends on the chosen specification and involves Brownian bridges and Brownian motion functionals.

Although Shin did not explicitly propose Dynamic OLS (DOLS), efficient residual estimation integrates naturally with the procedure. The DOLS estimator of Stock and Watson (1993), $y_t = \alpha + \beta' x_t + \sum_{j=-q}^{q} \delta_j \Delta x_{t-j} + u_t^*$, augments the regression with leads and lags of Δx_t to mitigate endogeneity and serial correlation, improving the asymptotic behavior of the residual-based statistic.

Miscellanea: Shin's procedure is one of the few classical residual-based cointegration tests that directly treats *cointegration* as the null hypothesis. This mirrors the philosophy of the KPSS stationarity test, where stationarity rather than nonstationarity is treated as the maintained hypothesis. The test emerged during a period of rapid development in residual-based cointegration testing during the early 1990s. Meanwhile, Stock and Watson's Dynamic OLS estimator became highly influential because it provided asymptotically efficient estimation using only a simple augmented regression without requiring a full system maximum-likelihood framework.

Reference(s): KPSS (1992); Stock & Watson (1993); Shin (1994)

Worked Example

Problem Statement An economist re-examines the commodity price–input cost pair using a two-step procedure: **Step 1** estimates the cointegrating vector via Stock and Watson's (1993) DOLS, which corrects for endogeneity and serial correlation by augmenting the static regression with leads and lags of Δx_t; **Step 2** applies Shin's (1994) C_μ statistic to the DOLS residuals to test the null of cointegration. See Engle-Granger data.

1. DOLS Regression ($q = 1, n = T - 2q - 1 = 17$)
Augment the static regression with one lead and one lag of Δx_t, using observations $t = 3, \dots, 19$:

$$y_t = \alpha + \beta x_t + \gamma_{-1}\Delta x_{t+1} + \gamma_0 \Delta x_t + \gamma_1 \Delta x_{t-1} + u_t^*$$

OLS gives: $\hat{\alpha} = 2.838, \quad \hat{\beta} = 1.498$

$$\hat{\gamma}_{-1} = -0.677, \quad \hat{\gamma}_0 = -0.389, \quad \hat{\gamma}_1 = -0.586$$

Compare with plain OLS ($T = 20$, no augmentation): $\hat{\beta}^{OLS} = 1.491$ vs $\hat{\beta}^{DOLS} = 1.500$.

2. DOLS Residuals ($t = 3, \dots, 19$)

$$\hat{u}_t^* = y_t - \hat{\alpha} - \hat{\beta} x_t - \hat{\gamma}_{-1}\Delta x_{t+1} - \hat{\gamma}_0 \Delta x_t - \hat{\gamma}_1 \Delta x_{t-1}$$

Selected residuals: $\hat{u}^*_{t=3} = +0.072$, $\hat{u}^*_{t=10} = +0.173$, $\hat{u}^*_{t=15} = -0.016$, $\hat{u}^*_{t=19} = -0.112$. Check: $\sum_t \hat{u}_t^* = 0.000$.

3. NW Long-Run Variance of $\hat{u}^*$ ($\ell = 2$)
Using the biased $1/n$ convention throughout:

$$\hat{\lambda}^2 = \hat{\gamma}_0^{NW} + \tfrac{4}{3}\hat{\gamma}_1^{NW} + \tfrac{2}{3}\hat{\gamma}_2^{NW} = 0.0091$$

where $\hat{\gamma}_0^{NW} = 0.0072$, $\hat{\gamma}_1^{NW} = 0.0016$, $\hat{\gamma}_2^{NW} = -0.0003$ (from DOLS residuals) are variance and auto-covariances.

4. C_μ Statistic
Let $i = 1, \dots, n = 17$ index the window and form partial sums $S_i = \sum_{k=1}^{i} \hat{u}^*_{t_k}$ (e.g. $S_1 = +0.072$, $S_8 = +0.124$, $S_{17} = 0.000$):

$$C_\mu = \frac{1}{n^2} \cdot \frac{\sum_{i=1}^{n} S_i^2}{\hat{\lambda}^2} = \frac{1}{289} \cdot \frac{0.207}{0.0091} = 0.079$$

5. Results and Decision
Critical values (Shin 1994, $m = 1$ regressor):

Statistic	Value	CV (5%)	Decision
C_μ	0.079	0.314	FTR H_0

We **fail to reject** H_0: the DOLS residuals are stationary, confirming cointegration. The long-run relationship is $y_t \approx 2.84 + 1.50 x_t$.

Practitioner Insight: This entry combines two distinct contributions. **Step 1** is from Stock–Watson (1993): DOLS augments the cointegrating regression with leads and lags of Δx_t to remove endogeneity and serial correlation, yielding asymptotically efficient estimates of β with standard Wald inference. In SW notation, negative j denote leads and positive j denote lags. Stock and Watson themselves did not propose a dedicated cointegration test, instead relying on procedures such as Engle–Granger or Johansen. **Step 2** follows Shin (1994), applying a KPSS stationarity framework to the DOLS residuals. Unlike Engle–Granger or Phillips–Ouliaris, which test the null of no cointegration, Shin's null is cointegration, so *small* C_μ values support a long-run relationship. The DOLS estimation window is $t = q + 1, \dots, T - q$ only when boundary differences exist; otherwise it shrinks by one, giving $t = 3, \dots, 19$ ($n = 17$) here. The partial sums returning to zero ($S_{17} = 0$) provide a useful arithmetic check since OLS residuals sum to zero. Despite the small sample, DOLS recovers $\hat{\beta} = 1.498$ close to the true value 1.5, and $C_\mu = 0.079$ lies comfortably below the 5% critical value, supporting cointegration.

Common Pitfalls

- **Misattributing the test statistic.** Stock and Watson (1993) propose the DOLS estimator only; the cointegration test (C, C_μ, C_τ) is due to Shin (1994). SW (1993) provide no cointegration critical values — use Shin (1994) tables.
- **Wrong DOLS window at the boundary.** The lag regressor Δx_{t-1} requires a pre-sample observation x_0. When none exists the window starts at $t = q + 2$, not $t = q + 1$, giving $n = T - 2q - 1$ rather than the textbook $T - 2q$.
- **Confusing lead/lag subscripts.** SW (1993) define γ_j as the coefficient on Δx_{t-j}, so negative j are leads and positive j are lags — reversing the convention used in some textbooks.
- **Using short-run instead of long-run variance.** Shin's test requires a long-run variance estimate in the denominator. Substituting the short-run residual variance inflates the statistic and distorts inference.
- **Ignoring the null reversal.** The Shin statistics (C, C_μ, C_τ) test the null of cointegration, not the null of no cointegration. Large values reject cointegration; small values support it. Applying EG/PO critical regions to these statistics leads to the wrong decision.

Park Added Variables Test

Variable-Addition Test for Cointegration

COINT.8 | **Category:** Cointegration | **Data:** Time Series | **Null:** Asymptotic Chi-Square

When to Use

- Directly test the null hypothesis of cointegration.
- Check residual stability through added trend functions.
- Complement Engle–Granger or Phillips–Ouliaris tests.
- Use after verifying the variables are I(1).
- Useful in single-equation cointegrating regressions.

Assumptions

- Variables are integrated of order one, I(1).
- Cointegrating relationship is linear in levels.
- Deterministic specification is correctly chosen.
- Added variables are excluded under cointegration.
- Long-run covariance is consistently estimated.

Test Statistic

$$H(p,q) = \frac{RSS_R - RSS_U}{\hat{\lambda}^2} \quad \Big| \quad \text{Under } H_0, \quad H(p,q) \xrightarrow{d} \chi^2(q-p).$$

Notation: RSS_R, RSS_U = RSS from restricted and unrestricted regressions; $\hat{\lambda}^2$ = long-run (Newey-West) variance of unrestricted residuals; p = highest maintained trend power; q = highest added trend power; $q - p$ = degrees of freedom

Plain English: Park's test asks if a proposed cointegrating regression is already stable, or if extra deterministic trend functions still explain systematic movement in the dependent variable. If the variables are truly cointegrated and the deterministic spec is adequate, the added trend functions should be redundant and have no additional explanatory power. A large $H(p,q)$ means the added variables still matter, suggesting the proposed long-run relationship is not a valid cointegrating relation.

⚠ Practitioner Warning

Park's Canonical Cointegrating Regression (CCR) is an efficient estimation method while the Added Variables test is a cointegration test. The null is *cointegration*, so large χ^2 rejects cointegration; failure to reject merely supports it. Results depend heavily on the maintained deterministic terms; omitting a required trend can spuriously reject H_0. Long-run variance (HAC estimator) must be estimated from the unrestricted residuals; bandwidth choice affects finite-sample behavior. Do not mix Park's test with Engle–Granger or Phillips–Ouliaris critical regions, which use the opposite null.

Hypotheses

- H_0: Added variables are irrelevant; cointegration holds.
- H_1: Added variables matter; no cointegration.

Decision Rule: Reject H_0 if $H(p,q) > \chi^2_{\alpha, q-p}$ or p-value $< \alpha$. Large values reject the null of cointegration.

Software References [Python, R, SAS]

Python	R	SAS
`Not Native (Manual)`	`Not Native (Manual)`	`Not Native (Manual)`

Additional Notes

Park's variable-addition idea exploits the behavior of spurious regressions with $I(1)$ variables. Under cointegration, the maintained regression captures the long-run equilibrium and the added deterministic terms become redundant. Under no cointegration, the added terms can absorb remaining nonstationary drift, producing a large χ^2. The $H(p,q)$ family is indexed by the polynomial degree of maintained trends (p) and added trends (q), with $q > p \geq 0$. Thus, $H(0,1)$ maintains only a constant and adds a linear trend, while $H(1,2)$ maintains a constant and linear trend and adds a quadratic term. Each added variable contributes one degree of freedom, so the asymptotic distribution is $\chi^2(q-p)$. Park's later CCR framework shares the same asymptotic machinery and can provide efficient estimation and valid χ^2 inference, but CCR itself is a separate estimation procedure differnt from the Added Variables test.

Miscellanea: Park's Added Variables test belongs to the small class of cointegration procedures that treat cointegration itself as H_0. Rather than testing residuals for a unit root, the procedure examines whether higher-order deterministic trends become asymptotically irrelevant once the long-run relation is correctly specified. Park's later CCR paper became far more widely known in applied econometrics, which is why the Added Variables test is often mistakenly conflated with CCR despite serving a fundamentally different purpose.

Reference(s): Park (1990); Park (1992); Park, Ouliaris & Choi (1988)

Park Added Variables Test

COINT.8

Worked Example

Problem Statement An economist tests cointegration between the commodity price (y_t) and input cost (x_t) series using Park's (1992) $H(p, q)$ Added Variables test. See Engle-Grainger for data.

1. The $H(p, q)$ Framework
The test compares two nested regressions. For $H(1, 2)$:

- **Restricted** (R): $y_t = \delta_0 + \beta x_t + \delta_1 t + e_t^R$
  Maintained: $\{1, x_t, t\}$ — needed under H_0.
- **Unrestricted** (U): $y_t = \delta_0 + \beta x_t + \delta_1 t + \gamma_2 t^2 + e_t^U$
  Added: $\{t^2\}$ — should be redundant if x_t spans the $I(1)$ space.

2. Restricted Regression
OLS of y_t on $[1, x_t, t]$:

$$\hat{\delta}_0 = -5.343,\ \hat{\beta} = 3.208,\ \hat{\delta}_1 = -0.859,\ RSS_R = 0.2375$$

Selected residuals: $\hat{e}_1^R = +0.063$, $\hat{e}_9^R = +0.206$, $\hat{e}_{17}^R = +0.248$, $\hat{e}_{20}^R = -0.103$.

3. Unrestricted Regression
OLS of y_t on $[1, x_t, t, t^2]$: $\hat{\delta}_0 = -5.407$, $\hat{\beta} = 3.207$

$$\hat{\delta}_1 = -0.841,\ \hat{\gamma}_2 = -0.000854\ (\approx 0),\quad RSS_U = 0.2247$$

The near-zero $\hat{\gamma}_2$ signals t^2 adds little; RSS barely falls.

4. NW Long-Run Variance of $\hat{e}^U$ ($\ell = 2$)

$$\hat{S}^2 = \hat{\sigma}_0 + \tfrac{4}{3}\hat{\sigma}_1 + \tfrac{2}{3}\hat{\sigma}_2 = 0.01142$$

where $\hat{\sigma}_0 = 0.0112$, $\hat{\sigma}_1 = 0.0024$, $\hat{\sigma}_2 = -0.0045$.
Here the unbiased autocovariance form is used, dividing each $\hat{\sigma}_k$ by the number of available lag-k pairs rather than by the effective sample size.

5. $H(1, 2)$ Statistic

$$H(1,2) = \frac{RSS_R - RSS_U}{\hat{S}^2} = 1.122$$

Under H_0 (cointegration), $H(1,2) \xrightarrow{d} \chi^2(q-p) = \chi^2(1)$.

6. Results and Decision Critical values $\chi^2(1)$:

Statistic	Value	CV (5%)	Decision
$H(1,2)$	1.122	3.841	FTR H_0

We **fail to reject** H_0. The added variable t^2 is redundant: x_t already spans the $I(1)$ space, confirming a stable long-run cointegrating relationship.

Practitioner Insight: Park's test is unusual in placing *cointegration as* H_0, so the critical region is the *right* tail — large $H(p, q)$ values reject cointegration, unlike EG/PO where sufficiently negative statistics reject the null of *no* cointegration. Model specification is critical: the maintained regressors must include all $I(1)$ components present in the data generating process. Using $H(0, 1)$ here — maintaining only $[1, x_t]$ and adding t — gives $H(0, 1) = 232.2$ and wrongly rejects cointegration, because x_t does not fully span the linear trend at $T = 20$. With $H(1, 2)$, the trend t is properly maintained alongside x_t, so only the genuinely redundant t^2 term is tested. The degrees of freedom equal $q - p$: adding one variable gives $\chi^2(1)$, adding two gives $\chi^2(2)$; the Park $H(p, q)$ family nests any desired polynomial order and can be extended to multivariate systems by replacing x_t with a vector of $I(1)$ regressors.

Common Pitfalls

- **Confusing the Added Variables test with CCR.** Park's Canonical Cointegrating Regression (CCR) is an efficient estimation procedure, whereas the Added Variables test is a specification-style cointegration test. The two papers share asymptotic machinery but serve different purposes.
- **Ignoring the null reversal.** Park's test treats cointegration as the null hypothesis. Large χ^2 values reject cointegration, while failure to reject merely supports it. Applying Engle–Granger or Phillips–Ouliaris critical regions leads to the wrong conclusion.
- **Misspecifying deterministic trends.** The maintained deterministic specification is crucial. Omitting a trend that belongs in the regression can spuriously reject cointegration, while adding unnecessary trends reduces power.
- **Misinterpreting the role of added variables.** The added polynomial trend terms are diagnostic devices, not structural regressors of substantive interest. Their purpose is to detect remaining nonstationary drift not captured by the proposed cointegrating relation.
- **Using the wrong long-run variance estimate.** The HAC long-run variance must be estimated from the unrestricted residuals. Bandwidth choice and kernel selection can materially affect finite-sample behavior and the resulting χ^2 statistic.

ARDL Bounds Test

Bounds Test for Long-Run Cointegration

COINT.9 | **Category:** Cointegration | **Data:** Time Series | **Null:** Tabled distribution

When to Use

- Test for long-run relationships among time-series variables.
- Analyze mixed $I(0)$ and $I(1)$ regressors jointly.
- Avoid pre-testing all variables for same integration order.
- Model short-run and long-run dynamics simultaneously.
- Apply cointegration testing in small-sample settings.

Assumptions

- No regressor is I(2) or higher — verify before applying.
- Correct lag orders selected for ARDL specification.
- Residuals are serially uncorrelated and homoskedastic.
- Long-run relationship is stable over the sample period.
- Large enough sample for asymptotic bounds to be valid.

Test Statistic

$$F = \frac{(RSS_R - RSS_U)/q}{RSS_U/(T-k)} \quad \Bigg| \quad \text{ARDL: } \Delta y_t = c + \underbrace{\pi_y y_{t-1} + \pi_x x_{t-1}}_{\text{level terms}} + \sum_{i=1}^{p-1} d_i \Delta y_{t-i} + \sum_{j=0}^{q-1} \delta_j \Delta x_{t-j} + \varepsilon_t.$$

Notation: RSS_R and RSS_U are restricted and unrestricted residual sums of squares, q is the number of restrictions (e.g., in the above ARDL, $\pi_y = \pi_x = 0$), T is sample size, and k is the number of estimated parameters (including intercept).

Plain English: The ARDL Bounds test determines whether a stable long-run equilibrium relationship exists among time-series variables. The procedure estimates an autoregressive distributed lag (ARDL) model and tests whether the lagged level variables jointly contribute to explaining the dependent variable. Unlike many classical cointegration procedures, the method accommodates a mix of stationary $I(0)$ and first-difference stationary $I(1)$ regressors without requiring that all variables share the same integration order.

⚠ Practitioner Warning

The Bounds procedure is invalid when any variable is integrated of order two or higher. Results can be highly sensitive to lag-length selection, deterministic trend specification, and structural breaks. The test indicates existence of cointegration but does not itself estimate the long-run coefficients reliably unless the ARDL model is properly specified. Small samples and residual autocorrelation may substantially distort inference.

Hypotheses

- H_0: $\pi_y = \pi_x = 0$ (No long-run relationship).
- H_1: $\pi_y \neq 0$ and/or $\pi_x \neq 0$ (cointegration exists).

Decision Rule: Reject H_0 if $F > I(1)$. Compare F-stat to Pesaran bounds $[I(0), I(1)]$. $F < I(0)$: FTR H_0.

Software References [Python, R, SAS]

Python	R	SAS
`statsmodels.tsa.ardl`	`ARDL::bounds_f_test`	`PROC AUTOREG`

Additional Notes

The Bounds framework compares the computed F statistic against two asymptotic critical-value sets: a lower bound assuming all regressors are $I(0)$ and an upper bound assuming all regressors are $I(1)$. If the test statistic exceeds the upper bound, cointegration is supported regardless of the exact integration mixture. If it falls below the lower bound, cointegration is not supported. Intermediate values produce inconclusive results. The unrestricted error-correction representation of an ARDL$(p, q_1, \ldots, q_k)$ model is commonly written as: $\Delta y_t = \alpha + \lambda y_{t-1} + \sum_{j=1}^{k} \delta_j x_{j,t-1} + \sum_{i=1}^{p-1} \phi_i \Delta y_{t-i} + \sum_{j=1}^{k} \sum_{m=0}^{q_j-1} \gamma_{jm} \Delta x_{j,t-m} + \varepsilon_t$. Cointegration corresponds to jointly testing whether the lagged level coefficients are zero.

Miscellanea: The ARDL Bounds framework was developed by Hashem Pesaran, Yongcheol Shin, and Richard Smith in 2001 as a practical alternative to traditional cointegration procedures such as Engle–Granger and Johansen tests. The method became especially popular in applied economics because it accommodates mixtures of $I(0)$ and $I(1)$ variables and performs reasonably well in relatively small samples.

Reference(s): Pesaran, Shin & Smith (2001); Narayan (2005); Enders (2015)

Worked Example

Problem Statement Twelve quarterly observations of real household income (x_t) and real consumption (y_t), both in index form, are recorded below. Fit an ARDL(1, 1) model and apply the Pesaran–Shin–Smith bounds test to determine at $\alpha = 0.05$ whether a long-run cointegrating relationship exists between consumption and income.

$H_0: \pi_y = 0$ and $\pi_x = 0$ (no long-run levels relationship); $H_1: \pi_y \neq 0$ and/or $\pi_x \neq 0$ (cointegration exists).

t	1	2	3	4	5	6	7	8	9	10	11	12
x_t	10	12	13	15	16	18	20	21	23	25	26	28
y_t	9	10	11	13	14	16	17	18	20	22	23	25
Δy_t		1	1	2	1	2	1	1	2	2	1	2
y_{t-1}		9	10	11	13	14	16	17	18	20	22	23
x_{t-1}		10	12	13	15	16	18	20	21	23	25	26
Δx_t		2	1	2	1	2	2	1	2	2	1	2

1. ARDL(1,1) Bounds Test Model

Unrestricted error-correction form of ARDL(1,1):

$$\Delta y_t = c + \pi_y y_{t-1} + \pi_x x_{t-1} + \delta \Delta x_t + u_t.$$

Cointegration $\Longrightarrow$ lagged levels are significant.

2. Unrestricted Model (OLS)

Regressing Δy_t on $(1, y_{t-1}, x_{t-1}, \Delta x_t)$:

$$\hat{c} = -0.8319, \quad \hat{\pi}_y = -0.8749,$$
$$\hat{\pi}_x = 0.8045, \quad \hat{\delta} = 0.9118.$$

$\text{RSS}_U = \boxed{0.6370}$, $\quad \hat{\sigma}^2 = RSS/(T-4) = \boxed{0.0796}$.

3. Restricted Model (OLS)

Under $H_0: \pi_y = \pi_x = 0$, model is $\Delta y_t = c + \delta \Delta x_t + u_t$.

$$\hat{\delta}_R = 5/7, \quad \hat{c}_R = 2/7, \quad \text{RSS}_R = \boxed{10/7}.$$

4. Bounds F-Statistic

With $q = 2$ restrictions and $k = 4$ unrestricted parameters:

$$F_{\text{PSS}} = \frac{(\text{RSS}_R - \text{RSS}_U)/q}{\text{RSS}_U/(T-k)} = \boxed{4.97}.$$

5. Decision

Use PSS (2001) bounds for $k = 1$, Case III (restricted constant):

α	$I(0)$ lower	$I(1)$ upper	Decision
10%	2.72	3.23	**Reject**
5%	3.62	4.16	**Reject** (4.97 > 4.16)
1%	5.13	5.85	FTR (4.97 < 5.13)

Since $F = 4.97$ exceeds the 5% upper $I(1)$ bound of 4.16, evidence supports cointegration at 5% regardless of whether x_t is $I(0)$ or $I(1)$. *Note:* PSS critical values are asymptotic; with $T = 12$, this example is illustrative.

6. Long-Run Relationship ($\pi_y y + \pi_x x = 0$)

From the unrestricted estimates: $\hat{\theta} = -\hat{\pi}_x/\hat{\pi}_y \approx 0.920$, giving estimated long-run consumption $y \approx 0.92x$.

Practitioner Insight: The ARDL bounds test is designed for situations where regressors may be a mixture of $I(0)$ and $I(1)$ variables, provided none are $I(2)$. Unlike Engle–Granger, it does not require all variables to be integrated of the same order. The test is based on the unrestricted ECM form of the ARDL model: under cointegration, the adjustment coefficient $\hat{\pi}_y$ is significantly negative, while $\hat{\pi}_x$ captures the long-run relationship. Because the asymptotic distribution depends on the unknown integration orders, the Wald statistic is compared against lower $I(0)$ and upper $I(1)$ critical bounds. Exceeding the upper bound, as here, gives conclusive evidence of cointegration; values between the bounds are inconclusive and usually require additional unit-root testing. Lag orders should be selected by criteria such as AIC or BIC before testing. For fully $I(1)$ multivariate systems, Johansen's trace and maximum-eigenvalue tests are standard.

Common Pitfalls

- **I(2) variables not screened out.** The bounds framework is invalid if any variable is I(2). Always run PP or KPSS pre-tests and remove I(2) series (typically via differencing) before applying the bounds test.
- **Wrong PSS case selected.** Cases I–V differ in intercept and trend placement. Using Case III bounds when the DGP has a restricted trend (Case IV or V) produces incorrect CVs. Match the case to the economic interpretation of the data.
- **Lag order chosen without residual checking.** Select p and q by AIC/BIC but always verify that ε_t passes Breusch-Godfrey. Serial correlation in residuals invalidates the F-statistic.
- **Using asymptotic bounds for small samples.** PSS (2001) asymptotic bounds are unreliable for $T < 80$. Use Narayan (2005) finite-sample bounds or bootstrap critical values for typical macroeconomic datasets.
- **Ignoring the inconclusive band.** An inconclusive result is not a soft rejection — it genuinely means the test cannot determine the answer. Do not report it as marginal evidence either way.

Pedroni Test (Panel)

Panel Cointegration Test with Heterogeneous Alternatives

COINT.10 | **Category:** Cointegration | **Data:** Panel Time Series | **Null:** Normal (asymptotic)

When to Use

- Test panel cointegration across cross-sections.
- Heterogeneous long-run panel relationships.
- Large-N, moderate-T panel-data analysis.
- Increase power by pooling cross-sectional information.

Assumptions

- Variables are integrated of order one.
- Cross-sectional units are independent.
- Cointegrating vectors may differ across panels.
- Residuals are stationary under cointegration.

Test Statistic

$$Z = \frac{\sqrt{N}(\hat{\theta}_N - \mu)}{\sigma} \xrightarrow{d} N(0,1), \quad \Big| \quad y_{i,t} = \alpha_i + \delta_i t + \beta_i' x_{i,t} + e_{i,t}, \quad \Big| \quad \textbf{Levels}: \hat{e}_{i,t} = \rho_i \hat{e}_{i,t-1} + u_{i,t}$$

Notation: $\hat{\theta}_N$ is pooled test statistic, μ, σ are mean/variance adjustments, N is panels. See additional notes

Plain English: Pedroni's panel cointegration tests extend residual-based cointegration testing to heterogeneous panel data by pooling evidence across multiple cross-sectional units. Each panel member may have different cointegrating vectors and short-run dynamics, allowing substantial heterogeneity across the panel. The tests are based on unit-root testing of the residuals from the hypothesized cointegrating regressions and converge to standard normal distributions after appropriate standardization.

⚠ Practitioner Warning

The seven statistics can yield conflicting conclusions, especially in small samples; no universal consensus exists on which statistic should be prioritized. Cross-sectional dependence (common shocks, spillovers, spatial correlation) violates the independence assumption and can severely distort inference; second-generation panel cointegration tests are preferable when such dependence is suspected. Unbalanced panels require modifications. Pre-testing for unit roots remains essential but is itself complicated by cross-sectional dependence.

Hypotheses

- H_0: No cointegration in any panel member ($\rho_i = 1, \forall i$).
- H_1: Cointegration in some members ($\rho_i < 1$ for some/all i).

Decision Rule: Reject H_0 if $|Z| > z_{\alpha/2}$ or p-value $< \alpha$. Rejection indicates cointegration evidence.

Software References [Python, R, SAS]

Python	R	SAS
`Not Native (Manual)`	`plm::pcdtest (related)`	`Not Native (Manual)`

Additional Notes

Pedroni proposes seven statistics divided into two categories. Within-dimension (panel) statistics — (1) panel v, (2) panel ρ, (3) panel PP, and (4) panel ADF — pool autoregressive coefficients across units. Between-dimension (group) statistics — (5) group ρ, (6) group PP, and (7) group ADF — average panel-level statistics and allow greater heterogeneity across panel units. The panel v statistic has a right-tail rejection region; all others reject in the left tail. All statistics standardized to $N(0,1)$: $Z = \sqrt{N}(\hat{\theta}_N - E[\hat{\theta}_N])/\sqrt{Var[\hat{\theta}_N]}$ where adjustment moments tabulated via simulation. The heterogeneous alternative allows $\beta_i \neq \beta_j$, making the framework realistic for economically diverse panel members. Kao (1999) proposed a related residual-based panel cointegration test under greater homogeneity assumptions. Westerlund (2007) later developed error-correction-based panel cointegration tests, while Banerjee–Carrion-i-Silvestre (2017) developed panel cointegration tests that account for cross-sectional dependence using Pesaran-style common correlated effects augmentation. Pedroni (2004) further extended the framework to multiple regressors and more general heterogeneous panels.

Miscellanea: Peter Pedroni (1999, 2004) developed panel cointegration tests that pool information across cross-sectional units while allowing heterogeneous relationships. By providing multiple test statistics with different properties, the approach enables robust inference in panel settings with limited time-series length. Exact formulas involve heavy notation and extensive asymptotic corrections, so only the general structure is shown here.

Reference(s): Pedroni (1999, 2004); Kao (1999)

Pedroni Test (Panel) — COINT.10

Worked Example

Problem Statement An economist tests whether a price–cost relationship holds as a stable cointegrating equilibrium across four regions ($N = 4, T = 10$). Data is provided at the end of this chapter.

$$\hat{e}_{i,t} = y_{i,t} - \hat{\alpha}_i - \hat{\beta}_i x_{i,t} \quad \Delta\hat{e}_{i,t} = \hat{\rho}_i \hat{e}_{i,t-1} + \hat{u}_{i,t} \quad \hat{\rho}_i = \frac{\sum_{t=2}^{10} \hat{e}_{i,t-1}\,\Delta\hat{e}_{i,t}}{\sum_{t=2}^{10} \hat{e}^2_{i,t-1}} \quad \hat{s}_i^2 = \frac{\sum_{t=2}^{10} \hat{u}^2_{i,t}}{T-2} \quad SE^2(\hat{\rho}_i) = \frac{\hat{s}_i^2}{\sum_t \hat{e}^2_{i,t-1}}$$

1. Step 1 — Individual Cointegrating Regressions
Estimate $y_{it} = \hat{\alpha}_i + \hat{\beta}_i x_{it} + \hat{e}_{it}$ separately for each unit:

Each unit has its own intercept and slope — heterogeneity is permitted. Residuals $\hat{e}_{it}$ oscillate around zero for all units.

	$\hat{\alpha}_i$	$\hat{\beta}_i$
U1	2.643	1.448
U2	3.445	1.376
U3	1.977	1.539
U4	4.104	1.311

2. Step 2 — Individual ADF on Residuals ($k = 0$)
ADF ($k = 0$, no constant) on each $\hat{e}_{it}$:

	$\hat{\rho}_i$	$\hat{s}_i^2$	SE_i	τ_i
U1	−1.836	0.036	0.156	−11.804
U2	−1.846	0.047	0.181	−10.181
U3	−1.845	0.058	0.197	−9.383
U4	−1.936	0.060	0.227	−8.543

3. Group-ADF Statistic (Between-Dimension, Stat 7)
Average the individual τ_i and standardize. Adjustment moments $\mu_\tau = -1.645$ and $\sigma_\tau^2 = 0.875$ are used for illustration; exact values for $m = 1$ are not tabulated in Pedroni (1999) and require simulation or software.

$$\bar{\tau} = \frac{1}{N}\sum_{i=1}^{N} \tau_i = -9.978, \quad Z_{\bar{\tau}} = \frac{\sqrt{N}\,(\bar{\tau} - \mu_\tau)}{\sigma_\tau} = -19.048$$

4. Panel-ADF Statistic (Within-Dimension, Stat 4)
Pool residuals across all units:

$$\hat{\rho}_P = \frac{\sum_i \sum_t \hat{e}_{i,t-1}\,\Delta\hat{e}_{it}}{\sum_i \sum_t \hat{e}^2_{i,t-1}} = \frac{-10.376}{5.573} = -1.862$$

$$\hat{s}_P^2 = 0.0461, \; SE_P = 0.0910, \; \tau_P = \frac{-1.862}{0.0909} = -20.475$$

$$Z_{\tau_P} = \frac{\sqrt{N}\,(\tau_P - \mu_\tau)}{\sigma_\tau} = -43.042$$

5. Interpretation

- **Decision:** Both statistics reject H_0 at $\alpha = 0.05$ (normal $CV = -1.645$). A stable long-run price–cost equilibrium holds in each of the four regions, with heterogeneous intercepts and slopes.
- **Practitioner Insight:** Pedroni provides seven statistics in two families. The *within-dimension* (Panel) statistics pool $\hat{\rho}$ across units—imposing a common adjustment speed under H_1 (analogous to LLC). The *between-dimension* (Group) statistics average individual τ_i—allowing heterogeneous speeds (analogous to IPS) and are preferred when adjustment differs across units (e.g. regional prices). All seven statistics are standardized to $N(0, 1)$ using unit-specific factors from Pedroni (1999, Table 2) that depend on T, regressors, and deterministic terms; using the wrong specification (e.g. trend vs. no-trend) yields incorrect p-values. The panel dimension substantially increases power against near-cointegrated alternatives compared with unit-by-unit single-equation tests

Common Pitfalls

- **Reporting only selected Pedroni statistics.** Pedroni proposed seven panel and group statistics to provide robustness across specifications. Conclusions should be based on the overall pattern rather than selectively chosen results.
- **Ignoring cross-sectional dependence.** Pedroni's asymptotic theory assumes cross-sectional independence across units. Strong common shocks or global factors can distort size and create spurious findings.
- **Using very small N or T.** The test relies on large-panel asymptotics with both dimensions reasonably large. Small samples can produce poor size and low power.
- **Misreading panel vs. group statistics.** Panel statistics assume homogeneous adjustment, while group statistics allow heterogeneity across units. Large disagreement between them often signals heterogeneous dynamics.
- **Misspecifying deterministic terms.** Pedroni permits models with no deterministic terms, fixed effects, or trends. Incorrect handling of trends or fixed effects can distort inference.

Westerlund Test (Panel)

Panel Cointegration Test Based on Error Correction

COINT.11 | **Category:** Cointegration | **Data:** Panel Time Series | **Null:** Tabled distribution

When to Use

- Test panel cointegration via error correction.
- Heterogeneous long-run panel relationships.
- Robust to cross-sectional dependence in panels.
- Financial or macroeconomic panel datasets.
- Preliminary step before panel ECM modeling.

Assumptions

- Variables are I(1) in each panel member.
- Error-correction dynamics are correctly specified.
- Cointegrating vectors may differ across panels.
- Panel dimensions are sufficiently large.
- Bootstrap often used under cross dependence.

Test Statistic

$$G_\tau = \frac{1}{N}\sum_{i=1}^{N}\frac{\hat{\alpha}_i}{SE(\hat{\alpha}_i)} \quad \Big| \quad G_\alpha = \frac{1}{N}\sum_{i=1}^{N}\frac{T\hat{\alpha}_i}{\hat{\alpha}_i(1)} \quad \Big| \quad P_\tau = \frac{\hat{\alpha}}{SE(\hat{\alpha})} \quad \Big| \quad P_\alpha = T\hat{\alpha}$$

Notation: $\hat{\alpha}_i$ is adjustment speed for unit i, group (G) vs. panel (P) statistics, τ is t-ratio, α is coefficient

Plain English: Westerlund's panel cointegration tests examine whether error correction mechanisms are present by testing the significance of adjustment speed parameters in panel error correction models. If cointegration exists, deviations from long-run equilibrium should trigger adjustments back toward equilibrium (negative α_i). The tests provide four statistics: two group statistics allowing heterogeneous adjustment speeds across panel members, and two panel statistics pooling under homogeneity. Unlike residual-based tests, this approach directly tests the defining feature of cointegration—error correction.

⚠ Practitioner Warning

Cross-sectional dependence common in panel data invalidates asymptotic distribution; bootstrap critical values strongly recommended. Four statistics can yield conflicting conclusions. Group statistics (G_τ, G_α) more robust to heterogeneity but potentially less powerful than panel statistics (P_τ, P_α). Requires specifying lag structure in error correction model; information criteria or sequential testing recommended. Unbalanced panels not directly accommodated in original formulation. Test assumes no structural breaks; Gregory-Hansen-type panel extensions needed for break robustness.

Hypotheses

- H_0: No cointegration (no error correction, $\alpha_i = 0, \forall i$).
- H_1: Cointegration exists ($\alpha_i < 0$ for some/all i).

Decision Rule: Reject H_0 if stat < CV or p-value < α. Panel statistics test $\alpha_i = \alpha < 0, \forall i$; group test $\alpha_i < 0$ for some i.

Software References [Python, R, SAS]

Python	R	SAS
`Not Native (Manual)`	`plm::pwtest (related)`	`Not Native (Manual)`

Additional Notes

Westerlund's test is based on the conditional error-correction model $y_{it} = \delta_i' d_t + \alpha_i y_{i,t-1} + \eta_i' x_{i,t-1} + \sum_{j=1}^{p_i} \alpha_{ij} y_{i,t-j} + \sum_{j=0}^{p_i} \gamma_{ij}' x_{i,t-j} + e_{it}$, where d_t contains deterministic components (none, intercept, or intercept plus trend), commonly $d_t = [1]$ or $d_t = [1, t]$. The key parameter is the adjustment coefficient α_i; cointegration implies $\alpha_i < 0$. The normalization factor $\alpha_i(1) = 1 - \sum_{j=1}^{p_i} \alpha_{ij}$ adjusts for short-run autoregressive lag augmentation. The four statistics are: G_τ, which averages the individual t-ratios for $\hat{\alpha}_i$; G_α, which averages normalized coefficient estimates; P_τ, which pools the standardized adjustment evidence across units; and P_α, which pools normalized coefficients. Short panels ($T < 20$) may have inadequate power. Westerlund and Edgerton (2007) extended the framework to allow structural breaks. Persyn and Westerlund (2008) developed a Stata implementation with bootstrap options. Banerjee–Carrion-i-Silvestre (2017) proposed panel cointegration methods that accommodate cross-sectional dependence through common correlated effects augmentation.

Miscellanea: Joakim Westerlund (2007) developed panel cointegration tests based on error-correction models, directly testing whether long-run equilibrium adjustments exist. The approach builds on the Engle–Granger representation and incorporates bootstrap methods to handle cross-sectional dependence, making it a widely used alternative to residual-based panel tests.

Reference(s): Westerlund (2007); Persyn & Westerlund (2008)

Worked Example

Problem Statement An economist tests for cointegration in a panel of four regions ($N = 4, T = 10$) using Westerlund's (2007) error-correction approach. H_0: $\alpha_i = 0$ for all i. Data is provided at the end of this chapter.

1. Individual ECM Regressions
For each unit, estimate

$$\Delta y_{it} = \delta_i + \alpha_i y_{i,t-1} + \gamma_i x_{i,t-1} + \theta_i \Delta x_{it} + e_{it}.$$

Here $n = T - 1 = 9$ and residual $df = 9 - 4 = 5$ per unit.

Unit	$\hat{\delta}_i$	$\hat{\alpha}_i$	$\hat{\gamma}_i$	$\hat{\theta}_i$
1	2.3736	−1.2574	1.8705	2.0748
2	0.0226	−0.2864	0.4091	2.6942
3	1.0399	−0.9578	1.5321	2.3951
4	6.9370	−1.7759	2.3361	1.4761

2. Unit-Specific Error-Correction Results

Unit	$\hat{\alpha}_i$	SE_i	τ_i	SSR_i
1	−1.2574	0.2517	−4.9953	0.0067
2	−0.2864	0.2407	−1.1900	0.0158
3	−0.9578	0.1071	−8.9470	0.0146
4	−1.7759	0.1639	−10.8323	0.0344

All four estimated adjustment (error correction) speeds are negative: Unit 4 (strongest) and Unit 2 (weakest).

3. Group Statistics
Since $\alpha_i(1) = 1$ (no lags – no Δy_{it-j}) for every unit,

$$G_\tau = -6.4912, \quad G_\alpha = -10.6939.$$

4. Panel Statistic Construction
For the panel statistics, partial out the nuisance regressors $[1, x_{i,t-1}, \Delta x_{it}]$ from both Δy_{it} and $y_{i,t-1}$. Let the residualized variables be $\widetilde{\Delta y}_{it}$ and $\tilde{y}_{i,t-1}$.

Unit	$\sum \tilde{y}^2_{i,t-1}$	$\sum \tilde{y}_{i,t-1}\widetilde{\Delta y}_{it}$
1	0.0211	−0.0265
2	0.0544	−0.0156
3	0.2550	−0.2443
4	0.2556	−0.4539
Total	0.5861	−0.7403

$$\hat{\alpha}_{\text{pool}} = \frac{-0.7403}{0.5861}$$

$$\hat{S}^2_N = \sum_{i=1}^{N} \frac{SSR_i}{5}$$

$$SE(\hat{\alpha}_{\text{pool}}) = \sqrt{\frac{\hat{S}^2_N}{0.5861}}$$

5. Panel Statistics

$$P_\tau = -16.1823, \quad P_\alpha = -12.6298.$$

6. Results and Decision
(μ, σ) from (Westerlund 2007 Table 1, Case 2, $K = 1$):

Stat	Raw	μ	σ	Z	Decision
G_τ	−6.4912	−1.7776	0.4492	−10.49	Reject
G_α	−10.6939	−7.1423	2.7218	−1.30	FTR
P_τ	−16.1823	−2.8952	0.9942	−13.36	Reject
P_α	−12.6298	−4.2303	2.2198	−3.78	Reject

Practitioner Insight: Westerlund's test is an error-correction test rather than a residual unit-root test: the key parameter is the adjustment coefficient α_i. A negative and statistically significant $\hat{\alpha}_i$ implies that deviations from the long-run equilibrium are corrected over time, providing direct evidence of cointegration. Westerlund's key innovation is to estimate α_i from a one-step ECM with $y_{i,t-1}$ and $x_{i,t-1}$ entered separately, avoiding the nuisance-parameter problems of two-step residual-based methods. When $p_i = 0$, the normalizing factor $\alpha_i(1) = 1 - \sum_{j=1}^{p_i} \alpha_{ij}$ equals 1, simplifying G_α and P_α to scaled sums of the estimated adjustment coefficients. The group statistics (G_τ, G_α) test the heterogeneous alternative that some units are cointegrated, whereas the panel statistics (P_τ, P_α) target the stronger alternative that the panel as a whole shares a common adjustment mechanism. Here all four units have negative adjustment speeds and three of the four Westerlund statistics are strongly negative, giving consistent evidence of panel cointegration.

Common Pitfalls

- **Using asymptotic CVs when cross-sectional dependence exists.** The Westerlund test assumes cross-sectional independence. If strong dependence is present, bootstrap CVs should be used; otherwise the test may over-reject.
- **Misinterpreting disagreement between group (G) and panel (P) statistics.** Group statistics allow heterogeneous adjustment across units, while panel statistics assume common adjustment dynamics. Differences between them can signal heterogeneous cointegration relationships across the panel.
- **Using inappropriate lag structures in the error-correction models.** Each cross-sectional unit may require a different lag length to capture short-run dynamics. Poor lag selection can introduce serial correlation or reduce test's power.
- **Applying the test to panels with very small T or N.** Reliable inference requires reasonably large time and cross-sectional dimensions. Small samples can lead to unstable estimates and weak test power.
- **Ignoring the integration order of the variables.** The Westerlund framework assumes variables are integrated of order one. Unit root tests should be conducted beforehand to verify the appropriate integration properties.

XV. COINT — Raw Data

Data used in selected worked examples are provided here to preserve the book's strict two-page-per-test format.

[COINT.10] Pedroni Test

Problem Statement An economist tests whether a price–cost relationship holds as a stable cointegrating equilibrium across four regions ($N = 4, T = 10$). Pedroni's test pools information across units, allowing heterogeneous intercepts and slopes while testing the *common* hypothesis of no cointegration. Individual-intercept specification, $k = 0$; two statistics shown: Group-ADF (between-dimension) and Panel-ADF (within-dimension).

		$t = 1$	$t = 2$	$t = 3$	$t = 4$	$t = 5$	$t = 6$	$t = 7$	$t = 8$	$t = 9$	$t = 10$
Unit 1	x	5.0	5.2	6.3	6.4	7.4	7.6	8.7	8.8	9.8	10.0
	y	10.3	9.9	12.1	11.4	13.7	13.3	15.5	14.8	17.3	17.0
Unit 2	x	8.0	8.1	9.3	9.5	10.4	10.5	11.7	11.9	12.8	12.9
	y	14.8	14.1	16.7	16.4	18.1	17.3	19.8	19.5	21.5	20.9
Unit 3	x	3.0	3.3	4.3	4.3	5.4	5.7	6.7	6.7	7.8	8.1
	y	7.0	6.7	8.9	7.9	10.7	10.6	12.8	12.0	14.3	14.0
Unit 4	x	6.0	6.0	7.3	7.6	8.4	8.4	9.7	10.0	10.8	10.8
	y	12.3	11.5	14.0	13.8	15.5	14.7	17.3	17.2	18.6	17.6

[COINT.11] Westerlund Test (Panel)

Problem Statement An economist tests for cointegration in a panel of four regions ($N = 4, T = 10$) using Westerlund's (2007) error-correction approach. Unlike residual-based tests (Pedroni), the Westerlund test directly estimates the speed of adjustment α_i in the ECM for each unit. H_0: $\alpha_i = 0$ for all i (no error correction, no cointegration). Two-step procedure, $k = 0$, individual intercepts.

		$t = 1$	$t = 2$	$t = 3$	$t = 4$	$t = 5$	$t = 6$	$t = 7$	$t = 8$	$t = 9$	$t = 10$
Unit 1	x	5.0	5.2	6.3	6.4	7.4	7.6	8.7	8.8	9.8	10.0
	y	10.0	9.6	11.9	11.3	13.5	13.1	15.5	14.9	17.1	16.7
Unit 2	x	8.0	8.1	9.3	9.5	10.4	10.5	11.7	11.9	12.8	12.9
	y	14.6	14.0	16.5	16.2	17.9	17.3	19.9	19.5	21.3	20.7
Unit 3	x	3.0	3.3	4.3	4.3	5.4	5.7	6.7	6.7	7.8	8.1
	y	6.9	6.7	8.7	8.0	10.6	10.5	12.6	11.8	14.5	14.3
Unit 4	x	6.0	6.0	7.3	7.6	8.4	8.4	9.7	10.0	10.8	10.8
	y	12.2	11.5	13.9	13.6	15.3	14.8	17.1	16.7	18.5	17.8

Part XVI

Time-Varying Volatility & Risk Dynamics

Detecting ARCH/GARCH effects and modeling conditional variance

Volatility and Conditional Heteroskedasticity

XVI. ARCH — A Conceptual Primer

What Are ARCH Effects?

A series $\{\varepsilon_t\}$ exhibits **autoregressive conditional heteroskedasticity** (ARCH) when its conditional variance is time-varying and depends on past squared innovations:

$$\sigma_t^2 = \text{Var}(\varepsilon_t \mid \mathcal{F}_{t-1}) \neq \text{const},$$

where $\mathcal{F}_{t-1}$ is the information set up to $t-1$. Writing $\varepsilon_t = \sigma_t z_t$ with $z_t \overset{\text{iid}}{\sim} (0, 1)$ separates the **volatility process** σ_t from the standardized shock z_t.

Critical distinction: ARCH residuals are *uncorrelated* ($\text{Cov}(\varepsilon_s, \varepsilon_t) = 0$, $s \neq t$) but *not independent*—dependence lives in the **second moment**, not the mean. A model passing standard autocorrelation tests (Ljung–Box on ε_t) can still be severely misspecified if ARCH is present.

Volatility Clustering

The defining empirical signature of ARCH is **volatility clustering**: large shocks tend to be followed by large shocks (of either sign), and calm periods by calm periods. This produces:

- **Positive autocorrelation** in ε_t^2 and $|\varepsilon_t|$ even when ε_t itself is uncorrelated.
- **Excess kurtosis** (fat tails) in the unconditional distribution of ε_t, even if z_t is Gaussian. The unconditional kurtosis of a GARCH(1,1) process is $3(1-(\alpha+\beta)^2)/(1-(\alpha+\beta)^2 - 2\alpha^2) > 3$.
- **Leptokurtic** marginal distributions, consistent with the heavy tails observed in financial returns.

Why It Matters

- **Incorrect standard errors:** OLS assumes $\text{Var}(\varepsilon_t) = \sigma^2$; ARCH violates this, producing inefficient estimates and misleading inference even when $\hat{\beta}$ is unbiased.
- **Risk and forecast:** constant-variance models underestimate risk during volatile periods and overestimate it during calm ones—directly affecting Value-at-Risk, option pricing, and portfolio allocation.
- **Model adequacy:** residual ARCH signals that the fitted model has not fully captured the data-generating process; forecasts will be suboptimal.

The ARCH(m) and GARCH(p, q) Models

ARCH(m): (Engle 1982)

$$\sigma_t^2 = \omega + \sum_{i=1}^{m} \alpha_i \varepsilon_{t-i}^2, \quad \omega > 0,\ \alpha_i \geq 0,\ \sum \alpha_i < 1.$$

Implies ε_t^2 follows an AR(m) with non-negative coefficients; this is the direct motivation for regressing $\hat{\varepsilon}_t^2$ on its own lags in Engle's LM test.

GARCH(p, q): Generalized ARCH (Bollerslev 1986)

$$\sigma_t^2 = \omega + \sum_{i=1}^{q} \alpha_i \varepsilon_{t-i}^2 + \sum_{j=1}^{p} \beta_j \sigma_{t-j}^2.$$

Parsimonious: GARCH(1,1) with $\alpha+\beta$ close to 1 captures the slow decay of volatility observed in practice. An ARCH(m) test significant at long lags signals that a GARCH model (not just ARCH) may be needed.

What Are We Testing?

$$H_0 : \alpha_1 = \alpha_2 = \cdots = \alpha_m = 0$$

(no ARCH effects; constant variance)

$$H_1 : \alpha_i \neq 0 \text{ for some } i$$

(conditional heteroskedasticity present)
All tests in this chapter are applied to **model residuals** $\hat{\varepsilon}_t$, not raw data. A correctly specified mean equation is a prerequisite: residual ARCH detected on poorly fitted mean models may reflect mean misspecification rather than genuine volatility clustering.

Engle's ARCH LM Test

The most widely used diagnostic. Regress squared residuals on m lagged values:

$$\hat{\varepsilon}_t^2 = \gamma_0 + \gamma_1 \hat{\varepsilon}_{t-1}^2 + \cdots + \gamma_m \hat{\varepsilon}_{t-m}^2 + v_t,$$

and form the Lagrange Multiplier statistic:

$$\text{LM} = T \cdot R^2 \xrightarrow{d} \chi_m^2 \quad \text{under } H_0,$$

where R^2 is from the auxiliary regression above. Equivalently, an F-version tests joint significance of $\gamma_1, \ldots, \gamma_m$. The LM form is preferred as it requires fitting only the null (OLS) model.

Lag selection: use $m = 5$ for daily data, $m = 12$ for monthly; in general $m \approx \ln(T)$. Report results for several values of m.

Ljung–Box Test on Squared Residuals (McLeod-Li)

The standard Ljung–Box statistic (see TSDG primer) applied to $\{\hat{\varepsilon}_t^2\}$ rather than $\{\hat{\varepsilon}_t\}$:

$$Q^*(h) = T(T+2) \sum_{k=1}^{h} \frac{r_k^2(\hat{\varepsilon}^2)}{T-k} \xrightarrow{d} \chi_h^2 \quad \text{under } H_0.$$

Tests whether squared residuals exhibit autocorrelation—a necessary condition for ARCH. **Complements** Engle's LM test: LM is optimal for ARCH alternatives; Ljung–Box on ε_t^2 is more general and also detects other forms of nonlinear dependence in the variance.

Parallel testing strategy:

- Ljung–Box on $\hat{\varepsilon}_t$: tests *mean* model adequacy.
- Ljung–Box on $\hat{\varepsilon}_t^2$: tests *variance* model adequacy.
- Both should be checked after any time series fit.

Engle–Ng Sign and Size Bias Tests

Standard ARCH and GARCH models impose a **symmetric** volatility response: a shock of $+\delta$ increases σ_t^2 by the same amount as $-\delta$. Financial markets routinely violate this—negative shocks (bad news) typically increase volatility more than positive shocks of equal magnitude (**leverage effect**).

Define $S_{t-1}^- = \mathbf{1}(\hat{\varepsilon}_{t-1} < 0)$ (indicator for negative past shocks) and $\hat{v}_t = \hat{z}_t^2 - 1$ (standardized variance residual). The three component tests and their joint form:

Test	Auxiliary regression
Sign bias	$\hat{v}_t = c_0 + c_1 S_{t-1}^- + e_t$; $c_1 \neq 0 \Rightarrow$ sign asymmetry.
Negative size bias	$\hat{v}_t = c_0 + c_1 S_{t-1}^- \hat{\varepsilon}_{t-1} + e_t$; magnitude of negative shocks matters.
Positive size bias	$\hat{v}_t = c_0 + c_1(1 - S_{t-1}^-)\hat{\varepsilon}_{t-1} + e_t$; magnitude of positive shocks matters.
Joint test	All three regressors; $TR^2 \sim \chi_3^2$ under H_0.

Rejection of sign bias $\Rightarrow$ consider GJR-GARCH or EGARCH. Rejection of size bias $\Rightarrow$ APARCH or nonlinear GARCH variants.

ICSS Test for Variance Breaks

While Engle's LM and McLeod–Li tests detect **dynamic** (ARCH-type) volatility clustering, the **Iterated Cumulative Sums of Squares** (ICSS) test (Inclan & Tiao 1994) detects **structural breaks** in the unconditional variance—abrupt, persistent shifts in the variance level, as opposed to autocorrelated fluctuations around a constant level.

Define the cumulative sum of squared residuals and its centerd ratio:

$$C_k = \sum_{t=1}^{k} \hat{\varepsilon}_t^2, \quad D_k = \frac{C_k}{C_T} - \frac{k}{T}, \quad k = 1, \ldots, T.$$

Under constant variance, $C_k/C_T \approx k/T$ so $D_k \approx 0$ throughout. A sustained departure signals a variance break. The test statistic

$$\text{IT} = \sqrt{\tfrac{T}{2}} \max_k |D_k|$$

converges to the supremum of a standard Brownian bridge; Inclan–Tiao (1994) tabulate critical values: 10% = 1.22, 5% = 1.36, 1% = 1.63. The break-point estimate is $k^* = \text{argmax}_k |D_k|$.

Iterated procedure: after locating k^*, re-apply the test to each sub-segment $[1, k^*]$ and $[k^*+1, T]$ until no further breaks are found, yielding all change-points.

Limitation: the raw-$\hat{\varepsilon}_t^2$ formulation has size distortion under ARCH/GARCH dependence; Sansó et al. (2004) correct this with a kernel long-run variance adjustment.

Why it matters for ARCH modeling: unmodelled variance breaks inflate GARCH persistence estimates, pushing $\hat{\alpha} + \hat{\beta}$ toward unity and suggesting spurious IGARCH behavior. ICSS should therefore be run *before* fitting GARCH models; detected breaks may warrant regime-specific GARCH estimation.

Diagnostic Workflow

1. Fit a mean model (ARMA or regression); obtain $\hat{\varepsilon}_t$.
2. **Check mean adequacy:** Ljung–Box on $\hat{\varepsilon}_t$ (not squared).
3. **Test for variance breaks:** ICSS test on $\hat{\varepsilon}_t^2$. If breaks are found, consider regime-specific modeling before proceeding.
4. **Test for ARCH:** Engle's LM and Ljung–Box on $\hat{\varepsilon}_t^2$.
5. If ARCH confirmed, **test for asymmetry:** Engle–Ng sign and size bias tests.
6. Fit a GARCH (or asymmetric) model; repeat steps 2, 4–5 on standardized residuals $\hat{z}_t = \hat{\varepsilon}_t/\hat{\sigma}_t$.

Common Pitfalls

- **Mean model misspecification:** residual ARCH in a poorly fitted mean model may be spurious; confirm the mean is well-specified first.
- **Confusing LB on ε_t and ε_t^2:** the former tests mean dependence; the latter tests variance dependence—both are needed.
- **Applying tests to raw returns:** apply to residuals from a fitted mean model, not raw data.
- **Ignoring asymmetry:** a symmetric GARCH on financial returns will fail the Engle–Ng test; misspecified variance models inflate risk estimates.
- **Over-relying on a single lag m:** ARCH LM is sensitive to m; report for multiple choices.
- **Not re-testing after GARCH fit:** standardized residuals $\hat{z}_t$ from a GARCH model should also pass all four tests; if not, the variance model is still misspecified.
- **Skipping the ICSS pre-test:** fitting GARCH without first testing for variance breaks can produce spuriously high persistence ($\hat{\alpha} + \hat{\beta} \approx 1$) driven by an unmodelled level shift rather than genuine GARCH dynamics.

Bottom Line

ARCH diagnostics reveal dependence in volatility, not the mean. Uncorrelated residuals do not imply independence when variance is time-varying. Run ICSS *first* to detect variance breaks before fitting GARCH; then test *both* $\hat{\varepsilon}_t$ and $\hat{\varepsilon}_t^2$ after every model fit; use Engle–Ng to determine whether asymmetric volatility models are needed; and always re-test standardized residuals after fitting GARCH.

ARCH Test (Engle's LM Test)

Lagrange Multiplier Test for Autoregressive Conditional Heteroskedasticity

ARCH.1 | **Category:** ARCH/GARCH | **Data:** Time Series | **Null:** Chi-square

When to Use

- Detect ARCH effects in time-series residuals.
- Assess volatility clustering after mean model fitting.
- Financial or economic data with time-varying variance.
- Diagnose conditional heteroskedasticity and assess need for ARCH/GARCH models.

Assumptions

- No severe model misspecification in mean equation.
- Residuals are serially uncorrelated (mean model adequate).
- Under H_0, residuals have constant conditional variance.
- Errors have finite fourth moments (for χ^2 approximation).
- Lag order q correctly specified for ARCH(q) testing.

Test Statistic

$$LM = T \cdot R^2 \approx \chi^2_q \text{ (under } H_0\text{)}. \qquad \text{Auxiliary Regression: } \hat{u}_t^2 = \alpha_0 + \sum_{i=1}^{q} \alpha_i \hat{u}_{t-i}^2 + \epsilon_t$$

Notation: $\hat{u}_t$ are residuals from mean equation, q is number of lags, R^2 from auxiliary regression with sample size T

Plain English: Engle's ARCH test examines whether past squared residuals help predict current squared residuals by regressing squared residuals on their lags. If conditional variance is constant, past squared residuals should have no predictive power for current variance. Significant coefficients indicate ARCH effects—volatility clustering where large shocks tend to be followed by large shocks (of either sign), and small by small. The test provides formal evidence for time-varying volatility requiring ARCH/GARCH modeling.

⚠ Practitioner Warning

Requires a correctly specified mean equation; misspecification can induce spurious ARCH effects. Lag choice q matters—use small lags or select via information criteria. Residual autocorrelation can distort size and power. Sensitive to outliers and heavy tails; large shocks may trigger false positives. Power is low for non-ARCH heteroskedasticity (e.g., structural breaks), and highly persistent volatility (e.g., IGARCH) may be difficult to detect with small q.

Hypotheses

- H_0: $\alpha_1 = \cdots = \alpha_q = 0$ (no ARCH effects).
- H_1: At least one $\alpha_i \neq 0$ (ARCH effects present).

Decision Rule: Reject H_0 if $LM > \chi^2_{\alpha,q}$ or p-value $< \alpha$. df = ARCH order q tested. For small n, use F formulation.

Software References [Python, R, SAS]

`statsmodels::.het_arch` `FinTS::ArchTest` `PROC AUTOREG (ARCHTEST)`

Additional Notes

The test is a score (Lagrange Multiplier) test for ARCH(q) effects in residuals. An equivalent formulation is TR^2, where T is the sample size used in the auxiliary regression. For testing nested hypotheses (e.g., ARCH(p) vs. ARCH(q), $p < q$), one may use **(A)**. Extensions: the Ljung–Box test (McLeod-Li) applied to squared residuals provides an alternative omnibus diagnostic for ARCH effects; robust ARCH tests (e.g., Lee-King-Strazicich (2003)) account for structural breaks. Relationship to volatility modeling: rejection indicates conditional heteroskedasticity, for which GARCH models are commonly used, often starting with GARCH(1,1) for parsimony. Engle's (1982) ARCH model is **(B)**, while GARCH extends this by including lagged conditional variances **(C)**.

$$\textbf{(A)}: F = \frac{(R_q^2 - R_p^2)/(q-p)}{(1-R_q^2)/(T-q-1)} \sim F_{q-p,\,T-q-1}, \quad \textbf{(B)}: \sigma_t^2 = \alpha_0 + \sum_{i=1}^{q} \alpha_i u_{t-i}^2, \quad \textbf{(C)}: \sigma_t^2 = \alpha_0 + \sum_{i=1}^{q} \alpha_i u_{t-i}^2 + \sum_{j=1}^{p} \beta_j \sigma_{t-j}^2.$$

Miscellanea: Robert Engle (1982) introduced ARCH models to capture time-varying volatility, motivated by the clustering of high and low variance observed in economic and financial data. By modeling variance as a function of past squared shocks, his work transformed volatility from a nuisance into a central feature of time series analysis, leading to widespread applications and a Nobel Prize in 2003.

Reference(s): Engle (1982); Bollerslev (1986)

ARCH Test (Engle's LM Test)

ARCH.1

Worked Example

Problem Statement Twenty weekly excess returns (%) on an equity index have been demeaned; the residuals e_t are given below. Informal inspection suggests two bursts of large absolute returns (weeks 4–6 and weeks 13–15). Apply Engle's ARCH LM test at $\alpha = 0.05$ to decide whether ARCH(1) effects are present. $H_0: \alpha_1 = 0$ (no ARCH(1) effects); $H_1: \alpha_1 \neq 0$.

t	1	2	3	4	5	6	7	8	9	10
e_t	0.5	−0.4	0.6	−1.5	1.6	−1.5	0.8	−0.5	0.4	−0.5
t	11	12	13	14	15	16	17	18	19	20
e_t	0.6	−0.4	1.7	−1.6	0.8	−0.5	0.4	−0.6	0.5	−0.4

1. Form Squared Residuals
Set $u_t = e_t^2$.

2. Auxiliary Regression
Regress u_t on $(1, u_{t-1})$ for $t = 2, \ldots, 20$ ($n = 19$ obs.):

$$u_t = \hat{\alpha}_0 + \hat{\alpha}_1 u_{t-1} + v_t. \qquad (q = 1)$$

OLS yields $\hat{\alpha}_0 = 0.426$, $\hat{\alpha}_1 = 0.512$, giving:

$$\text{ESS} = 4.834, \quad \text{RSS} = 13.710, \quad \text{TSS} = 18.545.$$

$$R^2 = \frac{\text{ESS}}{\text{TSS}} = \frac{4.834}{18.545} = 0.261.$$

3. ARCH LM Statistic
The LM statistic and its finite-sample F-form are:

$$LM = T \cdot R^2 = 19 \times 0.261 = 4.96 \sim \chi^2(1),$$

$$F = \frac{\text{ESS}/q}{\text{RSS}/(n-q-1)} = \frac{4.834/1}{13.710/17} = \frac{4.834}{0.807} = 5.99.$$

4. Decision

Form	Stat	CV (5%)	Decision
$\chi^2(1)$	4.96	3.841	**Reject**
$F(1, 17)$	5.99	4.451	**Reject**

$p = 0.026$. Reject H_0 at the 5% level: evidence of ARCH(1) effects. $\hat{\alpha}_1 = 0.51$ indicates moderate volatility persistence.

Practitioner Insight: Engle's ARCH LM test works by asking whether today's squared residual $u_t = e_t^2$ is predictable from yesterday's u_{t-1}. Under the null of no ARCH, u_t is serially uncorrelated and R^2 in the auxiliary regression should be near zero. The standard form of the test uses the LM statistic $T \cdot R^2 \sim \chi^2(q)$; the F-form shown here is preferable for small samples because the χ^2 approximation over-rejects when n is small — for $n \to \infty$ the two are equivalent via $q \cdot F \xrightarrow{d} \chi^2(q)$. The test extends to ARCH(q) by including q lags of u_t in the auxiliary regression, with df $(q, n-q-1)$; the Ljung–Box test applied to u_t (McLeod–Li test) is a closely related portmanteau alternative. A significant result identifies conditional heteroscedasticity but not the appropriate order; the partial autocorrelation function of u_t and information criteria on GARCH models should guide order selection. The two volatility clusters in the data (weeks 4–6 and weeks 13–15) are the feature the test is designed to detect: large squared residuals that persist for one or two periods before subsiding.

Common Pitfalls

- **Testing raw data instead of residuals from the mean equation.** The ARCH LM test applies to residuals, not the original series. Always estimate an appropriate mean model (e.g., AR/ARMA or regression) and test the resulting residuals $\hat{u}_t$.
- **Using arbitrary lag order q in the auxiliary regression.** The number of ARCH lags should reflect data frequency and volatility persistence. Testing several plausible values of q or using simple rules (e.g., $q \approx \ln T$) helps ensure results are not driven by a particular choice.
- **Ignoring heavy tails in the residual distribution.** The asymptotic χ^2 approximation assumes finite fourth moments. Very heavy-tailed data can distort the test size, so large kurtosis or extreme outliers should be examined before relying on the test.
- **Assuming rejection automatically implies a GARCH model.** Rejection indicates time-varying variance but does not identify the source. Volatility clustering, structural breaks, mean misspecification, or other forms of heteroskedasticity may produce similar results.
- **Relying solely on the test without examining squared residual diagnostics.** Plots of $\hat{u}_t^2$ and the ACF of squared residuals help identify volatility clustering and the appropriate model structure (e.g., ARCH vs. GARCH).

McLeod-Li Test

Portmanteau Test for ARCH Effects (Ljung-Box Application)

ARCH.2 | **Category:** ARCH/GARCH | **Data:** Time Series | **Null:** Chi-square

When to Use

- Detect ARCH effects using autocorrelation of squared residuals. Ljung-Box application on squared residuals.
- Assess volatility clustering after fitting a mean model.
- Screen multiple lag orders via a portmanteau statistic.
- Complement Engle's LM test for ARCH effects testing.

Assumptions

- No severe model misspecification in mean equation.
- Residuals are serially uncorrelated (mean model adequate).
- Under H_0, residuals have constant conditional variance.
- Errors have finite fourth moments (for χ^2 approximation).
- Large-sample χ^2 approximation is appropriate.

Test Statistic

$$Q(h) = T(T+2)\sum_{k=1}^{h} \frac{\hat{\rho}_k^2(\hat{u}_t)}{T-k}, \quad \text{where } \hat{\rho}_k = \frac{\sum_{t=k+1}^{T}(\hat{u}_t - \bar{u})(\hat{u}_{t-k} - \bar{u})}{\sum_{t=1}^{T}(\hat{u}_t - \bar{u})^2}. \quad \text{Under } H_0, Q(h) \sim \chi_h^2 \text{ (asymp.)}$$

Notation: $\hat{u}_t$ are squared residuals (e_t^2), $\hat{\rho}_k$ are autocorrelations of $\hat{u}_t$, h is maximum lag, T is sample size

Plain English: The McLeod-Li test (Ljung-Box test on squared residuals) detects ARCH effects by examining whether autocorrelations of squared residuals are jointly significant. Under constant conditional variance, squared residuals should behave like white noise with no autocorrelation. Significant autocorrelation in squared residuals indicates volatility clustering—current variance depends on past variance. This omnibus approach tests multiple lags simultaneously, potentially more powerful than testing specific ARCH orders when the true lag structure is unknown.

⚠ Practitioner Warning

Choice of lag h affects results; common choices use small fixed lags, $\sqrt{T}$, $\log T$, or inspection of the autocorrelogram of u_t. Too few lags may miss ARCH; too many reduce power. Requires correctly specified mean equation, as autocorrelation can confound results. Less powerful than Engle LM when the true ARCH order is known. Sensitive to fat tails and outliers. Does not identify ARCH order—use ACF/PACF of squared residuals or information criteria.

Hypotheses

- H_0: $\rho_1 = \cdots = \rho_h = 0$ (no ARCH effects present).
- H_1: At least one $\rho_k \neq 0$ (ARCH effects present).

Decision Rule: Reject H_0 if $Q(h) > \chi^2_{\alpha,h}$ or p-value $< \alpha$. df = maximum lag h tested.

Software References [Python, R, SAS]

Python	R	SAS
`statsmodels::acorr_ljungbox`	`Box.test(resid^2, type="Ljung")`	`Not Native (Manual)`

Additional Notes

The McLeod–Li test applies the Ljung–Box portmanteau statistic to squared residuals $u_t = \hat{e}_t^2$ to detect ARCH-type dependence. While the Ljung–Box test is traditionally used on residuals to assess autocorrelation in the mean equation, applying it to u_t targets serial dependence in the conditional variance. The test is closely related to Engle's LM test: both assess predictability in squared residuals, but differ in formulation. The Ljung–Box test uses sample autocorrelations $\hat{\rho}_k$, whereas the LM test uses a regression of u_t on its lagged values. The two approaches are asymptotically related but can differ in finite samples, especially when dependence is concentrated at specific lags. For diagnostic purposes, inspecting the ACF and PACF of u_t complements formal testing; significant spikes suggest candidate ARCH orders. Testing absolute residuals $|\hat{e}_t|$ can sometimes improve power against certain alternatives. After detecting ARCH effects, a GARCH model is typically fitted, and the test is reapplied to standardized residuals $\hat{e}_t/\hat{\sigma}_t$ to assess adequacy of the volatility specification.

Miscellanea: Following Engle's (1982) introduction of ARCH models, McLeod and Li (1983) showed that applying portmanteau tests to squared residuals provides a simple way to detect ARCH-type dependence. Its popularity stems from its simplicity and familiarity, making it a convenient complement to dedicated ARCH tests.

Reference(s): Ljung & Box (1978); McLeod & Li (1983)

McLeod-Li Test

ARCH.2

Worked Example

Problem Statement The same twenty demeaned weekly excess returns used in the Tsay example are reproduced below. Apply the McLeod–Li test with $h = 4$ lags at $\alpha = 0.05$ to test for ARCH effects. ρ_k is the k-th autocorrelation of $u_t = e_t^2$. $H_0: \rho_1 = \rho_2 = \rho_3 = \rho_4 = 0$; $H_1: \rho_k \neq 0$ for some $k \leq 4$.

t	1	2	3	4	5	6	7	8	9	10
e_t	0.5	−0.4	0.6	−1.5	1.6	−1.5	0.8	−0.5	0.4	−0.5
t	11	12	13	14	15	16	17	18	19	20
e_t	0.6	−0.4	1.7	−1.6	0.8	−0.5	0.4	−0.6	0.5	−0.4

1. Squared Residuals and Sample Mean
Set $u_t = e_t^2$. $\bar{u} = 0.846$.

2. Sample Autocorrelations of u_t
The denominator of $\hat{\rho}_k$ is $\sum(u_t - \bar{u})^2 = 18.918$.

k	Numerator of $\hat{\rho}_k$	$\hat{\rho}_k$	$\hat{\rho}_k^2/(T-k)$
1	9.459	0.5000	0.01316
2	−1.925	−0.1017	0.00057
3	−7.464	−0.3945	0.00916
4	−7.843	−0.4146	0.01074
		Sum	0.03363

3. McLeod–Li Statistic
The Ljung–Box statistic applied to u_t ($T = 20$):

$$Q(h) = T(T+2)\sum_{k=1}^{h}\frac{\hat{\rho}_k^2}{T-k} = 14.80.$$

4. Decision
Under H_0: $Q(h) \sim \chi^2(h) = \chi^2(4)$.

α	$\chi^2_\alpha(4)$	Decision
5%	9.488	**Reject** (14.80 > 9.488)
1%	13.277	**Reject** (14.80 > 13.277)

$p = 0.005$. Reject H_0 at both levels: the squared residuals exhibit significant autocorrelation, confirming ARCH effects.

Practitioner Insight: The McLeod–Li test is the Ljung–Box portmanteau test applied to $u_t = e_t^2$ rather than to e_t itself. Applying the test to e_t checks for residual serial correlation in the mean equation; applying it to u_t checks for serial correlation in squared residuals, a signature of ARCH-type conditional heteroskedasticity. Engle's LM test with $h = 1$ gives $LM = 4.96$, $p = 0.026$, while McLeod–Li with $h = 4$ gives $Q(4) = 14.80$, $p = 0.005$. The stronger rejection reflects substantial autocorrelation at lags 3 and 4 ($\hat{\rho}_3 = -0.39$, $\hat{\rho}_4 = -0.41$), which a single-lag test misses. McLeod–Li is closely related to Engle's LM test but is not equivalent. Engle's LM test regresses u_t jointly on h lagged values and forms $nR^2 \sim \chi^2_h$; with $h = 4$ this gives $nR^2 = 16(0.521) = 8.33$, $p = 0.080$, failing to reject at 5%. McLeod–Li instead accumulates the squared sample autocorrelations $\hat{\rho}_k^2$ one lag at a time using the Ljung–Box weight $T(T+2)/(T-k)$. The choice of lag truncation h is consequential: too few lags may miss higher-order ARCH structure, while too many lags waste degrees of freedom and reduce power. A significant McLeod–Li statistic motivates volatility modeling, often beginning with a parsimonious GARCH specification. Engle's LM loses h observations due to lags. Sample size: $n = T - h$.

Common Pitfalls

- **Using incorrect degrees of freedom.** For tests on squared residuals $\hat{e}_t^2$, the Ljung–Box statistic uses h degrees of freedom (not $h - p - q$). Adjustments for mean-model parameters, ARMA(p,q), apply only when testing residual levels, not variance dynamics.
- **Choosing inappropriate lag length h.** Too few lags may miss volatility persistence, while too many reduce power. Lag selection should reflect data frequency and sample size, and results should be checked for robustness across plausible values of h.
- **Testing squared residuals when the mean model is misspecified.** Serial correlation in $\hat{e}_t$ can appear as dependence in $\hat{u}_t$. Always ensure residuals from the mean equation are approximately white noise before testing for ARCH effects.
- **Assuming Ljung–Box and ARCH LM tests are identical.** Although asymptotically equivalent, the two tests can differ in finite samples or with heavy-tailed data. Comparing both diagnostics provides a useful robustness check.
- **Relying only on the omnibus statistic.** The Ljung–Box statistic indicates whether dependence exists but not where it occurs. Inspecting the ACF of $\hat{u}_t$ helps identify volatility persistence and possible ARCH/GARCH structure.

Engle-Ng Sign Bias Test

Leverage Effect Diagnostic for Asymmetric Volatility

ARCH.3 | **Category:** ARCH/Volatility Tests | **Data:** Time Series | **Null:** t / χ^2 distribution

When to Use

- Detect asymmetry (leverage effect) in conditional volatility.
- Diagnose misspecification after fitting ARCH/GARCH.
- Assess whether negative shocks impact volatility differently.
- Distinguish symmetric vs asymmetric volatility dynamics.
- Guide choice of asymmetric models (e.g., GJR, EGARCH).

Assumptions

- Mean & variance models are correctly specified under H_0.
- Standardized residuals z_t are serially uncorrelated.
- Under H_0, no asymmetry in conditional variance dynamics.
- Residuals have finite moments for model-based inference.
- No severe remaining ARCH effects in std. residuals.

Test Statistic

$$\hat{\varepsilon}_t^2 = \alpha + \beta_1 S_{t-1}^- + \beta_2 S_{t-1}^- \hat{\varepsilon}_{t-1} + \beta_3 S_{t-1}^+ \hat{\varepsilon}_{t-1} + v_t, \quad \text{Tests: } t_i = \hat{\beta}_i / SE(\hat{\beta}_i); \chi^2 = nR^2 \sim \chi_3^2, R^2 \text{from joint reg..}$$

Notation: $\hat{\varepsilon}_t$ = residuals, $S_{t-1}^- = I(\hat{\varepsilon}_{t-1} < 0)$, $S_{t-1}^+ = I(\hat{\varepsilon}_{t-1} \geq 0)$, Sign Bias($\beta_1$), -ve Size Bias($\beta_2$), +ve Size Bias($\beta_3$)

Plain English: The leverage effect (asymmetric volatility) reflects that bad news increases volatility more than equally sized good news. The Engle–Ng test decomposes this asymmetry into three parts: sign bias (do -ve shocks affect volatility differently than +ve ones?), -ve size bias (do larger -ve shocks have disproportionate effects?), and +ve size bias (same for +ve shocks). Each is tested via regressions of squared residuals on indicators and shock magnitudes. Significant sign bias suggests asymmetric models (e.g., EGARCH), while size bias points to threshold models (e.g., GJR-GARCH). A joint test provides an overall diagnostic. In essence, the test asks whether volatility responds differently to the direction and size of past shocks.

⚠ Practitioner Warning

Requires correctly specified mean equation; residual autocorrelation can distort results. Apply only after confirming ARCH effects. The three tests use overlapping information, so multiple rejections require caution—use the joint test for overall inference. Power is low with imbalanced signs (e.g., few -ve returns). The test detects asymmetry but does not specify or estimate the appropriate model; rejection indicates symmetric GARCH is inadequate, but model choice should be confirmed using likelihood-based criteria. Small n or highly imbalanced signs reduce reliability of asymmetry tests.

Hypotheses

- H_0: No sign or size bias (symmetric volatility).
- H_1: Sign or size bias present (asymmetric volatility).

Decision Rule: Reject H_0 if $|t| > t_{\alpha/2,n-2}$ or p-value $< \alpha$ (individual tests). Joint test: if $\chi^2 = nR^2 > \chi^2_{\alpha,3}$.

Software References [Python, R, SAS]

Python	R	SAS
`Not Native (Manual)`	`rugarch::signbias()`	`Not Native (Manual)`

Additional Notes

The test decomposes asymmetry into three forms. Sign bias captures constant differences between -ve and +ve shocks ($\Longrightarrow$ EGARCH). -ve size bias reflects threshold effects where large -ve shocks have disproportionate impact ($\Longrightarrow$ GJR-GARCH/TGARCH). +ve size bias, less common, indicates similar effects for +ve shocks. A joint test provides an omnibus diagnostic. The test is used before fitting asymmetric GARCH models: individual tests identify specific asymmetry, while the joint test guards against multiple testing. It complements broader nonlinear diagnostics (e.g., Tsay), trading generality for power in detecting leverage effects. Model choice follows patterns: sign bias $\rightarrow$ EGARCH; size bias $\rightarrow$ threshold GARCH. Final model selection requires likelihood-based criteria or information criteria.

Miscellanea: Robert Engle and Victor Ng (1993) developed sign and size bias tests to detect asymmetry in volatility (first noted by Fischer Black(1976)), motivated by the leverage effect (asymmetric volatility) observed in financial returns. Their approach provided a simple diagnostic for determining whether symmetric GARCH models were adequate or whether asymmetric specs. were needed. The tests remain widely used due to their simplicity and interpretability.

Reference(s): Engle & Ng (1993); Black (1976); Nelson (1991); Glosten, Jagannathan, & Runkle (1993); Tsay (1986)

Engle-Ng Sign Bias Test

ARCH.3

Worked Example

Problem Statement A symmetric GARCH(1,1) model with $\hat{\omega} = 0.30$, $\hat{\alpha} = 0.10$, $\hat{\beta} = 0.65$ has been fitted to 25 daily excess return residuals e_t. Apply the Engle–Ng Sign Bias test at $\alpha = 0.05$ to determine whether the symmetric GARCH specification is misspecified due to leverage effects. Data can be found at the end of this chapter.

1. Define the Indicator Variable
Set $S_t^- = \mathbf{1}(e_t < 0)$. For the regression ($t = 2, \ldots, 25$, $n = 24$), the lagged indicator S_{t-1}^-, the negative-size variable $S_{t-1}^- \cdot e_{t-1}$, and the positive-size variable $(1 - S_{t-1}^-) \cdot e_{t-1}$ are used.

3. Joint Test
Regress z_t^2 on all three predictors jointly. $R^2 = 0.414$, $n = 24$:

$$\xi = n \cdot R^2 = 24 \times 0.414 = 9.94.$$

Under H_0: $\xi \sim \chi^2(3)$.

α	$\chi^2_\alpha(3)$	Decision
5%	7.815	**Reject** (9.94 > 7.815)
1%	11.345	FTR (9.94 < 11.345)

$p = 0.019$. Reject H_0 at the 5% level.

2. Three Component Regressions
Each regresses z_t^2 on an intercept and one predictor ($n = 24$, $df = 22$).

Sign Bias Test: $z_t^2 = c_0 + c_1 S_{t-1}^- + v_t$

$$\hat{c}_1 = -1.644, \quad \widehat{\text{SE}} = 0.528, \quad t = -3.117.$$

Negative Size Bias Test: $z_t^2 = c_0 + c_1 S_{t-1}^- e_{t-1} + v_t$

$$\hat{c}_1 = 1.013, \quad \widehat{\text{SE}} = 0.392, \quad t = 2.584.$$

Positive Size Bias Test: $z_t^2 = c_0 + c_1(1 - S_{t-1}^-)e_{t-1} + v_t$

$$\hat{c}_1 = 0.340, \quad \widehat{\text{SE}} = 0.407, \quad t = 0.834.$$

Test	t-stat	df	p-value
Sign Bias	−3.117	22	0.005
Neg. Size Bias	2.584	22	0.017
Pos. Size Bias	0.834	22	0.413

Practitioner Insight: The Engle–Ng test identifies *which form* of asymmetry drives misspecification, not just its presence. A significant Sign Bias result ($t = -3.12$, $p = 0.005$) shows that the sign of the previous shock predicts current z_t^2: negative shocks produce higher volatility than positive shocks of the same size, the leverage effect (Black 1976). A significant Negative Size Bias ($t = 2.58$, $p = 0.017$) indicates that larger negative shocks further amplify z_t^2, while the insignificant Positive Size Bias ($t = 0.83$, $p = 0.413$) shows positive shocks have no additional predictive effect. This pattern—sign bias and negative size bias present, positive size bias absent—is the canonical signature of GJR-GARCH/TGARCH rather than EGARCH (which typically shows positive size bias). The joint statistic $\xi = nR^2 = 9.94$ tests all components simultaneously using $\chi^2(3)$, consistent with the LM framework. The test uses *standardized* residuals $z_t = e_t/\sqrt{h_t}$; using raw residuals would confound ARCH magnitude effects already captured by h_t with leverage asymmetry. Following McLeod–Li or Tsay, a significant Engle–Ng result refines the diagnosis from general volatility clustering to asymmetric clustering requiring leverage-augmented models.

Common Pitfalls

- **Confusing the Engle–Ng sign bias test with ARCH–LM or Ljung–Box tests on squared residuals.** ARCH–LM and related diagnostics test for the presence of conditional heteroskedasticity, whereas the Engle–Ng test evaluates whether volatility responds asymmetrically to positive and negative shocks.
- **Applying the test directly to raw returns rather than standardized residuals from a fitted GARCH model.** The Engle–Ng procedure assumes that mean dynamics and symmetric volatility clustering have already been filtered, so the test should be applied to standardized residuals from an estimated volatility model.
- **Using the test with small samples where asymmetric effects are difficult to detect.** Because leverage effects are often subtle, samples with only a few hundred observations provide limited power and may fail to identify genuine asymmetric volatility responses.
- **Assuming that a significant test automatically implies a specific asymmetric GARCH specification.** The test indicates the presence of asymmetry but does not determine the correct model form, so alternative specifications such as GJR-GARCH, EGARCH, or APARCH should be compared.
- **Interpreting non-rejection as evidence that volatility responds symmetrically to shocks.** Failure to reject the null may simply reflect low power or model misspecification, so the result should be interpreted alongside residual diagnostics and model comparisons.

ICSS Test

Iterated Cumulative Sums of Squares Test for Variance Shifts

ARCH.4 | **Category:** ARCH Tests | **Data:** Time Series | **Null:** Tabled (Brownian Bridge)

When to Use

- Detect sudden variance shifts in time-series data.
- Identify volatility regime changes in financial returns.
- Locate structural breaks in unconditional variance.
- Pre-screen series before fitting GARCH-type models.
- Analyze variance instability over long time periods.

Assumptions

- Observations are serially independent under the null.
- The variance is constant under H_0.
- Structural breaks occur abruptly rather than gradually.
- Large samples improve asymptotic approximations.
- Extreme heavy tails may distort test performance.

Test Statistic

$$ICSS = \max_{1 \le k < T} \sqrt{\frac{T}{2}} |D_k|. \quad C_k = \sum_{t=1}^{k} \varepsilon_t^2, \quad D_k = \frac{C_k}{C_T} - \frac{k}{T}, \quad k = 1, \dots, T, \quad C_T = \sum_{t=1}^{T} \varepsilon_t^2.$$

Notation: ε_t is the residual or return at time t, C_k is cumulative squared variation up to time k, and T is the sample size.

Plain English: The ICSS test detects sudden changes in variance by tracking cumulative sums of squared observations over time. Under constant variance, the cumulative variance process evolves approximately linearly. Structural breaks in volatility create noticeable departures from this linear pattern, allowing the procedure to identify candidate variance-shift points. The method is widely used in financial econometrics to detect volatility regime changes before fitting ARCH or GARCH models.

⚠ Practitioner Warning

The classical ICSS procedure is sensitive to heavy-tailed distributions and large outliers, which may generate false variance-break detections. Serial dependence and strong conditional heteroskedasticity can also distort the null distribution. The method detects abrupt variance changes rather than gradual volatility evolution, so smoothly changing variance processes may be poorly handled. Robust ICSS variants are often preferred for heavy-tailed financial-return data.

Hypotheses

- H_0: $\sigma_1^2 = \sigma_2^2 = \cdots = \sigma_T^2$ (Constant variance)
- H_1: At least one structural break in variance.

Decision Rule: Reject H_0 if $\max_k \sqrt{T/2}|D_k| > C_\alpha$, where C_α is the CV (Inclán & Tiao (1994)). A rejection identifies k as the point of maximal variance shift.

Software References [Python, R, SAS]

`arch::breaks` | `ICSS::ICSS` | `PROC AUTOREG / Custom Code`

Additional Notes

The ICSS algorithm iteratively partitions the series at detected breakpoints and reapplies the test within each segment until no further variance shifts are identified. The resulting break dates define volatility regimes that may subsequently be modeled using separate GARCH specifications or regime-switching frameworks. Under the null hypothesis of constant variance, $\sqrt{T/2}\,D_k$ converges weakly to a Brownian bridge process, so classical ICSS critical values are derived from the supremum distribution of the Brownian bridge. The standard ICSS procedure can substantially over-reject under ARCH/GARCH errors or heavy-tailed innovations; Sanso et al. (2004) proposed heteroscedasticity-robust modifications based on spectral density estimation. Detected variance breaks are often incorporated into GARCH models through intercept dummies or by estimating separate models within each variance regime.

Miscellanea: Inclán and Tiao introduced the ICSS procedure in 1994 while studying variance instability in financial-return series. The method became influential because it provided a simple nonparametric mechanism for locating volatility regime shifts without explicitly fitting complex GARCH structures. ICSS later inspired numerous robust and heavy-tail-adjusted variance-break procedures used in modern financial econometrics.

Reference(s): Inclán & Tiao (1994); Sansó, Aragó & Carrion (2004); Tsay (2010)

Worked Example

Problem Statement A financial analyst collects 20 daily return residuals ε_t (demeaned log-returns, %) for a stock. The first ten observations fall in a calm period and the second ten in a turbulent period. Apply the Inclan–Tiao ICSS test to determine at $\alpha = 0.05$ if a variance break is present, and locate the break if found. Data at the end of the chapter.
$H_0\colon \sigma_t^2 = \sigma^2\ \forall t$ (constant variance); $H_1:$ at least one variance break.

1. ICSS Statistic
For $k = 1, \ldots, T$

$$C_k = \sum_{t=1}^{k} \varepsilon_t^2, \qquad C_T = \sum_{t=1}^{T} \varepsilon_t^2, \qquad D_k = \frac{C_k}{C_T} - \frac{k}{T}.$$

The Inclan–Tiao statistic is $\text{IT} = \sqrt{T/2}\ \max_{1 \le k \le T} |D_k|$.

2. Cumulative Sums

$$C_k = \begin{cases} k, & 1 \le k \le 10, \\ 25k - 240, & 11 \le k \le 20, \end{cases} \qquad C_T = C_{20} = 260.$$

3. D_k Sequence
Substituting into $D_k = C_k/260 - k/20$:

$$D_k = \begin{cases} -3k/65, & 1 \le k \le 10, \\ 3(k-20)/65, & 11 \le k \le 20 \end{cases}$$

k	C_k	D_k	$\lvert D_k\rvert$	k	C_k	D_k	$\lvert D_k\rvert$
1	1	−3/65	0.046	11	35	−27/65	0.415
2	2	−6/65	0.092	12	60	−24/65	0.369
3	3	−9/65	0.138	13	85	−21/65	0.323
4	4	−12/65	0.185	14	110	−18/65	0.277
5	5	−3/13	0.231	15	135	−3/13	0.231
6	6	−18/65	0.277	16	160	−12/65	0.185
7	7	−21/65	0.323	17	185	−9/65	0.138
8	8	−24/65	0.369	18	210	−6/65	0.092
9	9	−27/65	0.415	19	235	−3/65	0.046
10	10	−6/13	0.462	20	260	0	0.000

4. Maximum and Break-Point Estimate

$$\max_k |D_k| = |D_{10}| = 6/13, \quad k^* = \operatorname{argmax}_k |D_k| = 10,$$

where k^* is the estimated break point.

5. ICSS Statistic
With $T = 20$:

$$\text{IT} = \sqrt{T/2}\,(6/13) \approx 1.460.$$

6. Decision
Inclan–Tiao (1994) critical values:

α	CV	Decision
10%	1.22	**Reject** (1.460 > 1.22)
5%	1.36	**Reject** (1.460 > 1.36)
1%	1.63	FTR (1.460 < 1.63)

Reject H_0 at 5%: evidence of a variance break at $k^* = 10$.

7. Variance Estimates
Splitting at $k^* = 10$:

$$\hat{\sigma}_1^2 = C_{10}/10 = 1, \quad \hat{\sigma}_2^2 = (C_{20} - C_{10})/10 = 25.$$

Volatility increases 5x after $t = 10$ ($\hat{\sigma}_2/\hat{\sigma}_1 = 5$), consistent with a market shock at the midpoint of the data.

Practitioner Insight: The ICSS test detects structural breaks in unconditional variance without pre-specifying the number or location of breaks. The centerd ratio D_k compares cumulative squared residuals with the proportion k/T expected under constant variance; sustained negative values indicate that early obs. contribute relatively little variation, consistent with a later high-variance regime. The statistic IT converges to the supremum of a Brownian bridge, from which critical values are obtained. A major limitation of the original ICSS procedure is over-rejection under ARCH/GARCH effects, since volatility clustering inflates $|D_k|$ even without true variance breaks; Sansó, Aragó & Carrion-i-Silvestre (2004) address this using long-run variance corrections. The ICSS algorithm applies binary segmentation recursively to detect multiple breaks. In practice, ICSS is often used before GARCH modeling because unmodeled variance shifts can artificially inflate persistence estimates and create the false appearance of IGARCH behavior.

Common Pitfalls

- **Using standard ICSS under ARCH effects.** Volatility clustering inflates $|D_k|$ and causes over-rejection even without true variance breaks. Use the Sansó et al. (2004) robust version when ARCH/GARCH dynamics are suspected.
- **Fitting one GARCH across variance regimes.** Unmodeled variance breaks inflate $\hat{\alpha} + \hat{\beta}$ toward unity and can falsely suggest IGARCH behavior. Refit GARCH within regimes or include break effects.
- **Applying ICSS to raw instead of filtered data.** Serial correlation in the mean causes systematic drift in D_k and spurious break detection. Remove mean dynamics first using demeaning or ARMA filtering.
- **Confusing unconditional and conditional variance.** ICSS detects abrupt shifts in unconditional variance, whereas ARCH tests target conditional volatility dynamics. Both may coexist and should be analyzed separately.

XVI. ARCH — Raw Data

Data used in selected worked examples are provided here to preserve the book's strict two-page-per-test format.

[ARCH.3] Engle-Ng Sign Bias Test

Problem Statement A symmetric GARCH(1,1) model with $\hat{\omega} = 0.30$, $\hat{\alpha} = 0.10$, $\hat{\beta} = 0.65$ has been fitted to 25 daily excess return residuals e_t. The fitted conditional variances h_t and standardized residuals $z_t = e_t/\sqrt{h_t}$ are tabulated below. Apply the Engle–Ng Sign Bias test at $\alpha = 0.05$ to determine whether the symmetric GARCH specification is misspecified due to leverage effects.

t	e_t	h_t	z_t^2	S_{t-1}^-	$S_{t-1}^- e_{t-1}$	$(1 - S_{t-1}^-)e_{t-1}$
1	−0.2	1.500	0.027			
2	0.6	1.279	0.281	1	−0.2	0.0
3	−0.9	1.167	0.694	0	0.0	0.6
4	−0.3	1.140	0.079	1	−0.9	0.0
5	0.6	1.050	0.343	1	−0.3	0.0
6	−1.4	1.018	1.925	0	0.0	0.6
7	−1.4	1.158	1.693	1	−1.4	0.0
8	−1.4	1.249	0.969	1	−1.4	0.0
9	1.1	1.233	3.927	0	0.0	1.1
10	−2.2	1.585	0.101	1	−2.2	0.0
11	0.4	1.346	1.456	0	0.0	0.4
12	1.4	1.371	3.858	0	0.0	1.4
13	1.5	1.720	1.308	0	0.0	2.3

t	e_t	h_t	z_t^2	S_{t-1}^-	$S_{t-1}^- e_{t-1}$	$(1 - S_{t-1}^-)e_{t-1}$
14	0.5	1.643	0.152	0	0.0	1.5
15	1.9	1.393	2.591	0	0.0	0.5
16	−0.4	1.566	0.102	0	0.0	1.9
17	−1.7	1.334	2.166	1	−0.4	0.0
18	0.4	1.456	0.110	1	−1.7	0.0
19	−1.7	1.263	2.289	0	0.0	0.4
20	0.4	1.410	0.114	1	−1.7	0.0
21	2.5	1.232	5.072	0	0.0	0.4
22	−1.5	1.726	1.304	0	0.0	2.5
23	−0.5	1.647	0.152	1	−1.5	0.0
24	0.7	1.395	0.351	1	−0.5	0.0
25	2.3	1.256	4.212	0	0.0	0.7

Define the Indicator Variable

Set $S_t^- = \mathbf{1}(e_t < 0)$. For the regression ($t = 2, \ldots, 25$, $n = 24$), the lagged indicator S_{t-1}^-, the negative-size variable $S_{t-1}^- \cdot e_{t-1}$, and the positive-size variable $(1 - S_{t-1}^-) \cdot e_{t-1}$ are shown above.

[ARCH.4] ICSS Test

Problem Statement A financial analyst collects 20 daily return residuals ε_t (demeaned log-returns, %) for a stock. The first ten observations fall in a calm period and the second ten in a turbulent period. Apply the Inclan–Tiao ICSS test to determine at $\alpha = 0.05$ whether a variance break is present, and locate the break if found.
$H_0\colon \sigma_t^2 = \sigma^2\ \forall t$ (constant variance); $H_1\colon$ at least one variance break.

t	1	2	3	4	5	6	7	8	9	10	11	12	13	14	15	16	17	18	19	20
ε_t^2	1	1	1	1	1	1	1	1	1	1	25	25	25	25	25	25	25	25	25	25

Part XVII

Stochastic Independence & Sequence Structure

Testing for randomness, runs, and serial independence

Randomness and Independence

XVII. RAND — A Conceptual Primer

What Is Randomness?

A sequence $\{x_1, x_2, \ldots, x_n\}$ is **random** (i.i.d.) if the observations are mutually independent and identically distributed. Randomness implies:

- No systematic trend (values do not consistently increase or decrease).
- No clustering (similar values do not group together).
- No serial dependence (knowing x_t gives no information about x_{t+1}).
- No cycles or oscillations.

Why test for randomness? Many statistical procedures assume i.i.d. data. Non-random sequences arise in time series (autocorrelation, trends), quality control (process drift), ecology (spatial clustering), and finance (volatility clustering). Detecting non-randomness is the first diagnostic step before modeling.

Critical distinction: Uncorrelated ≠ Independent. Zero linear autocorrelation ($\rho_k = 0$ for all $k > 0$) does not imply independence; nonlinear dependence (e.g. ARCH effects, chaotic dynamics) can produce sequences that are uncorrelated yet strongly dependent. Tests in this chapter range from detecting simple linear structure (runs, trend) to detecting general dependence (BDS).

Taxonomy of Departures from Randomness

Departure	Manifestation
Monotone trend	Values consistently increase or decrease; Mann–Kendall, Cox–Stuart detect.
Clustering	Similar values group together (positive autocorrelation); too few runs.
Oscillation	Values alternate (-ve autocorrelation); too many runs; high von Neumann ratio.
Serial dependence	x_t predicts x_{t+1} linearly; von Neumann, Bartels detect.
Nonlinearity in mean	Conditional mean depends on nonlinear functions of lags; Tsay (1986) detects; ARCH tests miss.
Nonlinear dependence	Dependence in higher moments or squared values; BDS detects; linear tests miss.

I. Runs-Based Tests

A **run** is a maximal unbroken subsequence of observations with a common property.

Wald–Wolfowitz runs test: For a binary sequence (e.g. values above/below the median coded as $+/-$), let R be the total number of runs. Under H_0 (randomness), with n_+ values above and n_- below the median:

$$E[R] = \frac{2n_+n_-}{n} + 1, \quad \mathrm{Var}(R) = \frac{2n_+n_-(2n_+n_- - n)}{n^2(n-1)}.$$

Standardized: $z = (R - E[R])/\sqrt{\mathrm{Var}(R)} \xrightarrow{d} \mathcal{N}(0,1)$ for large n. **Too few runs** ($z \ll 0$): clustering (positive autocorrelation). **Too many runs** ($z \gg 0$): oscillation (negative autocorrelation). For exact tables ($n \le 20$), compare R directly to the critical region.

Up-and-down runs test: Code each consecutive comparison as + (increase) or – (decrease): $s_i = \mathrm{sign}(x_{i+1} - x_i)$. Count runs of + and – in the sequence $s_1, s_2, \ldots, s_{n-1}$. Under randomness, the expected number of runs is $E[R] = (2n-1)/3$. Sensitive to oscillatory and clustered patterns that the median-based Wald–Wolfowitz test may miss.

Difference-sign test: Count $S = \#\{i : x_{i+1} > x_i\}$ (the number of positive differences). Under H_0: $E[S] = (n-1)/2$, $\mathrm{Var}(S) = (n+1)/12$. Standardized statistic is asymptotically $\mathcal{N}(0,1)$. The simplest directional test; equivalent to a sign test applied to successive differences.

II. Turning Points Test

A **turning point** occurs at position i if $x_{i-1} < x_i > x_{i+1}$ (local maximum) or $x_{i-1} > x_i < x_{i+1}$ (local minimum). Let T = total number of turning points in $x_2, x_3, \ldots, x_{n-1}$. Under randomness:

$$E[T] = \frac{2(n-2)}{3}, \quad \mathrm{Var}(T) = \frac{16n-29}{90}.$$

Standardized: $z = (T - E[T])/\sqrt{\mathrm{Var}(T)} \xrightarrow{d} \mathcal{N}(0,1)$. **Too few turning points**: trend or oscillation. **Too many turning points**: rapid alternation (high-frequency oscillation). The turning points test is particularly sensitive to periodic structure and is complementary to runs tests.

III. Trend Tests

Mann–Kendall test: A nonparametric test for a monotonic trend based on Kendall's τ applied to the temporal sequence. Define:

$$S = \sum_{i=1}^{n-1} \sum_{j=i+1}^{n} \mathrm{sgn}(x_j - x_i),$$

where $\mathrm{sgn}(\cdot) \in \{-1, 0, +1\}$. Under H_0 (no trend): $E[S] = 0$, $\mathrm{Var}(S) = n(n-1)(2n+5)/18$ (adjusted for ties). $z = S/\sqrt{\mathrm{Var}(S)} \xrightarrow{d} \mathcal{N}(0,1)$. $S > 0$: upward trend; $S < 0$: downward trend. Robust to non-normality, outliers, and missing values; the standard test for monotone trends in hydrology, climatology, and environmental monitoring. The **Sen slope** $\hat{\beta} = \mathrm{median}\{(x_j - x_i)/(j-i) : j > i\}$ estimates the trend magnitude (the Theil–Sen estimator applied to time-indexed data).

Cox–Stuart test: Splits the series into first and second halves; pairs x_i with $x_{i+n/2}$ and tests whether the paired differences $x_{i+n/2} - x_i$ are systematically positive (upward trend) or negative (downward trend) using the sign test on $\lfloor n/2 \rfloor$ pairs. $B \sim \text{Binomial}(m, 0.5)$ under H_0, where m is the number of non-zero differences. Simpler than Mann–Kendall; uses only sign, not magnitude—less powerful but more robust.

IV. Serial Dependence Tests

Von Neumann ratio (rank test):

$$\delta^2 = \frac{n \sum_{i=1}^{n-1} (x_{i+1} - x_i)^2}{(n-1) \sum_{i=1}^{n} (x_i - \bar{x})^2}.$$

Under H_0 (independence), $E[\delta^2] = 2n/(n-1) \approx 2$. $\delta^2 < 2$: positive serial correlation; $\delta^2 > 2$: negative serial correlation. Normalized version: $v = \delta^2/2 \in (0, 2)$; the ratio $v < 1$ signals positive autocorrelation. Based on raw values; sensitive to departure from normality.

Bartels test: The rank-based analog of the von Neumann ratio. Replace x_i with their ranks R_i:

$$RVN = \frac{\sum_{i=1}^{n-1} (R_{i+1} - R_i)^2}{\sum_{i=1}^{n} (R_i - \bar{R})^2}.$$

Distribution-free under H_0; more robust than von Neumann when the data are heavy-tailed or non-normal. Asymptotically normal with known mean and variance under H_0. The test of choice for serial independence without distributional assumptions.

V. BDS Test (Brock–Dechert–Scheinkman)

The BDS test detects **any form of dependence**— linear or nonlinear—making it a powerful omnibus test for i.i.d. against structural dependence.

Correlation integral: For embedding dimension m and tolerance ϵ, the correlation integral estimates the probability that two length-m trajectories are within ϵ of each other:

$$C_{m,n}(\epsilon) = \frac{2}{n_m(n_m - 1)} \sum_{t<s} \mathbf{1}\left(\|x_t^{(m)} - x_s^{(m)}\| < \epsilon\right),$$

where $x_t^{(m)} = (x_t, x_{t+1}, \ldots, x_{t+m-1})$ is the m-history and $n_m = n - m + 1$.

Under i.i.d., $C_{m,n}(\epsilon) = C_{1,n}(\epsilon)^m$ (the product of one-dimensional probabilities). The BDS statistic measures the standardized deviation:

$$W_{m,n}(\epsilon) = \sqrt{n_m} \frac{C_{m,n}(\epsilon) - C_{1,n}(\epsilon)^m}{\hat{\sigma}_m(\epsilon)} \xrightarrow{d} \mathcal{N}(0, 1).$$

Key properties:

- Detects ARCH effects, chaos, nonlinear dependence, and structural change—all invisible to linear autocorrelation tests.
- Applied to **residuals** after fitting a linear model; a significant BDS on residuals confirms remaining nonlinear structure.
- Choose $\epsilon \in [0.5s, 2s]$ where $s = \text{std}(x)$; use multiple values of m (typically 2–5).
- The test has power against a wide range of alternatives but does not identify the type of dependence.

VI. Tsay's Test for Nonlinearity (1986)

While BDS detects any form of dependence, Tsay's (1986) test targets specifically **nonlinearity in the conditional mean** — threshold effects, bilinear structure, or polynomial dynamics in the AR mean equation. The test extends Tukey's one-degree-of-freedom idea to the time series setting: if a linear AR(p) is adequate, no function of its lagged values should predict the residuals.

Procedure: Fit an AR(p) baseline; obtain residuals $\hat{e}_t$. Form all $q = p(p+1)/2$ cross-products of lagged observations $\mathbf{z}_t = \text{vech}(\mathbf{x}_t \mathbf{x}_t^\top)$, where $\mathbf{x}_t = (y_{t-1}, \ldots, y_{t-p})^\top$. Orthogonalize $\mathbf{z}_t$ against $\mathbf{x}_t$ to obtain residuals $\hat{\mathbf{v}}_t$ (purging linear dependence). Regress $\hat{e}_t$ on $\hat{\mathbf{v}}_t$ and test joint significance:

$$F = \frac{\text{SS}_{\text{reg}}/q}{\text{SS}_{\text{err}}/(n-q-1)} \sim F_{q,n-q-1} \quad \text{under } H_0.$$

Key distinctions:

- Tsay uses AR *residuals* on cross-products of lagged *observations* — targeting conditional mean nonlinearity.
- Engle's ARCH test uses *squared* residuals on lagged *squared* residuals — targeting conditional variance nonlinearity.
- The two tests are complementary, not equivalent; both can be significant simultaneously.
- A significant result indicates nonlinear structure but does not specify type (threshold, bilinear, etc.); use Tsay (1989) arranged autoregression as a follow-up to test for threshold structure specifically.

Test Selection Guide

Suspected departure	Recommended test
General randomness (binary/median)	Wald–Wolfowitz
Oscillation / clustering	Up-and-down runs
Directional change pattern	Difference-sign
Periodic / local extremes	Turning points
Monotone trend (nonparametric)	Mann–Kendall
Monotone trend (simple)	Cox–Stuart
Linear serial dependence	Von Neumann / Bartels
Nonlinear dependence	BDS
Nonlinearity in mean	Tsay (1986)

Common Pitfalls

- **Ignoring time ordering:** randomness tests are for sequences; applying them to cross-sectional data or after sorting is meaningless.
- **Uncorrelated ≠ random:** zero autocorrelation does not imply i.i.d.; always complement linear tests with BDS or squared- residual checks.
- **Multiple tests without adjustment:** running all tests simultaneously inflates Type I error; pre-specify the target departure or apply FDR correction.
- **Small samples:** asymptotic normal approximations for runs, turning points, and Mann–Kendall are unreliable for $n < 10$; use exact tables or permutation p-values.
- **Ties in Mann–Kendall:** ties in the series reduce power; apply the tie-adjusted variance formula.
- **BDS on raw data:** apply BDS to *model residuals* (after removing linear dependence); on raw financial returns it largely detects ARCH, not chaos.

Bottom Line

> Randomness tests detect hidden structure in sequences: trends, clustering, oscillation, or nonlinear dependence. Different tests target different departures—runs tests for sequence patterns, Mann–Kendall for monotone trends, von Neumann / Bartels for linear serial dependence, BDS for any form of dependence including nonlinear, and Tsay's test for nonlinearity specifically in the conditional mean. Always test residuals, not raw data, and remember that zero autocorrelation does not imply independence.

Wald-Wolfowitz Runs Test

Test for Randomness in a Binary Sequence

RAND.1 | **Category:** Randomness Tests | **Data:** Binary | **Null:** Exact (Runs) / Normal (asymptotic)

When to Use

- Assess randomness in a binary sequence (e.g., success/failure, above/below median).
- Detect clustering, trends, or dependence in ordered process.
- Analyze quality-control or sequential data where order is meaningful.

Assumptions

- Observations form an ordered binary sequence.
- Under H_0, the two symbols are randomly arranged in the sequence (conditional on the observed counts).
- For asymptotic inference, both symbol counts are large enough for the normal approximation to be reliable.

Test Statistic

$$z = \frac{R - E[R]}{\sqrt{\mathrm{Var}[R]}}. \quad E[R] = \frac{2n_1 n_2}{n} + 1, \quad \mathrm{Var}[R] = \frac{2n_1 n_2(2n_1 n_2 - n)}{n^2(n-1)}. \quad \text{Under } H_0,\ z \approx N(0,1)$$

Notation: R is observed runs, n_1, n_2 are counts of each outcome type, $n = n_1 + n_2$, run is a *maximal* uninterrupted sequence

Plain English: The Wald-Wolfowitz runs test evaluates randomness by counting runs—maximal sequences of identical outcomes. A sequence like AAABBAAB has 4 runs. Too few runs suggest clustering or trends (AAAABBBB has 2 runs); too many runs suggest alternation patterns (ABABAB has 6 runs). The test compares observed runs to expected under randomness, using normal approximation for inference. Significant deviations in either direction indicate non-random patterns.

⚠ Practitioner Warning

The exact null distribution is conditional on the observed counts n_1, n_2. The normal approximation may be poor when one or both counts are small, in which case an exact test is preferred. The test detects non-random ordering but does not identify the source: clustering, trend, periodicity, or other dependence can all produce unusual run counts. Two-sided tests are most common, though one-sided versions may be used for specific alternatives such as clustering or excessive alternation. A continuity correction, $z = \frac{|R-E[R]|-0.5}{\sqrt{\mathrm{Var}(R)}}$, is sometimes used to improve the normal approximation.

Hypotheses

- H_0: Outcome sequence is random.
- H_1: Outcome sequence is not random.

Decision Rule: Reject H_0 if $|z| > z_{\alpha/2}$ or p-value $< \alpha$ for two-tailed test. For small samaple, use exact distribution.

Software References [Python, R, SAS]

Python	R	SAS
`statsmodels.⋯.runs_test`	`randtests::runs.test`	`Not Native (Manual)`

Additional Notes

A *run* is a maximal uninterrupted sequence of identical symbols (e.g., 000|11|000|111 gives $R = 4$). Under randomness, $E[R] = 1 + \frac{2n_1 n_2}{n}$, with variance based on the hypergeometric structure. For small samples, use the exact distribution of R given n_1, n_2; for reasonably large n_1, n_2, the normal approximation is adequate. Fewer runs than expected ($R < E[R]$) indicate clustering; more runs ($R > E[R]$) suggest alternation. The two-sample Wald–Wolfowitz version pools observations, codes by group, and tests runs to detect distributional differences. The test is most sensitive to short-range dependence and is often used in quality control or sequence screening after dichotomizing observations (e.g., above/below a target or median).

Miscellanea: Abraham Wald and Jacob Wolfowitz introduced the runs test in 1940 as a simple nonparametric check of randomness in ordered binary data. Their key insight was that a process can have the correct average behavior yet still show non-random ordering, which the number of runs can reveal. The test became a staple of industrial quality control and sequential analysis.

Reference(s): Wald & Wolfowitz (1940); Gibbons & Chakraborti (2011)

Worked Example

Problem Statement A process engineer records whether each of $n = 16$ consecutive items off a production line is *long* (L, wait time above the median) or *short* (S, below the median). Test whether the ordering is consistent with randomness. H_0: sequence is random; H_1: non-random (too few or too many runs).
The resulting sequence is: $\underbrace{LLLLL}_{1}\,\underbrace{SSSS}_{2}\,\underbrace{LL}_{3}\,\underbrace{SSSS}_{4}\,\underbrace{L}_{5}$ with $n_L = 8$ long and $n_S = 8$ short items.

1. Run Count and Summary
A *run* is a maximal block of identical symbols. Reading left to right:

Run	Symbols	Length
1	L L L L L	5
2	S S S S	4
3	L L	2
4	S S S S	4
5	L	1

$R = 5, n_L = 8, n_S = 8, n = 16.$

2. Mean and Variance of R Under H_0

$$E[R] = 1 + \frac{2n_L n_S}{n} = 1 + \frac{2 \times 8 \times 8}{16} = 9$$

$$\mathrm{Var}[R] = \frac{2n_L n_S(2n_L n_S - n)}{n^2(n-1)} = \frac{2 \times 64 \times 112}{256 \times 15} = \frac{56}{15}$$

3. Normal Approximation

$$z = \frac{R - E[R]}{\sqrt{\mathrm{Var}[R]}} = \frac{5-9}{\sqrt{56/15}} \approx -2.070$$

Method	**Value**	**CV (5%)**	**Decision**
z (approx)	−2.070	±1.960	Reject*
Exact p	0.063	0.05	FTR

*Normal approx: $p = 0.038$; exact: $p = 0.063$.

4. Exact Test
$\binom{16}{8} = 12870$ equally likely arrangements. Exact probabilities summing over the lower tail:

$$P(R=2) = \tfrac{2}{12870},\ P(R=3) = \tfrac{14}{12870},\ P(R=4) = \tfrac{98}{12870},$$

$$P(R=5) = \tfrac{294}{12870},\quad P(R \le 5) = \tfrac{408}{12870} \approx 0.032$$

By the exact symmetry $P(R = r) = P(R = 18 - r)$ when $n_L = n_S$, the two-sided $p = 2 \times 0.032 \approx 0.063$. The exact 5% critical region is $R \le 4$ or $R \ge 14$; $R = 5$ falls outside it.

Practitioner Insight: The Wald–Wolfowitz runs test is the canonical test of sequence randomness, based solely on the pattern of signs after dichotomisation (not magnitudes). Two-sided rejection arises from either too few runs (clustering, often due to trend or autocorrelation) or too many runs (alternation, e.g., oscillatory error). Here $R = 5$ is well below the expected 9, indicating clustering (long follows long, short follows short). A key small-sample lesson: the normal approximation gives $z = -2.070$ ($p = 0.038$), but the exact test yields $p = 0.063$. For $n < 20$ or imbalanced groups, the normal approximation is anti-conservative; the exact test is preferred. The 5% lower critical value is $R = 4$, so the evidence is suggestive but not statistically significant. In practice, the dichotomization rule (e.g., median) and tie handling must be pre-specified; ties are typically dropped, reducing the effective sample size.

Common Pitfalls

- **Counting items instead of runs.** A run is a consecutive sequence of identical symbols. Count transitions between symbols (+1), not total observations.
- **Using normal approximation for small samples.** When n_1 or $n_2 < 10$, the normal approximation is unreliable. Use the exact runs distribution instead.
- **Over-interpreting rejection without context.** Strong autocorrelation naturally produces too few or too many runs; interpret results in the context of the underlying process rather than as evidence of a unique mechanism.
- **Choosing an arbitrary cutpoint.** For one-sample tests, dichotomize at the median (or a pre-specified threshold). Post-hoc cutpoints distort n_1, n_2 and invalidate inference.
- **Expecting it to detect all non-randomness.** The runs test detects clustering or alternation (first-order dependence) only; it may miss periodicity, nonlinear patterns, or variance shifts.

Up-and-Down Runs Test

Test for Randomness Using Directional Changes

RAND.2 | **Category:** Randomness Tests | **Data:** Continuous | **Null:** Normal (asymptotic)

When to Use

- Test for randomness or trend in continuous sequence data.
- Detect non-random ordering through patterns of increases and decreases.
- Quality control monitoring for process stability or drift.
- Nonparametric complement to autocorrelation-based tests.

Assumptions

- Data form a sequence of continuous or ordinal values.
- Under H_0, observations are independent and randomly ordered.
- No ties (or very few); ties must be handled explicitly.
- Sample size adequate for normal approximation ($n \geq 20$).

Test Statistic

$$z = \frac{U - E[U]}{\sqrt{\text{Var}[U]}}. \quad E[U] = \frac{2n-1}{3}, \quad \text{Var}[U] = \frac{16n-29}{90}. \quad \text{Under } H_0,\ z \approx N(0,1).$$

An up-run is a maximal sequence of increases; a down-run is a maximal sequence of decreases.

Notation: U is observed up-and-down runs, n is sequence length, run is a maximal sequence of identical directional signs

Plain English: The up-and-down runs test evaluates randomness by examining directional patterns—sequences of consecutive increases (up-runs) and decreases (down-runs). A random sequence produces a predictable number of such runs. Too few runs suggest persistence or trend; too many runs indicate excessive alternation. The test converts the data into directional signs (+/-), counts runs, and compares the result to its expected value under randomness using a normal approximation.

⚠ Practitioner Warning

Ties complicate analysis and must be handled (e.g., broken randomly or removed). Normal approximation requires $n \geq 20$; smaller samples require exact tables. The test does not identify the form of non-randomness (trend vs oscillation). It is sensitive to directional dependence but not all forms of serial dependence. It is less powerful than parametric methods (e.g., regression) for detecting specific trends. A continuity correction $z = \frac{|U-E[U]|-0.5}{\sqrt{\text{Var}[U]}}$ may improve approximation.

Hypotheses

- H_0: Sequence is random.
- H_1: Sequence is non-random (trends/persistence/swings).

Decision Rule: Reject H_0 if $|z| > z_{\alpha/2}$ or p-value $< \alpha$ for two-tailed test.

Software References [Python, R, SAS]

Python	R	SAS
`Not Native (Manual)`	`randtests::runs.test`	`Not Native (Manual)`

Additional Notes

Directional construction: from $x_1, \ldots, x_n$, define + if $x_{i+1} > x_i$ and − if $x_{i+1} < x_i$, yielding $n-1$ symbols. An up-and-down run is a maximal sequence of identical signs. Example: 3,1,4,1,5,9 → -, +, -, +, + with runs -|+|-|++ giving $U = 4$. Exact small-sample distributions were derived by Levene and Wolfowitz (1944), though normal approximation is common. Interpretation: $U < E[U]$ suggests trend or persistence; $U > E[U]$ indicates excessive alternation. The test is nonparametric and based only on direction (not magnitude), making it robust but less powerful than autocorrelation-based methods. Closely related to the turning-point test though the turning-point test focuses on local extrema rather than directional runs.

Miscellanea: Howard Levene and Jacob Wolfowitz introduced the up-and-down runs test in 1944 as a formal method for detecting non-random directional patterns in sequences of observations. They emphasized that tracking directional changes—not just absolute levels—could reveal structure that traditional summaries might miss. The test later gained adoption in sequential analysis and time-series applications. Barton and David (1957) provided extensive tables for exact inference, and Bradley's (1968) comprehensive nonparametric textbook helped popularize the method.

Reference(s): Levene & Wolfowitz (1944); Barton & David (1957); Bradley (1968)

Worked Example

Problem Statement A quality analyst records monthly defect counts for $n = 8$ consecutive months:

$$3,\ 6,\ 9,\ 12,\ 10,\ 8,\ 5,\ 2.$$

The up-and-down runs test assesses whether the sequence is consistent with random variation, or whether the pattern of rises and falls is too systematic. H_0: sequence is random (no trend); H_1: non-random (too few or too many direction changes).

Unlike the Wald–Wolfowitz test, which dichotomises a sequence into two symbols, the up-and-down test works directly on the signs of successive *differences* of a continuous sequence.

1. Difference Sequence and Run Count

Form $n - 1 = 7$ consecutive differences $d_i = x_{i+1} - x_i$ and encode each as U (up, $d_i > 0$) or D (down, $d_i < 0$):

i	1	2	3	4	5	6	7
$x_{i+1} - x_i$	+3	+3	+3	−2	−2	−3	−3
Sign	U	U	U	D	D	D	D

The sign sequence UUUDDDD contains

$$R = 2 \text{ runs: } \underbrace{\text{UUU}}_{1}\ \underbrace{\text{DDDD}}_{2}.$$

2. Mean and Variance Under H_0

The null moments depend on n only (the number of U's and D's is not fixed, unlike in Wald–Wolfowitz):

$$E[R] = \frac{2n-1}{3} = 5, \quad \text{Var}[R] = \frac{16n-29}{90} = \frac{11}{10}$$

3. Normal Approximation

$$z = \frac{R - E[R]}{\sqrt{\text{Var}[R]}} = \frac{2-5}{\sqrt{11/10}} \approx -2.860$$

Method	Value	CV (1%)	Decision
z (approx)	−2.860	±2.576	Reject H_0
Exact p	0.013	0.01	Reject H_0

Approx: $p = 0.004$; exact: $p = 0.013$. Both reject at 1%.

4. Exact Lower-Tail Probability

Out of $8! = 40320$ equally likely orderings:

$$P(R = 1) = \frac{2}{8!}, \qquad P(R = 2) = \frac{252}{8!}$$

$$P(R \le 2) = \frac{254}{8!} \approx 0.0063$$

Exact two-sided $p = 2 \times 0.0063 \approx 0.013$.

Practitioner Insight: The up-and-down runs test and the Wald–Wolfowitz test both use run counts but target different structures. Wald–Wolfowitz applies to a pre-labelled binary sequence (fixed n_1, n_2; hypergeometric null), whereas the up-and-down test uses the signs of successive differences in a continuous series, requiring no dichotomisation; the number of U/D signs is random, and the null depends only on n, with $E[R] = (2n-1)/3$ and $\text{Var}[R] = (16n-29)/90$. Here $E[R] = 5$ and $\text{Var}[R] = 11/10$ (for $n = 8$), and the observed $R = 2$ is far below expectation, consistent with a single trend reversal (one peak, two direction blocks). The normal approximation understates the exact p (0.004 vs 0.013) at small n; it becomes reliable for $n \ge 20$. Both methods reject at 1% here. Ties (equal consecutive values) are omitted, reducing the effective n.

Common Pitfalls

- **Confusing up-down with Wald–Wolfowitz runs.** They use different definitions. Wald–Wolfowitz (WW) counts category runs; up-down counts directional (increase/decrease) runs. Use WW for categorical randomness, up-down for trend/oscillation in ordered data.
- **Applying to unordered data.** Up-down runs require a natural sequence (time, space, production order). Arbitrary ordering makes results meaningless.
- **Mishandling ties.** Equal consecutive values must be treated consistently (continue prior direction or remove). Decide method before analysis.
- **Relying on the test without visualization.** The test detects non-random directional patterns but does not describe their form. Always plot the data to interpret trend or structure.
- **Equating significance with monotonic trend.** A significant result indicates deviation from randomness—not necessarily a steady upward or downward trend. Follow up to identify the specific pattern.

Turning Points Test

Test for Randomness Based on Local Extrema

RAND.3 | **Category:** Randomness Tests | **Data:** Continuous | **Null:** Normal (asymptotic)

When to Use

- Test randomness in sequential data using peaks and valleys
- Detect cycles, oscillations, or smooth trends in time series.
- Alternative to runs tests focusing on local extrema rather than directions.
- Process monitoring for drift or cyclical instability.

Assumptions

- Data form a sequence of continuous or ordinal values.
- Under H_0, observations are independent and identically distributed.
- Few or no ties are present (strict comparisons required).
- Sample size is large enough for normal approximation.

Test Statistic

$$z = \frac{T - E[T]}{\sqrt{\mathrm{Var}[T]}}, \quad \text{where } E[T] = \frac{2(n-2)}{3}, \quad \mathrm{Var}[T] = \frac{16n-29}{90}. \quad \text{Under } H_0,\ z \approx N(0,1).$$

Notation: T is observed turning points (local max/min), n is sequence length, peak is $x_{i-1} < x_i > x_{i+1}$, valley is $x_{i-1} > x_i < x_{i+1}$

Plain English: The turning points test examines randomness by counting peaks (local maxima) and valleys (local minima) in a sequence. A peak occurs when a value exceeds both neighbors; a valley when a value is less than both neighbors. Random sequences have expected number of turning points based on sequence length. Too few turning points suggest smooth trends or monotonicity; too many suggest rapid oscillation. The test provides nonparametric assessment of non-random ordering.

⚠ Practitioner Warning

Requires at least three observations for each turning point (x_{i-1}, x_i, x_{i+1}); first and last observations cannot be turning points. Ties complicate identification; typically broken randomly or observations excluded. For small samples, the normal approximation may be poor and exact distributions are preferred. Two-tailed tests are most common, though one-tailed versions may be used for specific alternatives such as smoothness or oscillation. Less powerful than parametric methods (regression) for linear trends. Cannot distinguish increasing from decreasing trends; detects departure from randomness generally. It is also sensitive to noise or measurement error that creates spurious local extrema.

Hypotheses

- H_0: Sequence is random (no serial dependence).
- H_1: Sequence is non-random (trends, cycles, or systematic patterns present).

Decision Rule: Reject H_0 if $|z| > z_{\alpha/2}$ or p-value $< \alpha$ for two-tailed test. Too few turning points ($z < 0$): smooth trends. Too many ($z > 0$): excessive oscillation.

Software References [Python, R, SAS]

Python	R	SAS
`Not Native (Manual)`	`randtests::turning.point.test`	`Not Native (Manual)`

Additional Notes

Exact small-sample distributions were tabulated by Wallis and Moore (1941). Fewer than expected turning points suggest unusually smooth movement or drift; more suggest oscillation or overcorrection. The test is closely related to the up-and-down runs test: both depend only on local order changes and share the same asymptotic variance formula, though they count different structures. It is simple to compute (single pass), nonparametric, and useful as a preliminary screen for non-random serial structure.

Miscellanea: Wallis and Moore (1941) introduced the turning point test while studying business cycles, aiming to quantify peaks and troughs in economic data. Because of its intuitive appeal—counting local highs and lows as a simple diagnostic for dependence and cyclical behavior—it later found use in time-series analysis and related applied settings. Early econometricians used turning points to identify business-cycle phases long before formal spectral methods were available.

Reference(s): Wallis & Moore (1941); Kendall & Stuart (1976)

Worked Example

Problem Statement An energy analyst records daily peak electricity demand (arbitrary units) for $n = 11$ consecutive days:

$$2,\ 4,\ 6,\ 8,\ 10,\ 8,\ 6,\ 4,\ 6,\ 8,\ 6.$$

The turning-points test assesses whether the number of local extrema is consistent with an i.i.d. sequence, or whether the series is too smooth (too few turning points, suggesting a trend or cycle) or too erratic (too many). H_0: sequence is random (i.i.d.); H_1: non-random (systematic pattern).

Unlike the up-and-down runs test, which counts *runs* in the sign sequence, the turning-points test counts *interior local maxima and minima* directly.

1. Identifying Turning Points
Position i $(2 \le i \le n-1)$ is a turning point if $x_i > \max(x_{i-1}, x_{i+1})$ (peak) or $x_i < \min(x_{i-1}, x_{i+1})$ (trough).

Day i	x_i	x_{i-1}	x_{i+1}	Status
2	4	2	6	– (rising)
3	6	4	8	– (rising)
4	8	6	10	– (rising)
5	10	8	8	**Peak**
6	8	10	6	– (falling)
7	6	8	4	– (falling)
8	4	6	6	**Trough**
9	6	4	8	– (rising)
10	8	6	6	**Peak**
11	6	8	–	endpoint

$T = 3$ turning points (two peaks, one trough).

2. Mean and Variance Under H_0

$$E[T] = \frac{2(n-2)}{3} = \frac{2\times 9}{3} = 6$$

$$\mathrm{Var}[T] = \frac{16n-29}{90} = \frac{176-29}{90} = \frac{49}{30}$$

$$\mathrm{SD}[T] = \sqrt{49/30} \approx 1.278$$

3. Normal Approximation

$$z = \frac{T - E[T]}{\mathrm{SD}[T]} = \frac{3-6}{7/\sqrt{30}} \approx -2.347$$

Method	**Value**	**CV (5%)**	**Decision**
z (approx)	−2.347	±1.960	Reject H_0
Exact p	0.042	0.05	Reject H_0

Approx: $p = 0.019$; exact: $p = 0.042$. Both reject at 5%, FTR at 1% (CV= ±2.576).

Practitioner Insight: The turning-points and up-and-down runs tests are equivalent: if R is the number of runs in the sign-of-differences sequence, then $T = R - 1$, yielding identical z-scores. The difference is interpretational: turning points count peaks and troughs directly in the series, making them more intuitive for time-series analysis. They share $\mathrm{Var} = (16n-29)/90$ but differ in mean, $E[T] = 2(n-2)/3$ and $E[R] = (2n-1)/3$ with $E[T] = E[R] - 1$. Here $n = 11$ gives $E[T] = 6$ and $\mathrm{Var}[T] = 49/30$. The observed $T = 3$ (half the expectation) indicates a series that is too smooth, consistent with a systematic oscillation rather than randomness. The exact $p = 0.042$ (obtained by enumeration over all $11! = 39916800$ orderings) exceeds the normal approximation ($p = 0.019$), illustrating small-sample anti-conservatism. Both reject at 5% but not at 1%.

Common Pitfalls

- **Counting endpoints as turning points.** Only interior observations ($i = 2, \ldots, n-1$) can be turning points. Endpoints lack two neighbors and must never be counted.
- **Using non-strict inequalities.** Turning points require strict comparisons: $x_i > x_{i-1}$ and $x_i > x_{i+1}$ (peak) or both $<$ (valley). Ties invalidate turning point status.
- **Confusing turning points with direction changes.** A turning point is a local extremum (greater/less than both neighbors), not merely a change in direction. Use the correct definition.
- **Applying to unordered data.** The test requires natural ordering (time, space, sequence). Arbitrary ordering makes results meaningless.
- **Interpreting significance without visualization.** A significant result indicates non-random pattern, not its form. Always plot the data to understand oscillation, trend, or structure.

Difference-Sign Test

Test for Randomness Based on Sign Patterns in First Differences

RAND.4 | **Category:** Randomness Tests | **Data:** Continuous | **Null:** Normal (asymptotic)

When to Use

- Test randomness or monotonic trend using signs of successive differences.
- Detect directional drift or non-random imbalance in ordered data.

Assumptions

- Observations are sequential (continuous or ordinal scale).
- Under H_0, the sequence is i.i.d. continuous (note: paired differences are dependent).
- Sample size large enough for normal approximation.

Test Statistic

$$z = \frac{S - E[S]}{\sqrt{\text{Var}[S]}}, \quad E[S] = \frac{n-1}{2}, \quad \text{Var}[S] = \frac{n+1}{12}. \qquad \text{Under } H_0,\ z \approx N(0,1).$$

Notation: S is number of positive differences, $d_i = x_{i+1} - x_i$ for $i = 1, \ldots, n-1$, n is the sequence length

Plain English: The difference-sign test examines randomness by computing consecutive differences and counting how many are positive versus negative. Under randomness with no trend, approximately half the differences should be positive, half negative. Systematic excess of positive differences indicates upward trend; excess negative differences indicates downward trend. The test provides simple nonparametric assessment of trend without assuming specific functional form.

⚠ Practitioner Warning

Zero differences (ties) typically excluded from analysis; reduces effective sample size. Assumes the observations are in random order under H_0; this is not the same as assuming the first differences are independent Bernoulli signs (not applicable here). Sensitive mainly to directional imbalance or drift; may miss oscillatory or more complex dependence patterns. Less powerful than parametric regression for detecting specific trends. Standard continuity correction (0.5 adjustment) can improve normal approximation. One-sided tests appropriate when trend direction hypothesized a priori.

Hypotheses

- H_0: The observations are in random order (i.i.d continuous).
- H_1: Evidence of trend or non-random ordering.

Decision Rule: Reject H_0 if $|z| > z_{\alpha/2}$ or p-value $< \alpha$ for two-tailed test.

Software References [Python, R, SAS]

Python	R	SAS
`Not Native (Manual)`	`Not Native (Manual)`	`Not Native (Manual)`

Additional Notes

The statistic counts how often the sequence increases from one observation to the next. Under an i.i.d. continuous null, successive indicators $I_i = \mathbb{I}(X_{i+1} > X_i)$ are not independent: adjacent terms share a common observation and are negatively correlated. This dependence reduces the variance from the binomial value $(n-1)/4$ to $(n+1)/12$. Refer to Technical Notes at the end of this chapter.

Proof: For an i.i.d. continuous sequence, let $S = \sum_{i=1}^{n-1} I_i$ with $I_i = \mathbb{I}(X_{i+1} > X_i)$. Then, under H_0, $E[I_i] = 1/2$ and $\text{Var}(I_i) = 1/4$. Non-adjacent indicators are independent, so only adjacent pairs contribute covariance. For adjacent terms, $E[I_i I_{i+1}] = P(X_{i+2} > X_{i+1} > X_i) = 1/6$, giving $\text{Cov}(I_i, I_{i+1}) = 1/6 - 1/4 = -1/12$. Hence, $\text{Var}(S) = (n-1)\frac{1}{4} + 2(n-2)\left(-\frac{1}{12}\right) = (n+1)/12$.

Miscellanea: The difference-sign test is frequently confused with the classical sign test of Dixon and Mood (1946), but the two arise in fundamentally different settings. The Dixon–Mood sign test applies to *paired differences* and assumes independence across pairs, leading to a binomial model with variance $(n-1)/4$. In contrast, the difference-sign test for a single ordered sequence, developed by Moore and Wallis (1943), is based on *signs of first differences*. These signs are **not independent**: adjacent differences share observations, inducing negative correlation. As a result, the correct variance is $(n+1)/12$, not $(n-1)/4$. This distinction reflects two different null hypotheses: random paired differences (Dixon–Mood's Sign test) vs. random ordering of observations (Moore–Wallis' Difference-Sign test).

Reference(s): Wallis & Moore (1943); Dixon & Mood (1946)

Difference-Sign Test RAND.4

Worked Example

Problem Statement A health economist records the monthly unemployment rate (percentage points) for $n = 11$ consecutive months:

$$10, 9, 8, 7, 8, 6, 5, 6, 4, 3, 2.$$

The difference-signs test counts the number of consecutive *increases* to test whether the sequence is consistent with a random ordering, or whether downward momentum is too persistent to arise by chance. H_0: sequence is random (i.i.d.); H_1: non-random (systematic trend).

1. Differences and Their Signs
Form $n - 1 = 10$ successive differences $d_i = x_{i+1} - x_i$ and encode each as U (up) or D (down); ties are dropped.

i	$d_i = x_{i+1} - x_i$	Sign
1	$9 - 10 = -1$	D
2	$8 - 9 = -1$	D
3	$7 - 8 = -1$	D
4	$8 - 7 = +1$	U
5	$6 - 8 = -2$	D
6	$5 - 6 = -1$	D
7	$6 - 5 = +1$	U
8	$4 - 6 = -2$	D
9	$3 - 4 = -1$	D
10	$2 - 3 = -1$	D

$S = 2$ +ve differences (U), $Q = 8$ -ve diff. (D), no ties.

2. Mean and Variance Under H_0
Consecutive differences d_i and d_{i+1} share the observation x_{i+1}, so they are negatively correlated under H_0. The correct null moments are:

$$E[S] = \frac{n-1}{2} = 5, \qquad \text{Var}[S] = \frac{n+1}{12} = 1$$

The variance equals *exactly* 1 for $n = 11$.

3. Test Statistic

$$z = \frac{S - E[S]}{\sqrt{\text{Var}[S]}} = \frac{2-5}{\sqrt{1}} = -3$$

Method	Value	CV (1%)	Decision
z (approx)	-3	± 2.576	Reject H_0
Exact p	0.008	0.01	Reject H_0

Approx: $p = 0.003$; exact: $p = 0.008$. Both reject at 1%.

4. Exact Lower-Tail Probability
Out of $11! = 39916800$ equally likely orderings, using Eulerian numbers $A(11, k)$: $P(S = 2) = 152637/11!$, $P(S \le 2) = 154674/11!$ By the symmetry $P(S = k) = P(S = 10 - k)$: $P(S \ge 8) = P(S \le 2)$, $p_{\text{exact}} = 309348/11! \approx 0.008$.

Practitioner Insight: The turning-points and up-and-down runs tests are equivalent: if R counts runs in the sign-of-differences sequence, then $T = R - 1$, giving identical z-scores. The distinction is interpretational: turning points count peaks and troughs directly, making them more intuitive. They share $\text{Var} = (16n - 29)/90$ but differ in mean, $E[T] = 2(n-2)/3$ and $E[R] = (2n-1)/3$. For $n = 11$, $E[T] = 6$ and $\text{Var}[T] = 49/30$; the observed $T = 3$ indicates an overly smooth, oscillatory series. The exact $p = 0.042$ exceeds the normal approximation ($p = 0.019$), showing small-sample anti-conservatism. Both reject at 5% but not at 1%. See Technical Notes (RAND) on Eulerian numbers.

Common Pitfalls

- **Using n instead of $n - 1$.** The binomial parameter is the number of *differences* ($n - 1$), not observations (n). Always verify: n observations $\to n - 1$ differences.
- **Interpreting as magnitude test.** The test examines the *frequency* of positive vs. negative changes, not their size. Large negative shifts can outweigh many small positives—magnitude requires regression or t-tests.
- **Handling ties inconsistently.** Zero differences must be treated systematically (typically excluded). Different tie rules change n and the test result—state the method used.
- **Applying to unordered data.** Requires natural sequential order (time, space, process). Arbitrary ordering makes the test meaningless.
- **Equating non-significance with randomness.** Failure to reject H_0 only indicates insufficient evidence of directional bias—not proof of randomness.

Mann-Kendall Test

Nonparametric Test for Monotonic Trend in Time Series

RAND.5 | **Category:** Randomness Tests | **Data:** Ordinal/Continuous | **Null:** Normal (asymptotic)

When to Use

- Detect monotonic trends (increasing or decreasing) in time series (environmental and longitudinal monitoring).
- Appropriate when exact functional form of the trend is unknown, non-normal data, or data contain outliers.

Assumptions

- Observations measured over time in natural sequence.
- Under null hypothesis, all orderings are equally likely.
- No specific distributional form required (nonparametric).
- Sample size large enough for normal approximation.

Test Statistic

$$z = \frac{S}{\sqrt{\text{Var}[S]}}, \qquad S = \sum_{i=1}^{n-1} \sum_{j=i+1}^{n} \text{sgn}(x_j - x_i), \qquad \text{Var}(S) = \frac{n(n-1)(2n+5)}{18} \text{ (no ties)}. \qquad \text{Under } H_0,\ z \approx N(0,1).$$

Notation: S is Kendall score statistic, sum over all pairs (i, j) with $i < j$, $\text{sgn}(x) = +1$ if $x > 0, -1$ if $x < 0, 0$ if $x = 0$

Plain English: The Mann–Kendall test detects monotonic trends by comparing all time-ordered pairs and counting the excess of concordant (increasing) over discordant (decreasing) pairs. A positive statistic indicates an upward trend; a negative statistic indicates a downward trend. The test is distribution-free and requires only ordered observations, making it robust to outliers and suitable for non-normal data common in environmental and longitudinal studies.

⚠ Practitioner Warning

Assumes independence; serial correlation inflates Type I error (Hamed–Rao adjustment available). Ties reduce power and require variance corrections. Detects monotonic trends only (not U-shaped or cyclical patterns) and does not estimate trend magnitude—use Sen's slope for effect size. Seasonal data require the seasonal variant. Very short series ($n < 10$) have limited power. Missing data must be handled appropriately.

Hypotheses

- H_0: Observations are i.i.d. with no monotonic trend.
- H_1: Monotonic trend exists.

Decision Rule: Reject H_0 if $|z| > z_{\alpha/2}$ or p-value $< \alpha$ for two-tailed test. +ve z: upward trend. -ve z: downward trend.

Software References [Python, R, SAS]

Python	R	SAS
`pyMannKendall.mk_test`	`Kendall::MannKendall`	`Not Native (Manual)`

Additional Notes

Kendall's S counts concordant minus discordant pairs over all $\binom{n}{2}$ ordered pairs ($i < j$): concordant if $x_i < x_j$, discordant if $x_i > x_j$, ties if $x_i = x_j$. Under H_0 (no monotonic trend), $E[S] = 0$. Variance with ties: $\text{Var}(S) = \frac{1}{18}[n(n-1)(2n+5) - \sum t(t-1)(2t+5)]$, where t denotes tied-group sizes. A continuity correction uses $z = (S - \text{sgn}(S))/\sqrt{\text{Var}(S)}$. Effect size: Kendall's $\tau = S/\binom{n}{2} \in [-1, 1]$. Trend magnitude: Sen's slope = median of all pairwise slopes $(x_j - x_i)/(j - i)$. Extensions: Seasonal Mann-Kendall (compute S within seasons, combine) and variance-adjusted versions (e.g., Hamed–Rao) for autocorrelation. Widely used in environmental and climate trend monitoring due to robustness to non-normality and outliers.

Miscellanea: Henry Mann (1945) proposed the statistic for testing monotonic trend; Maurice Kendall later connected it to rank correlation (τ) and popularized it in time-series settings. The test gained prominence in environmental regulation during the 1970s–80s, where skewed, censored, and outlier-prone pollutant data undermined parametric regression. Influential work by Hirsch and Slack (1984) established Mann–Kendall as a standard tool in water-quality analysis, and it remains widely recommended by environmental agencies for robust trend detection.

Reference(s): Mann (1945); Kendall (1975); Hirsch et al. (1984); Hamed & Rao (1998)

Worked Example

Problem Statement An environmental scientist records annual mean river discharge (m^3s^{-1}, rounded) at a gauging station over $n = 10$ consecutive years: 10, 5, 6, 7, 8, 9, 4, 3, 2, 1.

The Mann–Kendall test examines whether there is a monotone trend by comparing *all* $\binom{10}{2} = 45$ pairs of observations, not only consecutive differences. H_0: no trend (i.i.d.); H_1: monotone trend (upward or downward).

1. Concordant and Discordant Pairs

Pair (i, j) with $i < j$ is *concordant* if $x_j > x_i$ (later value higher) and *discordant* if $x_j < x_i$ (later value lower). Ties contribute 0. The 45 pairs split cleanly into four groups:

Group	Pairs	Count	Sign
A: x_1=10 vs $x_2 \ldots x_{10}$	9	9	D
B: $\{x_2, \ldots, x_6\}$ = (5–9) internal	$\binom{5}{2}$	10	C
C: $(x_2 \ldots x_6)$ vs $(x_7 \ldots x_{10})$	5×4	20	D
D: $\{x_7, \ldots, x_{10}\}$ = (4–1) internal	$\binom{4}{2}$	6	D
Total		**45**	

Group A: $x_1 = 10$ exceeds all nine later values.
Group B: values $5 < 6 < 7 < 8 < 9$ form a rising block; all $\binom{5}{2} = 10$ within-block pairs are concordant.
Group C: every value in $\{5, 6, 7, 8, 9\}$ exceeds every value in $\{4, 3, 2, 1\}$, so all 20 pairs discordant.
Group D: values $4 > 3 > 2 > 1$ form a falling block.

$$P = 10, \quad Q = 9 + 20 + 6 = 35, \quad S = P - Q = \mathbf{-25}.$$

2. Null Mean and Variance

Under H_0 with no ties:

$$E[S] = 0, \quad \text{Var}[S] = \frac{n(n-1)(2n+5)}{18} = 125$$

3. Test Statistic

Without continuity correction:

$$z = \frac{S}{\sqrt{\text{Var}[S]}} = \frac{-25}{\sqrt{125}} = -\sqrt{5} \approx -2.236$$

Kendall's $\tau = S/\binom{n}{2} = -25/45 \approx -0.556$.

Method	Value	CV (5%)	Decision
z (approx)	$-\sqrt{5}$	±1.960	Reject H_0
Exact p	0.029	0.05	Reject H_0

Approx: $p = 0.025$; exact: $p = 0.029$. Both reject at 5%, FTR at 1% (CV= ±2.576).

4. Exact Tail Probability

Out of $10! = 3628800$ equally likely orderings: $P(S \leq -25) = 51909/10!$, $p_{\text{exact}} = 103818/10! \approx 0.029$. The distribution is symmetric: $P(S = k) = P(S = -k)$.

Practitioner Insight: The Mann–Kendall and difference-signs tests both detect monotone trends, but use different information. The difference-signs test uses only the $n - 1$ consecutive differences, with $E[S] = (n-1)/2$ and $\text{Var}[S] = (n+1)/12$. Mann–Kendall uses all $\binom{n}{2}$ pairs, giving greater power for gradual trends while remaining nonparametric. Its statistic S equals unnormalised Kendall's τ: $\tau = S/\binom{n}{2} = -5/9$, indicating later values tend to be smaller. For $n = 10$, Var = 125, giving $z = -25/\sqrt{125} = -\sqrt{5}$ ($z^2 = 5$ exactly). The pair structure explains $P = 10$ and $Q = 35$: only the increasing block (positions 2–6) contributes concordant pairs ($\binom{5}{2} = 10$); all others are discordant, allowing an exact, transparent computation without enumerating all pairs.

Common Pitfalls

- **Ignoring seasonality.** Standard Mann–Kendall assumes no seasonal structure. For seasonal data (e.g., monthly), use the Seasonal Mann–Kendall test or deseasonalize before testing.
- **Equating significance with linear trend.** The test detects *monotonic* change, not necessarily linear form. A significant result may reflect exponential or other nonlinear trends.
- **Overlooking autocorrelation.** Independence is assumed. Positive serial correlation inflates Type I error. Check ACF and use variance-corrected or pre-whitened versions when needed.
- **Ignoring ties or censoring.** Ties reduce the variance of S and require tie corrections; censored values (e.g., below detection limits) need appropriate handling. Naively treating all values as distinct can misstate significance.
- **Reporting only the p-value.** Always accompany significance with Sen's slope (and confidence interval) to quantify magnitude of change.

Cox-Stuart Test

Simple Sign Test for Monotonic Trend in Time Series

RAND.6 | **Category:** Randomness Tests | **Data:** Ordinal/Continuous | **Null:** Binomial (exact)

When to Use

- Detect monotonic trends in time-ordered data with minimal computation.
- Quick preliminary or small-sample alternative to Mann–Kendall.
- Robust, distribution-free trend check with ordinal summary.

Assumptions

- Time-ordered sequence with $n \geq 6$.
- Under H_0, no systematic trend (early and late portions comparable).
- Observations can be paired into first and second halves.
- No distributional assumptions beyond ordinal measurement.

Test Statistic

$$z = \frac{S - E[S]}{\sqrt{\mathrm{Var}[S]}}, \quad E[S] = \frac{n_p}{2}, \quad \mathrm{Var}[S] = \frac{n_p}{4}, \quad n_p = \lfloor n/2 \rfloor, \quad \text{Under } H_0,\ S \sim \mathrm{Binomial}(n_p, 0.5).$$

Notation: S = #{pairs where late > early}, n_p is number of pairs, middle observation discarded if n odd

Plain English: The Cox–Stuart test splits a time series into two halves and pairs early observations with corresponding later ones (1 with $m+1$, 2 with $m+2$, etc.). For each pair, record whether the later value is larger or smaller. Under no trend, about half the pairs should increase. A significant excess of increases (or decreases) indicates an upward (or downward) monotonic trend. The procedure reduces trend detection to a simple sign test on paired differences.

⚠ Practitioner Warning

Less powerful than Mann–Kendall or regression since it uses only paired signs. Requires moderate sample size ($n \geq 6$, preferably $n \geq 10$). When n is odd, the middle observation is discarded; ties are usually excluded, reducing effective sample size (n_p). Detects only monotonic trends (not cycles or local patterns) and does not estimate slope magnitude. For large samples, a normal approximation with continuity correction may be used.

Hypotheses

- H_0: No monotonic trend ($P(\text{late} > \text{early}) = 0.5$).
- H_1: Monotonic trend exists ($P(\text{late} > \text{early}) \neq 0.5$).

Decision Rule: Reject H_0 if $p_{\text{exact}} = 2\min(P(X \leq S), P(X \geq S)) < \alpha$ where $X \sim \mathrm{Binomial}(n_p, 0.5)$. One-sided test uses single tail probability.

Software References [Python, R, SAS]

Python	R	SAS
`scipy.stats.binomtest`	`(manual with binom.test)`	`Not Native (Manual)`

Additional Notes

For a series $x_1, \ldots, x_n$, let $c = \lfloor n/2 \rfloor$ and form pairs $(x_1, x_{c+1}), (x_2, x_{c+2}), \ldots$; if n is odd, discard the middle value. Number of pairs $n_p = \lfloor n/2 \rfloor$. For each pair, record + if later > earlier, − if later < earlier (ties excluded), and let S be the number of + signs. The test is essentially a sign test comparing early vs. late halves. Compared to Wilcoxon signed-rank or Mann–Kendall, it ignores magnitudes (lower power) but is extremely simple, robust, and easy to compute by hand. For $n_p \geq 10$, use the normal approximation with continuity correction:

$$z = \frac{S - n_p/2 - 0.5\,\mathrm{sgn}(S - n_p/2)}{\sqrt{n_p/4}}.$$

Miscellanea: Introduced by Cox and Stuart (1955) as a hand-calculation alternative to more intensive trend tests, the method was designed for quick screening when computation was limited. Its divide-and-compare pairing scheme made trend detection feasible with only counting pluses and minuses. Though less powerful than Mann–Kendall, its simplicity ensured adoption in quality control and environmental monitoring—especially for preliminary, field-level assessments.

Reference(s): Cox & Stuart (1955); Gibbons & Chakraborti (2011)

Worked Example

Problem Statement A meteorologist records mean monthly temperature (°C) at a highland station over twelve consecutive months:

$$10,\ 9,\ 8,\ 7,\ 6,\ 5,\ 7,\ 6,\ 5,\ 4,\ 3,\ 2.$$

The Cox–Stuart test assesses whether there is a monotone trend by pairing each observation in the *first half* with the corresponding observation in the *second half*. H_0: no trend; H_1: monotone trend (upward or downward).

For $n = 12$ (even), the lag is $c = n/2 = 6$, producing $m = 6$ pairs (x_i, x_{i+6}) for $i = 1, \ldots, 6$.

1. Paired Differences and Signs

For each pair record the sign of $d_i = x_{i+6} - x_i$; ties are dropped (none here).

Pair i	x_i	x_{i+6}	d_i	Sign
1	10	7	−3	D
2	9	6	−3	D
3	8	5	−3	D
4	7	4	−3	D
5	6	3	−3	D
6	5	2	−3	D

All six second-half values lie exactly 3 °C below their first-half counterparts. $S = 0$ (upward signs), $m = 6$. Under H_0, $S \sim \text{Binomial}(m, \frac{1}{2})$.

2. Null Mean and Variance

$$E[S] = \frac{m}{2} = 3, \qquad \text{Var}[S] = \frac{m}{4} = \frac{3}{2}$$

3. Normal Approximation

$$z = \frac{S - E[S]}{\sqrt{\text{Var}[S]}} = \frac{0 - 3}{\sqrt{3/2}} = -\sqrt{6} \approx -2.449$$

Method	Value	CV (5%)	Decision
z (approx)	$-\sqrt{6}$	±1.960	Reject H_0
Exact p	0.031	0.05	Reject H_0

Approx: $p = 0.014$; exact: $p = 0.031$. Both reject at 5%, FTR at 1%.

4. Exact Probability

$S = 0$ is the most extreme possible outcome ($m = 6$): $P(S = 0) = \left(\frac{1}{2}\right)^6 = \frac{1}{64}$; $p_{\text{exact}} = 2 \times \frac{1}{64} \approx 0.031$.

Practitioner Insight: The Cox–Stuart test lies between the difference-signs and Mann–Kendall tests. The former uses $n - 1$ consecutive differences, the latter all $\binom{n}{2}$ pairs, while Cox–Stuart uses $m = n/2$ long-range pairs. By comparing observations $c = n/2$ apart, it is more sensitive to global trends and less affected by local oscillations. Under H_0, the m pairs are independent, giving an exact $\text{Binomial}(m, \frac{1}{2})$ test. Here $S = 0$ has probability $1/64$, with two-sided $p = 1/32 \approx 0.031$, rejecting at 5% but not 1%. The normal approximation ($z = -\sqrt{6}$, $p = 0.014$) is anti-conservative for small m. For odd n, the middle observation is dropped; ties are also removed, reducing m. The analog of Kendall's τ is $\hat{\tau}_{\text{CS}} = (P - Q)/m = (0 - 6)/6 = -1$, indicating perfect negative association between the two halves.

Common Pitfalls

- **Arbitrary pairing.** Pairs must preserve temporal order: observation i is compared with $i + m$ ($m = n/2$). Sorting or rearranging destroys the time structure and invalidates the test.
- **Using Cox–Stuart instead of Mann–Kendall unnecessarily.** Cox–Stuart uses only m paired comparisons, whereas Mann–Kendall uses all $\binom{n}{2}$ pairs and is more powerful. Prefer MK for formal analysis; Cox–Stuart is mainly pedagogical or for quick screening.
- **Over-interpreting non-significance.** Power is limited, especially for small m. A non-significant result does not prove absence of trend—only insufficient evidence.
- **Ignoring seasonality.** Seasonal or periodic patterns can produce systematic pair differences unrelated to trend. Remove seasonality or use seasonal methods before applying the test.
- **Mishandling odd n.** When n is odd, the middle observation is typically discarded. Be explicit and check sensitivity if the midpoint contains an anomaly.

Von Neumann's Ratio Test

Test for Randomness via Mean Square Successive Difference

RAND.7 | **Category:** Randomness Tests | **Data:** Continuous | **Null:** Normal (asymptotic)

When to Use

- Test randomness versus first-order serial correlation in ordered data.
- Screen for autocorrelation before fitting regressions.
- Monitor process stability or detect drift in quality control.

Assumptions

- Observations form a naturally ordered sequence.
- Under H_0, observations are independent with common mean and finite variance.
- Sample is large enough for asymptotic normality.

Test Statistic

$$z = \frac{\eta - E[\eta]}{\sqrt{\text{Var}[\eta]}}, \qquad \eta = \frac{n\sum_{i=1}^{n-1}(x_{i+1}-x_i)^2}{(n-1)\sum_{i=1}^{n}(x_i-\bar{x})^2} \text{ (unbiased version)}, \quad E[\eta] = \frac{2n}{(n-1)} \approx 2, \ \text{Var}[\eta] \approx \frac{4n^2(n-2)}{[(n+1)(n-1)^3]}.$$

Notation: η is von Neumann's ratio, x_i are actual observations (not ranks), $\bar{x}$ is sample mean, n is sample size

Plain English: Von Neumann's ratio compares the mean squared successive difference (local jumps) to the overall sample variance. Under independence, $\eta \approx 2$. $\eta < 2$ indicate positive serial correlation (smoothness, trend, clustering); $\eta > 2$ indicate negative serial correlation (alternation or oscillation). Because it uses raw values rather than ranks, the test is parametric and most powerful under normality.

⚠ Practitioner Warning

Assumes approximate normality; outliers or heavy tails can distort results (use the rank-based Bartels version if violated). The null relies on independence and distributional assumptions. Asymptotic normality may be inaccurate for $n < 30$. The test mainly detects first-order autocorrelation and cannot distinguish trend from stationary dependence. Durbin–Watson is preferred for regression residuals. Two-sided testing is standard; use one-sided only when direction is specified *a priori*.

Hypotheses

- H_0: Observations are serially independent.
- H_1: Serial correlation (+ve or -ve) is present.

Decision Rule: Reject H_0 if $|z| > z_{\alpha/2}$ or p-value $< \alpha$ for large n. For small n, table are available.

Software References [Python, R, SAS]

Python	R	SAS
`Not Native (Manual)`	`DescTools::vonNeumannTest`	`Not Native (Manual)`

Additional Notes

John von Neumann (1941) introduced the ratio of the mean square successive difference to the sample variance to test independence in sequential data. The numerator measures local variability; the denominator measures overall dispersion, yielding a scale-free statistic. Exact small-sample distributions under normality are complex (involving special functions), but large-sample normal approximations enable z-tests. Durbin and Watson (1950–51) later adapted this ratio to regression residuals, creating the Durbin–Watson statistic; the latter is essentially von Neumann's ratio with regression-specific critical values. Applications include detecting autocorrelation in time series, spatial sequences, and quality control monitoring. Sensitivity is greatest for first-order positive autocorrelation (trend or clustering). Caution: Some sources use the **original (biased) ratio** without the $n/(n-1)$ scaling, for which $E[\eta] = 2$ and the variance differs.

Miscellanea: John von Neumann, who introduced this statistic, was one of the 20th century's great polymaths, contributing to quantum mechanics, game theory, computing, and nuclear physics. During World War II and the Manhattan Project, he applied advanced mathematics to shock-wave analysis, ballistic trajectories, and nuclear implosion design. He also played a central role in developing Monte Carlo methods and early random number generators—making the testing of randomness, as in this statistic, both a theoretical and practical necessity. His career exemplifies how abstract statistical ideas and high-stakes engineering challenges emerged from the same mathematical foundations. von Neumann is often credited with formalizing the "round half to even" rule (banker's rounding), now widely used in computing and finance to reduce cumulative rounding bias.

Reference(s): von Neumann (1941); Durbin & Watson (1950, 1951); Hart (1942); Bingham & Nelson (1981)

Worked Example

Problem Statement Are 20 consecutive product weights random, or do they show a systematic pattern (trend/serial dependence)? H_0: observations are serially independent; H_1: serial correlation is present.

1. Data and Calculations

i	x_i	Δ_i	i	x_i	Δ_i
1	455		11	467	-4
2	462	7	12	473	6
3	458	-4	13	469	-4
4	465	7	14	475	6
5	461	-4	15	471	-4
6	467	6	16	477	6
7	463	-4	17	473	-4
8	469	6	18	479	6
9	465	-4	19	475	-4
10	471	6	20	481	6

$$\bar{x} = \frac{\sum_{i=1}^{20} x_i}{20} = 468.8$$

$$\sum_{i=2}^{20} \Delta_i^2 = 530$$

$$\sum_{i=1}^{20} (x_i - \bar{x})^2 = 925.2$$

2. von Neumann's statistic (unbiased version)
Use the standardized rank von Neumann ratio ($n = 20$)

$$\eta = \frac{20 \times 530}{19 \times 925.2} = 0.6030.$$

3. Normal approximation (large-sample)
Under H_0 (random order), $E[\eta] = 2.1053$ and

$$\text{Var}(\eta) = \frac{4n^2(n-2)}{(n+1)(n-1)^3} = 0.1999.$$

$$Z = \frac{0.6030 - 2.1053}{\sqrt{0.1999}} = -3.3597.$$

4. p-value and Decision
Two-sided: $p \approx 0.00078$ (one-sided lower-tail $p \approx 0.00039$).
Reject H_0 due to strong evidence of non-randomness.

Practitioner Insight: Von Neumann's ratio is unique in this chapter in using the *magnitudes* of successive differences rather than signs or ranks. As a result, small differences on a wide-ranging series yield a small ratio; here $\eta = 0.6030$, far below the null expectation $E[\text{VN}] = 2.1053$. For $n = 20$, Var $= 0.1999$, giving $z = - -3.3597$. Small η ($z \ll 0$) indicates positive serial correlation (successive values are similar), while large η ($z \gg 0$) signals negative correlation or oscillation. Unlike other RAND tests, η lacks a simple exact null; tabulated critical values exist (Hart, 1942), and the normal approximation is adequate for $n \geq 10$. Bingham & Nelson (1981) provided a small sample t approximation.

Common Pitfalls

- **Applying the test to clearly non-normal data.** For inference, the classical von Neumann ratio assumes normality. Heavy tails or outliers can distort both numerator and denominator, making η unreliable. Consider transformation or a rank-based alternative (e.g., Bartels).
- **Interpreting $\eta \approx 2$ as proof of independence.** $\eta \approx 2$ indicates no first-order linear serial correlation under the null—not complete independence. Nonlinear dependence, ARCH effects, or periodic structure may still be present.
- **Misinterpreting direction of departure.** $\eta < 2$ suggests positive serial correlation (trend or clustering), whereas $\eta > 2$ indicates negative correlation (oscillation or alternation). Always interpret magnitude and direction together.
- **Applying to non-stationary data.** Deterministic trends or structural shifts can artificially drive η below 2. Detrend the series or analyze residuals (e.g., via Durbin–Watson) before testing for additional serial dependence.
- **Using irregularly spaced observations.** The test assumes equally spaced sequential data. Unequal time gaps invalidate interpretation of successive differences.
- **Ignoring order dependence.** The statistic depends entirely on sequence order. Reordering observations changes η. Never apply to unordered or purely cross-sectional data.
- **Relying on asymptotics for small samples.** For small n, the normal approximation can be inaccurate. Prefer exact or permutation-based inference when feasible.
- **Applying directly to multivariate series.** The standard η is univariate. Multivariate dependence requires vector extensions or portmanteau-type tests rather than separate univariate applications.

Bingham & Nelson (1981)—*t* distribution based approximation for small sample.

$$\epsilon = \left(1 - \frac{n}{(n-1)}\frac{\eta}{2}\right), \qquad \tilde{\epsilon} = \epsilon\sqrt{\frac{n^2-1}{n^2-4}}, \qquad t = \sqrt{(n+1)}\frac{\tilde{\epsilon}}{1-\tilde{\epsilon}^2} \sim t_{n+1}.$$

Bartels Test

Rank-Based Test for Randomness Against Trend or Serial Correlation

RAND.8 | **Category:** Randomness Tests | **Data:** Ordinal/Continuous | **Null:** Normal (asymptotic)

When to Use

- Test randomness against trend or serial dependence in ordered data.
- Often more powerful than basic runs tests for smooth trend or serial-dependence alternatives.

Assumptions

- Data form an ordered sequence (temporal or spatial).
- Observations measurable on at least ordinal scale (ranking possible).
- Sample is large enough for reliable normal approximation.

Test Statistic

$$z = \frac{RVN - 2}{\sqrt{\mathrm{Var}[RVN]}}, \quad RVN = \frac{\sum_{i=1}^{n-1}(R_{i+1} - R_i)^2}{\sum_{i=1}^{n}(R_i - \bar{R})^2}, \quad E[RVN] = 2, \quad \mathrm{Var}[RVN] = \frac{4(n-2)(5n^2 - 2n - 9)}{5n(n+1)(n-1)^2} \approx \frac{4}{n}.$$

Notation: R_i are ranks of observations in sequence order, $\bar{R} = (n+1)/2$ is the average rank. Under H_0, $z \approx N(0,1)$.

Plain English: Bartels test converts observations to ranks and examines the mean squared successive difference of ranks, normalized by the total variance. The test statistic measures smoothness of the rank sequence. Under randomness, the statistic approximately equals 2. Values significantly below 2 indicate positive serial correlation or trend (smooth progression); values above 2 indicate negative correlation or oscillation. The test provides nonparametric assessment of sequence randomness with good power against many ordered alternatives.

⚠ Practitioner Warning

Ties in data require average rank assignment; many ties reduce power. Normal approximation adequate for $n \geq 10$; exact permutation test for smaller samples. Two-tailed test typical; one-tailed when specific alternative hypothesized. Cannot distinguish between trend and serial correlation; both yield $RVN < 2$. Does not identify specific autocorrelation structure or trend form. More complex computation than simple runs tests; requires ranking and successive differences. Sensitive to any departure from randomness; follow-up analysis needed to identify pattern type.

Hypotheses

- H_0: Sequence is random (rank orderings equally likely).
- H_1: Non-randomness present (trend, autocorrelation, or systematic patterns).

Decision Rule: Reject H_0 if $|z| > Z_{\alpha/2}$ or p-value $< \alpha$ for two-sided test. $RVN < 2$ (or $z < 0$): trend or +ve correlation. $RVN > 2$ (or $z > 0$): oscillation or -ve correlation.

Software References [Python, R, SAS]

Python	R	SAS
`Not Native (Manual)`	`randtests::bartels.rank.test`	`Not Native (Manual)`

Additional Notes

Bartels introduced a rank-based analog of the von Neumann ratio, yielding a distribution-free test with improved finite-sample behavior. Convert $x_1, \ldots, x_n$ to ranks $R_1, \ldots, R_n$ in original order and compute successive differences $d_i = R_{i+1} - R_i$. The statistic RVN is based on $\sum_{i=1}^{n-1} d_i^2$, normalized by the total rank variability. With no ties, $\sum_{i=1}^{n}(R_i - \bar{R})^2 = n(n^2-1)/12$. Under H_0, its mean and variance follow from the permutation distribution of ranks. Exact inference (small n) enumerates rank permutations; Monte Carlo resampling provides practical approximation. More powerful than basic runs tests and competitive with Mann–Kendall for monotonic trends, while also detecting non-monotonic serial dependence. In quality control, $RVN < 2$ signals positive autocorrelation or drift.

Miscellanea: Robert Bartels introduced the test in 1982 (*Biometrika*) as a rank-based refinement of the von Neumann ratio. By standardizing the statistic and clarifying its asymptotic behavior, he made it more practical for routine use. Its combination of simplicity, distribution-free robustness, and strong performance against many alternatives makes it an elegant bridge between runs tests and more specialized serial-dependence methods.

Reference(s): von Neumann (1941); Bartels (1982)

Worked Example

Problem Statement Are 20 consecutive product weights random, or do they show a systematic pattern (trend/serial dependence)? H_0: observations are serially independent; H_1: serial correlation is present.

Data (grams): $\{455, 462, 458, 465, 461, 467, 463, 469, 465, 471, 467, 473, 469, 475, 471, 477, 473, 479, 475, 481\}$.

1. Ranks (midranks for ties)
(Ties occur at 465, 467, 469, 471, 473, 475.)

i	x_i	R_i	Δ_i	i	x_i	R_i	Δ_i
1	455	1.0		11	467	8.5	-4
2	462	4.0	3	12	473	14.5	6
3	458	2.0	-2	13	469	10.5	-4
4	465	6.5	4.5	14	475	16.5	6
5	461	3.0	-3.5	15	471	12.5	-4
6	467	8.5	5.5	16	477	18.0	5.5
7	463	5.0	-3.5	17	473	14.5	-3.5
8	469	10.5	5.5	18	479	19.0	4.5
9	465	6.5	-4	19	475	16.5	-2.5
10	471	12.5	6	20	481	20.0	3.5

2. Successive rank differences
Let $\Delta_i = R_i - R_{i-1}$ for $i = 2, \ldots, 20$. $\bar{R} = (n+1)/2 = 10.5$

$$D = \sum_{i=2}^{20} \Delta_i^2 = 371.5, \quad SS = \sum_{i=1}^{20} (R_i - \bar{R})^2 = 662.0.$$

3. Bartels statistic
Use the standardized rank von Neumann ratio

$$RVN = \frac{D}{SS} = \frac{371.5}{662.0} = 0.5612.$$

4. Normal approximation (large-sample)
Under H_0 (random order), $E[RVN] = 2$ and

$$\text{Var}(RVN) = \frac{4(n-2)(5n^2 - 2n - 9)}{5n(n+1)(n-1)^2} = 0.1853.$$

$$Z = \frac{RVN - 2}{\sqrt{\text{Var}(RVN)}} = \frac{0.5612 - 2}{\sqrt{0.1853}} = -3.44.$$

5. p-value and Decision
Two-sided: $p \approx 0.0006$ (one-sided lower-tail $p \approx 0.0003$). Reject H_0 due to strong evidence of non-randomness.

Practitioner Insight: Bartels's test is the rank-based analog of von Neumann's ratio: it replaces x_i by ranks R_i and applies the same ratio. Assuming no rank ties, the denominator $\sum(R_i - \bar{R})^2 = n(n^2 - 1)/12$ is constant for all permutations, making the test exactly distribution-free under H_0, unlike VN which relies on normality asymptotically. Comparing expectations: $E[\text{RVN}] = 2$ vs. $E[\text{VN}] = 2.1053$. Comparig variances for $n = 20$: Bartel uses $\text{Var}[\text{RVN}] = 4(n-2)(5n^2-2n-9)/(5n(n+1)(n-1)^2) = 0.1853$ vs. von Neumann's $\text{Var}[\text{VN}] = 4n^2(n-2)/[(n+1)(n-1)^3] = 0.1999$. Since $RVN \ll 2$, the pattern is consistent with positive serial dependence / trend-like smoothness (successive ranks stay unusually close).

Common Pitfalls

- **Using Bartels when a specific alternative is known.** Bartels is a general test for sequence non-randomness. If a monotonic trend or oscillatory pattern is hypothesized *a priori*, a more targeted test (e.g., Mann–Kendall, runs, or turning points) is usually more powerful.
- **Treating non-significance as proof of randomness.** Failure to reject indicates insufficient evidence—not confirmation of randomness. Consider sample size, power, and visual diagnostics.
- **Not investigating significant results.** A significant test indicates departure from randomness but not the pattern type. Always plot the data and apply targeted follow-up tests.
- **Ignoring seasonality.** Seasonal structure can produce significance even when the remaining series is otherwise well behaved. Adjust for seasonality before interpreting the result.
- **Applying with very small samples.** Bartels has low power in small samples. Use caution or exact/permutation approaches for short sequences.

Tsay's Test

Omnibus Test for Nonlinearity in the Conditional Mean

RAND.9 | **Category:** Nonlinearity | **Data:** Continuous | **Null:** F-distribution

When to Use

- Suspect nonlinearity in conditional mean of a time series.
- Linear AR model fits poorly or residuals show patterns.
- Screen for threshold, bilinear, or polynomial dynamics.
- Use as omnibus pre-test before fitting TAR or STAR models.
- Exploratory analysis on new or poorly understood series.

Assumptions

- Series is stationary or has been stationarity-transformed.
- A linear AR(p) model has been fit as the null baseline.
- Lag order p is chosen before testing, not data-mined.
- Residuals are free of serial correlation under H_0.
- No structural breaks distorting the AR residuals.

Test Statistic

$$F = \frac{R^2/q}{(1-R^2)/(n-p-q-1)} \;\Big|\; \textbf{(Aux)}: \hat{e}_t = \beta_0 + \sum_{i=1}^{p} \beta_i y_{t-i} + \sum_{i=1}^{p}\sum_{j=i}^{p} \gamma_{ij}\, y_{t-i} y_{t-j} + \varepsilon_t \;\Big|\; \textbf{(Mean)}: y_t = \phi_0 + \sum_{i=1}^{p} \phi_i y_{t-i} + \hat{e}_t$$

Notation: Conditional mean model is AR(p); auxiliary regression tests quadratic nonlinear terms; R^2 is from the auxiliary regression; n is the sample size; under $H_0, F \sim F_{q,\,n-p-q-1}$, $q = p(p+1)/2$

Plain English: Tsay's test asks whether a linear AR model fully captures the dynamics of a time series, or whether the relationship between past and present values is nonlinear. After fitting a linear AR model, the residuals are regressed on squared and cross-product terms of the lagged observations. If those quadratic terms are jointly significant, the AR model leaves systematic structure unexplained, suggesting threshold effects, bilinear interactions, or other nonlinear mean dynamics.

⚠ Practitioner Warning

Squared and cross-product terms are often highly collinear, which can reduce power against specific alternatives. Lag order p must be chosen by the analyst; there is no standard rule, and results can be sensitive to this choice. In short series, even moderate p can strain degrees of freedom. Rejection indicates that some form of nonlinearity is present but does not pinpoint whether it is threshold, bilinear, or polynomial — follow-up with targeted tests or model fitting is always needed.

Hypotheses

- H_0**:** Linear AR model adequate (no 2nd order nonlinearity).
- H_1**:** Inadequate model; nonlinear in the conditional mean.

Decision Rule: Reject H_0 if $F > F_{\alpha,q,n-p-q-1}$ or p-value $< \alpha$. $AR(p) \Rightarrow q = p(p+1)/2$ nonlinear terms.

Software References [Python, R, SAS]

Python	R	SAS
`Not Native (Manual)`	`TSA::Tsay.test()`	`Not Native (Manual)`

Additional Notes

Tsay (1986) developed this procedure to detect general second-order nonlinearity in time-series models, including bilinear, threshold, and polynomial-type dynamics. After fitting a linear AR(p) model, the residuals are regressed on quadratic functions of the lagged observations: squared terms y_{t-i}^2 and cross-products $y_{t-i}y_{t-j}$ for $i < j$. Tsay orthogonalizes these regressors against the AR fitted values before running the auxiliary regression, ensuring the F-test is not contaminated by the linear component already captured. Under lag order p, the number of nonlinear regressors is $q = p(p+1)/2$; for example, $q = 6$ when $p = 3$. Subsets of terms may be tested separately — squares only or cross-products only — to diagnose the specific form of nonlinearity and guide selection among TAR, STAR, or bilinear follow-up models.

Miscellanea: Tsay (1986) published this test in *Biometrika*, a statistics journal better known for survival analysis and experimental design than for time-series econometrics. The appearance there signalled that nonlinear dynamics in autoregressive models were becoming a concern across disciplines, not just in economics. The paper predates the widespread adoption of TAR and STAR models and is one of the earliest proposals for a fully general omnibus screen for conditional mean nonlinearity.

Reference(s): Tsay (1986); Tong (1990); Teräsvirta (1994)

Worked Example

Problem Statement Twenty observations of a demeaned stationary series y_t are used to fit an AR(2) baseline model. Apply Tsay's F-test at $\alpha = 0.05$ to assess if the conditional mean contains nonlinear structure. H_0: AR(2) is adequate (linear cond. mean). H_1: nonlinear structure exists. Data at the end of the chapter.

1. Fit AR(2) and Obtain Residuals
Regress y_t on $(1, y_{t-1}, y_{t-2})$ for $t = 3, \dots, 20$ ($n = 18$ obs). OLS gives:

$$\hat{y}_t = 0.440 - 1.028\, y_{t-1} - 0.519\, y_{t-2}.$$

Residuals $\hat{e}_t = y_t - \hat{y}_t$:

t	$\hat{e}_t$	t	$\hat{e}_t$	t	$\hat{e}_t$
3	0.866	9	-0.750	15	-0.197
4	-0.163	10	0.765	16	-0.592
5	-0.140	11	-0.114	17	0.515
6	0.559	12	-0.189	18	-0.271
7	-0.656	13	0.908	19	-0.290
8	-0.812	14	0.003	20	0.557

2. Form Cross-Product Matrix
For $p = 2$, there are $q = p(p+1)/2 = 3$ distinct cross-products per observation: $\mathbf{z}_t = (y_{t-1}^2,\ y_{t-1}y_{t-2},\ y_{t-2}^2)^\top$. Selected rows of $\mathbf{Z}$:

t	y_{t-1}^2	$y_{t-1}y_{t-2}$	y_{t-2}^2
3	0.090	−0.240	0.640
4	1.440	−0.360	0.090
5	0.640	−0.960	1.440
20	0.090	−0.180	0.360

3. Orthogonalize Cross-Products
Regress each column of $\mathbf{Z}$ on the AR(2) regressors $\mathbf{X} = (1, y_{t-1}, y_{t-2})$ and collect the residuals $\hat{\mathbf{V}}$, representing the component of each cross-product unexplained by the linear AR structure. Selected rows of $\hat{\mathbf{V}}$ shown below:

t	$\hat{v}_{t,1}$	$\hat{v}_{t,2}$	$\hat{v}_{t,3}$
3	−0.680	0.270	−0.107
4	0.613	−0.464	−0.125
5	−0.123	−0.258	0.528
20	−0.306	0.219	−0.077

4. Auxiliary Regression and F-Statistic
Regress $\hat{e}_t$ on $\hat{\mathbf{V}}$ (no intercept; $\hat{e}_t$ is already orthogonal to $\mathbf{X}$): $\text{df}_1 = q = 3, \text{df}_2 = n - q - 1 = 14$

$$\text{SS}_{\text{reg}} = 3.111, \quad \text{SS}_{\text{err}} = 2.229, \quad F = \frac{\text{SS}_{\text{reg}}/3}{\text{SS}_{\text{err}}/14} = 6.51.$$

5. Decision
Under H_0: $F \sim F(3, 14)$.

α	$F_\alpha(3, 14)$	Decision
5%	3.344	**Reject** (6.51 > 3.344)
1%	5.564	**Reject** (6.51 > 5.564)

$p = 0.006$. Reject H_0: the AR(2) model is inadequate; nonlinear structure is present in the conditional mean.

Practitioner Insight: Tsay's test differs fundamentally from Engle's ARCH test. Engle regresses squared residuals on lagged squared residuals to detect nonlinearity in the conditional variance, whereas Tsay regresses AR residuals $\hat{e}_t$ on orthogonalized cross-products of lagged observations to detect nonlinearity in the conditional mean, such as threshold, bilinear, or polynomial structure. The orthogonalization step is crucial because it removes linear predictability already captured by the AR terms; without it, the test statistic would be inflated. A significant result indicates that some nonlinear function of the lags predicts the residuals, but it does not identify the form of nonlinearity. Tsay's arranged autoregression test is a common follow-up for detecting threshold effects and estimating the delay parameter.

Common Pitfalls

- **Confusing rejection with a specific form of nonlinearity.** Tsay's is omnibus — rejection signals 2nd-order nonlinearity but does not distinguish threshold, bilinear, or polynomial structure. Follow-up with targeted tests is required.
- **Choosing lag order p without justification.** Results can be sensitive to p, and there is no standard selection rule. Reporting results across a range of plausible values guards against cherry-picking a significant outcome.
- **Applying the test to a short series with large p.** The number of nonlinear regressors grows as $q = p(p+1)/2$, consuming degrees of freedom rapidly. With small n, even moderate p can make the F-test unreliable.
- **Testing without checking AR model's adequacy.** Tsay's auxiliary regression assumes the AR residuals are free of serial correlation. Remaining linear structure in the residuals inflates the F-statistic and produces spurious rejections.
- **Mistaking Tsay's for an ARCH test.** The test targets nonlinearity in the conditional mean, not the conditional variance. Using it as a substitute for Engle's LM test is a misapplication that can lead to incorrect model selection.

BDS Test

Test for Independence Based on Correlation Integral

RAND.10 | **Category:** Randomness Tests | **Data:** Continuous | **Null:** Normal (asymptotic)

When to Use

- Test for IID in time series data.
- Detect nonlinear dependence, deterministic chaos, or hidden structures (e.g: residuals). Diagnostic for model adequacy after fitting time series models (ARMA, GARCH).

Assumptions

- Data are a real-valued, stationary time series.
- Does not require normality or linearity.
- $n \geq 500$ for reliable normal approximation.

Test Statistic

$$W_{m,n}(\epsilon) = \sqrt{n}\frac{C_{m,n}(\epsilon) - [C_{1,n}(\epsilon)]^m}{\sigma_{m,n}(\epsilon)}, \quad C_{m,n}(\epsilon) = \frac{2}{n_m(n_m - 1)} \sum_{s<t} \mathbf{1}(\|X_s^m - X_t^m\| < \epsilon).$$

Under H_0, $W_{m,n}(\epsilon) \approx N(0,1)$. $X_t^m = (x_t, x_{t+1}, \ldots, x_{t+m-1})$ are m-histories, $n_m = n - m + 1$.

Notation: $C_{m,n}$ is correlation integral, ϵ is proximity tolerance, m is embedding dimension, $\sigma_{m,n}(\epsilon)$ is standard error

Plain English: The BDS test evaluates whether closeness of points in an m-dimensional embedding occurs as often as expected under independence. For IID data, the probability that two m-vectors are within distance ϵ should equal the product of one-dimensional probabilities. Deviations from this factorization imply dependence. The procedure embeds the series, counts nearby vector pairs, and compares to the independence benchmark. The test is omnibus: it rejects any violation of IID, including linear autocorrelation, nonlinear dependence, conditional heteroskedasticity, or structural breaks.

⚠ Practitioner Warning

Computational cost is $O(n^2)$, and reliable asymptotic inference typically requires large samples. The test assumes stationarity, so non-stationary series should be detrended, differenced, or otherwise transformed first. Rejection does not by itself imply chaos. In practice, BDS is most informative on residuals from an appropriate linear or volatility model.

Hypotheses

- H_0: Data are IID (no serial dependence, linear or nonlinear).
- H_1: Dependence structure present (autocorrelation, nonlinear dynamics, or deterministic patterns).

Decision Rule: Reject H_0 if $|W_{m,n}(\epsilon)| > z_{\alpha/2}$ or p-value $< \alpha$ for two-sided test. Positive values more common; indicate clustering or dependence.

Software References [Python, R, SAS]

`arch.unitroot.BDS`	`tseries::bds.test`	`PROC AUTOREG`

Additional Notes

Brock, Dechert, and Scheinkman (1987, 1996) proposed the BDS test to detect nonlinear dependence and possible low-dimensional chaos in time series. The method embeds the scalar series into m-dimensional delay vectors $X_t^m = (x_t, \ldots, x_{t+m-1})$ and computes the correlation integral $C_{m,n}(\epsilon)$—the proportion of vector pairs within distance ϵ (often using the maximum norm). Under independence, $C_{m,n}(\epsilon) \approx [C_{1,n}(\epsilon)]^m$; the BDS statistic measures standardized deviations from this relation. Typical choices are $m = 2$–5 and ϵ between 0.5σ and 2σ. Larger m increases power but requires larger n. Robust practice reports results across multiple (m, ϵ) combinations. Applications include testing model residuals (e.g., ARMA/GARCH), checking random number generators, and assessing nonlinear structure in economic and financial data. The test rejects IID behavior but does not identify the source (chaos, structural breaks, etc.).

Miscellanea: The BDS test emerged during the 1980s surge of interest in chaos theory. Brock, Dechert, and Scheinkman introduced it to examine whether economic time series exhibited deterministic chaos rather than purely stochastic behavior. Early applications sparked debate—significant results often reflected model misspecification rather than true chaos. Over time, BDS became a standard diagnostic tool for detecting residual dependence beyond linear models.

Reference(s): Brock et al. (1987, 1996); LeBaron (1994)

Worked Example

Research question. A financial analyst suspects daily stock returns exhibit nonlinear dependence (e.g., ARCH effects) that linear autocorrelation tests miss. We apply the BDS test to check the i.i.d. assumption. Choose BDS parameters: $m = 2$ (embedding dimension), $\epsilon = 1.0$ (on standardized scale, i.e., $1.0\,s$).
Data. Daily returns (in %; raw and standardized): $n = 10,\ \bar{x} = 0.20,\ s = 0.577$.

i	1	2	3	4	5	6	7	8	9	10
x_i	0.5	-0.3	0.8	-0.2	1.2	-0.5	0.4	-0.1	0.6	-0.4
z_i	0.52	-0.87	1.04	-0.69	1.73	-1.21	0.35	-0.52	0.69	-1.04

Due to space limitations, it is not practical to present all intermediate calculations for the BDS test here. The full step-by-step computation is provided in the next few pages.

$C_{1,n}(\epsilon)$ **for** $m = 1$.

$$C_{1,n}(\epsilon) = \frac{2 \times 18}{10 \times 9} = 0.400.$$

$C_{2,n_m}(\epsilon)$ **for** $m = 2$. Here, $n_m = n - m + 1$

$$C_{2,n_m}(\epsilon) = \frac{2 \times 11}{9 \times 8} = 0.306$$

Fourth-order integral $K(\epsilon)$.

$$K(\epsilon) = \frac{6 \times 7}{10 \times 9 \times \times 8 \times 7} = 0.00833.$$

Asymptotic variance.

$$V_2(\epsilon) = 0.0920$$

$$\sigma_{2,10}(1.0) = \sqrt{V_2(1.0)} \approx 0.303.$$

Test statistic:

$$W_{2,10}(1.0) = \frac{\sqrt{10} \times (0.306 - (0.400)^2)}{0.303} = 1.52$$

Critical value (two-tailed test, $\alpha = 0.05$): $z_{0.025} = 1.96$

p-value: $p = 2 \times P(Z > 1.52) = 2 \times 0.0640 = 0.1280$

Decision: Since $|1.52| < 1.96$ (or equivalently, $p = 0.1280 > 0.05$), we fail to reject H_0 at the 5% significance level.

Conclusion: Insufficient evidence to reject i.i.d. at 5% significance level. However, the conclusion is unreliable due to extremely small sample size.

Common Pitfalls

- **Using BDS with small samples.** BDS relies on asymptotic theory and requires large n (typically $n \geq 500$, preferably > 1000). Small samples produce unstable p-values and unreliable inference.
- **Equating significance with chaos.** A significant BDS test indicates deviation from IID—not chaos specifically. It may reflect linear autocorrelation, GARCH effects, structural breaks, or other dependence.
- **Testing only one** (m, ϵ) **combination.** Results depend on embedding dimension m and radius ϵ. Examine multiple combinations and report all; isolated significance may be spurious.
- **Failing to pre-whiten the series.** BDS detects any dependence. Remove linear structure first (e.g., ARMA), then apply BDS to residuals to test for remaining nonlinear dependence.
- **Stopping at significance.** BDS does not identify the source of dependence. Follow up with model diagnostics (e.g., GARCH, break tests) before drawing substantive conclusions.

Additional Formulas

$$K(\epsilon) = \frac{6}{n(n-1)(n-2)(n-3)} \sum_{1 \le i < j < k < \ell \le n} \prod_{q \in \{j,k,\ell\}} I(|x_i - x_q| < \epsilon)$$

$$V_m(\epsilon) = 4\left[K^m + 2\sum_{j=1}^{m-1} K^{m-j}C^{2j} + (m-1)^2 C^{2m} - m^2 K C^{2(m-1)}\right], \quad \text{where } C = C_{1,n}(\epsilon), \quad \sigma_{m,n}(\epsilon) = \sqrt{V_m(\epsilon)}$$

XVII. RAND — Detailed Worked Example: BDS Test on Stock Returns

Problem Statement A financial analyst suspects that daily stock returns exhibit nonlinear dependence (possibly ARCH effects) that linear autocorrelation tests fail to detect. We apply the BDS test to investigate. Choose BDS parameters: $m = 2$ (embedding dimension), $\epsilon = 1.0$ (on standardized scale, i.e., $1.0\,s$).

Data. Daily returns (in %; raw and standardized): $n = 10$, $\bar{x} = 0.20$, $s = 0.577$.

i	1	2	3	4	5	6	7	8	9	10
x_i	0.5	-0.3	0.8	-0.2	1.2	-0.5	0.4	-0.1	0.6	-0.4
z_i	0.52	-0.87	1.04	-0.69	1.73	-1.21	0.35	-0.52	0.69	-1.04

1. Compute $C_{1,n}(\epsilon)$ — Correlation Integral at $m = 1$.

Count pairs (z_i, z_j) with $i < j$ where $|z_i - z_j| < 1.0$. Total number of pairs: $\binom{10}{2} = 45$. Distance matrix:

	z_1	z_2	z_3	z_4	z_5	z_6	z_7	z_8	z_9	z_{10}
$z_1 = 0.52$	—	1.39	0.52	1.21	1.21	1.73	0.17	1.04	0.17	1.56
$z_2 = -0.87$		—	1.91	0.18	2.60	0.34	1.22	0.35	1.56	0.17
$z_3 = 1.04$			—	1.73	0.69	2.25	0.69	1.56	0.35	2.08
$z_4 = -0.69$				—	2.42	0.52	1.04	0.17	1.38	0.35
$z_5 = 1.73$					—	2.94	1.38	2.25	1.04	2.77
$z_6 = -1.21$						—	1.56	0.69	1.90	0.17
$z_7 = 0.35$							—	0.87	0.34	1.39
$z_8 = -0.52$								—	1.21	0.52
$z_9 = 0.69$									—	1.73
$z_{10} = -1.04$										—

Pairs within $\epsilon = 1.0$ are shown in red. Total pairs within ϵ: 18.

$$C_{1,10}(1.0) = \frac{2}{n(n-1)} \sum_{1 \le i < j \le n} \mathbf{1}\{|z_i - z_j| < \epsilon\} = \frac{2 \times 18}{10 \times 9} = \frac{34}{90} = 0.400$$

2. Construct m-Histories (Phase Space Embedding).

For $m = 2$, create 2-dimensional lag vectors: $n_m = n - m + 1 = 10 - 2 + 1 = 9$. The m-histories are:

$$X_1^2 = (z_1, z_2) = (0.52, -0.87), \quad X_2^2 = (z_2, z_3) = (-0.87, 1.04), \quad X_3^2 = (z_3, z_4) = (1.04, -0.69),$$
$$X_4^2 = (z_4, z_5) = (-0.69, 1.73), \quad X_5^2 = (z_5, z_6) = (1.73, -1.21), \quad X_6^2 = (z_6, z_7) = (-1.21, 0.35),$$
$$X_7^2 = (z_7, z_8) = (0.35, -0.52), \quad X_8^2 = (z_8, z_9) = (-0.52, 0.69), \quad X_9^2 = (z_9, z_{10}) = (0.69, -1.04)$$

3. Compute $C_{2,9}(\epsilon)$ — Correlation Integral at $m = 2$

Count pairs of 2-histories (X_s^2, X_t^2) with $s < t$ where:

$$\|X_s^2 - X_t^2\| = \max\{|z_s - z_t|, |z_{s+1} - z_{t+1}|\} < 1.0$$

Total number of pairs: $\binom{9}{2} = 36$

Example distance calculation:

Pair (X_1^2, X_2^2)**:** $X_1^2 = (0.52, -0.87), \quad X_2^2 = (-0.87, 1.04)$

$$\|X_1^2 - X_2^2\| = \max\{|0.52 - (-0.87)|, |-0.87 - 1.04|\} = \max\{1.39, 1.91\} = 1.91 > 1.0 \quad \text{(Not within } \epsilon)$$

Complete enumeration of all 36 pairs:

s	t	X_s^2	X_t^2	$\|X_s^2 - X_t^2\|$	s	t	X_s^2	X_t^2	$\|X_s^2 - X_t^2\|$
1	2	(0.52, −0.87)	(−0.87, 1.04)	1.91	3	7	(1.04, −0.69)	(0.35, −0.52)	0.69
1	3	(0.52, −0.87)	(1.04, −0.69)	0.52	3	8	(1.04, −0.69)	(−0.52, 0.69)	1.56
1	4	(0.52, −0.87)	(−0.69, 1.73)	2.60	3	9	(1.04, −0.69)	(0.69, −1.04)	0.35
1	5	(0.52, −0.87)	(1.73, −1.21)	1.21	4	5	(−0.69, 1.73)	(1.73, −1.21)	2.94
1	6	(0.52, −0.87)	(−1.21, 0.35)	1.73	4	6	(−0.69, 1.73)	(−1.21, 0.35)	1.38
1	7	(0.52, −0.87)	(0.35, −0.52)	0.35	4	7	(−0.69, 1.73)	(0.35, −0.52)	2.25
1	8	(0.52, −0.87)	(−0.52, 0.69)	1.56	4	8	(−0.69, 1.73)	(−0.52, 0.69)	1.04
1	9	(0.52, −0.87)	(0.69, −1.04)	0.17	4	9	(−0.69, 1.73)	(0.69, −1.04)	2.77
2	3	(−0.87, 1.04)	(1.04, −0.69)	1.91	5	6	(1.73, −1.21)	(−1.21, 0.35)	2.94
2	4	(−0.87, 1.04)	(−0.69, 1.73)	0.69	5	7	(1.73, −1.21)	(0.35, −0.52)	1.38
2	5	(−0.87, 1.04)	(1.73, −1.21)	2.60	5	8	(1.73, −1.21)	(−0.52, 0.69)	2.25
2	6	(−0.87, 1.04)	(−1.21, 0.35)	0.69	5	9	(1.73, −1.21)	(0.69, −1.04)	1.04
2	7	(−0.87, 1.04)	(0.35, −0.52)	1.56	6	7	(−1.21, 0.35)	(0.35, −0.52)	1.56
2	8	(−0.87, 1.04)	(−0.52, 0.69)	0.35	6	8	(−1.21, 0.35)	(−0.52, 0.69)	0.69
2	9	(−0.87, 1.04)	(0.69, −1.04)	2.08	6	9	(−1.21, 0.35)	(0.69, −1.04)	1.90
3	4	(1.04, −0.69)	(−0.69, 1.73)	2.42	7	8	(0.35, −0.52)	(−0.52, 0.69)	1.21
3	5	(1.04, −0.69)	(1.73, −1.21)	0.69	7	9	(0.35, −0.52)	(0.69, −1.04)	0.52
3	6	(1.04, −0.69)	(−1.21, 0.35)	2.25	8	9	(−0.52, 0.69)	(0.69, −1.04)	1.73

Pairs within $\epsilon = 1.0$ are shown in red. Total pairs within ϵ: 11.

$$C_{2,9}(1.0) = \frac{2}{n_m(n_m - 1)} \sum_{1 \le s < t \le n_m} \mathbf{1}\{\|X_s^2 - X_t^2\|_\infty < \epsilon\} = \frac{2 \times 11}{9 \times 8} = \frac{24}{72} = 0.306$$

4. Compute $K(\epsilon)$ — Fourth-Order Correlation Integral

Count 4-tuples (i, j, k, ℓ) with $i < j < k < \ell$ where all three pairs centered on x_i are within distance ϵ:

$$|z_i - z_j| < 1.0 \qquad |z_i - z_k| < 1.0 \qquad |z_i - z_\ell| < 1.0$$

Total number of 4-tuples: $\binom{10}{4} = 210$

Example check for 4-tuple $(1, 3, 7, 9)$**:**

Center on $i = 1$, check distances to $j = 3, k = 7, \ell = 9$:

$|z_1 - z_3| = |0.52 - 1.04| = 0.52 < 1.0$ ✓, $\quad |z_1 - z_7| = |0.52 - 0.35| = 0.17 < 1.0$ ✓, $\quad |z_1 - z_9| = |0.52 - 0.69| = 0.17 < 1.0$ ✓

This 4-tuple qualifies!

Complete enumeration finds the following qualifying 4-tuples:

- $(1, 3, 7, 9)$: Center z_1, distances $0.52, 0.17, 0.17$
- $(2, 4, 6, 8)$: Center z_2, distances $0.18, 0.34, 0.35$
- $(2, 4, 6, 10)$: Center z_2, distances $0.18, 0.34, 0.17$
- $(2, 4, 8, 10)$: Center z_2, distances $0.18, 0.35, 0.17$

- $(2, 6, 8, 10)$: Center z_2, distances $0.34, 0.35, 0.17$
- $(3, 5, 7, 9)$: Center z_3, distances $0.69, 0.69, 0.35$
- $(4, 6, 8, 10)$: Center z_4, distances $0.52, 0.17, 0.35$

Total qualifying 4-tuples: 7

$$K(1.0) = \frac{6}{n(n-1)(n-2)(n-3)} \sum_{1\le i<j<k<\ell\le n} \mathbf{1}\{|z_i - z_j| < \epsilon,\ |z_i - z_k| < \epsilon,\ |z_i - z_\ell| < \epsilon\} = \frac{6 \times 7}{10 \times 9 \times 8 \times 7} = \frac{18}{5040} = 0.00833$$

5. Compute Asymptotic Variance

For $m = 2$, the variance formula simplifies to:

$$V_2(\epsilon) = 4\left[K^2 - 2KC^2 + C^4\right] = 0.0920, \qquad \sigma_{2,10}(1.0) = \sqrt{0.0920} = 0.303$$

where $C = C_{1,10}(1.0) = 0.400$ and $K = K(1.0) = 0.00833$.

6. Compute BDS Test Statistic

Test statistic:

$$W_{2,10}(1.0) = \frac{\sqrt{n}\left[C_{2,n}(\epsilon) - C_{1,n}(\epsilon)^2\right]}{\sigma_{2,n}(\epsilon)} = \frac{\sqrt{10} \times (0.306 - (0.400)^2)}{0.303} = 1.52$$

Critical value (two-tailed test, $\alpha = 0.05$): $z_{0.025} = 1.96$

p-value: $p = 2 \times P(Z > 1.52) = 2 \times 0.0640 = 0.1280$

Decision: Since $|1.52| < 1.96$ (or equivalently, $p = 0.1280 > 0.05$), we fail to reject H_0 at the 5% significance level.

Conclusion: Insufficient evidence to reject i.i.d. at 5% significance level. However, the conclusion is unreliable due to extremely small sample size.

7. Summary of Calculations

Quantity	Value
Sample size (n)	10
Embedding dimension (m)	2
Distance threshold (ϵ)	1.0
Number of m-histories (n_m)	9
Pairs within ϵ at $m = 1$	18 out of 45
$C_{1,10}(1.0)$	0.400
Pairs within ϵ at $m = 2$	11 out of 36
$C_{2,9}(1.0)$	0.306
Qualifying 4-tuples	7 out of 210
$K(1.0)$	0.00833
Asymptotic variance $V_2(1.0)$	0.0920
Standard deviation $\sigma_{2,10}(1.0)$	0.303
BDS test statistic $W_{2,10}(1.0)$	1.52
Critical value ($\alpha = 0.05$)	1.96
p-value	0.1280
Decision	Fail to reject H_0

In practice: With a realistic sample size ($n \ge 500$), the analyst would:

1. Test multiple embedding dimensions ($m = 2, 3, 4, 5$)
2. Test multiple distance thresholds ($\epsilon = 0.5\sigma, 1.0\sigma, 1.5\sigma, 2.0\sigma$)
3. Examine consistency of rejections across the (m, ϵ) grid
4. Apply multiple testing corrections (Bonferroni or FDR)
5. Use optimized algorithms (fast BDS implementations) for computational efficiency

XVII. RAND — Raw Data

Data used in selected worked examples are provided here to preserve the book's strict two-page-per-test format.

[RAND.9] Tsay's Test

Problem Statement Twenty observations of a demeaned stationary series y_t are given below. Fit an AR(2) baseline and apply Tsay's (1986) F-test at $\alpha = 0.05$ to assess whether the conditional mean contains nonlinear structure.

t	1	2	3	4	5	6	7	8	9	10
y_t	0.8	−0.3	1.2	−0.8	0.5	0.9	−1.4	0.6	−0.2	1.1
t	11	12	13	14	15	16	17	18	19	20
y_t	−0.7	0.4	1.3	−1.1	0.7	−0.3	0.9	−0.6	0.3	1.0

H_0: AR(2) is adequate (linear conditional mean). H_1: nonlinear structure present.

Eulerian Numbers and the Exact Distribution of the Difference-Sign Test Statistic

The difference-sign test counts the number of positive first differences in a sequence—the number of positions i where $y_{i+1} > y_i$—and uses this count S as a test statistic for randomness. Under the null hypothesis of no serial dependence, all $n!$ orderings of the sequence are equally likely, and the exact distribution of S is governed by the *Eulerian numbers*, denoted $\langle{n \atop k}\rangle$. The Eulerian number $\langle{n \atop k}\rangle$ counts the number of permutations of $\{1, 2, \dots, n\}$ having exactly k *ascents*—positions where the next element exceeds the current one—and therefore $P(S = k) = \langle{n \atop k}\rangle/n!$ exactly. These numbers were studied by Leonhard Euler in the eighteenth century in the context of series expansions; they satisfy the recurrence

$$\left\langle{n \atop k}\right\rangle = (k+1)\left\langle{n-1 \atop k}\right\rangle + (n-k)\left\langle{n-1 \atop k-1}\right\rangle,$$

with $\langle{n \atop 0}\rangle = 1$ for all $n \geq 1$. The resulting distribution is symmetric about $(n-1)/2$, unimodal, and converges rapidly to normality, which is why the large-sample $N\left(\frac{n-1}{2}, \frac{n+1}{12}\right)$ approximation performs well even for moderate n.

What makes the Eulerian connection more than a combinatorial curiosity is that it provides an exact, closed-form null distribution requiring no distributional assumptions beyond exchangeability—a considerably weaker requirement than independence or normality. For small samples where the normal approximation may be unreliable, exact p-values can be computed directly from the Eulerian numbers, and tables for small n have been published for this purpose. The same Eulerian framework extends to the runs-up-and-down test and related nonparametric trend diagnostics, making it a unifying combinatorial structure across several tests (e.g: RAND, NPAR) of this compendium.

Moore and Wallis (1943): The Null Hypothesis and Variance Formula for the Difference-Sign Test

The difference-sign test was introduced by Geoffrey H. Moore and W. Allen Wallis in their 1943 *JASA* paper "Time Series Significance Tests Based on Signs of Differences." The test counts T, the number of positive first differences in a sequence of N observations, and compares it to its null distribution. A subtle but consequential choice underlies the test's construction: the variance of T under the null takes one of two forms depending on what the null hypothesis actually asserts. If the null is that the *observations themselves* are in random order—that is, the N values are exchangeable with no serial structure—then the correct variance is $(N+1)/12$. This is the formula derived from the Eulerian number distribution (see the preceding note) and treats all $N!$ orderings of the original values as equally probable. If instead the null is that the *first differences* $d_i = y_{i+1} - y_i$ are independent and identically distributed with $P(d_i > 0) = \frac{1}{2}$, then the T differences are Bernoulli trials, their sum has variance $N(1/2)(1/2)$, and adjusting for the fact that there are $N - 1$ differences gives variance $(N-1)/4$—a substantially larger quantity that treats each sign as independent of all others.

The distinction matters in practice because the two null hypotheses are not equivalent, and neither is a special case of the other. In a genuine time series setting, successive differences are not independent: a large positive value y_t inflates the chance of a negative difference $d_t = y_{t+1} - y_t$ and a positive difference $d_{t+1} = y_{t+2} - y_{t+1}$ through the shared term y_{t+1}, inducing negative lag-one autocorrelation in the difference sequence even under exchangeability of the original series. The variance $(N+1)/12$ correctly accounts for this induced dependence structure by deriving from the joint distribution of ascents under the exchangeability assumption, rather than treating signs as independent. Assuming independence of differences—and thus using $(N-1)/4$—overstates the variance and produces a conservative test that underrejects. Moore and Wallis's contribution was to recognize that for ordered time series data, the appropriate null is randomness of the original sequence, not independence of differences, and to derive the correct variance accordingly. This makes $(N+1)/12$ the standard formula for the difference-sign test as it appears in virtually all subsequent treatments of nonparametric time series analysis.

Moore and Wallis (1941): The Turning Points Test and Its Distinction from the Difference-Sign Test

The turning points test was introduced by W. Allen Wallis and Geoffrey H. Moore in their 1941 *JASA* paper "A Significance Test for Time Series Analysis" (Vol. 36, pp. 401–409), two years before their better-known 1943 paper on the difference-sign test. The test counts U, the number of *turning points* in a sequence of N observations, where position i is a turning point if $y_{i-1} < y_i > y_{i+1}$ (a local maximum) or $y_{i-1} > y_i < y_{i+1}$ (a local minimum). Under the null hypothesis that the observations are in random order (exchangeability), the expected number of turning points is $\mu_U = 2(N-2)/3$, and the variance is $\sigma_U^2 = (16N-29)/90$. Unlike the difference-sign test, there is no competing variance formula arising from an alternative independence assumption: the exchangeability null is the only natural framing for turning points, so the single-formula situation here contrasts directly with the two-formula ambiguity discussed in the preceding note on the difference-sign test.

The asymptotic distribution of the standardized statistic is $N(0, 1)$ under the same regularity conditions.

A persistent source of confusion in the applied literature is that both the turning points test and the difference-sign test are attributed to "Moore and Wallis," both are nonparametric randomness tests for ordered sequences, and both use sign-based counting — making citations frequently transposed across the two papers in textbooks and secondary sources. The R `trend` package's `wm.test` function correctly cites only the 1941 paper for its implementation of the turning points (phase-frequency) criterion, reflecting accurate attribution. However, in much of the applied literature the label "Moore–Wallis test" appears without qualification, and a reader encountering it cannot determine which test is meant without inspecting the test statistic being computed. The two tests are not interchangeable: the difference-sign test counts ascents and is primarily sensitive to monotone trend in the series, while the turning points test counts local extrema and is more sensitive to oscillatory or cyclical departures from randomness. A series with a strong upward trend will produce few turning points *and* many positive differences, so both tests would reject — but for different reasons and against different alternatives. Conversely, a highly oscillatory series alternating up and down produces a maximum number of turning points while yielding roughly equal numbers of positive and negative differences, so the turning points test would reject while the difference-sign test would not. Awareness of this power profile distinction is essential for choosing between the two tests and for correctly interpreting a significant result from either.

Von Neumann, Dickey–Fuller, and Bartels: A Connected Family of Ratio Tests

John von Neumann derived the ratio test bearing his name in a pair of papers in 1941 and 1942 (*Annals of Mathematical Statistics*), proposing the statistic

$$\delta = \frac{\frac{1}{N-1}\sum_{i=1}^{N-1}(y_{i+1}-y_i)^2}{\frac{1}{N}\sum_{i=1}^{N}(y_i-\bar{y})^2}$$

as a test of the null hypothesis that a sequence of observations is random against the alternative of positive serial autocorrelation. The numerator accumulates squared successive differences — large when adjacent values fluctuate unpredictably, small when the series moves smoothly — while the denominator is the sample variance. Under randomness, $\delta \approx 2$; values substantially below 2 indicate positive autocorrelation. Von Neumann derived the exact mean and variance of δ under normality and established its asymptotic distribution, making this one of the first rigorously derived test statistics for serial dependence in an ordered sequence. The Durbin–Watson statistic (1950, 1951), developed shortly afterward for regression residuals, is algebraically equivalent to δ up to a boundary correction and is best understood as a regression-context repackaging of the same underlying idea.

The connection to the Dickey–Fuller test (1979) is deeper than is generally acknowledged in econometrics pedagogy. Dickey and Fuller reframed the randomness-versus-autocorrelation question in an autoregressive model, testing $H_0\colon \rho = 1$ (unit root, non-stationary) against $H_1 : \rho < 1$ (stationary) via the t-ratio on $\hat{\rho}$ in the regression $\Delta y_t = (\rho - 1)y_{t-1} + \varepsilon_t$. Their genuine contribution was characterizing the non-standard limiting distribution of this t-ratio under the unit root null — a distribution that is not $N(0, 1)$ but depends on functionals of Brownian motion, requiring new critical values. However, the underlying diagnostic intuition — that small mean square successive differences relative to overall variance signal a smooth, autocorrelated, non-random series — is precisely von Neumann's. The Dickey–Fuller test is the unit-root reinterpretation of a question von Neumann had already posed and answered in a more general setting nearly four decades earlier. Practitioners who know Dickey–Fuller but not von Neumann are in the common position of knowing the application without knowing the origin.

R. J. Bartels (1982) extended von Neumann's ratio to a distribution-free setting by replacing the raw observations with their ranks before computing the statistic. Substituting ranks $R_i = \text{rank}(y_i)$ for y_i in δ yields a test whose null distribution does not depend on the underlying population — the normality assumption that von Neumann required is replaced by the weaker assumption of exchangeability. The rank transformation preserves the ratio's sensitivity to serial structure while making the test applicable to non-normal and heavy-tailed series. Bartels also showed that the rank-based statistic has competitive asymptotic relative efficiency against the original von Neumann ratio under normality, so little is lost by ranking even when normality holds. Together, the three tests — von Neumann (1941), Durbin–Watson (1950), and Bartels (1982) — form a coherent family of ratio-based diagnostics for serial dependence, each addressing the same core question under progressively weaker distributional assumptions.

Three Nonparametric Trend Tests: Difference-Sign, Mann–Kendall, and Cox–Stuart

The difference-sign test, the Mann–Kendall test, and the Cox–Stuart test all appear in this chapter as nonparametric procedures for detecting monotone trend in a sequence, and a reader could reasonably ask whether three tests are needed for what appears to be the same purpose. The answer lies in how much of the sequence each test actually uses. The difference-sign test compares only *adjacent* pairs: it counts the number of positions where $y_{i+1} > y_i$ and tests whether this count is consistent with randomness. Each observation participates in at most two comparisons. The Mann–Kendall test compares *all* $\binom{n}{2}$ pairs in the sequence, counting concordant pairs (where $y_j > y_i$ for $j > i$) against discordant pairs, and is directly equivalent to testing $H_0\colon \tau = 0$ using Kendall's rank correlation between the observations and their time indices. The Cox–Stuart test occupies a deliberate middle ground: it divides the sequence into an early third and a late third, forming $\lfloor n/3 \rfloor$ matched pairs $(y_i, y_{i+\lceil n/3 \rceil})$, and applies the sign test to those pairs only. It uses more information than the difference-sign test (comparing across the full span of the series rather than between neighbors) but far less than Mann–Kendall (examining only a small number of strategically chosen pairs rather than all possible ones).

This hierarchy of information use translates directly into a power ordering. Under most smooth monotone alternatives, Mann–Kendall is the most powerful of the three, Cox–Stuart is intermediate, and the difference-sign test is the weakest. The difference-sign test is not simply dominated, however: because it is sensitive to *local* consecutive movement, it can detect short-run directional runs that Mann–Kendall dilutes by averaging over the full pairwise structure. A series with a strong local trend confined to one segment, surrounded by flat or noisy sections, may produce a significant difference-sign result while Mann–Kendall, weighting all pairs equally, fails to reject. Conversely, a series with a weak but globally consistent trend — one that persistently, if slightly, increases over the full length of the record — will accumulate concordant pairs across the entire $\binom{n}{2}$ comparison set and give Mann–Kendall its best opportunity, while the difference-sign test, seeing only adjacent noise, may miss it entirely. Cox–Stuart is best understood as a screening tool: it is simpler to compute and explain than Mann–Kendall, sacrifices some power, but focuses specifically on whether the end of the series is systematically higher (or lower) than the beginning — a natural operationalization of what practitioners often mean informally by "trend."

A subtle distinction in the null hypotheses also deserves attention. The difference-sign and Cox–Stuart tests are framed as tests of *randomness* of the sequence: the null is that all orderings of the observations are equally likely (exchangeability), and trend is just one of many departures from randomness that could produce a significant result. The Mann–Kendall test is framed more specifically as a test of *monotone association* between the observations and time: the null is $\tau = 0$, with the alternative directional ($\tau > 0$ for upward trend, $\tau < 0$ for downward). In practice this distinction rarely affects conclusions, but it matters for interpretation: a non-rejection under Mann–Kendall specifically means no evidence of monotone rank association with time, while a non-rejection under the difference-sign test means only that the pattern of ups and downs is consistent with randomness — a weaker statement that does not rule out non-monotone structure. For most applied purposes where the question is simply "does this series trend upward or downward?", Mann–Kendall is the preferred choice on both power and interpretive grounds. The difference-sign test is most useful as a quick diagnostic for local directional consistency, and Cox–Stuart as a transparent, easy-to-explain alternative when the audience is unfamiliar with rank correlation methods.

Tsay's Nonlinearity Test: Two Papers, Two Procedures

Overview and the Confusion

Ruey Tsay published two related but distinct nonlinearity tests that are frequently conflated in the applied literature:

- **Tsay (1986)**, *Biometrika* 73(2), 461–466: a general nonlinearity test based on Tukey's one-degree-of-freedom idea, generalized to the time series setting. Tests whether any nonlinear structure is present.
- **Tsay (1989)**, *JASA* 84(405), 231–240: the *arranged autoregression* test, specifically designed to detect threshold (TAR/SETAR) nonlinearity and identify the threshold variable.

Most R implementations (`TSA::Tsay.test`, `NTS::Tsay`) implement the **1986** paper. The 1989 procedure is implemented separately as `@TSAYTEST` in RATS and via custom code in R. The two tests answer different questions and should not be substituted for each other.

Tsay (1986): General Nonlinearity Test

Motivation. Tukey (1949) proposed a one-degree-of-freedom test for nonadditivity in two-way tables by augmenting the model with a single quadratic term. Tsay (1986) generalises this idea to stationary time series: if a linear AR(p) model is correct, no function of its fitted values should predict the residuals. Adding cross-product terms of the lagged variables as regressors tests for any nonlinear structure consistent with a bilinear or polynomial alternative.

Procedure. Let $y_1, \ldots, y_T$ be a stationary time series.

Step 1. **Fit the linear AR(p) model.** Regress y_t on $(y_{t-1}, \ldots, y_{t-p})$ by OLS. Obtain residuals $\hat{e}_t$.

Step 2. **Form the cross-product matrix.** For $t = p+1, \ldots, T$, let $\mathbf{x}_t = (y_{t-1}, \ldots, y_{t-p})^\top$. Form all $q = p(p+1)/2$ distinct cross-products:

$$\mathbf{z}_t = \operatorname{vech}(\mathbf{x}_t \mathbf{x}_t^\top) = (y_{t-1}^2, y_{t-1}y_{t-2}, \ldots, y_{t-p}^2)^\top.$$

Step 3. **Regress cross-products on lags.** Regress each component of $\mathbf{z}_t$ on $\mathbf{x}_t$ and collect the residuals $\hat{\mathbf{v}}_t$ (the component of $\mathbf{z}_t$ orthogonal to the linear lags). This purges linear dependence from the cross-products.

Step 4. **Regress residuals on purged cross-products.** Regress $\hat{e}_t$ on $\hat{\mathbf{v}}_t$. Under H_0 (linearity), the AR residuals should be uncorrelated with any function of the lagged values, so $\hat{e}_t$ and $\hat{\mathbf{v}}_t$ should be orthogonal.

Step 5. **Compute the test statistic.**

$$F = \frac{\mathrm{SS}_{\mathrm{reg}}/q}{\mathrm{SS}_{\mathrm{err}}/(T-p-q-1)} \sim F_{q,T-p-q-1}$$

under H_0, where $\mathrm{SS}_{\mathrm{reg}}$ is the regression sum of squares from Step 4 and $\mathrm{SS}_{\mathrm{err}}$ is the residual sum of squares.

Null and alternative hypotheses.

$$H_0\colon \text{AR}(p) \text{ is adequate (linear).}$$

$$H_1\colon \text{nonlinear structure present.}$$

Reject H_0 when $F > F_{\alpha,q,T-p-q-1}$.

Key properties.

- Tests against a broad class of alternatives: bilinear, threshold, ARCH, polynomial — any deviation from linearity that manifests as correlation between residuals and cross-products of lags.
- With $q = p(p+1)/2$ numerator degrees of freedom, power is spread across many directions. For large p, the test can lose power against specific alternatives.
- Shown by Tsay (1986) to be more powerful than Keenan's (1985) test, which uses only a single quadratic term (1 df).
- **Not a threshold test.** A significant result indicates nonlinearity but gives no information about the type (threshold, bilinear, ARCH, etc.) or the delay parameter d.

Relationship to Engle's ARCH test. Engle's (1982) LM test for ARCH regresses *squared* residuals on squared lagged residuals. Tsay (1986) regresses OLS residuals on cross-products of lagged *observations*. Both detect nonlinearity but against different alternatives: Engle targets conditional heteroscedasticity in the error process; Tsay targets nonlinearity in the conditional mean. They are complementary, not equivalent.

Tsay (1989): Arranged Autoregression Test

Motivation. The 1986 test detects nonlinearity but cannot identify the threshold variable z_{t-d} or the delay parameter d that governs which regime the process is in. Tsay (1989) introduces the *arranged autoregression*, which sorts the data by the candidate threshold variable and uses recursive residuals to detect the structural change this sorting would expose if a true threshold existed.

Set-up. Assume a TAR model with p lags and a candidate threshold variable z_{t-d} (typically y_{t-d} for a SETAR, but can be any observable):

$$y_t = \begin{cases} \boldsymbol{\phi}_1^\top \mathbf{x}_t + \varepsilon_t & \text{if } z_{t-d} \le c \\ \boldsymbol{\phi}_2^\top \mathbf{x}_t + \varepsilon_t & \text{if } z_{t-d} > c, \end{cases}$$

where $\mathbf{x}_t = (1, y_{t-1}, \ldots, y_{t-p})^\top$ and c is the unknown threshold. The null is the linear AR(p): $\boldsymbol{\phi}_1 = \boldsymbol{\phi}_2$.

Procedure.

Step 1. **Fit the linear AR(p) baseline.** Estimate the AR(p) model on the full time-ordered sample to determine p.

Step 2. **Sort by the threshold variable.** Re-order the $T-p$ usable observations by the value of $z_{t-d} = y_{t-d}$ in ascending order. Denote the permuted index as $\pi(1), \pi(2), \ldots, \pi(T-p)$ where $z_{\pi(1)-d} \le z_{\pi(2)-d} \le \cdots \le z_{\pi(T-p)-d}$. *Note: the regressors $\mathbf{x}_{\pi(i)}$ retain their original alignment — only the order of entry into the recursive regression changes.*

Step 3. **Run recursive least squares (RLS) on the arranged data.** Add observations one at a time in the sorted order $\pi(1), \pi(2), \ldots$, updating the AR coefficient estimates after each addition. For observation $\pi(i)$, compute the *predictive (recursive) residual*:

$$\hat{e}^{\text{pred}}_{\pi(i)} = y_{\pi(i)} - \hat{\boldsymbol{\phi}}_{(i-1)}^\top \mathbf{x}_{\pi(i)},$$

where $\hat{\boldsymbol{\phi}}_{(i-1)}$ is the RLS estimate using only the first $i-1$ observations (in sorted order). The standardized predictive residual is $e^*_{\pi(i)} = \hat{e}^{\text{pred}}_{\pi(i)}/\hat{\sigma}_{(i-1)}\sqrt{1+h_{ii}}$, where h_{ii} is the leverage of $\mathbf{x}_{\pi(i)}$ given the current design matrix.

Step 4. **Test for correlation between predictive residuals and regressors.** Under H_0 (no threshold), the predictive residuals $\hat{e}^{\text{pred}}_{\pi(i)}$ are uncorrelated with $\mathbf{x}_{\pi(i)}$ even in the sorted order — the recursive estimates stabilize quickly and the residuals behave like white noise. Under H_1 (threshold at some c), observations below c and above c follow different coefficient vectors, so the recursive estimates will drift systematically as the threshold is crossed, producing correlated residuals. Regress the predictive residuals on the regressors:

$$\hat{e}^{\text{pred}}_{\pi(i)} = \boldsymbol{\gamma}^\top \mathbf{x}_{\pi(i)} + \eta_i, \quad i = p+1, \ldots, T-p.$$

Step 5. **Compute the F-statistic.** Test $H_0\colon \boldsymbol{\gamma} = \mathbf{0}$ (full coefficient vector including constant):

$$F = \frac{\mathrm{SS}_{\mathrm{reg}}/(p+1)}{\mathrm{SS}_{\mathrm{err}}/(T-2p-2)} \overset{\text{approx}}{\sim} F_{p+1,T-2p-2}$$

under H_0. Tsay (1989) shows this converges asymptotically to a χ^2_{p+1} after scaling, but the F approximation is recommended for finite samples. *Note: this tests the full coefficient vector including the intercept, which differs from the standard partial F-test.*

Null and alternative hypotheses.

$$H_0\colon \text{linear AR}(p) \text{ — no threshold.}$$

$$H_1\colon \text{TAR nonlinearity with threshold variable } z_{t-d}.$$

Reject when $F > F_{\alpha,p+1,T-2p-2}$.

Choosing the delay d. Repeat the procedure for $d = 1, 2, \ldots, d_{\max}$ and select $d^* = \arg\max_d F(d)$. The threshold variable with the largest F statistic is the most likely candidate. Tsay also recommends two graphical diagnostics:

- Scatter plot of standardized predictive residuals $e^*_{\pi(i)}$ versus the sorted threshold variable $z_{\pi(i)-d}$: structural breaks appear as abrupt shifts in the scatter.
- Scatter plot of individual RLS t-ratios for each AR coefficient versus the sorted threshold variable: instability in any coefficient signals the threshold location.

Key properties.

- Specifically designed for threshold alternatives; not a general nonlinearity test.
- Identifies both the threshold variable (via the delay d^* with largest F) and the approximate location of threshold values (via graphical diagnostics).
- More powerful than the 1986 test against TAR alternatives because it concentrates power in the threshold direction.
- Requires the baseline AR order p to be correctly specified; misspecification of p can distort the test.
- The nuisance parameter c is not identified under H_0 (the threshold only exists under H_1), so the asymptotic distribution of the F-statistic is non-standard. The $F_{p+1,T-2p-2}$ approximation is adequate for detecting the presence of nonlinearity, but the bootstrap (fixed-regressor) is preferred for accurate p-values.

Summary: Which Test to Use

	Tsay 1986	**Tsay 1989**
Alternative	General nonlinearity	Threshold (TAR/SETAR)
Mechanism	Cross-product regression	Arranged autoregression
Identifies d?	No	Yes
Identifies c?	No	Approximately
Numerator df	$p(p+1)/2$	$p+1$
R function	`TSA::Tsay.test`	Custom / `NTS::Tsay`

Use 1986 as a pre-test for any nonlinearity. If significant, use 1989 to test specifically for a threshold and identify the delay parameter.

References. Tsay (1986), *Biometrika* 73, 461–466; Tsay (1989), *JASA* 84, 231–240; Keenan (1985), *JASA* 80, 487–489; Engle (1982), *Econometrica* 50, 987–1007.

Part XVIII

Time-to-Event Modeling & Hazard Structure

Comparing survival curves and evaluating proportional hazards

Survival Analysis

XVIII. SURV — A Conceptual Primer

What Is Survival Analysis?

Survival analysis is a collection of methods for modeling the **time until an event of interest occurs**. Although the terminology originates in biomedical research (the event is often death), the framework is universal: equipment failure, loan default, employee attrition, time to recidivism, and customer churn are all survival-analytic problems.

The defining feature is that the **outcome is a duration**, not a static measurement. This creates two fundamental challenges absent from cross-sectional models:

- **Censoring:** the event may not occur before the study ends, so the exact event time is unknown for some subjects.
- **Time-varying risk:** the probability of the event occurring may change over time, requiring the hazard function rather than a simple probability.

How Survival Analysis Differs from Cross-Sectional Models

In a standard regression, the outcome Y_i is observed for all i. In survival analysis:

- The **outcome is the event time** T_i—a positive random variable with its own distribution.
- For censored subjects, T_i is only known to exceed some censoring time C_i.
- The **observed data** for subject i is the pair (Y_i, δ_i) where $Y_i = \min(T_i, C_i)$ and $\delta_i = \mathbf{1}(T_i \le C_i)$ is the event indicator.
- Standard regression applied to Y_i as if it were the outcome would produce **biased estimates** because censored observations are systematically excluded or misrepresented.
- The **risk set** $\mathcal{R}(t)$—subjects still under observation and event-free just before time t—is the fundamental quantity around which all test statistics are built.

Censoring

Right censoring (most common): the event has not occurred by the end of follow-up; we know $T_i > C_i$. Causes: administrative end-of-study, loss to follow-up, competing event, withdrawal.

Left censoring: the event occurred before the observation window opened; we know $T_i < C_i$. Example: a disease already present at study entry.

Interval censoring: the event occurred somewhere in (L_i, R_i) but the exact time is unknown. Example: quarterly clinic visits where infection status is checked periodically.

Informative vs. non-informative censoring:

- **Non-informative (random) censoring:** the censoring mechanism is independent of the event time: $T \perp C$. The censored subjects are representative of those still at risk; this is the standard assumption underlying all tests in this chapter.
- **Informative censoring:** T and C are correlated—subjects who drop out are systematically sicker (or healthier) than those remaining. All standard methods are **biased** under informative censoring; it must be addressed at the design stage, through sensitivity analyses, or via joint models for the event and dropout processes.

Truncation

Distinct from censoring. **Left truncation** (delayed entry): a subject is only included in the study if they survive long enough to be observed (e.g. subjects must reach a certain age to enter the study). Subjects who experienced the event before entry are never observed—the risk set at time t must exclude subjects not yet enrolled. Ignoring left truncation underestimates early hazard.

Right truncation: only subjects who have already experienced the event are sampled (e.g. disease registries). Rare; requires specialized methods.

Core Functions

Let $T \ge 0$ be the event time with density $f(t)$ and CDF $F(t) = P(T \le t)$.

Survival function:

$$S(t) = P(T > t) = 1 - F(t), \quad S(0) = 1, \quad S(\infty) = 0.$$

$S(t)$ is monotone non-increasing; a step function in the non-parametric setting.

Hazard function (hazard rate):

$$h(t) = \lim_{\Delta t \to 0} \frac{P(t \le T < t + \Delta t \mid T \ge t)}{\Delta t} = \frac{f(t)}{S(t)}.$$

$h(t)$ is the **instantaneous rate** of the event given survival to t; not a probability but a rate (events per unit time). Common shapes: constant (exponential), monotone (Weibull), unimodal (log-normal), bathtub.

Cumulative hazard function:

$$H(t) = \int_0^t h(u)\,du = -\log S(t), \quad S(t) = e^{-H(t)}.$$

$H(t)$ is used for model diagnostics and forms the basis of the Nelson–Aalen estimator.

Key relationships: $f(t) = h(t)S(t)$; $S(t) = \exp(-H(t))$; $h(t) = -S'(t)/S(t)$. Any one of $\{f, F, S, h, H\}$ determines all others.

Median survival: $t_{0.5}$ defined by $S(t_{0.5}) = 0.5$; preferred over the mean, which requires the entire tail of $S(t)$ and may be infinite.

Non-Parametric Estimation

Kaplan–Meier (KM) estimator: The product-limit estimator of $S(t)$:

$$\widehat{S}(t) = \prod_{i:t_i \le t} \left(1 - \frac{d_i}{n_i}\right),$$

where $t_1 < t_2 < \cdots$ are the ordered event times, d_i is the number of events at t_i, and $n_i = |\mathcal{R}(t_i)|$ is the risk-set size. The KM curve is a right-continuous step function that drops only at event times; censored observations leave the risk set without causing a step.

Greenwood's formula gives the variance of $\widehat{S}(t)$:

$$\widehat{\text{Var}}(\widehat{S}(t)) = \widehat{S}(t)^2 \sum_{i:t_i \le t} \frac{d_i}{n_i(n_i - d_i)}.$$

CIs are constructed on the $\log(-\log S)$ (complementary log-log) scale and back-transformed to ensure they lie in $[0, 1]$.

Nelson–Aalen estimator:

$$\widehat{H}(t) = \sum_{i:t_i \le t} \frac{d_i}{n_i}, \quad \widetilde{S}(t) = e^{-\widehat{H}(t)}.$$

Preferable to KM in small samples; forms the basis of cumulative hazard diagnostic plots.

Types of Survival Models

Class	Characteristics
Non-parametric	No distributional assumption; KM and Nelson–Aalen; no covariates; fully data-driven.
Semi-parametric	Unspecified baseline hazard $h_0(t)$; parametric covariate effects; Cox PH model is the standard.
Parametric	Fully specified distribution (exponential, Weibull, log-normal, Gamma, log-logistic); MLE estimation; stronger assumptions, more efficient if correct.
Accelerated failure time (AFT)	Models $\log T = \mathbf{x}^\top \boldsymbol{\beta} + \sigma\varepsilon$; covariates accelerate or decelerate time; alternative to PH when hazards cross.
Competing risks	Multiple mutually exclusive event types; cause-specific hazard and subdistribution hazard approaches; standard methods over-estimate event probabilities.

I. Comparing Survival Curves: Log-Rank Family

All tests in this family test H_0: identical survival functions across k groups by comparing observed and expected events at each event time. The general weighted statistic is:

$$\chi^2 = \frac{(\sum_t w(t)[d_{1t} - e_{1t}])^2}{\sum_t w(t)^2 v_{1t}} \mathrel{\dot\sim} \chi^2_{k-1},$$

where d_{1t} and e_{1t} are observed and expected events in group 1 at time t, v_{1t} is the hypergeometric variance, and $w(t)$ is a weight function.

Test	Weight $w(t)$ and sensitivity
Log-rank (Mantel-Cox)	$w(t) = 1$; equal weight to all event times; **most powerful under proportional hazards**.
Gehan–Breslow	$w(t) = n(t)$ (risk-set size); emphasises **early differences**; sensitive to early event times when risk sets are large.
Tarone–Ware	$w(t) = \sqrt{n(t)}$; intermediate between log-rank and Gehan; balanced early/late sensitivity.
Peto–Peto	$w(t) = \widehat{S}(t)$ from the pooled KM; robust to unequal censoring patterns across groups; emphasises early differences.
Fleming–Harrington	$w(t) = \widehat{S}(t)^\rho(1 - \widehat{S}(t))^\gamma$; $\rho = \gamma = 0$: log-rank; $\rho = 1, \gamma = 0$: early; $\rho = 0, \gamma = 1$: late; flexible family.

Choosing the weight:

- No prior knowledge $\Rightarrow$ log-rank.
- Hazards expected to diverge early (e.g. fast-acting treatment) $\Rightarrow$ Gehan–Breslow or Peto–Peto.
- Hazards expected to diverge late $\Rightarrow$ Fleming–Harrington with $\gamma > 0$.
- **Never choose the test after seeing the data**; pre-specify the weight based on subject-matter knowledge.
- **When hazards cross**, all log-rank family tests have low power (they test for systematic ordering); use alternatives such as the supremum test or restrict the time window.

II. The Cox Proportional Hazards Model

The Cox model specifies:

$$h(t|\mathbf{x}_i) = h_0(t) \exp(\mathbf{x}_i^\top \boldsymbol{\beta}),$$

where $h_0(t)$ is the unspecified **baseline hazard** (the hazard when all covariates are zero) and $\exp(\beta_j)$ is the **hazard ratio** (HR) associated with a one-unit increase in x_j.

Partial likelihood: Cox (1972) showed that $\boldsymbol{\beta}$ can be estimated without specifying $h_0(t)$ by maximizing the **partial likelihood**:

$$PL(\boldsymbol{\beta}) = \prod_{i:\delta_i=1} \frac{\exp(\mathbf{x}_i^\top \boldsymbol{\beta})}{\sum_{j \in \mathcal{R}(t_i)} \exp(\mathbf{x}_j^\top \boldsymbol{\beta})}.$$

At each event time, the denominator sums over all subjects still at risk. Tied event times require Breslow, Efron, or exact corrections.

Interpretation: $\exp(\hat{\beta}_j)$ = HR; a one-unit increase in x_j multiplies the hazard by $\exp(\hat{\beta}_j)$. HR > 1: increased risk; HR < 1: decreased risk; HR $= 1$: no effect. **The HR is assumed constant over time**—the proportional hazards assumption.

III. Testing in the Cox Model

Likelihood ratio test (LRT):

$$\Lambda = -2[\log PL(M_0) - \log PL(M_1)] \stackrel{\cdot}{\sim} \chi^2_q.$$

Tests $H_0 : \boldsymbol{\beta}_2 = \mathbf{0}$ (q restrictions). Most reliable in small samples; preferred over Wald.

Score (log-rank) test: Evaluates the score function of PL at $\boldsymbol{\beta} = \mathbf{0}$; requires only the null model. Equivalent to the log-rank test when there are no ties and covariates are group indicators—connecting the nonparametric and semi-parametric frameworks.

Wald test: $W = \hat{\beta}_j/\widehat{\text{se}}(\hat{\beta}_j) \sim \mathcal{N}(0,1)$; routinely reported in software output but **less reliable than LRT** for small n or large effects.

IV. The Proportional Hazards Assumption

The PH assumption requires that the HR between any two subjects is **constant over time**: $h(t|\mathbf{x}_1)/h(t|\mathbf{x}_2) = \text{const}$. Equivalently, the log-log survival curves $\log(-\log \widehat{S}(t|\mathbf{x}))$ vs. $\log t$ should be parallel.

Schoenfeld residuals: For each event time t_i and covariate j, the Schoenfeld residual is the difference between the observed covariate value and its risk-set weighted mean:

$$r^S_{ij} = x_{ij} - \frac{\sum_{k\in\mathcal{R}(t_i)} x_{kj}\exp(\hat{\beta}_j x_{kj})}{\sum_{k\in\mathcal{R}(t_i)} \exp(\hat{\beta}_j x_{kj})}.$$

Under PH, Schoenfeld residuals should be **uncorrelated with time**. Plotting them against t (or a function of t) reveals time-varying effects.

Grambsch–Therneau test: Formalises the Schoenfeld residual plot by testing the correlation between scaled Schoenfeld residuals and a function of time $g(t)$ (typically $\log t$ or the KM time transformation):

$$H_0 : \beta_j(t) = \beta_j \quad \text{(constant HR)}.$$

χ^2_p joint test (all covariates) and individual χ^2_1 tests per covariate. A significant result indicates the HR for that covariate is time-varying.

Remedies for PH violation:

- Stratify the Cox model by the offending variable (allows different $h_0(t)$ per stratum; no HR estimated for that variable).
- Include a time-by-covariate interaction: $\beta_j(t) = \beta_j + \gamma_j \log t$.
- Use an accelerated failure time (AFT) model which does not assume proportional hazards.
- Restrict analysis to a time window where PH holds.

V. Residual Diagnostics for the Cox Model

Cox–Snell residuals:

$$r^{CS}_i = \widehat{H}(Y_i|\mathbf{x}_i) = \hat{H}_0(Y_i)\exp(\mathbf{x}_i^\top\hat{\boldsymbol{\beta}}),$$

where $\hat{H}_0$ is the estimated baseline cumulative hazard. Under a correctly fitted model, r^{CS}_i should follow a unit exponential distribution, i.e. the KM curve of $\{r^{CS}_i\}$ should follow $\exp(-t)$. Deviations indicate overall model misspecification; plot $-\log \widehat{S}_{CS}(r)$ against r—a well-fitted model yields a straight line through the origin with slope 1.

Martingale residuals: $M_i = \delta_i - r^{CS}_i$; range $(-\infty, 1]$. Used to detect nonlinear covariate effects: plot M_i against a continuous covariate to suggest the functional form.

Deviance residuals: Normalized martingale residuals; approximately $\mathcal{N}(0,1)$ for a well-fitted model. Large positive values: subjects who experienced the event much sooner than expected; large negative values: those who survived much longer. Used to identify influential observations.

Score residuals and dfbeta: Measure the influence of each observation on $\hat{\boldsymbol{\beta}}$; analogous to DFBETAS in linear regression.

VI. Competing Risks

A subject faces **competing risks** when multiple mutually exclusive event types can occur, and the occurrence of one precludes the others (e.g. death from cancer vs. death from cardiovascular disease; discharge vs. in-hospital death).

Why standard methods fail: Applying the KM estimator to each cause separately, treating other-cause events as censored, **overestimates** the cause-specific probability because it assumes subjects censored by other causes remain at risk indefinitely—a counterfactual that cannot be realized.

Cause-specific hazard (CSH): $h_k(t)$ = instantaneous rate of cause k among those event-free. Cox models fitted separately for each cause (treating competing events as censored) give CSH hazard ratios—valid for aetiology questions (*what drives the cause-specific risk?*).

Subdistribution hazard (SDH): Defined via the **cumulative incidence function** (CIF): $F_k(t) = P(T \le t, \text{cause} = k)$. The SDH is $\tilde{h}_k(t) = -d\log(1 - F_k(t))/dt$; it treats competing-events subjects as remaining in the risk set with weight decreasing over time. The **Fine–Gray model** regresses on SDH; its coefficients directly describe effects on $F_k(t)$—the right quantity for *prognosis* (what is the probability a patient will experience cause k?).

CSH vs. SDH: A covariate can have opposite effects on the CSH and SDH of the same cause when competing risks are strong. **Use CSH** for aetiological inference; **use SDH/Fine–Gray** for absolute risk prediction. Always report the CIF, not the KM estimator, in the presence of competing risks.

Gray's test: Nonparametric k-sample test comparing CIFs across groups; the competing-risks analog of the log-rank test. Tests $H_0 : F_1^{(k)}(t) = F_2^{(k)}(t) = \cdots$ for cause k. Uses a weighted score statistic based on the subdistribution hazard; χ^2_{g-1} for g groups.

Parametric Survival Models

When the event-time distribution can be specified, parametric models offer efficiency gains and direct extrapolation beyond the follow-up period.

Distribution	Hazard shape and notes
Exponential	Constant $h(t) = \lambda$; memoryless; simplest baseline.
Weibull	$h(t) = \lambda p(\lambda t)^{p-1}$; monotone; $p > 1$ increasing, $p < 1$ decreasing; PH and AFT.
Log-normal	Unimodal hazard; AFT model; non-PH.
Log-logistic	Unimodal; closed-form $S(t)$; AFT; common for time-to-tumor-response.
Gompertz	$h(t) = \lambda e^{\gamma t}$; exponentially increasing; actuarial / demographic applications.

Model selection: use AIC/BIC; plot the empirical $-\log \widehat{S}(t)$ vs. t (exponential) or $\log(-\log \widehat{S}(t))$ vs. $\log t$ (Weibull) to assess linearity.

Diagnostic Workflow

1. **Plot KM curves** by group; assess censoring pattern; check for crossing curves.
2. **Compare groups** (log-rank or weighted variant pre-specified by expected hazard timing).
3. **Fit Cox model**; check convergence; interpret HRs with CIs.
4. **Test coefficients**: LRT (primary); Wald (routine output).
5. **Check PH**: Grambsch–Therneau test; Schoenfeld residual plots; log-log plots.
6. **Check overall fit**: Cox–Snell residuals; martingale residuals for nonlinear effects.
7. **Identify influential obs.**: deviance and score residuals; dfbeta plots.
8. **Competing risks?** Estimate CIF (not KM); Gray's test for group comparison; Fine–Gray model for covariate effects on $F_k(t)$.
9. **Consider parametric model** if extrapolation needed or hazard shape informative.

Common Pitfalls

- **Informative censoring:** standard methods assume $T \perp C$; dropout correlated with prognosis invalidates all results.
- **Ignoring left truncation:** subjects must survive to enter the study; omitting the delayed-entry adjustment underestimates early hazard.
- **Log-rank when hazards cross:** all log-rank family tests have near-zero power; use a two-sided test with restricted time window or Fleming–Harrington with $\rho = 0, \gamma = 1$.
- **Post-hoc weight selection:** choosing the weighted log-rank test after inspecting the data inflates Type I error; pre-specify.
- **PH assumption not tested:** always run Grambsch–Therneau; time-varying effects are common and materially change conclusions.
- **KM with competing risks:** using $1 - \widehat{S}(t)$ as the event probability overestimates the CIF; always use the CIF in competing-risk settings.
- **CSH vs. SDH confusion:** a predictor that reduces the CSH of cause k can increase $F_k(t)$ if it strongly reduces competing causes; report which model is used and why.
- **Cox with few events:** the rule of thumb is ≥ 10 events per predictor (EPV); below this, estimates are unstable and CIs are unreliable.

Bottom Line

Survival analysis is distinguished by time-to-event outcomes, censoring, and time-varying risk. Standard regression applied to censored outcomes is biased. Use KM for nonparametric description, log-rank for group comparison, and Cox for covariate adjustment—but always verify the proportional hazards assumption. In competing-risk settings, replace the KM estimator with the CIF and use Gray's test or the Fine–Gray model.

Log-Rank Test (Mantel-Cox)

Compare Survival Curves Between Groups

SURV.1 | **Category:** Survival Analysis | **Data:** Time-to-Event | **Null:** Chi-square

When to Use

- Compare survival curves of two or more groups.
- Assess treatment effects on time-to-event data.
- Right-censored survival data are present.
- Clinical trials or medical follow-up studies.
- Detect overall differences in survival experience.

Assumptions

- Observations are mutually independent.
- Censoring is noninformative and independent.
- Event times are correctly recorded.
- Groups are defined before follow-up begins.
- Hazards are approximately proportional.

Test Statistic

$$\chi^2_{LR} = \frac{\left(\sum_j w_j(d_{1j} - e_{1j})\right)^2}{\sum_j w_j^2 v_j}, \quad \text{where } w_j = 1, \quad e_{1j} = \frac{n_{1j} d_j}{n_j}, \quad v_j = \frac{n_{1j} n_{2j} d_j (n_j - d_j)}{n_j^2 (n_j - 1)}.$$

Under H_0, $\chi^2_{LR} \sim \chi^2_1$ (two groups) or χ^2_{K-1} (K groups) asymptotically.
Notation: n_j is total at risk at time j, d_{1j}, e_{1j} are observed/expected events, v_j is variance (hypergeometric formula)

Plain English: The log-rank test compares observed versus expected event counts across groups at each distinct event time. If survival curves are identical, events should be distributed across groups proportional to the number at risk in each group. The test sums differences between observed and expected events over all time points, weighting each equally. Large discrepancies between observed and expected counts indicate different survival experiences or hazard rates.

⚠ Practitioner Warning

Assumes PH; less powerful when hazards cross or converge over time. Use weighted log-rank tests (Gehan, Tarone-Ware, Peto) for non-proportional hazards. Ties in event times require special handling; common methods include Breslow, Efron, or exact likelihood. Small samples or few events reduce power and accuracy of chi-square approximation. Does not estimate magnitude of effect (hazard ratio)—use Cox regression for that. Multiple group comparisons (omnibus test) do not identify which specific groups differ—requires post-hoc pairwise tests with multiplicity adjustment.

Hypotheses

- H_0: Survival curves $S_i(t)$ identical across groups.
- H_1: At least one $S_i(t)$ is different.

Decision Rule: Reject H_0 if $\chi^2_{LR} > \chi^2_{\alpha,K-1}$ or p-value $< \alpha$.
For K groups, df = $K - 1$.

Software References [Python, R, SAS]

Python	R	SAS
`lifelines::logrank_test`	`survival::survdiff(rho=0)`	`PROC LIFETEST (STRATA; LOGRANK)`

Additional Notes

For $K > 2$ groups, test statistic becomes quadratic form: $\chi^2_{LR} = (\mathbf{O} - \mathbf{E})'\mathbf{V}^{-1}(\mathbf{O} - \mathbf{E})$ with $\mathbf{V}$ covariance matrix. Relationship to Cox regression: log-rank test equivalent to score test in Cox model with single group indicator. Stratified log-rank allows adjusting for confounders by computing test within strata and pooling. Weighted variants: Gehan-Breslow (early differences), Tarone-Ware (middle emphasis), Peto-Peto (modified Wilcoxon). Fleming-Harrington $G^{\rho,\gamma}$ family provides flexible weighting: $w(t) = \hat{S}(t)^\rho(1 - \hat{S}(t))^\gamma$.

Miscellanea: Nathan Mantel (1966) introduced the log-rank test, using an observed-minus-expected framework akin to Mantel–Haenszel methods. David Cox (1972) showed it is the score test from the proportional hazards model, linking nonparametric rank tests with semiparametric survival analysis. Richard Peto and others later popularized the test in clinical trials, especially in cancer research. Survival analysis is fundamentally driven by censoring: many event times are only partially observed, making ordinary comparisons of sample means inappropriate and motivating rank-based procedures such as the log-rank test.

Reference(s): Mantel (1966); Peto & Peto (1972); https://web.stanford.edu/~lutian/coursepdf/unit6.pdf

Log-Rank Test (Mantel-Cox)

Worked Example

Problem Statement A clinical researcher tests whether survival times differ between a treatment group ($n_1 = 6$) and a control group ($n_2 = 6$) following a medical intervention. H_0: identical survival curves; H_1: survival curves differ. Group 1: 4 events, 2 censored ($9^+, 20^+$). Group 2: 5 events, 1 censored (8^+). Under H_0, $LR \xrightarrow{d} \chi^2(1)$.

Subject	1	2	3	4	5	6
Time (Trt.)	6	9^+	12	16	20^+	25
Time (Ctrl.)	1	2	4	6	8^+	10

1. Event Table at Each Distinct Failure Time

At each t_j: n_{kj} = at risk in group k; d_{kj} = events; expected $e_{1j} = n_{1j}\,d_j/n_j$. Censored subjects leave the risk set after their recorded time.

Note: At $t_j = 12, 16, 25$ no control subjects remain at risk ($n_{2j} = 0$), so $e_{1j} = d_j$ and $v_j = 0$ — these events carry no discriminatory information.

t_j	n_{1j}	n_{2j}	n_j	d_{1j}	d_{2j}	d_j	e_{1j}	$d_{1j} - e_{1j}$	v_j
1	6	6	12	0	1	1	0.500	−0.500	0.250
2	6	5	11	0	1	1	0.545	−0.545	0.248
4	6	4	10	0	1	1	0.600	−0.600	0.240
6	6	3	9	1	1	2	1.333	−0.333	0.389
10	4	1	5	0	1	1	0.800	−0.800	0.160
12	4	0	4	1	0	1	1.000	0.000	0.000
16	3	0	3	1	0	1	1.000	0.000	0.000
25	1	0	1	1	0	1	1.000	0.000	0.000
							Total	**−2.779**	**1.287**

2. Variance at Each Event Time

Each v_j uses the hypergeometric variance formula:

$$v_j = \frac{n_{1j}\,n_{2j}\,d_j\,(n_j - d_j)}{n_j^2\,(n_j - 1)}$$

e.g. at $t_j = 6$: $v_6 = \frac{6 \cdot 3 \cdot 2 \cdot 7}{81 \cdot 8} = 0.389$.

3. Log-Rank Statistic

$$U = \sum_{j=1}^{D}(d_{1j} - e_{1j}), \quad V = \sum_{j=1}^{D} v_j, \quad LR = \frac{U^2}{V} = 6.001$$

4. Interpretation

- **Decision:** $LR = 6.001 > 3.841$, so we **reject** H_0. Group 1 saw fewer early failures ($U < 0$), indicating better survival.
- **Practitioner Insight:** The log-rank test weights event times equally, maximizing power under proportional hazards. The negative $U = -2.779$ indicates treatment subjects experienced fewer events than expected (better survival), consistently at early times. Right-censored observations contribute to the risk set n_{kj} until censoring, reducing at-risk counts but not d_{kj}. At late times ($t = 12, 16, 25$), $n_{2j} = 0$ and $v_j = 0$, so these events add no comparison information. For $K > 2$, the test generalises to a $(K-1)$-df chi-squared using the vector U_k and its covariance matrix.

Common Pitfalls

- **Using the log-rank test when survival curves cross.** The log-rank test is most powerful when hazards are proportional over time. When curves cross, weighted tests or restricted mean survival time comparisons may be more appropriate.
- **Ignoring informative censoring.** The test assumes censoring is independent of the event process. If censoring is related to prognosis or treatment response, the log-rank comparison may be biased.
- **Performing multiple pairwise log-rank tests without adjustment.** With more than two groups, the overall test evaluates equality of survival functions across all groups. Multiple pairwise comparisons require correction methods such as Bonferroni or Holm to control type I error.
- **Interpreting mean survival times with heavy censoring.** Censored observations prevent reliable estimation of avg. survival time. Reporting survival probabilities at fixed times or restricted mean survival time is often more informative.
- **Ignoring clustering or dependence among observations.** The log-rank test assumes independence of event times across individuals. Clustered, matched, or recurrent-event data require stratified or model-based survival methods.

Gehan-Breslow Test (Wilcoxon)

Weighted Log-Rank Test Emphasizing Early Differences

SURV.2 | **Category:** Survival Analysis | **Data:** Time-to-Event | **Null:** Chi-square

When to Use

- Compare survival curves of two or more groups.
- Assess treatment effects on time-to-event data.
- Right-censored survival data are present.
- Clinical trials or medical follow-up studies.
- Emphasize early differences in survival experience.

Assumptions

- Observations are mutually independent.
- Censoring is noninformative and independent.
- Event times are correctly recorded.
- Groups are defined before follow-up begins.
- Hazards are approximately proportional.

Test Statistic

$$\chi^2_{GB} = \frac{\left(\sum_j w_j(d_{1j} - e_{1j})\right)^2}{\sum_j w_j^2 v_j}, \quad \text{where } w_j = n_j, \quad e_{1j} = \frac{n_{1j}\, d_j}{n_j}, \quad v_j = \frac{n_{1j} n_{2j} d_j (n_j - d_j)}{n_j^2 (n_j - 1)}.$$

Under H_0, $\chi^2_{GB} \sim \chi^2_1$ (two groups) or χ^2_{K-1} (K groups) asymptotically.
Notation: n_j is total at risk at time j (weight), d_{1j}, e_{1j} are observed/expected events, v_j is variance (hypergeometric)

Plain English: The Gehan-Breslow test is a weighted version of the log-rank test that gives more weight to event times when more subjects are at risk, typically early in follow-up. By weighting each time point by the number at risk, the test emphasizes differences occurring when the study population is largest. This makes it more sensitive to early separation of survival curves but less sensitive to late differences when few subjects remain. The weighting addresses situations where early survival differences matter most clinically.

⚠ Practitioner Warning

More sensitive to early differences means less sensitive to late differences—can miss important late divergence in survival curves. Not optimal under proportional hazards (log-rank preferred in that case). Power depends on whether true differences occur early versus late in follow-up. Heavy censoring early on reduces effective weighting advantage. Test called "Wilcoxon" by analogy to rank-sum test but not equivalent to standard Wilcoxon for censored data. Multiple names (Gehan, Breslow, Wilcoxon) cause confusion; same test, different historical attributions.

Hypotheses

- H_0: Survival curves $S_i(t)$ identical across groups.
- H_1: Survival curves differ (during follow-up period).

Decision Rule: Reject H_0 if $\chi^2_{GB} > \chi^2_{\alpha,K-1}$ or p-value $< \alpha$. Emphasizes early survival differences.

Software References [Python, R, SAS]

`lifelines::multi..._logrank_test` `survival::survdiff(rho=1)` `PROC LIFETEST (WILCOXON)`

Additional Notes

Relationship to other weighted tests: Tarone-Ware uses $w_j = \sqrt{n_j}$ (intermediate between log-rank and Gehan-Breslow), Peto-Peto uses $w_j = \tilde{S}(t_j)$ where $\tilde{S}$ is Peto-Peto survival estimate. Fleming-Harrington $G^{\rho,\gamma}$ family unifies: $w_j = \hat{S}(t_{j-1})^{\rho}(1 - \hat{S}(t_{j-1}))^{\gamma}$. Gehan-Breslow is $G^{1,0}$, log-rank is $G^{0,0}$. Choice among tests should be pre-specified based on clinical hypothesis about timing of effects.

Miscellanea: Edmund Gehan developed his generalized Wilcoxon test in 1965 while at MD Anderson Cancer Center, extending rank tests to accommodate censored survival data. Norman Breslow independently derived a weighted log-rank formulation in 1970 and showed its close relationship to Gehan's procedure. The test is therefore commonly known as the Gehan–Breslow (or Breslow–Gehan) test. Although often treated as equivalent in practice, the original Gehan and Breslow formulations differ slightly in derivation and tie handling. The "Wilcoxon" label reflects its emphasis on early rank differences, analogous to the Wilcoxon rank-sum test.

Reference(s): Gehan (1965); Breslow (1970)

Worked Example

Problem Statement The same treatment ($n_1 = 6$) vs control ($n_2 = 6$) survival dataset from the Log-Rank example is re-analyzed using the Gehan-Breslow (Wilcoxon) test. The only modification: each event time t_j receives weight $w_j = n_j$ (total at risk), giving greater emphasis to early failures when the risk set is largest. H_0: $S_1(t) = S_2(t)$ for all t.

1. Weighted Event Table
Columns identical to Log-Rank; final two columns add the Gehan-Breslow weighting $w_j = n_j$:

t_j	n_j	d_{1j}	e_{1j}	$d_{1j} - e_{1j}$	v_j	w_j	$w_j(d-e)$	$w_j^2 v_j$
1	12	0	0.500	−0.500	0.250	12	−6.000	36.000
2	11	0	0.545	−0.545	0.248	11	−6.000	30.000
4	10	0	0.600	−0.600	0.240	10	−6.000	24.000
6	9	1	1.333	−0.333	0.389	9	−3.000	31.500
10	5	0	0.800	−0.800	0.160	5	−4.000	4.000
12	4	1	1.000	0.000	0.000	4	0.000	0.000
16	3	1	1.000	0.000	0.000	3	0.000	0.000
25	1	1	1.000	0.000	0.000	1	0.000	0.000
						Total	**−25.000**	**125.500**

2. Gehan-Breslow Statistic

$$U = \sum_j w_j (d_{1j} - e_{1j}) \quad V = \sum_j w_j^2 v_j,$$

$$GB = \frac{U^2}{V} \xrightarrow{d} \chi^2(1), \text{ under } H_0$$

3. Results and Decision

Test	Statistic	p-value	Weights	**Decision**
Log-Rank	6.001	0.014	$w_j = 1$	Reject H_0
Gehan-Breslow	4.980	0.026	$w_j = n_j$	Reject H_0

4. Interpretation

- **Decision:** $GB = 4.980 > 3.841$: we **reject** H_0. The treatment group survived significantly longer. Both tests agree on rejection; the log-rank is stronger here because the survival difference is pronounced at late event times ($t = 10$), which the Gehan-Breslow downweights relative to early times.

Practitioner Insight: The Gehan-Breslow statistic is identical in structure to the log-rank — the same e_{1j} and v_j — but multiplies each term by $w_j = n_j$ before summing. Because n_j decreases monotonically as follow-up progresses (subjects exit through events or censoring), this weighting scheme amplifies early differences and attenuates late ones. Here w_j ranges from 12 at $t = 1$ to 1 at $t = 25$, so the first three event times each contribute $w_j(d - e) = -6$ compared with -0.5 each in the unweighted log-rank. The practical choice between the two tests should be driven by where the hazard ratio is expected to be largest: Gehan-Breslow is preferred when treatment effects manifest early (e.g. acute toxicity studies); log-rank is preferred when effects emerge late or are proportional throughout follow-up. Neither test is uniformly more powerful — the relative performance depends entirely on the shape of the hazard difference over time.

Common Pitfalls

- **Using the Gehan–Breslow test when treatment effects occur late.** The Gehan–Breslow test places greater weight on early failures through the number at risk. When survival differences emerge primarily at later times, the log-rank test is usually more powerful.
- **Interpreting Gehan–Breslow results as evidence of overall survival differences.** Because early events receive the greatest weight, the test mainly reflects early survival differences. A significant result therefore indicates early divergence rather than a general difference over the entire follow-up period.
- **Ignoring the number at risk when interpreting weighted results.** The weighting scheme depends on the number of individuals at risk at each event time. Sharp declines in the risk set or heavy early censoring can substantially influence the test statistic.
- **Choosing the Gehan–Breslow test only after the log-rank test fails.** Selecting an alternative test post hoc increases the risk of data-driven inference. The weighting scheme should be specified in advance based on the expected timing of treatment effects.
- **Assuming the Gehan–Breslow test is uniformly more powerful than log-rank.** Power depends on when survival differences occur during follow-up. The Gehan–Breslow test favors early effects, whereas the log-rank test performs better when effects occur later.

Tarone-Ware Test

Weighted Log-Rank with Moderate Time Weighting

SURV.3 | **Category:** Survival Analysis | **Data:** Time-to-Event | **Null:** Chi-square

When to Use

- Compare survival curves of two or more groups.
- Assess treatment effects on time-to-event data.
- Right-censored survival data are present.
- Clinical trials or medical follow-up studies.
- Moderately emphasize early survival differences.

Assumptions

- Observations are mutually independent.
- Censoring is noninformative and independent.
- Event times are correctly recorded.
- Groups are defined before follow-up begins.
- Hazards are approximately proportional.

Test Statistic

$$\chi^2_{TW} = \frac{\left(\sum_j w_j(d_{1j} - e_{1j})\right)^2}{\sum_j w_j^2 v_j}, \quad \text{where } w_j = \sqrt{n_j}, \quad e_{1j} = \frac{n_{1j}\, d_j}{n_j}, \quad v_j = \frac{n_{1j} n_{2j} d_j (n_j - d_j)}{n_j^2 (n_j - 1)}.$$

Under H_0, $\chi^2_{TW} \sim \chi^2_1$ (two groups) or χ^2_{K-1} (K groups) asymptotically. Hypergeometric formula for v_j.

Notation: $w_j = \sqrt{n_j}$ weights each time point, n_j is number at risk, d_{1j}, e_{1j} are observed/expected events, v_j is variance

Plain English: The Tarone-Ware test uses moderate time weighting by taking the square root of the number at risk at each event time. This produces intermediate behavior between the log-rank test (equal emphasis throughout) and Gehan-Breslow test (heavy early emphasis). The square root weighting still gives more influence to early time points when more subjects are at risk, but less dramatically than Gehan-Breslow, providing balanced sensitivity across the follow-up period.

⚠ Practitioner Warning

Compromise nature means it may be less powerful than log-rank under proportional hazards or Gehan-Breslow for pure early differences. No clear clinical interpretation of "square root weighting"—chosen for mathematical properties, not clinical rationale. Less commonly used than log-rank or Gehan-Breslow, with less established practice regarding when to prefer it. Power properties intermediate between extremes but not always optimal. Some argue choosing among weighted tests is data-driven specification searching; prefer pre-specified primary test based on clinical hypothesis.

Hypotheses

- H_0: Survival curves $S_i(t)$ identical across groups.
- H_1: Survival curves differ at some point during follow-up.

Decision Rule: Reject H_0 if $\chi^2_{TW} > \chi^2_{\alpha,K-1}$ or p-value $< \alpha$. Balanced emphasis across time.

Software References [Python, R, SAS]

```
lifelines::multi..._logrank_test    survival::survdiff(rho=0.5)    PROC LIFETEST (STRATA; TARONE)
```

Additional Notes

Tarone and Ware (1977) proposed several weighted log-rank statistics, with $w_j = \sqrt{n_j}$ being most commonly implemented. The weighting scheme represents compromise between linear weight n_j (Gehan-Breslow) and constant weight $w_j = 1$ (log-rank). Relationship to variance stabilizing transformations: square root of counts is variance-stabilizing for Poisson, providing statistical motivation. Fleming-Harrington family $G^{\rho,\gamma}$ with $\rho = 0.5, \gamma = 0$ yields Tarone-Ware. Alternative Tarone-Ware formulations exist using different power weights; $\sqrt{n_j}$ most standard. In practice, researchers often report log-rank as primary test with Gehan-Breslow and Tarone-Ware as sensitivity analyses.

Miscellanea: Robert Tarone and James Ware introduced this test in 1977 to bridge the gap between the log-rank and Gehan–Breslow procedures, which can yield different conclusions when survival differences occur at different times. They proposed a family of weighted rank tests that place varying emphasis on early versus late events. The Tarone–Ware weight $w_j = \sqrt{n_j}$ provides a practical compromise between the equal weighting of the log-rank test and the stronger early-event emphasis of the Gehan–Breslow test.

Reference(s): Tarone & Ware (1977); Fleming & Harrington (1991)

Worked Example

Problem Statement The same treatment ($n_1 = 6$) vs control ($n_2 = 6$) survival dataset from the Log-Rank example is re-analyzed using the Tarone-Ware test. Weights $w_j = \sqrt{n_j}$ are intermediate between the Log-Rank ($w_j = 1$) and Gehan-Breslow ($w_j = n_j$), giving moderate emphasis to early event times without fully suppressing later ones.

1. Weighted Event Table

The columns E_{1j}, $d_{1j} - e_{1j}$, v_j are identical to the Log-Rank and Gehan-Breslow examples; only the weight column and its products change.

t_j	n_j	d_{1j}	e_{1j}	$d_{1j} - e_{1j}$	w_j	$w_j(d_{1j} - e_{1j})$	$n_j v_j$
1	12	0	0.500	−0.500	$\sqrt{12}$	−1.732	3.000
2	11	0	0.545	−0.545	$\sqrt{11}$	−1.809	2.727
4	10	0	0.600	−0.600	$\sqrt{10}$	−1.897	2.400
6	9	1	1.333	−0.333	3	−1.000	3.500
10	5	0	0.800	−0.800	$\sqrt{5}$	−1.789	0.800
12	4	1	1.000	0.000	2	0.000	0.000
16	3	1	1.000	0.000	$\sqrt{3}$	0.000	0.000
25	1	1	1.000	0.000	1	0.000	0.000
					Total	**−8.227**	**12.427**

2. Tarone-Ware Statistic

$$U = \sum_j \sqrt{n_j}\,(d_{1j} - e_{1j}), \quad V = \sum_j n_j v_j$$

$$TW = \frac{U^2}{V} \xrightarrow{d} \chi^2(1), \text{ under } H_0.$$

3. Results and Decision

Test	Statistic	p-value	Weights	**Decision**
Tarone-Ware	5.447	0.020	$\sqrt{n_j}$	Reject H_0
Log-Rank	6.001	0.014	$w_j = 1$	Reject H_0
Gehan-Breslow	4.980	0.026	$w_j = n_j$	Reject H_0

4. Interpretation

- **Decision:** $TW = 5.447 > 3.841$: we **reject** H_0 at 5%. The Tarone-Ware statistic is intermediate between Log-Rank and Gehan-Breslow, consistent with $\sqrt{n_j}$ weighting lying between 1 and n_j.

Practitioner Insight: These weighted log-rank tests form a natural family parameterised by the power of n_j: LR ($n_j^0 = 1$), TW ($n_j^{1/2}$), GB (n_j^1). The statistic ordering LR > TW > GB in this dataset arises because the largest single survival difference occurs at $t_j = 10$ when only 5 subjects remain at risk — a late signal that Gehan-Breslow downweights most aggressively ($w_{10} = 5$) and Log-Rank ignores ($w_{10} = 1$). Tarone-Ware's $w_{10} = \sqrt{5} \approx 2.24$ is the compromise. Note that $V_w = \sum n_j v_j$: the variance formula simplifies because $w_j^2 = n_j$, making V_w a simple weighted sum of the unweighted v_j values with weights equal to the risk set sizes. The Tarone-Ware test is most appropriate when the investigator has no strong prior about where the hazard difference is concentrated, and serves as a robust middle ground between the two extremes.

Common Pitfalls

- **Using the Tarone–Ware test as a default without considering clinical context.** The Tarone–Ware test applies intermediate weighting between log-rank and Gehan–Breslow. The choice of test should reflect the expected timing of treatment effects rather than convenience.
- **Interpreting similar results across weighted tests as validation of the chosen method.** When log-rank, Tarone–Ware, and Gehan–Breslow produce similar results, the weighting scheme likely has little influence. Agreement among tests usually indicates a strong or consistent effect across follow-up.
- **Assuming Tarone–Ware always has intermediate statistical power.** The weighting scheme lies between early- and late-weighted tests but power depends on the temporal pattern of effects. Under proportional hazards, the log-rank test often remains the most efficient.
- **Using Tarone–Ware to avoid specifying the expected timing of treatment effects.** The compromise weighting is appropriate when the timing of effects is genuinely uncertain. Whenever possible, the test choice should be justified using clinical or biological considerations.
- **Failing to specify which survival test is primary.** When multiple weighted tests are reported, the primary analysis should be declared in advance. Alternative tests should be treated as sensitivity analyses rather than post hoc selections.

Peto-Peto Test

Modified Wilcoxon with Peto-Peto Survival Weighting

SURV.4 | **Category:** Survival Analysis | **Data:** Time-to-Event | **Null:** Chi-square

When to Use

- Compare survival curves of two or more groups.
- Assess treatment effects on time-to-event data.
- Right-censored survival data are present.
- Clinical trials or medical follow-up studies.
- Emphasize early survival differences with censoring.

Assumptions

- Observations are mutually independent.
- Censoring is noninformative and independent.
- Event times are correctly recorded.
- Groups are defined before follow-up begins.
- Hazards are approximately proportional.

Test Statistic

$$\chi^2_{PP} = \frac{\left(\sum_j w_j(d_{1j} - e_{1j})\right)^2}{\sum_j w_j^2 v_j}, \quad \text{where } w_j = \tilde{S}(t_{j-1}), \quad e_{1j} = \frac{n_{1j}\, d_j}{n_j}, \quad v_j = \frac{n_{1j} n_{2j} d_j (n_j - d_j)}{n_j^2 (n_j - 1)}.$$

Under H_0, $\chi^2_{PP} \sim \chi^2_1$ (two groups) or χ^2_{K-1} (K groups) asymptotically.
Notation: $\tilde{S}(t)$ is Peto-Peto survival estimate, d_{1j}, e_{1j} are observed/expected events, v_j is variance

Plain English: The Peto-Peto test weights each event time by the estimated survival probability up to that point, calculated using a modified product-limit estimator. This weighting scheme gives more influence to earlier events (when survival probability near 1) and less to later events (when survival probability declining). Unlike Gehan-Breslow which uses number at risk, Peto-Peto uses survival probability, providing different behavior when censoring patterns vary between groups or when hazards are not proportional.

⚠ Practitioner Warning

More computationally intensive due to Peto–Peto survival estimates; implementations may vary slightly across software. Less commonly used than log-rank or Gehan–Breslow, with fewer established interpretation guidelines. Suboptimal under proportional hazards (log-rank preferred). Weighting by survival probability can be unintuitive—emphasizes periods with high survival rather than clinically relevant differences. Modern practice increasingly uses Fleming-Harrington family with explicit ρ, γ parameters rather than Peto-Peto, as Fleming-Harrington provides clearer control over weighting shape.

Hypotheses

- H_0: Survival curves $S_i(t)$ identical across groups.
- H_1: Survival curves differ (esp. middle follow-up period.

Decision Rule: Reject H_0 if $\chi^2_{PP} > \chi^2_{\alpha,K-1}$ or p-value $< \alpha$. Weighting emphasizes when survival high.

Software References [Python, R, SAS]

`lifelines::multi..._logrank_test` `survival::survdiff(rho=1)` `PROC LIFETEST; (STRATA; PETO)`

Additional Notes

Peto-Peto survival estimator $\tilde{S}(t)$ is modified Kaplan-Meier: $\tilde{S}(t_j) = \prod_{i=1}^{j}(1 - d_i/(n_i + 1))$ where modification adds 1 to denominator. This produces weights declining from near 1 early to near 0 late in follow-up. At time t_j, weight $w_j = \tilde{S}(t_{j-1})$ uses survival just before current event time. The test emphasizes differences occurring when most of the cohort still alive (middle period), less emphasis on very early (few events yet) or very late (few at risk). Peto-Peto approximates $G^{1,0}$ (Fleming-Harrington) but uses slightly different survival estimates.

Miscellanea: Richard Peto and Julian Peto (brothers) developed this test in 1972 while at Oxford University, contributing to the explosion of survival analysis methodology in the 1970s. Richard became particularly influential in medical statistics, leading landmark cancer trials and advocating for large simple trials. The test bears their family name, one of few sibling-attributed methods in statistics.

Reference(s): Peto & Peto (1972); Fleming & Harrington (1991)

Worked Example

Problem Statement Survival dataset from the Log-Rank example is re-analyzed using the Peto-Peto (mod. Wilcoxon) test. Its weight $w_j = \tilde{S}(t_j^-)$ is the *pooled* modified Kaplan-Meier estimate evaluated just before t_j, making weights data-driven and censoring-aware rather than mechanically proportional to n_j or $\sqrt{n_j}$. H_0: $S_1(t) = S_2(t)$ for all t.

t_j	n_j	d_j	$(1 - d_j/(n_j + 1))$	$\tilde{S}(t_j)$	$w_j = \tilde{S}(t_{j-1})$	$d_{1j} - e_{1j}$	v_j	$w_j(d_{1j} - e_{1j})$	$w_j^2 v_j$
1	12	1	12/13	0.923	1.000	-0.500	0.250	-0.500	0.250
2	11	1	11/12	0.846	0.923	-0.545	0.248	-0.503	0.211
4	10	1	10/11	0.769	0.846	-0.600	0.240	-0.508	0.172
6	9	2	8/10	0.615	0.769	-0.333	0.389	-0.256	0.230
10	5	1	5/6	0.513	0.615	-0.800	0.160	-0.492	0.061
12	4	1	4/5	0.410	0.513	0.000	0.000	0.000	0.000
16	3	1	3/4	0.308	0.410	0.000	0.000	0.000	0.000
25	1	1	1/2	0.154	0.308	0.000	0.000	0.000	0.000
							Total	**−2.259**	**0.924**

1. Pooled Modified KM Weights
Compute $\tilde{S}$ from the *pooled* sample using the Peto-Peto formula at each event time:

$$\tilde{S}(t_j) = \tilde{S}(t_{j-1}) \times (1 - d_j/(n_j + 1)), \qquad \tilde{S}(t_0) = 1$$

2. Peto-Peto Statistic Under H_0,

$$PP \xrightarrow{d} \chi^2(1), \quad PP = U^2/V = (-2.259)^2/0.924$$

3. Results and Four-Test Comparison

Test	Statistic	p	w_j	**Decision**
Peto-Peto	5.529	0.019	$\tilde{S}(t_j^-)$	Reject
Log-Rank	6.001	0.014	1	Reject
Tarone-Ware	5.447	0.020	$\sqrt{n_j}$	Reject
Gehan-Breslow	4.980	0.026	n_j	Reject

4. Interpretation

- **Statistical Decision:** $PP = 5.529 > 3.841$: we **reject** H_0 at 5%. All four tests agree; PP is between TW and GB.
- **Practitioner Insight:** The PP weight $\tilde{S}(t_j^-)$ differs from GB's n_j in using $n_j + 1$ in the denominator, ensuring the final event is not assigned zero weight — a small-sample correction. It is also censoring-aware: after heavy censoring, $\tilde{S}$ declines faster than n_j/n_1, down-weighting later, less reliable event times. The ordering LR > TW > PP > GB reflects this dataset's pattern, where the strongest signal at $t = 10$ is diluted as weights shift toward earlier times. In practice, Peto-Peto is often preferred since its KM-based weights adapt to censoring rather than relying on raw risk-set counts.

Common Pitfalls

- **Using the Peto–Peto test when treatment effects occur primarily late.** Because weights decline as survival probability decreases, late events receive less influence. The test may therefore have reduced power when survival differences emerge late in follow-up.
- **Assuming the Peto–Peto test is always less powerful than the log-rank test.** Power depends on when hazard differences occur. The Peto–Peto test can outperform log-rank when differences arise early or in mid follow-up, but may be less powerful for late effects.
- **Interpreting a significant Peto–Peto result as evidence of late effects.** The test emphasizes periods when survival is high; significant results typically reflect early or mid follow-up differences rather than late divergence.
- **Ignoring that weights are based on pooled survival estimates.** Weights use the pooled Kaplan–Meier estimator rather than group-specific curves, which may not represent either group well when survival differs substantially.
- **Confusing Peto–Peto with Gehan–Breslow weighting.** Both emphasize earlier events, but Peto–Peto adapts to censoring via survival probabilities, leading to different behavior when censoring patterns vary across groups.

Fleming-Harrington Test

Flexible Weighted Log-Rank with $G^{\rho,\gamma}$ Family

SURV.5 | **Category:** Survival Analysis | **Data:** Time-to-Event | **Null:** Chi-square

When to Use

- Compare survival curves of two or more groups.
- Assess treatment effects on time-to-event data.
- Right-censored survival data are present.
- Clinical trials or medical follow-up studies.
- Target early or late survival differences selectively.

Assumptions

- Observations are mutually independent.
- Censoring is noninformative and independent.
- Event times are correctly recorded.
- Groups are defined before follow-up begins.
- Hazards are approximately proportional.

Test Statistic

$$G_{\rho,\gamma} = \frac{\left(\sum_j w_j(d_{1j} - e_{1j})\right)^2}{\sum_j w_j^2 v_j}, \quad \text{where } w_j = \hat{S}(t_{j-1})^{\rho}(1 - \hat{S}(t_{j-1}))^{\gamma}, \quad e_{1j} = \frac{n_{1j}\, d_j}{n_j}, \quad v_j = \frac{n_{1j} n_{2j} d_j (n_j - d_j)}{n_j^2 (n_j - 1)}.$$

Under H_0, $G_{\rho,\gamma} \sim \chi_1^2$ (two groups) or χ_{K-1}^2 (K groups) asymptotically.

Notation: $\hat{S}(t)$ is Kaplan-Meier estimate, ρ weights early/late, γ weights across survival distribution

Plain English: The Fleming–Harrington test provides flexible weighting of survival comparisons through two non-negative parameters. The parameter ρ increases emphasis on early differences, while γ increases emphasis on late differences. The weight function w_j allows targeting specific phases where treatment effects are hypothesized.

⚠ Practitioner Warning

Requires pre-specifying ρ, γ before seeing data to maintain Type I error control. No consensus on optimal values for specific scenarios. Multiple testing across different (ρ, γ) pairs requires correction. Test is most powerful when weight pattern matches true difference pattern; misspecification reduces power. Parameters lacking direct clinical interpretation make results harder to communicate than "early differences" or "proportional hazards."

Hypotheses

- H_0: Survival curves $h_i(t)$ identical across groups.
- H_1: Survival curves differ (pattern matching (ρ, γ) spec.).

Decision Rule: Reject H_0 if $G_{\rho,\gamma} > \chi^2_{\alpha,K-1}$ or p-value $< \alpha$. Interpretation depends on (ρ, γ) chosen.

Software References [Python, R, SAS]

Python	R	SAS
`lifelines::multi..._logrank_test`	`EventPredInCure::FH_Test`	`Not Native (Manual)`

Additional Notes

Common parameter choices: $(0, 0)$ log-rank (equal weights), $(1, 0)$ early differences (Gehan-Breslow-like), $(0, 1)$ late differences, $(1, 1)$ middle period emphasis. For delayed treatment effects (immunotherapy, vaccines), use negative ρ to down-weight early non-differences. For diminishing effects, use positive γ to down-weight late convergence. Yang and Prentice (2005) proposed two-stage testing: first test with $(0, 0)$, if non-significant try alternative (ρ, γ) with adjusted critical value. MaxCombo test uses maximum of several Fleming-Harrington statistics with multiplicity correction for non-proportional hazards.

Miscellanea: Thomas Fleming and David Harrington developed this flexible testing framework in 1981 and expanded it in their influential 1991 textbook *Counting Processes and Survival Analysis*. They were motivated by increasing evidence that proportional hazards often violated in practice—immunotherapies showing delayed benefits, surgical interventions with early risks. Standard log-rank test missed non-proportional effects or had reduced power. Fleming and Harrington's $G_{\rho,\gamma}$ family provided principled way to target specific alternatives.

Reference(s): Fleming & Harrington (1981, 1991); Yang & Prentice (2005)

Worked Example

Problem Statement Survival dataset from the Log-Rank example is re-analyzed using the Fleming-Harrington $G(\rho, \gamma)$ family with $(\rho, \gamma) = (0, 1)$. Unlike all previous tests whose weights *decrease* over follow-up, $G(0, 1)$ weights $w_j = (1 - \hat{S}(t_j^-))^\gamma$ *increase* as the pooled survival falls — emphasizing late failures.

t_j	n_j	d_j	$\hat{S}(t_j)$	w_j	$d_{1j} - e_{1j}$	v_j	$w_j(d_{1j} - e_{1j})$	$w_j^2 v_j$
1	12	1	11/12 = 0.917	0.000	−0.500	0.250	0.000	0.000
2	11	1	5/6 = 0.833	0.083	−0.545	0.248	−0.045	0.002
4	10	1	3/4 = 0.750	0.167	−0.600	0.240	−0.100	0.007
6	9	2	7/12 = 0.583	0.250	−0.333	0.389	−0.083	0.024
10	5	1	7/15 = 0.467	0.417	−0.800	0.160	−0.333	0.028
12	4	1	7/20 = 0.350	0.533	0.000	0.000	0.000	0.000
16	3	1	7/30 = 0.233	0.650	0.000	0.000	0.000	0.000
25	1	1	0	0.767	0.000	0.000	0.000	0.000
						Total	**−0.562**	**0.061**

1. Pooled KM and G(0,1) Weights

Compute $\hat{S}$ from the *pooled* sample using the standard KM formula and then set $w_j = 1 - \hat{S}(t_{j-1})$:

$$\hat{S}(t_j) = \hat{S}(t_{j-1}) \times (n_j - d_j)/n_j, \qquad \hat{S}(t_0) = 1$$

2. Fleming-Harrington Statistic

Under H_0,

$$FH \xrightarrow{d} \chi^2(1), \quad FH = U_w^2/V_w = (-0.562)^2/0.061$$

3. Results and Four-Test Comparison

Test	Statistic	p	w_j	Decision
Log-Rank	6.001	0.014	1	Reject
FH G(1,0)	5.471	0.019	$\hat{S}$	Reject
FH G(0,1)	5.225	0.022	$1 - \hat{S}$	Reject
FH G(1,1)	4.996	0.025	$\hat{S}(1 - \hat{S})$	Reject
Gehan-Breslow	4.980	0.026	n_j	Reject

4. Interpretation

- **Decision:** $FH = 5.225 > 3.841$: we **reject** H_0 at 5% ($p = 0.022$). Despite zeroing out the earliest event time, the late-weighted statistic detects the survival difference.
- **Practitioner Insight:** The $G(\rho, \gamma)$ family unifies weighted log-rank tests via two parameters: ρ emphasizes early events (since $\hat{S}$ is large early) and γ emphasizes late events (as $1 - \hat{S}$ increases over time). Setting $(\rho, \gamma) = (0, 0)$ gives the log-rank; $G(1, 0)$ approximates Peto-Peto; $G(0, 1)$ uniquely up-weights late events and is suited for delayed treatment effects (e.g. immunotherapy). The ordering LR > G(1,0) > G(0,1) > G(1,1) ≈ GB reflects that the signal at $t = 10$ is moderate relative to early events—$G(0, 1)$ also up-weights later times with $v_j = 0$, limiting its gain over GB. All tests are nonparametric and require no distributional assumptions.

Common Pitfalls

- **Choosing (p, q) after examining the data.** The Fleming–Harrington test allows many weighting choices through (p, q). Selecting parameters post hoc inflates type I error and should be avoided.
- **Assuming larger p or q automatically increases power.** The parameters determine *where* the test places weight over time. Power improves only when the weighting aligns with the true timing of survival differences.
- **Ignoring how p and q jointly shape the weighting pattern.** Weights take the form $\hat{S}(t)^p(1 - \hat{S}(t))^q$, producing early-, middle-, or late-emphasis depending on the parameters. Understanding this interaction is essential for choosing weights consistent with the expected treatment effect.
- **Using extreme (p, q) values without clear justification.** Large parameter values concentrate weight on narrow portions of follow-up. Such choices increase variance and reduce robustness if the effect timing is misspecified.
- **Failing to declare which (p, q) specification is primary.** When multiple Fleming–Harrington tests are reported, the primary specification should be stated in advance. Alternative (p, q) choices should be treated as sensitivity analyses rather than post hoc selections.

Cox Proportional Hazards Score Test

Score Test for Covariate Effects in Survival Models

SURV.6 | **Category:** Survival Analysis | **Data:** Time-to-Event | **Null:** Chi-square

When to Use

- Test covariate effects in Cox survival models.
- Assess predictors of time-to-event outcomes.
- Right-censored survival data are present.
- Clinical, reliability, or follow-up studies.
- Significance testing without full model fitting.

Assumptions

- Hazards are proportional over time (PH assumption).
- Observations are mutually independent.
- Censoring is non-informative (indep. of x and survival).
- Covariate effects are log-linear on hazards.
- Event times are correctly recorded.

Test Statistic

$$\chi^2_{\text{score}} = \mathbf{U}'\mathcal{I}^{-1}\mathbf{U} \approx \chi^2_p, \text{ (under } H_0) \qquad U = \sum_{j=1}^{D} \left[\mathbf{z}_j - \bar{\mathbf{z}}_j\right], \qquad \mathcal{I} = \sum_{j=1}^{D} \mathbf{V}_j.$$

Notation: $\mathbf{z}_j$ is covariate vector for j-th event, $\bar{\mathbf{z}}_j$ is risk-set weighted mean, $\mathbf{V}_j$ is variance, p is number of covariates

Plain English: The Cox score test evaluates whether covariates significantly predict survival by examining the score function (the gradient of the partial log-likelihood) evaluated at the null hypothesis ($\boldsymbol{\beta} = \mathbf{0}$). If the covariates are unrelated to survival, the observed covariate values at each event time should, on average, match the expected values among subjects still at risk at that time. The test accumulates these observed-minus-expected differences across all event times and standardizes them using the information matrix. Large systematic departures indicate that the covariates influence the hazard rate, providing evidence against the null hypothesis of no covariate effect.

⚠ Practitioner Warning

Assumes proportional hazards; violations reduce power or inflate Type I error. Test requires fitting null model only (computational advantage) but provides no effect size estimates. Ties in event times require approximation (Breslow, Efron); different methods can yield different results. Equivalent to log-rank for single binary covariate, but not for continuous or multiple covariates. Does not handle time-varying covariates in standard form. Score test generally most conservative of three (Wald, LRT, score) in finite samples.

Hypotheses

- H_0: $\boldsymbol{\beta} = \mathbf{0}$ (covariates do not affect hazard).
- H_1: $\boldsymbol{\beta} \neq \mathbf{0}$ (covariates predict hazard, association exists).

Decision Rule: Reject H_0 if $\chi^2_{\text{score}} > \chi^2_{\alpha,p}$ or p-value $< \alpha$.

Software References [Python, R, SAS]

`lifelines.CoxPHFitter`	`survival::coxph`	`PROC PHREG`

Additional Notes

Score vector at null: $\mathbf{U}(\mathbf{0}) = \sum_j [\mathbf{z}_j - \bar{\mathbf{z}}_j]$ where $\mathbf{z}_j$ is covariate vector for individual experiencing j-th event, $\bar{\mathbf{z}}_j = \sum_{i \in R_j} \mathbf{z}_i / n_j$ is mean over risk set R_j. Information matrix: $\mathcal{I} = \sum_j \mathbf{V}_j$ where $\mathbf{V}_j$ is variance-covariance matrix of covariates in risk set j. The partial likelihood for Cox model: $PL(\boldsymbol{\beta}) = \prod_{j=1}^{D} \left[\exp(\mathbf{z}_j'\boldsymbol{\beta}) \div \sum_{i \in R_j} \exp(\mathbf{z}_i'\boldsymbol{\beta})\right]$ where product is over distinct event times. Score function: $\mathbf{U}(\boldsymbol{\beta}) = \partial \log PL / \partial \boldsymbol{\beta}$. Observed information: $\mathcal{I}(\boldsymbol{\beta}) = -\partial^2 \log PL / \partial \boldsymbol{\beta} \partial \boldsymbol{\beta}'$. Three tests: score (requires null model), Wald (requires full model), likelihood ratio (requires both); asymptotically equivalent but differ in finite samples.

Miscellanea: David Cox introduced the proportional hazards model in 1972 in a landmark *JRSS* paper, revolutionizing survival analysis. The score test emerged naturally from partial likelihood theory as computationally efficient testing approach. In the 1970s, fitting Cox models was computationally expensive; score test allowed hypothesis testing without parameter estimation, a significant practical advantage. Norman Breslow's work on computational methods and tie handling made Cox models practical.

Reference(s): Cox (1972); Breslow (1974)

Worked Example

Problem Statement A researcher tests whether a continuous biomarker z is associated with survival time in $N = 10$ patients. The Cox score test evaluates the partial likelihood score at $\hat{\beta} = 0$, asking: does the observed covariate value at each failure time exceed what chance alone would produce from the risk set? H_0: $\beta = 0$ (biomarker has no effect on hazard). 7 events, 3 censored (4^+, 8^+, 13^+).

Subject	1	2	3	4	5	6	7	8	9	10
Time t_i	2	3	4^+	5	6	8^+	9	11	13^+	15
Covariate z_i	7.0	8.0	4.0	6.5	5.0	3.0	6.0	3.5	5.0	4.5

1. Score and Information at $\hat{\beta} = 0$
At each t_j, define the risk set $\mathcal{R}_j = \{i : t_i \geq t_j\}$ and compute:

$$\bar{z}_j = \frac{\sum_{i \in \mathcal{R}_j} z_i}{n_j}, \quad u_j = z_{(j)} - \bar{z}_j \quad \text{Var}_j(z) = \frac{\sum_{i \in \mathcal{R}_j} z_i^2}{n_j} - \bar{z}_j^2$$

where $z_{(j)}$ is the covariate of the subject who failed at t_j. At $t_j = 2$: all 10 subjects at risk, $\bar{z} = 52.5/10 = 5.250$; failed subject has $z_{(j)} = 7.0$, so $u_j = 1.75$ (above avg.).

t_j	n_j	$z_{(j)}$	$\bar{z}_j$	u_j	Var_j
2	10	7.0	5.250	+1.750	2.313
3	9	8.0	5.056	+2.944	2.191
5	7	6.5	4.786	+1.714	1.347
6	6	5.0	4.500	+0.500	1.000
9	4	6.0	4.750	+1.250	0.813
11	3	3.5	4.333	−0.833	0.389
15	1	4.5	4.500	0.000	0.000
			Total	**+7.325**	**8.052**

2. Score Statistic Under H_0, $X^2 \xrightarrow{d} \chi^2(1)$.

$$U(0) = \sum u_j, \quad \mathcal{I}(0) = \sum \text{Var}_j(z),$$

$$X^2 = \frac{[U(0)]^2}{\mathcal{I}(0)} = \frac{(7.325)^2}{8.052} = 6.664$$

3. Interpretation

- **Decision:** $X^2 = 6.664 > 3.841$: we **reject** H_0. The biomarker z has a significant positive effect on the hazard — higher z is associated with earlier failure ($U(0) > 0$, failed subjects consistently above $\bar{z}_j$).

Practitioner Insight: The Cox score test evaluates the partial likelihood derivative $U(\beta) = \partial\ell/\partial\beta$ at $\beta = 0$, without ever fitting the Cox model. Each $u_j = z_{(j)} - \bar{z}_j$ measures how far the failing subject's covariate deviates from the risk-set mean — a positive $U(0)$ signals that high-z subjects tend to fail first. The information $\mathcal{I}(0) = \sum \text{Var}_j(z)$ is the sum of within-risk-set covariate variances: it is large when z is spread across the risk set, providing power to detect an association. Note that for a single binary covariate $z_i \in \{0, 1\}$ (group indicator), the score test reduces exactly to the log-rank test. For a continuous covariate the score test avoids arbitrary dichotomisation and uses the full covariate information. The score test, Wald test, and likelihood ratio test on β are asymptotically equivalent under H_0 but can differ in small samples; the score test is preferred when the model has not yet been fitted, as it requires only the null-hypothesis evaluation.

Common Pitfalls

- **Using the score test without checking the proportional hazards assumption.** The score test relies on the Cox model assumption that hazard ratios remain constant over time. Violation of proportional hazards can lead to misleading inference and reduced test power.
- **Expecting the score, Wald, and likelihood ratio tests to give identical results.** These tests are asymptotically equivalent but may differ in finite samples. Disagreement often reflects small samples, large effects, or instability in parameter estimates.
- **Assuming the score test is always identical to the log-rank test.** Equivalence holds only for a Cox model with a single binary covariate and consistent tie handling. With multiple or continuous covariates, the score test generalizes beyond the standard log-rank comparison.
- **Interpreting the score statistic as an estimate of the treatment effect.** The score test evaluates H_0 that the regression coefficient equals zero. Effect size and hazard ratios must be obtained from the fitted Cox model parameters.

Likelihood Ratio Test (Cox Model)

Test Covariate Effects via Partial Likelihood Comparison

SURV.7 | **Category:** Survival Analysis | **Data:** Time-to-Event | **Null:** Chi-square

When to Use

- Compare nested Cox proportional hazards models.
- Test significance of subset of covariates jointly.
- Assess overall model fit compared to null model.
- Standard approach for hypothesis testing in Cox regression.

Assumptions

- Proportional hazards assumption holds for both models.
- Models are nested (simpler is special case of complex).
- Censoring is non-informative (indep. of x and survival).
- Sample size adequate for asymptotic χ^2 approximation.

Test Statistic

$$LR = -2[\ell_0 - \ell_1] = -2\log\left(\frac{PL_0}{PL_1}\right), \quad \text{where } PL \text{ is partial likelihood.}$$

Under H_0, $LR \sim \chi^2_p$ asymptotically where p is difference in number of parameters.
Notation: ℓ_0 is log partial likelihood under null (restricted model), ℓ_1 under alternative (full model)

Plain English: The likelihood ratio test for Cox models compares the fit of two nested models by examining the ratio of their maximized partial likelihoods. If the additional covariates in the complex model genuinely improve prediction of survival, the partial likelihood will be substantially larger. The test statistic measures this improvement, doubling the difference in log partial likelihoods. Large values indicate the additional parameters significantly enhance model fit, justifying their inclusion.

⚠ Practitioner Warning

Requires fitting both null and alternative models (computational burden greater than score test). Assumes proportional hazards for both models; violations affect validity. Models must be truly nested with same subjects and event times. Tie handling method (Breslow, Efron, exact) must be consistent across models. Does not test PH assumption itself; use Schoenfeld residuals or time interaction tests. Generally more powerful than Wald test in finite samples but computationally more demanding than score test. Very large models relative to sample size can yield unreliable asymptotic approximation.

Hypotheses

- H_0: Restricted model adequate ($\boldsymbol{\beta}_{\text{extra}} = \mathbf{0}$).
- H_1: Full model needed ($\boldsymbol{\beta}_{\text{extra}} \neq \mathbf{0}$, extra covariates needed).

Decision Rule: Reject H_0 if $LR > \chi^2_{\alpha,p}$ or p-value $< \alpha$.

Software References [Python, R, SAS]

Python	R	SAS
`lifelines.CoxPHFitter`	`survival::anova.coxph`	`PROC PHREG`

Additional Notes

Partial likelihood for Cox model with covariates $\boldsymbol{\beta}$: $PL(\boldsymbol{\beta}) = \prod_{j=1}^{D} \frac{\exp(\mathbf{z}_j'\boldsymbol{\beta})}{\sum_{i\in R_j}\exp(\mathbf{z}_i'\boldsymbol{\beta})}$ where product is over D distinct event times, $\mathbf{z}_j$ is covariate vector for j-th event, R_j is risk set. Log partial likelihood: $\ell(\boldsymbol{\beta}) = \sum_j [\mathbf{z}_j'\boldsymbol{\beta} - \log(\sum_{i\in R_j}\exp(\mathbf{z}_i'\boldsymbol{\beta}))]$. The LRT compares nested models: full model with p_1 parameters versus restricted with $p_0 < p_1$ parameters, using $LR = -2(\ell_0 - \ell_1)$ with df $= p_1 - p_0$. Relationship to other tests: LRT, Wald, and score tests asymptotically equivalent but differ in finite samples. Ordering typically: Wald $\geq$ LRT $\geq$ Score, with LRT generally recommended for its better finite-sample properties. Information criteria use partial likelihood: AIC $= -2\ell + 2p$, BIC $= -2\ell + p\log(n)$. For testing proportional hazards assumption: fit model with time interactions $\boldsymbol{\beta}(t) = \boldsymbol{\beta} + \boldsymbol{\gamma} g(t)$ and test $\boldsymbol{\gamma} = \mathbf{0}$ using LRT.

Miscellanea: Cox's 1972 paper introduced partial likelihood, a revolutionary concept allowing inference without fully specifying baseline hazard function. The idea initially raised questions about whether it constituted a "true" likelihood, but subsequent asymptotic theory established its validity. Its computational simplicity and flexibility led to rapid adoption. Inference in Cox models is typically based on Wald, score, or partial likelihood ratio tests, all of which show good large-sample performance.

Reference(s): Cox (1972, 1975); Kalbfleisch & Prentice (2002)

Worked Example

Problem Statement The Cox partial likelihood ratio test is applied to the same biomarker–survival dataset as the score test, comparing the partial log-likelihood at $\hat{\beta}$ (the MLE) against its value at $\beta = 0$. Unlike the score test (which requires no model fitting) and the Wald test (which uses $\hat{\beta}/SE$), the LR test directly measures how much the partial likelihood improves when the covariate is included. H_0: $\beta = 0$; H_1: $\beta \neq 0$.

Subject	1	2	3	4	5	6	7	8	9	10
Time t_i	2	3	4^+	5	6	8^+	9	11	13^+	15
Covariate z_i	7.0	8.0	4.0	6.5	5.0	3.0	6.0	3.5	5.0	4.5

1. Cox Partial Log-Likelihood

$$\ell(\beta) = \sum_{j:\text{event}} \left[\beta z_{(j)} - \log \sum_{i \in \mathcal{R}_j} e^{\beta z_i} \right]$$

2. Evaluate $\ell(0)$
At $\beta = 0$: $e^{0 \cdot z_i} = 1$ for all i, so $\sum_{\mathcal{R}_j} e^{\beta z_i} = n_j$:

$$\ell(0) = \sum_j [0 - \log n_j] = -\sum_j \log n_j$$

3. Evaluate $\ell(\hat{\beta})$
The MLE $\hat{\beta} = 1.072$ (maximises ℓ, found numerically by solving $U(\beta) = 0$). Selected contributions:

t_j	n_j	$-\log n_j$	$z_{(j)}$	$\sum_{\mathcal{R}_j} e^{\hat{\beta} z_i}$	contrib
2	10	−2.303	7.0	9457.9	−1.654
3	9	−2.197	8.0	7648.8	−0.370
5	7	−1.946	6.5	2294.3	−0.773
6	6	−1.792	5.0	1235.6	−1.762
9	4	−1.386	6.0	998.5	−0.477
11	3	−1.099	3.5	378.9	−2.187
15	1	0.000	4.5	124.2	0.000
	Total	**−10.722**			**−7.224**

contrib = $1.072 \times 7.0 - \ln(9457.9) = -1.654$

4. LR Statistic Under H_0, $LR \xrightarrow{d} \chi^2(1)$.

$$LR = 2[\ell(\hat{\beta}) - \ell(0)] = 2[-7.224 - (-10.722)] = 6.997$$

5. Results and Decision

Test	Statistic	Value	p	**Decision**
LR	$2\Delta\ell$	6.997	0.008	Reject H_0
Score	$U^2/\mathcal{I}$	6.664	0.010	Reject H_0
Wald	$\hat{\beta}^2/\widehat{\text{Var}}$	4.756	0.029	Reject H_0

Practitioner Insight: The LR test compares the partial likelihood with and without β; the gain $\ell(\hat{\beta}) - \ell(0) = 3.499$ implies a likelihood ratio of $e^{3.499} \approx 33$. While LR, Wald, and score tests are asymptotically equivalent under H_0, they differ in small samples: the Wald test is typically smallest as it relies on local curvature at $\hat{\beta}$; the LR test is preferred since it uses the full likelihood; the score test lies between and avoids model fitting. The large sums $\sum e^{\hat{\beta} z_i}$ at $\hat{\beta} = 1.072$ (e.g. 9457.9 at $t = 2$) arise because $e^{1.072 \times 8} = e^{8.576} \approx 5288$ for the highest-z subject — this is normal and reflects the exponential hazard ratio parameterisation of the Cox model.

Common Pitfalls

- **Applying the likelihood ratio test to non-nested Cox models.** The likelihood ratio test is valid only when one model is a restricted version of the other. For non-nested models, model comparison should rely on criteria such as AIC, BIC, or cross-validation.
- **Using incorrect degrees of freedom for the test statistic.** The degrees of freedom equal the number of parameter restrictions imposed by the null model. This corresponds to the difference in the number of parameters between the full and restricted models.
- **Expecting likelihood ratio, Wald, and score tests to give identical results.** Although asymptotically equivalent, the three tests can differ noticeably in finite samples. Likelihood ratio tests are generally considered the most reliable when discrepancies occur.
- **Performing the test when the fitted models have not properly converged.** Likelihood ratio comparisons assume that both models have reached their maximum likelihood estimates. Convergence warnings or unstable parameter estimates can invalidate the test results.
- **Comparing models estimated on different samples or event definitions.** The likelihood ratio test requires both models to be fitted to the same dataset and event structure. Differences in sample size, censoring, or event definitions make the likelihoods incomparable.

Schoenfeld Residuals Test

Test Proportional Hazards Assumption in Cox Models

SURV.8 | **Category:** Survival Analysis | **Data:** Time-to-Event | **Null:** Chi-square

When to Use

- Test whether PH assumption holds in a Cox model.
- Diagnose time-varying covariate effects.
- Validate fitted Cox regression models.
- Right-censored survival data are present.
- Follow-up after fitting a Cox model.

Assumptions

- Cox proportional-hazards model is fitted.
- Observations are mutually independent.
- Censoring is noninformative and independent.
- Covariate effects are correctly specified.
- Residual trends reflect PH violations only.

Test Statistic

$$\chi^2_k = \left(\frac{\hat{b}_k}{SE(\hat{b}_k)}\right)^2 \sim \chi^2_1 \text{ under } H_0. \qquad r^*_{jk} = a_k + b_k\, g(t_j) + \varepsilon_j, \qquad r^S_{jk} = z_{jk} - \bar{z}^*_{jk}, \qquad r^*_{jk} = r^S_{jk}/v_{jk}$$

where r^S_{jk} and r^*_{jk} are the raw and scaled Schoenfeld residuals, v_{jk} is the weighted covariate variance in the risk set at t_j.
Notation: z_{jk} observed covariate of failing subject, $\bar{z}^*_{jk}$ weighted risk-set mean under $\hat{\boldsymbol{\beta}}$, $g(t)$ time function

Plain English: Schoenfeld residuals measure the difference between the observed covariate value of the failing subject at each event time and the value the Cox model predicted given the risk set. Under proportional hazards these residuals should be independent of time. The test regresses the scaled residuals on a chosen function of event time and tests whether the slope is zero; a significant slope indicates the hazard ratio is changing over time for that covariate.

⚠ Practitioner Warning

Test has low power with few informative events ($D < 30$–50, a rule of thumb) or when non-proportionality is mild. Event times with only one subject at risk ($n_j = 1$) must be excluded; including them can artificially deflate the slope and reverse the conclusion. Choice of $g(t)$ affects power; $\log t$ (R default in `cox.zph`) is more robust to late-event leverage than $g(t) = t$. Multiple testing issues arise when testing many covariates; consider Bonferroni or other corrections. Rejection indicates violation but not the remedy—options include stratification, time-varying coefficients, or alternative models.

Hypotheses

- H_0: Proportional hazards holds for covariate k ($\beta_k(t) = \beta_k$).
- H_1: Hazard ratio varies with time for covariate k.

Decision Rule: Reject H_0 if $\chi^2_k > \chi^2_{\alpha,1}$ or p-value $< \alpha$ for covariate k.

Software References [Python, R, SAS]

`lifelines.CoxPHFitter`	`survival::cox.zph`	`PROC PHREG`

Additional Notes

Schoenfeld residual for covariate k at event j: $r^S_{jk} = z_{jk} - E[Z_k \mid \mathcal{R}_j, \hat{\boldsymbol{\beta}}]$, where the expectation is the weighted mean over the risk set. Under PH, $E[r^S_{jk} \mid t_j] = 0$ for all j. Scaled residuals: $r^*_{jk} = r^S_{jk}/v_{jk}$, where $v_{jk} = \overline{z^2}^*_{jk} - (\bar{z}^*_{jk})^2$ is the weighted risk-set variance (Grambsch & Therneau, 1994). Degenerate event times ($n_j = 1$, so $v_{jk} = 0$) contribute $r_j = r^*_j = 0$ trivially and must be omitted from any regression or correlation. Common time functions: $g(t) = t$ (linear trend), $g(t) = \log(t)$ (log trend), $g(t) = \text{rank}(t)$ (Spearman), $g(t) = \hat{S}(t)$ (Kaplan–Meier). Global test: joint χ^2_p across all p covariates as in Grambsch–Therneau (1994), eq. (11). Remedies for violations: stratify on violating covariate, include time interaction $\beta_k g(t) x_k$, or use time-varying coefficient models.

Miscellanea: Unlike classical residuals (observed minus predicted outcome), Schoenfeld residuals are not defined on the response variable at all. Instead, they are computed for each covariate at each event time, measuring the difference between the observed covariate of the failing subject and its expected value given the risk set. In this sense they diagnose model fit in the covariate space rather than the outcome space—making them fundamentally different from residuals in standard regression.

Reference(s): Schoenfeld (1982); Grambsch & Therneau (1994)

Schoenfeld Residuals Test SURV.8

Worked Example

Problem Statement After fitting a Cox model to assess whether biomarker z affects survival, the analyst checks the proportional hazards assumption using Schoenfeld residuals. Use $\hat{\beta} = 0.522$.

$$W_j = \sum_{i \in \mathcal{R}_j} e^{\hat{\beta} z_i}, \quad \bar{z}^*_j = \frac{\sum_{i \in \mathcal{R}_j} z_i e^{\hat{\beta} z_i}}{W_j}, \quad \overline{z^2}^*_j = \frac{\sum_{i \in \mathcal{R}_j} z_i^2 e^{\hat{\beta} z_i}}{W_j}$$

Subject	1	2	3	4	5	6	7	8
Time t_i	2	4	6^+	7	9	12^+	14	18
Covariate z_i	7.0	8.0	4.0	6.0	5.5	3.0	4.0	6.5

1. Schoenfeld Residual at Each Event Time

At each t_j, compute the *weighted* risk-set mean using $\exp(\hat{\beta} z_i)$ weights, then subtract from the observed failure covariate (see formula above).

The raw and scaled residuals and risk-set information are:

$$r_j = z_{(j)} - \bar{z}^*_j, \qquad r^*_j = r_j \,/\, v_j, \qquad v_j = \overline{z^2}^*_j - (\bar{z}^*_j)^2$$

At $t_j = 2$: all 8 at risk, $W_2 = 195.4$, $\bar{z}^*_2 = 6.658$, $v_2 = 1.713$, $r_2 = 7.0 - 6.658 = 0.342$, $r^*_2 = 0.342/1.713 = 0.200$.

t_j	n_j	$z_{(j)}$	$\bar{z}^*_j$	$\overline{z^2}^*_j$	v_j	r_j	r^*_j
2	8	7.0	6.658	46.045	1.713	+0.342	+0.200
4	7	8.0	6.574	45.315	2.101	+1.426	+0.679
7	5	6.0	5.706	33.524	0.962	+0.294	+0.305
9	4	5.5	5.595	32.582	1.282	−0.095	−0.074
14	2	4.0	5.967	36.653	1.048	−1.967	−1.876
18	1	6.5	6.500	42.250	0.000	0.000	0.000

$\sum r_j \approx 0$ confirms $\hat{\beta}$ is the MLE. At $t_j = 18$: only 1 subject at risk, so $r_j = r^*_j = 0$ trivially—it carries no info and is **excluded** from the regression.

2. Regression of r^*_j on $g(t_j) = t_j$

Test for a linear time trend in the r^*_j ($t \neq 18, n = 5$):

$$r^*_j = a + b\,t_j + \varepsilon_j, \quad H_0 : b = 0$$

With $\bar{t} = 7.2$ and $\bar{r}^* = -0.153$:

$$\hat{b} = \frac{\sum(t_j - \bar{t})(r^*_j - \bar{r}^*)}{\sum(t_j - \bar{t})^2} = \frac{-16.162}{86.8} = -0.186$$

$$\hat{s}^2 = \frac{RSS}{n-2} = \frac{0.993}{3} = 0.331, \quad SE(\hat{b}) = \sqrt{\frac{\hat{s}^2}{S_{tt}}} = 0.062$$

$$\chi^2 = \left(\frac{\hat{b}}{SE(\hat{b})}\right)^2 = \left(\frac{-0.186}{0.062}\right)^2 = 9.094$$

3. Results and Decision

Statistic	Value	CV (5%)	Decision
χ^2	9.094	3.841	Reject H_0

4. Interpretation

- **Decision:** $\chi^2 = 9.094 > 3.841$: we **reject** H_0. Significant -ve time trend in the scaled Schoenfeld residuals.

Practitioner Insight: The Schoenfeld residual r_j is the difference between the observed covariate of the failing subject and its risk-set expectation at $\hat{\beta}$—positive if higher than expected, negative otherwise. Scaling by v^*_j standardises residuals for comparability across event times. The test regresses r^*_j on $g(t_j)$; here $g(t_j) = t_j$ tests for linear time variation, with alternatives such as $\log(t_j)$ or $\hat{S}_{\text{KM}}(t_j)$. A significant slope indicates a time-varying hazard ratio, violating proportional hazards. Incorrectly including, $r_j = 0$ deflates the slope estimate and inflates $SE(\hat{b})$ (here: χ^2 drops from 9.09 to 1.68, reversing the conclusion).The test is applied per covariate, with a global test combining them.

Common Pitfalls

- **Concluding proportional hazards holds from visual inspection alone.** Residual plots can appear randomly scattered even when subtle trends exist. Formal tests and smoothed trend lines should accompany visual inspection.
- **Testing residuals against only one time transformation.** Different transformations of time may reveal different types of PH violations. Examining multiple forms such as identity, log, or rank time improves diagnostic sensitivity.
- **Statistical significance with clinical importance.** Large samples may detect minor time variation that has little practical impact. Interpreting size of hazard ratio changes over time is essential for meaningful conclusions.
- **Testing when the number of events is very small.** Schoenfeld residuals are defined only at event times, so power depends on the number of observed events. With few events, both plots and formal tests may fail to detect real violations.
- **Using Schoenfeld residuals as a general Cox model diagnostic.** These residuals assess violations of the PH assumption only. Other model issues require different diagnostics such as martingale, deviance, or influence residuals.

Grambsch-Therneau Test

Global Test for Proportional Hazards Using Scaled Schoenfeld Residuals

SURV.9 | **Category:** Survival Analysis | **Data:** Time-to-Event | **Null:** Chi-square

When to Use

- Omnibus test for PH across all covariates simultaneously.
- Improved version of Schoenfeld test with better scaling and distributional properties.
- Standard diagnostic for Cox model PH assumption.

Assumptions

- Cox proportional hazards model has been fitted.
- Sufficient events ($D \geq 30$) for reliable asymptotic approx.
- Censoring is non-informative.
- $g(t)$ chosen appropriately for detecting departures.

Test Statistic

$$T(G) = \left(\textstyle\sum_k G_k \hat{r}_k\right)' D^{-1} \left(\textstyle\sum_k G_k \hat{r}_k\right), \qquad D = \textstyle\sum_k G_k \hat{V}_k G_k' - \left(\sum_k G_k \hat{V}_k\right)\left(\sum_k \hat{V}_k\right)^{-1}\left(\sum_k G_k \hat{V}_k\right)'.$$

Under H_0, $T(G) \sim \chi^2_p$ where p is number of covariates. $\hat{V}_k = V(\hat{\beta}, t_k)$ the weighted covariate variance in the risk set at t_k.
Notation: $\hat{r}_k$ and $r_k^* = \hat{V}_k^{-1}\hat{r}_k$ are the raw and scaled Schoenfeld residuals, $G_k = g(t_k)$ is the chosen time function

Plain English: The Grambsch-Therneau test improves upon the original Schoenfeld residuals test by using a properly weighted score statistic and a global form that combines evidence across all covariates. Scaling raw residuals by the inverse of the risk-set information matrix V_k converts them into estimates of the time-varying coefficient $\beta(t)$; if PH holds these should scatter randomly about a constant. The global test detects whether any covariate violates proportional hazards, providing omnibus assessment of model adequacy.

⚠ Practitioner Warning

Global test can reject even if only one covariate violates proportionality; examine individual covariate tests to identify culprits. Low power when violations are in opposite directions for different covariates (cancellation effect). Choice of time function $g(t)$ affects power; $g(t) = \log t$ (R default) is more robust to late-time leverage than $g(t) = t$. Test assumes model otherwise correctly specified (functional form, linearity). With many covariates, test becomes less interpretable—which covariate(s) violated? Graphical diagnostics essential for understanding nature of violations. $T(G)$ and the approximation $D\hat{\rho}^2$ are not identical; use the p-value from $T(G)$ as reported by software.

Hypotheses

- H_0: PH holds for all covariates ($\beta_k(t) = \beta_k$).
- H_1: At least one covariate has time-varying effect.

Decision Rule: Reject H_0 if $T(G) > \chi^2_{\alpha,p}$ or p-value $< \alpha$.

Software References [Python, R, SAS]

Python	R	SAS
`lifelines.CoxPHFitter`	`survival::cox.zph`	`PROC PHREG`

Additional Notes

Scaled Schoenfeld residuals: $r_k^* = \hat{V}_k^{-1}\hat{r}_k$, where $\hat{V}_k = V(\hat{\beta}, t_k)$ is the weighted covariate variance matrix in the risk set at event time t_k (GT 1994, §2). For graphical diagnostics, $r_k^* + \hat{\beta}$ estimates the time-varying coefficient $\beta(t_k)$; smoothing this against t_k (e.g. lowess, splines) visualises departures from PH. The formal score test $T(G)$ uses raw residuals $\hat{r}_k$ with information weights $\hat{V}_k$; the Pearson correlation $\hat{\rho} = \mathrm{cor}(r_k^*, g(t_k))$ reported by `cox.zph` is a diagnostic summary and $D\hat{\rho}^2 \approx T(G)$ only asymptotically. Common choices: $g(t) = t$ (linear), $g(t) = \log(t)$ (log-linear; R default), $g(t) = \mathrm{rank}(t)$ (rank-based), or $g(t) = \hat{S}(t)$ (KM-based). Individual covariate tests: $T_k \sim \chi^2_1$. Software reports both global and individual tests. Remedies include stratification, time interactions, or time-varying coefficient models.

Miscellanea: Grambsch and Therneau (1994) refined Schoenfeld residuals (1982) by introducing a formal score test framework and showing that scaled residuals estimate the time-varying coefficient $\beta(t)$, enabling both hypothesis testing and graphical diagnosis of non-proportionality. The method is widely implemented in statistical software and remains the standard tool for evaluating the PH assumption in Cox models.

Reference(s): Grambsch & Therneau (1994); Schoenfeld (1982)

Worked Example

Problem Statement A Cox model with biomarker covariate z is fitted to $N = 9$ patients (6 events, 3 censored: 5^+, 11^+, 22^+), yielding $\hat{\beta} = 0.200$. The Grambsch-Therneau (1994) test checks whether $\hat{\beta}$ is constant over time using $g(t_j) = \log t_j$ (the R default). H_0: $\beta(t) = \beta$ (PH holds); H_1: $\beta(t)$ varies with time.

Subject	1	2	3	4	5	6	7	8	9
Time t_i	3	5^+	7	9	11^+	13	16	19	22^+
Covariate z_i	6.0	4.0	7.5	5.0	3.5	4.5	5.5	4.0	6.5

1. Scaled Schoenfeld Residuals

At each event t_j, compute the weighted risk-set mean $\bar{z}_j^*$ and information v_j, then the raw residual $r_j = z_{(j)} - \bar{z}_j^*$ and the scaled residual $r_j^* = r_j / v_j$.

t_j	n_j	$z_{(j)}$	$\bar{z}_j^*$	v_j	r_j	$\log t_j$	r_j^*
3	9	6.0	5.491	1.671	+0.509	1.099	+0.305
7	7	7.5	5.570	1.832	+1.930	1.946	+1.053
9	6	5.0	5.033	1.017	−0.033	2.197	−0.032
13	4	4.5	5.313	0.945	−0.813	2.565	−0.860
16	3	5.5	5.537	0.974	−0.037	2.773	−0.038
19	2	4.0	5.556	1.469	−1.556	2.944	−1.059

$\sum r_j \approx 0$ confirms $\hat{\beta}$ is the MLE. Adding $\hat{\beta} = 0.200$ to each r_j^* gives the diagnostic plot values $\hat{\beta}(t_j)$ (not needed for the test itself).

2. Exact GT Score Test

Center $\tilde{g}_j = \log t_j - \bar{g}$ ($\bar{g} = 2.254$):

$$T_{GT} = \frac{(\sum_j r_j \tilde{g}_j)^2}{\sum_j v_j \tilde{g}_j^2} = \frac{(-2.527)^2}{3.461} = \mathbf{1.845}$$

$$p\text{-value} = P(\chi_1^2 > 1.845) = \mathbf{0.174}$$

3. Pearson Correlation (Diagnostic)

Compute $\hat{\rho} = \text{cor}(r_j^*, \log t_j)$:

$$\bar{g} = 2.254, \quad \bar{r}^* = -0.105$$

$$S_{gr} = \sum(\log t_j - \bar{g})(r_j^* - \bar{r}^*) = -1.693$$

$$S_{gg} = 2.275, \quad S_{rr} = 2.999$$

$$\hat{\rho} = \frac{-1.693}{\sqrt{2.275 \times 2.999}} = -0.648$$

Approximation: $D\hat{\rho}^2 = 6 \times 0.420 = 2.521$ ($p = 0.112$). This *differs* from T_{GT}; use T_{GT} for inference.

4. Interpretation

- **Decision:** $T_{GT} = 1.845 < 3.841$ (5% CV, χ_1^2): we **fail to reject** H_0 ($p = 0.174$). The correlation $\hat{\rho} = -0.648$ is moderate but not significant at $D = 6$. No evidence of a time-varying hazard ratio; the PH assumption is upheld.

Practitioner Insight: The formal GT statistic T_{GT} uses raw residuals r_j with information weights v_j in both numerator and denominator, and is the quantity whose p-value `cox.zph` reports. The Pearson correlation $\hat{\rho} = \text{cor}(r_j^*, g_j)$ is a useful diagnostic summary and $D\hat{\rho}^2$ is an unweighted approximation; the two agree asymptotically but can diverge in small samples (here: 1.845 vs. 2.521). For the diagnostic plot, $r_j^* + \hat{\beta}$ is plotted against t_j; adding $\hat{\beta}$ shifts the display to the $\beta(t)$ scale but does not affect any test quantity. The choice of g matters: t_j tests linear trends, $\log t_j$ (R default) logarithmic trends. Remedies when PH fails: stratification, $\beta(t) = \beta_0 + \beta_1 g(t)$ time-interaction model, or piecewise Cox.

Common Pitfalls

- **Confusing T_{GT} with $D\hat{\rho}^2$.** R's `cox.zph` reports both a χ^2 p-value (from T_{GT}) and a correlation $\hat{\rho}$. The approximation $D\hat{\rho}^2$ can differ meaningfully from T_{GT} in small samples; always use the reported p-value, not $D\hat{\rho}^2$, for inference.
- **Using only a single time transformation in the test.** Different time transformations emphasize different portions of follow-up. Evaluating multiple transformations can reveal violations that a single transformation may miss.
- **Relying only on the global test result.** The global statistic indicates whether any covariate violates the proportional hazards assumption. Individual covariate tests and residual plots are needed to identify which effects vary over time.
- **Treating borderline p-values as definitive conclusions.** Results near conventional thresholds often reflect limited power or modest deviations from PH. Visual diagnostics and clinical reasoning should accompany the formal test.
- **Ignoring violations because the overall Cox model remains significant.** A significant model does not guarantee that hazard ratios remain constant over time. Time-varying effects may still distort the notion of a constant hazard ratio.

Cox-Snell Residuals Test

Test Overall Model Fit via Cumulative Hazard

SURV.10 | **Category:** Survival Analysis | **Data:** Time-to-Event | **Null:** Visual/KS test

When to Use

- Assess overall goodness-of-fit for Cox PH model.
- Diagnose departures from model assumptions.
- Follow-up after fitting a survival model.
- Diagnostic for model specification beyond PH assumption.

Assumptions

- Cox PH (or other survival) model has been fitted.
- Model correctly specified under null hypothesis.
- Sufficient events to estimate cumulative hazard of residuals.
- Censoring is non-informative.

Test Statistic

$$r_i^{CS} = \hat{H}_0(t_i)\exp(\mathbf{x}_i'\hat{\boldsymbol{\beta}}), \quad \text{where } \hat{H}_0(t) = \sum_{j:t_j \le t} \frac{d_j}{\sum_{k \in R_j} \exp(\mathbf{x}_k'\hat{\boldsymbol{\beta}})}.$$

Under correct model, r_i^{CS} should follow unit exponential (censored). Plot $\hat{H}(r^{CS})$ vs. r^{CS}.

Notation: r_i^{CS} is Cox-Snell residual, $\hat{H}_0$ is estimated baseline cumulative hazard, unit slope indicates good fit

Plain English: Cox-Snell residuals transform observed survival times using the fitted model's cumulative hazard function. If the model fits well, these residuals should behave like a sample from a unit exponential distribution (for uncensored observations). The diagnostic plots cumulative hazard of residuals against the residuals themselves—good fit produces a straight line with slope one through the origin. Systematic deviations from this pattern indicate model misspecification, such as wrong functional form, omitted covariates, or violations of distributional assumptions.

⚠ Practitioner Warning

Primarily graphical diagnostic; formal tests (KS on residuals) have low power. Plot interpretation subjective—what constitutes "close to 45-degree line"? Residuals are censored when original observations censored, complicating interpretation. More useful for detecting gross misspecification than subtle violations. Does not identify specific problem when fit is poor (could be non-proportionality, non-linearity, omitted variables). Requires estimating baseline cumulative hazard, adding computation. Less commonly used than Schoenfeld residuals for proportional hazards specifically.

Hypotheses

- H_0: Model fits adequately ($r_i^{CS} \sim \text{Exp}(1)$).
- H_1: Model misspecified.

Decision Rule: Reject H_0 if $D > D_{\alpha,n}$ for KS test on r^{CS} vs. Exp(1). Plot of $\hat{H}(r^{CS})$ vs. r^{CS} must be 45° line.

Software References [Python, R, SAS]

`lifelines.CoxPHFitter`	`survival::coxph`	`PROC PHREG (OUTPUT CS)`

Additional Notes

Cox–Snell residuals: for a correctly specified model, the cumulative hazard $H(T)$ follows Exp(1), so $r_i^{CS} = \hat{H}(t_i|\mathbf{x}_i) = \hat{H}_0(t_i)\exp(\mathbf{x}_i'\hat{\boldsymbol{\beta}})$ should approximate unit exponential. In Cox models, $\hat{H}_0(t)$ is typically estimated via the Breslow estimator. Diagnostic: treat r_i^{CS} as survival times (with original censoring), estimate $\hat{H}(r^{CS})$ using KM or Nelson–Aalen, and plot against r^{CS}; a 45° line indicates good fit, while systematic deviations suggest misfit. Martingale residuals $M_i = \delta_i - r_i^{CS}$ help assess functional form, and deviance residuals provide a symmetrized, more normal alternative.

Miscellanea: David Cox and Dennis Snell developed these residuals in 1968, well before Cox's famous 1972 PH model. The residuals were originally proposed for parametric survival models (Weibull, exponential) to check distributional assumptions. Their application to Cox models came later as researchers sought goodness-of-fit diagnostics. The residuals' theoretical elegance—transforming survival times to unit exponential under correct model—appealed to mathematical statisticians. However, practical adoption lagged due to implementation complexity and preference for Schoenfeld residuals for PH checking specifically.

Reference(s): Cox & Snell (1968); Collett (2015)

Worked Example

Problem Statement The Cox PH model fitted to the biomarker–survival data from the Schoenfeld example ($\hat{\beta} = 0.522$, $N = 8$, 6 events) is assessed for overall goodness-of-fit using Cox-Snell residuals. H_0: model fits; H_1: model misspecified.

1. Breslow Baseline Cumulative Hazard

At each event time t_j, the Breslow increment is $\Delta\hat{H}_0(t_j) = d_j / \sum_{i \in \mathcal{R}_j} e^{\hat{\beta} z_i}$ (with $d_j = 1$ throughout):

t_j	n_j	$\sum_{\mathcal{R}_j} e^{\hat{\beta} z_i}$	$\hat{H}_0(t_j)$
2	8	195.355	0.0051
4	7	156.649	0.0115
7	5	83.320	0.0235
9	4	60.361	0.0401
14	2	37.888	0.0665
18	1	29.810	0.1000

2. Cox-Snell Residuals

$r_i = \hat{H}_0(t_i) \cdot e^{\hat{\beta} z_i}$ (use $\hat{H}_0$ at last event time $\le t_i$):

i	evt	z_i	$\hat{H}_0(t_i)$	$e^{\hat{\beta} z_i}$	r_i
1	Y	7.0	0.0051	38.705	0.198
2	Y	8.0	0.0115	65.252	0.751
3	N	4.0	0.0115	8.078	0.093
4	Y	6.0	0.0235	22.959	0.540
5	Y	5.5	0.0401	17.682	0.709
6	N	3.0	0.0401	4.792	0.192
7	Y	4.0	0.0665	8.078	0.537
8	Y	6.5	0.1000	29.810	2.981

3. KM of $\{r_i\}$ vs Exp(1)

Treat r_i as survival times with the original censoring indicators; compute KM and compare to $S(r) = e^{-r}$:

r_i	evt	n	$\hat{S}_{KM}(r)$	e^{-r}	diff
0.093	N	8	1.000	0.911	+0.089
0.192	N	7	1.000	0.825	+0.175
0.198	Y	6	0.833	0.820	+0.013
0.537	Y	5	0.667	0.585	+0.082
0.540	Y	4	0.500	0.583	−0.083
0.709	Y	3	0.333	0.492	−0.159
0.751	Y	2	0.167	0.472	−0.305
2.981	Y	1	0.000	0.051	−0.051

4. Results and Decision

$$D = \max_i |\hat{S}_{KM}(r_i) - e^{-r_i}| = 0.305$$

attained at $r_7 = 0.751$. Comparing to the Kolmogorov 5% critical value (K-S adj.= $(\sqrt{n} + 0.12 + 0.11/\sqrt{n})$) for $n = 8$:

Statistic	Value	CV (5%)	Decision
D	0.305	0.457	FTR H_0

$p > 0.05$. The KM curve tracks e^{-r} adequately; no evidence of model misspecification.

Practitioner Insight: The Cox–Snell residual $r_i = \hat{H}_0(t_i) e^{\hat{\beta} z_i}$ represents fitted cumulative hazard. Under H_0, the residuals behave like a censored Exp(1) sample, so the standard diagnostic is a plot of the estimated cumulative hazard (e.g., Nelson–Aalen) versus r, which should follow a 45° line. Tail deviations are common with small samples or heavy censoring and are not, by themselves, strong evidence of misspecification. The identity $\sum r_i = d$ is an arithmetic check only. The KS-type statistic D is reported as a descriptive measure of deviation from Exp(1). Because residuals are censored and model-based, standard KS critical values are not strictly valid; visual assessment remains primary.

Common Pitfalls

- **Expecting perfect alignment with the 45-degree reference line.** Even correctly specified models produce random scatter around the theoretical line. Interpretation should focus on systematic curvature (suggests misspecified covariate effects or omitted predictors) or persistent slope deviations rather than small random departures.
- **Using Cox-Snell residuals as the only model diagnostic.** The Cox–Snell plot evaluates overall model fit but does not test specific assumptions. Other diagnostics are needed to assess PH, functional form, and influential observations.
- **Ignoring the influence of censoring on the residual distribution.** Heavy censoring reduces the number of events contributing to the upper tail of the residual plot. Sparse late-event information can create apparent deviations even when the model is correctly specified.
- **Using Cox-Snell residuals to evaluate individual prediction errors.** These residuals assess population-level agreement between model and observed survival patterns. Individual prediction accuracy and influential observations require other residual diagnostics.

Gray's Test (Competing Risks)

Compare Cumulative Incidence Functions Between Groups

SURV.11 | **Category:** Survival Analysis | **Data:** Time-to-Event | **Null:** Chi-square

When to Use

- Compare cumulative incidence functions by group.
- Test treatment effects when multiple event types possible.
- Assess cause-specific event probabilities.
- Clinical or epidemiological follow-up studies.
- Alternative to log-rank under competing risks.

Assumptions

- Observations are mutually independent.
- Censoring is noninformative and independent.
- Causes of failure are correctly classified.
- Event times are correctly recorded.
- Competing risks are explicitly modeled.

Test Statistic

$$W = U_k^\top \hat{V}^{-1} U_k \dot{\sim} \chi^2_{K-1}, \quad U_k = \sum_j w(t_j)(d_{1j} - E_{1j}), \quad w(t) = \{1 - \hat{F}_k(t^-)\}^\rho \ (\rho = 0 \text{ standard})$$

Notation: $\hat{F}_k$: CIF; d_{1j}: observed cause-k events; E_{1j}: expected under H_0. For two groups, $W = U_k^2/\widehat{\text{Var}}(U_k) \sim \chi^2_1$.

Plain English: Gray's test compares cumulative incidence functions (CIFs) for a specific event type between groups when competing risks are present. Unlike Kaplan-Meier which treats competing events as censored, cumulative incidence properly accounts for the fact that subjects experiencing competing events can no longer experience the event of interest. The test weights differences in CIF at each event time by sample sizes, summing across all times. Large weighted differences indicate groups differ in probability of experiencing the specific event type.

⚠ Practitioner Warning

Log-rank test is inappropriate with competing risks, as treating competing events as censoring yields invalid inference. Gray's test compares cause-specific CIFs, not hazards (use Fine–Gray regression for effect estimation). Accurate event classification is critical; misclassification invalidates results. Power is limited with few events, and computation increases with many event times. Does not provide effect size. Multiple causes require separate tests with multiplicity adjustment.

Hypotheses

- H_0: $\text{CIF}_1(t,k) = \text{CIF}_2(t,k)$ for cause k.
- H_1: $\text{CIF}_1(t,k) \neq \text{CIF}_2(t,k)$ for cause k.

Decision Rule: Reject H_0 if $W > \chi^2_{\alpha,1}$ (two groups) or generalized statistic $> \chi^2_{\alpha,K-1}$ (K groups) or p-value $< \alpha$.

Software References [Python, R, SAS]

`lifelines.statistics...` | `cmprsk::cuminc` | `PROC LIFETEST (STRATA CROSSLOG)`

Additional Notes

Cumulative incidence for cause k: $CIF(t,k) = P(T \le t, \delta = k) = \int_0^t S(u^-)h_k(u)\,du$, combining overall survival $S(t)$ and cause-specific hazard $h_k(t)$. Estimated by $\hat{CIF}(t,k) = \sum_{j:t_j \le t} \frac{d_j(k)}{n_j} \prod_{i:t_i<t_j}(1 - d_i/n_i)$, where the product is the Kaplan–Meier estimate of overall survival. Gray (1988) proposed a weighted test (analogous to Fleming–Harrington) with weights w_j (commonly n_j), allowing emphasis on different time regions. Fine–Gray (1999) extends this to regression; Gray's test is the corresponding score test with a single group indicator. For multiple causes, test each separately with multiplicity adjustment (e.g., Bonferroni, Holm). Alternatives include the Pepe–Mori test for general K-sample settings.

Miscellanea: Robert Gray developed this test in 1988 while at Dana-Farber Cancer Institute, addressing a critical gap in survival analysis. Competing risks had been recognized since the 1960s (Chiang (1968), Gail (1975)), but no standard test existed for comparing cumulative incidence. Researchers inappropriately used log-rank tests treating competing events as censored, yielding biased results. Gray demonstrated how log-rank and his test could yield opposite conclusions when competing mortality was high. The test gained rapid acceptance, though Fine-Gray's 1999 regression extension eventually became more popular for its flexibility.

Reference(s): Gray (1988); Fine & Gray (1999)

Worked Example

Problem Statement An oncologist compares cumulative incidence of disease-specific death (cause 1) between a treatment group ($n_1 = 6$) and a control group ($n_2 = 6$) in the presence of death from other causes (cause 2, a *competing risk*).

Subject	1	2	3	4	5	6
Time (Trt)	2	4	6	9	13	17
Cause (Trt)	2	2	1	2	2	2
Time (Ctrl)	1	3	5	7	10	15
Cause (Ctrl)	1	1	1	1	2	1

1. The Subdistribution Risk Set

Gray's test replaces the ordinary risk set with the *subdistribution risk set* $\mathcal{R}_j^*$:

- Subjects not yet having *any* event ($t_i \geq t_j$): **included** as usual.
- Subjects who experienced a *competing* event ($c_i = 2$, $t_i < t_j$): **remain in** $\mathcal{R}_j^*$ (with weight 1 here; no censoring present).
- Subjects who experienced *cause-1* event ($c_i = 1$, $t_i < t_j$): **removed**.

2. Weighted Event Table (weight $w_j = 1$)

t_j	n_{1j}^*	n_{2j}^*	n_j^*	d_{1j}	d_{2j}	E_{1j}	$d_{1j} - E_{1j}$	V_j
1	6	6	12	0	1	0.500	−0.500	0.250
3	6	5	11	0	1	0.545	−0.545	0.248
5	6	4	10	0	1	0.600	−0.600	0.240
6	6	3	9	1	0	0.667	+0.333	0.222
7	5	3	8	0	1	0.625	−0.625	0.234
15	5	2	7	0	1	0.714	−0.714	0.204
						Total	**−2.651**	**1.399**

At $t_j = 3$: Group 1 has $n_{1j}^* = 6$ (no cause-1 exits yet; the cause-2 subject at $t = 2$ *stays*). Group 2 has $n_{2j}^* = 5$ (cause-1 subject at $t = 1$ removed).

3. Gray's Test Statistic

The statistic mirrors the log-rank but uses n_{gj}^*:

$$V_j = \frac{n_{1j}^* n_{2j}^* d_j (n_j^* - d_j)}{(n_j^*)^2 (n_j^* - 1)}, \qquad W = \frac{U^2}{V}$$

4. Results and Decision

Test	Statistic	p	Decision
Gray's (subdistribution)	5.026	0.025	Reject
Log-rank (cause-specific)	2.870	0.090	FTR

The CIF of cause-1 events is significantly lower in the treatment group.

5. Interpretation

- **Decision: Reject** H_0. The cause-specific log-rank misses this ($p = 0.090$) as it ignores the competing risk structure.

Practitioner Insight: Gray's test retains competing-event subjects in the subdistribution risk set. In Group 1, subjects experiencing other causes remain in $\mathcal{R}_j^*$, so n_{1j}^* stays elevated even when no subjects are at risk of cause 1. This reflects that, if competing causes were removed, these subjects would still be at risk—consistent with the CIF. In contrast, the cause-specific log-rank treats competing events as censoring, distorting early risk sets and understating CIF differences. With no censoring, the subdistribution weight equals 1; with right-censoring, it adjusts via the censoring KM ratio. Gray's test extends to $K > 2$ causes or $G > 2$ groups as a $(G - 1)$ df chi-squared using group-level score contributions.

Note: This example uses a simplified, textbook variance for Gray's test. Software (e.g., `cmprsk` in **R**) uses a score/covariance recursion, so results may differ slightly. Use software for final inference.

Common Pitfalls

- **Using the log-rank test while censoring competing events.** Treating competing events as censored observations violates the assumptions of standard survival methods. Gray's test should be used when competing risks are present because it properly accounts for the cumulative incidence of the event of interest.
- **Confusing cumulative incidence with cause-specific hazard.** Gray's test compares cumulative incidence functions, which represent the probability of experiencing an event by time t. Cause-specific hazards measure instantaneous risk among those still event-free and answer a different scientific question.
- **Applying competing risks methods when events are not mutually exclusive.** Competing risks analysis assumes that each individual experiences at most one type of event. If multiple event types can occur sequentially, a multi-state modeling framework is more appropriate.
- **Performing multiple pairwise Gray's tests without adjustment.** When more than two groups are compared, multiple pairwise tests inflate Type I error. Multiplicity adjustments such as Bonferroni or Holm procedures should be applied.

Fine-Gray Test

Test Covariate Effects on Subdistribution Hazard with Competing Risks

SURV.12 | **Category:** Survival Analysis | **Data:** Time-to-Event | **Null:** Chi-square

When to Use

- Model competing-risk outcomes with covariates.
- Assess effects on cumulative incidence functions.
- Competing risks are present in the data.
- Model subdistribution hazards when interest is in absolute risk (CIF–Cumulative Incidence Function).

Assumptions

- Observations are mutually independent.
- Censoring is noninformative and independent.
- Multiple causes of failure are correctly classified.
- Covariate effects are proportional over time.
- Subdistribution hazards are correctly specified.

Test Statistic

$$\text{Score: } \chi^2 = U(\hat{\beta})'\mathcal{I}^{-1}U(\hat{\beta}) \overset{H_0}{\sim} \chi^2_p \quad \Big| \quad \text{LR: } -2[\ell_0 - \ell_1], \quad \text{Wald: } \hat{\beta}'\mathcal{I}\hat{\beta} \quad \Big| \quad \lambda_k(t|\mathbf{x}) = \lambda_{0k}(t)\exp(\mathbf{x}'\boldsymbol{\beta}_k)$$

Notation: $\lambda_k(t|\mathbf{x})$ is subdistribution hazard, $\boldsymbol{\beta}_k$ are covariate effects on cause k, p is covariates, $\mathcal{I}$ is information matrix, $U(.)$ is the score vector, ℓ is modified partial likelihood for subdistribution

Plain English: The Fine-Gray model extends Cox regression to competing risks by modeling the subdistribution hazard. The model evaluates how covariates affect the probability of a specific event over time in the presence of competing risks. Instead of modeling instantaneous event rates, it directly targets cumulative incidence (absolute risk). Individuals who experience competing events are retained in a modified risk set, allowing estimated effects to correspond directly to differences in event probabilities across groups.

⚠ Practitioner Warning

The subdistribution hazard lacks a direct physical interpretation and uses a nonstandard risk set that includes individuals after competing events. Results differ from cause-specific models and should not be interpreted as effects on instantaneous risk. Violations of proportionality can bias estimates. Not suitable for studying underlying biological mechanisms—use cause-specific hazards instead.

Hypotheses

- H_0: $\boldsymbol{\beta}_k = \mathbf{0}$
- H_1: $\boldsymbol{\beta}_k \neq \mathbf{0}$

Decision Rule: Reject H_0 if Wald, LRT, or score statistic $> \chi^2_{\alpha,p}$ or p-value $< \alpha$.

Software References [Python, R, SAS]

`lifelines.CRCSplineFitter`	`cmprsk::crr`	`PROC PHREG`

Additional Notes

Subdistribution hazard: $\lambda_k(t) = \lim_{\Delta t\to 0} \frac{P(t\le T<t+\Delta t,\delta=k|T\ge t\cup(T<t,\delta\neq k))}{\Delta t}$, where the risk set includes those who have not yet experienced any event plus those who have already experienced a competing event. The risk set retains individuals with prior competing events (covariates frozen), handled via weighted partial likelihood. Fine and Gray (1999) established consistency and asymptotic normality. In contrast, cause-specific models treat competing events as censored and target event-specific hazards. Choice depends on goal: cumulative incidence (subdistribution) vs. etiologic rates (cause specific). With a single group indicator, the Fine–Gray score test reduces to Gray's test.

Miscellanea: Fine and Gray (1999, JASA) introduced the subdistribution hazards model to directly link covariate effects to cumulative incidence in competing risks settings. Although the modified risk set initially drew criticism for its lack of intuitive interpretation, the model gained wide acceptance because it provides clinically meaningful measures of absolute risk. For instance, if the event of interest is a heart attack and the competing event is death from other causes, the CIF gives the probability of having a heart attack by year 5, accounting for competing risks.

Reference(s): Fine & Gray (1999); Gray (1988)

Worked Example

Problem Statement Fine-Gray subdistribution hazard model is applied to $N = 10$ patients with continuous biomarker z, 5 cause-1 events (disease-specific death), and 5 cause-2 events (competing death). H_0: $\gamma = 0$; H_1: $\gamma \neq 0$.

1. Score Test at $\hat{\gamma} = 0$

The subdistribution risk set $\mathcal{R}_j^*$ retains competing-event subjects (cause 2, no censoring $\Rightarrow$ weight 1). At each cause-1 time t_j: $\bar{z}_j^* = \sum_{i\in\mathcal{R}_j^*} z_i / n_j^*$, $u_j = z_{(j)} - \bar{z}_j^*$.

t_j	n_j^*	Comp. stays	$z_{(j)}$	$\bar{z}_j^*$	u_j	v_j^*
3	10	{4.5}	7.5	5.100	+2.400	1.840
5	9	{4.5}	6.5	4.833	+1.667	1.333
8	8	{4.5, 3.0}	6.0	4.625	+1.375	1.109
12	7	{4.5, 3.0, 6.0}	4.5	4.429	+0.071	0.959
17	6	{4.5, 3.0, 6.0, 3.5}	5.5	4.417	+1.083	1.118

At $t_j = 8$: two cause-2 subjects (from $t = 2$ and $t = 6$) remain in $\mathcal{R}_8^*$, so $n_8^* = 8$ although only 6 subjects have not yet had *any* event.

$$U(0) = \sum_j u_j = 6.596, \qquad \mathcal{I}(0) = \sum_j v_j^* = 6.360$$

$$X^2 = \frac{[U(0)]^2}{\mathcal{I}(0)} = \frac{(6.596)^2}{6.360} = 6.842 \quad \Rightarrow \quad p = 0.009$$

2. Partial Log-Likelihood At $\gamma = 0$: $\ell(0) = -\sum_j \log n_j^*$:

$$\ell(\gamma) = \sum_{j:\,c=1}\left[\gamma z_{(j)} - \log\sum_{i\in\mathcal{R}_j^*} e^{\gamma z_i}\right]$$

t_j	n_j^*	$-\log n_j^*$
3	10	−2.303
5	9	−2.197
8	8	−2.079
12	7	−1.946
17	6	−1.792
	Total	**−10.317**

$\hat{\gamma} = 1.483$ (maximises ℓ, numerically); $\ell(\hat{\gamma}) = -6.347$.

$$LR = 2\big[-6.347 - (-10.317)\big] = 2 \times 3.970 = 7.940$$

3. Results and Three-Test Comparison

Test	Formula	Stat	p	Decision
Score	$U^2/\mathcal{I}$	6.842	0.009	Reject
LR	$2\Delta\ell$	7.940	0.005	Reject
Wald	$\hat{\gamma}^2/\widehat{\text{Var}}$	4.395	0.036	Reject

($\widehat{\text{SE}}(\hat{\gamma}) = 0.707$; all reject at 5%.)

4. Interpretation

- Higher z significantly increases the subdistribution hazard of cause 1 ($\hat{\gamma} = 1.483 > 0$). The CIF of cause-1 events rises faster for patients with elevated z values.

Practitioner Insight: Fine-Gray is the regression analog of Gray's test: instead of comparing CIFs nonparametrically, it models the subdistribution hazard via covariates. The partial likelihood mirrors Cox's but uses the subdistribution risk set $\mathcal{R}_j^*$, so $\sum e^{\gamma z_i}$ grows as competing-event subjects remain in the risk set. At $\gamma = 0$, $\ell(0) = -\sum \log n_j^*$, analogous to the Cox null. The usual ordering LR > Score > Wald holds when $\hat{\gamma} > 0$. Interpreting $e^{\hat{\gamma}} = e^{1.483} \approx 4.4$: a unit increase in z multiplies the rate of CIF increase by 4.4. This is a subdistribution hazard ratio—not cause-specific—so it directly reflects cumulative incidence. While lacking a simple biological interpretation, it models the clinically relevant probability of the event by time t. See next page for data.

Common Pitfalls

- **Subdistribution hazard ratio ≠ standard hazard ratio.** FG model estimates effects on the CIF rather than cause-specific hazard. Interpretation must focus on changes in event prob. over time rather than instantaneous rates.
- **Expecting FG and cause-specific Cox models to produce identical results.** They target different quantities and answer different questions. Differences arise because CIF depends on both the event of interest and competing events.
- **Applying the Fine–Gray model when competing events are rare or uninformative.** When competing risks occur infrequently or are unrelated to covariates, simpler cause-specific Cox models may suffice. Fine–Gray models are most useful when competing events substantially influence observed event probabilities.
- **Ignoring the subdistribution PH assumption.** The Fine–Gray model assumes that subdistribution hazard ratios are time-invariant. Violations can distort meaning and must be assessed using CIF plots or time-varying effects.
- **Overlooking the modified risk set used in FG.** Subjects who experience competing events remain in the weighted risk set for estimating the subdistribution hazard. Requires specialized software to properly handle weighted risk sets.

XVIII. SURV — Raw Data

Data used in selected worked examples are provided here to preserve the book's strict two-page-per-test format.

[SURV.12] Fine-Gray Test

Problem Statement Fine-Gray subdistribution hazard model is applied to $N = 10$ patients with continuous biomarker z, 5 cause-1 events (disease-specific death), and 5 cause-2 events (competing death). H_0: $\gamma = 0$; H_1: $\gamma \neq 0$.

Subject	1	2	3	4	5	6	7	8	9	10
Time t_i	2	3	5	6	8	10	12	15	17	20
Cause c_i	2	1	1	2	1	2	1	2	1	2
Covariate z_i	4.5	7.5	6.5	3.0	6.0	6.0	4.5	3.5	5.5	4.0

Kaplan–Meier Estimator (Heuristic Construction)

Let event times be ordered as $0 \leq t_1 \leq \cdots \leq t_n$. Define R_i as the number at risk just prior to t_i, d_i the number of events, and c_i the number censored at t_i.

i	R_i	d_i	c_i	$\hat{h}_i$	$\hat{S}(t_i)$
0	R_0	0	0	0	1.000
1	R_1	d_1	c_1	$\frac{d_1}{R_1}$	$(1 - \hat{h}_1)$
2	R_2	d_2	c_2	$\frac{d_2}{R_2}$	$\hat{S}(t_1)(1 - \hat{h}_2)$
3	R_3	d_3	c_3	$\frac{d_3}{R_3}$	$\hat{S}(t_2)(1 - \hat{h}_3)$
$\vdots$	$\vdots$	$\vdots$	$\vdots$	$\vdots$	$\vdots$
n	R_n	d_n	c_n	$\frac{d_n}{R_n}$	$\hat{S}(t_{n-1})(1 - \hat{h}_n)$

$$R_{i+1} = R_i - d_i - c_i$$

Core idea: Survival is built multiplicatively from conditional survival:

$$\hat{S}(t) = \prod_{t_i \leq t} \left(1 - \frac{d_i}{R_i}\right)$$

Greenwood Variance:

$$\widehat{\mathrm{Var}}[\hat{S}(t)] \approx \hat{S}(t)^2 \sum_{t_i \leq t} \frac{d_i}{R_i(R_i - d_i)}$$

Likelihood Insight:

$$\log L = \sum_i \left[d_i \log h_i + (R_i - d_i) \log(1 - h_i)\right], \qquad \hat{h}_i = \frac{d_i}{R_i} \text{ (estimated hazard rate).}$$

Substituting $\hat{h}_i$ into the discrete survival identity yields the Kaplan–Meier estimator.

Censoring reduces R_i but does not directly change $\hat{S}(t)$ at that time.

The MaxCombo Test: A Robust Combination of Weighted Log-Rank Statistics

Motivation and Setup

The standard log-rank test maximizes power when the hazard ratio is constant over time (proportional hazards), but loses power substantially when the treatment effect is delayed, diminishing, or otherwise non-proportional — scenarios increasingly common in immuno-oncology trials. The Fleming–Harrington $G(\rho,\gamma)$ family addresses this by applying a weight $w_j = \hat{S}(t_j^-)^\rho(1-\hat{S}(t_j^-))^\gamma$ at each event time, allowing early, late, or middle emphasis depending on the choice of (ρ,γ). The difficulty is that the optimal weight function depends on the (unknown) hazard ratio pattern.

The **MaxCombo test** (Roychoudhury et al. 2021; Lin et al. 2020) sidesteps this pre-specification problem by combining m FH statistics simultaneously and rejecting if *any one of them* is sufficiently large. The standard four-component version uses $G(0,0)$, $G(1,0)$, $G(0,1)$, and $G(1,1)$, providing robustness across proportional, early, delayed, and humped hazard patterns. Crucially, the p-value is not obtained from a standard normal table but from the multivariate normal (MVN) distribution of the component statistics, properly accounting for their correlation.

Data

Twelve patients are randomised to treatment ($n_1 = 6$) or control ($n_2 = 6$). Superscript + denotes right-censoring.

	Treatment			Control		
Time	6	9^+	12	1	2	4
	16	20^+	25	6	8^+	10

Treatment has 4 events (2 censored); control has 5 events (1 censored). The hypotheses are $H_0 : S_1(t) = S_2(t)$ for all t vs. $H_1 : S_1(t) > S_2(t)$ for some t (one-sided).

Base Event Table

At each distinct event time t_j we record the at-risk counts n_{kj}, events d_{kj}, expected events for Group 1 $E_{1j} = n_{1j}d_j/n_j$, and the hypergeometric variance $V_j = n_{1j}n_{2j}d_j(n_j - d_j)/[n_j^2(n_j-1)]$. These quantities are shared by all four FH components; only the weights differ.

Table 1: Base event table. Event times $t = 12, 16, 25$ have $n_{2j} = 0$ so $V_j = 0$ and contribute to no statistic.

t_j	n_{1j}	n_{2j}	d_j	E_{1j}	$d_{1j} - E_{1j}$	V_j
1	6	6	1	0.500	−0.500	0.250
2	6	5	1	0.545	−0.545	0.248
4	6	4	1	0.600	−0.600	0.240
6	6	3	2	1.333	−0.333	0.389
10	4	1	1	0.800	−0.800	0.160
12	4	0	1	1.000	0.000	0.000
16	3	0	1	1.000	0.000	0.000
25	1	0	1	1.000	0.000	0.000
				Sum	**−2.779**	**1.287**

The uniformly negative $d_{1j} - E_{1j}$ column (at informative times) confirms that treatment subjects consistently had fewer events than expected, a pattern visible across all event times where both groups had subjects at risk.

Pooled Kaplan–Meier Weights

The FH weight at time t_j is a function of $\hat{S}(t_j^-)$, the pooled KM survival estimate just before t_j. Table 2 gives $\hat{S}(t_j^-)$ and the three distinct non-trivial weight functions. Note that $G(0,0)$ has constant weight 1 (the log-rank), while $G(1,0)$ decreases, $G(0,1)$ increases, and $G(1,1)$ rises then falls, peaking near $\hat{S} = 0.5$ (around $t = 10$ where $\hat{S}(t^-) = 0.583$).

Table 2: Pooled KM survival and FH weights at each event time.

t_j	$\hat{S}(t_j^-)$	$w^{(0,0)}$	$w^{(1,0)}$	$w^{(0,1)}$	$w^{(1,1)}$
1	1.000	1.000	1.000	0.000	0.000
2	0.917	1.000	0.917	0.083	0.076
4	0.833	1.000	0.833	0.167	0.139
6	0.750	1.000	0.750	0.250	0.188
10	0.583	1.000	0.583	0.417	0.243
12	0.467	1.000	0.467	0.533	0.249
16	0.350	1.000	0.350	0.650	0.228
25	0.233	1.000	0.233	0.767	0.179

Computing the FH Component Statistics

For each component k, the weighted statistic, variance, and standardized Z-score are:

$$U_k = \sum_j w_{kj}(d_{1j} - E_{1j}), \quad V_k = \sum_j w_{kj}^2 V_j,$$

$$Z_k = U_k/\sqrt{V_k}.$$

$G(0,0)$ (the log-rank) and $G(0,1)$ are derived in full in the preceding chapter examples. Here we show $G(1,0)$ in detail to illustrate the general procedure; $G(1,1)$ follows identically.

Worked detail: $G(1,0)$

Table 3 applies the decreasing weight $w_j = \hat{S}(t_j^-)$ from Table 2. Three observations are worth noting. First, the first three rows contribute *exactly* −0.500 each to $U_{(1,0)}$ despite having different raw $d_{1j} - E_{1j}$ values — the decreasing KM weight exactly compensates the increasing expected-events fraction. Second, the final three rows again contribute zero because $V_j = 0$, regardless of the weight value. Third, the $w_j^2 V_j$ column shrinks faster than $w_j V_j$ because the weight is squared, so $G(1,0)$ has a notably smaller variance ($V_{(1,0)} = 0.898$) than the log-rank ($V_{(0,0)} = 1.287$).

Table 3: $G(1,0)$ weighted event table.

t_j	w_j	$d_{1j} - E_{1j}$	$w_j(d_{1j} - E_{1j})$	$w_j^2 V_j$
1	1.000	−0.500	−0.500	0.250
2	0.917	−0.545	−0.500	0.208
4	0.833	−0.600	−0.500	0.167
6	0.750	−0.333	−0.250	0.219
10	0.583	−0.800	−0.467	0.054
12–25	—	0.000	0.000	0.000
		Total	**−2.217**	**0.898**

$$Z_{(1,0)} = \frac{-2.217}{\sqrt{0.898}} = \frac{-2.217}{0.948} = -2.339.$$

The tables for $G(0,0)$, $G(0,1)$, and $G(1,1)$ are constructed identically, substituting the appropriate weights from Table 2. The results for all four components are collected in Table 4.

Table 4: Summary of all four FH component statistics. p values are one-sided.

Component	U_k	V_k	Z_k	p_k
$G(0,0)$ log-rank	−2.779	1.287	−2.450	0.007
$G(1,0)$ early	−2.217	0.898	−2.339	0.010
$G(0,1)$ late	−0.562	0.061	−2.286	0.011
$G(1,1)$ middle	−0.382	0.029	−2.235	0.013

All four Z-statistics are negative and similar in magnitude (ranging from −2.450 to −2.235), indicating consistent evidence of treatment benefit regardless of the temporal weighting. The slight weakening from $G(0,0)$ to $G(1,1)$ occurs because the only truly informative event times ($t = 1, 2, 4, 6, 10$) are all in the early-to-middle phase; late-emphasis weights pay a cost for also up-weighting $t = 12, 16, 25$ where $V_j = 0$ and no information is available.

The MaxCombo Combination

Statistic and null distribution

The MaxCombo statistic is simply the largest absolute Z-value:

$$z_{\max} = \max_k |Z_k| = |Z_{(0,0)}| = 2.450.$$

Under H_0, the vector $(Z_{(0,0)}, Z_{(1,0)}, Z_{(0,1)}, Z_{(1,1)})$ is asymptotically $N_4(\mathbf{0}, \hat{\boldsymbol{\Sigma}})$, so the one-sided p-value is:

$$p = 1 - \Phi_4(z_{\max}, \ldots, z_{\max}; \hat{\boldsymbol{\Sigma}}),$$

where Φ_4 is the four-dimensional normal CDF. This accounts for the multiplicity introduced by taking a maximum: a naive reading of $z_{\max}$ against a standard normal table would be anti-conservative because it ignores the other three tests.

Estimating $\hat{\boldsymbol{\Sigma}}$

The correlation between components k and l is estimated as:

$$\hat{\rho}_{kl} = \frac{\sum_j w_{kj} w_{lj} V_j}{\sqrt{V_k V_l}}.$$

The numerator $\widehat{\text{Cov}}(U_k, U_l) = \sum_j w_{kj} w_{lj} V_j$ is computed from the same per-event V_j values used in Table 1; the weight product $w_{kj} w_{lj}$ is the only quantity that changes across pairs.

Worked detail: $\hat{\rho}_{(0,0),(1,0)}$

For the pair $G(0,0)$–$G(1,0)$, the weight product is $1 \times \hat{S}(t_j^-) = \hat{S}(t_j^-)$, so the cross-covariance numerator is simply $\sum_j \hat{S}(t_j^-) V_j$:

$t = 1:$	$1.000 \times 0.250 = 0.250$
$t = 2:$	$0.917 \times 0.248 = 0.227$
$t = 4:$	$0.833 \times 0.240 = 0.200$
$t = 6:$	$0.750 \times 0.389 = 0.292$
$t = 10:$	$0.583 \times 0.160 = 0.093$
Total:	1.062

$$\hat{\rho}_{(0,0),(1,0)} = \frac{1.062}{\sqrt{1.287 \times 0.898}} = \frac{1.062}{1.075} = 0.988.$$

The remaining five pairs are computed in the same manner. A useful algebraic shortcut: the weight products for pairs $(0,0)$–$(1,1)$ and $(1,0)$–$(0,1)$ are both equal to $\hat{S}(1 - \hat{S})$, so these two cross-covariances are numerically identical ($= 0.164$).

Full correlation matrix

Table 5 summarises all six cross-covariances and correlations, yielding:

$$\hat{\boldsymbol{\Sigma}} = \begin{pmatrix} 1.000 & 0.988 & 0.805 & 0.846 \\ 0.988 & 1.000 & 0.704 & 0.757 \\ 0.805 & 0.704 & 1.000 & 0.989 \\ 0.846 & 0.757 & 0.989 & 1.000 \end{pmatrix}$$

(rows/cols: $G(0,0), G(1,0), G(0,1), G(1,1)$).

Table 5: Cross-covariances and estimated correlations for all six component pairs. Pairs $(0,0)$–$(1,1)$ and $(1,0)$–$(0,1)$ share the same covariance numerator by the algebraic identity $w_{(0,0)}w_{(1,1)} = w_{(1,0)}w_{(0,1)} = \hat{S}(1-\hat{S})$.

Pair (k,l)	$\sum_j w_{kj}w_{lj}V_j$	$\sqrt{V_k V_l}$	$\hat{\rho}_{kl}$
$(0,0)$–$(1,0)$	1.062	1.075	0.988
$(0,0)$–$(0,1)$	0.225	0.279	0.805
$(0,0)$–$(1,1)$	0.164	0.194	0.846
$(1,0)$–$(0,1)$	0.164	0.233	0.704
$(1,0)$–$(1,1)$	0.123	0.162	0.757
$(0,1)$–$(1,1)$	0.042	0.042	0.989

Two structural features stand out. First, adjacent pairs ($G(0,0)$–$G(1,0)$ and $G(0,1)$–$G(1,1)$) are nearly collinear ($\hat{\rho} > 0.98$): the early-weight pair and the late-weight pair each describe essentially the same one-dimensional signal within their respective temporal window. Second, all correlations are strongly positive ($\hat{\rho} > 0.70$), reflecting the fact that all four tests are measuring the same underlying treatment effect from four different vantage points.

Decision and Interpretation

The MaxCombo p-value is obtained by numerical integration of $\Phi_4(2.450, 2.450, 2.450, 2.450; \hat{\Sigma})$ using, for example, `pmvnorm` in R's `mvtnorm` package. Table 6 shows the final results.

Table 6: Component and MaxCombo results. All p-values are one-sided.

Test	Z_k or $z_{\max}$	p	$\alpha = 0.05$
$G(0,0)$ log-rank	-2.450	0.007	Reject
$G(1,0)$ early	-2.339	0.010	Reject
$G(0,1)$ late	-2.286	0.011	Reject
$G(1,1)$ middle	-2.235	0.013	Reject
MaxCombo	$z_{\max} = 2.450$	**0.014**	**Reject**

We reject H_0 at the 5% level: the treatment arm has significantly superior survival ($p = 0.014$).

The multiplicity penalty

The MaxCombo p-value (0.014) is larger than the most significant component ($p = 0.007$ from the log-rank), reflecting the implicit multiple-comparison adjustment for testing four statistics and selecting the largest. However, the penalty is modest: a Bonferroni correction would give $p_{\text{Bon}} = 4 \times 0.007 = 0.028$, roughly twice as large. The saving relative to Bonferroni arises directly from the high inter-component correlations: when $\hat{\rho} \approx 0.99$, the four tests carry almost no independent information, so selecting the maximum from four nearly identical statistics is only marginally more favourable than selecting from one. The MVN integration correctly accounts for this structure; Bonferroni does not.

When does MaxCombo add real value?

In this dataset all four components individually reject H_0, so MaxCombo offers no material power advantage over running the log-rank alone. The test's true benefit emerges under *non-proportional hazards*: if, for example, the hazard ratio were initially 1 (no early effect) and diverged only after month 6, then $G(0,0)$ might fail to reject while $G(0,1)$ would succeed. MaxCombo would correctly reject by taking the maximum, while a pre-specified log-rank would miss the signal entirely. The MVN p-value ensures this flexibility comes at only a modest, precisely quantified cost in the proportional-hazards case.

Computational Note

The MVN CDF for $m \geq 3$ has no closed form and requires numerical quadrature. Standard implementations: `pmvnorm()` in R's `mvtnorm` (Genz & Bretz 2009); Monte Carlo sampling from $N_4(\mathbf{0}, \hat{\Sigma})$ for a quick approximation; or `scipy.stats.multivariate_normal` in Python with `allow_singular=True` (needed here because $\hat{\Sigma}$ is near-singular, having a near-zero eigenvalue owing to the high correlations). The near-singularity is a dataset-specific artefact of having only five informative event times driving all four components; in larger trials with more diverse event times the correlation matrix will be less extreme.

Emerging Concerns: Mukhopadhyay et al. (*JAMA Oncology*, 2022) conducted a systematic review of immuno-oncology trials and found that MaxCombo and the standard log-rank test were largely concordant, questioning the practical advantage of MaxCombo in confirmatory settings. A related concern, elaborated by Magirr (2023), is that when survival curves cross, MaxCombo's down-weighting of early events can inflate the Type I error rate by effectively rewarding an early detrimental treatment effect—a behaviour absent from the standard log-rank test and a caution for any application where early-phase harm cannot be ruled out. In light of these emerging concerns, the author has placed this test in the Appendix rather than the main chapter; the treatment here remains faithful to the spirit of the compendium—explaining how the test works under the hood—so that readers can evaluate it on its merits as the literature continues to evolve.

XVIII. SURV — Survival Analysis Demystified

Key Concepts in Plain English

A reference for readers encountering survival analysis terminology for the first time. Each concept is anchored to an everyday situation before the formal meaning is given.

1. Survival Time and the Event

You are tracking how long light bulbs last until they burn out. In survival analysis, the **event** is the outcome of interest (death, failure, relapse, default) and the **survival time** is the elapsed time until that event occurs. "Survival" does not have to mean staying alive — it means not yet experiencing the event.

2. Censoring

Some bulbs are still working when the study ends, or were accidentally broken before burning out. A **censored** observation is one where the event had not yet occurred by the time the study closed or the subject was lost to follow-up. You know the subject survived *at least* that long — but not what happened afterward. Ignoring censoring and analyzing only complete cases discards information and biases results.

3. Survival Function $S(t)$

What fraction of bulbs are still working after 1,000 hours? The **survival function** $S(t)$ gives the probability of remaining event-free beyond time t. It starts at 1 (everyone is event-free at the beginning) and decreases toward 0 over time. The Kaplan–Meier curve is its nonparametric estimate.

4. Hazard Function $h(t)$

Given that a bulb has already lasted 1,000 hours, how likely is it to fail in the next minute? The **hazard function** measures the instantaneous rate of the event occurring *right now*, conditional on having survived to this point. It is a rate, not a probability — it can exceed 1. A bathtub-shaped hazard (high early, low in the middle, high again late) describes many mechanical and biological lifetimes.

5. Cumulative Hazard $H(t)$

How much total risk has a bulb accumulated over its first 2,000 hours? The **cumulative hazard** $H(t)$ sums up the hazard from the start until time t. It is directly linked to the survival function: $S(t) = e^{-H(t)}$. A steep climb in $H(t)$ signals a period of concentrated risk; a flat stretch means the event is rare in that interval.

6. Competing Risks

A retiree can stop working because of death, disability, or voluntary retirement — three outcomes that compete with each other. **Competing risks** arise when multiple distinct event types are possible and experiencing one prevents the others from ever occurring. Standard single-event methods break down here because removing one cause changes the probability of the rest.

7. Cause-Specific Hazard

Focus only on disability retirements, treating everyone who retired voluntarily or died as if they had simply left the study. The **cause-specific hazard** is the hazard for one particular cause, with individuals who experience a competing event treated as censored. It answers: "Among those still at risk, how fast is *this* cause claiming them?" It is the natural quantity for understanding the biology or mechanism of a specific cause.

8. Cumulative Incidence Function (CIF)

What is the actual probability that a retiree leaves due to disability by age 70, knowing that death could get there first? The **cumulative incidence function** gives the probability of failing from a specific cause by time t, properly accounting for the fact that competing events can foreclose it. Critically, in a competing-risks setting the CIF is *not* $1 - S(t)$; summing the CIFs for all causes gives the overall failure probability.

9. Sub-Hazard (Fine–Gray)

Imagine tracking disability risk while keeping already-deceased retirees "on the books" as still theoretically at risk. The **subdistribution hazard** (Fine–Gray) is the hazard that corresponds directly to the CIF. It is conceptually unusual — individuals who experienced a competing event remain in the risk set — but it has the practical advantage that a proportional sub-hazard model directly models the CIF itself. Use it when your scientific question is about the population-level probability of a specific outcome, not its underlying mechanism.

10. Proportional Hazards

Smokers have twice the failure rate of non-smokers at every point in time — not just on average, but consistently at age 50, at age 60, at age 70. The **proportional hazards** assumption states that the hazard ratio between any two covariate profiles is constant over time. The Cox model imposes this structure without specifying the shape of the baseline hazard. When the assumption fails — for example, if a treatment helps early but loses effect later — the log-rank test loses power and the Cox hazard ratio is a misleading average.

Part XIX

Spatial Dependence & Geographic Structure

Detecting spatial clustering, autocorrelation, and spatial heterogeneity

Spatial Statistics

XIX. SPAT — A Conceptual Primer

What Is Spatial Data?

Spatial data are observations $\{y_i\}$ indexed by location $\mathbf{s}_i$, where ordering in space carries information analogous to time-series ordering. The defining feature is that **nearby observations tend to be more similar than distant ones** (Tobler's First Law), violating the independence assumption of classical inference.

This chapter focuses on **areal (lattice) data**: the study region is partitioned into n discrete units (counties, census tracts, grid cells) and a single value y_i is observed for each. Other spatial data types include geostatistical (continuous surface) and point process data, but the tests in this chapter are designed for areal data.

The Spatial Weights Matrix W

The $n \times n$ **spatial weights matrix W** formalises neighborhood relationships: $w_{ij} > 0$ if units i and j are neighbors, $w_{ij} = 0$ otherwise, $w_{ii} = 0$ by convention.

Contiguity-based weights:

- **Rook:** $w_{ij} = 1$ if units share a common *edge* only.
- **Bishop:** $w_{ij} = 1$ if units share a *vertex*.
- **Queen:** $w_{ij} = 1$ if units share an edge *or vertex*; includes diagonals. Preferred for irregular polygons.
- **Higher-order:** $\mathbf{W}^{(2)}$, $\mathbf{W}^{(3)}$ extend to second- and third-order neighbors.

Distance-based weights:

- **k-nearest neighbors:** $w_{ij} = 1$ for the k closest units; each unit has exactly k neighbors.
- **Inverse distance:** $w_{ij} = 1/d_{ij}^{\alpha}$; continuous decay.
- **Distance band:** $w_{ij} = 1$ if $d_{ij} < d^*$.

Row standardization: $w_{ij}^* = w_{ij}/\sum_k w_{ik}$; each row sums to 1. The **spatial lag** $(\mathbf{Wy})_i = \sum_j w_{ij}^* y_j$ becomes the weighted average of neighbors of i. Define $S_0 = \sum_i \sum_j w_{ij}$ ($= n$ after row-standardization).

Results are sensitive to the choice of W. Always report the specification and check robustness.

Spatial Autocorrelation: Concepts

Positive (clustering): similar values near each other; **Negative** (dispersion): dissimilar values adjacent; **Spatial randomness**: no systematic pattern (null hypothesis for most tests).

Two inference frameworks:

- **Normality:** analytical variances; standardize to $\mathcal{N}(0, 1)$.
- **Randomization:** permutation distribution from randomly reassigning values to locations; more robust; recommended when normality is doubtful.

I. Global Spatial Autocorrelation Tests

Moran's I:

$$I = \frac{n}{S_0} \cdot \frac{\sum_i \sum_j w_{ij}(y_i - \bar{y})(y_j - \bar{y})}{\sum_i (y_i - \bar{y})^2}.$$

$E[I] = -1/(n-1) \approx 0$ under spatial randomness. Variance (normality):

$$\mathrm{Var}(I) = \frac{n^2(n-1)S_1 - n(n-1)S_2 - 2S_0^2}{(n+1)(n-1)^2 S_0^2},$$

where $S_1 = \frac{1}{2}\sum_{i,j}(w_{ij} + w_{ji})^2$ and $S_2 = \sum_i\left(\sum_j w_{ij} + \sum_j w_{ji}\right)^2$. Standardized: $z_I = (I - E[I])/\sqrt{\mathrm{Var}(I)} \xrightarrow{d} \mathcal{N}(0, 1)$.

Geometrically, I is the slope in a **Moran scatterplot** of **Wy** vs. **y** (standardized); the four quadrants identify spatial regimes HH, LL, HL, LH.

Geary's C:

$$C = \frac{(n-1)}{2S_0} \cdot \frac{\sum_i \sum_j w_{ij}(y_i - y_j)^2}{\sum_i (y_i - \bar{y})^2}.$$

$E[C] = 1$; $C < 1$: positive autocorrelation; $C > 1$: negative. Based on squared differences between neighbors; more sensitive to **local** dissimilarities than Moran's I.

Getis–Ord General G:

$$G(d) = \frac{\sum_i \sum_j w_{ij}(d)\, y_i y_j}{\sum_i \sum_j y_i y_j}, \quad i \neq j.$$

$G > E[G]$: clustering of **high values**; $G < E[G]$: clustering of **low values**. Requires $y_i > 0$; a directional capability Moran's I lacks.

II. Local Spatial Statistics (LISA)

Global tests reveal *whether* spatial autocorrelation exists; local statistics identify *where*.

Local Moran's I_i:

$$I_i = z_i \sum_j w_{ij}^* z_j, \quad z_i = (y_i - \bar{y})/s.$$

$\sum_i I_i \propto I$ (decomposes the global statistic). Moran scatterplot quadrants:

HH	High value, high-value neighbors: cluster
LL	Low value, low-value neighbors: cluster
HL	High value, low-value neighbors: outlier
LH	Low value, high-value neighbors: outlier

Getis–Ord G_i^*:

$$G_i^* = \frac{\sum_j w_{ij} y_j - \bar{y}\sum_j w_{ij}}{s\sqrt{\left[n\sum_j w_{ij}^2 - \left(\sum_j w_{ij}\right)^2\right]/(n-1)}},$$

$\sim \mathcal{N}(0,1)$. Large positive: **hotspot**; large negative: **coldspot**. Unlike LISA, G_i^* targets value concentration, not cluster/outlier type.

Multiple testing: n simultaneous tests inflate FWER; use conditional permutation p-values with FDR or Bonferroni correction. Uncorrected local patterns are exploratory only.

III. Join Count Test

For a **binary** variable (Black B / White W), count adjacent pairs:

$$\text{BB} = \tfrac{1}{2}\sum_i \sum_j w_{ij} x_i x_j,$$

where $x_i = 1$ if unit i is B. Under spatial randomness, $E[\text{BB}]$ and $\text{Var}(\text{BB})$ are derived under either **free sampling** (Bernoulli, $p = n_B/n$) or **non-free sampling** (hypergeometric, fixed n_B). Excess BB/WW joins $\Rightarrow$ positive autocorrelation; excess BW joins $\Rightarrow$ negative. Extends to $k > 2$ categories via all $(k^2+k)/2$ pair types.

IV. Spatial Regression Models and LM Tests

When spatial autocorrelation persists in OLS residuals, two structural misspecifications are possible.

Spatial Lag Model (SLM): $\mathbf{y} = \rho\mathbf{W}\mathbf{y} + \mathbf{X}\boldsymbol{\beta} + \boldsymbol{\varepsilon}$; $\mathbf{W}\mathbf{y}$ is endogenous; OLS biased; Spatial ML or IV required; global spillover via $(\mathbf{I} - \rho\mathbf{W})^{-1}$.

Spatial Error Model (SEM): $\mathbf{y} = \mathbf{X}\boldsymbol{\beta} + \mathbf{u}$, $\mathbf{u} = \lambda\mathbf{W}\mathbf{u} + \boldsymbol{\varepsilon}$; OLS unbiased but inefficient; SEs incorrect.

Let $\hat{\sigma}^2 = \hat{\boldsymbol{\varepsilon}}^\top\hat{\boldsymbol{\varepsilon}}/n$, $T = \text{tr}(\mathbf{W}^\top\mathbf{W} + \mathbf{W}^2)$, and $J = T + \hat{\sigma}^{-2}\hat{\boldsymbol{\varepsilon}}^\top\mathbf{W}\mathbf{X}(\mathbf{X}^\top\mathbf{X})^{-1}\mathbf{X}^\top\mathbf{W}^\top\hat{\boldsymbol{\varepsilon}}$.

Test	Statistic / null
LM-Lag	$(\hat{\boldsymbol{\varepsilon}}^\top\mathbf{W}\mathbf{y}/\hat{\sigma}^2)^2/J \sim \chi^2_1$; $H_0: \rho = 0$
LM-Error	$(\hat{\boldsymbol{\varepsilon}}^\top\mathbf{W}\hat{\boldsymbol{\varepsilon}}/\hat{\sigma}^2)^2/T \sim \chi^2_1$; $H_0: \lambda = 0$
Robust LM-Lag	Adjusts LM-Lag for local λ misspec.; tests $\rho = 0 \mid \lambda \neq 0$
Robust LM-Error	Adjusts LM Error for local ρ misspec.; tests $\lambda = 0 \mid \rho \neq 0$

Decision rule (Anselin–Florax): If only LM-Lag significant $\Rightarrow$ SLM. If only LM-Error significant $\Rightarrow$ SEM. If both significant, compare robust variants; prefer the model with the larger robust statistic.

Stationarity and Edge Effects

Second-order stationarity: mean μ and covariance $C(\mathbf{h})$ depend only on separation $\mathbf{h}$, not absolute location. **Isotropy:** C depends only on $\|\mathbf{h}\|$.

Edge effects: boundary units have fewer neighbors, distorting weight structure and inflating test statistics near the boundary. Torus correction and free-sampling assumptions partially mitigate this.

Diagnostic Workflow

1. Define **W**; document specification.
2. Map data; choropleth for visual inspection.
3. Global autocorrelation: Moran's I, Geary's C, Getis–Ord G.
4. Local patterns: LISA (I_i) and G_i^* with FDR correction.
5. Fit OLS; run LM-Lag and LM-Error; if both significant, use robust variants.
6. Sensitivity: repeat with alternative **W**.

Common Pitfalls

- **Arbitrary W:** results can change substantially; sensitivity analysis is essential.
- **No multiple testing for LISA:** n simultaneous tests produce spurious clusters; use conditional permutation with FDR.
- **OLS residual check first:** LM tests require a correctly specified mean model.
- **Edge effects ignored:** boundary units distort statistics.
- **Moran's I on binary data:** use Join Count test for categorical variables.
- **Skipping model selection:** Moran's I on OLS residuals alone cannot distinguish SLM from SEM; always use the LM sequence.

Bottom Line

Spatial statistics quantify dependence across locations; all inference depends critically on **W**. Start with Moran's I for global autocorrelation; use LISA and G_i^* to locate clusters with FDR correction; apply the Anselin LM sequence to choose between SLM and SEM. Always report **W** and check sensitivity.

Moran's I Test

Test for Global Spatial Autocorrelation

SPAT.1 | **Category:** Spatial Statistics | **Data:** Continuous | **Null:** Normal (asymptotic)

When to Use

- Detect global spatial autocorrelation.
- Assess clustering or dispersion across regions.
- Spatial residual diagnostics after regression.
- Geographic or lattice-based areal data.
- Test whether nearby values are more similar.

Assumptions

- Spatial weights matrix is specified a priori.
- Observations are indexed to fixed spatial units.
- Spatial randomness/exchangeability holds under H_0.
- Data are continuous or ordered numeric values.
- Normal or permutation approximation is appropriate.

Test Statistic

$$I = \frac{n}{\sum_i \sum_j w_{ij}} \cdot \frac{\sum_i \sum_j w_{ij}(x_i - \bar{x})(x_j - \bar{x})}{\sum_i (x_i - \bar{x})^2}, \quad z = \frac{I - E[I]}{\sqrt{\text{Var}[I]}}. \quad \text{Under } H_0, z \approx N(0,1).$$

Notation: w_{ij} are spatial weights, x_i are observed values, I is correlation-like but its exact range depends on **W**

Plain English: Moran's I measures the spatial autocorrelation between values at nearby locations weighted by their spatial proximity. I compares the product of deviations from the mean for neighboring locations to overall variance. Positive I values indicate clustering, while negative I values indicate dispersion, and values near zero suggest random spatial pattern. The test assesses whether observed spatial pattern differs significantly from random arrangement.

⚠ Practitioner Warning

Results critically depend on spatial weights matrix specification (contiguity, distance-based, k-nearest neighbors). Different weight matrices can yield different conclusions. Choice of weights should be theory-driven, not data-driven. Test detects global autocorrelation; may miss localized clusters (use LISA or Getis-Ord for local patterns). Assumes spatial structure correctly specified and stationary across study area. A single global statistic may mask spatial nonstationarity or localized clusters. Edge effects and irregular boundaries can affect results. Large samples detect trivial autocorrelation. Row-standardization of weights matrix affects interpretation.

Hypotheses

- H_0: Spatial randomness; $I = E[I]$.
- H_1: Spatial autocorrelation: $I \neq E[I]$; often $I > E[I]$.

Decision Rule: Reject H_0 if $|z| > Z_{\alpha/2}$ (two-tailed) or p-value $< \alpha$. $z > 0$: +ve autocorrelation and vice versa.

Software References [Python, R, SAS]

Python	R	SAS
`esda.Moran`	`spdep::moran.test`	`Not Native (Manual)`

Additional Notes

Spatial weights matrix **W**: element w_{ij} represents spatial relationship between locations i and j. Common specifications: binary contiguity ($w_{ij} = 1$ if neighbors, 0 otherwise), inverse distance ($w_{ij} = 1/d_{ij}^{\alpha}$), k-nearest neighbors ($w_{ij} = 1$ for k nearest, 0 otherwise). Row-standardization: divide each w_{ij} by row sum $\sum_j w_{ij}$ so weights sum to 1 for each location; under row-standardization, $S_0 = n$ (excluding isolated locations). Under spatial randomness: $E[I] = -1/(n-1) \approx 0$ for large n. Under common randomization assumptions, one widely used variance expression is: $\text{Var}[I] = [n^2 S_1 - nS_2 + 3S_0^2]/[S_0^2(n^2-1)] - E[I]^2$ where $S_0 = \sum_i \sum_j w_{ij}$, $S_1 = \frac{1}{2}\sum_i \sum_j (w_{ij} + w_{ji})^2$, $S_2 = \sum_i (\sum_j w_{ij} + \sum_j w_{ji})^2$. Permutation inference provides a robust alternative to asymptotic normality: randomly permute observed values across locations, recompute I, and compare the observed statistic with the permutation distribution. Cliff and Ord (1981) developed the classical asymptotic theory for Moran's I. Geary's C provides an alternative global measure of spatial autocorrelation.

Miscellanea: Patrick Moran (1950) introduced I as a measure of spatial autocorrelation, extending the idea of correlation to data indexed by spatial relationships. Later work by Cliff and Ord (1970s) established its statistical properties and broadened its application. The statistic is now widely used in spatial analysis to test whether observed geographic patterns reflect more than random variation.

Reference(s): Moran (1950); Cliff & Ord (1981)

Moran's I Test

SPAT.1

Worked Example

Problem Statement A city planner investigates whether high-value properties are spatially clustered across four adjacent districts ($n = 4$). A binary contiguity weight $w_{ij} = 1$ if districts share a border, 0 otherwise, arranged as a linear chain $1-2-3-4$. Property values are $x_1 = 2, x_2 = 3, x_3 = 4, x_4 = 5$ with mean $\bar{x} = 3.5$, giving deviations $z_i = (z_1, z_2, z_3, z_4) = (-1.5, -0.5, 0.5, 1.5)$.

1. Spatial Weights Matrix

$$W = \begin{pmatrix} 0 & 1 & 0 & 0 \\ 1 & 0 & 1 & 0 \\ 0 & 1 & 0 & 1 \\ 0 & 0 & 1 & 0 \end{pmatrix}$$

2. Geometry-Based Constants

- Row sums: $w_{1\cdot} = 1,\ w_{2\cdot} = 2,\ w_{3\cdot} = 2,\ w_{4\cdot} = 1$
- Col sums: $w_{\cdot 1} = 1,\ w_{\cdot 2} = 2,\ w_{\cdot 3} = 2,\ w_{\cdot 4} = 1$
- $S_0 = \sum_i \sum_j w_{ij} = 6$
- $S_1 = \sum_i \sum_j (w_{ij}^2 + w_{ij} w_{ji}) = 6 + 6 = 12$
- $S_2 = \sum_i (w_{i\cdot} + w_{\cdot i})^2 = 4 + 16 + 16 + 4 = 40$

3. Moran's I Statistic

Denominator:

$$\sum z_i^2 = (-1.5^2) + (-0.5^2) + 0.5^2 + 1.5^2 = 5.0$$

Numerator (sum only over $w_{ij} = 1$ pairs):

$$\sum \sum w_{ij} z_i z_j = 2(z_1 z_2 + z_2 z_3 + z_3 z_4) = 2.5$$

$$I = \frac{n}{S_0} \cdot \frac{\sum \sum w_{ij} z_i z_j}{\sum z_i^2} = \frac{4}{6} \cdot \frac{2.5}{5.0} = \mathbf{0.333}$$

4. Expected Value and Variance

Under spatial randomness:

$$E[I] = \frac{-1}{n-1} = \frac{-1}{3}$$

$$\mathrm{Var}[I] = \frac{n^2 S_1 - n S_2 + 3 S_0^2}{(n^2 - 1) S_0^2} - E[I]^2 = 0.148$$

5. Z-Statistic

$$Z = \frac{I - E[I]}{\sqrt{\mathrm{Var}[I]}} = 1.73$$

6. Two-sided test (any spatial autocorrelation)

$$H_0 : I - E[I] \quad \text{vs.} \quad H_a : I \neq E[I]$$

Critical value at $\alpha = 0.05$: $z_{0.025} = 1.96$.
Since $|1.73| < 1.96$, we fail to reject H_0 at the 5% level for a two-sided test.

7. One-sided test (positive spatial clustering)

$$H_0 : I = E[I] \quad \text{vs.} \quad H_a : I > E[I].$$

Critical value at $\alpha = 0.05$: $z_{0.05} = 1.645$.
Since $1.73 > 1.645$, the one-sided test rejects H_0, indicating evidence of positive spatial autocorrelation.

Conclusion: The observed Moran's $I = 0.333$ exceeds its expected value under randomness $E[I] = -1/3$, suggesting a positive clustering tendency: nearby districts tend to have similar property values. The conclusion, however, depends on the testing direction. A two-sided test does not reach significance at $\alpha = 0.05$, whereas a one-sided test focused specifically on positive clustering does. Because this toy example uses only $n = 4$ districts, the asymptotic normal approximation is unreliable and the result should be interpreted cautiously.

Common Pitfalls

- **Unjustified spatial weights.** Moran's I depends directly on the spatial weights matrix **W**. The choice of **W** should be theoretically justified, with sensitivity checks when appropriate.
- **Ignoring context when interpreting I.** The meaning of Moran's I depends on the spatial structure, sample size, and weights specification. Interpretation should consider both effect magnitude and statistical significance.
- **Equating positive I with true clustering.** Positive Moran's I indicates spatial association but not necessarily direct spatial interaction. Apparent clustering may reflect trends, covariates, or measurement processes.
- **Ignoring spatial trends.** Large-scale gradients can produce apparent autocorrelation even without local interaction. Trend effects should be modeled or removed before testing residual dependence.
- **Ignoring multiple testing.** Testing many variables inflates false-discovery risk. Multiple-testing adjustments or pre-specified hypotheses help maintain valid inference.

Geary's C Test

Test for Global Spatial Autocorrelation via Squared Differences

SPAT.2 | **Category:** Spatial Statistics | **Data:** Continuous | **Null:** Normal (asymptotic)

When to Use

- Detect global spatial autocorrelation.
- Sensitive to local spatial differences/discontinuities.
- Compare neighboring spatial observations.
- Spatial diagnostics for areal or lattice data.

Assumptions

- Spatial weights matrix is correctly specified.
- Spatially indexed interval/ratio-scale data.
- Spatial randomness/exchangeability holds under H_0.
- Normal or permutation approximation is appropriate.

Test Statistic

$$C = \frac{(n-1)\sum_i \sum_j w_{ij}(x_i - x_j)^2}{2\left(\sum_i \sum_j w_{ij}\right)\sum_i (x_i - \bar{x})^2}, \quad z = \frac{C - E[C]}{\sqrt{\mathrm{Var}[C]}}. \quad \text{Under } H_0, z \approx N(0,1)$$

Notation: w_{ij} are spatial weights, x_i are values, Geary's $C \in [0, 2]$ (common but can exceed bounds based on w_{ij})

Plain English: Geary's C measures spatial autocorrelation by examining squared differences between neighboring values, weighted by spatial proximity. The statistic compares the average squared difference among neighbors to overall variance. Values less than 1 indicate positive spatial autocorrelation (neighbors are similar), values greater than 1 indicate negative autocorrelation (neighbors are dissimilar), and values near 1 suggest spatial randomness. Unlike Moran's I which uses products of deviations, Geary's C uses squared differences, making it more sensitive to local dissimilarity.

⚠ Practitioner Warning

Inverse relationship to Moran's I: Geary's C decreases as autocorrelation increases (opposite of Moran). Results depend critically on spatial weights specification. More sensitive than Moran's I to differences in small neighborhoods but less powerful for detecting global trends. Can be influenced more by outliers due to squaring differences. Expected value is 1 regardless of sample size (Moran's $E[I] = -1/(n-1)$). Tests may disagree when local and global patterns differ. Edge effects and boundary conditions affect results. Permutation tests provide alternative to asymptotic approximation.

Hypotheses

- H_0: No spatial autocorrelation ($C = 1$).
- H_1: Spatial autocorrelation present.

Decision Rule: Reject H_0 if $|z| > Z_{\alpha/2}$ or p-value $< \alpha$.
$C < 1 \Rightarrow$ +ve autocorrelation; $C > 1 \Rightarrow$ -ve autocorrelation.

Software References [Python, R, SAS]

`esda.Geary`	`spdep::geary.test`	`(spatial analysis packages)`

Additional Notes

Relationship to Moran's I: both measure spatial autocorrelation but via different formulations. Moran uses $\sum w_{ij}(x_i - \bar{x})(x_j - \bar{x})$ (covariance-like), Geary uses $\sum w_{ij}(x_i - x_j)^2$ (variance of differences). For strong global autocorrelation, both typically agree (Moran positive, Geary < 1). For local clustering with weak global pattern, may disagree. Expected value: $E[C] = 1$ exactly under randomization. Variance: $\mathrm{Var}[C]$ complex function of spatial weights structure, similar complexity to Moran. Geary's C related to spatial variogram in geostatistics: semi-variance $\gamma(h) = \frac{1}{2N(h)} \sum (x_i - x_j)^2$ for pairs separated by distance h. Geary essentially standardized weighted average of semi-variances. Sensitivity differences: Geary more influenced by local irregularities; Moran by global trends. Getis (1991) showed Geary often has higher variance, hence lower power than Moran for same sample size under many alternatives.

Miscellanea: Robert Geary (1954) introduced Geary's C as an alternative measure of spatial autocorrelation, based on squared differences between neighboring observations. Compared to Moran's I, which captures global correlation, Geary's C is more sensitive to local variation. Later work by Cliff and Ord (1970s) clarified the relationship between the two, and although both are complementary, Moran's I is more widely used in practice. ($C < 1$: positive clustering, $C > 1$: negative dispersion)

Reference(s): Geary (1954); Cliff & Ord (1973, 1981)

Worked Example

Problem Statement An environmental scientist is measuring the "burn severity" index across 4 adjacent forest plots ($n = 4$) arranged in a linear transect. The scientist suspects that fire intensity is not randomly distributed but rather exhibits **positive spatial autocorrelation**, where adjacent plots show similar levels of damage.

1. Experimental Parameters

- **Burn Index Values (x_i):** $x_1 = 10, x_2 = 20, x_3 = 30, x_4 = 40$. **Mean ($\bar{x}$):** 25. **Sample Kurtosis (k):** 1.632.
- **Spatial Weights (W):** Binary contiguity (Linear: $1-2, 2-3, 3-4$) W is the same as in Moran's I example
- $S_0 = \sum\sum w_{ij} = 6$; $S_1 = \frac{1}{2}\sum\sum(w_{ij} + w_{ji})^2 = 12$; $S_2 = \sum(w_{i.} + w_{.i})^2 = 40$
- **Null Hypothesis (H_0):** $C = 1$ (Spatial randomness/no autocorrelation).

2. Calculation of the Denominator The denominator represents the total variation in the sample:

$$\sum_{i=1}^{n}(x_i - \bar{x})^2 = 500$$

3. Calculation of the Numerator For a linear chain, the pairs are (1,2), (2,3), and (3,4) (doubled for symmetry):

$$\sum\sum w_{ij}(x_i - x_j)^2 = 600$$

4. Final Statistic Calculation Given $S_0 = 6$:

$$C = \frac{(n-1)}{2S_0} \times \frac{Numerator}{Denominator} = 0.30$$

$$Z = \frac{C - E[C]}{\sqrt{Var[C]}} = \frac{0.30 - 1.00}{\sqrt{0.140}} = -1.87$$

5. Results and Interpretation

Metric	Value
Geary's C Statistic	0.30
Expected Value $E[C]$	1.00
Variance $Var[C]$	0.140
Interpretation	**+ve Spatial Autocorrelation**

6. Decision

For a one-sided test of positive spatial autocorrelation, H_a : $C < 1$, the critical value at $\alpha = 0.05$ is $z_{0.05} = -1.645$. Since $-1.87 < -1.645$, we reject H_0 and conclude evidence of positive spatial autocorrelation.

For a two-sided test of any spatial autocorrelation, $|Z| = 1.87 < 1.96$, so the result would not be significant at $\alpha = 0.05$.

Conclusion: For the one-sided alternative of positive spatial autocorrelation ($C < 1$), the observed statistic $C = 0.30$ is significantly below its expected value $E[C] = 1$. We therefore reject the null hypothesis of spatial randomness and conclude that neighboring plots exhibit similar burn severity levels (positive spatial clustering). Under a two-sided test, however, the result would not be significant at $\alpha = 0.05$. Variance is calculated using the formula given below with $k = 1.632$.

Common Pitfalls

- **Misinterpreting Geary's C.** Values below 1 indicate positive spatial autocorrelation, while values above 1 indicate dispersion. Interpretation should therefore be made relative to $E[C] = 1$.
- **Expecting agreement with Moran's I.** Geary's C emphasizes local differences, whereas Moran's I reflects broader spatial association. Disagreement between the two may indicate different spatial structures.
- **Comparing statistics using different weights.** Spatial autocorrelation measures should be compared using the same weights matrix. Different neighborhood definitions can yield substantially different results.
- **Over-interpreting small deviations from 1.** Minor departures from $C = 1$ may reflect normal sampling variability. Interpretation should emphasize significance and direction rather than small numeric differences.
- **Ignoring sensitivity to outliers.** Because Geary's C uses squared neighboring differences, extreme observations can strongly influence the statistic. Potential outliers should therefore be examined carefully.

Variance Formula

$$Var[C] = \frac{(n-1)S_1[n^2 - 3n + 3 - (n-1)k] - \frac{1}{4}(n-1)S_2[n^2 + 3n - 6 - (n^2 - n + 2)k] + S_0^2[n^2 - 3 - (n-1)^2 k]}{n(n-2)(n-3)S_0^2}$$

Note on Variance: The parameter k represents the sample kurtosis. Under the **Normality Assumption**, k is fixed at 3. Under the **Randomization Assumption**, k is estimated from the sample data, which accounts for non-normality and potential outliers in the spatial distribution.

Getis-Ord General G Test

Test for High-Value or Low-Value Clustering

SPAT.3 | **Category:** Spatial Statistics | **Data:** Continuous | **Null:** Normal (asymptotic)

When to Use

- Detect global clustering of high or low values.
- Assess spatial concentration of extremes globally.
- Complement Moran's I for intensity clustering.
- Spatial analysis of areal or lattice data.

Assumptions

- Spatial weights matrix is specified a priori.
- Values are non-negative and meaningfully ordered.
- Values are exchangeable under spatial randomness (H_0).
- Normal or permutation approximation is appropriate.

Test Statistic

$$G = \frac{\sum_i \sum_{j \neq i} w_{ij} x_i x_j}{\sum_i \sum_{j \neq i} x_i x_j}, \qquad z = \frac{G - E[G]}{\sqrt{\mathrm{Var}[G]}}. \qquad \text{Under } H_0, z \approx N(0,1).$$

Notation: w_{ij} are spatial weights, x_i are values

Plain English: The Getis–Ord General G statistic tests whether unusually high or unusually low values are spatially clustered. It compares the weighted sum of products among neighbouring values with the corresponding sum over all location pairs. A significantly positive standardized z value indicates clustering of high values, while a significantly negative z value indicates clustering of low values. Unlike Moran's I, which measures similarity of deviations from the mean, General G is directional and focuses on concentration of large or small magnitudes.

⚠ Practitioner Warning

Requires non-negative values; negative values invalidate interpretation. More sensitive to choice of spatial weights than Moran's I, especially distance band specifications. Does not detect all types of clustering—only concentration of high or low values. Can miss checkerboard patterns or clustering of intermediate values. Expected value and variance depend on specific weight matrix structure. Distance-based weights require choosing appropriate distance threshold. Test is global; use local Getis-Ord G_i^* to identify specific hotspot/coldspot locations. Permutation tests recommended for small samples.

Hypotheses

- H_0: No spatial concentration ($G = E[G]$, random pattern).
- H_1: Spatial concentration present.

Decision Rule: Reject H_0 if $|z| > Z_{\alpha/2}$ or p-value $< \alpha$. +ve significant z: high-value hotspot clustering and vice versa.

Software References [Python, R, SAS]

`esda.G`	`spdep::globalG.test`	`(spatial analysis packages)`

Additional Notes

Expected value and variance: $E[G] = \sum_{i \neq j} w_{ij}/(n(n-1))$, Var$[G]$ complex function involving moments of x and structure of $\mathbf{W}$. Binary weights simplification: when $w_{ij} \in \{0, 1\}$, formulas simplify considerably. Distance-based weights common: $w_{ij} = 1$ if $d_{ij} \leq d_0$, otherwise 0, where d_0 is distance threshold. Local version: Getis-Ord $G_i^* = \sum_j w_{ij} x_j / \sum_j x_j$ identifies specific hotspot/coldspot locations, widely used in crime analysis and epidemiology. Relationship to spatial statistics: General G emphasizes spatial concentration of large or small values; Moran's I emphasizes spatial similarity of deviations from the mean. Both may be significant, but they describe different aspects of spatial patterning. General G particularly useful for positive-valued phenomena: disease counts, housing prices, income levels. Ord and Getis (1995) extended to local statistics enabling hotspot mapping.

Miscellanea: Arthur Getis and Keith Ord (1992) introduced the General G statistic to detect clustering of high or low values, addressing a limitation of Moran's I, which identifies spatial autocorrelation without distinguishing cluster type. Their later local statistics (G_i, G_i^*) became widely used for hotspot analysis in GIS, particularly in crime mapping and public health.

Reference(s): Getis & Ord (1992); Ord & Getis (1995)

Getis-Ord General G Test

SPAT.3

Worked Example

Problem Statement A public health official is analyzing flu cases across 5 city sectors ($n = 5$). After performing local analysis, the official wants to determine the **global** nature of the spatial pattern. Specifically, they want to know if the map is dominated by **Hot Spots** (high-value clusters) or **Cold Spots** (low-value clusters). W is a 5x5 matrix with 1's and 0's.

1. Experimental Parameters

- **Values (x_i):** $x_1 = 10, x_2 = 12, x_3 = 45, x_4 = 38, x_5 = 15$. $\sum x_i = 120$; $\sum x_i^2 = 3938$.
- **Spatial Weights (W):** Binary contiguity (Linear chain: $1-2, 2-3, 3-4, 4-5$)
- **Number of Pairs ($j \neq i$):** For General G, we exclude self-influence ($w_{ii} = 0$).
- **Total Weight (W):** Total number of directed neighbor pairs is 8 ($1 \to 2, 2 \to 1, 2 \to 3, 3 \to 2, \ldots$). $W = 8$.

2. Calculation of the Observed G Statistic

$$G = \frac{\sum_{i=1}^{n} \sum_{j=1}^{n} w_{ij} x_i x_j}{\sum_{i=1}^{n} \sum_{j=1}^{n} x_i x_j}, \ j \neq i. \quad G_{obs} = \frac{5880}{10462} \approx 0.562$$

Numerator: Sum $x_i x_j$ for all $w_{ij} = 1$.

$$2[(x_1 x_2) + (x_2 x_3) + (x_3 x_4) + (x_4 x_5)] = 5880$$

Denominator: Sum $x_i x_j$ for all pairs $j \neq i$.

$$\sum_{i=1}^{n} \sum_{j \neq i}^{n} x_i x_j = (\sum x_i)^2 - \sum x_i^2 = 10462$$

3. Variance and Z-test

The expected value under randomness is

$$E[G] = \frac{W}{n(n-1)} = 0.400.$$

For this sample, exact permutation over the $5! = 120$ possible rearrangements gives $\sqrt{\text{Var}(G)} \approx 0.104$. Thus,

$$Z = \frac{G_{obs} - E[G]}{\sqrt{Var[G]}} = \frac{0.562 - 0.400}{0.104} = +1.56.$$

4. Significance and Decision

Metric	Value
Observed G	0.562
Expected $E[G]$	0.400
Z-statistic	+1.56
Critical Value ($\alpha = 0.05$)	1.96
Decision	**Fail to Reject H_0**

Conclusion: Since $Z = 1.56 < 1.96$, we fail to reject the null hypothesis of randomness. The positive Z-score suggests a tendency toward high-value clustering (Hot Spot Clustering), but the evidence is not statistically significant at $\alpha = 0.05$ If the Z-score had been negative and significant, it would have indicated a concentration of low values (**Cold Spot Clustering**).

$$W = \begin{pmatrix} 0 & 1 & 0 & 0 & 0 \\ 1 & 0 & 1 & 0 & 0 \\ 0 & 1 & 0 & 1 & 0 \\ 0 & 0 & 1 & 0 & 1 \\ 0 & 0 & 0 & 1 & 0 \end{pmatrix}$$

Common Pitfalls

- **Using General G with unsuitable values.** General G is most appropriate for non-negative variables because it is based on products of neighboring observations. Negative values or many zeros can distort clustering interpretation.
- **Confusing General G with local G_i^*.** General G tests whether high or low values cluster across the entire region. Local G_i^* statistics identify the specific locations driving clustering.
- **Ignoring the expected value of G.** The statistic is usually small because it is based on weighted-product ratios. Interpretation should rely on the expected value and standardized z-score.
- **Equating significance with meaningful hotspots.** Significant clustering may reflect covariates, boundaries, or extreme values rather than true hotspot processes. Spatial visualization and follow-up analysis are often needed.
- **Using raw counts instead of rates.** Clustering of counts may simply reflect population density rather than genuine spatial variation. Rates or standardized measures are usually more appropriate.

Local Moran's I (LISA)

Local Indicator of Spatial Association

SPAT.4 | **Category:** Spatial Statistics | **Data:** Continuous | **Null:** Permutation test

When to Use

- Detect local spatial autocorrelation patterns across regions.
- Identify local clusters, outliers, hot/cold spots.
- Explore regional heterogeneity in spatial relationships.
- Follow-up analysis after significant global spatial tests.

Assumptions

- Spatial weights matrix is correctly specified.
- Spatially indexed interval/ratio-scale data.
- Values are exchangeable under spatial randomness (H_0).
- Multiple-testing adjustment may be needed.

Test Statistic

$$I_i = \frac{x_i - \bar{x}}{s^2} \sum_j w_{ij}(x_j - \bar{x}), \quad \text{where } s^2 = \frac{\sum_i (x_i - \bar{x})^2}{n}$$

(Note: Divided by n and not $n-1$).

Notation: I_i is local Moran for location i, w_{ij} are spatial weights for i's neighbors, s^2 is scaled variance

Plain English: Local Moran's I measures, for each location, whether its value is similar to or different from nearby values. Positive I_i with high value indicates a high-value location surrounded by high-value neighbors (hotspot); positive I_i with low value indicates low-low cluster (coldspot). Negative I_i indicates spatial outliers—high values surrounded by low neighbors or vice versa. Permutation tests establish significance for each location.

⚠ Practitioner Warning

Exact scaling (s^2 with n or $n-1$) may vary slightly across software. Multiple testing issue: testing all n locations inflates family-wise error rate. Consider False Discovery Rate (FDR) or Bonferroni correction for controlling Type I error. Permutation tests computationally intensive with many locations. Choice of spatial weights critically affects which locations flagged as significant. Edge effects: boundary locations have fewer neighbors, reducing power. Significance depends on number of permutations (typically 999 or 9,999). Does not account for global spatial trend; may detect spurious local clusters in presence of unmodeled trend.

Hypotheses

- H_0: No local spatial association at location i.
- H_1: Significant local spatial association.

Decision Rule: Reject H_0 if p-value $< \alpha$. Classify: HH (high-high), LL (low-low), HL (high-low), LH (low-high).

Software References [Python, R, SAS]

Python	R	SAS
`esda.Moran_Local`	`spdep::localmoran`	`(GIS spatial statistics tools)`

Additional Notes

Average of local statistics equals global: $\sum_i I_i/n = I$ (global Moran's I), establishing mathematical relationship. Four quadrants in Moran scatterplot: HH (high value, high neighbors, positive I_i), LL (low value, low neighbors, positive I_i), HL (high value, low neighbors, negative I_i), LH (low value, high neighbors, negative I_i). Anselin (1995) formalized LISA framework with five criteria: local statistic for each location, scaled sum gives global statistic, expected value calculable, variance derivable, inference via permutation. Conditional permutation: fix focal location's value, permute only neighbor values (more conservative). Bonferroni correction: $\alpha_i = \alpha/n$ controls family-wise error but very conservative. False Discovery Rate: controls expected proportion of false positives among rejections, less conservative. Getis-Ord G_i^* provides alternative local statistic. Cluster mapping: LISA maps show spatial distribution of significant clusters and outliers, widely used in disease surveillance and crime analysis.

Miscellanea: Luc Anselin (1995) introduced the LISA framework (Local Indicators of Spatial Association), formalizing local spatial statistics and establishing criteria for their interpretation. Unlike global measures such as Moran's I, LISA identifies where clustering occurs, using permutation-based inference to assess significance. It is widely used in spatial analysis across multiple disciplines.

Reference(s): Anselin (1995); Ord & Getis (1995)

Worked Example

Problem Statement A city auditor is examining crime rates across 5 administrative districts ($n = 5$). While the Global Moran's I shows moderate clustering, the auditor wants to perform a **LISA** analysis on **District 2 (Target Cell)** to determine if it is a "Spatial Outlier" (a high-crime district surrounded by low-crime neighbors).

1. Experimental Parameters

- **Crime Rates (x_i):** $x_1 = 5, x_2 = 50, x_3 = 8, x_4 = 12, x_5 = 7$.
- **Mean ($\bar{x}$):** 16.4.
- **Scaled Variance term (s^2):** 287.44 (divided by n and not the usual $n - 1$)
- **Spatial Weights (W):** Binary contiguity. District 2 has neighbors Districts 1 and 3 ($w_{21} = 1, w_{23} = 1$).

2. Calculation of the Local I_i Statistic
The formula for the Local Moran's I for a specific location i is:

$$I_i = \frac{(x_i - \bar{x})}{s^2} \sum_{j=1, j\neq i}^{n} w_{ij}(x_j - \bar{x})$$

a. Deviation of the target district (z_i):

$$z_2 = (x_2 - \bar{x}) = 50 - 16.4 = 33.6$$

b. Sum of weighted deviations of neighbors:
District 2's neighbors are 1 and 3.

$$\sum_{j\neq 2} w_{2j}(x_j - \bar{x}) = (x_1 - 16.4) + (x_3 - 16.4) = -19.8$$

c. Final I_2 Calculation:

$$I_2 = \frac{33.6}{287.44} \times (-19.8) = 0.1177 \times (-19.8) = -2.32$$

3. Expected Value Under Conditional Randomization
Holding the value at location i fixed and randomly permuting the remaining observations gives

$$E[I_2] = -\frac{\sum_{j=1, j\neq 2}^{n} w_{2j}}{n-1} = -\frac{w_{2.}}{n-1} = \frac{2}{4} = -0.5$$

4. Results and Interpretation

Metric	Value
Local Moran's I_2	-2.32
Expected Value $E[I_2]$	-0.5 (for $n = 5$)
Pattern Type	**High-Low Outlier (HL)**

Conclusion: The **strongly negative** I_2 value indicates spatial dissimilarity: District 2 differs markedly from its neighboring districts and is classified as a **High-Low (HL) spatial outlier**.

Note: In practice, LISA significance is usually assessed using permutation tests because asymptotic approximations can be unreliable for very small samples.

Common Pitfalls

- **Using LISA without a global test.** Local Moran statistics identify clustering at specific locations but do not establish overall spatial autocorrelation. Testing global Moran's I first is often advisable.
- **Ignoring multiple testing.** LISA performs a separate hypothesis test for each spatial unit. FDR or related adjustments help reduce false positives.
- **Confusing clusters and outliers.** High-high and low-low patterns indicate clusters, whereas high-low and low-high patterns indicate spatial outliers. These require different interpretations.
- **Equating significance with strength.** Significant local statistics may still have small magnitudes. Interpretation should consider both effect size and significance.
- **Ignoring boundary effects.** Edge locations often have fewer neighbors, producing less stable local statistics. Boundary structure should therefore be considered during interpretation.

Getis-Ord G_i^* Test

Local Hotspot and Coldspot Analysis

SPAT.5 | **Category:** Spatial Statistics | **Data:** Continuous | **Null:** Normal (asymptotic)

When to Use

- Identify local hot spots and cold spots in data.
- Detect localized clustering of unusually extreme values.
- Explore local spatial concentration patterns across regions.
- Follow-up analysis after significant global spatial tests.

Assumptions

- Spatial weights matrix is correctly specified.
- Spatially indexed interval/ratio-scale data.
- Values are exchangeable under spatial randomness (H_0).
- Multiple-testing adjustment may be needed.

Test Statistic

$$G_i^* = \frac{\sum_j w_{ij}x_j - \bar{x}\sum_j w_{ij}}{D_i}, \quad \text{where } D_i = s\sqrt{\frac{n\sum_j w_{ij}^2 - (\sum_j w_{ij})^2}{n-1}}, \quad \bar{x} = \frac{\sum_j x_j}{n}, \quad s^2 = \frac{\sum_j x_j^2}{n} - \bar{x}^2.$$

Notation: x_j are measurements, n is the sample size, w_{ij} includes $w_{ii} = 1$. Under $H_0, G_i^* \approx N(0,1)$.

Plain English: The Getis-Ord G_i^* statistic identifies locations where high or low values cluster by comparing the local sum of values (location plus neighbors) to the global mean. Positive G_i^* values indicate hotspots where a location and its neighbors have high values; negative G_i^* values indicate coldspots where values are low. The statistic is a standardized z-score, making interpretation straightforward: larger absolute values indicate stronger spatial concentration. Unlike Local Moran which identifies outliers as well, G_i^* focuses specifically on high-value and low-value clustering.

⚠ Practitioner Warning

Multiple testing arises because all n locations are tested; consider FDR, Bonferroni, or clearly labeled exploratory maps. Results highly sensitive to spatial weights specification, especially distance threshold. Edge effects: locations near study boundaries may have incomplete neighborhoods. Does not identify spatial outliers (high-low or low-high patterns)—use Local Moran's I for that. Assumes positive values; negative values require transformation. Distance bands should be chosen based on theory or exploratory analysis, not p-value optimization. Permutation tests are useful when normal approximation is questionable.

Hypotheses

- H_0: No local high/low concentration at location i.
- H_1: Significant local high- or low-value concentration.

Decision Rule: Reject H_0 if $|G_i^*| > z_{\alpha/2}$ or p-value $< \alpha$. Higher (+ve) z: hotspot; lower (-ve) z: coldspot.

Software References [Python, R, SAS]

Python	R	SAS
`esda.getisord.G_Local`	`spdep::localG`	`Not Native (Manual)`

Additional Notes

Difference between G_i and G_i^*: G_i excludes focal location from its own calculation ($w_{ii} = 0$), while G_i^* includes it ($w_{ii} = 1$). Most applications use G_i^* because including focal location increases power and interpretability. Distance-based weights common: $w_{ij} = 1$ if $d_{ij} \leq d_0$, otherwise 0. Fixed distance band ensures all locations have comparable neighborhood definitions. Adaptive distance (k-nearest neighbors) ensures all locations have same number of neighbors but varies in spatial extent. Standardization: row-standardized weights sum to 1 for each location, binary weights sum to number of neighbors. Interpretation of z-scores: $|z| > 1.96$ (95% confidence), $|z| > 2.58$ (99% confidence). Higher absolute z indicates stronger clustering. Cluster mapping: color-code by z-score ranges (e.g., dark red for $z > 2.58$ hotspots, dark blue for $z < -2.58$ coldspots). Widely implemented in GIS software for crime analysis, real estate, epidemiology. Ord and Getis (1995) developed local statistics; Getis and Ord (2010) provided computational refinements.

Miscellanea: Keith Ord and Arthur Getis developed local statistics (G_i, G_i^*) as extensions of their General G to identify spatial hotspots and coldspots. The G_i^* version includes the focal observation (i), distinguishing it from G_i. These statistics are widely used in GIS-based hotspot analysis, particularly in applications such as crime mapping and public health.

Reference(s): Ord & Getis (1995); Getis & Ord (2010)

Getis-Ord G_i^* Test

SPAT.5

Worked Example

Problem Statement A public health official is examining the number of flu cases across 5 city sectors ($n = 5$). While the global Moran's I suggests some clustering, the official needs to identify if **Sector 3 (Target Cell)** is a statistically significant **Hot Spot**. This would justify the deployment of a mobile vaccination clinic to that specific neighborhood and its immediate neighbors.

1. Experimental Parameters

- **Values (x_i):** $x_1 = 10, x_2 = 12, x_3 = 45, x_4 = 38, x_5 = 15$. Global mean: $\bar{x} = 24$; standard deviation: $s : 14.55$
- **Spatial Weights (W):** A contiguity matrix where Sector 3 is adjacent to Sectors 2 and 4. In G_i^*, the target cell is included in its own neighborhood ($w_{33} = 1$).
- **Neighborhood (j):** For $i = 3, w_{32} = 1, w_{33} = 1, w_{34} = 1$. All other $w_{3j} = 0$. $\sum w_{3j} = 3$ and $\sum w_{3j}^2 = 3$. $\sum w_{3j}x_j = x_2 + x_3 + x_4 = 95$.

2. Calculation of the G_i^* Statistic The G_i^* formula for Sector 3 is:

$$G_i^* = \frac{\sum_{j=1}^{n} w_{ij}x_j - \bar{x}\sum_{j=1}^{n} w_{ij}}{s\sqrt{\frac{n\sum w_{ij}^2-(\sum w_{ij})^2}{n-1}}} = \frac{23}{17.82} = +1.29$$

Numerator:

$$\sum_{j=1}^{n} w_{3j}x_j - \bar{x}\sum_{j=1}^{n} w_{3j} = 95 - (24 \times 3) = 23$$

Denominator:

$$s\sqrt{\frac{n\sum w_{3j}^2 - (\sum w_{3j})^2}{n-1}} = 14.55\times\sqrt{\frac{5(3)-(3)^2}{4}} \approx 17.82$$

3. Significance and Decision The G_i^* statistic is already a Z-score.

Metric	Value
Calculated G_i^* (Z-score)	+1.29
Confidence Level	90% (Critical $Z = 1.645$)
P-value	0.098
Decision	**Fail to Reject H_0**

Conclusion: Although Sector 3 has the highest value (45) and is adjacent to another high-value sector (38), the standardized G_i^* score of +1.29 does not exceed the one-sided critical value of 1.645. Thus, there is insufficient evidence at the 10% significance level to classify Sector 3 as a statistically significant hotspot.

Common Pitfalls

- **Confusing G_i and G_i^*.** The G_i statistic excludes the focal location, whereas G_i^* includes it. Most hotspot analyses therefore use G_i^* for neighborhood clustering.
- **Ignoring multiple testing.** Local hotspot analysis performs a separate test for each spatial unit. FDR or related adjustments help reduce false positives across many locations.
- **Using unsuitable variables.** G_i^* is most appropriate for positive variables such as counts, rates, or prices. Negative values or many zeros can distort hotspot interpretation.
- **Treating all hotspots equally.** The statistic produces standardized z-scores whose magnitudes reflect clustering strength. Larger values indicate stronger hotspot evidence.
- **Assuming hotspots are causal.** The test detects spatial concentration but does not explain its mechanism. Additional modeling is needed to interpret causes of clustering.

LM Test for Spatial Lag

Lagrange Multiplier Test for Spatially Lagged Dependent Variable

SPAT.6 | **Category:** Spatial Statistics | **Data:** Continuous | **Null:** Chi-square

When to Use

- Detect spatial lag dependence in regression models.
- Assess possible spillover effects across neighboring regions.
- Diagnose omitted spatial interaction effects in residuals.
- Perform spatial regression residual diagnostics.
- Choose between OLS and spatial lag specifications.

Assumptions

- OLS regression model has been fitted.
- Spatial weights matrix is specified a priori.
- Linear mean structure is correctly specified.
- Errors are homoskedastic under H_0.
- LM statistic is asymptotically χ^2_1 under H_0.

Test Statistic

$$LM_{\text{lag}} = \frac{[\mathbf{e}'\mathbf{Wy}/\hat{\sigma}^2]^2}{D} \sim \chi^2_1 \text{ (under } H_0), \quad \text{where } D = [\frac{(\mathbf{WX}\hat{\beta})'\mathbf{M}(\mathbf{WX}\hat{\beta})}{\hat{\sigma}^2} + \text{tr}(\mathbf{W}'\mathbf{W} + \mathbf{W}^2)], \ \hat{\sigma}^2 = \frac{\mathbf{e}'\mathbf{e}}{n}.$$

Notation: $\mathbf{e}$ are OLS residuals, $\mathbf{Wy}$ is spatially lagged dependent variable, $\mathbf{M} = \mathbf{I} - \mathbf{X}(\mathbf{X}'\mathbf{X})^{-1}\mathbf{X}'$

Plain English: The LM test for spatial lag evaluates whether the dependent variable exhibits spatial dependence by testing if neighbors' values of the dependent variable help predict the focal location's value. The test examines the correlation between OLS residuals and the spatially lagged dependent variable **Wy**. Significant results indicate a spatial autoregressive (SAR) model is more appropriate than standard OLS, where y_i depends on both covariates and neighbors' y values.

⚠ Practitioner Warning

Results depend critically on spatial weights matrix specification. Misspecified **W** can lead to incorrect conclusions. Test assumes homoskedastic errors; heteroskedasticity affects validity. Does not distinguish between spatial lag and spatial error models when both present—use robust LM tests. Power decreases with weak spatial weights or dispersed spatial patterns. Large samples can detect trivial spatial dependence. Always compare with LM test for spatial error to diagnose specification. If both significant, use robust versions to determine primary source.

Hypotheses

- H_0: $\rho = 0$ (no spatial lag dependence, OLS appropriate).
- H_1: $\rho \neq 0$ (spatial lag dependence present).

Decision Rule: Reject H_0 if $LM_{\text{lag}} > \chi^2_{\alpha,1}$ or p-value $< \alpha$. Rejection suggests spatial autoregressive model (SAR).

Software References [Python, R, SAS]

`spreg (PySAL)`	`spatialreg::lm.LMtests`	`PROC SPATIALREG`

Additional Notes

Spatial autoregressive (SAR) model: $\mathbf{y} = \rho\mathbf{Wy} + \mathbf{X}\boldsymbol{\beta} + \boldsymbol{\epsilon}$ where ρ is spatial autoregressive parameter, which measures dependence in neighboring outcome values. Positive ρ is often interpreted as spatial spillover, but causal interpretation requires substantive theory and careful identification. SAR estimation requires maximum likelihood, generalized method of moments, or instrumental variables rather than ordinary OLS. Anselin (1988) developed LM diagnostics for spatial dependence using OLS residuals. Robust LM test for lag: LM^*_{lag} adjusts for spatial error dependence. Decision tree: if both LM_{lag} and LM_{error} significant, use robust versions; whichever robust test is significant indicates primary specification. Spatial Durbin model (SDM) includes both **Wy** and **WX**: $\mathbf{y} = \rho\mathbf{Wy} + \mathbf{X}\boldsymbol{\beta} + \mathbf{WX}\boldsymbol{\gamma} + \boldsymbol{\epsilon}$; it nests SAR when $\boldsymbol{\gamma} = 0$ and is related to SEM under common-factor restrictions. LeSage and Pace (2009) recommend SDM as starting point.

Miscellanea: Luc Anselin (1988) developed Lagrange Multiplier tests for detecting spatial dependence in regression models, providing a practical framework for distinguishing between spatial lag and spatial error processes. These tests use OLS residuals and avoid fitting alternative spatial models, making them computationally efficient. They remain a standard diagnostic tool in spatial econometrics.

Reference(s): Anselin (1988); Anselin et al. (1996)

LM Test for Spatial Lag

SPAT.6

Worked Example

Problem Statement An economist studies the **Innovation Score** (0–100) of four ($n = 4$) geographically connected tech hubs to determine whether innovation is explained by **R&D Expenditure** (millions), while testing for a **spatial lag effect** — where a hub's innovation may be influenced by neighbouring hubs through knowledge spillovers. Hubs are arranged in a linear chain $A-B-C-D$.

Data

Hub	Inno. Score (y)	R&D (x)	Neighbours
A	50	10	B
B	60	20	A, C
C	85	30	B, D
D	90	45	C

1. Summary

- OLS Regression: $y = 39.112 + 1.224x$.
- Residuals (e'): $(-1.36, -3.60, 9.16, -4.21)$
- Lagged Dep. Var. $(Wy)'$: $(60, 67.5, 75, 85)$
- Residual Variance ($\hat{\sigma}^2$): $e'e/n = 29.09$
- Trace Constant (T): $tr(W'W + W^2) = 5.50$
- Scaling Constant (D): 7.95

2. Calculating Test Statistic

$$\text{Score} = \frac{e'Wy}{\hat{\sigma}^2} = \frac{5.257}{29.09} = 0.181,$$

$$LM_{Lag} = \frac{\text{Score}^2}{D} = 0.00411$$

Weights Matrix (W) & M Matrix

$$W = \begin{pmatrix} 0 & 1 & 0 & 0 \\ 0.5 & 0 & 0.5 & 0 \\ 0 & 0.5 & 0 & 0.5 \\ 0 & 0 & 1 & 0 \end{pmatrix} \quad M = \begin{pmatrix} 0.36 & -0.40 & -0.16 & 0.21 \\ -0.40 & 0.69 & -0.21 & -0.07 \\ -0.16 & -0.21 & 0.73 & -0.36 \\ 0.21 & -0.07 & -0.36 & 0.22 \end{pmatrix}$$

H_0: No spatial lag ($\rho = 0$).
H_a: Significant spatial lag is present $\rho \neq 0$.

3. Significance and Decision The LM_{Lag} statistic follows a χ^2 distribution with $df = 1$.

Metric	Value
Calculated LM_{Lag}	0.00411
Critical $\chi^2_{0.05,1}$	3.841
P-value	0.949
Decision	**Fail to Reject H_0**

Conclusion: Since $0.00411 < 3.841$, we fail to reject H_0. There is no statistically significant evidence of spatial lag. The current OLS model is sufficient, and there's no need to transition to a Spatial Autoregressive (SAR) model.

Note: This worked example uses only four observations for pedagogical illustration. In practice, LM spatial diagnostics rely on large-sample asymptotic χ^2 approximations, which are not reliable with such a small sample size.

Common Pitfalls

- **Skipping the baseline OLS model.** The LM-Lag test is applied to residuals from an initial OLS regression. Always estimate and report the OLS model first.
- **Using only LM-Lag.** Spatial lag and spatial error dependence can both produce significant LM statistics. LM-Lag and LM-Error should therefore be evaluated together.
- **Assuming causal spillovers.** A significant LM-Lag statistic indicates spatial association, not necessarily causal influence among neighbors. Further modeling and theory are needed.
- **Ignoring omitted spatial variables.** Spatial autocorrelation in residuals may reflect missing explanatory variables rather than true spatial interaction. Additional controls should be considered.
- **Using unsuitable spatial weights.** The LM-Lag statistic depends directly on the spatial weights matrix **W**. The choice of **W** should be theoretically justified and checked.

Scaling Factor Calculation

Residual Maker (matrix): $M = I - X(X'X)^{-1}X'$, **Trace Constant:** $T = \text{tr}(W'W + W^2)$, **Scaling Constant:** $D = \frac{(W\hat{y})'M(W\hat{y})}{\hat{\sigma}^2} + T$.

LM Test for Spatial Error

Lagrange Multiplier Test for Spatial Error Autocorrelation

SPAT.7 | **Category:** Spatial Statistics | **Data:** Continuous | **Null:** Chi-square

When to Use

- Detect spatial error dependence in regression.
- Diagnose omitted spatially correlated shocks.
- Assess residual spatial autocorrelation.
- Spatial regression residual diagnostics.
- Choose between OLS and spatial error models.

Assumptions

- OLS regression model has been fitted.
- Spatial weights matrix is specified a priori.
- Linear mean structure is correctly specified.
- Errors are homoskedastic under H_0.
- LM statistic is asymptotically χ^2_1 under H_0.

Test Statistic

$$LM_{\text{error}} = \frac{[\mathbf{e}'\mathbf{W}\mathbf{e}/\hat{\sigma}^2]^2}{\text{tr}(\mathbf{W}'\mathbf{W} + \mathbf{W}^2)} \sim \chi^2_1 \text{ (under } H_0), \quad \text{where } \hat{\sigma}^2 = \mathbf{e}'\mathbf{e}/n.$$

Notation: $\mathbf{e}$ are OLS residuals, $\mathbf{We}$ is spatially lagged residuals, tr is matrix trace

Plain English: The LM test for spatial error evaluates whether regression residuals exhibit spatial autocorrelation by testing the correlation between residuals and their spatial lag $\mathbf{We}$. The test examines whether errors at nearby locations are correlated, violating OLS independence assumption. Significant results indicate a spatial error model (SEM) is more appropriate, where errors follow a spatial AR process. This differs from spatial lag model where the dependent variable itself exhibits spatial dependence.

⚠ Practitioner Warning

Results critically depend on spatial weights matrix specification. Misspecified $\mathbf{W}$ leads to incorrect diagnosis. Assumes homoskedastic errors; heteroskedasticity affects test validity. Cannot distinguish between spatial lag and spatial error when both present—use robust LM tests. The test has power against spatial error but also against omitted spatially correlated variables. If both LM lag and LM error significant, robust versions needed for correct specification. Edge effects and boundary conditions can affect results.

Hypotheses

- H_0: $\lambda = 0$ (no spatial error autocorrelation).
- H_1: $\lambda \neq 0$ (spatial error autocorrelation present).

Decision Rule: Reject H_0 if $LM_{\text{error}} > \chi^2_{\alpha,1}$ or p-value $< \alpha$. Rejection suggests spatial error model (SEM).

Software References [Python, R, SAS]

`spreg (PySAL)`	`spatialreg::lm.LMtests`	`PROC SPATIALREG`

Additional Notes

Spatial error model (SEM): $\mathbf{y} = \mathbf{X}\beta + \epsilon$ where $\epsilon = \lambda\mathbf{W}\epsilon + \mathbf{u}$ and $\mathbf{u} \sim N(\mathbf{0}, \sigma^2\mathbf{I})$. Parameter λ measures spatial error dependence; $\lambda > 0$ indicates positive autocorrelation in errors. Reduced form: $\epsilon = (\mathbf{I} - \lambda\mathbf{W})^{-1}\mathbf{u}$, creating spatial correlation structure. Robust LM test for error: LM^*_{error} adjusts for potential spatial lag. Decision tree for specification: (1) Estimate OLS, (2) Test LM_{lag} and LM_{error}, (3) If both significant, use robust versions, (4) Whichever robust test significant indicates appropriate model. Interpretation: spatial error often reflects unobserved spatially correlated variables or measurement error with spatial structure. Spatial lag indicates substantive spatial spillovers. Anselin and Bera (1998) provided comprehensive testing framework.

Miscellanea: Luc Anselin (1988) developed LM tests for both spatial lag and spatial error dependence, providing a practical framework for diagnosing spatial model misspecification using OLS residuals. Earlier spatial error models existed, but formal testing procedures were limited. Later extensions introduced robust LM tests to handle cases where both forms of dependence appear present, and these diagnostics remain widely used in spatial econometrics.

Reference(s): Anselin (1988); Anselin et al. (1996)

LM Test for Spatial Error

SPAT.7

Worked Example

Problem Statement An economist studies the **Innovation Score** (0–100) of four ($n = 4$) geographically connected tech hubs to determine whether innovation is explained by **R&D Expenditure** (millions), while testing for a **spatial error autocorrelation** — where a hub's innovation may be influenced by neighbouring hubs through knowledge spillovers. Hubs are arranged in a linear chain $A-B-C-D$.

Data

Hub	Inno. Score (y)	R&D (x)	Neighbours
A	50	10	B
B	60	20	A, C
C	85	30	B, D
D	90	45	C

Weights Matrix (W) & M Matrix

$$W = \begin{pmatrix} 0 & 1 & 0 & 0 \\ 0.5 & 0 & 0.5 & 0 \\ 0 & 0.5 & 0 & 0.5 \\ 0 & 0 & 1 & 0 \end{pmatrix} \quad M = \begin{pmatrix} 0.36 & -0.40 & -0.16 & 0.21 \\ -0.40 & 0.69 & -0.21 & -0.07 \\ -0.16 & -0.21 & 0.73 & -0.36 \\ 0.21 & -0.07 & -0.36 & 0.22 \end{pmatrix}$$

H_0: No spatial error autocorrelation ($\lambda = 0$).
H_a: Significant spatial autocorrelation is present $\lambda \neq 0$.

1. Summary

- OLS Regression: $y = 39.112 + 1.224x$.
- Residuals (e'): $(-1.36, -3.60, 9.16, -4.21)$
- Spatial Error Score ($e'We$): -83.419
- Residual Variance ($\hat{\sigma}^2$): $e'e/n = 29.09$
- Trace Constant (T): $tr(W'W + W^2) = 5.50$

2. Calculating Test Statistic

$$\text{Score} = \frac{e'We}{\hat{\sigma}^2} = \frac{-83.419}{29.09} = -2.868,$$

$$LM_{Err} = \frac{\text{Score}^2}{T} = 1.495$$

3. Significance and Decision The LM_{Err} statistic follows a χ^2 distribution with $df = 1$.

Metric	Value
Calculated LM_{Err}	1.495
Critical $\chi^2_{0.05,1}$	3.841
P-value	0.2214
Decision	**Fail to Reject H_0**

Conclusion: Since $1.495 < 3.841$, we fail to reject H_0. There is no statistically significant evidence of spatial error autocorrelation. The OLS residuals are spatially random.

Note: This worked example uses only four observations for pedagogical illustration. In practice, LM spatial diagnostics rely on large-sample asymptotic χ^2 approximations, which are not reliable with such a small sample size.

Common Pitfalls

- **Treating spatial error dependence as substantive.** Spatial error autocorrelation often reflects omitted spatially patterned variables rather than a true spatial process. Additional covariates should therefore be considered first.
- **Choosing a model without LM comparison.** LM-Lag and LM-Error diagnostics should both be examined because either may become significant under misspecification. Robust LM tests help distinguish them.
- **Confusing lag and error models.** Spatial lag models imply dependence among neighboring outcomes, whereas spatial error models capture correlation among unobserved disturbances.
- **Expecting major coefficient changes.** Spatial error models mainly affect coefficient variances and standard errors rather than point estimates themselves. Inference often changes more than estimates.
- **Assuming autocorrelation is eliminated.** Residual spatial dependence may persist if covariates or spatial structure remain inadequate. Post-estimation diagnostics should therefore still be checked.

Why quadratic form $e'We$ is -ve?

The quadratic form $e'We$ need not be nonnegative (≥ 0) because the spatial weights matrix W is not positive definite. In fact, W is not even symmetric (due to row standardization), and its eigenvalues include negative values. A quadratic form $x'Ax$ is guaranteed to be nonnegative only when A is positive semidefinite. Since W does not satisfy this property, the value of $e'We$ may be either positive or negative; therefore obtaining a negative value here is not surprising.

Robust LM Test (Spatial Lag)

Robust Lagrange Multiplier Test for Spatial Lag Conditional on Spatial Error

SPAT.8 | **Category:** Spatial Statistics | **Data:** Continuous | **Null:** Chi-square

When to Use

- Confirm spatial lag after controlling error dependence.
- Distinguish lag from spatial error effects.
- Resolve ambiguity in standard LM diagnostics.
- Spatial regression model selection.
- Follow-up after significant LM lag/error tests.

Assumptions

- OLS regression model has been fitted.
- Spatial weights matrix is specified a priori.
- Linear mean structure is correctly specified.
- Errors are homoskedastic under H_0.
- Robust LM statistic is asymptotically χ_1^2 under H_0.

Test Statistic

$$LM^*_{\text{lag}} = \frac{\left[\mathbf{e}'\mathbf{W}\mathbf{y}/\hat{\sigma}^2 - \mathbf{e}'\mathbf{W}\mathbf{e}/\hat{\sigma}^2\right]^2}{D - T} \sim \chi_1^2 \text{ (under } H_0), \quad T = \text{tr}(\mathbf{W}'\mathbf{W} + \mathbf{W}^2).$$

Notation: $\mathbf{e}$ are OLS residuals, $\mathbf{Wy}$ is spatially lagged dependent variable, D from spatial lag test (SPAT.6)

Plain English: The robust LM test for spatial lag adjusts the standard spatial lag test by accounting for potential spatial error autocorrelation. When both spatial lag and spatial error dependencies may be present, the standard LM tests can both be significant, making specification ambiguous. The robust test isolates the spatial lag component by conditioning on spatial error, providing clearer guidance. If robust LM lag is significant but robust LM error is not, spatial lag model is appropriate.

⚠ Practitioner Warning

Requires both standard LM lag and LM error tests to be conducted first for proper interpretation. Decision rule requires examining all four tests (LM lag, LM error, robust LM lag, robust LM error). Lower power than standard LM tests when only one type of dependence present. Assumes homoskedastic errors; violations affect validity. Results still depend on spatial weights specification. If both robust tests significant, consider spatial Durbin model (SDM) or general nesting spatial (GNS) model. Does not test for spatial heterogeneity or structural instability.

Hypotheses

- H_0: $\rho = 0$ conditional on potential spatial error.
- H_1: $\rho \neq 0$ (spatial lag dependence present).

Decision Rule: Reject H_0 if $LM^*_{\text{lag}} > \chi^2_{\alpha,1}$ or p-value $< \alpha$. If both standard tests significant, compare robust versions.

Software References [Python, R, SAS]

`spreg (PySAL)`	`spatialreg::lm.LMtests`	`PROC SPATIALREG`

Additional Notes

Decision tree for spatial model selection: (1) Estimate OLS regression, obtain residuals. (2) Compute standard LM_{lag} and LM_{error}. (3) If only one significant, use corresponding model (SAR or SEM). (4) If both significant, compute robust versions LM^*_{lag} and LM^*_{error}. (5) If only robust lag significant, use SAR model. (6) If only robust error significant, use SEM model. (7) If both robust tests significant, use spatial Durbin model (SDM). Anselin et al. (1996) developed robust LM tests addressing the problem that standard tests confound different types of spatial dependence. The robust tests use orthogonalized components, isolating each type of dependence. Mathematical relationship: $LM^*_{\text{lag}} = LM_{\text{lag}} - LM_{\text{error}}$ approximately under certain conditions. Interpretation: robust lag test asks whether spatial lag improves fit beyond what spatial error already explains. Spatial Durbin model: $\mathbf{y} = \rho\mathbf{W}\mathbf{y} + \mathbf{X}\boldsymbol{\beta} + \mathbf{W}\mathbf{X}\boldsymbol{\gamma} + \boldsymbol{\epsilon}$ includes both spatial lag and spatially lagged covariates.

Miscellanea: Luc Anselin, Anil Bera, Raymond Florax, and Mann Yoon (1996) developed robust LM tests to address cases where standard LM tests for spatial lag and error dependence are both significant. By adjusting for the alternative form of dependence, these tests improve model selection between spatial lag and spatial error specifications. They are now commonly used alongside standard LM diagnostics in spatial econometrics.

Reference(s): Anselin et al. (1996); Anselin (1988)

Robust LM Test (Spatial Lag) SPAT.8

Worked Example

Problem Statement An economist studies the **Innovation Score** (0–100) of four ($n = 4$) geographically connected tech hubs to determine whether innovation is explained by **R&D Expenditure** (millions), while testing for a **spatial lag effect** — where a hub's innovation may be influenced by neighbouring hubs through knowledge spillovers. Hubs are arranged in a linear chain $A-B-C-D$.

Data

Hub	Inno. Score (y)	R&D (x)	Neighbours
A	50	10	B
B	60	20	A, C
C	85	30	B, D
D	90	45	C

Weights Matrix (W) & M Matrix

$$W = \begin{pmatrix} 0 & 1 & 0 & 0 \\ 0.5 & 0 & 0.5 & 0 \\ 0 & 0.5 & 0 & 0.5 \\ 0 & 0 & 1 & 0 \end{pmatrix} \quad M = \begin{pmatrix} 0.36 & -0.40 & -0.16 & 0.21 \\ -0.40 & 0.69 & -0.21 & -0.07 \\ -0.16 & -0.21 & 0.73 & -0.36 \\ 0.21 & -0.07 & -0.36 & 0.22 \end{pmatrix}$$

H_0: $\rho = 0$ conditional on potential spatial error.
H_a: $\rho \neq 0$ conditional on potential spatial error.

1. Summary

- OLS Regression: $y = 39.112 + 1.224x$.
- Residuals (e'): $(-1.36, -3.60, 9.16, -4.21)$
- Spatial Lag Score ($e'Wy$): 5.257
- Spatial Error Score ($e'We$): -83.419
- Residual Variance ($\hat{\sigma}^2$): $e'e/n = 29.09$
- Trace Constant (T): $tr(W'W + W^2) = 5.50$
- Scaling Constant (D): 7.95

2. Calculating Test Statistic

$$\text{Score} = \frac{e'Wy - e'We}{\hat{\sigma}^2} = \frac{5.257 + 83.419}{29.09} = 3.048,$$

$$LM^*_{\text{lag}} = \frac{\text{Score}^2}{D - T} = 3.79$$

3. Significance and Decision The LM^*_{lag} statistic follows a χ^2 distribution with $df = 1$.

Metric	Value
Calculated LM^*_{lag}	3.79
Critical $\chi^2_{0.05,1}$	3.841
P-value	0.0515
Decision	**Fail to Reject H_0**

Conclusion: Since $3.79 < 3.841$, we failed to reject H_0. After adjusting for possible spatial error dependence, the robust LM-lag statistic is borderline but does not exceed the 5% chi-square critical value. Thus, this toy example does not provide strong evidence requiring a spatial lag specification.

Note: This worked example uses only four observations for pedagogical illustration. In practice, LM spatial diagnostics rely on large-sample asymptotic χ^2 approximations, which are not reliable with such a small sample size.

Common Pitfalls

- **Using robust LM tests unnecessarily.** Robust diagnostics are mainly used when both standard LM tests indicate spatial dependence. If only one standard LM statistic is significant, the corresponding model is often sufficient.
- **Interpreting RLM-Lag alone.** Robust LM tests are comparative diagnostics for competing spatial processes. Model selection should therefore compare both RLM-Lag and RLM-Error together.
- **Replacing standard LM tests with robust versions.** Standard LM tests determine whether spatial dependence exists initially. Robust LM tests then help distinguish lag from spatial error dependence.
- **Ignoring cases where both robust tests are significant.** Significant RLM-Lag and RLM-Error statistics may indicate more complex spatial structure. Broader models such as spatial Durbin specifications may then be needed.
- **Assuming dependence is fully removed.** Selecting a spatial lag or error model does not guarantee residual independence. Post-estimation diagnostics should still be used to assess remaining autocorrelation.

Scaling Factor Calculation

Residual Maker (matrix): $M = I - X(X'X)^{-1}X'$, **Trace Constant:** $T = \text{tr}(W'W + W^2)$, **Scaling Constant:** $D = \frac{(W\hat{y})'M(W\hat{y})}{\hat{\sigma}^2} + T$.

Robust LM Test (Spatial Error)

Robust Lagrange Multiplier Test for Spatial Error Conditional on Spatial Lag

SPAT.9 | **Category:** Spatial Statistics | **Data:** Continuous | **Null:** Chi-square

When to Use

- Distinguish spatial error from lag dependence.
- Resolve ambiguity in standard LM diagnostics.
- Spatial regression model selection.
- Follow-up after significant LM lag/error tests.

Assumptions

- Spatial weights matrix is correctly specified.
- Linear regression model is correctly specified.
- Observations are spatially indexed.
- Errors are independent under H_0 and approx. normal.

Test Statistic

$$LM^*_{\text{error}} = \frac{\left[\mathbf{e}'\mathbf{W}\mathbf{e}/\hat{\sigma}^2 - (T/D)\mathbf{e}'\mathbf{W}\mathbf{y}/\hat{\sigma}^2\right]^2}{T\,(1-(T/D))} \approx \chi^2_1 \text{ (under } H_0), \quad T = \text{tr}(\mathbf{W}'\mathbf{W}+\mathbf{W}^2),\ D = T + (\mathbf{W}\hat{\mathbf{y}})'\mathbf{M}(\mathbf{W}\hat{\mathbf{y}})/\hat{\sigma}^2$$

Notation: $\mathbf{e}$ are OLS residuals, $\mathbf{Wy}$ is spatially lagged dependent variable, $\mathbf{M}$ the OLS residual maker ($\perp$ Hat matrix).

Plain English: The robust LM test for spatial error adjusts the standard spatial error test by accounting for potential spatial lag dependence in the dependent variable. When both spatial error and spatial lag may be present, the standard LM tests can both reject, creating specification uncertainty. The robust test isolates the spatial error component by conditioning on spatial lag, providing clearer diagnostic information. If robust LM error is significant but robust LM lag is not, spatial error model is appropriate.

⚠ Practitioner Warning

Requires all four LM statistics (standard lag/error, robust lag/error) for proper interpretation. Lower power than standard LM error test when only spatial error present. Assumes homoskedastic errors; violations affect validity. Still dependent on spatial weights matrix specification. If both robust tests significant, simple SAR or SEM inadequate—consider spatial Durbin or general nesting spatial model. Does not identify source of spatial error (omitted variables vs. measurement error). Interpretation requires understanding distinction between spatial lag and spatial error processes.

Hypotheses

- H_0: $\lambda = 0$ conditional on potential spatial lag.
- H_1: $\lambda \neq 0$ (spatial error autocorrelation present).

Decision Rule: Reject H_0 if $LM^*_{\text{error}} > \chi^2_{\alpha,1}$ or p-value $< \alpha$.

Software References [Python, R, SAS]

Python	R	SAS
`spreg (PySAL)`	`spatialreg::lm.LMtests`	`PROC SPATIALREG`

Additional Notes

Complete decision framework: (1) OLS regression yields residuals $\mathbf{e}$. (2) Compute LM_{lag} and LM_{error}. (3) If neither significant: OLS appropriate (no spatial dependence). (4) If only LM_{lag} significant: use spatial autoregressive (SAR) model. (5) If only LM_{error} significant: use spatial error model (SEM). (6) If both significant: compute LM^*_{lag} and LM^*_{error}. (7) If only LM^*_{lag} significant: use SAR. (8) If only LM^*_{error} significant: use SEM. (9) If both robust tests significant: use spatial Durbin model (SDM) including $\rho\mathbf{Wy}$ and $\mathbf{WX}\gamma$. Interpretation difference: spatial error indicates nuisance dependence (correlated omitted variables), spatial lag indicates substantive spillover effects. LeSage and Pace (2009) argue SDM should be default, nesting both SAR and SEM. Robust tests help choose among nested specifications. Florax et al. (2003) provided empirical guidance on test performance.

Miscellanea: The robust LM error test, introduced by Anselin et al. (1996), complements standard and robust lag diagnostics in spatial model specification. By adjusting for alternative forms of spatial dependence, these tests help distinguish between spatial lag and spatial error processes. In practice, both standard and robust LM tests are often used together to guide model selection.

Reference(s): Anselin et al. (1996); Florax et al. (2003)

Worked Example

Problem Statement An economist studies the **Innovation Score** (0–100) of four ($n = 4$) geographically connected tech hubs to determine whether innovation is explained by **R&D Expenditure** (millions), while testing for a **spatial error autocorrelation** — where a hub's innovation may be influenced by neighbouring hubs through knowledge spillovers. Hubs are arranged in a linear chain $A-B-C-D$.

Data

Hub	Inno. Score (y)	R&D (x)	Neighbours
A	50	10	B
B	60	20	A, C
C	85	30	B, D
D	90	45	C

1. Summary

- OLS Regression: $y = 39.112 + 1.224x$.
- Residuals (e'): $(-1.36, -3.60, 9.16, -4.21)$
- Spatial Lag Score ($e'Wy$): 5.257
- Spatial Error Score ($e'We$): −83.419
- Residual Variance ($\hat{\sigma}^2$): $e'e/n = 29.09$
- Trace Constant (T): $tr(W'W + W^2) = 5.50$
- Scaling Constant (D): 7.95

2. Calculating Test Statistic

$$\text{Score} = \frac{e'We - (T/D)e'Wy}{\hat{\sigma}^2} = -2.993,$$

$$LM^*_{\text{Err}} = \frac{\text{Score}^2}{T(1 - T/D)} = 5.284$$

Weights Matrix (W) & M Matrix

$$W = \begin{pmatrix} 0 & 1 & 0 & 0 \\ 0.5 & 0 & 0.5 & 0 \\ 0 & 0.5 & 0 & 0.5 \\ 0 & 0 & 1 & 0 \end{pmatrix} \quad M = \begin{pmatrix} 0.36 & -0.40 & -0.16 & 0.21 \\ -0.40 & 0.69 & -0.21 & -0.07 \\ -0.16 & -0.21 & 0.73 & -0.36 \\ 0.21 & -0.07 & -0.36 & 0.22 \end{pmatrix}$$

H_0: $\lambda = 0$ conditional on potential spatial lag.
H_a: $\lambda \neq 0$ conditional on potential spatial lag.

3. Significance and Decision The LM^*_{Err} statistic follows a χ^2 distribution with $df = 1$.

Metric	Value
Calculated LM^*_{Err}	5.284
Critical $\chi^2_{0.05,1}$	3.841
P-value	0.0215
Decision	**Reject H_0**

Conclusion: Since $5.284 > 3.841$, we reject H_0. After adjusting for possible spatial lag dependence, the robust LM-error statistic exceeds the 5% critical value. In this toy calculation, the robust diagnostic points more toward spatial error dependence than spatial lag dependence; however, because $n = 4$ and the standard LM tests were not significant, this result should be treated as illustrative rather than as reliable model selection evidence.

Note. This worked example uses only four observations for pedagogical illustration. In practice, LM spatial diagnostics rely on large-sample asymptotic χ^2 approximations, which are not reliable with such a small sample size.

Common Pitfalls

- **Choosing a model from RLM-Error alone.** Robust LM diagnostics should be interpreted comparatively rather than in isolation. RLM-Error should therefore be evaluated alongside RLM-Lag.
- **Treating RLM-Error as proof of omitted variables.** Significant spatial error dependence may reflect omitted covariates, spatial shocks, or measurement processes rather than a specific missing variable.
- **Ignoring ambiguous robust diagnostics.** RLM-Lag and RLM-Error may both be significant with similar magnitudes. Broader specifications such as spatial Durbin models, and sensitivity checks may then be needed.
- **Assuming non-significance means no dependence.** A non-significant RLM-Error statistic only suggests spatial error dependence is not the dominant process after accounting for lag effects.
- **Reporting only robust LM tests.** Standard LM tests establish whether spatial dependence exists initially. Robust tests should be reported alongside them to clarify the dominant process.

Join Count Test

Test Spatial Autocorrelation for Binary Data

SPAT.10 | **Category:** Spatial Statistics | **Data:** Binary | **Null:** Normal (asymptotic)

When to Use

- Detect spatial clustering of binary categories.
- Analyze adjacency of similar spatial regions.
- Test spatial randomness in lattice data.
- Spatial epidemiology or segregation studies.

Assumptions

- Data are binary/categorical and spatially indexed.
- Spatial weights matrix is correctly specified.
- Spatial randomness holds under H_0.
- Join counts follow asymptotic normality.

Test Statistic

$$BB = \frac{1}{2}\sum_i \sum_j w_{ij}x_i x_j, \quad z_{BB} = \frac{BB - E[BB]}{\sqrt{\text{Var}[BB]}} \sim N(0,1) \text{ (under } H_0), \quad \text{for black-black joins.}$$

Notation: w_{ij} are adjacency weights (typically binary), $x_i \in \{0,1\}$, BB black-black joins. Likewise for other joins

Plain English: The join count test counts the number of times adjacent spatial units share the same category (same-color joins) versus different categories (different-color joins). For binary data coded black/white, the test computes black-black (BB), white-white (WW), and black-white (BW) joins. Under spatial randomness, join counts follow expected values based on proportions and neighbor structure. Excess same-color joins indicate positive spatial autocorrelation (clustering); excess different-color joins indicate negative autocorrelation (checkerboard pattern).

⚠ Practitioner Warning

Limited to binary or small-number categorical data; Moran's I preferred for continuous data. Results depend on adjacency structure definition (rook vs. queen contiguity). Assumes *free sampling* (observations randomly assigned to locations); alternative is *nonfree sampling* (locations sampled from larger area). Expected values and variances differ between sampling schemes. Low power with sparse spatial arrangements or small n. Does not identify location of clusters—only tests global pattern. Multiple categories require multiple tests with potential multiplicity issues. Edge effects affect locations with fewer neighbors.

Hypotheses

- H_0: Spatial randomness (joins distributed randomly).
- H_1: Spatial autocorrelation present (clustering/dispersion).

Decision Rule: Reject H_0 if $|z_{BB}| > Z_{\alpha/2}$, $|z_{WW}| > Z_{\alpha/2}$, or $|z_{BW}| > Z_{\alpha/2}$.

Software References [Python, R, SAS]

Python	R	SAS
`esda.Join_Counts`	`spdep::joincount.test`	`Not Native (Manual)`

Additional Notes

Expected values under free sampling (randomization assumption): $E[BB] = \frac{W}{2} \cdot \frac{n_B(n_B-1)}{n(n-1)}$ where $W = \sum_i \sum_j w_{ij}$ is total joins, n_B is black observations. Similarly $E[WW] = \frac{W}{2} \cdot \frac{n_W(n_W-1)}{n(n-1)}$, $E[BW] = W \cdot \frac{n_B n_W}{n(n-1)}$. Variances computed from moments of join distribution. Nonfree sampling assumes locations fixed, categories sampled: $E[BB] = \frac{W}{2} \cdot p_B^2$ where $p_B = n_B/n$, with different variance formulas Contiguity definitions: rook (share edge), queen (share edge or vertex), distance-based. Test generalizes to k categories: compute joins for all category pairs. Moran's I can be applied to binary data coded 0/1 as special case. Historical context: join count statistics predate Moran's I, developed in 1940s-1950s for agricultural and ecological applications. Cliff and Ord (1981) formalized distributional theory. Modern applications: disease mapping (infected/not infected), land use (urban/rural), political geography (voting patterns).

Miscellanea: Join count tests are among the earliest methods for analyzing spatial patterns in binary data, widely used in ecology and agricultural studies. Based on counting adjacent pairs (e.g., black–black or black–white), they provide an intuitive measure of spatial clustering. Later formalized by Cliff and Ord, the method remains valued for its simplicity and interpretability.

Reference(s): Cliff & Ord (1981); Krishna Iyer (1949)

Worked Example

Problem Statement A marine biologist records coral presence (B = 1) or absence (W = 0) across $n = 10$ contiguous linear reef plots arranged as a linear chain. The objective is to test whether presences cluster spatially or are randomly distributed. Sample summary: $BB_{\text{obs}}(n_B) = 3,\ WW_{\text{obs}}(n_W) = 2,\ J = 9$ adjacent pairs (joins).

H_0: Random arrangement (no spatial autocorrelation). H_a: Similar values cluster (positive autocorrelation).

Plot i [$\mathbf{x_i}$]	1 [**1**]	2 [**1**]	3 [**1**]	4 [**0**]	5 [**0**]	6 [**1**]	7 [**0**]	8 [**0**]	9 [**1**]	10 [**1**]
Join ($J = 9$)	$1 \to 2$	$2 \to 3$	$3 \to 4$	$4 \to 5$	$5 \to 6$	$6 \to 7$	$7 \to 8$	$8 \to 9$	$9 \to 10$	*
Type	BB	BB	BW	WW	BW	BW	WW	BW	BB	*

1. Structural Constants

- Degrees: endpoints $r = 1$, interior plots $r = 2$
- $k = \frac{1}{2}\sum r_i(r_i - 1) = \frac{1}{2}(0 + 2 \times 8 + 0) = 8$
- $M = \frac{J(J-1)}{2} - k = 36 - 8 = 28$

2. Expected Values and Variances under H_0

$$E[BB] = 3.0,\ E[WW] = 1.2,\ E[BW] = 4.8$$

$\text{Var}(BB) = 0.667,\ \text{Var}(WW) = 0.560,\ \text{Var}(BW)_{\text{exact}} = 2.027$

3. z-Statistics and p-Values

	Obs	$E[\cdot]$	z	p (approx.)	p (Exact)
BB	3	3.0	0.000	0.500	0.738
WW	2	1.2	+1.069	0.143	0.333
BW	4	4.8	−0.562	0.287	0.405

Exact permutation p-values ($\binom{10}{6}$ arrangements).

Conclusion: All z-statistics are well below $z_{0.05} = 1.645$ and all p-values (approximate and exact) exceed $\alpha = 0.05$; we **fail to reject** H_0 for every join type. The observed pattern is consistent with a random arrangement of 6 presences in 10 plots. For small n, the normal approximation may be unreliable—here the exact permutation p-values are notably larger (e.g. $p_{WW}: 0.143 \to 0.333$) and should be preferred whenever $n < 30$.

Common Pitfalls

- **Ignoring multiplicity across join types.** Separate tests for BB, WW, and BW joins increase false-positive risk. Multiplicity adjustments help maintain valid inference across several join comparisons.
- **Using row-standardized weights.** Join count statistics are based on counts of neighboring pairs rather than weighted averages. Binary adjacency weights should therefore be used.
- **Expecting joins to equal sample size.** Join counts measure neighboring pairs, not individual observations. The total number of joins depends on the spatial weights structure.
- **Treating non-significance as no structure.** Failure to reject randomness does not prove spatial independence. Weak clustering or limited sample size may still obscure meaningful spatial patterns.
- **Ignoring many-category complexity.** With multiple categories, the number of possible join types increases rapidly. Simultaneous testing therefore requires careful interpretation and adjustment.

Formulas: Expected Values and Variances

Variance formulas are complex. Thus, permutation-based estimates are more common. Here, $n_B^{(r)}$ is the falling factorial $n_B(n_B - 1)\cdots(n_B - r + 1)$.

$$E[BB] = \frac{J\, n_B(n_B - 1)}{n(n-1)}, \qquad E[WW] = \frac{J\, n_W(n_W - 1)}{n(n-1)}, \qquad E[BW] = \frac{2J\, n_B n_W}{n(n-1)},$$

$$\text{Var}(BB) = E[BB] + \frac{2k\, n_B^{(3)}}{n^{(3)}} + \frac{2M\, n_B^{(4)}}{n^{(4)}} - [E(BB)]^2,$$

$$\text{Var}(WW) = E[WW] + \frac{2k\, n_W^{(3)}}{n^{(3)}} + \frac{2M\, n_W^{(4)}}{n^{(4)}} - [E(WW)]^2,$$

$\text{Var}(BW)$ is complicated; exact permutations are used above

PART XX

Statistical Process Stability & Capability

Monitoring variation, detecting shifts, and ensuring quality compliance

QUALITY CONTROL AND PROCESS MONITORING

XX. QUAL — A Conceptual Primer

Statistical Process Control (SPC)

SPC monitors a process over time to distinguish **common cause** variation (inherent, random, always present in a stable process) from **special cause** variation (systematic, assignable, indicating a process change).

- **In-control (stable):** only common cause variation; the process is predictable.
- **Out-of-control:** special causes are present; the process has shifted in mean, variance, or distribution.

Two types of errors in SPC:

- **False alarm (Type I):** signal when the process is in-control. Shewhart 3σ limits give $P(\text{false alarm}) \approx 0.0027$ per point.
- **Missed detection (Type II):** no signal when the process is out-of-control. Quantified via the **average run length (ARL)**: the expected number of observations until a signal is raised.

ARL_0 (in-control ARL): the expected run length before a false alarm; for Shewhart 3σ limits, $\text{ARL}_0 = 370$. ARL_1 (out-of-control ARL): the expected delay until a true shift is detected; smaller is better. All SPC methods balance ARL_0 and ARL_1.

I. Shewhart Control Charts

The Shewhart chart plots a statistic Q_t (sample mean, range, proportion, count) against time with:

$$\text{UCL} = \mu_Q + k\sigma_Q, \quad \text{LCL} = \mu_Q - k\sigma_Q,$$

where $k = 3$ gives the standard 3σ limits. A point outside the limits signals a potential special cause.

Key Shewhart charts:

Chart	Monitors
$\bar{X}$	Sample means; detects location shifts.
R	Sample ranges; monitors spread.
S	Sample SDs; more efficient than R.
p	Proportion nonconforming; Binomial.
np	Count nonconforming; Binomial (n fixed).
c	Count of defects per item; Poisson.
u	Defects per unit; Poisson (n variable).

Shewhart charts are most powerful for large sudden shifts ($\geq 1.5\sigma$); they have poor ARL_1 for small persistent shifts, motivating CUSUM and EWMA.

II. Pattern Recognition Rules

Shewhart's one-point-beyond-3σ rule is insensitive to trends and runs. Supplementary rules detect non-random patterns within the control limits.

Western Electric (WECO) rules (flagging requires any one of):

- 1 point beyond 3σ limits.
- 2 of 3 consecutive points beyond 2σ (same side).
- 4 of 5 consecutive points beyond 1σ (same side).
- 8 consecutive points on the same side of the centerline.

Nelson rules extend WECO with:

- 6 consecutive points steadily increasing or decreasing (trend).
- 15 consecutive points within 1σ (stratification / reduced variance).
- 14 consecutive points alternating up and down (over-control).
- 8 consecutive points $> 1\sigma$ from centerline, alternating sides (mixture).

Warning: using all rules simultaneously inflates the false-alarm rate substantially below $\text{ARL}_0 = 370$. Apply rules selectively based on the process failure modes expected.

III. CUSUM Chart

The CUSUM (**cumulative sum**) chart accumulates deviations from the target μ_0, making it sensitive to small persistent shifts.

Two-sided tabular CUSUM:

$$C_t^+ = \max\left[0,\ x_t - (\mu_0 + K) + C_{t-1}^+\right],$$

$$C_t^- = \max\left[0,\ (\mu_0 - K) - x_t + C_{t-1}^-\right],$$

starting at $C_0^+ = C_0^- = 0$. Signal when $C_t^+ > H$ or $C_t^- > H$.

Parameter choice:

- $K = \delta/2$ where δ is the minimum shift to detect (in σ units); typically $K = 0.5$ for a 1σ shift.
- $H = h\sigma$; typically $h = 4$ or 5 for $\text{ARL}_0 \approx 370$.

CUSUM has $\text{ARL}_1 \approx 10.4$ for a 1σ shift vs. $\text{ARL}_1 \approx 43.9$ for Shewhart—four times faster detection.

IV. EWMA Chart

The **exponentially weighted moving average** applies geometrically decreasing weights to past observations:

$$Z_t = \lambda x_t + (1 - \lambda)Z_{t-1}, \quad Z_0 = \mu_0,$$

with $\lambda \in (0, 1]$. The control limits are:

$$\text{UCL/LCL} = \mu_0 \pm L\sigma\sqrt{\frac{\lambda}{2-\lambda}[1 - (1-\lambda)^{2t}]},$$

converging to steady-state width $L\sigma\sqrt{\lambda/(2-\lambda)}$.

Parameter choice:

- Small λ (0.05–0.10): heavy weighting of past; good for small shifts.
- Large λ ($\rightarrow 1$): reduces to Shewhart.
- $L = 3$ (standard); $L = 2.7$ with $\lambda = 0.1$ gives $\text{ARL}_0 \approx 500$.

EWMA and CUSUM have comparable performance for small shifts; EWMA is simpler to implement and interpret.

V. Moving Average Control Chart

The moving average chart plots $\bar{X}_t^{(w)} = (x_t + x_{t-1} + \cdots + x_{t-w+1})/w$ over a window of width w. Control limits: $\mu_0 \pm 3\sigma/\sqrt{w}$ (steady state). Smooths noise but introduces **lag in detection**: a shift at time τ is not fully reflected until time $\tau + w - 1$. Best for smoothing out measurement noise when process dynamics are slow.

VI. Change-Point Detection

Change-point methods estimate *when* a shift occurred, rather than raising a sequential alarm.

Pettitt's test: A nonparametric test for a single change point in the mean. Uses the Mann–Whitney statistic applied to all possible split points τ:

$$U_{\tau,n} = \sum_{i=1}^{\tau} \sum_{j=\tau+1}^{n} \operatorname{sgn}(x_j - x_i).$$

The test statistic $K_n = \max_\tau |U_{\tau,n}|$ and its p-value are computed from tables or approximations. The estimated change point is $\hat{\tau} = \arg\max_\tau |U_{\tau,n}|$. Nonparametric; robust to outliers and non-normality; limited to a **single** change point.

Shiryaev–Roberts (SR) procedure: A sequential Bayesian-flavored detection scheme with a recursive statistic:

$$R_t = (1 + R_{t-1}) \cdot \frac{p(x_t|\mu_1)}{p(x_t|\mu_0)}, \qquad R_0 = 0.$$

Signal when $R_t > A$ (threshold chosen for target ARL_0). Unlike CUSUM, SR accumulates evidence from all possible change points equally; it is nearly minimax optimal for detecting the timing of a shift and has slightly better ARL_1 than CUSUM when the shift time is unknown.

VII. Process Capability

Capability indices measure whether a stable, in-control process meets specification limits (USL and LSL). **First verify the process is in-control** (via control charts) before computing capability indices; applying them to an out-of-control process is meaningless.

$$C_p = \frac{\text{USL} - \text{LSL}}{6\sigma}, \qquad C_{pk} = \min\left(\frac{\text{USL} - \mu}{3\sigma}, \frac{\mu - \text{LSL}}{3\sigma}\right).$$

C_p measures **potential capability** (spread relative to tolerance); C_{pk} measures **actual capability** (accounts for centering). Always $C_{pk} \le C_p$; equality only when $\mu = (\text{USL} + \text{LSL})/2$.

Interpretation:

C_{pk}	**Capability level**
< 1.00	Inadequate; process produces nonconformances
1.00–1.33	Marginally capable; improvement needed
1.33–1.67	Capable; industry-standard minimum
≥ 1.67	Highly capable; Six Sigma target = 2.0

Extended capability indices:

- C_{pm} (Taguchi index): penalizes deviation from the target value T, not just the specification center: $C_{pm} = C_p/\sqrt{1 + [(\mu - T)/\sigma]^2}$.
- C_{pmk}: combines C_{pk} and C_{pm}.
- **PPM** (parts per million defective): for a centered normal process, $\text{PPM} = 2 \times 10^6 \Phi(-3C_p)$.

ARL Comparison of Key Methods

Method	ARL_0	ARL_1 (1σ shift)
Shewhart (3σ)	370	43.9
CUSUM ($K = 0.5, h = 5$)	465	10.4
EWMA ($\lambda = 0.1, L = 2.7$)	500	10.3
Moving average ($w = 5$)	≈370	≈20

Common Pitfalls

- **Applying all pattern rules:** using all WECO and Nelson rules simultaneously dramatically inflates false-alarm rates; select rules based on expected failure modes.
- **Autocorrelated data:** standard control chart limits assume i.i.d. observations; autocorrelation inflates false alarms; use ARIMA residuals or special-cause charts for autocorrelated processes.
- **Capability before stability:** C_p and C_{pk} are meaningless if the process is not yet in-control; always check charts first.
- **C_p without C_{pk}:** $C_p = 2.0$ with a severely off-center process still produces nonconformances; always report both.
- **Ignoring measurement system error:** gauge $R\&R$ studies separate process variation from measurement variation; poor gauge inflates apparent process variability.
- **Using Pettitt's for multiple breaks:** Pettitt's test is designed for a single change point; use binary segmentation or PELT for multiple breaks.

Bottom Line

Quality control distinguishes common cause from special cause variation. Use Shewhart charts for immediate large shifts; CUSUM or EWMA for small persistent shifts; pattern rules selectively for trends and runs; Pettitt's or Shiryaev–Roberts to identify *when* change occurred; and C_{pk} (after confirming stability) to assess whether the process meets specifications.

Pattern Recognition Rules (Western Electric & Nelson)

Supplementary Tests for Non-Random Patterns on Shewhart Control Charts

QUAL.1 | **Category:** Quality Control Tests | **Data:** Continuous | **Null:** Zone-based patterns

When to Use

- Detect non-random patterns within Shewhart control limits.
- Identify runs, trends, cycles, or stratification indicating special causes.
- Supplement 3σ limits for earlier detection of instability.

Assumptions

- Data are plotted on a Shewhart chart with established limits from in-control (Phase I) data.
- Observations are independent (or form rational subgroups).
- Process is approximately normal under stable conditions.

Test Statistic

R1: 1 or more points in Zone A^+ (beyond 3σ).
R2: 2 out of 3 consecutive in Zone A^+ (same side).
R3: 4 out of 5 consecutive in Zone B^+ (same side).
R4: 8 consecutive on one side of CL (centerline).
R5: 6 consecutive ↑ or ↓ (trend).
R6: 15 consecutive in Zone C (either side).
R7: 14 consecutive alternating ↑↓.
R8: 8 consecutive, both sides, none in Zone C.

Notation: Zone C: $|z| \leq 1$ (within 1σ), B: $1 < |z| \leq 2$ (between 1σ and 2σ), A: $2 < |z| \leq 3$ (between 2σ and 3σ)

Plain English: Pattern rules divide the control chart into six horizontal zones (three above and three below the centerline) and look for non-random patterns. Zone C reflects typical variation, Zone B moderate deviations, and Zone A large deviations within the three-sigma limits. The rules detect issues by tracking sequences of points within zones or specific patterns. A stable process produces points scattered across zones without systematic structure. Violations indicate a process change, even when all points remain within control limits. If process parameters shift during Phase I, zone boundaries must be recalculated.

⚠ Practitioner Warning

Applying all eight rules inflates the false alarm rate (often in the range ≈5–10% vs. 0.27% for three-sigma limits). Individual rules have low false alarm probabilities, but combined use without adjustment yields frequent signals. In practice, use a subset (often R1–R4) or tailor rules to expected failure modes. The rules assume normality; non-normal data—especially for R6 and R7—can inflate false alarms. Rules flag patterns, not causes—each signal requires investigation. Autocorrelation violates independence and can spuriously trigger run-based rules (especially R4 & R5).

Hypotheses

- H_0: Process exhibits only common-cause variation.
- H_1: Special-cause variation present.

Decision Rule: Any pattern rule is violated. Investigate assignable cause and take corrective action.

Software References [Python, R, SAS]

Python	R	SAS
`Not Native (Manual)`	`qcc::qcc()`	`PROC SHEWHART`

Additional Notes

The Western Electric rules were formalized in the 1956 *Statistical Quality Control Handbook* (Bell Labs), extending Shewhart's three-sigma rule (Rule 1) to detect earlier signs of process deterioration via runs, trends, and clustering. Rules 2–4 target shifts of decreasing magnitude: $\sim 2\sigma$, $\sim 1\sigma$, and small sustained mean shifts; the A/B/C zone framework became standard terminology. Nelson (1984) added four rules: Rule 5 (monotonic trends), Rule 6 (over-centering/stratification), Rule 7 (rapid alternation, often due to over-control or measurement error), and Rule 8 (mixture of process levels). These eight rules became standard across manufacturing (AIAG, AS9100, ISO) and are widely implemented in software. Under normality, individual false alarm probabilities are small (Rule 1: 0.0027; others smaller or $\sim$ 0.001–0.01). Applying all rules simultaneously inflates the overall false alarm rate to roughly 5–10% per point, reducing in-control run length.

Miscellanea: The Western Electric (1956) rules were developed to detect non-random patterns in control charts and became widely used in industrial quality control. Nelson (1984) later proposed additional rules to improve sensitivity to trends and other patterns, though at the cost of increased false alarms. In practice, these rules are often applied selectively based on the process and monitoring objectives.

Reference(s): Western Electric (1956); Nelson (1984, 1985); Champ & Woodall (1987); Montgomery (2019)

Worked Example

Problem Statement A laboratory monitors the concentration of a reagent. The process is stable with a target mean (μ) of 50.0 mg/L and a standard deviation (σ) of 2.0 mg/L.

We evaluate 7 consecutive points to determine if the process has shifted, despite all points being within the 3σ Upper Control Limit ($UCL = 56.0$). We divide the control chart into three zones: **Zone A:** Between 2σ and 3σ (54.0 to 56.0); **Zone B:** Between 1σ and 2σ (52.0 to 54.0); **Zone C:** Between the mean and 1σ (50.0 to 52.0).

	1	2	3	4	5	6	7
x_t	51.2	52.5	53.1	52.8	54.5	55.2	54.8
Deviation from Mean	$+0.6\sigma$	$+1.25\sigma$	$+1.55\sigma$	$+1.4\sigma$	$+2.25\sigma$	$+2.6\sigma$	$+2.4\sigma$
Zone	C	B	B	B	A	A	A

1. Testing Western Electric Rule 2 (Zone A) Rule: 2 out of 3 consecutive points fall in Zone A or beyond ($> 2\sigma$).

- **Observations:** x_5, x_6, x_7.
- **Result:** All three points are in Zone A.
- **Decision: VIOLATION.** The process is out of control.

2. Testing Western Electric Rule 3 (Zone B) Rule: 4 out of 5 consecutive points fall in Zone B or beyond ($> 1\sigma$).

- **Observations:** $x_2, \cdots, x_6$.
- **Result:** 5 out of 5 points are in Zone B or A.
- **Decision: VIOLATION.**

3. Results Summary

Rule Number	Description	Status
R1 (1 pt $> 3\sigma$)	Single point extreme	Pass
R2 (2 of 3 $> 2\sigma$)	Shift in mean	**Fail** ($t = 6$)
R3 (4 of 5 $> 1\sigma$)	Small shift in mean	**Fail** ($t = 5$)
R4 (8 of 8 on one side)	Persistent bias	Pending

Note: R4 is pending, as only 7 data points are available.

4. Interpretation

- **Statistical Insight:** Although no single point exceeded 56.0 mg/L, the clusters in Zone B and Zone A are statistically improbable under a random normal distribution ($p < 0.005$).
- **Practitioner Action:** Stop the process. The "run" of points far from the mean indicates a significant upward shift has occurred, likely due to a change in raw material or sensor calibration.

Common Pitfalls

- **Confusing Western Electric rules with Nelson rules or treating them as identical.** Western Electric defines four original pattern rules, while Nelson extends them to eight rules detecting additional patterns such as trends and stratification; always specify which rule set is used.
- **Applying many pattern rules simultaneously without considering inflated false alarm rates.** Each rule has its own false alarm probability, and applying several rules together substantially increases the overall rate of false signals even when the process is in control.
- **Applying pattern recognition rules without verifying distributional assumptions.** The zone probabilities underlying the rules assume approximately normal and independent observations; strong skewness, heavy tails, or autocorrelation can dramatically distort false alarm rates.
- **Using pattern rules without proper rational subgrouping.** Subgroups must represent homogeneous production conditions so that patterns across subgroups reflect temporal process changes rather than mixing of machines, operators, or materials.
- **Interpreting rule violations as causal process explanations.** A rule signal indicates an unusual statistical pattern in the chart but does not by itself identify the underlying physical cause or mechanism.

Attribute Control Chart Tests

Control Charts for Discrete Count and Proportion Data

QUAL.2 | **Category:** Quality Control | **Data:** Binary/Count | **Null:** Binomial/Poisson

When to Use

- Monitor attribute data: proportions (p, np charts) or counts (c, u charts).
- Track defects/nonconformities when data are discrete (pass/fail or counts).
- Use when continuous measurement is infeasible but classification is possible.

Assumptions

- Data follow binomial (p, np) or Poisson (c, u) models.
- Samples are independent over time.
- Process is stable with constant probability or rate when in control.
- Sample sizes support normal approximation ($np \geq 5$, $n(1-p) \geq 5$ for p-chart; (approximate guideline).

Test Statistic

p-chart: $p_i = \frac{x_i}{n_i}$, $\text{UCL} = \bar{p}+3\text{SE}_i$, $\text{LCL} = \bar{p}-3\text{SE}_i$, $\text{SE}_i = \sqrt{\bar{p}(1-\bar{p})/n_i}$; c-chart: $\text{UCL} = \bar{c}+3\sqrt{\bar{c}}$, $\text{LCL} = \bar{c}-3\sqrt{\bar{c}}$.

Notation: p_i is proportion nonconforming, c_i is defect count, $\bar{p}, \bar{c}$ are process averages (if LCL < 0, set LCL=0)

Plain English: Attribute control charts monitor discrete data like proportions of defective items or counts of defects. The p-chart tracks proportion nonconforming in samples, using binomial distribution for control limits. The np-chart plots number nonconforming directly. The c-chart monitors total defects per unit assuming Poisson distribution with constant sample size. The u-chart tracks defects per unit when sample sizes vary. Control limits based on binomial or Poisson variance formulas.

⚠ Practitioner Warning

Requires adequate sample sizes for normal approx; small n or extreme p violate approx. Zero lower control limit common when $\bar{p}$ or $\bar{c}$ small; use exact limits or probability limits. Variable sample sizes complicate p-chart and u-chart; use varying limits or standardized chart. Overdispersion (variance exceeds binomial/Poisson) indicates additional variation sources; Laney $p-$ or $u-$ charts adjust. Assumes constant process probability or rate; changing p or λ invalidates limits. Not sensitive to small shifts; supplementary runs rules sometimes applied. Integer constraints: plotted points must be discrete but limits are continuous. Seasonal or systematic patterns require stratification or specialized methods.

Hypotheses

- H_0: Process proportion/rate constant at centerline value.
- H_1: Process proportion/rate shifted from centerline.

Decision Rule: Point outside 3σ control limits. Pattern recognition rules sometimes applied.

Software References [Python, R, SAS]

Python	R	SAS
`Not Native (Manual)`	`qcc::qcc`	`PROC SHEWHART`

Additional Notes

Four attribute charts: p-chart (proportion nonconforming) with $p_i = x_i/n_i$ and limits $\bar{p} \pm 3\sqrt{\bar{p}(1-\bar{p})/n_i}$ (binomial); np-chart (count nonconforming) with limits $n\bar{p} \pm 3\sqrt{n\bar{p}(1-\bar{p})}$, requiring constant n; c-chart (defect counts) with limits $\bar{c} \pm 3\sqrt{\bar{c}}$ (Poisson, fixed inspection unit); u-chart (defects per unit) with $u_i = c_i/n_i$ and limits $\bar{u} \pm 3\sqrt{\bar{u}/n_i}$. For varying sample sizes, use sample-specific limits or standardized charts $z_i = (p_i - \bar{p})/\sqrt{\bar{p}(1-\bar{p})/n_i}$ with limits ± 3. Overdispersion ($s_p^2 > \bar{p}(1-\bar{p})/n$, $s_c^2 > \bar{c}$) indicates extra-binomial/Poisson variation; Laney adjustments correct this. For small samples or extreme probabilities, exact binomial or Poisson limits are preferred. Variance-stabilizing transforms (arcsin for proportions, square root for counts) may be applied.

Miscellanea: Walter Shewhart (1920s–1930s) developed control charts for both continuous (variables) and discrete (attribute) data. Attribute charts, such as p-charts and c-charts, monitor proportions or counts of defects and are widely used when quality is measured in categorical terms rather than continuous measurements.

Reference(s): Shewhart (1931); Duncan (1986); Montgomery (2019)

Worked Example

Problem Statement An electronics manufacturer monitors the assembly of printed circuit boards (PCBs) to ensure soldering integrity. The quality control department uses a **p-chart** to track the proportion of boards containing at least one soldering defect.

1. Process Parameters A stable process has an average fraction defective ($\bar{p}$) based on historical data. We test a new subgroup of size $n = 100$.

- **Average Proportion Defective ($\bar{p}$):** 0.02
- **Subgroup Size (n):** 100
- **Observed Defectives (x):** 6
- **Observed Proportion ($\hat{p}$):** $6/100 = 0.06$

2. Calculation of Control Limits The standard error for the proportion is calculated as:

$$\text{Standard error:} \quad \sigma_p = \sqrt{\frac{0.02(1-0.02)}{100}} = 0.014$$

The Upper Control Limit (UCL) is:

$$UCL = \bar{p} + 3\sigma_p = 0.02 + 3(0.014) = 0.062$$

The Lower Control Limit (LCL) is:

$$LCL = \bar{p} - 3\sigma_p = 0.02 - 0.042 \rightarrow \text{Set to } 0$$

3. Results and Decision

Metric	Value
Observed Proportion ($\hat{p}$)	0.060
Upper Control Limit (UCL)	0.062
Center Line (CL)	0.020
Status	**In Control (Marginal)**

4. Interpretation

- **Statistical Decision:** Since $0.060 < 0.062$, the point is technically within the 3σ limits.
- **Practitioner Insight:** While the process is "In Control," an observation of 6% defects represents a threefold increase relative to the baseline rate. Under Western Electric Rules, a single point so close to the UCL, or a run of points above the CL, would trigger a "Warning" to investigate the soldering machines before the process officially fails.

Common Pitfalls

- **Using an attribute chart when continuous measurements are available.** Attribute charts reduce measurements to binary outcomes (e.g., pass/fail). When actual measurements are available, variables charts typically provide greater sensitivity to process shifts.
- **Applying normal approximation with inadequate expected counts.** Control limits for attribute charts rely on approximate normality when expected counts are sufficiently large (e.g., np, $n(1-p)$). Very low defect probabilities or small sample sizes can invalidate these approximations.
- **Confusing defective units with defect counts.** A defective unit represents a single binary outcome, whereas defects represent multiple possible flaws per unit. Different control charts are designed for these two situations (e.g., p/np vs. c/u charts).
- **Using fixed control limits when sample sizes vary substantially.** For charts such as the p-chart or u-chart, control limits depend directly on the sample size. When sample sizes vary, limits must be recalculated for each sample or plotted as varying limits.
- **Ignoring unusually low defect rates.** Points falling below the lower control limit also indicate unusual process behavior (e.g., process changes, inspection differences, or data issues). Both unusually high and unusually low defect counts may signal special causes requiring investigation.
- **Assuming independence when data are clustered or autocorrelated.** Attribute charts assume independent observations; clustering (e.g., by batch, operator, or time) can inflate variability and produce misleading signals. Appropriate subgrouping or alternative models may be required.

Moving Average Control Chart Test

Control Chart Based on Sequence of Moving Averages

QUAL.3 | **Category:** Quality Control | **Data:** Continuous | **Null:** Normal (moving average)

When to Use

- Smooth noisy data for easier visual interpretation.
- Detect moderate process shifts with simple calculation.
- When equal weighting of recent observations desired (vs. EWMA's decay).

Assumptions

- Process measurements approximately normally distributed.
- Raw data are independent (autocorrelation minimal).
- Target mean μ_0 and standard deviation σ known.
- Window size w chosen (typically 3-10 observations).

Test Statistic

$$\bar{X}_i = \frac{1}{w}\sum_{j=0}^{w-1} x_{i-j}, \quad \text{UCL} = \mu_0 + 3\frac{\sigma}{\sqrt{w}}, \quad \text{LCL} = \mu_0 - 3\frac{\sigma}{\sqrt{w}} \quad \text{(assuming } \sigma \text{ known or estimated).}$$

Notation: $\bar{X}_i$ is moving average at time i, w is span (window width), limits based on $\sigma/\sqrt{w}$

Plain English: Moving average control charts plot the average of the most recent w observations rather than individual points. Each plotted point represents the mean of a sliding window of the most recent w observations. As new observations arrive, the oldest observation drops out and the newest enters, creating a moving window. The averaging reduces random variability, making systematic shifts easier to detect. Control limits are narrower than Shewhart charts by factor $\sqrt{w}$, reflecting reduced variability of averages compared to individual observations.

⚠ Practitioner Warning

Introduces autocorrelation even when original data independent; consecutive moving averages share $w-1$ observations. Autocorrelation invalidates standard control limit theory; actual ARL differs from theoretical. Less sensitive than CUSUM or EWMA for small shifts; mainly useful for moderate shifts. Delay in detection: shift not reflected until enough shifted observations enter the window. Larger w increases smoothing but increases detection delay and autocorrelation. Assumes constant variance; non-stationary variance complicates interpretation. Not optimal for detecting any specific shift size; EWMA generally preferred. Startup period: first $w-1$ points cannot be plotted using full window.

Hypotheses

- H_0: Process mean on target ($\mu = \mu_0$, in control).
- H_1: Process mean shifted from target ($\mu \neq \mu_0$).

Decision Rule: Reject H_0 if $\bar{X}_i >$ UCL or $\bar{X}_i <$ LCL.
Standard limits: $\mu_0 \pm 3\sigma/\sqrt{w}$ for 3-sigma control.

Software References [Python, R, SAS]

Python	R	SAS
`(pandas.rolling or manual)`	`qcc (with moving average)`	`Not Native (Manual)`

Additional Notes

Moving average charts smooth data via $\bar{X}_i = \frac{1}{w}\sum_{j=0}^{w-1} x_{i-j}$, with variance σ^2/w and control limits $\pm 3\sigma/\sqrt{w}$ under independence. However, overlapping windows induce autocorrelation (corr $\approx (w-1)/w$), invalidating standard limits. Larger w increases smoothing but delays detection. EWMA provides a more efficient alternative by weighting recent data more heavily without fixed windows. Moving averages remain useful for simple smoothing and visualization.

Miscellanea: Moving averages were among the earliest tools used to smooth noisy data, with applications in astronomy and later widespread adoption in industrial quality control. While simple and intuitive for revealing trends, they introduce dependence between observations and can delay detection of process shifts, limiting their effectiveness for formal statistical monitoring. These limitations—particularly induced autocorrelation and sluggish response—motivated the development of more refined methods such as EWMA and CUSUM. From a signal-processing perspective, the moving average acts as a low-pass filter, removing high-frequency noise while preserving slower trends, which explains both its smoothing benefit and its delayed response to abrupt changes.

Reference(s): Roberts (1959); Montgomery (2019)

Worked Example

Problem Statement A process with $\mu_0 = 1000$ and $\sigma = 5$ is monitored using an MA chart with a window size $w = 3$ and control factor $L = 3$. We test for a mean shift in filling volumes.

Time (t)	1	2	3	4	5
x_t	1002	1005	1008	1006	1009
M_t	1002.0	1003.5	1005.0	1006.3	1007.67
UCL $(1000 + 15/\sqrt{n})$	1015.0	1010.6	1008.7	1008.7	1008.7

1. Calculation of the Moving Average Statistic The moving average M_t at time $t = 5$ is the mean of the three most recent observations:

$$M_5 = \frac{x_5 + x_4 + x_3}{3} = \frac{1009 + 1006 + 1008}{3} = 1007.67$$

2. Control Limit Calculation For $t \geq w$, the control limits are constant:

$$UCL = \mu_0 + \frac{L\sigma}{\sqrt{w}} = 1008.66$$

$$LCL = 1000 - 8.66 = 991.34$$

3. Results

Metric	Value
Target Mean (μ_0)	1000.00
Current MA Value (M_5)	1007.67
Upper Control Limit (UCL)	1008.66
Lower Control Limit (LCL)	991.34

Note: The moving average chart uses *rectangular weights*, assigning equal weight within a fixed window and zero weight outside, leading to abrupt information loss when observations leave the window. EWMA instead applies exponentially decaying weights, providing smoother memory and better sensitivity to small shifts.

4. Decision and Interpretation

- **Decision:** Since 1007.67 < 1008.66, the process is "In Control."
- **Interpretation:** While the most recent raw observation (1009) was quite high, the moving average stabilizes the data. However, the M_t value is very close to the UCL, suggesting a potential upward drift that warrants close observation.
- **Practitioner Insight:** The MA chart is less computationally intensive than the EWMA chart but can suffer from "cycling" or "slumping" due to abrupt dropping of older observations (rectangular weighting)..

Common Pitfalls

- **Choosing the window size without considering detection goals.** The window size determines the trade-off between noise reduction and responsiveness to process shifts. Larger windows smooth variability but delay detection, while smaller windows respond more quickly to changes.
- **Ignoring the inherent detection lag introduced by averaging.** A moving average reflects information from several past observations rather than the current value alone. Consequently, signals typically occur several observations after the actual process shift.
- **Treating successive moving averages as independent observations.** Adjacent moving average values share many of the same data points and are therefore strongly correlated. Runs-based rules, which are designed for independent observations, may therefore give misleading signals.
- **Applying moving averages to data with strong autocorrelation.** If the original process data are autocorrelated, averaging can amplify dependence among observations. Modeling the time-series structure or analyzing residuals may provide more reliable monitoring.
- **Comparing moving average charts directly with Shewhart charts using simple ARL values.** Average run length calculations for moving averages must account for the correlation among successive statistics. Direct comparisons with Shewhart charts can therefore be misleading without proper adjustment.

CUSUM Test

Cumulative Sum Control Chart for Detecting Small Process Shifts

QUAL.4 | **Category:** Quality Control | **Data:** Continuous | **Null:** Normal (tabular)

When to Use

- Detect small persistent shifts in process mean (≈ 0.5–2σ).
- Early detection desired than Shewhart charts provide.
- Monitor processes where small shifts are critical.
- Superior to control chart zone tests for sustained shifts.

Assumptions

- Process measurements approximately normally distributed.
- Raw data are independent (autocorrelation minimal).
- Target mean μ_0 and standard deviation σ known.
- CUSUM parameters (k and h) are specified.

Test Statistic

$$C_i^+ = \max(0, C_{i-1}^+ + x_i - (\mu_0 + k\sigma)), \quad C_i^- = \max(0, C_{i-1}^- - x_i + (\mu_0 - k\sigma)).$$

Notation: C^+, C^- are cumulative sums for upward/downward shifts, k is reference value, h is decision interval

Plain English: CUSUM accumulates deviations of observations from target, resetting to zero when sum becomes negative. Separate CUSUMs track upward and downward shifts. If observations consistently above target (even slightly), the positive CUSUM grows; if below target, negative CUSUM grows. When CUSUM exceeds decision interval h, the chart signals an out-of-control condition. Unlike Shewhart charts examining each point independently, CUSUM integrates information over time, making it highly sensitive to persistent small shifts that Shewhart charts might miss.

⚠ Practitioner Warning

Requires specifying k (typically 0.5 for 1) and h (controls false alarms). Designed for a specific shift size; less sensitive to large, sudden shifts than Shewhart charts. Autocorrelation degrades performance—use modified CUSUM or alternatives for correlated data. Performance depends on accurate σ estimation; misspecification reduces effectiveness. Interpretation is less intuitive and does not directly quantify the magnitude of the shift—follow-up analysis is required. Tabular CUSUM is standard; Fast inital response, FIR, (reset/headstart) can improve responsiveness.

Hypotheses

- H_0: Process mean on target ($\mu = \mu_0$, in control).
- H_1: Process mean shifted from target ($\mu \neq \mu_0$).

Decision Rule: Reject H_0 if $C_i^+ > h\sigma$ (upward shift) or $C_i^- > h\sigma$ (downward).

Software References [Python, R, SAS]

`Not Native (Manual)`	`qcc::cusum`	`PROC CUSUM`

Additional Notes

The tabular form initializes $C_0^+ = C_0^- = 0$ and updates $C_i^+ = \max(0, C_{i-1}^+ + x_i - (\mu_0 + k\sigma))$, $C_i^- = \max(0, C_{i-1}^- - x_i + (\mu_0 - k\sigma))$. The reference value k (typically half the target shift) controls sensitivity; e.g., $k = 0.5$ is optimal for detecting a 1σ shift. The decision interval h governs false alarms: larger h increases ARL_0 (In-control Average Run Length) but delays detection. Design selects (k, h) to achieve desired ARL_0 while minimizing ARL_1 (Out-of-control Average Run Length); e.g., $(0.5, 4) \Rightarrow \text{ARL}_0 \approx 168$, $(0.5, 5) \Rightarrow \text{ARL}_0 \approx 465$. Variants include fast initial response (initialize at $h/2$), V-mask (graphical form), and combined Shewhart–CUSUM schemes (CUSUM for small shifts, Shewhart for large). Extensions handle Poisson/count data and residual monitoring in regression/time series. Compared to EWMA, CUSUM is optimal for a specified shift size, while EWMA is more robust across a range of shifts.

Miscellanea: CUSUM (short for *CUmulative SUM*) was introduced by E. S. Page (1954) at the University of Cambridge and published in *Biometrika* as a sequential method for detecting small, persistent shifts in process mean. Its cumulative structure—rooted in likelihood-ratio principles—gives it strong sensitivity to gradual drift compared to Shewhart charts. George Barnard later developed the V-mask (1959), enabling practical visual implementation. Because it accumulates evidence over time, CUSUM is often described as having "memory," making it one of the earliest and most influential tools in sequential change detection, with applications extending beyond quality control to fields such as medicine and finance.

Reference(s): Page (1954); Lucas & Crosier (1982); Hawkins & Olwell (1998)

Worked Example

Problem Statement A quality control engineer monitors pH levels with a target $\mu_0 = 7.00$ and $\sigma = 0.10$. The CUSUM test is applied to detect a small upward shift. The parameters are set to $k = 0.5$ (or $k\sigma = 0.05$) and the decision interval $h = 5$ (or $h\sigma = 0.50$).

Time (t)	0	1	2	3	4	5	6	7	8
x_t	-	7.02	7.05	7.08	7.10	7.12	7.11	7.15	7.14
$x_t - (7.00 + 0.05)$	-	-0.03	0.00	0.03	0.05	0.07	0.06	0.10	0.09
C_t^+	0	0	0	0.03	0.08	0.15	0.21	0.31	0.40

1. Recursive Calculation The upper CUSUM (C_i^+) is calculated as: $C_i^+ = \max[0, C_{i-1}^+ + x_i - (\mu_0 + k\sigma)]$. At $t = 8$, the cumulative sum reaches $0.40 (= 0.09 + 0.31)$. While no single point exceeded the Shewhart upper control limit ($UCL = 7.30$), the cumulative sum is rapidly approaching the threshold $h = 0.50$.

2. Results

Metric	Value
Target Mean (μ_0)	7.00
Reference Value ($k\sigma$)	0.05
Current CUSUM Value (C_8^+)	0.40
Decision Limit ($h\sigma$)	0.50

3. Decision and Interpretation

- **Decision:** The process is not yet "officially" out-of-control ($0.40 < 0.50$), but the persistent upward trend is highly evident.
- **Interpretation:** Even though every individual pH reading was "legal" (within 3σ), the CUSUM test reveals a systematic drift. If the next measurement is > 7.15, the CUSUM will exceed 0.50, triggering a signal.
- **Practitioner Insight:** CUSUM is the "memory" test. It is superior to the Western Electric Rules for detecting very small shifts because it uses the exact magnitude of every deviation rather than just counting points in zones.

Common Pitfalls

- **Starting a CUSUM chart with an incorrect baseline mean.** CUSUM monitoring assumes the reference value reflects the stable in-control process mean. An incorrect baseline can cause immediate accumulation and spurious signals.
- **Using CUSUM primarily to detect large process shifts.** CUSUM charts are designed to detect small and persistent shifts in the process mean. Large sudden changes are typically detected more quickly by traditional Shewhart charts.
- **Failing to reset the cumulative sum after correcting a detected shift.** Once a special cause is identified and corrected, the cumulative statistic should be restarted. Continuing from a large accumulated value can delay detection of subsequent shifts.
- **Interpreting the CUSUM plot as if it represents individual observations.** CUSUM values represent cumulative deviations from the target rather than the raw measurements. Individual observations must be examined separately to understand the source of a signal.
- **Applying standard CUSUM procedures to autocorrelated data.** CUSUM assumes successive observations are approximately independent. Autocorrelation can produce systematic drift in the cumulative statistic and increase false alarms.
- **Choosing the reference value k and decision interval h without calibration.** The performance of a CUSUM chart depends critically on the choice of k (target shift) and h (decision threshold). Poorly chosen parameters can lead to excessive false alarms or slow detection; calibration using ARL targets is essential.

EWMA Test

Exponentially Weighted Moving Average Control Chart

QUAL.5 | **Category:** Quality Control | **Data:** Continuous | **Null:** Normal (asymptotic)

When to Use

- Detect small to moderate process shifts (0.5 – 2σ range).
- When process has slight autocorrelation (EWMA more robust than CUSUM).
- Balance between Shewhart (no memory) and CUSUM (full memory).

Assumptions

- Observations sequential (time-ordered) and approximately independent.
- Process measurements approximately normally distributed.
- Target mean μ_0 and standard deviation σ established.
- Smoothing parameter λ chosen (typically 0.05-0.25).

Test Statistic

$$Z_i = \lambda x_i + (1-\lambda)Z_{i-1}, \quad \text{UCL} = \mu_0 + L\sigma\sqrt{\frac{\lambda}{2-\lambda}}, \quad \text{LCL} = \mu_0 - L\sigma\sqrt{\frac{\lambda}{2-\lambda}}.$$

Notation: Z_i is EWMA statistic, λ is smoothing constant, L is width of control limits (typically 3), $Z_0 = \mu_0$ (or target mean)

Plain English: EWMA forms a weighted average where the current observation receives weight λ, the previous EWMA receives weight $(1-\lambda)$, with older observations receiving exponentially decreasing weights. Small λ (e.g., 0.05) gives smooth, slow-responding chart sensitive to small sustained shifts. Large λ (e.g., 0.40) gives more weight to recent data, approaching Shewhart chart behavior. The chart signals when EWMA exceeds control limits calculated to achieve desired false alarm rate. EWMA naturally downweights outliers and old information, focusing on recent process behavior.

⚠ Practitioner Warning

Requires choosing smoothing parameter λ; different values optimize for different shift sizes. Control limit width L (typically 2.7-3.0) determines false alarm rate; must be chosen with λ. Assumes normality for control limits; violations affect false alarm rates. Autocorrelation complicates control limit calculation; may require modified limits or alternative methods. Less sensitive to large sudden shifts than Shewhart charts; combined Shewhart-EWMA addresses this. Startup period needed for EWMA to converge; time-varying limits for initial observations improve performance. Does not directly indicate shift timing; requires additional analysis. Popular but sometimes overused; CUSUM theoretically superior for a specified shift size. *Standard:* $\lambda = 0.2$, $L = 3$ *for moderate shift sensitivity;* $\lambda = 0.05$, $L = 2.7$ *for small shifts.*

Hypotheses

- H_0: Process mean on target ($\mu = \mu_0$, in control).
- H_1: Process mean shifted from target ($\mu \neq \mu_0$).

Decision Rule: Reject H_0 if $Z_i > \text{UCL}$ or $Z_i < \text{LCL}$.

Software References [Python, R, SAS]

Python	R	SAS
`(pandas.ewm or manual)`	`qcc::ewma`	`PROC ANOM; ewma;`

Additional Notes

EWMA monitors a weighted average: $Z_i = \lambda x_i + (1-\lambda)Z_{i-1}$, giving exponentially decreasing weights to past data. Initialize with $Z_0 = \mu_0$ or $\bar{x}$. The steady-state variance is $\frac{\lambda}{2-\lambda}\sigma^2$, with smaller variance early on. Smaller λ (0.05–0.10) targets small shifts; larger λ (0.20–0.30) responds faster to moderate shifts; $\lambda = 1$ reduces to Shewhart. Control limits depend on λ and desired ARL. EWMA is comparable to CUSUM for small shifts but more robust and easier to interpret. It can be combined with Shewhart limits to detect both small and large changes and is closely related to exponential smoothing in time series.

Miscellanea: The EWMA chart was introduced by Roberts (1959) as a simple alternative to CUSUM for detecting small shifts in process mean. Its recursive form makes it mathematically equivalent to exponential smoothing used in time series forecasting, linking quality control with classical forecasting methods. From a signal-processing perspective, EWMA acts as a first-order low-pass filter, weighting recent observations more heavily while gradually discounting older data.

Reference(s): Roberts (1959); Hunter (1986); Lucas & Saccucci (1990)

Worked Example

Problem Statement A process with $\mu_0 = 50.0$ and $\sigma = 2.0$ is monitored using an EWMA chart ($\lambda = 0.2, L = 3$). We test if the process has shifted from its target mean.

Time (t)	0	1	2	3	4	5
x_t	-	51.0	50.5	52.0	51.5	53.0
Z_t	50.00	50.20	50.26	50.61	50.79	51.23

1. Calculation of the EWMA Statistic The recursive formula for the EWMA statistic is:

$$Z_t = \lambda x_t + (1 - \lambda)Z_{t-1}$$

Starting with $Z_0 = 50.0$, the final value at $t = 5$ is $Z_5 = 51.23 (= 0.2 * 53.0 + 0.8 * 50.79)$.

2. Control Limit Calculation (Steady State) As t increases, the control limits converge to:

$$UCL = \mu_0 + L\sigma\sqrt{\frac{\lambda}{2-\lambda}} = 52.0$$

$$LCL = 50.0 - 2.0 = 48.0$$

3. Results

Metric	Value
Target Mean (μ_0)	50.00
Final EWMA Statistic (Z_5)	51.23
Upper Control Limit (UCL)	52.00
Lower Control Limit (LCL)	48.00

Variance Derivation
1) $Z_i = \lambda X_i + (1 - \lambda)Z_{i-1}$.
2) $E[X_i] = \mu_0$, $\text{Var}(X_i) = \sigma^2$, and $Z_0 = \mu_0$.
3) $Z_i = \sum_{j=0}^{i-1} \lambda(1-\lambda)^j X_{i-j} + (1-\lambda)^i Z_0$.
4) Independence implies

$$\text{Var}(Z_i) = \sigma^2\lambda^2 \sum_{j=0}^{i-1}(1-\lambda)^{2j} = \sigma^2 \frac{\lambda}{2-\lambda}\left[1 - (1-\lambda)^{2i}\right],$$

which converges to $\sigma^2 \frac{\lambda}{2-\lambda}$ as $i \to \infty$.

4. Decision and Interpretation

- **Decision:** Since $51.23 < 52.00$, the process is technically "In Control."
- **Interpretation:** Although no points have breached the limits, the EWMA value has climbed steadily from 50.0 to 51.23. This indicates a persistent upward drift.
- **Practitioner Insight:** A standard X-bar chart would have $UCL = 56.0$. All of these raw points (max 53.0) are far below the Shewhart limit. The EWMA chart catches the drift much sooner because its "effective" UCL is only 52.0 due to the smoothing parameter.

Common Pitfalls

- **Using asymptotic control limits from the first observation.** EWMA control limits gradually widen as the number of observations increases. Applying asymptotic limits too early can reduce sensitivity to shifts during the initial monitoring period.
- **Selecting the smoothing parameter without considering the target shift size.** The choice of λ determines how quickly the chart responds to changes in the process mean. Smaller values emphasize long-term memory and are more sensitive to small persistent shifts.
- **Assuming EWMA responds equally well to all types of shifts.** EWMA is most effective for small to moderate persistent shifts. Very large or abrupt changes may be detected more quickly by Shewhart charts.
- **Failing to reset the statistic after correcting a detected shift.** Once a signal is investigated and the process is restored, the EWMA statistic should be restarted at the target mean. Continuing from an extreme value may delay detection of subsequent changes.
- **Confusing EWMA with a simple moving average.** A simple moving average assigns equal weights to a fixed number of recent observations. EWMA instead assigns exponentially decreasing weights so that all past observations influence the statistic.
- **Applying EWMA without evaluating autocorrelation in the data.** EWMA charts assume approximately independent observations. Strong autocorrelation can cause persistent trends in the statistic and increase false alarm rates.

Shiryaev-Roberts Test

Optimal Sequential Change-Point Detection via Likelihood Ratios

QUAL.6 | **Category:** Quality Control | **Data:** Continuous | **Null:** Likelihood ratio

When to Use

- Sequential detection of an unknown change point with minimal delay.
- Alternative to CUSUM for unknown change-point scenarios. Benchmark procedure for evaluating sequential methods.

Assumptions

- Pre- and post-change distributions (f_0, f_1) are known.
- Change occurs at an unknown time (possibly never).
- Observations are independent conditional on the change point.

Test Statistic

$$R_n = \sum_{k=1}^{n} \prod_{i=k}^{n} \frac{f_1(x_i)}{f_0(x_i)} \quad \text{(likelihood ratios assuming change at time } k\text{)}, \quad \text{or recursively: } R_n = (1 + R_{n-1})\frac{f_1(x_n)}{f_0(x_n)}.$$

Notation: R_n is Shiryaev-Roberts statistic (with $R_0 = 0$), f_1/f_0 is likelihood ratio, sum over hypothesized change times

Plain English: The Shiryaev-Roberts procedure evaluates the likelihood that a change occurred at each possible past time point, assuming it has not been detected yet, and sums these likelihoods. At each observation, it calculates the likelihood ratio for all scenarios where change happened at time $1, 2, \dots, n$ and accumulates evidence. When accumulated evidence exceeds a threshold, the procedure signals a change. The method is asymptotically optimal, minimizing detection delay for given false alarm rate when change time is unknown.

⚠ Practitioner Warning

Requires specifying pre-change and post-change distributions explicitly; misspecification reduces performance. Computationally more intensive than CUSUM; recursive formula helps but still demanding. Threshold selection challenging; no simple relationship to ARL like CUSUM h parameter. Optimality results asymptotic (large threshold); finite-sample performance may differ. Assumes change permanent (not temporary); not suited for transient shifts. Less commonly used in practice than CUSUM or EWMA; limited software implementation. Theoretical benchmark more than practical tool.

Hypotheses

- H_0: No change occurred yet ($\tau > n$, τ: unknown change t).
- H_1: Change occurred ($\tau \leq n$, process shifted from f_0 to f_1).

Decision Rule: Reject H_0 if $R_n > h$. Larger h reduces false alarms but increases detection delay.

Software References [Python, R, SAS]

Python	R	SAS
`(specialized implementations)`	`(research code available)`	`Not Native (Manual)`

Additional Notes

Shiryaev derived it via posterior odds under a geometric prior on the change point, while Roberts interpreted it as the cumulative sum of likelihood ratios over all possible change times. The recursion is $R_n = (1 + R_{n-1})\Lambda_n$, with $\Lambda_n = f_1(x_n)/f_0(x_n)$ and $R_0 = 0$. For normal models $N(\mu_0, \sigma^2)$ vs. $N(\mu_1, \sigma^2)$, $\log \Lambda_n$ reduces to CUSUM-like increments. Pollak (1985) showed that the Shiryaev-Roberts procedure is asymptotically optimal, minimizing worst-case detection delay for a given false alarm rate as $h \to \infty$. Unlike CUSUM, which assumes a most recent change, it averages over all possible change points; for large thresholds, their performance is nearly identical. Practical variants include window-limited versions (restricting to recent w points) and headstarts ($R_0 = r > 0$) for faster initial response. Threshold selection requires simulation or asymptotic approximations to achieve a target ARL_0. Although theoretically superior, its complexity limited early adoption relative to CUSUM. Recent advances have renewed interest due to improved computational feasibility and strong optimality.

Miscellanea: Albert Shiryaev (1960s) developed Bayesian change-point detection methods based on optimal stopping theory. The related Shiryaev–Roberts procedure, later studied by Pollak (1985), has strong optimality properties in sequential detection. Despite this, simpler methods such as CUSUM are often preferred in practice due to ease of implementation.

Reference(s): Shiryaev (1963); Roberts (1966); Pollak (1985)

Worked Example

Problem Statement We monitor a Gaussian process with $\mu_0 = 5.0$ and $\sigma = 1.0$. We aim to detect a shift to $\mu_1 = 6.0$. The decision threshold is set to $A = 20$ to maintain a low false alarm rate.

1. Process Parameters and Data We define the "In-Control" (H_0) and "Out-of-Control" (H_1) states:

- **H**$_0$: $\mu_0 = 5.0, \sigma = 1.0$ vs. **H**$_1$: $\mu_1 = 6.0$ (a 1σ shift)
- **Data (x):** $\{5.1, 4.9, 5.8, 6.2, 6.5\}$
- **Threshold (A):** 20.0

2. The Log-Likelihood Ratio (LLR) For normal distributions with equal variance, the Likelihood Ratio (L_i) is:

$$L_i = \exp\left(\frac{\mu_1 - \mu_0}{\sigma^2}\left(x_i - \frac{\mu_0 + \mu_1}{2}\right)\right) = \exp(x_i - 5.5)$$

3. Likelihood Ratio and Recursion The Shiryaev-Roberts statistic R_n is updated using the current likelihood ratio L_n:

$$R_n = (1 + R_{n-1})\frac{f(x_n|\mu_1)}{f(x_n|\mu_0)}$$

Starting from $R_0 = 0$, the statistic accumulates evidence of a shift. At $t = 5$ with $x_5 = 6.5$, the likelihood ratio is $L_5 = e^{6.5-5.5} = 2.718$. The update is:

$$R_5 = (1 + 7.22) \times 2.72 = 22.35$$

n	0	1	2	3	4	5
x_n		5.1	4.9	5.8	6.2	6.5
L_n		0.67	0.55	1.35	2.01	2.72
R_n	0	0.67	0.92	2.59	7.22	22.35

4. Results and Decision

Metric	Value
Target Shift ($\Delta\mu$)	1.0
Current SR Statistic (R_5)	22.35
Decision Threshold (A)	20.00
Status	**ALARM: Shift Detected**

5. Decision and Interpretation

- **Decision:** Since $R_5 = 22.36 > 20$, the null hypothesis of process stability is rejected at time $t = 5$.
- **Interpretation:** The SR test signaled a shift immediately following the third consecutive observation above the mid-point (5.5).
- **Practitioner Insight:** The SR test is often preferred over CUSUM in "steady-state" monitoring where the change could happen at any time. It has been mathematically proven to be quasi-optimal for minimizing the expected delay.

Common Pitfalls

- **Using the basic Shiryaev–Roberts recursion without an appropriate initialization.** The standard recursion can drift upward even when the process remains in control. Practical implementations often use a headstart (SR-r) or stationary variant to control long-run behavior and reduce start-up bias.
- **Selecting the decision threshold without proper calibration.** The relationship between the threshold and the in-control average run length (ARL) generally lacks a simple closed-form expression. Simulation or specialized tables are therefore used to determine thresholds that achieve the desired false alarm rate.
- **Failing to specify the anticipated magnitude of the change.** The likelihood ratio in the Shiryaev–Roberts statistic depends on the assumed post-change distribution. Detection performance is best when the specified change size matches the shift likely to occur.
- **Comparing Shiryaev–Roberts performance directly with CUSUM without considering start-up behavior.** The procedures differ in how evidence accumulates at the start of monitoring. Performance comparisons therefore depend on assumptions about the timing of the change-point.
- **Applying the procedure without understanding its modeling assumptions.** The Shiryaev–Roberts framework assumes correctly specified pre- and post-change distributions. Misspecification can substantially degrade detection performance.
- **Ignoring sensitivity to parameter misspecification.** The Shiryaev–Roberts procedure is highly sensitive to the assumed post-change parameters (e.g., mean shift or variance change). Poor specification can lead to delayed detection or inflated false alarm rates.

Change-Point Detection Test (Pettitt's)

Nonparametric Test for Single Abrupt Change in Time Series

QUAL.7 | **Category:** Quality Control | **Data:** Continuous/Ordinal | **Null:** Mann-Whitney U

When to Use

- Detect a single abrupt change-point in the central tendency (median/location) of a time-ordered series.
- Nonparametric alternative not requiring normality.
- Applications such climate data with potential regime shifts.

Assumptions

- Single time series with atmost one potential change point.
- Data are independent (no strong serial dependence).
- Change manifests as shift in location, not variance.
- Data are continuous or ordinal (few or no ties)

Test Statistic

$$K_T = \max_{1 \le t < T} |U_t|, \quad U_t = \sum_{i=1}^{t} \sum_{j=t+1}^{T} \operatorname{sgn}(x_i - x_j).$$

Notation: K_T is Pettitt statistic, $U_{t,T}$ is Mann-Whitney U at potential change point t, T is series length

Plain English: Pettitt's test identifies change points by computing Mann-Whitney U statistics comparing observations before and after each potential change point. At each time t, the test compares the first t observations against the remaining $T - t$ observations using rank-based comparison. The test statistic is the maximum absolute value of these comparisons across all possible change points. If this maximum exceeds a critical value, a change point is declared, and its location is where the maximum occurred.

⚠ Practitioner Warning

Detects at most one change point; multiple changes require segmentation or alternative methods. Assumes change is abrupt shift in location; gradual trends not detected. Less powerful than parametric methods (CUSUM, likelihood ratio) when assumptions hold. P-values require approximation or simulation; exact distribution complex. Change point location estimate can be imprecise with small shifts or high variability. Does not test for specific change magnitude, only whether change occurred. Autocorrelation affects test validity; modified versions needed for dependent data. False detection possible with natural variability appearing as shift.

Hypotheses

- H_0: No change point (location constant).
- H_1: Change point exists (location shifts at unknown t^*).

Decision Rule: Reject H_0 if $K_T > \text{CV}$ from tables/simulation.
Approx. p-value: $p \approx 2\exp(-6K_T^2/(T^3 + T^2))$ for large T.

Software References [Python, R, SAS]

Python	R	SAS
`(pyhomogeneity.pettitt_test)`	`trend::pettitt.test`	`Not Native (Manual)`

Additional Notes

Distribution under null: exact distribution complicated; Pettitt derived asymptotic approximation. Critical values: tabulated for common significance levels and sample sizes; simulation provides exact values. Multiple testing adjustment: searching all $T - 1$ split points introduces a multiple testing issue, accounted for in critical value distribution. Alternative formulations: two-sided version uses $|U_{t,T}|$; one-sided versions test specific direction of shift. Confidence interval for change point: bootstrap or likelihood-based methods provide uncertainty quantification. Extensions: Lombard's test allows multiple change points; binary segmentation recursively applies Pettitt's test. Comparison to CUSUM: Pettitt retrospective (analyzes complete historical data), CUSUM prospective (monitors ongoing process).

Miscellanea: A. N. Pettitt (1979) developed a nonparametric test for detecting a single change-point based on rank statistics. The method evaluates all possible split points and identifies the most significant change without assuming a specific distribution, making it widely used in environmental and time series analysis.

Reference(s): Pettitt (1979); Kundzewicz & Robson (2004)

Worked Example

Problem Statement A time series of 10 (T) observations is analyzed to detect a shift in the mean. The null hypothesis H_0 states the observations are identically distributed; H_a states a change-point exists at time k.

Year (t)	1	2	3	4	5	6	7	8	9	10
Rain (x_t)	550	580	540	560	**710**	**740**	**700**	**730**	**720**	**750**
Rank	2	4	1	3	6	9	5	8	7	10

Values in bold ($t = 5, \cdots, 10$) appear markedly higher than the first four observations; a possible upward shift in mean rainfall around $t = 4$ or $t = 5$.

1. Rank and Summation Analysis By comparing the values before and after each split point k, we generate a sequence of D_k values. We calculate the statistic U_k using the recursive formula. With no ties, $\boxed{D_k = 2 \cdot \text{Rank}_k - (T + 1)}$.

- At $k = 2$: $x_2 = 580$ exceeds 3 x's but is less than 6 others. So, $D_2 = +3 - 6 = -3$.
- At $k = 5$: likewise for $x_5 = 710$, $D_5 = +5 - 4 = +1$.

The absolute values of U_k for this series are calculated as follows:

Split Point (k)	0	1	2	3	**4**	5	6	7	8	9	10
$D_k = \sum_{j=1}^{n} \text{sgn}(x_k - x_j)$		-7	-3	-9	-5	+1	+7	-1	+5	+3	+9
$U_k = U_{k-1} + D_k$	0	-7	-10	-19	-24	-23	-16	-17	-12	-9	0
$\lvert U_k \rvert$	0	7	10	19	**24**	23	16	17	12	9	0

The maximum value is $K_n = \max |U_k| = 24$, occurring at $k = 4$.

2. Statistical Significance The p-value is approximated by ($K_T = 24$ and $T = 10$):

$$p \approx 2\exp\left(\frac{-6K_T^2}{T^3 + T^2}\right) = 2\exp(-3.14) \approx 0.086$$

3. Results

Metric	**Value**
Test Statistic (K_T)	24
Estimated Change Point (k)	Year 4
Approximate P-value	0.086

4. Decision and Interpretation

- **Decision:** Though the change is visually striking, at the 5% level, we do not have enough data points (T = 10) to confirm the shift is statistically significant.
- **Interpretation:** Year 4 is the most likely location of a regime shift, but the small sample size limits the power of the test.
- **Practitioner Insight:** Pettitt's test is sensitive to changes in the middle of a series but loses power near the beginning or end ($k = 1$ or $k = T - 1$). It is widely used in hydrology to detect the impact of dam construction or land-use changes on river flow.

Common Pitfalls

- **Applying Pettitt's test when multiple change-points may exist.** Pettitt's procedure is designed to detect a single structural change in the series. When several changes occur, the test typically identifies only the most prominent one.
- **Interpreting the estimated change-point as an exact time of change.** The test identifies the most likely location of a shift rather than a precise instant. Actual changes may occur gradually or within a range of nearby observations.
- **Using the test on strongly autocorrelated data without adjustment.** Pettitt's test assumes independence among observations in the series. Autocorrelation can create persistent runs that mimic structural changes.
- **Equating statistical detection with practical importance.** Large samples may reveal statistically significant changes that are negligible in magnitude. Evaluating the size and practical relevance of the shift is therefore essential.
- **Using Pettitt's test for real-time monitoring.** The method is primarily intended for retrospective analysis of completed data sets. Sequential monitoring typically relies on procedures such as CUSUM or EWMA charts.

Process Capability Test (Cp, Cpk)

Indices Measuring Process Performance Against Specifications

QUAL.8 | **Category:** Quality Control | **Data:** Continuous | **Null:** Normal (capability)

When to Use

- Assess if process can consistently meet specification limits.
- Compare process variability to tolerance width.
- Quantify process centering and spread relative to customer requirements.

Assumptions

- Process in statistical control (stable, no special causes).
- Process measurements approximately normally distributed.
- Spec limits (USL, LSL) set by customer requirements.
- Process mean μ and std dev σ estimated from stable data.

Test Statistic

$$C_p = \frac{\text{USL} - \text{LSL}}{6\sigma}, \quad C_{pk} = \min\left(\frac{\text{USL} - \mu}{3\sigma}, \frac{\mu - \text{LSL}}{3\sigma}\right).$$ C_p = potential capability; C_{pk} = actual capability.

Notation: USL/LSL are upper/lower specification limits, μ is process mean, σ is within-process standard deviation

Plain English: Process capability indices quantify how well a process can meet specifications. C_p compares specification width to process spread (6σ), measuring potential capability if process perfectly centered. C_{pk} accounts for centering by measuring distance from mean to nearest specification limit in standard deviation units. $C_p \geq 1.33$ indicates capable process (specification width exceeds natural process spread). $C_{pk} \geq 1.33$ indicates process both capable and adequately centered. Values below 1.0 indicate process cannot consistently meet specifications.

⚠ Practitioner Warning

Assumes normality; heavy tails or skewness invalidate 6σ spread estimate. Requires stable process; capability meaningless if process out of control. Short-term vs. long-term capability: indices typically use within-subgroup variability (short-term); long-term may differ. C_p and C_{pk} address different aspects; both needed for complete assessment. Does not indicate where defects occur (high or low end); use both indices and histograms. Sampling variability: indices are estimates; confidence intervals or hypothesis tests available. One-sided specifications require modified indices (C_{pu}, C_{pl}). Industry-specific benchmarks vary; 1.33 common but some industries require 1.67 or 2.0.

Hypotheses

- H_0: Process capability adequate ($C_p \geq 1.33$, $C_{pk} \geq 1.33$).
- H_1: Process capability inadequate (below thresholds).

Decision Rule: Reject H_0 if $C_{pk} < 1.33$. Large ($C_p - C_{pk}$) indicates centering problem. Use industry-specific limits

Software References [Python, R, SAS]

```
(scipy.stats or manual)        qcc::processCapability        PROC CAPABILITY;
```

Additional Notes

C_p measures potential capability: $C_p = 1$ implies the specification width equals 6σ (corresponding to 99.73% coverage under normality if centered), $C_p = 1.33 \approx 99.99\%$, and $C_p = 2$ corresponds to Six Sigma capability. C_{pk} measures actual capability by accounting for centering: $C_{pk} = C_p$ if centered, while $C_{pk} < C_p$ indicates off-center processes. The Taguchi index $C_{pm} = \frac{\text{USL}-\text{LSL}}{6\sqrt{\sigma^2+(\mu-T)^2}}$ penalizes deviation from target T, even within specifications. Since C_p and C_{pk} are estimates, uncertainty can be assessed via bootstrap or asymptotic intervals, and capability can be assessed using confidence intervals or hypothesis tests such as $H_0 : C_{pk} \geq c_0$ vs. $H_1 : C_{pk} < c_0$. For non-normal data, use transformations (e.g., Box-Cox) or percentile-based methods. Distinguish short-term (σ_{ST}) vs. long-term (σ_{LT}) variation; P_p and P_{pk} use long-term σ.

Miscellanea: Process capability indices such as C_p and C_{pk} became widely used in the 1980s with the rise of quality management practices. Unlike control charts, which assess stability, these indices measure how well a stable process meets specification limits. Despite their popularity, capability indices depend on correct specifications and do not alone guarantee product quality.

Reference(s): Kane (1986); Kotz & Johnson (2002); Montgomery (2019)

Worked Example

Problem Statement A high-performance engine manufacturer requires piston rings with a specific outer diameter to ensure a proper seal within the cylinder. The design engineers have established a target diameter of 74.00 mm, with an allowable tolerance of ±0.05 mm.

1. Parameters and Statistics We evaluate a piston ring process with the following engineering specifications and observed data:

- **Specifications:** $USL = 74.05$, $LSL = 73.95$
- **Process Mean:** $\bar{x} = 74.02$
- **Process Standard Deviation:** $\sigma = 0.01$

2. Calculation of C_p The C_p index measures the process potential if it were perfectly centered:

$$C_p = \frac{USL - LSL}{6\sigma} = \frac{74.05 - 73.95}{6(0.01)} \approx 1.67$$

3. Calculation of C_{pk} The C_{pk} index accounts for the mean shift of +0.02 from the target:

$$C_{pu} = \frac{USL - \bar{x}}{3\sigma} = \frac{74.05 - 74.02}{0.03} = 1.00$$

$$C_{pl} = \frac{\bar{x} - LSL}{3\sigma} = \frac{74.02 - 73.95}{0.03} = 2.33$$

$$C_{pk} = \min(C_{pu}, C_{pl}) = 1.00$$

4. Results and Decision

Metric	Value
Potential Capability (C_p)	1.67 (Excellent)
Actual Capability (C_{pk})	1.00 (Marginal)
Status	**Requires Centering**

5. Interpretation

- **Statistical Insight:** The process has the potential for high capability (consistent with Six Sigma benchmarks, $C_p > 1.33$), but because the mean is shifted toward the Upper Specification Limit, the actual performance is lower ($C_{pk} = 1.00$).
- **Decision:** The process is "capable" but has no margin for error. An operator should adjust the machine to shift the mean from 74.02 back down to the target of 74.00.

Common Pitfalls

- **Estimating capability for a process that is not in statistical control.** Capability indices describe the performance of a stable process operating under common-cause variation. Control charts should therefore confirm process stability before computing capability measures.
- **Applying normal-based capability indices to non-normal data.** Indices such as C_p and C_{pk} assume approximate normality. When data are skewed or heavy-tailed, transformations (e.g., Box–Cox) or distribution-specific capability methods should be used.
- **Confusing specification limits with control limits.** Specification limits are externally defined requirements (design or customer), whereas control limits reflect inherent process variation. These serve fundamentally different purposes and should not be interchanged.
- **Reporting capability indices without acknowledging sampling uncertainty.** Capability estimates can vary substantially, especially with small samples. Confidence intervals or resampling methods provide useful context for assessing reliability.
- **Improving capability indices by changing specification limits.** Capability analysis evaluates how well a process meets fixed specifications. Adjusting specification limits may change the index numerically but does not improve the underlying process performance.
- **Ignoring process centering when interpreting capability.** High C_p indicates low variability but does not guarantee the process is centered within specifications. Indices such as C_{pk} or C_{pm} should be used to assess both spread and centering.
- **Using short-term data to estimate long-term capability.** Short-term capability (within-subgroup variation) may overstate process performance. Long-term capability should incorporate all sources of variation, including shifts and drifts over time.

Part XXI

Comparability & Threshold-Based Inference

Establishing equivalence or verifying non-inferiority relative to benchmarks

Equivalence and Non-Inferiority

XXI. EQVL — A Conceptual Primer

The Fundamental Shift in Logic

Classical hypothesis tests ask: *Is there a difference?* The null hypothesis of equality is the default, and we seek evidence to reject it.

Equivalence and noninferiority tests reverse this logic: they ask *Is the difference small enough to be practically irrelevant?* The null hypothesis now asserts a **meaningful difference**, and the burden of proof is to demonstrate similarity.

This distinction has profound consequences:

- A **non-significant traditional test** does not establish equivalence; it merely fails to find a difference. With small n, a test can be underpowered to detect even large differences.
- Equivalence requires **sufficient power** and a **pre-specified margin** Δ. Declaring equivalence without these is scientifically invalid.
- Applications: bioequivalence (generic vs. brand-name drugs), clinical non-inferiority trials, regulatory submissions, A/B testing, analytical method validation.

The Equivalence Margin Δ

$\Delta > 0$ is the **maximum clinically or practically acceptable difference**—the largest difference that would not matter for the application. It must be:

- **Pre-specified before data collection**, not chosen after seeing the data.
- **Substantively justified** from clinical, regulatory, or scientific grounds—not chosen to make the test easy to pass.
- **Expressed on the scale of the outcome** (absolute difference, ratio, or log-ratio depending on context).

In bioequivalence, regulatory guidance (e.g. FDA) typically sets Δ as 20% on the log scale (80–125% ratio bounds for AUC and $C_{\max}$). For clinical non-inferiority, Δ is often set as a fraction of the active control's established effect size (the "M2 criterion").

Hypothesis Structures

Let $\delta = \mu_1 - \mu_2$ (or a ratio/difference of interest).

Equivalence:

$$H_0 : |\delta| \geq \Delta \quad \text{vs} \quad H_1 : |\delta| < \Delta.$$

Reject $H_0 \Rightarrow$ conclude equivalence.

Non-inferiority (new vs. control; higher = better):

$$H_0 : \delta \leq -\Delta \quad \text{vs} \quad H_1 : \delta > -\Delta.$$

Reject $H_0 \Rightarrow$ new treatment is not worse than control by more than Δ.

Superiority-by-margin (new vs. control):

$$H_0 : \delta \leq \Delta \quad \text{vs} \quad H_1 : \delta > \Delta.$$

Reject $H_0 \Rightarrow$ new treatment is better than control by at least Δ.

Relationship among the three: All three involve a margin Δ. Non-inferiority seeks to exclude the "unacceptably worse" region ($< -\Delta$). Superiority-by-margin is a more stringent form of superiority requiring the effect to exceed a clinically meaningful threshold $\Delta > 0$. Equivalence imposes the narrowest acceptable interval, requiring the effect to remain within $[-\Delta, +\Delta]$.

Logical implications:

Under the **same margin** and direction convention:

$$\text{Equivalence} \Rightarrow \text{Non-inferiority},$$

$$\text{Superiority-by-margin} \Rightarrow \text{Superiority} \Rightarrow \text{Non-inferiority}.$$

Equivalence and superiority-by-margin are mutually exclusive (*important: using the same margin*).

The TOST Procedure

The **Two One-Sided Tests** procedure (Schuirmann 1987) decomposes the equivalence null into two one-sided hypotheses and rejects both:

$$H_{01} : \delta \leq -\Delta \quad \text{and} \quad H_{02} : \delta \geq +\Delta.$$

Reject equivalence H_0 if *both* H_{01} and H_{02} are rejected at level α. Each one-sided test uses a t-statistic:

$$t_1 = \frac{\hat{\delta} + \Delta}{\widehat{\text{se}}(\hat{\delta})}, \quad t_2 = \frac{\hat{\delta} - \Delta}{\widehat{\text{se}}(\hat{\delta})}.$$

Reject H_{01} if $t_1 > t_{\alpha,\nu}$; reject H_{02} if $t_2 < -t_{\alpha,\nu}$. Overall Type I error is controlled at α (not 2α) because the two tests are positively correlated.

Confidence interval equivalence: TOST at level α is exactly equivalent to checking whether the $(1 - 2\alpha)$ CI for δ falls entirely within $(-\Delta, +\Delta)$:

$$\hat{\delta} \pm t_{\alpha,\nu} \cdot \widehat{\text{se}}(\hat{\delta}) \subset (-\Delta, +\Delta).$$

For $\alpha = 0.05$, this is a 90% CI, *not* 95%. This is the most intuitive representation and is required in regulatory submissions.

Paired TOST: Apply TOST to the paired differences $d_i = y_{1i} - y_{2i}$; the SE is $s_d/\sqrt{n}$ and df $= n - 1$.

Schuirmann's independent-samples TOST: Use the two-sample t-statistic with the pooled or Welch SE depending on whether variances are equal.

Confidence Interval Framework

The CI approach unifies all four test types:

Conclusion	CI condition
Equivalence	$(1-2\alpha)$ CI $\subset (-\Delta, +\Delta)$
Non-inferiority	Lower bound of $(1-2\alpha)$ CI $> -\Delta$
Superiority-by-margin	Lower bound of $(1-2\alpha)$ CI $> +\Delta$
Traditional superiority	$(1-2\alpha)$ CI excludes 0

The 90% CI is not a mistake: using $\alpha = 0.05$ for each one-sided test corresponds to a 90% (not 95%) two-sided interval. Presenting a 95% CI and checking whether it falls inside $(-\Delta, +\Delta)$ is a *different* (more conservative) test.

Anderson–Hauck Test

An alternative parametric test for equivalence; instead of decomposing into two one-sided tests, it directly tests $H_0 : |\delta| \geq \Delta$ using a test statistic based on the t-distribution with a non-centrality parameter approach. Compared to TOST:

- May have higher power than TOST when the true difference is near zero, but lower power near the equivalence boundaries.
- Produces a single test statistic and p-value rather than two separate one-sided tests.
- Less widely used and not universally accepted by regulatory agencies; TOST remains the standard.

Equivalence Test for Proportions

For binary outcomes with proportions p_1 and p_2:

Absolute difference: $H_0 : |p_1 - p_2| \geq \Delta$ vs. $H_1 : |p_1 - p_2| < \Delta$. The test statistic uses separate SE under the null boundary $|p_1 - p_2| = \Delta$.

Relative difference / ratio: Equivalence is often stated as a risk ratio or odds ratio within bounds $[1/\theta, \theta]$; log-transform and apply TOST on the log scale.

Regulatory bioequivalence for proportions: For pharmacokinetic endpoints that are ratios, the FDA 80–125% bounds correspond to $(-\log(1.25), +\log(1.25))$ on the log scale.

Mann–Whitney Equivalence Test

A nonparametric equivalence test based on the probability of stochastic equivalence:

$$H_0 : \left|P(X_1 < X_2) + \tfrac{1}{2}P(X_1 = X_2) - 0.5\right| \geq \epsilon$$
$$H_1 : \left|P(X_1 < X_2) + \tfrac{1}{2}P(X_1 = X_2) - 0.5\right| < \epsilon$$

where ϵ is a pre-specified margin on the probability scale (e.g. $\epsilon = 0.1$ corresponds to $P(X_1 < X_2) \in [0.4, 0.6]$). The test statistic is the Mann–Whitney U statistic (estimating $P(X_1 < X_2)$) with bootstrap or asymptotic CIs.

- Does not assume normality or a specific distributional form.
- The margin ϵ must be specified on the probability scale, which is more interpretable for ordinal outcomes.
- Less powerful than the t-based TOST under normality; preferable for skewed or ordinal data.

Power and Sample Size

Equivalence tests often require **larger samples** than traditional superiority tests for the same α and power. For the independent-samples TOST:

$$n \approx \frac{2\sigma^2(z_\alpha + z_\beta)^2}{(\Delta - |\delta_{\text{true}}|)^2},$$

where δ_{true} is the assumed true difference (often 0 for symmetric equivalence), z_α and z_β are the one-sided normal quantiles. Key observations:

- Power decreases as $|\delta_{\text{true}}|$ approaches Δ; maximum power when $\delta_{\text{true}} = 0$.
- A larger margin Δ requires smaller n but is harder to justify scientifically.
- At 80% power and $\alpha = 0.05$: use $z_\alpha = 1.645$ and $z_\beta = 0.842$.

Common Pitfalls

- **Non-significance $\neq$ equivalence:** a $p > 0.05$ from a traditional test does not establish equivalence; the test may simply be underpowered.
- **Post-hoc margin:** choosing Δ after seeing the data to make the test pass is scientifically invalid; Δ must be pre-specified.
- **95% CI instead of 90%:** using a 95% CI for a TOST at $\alpha = 0.05$ applies a stricter test than intended; use 90%.
- **Insufficient power:** equivalence requires power to detect the margin; a study designed for a superiority test will almost always be underpowered for equivalence.
- **Confusing non-inferiority and equivalence:** non-inferiority is a one-sided claim; equivalence is two-sided; they use different hypothesis structures and margins.
- **Ignoring the direction convention:** in non-inferiority, "higher is better" and "lower is better" lead to opposite signs for Δ; always state the direction explicitly.
- **Switching from non-inferiority to superiority post hoc:** a non-inferiority trial that incidentally shows superiority must pre-plan the superiority analysis or apply multiplicity correction.

Bottom Line

Equivalence and noninferiority tests demonstrate that differences are small enough to be practically irrelevant—not merely statistically insignificant. Pre-specify Δ before data collection, use TOST (check whether the 90% CI lies within the margin), and ensure adequate power. Never claim equivalence from a non-significant traditional test.

TOST for Independent Samples (Schuirmann's Test)

Two One-Sided Tests for Equivalence Between Independent Groups

EQVL.1 | **Category:** Equivalence Testing | **Data:** Continuous | **Null:** t-distribution

When to Use

- Demonstrate equivalence between two independent groups within predefined margins. Show differences are practically negligible rather than merely non-significant.
- Bioequivalence, comparability, or non-inferiority-type studies (regulatory/clinical settings).

Assumptions

- Data are at least interval-scale.
- Two groups are independent with independent observations.
- Group distributions are approximately normal.
- Equivalence margin $[-\Delta, \Delta]$ specified a priori.
- Equal (pooled) or unequal (Welch) variances allowed.

Test Statistic

$$t_L = \frac{(\bar{x}_1 - \bar{x}_2) - (-\Delta)}{\text{SE}}, \quad t_U = \frac{(\bar{x}_1 - \bar{x}_2) - \Delta}{\text{SE}}, \quad s_p^2 = \frac{(n_1 - 1)s_1^2 + (n_2 - 1)s_2^2}{n_1 + n_2 - 2}, \quad \text{SE} = s_p\sqrt{1/n_1 + 1/n_2}.$$

Notation: $\bar{x}_i$ are group means, s_p is pooled SD, Δ is equivalence margin (1-sided), $df = n_1 + n_2 - 2$ (pooled)

Plain English: TOST for independent samples establishes equivalence between two groups by testing whether the mean difference lies within pre-specified equivalence bounds. Two one-sided tests are conducted: one checks if the difference exceeds the lower bound, the other checks if it falls below the upper bound. When both tests reject their respective null hypotheses, equivalence is established—the groups differ by less than the practically important amount Δ in either direction.

⚠ Practitioner Warning

Equivalence margin Δ must be scientifically justified and pre-specified; cannot be chosen based on observed data. Equal variance assumption affects pooled version; use Welch correction if violated. Normality assumption more critical with small samples; robust alternatives exist for non-normal data. Cannot demonstrate exact equality, only practical equivalence within margins. Confidence interval approach (90% CI) provides equivalent inference with more familiar interpretation. Sample size requirements larger than for superiority testing. Balanced designs ($n_1 = n_2$) maximize power.

Hypotheses

- H_0: $|\mu_1 - \mu_2| \geq \Delta$, non-equivalence.
- H_1: $|\mu_1 - \mu_2| < \Delta$, practical equivalence.

Decision Rule: Reject H_0 if $t_L > t_{1-\alpha,df}$ AND $t_U < -t_{1-\alpha,df}$. Or, reject if $\text{CI}_{1-2\alpha}(\mu_1 - \mu_2) \subset (-\Delta, +\Delta)$.

Software References [Python, R, SAS]

Python	R	SAS
`scipy.stats.ttest_ind + manual`	`equivalence::tost`	`PROC TTEST`

Additional Notes

Two one-sided hypotheses are tested: $H_{01} : \mu_1 - \mu_2 \leq -\Delta$ (group 1 not unacceptably lower) and $H_{02} : \mu_1 - \mu_2 \geq \Delta$ (group 1 not unacceptably higher). Both must be rejected to conclude $-\Delta < \mu_1 - \mu_2 < \Delta$. For unequal variances, the Welch version uses **(A)** as the standard error with Welch–Satterthwaite degrees of freedom **(B)**. Equivalently, construct the $(1 - 2\alpha)$ CI for $\mu_1 - \mu_2$; if the entire interval lies within $[-\Delta, \Delta]$, conclude equivalence. For $\alpha = 0.05$, this corresponds to a 90% CI. Approximate equal-group sample size is **(C)**, where Δ is the assumed true difference and β is Type II error rate. Power increases as the true difference approaches zero and as Δ increases. Regulatory bioequivalence studies commonly use TOST with 90% CIs on log-transformed pharmacokinetic endpoints.

$$\textbf{(A)}\ \sqrt{s_1^2/n_1 + s_2^2/n_2}, \quad \textbf{(B)}\ df = \frac{(s_1^2/n_1 + s_2^2/n_2)^2}{(s_1^2/n_1)^2/(n_1 - 1) + (s_2^2/n_2)^2/(n_2 - 1)}, \quad \textbf{(C)}\ n = \frac{2(z_{1-\alpha} + z_{1-\beta})^2\sigma^2}{(\Delta - |\delta|)^2}.$$

Miscellanea: TOST, introduced by Donald Schuirmann (1987), applies to both paired and independent (parallel-group) designs commonly used in clinical trials. It became central to regulatory bioequivalence and non-inferiority testing, providing a framework to demonstrate practical equivalence within predefined margins rather than relying on traditional tests of difference.

Reference(s): Schuirmann (1987); Wellek (2010); US FDA (2001)

Worked Example

Problem Statement Two independent laboratories each measure the same analyte concentration (mg/L) on ten samples. Test whether the labs are equivalent to within $\Delta = 5$ mg/L at $\alpha = 0.05$ using the pooled two-sample TOST. $H_0: |\mu_1 - \mu_2| \geq \Delta$; $H_1: |\mu_1 - \mu_2| < \Delta$. Reference distribution for both t-tests is $t(n_1 + n_2 - 2) = t(18)$.

Lab 1	93.7	95.9	97.2	98.4	100.5	101.9	104.1	104.8	106.8	107.0
Lab 2	94.4	99.3	100.0	100.8	101.0	101.3	102.3	102.4	107.7	109.4

1. Group Summary Statistics

	n	$\bar{y}$	s^2
Lab 1	10	101.03	21.693
Lab 2	10	101.86	17.698

2. Pooled Standard Error

$$s_p = \sqrt{\frac{(n_1 - 1)s_1^2 + (n_2 - 1)s_2^2}{n_1 + n_2 - 2}} = 4.438.$$

$$\widehat{\text{SE}} = s_p\sqrt{1/n_1 + 1/n_2} = 1.985.$$

3. Two One-Sided Test Statistics

$\bar{y}_1 - \bar{y}_2 = -0.830.$
Lower $(H_{01}: \mu_1 - \mu_2 \leq -5)$ & *Upper* $(H_{02}: \mu_1 - \mu_2 \geq +5)$:

$$t_1 = \frac{(\bar{y}_1 - \bar{y}_2) + \Delta}{\widehat{\text{SE}}} = 2.101, \quad t_2 = \frac{(\bar{y}_1 - \bar{y}_2) - \Delta}{\widehat{\text{SE}}} = -2.937.$$

4. Decision via t-Statistics

Critical value (one-sided, $\alpha = 0.05$): $t_{0.05}(18) = 1.734$.
Conclude equivalence if and only if:

$$t_1 > 1.734 \quad \text{AND} \quad t_2 < -1.734.$$

Test	Statistic	Critical	
$H_{01}: \mu_1 - \mu_2 \leq -5$	$t_1 = 2.101$	> 1.734	Reject
$H_{02}: \mu_1 - \mu_2 \geq +5$	$t_2 = -2.937$	< -1.734	Reject

$p_1 = 0.025$, $p_2 = 0.004$, $p_{\text{TOST}} = \max(p_1, p_2) = 0.025$.
Both nulls rejected: the labs are declared equivalent.

5. Equivalent CI Approach

$$90\% \text{ CI} = -0.830 \pm 1.734 \times 1.985 = [-4.27,\ 2.61].$$

Since $[-4.27, 2.61] \subset (-5, 5)$, equivalence is confirmed. The ordinary two-sample t-test gives $t(18) = -0.418$, $p = 0.681$: it fails to reject $H_0: \mu_1 = \mu_2$, but this is irrelevant to equivalence.

Practitioner Insight: The two-sample TOST (Schuirmann 1987) extends the paired procedure to independent groups by replacing the paired difference and its standard error with the two-sample mean difference and its pooled (or Welch) standard error. The pooled version assumes equal variances, which is reasonable here ($s_1^2 = 21.7$, $s_2^2 = 17.7$); Welch's version is preferred when variances differ substantially. The lower test is the binding constraint: $t_1 = 2.101$ only modestly exceeds the critical value, while the upper test is easily satisfied because the observed difference (−0.83) lies closer to the lower equivalence bound −5 than to +5. The 90% CI therefore provides a clearer interpretation by showing how close the estimate lies to each boundary. Sample-size planning follows the usual one-sided t-test framework but depends on the gap $\Delta - |\mu_1 - \mu_2|$; wider margins or smaller variances require fewer observations. As in the paired case, a non-significant ordinary t-test ($p = 0.681$ here) does not establish equivalence without pre-specified margins and adequate power.

Common Pitfalls

- **Using pooled variance when variances are unequal.** The pooled test assumes equal population variances. When variances differ substantially, Welch's version gives more reliable inference.
- **Choosing Δ without considering the measurement scale.** Equivalence margins may be absolute or relative depending on context. Relative margins often require analysis on a transformed scale such as logarithms.
- **Using superiority-based sample size calculations.** Equivalence studies require enough precision for the confidence interval to fall within the equivalence bounds. Studies powered for superiority may therefore be underpowered for equivalence.
- **Confusing equivalence with non-inferiority.** Equivalence tests whether differences remain within bounds on both sides of zero. Non-inferiority tests only whether a treatment is not worse than a specified margin.
- **Treating overlapping 95% CIs as evidence of equivalence.** Overlapping conventional intervals only indicate weak evidence of a difference. Equivalence requires the appropriate CI to lie entirely within the predefined equivalence region.

TOST (Paired Data)

Two One-Sided Tests for Equivalence in Paired Samples

EQVL.2 | **Category:** Equivalence Testing | **Data:** Continuous | **Null:** t-distribution

When to Use

- Demonstrate equivalence of paired measurements within predefined practical margins.
- Bioequivalence, crossover, method-comparison, or other repeated-measures studies (regulatory/clinical settings).
- Require evidence of equivalence rather than "no difference."

Assumptions

- Data are at least interval-scale.
- Paired observations are meaningfully matched.
- Paired differences are independent across subjects.
- Paired differences are approximately normally distributed.
- Equivalence margins are specified *a priori*.

Test Statistic

$$t_L = \frac{\bar{d} - (-\Delta)}{s_d/\sqrt{n}}, \quad t_U = \frac{\bar{d} - \Delta}{s_d/\sqrt{n}}, \quad \text{where } \bar{d} = \frac{\sum d_i}{n}, \quad s_d^2 = \frac{\sum (d_i - \bar{d})^2}{n-1}.$$

Notation: d_i are paired differences, $\bar{d}$ is mean difference, Δ is equivalence margin (one-sided)

Plain English: TOST for paired data tests equivalence by conducting two one-sided tests simultaneously: one testing whether the mean difference exceeds the lower equivalence bound, another testing whether it falls below the upper bound. Equivalence is demonstrated when both tests reject, establishing the mean difference lies within $[-\Delta, \Delta]$. This provides positive statistical evidence that treatments are equivalent within practical limits, not merely "not significantly different."

⚠ Practitioner Warning

Equivalence margin Δ must be pre-specified and scientifically justified; post-hoc selection invalidates test. Margin choice critical—too wide makes equivalence trivial, too narrow makes it unattainable. Assumes normality of differences; violations affect validity with small samples. Cannot prove exact equivalence (mean difference = 0), only practical equivalence within margins. The 90% confidence interval for the paired mean difference is equivalent to TOST but usually more interpretable. Failure to reject a traditional paired *t*-test does not establish equivalence; absence of evidence and evidence of equivalence are fundamentally different conclusions.

Hypotheses

- H_0: $|\mu_d| \geq \Delta$, non-equivalence.
- H_1: $|\mu_d| < \Delta$, practical equivalence.

Decision Rule: Reject H_0 if $t_L > t_{1-\alpha,n-1}$ AND $t_U < -t_{1-\alpha,n-1}$. Equivalently, reject if $\mathrm{CI}_{1-2\alpha}(\mu_d) \subset (-\Delta, +\Delta)$.

Software References [Python, R, SAS]

Python	R	SAS
`scipy.stats.ttest_rel + manual`	`equivalence::tost`	`Not Native (Manual)`

Additional Notes

The two one-sided hypotheses: $H_{01}: \mu_d \leq -\Delta$ tested via t_L, and $H_{02}: \mu_d \geq \Delta$ tested via t_U. Reject both to establish $-\Delta < \mu_d < \Delta$. Confidence interval equivalence: $(1-2\alpha)$ CI for μ_d corresponds to TOST at level α. For $\alpha = 0.05$, use 90% CI; if entirely within $[-\Delta, \Delta]$, equivalence established. Relationship to traditional testing: TOST reverses burden of proof—must demonstrate equivalence rather than non-difference. Sample size: power analysis requires specifying true difference, equivalence margin, and desired power. Larger n needed for equivalence than superiority testing. Regulatory guidance: FDA requires 90% CI approach for bioequivalence (equivalent to TOST $\alpha = 0.05$). Equivalence margins typically 80-125% of reference for bioequivalence, or minimal clinically important difference (MCID) in clinical trials.

Miscellanea: Donald Schuirmann (1987) introduced the TOST procedure for equivalence testing, motivated by bioequivalence studies in drug regulation. Unlike traditional tests, TOST assesses whether differences lie within predefined equivalence bounds, shifting focus from exact equality to practical equivalence. This framework became central to regulatory evaluation of generic drugs, including the widely used 80–125% margin.

Reference(s): Schuirmann (1987); Wellek (2010)

Worked Example

Problem Statement Systolic blood pressure (mmHg) is measured on twelve patients using two devices: a reference sphygmomanometer (Method A) and a new automated monitor (Method B). Measurements are paired by patient. Test equivalence with margin $\Delta = 10\,\text{mmHg}$ at $\alpha = 0.05$. $H_0: |\mu_d| \geq \Delta$; $H_1: |\mu_d| < \Delta$.

i	1	2	3	4	5	6	7	8	9	10	11	12
A_i	123	127	131	134	137	139	142	144	147	150	153	156
B_i	127	130	132	134	136	138	140	141	143	145	147	149
d_i	−4	−3	−1	0	1	1	2	3	4	5	6	7

1. Summary Statistics for $d_i = A_i - B_i$

$$\bar{d} = 1.750, \qquad s_d = \sqrt{\frac{\sum(d_i - \bar{d})^2}{n-1}} = 3.441,$$

$$\widehat{\text{SE}} = \frac{s_d}{\sqrt{n}} = 0.993.$$

2. Two One-Sided Test Statistics

Lower ($H_{01}: \mu_d \leq -\Delta$):

$$t_1 = \frac{\bar{d} + \Delta}{\widehat{\text{SE}}} = 11.83.$$

Upper ($H_{02}: \mu_d \geq +\Delta$):

$$t_2 = \frac{\bar{d} - \Delta}{\widehat{\text{SE}}} = -8.31.$$

Both statistics are referred to $t(n-1) = t(11)$.

3. Decision via t-Statistics

Critical value (one-sided, $\alpha = 0.05$): $t_{0.05}(11) = 1.796$.
Conclude equivalence if and only if:

$$t_1 > 1.796 \quad \text{AND} \quad t_2 < -1.796.$$

Test	Statistic	Critical	Decision
H_{01}: $\mu_d \leq -10$	$t_1 = 11.83$	> 1.796	Reject
H_{02}: $\mu_d \geq +10$	$t_2 = -8.31$	< -1.796	Reject

$p_{\text{TOST}} = \max(p_1, p_2) < 0.001$.

Both H_{01} and H_{02} are rejected: the two devices are declared equivalent.

4. Equivalent CI Approach

The same conclusion follows from the 90% CI for μ_d:

$$90\%\ \text{CI} = \bar{d} \pm t_{0.05}(11) \times \widehat{\text{SE}} = [-0.03,\ 3.53].$$

As $[-0.03, 3.53] \subset (-10, 10) \implies$, equivalence is confirmed.

Practitioner Insight: The TOST procedure (Schuirmann 1987; Westlake 1981) reverses the usual testing logic by treating non-equivalence ($|\mu_d| \geq \Delta$) as the null and equivalence as the alternative. Rejecting both one-sided nulls at level α is equivalent to showing that the $100(1-2\alpha)\%$ CI lies entirely within $(-\Delta, +\Delta)$; here the 90% CI $[-0.03, 3.53]$ remains well inside the equivalence margins. The margin $\Delta = 10\,\text{mmHg}$ reflects a clinically unimportant difference in blood-pressure monitoring. Unlike TOST, the ordinary two-sided t-test asks whether $\mu_d = 0$ and a non-significant result ($p = 0.106$ here) does *not* establish equivalence; absence of evidence is not evidence of absence. The equivalence margin must be specified *a priori*; post-hoc widening inflates Type I error. No multiplicity correction is required because equivalence is concluded only when both one-sided null hypotheses are rejected simultaneously. The independent-samples version replaces $\bar{d}$ and $s_d/\sqrt{n}$ with the two-sample mean difference and its pooled or Welch standard error.

Common Pitfalls

- **Treating a non-significant paired t-test as evidence of equivalence.** Failing to reject zero difference does not establish equivalence. Equivalence requires the confidence interval to lie entirely within the predefined bounds.
- **Choosing Δ after seeing the data.** The equivalence margin must be specified *a priori* to avoid data-driven conclusions. Margins should be justified clinically or from prior evidence.
- **Testing only one equivalence boundary.** Equivalence requires rejecting both one-sided null hypotheses. Demonstrating only an upper or lower bound is insufficient.
- **Using the wrong confidence interval level.** At $\alpha = 0.05$, TOST corresponds to a 90% confidence interval, not 95%. Using 95% intervals imposes a stricter and less powerful criterion.
- **Equating statistical and clinical equivalence.** Equivalence conclusions depend entirely on the chosen margin. Poorly justified or overly wide margins may lack practical meaning.

Non-Inferiority Test

One-Sided Test for Demonstrating Not Worse Than Reference

EQVL.3 | **Category:** Equivalence Testing | **Data:** Continuous | **Null:** t-distribution

When to Use

- Demonstrate non-inferiority relative to a reference treatment using a predefined clinical or regulatory margin.
- Compare new treatments offering lower cost, risk, or greater convenience.
- One-sided inference where superiority is unnecessary.
- Trials where placebo control is unethical or impractical.

Assumptions

- Treatment groups are independent.
- Observations within groups are independent.
- Outcome distributions satisfy assumptions of the underlying test.
- Non-inferiority margin is specified *a priori*.
- Assay sensitivity and constancy assumptions are plausible.

Test Statistic

$$t = \frac{(\bar{x}_T - \bar{x}_R) - (-\Delta)}{s_p\sqrt{1/n_T + 1/n_R}}, \quad \text{where } s_p^2 = \frac{(n_T - 1)s_T^2 + (n_R - 1)s_R^2}{n_T + n_R - 2}. \quad \text{Under } H_0 \text{ at the boundary, } t \sim t_{n_T+n_R-2}.$$

Notation: $\bar{x}_T, \bar{x}_R$ are test/reference means, Δ is non-inferiority margin (positive), s_p is pooled SD

Plain English: The non-inferiority test evaluates whether a test treatment performs no worse than a reference treatment by more than a pre-specified clinically acceptable margin. Unlike equivalence testing which requires similarity in both directions, non-inferiority only requires the test treatment not be substantially worse. The test rejects H_0 that the test treatment is inferior by at least Δ, concluding that any disadvantage relative to the reference is smaller than the clinically irrelevant threshold.

⚠ Practitioner Warning

Non-inferiority margin Δ must be pre-specified and clinically justified; choosing it too large makes non-inferiority trivial, while choosing it too small may make the study infeasible. Outcome direction must be defined correctly (reverse inequalities for adverse outcomes). Non-inferiority alone does not imply equivalence or superiority, though superiority may also be concluded if the lower CI bound exceeds 0. Intention-to-treat (ITT) analyses may bias toward non-inferiority through treatment dilution, whereas per-protocol analyses may introduce selection bias; regulators often recommend both alongside sensitivity analyses for missing data.

Hypotheses

- H_0: $\mu_T - \mu_R \leq -\Delta$, clinically worse than reference.
- H_1: $\mu_T - \mu_R > -\Delta$, not clinically worse.

Decision Rule: Reject H_0 if $t > t_{1-\alpha,df}$. Or, lower bound of $(1 - 2\alpha)$ CI for $(\mu_T - \mu_R)$ exceeds $-\Delta$.

Software References [Python, R, SAS]

```
statsmodels.stats.weightstats      TrialSize::OneArmNonInf      PROC TTEST; tost; sided=1;
```

Additional Notes

Non-inferiority uses one of the two one-sided tests employed in TOST equivalence testing: $H_0 : \mu_T - \mu_R \leq -\Delta$ vs. $H_1 : \mu_T - \mu_R > -\Delta$. Using the CI approach, non-inferiority is concluded if the lower bound of a $(1 - 2\alpha)$ CI (e.g., 90% for $\alpha = 0.05$) exceeds $-\Delta$. Welch versions accommodate unequal variances using adjusted standard errors and Welch–Satterthwaite degrees of freedom. Margin selection is often based on preserving a fraction of the active-control effect or the minimal clinically important difference, as recommended by FDA, EMA, and ICH E10 guidance. Required sample sizes are usually larger than for superiority trials.

Miscellanea: Non-inferiority testing developed in the 1980s–1990s as ethical concerns limited placebo-controlled trials when effective treatments existed. Active-controlled designs compare new treatments to established ones, focusing on whether they are not unacceptably worse within a predefined margin. The ICH E10 guidance (2000) formalized this framework, emphasizing margin justification and concerns such as biocreep. Repeated non-inferiority comparisons against previously approved treatments can lead to "biocreep," where small accepted losses of efficacy compound across successive generations of therapies.

Reference(s): ICH E10 (2000); FDA (2016); Pigeot et al. (2003)

Worked Example

Problem Statement A new drug is compared to a standard treatment in a randomized trial ($n_1 = n_2 = 20$). Higher efficacy scores are better. The new drug is acceptable if its mean is no more than $\Delta = 5$ points below the standard. Test non-inferiority at $\alpha = 0.05$. $H_0: \mu_{\text{new}} - \mu_{\text{std}} \le -\Delta$; $H_1: \mu_{\text{new}} - \mu_{\text{std}} > -\Delta$. Under H_0: $t_{\text{NI}} \sim t(n_1 + n_2 - 2) = t(38)$.

New	68.8	70.8	72.1	74.7	75.4	75.8	76.5	77.2	77.5	77.5
	79.0	80.8	81.5	83.1	83.3	83.6	84.1	88.3	89.2	92.2
Std	69.2	70.2	70.7	75.9	77.1	77.8	79.0	79.6	79.8	80.9
	81.6	81.9	81.9	82.5	83.0	84.4	84.7	87.9	89.0	94.6

1. Group Summary Statistics

	n	$\bar{y}$	s
New	20	79.57	6.176
Std	20	80.59	6.272

2. Pooled Standard Error

$$s_p = \sqrt{\frac{19 \times 6.176^2 + 19 \times 6.272^2}{38}} = 6.224.$$

$$\widehat{\text{SE}} = s_p\sqrt{1/n_1 + 1/n_2} = 1.968.$$

3. Non-Inferiority Test Statistic

$$\bar{y}_{\text{new}} - \bar{y}_{\text{std}} = -1.02, \quad t_{\text{NI}} = \frac{(\bar{y}_{\text{new}} - \bar{y}_{\text{std}}) + \Delta}{\widehat{\text{SE}}} = 2.024.$$

4. Decision via Test Statistic

Critical value (one-sided, $\alpha = 0.05$): $t_{0.05}(38) = 1.686$.

α	$t_{0.05}(38)$	Decision
5%	1.686	**Reject** (2.024 > 1.686)
1%	2.429	FTR (2.024 < 2.429)

$p_{\text{NI}} = 0.025$. **Non-inferiority of the new drug is established at $\alpha = 0.05$.**

5. Equivalent CI Approach

The one-sided 95% lower confidence bound equals the lower limit of the 90% CI:

$$\bar{y}_{\text{new}} - \bar{y}_{\text{std}} - t_{0.05}(38) \times \widehat{\text{SE}} = -4.33.$$

Since $-4.33 > -5$, the lower bound lies above $-\Delta$: non-inferiority is confirmed.

Contrast: the ordinary two-sided t-test gives $t(38) = -0.516$, $p = 0.609$ (FTR); and the 95% CI $[-5.00, 2.97]$ touches the boundary, illustrating that it uses the wrong criterion for non-inferiority.

Practitioner Insight: Non-inferiority testing (Blackwelder 1982; ICH E10 2001) asks whether a new treatment is no more than Δ units worse than the standard, leading to the one-sided hypothesis $H_0: \mu_{\text{new}} - \mu_{\text{std}} \le -\Delta$ vs $H_1: \mu_{\text{new}} - \mu_{\text{std}} > -\Delta$. The test shifts the observed difference by $+\Delta$, recentering the reference distribution at the non-inferiority boundary rather than at zero. The correct interval is therefore a 90% CI (equivalently, a one-sided 95% lower bound): non-inferiority is established if the lower bound exceeds $-\Delta$. Here the 90% CI $[-4.33, 2.30]$ confirms non-inferiority because it remains 0.67 points above the -5 margin; using a 95% two-sided CI would incorrectly fail the criterion. The margin Δ must be specified *a priori* based on the largest clinically acceptable loss of efficacy. Unlike equivalence testing, which requires both upper and lower bounds to hold, non-inferiority requires only one; a treatment declared non-inferior may still be substantially superior. *Note:* Because H_0 is composite, distributional results "under H_0" are evaluated at the boundary of H_0.

Common Pitfalls

- **Choosing an overly large margin.** The non-inferiority margin must be smaller than the historical control effect versus placebo. Otherwise an ineffective treatment could incorrectly appear non-inferior.
- **Using the wrong confidence interval.** Non-inferiority is a one-sided hypothesis and requires a one-sided confidence bound or equivalent interpretation. Using the wrong confidence level can lead to incorrect conclusions.
- **Confusing non-inferiority with equivalence.** Non-inferiority only shows the treatment is not unacceptably worse than the control. Equivalence requires bounds to hold on both sides of zero.
- **Ignoring the constancy assumption.** Non-inferiority trials assume the control retains the same effect seen in earlier studies. Changes in populations or study conditions can invalidate this assumption.
- **Switching to non-inferiority post-hoc.** Non-inferiority margins and hypotheses must be specified before data collection. Post-hoc switching invalidates inference and is generally unacceptable to regulators.

Superiority-by-Margin Test

Test for Exceeding Clinically Meaningful Improvement Threshold

EQVL.4 | **Category:** Equivalence Testing | **Data:** Continuous | **Null:** t-distribution

When to Use

- To prove treatment is better by a specific margin.
- Use when small improvements lack clinical significance.
- When new interventions must justify higher costs.
- Required when a minimum effect size is strictly mandated.
- Used for comparing active drugs against known controls.

Assumptions

- Superiority margin Δ specified a priori.
- Independent observations within and between groups.
- Outcome approximately normal or sample sizes large.
- Appropriate standard error estimate for the design used.
- Margin reflects clinically meaningful improvement.

Test Statistic

$$t = \frac{(\bar{x}_T - \bar{x}_R) - \Delta}{s_p\sqrt{1/n_T + 1/n_R}}, \quad \text{where } s_p^2 = \frac{(n_T - 1)s_T^2 + (n_R - 1)s_R^2}{n_T + n_R - 2}. \qquad \text{Under } H_0 \text{ at the boundary, } t \sim t_{n_T+n_R-2}.$$

Notation: $\bar{x}_T, \bar{x}_R$ are test/reference means, Δ is superiority margin (positive), s_p is pooled SD (for equal σ^2)

Plain English: The superiority-by-margin test evaluates whether a test treatment exceeds a reference treatment by at least a pre-specified clinically meaningful amount. Unlike standard superiority testing which only requires any detectable difference, this test demands the difference exceed a threshold representing minimal clinical importance. The test rejects the null hypothesis that the improvement is less than Δ, concluding not just statistical superiority but clinically relevant superiority.

⚠ Practitioner Warning

Superiority margin Δ must be pre-specified and clinically justified; typically set to minimal clinically important difference (MCID). Larger margin than statistical superiority test requires larger sample size for same power. Assumes higher values better; reverse signs for adverse outcomes or costs. Pooled variance assumes homogeneity; use Welch correction if violated. Does not test equivalence or non-inferiority, only whether improvement exceeds threshold. If the test rejects, the observed improvement exceeds the pre-specified clinically meaningful threshold; if fails to reject, cannot conclude meaningful superiority even if statistically significant by standard test. Conservative approach balancing statistical and clinical evidence.

Hypotheses

- H_0: $\mu_T - \mu_R \leq \Delta$, not clinically superior.
- H_1: $\mu_T - \mu_R > \Delta$, clinically meaningful superiority.

Decision Rule: Reject H_0 if $t > t_{1-\alpha,df}$. Or, lower bound of $(1 - 2\alpha)$ CI for $(\mu_T - \mu_R)$ exceeds Δ.

Software References [Python, R, SAS]

Python	R	SAS
`scipy.stats.ttest_ind + manual`	`Not Native (Manual)`	`PROC TTEST; margin=; sided=u;`

Additional Notes

Superiority-by-margin testing addresses the limitation that statistical significance may not imply clinical relevance. Standard test: $H_0 : \mu_T = \mu_R$ vs. $H_1 : \mu_T > \mu_R$; margin test: $H_0 : \mu_T - \mu_R \leq \Delta$ vs. $H_1 : \mu_T - \mu_R > \Delta$, where Δ is the minimal clinically important difference (MCID). Using the CI approach, if the lower bound of a $(1 - 2\alpha)$ CI (e.g., 90% for $\alpha = 0.05$) exceeds Δ, meaningful superiority is established. The method parallels non-inferiority testing but with a positive margin. Larger sample sizes are typically required, and Welch versions handle unequal variances. It helps identify cases where statistical significance reflects trivial effects rather than clinically important differences.

Miscellanea: Clinical research increasingly recognized that statistical significance does not imply clinical relevance, especially in large studies where small effects can achieve low p-values. This led to greater emphasis on clinically meaningful differences, including margin-based interpretations and minimal clinically important differences (MCID). Guidelines such as CONSORT and regulatory agencies now stress reporting effect sizes alongside their practical significance.

Reference(s): Jaeschke et al. (1989); Cook (2008); FDA (2019)

Worked Example

Problem Statement A new training program is evaluated against a control condition ($n_1 = n_2 = 12$). Higher scores indicate better performance. The program is worth adopting only if it improves scores by more than $\Delta = 5$ points (the minimum clinically important difference, MCID). Test superiority by margin at $\alpha = 0.05$.
$H_0: \mu_{new} - \mu_{ctrl} \le \Delta$; $H_1: \mu_{new} - \mu_{ctrl} > \Delta$. Let $\mu_d = \mu_{new} - \mu_{ctrl}$. Under H_0 at the boundary: $t_{SM} \sim t(n_1 + n_2 - 2) = t(22)$.

New	75.5	81.2	82.7	85.6	85.9	87.7	88.3	90.8	91.0	92.4	93.3	98.9
Ctrl	62.9	69.2	71.8	74.2	75.4	77.6	80.4	81.2	82.6	83.3	84.3	84.4

1. Group Summary Statistics

	n	$\bar{y}$	s
New	12	87.775	6.204
Ctrl	12	77.275	6.775

$\bar{y}_{new} - \bar{y}_{ctrl} = 10.500.$

2. Pooled Standard Error

$$s_p = \sqrt{\frac{11 \times 6.204^2 + 11 \times 6.775^2}{22}} = 6.496.$$

$$\widehat{SE} = s_p\sqrt{1/n_1 + 1/n_2} = 2.652.$$

3. Superiority-by-Margin Statistic

$$t_{SM} = \frac{(\bar{y}_{new} - \bar{y}_{ctrl}) - \Delta}{\widehat{SE}} = 2.074.$$

Practitioner Insight: Superiority-by-margin (clinical superiority) tests whether a treatment exceeds a reference by a clinically meaningful amount Δ, not merely by any positive amount. The null hypothesis $H_0: \mu_{new} - \mu_{ctrl} \le \Delta$ shifts the reference point from zero to the superiority margin, so the test statistic $t_{SM} = (\bar{\Delta} - \Delta)/\widehat{SE}$ requires stronger evidence than an ordinary superiority test. Here the observed difference (10.5 points) exceeds the 5-point margin, but the 90% CI lower bound (5.95) only narrowly clears the threshold: ordinary superiority is highly significant ($p < 0.001$), whereas superiority beyond $\Delta = 5$ is only marginally established ($p = 0.025$). The margin must be specified *a priori* based on practical or clinical importance; post-hoc selection inflates Type I error. Unlike equivalence testing, which requires both upper and lower bounds to hold, superiority-by-margin requires only evidence that the treatment exceeds the clinically important threshold.

4. Decision via Test Statistic

Critical value (one-sided, $\alpha = 0.05$): $t_{0.05}(22) = 1.717$.

α	$t_{0.05}(22)$	Decision
5%	1.717	**Reject** (2.074 > 1.717)
1%	2.508	FTR (2.074 < 2.508)

$p_{SM} = 0.025$. **Superiority by margin is established at** $\alpha = 0.05$.

5. Equivalent CI Approach

90% CI = $10.500 \pm 1.717 \times 2.652 = [5.946,\ 15.054]$.

Since $5.946 > \Delta = 5$, the lower bound of the 90% CI exceeds the margin: superiority by margin is confirmed.

6. Comparison Across Tests

Hypothesis	t	p
Sup. by margin ($H_0: \mu_d \le 5$)	2.074	0.025
Plain superiority ($H_0: \mu_d \le 0$)	3.960	< 0.001
Two-sided test ($H_0: \mu_d = 0$)	3.960	0.001

Plain superiority is easily established ($p < 0.001$); the more demanding superiority-by-margin test just passes at $p = 0.025$.

Common Pitfalls

- **Choosing a margin for statistical convenience.** The superiority margin should reflect the minimal clinically important difference. Margins chosen only to ensure significance weaken the clinical interpretation.
- **Confusing significance with clinical superiority.** Statistical significance only shows evidence of improvement over zero. Superiority-by-margin additionally requires exceeding the pre-specified clinical threshold.
- **Applying margin testing post-hoc.** The superiority margin and testing strategy must be specified before data collection. Post-hoc testing introduces bias and weakens the validity of the conclusion.
- **Ignoring increased sample-size requirements.** Demonstrating superiority beyond a margin usually requires greater precision than testing against zero. Sample-size planning should therefore be margin-based.
- **Confusing superiority-by-margin with non-inferiority.** Both methods use margins but test different directions of treatment effect. Superiority-by-margin shows clinically meaningful improvement, whereas non-inferiority shows a treatment is not unacceptably worse.

Equivalence Test for Proportions

Two One-Sided Tests for Equivalence of Binomial Proportions

EQVL.5 | **Category:** Equivalence Testing | **Data:** Binary | **Null:** Normal (asymptotic)

When to Use

- Demonstrate equivalence of proportions between two groups (proportion differences are practically negligible).
- Bioequivalence, comparability, or device-validation studies (regulatory or clinical settings requiring equivalence).
- Binary outcomes with independent groups or treatments.

Assumptions

- Observations independent within and between groups.
- Independent samples from binomial populations.
- Equivalence margin $[-\Delta, \Delta]$ specified a priori.
- Adequate sample size for normal approximation ($np, n(1-p) \gtrsim 5$).

Test Statistic

$$z_L = \frac{(\hat{p}_1 - \hat{p}_2) - (-\Delta)}{\sqrt{\hat{p}_1(1-\hat{p}_1)/n_1 + \hat{p}_2(1-\hat{p}_2)/n_2}}, \quad z_U = \frac{(\hat{p}_1 - \hat{p}_2) - \Delta}{\sqrt{\hat{p}_1(1-\hat{p}_1)/n_1 + \hat{p}_2(1-\hat{p}_2)/n_2}}. \quad \text{Under } H_0, z_L, z_U \approx N(0,1)$$

Notation: $\hat{p}_i = x_i/n_i$ are sample proportions, Δ is equivalence margin on difference scale, separate variance estimates

Plain English: This test for proportions extends TOST to binary outcomes by testing whether the difference in success rates falls within pre-specified equivalence bounds. Two one-sided tests evaluate the lower and upper margins: one tests whether proportion difference is not too low, the other tests whether not too high. When both tests reject, the difference in proportions is established to lie within the equivalence region, demonstrating practical similarity in success rates between groups.

⚠ Practitioner Warning

Requires adequate sample sizes for normal approximation; exact methods exist for small samples. Equivalence margin Δ must be pre-specified and justified; typical values 0.10-0.20 depending on context. Test can be formulated on difference scale (additive), ratio scale (relative risk), or odds ratio scale; results may differ. Standard error uses separate variances (Wald approach); alternatives include pooled variance or score test approaches. Very unbalanced sample sizes reduce power. Confidence interval for difference provides equivalent inference: if 90% CI entirely within $[-\Delta, \Delta]$, equivalence shown for $\alpha = 0.05$. Continuity correction sometimes applied but generally not recommended for equivalence testing.

Hypotheses

- H_0: $|p_1 - p_2| \geq \Delta$, non-equivalence.
- H_1: $|p_1 - p_2| < \Delta$, equivalence.

Decision Rule: Reject H_0 if $z_L > Z_{1-\alpha}$ AND $z_U < -Z_{1-\alpha}$. Or, 90% CI for $(p_1 - p_2)$ entirely within $[-\Delta, \Delta]$ at $\alpha = 5\%$.

Software References [Python, R, SAS]

```
statsmodels.stats.proportion     TrialSize::TwoSampleProportion.NIS PROC FREQ; test equiv;
```

Additional Notes

Equivalence can be formulated on different scales: difference, ratio (relative risk), or odds ratio, each with context-dependent interpretation. Transformations (log, logit) improve normal approximation and respect parameter bounds. The Farrington–Manning score test improves Type I error control over Wald methods, while exact methods (e.g., Barnard) are preferred for small samples. Sample size planning depends on both proportions, margin, and power, with balanced designs maximizing efficiency. Non-inferiority tests only one margin. Stratified analyses use Mantel–Haenszel methods. For rare outcomes ($p < 0.05$), exact methods are preferred over asymptotic approximations.

Miscellanea: Equivalence testing for proportions developed alongside methods for continuous outcomes, with major applications in vaccine and oncology trials. A central issue is the choice of scale—risk difference, risk ratio, or odds ratio—since each leads to different interpretations. Regulatory guidance emphasizes selecting clinically meaningful margins, and analyses may consider multiple scales to ensure robustness. *Note:* Since H_0 is composite, distributional results are evaluated at the boundary of H_0.

Reference(s): Farrington & Manning (1990); Miettinen & Nurminen (1985)

Worked Example

Problem Statement Two formulations of a vaccine are tested in independent groups of $n_1 = n_2 = 100$ participants. Formulation A produces 83 seroconversions; Formulation B produces 85. Test whether the seroconversion rates are equivalent to within $\delta = 0.12$ at $\alpha = 0.05$. Let $\Delta p = p_1 - p_2$. $H_0: |\Delta p| \geq \delta$; $H_1: |\Delta p| < \delta$.

1. Observed Proportions and Difference

$$\hat{p}_1 = \frac{83}{100}, \quad \hat{p}_2 = \frac{85}{100}, \quad \hat{p}_1 - \hat{p}_2 = -0.020.$$

2. Standard Error (Wald, Unpooled)

$$\widehat{\mathrm{SE}} = \sqrt{\frac{\hat{p}_1(1-\hat{p}_1)}{n_1} + \frac{\hat{p}_2(1-\hat{p}_2)}{n_2}} = 0.0518.$$

3. Two One-Sided Test Statistics

Critical value: $z_{0.05} = 1.645$.

Lower test ($H_{01}: p_1 - p_2 \leq -0.12$):

$$z_1 = \frac{(\hat{p}_1 - \hat{p}_2) + \delta}{\widehat{\mathrm{SE}}} = \frac{-0.020 + 0.12}{0.0518} = 1.930.$$

Upper test ($H_{02}: p_1 - p_2 \geq +0.12$):

$$z_2 = \frac{(\hat{p}_1 - \hat{p}_2) - \delta}{\widehat{\mathrm{SE}}} = \frac{-0.020 - 0.12}{0.0518} = -2.701.$$

Practitioner Insight: Equivalence testing for two proportions follows the same TOST logic as for means: equivalence is declared when both one-sided z-tests reject, equivalently when the 90% CI for $p_1 - p_2$ lies entirely within $(-\delta, +\delta)$. The unpooled Wald standard error, $\widehat{\mathrm{SE}} = \sqrt{\hat{p}_1(1-\hat{p}_1)/n_1 + \hat{p}_2(1-\hat{p}_2)/n_2}$, is used because the hypotheses are evaluated at the equivalence boundaries rather than equality ($p_1 = p_2$); pooled SEs are appropriate under equality ($p_1 = p_2$). Here the lower bound is binding: the 90% CI lower limit -0.105 clears the margin -0.12 by only 0.015, reflecting that $\hat{p}_1 - \hat{p}_2 = -0.020$ lies closer to the lower boundary. The equivalence margin $\delta = 0.12$ must be specified *a priori* from clinical or regulatory considerations. A non-significant ordinary two-sided z-test ($p = 0.699$) does not establish equivalence because it does not define what counts as a negligible difference. For small samples or extreme proportions, exact or score-based methods are preferred over the Wald approximation. Related equivalence procedures also apply to log risk ratios and odds ratios, including the standard 80–125% bioequivalence limits on the log scale.

4. Decision via Test Statistics

Test	Stat	Critical	
H_{01}: $\Delta p \leq -0.12$	$z_1 = 1.930$	> 1.645	Reject
H_{02}: $\Delta p \geq +0.12$	$z_2 = -2.701$	< -1.645	Reject

$p_1 = 1 - \Phi(1.930) = 0.027$; $p_2 = \Phi(-2.701) = 0.004$; $p_{\mathrm{TOST}} = \max(0.027, 0.004) = 0.027$.

Both nulls rejected: the formulations are declared equivalent.

5. Equivalent CI Approach

$$90\%\ \mathrm{CI} = -0.020 \pm 1.645 \times 0.0518 = [-0.105, 0.065].$$

Since $[-0.105, 0.065] \subset (-0.12, 0.12)$, equivalence is confirmed. The lower bound -0.105 is 0.015 above the margin -0.12: the lower test is the binding constraint.

6. Comparison

Test	Statistic	p
TOST equivalence ($\delta = 0.12$)	$z_1 = 1.930$	0.027
Ordinary two-sided z	$z = -0.386$	0.699

The ordinary z-test of $H_0: p_1 = p_2$ gives $p = 0.699$: it fails to reject equality, but this is not evidence of equivalence.

Common Pitfalls

- **Pooling variances in equivalence tests.** Standard two-proportion tests pool variances under equal proportions. Equivalence testing instead uses separate variance estimates because the null allows differences between proportions.
- **Using relative instead of absolute margins.** Equivalence margins for proportions are usually defined as absolute probability differences. Relative margins typically require log-risk ratio or odds-ratio formulations.
- **Ignoring baseline proportions in sample-size planning.** Proportion variance depends on $p(1-p)$ and changes across the probability scale. Sample-size calculations should therefore use realistic baseline values.
- **Using normal approximations with sparse data.** Large-sample methods require adequate expected counts. Exact or score-based methods are preferable for small samples or extreme proportions.
- **Confusing equivalence with non-inferiority.** Equivalence requires the difference to remain within two bounds. Non-inferiority evaluates only a single lower bound.

Mann-Whitney Equivalence Test

Distribution-Free Equivalence Test Based on Rank Statistics

EQVL.6 | **Category:** Equivalence Testing | **Data:** Ordinal/Continuous | **Null:** Permutation/Bootstrap

When to Use

- Demonstrate equivalence when normality is questionable.
- Test equivalence for ordinal or skewed continuous data.
- Distribution-free alternative to parametric TOST.

Assumptions

- Independent random samples from two groups.
- Data at least ordinal (rankable).
- Equivalence margins defined on an appropriate scale.

Test Statistic

$$U = \sum_{i=1}^{n_1} \sum_{j=1}^{n_2} \left[\mathbf{1}(X_i < Y_j) + \tfrac{1}{2}\mathbf{1}(X_i = Y_j) \right], \quad P = \frac{U}{n_1 n_2} = P(X < Y) + \tfrac{1}{2}P(X = Y) \text{ (probability of superiority)}$$

Notation: X_i and Y_j are actual data, n_1 and n_2 are sample sizes, Δ_L, Δ_U are equivalence bounds on probability scale

Plain English: The Mann-Whitney equivalence test extends distribution-free rank-based testing to equivalence framework by examining whether the probability that a randomly selected observation from one group exceeds one from the other group falls within equivalence bounds. Rather than testing location shifts under normality, the test evaluates stochastic equivalence using ranks. Bootstrap or permutation methods construct confidence intervals for the probability parameter, and equivalence is established when this interval lies entirely within pre-specified margins.

⚠ Practitioner Warning

Equivalence margins must be specified on probability scale: 0.50 indicates identical distributions, values near 0.50 indicate similarity. Typical equivalence region might be [0.40, 0.60] but depends on context. No simple parametric critical values; requires computational methods (bootstrap, permutation). Interpretation differs from location parameter equivalence; evaluates stochastic equivalence through relative rankings rather than mean differences. Ties complicate inference; mid-rank or randomization methods needed. Sample size requirements larger than parametric tests for equivalent power. Does not require normality or equal variances, though interpretation is simplest when group distributions have similar shapes. Alternative formulations use Hodges-Lehmann estimator of location shift with confidence intervals.

Hypotheses

- H_0: $P(X < Y) + \frac{1}{2}P(X = Y) \notin [\Delta_L, \Delta_U]$.
- H_1: $P(X < Y) + \frac{1}{2}P(X = Y) \in [\Delta_L, \Delta_U]$.

Decision Rule: Reject H_0 if 90% CI for $P(X < Y) \subset [\Delta_L, \Delta_U]$. Bootstrap or permutation-based CI.

Software References [Python, R, SAS]

Python	R	SAS
`scipy.stats.mannwhitneyu`	`Not Native (Manual)`	`Not Native (Manual)`

Additional Notes

The Mann–Whitney statistic U counts pairs where $X < Y$, and $P = U/(n_1 n_2)$ estimates the probability a random Y exceeds X; under identical distributions, $P = 0.50$. Equivalence bounds are typically symmetric: $[0.50 - \Delta, 0.50 + \Delta]$.
Inference can use bootstrap (percentile or BCa) or permutation methods. The Hodges–Lehmann estimator ($\hat{\theta}_{HL} = \text{median}(X_i - Y_j)$ for $1 \le i \le m$, $1 \le j \le n$) provides an alternative location-shift formulation of equivalence. Brunner–Munzel extends to unequal variances. Ties use mid-ranks, and exact permutation tests are available for small samples. While less powerful than parametric methods under normality, the approach is robust and well-suited for ordinal data.

Miscellanea: Mann and Whitney (1947) introduced a rank-based, distribution-free alternative to the t-test. Extensions to equivalence testing developed later, allowing assessment of similarity without strong distributional assumptions. These approaches often use the probability of superiority $P(X < Y)$, though specifying meaningful equivalence margins on this scale can be less intuitive than for mean differences.

Reference(s): Mann & Whitney (1947); Wellek (2010); Hodges & Lehmann (1963)

Worked Example

Problem Statement Two independent groups each provide ten pain scores (0–40 scale). Test stochastic equivalence with margin $\Delta = 0.30$ at $\alpha = 0.05$: the two groups are equivalent if $|P(X_1 < X_2) - 0.5| < 0.30$, i.e. $\theta \in (0.20, 0.80)$.
$H_0: |\theta - 0.5| \geq 0.30$; $H_1: |\theta - 0.5| < 0.30$, where $\theta = P(X_1 < X_2) + \frac{1}{2}P(X_1 = X_2)$.

1. Combined and Within-Group Ranks
Rank all $N = 20$ obs. jointly (R_{ij}) and within each group separately (r_{ij}). $\bar{R}_1 = 10.7$; $\bar{R}_2 = 10.3$; $\bar{r}_1 = \bar{r}_2 = 5.5$.

Group 1			Group 2		
x	R	r	y	R	r
5	4	1	1	1	1
7	5	2	2	2	2
8	6	3	3	3	3
9	7	4	11	8	4
12	9	5	17	11	5
16	10	6	21	12	6
22	13	7	25	15	7
24	14	8	27	16	8
38	19	9	29	17	9
39	20	10	32	18	10

2. Estimate θ via the Mann–Whitney U
Count the number of pairs (x_i, y_j) with $x_i < y_j$:

Obs	x_i	less than k of the y's
x_1–x_4	5, 7, 8, 9	7 each
x_5–x_6	12, 16	6 each
x_7–x_8	22, 24	4 each
x_9–x_{10}	38, 39	0 each

No ties exist between groups ($T = 0$), so:
$U = 4\times7+2\times6+2\times4+2\times0 = 48$, $\hat{\theta} = U/(n_1 n_2) = 0.48$.

3. Brunner–Munzel Variance Estimators

$$S_j^2 = [\textstyle\sum_i (R_{ij} - r_{ij} - \bar{R}_j + \bar{r}_j)^2]/(n_j - 1).$$

$$S_1^2 = 7.733, \qquad S_2^2 = 12.622.$$

4. Standard Error and TOST Statistics

$$\widehat{\text{SE}} = \sqrt{(S_1^2/n_2 + S_2^2/n_1)/(n_1 n_2)} = 0.143.$$

$$z_1 = \frac{\hat{\theta} - (0.5 - \Delta)}{\widehat{\text{SE}}} = 1.962, \quad z_2 = \frac{\hat{\theta} - (0.5 + \Delta)}{\widehat{\text{SE}}} = -2.243.$$

5. Decision ($z_{0.05} = 1.645$)

Test	Stat	Critical	
H_{01}: $\theta \leq 0.20$	$z_1 = 1.962$	> 1.645	Reject
H_{02}: $\theta \geq 0.80$	$z_2 = -2.243$	< -1.645	Reject

$p_{\text{TOST}} = \max(0.025, 0.012) = 0.025$. 90% CI for θ: $0.480 \pm 1.645 \times 0.143 = [\mathbf{0.245}, \mathbf{0.715}] \subset (0.20, 0.80)$.

Stochastic equivalence established.

Practitioner Insight: The Mann–Whitney equivalence test evaluates the stochastic superiority parameter θ, with $\theta = 0.5$ representing stochastic equality and the null hypothesis asserting non-equivalence, $H_0: |\theta - 0.5| \geq \Delta$. The estimator $\hat{\theta} = U/(n_1 n_2)$ is based on the MW U statistic and requires no distributional shape assumptions. Standard errors use BM variance estimators S_j^2, which measure how strongly the groups intermingle in the combined ranking; here $S_2^2 > S_1^2$ because group 2 spans the ranking more widely. The lower boundary is binding: $\hat{\theta} = 0.480$ lies closer to 0.20 than 0.80, and the 90% CI lower bound 0.245 clears the margin by only 0.045. As in all equivalence settings, a non-significant ordinary MW test ($p = 0.889$ here) does not establish equivalence because it lacks a pre-specified equivalence margin. The nonparametric approach is preferable for ordinal data or unequal distributional shapes, whereas parametric TOST procedures are more powerful under pure location shifts.

Common Pitfalls

- **Mixing probability and location margins.** Equivalence margins may be defined on the probability scale $P(X_1 > X_2)$ or on a location measure such as the Hodges–Lehmann estimator. Pre-specified margin must match the inferential scale.
- **Treating Mann–Whitney as a median test.** The test evaluates stochastic ordering, not median equality in general. Median interpretation requires the location-shift assumption.
- **Ignoring tie corrections.** Ties change the variance of the Mann–Whitney statistic. Omitting tie adjustments can distort standard errors and confidence intervals.
- **Using nonparametric methods unnecessarily.** Nonparametric tests are robust but may lose power under near-normality. Parametric equivalence tests can provide more efficient inference when assumptions are reasonable.

Anderson–Hauck Test

Single-Statistic Equivalence Test via Noncentral t Approximation

EQVL.7 | **Category:** Equivalence Testing | **Data:** Continuous | **Null:** Noncentral t (approx. central t)

When to Use

- To test equivalence between two treatments or formulations.
- Alternative to TOST for continuous parametric outcomes.
- When a single equivalence p-value is preferred.
- Bioequivalence and comparative pharmacokinetic studies.
- Primarily of historical or methodological interest.

Assumptions

- Independent samples or paired crossover observations.
- Continuous outcome approximately normally distributed.
- Equivalence margin $[-\Delta, \Delta]$ specified a priori.
- Appropriate standard error estimate for the design used.
- Noncentral t approximation reasonably accurate.

Test Statistic

$$t_{AH} = \frac{\bar{x}_1 - \bar{x}_2}{SE}, \quad SE = s_p\sqrt{\tfrac{1}{n_1} + \tfrac{1}{n_2}}, \quad \lambda = \frac{\Delta}{SE}. \qquad p\text{-value} = P\Big(-\tau < t_\nu(\lambda) < \tau \,\Big|\, \lambda\Big), \quad \nu = n_1 + n_2 - 2.$$

Notation: t_{AH} is observed t-statistic, λ is noncentrality parameter at the boundary, τ is critical value at α

Plain English: The Anderson–Hauck test evaluates equivalence using a single t-statistic rather than two separate one-sided tests as in TOST. It asks whether the observed treatment difference is sufficiently close to zero relative to the equivalence boundary $\pm\Delta$. The method compares the observed statistic with the distribution expected when the true difference lies exactly at an equivalence limit, using a noncentral t calibration. Equivalence is declared when the observed statistic falls sufficiently near zero. Unlike TOST, which requires two separate hypothesis tests and has a direct confidence-interval interpretation, Anderson–Hauck produces a single p-value and decision.

⚠ Practitioner Warning

The Anderson–Hauck procedure is calibrated using the noncentral t distribution at the equivalence boundary; simple central t approximations may be inaccurate in some settings. Unlike TOST, the method does not correspond directly to checking whether a $(1 - 2\alpha)$ confidence interval lies within $[-\Delta, \Delta]$. Regulatory agencies generally require TOST for bioequivalence submissions, so it is rarely used in practice. Relative to TOST, it may have higher power near zero difference but lower power near the equivalence boundaries. Dedicated software implementations are uncommon.

Hypotheses

- H_0: Non-equivalence: $|\mu_1 - \mu_2| \geq \Delta$.
- H_1: Equivalence: $|\mu_1 - \mu_2| < \Delta$.

Decision Rule: Reject H_0 if $-\tau < t_{AH} < \tau$ where τ satisfies $P(-\tau < t_\nu(\lambda) < \tau) = \alpha$ with $\lambda = \Delta/SE$.

Software References [Python, R, SAS]

`Not Native (Manual)`	`Not Native (Manual)`	`Not Native (Manual)`

Additional Notes

The test begins with the noncentral t statistic $T^* = ((\bar{x}_1 - \bar{x}_2) - \mu)/SE$ where under H_0 the true difference μ lies at the boundary $\pm\Delta$. Since the exact noncentral distribution depends on the unknown σ, Anderson and Hauck (1983) proposed approximating T^* by a central t_ν plus the estimated noncentrality $\hat{\lambda} = \Delta/SE$, finding this approximation to be accurate in simulation. The rejection region is the symmetric interval $(-\tau, \tau)$ where τ is determined so that the test size is α at the boundary. Compared with TOST, the Anderson–Hauck method tends to have higher power when the true difference is very near zero but slightly lower power near the equivalence limits; neither procedure uniformly dominates the other. Extensions to paired and crossover designs replace SE with the appropriate within-subject standard error.

Miscellanea: Sharon Anderson and Walter Hauck introduced their equivalence-testing procedure in 1983 during the early development of bioequivalence methodology. Their method used equivalence hypotheses (H_0: non-equivalence, H_1: equivalence) rather than the traditional "no difference" framework. Although theoretically elegant, it was later eclipsed by Schuirmann's TOST procedure, whose direct confidence-interval interpretation proved more attractive to regulators and practitioners.

Reference(s): Anderson & Hauck (1983); Hauck & Anderson (1984); Schuirmann (1987); Berger & Hsu (1996)

Worked Example

Problem Statement Two analytical instruments measure the concentration (mg/L) of the same analyte on ten paired samples. Test whether the instruments are equivalent to within $\Delta = 5$ mg/L at $\alpha = 0.05$ using the Anderson–Hauck test. $H_0: |\mu_d| \geq \Delta$; $H_1: |\mu_d| < \Delta$. Under the central-t approximation, $t^* \sim t(n-1) = t(9)$ under H_0 at the boundary.

i	1	2	3	4	5	6	7	8	9	10
Instr. A	110.2	104.2	102.7	104.1	102.6	96.3	105.6	93.0	107.7	105.3
Instr. B	106.8	98.1	100.1	101.6	96.8	100.0	100.0	93.0	104.1	102.4
$d_i = A - B$	3.4	6.1	2.6	2.5	5.8	−3.7	5.6	0.0	3.6	2.9

1. Difference Statistics

$$\bar{d} = 2.880,\ s_d = \sqrt{\frac{\sum(d_i - \bar{d})^2}{n-1}} = 2.961.$$

$$\widehat{\text{SE}} = \frac{s_d}{\sqrt{n}} = 0.936.$$

2. Anderson–Hauck Test Statistic

The AH test statistic and the noncentrality parameter at the boundary:

$$t_{\text{AH}} = \frac{|\bar{d}|}{\widehat{\text{SE}}} = 3.076, \quad \lambda = \frac{\Delta}{\widehat{\text{SE}}} = 5.340.$$

Under the central-t approximation, shift t_{AH} by $-\lambda$ to obtain the approximate pivot: $t^* = t_{AH} - \lambda = -2.264$.

3. Decision

Reject H_0 if $t^* < -t_\alpha(9)$.
Critical value (one-sided, $\alpha = 0.05$): $t_{0.05}(9) = 1.833$.

α	$-t_{0.05}(9)$	Decision
5%	−1.833	**Reject** (−2.264 < −1.833)
1%	−2.821	FTR (−2.264 > −2.821)

$p_{\text{AH}} = F_t(-2.264,\ 9) = 0.025$.
Equivalence of the two instruments is established at $\alpha = 0.05$.

4. Comparison with TOST

The TOST binding statistic is: $t_2 = (\bar{d} - \Delta)/\widehat{\text{SE}} = -2.264$. The tests differ in motivation, construction, and power profile; numerical equivalence here is a consequence of the central-t approximation and symmetric bounds.

The 90% CI $d \pm t_{0.05}(9) \times SE$: $[1.164,\ 4.596]$.
Since CI $\subset (-5, 5)$, the TOST/CI criterion also confirms equivalence. This CI is the TOST confidence interval; AH does not have a direct CI interpretation.

Practitioner Insight: The Anderson–Hauck test begins with the natural t-statistic $t_{AH} = \bar{d}/SE = 3.076$ and asks: how probable is this observation under the null hypothesis that the true difference equals the boundary $\pm\Delta$? Because the null distribution is a noncentral t with noncentrality $\lambda = \Delta/SE = 5.34$, the observed $t_{AH} = 3.076$ lies well to the left of the boundary distribution's center, indicating the data are consistent with equivalence. Anderson and Hauck proposed approximating the noncentral t by a central t shifted by $\hat{\lambda}$, yielding the pivotal quantity $t^* = t_{AH} - \lambda = (\bar{d} - \Delta)/SE$, which is referred to $t(n-1)$ as a one-sided lower-tail test. Numerically $t^* = -2.264$ equals the TOST binding statistic, so both tests give $p = 0.025$ here. This numerical agreement under the central-t approximation and symmetric bounds is a known result (Berger & Hsu 1996), but the two procedures have different power profiles: AH has higher power when the true difference is near zero (low noncentrality), while TOST has slightly higher power at moderate effect sizes. The 90% CI $[1.16, 4.60]$ confirms equivalence via the TOST/CI criterion; AH itself does not have a direct CI interpretation; not accepted as a standalone FDA/EMA procedure.

Common Pitfalls

- **Assuming equivalence to TOST.** Anderson–Hauck and TOST are related but not identical procedures and may differ in power and interpretation.
- **Using central t approximations blindly.** The method is calibrated using the noncentral t distribution; simple central approximations may be inaccurate in some settings.
- **Interpreting the test through CI inclusion.** Unlike TOST, Anderson–Hauck does not correspond directly to checking whether a confidence interval lies within the equivalence region.
- **Using inappropriate equivalence margins.** The equivalence margin should reflect scientific or clinical relevance for the specific study design and outcome scale.
- **Choosing Δ after viewing the data.** Equivalence margins must be specified before analysis; post-hoc adjustment invalidates the inference.

Part XXII

Heavy-Tailed & Outlier-Resistant Inference

Testing using methods stable under non-normality and contamination

Robust and Resistant Statistics

XXII. ROBU — A Conceptual Primer

What Is Robust Statistics?

Robust methods remain reliable under moderate departures from classical assumptions—particularly contamination by outliers, heavy-tailed distributions, and heteroscedasticity. Classical estimators such as the sample mean and OLS are optimal under Gaussian assumptions but can be severely distorted by even a single extreme observation.

The core trade-off: robust methods accept a small efficiency loss under ideal (Gaussian) conditions in exchange for dramatically better behavior when assumptions are violated. Key sensitivity measures — breakdown point, influence function, and asymptotic relative efficiency (ARE) — are defined below.

Sensitivity Measures

Breakdown point ε^*: The smallest fraction of observations that can be replaced by arbitrary values to make the estimator unbounded. The sample mean has $\varepsilon^* = 1/n \approx 0$; the sample median has $\varepsilon^* = 0.5$, the maximum achievable for a location estimator. The trimmed mean with proportion α has $\varepsilon^* = \alpha$.

Influence function (IF): Measures the effect of an infinitesimal contamination at point x:

$$\mathrm{IF}(x;T,F) = \lim_{\epsilon\to 0}\frac{T((1-\epsilon)F+\epsilon\delta_x)-T(F)}{\epsilon}.$$

Here, T is the statistical functional (estimator), F is the underlying population distribution, and δ_x is a point mass placing all probability at x. Bounded IF $\Rightarrow$ robust to extreme observations; unbounded IF $\Rightarrow$ highly sensitive to outliers.

Gross error sensitivity $\gamma^* = \sup_x |\mathrm{IF}(x;T,F)|$: the worst-case effect of a single outlier. The sample mean has unbounded IF ($\gamma^* = \infty$); trimmed means (an L-estimator) and M-estimators have bounded IF, which is the key robustness property.

Asymptotic relative efficiency (ARE): Ratio of the asymptotic variance of the classical estimator to that of the robust estimator under the same model. ARE > 1 means the robust method is more efficient; ARE < 1 means a small cost is paid. The trimmed mean with $\alpha = 0.20$ has ARE ≈ 0.91 vs. the mean under normality — a modest 9% cost for substantial outlier protection.

Robust Location: Trimmed Means and Winsorization

Trimmed mean $\bar{x}_t$: Remove the $\lfloor g = \alpha n\rfloor$ smallest and g largest observations, then average the remaining $h = n - 2g$ values:

$$\bar{x}_t = \frac{1}{h}\sum_{i=g+1}^{n-g} x_{(i)}.$$

The trimming proportion $\alpha \in [0, 0.5)$; common choices are $\alpha = 0.10$ or 0.20. The trimmed mean has breakdown point α and is asymptotically normal with variance σ_t^2/h.

Winsorized mean and variance: Rather than discarding extremes, **Winsorization** replaces them with the nearest retained value:

$$x_{(i)}^W = \begin{cases} x_{(g+1)} & i \le g \\ x_{(i)} & g < i \le n-g \\ x_{(n-g)} & i > n-g \end{cases}$$

The Winsorized mean and variance are:

$$\bar{x}_W = \frac{1}{n}\sum_{i=1}^{n} x_{(i)}^W, \quad S_W^2 = \frac{\sum_{i=1}^n (x_{(i)}^W - \bar{x}_W)^2}{n-1}.$$

The variance of the trimmed mean is estimated as $\widehat{\mathrm{Var}}(\bar{x}_t) = (n-1)S_W^2/h(h-1)$, where $h = n - 2g$ is the effective sample size.

Trimming proportion choice: $\alpha = 0.20$ provides good protection against heavy tails with modest efficiency loss. $\alpha = 0$ recovers the sample mean; $\alpha \to 0.5$ recovers the median. Over-trimming discards valid data; pre-specify α before analysis.

I. Yuen's Test (One- and Two-Sample)

Yuen (1974) replaced the sample mean with the trimmed mean in Welch's t-test structure.

One-sample Yuen test: $H_0 : \mu_t = \mu_0$ (population trimmed mean):

$$T = \frac{\bar{x}_t - \mu_0}{\sqrt{(n-1)S_W^2/(h(h-1))}} \mathrel{\dot\sim} t_{h-1}.$$

Two-sample Yuen test (Yuen-Welch): $H_0 : \mu_{t1} = \mu_{t2}$:

$$T = \frac{\bar{x}_{t1} - \bar{x}_{t2}}{\sqrt{d_1 + d_2}}, \quad d_j = \frac{(n_j - 1)S_{Wj}^2}{h_j(h_j - 1)},$$

with Welch–Satterthwaite degrees of freedom:

$$\nu = \frac{(d_1 + d_2)^2}{d_1^2/(h_1 - 1) + d_2^2/(h_2 - 1)}.$$

Yuen's test is robust to non-normality and outliers while accounting for unequal variances—a robust alternative to both the pooled and Welch t-tests. It does not assume equal trimming proportions but equal α in both groups is conventional.

II. Trimmed Mean ANOVA and Winsorized Variance Test

Trimmed mean one-way ANOVA: Extends Yuen's two-sample test to $J \ge 2$ groups. Let $\bar{x}_{tj}, S_{Wj}^2, h_j, n_j$ denote the trimmed mean, Winsorized variance, effective and actual sample sizes for group j. Define $d_j = (n_j - 1)S_{Wj}^2/h_j(h_j - 1)$, $c_j = 1/d_j$, and $C = \sum_j c_j$. The test statistic is:

$$F_t = \frac{\sum_j c_j(\bar{x}_{tj} - \tilde{x})^2 / (J-1)}{1 + \frac{2(J-2)}{J^2-1} \sum_j \frac{(1-c_j/C)^2}{h_j - 1}},$$

where $\tilde{x} = \sum_j c_j \bar{x}_{tj}/C$ is the weighted grand trimmed mean. Under H_0, $F_t \stackrel{\cdot}{\sim} F_{k-1,\nu}$ with Welch–James denominator df ν (Johansen's approximation).

Winsorized variance test: Tests $H_0\colon \sigma^2_{W1} = \cdots = \sigma^2_{WJ}$ using the Winsorized variances S^2_{Wj} in place of raw sample variances. The Levene-type statistic is:

$$W_{\text{Rob}} = \frac{\sum_j n_j(\bar{z}_j - \bar{z})^2 / (J-1)}{\sum_j \sum_i (z_{ij} - \bar{z}_j)^2 / (N-J)},$$

where $z_{ij} = |x^W_{ij} - \bar{x}_{Wj}|$ are absolute Winsorized deviations; $\bar{z}_j$ and $\bar{z}$ are group and grand means of z_{ij}. Under H_0, $W_{\text{Rob}} \stackrel{\cdot}{\sim} F_{J-1,N-J}$. Often more stable than Bartlett or Levene under heavy tails.

III. Welch–James Test

The Welch–James (James 1951; Johansen 1980) procedure tests $H_0\colon \mu_1 = \cdots = \mu_J$ under heteroscedasticity *and* moderate non-normality. Define weights $w_j = n_j/s^2_j$, $W = \sum_j w_j$, and weighted grand mean $\bar{y}_w = \sum_j w_j \bar{y}_j / W$. The test statistic is:

$$F^* = \frac{\sum_j w_j(\bar{y}_j - \bar{y}_w)^2 / (J-1)}{1 + \frac{2(J-2)}{J^2-1} \sum_j \frac{(1-w_j/W)^2}{n_j - 1}},$$

which is compared to $F_{k-1,\nu}$ with Satterthwaite df:

$$\nu = \frac{J^2 - 1}{3 \sum_j \frac{(1-w_j/W)^2}{n_j - 1}}.$$

The correction term in the denominator provides better Type I error control than standard Welch ANOVA under skewed or heavy-tailed distributions, particularly for small unbalanced groups. With trimmed means and Winsorized variances in place of $\bar{y}_j$ and s^2_j, this becomes the trimmed mean ANOVA of Section II.

IV. Brunner–Munzel Test

A nonparametric test for the stochastic equality $H_0 : P(X_1 > X_2) = 0.5$, similar to the Mann–Whitney test but **valid under unequal distributions and unequal variances**—a limitation of the Mann–Whitney when distributions differ in shape.

The Brunner–Munzel statistic is based on the relative effect $\hat{p} = P(X_1 > X_2) + \frac{1}{2}P(X_1 = X_2)$ (estimated from ranks) with a variance estimate that does not assume equal distributions:

$$W_{BM} = \frac{\sqrt{n_1 n_2 / N}\,(\hat{p} - 0.5)}{\hat{\sigma}_{BM}} \stackrel{\cdot}{\sim} t_\nu,$$

where

$$\hat{\sigma}_{BM} = \sqrt{n_1 S^2_2 + n_2 S^2_1},$$

and

$$S^2_j = \frac{1}{n_j - 1} \sum_i \left(R_{ij} - r_{ij} - \bar{R}_j + \bar{r}_j\right)^2.$$

Here R_{ij} are pooled ranks, r_{ij} are within-group ranks, and $\bar{R}_j, \bar{r}_j$ are their group means. Degrees of freedom ν are estimated by a Welch–Satterthwaite-type formula. Unlike Mann–Whitney, the BM test maintains nominal Type I error when variances differ substantially—a common setting where Mann–Whitney over-rejects.

V. Percentile Bootstrap Test

The bootstrap is the natural inferential companion for robust estimators whose sampling distributions have no clean closed form — trimmed means, M-estimators, and Winsorized statistics all fall into this category. For any robust test statistic T (e.g. a trimmed mean difference):

1. Draw B bootstrap samples $\mathbf{x}^{*(b)}$ (with replacement).
2. Compute the robust statistic $T^{*(b)}$ for each resample.
3. Center the bootstrap distribution under H_0 by shifting: use $T^{*(b)} - \hat{T}$ and compare to the observed $\hat{T} - \mu_0$.
4. The p-value is the proportion of resamples at least as extreme as observed.

A $(1-\alpha)$ bootstrap **percentile CI** uses quantiles $[T^*_{(\alpha/2)}, T^*_{(1-\alpha/2)}]$ directly. See PERM.6–7 for bootstrap hypothesis tests and the four CI types (percentile, basic, normal, BCa) in a general resampling context. For robust estimators specifically, $B \geq 1999$ resamples and $n \geq 20$ are recommended; BCa CIs are preferred over percentile CIs when the bootstrap distribution is skewed.

VI. Huber's M-Estimator

M-estimators generalize the MLE by replacing the squared loss with a robust loss function ρ, minimizing $\sum_i \rho(y_i - \mu)$, or equivalently solving $\sum_i \psi(y_i - \mu) = 0$ where $\psi = \rho'$.

Huber's ψ:

$$\psi_k(u) = \begin{cases} u & |u| \leq k \\ k\,\text{sign}(u) & |u| > k \end{cases}$$

Behaves like the mean for small residuals and like the median for large residuals; the tuning constant k (typically $k = 1.345$ for 95% efficiency at Gaussian) controls the transition. The asymptotic variance of $\hat{\mu}_M$ is $\sigma^2 \int \psi^2\, dF / \left(\int \psi'\, dF\right)^2$ (sandwich form).

Huber's M-test: Tests $H_0 : \mu = \mu_0$ using the M-estimator as the test statistic with its asymptotic normal distribution. More efficient than the sign test and more robust than the

t-test; appropriate when the error distribution is symmetric but heavier-tailed than Gaussian.

VII. Theil–Sen Slope Estimator and Test

The Theil–Sen estimator is the **median of all pairwise slopes**:

$$\hat{\beta}_{\text{TS}} = \text{median}\left\{\frac{y_j - y_i}{x_j - x_i} : i < j,\ x_i \neq x_j\right\}.$$

- **Breakdown point** $\varepsilon^* \approx 0.29$; resistant to $\sim 29\%$ outlier contamination in y.
- **ARE** $= 3/\pi \approx 0.955$ vs. OLS under Gaussian errors; no efficiency loss for heavy-tailed errors.
- **Confidence interval:** based on the distribution of pairwise slopes (Kendall's τ inversion); no normality assumption required.
- **Intercept** estimated as $\hat{\alpha} = \text{median}(y_i - \hat{\beta}_{\text{TS}} x_i)$.
- Particularly popular in **trend detection** (climate, hydrology) where outlying observations are common.

Choosing Among Robust Methods

Setting	Recommended
1 or 2 group means, outliers	Yuen's test
k group means, outliers	Trimmed ANOVA
k groups, heavy tails + unequal var	Welch–James
2 groups, unequal σ^2 and shapes	Brunner–Munzel
Any test, distributional uncertainty	Percentile bootstrap
Location M-estimation	Huber's M-estimator
Simple regression, outliers in y	Theil–Sen

Common Pitfalls

- **Over-trimming:** removing too large an α fraction discards valid signal; pre-specify α and justify it.
- **Applying robust methods unnecessarily:** when data are truly Gaussian with no outliers, classical tests are slightly more efficient; use robust methods when there is diagnostic evidence of violation.
- **Interpreting trimmed-mean tests as median tests:** the trimmed mean is not the median (unless $\alpha = 0.5$); it is a different estimand.
- **Bootstrap with small n:** the bootstrap requires a representative sample; below $n \approx 20$ the resampling distribution is unreliable.
- **Confusing breakdown point and efficiency:** high breakdown point (e.g. median) does not imply high efficiency; Huber's M-estimator and trimmed mean balance both.
- **Post-hoc trimming:** choosing α after inspecting the data inflates Type I error; always pre-specify.

Bottom Line

Robust methods protect inference by limiting the influence of extreme observations. Trimmed means and Yuen's test handle outliers while preserving approximate normality of the test statistic; Brunner–Munzel handles unequal variances and distributions simultaneously; bootstrap provides distribution-free inference; Theil–Sen provides outlier-resistant regression. The efficiency cost relative to classical methods is small; the protection against assumption violations is substantial.

Yuen's Test (One-Sample)

Robust Test for Trimmed Mean Based on One Sample

ROBU.1 | **Category:** Robust Methods | **Data:** Continuous | **Null:** t-distribution (approximate)

When to Use

- Test population trimmed mean against hypothesized value with outlier robustness.
- Data contain extreme observations or mild contamination.
- Alternative to one-sample t-test with outliers or heavy tails.
- Mean-based inference is desired with less outlier sensitivity.

Assumptions

- Independent random sample from continuous distribution.
- Distribution approximately symmetric after trimming.
- Trimming proportion γ is pre-specified.
- Sample size adequate after trimming ($n(1-2\gamma) \geq 20$ recommended).

Test Statistic

$$t_Y = \frac{\bar{y}_t - \mu_0}{\sqrt{(n-1)S_W^2/(h(h-1))}} \approx t_{h-1} \text{ (under } H_0), \quad \text{where } \bar{y}_t = \frac{1}{h}\sum_{i=g+1}^{n-g} y_{(i)}, \; h = n-2g, \; g = \lfloor n\gamma \rfloor, \gamma = \text{trim \%}.$$

Notation: $\bar{y}_t$ is trimmed mean, $y_{(i)}$ are ordered values, h is eff. sample size, S_W^2 is Winsorized variance, t for trimmed

Plain English: Yuen's one-sample test evaluates whether a population trimmed mean equals a hypothesized value by removing a specified percentage of extreme observations from each tail before computing the mean. The test uses Winsorized variance—replacing trimmed values with the most extreme remaining values—to estimate standard error. This provides robustness to outliers while maintaining better power than median-based tests. The trimmed mean balances resistance to extreme values with efficiency for the bulk of the distribution.

⚠ Practitioner Warning

Trimming proportion γ must be chosen a priori (commonly 10%, 20%, or 25%); higher trimming increases robustness but reduces efficiency. Assumes approximate symmetry after trimming. Winsorized variance may underestimate variability for heavy tails. Effective sample size is reduced ($df = n - 2g - 1$), limiting use in small samples. The test targets the trimmed mean, not the usual population mean. Bootstrap methods offer distribution-free alternatives.

Hypotheses

- H_0: $\mu_t = \mu_0$ (μ_t is the population trimmed mean)
- H_1: $\mu_t \neq \mu_0$

Decision Rule: Reject H_0 if $|t_Y| > t_{\alpha/2,h-1}$ or p-value $< \alpha$. Use t-distribution with $h-1$ degrees of freedom.

Software References [Python, R, SAS]

Python	R	SAS
`scipy.stats.trim_mean + manual`	`WRS2::yuend (adapted)`	`Not Native (Manual)`

Additional Notes

Yuen's test provides robust inference for the trimmed mean under departures from normality while retaining good power. Heavier tails generally favor more trimming, often around 20–25%, whereas lighter tails may require less, around 10–15%. A confidence interval is $\bar{y}_t \pm t_{\alpha/2,h-1}\sqrt{(n-1)S_W^2/[h(h-1)]}$, using the Winsorized variance S_W^2 and effective sample size $h = n - 2g$. The trimmed mean is a robust L-estimator (linear combination of order statistics), not a Huber M-estimator (minimizer of Huber's piecewise quadratic-linear loss); it removes fixed tail proportions rather than smoothly downweighting residuals. Bootstrap percentile intervals may also be obtained by resampling and recomputing the trimmed mean. Compared with median tests, trimmed-mean tests often have higher power under contaminated normal models and approach ordinary t-test efficiency under clean normal data.

Miscellanea: Yuen's (1974) Biometrika paper introduced a two-sample trimmed-mean test with rigorous sampling theory. Building on earlier descriptive use of trimming, it emerged within the 1970s robust statistics movement (Tukey, Huber, Hampel). By balancing efficiency and robustness, the method became widely used in psychology and the social sciences, where outliers and heavy-tailed data are common.

Reference(s): Yuen (1974); Wilcox (2017); https://www.medcalc.org/en/manual/note-trimmedmean.php

Worked Example

Problem Statement Twelve ($n = 12$) participants complete a cognitive task; response times (ms) are recorded. Two participants were distracted (unusually fast) and two had technical issues (unusually slow), producing a contaminated sample. Test $H_0: \mu_t = 400$ ms at $\alpha = 0.05$ using Yuen's test with trimming proportion $\gamma = 0.20$. $H_0: \mu_t = 400$; $H_1: \mu_t \neq 400$.

1. Trimming Parameters

$$g = \lfloor \gamma n \rfloor = \lfloor 0.20 \times 12 \rfloor = 2, \qquad h = n - 2g = 8.$$

Trim the $g = 2$ smallest and $g = 2$ largest observations.

2. Trimmed Mean

Retain positions 3–10: $\bar{y}_t = 416.250$.

3. Winsorized Sample

Replace the $g = 2$ lower and upper trimmed values with $y_{(3)} = 398$ and $y_{(10)} = 423$, respectively:

i	y_i	w_i	$w_i - \bar{w}$	$(w_i - \bar{w})^2$
1	152	398	−16.333	266.78
2	262	398	−16.333	266.78
3	398	398	−16.333	266.78
4	410	410	−4.333	18.78
5	416	416	1.667	2.78
6	419	419	4.667	21.78
7	420	420	5.667	32.11
8	421	421	6.667	44.44
9	423	423	8.667	75.11
10	423	423	8.667	75.11
11	667	423	8.667	75.11
12	826	423	8.667	75.11

$$\bar{w} = 414.333, \quad S_W^2 = \frac{\sum (w_i - \bar{w})^2}{n \;\; 1} = 110.97.$$

4. Standard Error of Trimmed Mean

$$\widehat{\text{SE}} = \sqrt{(n-1)S_W^2/(h(h-1))} = \sqrt{1220.67/(8 \times 7)} = 4.669.$$

5. Yuen Test Statistic

$$t = (\bar{y}_t - \mu_0)/\widehat{\text{SE}} = (416.250 - 400)/4.669 = 3.481.$$

Under H_0: $t \sim t(h-1) = t(7)$.

6. Decision

α	$t_{\alpha/2}(7)$	Decision
5%	2.365	**Reject** (3.481 > 2.365)
1%	3.500	FTR (3.481 < 3.500)

$p = 0.010 \Rightarrow$ reject H_0. By contrast, the usual t-test gives $\bar{y} = 436.4$, $SE = 49.2$, $t = 0.740$, $p = 0.475$—a failure to reject due to four outliers inflating the sample variance 8x relative to S_W^2.

Practitioner Insight: Yuen's test is a robust one-sample location test that preserves the interpretability of the t-test while reducing the influence of outliers on the variance estimate. Its key mechanism is the Winsorized variance S_W^2: extreme observations such as 152, 262, 667, and 826 are replaced by adjacent order statistics ($y_{(3)} = 398$, $y_{(10)} = 423$), reducing variability from the ordinary $SE = 49.2$ to $SE = 4.669$. The estimand is the γ-trimmed population mean, so the null $\mu_t = 400$ refers specifically to this trimmed location parameter. The approximate $t(h-1)$ distribution requires only mild assumptions on the untrimmed tails and remains valid under heavy-tailed or skewed distributions where the ordinary t-test may be poorly sized. A common default is $\gamma = 0.20$; smaller values retain more efficiency under near-normality, while larger values increase robustness. Unlike the Wilcoxon signed-rank test, Yuen's test uses the magnitudes of the retained observations and yields a directly interpretable effect size ($\bar{y}_t - \mu_0 = 16.25$ ms); compared with the sign test, it typically has greater power when the central distribution is approximately symmetric.

Common Pitfalls

- **Choosing the trim proportion after inspecting the data.** The trimming level γ must be pre-specified; selecting γ to exclude observed outliers inflates Type I error. Choose γ a priori (commonly 0.20) and apply it mechanically.
- **Using the original sample df instead of trimmed df.** Critical values must use the effective sample size after trimming. With $h = n(1 - 2\gamma)$, the correct degrees of freedom are $df = h - 1$, not $n - 1$.
- **Confusing the trimmed mean with the median or truncated mean.** Yuen's test evaluates a *trimmed mean* (fixed proportion removed from each tail). This differs from the median and from truncation rules based on value thresholds.
- **Computing variance from the trimmed sample instead of Winsorizing.** Yuen's statistic requires the *Winsorized variance*; simply computing the variance of the trimmed data underestimates variability.
- **Using Yuen's test for median inference.** Yuen's test targets a robust mean under approximate symmetry. For heavily skewed distributions where interest centers on the median, use sign or Wilcoxon tests instead.

Yuen's Test (Two-Sample)

Robust Comparison of Trimmed Means Between Independent Groups

ROBU.2 | **Category:** Robust Methods | **Data:** Continuous | **Null:** t-distribution (Welch-type)

When to Use

- Compare two independent groups in the presence of outliers or heavy-tailed distributions, with possible contamination and unequal variances.
- Robust alternative to Welch's two-sample t-test.
- Mean-based inference is desired with less outlier sensitivity.

Assumptions

- Independent samples from two continuous distributions.
- Distributions approximately symmetric after trimming in each group.
- Trimming percentage γ specified (same for both groups).
- Eff. sample sizes adequate after trimming ($h_1, h_2 \geq 20$).

Test Statistic

$$t_Y = \frac{\bar{y}_{t1} - \bar{y}_{t2}}{\sqrt{d_1 + d_2}} \approx t_{df} \text{ (under } H_0), \quad \widehat{\text{SE}} = \sqrt{d_{t1} + d_{t2}}, \quad df = \frac{(d_1 + d_2)^2}{d_1^2/(h_1 - 1) + d_2^2/(h_2 - 1)} \text{(Welch-Satterthwaite)}.$$

Notation: $\bar{x}_{ti}$ = trimmed means, d_i = variances of $\bar{y}_{ti}$, $h_i = n_i - 2g_i$ (eff. sample sizes)

Plain English: Yuen's two-sample test compares trimmed means between independent groups by removing specified percentages of extreme observations from each tail of both samples. The test uses separate Winsorized variance estimates for each group, avoiding the equal variance assumption, and applies Welch-Satterthwaite df adjustment. This provides simultaneous robustness to outliers and heteroscedasticity, combining advantages of trimmed means with Welch's approach to unequal variances.

⚠ Practitioner Warning

Trimming percentage γ must be pre-specified; typical values 10%, 20%, or 25%. Same trimming percentage used for both groups unless strong rationale for different amounts. Assumes approximate symmetry in central portions of distributions; asymmetry affects validity. Effective sample sizes reduced after trimming: $h_i = n_i - 2\lfloor n_i\gamma \rfloor$. Very unequal sample sizes combined with different variances reduce power. Tests difference in trimmed means, not population means; interpretation differs. Degrees of freedom can be non-integer; software handles automatically. Bootstrap methods provide distribution-free alternative.

Hypotheses

- H_0: $\mu_{t1} = \mu_{t2}$ (population trimmed means are equal).
- H_1: $\mu_{t1} \neq \mu_{t2}$ (population trimmed means are different).

Decision Rule: Reject H_0 if $|t_Y| > t_{\alpha/2,df}$ or p-value $< \alpha$. Use Welch-Satterthwaite degrees of freedom.

Software References [Python, R, SAS]

Python	R	SAS
`scipy.stats.trim_mean + manual`	`WRS2::yuen`	`Not Native (Manual)`

Additional Notes

Yuen and Dixon (1973) developed the two-sample version, extending Yuen's (1974) one-sample approach. Confidence interval: $(\bar{y}_{t1} - \bar{y}_{t2}) \pm t_{\alpha/2,df} \cdot \widehat{\text{SE}}$ estimates difference in trimmed means. Power properties: more powerful than Wilcoxon-Mann-Whitney for symmetric contaminated distributions, nearly as efficient as t-test under normality. Trimming reduces influence of outliers but maintains higher breakdown point than M-estimators. Bootstrap percentile-t method: resample each group separately, compute test statistic, approximate null distribution empirically. The test combines Yuen's trimmed mean framework with Welch's heteroscedastic approach, providing dual robustness.

Miscellanea: Yuen's trimmed-mean tests were introduced in two 1974 papers: the two-sample version (appeared first) in *Biometrika* and the one-sample version in *Communications in Statistics*. Her work provided rigorous inference for trimmed means, extending earlier descriptive use and offering a balance between robustness and efficiency. The methods became widely used in psychology and the social sciences, and were later popularized through applied work by Rand Wilcox and implementations such as the WRS2 R package.

Reference(s): Yuen & Dixon (1973); Yuen (1974); Wilcox (2017); https://www.medcalc.org/en/manual/note-trimmedmean.php

Worked Example

Problem Statement Ten participants in a control group and ten in a treatment group complete a cognitive task; response times (ms) are recorded. Both groups contain two anomalously fast and two anomalously slow observations. Test $H_0: \mu_{t1} = \mu_{t2}$; $H_1: \mu_{t1} \neq \mu_{t2}$ at $\alpha = 0.05$ using Yuen's two-sample test with $\gamma = 0.20$.

Group 1	97	141	366	368	376	386	394	399	752	906
Group 2	79	80	337	343	347	357	363	364	680	764

1. Trimming Parameters
Both groups: $n_j = 10$, $\gamma = 0.20$, $g_j = \lfloor 0.20 \times 10 \rfloor = 2$, $h_j = 10 - 4 = 6$.

2. Trimmed Means
Drop the 2 smallest and 2 largest from each group.

$$\bar{y}_{t1} = 2289/6 = 381.500, \quad \bar{y}_{t2} = 2111/6 = 351.833.$$

3. Winsorized Samples and Variances
Group 1: replace bottom 2 with $y_{(3)} = 366$, top 2 with $y_{(8)} = 399$.

$$\bar{w}_1 = 3819/10 = 381.9, \quad S^2_{W1} = 2026.9/9 = 225.211.$$

Group 2: replace bottom 2 with $y_{(3)} = 337$, top 2 with $y_{(8)} = 364$.

$$\bar{w}_2 = 3513/10 = 351.3, \quad S^2_{W2} = 1354.1/9 = 150.456.$$

4. Variances of Trimmed Means

$$d_j = \frac{(n_j - 1)S^2_{Wj}}{h_j(h_j - 1)}, \quad d_1 = 67.563, \quad d_2 = 45.137.$$

5. Standard Error and Test Statistic

$$\widehat{\text{SE}} = \sqrt{d_1 + d_2} = 10.616.$$

$$t = \frac{\bar{y}_{t1} - \bar{y}_{t2}}{\widehat{\text{SE}}} = \frac{381.500 - 351.833}{10.616} = 2.794.$$

6. Welch–Satterthwaite Degrees of Freedom

$$\nu = \frac{(d_1 + d_2)^2}{d_1^2/(h_1 - 1) + d_2^2/(h_2 - 1)} = 9.62.$$

7. Decision
Under H_0: $t \sim t(\nu) = t(9.62)$.

α	$t_{\alpha/2}(9.62)$	Decision
5%	2.240	**Reject** (2.794 > 2.240)
1%	3.192	FTR (2.794 < 3.192)

$p = 0.020$. Ordinary Welch t-test: $\bar{y}_1 = 418.5$, $\bar{y}_2 = 371.4$, $t = 0.456$, $p = 0.654$—fails to reject because four outliers per group inflate the pooled variance.

Practitioner Insight: Yuen's two-sample test extends the one-sample procedure to two independent groups by trimming and Winsorizing within each group and combining the resulting scaled variance terms d_j in a Welch-style standard error. The Welch–Satterthwaite df are applied to the Winsorized variances rather than the ordinary sample variances, allowing unequal post-Winsorization variability across groups. Here the four outliers per group inflate the ordinary Welch standard error to $\widehat{\text{SE}}_{\text{Welch}} = 103.3$, whereas Yuen's procedure reduces it to $\widehat{\text{SE}}_{\text{Yuen}} = 10.6$, changing a non-significant $p = 0.654$ into a significant $p = 0.020$. The estimand is the difference in γ-trimmed population means, $\mu_{t1} - \mu_{t2} = 29.7$ ms, not the difference in ordinary means. Unlike the Mann–Whitney–Wilcoxon test, Yuen's test yields a directly interpretable effect size in the original units. The test assumes approximate symmetry in the trimmed portion but not normality and remains valid under heteroskedasticity because the d_j terms are computed separately. For $J > 2$ groups the natural extension is the Yuen-Welch one-way ANOVA test, analogous to Welch's ANOVA but with trimmed means and Winsorized variances.

Common Pitfalls

- **Pooling Winsorized variances.** Yuen's two-sample test uses a Welch-type formulation with *separate* Winsorized variances. Pooling variances destroys robustness to heteroscedasticity and defeats the purpose of the test.
- **Trimming the same number of observations in both groups.** Trimming must be proportional (γ), not a fixed count. Compute $g_i = \lfloor n_i \gamma \rfloor$ separately for each group so both samples are trimmed at the same proportion.
- **Using incorrect degrees of freedom.** The test requires Welch–Satterthwaite df based on the *trimmed* sample sizes $h_i = n_i(1 - 2\gamma)$, not the original n_i or pooled $n_1 + n_2 - 2$.
- **Applying the independent-sample test to paired data.** When observations are paired (before–after or matched designs), analyze pairwise differences and apply a one-sample trimmed test instead of treating samples as independent.
- **Using the test for strongly skewed distributions.** Yuen's test targets a trimmed mean under approximate symmetry. For highly skewed data, consider transformations or median/quantile-based tests instead.

Trimmed Mean ANOVA

Robust One-Way Analysis of Variance for Trimmed Means

ROBU.3 | **Category:** Robust Methods | **Data:** Continuous | **Null:** F-distribution (approximate)

When to Use

- Compare multiple group locations under outliers, heavy tails, unequal variances, or moderate non-normality.
- Mean-based inference is desired with less outlier influence.
- Robust option for one-way ANOVA under contamination.

Assumptions

- Independent groups with independent interval-scale data.
- Group distributions are approximately symmetric.
- Trimming proportion is specified *a priori*.
- Effective sample sizes remain adequate after trimming.

Test Statistic

$$F_t = \frac{\sum c_j(\bar{x}_{tj} - \bar{x}_{t.})^2/(J-1)}{1 + \frac{2(J-2)}{J^2-1}\Lambda} \approx F \text{ (under } H_0). \quad \text{Here } c_j = \frac{1}{d_j}, \; C = \textstyle\sum_j c_j, \quad \bar{x}_{t.} = \frac{\sum_j c_j \bar{x}_{tj}}{C}, \; \Lambda = \sum \frac{(1-c_j/C)^2}{h_j - 1}.$$

Notation: $\bar{x}_{tj}$ = group trimmed means, $h_j = n_j - 2g_j$ = effective sample sizes, d_j = variances of trimmed means

Plain English: Trimmed mean ANOVA extends one-way ANOVA to robust framework by computing trimmed means within each group after removing specified percentages of extreme values. The test uses Winsorized variances allowing heteroscedasticity across groups, similar to Welch's approach for two groups extended to multiple groups. Between-group variation in trimmed means is compared to within-group Winsorized variation, providing robustness to outliers while maintaining reasonable power for central portions of distributions.

⚠ Practitioner Warning

Trimming percentage γ must be pre-specified; typical values 10%, 20%, or 25%. Same percentage used across all groups unless strong justification for differences. Assumes approximate symmetry in central portions within groups; violations affect validity. Effective sample sizes reduced: $h_j = n_j - 2\lfloor n_j\gamma \rfloor$ for each group. Very unequal sample sizes combined with heteroscedasticity reduce power. Tests equality of trimmed means, not population means. Critical values and degrees of freedom complex; software implementation essential.

Hypotheses

- H_0: All population trimmed means (μ_{ti}) are equal.
- H_1: At least one μ_{ti} differs from others.

Decision Rule: Reject H_0 if $F_t > F_{\alpha,df_1,df_2}$ or p-value $< \alpha$. Use $df_1 = J - 1$, $df_2 = (J^2 - 1)/(3\Lambda)$.

Software References [Python, R, SAS]

Python	R	SAS
`Not Native (Manual)`	`WRS2::t1way`	`Not Native (Manual)`

Additional Notes

The Welch–James trimmed ANOVA uses numerator $df_1 = J - 1$ and approximate denominator $df_2 = (J^2 - 1)/(3\Lambda)$, where $c_j = 1/d_j$ are inverse-variance weights and $d_j = (n_j - 1)S^2_{Wj}/[h_j(h_j - 1)]$ estimates the variance of the trimmed mean. Here $h_j = n_j - 2g_j$ is the effective sample size after trimming and S^2_{Wj} is the Winsorized variance. Studied extensively by Wilcox, trimmed-mean ANOVA often has substantially higher power than Kruskal–Wallis under contamination while retaining good efficiency near classical ANOVA under normality. Post-hoc comparisons typically use Yuen's two-sample test with multiplicity adjustments (Bonferroni, Holm, etc.) or bootstrap procedures that resample within groups. Standard trimming removes equal proportions from both tails, though adaptive schemes exist. Trimmed means are robust L-estimators (linear combination of order statistics) based on order statistics and are especially useful for heavy-tailed or contaminated data; under extreme asymmetry, median-based methods may sometimes be preferable.

Miscellanea: Trimmed-mean ANOVA developed as part of the robust statistics literature, extending earlier two-sample methods (e.g., Yuen, 1974) to multiple groups. The approach provides resistance to outliers while maintaining good efficiency, offering an alternative to classical ANOVA under non-normality or unequal variances. Wilcox's extensive applied work popularized these methods in practice.

Reference(s): Yuen & Dixon (1973); Wilcox (1995, 2017)

Worked Example

Problem Statement Thirty participants are assigned equally to three ($J = 3$) experimental conditions (A, B, C); response times (ms) are recorded. Each group of ten ($n_j = 10$) contains two ($g_j = 2$) anomalously fast and two anomalously slow observations. Test $H_0: \mu_{tA} = \mu_{tB} = \mu_{tC}$ at $\alpha = 0.05$ using the trimmed-mean one-way ANOVA with $\gamma = 0.20$.

A	119	148	338	351	363	365	368	372	864	938
B	138	159	348	367	375	381	382	385	726	805
C	102	150	386	387	389	396	399	414	755	772

1. Group Statistics ($n_j = 10$, $g_j = 2$, $h_j = 6$)

Group j	$\bar{y}_{tj}$	$\bar{w}_j$	S^2_{Wj}	d_j
A	359.500	357.700	223.344	67.003
B	373.000	370.400	269.378	80.813
C	395.167	397.100	155.433	46.630

2. Weights and Weighted Grand Trimmed Mean

$c_j = 1/d_j$: $c_A = 0.01493$, $c_B = 0.01237$, $c_C = 0.02145$.

$$C = c_A + c_B + c_C = 0.04875.$$

$$\bar{y}_{tw} = (c_A\bar{y}_{tA} + c_B\bar{y}_{tB} + c_C\bar{y}_{tC})/C = 378.619.$$

$$\Lambda = \sum_j \frac{(1 - c_j/C)^2}{h_j - 1} = 0.2704,$$

3. Welch F-Statistic

$$\text{Num} = \frac{\sum_j c_j(\bar{y}_{tj} - \bar{y}_{tw})^2}{J - 1} = \frac{11.719}{2} = 5.859.$$

$$\text{Den} = 1 + \frac{2(J-2)}{J^2 - 1}\Lambda = 1.068.$$

$$F = \frac{\text{Num}}{\text{Den}} = 5.488, \quad \nu_e = \frac{J^2 - 1}{3\Lambda} = 9.86.$$

4. Decision

Under H_0: $F \sim F(J - 1, \nu_e) = F(2, 9.86)$.

α	$F_\alpha(2, 9.86)$	Decision
5%	4.122	**Reject** (5.488 > 4.122)
1%	7.615	FTR (5.488 < 7.615)

$p = 0.025 \Rightarrow$ reject H_0.

Ordinary one-way ANOVA: $F(2, 27) = 0.012$, $p = 0.988$—completely fails to detect the difference.

Practitioner Insight: The trimmed-mean one-way ANOVA (Welch 1951; Yuen 1974; Wilcox 2012) extends Yuen's two-sample test to J groups. The ordinary ANOVA fails here ($p = 0.988$) because four outliers per group inflate the within-group variances and obscure the true trimmed-mean differences (359.5, 373.0, 395.2 ms). After trimming, the Winsorized variances ($S^2_{WA} = 223$, $S^2_{WB} = 269$, $S^2_{WC} = 155$) are moderate and similar across groups, so the weights c_j are also similar; the weighted grand mean $\bar{y}_{tw} = 378.6$ is close to the simple average of the three trimmed means (375.9). The denominator correction Den = 1.068 accounts for residual heteroscedasticity in the weights; the effective degrees of freedom $\nu_e = 9.86$ are substantially below the ordinary $n - J = 27$. For $J = 2$, the omnibus statistic is equivalent to Yuen's two-sample t^2 test. Post-hoc pairwise comparisons use Yuen's two-sample test on each pair with a Bonferroni or other multiplicity correction.

Notation: $\bar{y}_{tj}$ = Trimmed means, $\bar{w}_j$ = Winsorized means, S^2_{Wj} = Winsorized variances, d_j = Variances (est.) of the trimmed means.

Common Pitfalls

- **Using classical ANOVA degrees of freedom.** Trimmed ANOVA relies on Welch-type degrees of freedom that account for unequal variances across groups. Using the classical denominator df ($N - J$) assumes homoscedasticity and produces incorrect inference.
- **Pooling Winsorized variances across groups.** The method is designed to handle heteroscedasticity by using group-specific Winsorized variances. Pooling variances removes the robustness advantage and can distort test results.
- **Ignoring multiplicity in post-hoc comparisons.** A significant omnibus result only indicates that some group differences exist. Pairwise trimmed comparisons require appropriate multiple-comparison adjustments.
- **Discarding data to force equal sample sizes.** Trimmed ANOVA naturally accommodates unequal group sizes through weighting. Artificially balancing groups wastes data and reduces statistical power.
- **Using the test with dissimilar group shapes.** The procedure assumes roughly similar distributional shapes across groups. Substantial differences in skewness or shape can make trimmed means difficult to interpret.

Winsorized Variance Test

Robust Test for Homogeneity of Variance Using Winsorized Estimates

ROBU.4 | **Category:** Robust Methods | **Data:** Continuous | **Null:** Chi-square (approximate)

When to Use

- Compare variances robustly in the presence of outliers.
- Assess scale differences under heavy-tailed distributions.
- Robust alternative to Bartlett's variance test.
- Variance inference is needed without discarding data.
- Prelim. scale assessment before robust mean comparisons.

Assumptions

- Independent groups with at least interval-scale observations that are independent.
- Trimming/Winsorization proportion is specified *a priori*.
- Approximate symmetry after Winsorization.
- Adequate sample sizes after Winsorization ($n_j \gtrsim 20$).

Test Statistic

$$F_w = S^2_{w,max} \; / \; S^2_{w,min}, \qquad \text{largest to smallest ratio is used by default}$$

Notation: $S^2_{w,*}$ is Winsorized variance. For 2+ groups: various approaches (Brown-Forsythe style, or ANOVA on abs. deviations from Winsorized median)

Plain English: The Winsorized variance test evaluates homogeneity of variances by replacing extreme values in each tail with the most extreme remaining values (Winsorization), computing variances from Winsorized samples, then testing whether these robust variance estimates differ across groups. Log-transformation of variances stabilizes the test statistic, and a correction factor improves chi-square approximation. This provides resistance to outliers that can severely distort traditional variance homogeneity tests. The Winsorized variance test is a robust analogue of classical variance-ratio tests.

⚠ Practitioner Warning

Winsorization percentage γ must be pre-specified; typical values 10%, 20%, or 25%. Same percentage used across all groups. Assumes approximate symmetry within groups; asymmetry affects Winsorized variance interpretation. Effective sample sizes reduced after Winsorization: $h_j = n_j - 2\lfloor n_j\gamma \rfloor$. Very small samples may have inadequate effective size. Tests equality of Winsorized variances, not population variances; interpretation differs. Less powerful than Levene's or Brown-Forsythe under normality but more robust to outliers. Bootstrap methods provide distribution-free alternative. Rejection indicates heteroscedasticity requiring robust procedures (Welch-type tests, separate variance models).

Hypotheses

- H_0: Winsorized variances (σ^2_{Wi}) equal across groups.
- H_1: At least one Winsorized variance differs from others.

Decision Rule: Reject H_0 if $F_w > F_{\alpha/2,h_1-1,h_2-1}$ or p-value $< \alpha$. Here $h_i = n_i(1-2\gamma)$ are eff. sample sizes.

Software References [Python, R, SAS]

Python	R	SAS
`Not Native (Manual)`	`WRS2::winvar`	`Not Native (Manual)`

Additional Notes

Winsorization replaces the lowest and highest $g_j = \lfloor n_j\gamma \rfloor$ observations in group j with the nearest retained values before computing the Winsorized variance S^2_{Wj}. Unlike trimming, which removes observations entirely, Winsorization preserves the sample size while limiting the influence of extreme values. The resulting variance ratio $F_W = S^2_{W1}/S^2_{W2}$ is compared with an approximate $F(h_1-1, h_2-1)$ distribution, where $h_j = n_j - 2g_j$ are the effective trimmed sample sizes. The method is conceptually analogous to the classical variance-ratio F-test but is substantially more robust to outliers and heavy tails. Alternatives include Levene, Brown–Forsythe, O'Brien (transforms data, then applies ANOVA to compare variances), and bootstrap-based procedures, which differ in robustness and sensitivity to departures from normality.

Miscellanea: Winsorization, named after Charles P. Winsor, replaces extreme values with less extreme ones to reduce outlier influence while retaining all observations. The approach became associated with the development of robust statistics, offering a practical alternative to trimming by limiting, rather than discarding, extreme data points.

Reference(s): Winsor (1946); Wilcox (2017)

Worked Example

Problem Statement Two independent groups of twelve participants ($n_i = 12$) each complete a reaction-time task (ms). Both groups contain two anomalously fast and two anomalously slow observations. Test $H_0: \sigma^2_{W1} = \sigma^2_{W2}$ (equal Winsorized variances) vs $H_1: \sigma^2_{W1} \neq \sigma^2_{W2}$ at $\alpha = 0.05$ using $\gamma = 0.20$.

Group 1	100	132	335	352	356	397	412	421	430	434	841	898
Group 2	104	146	385	390	392	393	396	413	418	425	813	864

1. Winsorization

$n = 12, \gamma = 0.20, g = \lfloor 0.20 \times 12 \rfloor = 2, h = n - 2g = 8.$

Group 1: replace bottom 2 with $y_{(3)} = 335$, top 2 with $y_{(10)} = 434$:

$$\bar{w}_1 = \frac{4675}{12} = 389.583,$$

$$S^2_{W1} = \frac{\sum(w_{1i} - \bar{w}_1)^2}{n_1 - 1} = \frac{20574.92}{11} = 1870.45.$$

Group 2: replace bottom 2 with $y_{(3)} = 385$, top 2 with $y_{(10)} = 425$:

$$\bar{w}_2 = \frac{4832}{12} = 402.667,$$

$$S^2_{W2} = \frac{\sum(w_{2i} - \bar{w}_2)^2}{n_2 - 1} = \frac{3186.67}{11} = 289.70.$$

2. Test Statistic

$$F_W = \frac{S^2_{W1}}{S^2_{W2}} = \frac{1870.45}{289.70} = 6.457.$$

Under H_0: $F_W \approx F(h-1, h-1) = F(7,7)$.

3. Decision

α	$F_{\alpha/2}(7,7)$	Decision
5%	4.995	**Reject** (6.457 > 4.995)
1%	8.885	FTR (6.457 < 8.885)

$p = 0.025$. Ordinary F-test: $s_1^2 = 55229$, $s_2^2 = 48181$, $F = 1.146$, $p = 0.825$—cannot detect the difference in spread because the outliers inflate both group variances equally.

Practitioner Insight: The Winsorized variance test compares group spreads using the Winsorized variances S^2_{Wj} and evaluates the ratio $F_W = S^2_{W1}/S^2_{W2}$ against an $F(h_1 - 1, h_2 - 1)$ distribution, where $h_j = n_j - 2g_j$ are the effective trimmed sample sizes. Here both groups contain structurally similar outliers, so the ordinary sample variances are inflated by nearly equal amounts ($s_1^2 = 55229$ vs $s_2^2 = 48181$, ratio 1.15), masking the true difference in core spread. After Winsorization, the central observations reveal much greater variability in Group 1 ($S^2_{W1} = 1870$) than Group 2 ($S^2_{W2} = 290$), giving a ratio of 6.46. Winsorized variances estimate a robust scale parameter and are far less sensitive to tail contamination than ordinary variances. Because trimming reduces the effective sample size ($h_j - 1 = 7$ instead of 11), the test sacrifices some power under exact normality but performs substantially better under heavy-tailed or contaminated distributions. The procedure is the variance analog of Yuen's trimmed-mean test; for more than two groups, robust Levene-type tests (e.g., Brown–Forsythe using trimmed means or medians) are generally preferred because they provide more stable and interpretable inference on group dispersion.

Transformation used in O'Brien's test followed by ANOVA: $r_{ij} = [(n_j - 1.5)n_j(x_{ij} - \bar{x}_j)^2 - 0.5s_j^2(n_j - 1)] \,/\, [(n_j - 1)(n_j - 2)]$.

Common Pitfalls

- **Using the test to decide between pooled and Welch-type methods.** Winsorized variance tests are descriptive diagnostics, not gatekeepers for choosing analysis methods. Robust procedures such as Yuen–Welch remain valid whether variances are equal or not.
- **Equating equal Winsorized variances with equal raw variances.** Equality of s^2_w reflects similar *typical* variability after downweighting extremes; raw variances may still differ if groups contain different outlier structures.
- **Using incorrect degrees of freedom.** Degrees of freedom must reflect the effective trimmed sample size $h = n(1 - 2\gamma)$, giving $df = h - 1$, not $n - 1$.
- **Applying on strongly skewed data.** Winsorization assumes approximate symmetry; severe skewness can distort interpretation of the variance comparison. Consider transformation or quantile-based variability measures instead.
- **Testing Winsorized variances but analyzing with raw variances.** If variability is assessed using Winsorized statistics, subsequent inference should also rely on Winsorized quantities to maintain a consistent robust framework.

Welch-James Test

Heteroscedastic One-Way ANOVA Without Equal Variance Assumption

ROBU.5 | **Category:** Robust Methods | **Data:** Continuous | **Null:** F-distribution (approximate)

When to Use

- Compare multiple group means under unequal variances.
- Robust alternative to one-way ANOVA when group variances and sample sizes are unequal.
- Mean-based inference is desired despite variance heterogeneity and moderate departures from normality.

Assumptions

- Groups are independent and possibly unbalanced.
- Observations within groups are independent.
- Response variable is approximately continuous.
- Group distributions are approximately normal.
- Variances may differ across groups.

Test Statistic

$$F_{WJ} = \frac{\sum_{j=1}^{J} w_j(\bar{x}_j - \bar{x}_w)^2/(J-1)}{1 + \frac{2(J-2)}{J^2-1}\sum_{j=1}^{J}\frac{(1-w_j/W)^2}{n_j-1}}, \quad w_j = \frac{n_j}{s_j^2}, \quad W = \sum_{j=1}^{J} w_j, \quad \bar{x}_w = \frac{\sum w_j\bar{x}_j}{W}. \quad \text{Under } H_0,\ F_{WJ} \approx F \text{ dist.}$$

Notation: w_j are precision weights (inv. variance), $\bar{x}_w$ is weighted grand mean, J is number of groups, n_j are sample sizes

Plain English: The Welch-James test extends Welch's approach for unequal variances to multiple groups by using precision weights (inverse variances) to weight group means. Groups with smaller variances receive more weight in computing the grand mean and test statistic. The denominator adjusts for heteroscedasticity using a correction factor based on the weighted structure. This provides valid inference when variances differ across groups without requiring transformation or pooling.

⚠ Practitioner Warning

Welch–James is reasonably robust to mild non-normality, but severe skewness or heavy tails—especially with small, unequal group sizes—can affect accuracy. Unequal sample sizes combined with highly disparate variances may substantially reduce power and destabilize variance estimates. The denominator degrees of freedom use a complex approximation based on group variances and sample sizes. Although slightly less powerful than standard ANOVA when variances are truly equal, Welch–James is generally preferred under heteroskedasticity. Post-hoc comparisons should use Welch-type pairwise procedures with multiplicity correction. Bootstrap methods provide useful alternatives when distributional assumptions are doubtful.

Hypotheses

- H_0: All population means equal ($\mu_1 = \mu_2 = \cdots = \mu_J$).
- H_1: At least one population mean differs from others.

Decision Rule: Reject H_0 if $F_{WJ} > F_{\alpha,df_1,df_2}$ or p-value $< \alpha$. $df_1 = J-1$; df_2 computed via Welch-James approx.

Software References [Python, R, SAS]

Python	R	SAS
`Not Native (Manual)`	`onewaytests::welch.test`	`Not Native (Manual)`

Additional Notes

Welch–James test uses precision weighting, giving more influence to groups with smaller variances. A correction factor and approximate degrees of freedom, $df_2 = (J^2-1)/[3\sum_j(1-w_j/W)^2/(n_j-1)]$, adjust for heteroscedasticity. For two groups, it reduces to Welch's t-test. Welch–James maintains proper Type I error under unequal variances, unlike standard ANOVA, whose behavior depends on the variance–sample size relationship. Alternatives include Brown–Forsythe (median-based) for robustness to non-normality. Post-hoc comparisons use Games–Howell, and trimmed-mean extensions combine this framework with Yuen's methods for added robustness.

Miscellanea: Bernard Welch (1951) and G.S. James (1951) independently developed extensions of ANOVA for unequal variances, building on Welch's earlier two-sample work. Their methods use weighted means and approximate F distributions to provide valid inference under heteroscedasticity. Although computationally demanding at the time, modern computing has made these approaches more accessible, and they are now widely recommended when variance equality is questionable.

Reference(s): Welch (1951); James (1951); Brown & Forsythe (1974)

Worked Example

Problem Statement Participants from three ($J = 3$) independent groups (A, B, C) complete a cognitive task; scores are recorded. The groups have unequal sizes and markedly unequal spreads. Apply the Welch–James test at $\alpha = 0.05$ to test for differences in group means without assuming equal variances. H_0: $\mu_A = \mu_B = \mu_C$; H_1: at least one mean differs.

A ($n = 4$)	48.2	49.0	51.9	55.6								
B ($n = 8$)	37.5	55.3	58.1	60.9	61.3	64.4	69.6	74.2				
C ($n = 12$)	30.8	41.9	47.2	61.2	61.5	62.4	72.7	76.3	82.7	87.1	94.8	112.6

1. Group Means and Variances

	n_j	$\bar{y}_j$	s_j^2
A	4	51.175	11.229
B	8	60.163	121.057
C	12	69.267	543.415

Group C has variance $543/11 \approx 48$ times Group A's, with $n_C = 12$ observations; its large $\text{SS}_{\text{within}}$ dominates the pooled error in the ordinary ANOVA.

2. Welch Weights

$w_j = n_j/s_j^2$: $w_A = 0.3562$, $w_B = 0.0661$, $w_C = 0.0221$.

$$W = w_A + w_B + w_C = 0.4444.$$

3. Weighted Grand Mean

$$\bar{y}_w = \frac{w_A\bar{y}_A + w_B\bar{y}_B + w_C\bar{y}_C}{W} = 53.411.$$

Group A receives 80.2% of the total weight; the weighted grand mean of 53.4 is pulled far towards Group A's mean of 51.2, well below the simple average of 60.2.

4. Welch–James Statistic

Heteroscedasticity correction:

$$A = \sum_j \frac{(1 - w_j/W)^2}{n_j - 1} = 0.199.$$

$$\text{Num} = \frac{\sum_j w_j(\bar{y}_j - \bar{y}_w)^2}{J-1} = 5.172, \quad \text{Den} = 1+\frac{2(J-2)}{J^2-1}A = 1.05.$$

$$F_{WJ} = \frac{5.172}{1.050} = 4.928, \quad \nu_e = \frac{J^2-1}{3A} = 13.42.$$

5. Decision

Under H_0: $F_{WJ} \sim F(J-1, \nu_e) = F(2, 13.42)$.

α	$F_\alpha(2, 13.42)$	Decision
5%	3.776	**Reject** (4.928 > 3.776)
1%	6.619	FTR (4.928 < 6.619)

$p = 0.025$. Ordinary one-way ANOVA: $F(2, 21) = 1.675$, $p = 0.212$—fails to reject because Group C's large variance (with $n = 12$ observations) inflates SS_W to 6858.7, swamping the between-group signal of 1093.9.

Practitioner Insight: The Welch–James test (Welch 1951; James 1951) extends Welch's two-sample t-test to J groups by weighting each group according to the precision of its mean, $w_j = n_j/s_j^2$. Ordinary ANOVA fails here because Group C ($s_C^2 = 543$) contributes 87% of the pooled within-group SS, so the pooled error is dominated by the noisiest group and masks genuine mean differences. Welch–James instead gives Group C only 5% of the total weight while Group A—the smallest but most precise group—receives 80%, making the test more sensitive to location differences under heteroscedasticity. The denominator correction moderates the statistic for residual heteroscedasticity, while the reduced effective degrees of freedom ($\nu_e = 13.42 < n - J = 21$) reflects the penalty for estimating separate variances. Unlike trimmed-mean ANOVA, which uses trimmed means and Winsorized variances for contaminated or heavy-tailed data, Welch–James uses ordinary means and variances and is intended for unequal variances without severe outliers. For $J = 2$, the statistic reduces to Welch's two-sample t^2 with Welch–Satterthwaite degrees of freedom.

Note: When the trimming proportion $\gamma = 0$, the trimmed-mean ANOVA reduces to the classical Welch–James heteroskedastic ANOVA.

Common Pitfalls

- **Omitting the second-order correction.** The Welch–James requires a correction for uncertainty in variance estimation. Dropping this adjustment produces a liberal test, especially with small samples or strong heteroscedasticity.
- **Using the asymptotic χ^2 in small samples.** The χ^2_{J-1} approximation works only when effective sample sizes are large. With small or unequal group sizes, bootstrap or F-type approximations provide more reliable inference.
- **Assuming higher power than trimmed ANOVA.** WJ prioritizes Type I error control under heteroscedasticity; its correction can make it slightly conservative when σ_i^2 are similar. Its advantage is validity, not guaranteed higher power.
- **Applying the test when groups differ in shape.** Like other trimmed-mean methods, Welch–James assumes approximate symmetry within groups. Severe skewness or differing distributional shapes across groups can undermine interpretation.

Brunner-Munzel Test

Generalized Rank-Based Test for Stochastic Dominance

ROBU.6 | **Category:** Robust Methods | **Data:** Ordinal/Continuous | **Null:** t-distribution (approximate)

When to Use

- Compare two independent groups under unequal variances or shapes. Robust alternative to Mann–Whitney.
- Test stochastic dominance rather than mean differences.
- Small to moderate samples with non-normal distributions.

Assumptions

- Independent observations from two groups with at least ordinal-scale measurements.
- Response variable permits meaningful ranking.
- Population distributions may differ in variance or shape.

Test Statistic

$$t_{BM} = \frac{\sqrt{n_1 n_2/(n_1+n_2)}\,(\bar{R}_1 - \bar{R}_2)}{\sqrt{n_1 S_2^2 + n_2 S_1^2}} \approx t_{df} \text{ (under } H_0\text{)}. \quad \hat{p} = \frac{\bar{R}_1 - \bar{R}_2}{n_1 + n_2} + \frac{1}{2}, \; S_j^2 = \frac{1}{n_j - 1} \sum_i (R_{ij} - r_{ij} - \bar{R}_j + \bar{r}_j)^2.$$

Notation: $\bar{R}_j$ =mean pooled rank in group j; S_j^2 =within-group rank variance; n_j =sample sizes; df from Welch formula.

Plain English: The Brunner-Munzel test evaluates whether observations from one group tend to be larger than those from another by estimating the probability that a randomly selected observation from the first group exceeds one from the second. Unlike Mann-Whitney which assumes equal distributions under the null hypothesis, Brunner-Munzel tests whether this probability equals 0.5 without assuming identical shapes or variances. The test uses separate variance estimates for each group and Welch-type degrees of freedom, providing robustness to heteroscedasticity.

⚠ Practitioner Warning

Does not test for location shift specifically; tests general stochastic superiority. Requires different interpretation than Mann-Whitney—tests $P(X > Y) = 0.5$ not identical distributions. Ties handled via mid-rank assignment but reduce power. Very small sample sizes ($n_i < 10$) make variance estimates unstable; exact methods preferred. Degrees of freedom can be non-integer; software handles automatically. More conservative than Mann-Whitney when equal variance holds but appropriate when violated. Does not assume equal shapes; can detect differences in any distributional aspect. Permutation tests provide alternative avoiding asymptotic approximation.

Hypotheses

- H_0: $P(X > Y) = 0.5$ (no stochastic dominance).
- H_1: $P(X > Y) \neq 0.5$ (stochastic dominance).

Decision Rule: Reject H_0 if $|t_{BM}| > t_{\alpha/2,df}$ or p-value $< \alpha$. Positive $t_{BM} \Rightarrow$ group 1 tends higher and vice versa.

Software References [Python, R, SAS]

```
scipy.stats.brunnermunzel        brunnermunzel::brunnermunzel.test  Not Native (Manual)
```

Additional Notes

Probability of superiority: $p = P(X > Y) + 0.5P(X = Y)$; $p = 0.5$ indicates no stochastic dominance, $p > 0.5$ favors group 1. Variance is estimated using ranks from the combined sample, with group-specific rank variances. Inference uses Welch–Satterthwaite degrees of freedom applied to these variances. Brunner and Munzel (2000) showed the test maintains correct Type I error under heteroscedasticity, unlike Mann–Whitney. It is particularly useful when groups differ in spread or shape; for pure location shifts with equal variances, Mann–Whitney is more efficient. The estimate $\hat{p}$ provides an interpretable effect size. Refer to Neubert and Brunner (2007) for extensions.

Miscellanea: Brunner and Munzel (2000) developed a nonparametric test addressing limitations of the Mann–Whitney test under unequal variances or distribution shapes. It tests stochastic dominance—whether one group tends to yield larger values—without requiring identical distributions. Often viewed as a nonparametric analog of Welch's t-test, it provides more reliable inference under heteroscedasticity.

Reference(s): Brunner & Munzel (2000); Neubert & Brunner (2007)

Worked Example

Problem Statement A control group ($n_1 = 9$) and a treatment group ($n_2 = 7$) each produce scores on a performance measure. The groups may differ in shape as well as location. Apply the Brunner–Munzel test at $\alpha = 0.05$ to test stochastic equality using the relative effect size. $H_0: P(X_1 < X_2) + \frac{1}{2}P(X_1 = X_2) = \frac{1}{2}$; $H_1: P(X_1 < X_2) + \frac{1}{2}P(X_1 = X_2) \neq \frac{1}{2}$.

1. Ranks
Rank all $N = 16$ observations jointly (combined ranks R_{ij}) and within each group separately (within-group ranks r_{ij}).

Group 1				Group 2			
x	R_{i1}	r_{i1}	$R_{i1} - r_{i1}$	y	R_{i2}	r_{i2}	$R_{i2} - r_{i2}$
1	1	1	0	11	6	1	5
2	2	2	0	22	10	2	8
6	3	3	0	25	12	3	9
8	4	4	0	27	13	4	9
9	5	5	0	28	14	5	9
12	7	6	1	29	15	6	9
14	8	7	1	34	16	7	9
19	9	8	1				
23	11	9	2				

The column $R_{ij} - r_{ij}$ counts how many observations from the other group lie below each observation.

2. Mean Ranks

$$\bar{R}_1 = 5.556, \quad \bar{R}_2 = 12.286; \quad \bar{r}_1 = 5, \quad \bar{r}_2 = 4.$$

3. Within-Group Variance Estimators

$$S_j^2 = \frac{1}{n_j - 1} \sum_i (R_{ij} - r_{ij} - \bar{R}_j + \bar{r}_j)^2.$$

$$S_1^2 = \frac{5(0.309) + 3(0.197) + 2.086}{8} = 0.528.$$

$$S_2^2 = \frac{10.796 + 0.082 + 4 \times 0.510}{6} = 2.238.$$

4. Test Statistic (Altered form using $\hat{p}$ also exists)

$$t_{BM} = \frac{\sqrt{n_1 n_2/(n_1 + n_2)}\,(\bar{R}_1 - \bar{R}_2)}{\sqrt{n_1 S_2^2 + n_2 S_1^2}} = -2.735.$$

5. Degrees of Freedom

$$\nu = \frac{(n_1 S_2^2 + n_2 S_1^2)^2}{(n_1 S_2^2)^2/(n_2 - 1) + (n_2 S_1^2)^2/(n_1 - 1)} = 8.20.$$

6. Decision
Under H_0: $t_{BM} \sim t(\nu) = t(8.20)$.

α	$t_{\alpha/2}(8.20)$	Decision
5%	2.296	**Reject** (2.735 > 2.296)
1%	3.332	FTR (2.735 < 3.332)

$p = 0.025$. Estimated relative effect: $\hat{p} = (\bar{R}_2 - \bar{r}_2) / n_1 = (12.286 - 4)/9 = 0.921$, i.e. there is an estimated 92% probability that a Group 2 score exceeds a Group 1 score.

Practitioner Insight: The Brunner–Munzel test tests stochastic equality without requiring the two groups to have the same distributional shape, unlike the Mann–Whitney–Wilcoxon test. The key quantities are the within-group variance estimators S_j^2, which measure how strongly observations from the two groups intersperse in the combined ranking. Here Group 2 has much larger rank variability ($S_2^2 = 2.24$ vs $S_1^2 = 0.53$), indicating greater interspersion with Group 1. The degrees of freedom $\nu = 8.20$ arise from a Welch–Satterthwaite approx. applied to S_j^2, making the BM test more conservative than MW when group shapes are similar (here $p = 0.025$ versus 0.003), but properly sized when shapes differ. The natural effect size is the relative effect $\hat{p} = 0.921$, which estimates $P(X_1 < X_2) + \frac{1}{2}P(X_1 = X_2)$, the probability that a randomly chosen Group 2 score exceeds a randomly chosen Group 1 score, with ties receiving half credit. Unlike the Hodges–Lehmann estimator, $\hat{p}$ remains valid under unequal σ_i^2's, skewness, or multimodal distributions.

Common Pitfalls

- **Interpreting $\hat{p}$ as a median or location difference.** The BM parameter $\hat{p}$ is the probability that an observation from one group exceeds an observation from the other. It is a relative effect measure, not a difference in medians or means.
- **Using Mann–Whitney when group shapes differ.** The MW test assumes a location shift with similar distributional shapes. When variances or shapes differ substantially, the Brunner–Munzel test provides a more reliable alternative.
- **Applying the test to paired or dependent samples.** The Brunner–Munzel test assumes independent observations between groups. Paired or repeated-measures data require methods designed for dependent samples.
- **Reporting only the p-value without the effect size.** The relative effect $\hat{p}$ provides a natural and interpretable measure of treatment difference. Reporting $\hat{p}$ with a CI $[\hat{p} \pm t_{\alpha/2,\nu} \cdot \text{se}(\hat{p})]$ conveys both magnitude and uncertainty.

Percentile Bootstrap Test

Distribution-Free Hypothesis Test via Resampling

ROBU.7 | **Category:** Robust Methods | **Data:** Any | **Null:** Empirical (bootstrap)

When to Use

- Classical parametric approximations may be unreliable.
- Sampling distribution is analytically difficult to derive.
- Construct robust confidence intervals for medians, trimmed means, or other complex statistics without strong distributional assumptions (e.g: unknown or non-normal data).

Assumptions

- Sample is representative of the population.
- Resampling scheme matches the data structure.
- Adequate sample size for resampling ($n \gtrsim 20$).
- Statistic is well-defined and computable on resamples.
- Sufficient bootstrap replications ($B \gtrsim 1000$).

Test Statistic

$$\hat{\theta}^* = g(\mathbf{X}^*). \quad \text{Construct empirical distribution of } \hat{\theta}^*.$$

Notation: $\hat{\theta}^*$ is bootstrap statistic, $\mathbf{X}^*$ is bootstrap sample, B is replications, percentile CI=$[\hat{\theta}^*_{\alpha/2}, \hat{\theta}^*_{1-\alpha/2}]$

Plain English: The percentile bootstrap generates an empirical sampling distribution by repeatedly resampling with replacement from the observed data and recomputing the statistic of interest. The resulting bootstrap distribution approximates the true sampling distribution without parametric assumptions. Hypothesis tests are performed by checking whether the null value lies within the bootstrap percentile CI, providing distribution-free inference for virtually any computable statistic.

⚠ Practitioner Warning

Requires sufficient bootstrap replications; too few ($B < 1000$) yields unstable percentiles. Cannot correct biased or unrepresentative samples. Small original samples ($n < 20$) may not capture population variability adequately. Percentile method can be biased with skewed distributions; bias-corrected and accelerated (BCa) bootstrap improves coverage. Computationally intensive for complex statistics or large datasets. Does not work well for discrete statistics or extreme quantiles. Assumes independence; clustered or dependent data require block bootstrap.

Hypotheses

- H_0: $\theta = \theta_0$ (parameter equals null value).
- H_1: $\theta \neq \theta_0$ (parameter differs from null value).

Decision Rule: Reject H_0 if θ_0 falls outside $(1-\alpha)$ percentile bootstrap CI: $[\hat{\theta}^*_{\alpha/2}, \hat{\theta}^*_{1-\alpha/2}]$.

Software References [Python, R, SAS]

Python	R	SAS
`scipy.stats.bootstrap`	`boot::boot`	`Not Native (Manual)`

Additional Notes

Bootstrap procedure: (1) Draw n observations with replacement from original sample $\mathbf{X} = \{x_1, \ldots, x_n\}$ to create bootstrap sample $\mathbf{X}^*_b$. (2) Compute statistic $\hat{\theta}^*_b = g(\mathbf{X}^*_b)$. (3) Repeat B times to obtain $\{\hat{\theta}^*_1, \ldots, \hat{\theta}^*_B\}$. (4) Order bootstrap statistics and extract percentiles. Percentile confidence interval: $(1-\alpha)$ CI is $[Q_{\alpha/2}, Q_{1-\alpha/2}]$ where Q_p is p-th quantile of $\{\hat{\theta}^*_b\}$. Bias-corrected and accelerated (BCa) bootstrap: adjusts for bias and skewness, improving coverage accuracy. Studentized bootstrap: compute $t^*_b = (\hat{\theta}^*_b - \hat{\theta})/SE^*_b$ where SE^*_b estimated from nested bootstrap; generally more accurate but computationally intensive. Bootstrap-t hybrid: use bootstrap to estimate standard error, construct t-statistic, compare to bootstrap-t distribution. Hypothesis testing: two-tailed test rejects if null value outside CI; one-tailed tests use appropriate tail. Bootstrap particularly valuable for median, trimmed mean, correlation coefficients, regression coefficients under heteroscedasticity, and complex model parameters. The Davison–Hinkley plus-one correction estimates bootstrap tail probabilities as $(r+1)/(B+1)$, avoiding zero p-values and reducing small-sample bias.

Miscellanea: Bradley Efron introduced the bootstrap in 1979, revolutionizing statistical inference with distribution-free alternative to parametric methods. The name "bootstrap" references pulling oneself up by bootstraps—generating sampling distribution from sample itself. Its applicability to virtually any computable statistic—without deriving theoretical distribution—transformed applied statistics.

Reference(s): Efron (1979); Efron & Tibshirani (1993)

Worked Example

Problem Statement Twelve ($n = 12$) participants complete a processing-speed task; scores are right-skewed and normality cannot be assumed. Use the percentile bootstrap with $B = 1{,}000$ resamples and $\gamma = 0.20$ trimming to test $H_0: \mu_t = 400$ at $\alpha = 0.05$. $H_0: \mu_t = 400$; $H_1: \mu_t \neq 400$. Use Davison–Hinkley plus-one correction.

i	1	2	3	4	5	6	7	8	9	10	11	12
y_i	395	398	412	417	445	450	461	471	493	533	549	559

1. Observed Trimmed Mean

$g = \lfloor n\gamma \rfloor = 2$, $h = n - 2g = 8$. Drop the two smallest (395, 398) and two largest (549, 559). Compute $\hat{\theta} = \bar{y}_t = 460.25$.

2. Bootstrap Procedure

Draw $B = 1{,}000$ bootstrap samples of size $n = 12$ with replacement and compute the trimmed mean $\bar{y}_t^*$ (dropping the two smallest and largest values) for each. Eight representative values are

481.50, 445.12, 513.88, 461.62, 466.62, 453.12, 456.00, and 479.75.

All exceed 400; across all resamples, $\bar{y}_t^*$ ranges from 416.75 to 527.50, with none at or below $\mu_t = 400$.

3. Bootstrap Distribution and CI

After sorting all $B = 1{,}000$ bootstrap trimmed means $\bar{y}^*_{t(1)} \leq \cdots \leq \bar{y}^*_{t(1000)}$:

Percentile	Rank	Value
2.5th	25	424.75
50th	500	460.50
97.5th	975	504.62

The 95% percentile bootstrap CI is:

$$[\hat{\theta}^*_{(0.025)}, \hat{\theta}^*_{(0.975)}] = [424.75,\ 504.62].$$

4. Decision

$$\mu_0 = 400 \notin [424.75,\ 504.62].$$

Reject H_0 at $\alpha = 0.05$: the bootstrap CI lies entirely above $\mu_t = 400$. As no bootstrap resample yielded $\bar{y}_t^* \leq 400$, the two-sided bootstrap p-value is $< 2/1001 = 0.002$.

5. Comparison

Method	Statistic	p
Ordinary t	$t(11) = 3.93$	0.002
Yuen t	$t(7) = 2.79$	<0.027
Bootstrap	CI= [424.8, 504.6]	<0.002

All three methods reject; the bootstrap CI requires no distributional assumption.

Practitioner Insight: The percentile bootstrap (Efron 1979; Efron & Tibshirani 1993) combined with the trimmed mean provides two layers of robustness. Trimming removes the influence of the skewed upper tail by discarding the two largest observations (549, 559), so inference targets the population trimmed mean μ_t rather than the ordinary mean μ. The bootstrap then estimates the sampling distribution directly from the empirical distribution of $\bar{y}_t^*$, avoiding the approximate $t(h-1)$ pivot used in Yuen's test. Here Yuen's test gives $t(7) = 2.79$, $p = 0.027$, with $\widehat{\text{SE}} = 21.57$ and a 95% CI of [409.2, 511.3] (width 102.0), whereas the bootstrap CI is narrower ([424.8, 504.6], width 79.9) because the asymmetric resample distribution captures the skewed sampling behavior more efficiently than the symmetric t approximation. Unlike pivot-based methods, the same bootstrap algorithm applies broadly to medians, correlations, regression coefficients, and other statistics without requiring new distributional derivations. A limitation is that percentile intervals can undercover in small skewed samples; the BC$_a$ bootstrap improves this. Typical choices: $B = 1{,}000$ for CI and larger odd values such as 4,999 or 9,999 for more accurate two-sided p-values.

Common Pitfalls

- **Resampling without centering under H_0.** For hypothesis testing, bootstrap samples must reflect H_0 condition. Data should be centered or adjusted so the bootstrap resampling represents the null distribution being tested. Refer to PERM.
- **Using too few bootstrap resamples.** Reliable inference requires a sufficiently large number of bootstrap replications. Values such as $B \geq 1000$ are typically needed to obtain stable p-values and interval estimates.
- **Applying the standard bootstrap to dependent data.** The basic bootstrap assumes independent observations. Time-series, spatial, or clustered data require specialized methods such as block or cluster bootstrap procedures.
- **Bootstrapping with extremely small samples.** Bootstrap performance depends on the sample providing a reasonable approximation to the population. With very small n, the resampled distribution may be unstable and inference unreliable.
- **Treating percentile intervals as universally accurate.** Percentile bootstrap intervals can perform poorly when the sampling distribution is skewed or biased. More refined methods (e.g., BCa intervals) may provide improved accuracy.

Huber's M-Test

Robust Test Based on M-Estimator of Location

ROBU.8 | **Category:** Robust Methods | **Data:** Continuous | **Null:** t-distribution (approximate)

When to Use

- Robustly estimate central tendency by downweighting, rather than discarding, extreme observations.
- Robust alternative to the sample mean under contamination.
- Data exhibit moderate heavy tails or influential points.
- Regression or location estimation requiring stable inference.

Assumptions

- Independent random sample from continuous distribution.
- Majority of observations arise from a common distribution.
- Tuning constant is specified appropriately.
- Outliers represent contamination rather than dominant structure.

Test Statistic

$$t_M = \frac{\hat{\mu}_M - \mu_0}{\hat{\sigma}_M/\sqrt{n}}, \quad \hat{\mu}_M : \sum_{i=1}^{n} \psi\left(\frac{x_i - \hat{\mu}_M}{\hat{\sigma}}\right) = 0, \quad \psi(u) = \max(-k, \min(k, u)). \quad \text{Under } H_0, t_M \approx t_{n-1}.$$

Notation: $\hat{\mu}_M$ is Huber M-estimate, ψ is Huber influence function, k is tuning constant, $\hat{\sigma}$ is MAD or similar

Plain English: Huber's M-test uses an M-estimator of location that smoothly downweights extreme observations rather than completely removing them. The estimator minimizes a loss function that is quadratic for small deviations but linear for large deviations, providing a compromise between efficiency and robustness. The test statistic compares the M-estimate to a hypothesized value using a robust standard error, offering protection against outliers while maintaining good power under normality.

⚠ Practitioner Warning

Tuning constant k must be chosen a priori; $k = 1.345$ balances robustness and ≈95% efficiency under normality. Requires iterative estimation and may face convergence issues with extreme outliers. Assumes approximate symmetry; results depend on the scale estimate (often MAD). Inference may be less accurate in small samples, and estimates can be unstable under heavy contamination. Less robust than trimmed means. Bootstrap methods offer a distribution-free alternative.

Hypotheses

- H_0: $\mu_M = \mu_0$
- H_1: $\mu_M \neq \mu_0$

Decision Rule: Reject H_0 if $|t_M| > t_{\alpha/2,n-1}$ or p-value $< \alpha$. Use approximate t-distribution or bootstrap p-values.

Software References [Python, R, SAS]

Python	R	SAS
`statsmodels::RLM`	`MASS::rlm`	`Not Native (Manual)`

Additional Notes

Huber loss is $\rho(u) = u^2/2$ for $|u| \leq k$ and $\rho(u) = k|u| - k^2/2$ for $|u| > k$; hence $\psi(u) = \rho'(u) = u$ for $|u| \leq k$ and $k\,\text{sign}(u)$ otherwise, so extreme observations are downweighted rather than discarded. IRLS initializes with a robust location, often the median, and iterates $\hat{\mu}^{(t+1)} = \sum_i w_i^{(t)} x_i / \sum_i w_i^{(t)}$, where $u_i = (x_i - \hat{\mu}^{(t)})/\hat{\sigma}$ and $w_i = \psi(u_i)/u_i$ with $w_i = 1$ if $u_i = 0$. Scale may be fixed using the normal consistent MAD, $\hat{\sigma} = 1.4826\,\text{median}|x_i - \text{median}(x)|$, or updated jointly with location. The usual normal-theory tuning constant is $k = 1.345$; smaller k gives more robustness but lower efficiency, while larger k approaches the ordinary mean. A sandwich standard error is based on $\widehat{\text{Var}}(\hat{\mu}_M) = \hat{\sigma}^2 \sum_i \psi^2(u_i)/[\sum_i \psi'(u_i)]^2$.

Miscellanea: Peter Huber developed M-estimation theory in 1960s-1970s. His 1964 paper "Robust Estimation of a Location Parameter" in *Annals of Mathematical Statistics* became seminal work. Huber sought procedures balancing efficiency under ideal conditions with robustness under contamination. The Huber loss function—quadratic in center, linear in tails—arose from minimax theory, minimizing maximum asymptotic variance over neighborhoods of normal distribution.

Reference(s): Huber (1964, 1981); Maronna et al. (2006)

Worked Example

Problem Statement Twelve response times (ms) are recorded, including two large outliers. Test $H_0: \mu = 400$ at $\alpha = 0.05$ using Huber's M-test with tuning constant $k = 1.345$. $H_0: \mu = 400$; $H_1: \mu \neq 400$.

1. Robust Scale Estimate

$\tilde{y} = \text{median}(y) = 440.5$, $\text{MAD} = \text{median}|y_i - \tilde{y}| = 23.5$.

$$s = \text{MAD}/0.6745 = 34.841.$$

2. Standardized Residuals and Huber Scores

Compute $r_i = (y_i - \mu_0)/s$ and apply Huber's $\psi_k(r)$: pass through unchanged if $|r| \leq k = 1.345$, cap at $\pm k$ otherwise.

i	y_i	r_i	$\psi(r_i)$	$\psi'(r_i)$
1	410	0.287	0.287	1
2	414	0.402	0.402	1
3	414	0.402	0.402	1
4	415	0.431	0.431	1
5	419	0.545	0.545	1
6	440	1.148	1.148	1
7	441	1.177	1.177	1
8	442	1.206	1.206	1
9	443	1.234	1.234	1
10	444	1.263	1.263	1
11	855	13.060	**1.345**	0
12	895	14.208	**1.345**	0

3. Test Statistic

Observations 11–12 exceed $k = 1.345$ and are capped; all others pass through unchanged.

$$\bar{\psi} = \frac{\sum_{i=1}^{12} \psi(r_i)}{12} = 0.899, \quad A^2 = \frac{\sum_{i=1}^{12} \psi(r_i)^2}{12} = 0.982.$$

$$B = \frac{\sum_{i=1}^{12} \psi'(r_i)}{12} = 0.833, \quad t_H = \frac{\sqrt{n}\,\bar{\psi}\,B}{\sqrt{A^2}} = 2.618.$$

4. Decision

Under H_0: $t_H \sim t(n-1) = t(11)$ approximately.

α	$t_{\alpha/2}(11)$	Decision
5%	2.201	**Reject** (2.618 > 2.201)
1%	3.106	FTR (2.618 < 3.106)

$p = 0.024$. The Huber M-estimator converges to $\hat{\mu} = 437.6$. By contrast, the ordinary t-test gives $\bar{y} = 502.7$, $s = 174.6$, $t(11) = 2.037$, $p = 0.066$—fails to reject as the two outliers inflate the std. dev. from ≈ 15 (core data) to 174.6.

Practitioner Insight: Huber's M-estimator (Huber 1964) replaces the least-squares criterion $\sum(y_i - \mu)^2$ with $\sum \rho_k(y_i - \mu)$, where ρ_k grows quadratically near the center but only linearly in the tails, limiting the influence of extreme observations. The corresponding score function $\psi_k = \rho'_k$ caps the contribution of large residuals: observations 11–12 have residuals 13.1 and 14.2, but with $k = 1.345$ each contributes at most 1.345 rather than its full magnitude as in the ordinary t-test. The statistic uses sandwich-like quantities $A^2 = \overline{\psi^2}$ and $B = \overline{\psi'}$, where B reflects the fraction of observations within the central linear region. The standard choice $k = 1.345$ gives about 95% asymptotic efficiency under normality while remaining robust to outliers. Unlike Yuen's trimmed-mean test, Huber's method does not discard observations but downweights them continuously. The resulting estimate $\hat{\mu} = 437.6$ lies between the trimmed mean (408.6) and the ordinary mean (502.7), reflecting partial influence from the high outliers. Compared with Yuen's test, the M-test is especially natural when contamination is primarily one-sided and retaining the full sample is desirable.

Huber's M-estimator is obtained numerically (e.g., via Excel Solver) by solving $\hat{\mu}_M : \sum_{i=1}^{n} \psi\left(\frac{y_i - \hat{\mu}_M}{s}\right) = 0$. Here, $\hat{\mu}_M = 437.6$.

Common Pitfalls

- **Post-hoc selection of tuning constant.** The robustness constant k must be specified in advance based on desired efficiency and robustness. Selecting k after identifying outliers introduces bias and invalidates inference.
- **Applying Huber M-estimation to strongly skewed data.** Huber's estimator is designed for symmetric distributions with outliers. For highly skewed data, transformations or quantile-based methods may be more appropriate.
- **Failing to scale residuals using a robust scale estimate.** Residuals must be standardized with a robust scale measure (e.g., MAD). Using the sample standard deviation can mask outliers and undermine robustness.
- **Stopping the iterative algorithm too early.** Huber estimation requires iterative reweighting until convergence. Premature termination can produce unstable estimates and incorrect inference.
- **Using asymptotic normal inference with very small n.** The large-sample normal approximation may be unreliable when sample sizes are small. Bootstrap or small-sample adjustments provide more reliable inference in such cases.

Theil-Sen Slope Test

Robust Nonparametric Test for Regression Slope

ROBU.9 | **Category:** Robust Methods | **Data:** Continuous | **Null:** Normal (asymptotic)

When to Use

- Robust alternative to least squares regression slope test.
- Situations where outliers or heteroscedasticity compromise OLS inference.
- Slope estimation requiring no dist. assumptions.

Assumptions

- Observations are independent pairs (x_i, y_i).
- Relationship between x and y approximately linear.
- Errors independent of x values (need not be normal).
- Sample size adequate for asymptotic approx. ($n \geq 30$).

Test Statistic

$$\hat{\beta}_{TS} = \text{median}\left\{\frac{y_j - y_i}{x_j - x_i} : i < j, x_i \neq x_j\right\}, \quad z = \frac{\hat{\beta}_{TS} - \beta_0}{\hat{\sigma}_\beta}. \qquad \text{Under } H_0, z \approx N(0,1).$$

Notation: $\hat{\beta}_{TS}$ is Theil-Sen slope estimator, median over all $\binom{n}{2}$ pairs, $\hat{\sigma}_\beta$ is robust SE

Plain English: The Theil-Sen slope test estimates the regression slope as the median of all pairwise slopes between data points. This median-based approach provides resistance to outliers since extreme pairwise slopes affect the median minimally. The test evaluates whether this robust slope estimate differs significantly from a hypothesized value. Standard errors can be estimated via bootstrap or asymptotic formulas, providing valid inference without normality assumptions.

⚠ Practitioner Warning

Computationally intensive for large samples; $\binom{n}{2}$ slopes computed and sorted. Tied x values create undefined slopes (division by zero); exclude such pairs. Does not provide robust intercept simultaneously; Sen's extension estimates intercept separately. Standard error estimation requires bootstrap or complex asymptotic formulas. Confidence intervals often wider than OLS when data truly normal. Assumes linear relationship; robust to outliers but not to nonlinearity. Multiple predictors require separate treatment or multivariate extensions. Breakdown point 29%; resistant to substantial contamination but not arbitrarily high.

Hypotheses

- H_0: $\beta = \beta_0$
- H_1: $\beta \neq \beta_0$

Decision Rule: Reject H_0 if $|z| > Z_{\alpha/2}$ or p-value $< \alpha$. Or, use bootstrap percentile CI; reject if β_0 outside CI.

Software References [Python, R, SAS]

Python	R	SAS
`scipy.stats.theilslopes`	`mblm::mblm`	`Not Native (Manual)`

Additional Notes

Theil (1950) introduced the median-of-pairwise-slopes estimator for simple linear regression, and Sen (1968) extended it to handle ties and provided distributional theory. For all pairs (x_i, y_i) and (x_j, y_j) with $i < j$ and $x_i \neq x_j$, compute slopes $m_{ij} = (y_j - y_i)/(x_j - x_i)$; the Theil–Sen slope is the median of these slopes. The intercept is estimated as the median of $y_i - \hat{\beta}_{TS} x_i$. The estimator is robust, with a breakdown point of about 29%, and has roughly 64% efficiency under normality while outperforming OLS under heavy-tailed contamination. Asymptotic normality and variance expressions follow from Sen's theory, though bootstrap intervals are common in practice. The repeated-median estimator (Siegel 1982) attains a 50% breakdown point but is more computationally intensive. Theil–Sen is closely related to Kendall's tau, and testing slope = 0 corresponds to testing tau = 0. Applications include environmental trend detection and other settings requiring robustness to outliers.

Miscellanea: Henri Theil (1950) introduced the median-of-pairwise-slopes estimator for trend estimation, and Pranab K. Sen (1968) independently extended the method to general regression settings, including treatment of tied x-values and asymptotic theory. By the 1980s the dual attribution "Theil–Sen" had become standard. The estimator's robustness to outliers and its connection to Kendall's tau contributed to its adoption in fields, where non-normal errors and extreme observations are common.

Reference(s): Theil (1950); Sen (1968); Wilcox (2017)

Worked Example

Problem Statement Eight bivariate observations are collected (x = study hours, y = test score). Observation 4 is an outlier: a student who scored unusually high relative to the linear trend. Use Theil–Sen method to test $H_0: \beta_1 = 0$ vs. $H_1: \beta_1 \neq 0$, at $\alpha = 0.05$.

i	1	2	3	4	5	6	7	8
x_i	1	2	3	4	5	6	7	8
y_i	2.1	2.9	3.3	8.4	4.9	4.7	5.2	5.8

1. All Pairwise Slopes
Compute $d_{ij} = (y_j - y_i)/(x_j - x_i)$ for all $M = \binom{8}{2}$ pairs.

2. Theil–Sen Slope
Sort all 28 slopes; slopes marked $*$ involve obs 4 (the outlier); for even M, take the average of ranks $M/2$ and $M/2+1$:

Rank	$d_{(k)}$	(i,j)	Rank	$d_{(k)}$	(i,j)
1	−3.500	(4,5) *	15	0.500	(3,8)
2	−1.850	(4,6) *	16	0.517	(1,7)
3	−1.067	(4,7) *	17	0.520	(1,6)
4	−0.650	(4,8) *	18	0.529	(1,8)
5	−0.200	(5,6)	19	0.550	(6,8)
6	0.150	(5,7)	20	0.600	(1,3)
7	0.300	(5,8)	21	0.600	(7,8)
8	0.400	(2,3)	22	0.667	(2,5)
9	0.450	(2,6)	23	0.700	(1,5)
10	0.460	(2,7)	24	0.800	(1,2)
11	0.467	(3,6)	25	0.800	(3,5)
12	0.475	(3,7)	26	2.100	(1,4) *
13	0.483	(2,8)	27	2.750	(2,4) *
14	0.500	(6,7)	28	5.100	(3,4) *

$$\hat{\beta}_1 = (d_{(14)} + d_{(15)})/2 = 0.500.$$

$$\hat{\beta}_0 = \text{med}(y) - \hat{\beta}_1\,\text{med}(x) = 4.800 - 0.500 \times 4.500 = 2.550.$$

3. Confidence Interval (Kendall's S Variance formula)

$$\text{Var}(S) = n(n-1)(2n+5)/18 = 65.33.$$

$$C_L = \frac{M - z_{0.025}\sqrt{65.33}}{2} = \frac{28 - 1.960 \times 8.083}{2} = 6.08.$$

$$C_U = M + 1 - C_L = 22.92.$$

The 95% CI uses $d_{(\lfloor C_L \rfloor)} = d_{(6)}$ and $d_{(\lceil C_U \rceil)} = d_{(23)}$:

$$95\%\ \text{CI} = [0.150,\ 0.700].$$

4. Decision
$0 \notin [0.150, 0.700] \Rightarrow$ **reject** H_0. OLS gives $\hat{\beta}_1 = 0.454$, $t(6) = 1.674$, $p = 0.145$—fails to reject because obs 4 creates large residual noise; its 95% CI [−0.209, 1.117] includes zero.

Practitioner Insight: The Theil–Sen estimator is the median of all $\binom{n}{2}$ pairwise slopes d_{ij}. The outlier at obs 4 generates several extreme +ve and -ve slopes, but these affect only the tails of the ordered slope distribution and cannot move the median; both middle ranks equal 0.5, giving a Theil–Sen slope of exactly 0.5 regardless of the outlier's magnitude. In contrast, the OLS slope is a weighted average of all slopes and is pulled toward the contaminated values, inflating the residual variance from about 0.1 (without the outlier) to 6.0 (with it) and producing a wide CI that includes zero. The Theil–Sen estimator has a breakdown point of $1 - 1/\sqrt{2} \approx 29\%$, meaning nearly one-third of observations can be arbitrarily corrupted without destroying the estimate, compared with 0% for OLS. Inference is based on the Hollander–Wolfe z approx. to Kendall's S, which also underlies Kendall's τ test. Extensions to multiple regression include Siegel's repeated-median estimator and MM-estimation, though the simple-regression case retains the exact median interpretation.

Common Pitfalls

- **Applying the method when x values contain ties.** The Theil–Sen slope is based on pairwise ratios $(y_j - y_i)/(x_j - x_i)$ and therefore requires distinct x values. Repeated x values must be aggregated or alternative methods considered.
- **Using Theil–Sen for clearly nonlinear patterns.** Theil–Sen summarizes the median of pairwise slopes and assumes an approx. linear or monotonic trend. Strong curvature or threshold behavior can produce misleading estimates.
- **Ignoring the computational cost for large samples.** The method requires computing $\binom{n}{2}$ pairwise slopes, which grows quadratically with sample size. Large datasets may require efficient algorithms or approximate implementations.
- **Applying the test to strongly autocorrelated time-series data.** The method assumes independent observations. Serial correlation can underestimate uncertainty and invalidate standard inference.
- **Reporting only the slope estimate without uncertainty measures.** The Theil–Sen estimate should be accompanied by a confidence interval and significance assessment. Point estimates alone provide no indication of statistical uncertainty.

XXII. ROBU — ANOVA-Type Statistic (ATS)

The author encountered the ATS in the late stages of writing this compendium and, given the time constraints of publication, was unable to prepare the full set of materials required for a main-chapter entry. Rather than omit it entirely, he has chosen to present it here in the spirit that defines the rest of this book: an explanation of how the test works under the hood, grounded in a complete numerical example. The ATS is a theoretically sound and practically important procedure with no credible challenges to its validity; the appendix placement reflects only a matter of timing, not merit.

What the ATS Does

The ANOVA-Type Statistic (Brunner, Domhof, and Langer 2002) is a nonparametric test for comparing groups in factorial or repeated measures designs. It makes no distributional assumption and permits heteroskedasticity across groups. So, the ATS is distribution-free in spirit and does not require normality, but it still relies on independent sampling, correct design specification, and large-sample or Box-type approximation logic. Instead of comparing means, it tests whether the **relative treatment effects (RTEs)** — $\hat{p}_i = \bar{R}_i/N - 1/(2N)$, where $\bar{R}_i$ is the mean mid-rank of group i across all N observations — differ across groups. An RTE near 0.5 indicates the group is average; an RTE near 1 indicates the group systematically outranks the others. Because RTEs are dimensionless probabilities, they are interpretable regardless of the original scale.

The test statistic is $F_{\text{ATS}} = Q_A/[(k-1)\,\widehat{\text{tr}}(MS)]$, where $Q_A = N\sum_i n_i(\hat{p}_i - \frac{1}{2})^2$ is the weighted sum of squared deviations of the RTEs from $\frac{1}{2}$, and $\widehat{\text{tr}}(MS)$ is a variance-based scaling derived from within-group rank variances. The denominator degrees of freedom are estimated by a Box-type approximation: $\hat{f} = [\widehat{\text{tr}}(MS)]^2/\widehat{\text{tr}}[(MS)^2]$.

Worked Example

Context. Three groups of eight patients each receive Drug A, Drug B, or Placebo. Pain relief is scored on an ordinal 1–12 scale (higher is better). Scores are skewed and variances differ across groups. Group means are Drug A = 5.9, Drug B = 3.2, Placebo = 8.9.

	Drug A	Drug B	Placebo
Observations	3, 4, 5, 5, 6, 7, 8, 9	1, 2, 2, 3, 3, 4, 5, 6	6, 7, 8, 9, 9, 10, 11, 12
n_i	8	8	8
Mean mid-rank $\bar{R}_i$	12.31	5.81	19.38
RTE $\hat{p}_i$	0.492	0.221	0.787
Within-rank var V_i	26.28	16.92	14.48
$c_i = n_i V_i/N^2$	0.3650	0.2351	0.2011

Step 1: Global mid-ranks ($N = 24$). Pool all 24 observations, assign average ranks for ties. Assign ranks back to each group.

Step 2: RTEs. $\hat{p}_i = \bar{R}_i/N - 1/(2N)$. For Drug A: $12.3125/24 - 1/48 = 0.492$. For Drug B: $5.8125/24 - 1/48 = 0.221$. For Placebo: $19.375/24 - 1/48 = 0.787$.

Step 3: Q_A**.**

$$Q_A = N\sum_{i=1}^{3} n_i(\hat{p}_i - \tfrac{1}{2})^2 = 24\,[8(0.492-0.5)^2 + 8(0.221-0.5)^2 + 8(0.787-0.5)^2] = 30.67$$

Step 4: Construct M, S, and MS. $M = I_3 - \frac{1}{3}J_3$ is the 3×3 centering matrix (projects deviations from the grand mean); $S = \text{diag}(c_1, c_2, c_3)$ with $c_i = n_i V_i/N^2$ where V_i is the within-group sample variance of mid-ranks:

$$c_1 = \frac{8\times 26.28}{576} = 0.3650, \quad c_2 = \frac{8\times 16.92}{576} = 0.2351, \quad c_3 = \frac{8\times 14.48}{576} = 0.2011.$$

The product MS has $(MS)_{ij} = M_{ij}\cdot c_j$, i.e. each column j of M is scaled by c_j:

$$MS = \begin{pmatrix} \frac{2}{3}c_1 & -\frac{1}{3}c_2 & -\frac{1}{3}c_3 \\ -\frac{1}{3}c_1 & \frac{2}{3}c_2 & -\frac{1}{3}c_3 \\ -\frac{1}{3}c_1 & -\frac{1}{3}c_2 & \frac{2}{3}c_3 \end{pmatrix} = \begin{pmatrix} 0.2433 & -0.0784 & -0.0670 \\ -0.1217 & 0.1567 & -0.0670 \\ -0.1217 & -0.0784 & 0.1341 \end{pmatrix}$$

$\widehat{\text{tr}}(MS)$ is simply the sum of the diagonal entries, which always reduces to $\frac{2}{3}\sum_i c_i$:

$$\widehat{\text{tr}}(MS) = 0.2433 + 0.1567 + 0.1341 = \tfrac{2}{3} \times 0.8012 = 0.5341.$$

Step 5: Box degrees of freedom. $\widehat{\text{tr}}[(MS)^2]$ is the trace of the squared matrix, computed as the sum of diagonal entries of $(MS)^2$:

$$(MS)^2_{11} = 0.2433(0.2433) + (-0.0784)(-0.1217) + (-0.0670)(-0.1217) = 0.0769$$
$$(MS)^2_{22} = (-0.1217)(-0.0784) + 0.1567(0.1567) + (-0.0670)(-0.0784) = 0.0394$$
$$(MS)^2_{33} = (-0.1217)(-0.0670) + (-0.0784)(-0.0670) + 0.1341(0.1341) = 0.0315$$

$$\widehat{\text{tr}}[(MS)^2] = 0.0769 + 0.0394 + 0.0315 = 0.1478, \qquad \hat{f} = \frac{(0.5341)^2}{0.1478} = 1.93.$$

Step 6: ATS F-statistic and decision.

$$F_{\text{ATS}} = \frac{Q_A}{(k-1)\,\widehat{\text{tr}}(MS)} = \frac{30.67}{2 \times 0.5341} = 28.71 \sim F(2,\ 1.93) \text{ approx.}$$

Critical value $F_{2,\,1.93,\,0.05} \approx 20.49$. Since $28.71 > 20.49$, **reject** H_0. $p \approx 0.037$. The RTEs confirm the same practical ordering as the raw means, but the ATS reaches this conclusion through rank-based treatment effects rather than relying on normality or equal-variance assumptions.

Software. R: `nparLD` package (`f1.ld.f1()` for one-way, `f1.ld.f2()` for two-factor designs). Python: no native implementation; use R via `rpy2`. SAS: no official `PROC`; the `%Brunner` macro provides an approximation.

Primary references:

- Brunner, E., Domhof, S., & Langer, F. (2002). *Nonparametric Analysis of Longitudinal Data in Factorial Experiments*. Wiley Series in Probability and Statistics. John Wiley & Sons.
- Box, G. E. P. (1954). Some theorems on quadratic forms applied in the study of analysis of variance problems, I. Effect of inequality of variance in the one-way classification. *The Annals of Mathematical Statistics*, **25**(2), 290–302.

XXII. ROBU — Aligned Rank Transform (ART)

As with the ATS, the author encountered ART in the late stages of writing and was unable to include it in the main chapter. The under-the-hood treatment below follows the same spirit as the rest of the compendium. Unlike the ATS, however, ART carries a specific empirical caveat: Elkin, Kay, Higgins, and Wobbrock (Behavior Research Methods, 2021) demonstrated through simulation that ART can produce severely inflated Type I error rates for ***interaction effects when non-zero main effects are present*** *— precisely the most common use case in factorial designs. The ART-C contrast procedure (same paper) largely addresses this for pairwise comparisons, but readers should be aware that the interaction F-statistic from a standard ART analysis may over-reject in practice. This concern is analogous to the MaxCombo discussion in the SURV chapter appendix.*

What ART Does

The Aligned Rank Transform (Higgins, Blair & Tashtoush 1990; Wobbrock et al. 2011) extends nonparametric rank-based testing to **factorial designs with interaction terms**. A direct rank transform (simply replacing every observation by its global rank and running a standard ANOVA) is known to produce badly distorted Type I error for interactions. ART's solution is to first **align** the data for each effect by stripping out all other effects, and only then rank.

For a 2×2 factorial design with factors A and B, the procedure produces three separate aligned-and-ranked datasets — one for the main effect of A, one for B, and one for the $A \times B$ interaction. Each dataset is analysed by a standard parametric ANOVA. A useful diagnostic is that the alignment should remove all effects except the one currently being tested. In practice, ANOVA applied to the aligned (but not yet ranked) responses should show negligible sums of squares and F-statistics for the nuisance effects.

Alignment Formulas

Let $\bar{y}_{ij}$ be the cell mean, $\bar{y}_{i\cdot}$ and $\bar{y}_{\cdot j}$ the marginal means, and $\bar{y}_{\cdot\cdot}$ the grand mean. For observation y_{ijk} in cell (i, j):

$$\tilde{y}^{A}_{ijk} = y_{ijk} - \bar{y}_{ij} + \bar{y}_{i\cdot} \qquad \text{(align for main effect of } A)$$
$$\tilde{y}^{B}_{ijk} = y_{ijk} - \bar{y}_{ij} + \bar{y}_{\cdot j} \qquad \text{(align for main effect of } B)$$
$$\tilde{y}^{AB}_{ijk} = y_{ijk} - \bar{y}_{ij} + (\bar{y}_{ij} - \bar{y}_{i\cdot} - \bar{y}_{\cdot j} + \bar{y}_{\cdot\cdot}) \qquad \text{(align for interaction)}$$

Each aligned dataset is then ranked globally (mid-ranks for ties), and a standard F-test is applied.

Worked Example

Context. A 2×2 between-subjects design: Factor A = Treatment (Drug vs. Placebo) and Factor B = Dose (Low vs. High). $n = 5$ per cell ($N = 20$). Response: ordinal pain relief score (1–10, higher is better).

	Low Dose (B1)	High Dose (B2)	Marginal mean
Drug (A1)	3, 4, 5, 5, 6 ($\bar{y} = 4.6$)	5, 6, 7, 7, 8 ($\bar{y} = 6.6$)	5.6
Placebo (A2)	2, 3, 3, 4, 5 ($\bar{y} = 3.4$)	3, 4, 4, 5, 6 ($\bar{y} = 4.4$)	3.9
Marginal mean	4.0	5.5	Grand: 4.75

Step 1: Align for Factor A (Treatment). Subtract cell mean, add marginal mean of A: $\tilde{y}^{A} = y - \bar{y}_{ij} + \bar{y}_{i\cdot}$. Both Drug cells add 5.6; both Placebo cells add 3.9. Rank the 20 aligned values globally (mid-ranks for ties):

Cell	Aligned values	Global ranks	Mean rank
A1B1 (Drug/Low)	4.0, 5.0, 6.0, 6.0, 7.0	7.5, 11.5, 16.5, 16.5, 19.5	14.3
A1B2 (Drug/High)	4.0, 5.0, 6.0, 6.0, 7.0	7.5, 11.5, 16.5, 16.5, 19.5	14.3
A2B1 (Placebo/Low)	2.5, 3.5, 3.5, 4.5, 5.5	1.5, 4.5, 4.5, 9.5, 13.5	6.7
A2B2 (Placebo/High)	2.5, 3.5, 3.5, 4.5, 5.5	1.5, 4.5, 4.5, 9.5, 13.5	6.7

Drug combined mean rank = 14.3; Placebo = 6.7. One-way ANOVA: $SS_{between} = 10(14.3 - 10.5)^2 + 10(6.7 - 10.5)^2 = 288.8$, $SS_{within} = 367.7$, $F(1, 18) = 288.8/20.43 = 14.14$, $p = 0.0014$.

Step 2: Align for Factor *B* (Dose). $\tilde{y}^B = y - \bar{y}_{ij} + \bar{y}_{\cdot j}$: Low cells add 4.0, High cells add 5.5. After global ranking, Low combined mean rank = 7.1, High = 13.9. One-way ANOVA: $F(1, 18) = 9.64$, $p = 0.0061$.

Step 3: Align for Interaction $A \times B$. $\tilde{y}^{AB} = y - \bar{y}_{ij} + (\bar{y}_{ij} - \bar{y}_{i\cdot} - \bar{y}_{\cdot j} + \bar{y}_{\cdot\cdot})$, i.e. subtract cell mean then add the cell's interaction component:

Cell	Interaction component	Aligned values	Cell mean rank
A1B1	4.6 − 5.6 − 4.0 + 4.75 = −0.25	−1.85, −0.85, 0.15, 0.15, 1.15	9.2
A1B2	6.6 − 5.6 − 5.5 + 4.75 = +0.25	−1.35, −0.35, 0.65, 0.65, 1.65	11.8
A2B1	3.4 − 3.9 − 4.0 + 4.75 = +0.25	−1.15, −0.15, −0.15, 0.85, 1.85	11.8
A2B2	4.4 − 3.9 − 5.5 + 4.75 = −0.25	−1.65, −0.65, −0.65, 0.35, 1.35	9.2

All four marginal means of the interaction-aligned ranks equal 10.5 (grand mean), so $SS_A = SS_B = 0$ in the two-way ANOVA on these ranks — the interaction *SS* is isolated cleanly. The cell mean ranks $\{9.2, 11.8, 11.8, 9.2\}$ give:

$$SS_{int} = 5\sum_{cells}(\bar{r}_{ij} - 10.5)^2 = 5\left[2(9.2 - 10.5)^2 + 2(11.8 - 10.5)^2\right] = 5 \times 6.76 = 33.8$$

$SS_{error} = 629.2$ (within-cell rank variation), $df_{error} = 16$.

$$F(1, 16) = \frac{33.8/1}{629.2/16} = \frac{33.8}{39.33} = 0.86, \quad p = 0.368.$$

Summary:

Effect	F	p	Decision
Treatment (A)	14.14	0.001	Reject H_0: Drug > Placebo
Dose (B)	9.64	0.006	Reject H_0: High > Low
$A \times B$	0.86	0.368	Fail to reject: no interaction

Both main effects are significant. The interaction is not, suggesting that the treatment effect is reasonably consistent across dose levels.

Software. R: `ARTool` package (`art()` for alignment, `anova(art(...))` for *F*-tests, `art.con()` for ART-C contrasts). Python: `pyARTool` (limited maintenance; verify version). SAS: no official implementation.

Primary references:

- Higgins, J. J., Blair, R. C., & Tashtoush, S. (1990). The aligned rank transform procedure. *Proceedings of the Kansas State University Conference on Applied Statistics in Agriculture*, 185–195.
- Wobbrock, J. O., Findlater, L., Gergle, D., & Higgins, J. J. (2011). The Aligned Rank Transform for nonparametric factorial analyses using only ANOVA procedures. *Proceedings of the ACM Conference on Human Factors in Computing Systems (CHI '11)*, 143–146.
- Elkin, L. A., Kay, M., Higgins, J. J., & Wobbrock, J. O. (2021). An Aligned Rank Transform procedure for multifactor contrast tests. *Proceedings of the ACM Symposium on User Interface Software and Technology (UIST '21)*, 126–141.

Part XXIII

Exact Inference via Resampling

Evaluating hypotheses using data-driven null distributions

Permutation and Resampling

XXIII. PERM — A Conceptual Primer

What Are Resampling Methods?

Resampling methods build inferential distributions directly from the observed data rather than relying on parametric assumptions (normality, equal variances, known distributions). They approximate the sampling distribution of a test statistic by repeated manipulation of the original data.

Three distinct approaches underpin the tests in this chapter:

Method	Core mechanism
Permutation	Shuffle group labels (without replacement) under the null hypothesis of exchangeability; exact p-value over all possible rearrangements.
Randomization	Uses the actual experimental assignment mechanism; valid inference derived from the design itself, not a population model.
Bootstrap	Resample observations *with replacement* from the data; approximates the sampling distribution of any statistic.

Exchangeability: The Foundational Assumption

The key assumption for permutation and randomization tests is **exchangeability**: under H_0, the observed values could have been assigned to any unit (or group) with equal probability. Formally, the joint distribution of $(y_1, y_2, \dots, y_n)$ is unchanged under any permutation of the indices.

- In a **two-sample** test, exchangeability holds under H_0: the two populations have identical distributions; group labels become arbitrary.
- In a **randomized experiment**, exchangeability is guaranteed by design—the assignment mechanism is known and the permutation distribution is exact without any population assumption.
- Exchangeability **does not hold** for clustered, hierarchical, or time-series data without modification; naive permutation inflates Type I error in these settings.

The General Permutation Procedure

1. Compute the observed test statistic T_{obs} from the data.
2. Under H_0, generate the **reference (permutation) distribution** by reassigning labels across all (or a random subset of) permutations and computing $T^{*(b)}$ for $b = 1, \dots, B$.
3. The exact p-value is:

$$p = \frac{\#\{T^{*(b)} \geq T_{\text{obs}}\} + 1}{B + 1},$$

where the $+1$ in numerator and denominator counts the observed statistic itself; this ensures $p \geq 1/(B+1)$ and controls Type I error exactly.

4. Reject H_0 if $p \leq \alpha$.

Exact vs. Monte Carlo permutation: The exact test enumerates all $n!/(n_1!n_2!)$ permutations; feasible only for small n. Monte Carlo permutation (sampling B random permutations) approximates the exact p-value with precision $\sim 1/\sqrt{B}$; use $B \geq 999$ for $\alpha = 0.05$ and $B \geq 9999$ for $\alpha = 0.01$.

I. Permutation t-Test

The permutation analog of the two-sample t-test. The test statistic is the observed mean difference $T_{\text{obs}} = \bar{y}_1 - \bar{y}_2$ (or equivalently the t-statistic). Under H_0, all $\binom{n}{n_1}$ assignments of n observations to two groups of sizes n_1 and n_2 are equally likely.

Key properties:

- Exact Type I error control under exchangeability; no normality assumption required.
- Asymptotically equivalent to the t-test under normality (same limiting power).
- More powerful than the Wilcoxon rank-sum test when the data are genuinely Gaussian but equally powerful under broad alternatives.
- For paired data, permute the signs of the differences $d_i = y_{1i} - y_{2i}$ (2^n sign assignments under H_0).

II. Permutation ANOVA

Extends the permutation t-test to $k \geq 2$ groups. The test statistic is typically the F-ratio $F = \text{MS}_\text{B}/\text{MS}_\text{W}$; under H_0, all $n!/(n_1! \cdots n_k!)$ assignments of observations to groups are equally likely.

Advantages over classical ANOVA: No normality or homoscedasticity assumption; the F-statistic is used as a summary but the p-value comes from the permutation distribution, not the $F_{k-1,N-k}$ table. Valid for small, unbalanced groups where the asymptotic F approximation is poor.

Post-hoc comparisons: permutation versions of pairwise t-tests with max-statistic correction (permute once, compare all pairs) or step-down procedures control the FWER without Bonferroni conservatism.

III. Permutation Test for Correlation and Mantel Test

Permutation correlation test: Tests $H_0 : \rho = 0$ by permuting one variable $(y_1, \dots, y_n)$ while holding the other $(x_1, \dots, x_n)$ fixed. Equivalent to permuting the pairing of (x_i, y_i). The permutation p-value is the proportion of permuted correlations r^* exceeding $|r_{\text{obs}}|$. Valid for Pearson, Spearman, or Kendall without distributional assumptions.

Mantel test: A matrix permutation test for association between two $n \times n$ distance (or dissimilarity) matrices $\mathbf{A}$ and $\mathbf{B}$:

$$M = \sum_{i<j} A_{ij} B_{ij}.$$

Under H_0: no relationship between the matrices; generate the reference distribution by permuting the rows and columns of **A** (or **B**) simultaneously and recomputing M.

The **partial Mantel test** controls for a third matrix, analogous to partial correlation.

IV. Randomization Test

The randomization test is philosophically distinct from the permutation test: it derives validity from the **actual assignment mechanism** of a randomized experiment rather than a population model.

Under Fisher's randomization inference, the only randomness is the assignment of units to treatments. The reference distribution consists of all possible reassignments consistent with the experimental design:

- For completely randomized designs: all $\binom{n}{n_1}$ label permutations.
- For block designs: permute only within blocks.
- For matched-pairs: permute signs of d_i.

The null hypothesis is the **sharp null** of no treatment effect on any unit (Fisher's null), not merely equal means. Exact p-values require no distributional assumption; valid for any sample size.

V. Bootstrap Hypothesis Test and CIs

Unlike permutation tests, the bootstrap generates the null distribution by **resampling with replacement** from a version of the data consistent with H_0.

Bootstrap hypothesis test: For $H_0 : \theta = \theta_0$, center the data at θ_0 before resampling, compute $T^{*(b)}$ from each bootstrap sample, and compare T_{obs} to the centered bootstrap distribution.

Bootstrap confidence intervals: Draw B bootstrap samples $\mathbf{y}^{*(b)}$; compute $\hat{\theta}^{*(b)}$ for each.

CI type	Method and notes
Percentile	$[\hat{\theta}^*_{(\alpha/2)}, \hat{\theta}^*_{(1-\alpha/2)}]$; simplest; can be biased for skewed statistics.
Basic (pivot)	$[2\hat{\theta} - \hat{\theta}^*_{(1-\alpha/2)}, 2\hat{\theta} - \hat{\theta}^*_{(\alpha/2)}]$; corrects bias.
BCa	Bias-corrected and accelerated; corrects for both bias and skewness in the bootstrap distribution; recommended for small samples.
Studentised	Rescales by bootstrap SE; better coverage for smooth statistics; requires double bootstrap.

Practical guidelines: $B \geq 999$ for p-values; $B \geq 1999$ for 95% CIs; $B \geq 9999$ for precise tail probabilities. BCa is the default recommendation; percentile is acceptable for large n and symmetric statistics.

Permutation vs. Bootstrap: Key Distinctions

	Permutation	Bootstrap
Sampling	Without replacement	With replacement
H_0 used?	Yes (labels shuffled)	No (data centered)
Exactness	Exact (under exch.)	Approximate
Best for	Hypothesis tests	CIs and estimation
Assumption	Exchangeability	i.i.d. (or weaker)
Handles clusters?	With modification	With stratification

Handling Dependence: Block and Restricted Permutations

Standard permutation tests assume i.i.d. or exchangeable observations. For structured data:

- **Clustered data:** permute whole clusters (not individual observations) to preserve within-cluster dependence.
- **Block designs:** permute within blocks only; the Mantel–Haenszel analog for permutation tests.
- **Time series:** use block bootstrap (resample contiguous blocks) or stationary bootstrap to preserve serial correlation.
- **Spatial data:** Mantel test accounts for spatial structure via the distance matrix formulation.

Common Pitfalls

- **Violating exchangeability:** permuting individual observations in clustered or longitudinal data inflates Type I error; permute at the cluster level.
- **Too few resamples:** $B = 99$ is insufficient for $\alpha = 0.05$; use $B \geq 999$ as a minimum.
- **Wrong centring for bootstrap tests:** the bootstrap data must be shifted to be consistent with H_0; uncentered resampling produces an incorrect null distribution.
- **Confusing permutation p-values with bootstrap CIs:** permutation is for H_0-based testing; bootstrap is for estimation-based inference; they answer different questions.
- **Using percentile CI for skewed statistics:** the percentile CI is biased for skewed estimators; use BCa instead.
- **Ignoring the +1 correction:** omitting the +1 from the p-value formula can produce $p = 0$, which is not achievable by a valid test.

Bottom Line

Permutation and bootstrap methods derive inference directly from the data, replacing distributional assumptions with exchangeability or i.i.d. sampling. Permutation tests are exact under exchangeability; bootstrap approximates sampling distributions for CIs and estimation. Always respect the dependence structure of the data when designing the resampling scheme, and use $B \geq 999$ resamples.

Permutation t-Test

Distribution-Free Alternative to Student's t-Test via Permutation

PERM.1 | **Category:** Permutation Tests | **Data:** Continuous | **Null:** Empirical (permutation)

When to Use

- Test mean difference between two groups without normality.
- Small-sample settings where parametric t-test approximations may be unreliable.
- Situations with skewed or heavy-tailed data where robustness is desired.

Assumptions

- Observations are independent within and between groups.
- Under H_0, observations are exchangeable across groups.
- Normality or variance homogeneity not required.
- Data are measured on at least an interval scale (for mean-based statistics).

Test Statistic

$$t_{\text{obs}} = \frac{\bar{x}_1 - \bar{x}_2}{s_p\sqrt{1/n_1 + 1/n_2}} \text{ (for equal variances),} \quad \text{compare to distribution of } t^{(b)} \text{ under random reassignments.}$$

Notation: t_{obs} is observed t-statistic, $t^{(b)}$ are permutation statistics, $b = 1, \ldots, B$ permutations; applicable for other variants of t tests as well

Plain English: The permutation t–test forms a null distribution by repeatedly reallocating observations to groups under the assumption that, if H_0 is true, all group assignments are equally likely. For each reassignment the t–statistic is recomputed, and the observed statistic is compared against this permutation distribution. Enumerating all permutations yields an exact test; random sampling provides an accurate Monte Carlo approximation when enough replications are used.

⚠ Practitioner Warning

Complete enumeration requires $\binom{n_1+n_2}{n_1}$ permutations and is infeasible for large samples. Monte Carlo versions need sufficiently many replications ($B \geq 1000$, preferably ≥ 10000). Validity rests on exchangeability; differences in variance or distributional shape can violate exchangeability under H_0. When distributions may differ, the test detects any distributional change—not only mean shifts. Very small or unequal sample sizes reduce the number of unique permutations and limit resolution. Confidence intervals are not obtained directly and require test inversion.

Hypotheses

- H_0: Means are equal across groups (exchangeability).
- H_1: Means differ between groups.

Decision Rule: Reject H_0 if $p_{\text{perm}} = \frac{\#\{|t^{(b)}| \geq |t_{\text{obs}}|\}+1}{B+1} < \alpha$."+1" not required for full enumeration (exact test).

Software References [Python, R, SAS]

```
scipy.stats.permutation_test      coin::independence_test      PROC NPAR1WAY; exact;
```

Additional Notes

Choice of test statistic: any statistic valid (difference in means, t-statistic, Welch statistic); different statistics have different power. Paired designs permute signs of within-pair differences; stratified designs restrict permutations within strata. When exchangeability holds, the test provides exact Type I error control, matches the t–test under normality, and can outperform it under skewed or heavy-tailed distributions. Confidence intervals arise via test inversion by identifying null values not rejected by the permutation test. The "+1" in the decision rule includes the observed statistic as one of the permutations, ensuring a valid, nonzero p-value and preventing downward bias in finite samples.

Miscellanea: Ronald A. Fisher introduced permutation testing in *The Design of Experiments* (1935), using the tea-tasting example to illustrate exact inference from experimental design. Early critics noted that exhaustive enumeration becomes infeasible for large n; Fisher acknowledged this but emphasized the principle's validity. For decades, permutation tests were limited in practice by computational constraints. With modern computing, they became routine by the 1990s–2000s, and high-performance computing now enables large-scale permutation inference via parallelization. Efron (1979) later noted their close conceptual connection to the bootstrap, both relying on resampling ideas. The commonly used "+1" correction for Monte Carlo p-values is known as the Davison–Hinkley adjustment.

Reference(s): Fisher (1935); Good (2005); Ernst (2004)

Worked Example

Problem Statement A marine biologist compares the bioluminescence intensity (cd/m^2) of two jellyfish species (A and B). Due to the small sample size ($n_1 = n_2 = 5$) and unknown population distribution, a Permutation t-test is used to test $H_0 : \mu_A = \mu_B$ against $H_a : \mu_B > \mu_A$.
Data Summary

Species	**Intensity Observations** (cd/m^2)	n	**Mean** ($\bar{x}$)	$S^2 = \sum(x_i - \bar{x})^2$
Species A	$\{1.2, 1.5, 0.9, 1.8, 2.1\}$	$n_A = 5$	1.50	0.90
Species B	$\{2.5, 1.9, 3.2, 2.8, 2.1\}$	$n_B = 5$	2.50	1.10
Difference	$\Delta_{obs} = \bar{x}_B - \bar{x}_A$		**1.00**	

1. Calculation of Observed t-statistic (t_{obs}) We first compute the t-statistic for the original data. Using a pooled standard error:

$$s_{\text{pooled}} = \sqrt{\frac{0.90 + 1.10}{n_A + n_B - 2}} = 0.50, \quad SE = s_{\text{pooled}}\sqrt{\frac{1}{n_A} + \frac{1}{n_B}} = 0.316, \quad t_{obs} = \frac{\bar{x}_B - \bar{x}_A}{SE} = \frac{1.00}{0.316} = 3.162$$

2. Exact Permutation Results Under the null hypothesis of exchangeability, all $\binom{10}{5} = 252$ distinct label permutations were enumerated. For each permutation, a test statistic t^* was computed to obtain the exact null distribution. Let K denote the number of permutations for which $t^* \geq t_{obs}$, where $t_{obs} = 3.162$.

Metric	**Value**
Observed t-statistic (t_{obs})	3.162
Number of Permutations (M)	252
Count, K ($t^* \geq 3.162$)	3 (including t_{obs})
Exact p**-value**	$3/252$ =**0.0119**

3. Decision and Interpretation

- **Decision:** Since $p = 0.0119 < 0.05$, we **Reject the Null Hypothesis**. Since all $\binom{10}{5} = 252$ permutations were enumerated, the p-value is computed exactly as $p = K/252$. Davison-Hinkley adjustment is not required.
- **Interpretation:** The difference in bioluminescence is highly significant. Species B is statistically brighter than Species A.
- **Practitioner Insight:** The permutation p-value (0.0119) provides the exact probability of observing such a large difference by chance, given this specific set of 10 specimens. It avoids reliance on the t-distribution, which may be inaccurate for small or non-normal samples.

Common Pitfalls

- **Assuming exchangeability when it fails.** Permutation tests require exchangeable observations under H_0; fixed group attributes (e.g., sex, age) break exchangeability and invalidate random relabeling.
- **Using too few Monte Carlo permutations.** Small B yields unstable p–values; Monte Carlo error is $O(1/\sqrt{B})$, so reliable inference typically requires $B \geq 10{,}000$.
- **Permuting paired samples as if independent.** Paired designs need sign-flip permutations on within-pair differences; shuffling group labels discards the pairing structure.
- **Expecting higher power than the t–test.** Permutation tests trade a slight loss of efficiency under normality for robustness; gains in power occur mainly with skewed or heavy-tailed data.
- **Not distinguishing exact from Monte Carlo p–values.** Permutation results must state whether all permutations were enumerated or a Monte Carlo approximation was used, and report B when applicable.

Permutation ANOVA

Randomization Test for Comparing Multiple Group Means

PERM.2 | **Category:** Permutation Tests | **Data:** Continuous | **Null:** Empirical (permutation)

When to Use

- Compare multiple groups ($J \geq 2$) without normality or equal variance assumptions.
- Small samples per group where asymptotic ANOVA unreliable.
- Non-normal data (skewed, heavy-tailed, discrete) requiring exact inference.

Assumptions

- J independent groups ($J \geq 2$, typically ≥ 3).
- Exchangeability under null hypothesis (observations can be randomly reassigned to groups).
- Does NOT require normality, homoscedasticity, or specific distribution shape.
- Sample sizes can be unequal and small.

Test Statistic

$$F_{\text{obs}} = \frac{\sum_{j=1}^{J} n_j(\bar{x}_j - \bar{x})^2/(J-1)}{\sum_{j=1}^{J}\sum_{i=1}^{n_j}(x_{ij} - \bar{x}_j)^2/(N-J)}, \quad \text{compare to permutation distribution of } F^{(b)}.$$

Notation: F_{obs} is observed F-statistic, $F^{(b)}$ are permutation F-values, J is groups, N is total sample size

Plain English: Permutation ANOVA generates a null distribution by randomly reassigning observations to groups and computing the F-statistic for each reassignment. Under the null hypothesis of no group differences, all possible group assignments are equally likely. The observed F-statistic is compared to this empirical permutation distribution to assess significance.

⚠ Practitioner Warning

Complete permutation requires $\frac{N!}{n_1!n_2!\cdots n_J!}$ evaluations; computationally prohibitive except small samples. Monte Carlo approximation requires adequate replications ($B \geq 1000$). Assumes exchangeability under null; violated if groups differ in variance or shape under null. Tests equality of entire distributions, not just means. Post-hoc pairwise comparisons require permutation tests for each pair with multiplicity correction. Very unequal sample sizes reduce number of unique permutations. Does not identify which specific groups differ when null rejected.

Hypotheses

- H_0: All group distributions identical.
- H_1: At least one group distribution differs from others.

Decision Rule: Reject H_0 if $p_{\text{perm}} = \frac{\#\{F^{(b)} \geq F_{\text{obs}}\}+1}{B+1} < \alpha$.The +1 adjustment prevents zero p-values.

Software References [Python, R, SAS]

```
scipy.stats.permutation_test      coin::oneway_test      PROC NPAR1WAY; exact;
```

Additional Notes

Permutation ANOVA pools all N observations and repeatedly reassigns them into J groups of fixed sizes $(n_1, \ldots, n_J)$, computing the chosen test statistic (typically the F-statistic) each time. Alternatives to the classical F include Welch-type statistics (heteroscedasticity), rank-based choices (Kruskal–Wallis), or trimmed/robust statistics. Confidence intervals via inversion are computationally demanding, so analyses typically report p-values and effect sizes. Power is comparable to parametric ANOVA under normality and superior when assumptions fail. PERMANOVA (Anderson 2001; Anderson–ter Braak 2003) generalizes this framework to multivariate responses.

Miscellanea: Permutation ANOVA extends Fisher's randomization principle to multi-group testing, but only became practical with modern computing. Dwass (1957) established permutation theory for multiple comparisons; Edgington (1969–1995) strongly promoted randomization tests as exact, assumption-free alternatives. By the 1990s, computational advances and work such as Manly's text *Randomization, Bootstrap and Monte Carlo Methods* popularized permutation methods across ecology, neuroimaging, and genomics. Especially valuable with messy, non-normal field data, freeing analysts from questionable distributional assumptions.

Reference(s): Fisher (1935); Edgington (1995); Good (2005)

Worked Example

Problem Statement An analyst evaluates user engagement (seconds) across three mobile app UI colors. Due to the presence of extreme outliers and small, unequal group sizes, a Permutation ANOVA is used to determine if the mean engagement time differs significantly across groups without assuming normality.

Data Summary

Group	Engagement Times (sec)	n	Mean ($\bar{x}$)	Var (s^2)
Blue	$\{12, 15, 11, 45\}$	4	20.75	264.25
Green	$\{18, 22, 20, 25, 21\}$	5	21.20	6.70
Red	$\{45, 50, 48, 60\}$	4	50.75	42.25
Total	$N = 13$	13	$\bar{x}_G$=30.15	

$$SS_{\text{Between}} = \sum_{j=1}^{k} n_j(\bar{x}_j - \bar{x}_G)^2 = 2451.39$$

$$SS_{\text{Within}} = \sum_{j=1}^{k} (n_j - 1)s_j^2 = 946.30$$

1. Calculation of Observed Statistic (F_{obs}) Using the standard ANOVA formulas for $SS_{Between}$ and SS_{Within}:

$$F_{obs} = \frac{MS_{Between}}{MS_{Within}} = \frac{2451.39/2}{946.3/10} = 12.9525$$

2. Monte Carlo Permutation Results The group labels were randomly permuted $M = 5{,}000$ times (with repetition). In each iteration, the F^* statistic was recalculated to construct the empirical null distribution.

Metric	Value
Observed F-statistic (F_{obs})	12.9525
Number of Permutations (M)	5,000
Count ($F^* \geq 12.9525$)	59
Empirical (+1 adj.) p-value	**0.006**

3. Decision and Interpretation

- **Decision:** Since $p = 0.006 < 0.05$, we **Reject the Null Hypothesis**.
- **Interpretation:** There is a statistically significant difference in user engagement across the three colors.
- **Practitioner Insight:** Both methods use the same ANOVA table and F statistic; they differ only in how the p-value is computed—F distribution vs. permutation (exact or Monte Carlo). The classical ANOVA relies on normality and equal variance, while the permutation test does not. In this example, both methods reject ($p_{\text{perm}} = 0.006$, $p_{\text{ANOVA}} = 0.0017$), indicating that the effect is strong enough to be detected even under assumption violations. When the two agree, inference is robust; when they differ, the discrepancy is often driven by those assumptions rather than the data itself.

Conclusion: The Red CTA button appears to drive significantly higher engagement than the Blue or Green alternatives.

Common Pitfalls

- **Assuming exchangeability when it fails.** Permutation tests require exchangeable observations under H_0; fixed group attributes (e.g., sex, age) break exchangeability and invalidate random relabeling.
- **Using too few Monte Carlo permutations.** Small B yields unstable p–values; Monte Carlo error is $O(1/\sqrt{B})$, so reliable inference typically requires $B \geq 10{,}000$.
- **Permuting paired samples as if independent.** Paired designs need sign-flip permutations on within-pair differences; shuffling group labels discards the pairing structure.
- **Expecting higher power than the t–test.** Permutation tests trade a slight loss of efficiency under normality for robustness; gains in power occur mainly with skewed or heavy-tailed data.
- **Not distinguishing exact from Monte Carlo p–values.** Permutation results must state whether all permutations were enumerated or a Monte Carlo approximation was used, and report B when applicable.

Permutation Test for Correlation

Distribution-Free Test of Association Between Variables

PERM.3 | **Category:** Permutation Tests | **Data:** Continuous | **Null:** Empirical (permutation)

When to Use

- Test association between two variables without distributional assumptions.
- Small samples where standard tests may be unreliable.
- Data with outliers or non-normal marginal distributions.
- Validate or cross-check parametric correlation test results.
- Inference when exact sampling distribution is unknown.

Assumptions

- Observations are independent pairs (X_i, Y_i).
- Under H_0, X and Y are exchangeable (no association).
- Test statistic computed consistently across permutations.
- All (or sampled) permutations are equally likely under H_0.
- Permutations preserve one variable while shuffling the other.

Test Statistic

$$r_{\text{obs}} = \frac{\sum_{i=1}^{n}(x_i - \bar{x})(y_i - \bar{y})}{\sqrt{\sum_{i=1}^{n}(x_i - \bar{x})^2 \sum_{i=1}^{n}(y_i - \bar{y})^2}}, \quad \text{compare to } \{r^{(b)}\}_{b=1}^{B} \text{ under permutations.}$$

Notation: r_{obs} is observed correlation, $r^{(b)}$ are permutation correlations, permute y while holding x fixed

Plain English: The permutation test for correlation generates a null distribution by randomly shuffling one variable (typically y) while keeping the other (x) fixed, computing the correlation for each permutation. Under the null hypothesis of no association, all permutations of y values are equally likely. The observed correlation is compared to this permutation distribution to assess significance. This provides exact or approximate inference without requiring bivariate normality.

⚠ Practitioner Warning

Full enumeration requires $n!$ permutations—impractical for $n > 10$—so Monte Carlo sampling ($B \geq 1000$) is used. The test assesses independence, not just linear correlation; rank-based versions (Spearman, Kendall) detect nonlinear monotonic associations and handle outliers. Does not yield confidence intervals directly (requires test inversion). Very small samples limit the number of unique permutations and reduce test resolution.

Hypotheses

- H_0: No association between variables ($\rho = 0$).
- H_1: Association exists between variables ($\rho \neq 0$).

Decision Rule: Reject H_0 if $p_{\text{perm}} = \frac{\#\{|r^{(b)}| \geq |r_{\text{obs}}|\}+1}{B+1} < \alpha$.The +1 adjustment prevents zero p-values.

Software References [Python, R, SAS]

```
scipy.stats.permutation_test      coin::independence_test      PROC NPAR1WAY; exact;
```

Additional Notes

Permutation mechanism: fix the x values in observed order and randomly permute the y values; compute the correlation statistic for each permutation. Under H_0 of independence, all y-orderings are equally likely, so the permutation distribution of the statistic is valid. Choice of statistic: Pearson r (linear association), Spearman ρ_s (monotonic, robust), Kendall τ (concordance, handles ties). Rank-based versions permute ranks rather than raw data. Power: Pearson-based permutation test matches classical test under normality and outperforms it with heavy tails; rank-based statistics trade some efficiency for robustness. Extensions include partial correlation via residual permutation and the Mantel test for matrix correlation. Confidence intervals can be constructed by test inversion, though computationally intensive.

Miscellanea: Permutation tests for correlation trace to Pitman (1937–1938) and early randomization ideas but became practical only with modern computing. They gained prominence in ecology, where normality rarely holds, and later in spatial genetics through the Mantel test (1967). Permutation testing liberated correlation analysis from strict normality assumptions, enabling reliable inference in fields with messy, highly non-normal data.

Reference(s): Fisher (1935); Mantel (1967); Good (2005)

Permutation Test for Correlation	PERM.3

Worked Example

Problem Statement An analyst examines the relationship between Loan Officer seniority (Years) and recovery rates (%). Due to a small sample size ($n = 8$) and concerns regarding the normality of recovery data, a Permutation Test is used to assess the significance of the correlation. We test $H_0 : \rho = 0$ vs. $H_a : \rho > 0$.

Data Summary

Officer	1	2	3	4	5	6	7	8
Years (X)	2	3	5	6	8	10	12	15
Recovery % (Y)	15	18	22	20	28	25	35	32

1. Observed Statistic The Pearson product-moment correlation coefficient for the original pairing is:

$$r_{obs} = \frac{\sum(x_i - \bar{x})(y_i - \bar{y})}{\sqrt{\sum(x_i - \bar{x})^2 \sum(y_i - \bar{y})^2}} = 0.9186$$

2. Permutation Distribution Results The Y vector was shuffled 10,000 times, while the X vector remained fixed. The correlation r^* was recalculated for each iteration to construct the null distribution.

Metric	**Value**
Observed Correlation (r_{obs})	0.9186
Number of Permutations (M)	10,000
Count ($r^* \geq 0.9186$)	9
Empirical p-value (+1 adjusted)	**0.001**

3. Decision and Interpretation

- **Decision:** Since $p = 0.001 < 0.05$, we **Reject the Null Hypothesis**.
- **Interpretation:** There is a statistically significant positive correlation between officer seniority and recovery rates.
- **Practitioner Insight:** Unlike a standard t-test for correlation, this p-value does not rely on the assumption that X and Y follow a bivariate normal distribution. The low p-value suggests that the strong observed relationship (0.9186) is highly unlikely to have occurred by random chance.

Common Pitfalls

- **Permuting both variables.** Only one variable (typically y) should be permuted while x remains fixed; permuting both destroys the null structure and yields meaningless inference.
- **Relying on Pearson r for non-monotonic patterns.** Permutation tests detect structure captured by the chosen statistic (e.g: linear association by Pearson r). For nonlinear or U-shaped relationships, use alternative statistics (e.g., distance correlation, MIC) or visualize/transform the data first.
- **Ignoring dependence or clustering.** Permutation correlation assumes independent pairs. With repeated measures or clustered data, use restricted schemes (permute whole clusters or permute within subjects); unrestricted permutation inflates Type I error.
- **Using the wrong correlation measure.** Choose the statistic to match the data: Pearson for linear/continuous, Spearman or Kendall for ordinal data or outliers. The permutation framework works with any correlation measure, but the measure must reflect the relationship of interest.
- **Interpreting correlation as causation.** A significant permutation test confirms association, not causality. Confounding variables or common drivers can still explain the relationship; causal conclusions require experimental or causal-model designs.

Mantel Test

Permutation Test for Association Between Distance Matrices

PERM.4 | **Category:** Permutation Tests | **Data:** Continuous | **Null:** Empirical (permutation)

When to Use

- Test correlation between two distance or dissimilarity matrices.
- Assess association between spatial and genetic distances.
- Detect relationships between ecological, geographic, or network proximity measures.

Assumptions

- Two symmetric distance or dissimilarity matrices of same dimension.
- Observations are units with defined pairwise relationships.
- Matrix elements not independent (permutation accounts for dependence structure).

Test Statistic

$$Z_{\text{obs}} = \sum_{i<j} X_{ij} Y_{ij}, \quad \text{or standardized } r_M = \frac{\sum_{i<j}(X_{ij} - \bar{X})(Y_{ij} - \bar{Y})}{\sqrt{\sum_{i<j}(X_{ij} - \bar{X})^2 \sum_{i<j}(Y_{ij} - \bar{Y})^2}}.$$

Notation: X_{ij}, Y_{ij} are elements of distance matrices, sum over unique pairs, r_M is Mantel correlation

Plain English: The Mantel test evaluates whether two distance matrices are associated by computing a correlation (or summed cross-products) between corresponding entries. Because matrix elements share objects and are therefore not independent, ordinary correlation tests fail. A valid Mantel test permutes row/column labels of one matrix simultaneously, recomputes the statistic for each permutation, and compares the observed value to this reference distribution.

⚠ Practitioner Warning

Permutation must preserve matrix structure—permute indices jointly, not individual cells. Requires symmetric distance/dissimilarity matrices; asymmetric cases need specialized methods. Power is limited for small n and results may be confounded by third variables. Partial Mantel variants exist but their interpretation is debated. Sparse or zero-inflated matrices can distort correlations. Monte Carlo permutation requires sufficient replicates ($B \geq 1000$, ideally $> 10{,}000$).

Hypotheses

- H_0: Distance matrices are independent ($\rho_M = 0$).
- H_1: Distance matrices are associated ($\rho_M \neq 0$).

Decision Rule: Reject H_0 if $p_{\text{perm}} = \frac{\#\{|Z^{(b)}| \geq |Z_{\text{obs}}|\}+1}{B+1} < \alpha$.The +1 adjustment prevents zero p-values.

Software References [Python, R, SAS]

Python	R	SAS
`scipy.spatial.distance + manual`	`vegan::mantel`	`Not Native (Manual)`

Additional Notes

Matrix structure: for n objects, $\mathbf{X}$ and $\mathbf{Y}$ are $n \times n$ symmetric matrices with fixed diagonals (0 for distances, 1 for similarities). Test statistic: Mantel's original $Z = \sum_{i<j} X_{ij} Y_{ij}$; standardized form r_M is the Pearson correlation between vectorized lower triangles. Permutation scheme: randomly permute object labels π, apply to both rows and columns of one matrix ($Y^*_{ij} = Y_{\pi(i)\pi(j)}$), recompute Z^* or r^*_M, and compare to the permutation distribution. Complete enumeration is infeasible; Monte Carlo sampling ($B \geq 1000$) is standard. Partial Mantel: regresses out a third matrix—widely used but criticized for inflated Type I error (Legendre et al. 2015). Applications include isolation by distance (genetic vs. geographic distance), community ecology (composition vs. environment), and network analysis (proximity vs. attribute similarity). Standardized r_M provides effect-size interpretation. Multiple regression on distance matrices extends Mantel to multiple predictors.

Miscellanea: Nathan Mantel introduced the test in 1967 at NIH while analyzing spatial clustering of disease cases, recognizing that pairwise distances violate independence assumptions of standard correlation tests. Early work focused on epidemiology; later adoption in ecology and genetics followed through the 1980s–1990s. Sokal and Rohlf helped popularize it. Partial Mantel tests later drew criticism for reliability issues, but the classic Mantel test remains a core tool for assessing matrix association.

Reference(s): Mantel (1967); Legendre et al. (2015); Sokal & Rohlf (1995)

Worked Example

Problem Statement An operational risk analyst investigates whether geographic proximity between five regional hubs (Matrix A) correlates with the dissimilarity of their risk loss profiles (Matrix B). A significant positive correlation would indicate that offices closer to one another share similar risk characteristics.
Data Summary The analysis utilizes two symmetric 5 × 5 distance matrices. The lower triangular values for both are vectorized for correlation analysis.

Element	2,1	3,1	3,2	4,1	4,2	4,3	5,1	5,2	5,3	5,4
Geo Dist (A)	15	40	25	100	85	60	150	140	120	55
Risk Diss (B)	0.2	0.5	0.3	0.8	0.9	0.7	0.9	0.8	0.9	0.4

1. Observed Mantel Statistic (r_M) The observed correlation between the two vectors is calculated using the standard Pearson product-moment formula:

$$r_M = \frac{\sum(A_i - \bar{A})(B_i - \bar{B})}{\sqrt{\sum(A_i - \bar{A})^2 \sum(B_i - \bar{B})^2}} = 0.876$$

2. Permutation Results Because distance matrix entries are interdependent, we determine significance by permuting the rows and columns of Matrix A 5,000 times to construct a null distribution.

Metric	Value
Observed Mantel Correlation (r_M)	0.876
Permutation Iterations (M)	5,000
Count ($r_M^* \geq 0.876$)	44
Empirical p-value (+1 adjusted)	**0.009**

3. Decision and Interpretation

- **Decision:** Since $p = 0.009 < 0.05$, we **Reject the Null Hypothesis.**
- **Interpretation:** There is a strong, statistically significant positive correlation between geographic distance and risk profile dissimilarity.
- **Practitioner Insight:** Offices that are geographically closer tend to have more similar risk profiles, suggesting that location is a key driver of operational loss patterns.

Common Pitfalls

- **Permuting rows and columns independently.** Always apply the *same* permutation to rows and columns of one matrix; independent permutations break symmetry and invalidate the distance structure.
- **Including diagonal elements in calculations.** Use only the upper (or lower) triangle of each matrix. Diagonal entries are structural zeros and artificially inflate sample size if included.
- **Treating pairwise distances as independent observations.** Although there are $\binom{n}{2}$ distances, they are not independent. Effective sample size is roughly n objects, so p-values should be interpreted cautiously and effect size (r_M) emphasized.
- **Using Mantel when simpler regression is appropriate.** Mantel is for correlating *two distance matrices*. When relating a single response variable to predictors, conventional (possibly spatial) regression is more suitable.
- **Misinterpreting significance under spatial autocorrelation.** Mantel detects association but cannot distinguish environmental effects from shared spatial structure. Use partial Mantel tests or spatial models for clearer interpretation.

Randomization Test

Inference Based on Random Treatment Assignment in Experiments

PERM.5 | **Category:** Permutation Tests | **Data:** Any | **Null:** Empirical (randomization)

When to Use

- Test treatment effects in randomized experiments.
- Small samples where asymptotic tests are unreliable.
- Assumption-light inference (no normality/homogeneity).
- Exact, design-based inference from known randomization mechanism.

Assumptions

- Sharp null: $Y_i(1) = Y_i(0)$ for all units.
- No interference between units (SUTVA holds).
- Test statistic computed consistently across all assignments.
- Treatment assignment is randomized with known probabilities (e.g., equal or stratified).

Test Statistic

$$T_{\text{obs}} = g(Y_1^{\text{obs}}, Y_0^{\text{obs}}, \mathbf{Z}), \quad \text{compare to } \{T^{(b)}\}_{b=1}^{B} \text{ under alternative assignments } \mathbf{Z}^{(b)}. \text{ (distribution from randomization)}$$

Notation: T_{obs} is observed (computable) statistic, Y_1, Y_0 are potential outcomes, $\mathbf{Z}$ is assignment vector

Plain English: Randomization tests base inference entirely on the experiment's assignment mechanism. Under Fisher's sharp null (no unit-level treatment effects), all assignments consistent with the design form the reference distribution of the test statistic. Comparing the observed statistic to this distribution yields exact, assumption-free inference rooted in the experimental design, making randomization tests a cornerstone of causal inference in randomized studies.

⚠ Practitioner Warning

Valid only when treatments were truly randomized; observational studies cannot justify randomization inference. Fisher's sharp null is stronger than the weak null of zero average effect and cannot test the latter without extra assumptions. Complete enumeration may be infeasible, requiring Monte Carlo sampling with adequate replications ($B \geq 1000$). Complex designs (clustered, stratified, sequential) require matching the permutation structure.

Hypotheses

- H_0: Sharp null: treatment has zero effect for every unit.
- H_1: Treatment has non-zero effect for at least one unit.

Decision Rule: Reject H_0 if $p_{\text{rand}} = \frac{\#\{|T^{(b)}| \geq |T_{\text{obs}}|\}+1}{B+1} < \alpha$.The +1 adjustment prevents zero p-values.

Software References [Python, R, SAS]

Python	R	SAS
`scipy.stats.permutation_test`	`ri2::conduct_ri`	`Not Native (Manual)`

Additional Notes

Fisher's sharp null $H_0 : Y_i(1) = Y_i(0)$ treats outcomes as fixed, with randomness arising only from assignment. The randomization distribution is obtained by enumerating or sampling all assignments consistent with the design (e.g., $\binom{N}{N_1}$ under complete randomization; permute within strata or clusters as appropriate) and recomputing the test statistic. Any statistic (mean difference, regression coefficient, rank-based) is valid; choice affects power, not validity. Randomization inference provides exact finite-sample Type I error control without distributional assumptions. Fisher's sharp null differs from Neyman's weak null (zero average effect). Confidence intervals follow by inverting tests over constant additive effects. Modern developments (Rosenbaum; Imbens–Rubin) extend and formalize this design based framework for causal inference.

Miscellanea: Fisher established randomization-based inference in *The Design of Experiments* (1935), emphasizing that validity comes from the random assignment actually performed; his "lady tasting tea" example illustrated exact inference with very small samples. Neyman (1930s) challenged the sharp null, advocating inference on average effects—forming a debate foundational to causal inference. For decades, randomization tests were conceptually important but computationally limited; advances in the 1960s–1970s made them practical. Rubin's potential-outcomes framework later unified Fisher's ideas into a formal causal model. Today, randomization tests are viewed as a gold standard for experimental inference: exact, assumption-light, and design-driven.

Reference(s): Fisher (1935); Rosenbaum (2002); Imbens & Rubin (2015)

Worked Example

Problem Statement A fintech firm conducts a randomized experiment to see if a New UI (Group B) reduces application time compared to the Old UI (Group A). Ten users are randomly assigned to the two groups. Because the sample is small and the randomization process is known, an exact Randomization Test is performed.

Data Summary

Group	Completion Times (min)	Mean ($\bar{x}$)
Control (A)	{12, 15, 11, 14, 18}	14.0
Treatment (B)	{8, 12, 10, 9, 11}	10.0
Difference	$\Delta_{obs} = \bar{x}_A - \bar{x}_B$	**4.0**

1. Randomization Procedure Under H_0, the completion time for each participant is fixed (data is not random). The only source of variation is the random assignment. There are $\binom{10}{5} = 252$ possible ways these 10 users could have been partitioned into two groups of five.
2. Exact Distribution and p-value We evaluate all 252 possible partitions to determine the number of assignments where the mean difference Δ^* is greater than or equal to the observed difference of 4.0.

Metric	Value
Observed Difference (Δ_{obs})	4.0
Total Possible Assignments ($N!/(n_1!n_2!)$)	252
Count ($\Delta^* \geq 4.0$)	5
Exact p-value	**0.01984**

3. Decision and Interpretation

- **Decision:** Since $p = 0.01984 < 0.05$, we **Reject the Null Hypothesis.**
- **Interpretation:** The results are statistically significant. The probability of observing a difference of 4 minutes or more purely by the luck of the draw in random assignment is only 1.98%.
- **Practical Application:** The firm can conclude that the New UI layout effectively reduces user friction and speeds up the application process.

Common Pitfalls

- **Using randomization tests for observational data.** Valid only when treatment was truly randomized; observational comparisons lack known assignment probabilities and require confounding-adjustment methods instead.
- **Confusing Fisher's sharp null with Neyman's weak null.** Randomization tests target the sharp null of no unit-level effects, not the weaker hypothesis of zero average effect, which cannot be tested exactly without further assumptions.
- **Ignoring the actual randomization scheme.** The reference distribution must reflect all assignments allowed by the design (complete, blocked, stratified, clustered), not arbitrary permutations.
- **Applying randomization tests with non-random missing data.** Dropouts that depend on treatment or outcome undermine validity; imputation or sensitivity analyses are required rather than simple complete-case analysis.
- **Reporting results without describing the randomization mechanism.** Validity depends on the assignment procedure, so reports must specify how units were randomized and whether the test used exact or Monte Carlo enumeration.

Sharp Null vs. Weak Null Hypothesis

The **Fisherian (sharp) null** hypothesis ($H_0 : Y_i(1) = Y_i(0)$ for all i) assumes the treatment effect is zero for *every* individual, enabling exact permutation tests. In contrast, the **Neyman (weak) null** hypothesis ($H_0 : E[Y_i(1) - Y_i(0)] = 0$) only assumes the *average* treatment effect is zero, allowing for individual-level variation (heterogeneity).

Bootstrap Hypothesis Test

Resampling-Based Test via Bootstrap Null Distribution

PERM.6 | **Category:** Permutation Tests | **Data:** Any | **Null:** Empirical (bootstrap)

When to Use

- Test hypotheses when parametric distributions unknown or intractable.
- Construct null distribution for complex/non-standard statistics (medians, ratios, trimmed means) without closed-form.
- Alternative to asymptotic approximations in small or moderate samples. Non-normal or heteroskedastic data settings.

Assumptions

- Sample is random and representative of population.
- i.i.d observations (or known dependence structure for block bootstrap). Sample is representative of the population.
- Bootstrap samples drawn with replacement from null-adjusted data. Statistic is computable and stable.
- Adequate replications ($B \geq 1000$, preferably ≥ 10000).

Test Statistic

$$\hat{\theta}_{\text{obs}} = g(\mathbf{X}), \quad \text{generate } \{\hat{\theta}_b^*\}_{b=1}^B \text{ from null distribution } F_0, \quad p_b = \frac{\#\{|\hat{\theta}_b^* - \theta_0| \geq |\hat{\theta}_{\text{obs}} - \theta_0|\} + 1}{B+1} \text{ (2-sided).}$$

Notation: $\hat{\theta}_{\text{obs}}$ is observed statistic, $\hat{\theta}_b^*$ are bootstrap statistics under null, θ_0 is null value; applies to 1-sided as well

Plain English: Bootstrap hypothesis testing generates a null distribution by resampling from data that have been adjusted to satisfy the null hypothesis. For testing a parameter value, center the data at the null value; for comparing groups, pool and resample assuming no difference. The observed statistic is compared to this bootstrap null distribution. Unlike permutation tests which rely on exchangeability, bootstrap tests explicitly construct the null distribution through resampling.

⚠ Practitioner Warning

Requires adjusting data to satisfy null hypothesis before bootstrapping; improper adjustment invalidates test. Different from bootstrap confidence interval—CI approach inverts acceptance regions, hypothesis test uses null-centered bootstrap. Adequate replications critical; too few produce unstable p-values. Small original samples ($n < 20$) may not capture population variability adequately. Does not work well for testing at distribution boundaries (e.g., variance = 0). Parametric bootstrap alternative when null distribution form known. Computationally intensive compared to analytical tests.

Hypotheses

- H_0: $\theta = \theta_0$ (parameter equals null value).
- H_1: $\theta \neq \theta_0$ (parameter differs from null value).

Decision Rule: Reject H_0 if $p_b < \alpha$.The +1 adjustment prevents zero p-values and preserves proper Type I error.

Software References [Python, R, SAS]

Python	R	SAS
`scipy.stats.bootstrap`	`boot::boot`	`Not Native (Manual)`

Additional Notes

Null-centered bootstrap: to test $H_0 : \mu = \mu_0$, recenter data as $x_i^* = x_i - \bar{x} + \mu_0$, resample with replacement, compute $\bar{x}_b^*$, and compare $|\bar{x}_{\text{obs}} - \mu_0|$ with $\{|\bar{x}_b^* - \mu_0|\}$. Two-sample tests: pool data under the null and bootstrap differences in means. Regression tests: bootstrap residuals from the reduced model (e.g., $\beta_j = 0$). Studentized bootstrap uses bootstrap t-statistics for improved finite sample accuracy. Parametric bootstrap draws samples from the model implied by the null (e.g. normal with μ_0, σ^2). Relation to permutation: bootstrap accommodates complex nulls, while permutation is exact when exchangeability holds. Higher-order refinements (Hall and Wilson 1991; Davidson and MacKinnon 1999) show bootstrap tests can outperform standard asymptotics in small samples.

Miscellanea: Bootstrap began with Efron (1979) for interval estimation; its extension to hypothesis testing grew through 1980s work by Hall, Beran, and others. Hall emphasized its robustness: when classical assumptions hold, bootstrap matches parametric tests; when they fail, bootstrap still works. The key insight—recenter or resimulate data so they reflect the null—enabled testing complex functionals and models without deriving new theory.

Reference(s): Efron (1979); Hall & Wilson (1991); Davison & Hinkley (1997)

Bootstrap Hypothesis Test **PERM.6**

Worked Example

Problem Statement An analyst evaluates if Model 2 has a significantly lower Mean Absolute Error (MAE) than Model 1. Given the non-normal nature of error metrics, a bootstrap hypothesis test is utilized. We test $H_0 : \mu_1 - \mu_2 = 0$ against $H_a : \mu_1 - \mu_2 > 0$ (one-sided).

Data Summary

Group	MAE Observations	Mean ($\bar{x}$)
Model 1 ($n_1 = 10$)	{2.4, 3.1, 2.8, 4.5, 3.9, 2.7, 3.3, 4.1, 5.0, 3.2}	3.50
Model 2 ($n_2 = 10$)	{2.1, 2.5, 2.3, 3.2, 2.8, 2.2, 2.4, 3.5, 2.9, 2.1}	2.60
Difference	$\Delta_{obs} = \bar{x}_1 - \bar{x}_2$	**0.90**

1. Bootstrap Null Distribution Construction To test the null hypothesis, the samples are shifted so their means both equal the pooled mean ($\bar{x}_p = 3.05$):

- $z_{1,i} = x_{1,i} - 3.50 + 3.05$
- $z_{2,i} = x_{2,i} - 2.60 + 3.05$

We then draw $B = 5,000$ samples with replacement from z_1 and z_2 to calculate the distribution of $\Delta^* = \bar{z}_1^* - \bar{z}_2^*$.

2. Results and p-value The empirical p-value is calculated as the proportion of bootstrap differences exceeding the observed difference.

Metric	Value
Observed Statistic (Δ_{obs})	0.9000
Bootstrap Iterations (B)	5,000
Count ($\Delta^* \geq \Delta_{obs}$)	9
Empirical p-value (+1 adjusted)	**0.002**

3. Decision and Interpretation

- **Decision:** Since $p = 0.002 < 0.05$, we **Reject the Null Hypothesis**.
- **Interpretation:** There is strong evidence to suggest that Model 2 provides a statistically significant improvement in accuracy over Model 1.
- **Practitioner Insight:** This test does not assume that the underlying error distributions are normal, making the conclusion more reliable than a standard t-test.

Common Pitfalls

- **Resampling original data instead of null-centered data.** Bootstrap tests must generate samples under H_0; resampling the raw data centers the distribution at the estimate, not the null, producing an invalid test.
- **Using too few bootstrap samples.** Bootstrap p-values have Monte Carlo error ($SE \approx 1/\sqrt{B}$). Hypothesis testing requires large B; use $B \geq 10{,}000$ for stable decisions.
- **Confusing bootstrap tests with bootstrap CIs.** CI methods resample the original data; hypothesis tests resample data adjusted to satisfy the null. Mixing the two leads to incorrect p-values.
- **Ignoring dependence in the data.** Standard bootstrap assumes independent observations. Time series, spatial data, or clusters require block or dependent-resampling schemes to preserve structure.
- **Not reporting B or Monte Carlo uncertainty.** Bootstrap p-values depend on the number of resamples. Always report B (and ideally the bootstrap SE) so readers can assess the reliability of the result.

Bootstrap Confidence Interval Test

Hypothesis Testing via Inversion of Bootstrap Confidence Intervals

PERM.7 | **Category:** Permutation Tests | **Data:** Any | **Null:** Empirical (bootstrap)

When to Use

- Construct intervals when the sampling distribution is unknown or complex, especially when a unified estimate–interval–test framework is desired.
- Small to moderate samples with unreliable asymptotics.
- Non-standard statistics lacking parametric interval theory.
- Non-normal or skewed data where classical intervals fail.

Assumptions

- Sample is random and representative of the population.
- Observations are i.i.d. or dependence is correctly modeled.
- Resampling with replacement approximates the true sampling process.
- Statistic is stable and sufficiently smooth for bootstrap.
- Adequate replications ($B \geq 1000$, preferably ≥ 10000).

Test Statistic

$\mathrm{CI}_{1-\alpha} = [\hat{\theta}^*_{(B\alpha/2)}, \hat{\theta}^*_{(B(1-\alpha/2))}]$ (percentile method), where $\hat{\theta}^*_{(p)}$ is the p-th quantile of the bootstrap distribution

Notation: $\hat{\theta}^*_b$ is statistic from b-th bootstrap sample, B is number of bootstrap replications

Plain English: Bootstrap confidence–interval testing builds a CI from percentiles of many resampled datasets and rejects the null when its value lies outside this interval. By resampling the observed data rather than relying on theoretical distributions, the method adapts automatically to the statistic's empirical variability and remains useful even when no parametric formula is available.

⚠ Practitioner Warning

Accuracy depends on the bootstrap mimicking the true sampling distribution, which can fail with very small samples or non-smooth (not differentiable or discontinuous jumps with small data perturbations) statistics. Percentile intervals perform poorly under strong skewness; BCa corrections are preferred. CI-based tests resample the original data (not null-centered), unlike bootstrap hypothesis tests. One-sided tests use one-sided bounds. BCa and Studentized (bootstrap-t) require extra resampling or jackknife steps, increasing computation.

Hypotheses

- H_0: $\theta = \theta_0$ (parameter equals specified null value).
- H_1: $\theta \neq \theta_0$ (two-sided), or $\theta > \theta_0$ or $\theta < \theta_0$ (one-sided).

Decision Rule: Reject H_0 if $\theta_0 \notin [\hat{\theta}^*_{\alpha/2}, \hat{\theta}^*_{1-\alpha/2}]$. For one-sided tests, use $[\hat{\theta}^*_{\alpha}, \infty)$ or $(-\infty, \hat{\theta}^*_{1-\alpha}]$.

Software References [Python, R, SAS]

`scipy::bootstrap`	`boot::boot.ci`	`(Manual implementation)`

Additional Notes

Percentile intervals use ordered bootstrap statistics: for a 95% CI with B=1000, the 25th and 975th values form the endpoints. CI–test duality follows from the definition of coverage probability and holds exactly for pivots but only approximately for general statistics, improving with larger n. The BCa method (Efron, 1987) adjusts percentile endpoints using a bias term z_0 and an acceleration term a, with corrected quantiles α_1 and α_2. BCa achieves $O(n^{-1})$ coverage error versus $O(n^{-1/2})$ for percentile intervals (Hall, 1988). BCa offers second-order accuracy (DiCiccio–Efron, 1996) with modest overhead (requires z_0 and a). Studentized (bootstrap-t) intervals standardize each bootstrap statistic using its bootstrap or jackknife SE, improving accuracy for skewed statistics at higher computational cost.

Miscellanea: Efron introduced the bootstrap in 1979, and his 1982 monograph helped establish confidence interval construction as a central application. The BCa method arose from the observation that simple percentile intervals can exhibit bias and poor coverage, especially in skewed settings. "Bias-corrected and accelerated" reflects adjustments for median bias and skewness, though the acceleration term is often viewed as less intuitive in practice. The bootstrap CI's dual role—providing both interval estimates and hypothesis tests via inversion—has made it especially valuable in applied work.

Reference(s): Efron (1979, 1982, 1987); Hall (1988); DiCiccio & Efron (1996)

Worked Example

Problem Statement A risk analyst is estimating the mean recovery rate (%) for a portfolio of defaulted commercial loans. Given the typically skewed nature of recovery data and a small sample size, the Bootstrap method is employed to construct robust 95% confidence intervals. Original Sample ($n = 15$): $\mathcal{X} = \{12, 18, 22, 15, 25, 30, 10, 20, 28, 14, 19, 21, 24, 17, 26\}$.

1. Summary Statistics and Bootstrap Parameters

Metric	Value	Metric	Value	Metric	Value
Sample Mean ($\hat{\theta}$)	20.067	Bootstrap Mean ($\bar{\theta}^*$)	20.058	$\#\{\hat{\theta}^*(b) < \hat{\theta}\}$	1036
Sample Std. Dev (s)	5.885	Bootstrap Std. Error (se_{boot})	1.442	Bias-Correction (z_0)	0.045
Significance Level (α)	0.05	Critical Value ($z_{1-\alpha/2}$)	1.960	Acceleration (a)	-0.012
Bootstrap Samples (B)	2000	Mean of Jackknife Means ($\bar{\theta}_{(\cdot)}$)	20.067		

2. Confidence Intervals For the percentile-based methods, let $\hat{\theta}^*_{(q)}$ denote the qth quantile of the bootstrap distribution.

A. Normal Approximation (SE)

Formula: $CI = \hat{\theta} \pm z_{1-\alpha/2} \cdot se_{boot}$

$CI = 20.067 \pm 1.96(1.442)$

Result: $[17.24, 22.89]$

B. Basic (Reverse Percentile)

Formula: $CI = [2\hat{\theta} - \hat{\theta}^*_{(1-\alpha/2)}, 2\hat{\theta} - \hat{\theta}^*_{(\alpha/2)}]$

$CI = [2(20.067) - 23.07, 2(20.067) - 17.33]$

Result: $[17.06, 22.80]$

C. Percentile Bootstrap

Formula: $CI = [\hat{\theta}^*_{(\alpha/2)}, \hat{\theta}^*_{(1-\alpha/2)}]$

$CI = [17.33, 23.07]$

Result: $[17.33, 23.07]$

D. BCa (Bias-Corrected Accelerated)

Formula: $CI = [\hat{\theta}^*_{(\alpha_1)}, \hat{\theta}^*_{(\alpha_2)}]$

Adjusting for z_0, a: $\alpha_1 \approx 0.028, \alpha_2 \approx 0.977$

Result: $[17.47, 23.20]$

Practitioner Insight The **Normal Approximation** is symmetric but may underestimate the tail risk if the distribution is non-normal. The **BCa interval** is the most robust, as it adjusts for the slight right-skew in the recovery data. We are 95% confident that the true population mean recovery rate falls between approximately 17.47% and 23.2% (BCa).

Common Pitfalls

- **Resampling paired data incorrectly.** Paired or matched observations must be resampled as units; resampling x and y independently destroys the dependence structure the bootstrap is meant to preserve.
- **Using percentile intervals for skewed statistics.** Percentile CIs assume symmetry of the bootstrap distribution; for skewed estimators (ratios, variances, correlations near ± 1), BCa intervals provide markedly better finite-sample coverage.
- **Misinterpreting confidence intervals.** Bootstrap CIs describe uncertainty in a population parameter, not the distribution of observations; they should never be interpreted as capturing "95% of the data."
- **Relying on too few bootstrap replications.** Quantile estimates are unstable when B is small; stable inference typically requires $B \geq 1000$ for exploration and $B \geq 5000$–10000 for final reporting.
- **Claiming exact coverage without qualification.** Bootstrap intervals achieve only approximate nominal coverage, with accuracy affected by sample size, skewness, and boundary parameters; results should be reported as "nominal 95% CI."

BCa Interval Constants: Bias-Correction (z_0), Acceleration (a), and Corrected Quantiles (α_1, α_2)

Bias-Correction (z_0) measures the median bias of the bootstrap distribution, while **acceleration** (a) accounts for the rate of change of the standard error. It is typically estimated using the jackknife values $\hat{\theta}_{(i)}$. Φ^{-1} is the inverse CDF of the standard normal distribution, $\hat{\theta}$ is the original estimate, and $\hat{\theta}_{(\cdot)} = \frac{1}{n}\sum \hat{\theta}_{(i)}$.

$$z_0 = \Phi^{-1}\left(\frac{\#\{\hat{\theta}^*(b) < \hat{\theta}\}}{B}\right), \text{ and } a = \frac{\sum_{i=1}^{n}(\hat{\theta}_{(\cdot)} - \hat{\theta}_{(i)})^3}{6\left[\sum_{i=1}^{n}(\hat{\theta}_{(\cdot)} - \hat{\theta}_{(i)})^2\right]^{3/2}}; \quad \alpha_1 = \Phi\left(z_0 + \frac{z_0 + z_{\alpha/2}}{1 - a(z_0 + z_{\alpha/2})}\right), \text{ and } \alpha_2 = \Phi\left(z_0 + \frac{z_0 + z_{1-\alpha/2}}{1 - a(z_0 + z_{1-\alpha/2})}\right).$$

XXIII. PERM — Detailed Worked Example: Bootstrap Confidence Intervals

Background and Motivation: Bootstrap confidence intervals provide a distribution-free method for quantifying uncertainty about a population parameter when the theoretical sampling distribution is unknown or difficult to derive. Unlike parametric methods that assume a specific distribution (e.g., normal), bootstrap methods use resampling from the observed data to empirically approximate the sampling distribution.

Problem Statement A psychologist measures reaction times (in milliseconds) for 10 participants performing a cognitive task. The researcher wants to estimate the population median reaction time with a 95% confidence interval. Since reaction times are often right-skewed, the normal-theory confidence interval may be inappropriate. The reaction times (ms) are:

$$\mathbf{x} = [245, 280, 310, 265, 420, 290, 305, 275, 315, 380]$$

Sample size: $n = 10$. $\text{median}(\mathbf{x}) = \hat{\theta} = \frac{290+305}{2} = 297.5$ ms, which is the point estimate. Now we need to quantify the uncertainty around this estimate.

Generate Bootstrap Samples: The bootstrap method generates new samples by **sampling with replacement** from the observed data. Each bootstrap sample has the same size as the original ($n = 10$), but some observations may appear multiple times while others may not appear at all.

Number of bootstrap samples: $B = 20$ (for illustration)

b	θ_b^*	$\theta_b^* < \hat{\theta}$	Bootstrap Sample (unsorted)
1	285.0	1	[280, 245, 290, 280, 305, 420, 275, 290, 310, 265]
2	297.5	0	[315, 420, 280, 305, 245, 280, 380, 290, 315, 265]
3	285.0	1	[265, 290, 275, 310, 245, 305, 280, 275, 290, 380]
4	297.5	0	[420, 305, 290, 315, 280, 245, 310, 275, 420, 265]
5	290.0	1	[280, 310, 290, 380, 245, 275, 305, 290, 265, 315]
6	280.0	1	[275, 245, 280, 290, 305, 265, 280, 310, 290, 275]
7	297.5	0	[310, 315, 305, 280, 305, 290, 420, 275, 265, 290]
8	290.0	1	[290, 305, 280, 310, 290, 275, 315, 245, 305, 280]
9	310.0	0	[315, 310, 380, 305, 310, 290, 315, 305, 290, 420]
10	280.0	1	[265, 275, 280, 290, 245, 305, 310, 280, 315, 275]
11	297.5	0	[305, 290, 315, 280, 310, 275, 305, 265, 420, 245]
12	290.0	1	[290, 280, 305, 245, 310, 275, 290, 315, 265, 305]
13	297.5	0	[305, 315, 310, 290, 280, 305, 420, 290, 245, 275]
14	275.0	1	[245, 265, 280, 275, 290, 280, 265, 275, 310, 245]
15	297.5	0	[305, 290, 315, 280, 310, 275, 305, 380, 265, 245]
16	307.5	0	[315, 310, 380, 420, 305, 290, 310, 280, 245, 275]
17	277.5	1	[275, 280, 265, 290, 245, 310, 275, 280, 305, 265]
18	290.0	1	[290, 305, 315, 280, 310, 275, 290, 305, 245, 265]
19	285.0	1	[280, 290, 305, 275, 310, 265, 290, 315, 280, 245]
20	305.0	0	[310, 315, 305, 380, 290, 305, 280, 420, 275, 265]

This distribution approximates the sampling distribution of the median.

Sorted bootstrap distribution:

$$\theta^*_{\text{sorted}} = [275.0, 277.5, 280.0, 280.0, 285.0, 285.0, 285.0, 290.0, 290.0, 290.0, 290.0, 297.5, 297.5, 297.5, 297.5, 297.5, 297.5, 305.0, 307.5, 310.0]$$

Bootstrap mean

$$\bar{\theta}^* = \frac{1}{20}\sum_{b=1}^{20} \theta_b^* = 291.75$$

Bootstrap standard error

$$\text{SE}_{\text{boot}} = \sqrt{\frac{1}{B-1}\sum_{b=1}^{B}(\theta_b^* - \bar{\theta}^*)^2} = 9.904$$

Bias-Correction Factor z_0

$$z_0 = \Phi^{-1}\left(\frac{\#\{\theta_b^* < \hat{\theta}\}}{B}\right) = \Phi^{-1}\left(\frac{11}{20}\right) = 0.1257$$

Acceleration Factor a (using jackknife)

$$a = \frac{\sum_{i=1}^{n}(\bar{\theta}_{(\cdot)} - \theta_{(i)})^3}{6\left[\sum_{i=1}^{n}(\bar{\theta}_{(\cdot)} - \theta_{(i)})^2\right]^{3/2}}$$

where $\theta_{(i)}$ is the median computed from data with i-th observation deleted.

Method 1: Percentile Bootstrap Confidence Interval

The percentile method uses the quantiles of the bootstrap distribution directly as confidence limits.

Theory

For a $(1-\alpha) \times 100\%$ confidence interval, find the $\alpha/2$ and $1-\alpha/2$ quantiles of the bootstrap distribution.

For 95% CI: $\alpha = 0.05$, so we need the 2.5th and 97.5th percentiles.

Calculation

With $B = 20$ bootstrap samples:

Lower index: $\lceil B \times 0.025 \rceil = 1$, Upper index: $\lfloor B \times 0.975 \rfloor = 19$

From the sorted bootstrap distribution:

$$\theta^*_{(1)} = 275.0, \ \theta^*_{(19)} = 307.5$$

95% Percentile Bootstrap Confidence Interval:

$$\text{CI}_{\text{percentile}} = [275.0, 307.5] \text{ ms}$$

Interpretation

We are 95% confident that the population median reaction time lies between 275.0 ms and 307.5 ms.

Note: With only $B = 20$ samples, this interval is unstable. In practice, use $B \geq 1000$ for reliable percentile intervals.

Method 2: Basic Bootstrap Confidence Interval

The basic (or reverse percentile) method uses the empirical quantiles of $(\hat{\theta} - \theta^*)$ to construct the interval.

Theory

The basic bootstrap interval is:

$$[\hat{\theta} - (\theta^*_{(1-\alpha/2)} - \hat{\theta}), \hat{\theta} - (\theta^*_{(\alpha/2)} - \hat{\theta})]$$

Simplifying: $[2\hat{\theta} - \theta^*_{(1-\alpha/2)}, 2\hat{\theta} - \theta^*_{(\alpha/2)}]$.

Calculation

Original estimate: $\hat{\theta} = 297.5$

Bootstrap quantiles: $\theta^*_{(0.025)} = 275.0, \theta^*_{(0.975)} = 307.5$

95% Basic Bootstrap Confidence Interval:

$$\text{CI}_{\text{basic}} = [287.5, 320] \text{ ms}$$

Interpretation

This interval reflects the distribution of $\hat{\theta} - \theta^*$ around the original estimate. It's particularly useful when the bootstrap distribution is asymmetric.

Method 3: Standard Error (Normal Approximation) Method

This method assumes the bootstrap distribution is approximately normal.

Theory

The standard error method uses:

$$\text{CI} = \hat{\theta} \pm z_{\alpha/2} \cdot \text{SE}_{\text{boot}}$$

where SE_{boot} is the standard deviation of the bootstrap distribution. For 95% CI: $z_{0.025} = 1.96$.

95% Standard Error Bootstrap Confidence Interval:

$$\text{CI}_{\text{SE}} = 291.75 \pm 1.96 \times 9.904 = [273.34, 312.16] \text{ ms}$$

Interpretation

This interval assumes normality of the bootstrap distribution. It's simpler to compute but may be less accurate if the bootstrap distribution is skewed.

Method 4: Bias-Corrected and Accelerated (BCa) Bootstrap

The BCa method corrects for bias and skewness in the bootstrap distribution, providing more accurate coverage.

Theory

The BCa interval uses adjusted percentiles: $\text{CI}_{\text{BCa}} = [\theta^*_{(\alpha_1)}, \theta^*_{(\alpha_2)}]$, where:

$$\alpha_1 = \Phi\left(z_0 + \frac{z_0 + z_{\alpha/2}}{1 - a(z_0 + z_{\alpha/2})}\right),$$
$$\alpha_2 = \Phi\left(z_0 + \frac{z_0 + z_{1-\alpha/2}}{1 - a(z_0 + z_{1-\alpha/2})}\right)$$

Φ is the standard normal CDF, z_0 is the bias-correction factor, and a is the acceleration factor.

Calculation

Step 1: Bias-Correction Factor z_0

$$z_0 = \Phi^{-1}\left(\frac{\#\{\theta^*_b < \hat{\theta}\}}{B}\right) = 0.1257$$

Step 2: Acceleration Factor a

The acceleration is estimated using jackknife:

$$a = \frac{\sum_{i=1}^{n}(\bar{\theta}_{(\cdot)} - \theta_{(i)})^3}{6\left[\sum_{i=1}^{n}(\bar{\theta}_{(\cdot)} - \theta_{(i)})^2\right]^{3/2}}$$

where $\theta_{(i)}$ is the median computed from data with i-th observation deleted.

Jackknife samples:

i	Data with x_i removed	$\theta_{(i)}$
1	[280, 310, 265, 420, 290, 305, 275, 315, 380]	305
2	[245, 310, 265, 420, 290, 305, 275, 315, 380]	305
3	[245, 280, 265, 420, 290, 305, 275, 315, 380]	290
4	[245, 280, 310, 420, 290, 305, 275, 315, 380]	305
5	[245, 280, 310, 265, 290, 305, 275, 315, 380]	290
6	[245, 280, 310, 265, 420, 305, 275, 315, 380]	305
7	[245, 280, 310, 265, 420, 290, 275, 315, 380]	290
8	[245, 280, 310, 265, 420, 290, 305, 315, 380]	305
9	[245, 280, 310, 265, 420, 290, 305, 275, 380]	290
10	[245, 280, 310, 265, 420, 290, 305, 275, 315]	290

Jackknife mean: $\bar{\theta}_{(\cdot)} = 297.50$. Squared and Cubed deviations:

$$\sum(\bar{\theta}_{(\cdot)} - \theta_{(i)})^2 = 475.625, \quad \sum(\bar{\theta}_{(\cdot)} - \theta_{(i)})^3 = 371.25$$

$$a = \frac{371.25}{6 \times (475.625)^{3/2}} = 0.00596$$

Step 3: Adjusted Percentiles

For 95% CI: $z_{\alpha/2} = -1.96$, $z_{1-\alpha/2} = 1.96$

Lower & upper percentiles:

$$\alpha_1 = 0.0456, \quad \alpha_2 = 0.9874$$

Step 4: Extract Quantiles

From the sorted bootstrap distribution of 20 values:

Index for 2.63th percentile: $20 \times 0.0456 \approx 1$

$$\theta^*_{(\alpha_1)} = \theta^*_{(1)} = 275.0$$

Index for 97.63th percentile: $20 \times 0.9874 \approx 20$

$$\theta^*_{(\alpha_2)} = \theta^*_{(20)} = 310.0$$

95% BCa Bootstrap Confidence Interval:

$$\text{CI}_{\text{BCa}} = [280.0, 310.0] \text{ ms}$$

Interpretation

The BCa interval adjusts for bias and skewness. With $z_0 \approx 0$ (no bias) and small a (minimal skewness), the adjustments are modest in this example.

Comparison of All Four Methods

Method	95% CI	Width
Percentile	[275.0, 307.5]	32.5
Basic	[287.5, 320.0]	32.5
Standard Error	[273.34, 312.16]	38.82
BCa	[275.0, 310.0]	35.0

Observations:

- All intervals overlap substantially, suggesting reasonable agreement
- Percentile and basic intervals have equal width but different locations
- Standard error interval is widest (assumes normality, may be conservative)
- BCa interval is slightly wider than percentile (adjusts for skewness)
- With only $B = 20$ bootstrap samples, all intervals are unstable and should not be trusted

Effect of Sample Size: Comparison with Normal-Theory Interval

For comparison, the normal-theory confidence interval for the mean (not median):

Sample mean: $\bar{x} = 308.5$ ms

Sample standard deviation: $s = 55.87$ ms

Standard error: $\text{SE} = s/\sqrt{n} = 55.87/\sqrt{10} = 17.67$ ms

95% CI for mean: $308.5 \pm 1.96 \times 17.67 = 308.5 \pm 34.6 = [273.9, 343.1]$ ms

Note: This is for the *mean*, not the median. There is no simple closed-form confidence interval for the median under normality, which is exactly why bootstrap methods are valuable!

Bootstrap Principles and Key Concepts

1. The Bootstrap Assumption

The bootstrap assumes that the **empirical distribution** (the observed sample) is a good approximation to the true population distribution. We treat the sample as a "pseudo-population" and resample from it.

Analogy:

- Population → Sample: Real sampling (what nature did)
- Sample → Bootstrap samples: Simulated resampling (what we do)

2. Sampling With Replacement

Why with replacement?

If we sampled without replacement, every bootstrap sample would be identical to the original (just reordered). Sampling with replacement allows variation between bootstrap samples, mimicking the sampling variability we'd see if we could repeatedly sample from the population.

3. Number of Bootstrap Samples (B)

Guidelines:

- Exploratory analysis: $B = 200$–500
- Standard confidence intervals: $B = 1{,}000$–$2{,}000$
- BCa intervals or hypothesis tests: $B = 2{,}000$–$10{,}000$
- Publication-quality work: $B = 10{,}000+$

Our example used $B = 20$ only for pedagogical purposes to show all calculations explicitly.

4. Bootstrap Standard Error

The standard deviation of the bootstrap distribution estimates the standard error of the statistic:

$$\text{SE}_{\text{boot}}(\hat{\theta}) = \sqrt{\frac{1}{B-1}\sum_{b=1}^{B}(\theta_b^* - \bar{\theta}^*)^2}$$

This is the plug-in estimate of $\text{SE}(\hat{\theta})$ and can be used even when the theoretical standard error is unknown or intractable.

5. What Can Be Bootstrapped?

Works well for:

- Means, medians, trimmed means
- Variances, standard deviations
- Correlation coefficients, regression coefficients
- Quantiles, interquartile ranges
- Ratios, differences
- Complex statistics with unknown sampling distributions

Doesn't work well for:

- Extreme order statistics (minimum, maximum) — underestimates variability
- Statistics dependent on sample size (e.g., range)
- When original sample size is very small ($n < 20$)
- Heavy-tailed distributions (variance may not exist)

Advantages of Bootstrap Methods

1. **Distribution-free:** No assumption about population distribution (normal, exponential, etc.)
2. **Versatile:** Works for complex statistics where theoretical formulas are unavailable
3. **Automatic:** No need to derive sampling distributions or standard errors mathematically
4. **Honest uncertainty:** Captures asymmetry and skewness in sampling distribution
5. **Minimal assumptions:** Only requires that sample is representative of population

Limitations and Cautions

1. **Computational intensity:** Requires many resamples ($B \geq 1000$)
2. **Random variability:** Different runs give slightly different intervals (set random seed for reproducibility)
3. **Small sample sizes:** Bootstrap can fail with $n < 20$; empirical distribution is poor approximation to true distribution
4. **Dependence structures:** Standard bootstrap assumes independence; time series or clustered data require specialized methods (block bootstrap, cluster bootstrap)
5. **Not magic:** Bootstrap cannot extract information that isn't in the data; if sample is unrepresentative, bootstrap will be too
6. **Extreme statistics:** Poor performance for minima, maxima, or statistics sensitive to extreme values

Practical Recommendations

1. **Use BCa for publication:** BCa intervals have best theoretical properties (accurate coverage, transformation-invariant)
2. **Check bootstrap distribution:** Plot histogram of $\{\theta_b^*\}$ to assess normality, bias, skewness
3. **Use sufficient B:** At least $B = 1{,}000$ for confidence intervals; $B = 10{,}000$ is safer
4. **Set random seed:** For reproducibility in publications
5. **Compare methods:** If percentile, basic, SE, and BCa intervals differ substantially, investigate further
6. **Validate assumptions:** Bootstrap assumes sample is representative; check for outliers, data quality issues
7. **Consider alternatives:** For small samples ($n < 20$), permutation tests or exact methods may be better

XXIII. PERM — Technical Notes

Permutation Tests vs. Randomization Tests

The Core Distinction Permutation tests and randomization tests share identical mechanics — both construct a reference distribution by relabelling observations — but they rest on fundamentally different inferential foundations.

	Permutation	Randomization
Justification	Exchangeability of observations (distributional)	Random assignment by experimenter (design-based)
Requires random sample?	Yes, in principle	No
Scope of inference	Population from which units were sampled	The specific units in the experiment
Null hypothesis	Groups drawn from same distribution	Treatment has no effect on any unit

Permutation Test: Distributional Justification The permutation test is valid because, under H_0, the observations are **exchangeable** — any relabelling of group membership is equally probable. Exchangeability requires the observations to be a random sample from a common population (or two populations with the same distribution under H_0).

Example. A biologist measures enzyme activity in $n_1 = 10$ treated fish and $n_2 = 10$ control fish, both randomly sampled from a lake population. Under $H_0\colon \mu_1 = \mu_2$, the treatment labels are arbitrary — any assignment of the 20 measurements to two groups of 10 is equally likely. The reference distribution of the t-statistic is constructed over all $\binom{20}{10} = 184{,}756$ relabellings. Inference generalizes to the lake population because the fish were randomly sampled.

Randomization Test: Design-Based Justification The randomization test is valid because the experimenter **physically randomized** units to treatments. No population and no random sampling are assumed. The reference distribution is the set of all treatment assignments that could have been made at the design stage. Inference is restricted to the specific units in the experiment — no generalization to a wider population is licensed by the test alone.

Example. Fisher's tea-tasting experiment (1935): a lady claims to detect whether milk or tea was poured first. Eight cups are prepared — four of each kind — and she is asked to identify them. There is no population of cups or of ladies. The experimenter randomly assigned the order of the eight cups. Under H_0 (no ability to discriminate), all $\binom{8}{4} = 70$ assignments are equally likely. The lady correctly identifies all eight: $p = 1/70 \approx 0.014$. The conclusion is about *this lady* and *these cups* — not about any population.

The Sharp Null Hypothesis A subtle but important difference lies in the precise null hypothesis. The permutation test null is **distributional**: $H_0\colon F_1 = F_2$ (same distribution). The randomization test null is the **sharp null** (Fisher 1935): $H_0\colon y_i(1) = y_i(0)$ for every unit i — the treatment has *literally no effect on any individual unit*. The sharp null is strictly stronger: it rules out unit-level effects even when population means are equal. This distinction matters in the potential outcomes framework (Rubin causal model) where treatment effects may vary across units.

When the Two Coincide In a randomized controlled trial where units are also a random sample from a population, both justifications apply simultaneously. The permutation test is valid via exchangeability; the randomization test is valid via the design. They produce identical p-values. In practice this is the most common setting, which is why the terms are used interchangeably in much of the applied literature. The distinction matters when *only one* justification holds:

- **Randomized experiment, convenience sample** (e.g. a clinical trial at one hospital): randomization justification holds; permutation exchangeability is questionable. Use the randomization test; restrict inference to these patients.
- **Observational study, random sample** (e.g. survey data with no random assignment): permutation exchangeability holds under H_0; randomization justification does not. Use the permutation test; generalize to the population.
- **Observational study, convenience sample**: neither justification holds cleanly; interpret with caution regardless of which relabelling test is applied.

Practical Implication for This Chapter The permutation t-test (PERM.01), permutation ANOVA (PERM.02), and permutation correlation test (PERM.03) rely on exchangeability and target population-level inference. The Mantel test (PERM.04) similarly requires exchangeability of distance matrices. The randomization test (PERM.05) is explicitly design-based and appropriate when random assignment is known but the sample is not necessarily representative. In randomized experiments with representative samples, either framework applies and the p-values are numerically identical.

Part XXIV

Influential Observations & Anomalies

Identifying unusual data points and assessing their statistical impact

Outlier Detection

XXIV. OUTL — A Conceptual Primer

What Is an Outlier?

An **outlier** is an observation that deviates markedly from the bulk of the data under the assumed model. This definition is inherently **model-dependent**: an observation is an outlier relative to a specific distributional assumption. Under normality, a value at 4σ is extreme; under a Cauchy distribution, it is unremarkable.

Outliers arise from three distinct sources:

- **Measurement or recording errors:** transcription mistakes, instrument malfunction. Should be corrected or removed.
- **Sampling from a different population:** a contaminated observation from a different distribution (e.g. a wrong subject included). Should be removed with justification.
- **Genuine extreme events:** rare but valid observations from the target population. **Should not be removed**; they carry real information.

Distinguishing among these requires domain knowledge, not statistics alone.

Why Outliers Matter

Outliers disproportionately influence:

- The sample mean and variance ($\varepsilon^* = 1/n$ for both).
- OLS regression coefficients and residuals.
- Pearson correlation.
- Classical hypothesis tests (t, F, χ^2).

A single outlier can shift the mean, inflate s^2, and mask or create apparent significance. Detecting outliers before analysis—or using robust alternatives (see ROBU chapter)—protects against these distortions.

Masking and Swamping

Two pathological phenomena complicate outlier detection:

Masking: multiple outliers in the same direction inflate $\bar{x}$ and s, pulling the test statistic toward the center and making individual outliers *appear* non-extreme. Single-outlier tests (Grubbs, Dixon) fail in this setting.

Swamping: the presence of outliers distorts $\bar{x}$ and s so severely that genuine inliers are flagged as outliers.

Both phenomena are avoided by **simultaneous** multiple-outlier methods (Rosner, Tietjen–Moore) or **robust estimators** of center and scale (Iglewicz–Hoaglin, MCD-based Mahalanobis).

The Z-Score and Its Limitations

The classical outlier measure is the standardized z-score:

$$z_i = \frac{x_i - \bar{x}}{s}.$$

$|z_i| > 3$ is a common rule of thumb (covers 99.7% of a Gaussian distribution). **Fundamental limitation:** both $\bar{x}$ and s are sensitive to the very outliers being detected; large outliers inflate s, compressing z_i and making detection harder. This is why robust alternatives to $\bar{x}$ and s are essential.

I. Single Outlier Tests (Univariate, Normal)

All tests below assume normality and test the most extreme observation.

Grubbs' test: Tests H_0: no outlier in $\{x_1, \ldots, x_n\}$ from $\mathcal{N}(\mu, \sigma^2)$. The statistic is the maximum absolute z-score:

$$G = \frac{\max_i |x_i - \bar{x}|}{s}.$$

Critical values are tabulated; $G \sim t_{n-2}$ transformation is used. Apply only once per dataset; applying sequentially after removing suspects inflates Type I error without correction.

Dixon's Q test: Designed for **small samples** (n = 3–30). For the suspected maximum outlier $x_{(n)}$:

$$Q = \frac{x_{(n)} - x_{(n-1)}}{x_{(n)} - x_{(1)}},$$

the ratio of the gap to the range. Different formulas apply for small, medium, and large n (using the range of the full sample or subsets). Critical values depend on n; exact tables required. Assumes normal distribution; sensitive to a single extreme value.

Chauvenet's criterion: Flag observation x_i as a potential outlier if the expected number of observations at least as extreme as x_i in n trials is < 0.5:

$$n \cdot 2\,[1 - \Phi(|z_i|)] < 0.5 \;\Rightarrow\; |z_i| > \Phi^{-1}(1 - 1/(4n)).$$

Threshold increases with n; heuristic rather than formal hypothesis test; not recommended for $n > 30$.

Thompson Tau test: Uses a threshold $\tau = t_{\alpha/2,n-2}(n-1)/\sqrt{n(n-2+t^2_{\alpha/2,n-2})}$ and flags x_i if $|x_i - \bar{x}|/s > \tau$. Related to Grubbs but uses a different critical-value derivation; iterative application requires α-adjustment.

Note: Thompson's Tau test was intentionally excluded. It is closely related to Grubbs' test through a simple transformation and typically produces identical conclusions regarding outlier detection.

II. Multiple Outlier Tests

Rosner's generalized ESD test: Tests for up to r outliers simultaneously, where r is pre-specified. Computes r Grubbs-type statistics $G_1 \geq G_2 \geq \cdots \geq G_r$ by sequentially removing the most extreme observation:

$$G_i = \frac{\max_j |x_j - \bar{x}_{(i)}|}{s_{(i)}},$$

where $\bar{x}_{(i)}$ and $s_{(i)}$ are computed with previous suspects removed. The number of outliers is the largest k such that $G_k > \lambda_k$, where λ_k uses a t_{n-k-1} critical value. **Avoids masking and swamping**; the recommended default for multiple-outlier detection under normality.

Tietjen–Moore test: Tests for exactly k outliers (pre-specified). Computes a likelihood-ratio-type statistic comparing the sum of squared deviations with k suspects removed vs. the full dataset. More powerful than Rosner when the exact number of outliers is known; less flexible.

Peirce's criterion: A classical (pre-computer) criterion for removing outliers based on maximizing the product of the probability of the remaining observations. Uses a tabulated threshold for R (ratio of the deviation to s) depending on n and the number of suspected outliers. Historically important but not easily extended to modern settings.

III. Robust Outlier Detection

Iglewicz–Hoaglin modified Z-score: Replaces mean and standard deviation with robust alternatives:

$$M_i = \frac{0.6745\,(x_i - \tilde{x})}{\text{MAD}},$$

where $\tilde{x}$ is the sample median and MAD = median($|x_i - \tilde{x}|$). The factor $0.6745 = \Phi^{-1}(0.75)$ makes MAD a consistent estimator of σ under normality. Flag $|M_i| > 3.5$ as a potential outlier (Iglewicz and Hoaglin's recommended threshold).

Advantages over classical z-score: M_i is resistant to masking (the median and MAD have breakdown point $\varepsilon^* = 0.5$); computed without the influence of the very observations being tested. Suitable for skewed distributions and small samples.

IV. Multivariate Outlier Detection

In multivariate data ($p \geq 2$ variables), a univariate outlier on any single variable is detectable by marginal tests. However, an observation can be **a multivariate outlier without being a univariate outlier on any single variable**—it lies far from the multivariate center in the joint space.

Mahalanobis distance:

$$D_i^2 = (\mathbf{x}_i - \bar{\mathbf{x}})^\top \mathbf{S}^{-1} (\mathbf{x}_i - \bar{\mathbf{x}}),$$

where $\bar{\mathbf{x}}$ and $\mathbf{S}$ are the sample mean vector and covariance matrix. Under multivariate normality, $D_i^2 \xrightarrow{d} \chi_p^2$. Flag $D_i^2 > \chi^2_{p,0.975}$ as a potential outlier (two-sided, $\alpha = 0.05$ per observation; Bonferroni-corrected threshold: $\chi^2_{p,1-0.025/n}$).

Classical vs. robust Mahalanobis: Like the z-score, $\bar{\mathbf{x}}$ and $\mathbf{S}$ are influenced by the outliers being detected. The **minimum covariance determinant (MCD)** estimator finds the subset of $h = \lfloor (n + p + 1)/2 \rfloor$ observations with the smallest determinant of the sample covariance; $\bar{\mathbf{x}}_{\text{MCD}}$ and $\mathbf{S}_{\text{MCD}}$ computed from this subset are highly robust ($\varepsilon^* \approx 0.5$). The robust Mahalanobis distance $D^2_{i,\text{MCD}} = (\mathbf{x}_i - \bar{\mathbf{x}}_{\text{MCD}})^\top \mathbf{S}^{-1}_{\text{MCD}} (\mathbf{x}_i - \bar{\mathbf{x}}_{\text{MCD}})$ avoids masking in multivariate settings.

Test Selection Guide

Setting	Recommended test
1 outlier, normal, $n \geq 7$	Grubbs
1 outlier, normal, $n \leq 30$	Dixon's Q
Up to r outliers, normal	Rosner (ESD)
Exactly k outliers, normal	Tietjen–Moore
Robust, distribution-free	Iglewicz–Hoaglin
Multivariate, classical	Mahalanobis D^2
Multivariate, robust	MCD-based D^2

Common Pitfalls

- **Applying single-outlier tests when multiple outliers are present:** masking leads to missed detection; use Rosner or robust methods.
- **Removing without investigation:** outliers may be the most scientifically important observations; always investigate the cause before removal.
- **Violating normality assumptions:** Grubbs, Dixon, and Rosner assume normality; use Iglewicz–Hoaglin or robust Mahalanobis for non-Gaussian data.
- **Iterative application without correction:** applying Grubbs sequentially without adjusting α (as Rosner does internally) inflates Type I error.
- **Using classical Mahalanobis in multivariate data:** the swamping and masking problems apply in p dimensions; use MCD.
- **No multiplicity adjustment for marginal checks:** flagging observations across p variables simultaneously requires Bonferroni or FDR correction.

Bottom Line

Outlier detection is model-dependent: an observation is an outlier relative to an assumed distribution. Distinguish between errors (remove), sampling contamination (remove with justification), and genuine extremes (retain). Use Rosner's test for multiple suspected outliers, Iglewicz–Hoaglin for robust univariate detection, and MCD-based Mahalanobis for multivariate data. Never remove outliers without investigating their cause.

Grubbs' Test

Test for Single Outlier in Univariate Normal Data

OUTL.1 | **Category:** Outlier Detection | **Data:** Continuous | **Null:** Grubbs' distribution (t related)

When to Use

- Detect a single extreme outlier in a univariate dataset.
- Test if the sample maximum or minimum is discordant.
- Quality control aberrant or erroneous measurements.
- Confirm a suspected extreme value before removal.
- Visual outlier detection is inconclusive (small sample).

Assumptions

- Data are approximately normally distributed (critical).
- Observations are independent and identically distributed.
- At most one outlier is present per test (test sequentially).
- Sample size is adequate ($n \geq 7$, preferably $n \geq 10$).
- Data are continuous (interval or ratio scale).

Test Statistic

$$G = \frac{\max_i |x_i - \bar{x}|}{s} = \frac{\max(x_{\max} - \bar{x}, \bar{x} - x_{\min})}{s}, \quad \text{where } \bar{x} = \frac{\sum x_i}{n}, \quad s = \sqrt{\frac{\sum (x_i - \bar{x})^2}{n-1}}.$$

Notation: G is Grubbs' statistic, $x_{\max}, x_{\min}$ are extreme values, $\bar{x}$ is sample mean, s is sample standard deviation

Plain English: Grubbs' test detects a single extreme observation by computing how many standard deviations it lies from the mean. If this maximum absolute z-score exceeds a sample-size–adjusted critical value, the observation is flagged as an outlier. The procedure controls Type I error when evaluating the most extreme value in the sample.

⚠ Practitioner Warning

Assumes normality; departures weaken validity and power. The test detects only one outlier per application, and repeated use inflates error rates. Multiple outliers may mask each other. Masking occurs when outliers distort the mean and standard deviation, hiding their own extremeness. Critical values depend on sample size and must be obtained from tables or formulas. The test does not determine whether an outlier is erroneous or genuine; subject-matter judgment is required. One-sided versions test only the maximum or minimum.

Hypotheses

- H_0: No outliers present ($x_i \sim$ normal distribution).
- H_1: Extreme observation is outlier (contamination).

Decision Rule: Reject H_0 if $G > G_{\text{crit}}(\alpha, n)$ where critical value from Grubbs' table.

Software References [Python, R, SAS]

Python	R	SAS
`scipy.stats (via outliers pkg)`	`outliers::grubbs.test`	`Not Native (Manual)`

Additional Notes

Grubbs' test provides a formal test for a *single* outlier using the statistic $G = \max_i |x_i - \bar{x}|/s$, the largest standardized deviation from the mean. One-sided variants test only the maximum or minimum; the two-sided form (default) tests the most extreme observation. Critical values are derived from the Student's t-distribution under normality, with the significance level adjusted to $\alpha/(2n)$ to account for selecting the most extreme of n observations. This Bonferroni-style correction makes the test conservative. The method is sensitive to multiple outliers. *Masking* occurs when outliers inflate s and reduce G, hiding true outliers; *swamping* occurs when legitimate observations are flagged due to other extremes. Iterative application is not theoretically justified and inflates the family-wise error rate (FWER). For multiple outliers, the Generalized ESD test (Rosner, 1983) extends Grubbs' approach while controlling error rates. Simpler alternatives, such as Dixon's Q-test, are sometimes used for very small samples. See Technical Notes for derivation of the t distribution relationship.

Miscellanea: Frank E. Grubbs developed his outlier test while working at the U.S. Army Ballistic Research Laboratory, motivated by the need to identify aberrant measurements in experimental data. Earlier practice often relied on informal judgment to discard extreme values; Grubbs' work provided a formal statistical criterion with a controlled error rate. His original contributions appeared in the 1950s, with a later synthesis in a 1969 *Technometrics* paper. Despite its simplicity, Grubbs' test remains widely used in laboratory and quality-control settings, where detecting a single aberrant measurement is often the primary objective.

Reference(s): Grubbs (1950, 1969); Barnett & Lewis (1994)

Worked Example

Problem Statement A quality engineer records six instrument readings from a production line:

$$x = (1,\ 2,\ 3,\ 4,\ 5,\ 15).$$

The value 15 appears far above the rest. Apply the Grubbs (1950) test to determine whether 15 is a statistically significant outlier, assuming the underlying population is normal. H_0: no outlier (all observations from the same normal population) vs. H_1: the most extreme observation is an outlier.

1. Sample Mean and Standard Deviation

$$\bar{x} = 5, \qquad s^2 = 26 \Rightarrow s = \sqrt{26}.$$

2. Grubbs Statistic

The Grubbs statistic is the maximum absolute standardized deviation:

$$G = \frac{\max_i |x_i - \bar{x}|}{s} = \frac{|15 - 5|}{\sqrt{26}} = \frac{10}{\sqrt{26}} \approx 1.961.$$

3. Critical Value and Decision

From the Grubbs (1969) table, the critical value at $\alpha = 0.05$ for $n = 6$ is $G_{\text{crit}} = 1.887$.

$$G = 1.961 > G_{\text{crit}} = 1.887 \implies \textbf{reject } H_0.$$

The value $x = 15$ is a statistically significant outlier at the 5% level.

4. Equivalent t-Formulation

Inverting the G_{crit} formula algebraically gives the equivalent t-statistic. Starting from $G_{\text{crit}} = \frac{n-1}{\sqrt{n}}\sqrt{t_c^2/(n-2+t_c^2)}$ and solving for t_c:

$$t^2 = \frac{n\,G^2\,(n-2)}{(n-1)^2 - n\,G^2} = 48, \Rightarrow t \approx 6.928, \ \nu = n-2 = 4 \text{ df.}$$

$$t \approx 6.928 > t_{0.05/12,\,4} = 4.851 \implies \textbf{reject} \checkmark$$

consistent with $G > G_{\text{crit}}$.

	Value	CV (5%)	Decision
G	1.961	1.887	Reject H_0
t (equivalent)	6.928	4.851	Reject H_0

Practitioner Insight: The Grubbs (1950) test—also called the extreme Studentised deviate (ESD)—assesses whether the single most extreme observation is inconsistent with the rest under normality. The statistic $G = \max_i |x_i - \bar{x}|/s$ compares the largest standardized deviation to a critical value derived via a t-distribution transformation. For multiple outliers, Rosner's generalized ESD extends this approach; nonparametric alternatives include the Hampel identifier and Tukey's IQR fence.

Common Pitfalls

- **Removing values without formal testing.** Never delete data because it "looks wrong." Apply a formal test, investigate the cause, and document decisions. If unexplained, report results with and without the observation.
- **Repeated application without adjustment.** Sequential Grubbs testing inflates Type I error. For multiple suspected outliers, adjust α or use procedures designed for multiple detection (e.g., Rosner/Generalized ESD).
- **Applying to non-normal data.** Grubbs assumes normality; violations distort size and power. Check normality first (See `NORM` chapter) or transform/use robust alternatives when data are skewed.
- **Confusing outliers with influential regression points.** Grubbs detects univariate extremes only. Leverage or influential observations in regression require model-based diagnostics (e.g., Cook's distance).
- **Failing to report outlier handling.** Transparency is essential—state the test used, α, identified values, investigation, and sensitivity of results to removal.

Relationship to Thompson Tau.

Thompson (1935) proposed an equivalent procedure—predating Grubbs by fifteen years—that uses the same test statistic $G = |x_i - \bar{x}|/s$ but compares it against a critical value derived from the uncorrected t-distribution: $\tau = t_{\alpha/2,n-2}(n-1)/\sqrt{n(n-2+t^2_{\alpha/2,n-2})}$, without the Bonferroni adjustment for the n implicit comparisons (i.e., using $\alpha/2$ instead of $\alpha/(2n)$). Thompson Tau is therefore slightly more liberal than Grubbs — it rejects more readily — and the two converge numerically for large n. Thompson's procedure survives in some engineering and quality control standards (e.g. ASTM E178) under the name "Thompson Tau test"; it is methodologically identical to Grubbs and should be treated as such when encountered in practice.

Dixon's Q Test

Simple Outlier Test Based on Range Ratios for Small Samples

OUTL.2 | **Category:** Outlier Detection | **Data:** Continuous | **Null:** Dixon Q distribution

When to Use

- Detect a single outlier in small samples ($3 \leq n \leq 30$).
- Quick screening for extreme values in ordered data.
- Quality control settings with limited or sparse observations.
- Computational simplicity is preferred over max. power.
- Simple alternative to Grubbs' test for small datasets.

Assumptions

- Data are approximately normally distributed.
- Observations are independent and identically distributed.
- At most one outlier is present per test.
- Data are continuous and can be meaningfully ordered.
- No strong skewness or multiple extreme values present.

Test Statistic

$$Q = \frac{\text{gap}}{\text{range}}; \quad Q_{10} = \frac{x_{(n)} - x_{(n-1)}}{x_{(n)} - x_{(1)}}, \quad Q_{11} = \frac{x_{(n)} - x_{(n-1)}}{x_{(n)} - x_{(2)}} \quad Q_{21} = \frac{x_{(n)} - x_{(n-2)}}{x_{(n)} - x_{(2)}} \quad Q_{22} = \frac{x_{(n)} - x_{(n-2)}}{x_{(n)} - x_{(3)}}.$$

Notation: $x_{(i)}$ are ordered observations, gap is distance from extreme to nearest neighbor, range depends on variant

Plain English: Dixon's Q test detects a single outlier by comparing the gap between an extreme value and its nearest neighbor to the sample range. If this ratio is large, the extreme observation may be discordant. The method requires only ordering the data and computing a simple ratio, making it convenient for small samples without specialized software.

⚠ Practitioner Warning

Different Q variants apply depending on sample size and which extreme is tested; correct critical-value tables must be used. The test is limited to small samples ($n \leq 30$), assumes normality, and detects only one outlier at a time. Multiple outliers may mask each other, and repeated testing inflates Type I error. It's simpler but less powerful than Grubbs' test.

Hypotheses

- H_0: No outliers present ($x_i \sim$ normal distribution).
- H_1: Extreme observation is outlier (contamination).

Decision Rule: Reject H_0 if $Q > Q_{\text{crit}}(\alpha, n)$ where critical value (differs by Q variant and n) from Dixon tables.

Software References [Python, R, SAS]

Python	R	SAS
`scipy.stats (via outliers pkg)`	`outliers::dixon.test`	`Not Native (Manual)`

Additional Notes

Dixon's framework comprises a family of related statistics ($Q_{10}, Q_{11}, Q_{12}, Q_{20}, Q_{21}, Q_{22}$), all constructed as ratios of a gap involving a suspected extreme value to an appropriate range or subrange. The purpose of these variants is to mitigate the influence of additional outliers that may distort the full range. The simplest form, Q_{10} (the Q test), compares the nearest-neighbor gap to the total range and is appropriate when only a single outlier is suspected in very small samples. As sample size increases, more robust ratios are preferred: Q_{11} reduces sensitivity to a possible outlier on the opposite end, while Q_{21} and Q_{22} progressively exclude observations from both ends of the sample to guard against masking by multiple extremes. Although Q_{12} and Q_{20} are part of Dixon's original formulation, they are not generally recommended, as alternative ratios provide better overall performance. In practice, a simple progression is followed based on sample size: use Q_{10} for $3 \leq n \leq 7$, Q_{11} for $8 \leq n \leq 10$, Q_{21} for $11 \leq n \leq 13$, and Q_{22} for larger samples. This sequence reflects an increasing emphasis on robustness, trading simplicity for protection against multiple or competing outliers. More information in Technical Notes.

Miscellanea: Wilfrid Dixon's range-based approach emerged from a broader comparison of outlier detection methods and gained prominence because of its simplicity and strong performance in small samples. The Q_{10} statistic became widely known as the "Q test," a name that persists in most analytical chemistry texts. Over time, inconsistencies and errors in published critical values—particularly at the commonly used 95% confidence level—led to widespread misuse, prompting Rorabacher's (1991) influential corrections. The test's enduring popularity reflects the realities of laboratory practice, where small datasets are common, even as statisticians such as Deming cautioned against rejecting observations based solely on statistical criteria.

Reference(s): Dixon (1950, 1951); Dean & Dixon (1951); Rorabacher (1991)

Worked Example

Problem Statement An analytical chemist records five replicate concentration measurements (mg/L):

$$\text{Sample (order statistics): } x_{(1)} = 1, \quad x_{(2)} = 10, \quad x_{(3)} = 11, \quad x_{(4)} = 12, \quad x_{(5)} = 13.$$

The value 1 appears far below the rest. Apply Dixon's Q test to determine whether 1 is a statistically significant outlier.

$$H_0: \text{no outlier (all from same population)}; \quad H_1: \text{the most extreme value is an outlier.}$$

1. Compute the Q Statistic
The Q statistic is the ratio of the *gap* (distance from the suspect value to its nearest neighbor) to the *range* (total spread of the sample): Q_{10} formula is used here

Testing $x_{(1)} = 1$ as a lower outlier:

$$Q_{\text{min}} = \frac{x_{(2)} - x_{(1)}}{x_{(n)} - x_{(1)}} = \frac{10 - 1}{13 - 1} = 0.750.$$

Testing $x_{(5)} = 13$ as an upper outlier:

$$Q_{\text{max}} = \frac{x_{(n)} - x_{(n-1)}}{x_{(n)} - x_{(1)}} = \frac{13 - 12}{13 - 1} \approx 0.083.$$

The lower value is far more extreme: $Q = Q_{\text{min}} = 3/4$.

2. Critical Values and Decision
Critical values for $n = 5$ (Rorabacher 1991, two-sided):

α	10%	**5%**	1%
Q_{crit}	0.642	**0.710**	0.821

$$Q = 0.750 > Q_{\text{crit}}(5\%) = 0.710 \implies \textbf{reject } H_0 \text{ at } 5\%.$$

$$Q = 0.750 < Q_{\text{crit}}(1\%) = 0.821 \implies \text{FTR at } 1\%.$$

The value $x = 1$ is a significant outlier at the 5% level but not at the 1% level. Evidence is moderate.

3. Grubbs Comparison
For reference, the Grubbs statistic on the same data:

$$\bar{x} = 9.4, \quad s \approx 4.827, \quad G = \frac{|1 - 9.4|}{4.827} \approx 1.740.$$

$G_{\text{crit}}(5\%, n{=}5) = 1.715$: also rejects at 5%. The two tests agree; Dixon required no mean or variance.

Test	Stat	CV (5%)	Decision
Dixon	$Q = 0.750$	0.710	Reject
Grubbs	$G \approx 1.740$	1.715	Reject

Practitioner Insight: Dixon's (1950) Q test is a pure order-statistic method, and avoids using the sample mean and variance. This makes it ideal for small samples ($n \leq 10$) and quick hand calculation. Q has a simple geometric meaning: it measures what fraction of the total spread is due to the step from the suspected outlier to its nearest neighbor. If that step dominates the range (e.g., 0.75 here), the value is unusually isolated. Dixon's test is less powerful than Grubbs for $n \geq 8$ because it discards interior information, though both agree in this example. Critical values depend on formulation: Dean and Dixon (1951) provide one-sided values, while Rorabacher (1991) gives corrected two-sided values. Unlike Grubbs, Q has no simple t-distribution link; its critical values come from the exact distribution of the gap-to-range ratio under normality.

Common Pitfalls

- **Using the wrong Q variant.** Each sample size requires a specific Q formula and matching critical values. Using the wrong variant invalidates the test. Always match n to the correct table or use reliable software.
- **Testing both extremes without adjustment.** Dixon is one-sided. Testing minimum and maximum separately doubles Type I error. Pre-specify the extreme or adjust α (e.g., Bonferroni).
- **Applying to large samples.** Designed for small samples ($n \leq 30$; best for $n \leq 10$). For moderate or large n, Grubbs' test is generally more appropriate and powerful.
- **Using on non-normal data.** Assumes approximate normality. Severe skewness inflates false positives. Check distribution or transform before applying.
- **Failing to document decisions.** Report the Q variant used, test statistic, critical value, flagged observation, investigation, and impact on results. Never remove values without documentation.

Chauvenet's Criterion

Classical Outlier Rejection Based on Probability Threshold

OUTL.3 | **Category:** Outlier Detection | **Data:** Continuous | **Null:** Normal distribution

When to Use

- Detect outliers using a probability-based rejection rule.
- Small to moderate samples where normality is plausible.
- Quick screening before formal modeling or testing.
- Situations favoring simple rules over formal statistical tests.
- Laboratory or measurement data with expected random variation.

Assumptions

- Data are approximately normally distributed.
- Observations are independent and identically distributed.
- Outliers are rare under the assumed distribution.
- Data are continuous and can be meaningfully ordered.
- Sample represents a common population except possible outliers.

Test Statistic

$$z_i = \frac{|x_i - \bar{x}|}{s}, \quad \text{reject if } P(|Z| \geq z_i) < \frac{1}{2n}, \quad \text{i.e., } z_i > z_{1-1/(4n)}.$$

Notation: z_i is standardized deviation, $\bar{x}$ is sample mean, s is sample standard deviation, n is sample size

Plain English: Chauvenet's criterion labels an observation an outlier if its two-tailed normal probability is less than $1/(2n)$—i.e., fewer than half an expected occurrence in a sample of size n. Each value is standardized, its tail probability computed, and those beyond the threshold are rejected. It provides a fixed probabilistic screening rule rather than a formal hypothesis test.

⚠ Practitioner Warning

Not a formal test and does not control Type I error rigorously. The $1/(2n)$ cutoff is arbitrary. Assumes normality and uses sample mean and standard deviation (including the suspected value), creating circularity and potential masking. Iterative removal inflates rejection rates. Modern practice prefers Grubbs', Dixon's, or Rosner's tests for valid inference.

Hypotheses

- H_0: Observation consistent with normal distribution.
- H_1: Observation is outlier (too extreme to occur by chance).

Decision Rule: Reject H_0 if observation $z_i > z_{1-1/(4n)}$ where z_p is standard normal quantile.

Software References [Python, R, SAS]

`scipy.stats.norm.ppf`	`Not Native (Manual)`	`Not Native (Manual)`

Additional Notes

William Chauvenet (1863) introduced this rule in *Manual of Spherical and Practical Astronomy* for screening astronomical observations. An observation is rejected if its two-tailed probability under normality is less than $1/(2n)$—interpreted as fewer than half an expected occurrence in a sample of size n. The cutoff depends on n (e.g., $z \approx 1.96$ for $n = 10$, 2.58 for $n = 50$, 2.81 for $n = 100$) and increases slowly with sample size. Because mean and variance are computed from the full sample, true outliers can mask themselves. Iterative application inflates error rates and lacks theoretical justification. Modern tests (Grubbs, Dixon, Rosner) provide formal Type I error control, but Chauvenet remains pedagogically useful and appears in some engineering standards.

Miscellanea: William Chauvenet, a 19th-century American astronomer and founder of Washington University's mathematics department, proposed this rule to standardize rejection of faulty astronomical measurements. He later became the second chancellor of the university. Observers faced atmospheric distortions, instrument defects, and recording mistakes, and needed a consistent rejection guideline. While modern practitioners often view such fixed cutoffs as arbitrary, the criterion persisted in physics and engineering education as a historical bridge between intuition and formal hypothesis testing.

Reference(s): Chauvenet (1863); Barnett & Lewis (1994)

Worked Example

Problem Statement A physics student records five independent measurements of a calibration constant: $x = (10, 10, 10, 8, 17)$. The reading 17 appears suspiciously high. Apply Chauvenet's criterion to determine whether any observation should be flagged for rejection, and compare with Dixon's and Grubbs' tests on the same data.

1. Sample Mean and Variance

$$\bar{x} = 11, \quad s^2 = 12.$$

2. Chauvenet's Criterion

Flag x_i if the expected number of observations at least as extreme is less than $1/2$:

$$z_i = \frac{|x_i - 11|}{\sqrt{12}}, \quad E_i = n \cdot 2(1 - \Phi(z_i)) < \tfrac{1}{2}.$$

x_i	z_i	E_i	Flag?
10	0.289	$5 \times 0.773 = 3.86$	No
8	0.866	$5 \times 0.386 = 1.93$	No
17	$\sqrt{3} \approx 1.732$	$5 \times 0.083 = \mathbf{0.416}$	**Yes**

For $x = 17$:

$$E_{17} = 5 \times 2(1 - \Phi(\sqrt{3})) = 0.416 < \tfrac{1}{2}.$$

Flag $x = 17$: fewer than half an observation is expected this far from the mean by chance.

3. Critical-z Formulation

Equivalently, flag if $z_i > z_{\text{crit}}$ where:

$$z_{\text{crit}} = \Phi^{-1}\left(1 - \tfrac{1}{4n}\right) = \Phi^{-1}\left(\tfrac{19}{20}\right) = 1.645.$$

$z_{17} \approx 1.732 > 1.645$: flag. $\quad z_8 \approx 0.866 < 1.645$: keep.

For $n = 5$, the Chauvenet threshold $1 - 1/(4n) = 0.95$ gives $z_{\text{crit}} = 1.645$, the one-sided 5% normal cutoff (equivalently, the two-sided 10% cutoff per observation).

4. Comparison with Dixon and Grubbs

Test	Statistic	CV (5%)	Decision
Chauvenet	0.416	< 0.5	Reject
Grubbs	1.732	1.715	Reject
Dixon	0.778	0.710	Reject

All three tests flag $x = 17$. Chauvenet and Grubbs share the same test statistic $\sqrt{3}$ on this data (both use $z = |x - \bar{x}|/s$); they differ only in how the threshold is set.

Practitioner Insight: Chauvenet's criterion (1863) is one of the earliest widely used outlier rules, based on probability rather than formal hypothesis testing. It flags an observation if fewer than half an observation would be expected at least as extreme under normality, i.e., $E < 1/2$, giving $z_{\text{crit}} = \Phi^{-1}(1 - 1/(4n))$, which increases with n. Unlike Grubbs or Dixon, Chauvenet does not control a fixed significance level; the $E < 1/2$ rule is a convention. For $n = 5$, the cutoff is $z_{\text{crit}} = 1.645$, equivalent to a two-sided 10% cutoff per observation. Both Chauvenet and Grubbs agree here, but repeated application of Chauvenet's rule can inflate false positives and should be avoided.

Common Pitfalls

- **Treating it as a formal hypothesis test.** Chauvenet's criterion is a screening rule, not a test with controlled α. Do not claim statistical significance based on it alone; confirm with a formal test (e.g., Grubbs) when inference matters.
- **Iterating excessively.** Repeated recalculation and removal can eliminate valid observations and inflate rejection rates. If multiple outliers are suspected, use Rosner's test instead.
- **Using the $1/(2n)$ cutoff dogmatically.** The 0.5 threshold is arbitrary and context-dependent. Document the rule used and avoid treating it as theoretically justified.
- **Ignoring normality.** The rule relies on normal tail probabilities. Apply only after checking approximate normality or transforming skewed data.
- **Preferring it over modern methods for formal research.** Chauvenet lacks rigorous error control. For publication or regulatory work, use Grubbs, Dixon, or Rosner instead.

Critical z-values by sample size

Sample size, n	5	10	15	20	25	30	35	40	45	50	75	100
Critical $\lvert z \rvert$	1.64	1.96	2.13	2.24	2.33	2.39	2.45	2.50	2.54	2.58	2.71	2.81

Rosner's Test

Generalized ESD Test for Multiple Outliers

OUTL.4 | **Category:** Outlier Detection | **Data:** Continuous | **Null:** t-distribution (related)

When to Use

- Detect multiple outliers (pre-chosen k) in univariate data.
- Appropriate when several extremes are suspected and masking is possible.
- Extension of Grubbs' test with controlled family-wise error.

Assumptions

- Observations are i.i.d. normal in the absence of outliers.
- Upper bound k specified in advance ($k < n/2$).
- Sample size adequate for reliable inference ($n \geq 25$ recommended).

Test Statistic

$$R_i = \frac{\max_j |x_j - \bar{x}_i|}{s_i}, \quad i = 1, \ldots, k, \quad \lambda_i = \frac{(n-i)t_{p,n-i-1}}{\sqrt{(n-i-1+t^2_{p,n-i-1})(n-i+1)}}.$$

Sequentially remove most extreme, recompute. Compare $\{R_i\}$ to $\{\lambda_i\}$ to determine number of outliers.
Notation: R_i is i-th test statistic, $\bar{x}_i, s_i$ computed after removing $i-1$ extremes, λ_i is critical value, $p = \alpha/(2(n-i+1))$

Plain English: Rosner's test identifies multiple outliers by sequentially removing the most extreme observation and recomputing the statistic, up to a user-specified maximum k. At each step, the extremeness of the remaining value is evaluated using critical values that control family-wise error. The final number of outliers is determined by the last step at which the test statistic exceeds its critical value.

⚠ Practitioner Warning

Requires specifying an upper bound k in advance ($k < n/2$). Assumes normality of the non-outlying data; violations reduce validity and power. If more than k outliers exist, only k will be detected. Performance is limited in small samples ($n < 25$), and results depend on the sequential removal process.

Hypotheses

- H_0: No outliers present in dataset ($x_i \sim$ normal distribution).
- H_1: Up to k outliers present (some discordant from rest).

Decision Rule: Reject H_0 if Number of outliers = largest i for which $R_i > \lambda_i$. If all $R_i \leq \lambda_i$, conclude no outliers.

Software References [Python, R, SAS]

Python	R	SAS
`Not Native (Manual)`	`EnvStats::rosnerTest`	`Not Native (Manual)`

Additional Notes

Rosner (1983) introduced the generalized Extreme Studentized Deviate (ESD) test to address masking in multiple outlier detection. The procedure sequentially removes the most extreme value, recomputes statistics, and forms a sequence $R_1, \ldots, R_k$, where R_i is calculated after removing $i-1$ observations. Each R_i is compared to a critical value λ_i that adjusts for multiple testing. If $R_1 > \lambda_1, \ldots, R_r > \lambda_r$ but $R_{r+1} \leq \lambda_{r+1}$, then r outliers are declared. The user specifies an upper bound k; too small may miss outliers, too large reduces power (typically $k \ll n/2$). The method controls the family-wise error rate across all tests. Tietjen–Moore is an alternative when the exact number of outliers is known.

Miscellanea: Bernard Rosner developed the generalized ESD test in 1983 at the Harvard School of Public Health after observing that single-outlier tests failed when multiple aberrant values masked each other. Rosner's procedure replaced informal "rules of thumb"—such as the 3σ threshold—with a formal framework for rigorous error control. By enabling the simultaneous detection of up to k outliers while maintaining a stable Type I error rate, it transformed outlier identification from an arbitrary exercise into a mathematically grounded discipline. The method quickly gained adoption in medical, environmental, and laboratory research and was later incorporated into EPA guidance for environmental data analysis.

Reference(s): Rosner (1983); Barnett & Lewis (1994)

Worked Example

Problem Statement Eight laboratory measurements are recorded: $x = (2, 3, 4, 5, 6, 7, 15, 21)$. Both 15 and 21 are suspected outliers, but the masking problem prevents a single Grubbs test from detecting either. Apply Rosner's generalized ESD test with $k = 2$ (test for up to two outliers) at $\alpha = 0.05$.

1. Step $i = 1$: Test on Full Sample ($m = 8$)

$$\bar{x}_1 = 7.875, \quad s_1^2 = 353/8.$$

For the most extreme observation: $x = 21$,

$$R_1 = \frac{|21 - 7.875|}{\sqrt{353/8}} \approx 1.976.$$

Critical value with $p = \alpha/(2m) = 0.05/16$:

$$\lambda_1 = \frac{m-1}{\sqrt{m}}\sqrt{\frac{t^2_{p,m-2}}{m-2+t^2_{p,m-2}}} = 2.127.$$

$R_1 = 1.976 < \lambda_1 = 2.127$: **do not reject at step 1.**

The two moderate outliers inflate $\bar{x}_1$ and s_1 simultaneously, masking both: neither looks extreme relative to the distorted spread. A single Grubbs test ($G = R_1 \approx 1.976$, $G_{crit} = 2.127$) would also fail to detect any outlier.

2. Step $i = 2$: Test After Removing $x = 21$ ($m = 7$)
Remaining data: $\{2, 3, 4, 5, 6, 7, 15\}$.

$$\bar{x}_2 = 6, \qquad s_2^2 = 56/3.$$

For the most extreme: $x = 15$,

$$R_2 = \frac{|15 - 6|}{\sqrt{56/3}} \approx 2.083.$$

Critical value with $p = \alpha/(2m) = 0.05/14$:

$$\lambda_2 = 2.020.$$

$R_2 = 2.083 > \lambda_2 = 2.020$: **reject at step 2.**

3. Decision Rule
The largest i for which $R_i > \lambda_i$ is $i = 2$.

$\Rightarrow$ **Two outliers** : $x = 21$ and $x = 15$.

Step	m	R_i	λ_i	$R_i > \lambda_i$?	
1	8	1.976	2.127	No	(masking)
2	7	2.083	2.020	**Yes**	outlier

Practitioner Insight: Rosner's (1975, 1983) generalized ESD test addresses the *masking problem*: multiple outliers inflate s and pull $\bar{x}$, making none appear extreme under a single Grubbs test. The remedy is sequential removal with Bonferroni-adjusted cutoffs. At step i, compute R_i on the reduced sample of size $m = n - i + 1$ and compare to λ_i using $p = \alpha/(2m)$ to control the family-wise error. The decision is based on the *largest* i with $R_i > \lambda_i$, avoiding *swamping* (over-rejection). Here, 15 and 21 initially mask each other ($R_1 < \lambda_1$), but removing 21 reveals 15 as significant ($R_2 > \lambda_2$). Each λ_i equals the Grubbs critical value for the reduced sample, so Rosner is effectively sequential Grubbs testing. The parameter k must be chosen in advance; selecting a conservative upper bound balances missed detections and loss of power.

Common Pitfalls

- **Choosing k post hoc.** The upper bound k must be specified before analysis. Selecting k after inspecting the data invalidates error control and inflates Type I error.
- **Using with small samples.** Rosner's test relies on large-sample approximations and is not recommended for $n < 25$ (prefer $n \geq 50$). Use Dixon or Grubbs for small samples.
- **Setting k too large.** Excessive k reduces power and over-adjusts for multiple testing. Choose a conservative bound (typically $k \leq 0.1n$ and $k < n/2$).
- **Removing values without investigation.** The test flags statistical outliers but does not determine cause. Investigate and document before removal; some outliers may represent real phenomena.
- **Equating non-detection with absence.** Failure to detect outliers does not prove none exist; limited power or an overly small k may mask them.

Tietjen-Moore Test

Test for Exactly k Outliers in Normal Data

OUTL.5 | **Category:** Outlier Detection | **Data:** Continuous | **Null:** Permutation/Tables

When to Use

- Detect multiple outliers simultaneously in a univariate dataset.
- Data may exhibit masking due to multiple outliers.
- Situations requiring a formal test for exactly k outliers.
- Preferable over sequential single-outlier tests (e.g., Grubbs).

Assumptions

- Data are approximately normal (excluding outliers).
- Observations are independent and identically distributed.
- Exact number of outliers (k) is specified a priori.
- Outliers arise from a different distribution than main data.
- Sample size is sufficient relative to k (e.g., $n \geq 3k + 3$).

Test Statistic

$$L_k = \frac{\sum_{i \in S}(x_i - \bar{x}_S)^2}{\sum_{i=1}^{n}(x_i - \bar{x})^2}.$$

Use when outliers are one-sided; use E_k if both sides (See Technical Notes).

Notation: L_k is ratio of variances, S is subset excluding k most discordant observations, $\bar{x}_S$ is mean of subset

Plain English: The Tietjen–Moore test determines whether exactly k observations are outliers by comparing the variance of the most homogeneous $n - k$ subset to the total variance. If removing k values substantially reduces variance, the statistic L_k becomes small. The test evaluates whether this reduction is greater than expected by chance under no outliers.

⚠ Practitioner Warning

The exact k must be specified a priori; misspecifying k invalidates inference. Less robust than generalized ESD when k is uncertain. Critical values depend on both n and k and are limited in tabulated form (permutation methods often used). The test does not automatically identify which observations are outliers and assumes normality of the uncontaminated data. Testing multiple values of k inflates Type I error.

Hypotheses

- H_0: No outliers present in dataset ($x_i \sim$ normal distribution).
- H_1: Exactly k outliers present (from different distribution).

Decision Rule: Reject H_0 if $L_k < L_{\text{crit}}(\alpha, n, k)$ from tables or permutation distribution. Small $L_k \Rightarrow k$ outliers.

Software References [Python, R, SAS]

`Not Native (Manual)`	`Not Native (Manual)`	`Not Native (Manual)`

Additional Notes

Tietjen and Moore (1972) proposed this test for a *known* number of outliers k. The procedure identifies the subset of $n - k$ observations with minimum variance, computes E_k from that subset, and compares it to tabulated critical values (depending on n and k). Exact minimization requires evaluating $\binom{n}{k}$ subsets; sequential or heuristic algorithms are often used in practice. A permutation approach provides flexible p-values. When $k = 1$, the test reduces to a Grubbs-type statistic. It is more powerful than generalized ESD when k is correctly specified, but performs poorly if k is misspecified. Observations excluded from the optimal subset are declared outliers.

Miscellanea Gary L. Tietjen and Roger H. Moore (1972) introduced this test in *Technometrics* as "Some Grubbs-Type Statistics for the Detection of Several Outliers," explicitly extending Grubbs' single-outlier test. For $k = 1$, the methods are equivalent. The test was motivated by the masking effect, where multiple outliers hide each other by inflating variance; by testing k observations simultaneously, Tietjen–Moore addresses this limitation of single-outlier methods. Its statistics (L_k and E_k) are based on the reduction in variance after removing the k most extreme observations. Critical values are obtained via simulation rather than closed-form distributions. The method is highly powerful when k is correctly specified, but performance deteriorates when k is misspecified. Because the exact number of outliers is rarely known in practice, more flexible procedures such as Rosner's generalized ESD test are often preferred, leaving Tietjen–Moore as an elegant but specialized tool.

Reference(s): Tietjen & Moore (1972); Barnett & Lewis (1994)

Worked Example

Problem Statement Eight chemical assay measurements are recorded: $x = (1, 2, 3, 4, 5, 6, 14, 21)$. Two values (14 and 21) are suspected to be outliers. Apply the Tietjen–Moore test with $k = 2$ at $\alpha = 0.05$, then compare with the wrong specification $k = 1$ to illustrate the masking effect. H_0: no outliers; H_1: exactly two outliers present.

1. Overall Mean and Total Sum of Squares: $k = 0$

$$\bar{x}_{[k=0]} = 7, \qquad S^2_{[k=0]} = \sum_{i=1}^{8}(x_i - \bar{x}_{[k=0]})^2 = 336.$$

2. Remove the $k = 2$ Most Extreme Observations
The two largest absolute deviations are $x = 21$ (rank 1) and $x = 14$ (rank 2). Retain the remaining $n - k = 6$ observations: $\{1, 2, 3, 4, 5, 6\}$.

$$\bar{x}_{[k=2]} = 3.5, \qquad S^2_{[k=2]} = \sum_{i=1}^{6}(x_i - \bar{x}_{[k=2]})^2 = 17.5.$$

3. Test Statistic and Decision

$$L_2 = \frac{S^2_{[k=2]}}{S^2_{[k=0]}} = \frac{17.5}{336} \approx 0.052.$$

Critical value $c(n{=}8, k{=}2, \alpha{=}5\%) = 0.146$ (Tietjen–Moore 1972, Table 1(c); lower-tail 5th percentile of L_2 under H_0).

$$L_2 = 0.052 < c = 0.146 \implies \textbf{reject } H_0.$$

Both $x = 14$ and $x = 21$ are identified as outliers.

When suspected outliers are on both sides, use E_k statistic.

4. Effect of Wrong Specification: $k = 1$
If only $k = 1$ is specified, the test removes only $x = 21$:

$$\bar{x}_{[k=1]} = 5, \qquad S^2_{[k=1]} = \sum_{i=1}^{7}(x_i - \bar{x}_{[k=1]})^2 = 112.$$

$$L_1 = \frac{S^2_{[k=1]}}{S^2_{[k=0]}} = \frac{112}{336} \approx 0.333.$$

$c(n{=}8, k{=}1, \alpha{=}5\%) = 0.326$; $L_1 = 0.333 > 0.326$: **FTR** — $x = 14$ is missed entirely.

k	L_k	c	**Decision**
2	0.052	0.146	Reject ($k = 2$ outliers)
1	0.333	0.326	FTR (misses $x = 14$)

Practitioner Insight: The Tietjen–Moore (1972) test generalizes Grubbs to *exactly* k outliers by comparing the retained sum of squares to the total: $L_k = S^2_{[k]}/S^2_{[k=0]} \in [0, 1]$. Under H_0, $L_k \approx (n-k-1)/(n-1)$; outliers inflate $S^2_{[k=0]}$ while leaving $S^2_{[k]}$ unchanged, driving L_k toward zero. Critical values are lower-tail α quantiles obtained via simulation or tables. Unlike Rosner (which tests for *up to* k outliers), Tietjen–Moore requires specifying *exactly* k in advance: under-specifying causes masking, over-specifying causes swamping. In this example, $k = 1$ fails to detect both outliers (masking), while $k = 2$ yields a small L_k and rejects.

Common Pitfalls

- **Choosing k post hoc.** The exact number k must be specified before analysis. Trying multiple values of k inflates Type I error and invalidates inference.
- **Using when k is uncertain.** Tietjen–Moore requires the exact number of outliers. If only an upper bound is known, use Rosner's test instead.
- **Applying to very small samples.** Adequate sample size is required (generally $n \geq 10$ and $n \geq 3k + 3$). Very small samples yield unstable variance estimates.
- **Not pre-specifying tail direction.** Specify whether testing upper-tail, lower-tail, or two-sided outliers. Default practice removes the k largest absolute deviations.
- **Using incorrect or mismatched critical values.** Critical values depend on (n, k, α). If unavailable in tables, obtain them via simulation; never substitute from another setting.

Peirce's Criterion

Maximum Likelihood-Based Outlier Rejection Criterion

OUTL.6 | **Category:** Outlier Detection | **Data:** Continuous | **Null:** Normal distribution

When to Use

- Detect multiple outliers using a likelihood-based rejection rule.
- When number of suspected outliers is specified in advance.
- Situations favoring principled rejection over ad hoc rules.
- Iterative screening where several extremes may distort estimates.

Assumptions

- Data are approximately normal (excluding outliers).
- Observations are independent from a common population.
- Number of outliers (k) is specified or evaluated.
- Outliers arise from a different or higher-variance process.
- Sample size is sufficient relative to number of outliers ($n > k + 2$).

Test Statistic

$$R = \frac{x_{\max}}{n\sigma}, \quad \text{reject if } |x_i - \bar{x}| > R\sigma, \quad \text{where } R \text{ from Peirce's tables.}$$

Notation: R is Peirce's criterion ratio, $\bar{x}$ is sample mean, and σ is population or sample standard deviation.

Plain English: Peirce's criterion sets a maximum allowable deviation based on sample size n and specified outlier count k. Derived from a maximum likelihood framework, it rejects observations whose standardized deviations exceed a tabulated critical ratio $R(n, k)$. The goal is to remove values whose exclusion maximizes the likelihood of the remaining data under normality.

⚠ Practitioner Warning

The exact number of outliers k must be fixed in advance; misspecification invalidates results. The method predates modern hypothesis testing and does not provide formal Type I error control. Tables are limited and may require interpolation. Using sample mean and variance introduces circularity and masking. Modern tests (Grubbs, Rosner, Tietjen–Moore) are generally preferred for rigorous inference.

Hypotheses

- H_0: k observations are consistent with normal distribution.
- H_1: k observations are outliers.

Decision Rule: Reject H_0 if observations with $|x_i - \bar{x}| > R(\alpha, n, k) \cdot \sigma$ where R from Peirce's tables.

Software References [Python, R, SAS]

Python	R	SAS
`Not Native (Manual)`	`(limited implementation)`	`Not Native (Manual)`

Additional Notes

Benjamin Peirce (1852) developed this maximum-likelihood–based rejection rule while working at the Harvard University. For n observations with k specified outliers and p estimated parameters, Gould (1855) provided formulas and tables for computing the Peirce rejection threshold. Solving the governing equations requires numerical methods, which led Peirce and Gould to publish extensive tabulations for practical use. Observations with standardized residuals exceeding the critical ratio $R(n, k)$ are rejected. Unlike Chauvenet's rule, Peirce's method requires fixing k in advance and derives the rejection threshold from likelihood principles rather than an arbitrary probability cutoff. Although historically influential in astronomy and geodesy, it does not provide modern Type I error guarantees, and contemporary methods are generally preferred. Ross (2003) later revisited the method using modern computational tools.

Miscellanea: Benjamin Peirce—mathematician, philosopher, and scientist—proposed this criterion in 1852 in the context of observational astronomy. Well before Fisher's formal development of likelihood theory, Peirce used probability-based reasoning to derive an objective rule for rejecting doubtful observations. His approach was notable for replacing arbitrary cutoffs with a mathematically justified decision rule. The method gained practical importance after Benjamin A. Gould published tables to facilitate its use and became widely applied in 19th-century astronomical data reduction.

Reference(s): Peirce (1852); Gould (1855); Ross (2003)

Worked Example

Problem Statement An astronomer records seven independent measurements of a stellar parallax (milliarcseconds): $x = (2, 3, 4, 5, 6, 7, 15)$. The reading 15 appears anomalously large. Apply Peirce's criterion (1852) with $m = 1$ suspected outlier, and compare with Chauvenet's criterion and the Grubbs test.

1. Sample Mean and Standard Deviation

$$\bar{x} = 6, \qquad s^2 = 56/3, \qquad s \approx 4.320$$

2. Most Extreme Observation
The largest absolute deviation is $|15 - 6| = 9$, so $x = 15$ is the suspect.

3. Peirce's Criterion
From Peirce's table (Ross 2003, Table 1) with $n = 7$ and $m = 1$ suspected outlier: $R = 1.693$, and $R \cdot s = 7.315$.

Flag any observation with $|x_i - \bar{x}| > R\,s$:

$$|15 - 6| = 9 > 7.315 \implies \textbf{flag}\ x = 15.$$

Equivalently, compute the standardized deviate $z_{\text{obs}} = 9/\sqrt{56/3} = \sqrt{243/56} \approx 2.083$ and compare to $R = 1.693$:

$$z_{\text{obs}} = 2.083 > R = 1.693.$$

4. Three-Way Comparison

Method	Threshold	Value	Obs	Flag?
Peirce	$R = 1.693$	$z_{\text{obs}} = 2.083$	>	Yes
Chauvenet	$z = 1.803$	$z_{\text{obs}} = 2.083$	>	Yes
Grubbs	$G_c = 2.020$	$G = 2.083$	>	Yes

All three methods agree: $x = 15$ is a significant outlier. Peirce's threshold ($R = 1.693$) is the most permissive (smallest), followed by Chauvenet ($z = 1.803$), then Grubbs ($G_c = 2.020$); here all three are exceeded by the same statistic $\sqrt{243/56} \approx 2.083$.

Practitioner Insight: Peirce's criterion (1852), predating Chauvenet, is the earliest formal outlier rule based on a maximum-likelihood argument. For n observations and m suspected outliers, it determines a threshold R by solving a transcendental equation that effectively compares the likelihood of all points arising from one normal population versus m arising from a different one. Peirce's cutoff is more aggressive than Chauvenet's, with $R < z_{\text{Chauvenet}}$, reflecting its astronomical origins where retaining bad measurements distorted estimates. For example, at $n = 7$, $R_{\text{Peirce}} = 1.693 < z_{\text{Chauvenet}} = 1.803 < G_{\text{Grubbs}} = 2.020$, so Peirce may flag observations that others retain. The procedure is iterative: remove flagged values, recompute $\bar{x}$ and s, and reapply until no further outliers are detected. The parameter m specifies how many outliers are tested simultaneously, with larger m yielding higher thresholds. This likelihood-based framework makes Peirce's criterion an early precursor to modern multiple-outlier tests.

Common Pitfalls

- **Using it for modern data analysis.** Peirce's criterion is historically important but impractical today. Modern tests (Grubbs, Rosner, Tietjen–Moore) are computationally simpler and provide clearer inferential properties.
- **Assuming maximum likelihood implies superiority.** Although derived from a likelihood argument, Peirce's method is not inherently better than modern procedures, which have stronger distributional results and error-control properties.
- **Misattributing authorship.** The criterion was developed by Benjamin Peirce (1852), not his son Charles Sanders Peirce, the philosopher.
- **Overlooking practical simplicity.** Peirce's approach was theoretically elegant but computationally demanding; simpler rules like Chauvenet gained wider adoption due to usability.
- **Dismissing it as obsolete.** Though superseded, Peirce's work was foundational and helped establish principled outlier detection in the 19th century.

Iglewicz and Hoaglin's Test

Modified Z-Score Test Using Median Absolute Deviation

OUTL.7 | **Category:** Outlier Detection | **Data:** Continuous | **Null:** Asymptotic (MAD)

When to Use

- Detect outliers robustly in skewed or heavy-tailed data.
- When mean/SD-based tests (e.g., Grubbs) are unreliable.
- Data contain multiple outliers or potential masking effects.
- Initial screen without strict distributional assumptions.
- Prefer median/MAD over mean and standard deviation.

Assumptions

- Observations are independent.
- Data are univariate and measured on a continuous scale.
- Distribution is roughly symmetric about the median.
- Sample size is adequate for reliable MAD estimation (e.g., $n \geq 10$).

Test Statistic

$$M_i = \frac{0.6745(x_i - \text{median})}{\text{MAD}}, \quad \text{where MAD} = \text{median}|x_i - \text{median}|.$$

Notation: M_i is modified z-score, MAD is median absolute deviation, 0.6745 is normalization constant

Plain English: Iglewicz and Hoaglin's test replaces the classical z-score (using mean and standard deviation) with a modified z-score using median and median absolute deviation. The median and MAD are robust to outliers, preventing outliers from inflating the scale estimate and masking their own extremeness. The modified z-score measures how many "robust standard deviations" an observation lies from the median. Values with $|M_i| > 3.5$ are commonly flagged as outliers.

⚠ Practitioner Warning

Threshold of 3.5 is rule of thumb, not rigorously derived; lacks formal Type I error control. Different thresholds (2.5, 3.0, 4.0) yield different results; choice somewhat arbitrary. Assumes approximate symmetry; strongly skewed distributions problematic. MAD can be zero or very small with clustered data, causing undefined or extreme M_i. Factor 0.6745 assumes normality for consistency with standard deviation; not appropriate for all distributions. Multiple outliers on same side can still bias median and MAD. Not a formal hypothesis test; use for screening, not definitive inference. Post-hoc examination of flagged values essential.

Hypotheses

- H_0: Observation is consistent with bulk of data.
- H_1: Observation is an outlier.

Decision Rule: Reject H_0 if $|M_i| > 3.5$ (or chosen threshold). Other choices: $|M_i| > 2.5$ (liberal), $|M_i| > 4.0$ (conservative).

Software References [Python, R, SAS]

Python	R	SAS
`scipy.stats.median_abs_deviation`	`Not Native (Manual)`	`Not Native (Manual)`

Additional Notes

Iglewicz and Hoaglin (1993) popularized the modified z-score in *How to Detect and Handle Outliers* (ASQC). The method uses the median absolute deviation, $\text{MAD} = \text{median}|x_i - \text{median}(x)|$, a robust scale estimator with 50% breakdown point. The constant 0.6745 makes MAD/0.6745 consistent for σ under normality. Modified z-scores behave like classical z-scores but resist masking since median and MAD are not inflated by extremes. The common cutoff $|M_i| > 3.5$ (heuristic) roughly corresponds to a very small two-tailed probability under normality. The method is computationally simple and widely used in exploratory analysis and data-cleaning workflows. More efficient robust scales (e.g., Q_n, S_n), proposed by Rousseeuw & Croux (1993), exist but are less simple.

Miscellanea: Boris Iglewicz and David Hoaglin popularized the modified z-score in the early 1990s as a simple, robust alternative to classical outlier rules. Hoaglin, influenced by Tukey's exploratory data analysis philosophy, emphasized practical, interpretable methods over strict optimality. The commonly used cutoff $|M_i| > 3.5$ is a pragmatic guideline rather than a formally derived threshold, reflecting empirical experience under normality. The method has since become widely used in modern data analysis workflows due to its simplicity, speed, and resistance to contamination.

Reference(s): Iglewicz & Hoaglin (1993); Rousseeuw & Croux (1993)

Worked Example

Problem Statement Nine replicate measurements are sorted and recorded: $x = (1, 2, 3, 4, 5, 6, 7, 10, 16)$. The values 10 and 16 are moderately elevated. Apply the Iglewicz–Hoaglin modified Z-score test with threshold 3.5, and compare with the Grubbs test to demonstrate the masking resistance of the robust method.

1. Median and Median Absolute Deviation
The median is the 5th value: $\tilde{x} = 5$. Absolute deviations $|x_i - 5|$, sorted:

$\lvert x_i - 5\rvert$	4	3	2	1	0	1	2	5	11
Sorted	0	1	1	2	**2**	3	4	5	11

The MAD is the median (5th value) of the sorted absolute deviations: MAD = 2.

2. Modified Z-Scores
Only highest three values shown

$$M_i = \frac{0.6745\,(x_i - \tilde{x})}{\text{MAD}} = \frac{0.6745\,(x_i - 5)}{2}.$$

x_i	$x_i - 5$	M_i	
7	+2	0.674	
10	+5	1.686	
16	+11	**3.710**	← **flag**

3. Decision
Flag any observation with $|M_i| > 3.5$:

$$M_{16} = \frac{0.6745 \times 11}{2} = 3.710 > 3.5 \implies \textbf{flag}\ x = 16.$$

All other $|M_i| \leq 1.686 < 3.5$: no further flags.

4. Grubbs Test on the Same Data
The Grubbs test uses the mean and standard deviation:

$$\bar{x} = 6, \quad s^2 = 21.5, \quad G = \frac{|16 - 6|}{s} \approx 2.157.$$

Critical value $G_{\text{crit}}(n{=}9, 5\%) = 2.215$:

$$G = 2.157 < 2.215 \implies \textbf{FTR}\ \text{(misses } x = 16\text{!)}$$

Method	**Stat**	**Threshold**	>?	**Decision**
IH	$M = 3.710$	3.5	Yes	Flag $x = 16$
Grubbs	$G = 2.157$	2.215	No	FTR

The robust method detects the outlier that the classical method misses.

Practitioner Insight: The Iglewicz–Hoaglin (1993) modified Z-score replaces $\bar{x}$ and s with the median $\tilde{x}$ and MAD $= \text{median}_i|x_i - \tilde{x}|$, yielding robust location and scale estimators. The median has a 50% breakdown point (resistant to up to half contaminated data), whereas s has 0% (one outlier can distort it arbitrarily). The scaling factor $0.6745 = \Phi^{-1}(3/4)$ ensures MAD/0.6745 $\approx \sigma$ under normality, so the modified score behaves like a classical Z-score for clean data but remains stable under contamination. In this example, the pair $\{10, 16\}$ inflates s^2 and masks detection in Grubbs, while the median and MAD remain unaffected, yielding $M_{16} = 3.710$. The threshold 3.5 is simulation-based (not tied to a formal α), with 2.5 sometimes used for higher sensitivity. Unlike parametric OUTL tests, this method provides no p-value; the cutoff acts as a heuristic rather than a distribution-based critical value.

Note: $\Phi^{-1}(0.75) = 0.6475$, where $\Phi(.)$ is the standard normal CDF. For normal data, MAD = $0.6475s$, where s is the standard deviation. In this case, M_i coincides with the usual z-score.

Common Pitfalls

- **Omitting the 0.6745 scaling constant.** The modified z-score must include the 0.6745 factor. Without it, the scale is incorrect and the 3.5 cutoff is invalid.
- **Using 3.5 as a rigid cutoff.** The 3.5 threshold is a guideline, not a universal rule. Adjust based on context, sample size, and tolerance for false positives, and document the choice.
- **Treating it as a formal hypothesis test.** Modified z-scores are screening tools and do not provide α levels or exact Type I error control. Use formal tests (e.g., Grubbs, Rosner) when inferential guarantees are required.
- **Assuming MAD always dominates SD.** MAD is more robust but less efficient under clean normal data. Choose based on contamination risk and analysis goals.
- **Removing flagged values automatically.** A large modified z-score signals investigation, not deletion. Verify cause before excluding observations.

Mahalanobis Distance Test

Multivariate Outlier Detection Based on Statistical Distance

OUTL.8 | **Category:** Outlier Detection | **Data:** Continuous (Multivariate) | **Null:** Chi-square or F

When to Use

- Detect multivariate outliers accounting for covariance.
- Identify observations unusual in joint (not marginal) space.
- Multivariate data where univariate checks miss anomalies.
- Screening before multivariate modeling (PCA, regression).
- QC monitoring involving correlated measurements.

Assumptions

- Observations are i.i.d
- Data are approximately multivariate normal.
- Variables are continuous; covariance handles scaling.
- Covariance matrix is nonsingular (invertible).
- $n \gg p$ for stable covariance estimation.

Test Statistic

$$D_i^2 = (\mathbf{x}_i - \bar{\mathbf{x}})'\mathbf{S}^{-1}(\mathbf{x}_i - \bar{\mathbf{x}}). \quad \text{Under } H_0, \ \frac{(n-p-1)pD_i^2}{n(n-p)} \sim F_{p,n-p-1} \text{ (large sample)}, \ D_i^2 \sim \chi_p^2 \text{ (small sample)}.$$

Notation: D_i^2 is squared Mahalanobis distance, $\mathbf{x}_i$ is observation vector, $\bar{\mathbf{x}}$ is $p \times 1$ mean vector, $\mathbf{S}$ is covariance matrix

Plain English: Mahalanobis distance quantifies how far an observation lies from the multivariate center while accounting for scale and correlations among variables. Unlike Euclidean distance, it uses the covariance structure to weight directions appropriately. Large squared distances are compared to χ^2 (or F) critical values to flag multivariate outliers.

⚠ Practitioner Warning

Requires reliable covariance estimation; small n relative to p produces unstable results. Highly sensitive to masking and swamping when multiple outliers distort the covariance matrix. Assumes multivariate normality. In high dimensions, computation and multiple-testing concerns arise; robust covariance estimators (e.g., MCD) are often preferred.

Hypotheses

- H_0: Suspect observation is not outlier.
- H_1: Suspect observation is a multivariate outlier.

Decision Rule: Reject H_0 if $D_i^2 > \chi^2_{1-\alpha,p}$ (large sample) or $(n-p-1)pD_i^2/(n(n-p)) > F_{1-\alpha,p,n-p-1}$ (exact).

Software References [Python, R, SAS]

Python	R	SAS
`scipy.spatial.distance.mahalanobis`	`Not Native (Manual)`	`PROC IML`

Additional Notes

Mahalanobis (1936) introduced this distance for multivariate analysis, originally motivated by anthropometric studies in India, making it one of the earliest multivariate tools in statistics. For data $\mathbf{x}_1, \dots, \mathbf{x}_n$, formulas to compute the mean vector $\bar{\mathbf{x}}$, and covariance matrix $\mathbf{S}$ are:

$$\bar{\mathbf{x}} = \frac{1}{n}\sum_i \mathbf{x}_i \qquad \mathbf{S} = \frac{1}{n-1}\sum_i (\mathbf{x}_i - \bar{\mathbf{x}})(\mathbf{x}_i - \bar{\mathbf{x}})'.$$

Under multivariate normality, $D_i^2 \sim \chi_p^2$ asymptotically (finite-sample results use an F distribution). The distance is scale-invariant and accounts for correlations; when variables are uncorrelated it reduces to standardized Euclidean distance. A key feature is that an observation may appear unremarkable in each variable individually yet be extreme in the joint space; Mahalanobis distance detects such "hidden" outliers by incorporating the covariance structure. The method is closely related to principal component analysis: in the PCA coordinate system, the distance becomes a sum of squared, variance-standardized scores along orthogonal axes. Because outliers can inflate $\mathbf{S}$ (masking), the classical version may fail in contaminated samples; robust covariance estimators such as minimum covariance determinant (MCD) or minimum volume ellipsoid (MVE) are therefore often preferred. Common practice flags observations exceeding a high χ_p^2 percentile, with multiplicity adjustments when testing all n points.

Miscellanea: Mahalanobis developed this measure while studying anthropometric classification, recognizing that correlated measurements (e.g., skull length and width) required covariance-based weighting. What began as a tool for population studies became foundational in multivariate statistics, influencing discrimination analysis, quality control, fraud detection, and modern anomaly detection.

Reference(s): Mahalanobis (1936); Rousseeuw & van Zomeren (1990)

Worked Example

Problem Statement Ten bivariate observations (x, y) are recorded:

$$(1,1), (2,2), (3,3), (4,4), (5,5), (6,6), (7,7), (8,8), (9,9), (2,9).$$

The first nine lie on the line $y = x$; the tenth shares $x = 2$ with the second point and $y = 9$ with the ninth, yet breaks the strong positive correlation. Apply the Mahalanobis distance test at $\alpha = 0.05$ ($p = 2$ variables) and compare with univariate Grubbs tests on x and y separately.

1. Sample Mean Vector & Covariance Matrix

$$\bar{\mathbf{x}} = \begin{pmatrix} 4.7 \\ 5.4 \end{pmatrix}, \qquad \mathbf{S} = \begin{pmatrix} 7.567 & 5.467 \\ 5.467 & 8.267 \end{pmatrix}.$$

2. Inverse Covariance Matrix

$$\mathbf{S}^{-1} = \begin{pmatrix} 0.253 & -0.167 \\ -0.167 & 0.232 \end{pmatrix}.$$

3. Mahalanobis Distance for $(2, 9)$

$$\mathbf{x}_{10} - \bar{\mathbf{x}} = \begin{pmatrix} -2.7 \\ 3.6 \end{pmatrix}, \; D^2_{10} = (\mathbf{x}_{10} - \bar{\mathbf{x}})' \mathbf{S}^{-1} (\mathbf{x}_{10} - \bar{\mathbf{x}}) = 8.1.$$

4. Decision

Under H_0: $D_i^2 \overset{\cdot}{\sim} \chi^2(p)$ for large n. With $p = 2$: $\chi^2_{2,0.95} = 5.991$.

$$D^2_{10} = 8.1 > 5.991 \implies \textbf{flag } (2, 9).$$

All other points have $D_i^2 \le 2.5 < 5.991$: no false flags.

5. Comparison: Univariate Grubbs on x and y Separately

	$\mathbf{x}_{10} - \bar{\mathbf{x}}$	s	G	vs $G_{\text{crit}}(n{=}10) = 2.290$
x-variable	-2.7	$\sqrt{7.567} \approx 2.751$	0.982	FTR
y-variable	3.6	$\sqrt{8.267} \approx 2.875$	1.252	FTR
Mahalanobis		$D^2_{10} = 8.1$		**Flag** ✓

Observation $(2, 9)$ is not a marginal outlier on x alone ($G = 0.98$) nor on y alone ($G = 1.25$), yet it is a clear multivariate outlier ($D^2 = 8.1 > 5.991$) because it violates the strong positive correlation of the remaining nine points.

Practitioner Insight: The Mahalanobis distance $D_i^2 = (\mathbf{x}_i - \bar{\mathbf{x}})^\top S^{-1} (\mathbf{x}_i - \bar{\mathbf{x}})$ generalises the squared z-score to p dimensions by accounting for variable scale and correlation. A point may appear typical marginally yet lie in a low-density joint region: here $x = 2$ and $y = 9$ are each common, but $(2, 9)$ contradicts the pattern $y \approx x$ and is far from the data cloud after rotation and scaling into principal axes. Under multivariate normality, $D_i^2 \overset{\cdot}{\sim} \chi^2(p)$ approximately for large n, providing threshold-based detection (e.g., $\chi^2_{2,0.95} = 5.991$). The example is constructed so that the outlier has $D^2 = 8.1$ while all other points lie well below the cutoff. A key limitation is masking: the sample covariance S is itself influenced by outliers, potentially reducing D^2; robust alternatives replace S with estimators such as the MCD.

Common Pitfalls

- **Using too small a sample ($n < 3p$).** Mahalanobis distance requires stable covariance estimation. If n is not substantially larger than p (preferably $n \ge 5p$), $\mathbf{S}$ is unstable or non-invertible, leading to unreliable distances.
- **Ignoring multicollinearity.** Highly correlated or redundant variables can make $\mathbf{S}$ nearly singular, producing unstable inverses. Check correlations or condition number; remove redundant variables or use PCA/regularization if necessary.
- **Relying on classical estimates with multiple outliers.** Outliers inflate the mean and covariance (masking effect), reducing their own distances. When contamination is suspected, use robust covariance estimators (e.g., MCD).
- **Failing to adjust for multiple testing.** Testing all n observations at level α inflates false positives. Apply Bonferroni or other familywise error corrections when declaring statistical significance.
- **Including categorical or discrete variables.** Mahalanobis distance assumes continuous variables and multivariate normality. Do not apply it directly to binary or categorical data.

Grubbs' Test and Its Relationship to the t-Distribution

The Grubbs statistic for testing whether the maximum observation x_n is an outlier in a sample of size n is:

$$G = \frac{x_n - \bar{x}}{s},$$

where $\bar{x}$ and s are computed from all n observations. Although G resembles a z-score, it is not t-distributed; its connection to the t-distribution arises through an exact algebraic transformation.

Deriving the Relationship. Suppose $x_1, \dots, x_n \overset{\text{iid}}{\sim} N(\mu, \sigma^2)$. For any fixed observation x_i, consider the statistic

$$T_i = \frac{\sqrt{n-2}\,(x_i - \bar{x})}{\sqrt{(n-1)s^2 - (x_i - \bar{x})^2}},$$

which follows a t_{n-2} distribution under H_0. This statistic serves as an intermediate bridge linking G to the t-distribution.

Solving this relationship for

$$G_i = \frac{|x_i - \bar{x}|}{s}, \quad \text{gives} \quad T_i^2 = \frac{(n-2)G_i^2}{(n-1) - G_i^2},$$

and hence,

$$G_i^2 = \frac{(n-1)T_i^2}{n-2+T_i^2}, \qquad G_i = \frac{n-1}{\sqrt{n}}\sqrt{\frac{T_i^2}{n-2+T_i^2}}.$$

For Grubbs' test, we consider the largest absolute deviation, so the critical value is based on a Bonferroni-adjusted t-quantile

$$t_{\alpha/(2n),\,n-2}.$$

Therefore,

$$G_{\text{crit}} = \frac{n-1}{\sqrt{n}}\sqrt{\frac{t^2_{\alpha/(2n),\,n-2}}{n-2+t^2_{\alpha/(2n),\,n-2}}}.$$

This shows that Grubbs' critical values are obtained by transforming a t_{n-2} statistic into the scale of the maximum standardized deviation, rather than from a direct t-distribution for G itself.

Degrees of Freedom. The appearance of $n-2$ degrees of freedom arises from the structure of the associated t-statistic, which effectively depends on a variance estimate with $n-2$ degrees of freedom.

Note on Two-Sided Testing. The factor $\alpha/(2n)$ controls the familywise error rate, accounting for the fact that any one of the n observations could be the extreme value. This Bonferroni adjustment becomes increasingly conservative as n grows.

References. Grubbs (1969); Stefansky (1972).

Dixon's Q Test: Complete Statistic Reference

The main test entry (OUTL.2) presents Q_{10}, the classical single-outlier statistic for small samples. Rorabacher (1991) systematized the full family of six statistics, each designed for a specific sample size range and number of suspected outliers. The data are assumed sorted in ascending order: $x_1 \le x_2 \le \cdots \le x_n$. For each statistic, the left-end formula (testing x_1 as the low outlier) is given first; the right-end formula (testing x_n as the high outlier) follows in parentheses. The numerator always measures the gap between the suspect value and its nearest non-suspect neighbor; the denominator is the range of the non-suspect portion of the sample.

Stat.	Range of n	Formula		
Q_{10}	3–7	$\frac{x_2 - x_1}{x_n - x_1}$	or	$\frac{x_n - x_{n-1}}{x_n - x_1}$
Q_{11}	8–10	$\frac{x_2 - x_1}{x_{n-1} - x_1}$	or	$\frac{x_n - x_{n-1}}{x_n - x_2}$
Q_{12}	11–13	$\frac{x_2 - x_1}{x_{n-2} - x_1}$	or	$\frac{x_n - x_{n-1}}{x_n - x_3}$
Q_{20}	14–17	$\frac{x_3 - x_1}{x_n - x_1}$	or	$\frac{x_n - x_{n-2}}{x_n - x_1}$
Q_{21}	18–24	$\frac{x_3 - x_1}{x_{n-1} - x_1}$	or	$\frac{x_n - x_{n-2}}{x_n - x_2}$
Q_{22}	25–30	$\frac{x_3 - x_1}{x_{n-2} - x_1}$	or	$\frac{x_n - x_{n-2}}{x_n - x_3}$

The subscript notation Q_{ij} encodes the structure of each statistic: i is the number of values excluded from the numerator end of the sample (i.e. how many near-neighbors of the suspect value are skipped), and j is the number of values excluded from the opposite end of the denominator range (i.e. how many extreme values on the far side are removed from the range). Thus Q_{10} skips no neighbors and excludes nothing from the range; Q_{22} skips the nearest neighbor and excludes two values from the far end — the most conservative statistic in the family, appropriate for the largest samples where the range is otherwise dominated by the bulk of the data.

Select the statistic whose n-range bracket contains the actual sample size; apply the left-end formula to test the minimum x_1 and the right-end formula to test the maximum x_n. Critical values are tabulated in Rorabacher (1991, Table 1).

Reference. Rorabacher (1991).

Peirce's Criterion: Iterative Calculation Procedure

Peirce's criterion requires an iterative procedure to determine the squared threshold x^2, beyond which a deviation is flagged as an outlier. The inputs are N (total observations), n (number of suspected outliers), and m (number of model unknowns; typically $m = 1$ for a simple mean). The procedure follows Gould's (1855) reformulation of Peirce's original equations.

Step 1. Compute Q from Gould's Equation B:

$$Q = \frac{n^{n/N}(N-n)^{(N-n)/N}}{N}.$$

Step 2. Initialize $R_{\text{old}} = 0$ and $R_{\text{new}} = 1$.

Step 3. While $|R_{\text{new}} - R_{\text{old}}| > \varepsilon$ (e.g. $\varepsilon = 10^{-6}$), iterate:

(a) Set $R_{\text{old}} \leftarrow R_{\text{new}}$. Compute Λ from Gould's Equation A′:

$$\Lambda = \left(\frac{Q^N}{R_{\text{old}}^n}\right)^{1/(N-n)}.$$

(b) Compute x^2 from Gould's Equation C:

$$x^2 = 1 + \frac{N-m-n}{n}\left(1-\Lambda^2\right).$$

If $x^2 \leq 0$, set $x^2 = 0$ and stop.

(c) Update R via Gould's Equation D:

$$R_{\text{new}} = e^{(x^2-1)/2}\,\text{erfc}\left(\frac{x}{\sqrt{2}}\right).$$

Step 4. The converged x^2 is the squared threshold. An observation is flagged as an outlier if its squared deviation from the mean exceeds $x^2 \cdot s^2$, where s^2 is the sample variance.

The erfc function is the complementary error function $\text{erfc}(z) = 1 - \text{erf}(z)$, available natively in R (`pracma::erfc`), Python (`scipy.special.erfc`), and most numerical libraries. Convergence is typically achieved in fewer than twenty iterations.

Reference. Gould (1855).

Tietjen–Moore Test: The L_k and E_k Statistics

Tietjen and Moore (1972) proposed two distinct statistics to handle different patterns of multiple outliers in a sample of size n. Let $y_1 \leq y_2 \leq \cdots \leq y_n$ denote the ordered observations and $\bar{y}$ the full sample mean.

The L_k statistic addresses the situation where the k suspected outliers all lie on the *same side* of the bulk — either all k largest or all k smallest values. Define $\bar{y}_k$ as the mean of the $(n-k)$ non-suspect observations. Then:

$$L_k = \frac{\sum_{i=1}^{n-k}(y_i - \bar{y}_k)^2}{\sum_{i=1}^{n}(y_i - \bar{y})^2}.$$

For the k *smallest* suspects, the numerator sums over the $(n-k)$ largest observations; for the k *largest* suspects, the numerator sums over the $(n-k)$ smallest. Small values of L_k indicate that removing the k suspect values disproportionately reduces the sum of squares, evidencing outliers. L_1 is algebraically equal to Grubbs' S_n^2/S^2 statistic for a single outlier.

The E_k statistic addresses the situation where the k suspected outliers may lie on *both sides* of the bulk simultaneously — some too large, some too small. Relabel the observations as $z_1, \ldots, z_n$ ordered by *absolute residual* from $\bar{x}$, so that z_1 is closest to the mean and z_n is farthest. Define $\bar{z}_k$ as the mean of the k least extreme observations. Then:

$$E_k = \frac{\sum_{i=1}^{n-k}(z_i - \bar{z}_k)^2}{\sum_{i=1}^{n}(z_i - \bar{z})^2}.$$

Small values of E_k signal that the k most extreme observations (regardless of direction) are outliers. E_1 is equivalent to L_1 at significance level 2α, since a single extreme value may be either large or small.

The key distinction: use L_k when all suspected outliers are expected on one side (e.g. all high or all low); use E_k when outliers may appear on both sides. Both statistics require the number of suspected outliers k to be specified in advance — over- or under-specifying k can produce incorrect conclusions, as the authors illustrate with their Example 3 in the original paper. Critical values for both statistics at $\alpha = 0.01, 0.025, 0.05, 0.10$ are tabulated in Tietjen and Moore (1972, Tables I and II) for $n = 3$–50 and $k = 1$–10.

A convenient mnemonic: E for either side.

Minimum Covariance Determinant (MCD)

The **Minimum Covariance Determinant** estimator (Rousseeuw 1984) is a highly robust alternative to the classical sample mean and covariance matrix for multivariate data. Its purpose is to produce location and scatter estimates that are resistant to the influence of outliers, which the classical estimators $\bar{\mathbf{x}}$ and $\mathbf{S}$ are not.

The core idea. Among all subsets of h observations drawn from the full sample of n, find the subset H^* whose sample covariance matrix has the smallest determinant:

$$H^* = \arg\min_{|H|=h} \det(\mathbf{S}_H),$$

where $\mathbf{S}_H$ is the covariance matrix computed from the h observations in subset H. The MCD location and scatter estimates are then:

$$\hat{\mu}_{\text{MCD}} = \bar{\mathbf{x}}_{H^*}, \qquad \hat{\mathbf{\Sigma}}_{\text{MCD}} = c_h \, \mathbf{S}_{H^*},$$

where c_h is a consistency factor that corrects for the bias introduced by selecting only h observations. The subset size h is typically chosen as $h = \lfloor (n + p + 1)/2 \rfloor$, giving a **breakdown point** of approximately 50% — the estimator remains bounded even if up to half the observations are replaced by arbitrary values.

Connection to outlier detection. The MCD estimates are used to compute robust Mahalanobis distances for each observation:

$$d_i^{\text{MCD}} = \sqrt{(\mathbf{x}_i - \hat{\mu}_{\text{MCD}})^\top \hat{\mathbf{\Sigma}}_{\text{MCD}}^{-1} (\mathbf{x}_i - \hat{\mu}_{\text{MCD}})}.$$

Under multivariate normality, $(d_i^{\text{MCD}})^2 \approx \chi^2_p$, so observations with d_i^{MCD} exceeding $\sqrt{\chi^2_{p,0.975}}$ are flagged as outliers. Because the MCD estimates are not distorted by the outliers themselves, this procedure avoids the *masking* problem that afflicts classical Mahalanobis distances — outliers can inflate $\mathbf{S}$ and pull $\bar{\mathbf{x}}$, causing their own distances to appear small. The Iglewicz–Hoaglin test (OUTL.07) applies the same resistant philosophy in the univariate case; MCD is its multivariate counterpart.

Computation. Exact enumeration of all $\binom{n}{h}$ subsets is infeasible for moderate n. The **FAST-MCD** algorithm (Rousseeuw & Van Driessen 1999) uses concentration steps — iteratively expanding a random initial subset toward lower determinant — to find a near-optimal solution efficiently. Available as `robustbase::covMcd()` in R and `sklearn.covariance.MinCovDet` in Python.

Reference. Rousseeuw (1984); Rousseeuw & Van Driessen (1999).

Minimum Volume Ellipsoid (MVE)

The **Minimum Volume Ellipsoid** estimator (Rousseeuw 1985) is a robust location and scatter estimator that predates MCD and shares the same motivation: produce estimates of μ and $\mathbf{\Sigma}$ that are resistant to outliers in multivariate data.

The core idea. Among all ellipsoids of the form $\mathcal{E}(\mu, \mathbf{C}) = \{\mathbf{x} : (\mathbf{x} - \mu)^\top \mathbf{C}^{-1} (\mathbf{x} - \mu) \le 1\}$ that contain at least h of the n observations, find the one with smallest volume:

$$(\hat{\mu}_{\text{MVE}}, \hat{\mathbf{C}}_{\text{MVE}}) = \arg\min \operatorname{vol}(\mathcal{E}(\mu, \mathbf{C})).$$

Since $\operatorname{vol}(\mathcal{E}) \propto [\det(\mathbf{C})]^{1/2}$, minimizing volume is equivalent to minimizing the determinant of $\mathbf{C}$ subject to the ellipsoid covering at least h points. The scatter estimate is $\hat{\mathbf{\Sigma}}_{\text{MVE}} = c_h \hat{\mathbf{C}}_{\text{MVE}}$ after a consistency correction c_h. As with MCD, $h \approx (n + p + 1)/2$ gives a breakdown point near 50%.

MVE vs. MCD. Both estimators solve related but distinct optimization problems. MVE minimises the volume of an enclosing ellipsoid; MCD minimises the determinant of the covariance matrix of a selected subset. In practice MCD is generally preferred: it has a faster exact algorithm (FAST-MCD), better finite-sample statistical efficiency, and its objective — the covariance determinant of the retained subset — has a more direct statistical interpretation than the geometric volume of an enclosing ellipsoid. MVE can produce slightly different outlier classifications than MCD and has lower asymptotic efficiency, but it is conceptually simpler and remains in use in some robust regression and data-depth contexts.

Robust Mahalanobis distances based on MVE are computed identically to those for MCD, substituting $\hat{\mu}_{\text{MVE}}$ and $\hat{\mathbf{\Sigma}}_{\text{MVE}}$, and referred to the same χ^2_p reference distribution. Available via `robustbase::covMve()` in R.

Reference. Rousseeuw (1985); Rousseeuw & Leroy (1987).

PART XXV

Posterior Evidence & Probabilistic Inference

Evaluating hypotheses using Bayes factors and posterior probabilities

BAYESIAN ANALYSIS

XXV. BAYES — A Conceptual Primer

The Bayesian Paradigm

Bayesian inference treats unknown parameters $\boldsymbol{\theta}$ as random variables with a **prior distribution** $p(\boldsymbol{\theta})$ encoding beliefs before seeing the data. After observing $\mathbf{y}$, beliefs are updated via Bayes' theorem:

$$\underbrace{p(\boldsymbol{\theta}|\mathbf{y})}_{\text{posterior}} \propto \underbrace{p(\mathbf{y}|\boldsymbol{\theta})}_{\text{likelihood}} \times \underbrace{p(\boldsymbol{\theta})}_{\text{prior}} .$$

The normalizing constant $p(\mathbf{y}) = \int p(\mathbf{y}|\boldsymbol{\theta})p(\boldsymbol{\theta})\,d\boldsymbol{\theta}$ is the **marginal likelihood** (model evidence).

Key difference from frequentist inference: Parameters have probability distributions; data are fixed. Conclusions are direct probability statements about $\boldsymbol{\theta}$ given the observed $\mathbf{y}$—not about hypothetical repeated samples.

Prior Distributions

Informative priors: Encode genuine prior knowledge (e.g. previous studies, expert opinion). **Conjugate priors** yield closed-form posteriors: Beta–Binomial, Gamma–Poisson, Normal–Normal (canonical pairs).

Weakly informative priors: Regularize estimation without strongly constraining the posterior. Recommended defaults: half-Normal or half-Cauchy on standard deviations; Cauchy(0, 2.5) on regression coefficients (Gelman et al.).

Non-informative (reference) priors: Jeffreys prior $p(\boldsymbol{\theta}) \propto |\mathcal{I}(\boldsymbol{\theta})|^{1/2}$ is invariant to reparametrization; the reference prior for single-parameter models. Flat priors $p(\theta) \propto 1$ are improper; valid only if the posterior is proper.

Prior sensitivity: Always check whether conclusions change substantially under alternative reasonable priors. Bayes factors are particularly sensitive to prior specification (Bartlett's paradox): diffuse priors can produce arbitrarily small Bayes factors even when the data strongly support the alternative.

Posterior Summaries *Posterior mean* $E[\theta|\mathbf{y}]$ (minimises posterior expected squared error); *posterior median* (minimises posterior expected absolute error; robust to skewness); *MAP* (mode of $p(\theta|\mathbf{y})$, equivalent to penalized MLE with $-\log p(\theta)$ penalty); *HDI* (shortest interval containing $1-\alpha$ posterior mass); *equal-tailed CI* $[\theta_{\alpha/2}, \theta_{1-\alpha/2}]$ (simpler, but less natural for skewed posteriors).

Credible interval vs. confidence interval: A 95% HDI contains the true θ with probability 0.95 *given the observed data*—a direct probability statement about θ. A frequentist 95% CI contains θ in 95% of repeated samples—a statement about the procedure, not the specific interval.

I. Bayes Factor (BF_{10})

The Bayes factor quantifies the relative evidence for M_1 over M_0 as the ratio of marginal likelihoods:

$$\mathrm{BF}_{10} = \frac{p(\mathbf{y}|M_1)}{p(\mathbf{y}|M_0)} = \frac{\int p(\mathbf{y}|\boldsymbol{\theta}_1, M_1)p(\boldsymbol{\theta}_1|M_1)\,d\boldsymbol{\theta}_1}{\int p(\mathbf{y}|\boldsymbol{\theta}_0, M_0)p(\boldsymbol{\theta}_0|M_0)\,d\boldsymbol{\theta}_0}.$$

$\mathrm{BF}_{01} = 1/\mathrm{BF}_{10}$ gives evidence for M_0. Bayes factors are **symmetric and transitive**: $\mathrm{BF}_{12} = \mathrm{BF}_{10}/\mathrm{BF}_{20}$.

Interpretation scales: Two scales are in common use; neither should be applied as a hard threshold.

BF_{10}	**Jeffreys (1961)**	**Kass–Raftery (1995)**
1–3	Anecdotal	Bare mention
3–10	Moderate	Positive
10–20	Strong	
20–30	Strong	Strong
30–100	Very strong	
100–150	Decisive	Strong
≥ 150	Decisive	Very strong

Kass and Raftery raised the bar for "strong" evidence, reflecting concern that Jeffreys' thresholds are too lenient. Both scales apply symmetrically to BF_{01} when assessing evidence for M_0.

BIC approximation: For large n, $\log \mathrm{BF}_{10} \approx \frac{1}{2}(\mathrm{BIC}_0 - \mathrm{BIC}_1)$; connects Bayesian model selection to penalized likelihood without specifying priors.

Lindley's paradox: In large samples, a frequentist test may reject H_0 ($p < 0.05$) while the Bayes factor simultaneously favours H_0. This is not a contradiction: p-values measure compatibility of the data with H_0 alone, while Bayes factors compare predictive adequacy of both H_0 and H_1. When H_1 carries a diffuse prior, the data may be almost as probable under H_0 as under H_1, favouring the simpler model despite a small p-value. The paradox is a principled reason to prefer Bayes factors in large-sample settings.

II. Savage–Dickey Density Ratio

For **nested models** where $M_0 : \theta = \theta_0$ is a special case of M_1, the Bayes factor simplifies to the ratio of the posterior to prior density at the null value:

$$\mathrm{BF}_{01} = \frac{p(\theta_0|\mathbf{y}, M_1)}{p(\theta_0|M_1)}.$$

Computationally convenient: requires only the posterior under M_1, not the marginal likelihood. **Requires** the prior under M_0 to be the marginal of the prior under M_1 at $\theta = \theta_0$ (the Savage–Dickey condition). Commonly used for point-null hypotheses in Bayesian t-tests, correlation tests, and regression coefficient tests. Graphically, if the posterior is lower than the prior at θ_0, the data have moved probability mass away from θ_0, favouring M_1.

III. Bayesian Model Comparison and Averaging

When multiple models $M_1, \ldots, M_K$ are plausible, compare via posterior model probabilities:

$$P(M_k|\mathbf{y}) \propto p(\mathbf{y}|M_k)P(M_k).$$

Bayesian model averaging (BMA) weights parameter estimates or predictions across models: $\hat{\theta}_{\text{BMA}} = \sum_k P(M_k|\mathbf{y})\hat{\theta}_k$, propagating model uncertainty into inference.

Information criteria approximations:

- **BIC** (consistent; assumes $n \to \infty$; no prior needed).
- **WAIC** (Watanabe–Akaike): fully Bayesian; targets predictive accuracy; sum of pointwise log predictive densities minus a penalty for posterior variance.
- **LOO-CV** (leave-one-out cross-validation via PSIS): gold standard for predictive comparison; implemented efficiently using importance sampling.

IV. Bayesian Analogs of Classical Tests

Each test uses the Bayes factor for model comparison and the posterior distribution for parameter inference, replacing p-values and confidence intervals.

Default priors: the JZS prior. The **Jeffreys–Zellner–Siow (JZS) prior** is the standard default for Bayesian t-tests and linear regression. It places Zellner's g-prior on coefficients, $\boldsymbol{\beta} \mid g, \sigma^2 \sim N(\mathbf{0}, g\sigma^2(\mathbf{X}'\mathbf{X})^{-1})$, with hyperprior $g \sim \text{Inv-Gamma}(\frac{1}{2}, \frac{n}{2})$. Integrating out g yields a Cauchy marginal on the standardized effect size $\delta = \beta/\sigma$, giving a heavy-tailed, scale-invariant prior that remains proper. The scale parameter r (default $r = \sqrt{2}/2$ in `BayesFactor`) should always be reported alongside any Bayes factor.

Bayesian t-test: Tests $H_0 : \mu_1 = \mu_2$ using a Cauchy prior on the effect size (Rouder et al. 2009); default scale $r = \sqrt{2}/2$. Computes BF_{10} for the difference and the posterior distribution of Cohen's d.

Bayesian ANOVA: Compares models with and without group effects using BF for each term; uses default g-priors (Zellner–Siow). Can quantify evidence for the null (no group effect) as well as the alternative— an advantage over classical ANOVA.

Bayesian correlation test: Places a stretched beta prior on $\rho \in (-1, 1)$, $\pi(\rho) \propto (1-\rho^2)^{\alpha-1}$ (Wetzels & Wagenmakers 2012; Ly et al.), with $\alpha = 1$ giving a uniform prior over $(-1, 1)$. Computes BF_{10} vs. $H_0 : \rho = 0$ via the Savage–Dickey ratio; reports posterior distribution of ρ. The key advantage over classical tests: the Bayes factor can actively support $H_0\colon \rho = 0$ (independence), not merely fail to reject it.

Bayesian linear regression: Uses g-priors or Zellner–Siow priors on $\boldsymbol{\beta}$; Bayes factors for variable inclusion; posterior inclusion probabilities (PIP) summarize evidence for each predictor.

Bayesian proportion test: Places a Beta(a, b) prior on p; posterior is Beta($a + x$, $b + n - x$) (conjugate); BF_{10} vs. $H_0 : p = p_0$ via Savage–Dickey.

V. HDI + ROPE: Practical Significance

The **HDI + ROPE** framework combines a credible interval with a **Region of Practical Equivalence** (ROPE) to make practical decisions:

ROPE = $(-\epsilon, +\epsilon)$: the range of parameter values that are practically equivalent to zero (or to a reference value). Typical choice: $\epsilon = 0.1$ in Cohen's d units (one-tenth of a small effect).

Decision rules:

HDI–ROPE	**Decision**
HDI $\cap$ ROPE $= \varnothing$	Reject null; meaningful effect
HDI $\subseteq$ ROPE	Accept null; practical equivalence
HDI $\cap$ ROPE $\neq \varnothing$	Undecided; more data needed

The HDI + ROPE approach separates **statistical significance** (is the effect non-zero?) from **practical significance** (is the effect large enough to matter?). It is the Bayesian analog of the TOST equivalence procedure (see EQVL chapter) but operates on the full posterior rather than a single point estimate.

VI. Posterior Predictive p-Value (PPP)

The PPP is a Bayesian model-checking device. Generate replicated datasets $\mathbf{y}^{\text{rep}}$ from the **posterior predictive distribution**:

$$p(\mathbf{y}^{\text{rep}}|\mathbf{y}) = \int p(\mathbf{y}^{\text{rep}}|\boldsymbol{\theta})p(\boldsymbol{\theta}|\mathbf{y})\, d\boldsymbol{\theta}.$$

For a discrepancy measure T (e.g. skewness, maximum, χ^2 statistic):

$$\text{PPP} = P(T(\mathbf{y}^{\text{rep}}) \geq T(\mathbf{y})|\mathbf{y}).$$

PPP near 0 or 1 signals model misfit; PPP near 0.5 indicates good model fit for that discrepancy.

Limitations:

- PPP is **not a p-value** in the frequentist sense; it uses the data twice (to fit the model and to compute the test), making it conservative.
- Expected PPP for a well-specified model is 0.5, not 0.05; do not apply an α threshold.
- Choose T to target the specific aspect of fit of concern; no single T is universally informative.

Computation

Closed-form posteriors exist only for conjugate models. The general case requires:

- **MCMC:** Metropolis–Hastings, Gibbs, Hamiltonian Monte Carlo (HMC; Stan). Convergence: $\hat{R} < 1.01$; ESS > 400.
- **Variational Bayes:** fast approximation; biased toward underestimating posterior variance.
- **Laplace approximation:** Gaussian centered at MAP; adequate for unimodal posteriors with large n.
- **Bridge sampling:** accurate marginal likelihood computation for Bayes factors; substantially more stable than the harmonic mean estimator; default in R package `bridgesampling`.

Common Pitfalls

- **Bartlett's paradox:** diffuse priors on model parameters make $p(\mathbf{y}|M_1)$ very small; Bayes factors are undefined with improper priors and arbitrarily small with very diffuse ones. Use default priors (Cauchy, Zellner–Siow) or unit-information priors.
- **Treating HDI as CI:** a 95% HDI is *not* equivalent to a 95% CI; they agree only when the posterior is symmetric and unimodal.
- $\mathrm{BF}_{10} > 3$ **as definitive:** "moderate" evidence is not conclusive; Bayes factors are continuous evidence measures, not binary decisions.
- **No sensitivity analysis:** always check whether BF_{10} changes substantially with alternative priors.
- **PPP near 0.5 as proof of fit:** PPP is conservative; use multiple T statistics targeting different potential failure modes.
- **MCMC convergence not checked:** all Bayesian summaries from MCMC are invalid if chains have not converged; inspect $\hat{R}$ and trace plots.

Bottom Line

> Bayesian analysis provides a coherent framework for updating beliefs and comparing models. Use Bayes factors and Savage–Dickey for model comparison; HDI + ROPE for practical decisions that distinguish statistical from practical significance; posterior predictive checks for model validation. Specify priors carefully—the JZS prior is the recommended default for t-tests and regression—and always conduct sensitivity analysis; Bayes factors are especially sensitive to prior specification.

Bayes Factor Test (BF_{10})

Bayesian Evidence Ratio for Model Comparison

BAYES.1 | **Category:** Bayesian Methods | **Data:** Any | **Null:** Posterior distribution

When to Use

- Quantify strength of evidence without arbitrary p-value thresholds. Situations requiring evidence for H_0, not just failure to reject.
- Sequential or accumulating data analysis (Bayesian update).
- Comparing non-nested or competing models.

Assumptions

- Likelihood functions for data, and priors for parameters are correctly specified under both hypotheses.
- Priors reasonably reflect substantive knowledge or are justified as defaults.
- Marginal likelihoods are computable.

Test Statistic

$$BF_{10} = \frac{P(\text{Data}|H_1)}{P(\text{Data}|H_0)} = \frac{\int P(\text{Data}|\theta, H_1)\,\pi(\theta|H_1)\,d\theta}{\int P(\text{Data}|\theta, H_0)\,\pi(\theta|H_0)\,d\theta} \quad \text{(Ratio of marginal likelihoods).}$$

Notation: $P(\text{Data} \mid H)$ is the marginal likelihood under H; $\pi(\theta \mid H)$ is the prior distribution for parameters under H

Plain English: The Bayes factor quantifies the relative evidence for two competing hypotheses by comparing how well each predicts the observed data, averaged over plausible parameter values. It answers: "How much more likely are the data under the alternative hypothesis compared to the null?" Unlike p-values which assume the null is true, Bayes factors weigh evidence for both hypotheses. Values much greater than 1 favor the alternative; values much less than 1 favor the null; values near 1 indicate data uninformative.

⚠ Practitioner Warning

Results depend critically on prior specifications; different priors can yield different Bayes factors. Default priors (e.g., Cauchy, JZS) are convenient but not universally appropriate. Interpretation scales (e.g. Jeffreys) are conventional, not strict thresholds. Bayes factors do not directly provide effect size; use posterior distributions for that. Computation can be challenging in complex models, often requiring approximations. Bayes factors are not posterior probabilities unless prior odds are specified. Vague priors under H_1 can inflate evidence for H_0 (Jeffreys–Lindley paradox).

Hypotheses

- H_0: Null model (e.g., no effect, $\theta = 0$).
- H_1: Alternative model (e.g., $\theta \neq 0$ with prior $\theta \sim \pi(\theta)$).

Decision Rule: Interpret via Jeffreys' scale or modern Kass-Raftery scale, but no universal cutoff applies.

Software References [Python, R, SAS]

Python	R	SAS
`pingouin.bayesfactor_ttest`	`BayesFactor::extractBF`	`Not Native (Manual)`

Additional Notes

Jeffreys (1961) formalized Bayes factors as a Bayesian alternative to hypothesis testing. The marginal likelihood $P(\text{Data} \mid H) = \int P(\text{Data} \mid \theta, H)P(\theta \mid H)\,d\theta$ integrates over parameters, naturally penalizing complex models (Occam's razor). Interpretation (Jeffreys scale): $BF_{10} = 1{-}3$ (anecdotal), $3{-}10$ (moderate), $10{-}30$ (strong), $30{-}100$ (very strong), > 100 (decisive); $BF_{01} = 1/BF_{10}$ gives evidence for H_0. Common setup: point null $H_0 : \theta = \theta_0$ vs. composite $H_1 : \theta \sim p(\theta)$. Default priors (e.g., JZS prior for t tests) use a Cauchy prior on standardized effect size. When H_0 is nested, the Savage–Dickey ratio simplifies computation; otherwise, numerical methods such as bridge sampling are used. Advantages: allows evidence for H_0, provides continuous evidence, incorporates prior information, and updates naturally with sequential data. Limitations: depends on prior choice, can be computationally intensive, and requires careful interpretation.

Miscellanea: Jeffreys (1930s) introduced Bayes factors as a formal alternative to significance testing, later codified in *Theory of Probability* (1961). A key philosophical shift: instead of asking "Is the null unlikely?", Bayes factors ask "Which model predicts the data better?" — a predictive competition rather than a rejection framework. Their use has grown with advances in computation.

Reference(s): Jeffreys (1961); Kass & Raftery (1995); Rouder et al. (2009)

Bayes Factor Test (BF_{10}) BAYES.1

Worked Example

Problem Statement A quality engineer tests whether a manufacturing process produces defective items at the benchmark rate $\theta_0 = \frac{1}{2}$ (binomial success = defect). In $n = 10$ independent inspections, $x = 9$ defects are found. Prior knowledge about the true rate is vague, so the engineer assigns $\theta \sim \text{Uniform}(0, 1)$ under H_1. Test H_0: $\theta = \frac{1}{2}$ vs H_1: $\theta \sim \text{Uniform}(0, 1)$ by computing the Bayes factor $\text{BF}_{10} = p(\text{data} \mid H_1)/p(\text{data} \mid H_0)$ directly from the two marginal likelihoods.

1. Marginal Likelihood Under H_0
Under H_0 the probability of the data is fixed:

$$p(x{=}9 \mid H_0) = \binom{10}{9}\left(\tfrac{1}{2}\right)^{10} = \frac{5}{512}.$$

2. Marginal Likelihood Under H_1
Average the binomial likelihood over the uniform prior:

$$p(x{=}9 \mid H_1) = \int_0^1 \binom{10}{9}\theta^9(1-\theta)\cdot 1\, d\theta = \frac{1}{11}.$$

From the properties of the Beta function:

$$\boxed{\int_0^1 \binom{n}{k}\theta^k(1-\theta)^{n-k}\, d\theta = \frac{1}{n+1}}$$

3. Bayes Factor

$$\text{BF}_{10} = \frac{p(x{=}9 \mid H_1)}{p(x{=}9 \mid H_0)} = \frac{1/11}{5/512} = \frac{512}{55} \approx 9.31$$

	Value	Jeffreys scale
$\text{BF}_{10} = 512/55$	9.31	Moderate–strong for H_1
$\text{BF}_{01} = 55/512$	0.107	H_0 disfavored

4. Posterior Under H_1
Conjugate update: $\theta \mid \text{data}, H_1 \sim \text{Beta}(10, 2)$, with posterior mean $10/12 = 5/6 \approx 0.833$.

5. Exact Binomial Test
Exact two-sided binomial p-value. Reject H_0 at $\alpha = 5\%$.

$$p = \frac{2(C(10,9) + C(10,10))}{2^{10}} = \frac{11}{512} \approx 0.021.$$

6. Frequentist Comparison
Here both approaches support H_1: $\text{BF}_{10} \approx 9.3$ (moderate–strong), and the exact binomial test gives $p \approx 0.021$ (significant).

Practitioner Insight: BF provides a coherent Bayesian measure of evidence, updating prior odds to posterior odds via $\text{Odds}(H_1 \mid \text{data}) = \text{BF}_{10} \times \text{Odds}(H_1)$. With equal priors, $\text{BF}_{10} = 512/55 \approx 9.3$ implies H_1 is about nine times more credible than H_0 after observing 9 defects in 10 trials. Using a uniform prior Beta(1, 1) yields the posterior Beta(10, 2) by conjugacy, and simplifies the marginal likelihood under H_1 to $1/(n+1)$, making computation straightforward. On standard scales, $\text{BF}_{10} \approx 9.3$ indicates moderate to strong evidence. The two-sided p-value $11/512 \approx 0.021$ also rejects H_0, so both frameworks agree here, though not always: with small n or borderline data, BFs may remain near 1 while p-values cross 0.05 (Jeffreys–Lindley paradox). The distinction is fundamental: a p-value assesses extremeness under H_0, whereas the Bayes factor compares how well the data support H_1 relative to H_0, averaging over the prior.

Common Pitfalls

- **Interpreting the Bayes factor as the probability that a hypothesis is true.** BF_{10} is an evidence ratio, not a posterior probability. Posterior probability requires prior odds: Posterior Odds = $BF_{10} \times$ Prior Odds.
- **Using default priors without examining their implications.** Default priors (e.g., Cauchy priors in many implementations) encode assumptions about plausible effect sizes. Results can change meaningfully with different priors, so sensitivity checks and clear reporting are essential.
- **Using overly diffuse priors under H_1 (Jeffreys–Lindley effect).** Very broad priors spread probability mass over implausible parameter values, which can reduce the marginal likelihood under H_1 and artificially favor H_0. Careful prior specification is essential to avoid misleading evidence.
- **Equating Bayes factor thresholds with p-value cutoffs.** Bayes factors quantify relative model evidence, whereas p-values measure tail extremeness under H_0. They do not correspond directly to thresholds such as $p < 0.05$, and the two approaches may disagree.
- **Ignoring computational reliability in complex models.** BFs require estimating marginal likelihoods, which can be numerically unstable. Verify convergence and, when possible, confirm results using multiple estimation approaches.
- **Failing to report evidence for H_0 when $BF_{10} < 1$.** When the Bayes factor favors the null, reporting $BF_{01} = 1/BF_{10}$ clarifies interpretation by expressing the evidence ratio directly for H_0.

Savage-Dickey Density Ratio Test

Simplified Bayes Factor via Posterior-Prior Ratio at Null Value

BAYES.2 | **Category:** Bayesian Methods | **Data:** Any | **Null:** Posterior distribution

When to Use

- Compute Bayes factors without direct marginal likelihood integration.
- Situations where H_0 is nested within H_1 (point null).
- When posterior samples under H_1 are already available (e.g., via MCMC).

Assumptions

- Null hypothesis specifies point value for θ: $H_0 : \theta = \theta_0$.
- H_1 assigns continuous prior distribution to θ.
- H_0 is nested within H_1 under the same parameterization.
- Prior density $p(\theta_0 \mid H_1)$ is well-defined and nonzero.
- Posterior distribution under H_1 available or computable.

Test Statistic

$$BF_{01} = \frac{p(\theta_0|\text{Data}, H_1)}{p(\theta_0|H_1)}, \quad BF_{10} = \frac{1}{BF_{01}} = \frac{p(\theta_0|H_1)}{p(\theta_0|\text{Data}, H_1)} \quad \text{(only densities required at single point).}$$

Notation: BF_{01} favors null, $p(\theta_0|\text{Data}, H_1)$ is posterior density at θ_0, $p(\theta_0|H_1)$ is prior density at θ_0

Plain English: The Savage-Dickey density ratio simplifies Bayes factor computation by comparing the posterior and prior densities of a parameter at the null hypothesis value. If the data support the null, the posterior density at θ_0 increases relative to the prior; if the data support the alternative, the posterior density at θ_0 decreases. The ratio directly yields the Bayes factor without computing marginal likelihoods. This works because when the null is nested, the complex integrals in the Bayes factor collapse to a simple ratio.

⚠ Practitioner Warning

Only valid when null hypothesis nested in alternative (point null within continuous prior). Requires continuous prior at null value; improper priors or priors with zero density at θ_0 invalid. Prior density at exact null value can be sensitive to prior specification; results depend on prior choice. Posterior and prior densities must be evaluated at exactly the same point. Computational precision matters; numerical approximations to densities affect ratio. Does not work for composite nulls or non-nested hypotheses. Interpretation follows standard Bayes factor scale (Jeffreys).

Hypotheses

- H_0: Point null hypothesis ($H_0 : \theta = \theta_0$, typically $\theta_0 = 0$).
- H_1: Composite hypothesis ($H_1 : \theta \sim p(\theta)$).

Decision Rule: Interpret via Jeffreys' scale or modern Kass-Raftery scale, but no universal cutoff applies.

Software References [Python, R, SAS]

Python	R	SAS
`pingouin.bayesfactor_ttest`	`BayesFactor::extractBF`	`Not Native (Manual)`

Additional Notes

When $H_0 : \theta = \theta_0$ is nested in H_1, the Bayes factor simplifies to

$$BF_{01} = \frac{p(\theta_0 \mid \text{Data}, H_1)}{p(\theta_0 \mid H_1)},$$

avoiding direct computation of marginal likelihoods. Computation: fit the model under H_1, evaluate the posterior density at θ_0, evaluate the prior density at θ_0, and take their ratio. With MCMC, the posterior density is typically estimated via kernel density methods. The result depends on the prior: low prior density at θ_0 can favor H_0 even with small posterior mass. The method extends to multi-parameter settings using joint densities. Its simplicity and compatibility with MCMC make it widely used despite sensitivity to prior specification.

Miscellanea: Dickey (1971) formalized the Savage–Dickey density ratio for computing Bayes factors in nested models, building on earlier ideas attributed to Savage. The result shows that, under certain conditions, Bayes factors can be obtained as a ratio of posterior to prior densities at the null value, avoiding direct computation of marginal likelihoods. The method has become especially useful with modern Bayesian computation.

Reference(s): Dickey (1971); Verdinelli & Wasserman (1995); Wagenmakers et al. (2010)

Worked Example

Problem Statement A researcher tests whether a short training intervention has *zero* mean effect on a standardized performance score. Four independent observations are modeled as $X_i \sim \mathrm{N}(\mu, \sigma^2{=}1)$, yielding $\bar{x} = 1$ (one standard deviation above the null). Under H_1 the effect μ is assigned a N(0, 1) prior (centered on the null, unit variance, reflecting genuine uncertainty in both directions). Test

$$H_0\colon \mu = 0 \quad \text{vs} \quad H_1\colon \mu \sim \mathrm{N}(0, \tau^2{=}1)$$

using the Savage–Dickey density ratio (SDDR), which gives BF_{01} as a ratio of two density evaluations at the null point $\mu_0 = 0$, with no integration required.

1. Update the Prior to the Posterior Under H_1

The conjugate posterior for μ given $\bar{x} = 1$, $n = 4$, $\sigma^2 = 1$, $\tau^2 = 1$ is $\mu \mid \text{data}, H_1 \sim \mathrm{N}(\mu_n, \sigma_n^2)$ with:

$$\sigma_n^2 = \frac{1}{\frac{1}{\tau^2} + \frac{n}{\sigma^2}} = \frac{1}{1+4} = \frac{1}{5}$$

$$\mu_n = \sigma_n^2\left(\frac{\mu_0}{\tau^2} + \frac{n\bar{x}}{\sigma^2}\right) = \frac{1}{5}(0+4) = \frac{4}{5}$$

The posterior has shifted from the prior mean 0 toward $\bar{x} = 1$.

2. Evaluate Prior at the Null

$$\pi(\mu_0 \mid H_1) = \mathrm{N}(0;\, 0,\, 1) = \frac{1}{\sqrt{2\pi}} \approx 0.3989$$

Notation: $\mathrm{N}(x;\, \mu, \sigma^2)$ denotes the *normal density evaluated at* x for a normal distribution with mean μ and variance σ^2. Thus, N(0; 0, 1) means the value of the $\mathcal{N}(0, 1)$ density at the point $x = 0$.

3. Evaluate Posterior at the Null

$$\pi(\mu_0 \mid \text{data}, H_1) = \mathrm{N}\left(0;\, \tfrac{4}{5},\, \tfrac{1}{5}\right)$$

$$= \sqrt{\frac{5}{2\pi}}\exp\left(-\frac{1}{2}\cdot\frac{(4/5)^2}{1/5}\right) = \sqrt{\frac{5}{2\pi}}\, e^{-8/5} \approx 0.1801$$

4. Savage–Dickey Ratio

$$\mathrm{BF}_{01} = \frac{\pi(\mu_0 \mid \text{data}, H_1)}{\pi(\mu_0 \mid H_1)} = \frac{\sqrt{5/(2\pi)}\, e^{-8/5}}{1/\sqrt{2\pi}} = \sqrt{5}\, e^{-8/5}$$

$$\approx 2.236 \times 0.202 = \mathbf{0.451}$$

	Value	Interp.
$\mathrm{BF}_{01} = \sqrt{5}\, e^{-8/5}$	0.451	H_0 disfavoured
$\mathrm{BF}_{10} = 1/\mathrm{BF}_{01}$	2.22	Anecdotal vs H_0

Frequentist z-test: $z = \bar{x}/(\sigma/\sqrt{n}) = 2$, $p = 0.046$ (rejects H_0 at 5%).

Practitioner Insight: The SDDR reduces Bayes factor computation to a single density ratio, avoiding marginal-likelihood integration. Here, $\mathrm{BF}_{01} = \sqrt{5}\, e^{-8/5}$ arises from cancellation of normalizing constants, leaving a precision ratio $\tau/\sigma_n = \sqrt{5}$ and Gaussian discrepancy $e^{-8/5}$. Validity requires a *point null nested within* H_1, a proper prior, and no additional nuisance parameters affecting the null restriction. The example illustrates the Jeffreys–Lindley paradox: a frequentist rejection ($p = 0.046$) contrasts with only anecdotal Bayesian evidence ($\mathrm{BF}_{10} \approx 2.2$), reflecting small n and data consistent with moderate effects under H_1. As n increases, BF_{10} grows while p decreases, resolving the tension. Equivalently, the SDDR can be expressed as a ratio of marginal likelihoods, $\mathrm{BF}_{01} = \mathrm{N}(1; 0, \frac{1}{4})/\mathrm{N}(1; 0, \frac{5}{4})$, yielding the same result.

Common Pitfalls

- **Using improper or flat priors.** Savage–Dickey requires a proper prior with finite density at the null value θ_0. Improper priors (e.g., flat on $\mathbb{R}$) make $\pi(\theta_0)$ undefined, rendering the ratio meaningless.
- **Ignoring parameterization sensitivity.** The Savage–Dickey ratio depends on the chosen parameterization. Reparameterizing (e.g., $\theta \to \phi = g(\theta)$) changes the prior and posterior densities at the null, potentially altering the Bayes factor. Results are valid only under a consistent and justified parameterization.
- **Applying the method to non-nested models.** The test works only when H_0 is a restriction of H_1 (e.g., $\theta = \theta_0$ within a broader model). For non-nested models, Bayes factors must be computed using full marginal likelihood methods.
- **Unstable density estimation at the null value.** Estimating the posterior density at a single point from MCMC samples can be sensitive to bandwidth choice and local sample density. Poor estimation near θ_0 can distort the Bayes factor.
- **Failing to specify which Bayes factor is reported.** Savage–Dickey naturally yields BF_{01} (evidence for H_0). Always state the direction clearly or report the reciprocal so the value exceeding 1 corresponds to the supported hypothesis.
- **Using the method for composite or interval null hypotheses.** Savage–Dickey applies to point nulls only ($\theta = \theta_0$). Interval or composite nulls require alternative Bayes factor formulations.

Bayesian Model Comparison

Model Selection via Marginal Likelihoods and Posterior Probabilities

BAYES.3 | **Category:** Bayesian Methods | **Data:** Any | **Null:** Posterior model probabilities

When to Use

- Compare and rank multiple competing models.
- Quantify relative evidence across a candidate model set.
- Incorporate model uncertainty via model averaging.
- Situations with multiple scientifically plausible models.

Assumptions

- Finite set of candidate models $\{M_i, i = 1, ..., k\}$ specified.
- Prior probabilities $P(M_k)$ are assigned, $\sum_k P(M_k) = 1$.
- Each model has a well-defined likelihood and prior over parameters. Marginals $P(\text{Data} \mid M_k)$ are computable.

Test Statistic

$$P(M_k|\text{Data}) = \frac{P(\text{Data}|M_k)P(M_k)}{\sum_{j=1}^{K} P(\text{Data}|M_j)P(M_j)}, \quad BF_{kj} = \frac{P(\text{Data}|M_k)}{P(\text{Data}|M_j)}.$$

Notation: $P(M_k|\text{Data})$ is posterior probability of model k, $P(\text{Data}|M_k)$ is marginal likelihood, BF_{kj} is Bayes factor

Plain English: Bayesian model comparison computes the posterior probability of each candidate model given the data by weighting each model's marginal likelihood by its prior probability and normalizing. The marginal likelihood measures how well the model predicts the data, averaged over parameter uncertainty and automatically penalizing complexity (Occam's razor). Models with higher posterior probabilities fit the data better while remaining parsimonious. Pairwise Bayes factors quantify relative evidence between specific models.

⚠ Practitioner Warning

Requires specifying prior probabilities for models; equal priors $(1/K)$ common default but not always appropriate. Marginal likelihood computation challenging; various approximations (BIC, Laplace, bridge sampling) vary in accuracy. Results sensitive to parameter priors within each model, not just model priors. Assumes true model in candidate set; if all models misspecified, selects "least bad" not "correct." Posterior probabilities can concentrate on single model even when multiple models similarly plausible. Does not account for models not considered.

Hypotheses

- H_0: Candidate models $\{M_1, \ldots, M_K\}$ with priors $P(M_k)$.
- H_1: Posterior model probabilities $P(M_k \mid \text{Data}) \forall k$.

Decision Rule: Interpret $P(M_k \mid \text{Data})$ as relative support across models. Use BF_{kj} for pairwise comparisons.

Software References [Python, R, SAS]

`(bridgesampling, pymc)` `BayesFactor::lmBF` `(PROC MCMC with DIC/WAIC)`

Additional Notes

Model priors reflect beliefs about plausibility: equal priors $P(M_k) = 1/K$ assume no preference, while informative priors can favor simpler or theory-driven models. The marginal likelihood $P(\text{Data} \mid M_k) = \int P(\text{Data} \mid \theta_k, M_k)P(\theta_k \mid M_k)\, d\theta_k$ averages likelihood over the prior, naturally penalizing complex models (Occam's razor). Computation methods include Laplace approximation (fast), bridge sampling (accurate), and path sampling (gold standard but costly); the harmonic mean estimator is unstable (infinite-variance behavior) and not recommended. The BIC provides a convenient approximation. When multiple models are plausible, Bayesian model averaging combines them via posterior weights: $\bar{\theta} = \sum_k \theta_k P(M_k \mid \text{Data})$, with predictions similarly averaged. Modern alternatives such as PSIS-LOO (Pareto Smoothed Importance Sampling-Leave One Out) offer predictive evaluation without direct marginal likelihood computation.

Miscellanea: Bayesian model comparison originates from Jeffreys' development of Bayes factors in the mid-20th century. Practical use was limited by the difficulty of computing marginal likelihoods, but advances in MCMC and related methods led to effective estimation techniques, including contributions by Chib (1995) and later bridge sampling approaches. Modern Bayesian practice emphasizes model uncertainty and often incorporates model averaging based on evidence.

Reference(s): Jeffreys (1961); Kass & Raftery (1995); Hoeting et al. (1999); Gronau et al. (2017)

Worked Example

Problem Statement A server records the number of connection requests arriving in each of $n = 4$ successive one-minute slots: $\mathbf{x} = (0, 0, 2, 2)$. The analyst entertains two models:

- M_0: constant rate — all four slots share a single Poisson rate λ, with prior $\lambda \sim \text{Gamma}(1, 1) = \text{Exp}(1)$;
- M_1: step change — slots 1–2 have rate λ_1 and slots 3–4 have rate λ_2, with independent priors $\lambda_1, \lambda_2 \sim \text{Exp}(1)$.

Compute the Bayes factor BF_{10} and posterior model probabilities under equal prior weights $P(M_0) = P(M_1) = \frac{1}{2}$.

1. Marginal Likelihood Under M_0
Average the Poisson likelihood over the Exp(1) prior:

$$p(\mathbf{x} \mid M_0) = \frac{1}{0!\,0!\,2!\,2!} \int_0^\infty \lambda^S e^{-(n+1)\lambda}\, d\lambda = \frac{6}{3125}.$$

where $S = \sum x_i = 4$, $n = 4$.

2. Marginal Likelihood Under M_1
The two halves are independent given M_1:

$$p(\mathbf{x} \mid M_1) = p(x_1, x_2 \mid M_1) \times p(x_3, x_4 \mid M_1).$$

First half ($x_1{=}x_2{=}0$, $S_1{=}0$):

$$\int_0^\infty e^{-2\lambda_1} \cdot e^{-\lambda_1}\, d\lambda_1 = \int_0^\infty e^{-3\lambda_1}\, d\lambda_1 = \frac{1}{3}.$$

Second half ($x_3{=}x_4{=}2$, $S_2{=}4$):

$$\frac{1}{2!\,2!} \int_0^\infty \lambda_2^4 e^{-3\lambda_2}\, d\lambda_2 = \frac{\Gamma(5)}{4 \times 3^5} = \frac{2}{81}.$$

Hence

$$p(\mathbf{x} \mid M_1) = \frac{1}{3} \times \frac{2}{81} = \frac{2}{243}.$$

3. Bayes Factor and Posterior Probabilities

$$\text{BF}_{10} = \frac{p(\mathbf{x} \mid M_1)}{p(\mathbf{x} \mid M_0)} = \frac{2/243}{6/3125} = \frac{3125}{729} \approx 4.29.$$

With equal prior odds ($P(M_0) = P(M_1) = \frac{1}{2}$), posterior odds equal BF_{10}:

$$P(M_1 \mid \mathbf{x}) = \frac{\text{BF}_{10}}{1 + \text{BF}_{10}} = \frac{3125}{3854} \approx 0.811.$$

$$P(M_0 \mid \mathbf{x}) = 729/3854 \approx 0.189.$$

4. Posterior Rate Estimates
Updating each Exp(1) prior by conjugacy:

Model	Parameter	Posterior	$E[\lambda \mid \mathbf{x}]$
M_0	λ	Gamma(5, 5)	1
M_1	λ_1	Gamma(1, 3)	1/3
M_1	λ_2	Gamma(5, 3)	5/3

Under M_1, the estimated rates 1/3 and 5/3 straddle the M_0 estimate of 1, confirming the changepoint interpretation.

Practitioner Insight: Unlike the point-null Bayes factor, Bayesian model comparison here integrates over parameters in *both* models and yields posterior model probabilities $P(M_k \mid \mathbf{x})$. Using Poisson–Gamma conjugacy, the marginal likelihood for n observations with sum S is $\beta^\alpha \Gamma(S + \alpha)/[\Gamma(\alpha)(n + \beta)^{S+\alpha} \prod x_i!]$; with $\alpha = \beta = 1$, this reduces to exact arithmetic, giving $p(\mathbf{x} \mid M_0) = 6/3125$ and $p(\mathbf{x} \mid M_1) = 2/243$. The resulting Bayes factor $3125/729 \approx 4.29$ indicates "positive" evidence for M_1. Despite having more parameters, M_1 is favored by *Occam's razor* because the data show a sharp contrast between segments; with weaker contrast, the advantage diminishes. Posterior means $(S_k + 1)/(n_k + 1)$ yield 1/3 and 5/3, symmetric about the single-rate estimate 1, reflecting the balanced design and prior.

Common Pitfalls

- **Expanding the candidate model set after inspecting the data.** Posterior model probabilities assume the candidate models and their priors were specified *a priori*. Adding models after examining the data invalidates the original prior model probabilities and the resulting inference.
- **Treating BIC as an exact marginal likelihood.** BIC approximates the log marginal likelihood only asymptotically. With small samples or complex models, it can diverge from the true Bayes factor; more accurate methods (e.g., bridge sampling or cross-validation) may be required.
- **Ignoring model uncertainty in predictions.** When several models have comparable posterior support, selecting a single "best" model discards useful information. Bayesian model averaging weights predictions by posterior model probabilities to account for uncertainty.
- **Comparing models fitted to different datasets.** Bayesian model comparison requires identical observations across models. Differences in sample size, missing-data handling, or preprocessing make marginal likelihoods incomparable.
- **Assuming the true model must be among the candidates.** Posterior probabilities rank models within the candidate set but do not guarantee that any model is correct. Interpret results as selecting the best among the considered models.

Bayesian t-Test

Bayesian Alternative to Classical t-Test with Bayes Factors

BAYES.4 | **Category:** Bayesian Methods | **Data:** Continuous | **Null:** Posterior distribution

When to Use

- Compare means between groups or against hypothesized value using Bayesian inference.
- Quantify evidence for H_0 as well as H_1.
- Obtain credible intervals for effect sizes.
- Monitor evidence sequentially without fixed sample size.

Assumptions

- Data approximately normally distributed within groups.
- Independent observations (a. one-sample, b. two-sample or c. paired). Equal variances for two-sample version (or use Behrens-Fisher alternative).
- Prior distribution specified for effect size (e.g: Cauchy/JZS).

Test Statistic

$$BF_{10} = \frac{P(\text{Data}|H_1, r)}{P(\text{Data}|H_0)}, \quad \delta \sim \text{Cauchy}(0, r), \quad r_{\text{default}} = \sqrt{2}/2.$$

Notation: BF_{10} is Bayes factor for alternative, δ is standardized effect size, r is scale parameter of Cauchy prior

Plain English: The Bayesian t-test replaces classical t-test's p-value with a Bayes factor quantifying relative evidence for the alternative hypothesis (mean difference exists) versus null hypothesis (no difference). The test uses a default prior on effect size—typically Cauchy distribution centered at zero—representing plausible effect magnitudes before seeing data. After observing data, the Bayes factor indicates how much more likely the data are under the alternative compared to the null, providing continuous evidence rather than binary reject/fail-to-reject decisions.

⚠ Practitioner Warning

Results depend on prior specification, especially scale parameter r; sensitivity analysis recommended. Default Cauchy prior (Rouder et al. 2009) widely used but not universally appropriate. Assumes normality; robust alternatives may be needed for non-normal data. Does not directly provide effect size estimate; use posterior distribution for that. Bayes factor interpretation requires understanding; not equivalent to posterior probability without equal prior odds. For very large samples, Bayes factors can strongly favor alternative even with tiny effect sizes, while diffuse priors can sometimes favor H_0 (Jeffreys–Lindley effect).

Hypotheses

- H_0: No effect ([a].$\mu = \mu_0$; [b].$\mu_1 = \mu_2$; [c].$\mu_d = 0$.).
- H_1: Effect exists ([a].$\mu \neq \mu_0$; [b].$\mu_1 \neq \mu_2$; [c].$\mu_d \neq 0$).

Decision Rule: Interpret via Jeffreys' scale or modern Kass-Raftery scale. No universal cutoff applies.

Software References [Python, R, SAS]

`pingouin.bayesfactor_ttest`	`BayesFactor::ttestBF`	`PROC MCMC (Manual)`

Additional Notes

Default Bayesian t-tests using the JZS prior, placing a Cauchy$(0, r)$ prior on the standardized effect size $\delta = (\mu_1 - \mu_2)/\sigma$ with default $r = \sqrt{2}/2$. The framework applies to one-sample, two-sample, and paired designs (via differences). The Cauchy prior is centered at zero (favoring small effects) but heavy-tailed, allowing large effects and maintaining scale invariance. Bayes factors are computed using the familiar t-statistic, but incorporate prior information to quantify evidence for both H_0 and H_1. The approach yields a full posterior for δ, enabling credible intervals for effect size. It also supports sequential data collection and directional hypotheses through appropriate prior restrictions. Compared to classical t-tests, the Bayesian version provides direct evidence measures rather than binary decisions.

Miscellanea: Rouder et al. (2009) introduced default Bayesian t-tests using the JZS prior, enabling routine computation of Bayes factors from standard t-statistics. A key insight: the Bayesian t-test uses the *same* t-statistic as the classical test, but interprets it differently—rather than measuring extremeness under H_0, it evaluates how well competing hypotheses predict the data. The Cauchy prior's heavy tails ensure that large effects remain plausible while preserving scale invariance, reflecting Jeffreys' original design principles.

Reference(s): Rouder et al. (2009); Jeffreys (1961); Wetzels et al. (2011)

Worked Example

Problem Statement Nine participants in a cognitive training study score 0, 0, 0, 2, 2, 2, 4, 4, 4 on a standardized test (population mean under no effect: $\mu_0 = 0$). Both the true mean μ and true variance σ^2 are unknown. Apply the Bayesian t-test of $H_0: \mu = 0$ against $H_1: \mu \neq 0$, using the Jeffreys–Zellner–Siow (JZS) Cauchy prior on the standardized effect size $\delta = \mu/\sigma$, and compare the result with the classical one-sample t-test. Assume equal model priors ($P(M_0) = P(M_1) = \frac{1}{2}$).

1. Sufficient Statistics

$$n = 9, \quad \bar{x} = 2, \quad s^2 = 2, \quad \text{SE} = \frac{s}{\sqrt{n}} = \frac{1}{\sqrt{3}}$$

2. Test Statistic

$$t = \frac{\bar{x} - \mu_0}{\text{SE}} = \frac{2}{1/\sqrt{3}} = 2\sqrt{3} \approx 3.464, \quad \nu = n - 1 = 8$$

Note $t^2 = 12$ and $t^2/\nu = 3/2$ exactly.

3. Frequentist p-value

$$p = 2\,P\Big(t_{(8)} > 2\sqrt{3}\Big) \approx 0.0085.$$

Reject H_0 at 1%. The 95% confidence interval for μ is (0.669, 3.331), which excludes 0.

Notes:

$$P(H_1 \mid \text{data}) = \frac{\text{BF}_{10}}{(1 + \text{BF}_{10})} = 0.881$$

(equal priors)

4. JZS Prior and Bayes Factor

The JZS prior places a standard Cauchy distribution on $\delta = \mu/\sigma$, equivalent to a scale mixture of normals: $\delta \mid g \sim \text{N}(0, g)$, $g \sim \text{Inv-}\chi^2(1)$. With Jeffreys prior $\pi(\mu_0, \sigma) \propto 1/\sigma$ on the shared nuisance parameters, the Bayes factor is (Rouder et al. 2009):

$$\text{BF}_{10} = \left(1 + \frac{t^2}{\nu}\right)^{n/2} \int_0^\infty h(g)\,\mathrm{d}g$$

where

$$h(g) = \frac{(1+ng)^{-1/2}\left(1 + \frac{t^2}{(1+ng)\,\nu}\right)^{-n/2}}{(2\pi g^3)^{1/2}\, e^{1/(2g)}}.$$

Substituting $t^2 = 12$, $n = 9$, $\nu = 8$: Numerical integration gives $\int_0^\infty h(g)\,\mathrm{d}g \approx 0.1198$, and $(1 + 3/2)^{9/2} \approx 61.76$, so: $\text{BF}_{10} \approx 7.40$.

	Value	Scale
BF_{10}	7.40	Positive evidence
$P(H_1 \mid \text{data})$	0.881	(equal priors)

Practitioner Insight: The Bayesian t-test integrates out the unknown variance σ^2 and places a prior on the standardized effect size $\delta = \mu/\sigma$, yielding scale-invariant inference. The default JZS prior $\delta \sim \text{Cauchy}(0, 1)$ is symmetric and heavy-tailed, allowing both small and large effects while remaining centered at zero. The Bayes factor requires one-dimensional numerical integration over g, typically peaking at moderate effect sizes. In this example ($\bar{x} = 2$, $s = \sqrt{3}$, $t = 2\sqrt{3}$), both frameworks agree: the frequentist test rejects ($p = 0.0085$) and $\text{BF}_{10} \approx 7.4$ indicates "positive" evidence. This contrasts with the Jeffreys–Lindley paradox seen in simpler settings. The posterior credible interval for μ matches the frequentist t-interval, reflecting their common pivot, though the Bayes factor additionally depends on the prior on δ.

Common Pitfalls

- **Using default priors without considering realistic effect sizes.** Default implementations use a Cauchy prior centered at zero, with scale r controlling plausible effect sizes. If typical effects are much smaller or larger, the default prior may be inappropriate; sensitivity analysis or informed priors should be used.
- **Equating Bayes factor thresholds with p-value cutoffs.** Bayes factors quantify relative evidence between models, not long-run error probabilities. They do not map directly to thresholds such as $p < 0.05$, and the two approaches may disagree (Jeffreys–Lindley paradox).
- **Interpreting BF_{10} as the probability that H_1 is true.** The Bayes factor is an evidence ratio, not a posterior probability. Posterior probabilities require prior odds through Posterior Odds = $BF_{10} \times$ Prior Odds.
- **Ignoring distributional assumptions in small samples.** The Bayesian t-test still assumes approximate normality of the data (or large-sample robustness). Outliers or heavy skew can distort inference unless robust models or transformations are used.
- **Treating paired observations as independent samples.** For matched or repeated-measures designs, analyze within-pair differences using a paired Bayesian t-test rather than two independent samples.

Bayesian ANOVA

Bayesian Analysis of Variance with Model Comparison

BAYES.5 | **Category:** Bayesian Methods | **Data:** Continuous | **Null:** Posterior distribution

When to Use

- Compare group means within a Bayesian framework.
- Perform model comparison across nested factor structures.
- Quantify evidence for main effects and interactions.
- Obtain posterior estimates for effects and effect sizes.

Assumptions

- Observations are independent within and across groups.
- Data are approximately normal within groups.
- Priors specified for effects and variance components.
- Homogeneity of variance (or modeled appropriately).

Test Statistic

$$BF_{10} = \frac{P(\text{Data}|M_{\text{effects}})}{P(\text{Data}|M_{\text{null}})}, \quad \delta_i \sim \text{Cauchy}(0, r), \quad r_{\text{default}} = 1/2.$$

Notation: BF_{10} is Bayes factor for effects model, δ_i are standardized group deviations, r is Cauchy scale

Plain English: Bayesian ANOVA compares models with and without group effects by computing Bayes factors, quantifying evidence for mean differences across groups. The null model assumes all groups share a common mean; the alternative model allows groups to differ with effect sizes governed by prior distributions. Unlike classical ANOVA's F-test, Bayesian ANOVA provides continuous evidence quantification, can support the null hypothesis, and yields full posterior distributions for group means and effect sizes enabling probabilistic statements about parameters.

⚠ Practitioner Warning

Results depend on prior specifications for effect sizes; sensitivity analysis recommended. Default priors (Rouder et al. 2012) designed for balanced designs; unbalanced data may require adjustment. Assumes normality within groups; violations may affect inference unless robust models are used. For factorial designs, must specify priors for main effects and interactions separately. Model comparison among multiple models (main effects only, interactions, etc.) requires careful interpretation. Large number of groups increases computational burden. Does not automatically identify which specific groups differ; requires post-hoc comparisons.

Hypotheses

- H_0: All group means are equal ($\mu_1 = \mu_2 = \cdots = \mu_k$).
- H_1: At least one μ_i differs, effect size $\delta_i \sim \text{Cauchy}(0, r)$.

Decision Rule: Interpret Bayes factors as continuous measures of evidence across competing models.

Software References [Python, R, SAS]

Python	R	SAS
`(pymc, numpyro, arviz)`	`BayesFactor::anovaBF`	`PROC MCMC (Manual)`

Additional Notes

Rouder et al. (2012) developed default Bayesian ANOVA using a random-effects formulation with JZS priors. The null model assumes $y_{ij} = \mu + \epsilon_{ij}$, while the effects model adds group effects $\alpha_i = \delta_i \sigma$ with $\delta_i \sim \text{Cauchy}(0, r)$ and a sum-to-zero constraint for identifiability. A default scale $r = 1/2$ is used for ANOVA. For factorial designs, priors are specified for main effects and interactions, and Bayes factors are computed across competing models (null, main effects, interactions) to rank them by posterior evidence. The approach yields full posterior distributions for parameters, enabling credible intervals and effect size summaries. Extensions include post-hoc comparisons via Bayesian t-tests, inclusion Bayes factors for assessing factors, and repeated-measures ANOVA. The framework naturally accommodates missing or unbalanced data.

Miscellanea: Morey and Rouder (2011–2012) developed default Bayesian ANOVA methods, extending Bayes factor approaches to factorial designs using principled priors. A key insight is that the same JZS prior used in Bayesian t-tests generalizes naturally to multi-group settings, enabling coherent model comparison across main effects and interactions. Their work, along with accessible software implementations, contributed to increased use of Bayesian model comparison in psychology and related fields.

Reference(s): Rouder et al. (2012); Morey & Rouder (2015)

Worked Example

Problem Statement Three teaching methods (A, B, C) are each applied to $m = 3$ students. Post-test scores are:

$$A: -1,\ 0,\ 1; \qquad B: 1,\ 2,\ 3; \qquad C: 3,\ 4,\ 5.$$

Test whether teaching method affects mean score, comparing: M_0: no group effect, all means equal; M_1: group means differ.

Use the Rouder et al. (2012) JZS Bayes factor with the unit-information g-prior ($g = n = 9$) on standardized effect sizes, and Jeffreys prior $\pi(\mu, \sigma) \propto 1/\sigma$ on shared nuisance parameters. Assume equal model priors ($P(M_0) = P(M_1) = \frac{1}{2}$).

1. ANOVA Summary Statistics

Group	$\bar{x}_j$	$\sum_i (x_{ij} - \bar{x}_j)^2$
A $(-1, 0, 1)$	0	$1 + 0 + 1 = 2$
B $(1, 2, 3)$	2	$1 + 0 + 1 = 2$
C $(3, 4, 5)$	4	$1 + 0 + 1 = 2$

$$\bar{x} = 2, \quad SS_b = 24, \quad SS_w = 6, \quad SS_T = 30.$$

2. Classical F-test

$$F = \frac{SS_b/(k-1)}{SS_w/(n-k)} = \frac{24/2}{6/6} = 12, \quad (2, 6) \text{ df.}$$

$p = 0.008$; reject H_0 at 1%.

3. Effect-Size Coefficient

$$R^2 = \frac{SS_b}{SS_T} = \frac{24}{30} = \frac{4}{5}.$$

4. g-Prior Bayes Factor

With $g = n = 9$, the closed-form BF is (Rouder et al. 2012):

$$\mathrm{BF}_{10}(g) = (1 + g)^{-(k-1)/2} \left(1 - \frac{g R^2}{1 + g}\right)^{-(n-1)/2}$$

Substituting $g = 9$, $k = 3$, $n = 9$, $R^2 = 4/5$:

$$\mathrm{BF}_{10}(g) = \frac{1}{10} \left(\frac{25}{7}\right)^4 \approx 16.3.$$

	Value	Interpretation
F-statistic	12.00	—
p-value	0.008	(reject at 1%)
BF_{10}	16.27	Strong evidence
$P(M_1 \mid \text{data})$	0.942	(equal priors)

$P(M_1 \mid \text{data}) = \mathrm{BF}_{10}/(1 + \mathrm{BF}_{10}) = 0.942$

Practitioner Insight: Bayesian ANOVA extends the Bayesian t-test to $k \geq 3$ groups, placing a g-prior on group effects $\alpha \mid g, \sigma \sim \mathrm{N}(\mathbf{0}, g\sigma^2\mathbf{H})$, where $\mathbf{H}$ encodes contrasts. With a fixed unit-information prior $g = n$, the marginal likelihood is closed-form and the Bayes factor depends on the F-statistic via $R^2 = (k-1)F/((k-1)F + n - k)$. In this example, $R^2 = 4/5$ yields $\mathrm{BF}_{10} = (1/10)(25/7)^4 \approx 16.3$, indicating "strong" evidence and agreeing with the frequentist result ($p = 0.008$). Relative to the t-test, R^2 generalizes $t^2/(t^2 + \nu)$, while the factor $(1 + g)^{-(k-1)/2}$ penalizes additional parameters, implementing Occam's razor. In practice, JZS priors require one-dimensional numerical integration (e.g., `BayesFactor`), while the closed-form g-prior highlights the structure of the computation.

Common Pitfalls

- **Testing all possible factorial models without prior specification.** Exhaustively fitting every combination of factors (e.g., 2^k models) after seeing the data introduces selection bias. Define theory-driven models in advance or use principled model averaging.
- **Ignoring heterogeneity of variances across groups.** Standard Bayesian ANOVA formulations assume a common error variance. Large differences can distort inference, requiring transformations or models allowing unequal variances.
- **Treating $BF_{10} > 1$ as decisive evidence for group differences.** A Bayes factor slightly above 1 indicates only weak evidence. Interpret results on a graded scale (e.g., anecdotal, moderate, strong) rather than using a binary threshold.
- **Ignoring repeated-measures structure.** When the same subjects appear in multiple conditions, independence assumptions fail. Use a repeated-measures Bayesian ANOVA that models subject-level variation.
- **Ignoring sensitivity to priors.** Bayes factors depend on prior distributions for main effects and interactions. Diffuse priors can penalize complex models, while concentrated priors can favor them; sensitivity analysis is essential.
- **Stopping data collection when a desired Bayes factor appears.** Sequential monitoring is valid in Bayesian analysis, but stopping rules should be pre-specified. Stopping only when evidence favors one hypothesis introduces bias.

Bayesian Correlation Test

Bayesian Inference for Pearson Correlation Coefficient

BAYES.6 | **Category:** Bayesian Methods | **Data:** Continuous (Bivariate) | **Null:** Posterior distribution

When to Use

- Test association between two continuous variables in a Bayesian framework.
- Quantify evidence for both presence and absence of correlation ($\rho = 0$).
- Obtain posterior estimates and credible intervals for ρ.

Assumptions

- Paired observations are independent.
- Data are approximately bivariate normal (or appropriately transformed).
- A prior distribution is specified for ρ on $[-1, 1]$.
- Sample size is adequate for stable posterior estimation.

Test Statistic

$$BF_{10} = \frac{P(\text{Data}|H_1)}{P(\text{Data}|H_0)}, \quad \rho \sim \text{Beta}^*(\kappa) \text{ on } [-1, 1] \text{ (stretched Beta distribution, } \rho = 2x - 1).$$

Notation: BF_{10} compares models with and without correlation; ρ is the population correlation

Plain English: Bayesian correlation test quantifies evidence for association between variables by comparing models with and without correlation. The null hypothesis specifies zero correlation; the alternative places a prior distribution on correlation (typically stretched (κ) beta distribution on $[-1, 1]$). The Bayes factor indicates how much more likely the data are under the alternative compared to null. Additionally, the full posterior distribution of correlation provides credible intervals and probability statements about correlation strength.

⚠ Practitioner Warning

Results depend on prior specification for correlation under alternative; default priors available but not universally appropriate. Assumes bivariate normality; violations affect inference though robust alternatives exist using Spearman's ρ. Prior on correlation coefficient must have support on $[-1, 1]$. Very large samples can yield strong Bayes factors for trivially small correlations; examine posterior distribution and effect size. Does not establish causation; correlation measures association only. Outliers can substantially influence results; diagnostic checks essential.

Hypotheses

- H_0: $\rho = 0$
- H_1: $\rho \sim \text{Beta}^*(\kappa)$, stretched Beta* on $[-1, 1]$.

Decision Rule: Interpret via Jeffreys' scale or modern Kass-Raftery scale.. No universal cutoff applies.

Software References [Python, R, SAS]

Python	R	SAS
`(pingouin, pymc)`	`BayesFactor::correlationBF`	`PROC MCMC (Manual)`

Additional Notes

Wetzels and Wagenmakers (2012) proposed a default Bayesian correlation test using a stretched beta prior on $\rho \in [-1, 1]$. The default $\kappa = 1$ yields a uniform prior, while larger κ concentrates mass near zero, reflecting skepticism about strong correlations. Using the Savage–Dickey ratio, $BF_{01} = p(\rho = 0 \mid \text{Data})/p(\rho = 0)$, Bayes factors can be computed from the posterior and prior densities at zero. Posterior inference provides full distributions and credible intervals for ρ. Extensions include one-sided tests, robust rank-based (Spearman) versions, and partial correlations controlling for covariates. The method is especially useful in exploratory settings where evidence for no correlation is scientifically meaningful.

Miscellanea: Wetzels and Wagenmakers (2012) demonstrated that Bayes factors can provide evidence for the absence of a relationship, not just detect its presence. This work highlighted a key limitation of classical significance testing and contributed to the growing use of Bayesian methods in applied research. Unlike classical methods that rely on Fisher's z-transformation, the Bayesian approach models ρ directly with priors defined on $[-1, 1]$, ensuring valid inference within the natural bounds of correlation. The need to define priors on a bounded parameter highlights a key difference from mean-based tests, where parameters are unbounded.

Reference(s): Wetzels & Wagenmakers (2012); Ly et al. (2016)

Worked Example

Problem Statement A researcher records a predictor variable x and outcome y for $n = 5$ participants:

$$x: 1, 2, 3, 4, 5; \qquad y: 1, 2, 3, 5, 4.$$

Test whether x and y are correlated in the population. $H_0: \rho = 0$ (no correlation); $H_1: \rho \neq 0$, with prior $\rho \sim \text{Uniform}(-1, 1)$.

1. Sample Correlation
With $\bar{x} = \bar{y} = 3$:

$$\sum(x_i - \bar{x})^2 = 10, \quad \sum(y_i - \bar{y})^2 = 10, \quad \sum(x_i - \bar{x})(y_i - \bar{y}) = 9$$

$$r = \frac{\sum(x_i - \bar{x})(y_i - \bar{y})}{\sqrt{\sum(x_i - \bar{x})^2 \cdot \sum(y_i - \bar{y})^2}} = \frac{9}{10}$$

2. Frequentist t-test

$$t = \frac{r\sqrt{n-2}}{\sqrt{1-r^2}} = \frac{9\sqrt{3}}{\sqrt{19}} \approx 3.576, \quad \nu = 3 \text{ df}$$

$t^2 = 243/19$; $p = 0.037$; reject H_0 at 5%.

Practitioner Insight: The Bayesian correlation test is unique in this chapter in having a bounded parameter space $\rho \in (-1, 1)$, making the uniform prior Uniform$(-1, 1)$ natural and proper. The likelihood depends nonlinearly on ρ via a hypergeometric function (${}_2F_1\left(\frac{1}{2}, \frac{1}{2}; \frac{2n-1}{2}; \frac{1+r\rho}{2}\right)$), so the Bayes factor requires one-dimensional numerical integration. This example illustrates the Jeffreys–Lindley paradox: the frequentist test rejects ($p = 0.037$), yet $\text{BF}_{10} \approx 2.95$ provides only anecdotal evidence, with $P(H_1 \mid \text{data}) \approx 0.747$. The discrepancy is driven by small $n = 5$, where the data are consistent with a wide range of ρ values, diluting evidence under the prior. Although $r = 0.9$, the posterior mean shrinks to about 0.63, reflecting this prior-driven moderation.

3. Bayesian BF via Marginal Likelihood Ratio
With uniform prior $\pi(\rho) = \frac{1}{2}$ on $(-1, 1)$ (Ly et al. 2016), the Bayes factor is:

$$\text{BF}_{10} = \frac{\frac{1}{2}\int_{-1}^{1} f(r \mid \rho, n)\, d\rho}{f(r \mid 0, n)}$$

where $f(r \mid \rho, n)$ is the bivariate-normal correlation density:

$$f(r \mid \rho, n) \propto \frac{(1-\rho^2)^{(n-1)/2}}{(1-r\rho)^{n-3/2}}\, {}_2F_1\left(\tfrac{1}{2}, \tfrac{1}{2}; \tfrac{2n-1}{2}; \tfrac{1+r\rho}{2}\right).$$

At $\rho = 0$: $f(r \mid 0, n) \propto (1 - r^2)^{(n-4)/2}$.

Substituting $r = 9/10$, $n = 5$ and evaluating the integral numerically:

$$\text{BF}_{10} \approx 2.95.$$

	Value	Interpretation
r	$9/10$	(exact)
t	$9\sqrt{3}/\sqrt{19}$	≈ 3.576, df= 3
p	0.037	Reject at 5%
BF_{10}	2.95	Anecdotal for H_1
$P(H_1 \mid \text{data})$	0.747	(equal priors)
Posterior mean $\hat{\rho}$	0.630	(under H_1)

Common Pitfalls

- **Specifying priors directly on ρ without considering Fisher's transformation.** Because $\rho \in [-1, 1]$, priors placed directly on ρ can behave poorly near the boundaries. Many implementations place priors on Fisher's $z = \tanh^{-1}(\rho)$, which is unbounded and easier to model.
- **Ignoring the bivariate normality assumption.** Bayesian Pearson correlation assumes approximate bivariate normality. Strong skewness, nonlinear relationships, or outliers can distort inference unless transformations or robust correlation measures are used.
- **Interpreting correlation as evidence of causation.** A Bayesian correlation quantifies association between variables, not causal direction. Confounding variables or common causes may generate strong correlations without causal linkage.
- **Relying on default priors with very small samples.** When n is small, the prior can dominate the posterior and strongly influence the Bayes factor. Sensitivity analysis or informed priors are recommended when data are limited.
- **Assuming Bayesian correlation is robust to outliers.** Bayesian Pearson correlation uses the same underlying statistic as classical r and remains sensitive to extreme observations. Robust or rank-based alternatives may be preferable when outliers are present.

Bayesian Linear Regression Test

Bayesian Inference for Regression Coefficients with Model Comparison

BAYES.7 | **Category:** Bayesian Methods | **Data:** Continuous | **Null:** Posterior distribution

When to Use

- Evaluate regression effects using a Bayesian framework.
- Compare nested models via Bayes factors.
- Obtain posterior estimates and credible intervals for coefficients and predictions.

Assumptions

- Linear relationship between predictors and response.
- Errors and independent and approximately normal.
- Observations are independent.
- Priors specified for coefficients and error variance.

Test Statistic

$$BF_{10} = \frac{P(\text{Data}|M_{\text{full}})}{P(\text{Data}|M_{\text{reduced}})}, \quad \beta_j \sim \text{Cauchy}(0, r\sigma_y), \quad r_{\text{default}} = \sqrt{2}/2.$$

Notation: BF_{10} is Bayes factor for inclusion, β_j is regression coefficient, σ_y is response SD, r is scale parameter

Plain English: Bayesian linear regression tests predictor importance by comparing models with and without specific coefficients using Bayes factors. The test places prior distributions on regression coefficients (typically Cauchy priors scaled by response variable standard deviation) representing plausible effect sizes. After observing data, Bayes factors quantify evidence for including each predictor. Additionally, full posterior distributions for coefficients provide credible intervals and enable probabilistic statements about effect magnitudes and predictions.

⚠ Practitioner Warning

Results depend on prior specifications; default priors (Liang et al. 2008, Rouder & Morey 2012) widely used but not universally appropriate. Assumes linearity; nonlinear relationships require different models. Multicollinearity affects posterior distributions and model comparison. Prior on coefficients scaled by response SD; changes to response scale affect results. Exploring many predictor subsets (2^p models) can introduce multiplicity issues; careful model specification or regularization is needed. Does not establish causation; observational data require careful interpretation. Posterior predictive checks essential for model validation.

Hypotheses

- H_0: Reduced model: $\beta_j = 0$.
- H_1: Full model: includes $\beta_j \sim \text{Cauchy}(0, r\sigma_y)$.

Decision Rule: Interpret via Jeffreys' scale or modern Kass-Raftery scale. No universal cutoff applies.

Software References [Python, R, SAS]

```
(bambi, pymc, arviz)          BayesFactor::lmBF, regressionBF    PROC MCMC
```

Additional Notes

The regression model $y_i = \beta_0 + \sum_{j=1}^{p} \beta_j x_{ij} + \epsilon_i$ assumes $\epsilon_i \sim N(0, \sigma^2)$, with diffuse priors (large variance to minimize *prior* influence) on β_0, Cauchy priors on standardized slopes, and weakly informative priors on σ^2. Model comparison considers subsets of predictors via Bayes factors. Inclusion Bayes factors assess evidence for each predictor by averaging over models including versus excluding it. Posterior model probabilities combine likelihood and model priors, enabling Bayesian model averaging of coefficients. Posterior inference provides full distributions, credible intervals, and predictions. Compared to classical regression, Bayesian methods quantify uncertainty more fully, allow evidence for null effects, and naturally incorporate model uncertainty.

Miscellanea: Bayesian regression builds on early work by Jeffreys and has advanced through modern developments in prior specification and computation. Liang et al. (2008) introduced mixtures of g-priors for variable selection, addressing limitations of earlier approaches such as Bartlett's paradox (vague priors favoring simpler models). Bayesian regression emphasizes estimation and uncertainty, providing full posterior distributions for model parameters.

Reference(s): Liang et al. (2008); Rouder & Morey (2012)

Worked Example

Problem Statement A researcher records the number of hours studied (x) and the test score (y, scaled) for five students: $\{(1,2), (2,3), (3,3), (4,3), (5,4)\}$.

Compare two models: M_0: $y_i = \beta_0 + \varepsilon_i$ (intercept only, no slope); M_1: $y_i = \beta_0 + \beta_1 x_i + \varepsilon_i$ (linear effect of study hours). Use the unit-information g-prior $\beta_1 \mid \sigma^2 \sim \mathrm{N}(0, g\sigma^2(\mathbf{X}'\mathbf{X})^{-1})$ with $g = n = 5$, and Jeffreys prior on (β_0, σ^2).

1. OLS and Sums of Squares

With $\bar{x} = \bar{y} = 3$:

$$\sum(x_i-\bar{x})^2 = 10, \qquad \mathrm{SS}_T = \sum(y_i-\bar{y})^2 = 2,$$
$$\sum(x_i-\bar{x})(y_i-\bar{y}) = 4$$

OLS slope, and intercept:

$$\hat{\beta}_1 = \frac{\sum(x_i-\bar{x})(y_i-\bar{y})}{\sum(x_i-\bar{x})^2} = \frac{2}{5}, \quad \hat{\beta}_0 = \bar{y} - \hat{\beta}_1\bar{x} = \frac{9}{5}.$$

Sums of squares:

$$\mathrm{SS}_{\mathrm{reg}} = \hat{\beta}_1^2 \sum(x_i-\bar{x})^2 = \frac{8}{5}, \quad \mathrm{SS}_{\mathrm{res}} = \mathrm{SS}_T - \mathrm{SS}_{\mathrm{reg}} = \frac{2}{5}.$$

2. R^2 and F-statistic

$$R^2 = \frac{\mathrm{SS}_{\mathrm{reg}}}{\mathrm{SS}_T} = \frac{4}{5}, \quad F = \frac{\mathrm{SS}_{\mathrm{reg}}/1}{\mathrm{SS}_{\mathrm{res}}/(n-2)} = 12, \quad (1,3)\ \mathrm{df}$$

$p = P(F_{1,3} > 12) = 0.040$; reject H_0 at 5%.

3. g-Prior Bayes Factor

The closed-form BF under the g-prior is:

$$\mathrm{BF}_{10}(g) = (1+g)^{(n-2)/2} \cdot (1 + g(1-R^2))^{-(n-1)/2}$$

With $g = 5$, $R^2 = 4/5$:

$$\mathrm{BF}_{10}(g) = 6^{3/2} \cdot \left(1 + 5 \cdot \tfrac{1}{5}\right)^{-2} = \frac{3\sqrt{6}}{2} \approx 3.67$$

	Value	Interp.
R^2	4/5	(exact)
F	12	df= (1, 3)
p	0.040	Reject at 5%
BF_{10}	$3\sqrt{6}/2 \approx 3.67$	Positive for M_1
$P(M_1 \mid \text{data})$	0.786	(equal priors)

$P(M_1 \mid \text{data}) = \mathrm{BF}_{10}/(1 + \mathrm{BF}_{10}) = 0.786$

Practitioner Insight: Bayesian linear regression uses Zellner's g-prior $\beta_1 \mid \sigma^2 \sim \mathrm{N}(0, g\sigma^2(\mathbf{X}'\mathbf{X})^{-1})$, which is scale-equivariant and leaves the BF unchanged under rescaling of x. With the unit-information choice $g = n$, the Bayes factor reduces to a function of R^2 (or F), computable from standard regression output. In this example, $R^2 = 4/5$ yields a closed-form $\mathrm{BF}_{10} = 3\sqrt{6}/2 \approx 3.67$, indicating modest evidence for a nonzero slope, consistent with $p = 0.040$. The g-prior also implements Occam's razor via a penalty factor $(1+g)^{-(n-2)/2}$, limiting evidence from finite data. Under equal priors, $\mathrm{BF}_{10} \approx 3.67$ corresponds to about 79% posterior probability for M_1.

Common Pitfalls

- **Ignoring the impact of the g-prior scale on Bayes factors.** The choice of g in Zellner's g-prior controls prior variance for regression coefficients and can strongly influence Bayes factors. Always report the prior specification and examine sensitivity to common choices (e.g., $g = n$, hyper-g).
- **Confusing inclusion Bayes factors with model Bayes factors.** Inclusion Bayes factors assess evidence for a predictor across many models, while model Bayes factors compare specific model pairs. Interpreting one as the other can lead to incorrect conclusions.
- **Ignoring multicollinearity among predictors.** Highly correlated predictors produce unstable coefficient posteriors and misleading model comparisons. Inspect correlation matrices or VIF diagnostics and remove or combine redundant predictors.
- **Averaging regression coefficients across different models.** Regression coefficients change meaning depending on which predictors are included. Model averaging is appropriate for predictions, but coefficients should usually be interpreted within each model.
- **Assuming Bayesian regression is robust to outliers.** Standard Bayesian regression typically uses normal error models and remains sensitive to extreme observations. Robust alternatives (e.g., heavy-tailed likelihoods) or diagnostic checks may be necessary.

Bayesian Proportion Test

Bayesian Inference for Binomial Proportions with Bayes Factors

BAYES.8 | **Category:** Bayesian Methods | **Data:** Binary | **Null:** Posterior distribution

When to Use

- Test a proportion against a benchmark (e.g., $p = p_0$).
- Compare two-sample proportions in a Bayesian framework.
- Quantify evidence for or against specific values.
- Obtain posterior estimates and credible intervals for p.

Assumptions

- Independent Bernoulli trials with constant success probability.
- Prior specified for p on $[0, 1]$.
- Correct likelihood specification (binomial sampling).

Test Statistic

$$BF_{10} = \frac{P(x \mid H_1)}{P(x \mid H_0)}, \quad P(x \mid H_1) = \binom{n}{x} \frac{B(\alpha + x, \beta + n - x)}{B(\alpha, \beta)}, \quad P(x \mid H_0) = \binom{n}{x} p_0^x (1 - p_0)^{n-x}.$$

Notation: BF_{10} is Bayes factor for alternative, x is successes, p is proportion, α, β are beta parameters

Plain English: Bayesian proportion test compares hypotheses about binomial proportions by computing Bayes factors. For testing against a specific value (e.g., $p = 0.5$), the null hypothesis fixes the proportion while the alternative assigns a prior distribution (typically beta distribution) representing plausible values. The Bayes factor quantifies how much more likely the observed data are under the alternative compared to the null. The posterior distribution provides credible intervals and probability statements about the proportion's true value.

⚠ Practitioner Warning

Results depend on choice of prior distribution; uniform Beta(1,1) common but assumes all proportions equally likely a priori. For p near 0 or 1, prior choice can strongly affect inference. Informative priors appropriate when prior knowledge exists; elicitation methods help translate expertise to beta parameters. For two-proportion comparison, requires specifying priors for both proportions. Small sample sizes yield wide credible intervals; prior influence substantial. Does not address dependency between trials; clustered data require hierarchical models. Savage-Dickey ratio simplifies computation when testing point null nested in continuous prior.

Hypotheses

- H_0**:** Point null: $p = p_0$ (often 0.5).
- H_1**:** $p \sim \text{Beta}(\alpha, \beta)$.

Decision Rule: Interpret via Jeffreys' scale or modern Kass-Raftery scale. No universal cutoff applies.

Software References [Python, R, SAS]

```
(scipy.stats, pymc, arviz)        BayesFactor::proportionBF        PROC MCMC (Manual)
```

Additional Notes

With a Beta(α, β) prior and binomial likelihood, the posterior is Beta$(\alpha + x, \beta + n - x)$. The posterior mean $\hat{p} = (\alpha + x)(\alpha + \beta + n)$ blends prior information with the sample proportion. Credible intervals are obtained directly from the Beta posterior. Common priors include Beta(1,1) (uniform), Jeffreys prior Beta(0.5,0.5), and the improper (non-normalizable) Haldane prior Beta(0,0). Bayes factors for testing $H_0 : p = p_0$ compare the predictive under the prior to the likelihood at p_0. Extensions include two-proportion comparisons via independent Beta priors, sequential analysis without penalty for optional stopping, and equivalence testing using posterior probabilities over a specified interval. Prior elicitation can incorporate expert knowledge when available.

Miscellanea: Bayes's 1763 essay (published posthumously by Richard Price) introduced inference for binomial proportions, laying the foundation of Bayesian statistics. Laplace later popularized the approach using uniform priors, leading to the *rule of succession*, where the posterior mean is $(x + 1)/(n + 2)$—avoiding extreme estimates of 0 or 1. Laplace famously applied this reasoning to estimate the probability that the sun will rise tomorrow—very close to 1, yet never exactly equal to 1. These early ideas established Bayesian methods for proportion estimation that remain widely used today.

Reference(s): Bayes (1763); Agresti & Coull (1998); Gronau et al. (2020)

Worked Example

Problem Statement A clinical trial tests whether a new drug achieves a success rate greater than the historical benchmark of $\theta_0 = 1/3$. In a pilot study, $x = 5$ successes are recorded in $n = 6$ independent trials. The analyst's prior, $\theta \sim \text{Beta}(1,2)$, reflects vague but coherent skepticism: its mean equals the benchmark $1/(1+2) = 1/3 = \theta_0$, so the prior is centered *at* the null value rather than away from it. Test H_0: $\theta = \frac{1}{3}$ vs H_1: $\theta \sim \text{Beta}(1,2)$.

1. Marginal Likelihood Under H_0

$$p(x{=}5 \mid H_0) = \binom{6}{5}\left(\frac{1}{3}\right)^5\left(\frac{2}{3}\right)^1 = \frac{4}{243}$$

2. Marginal Likelihood Under H_1
Average the binomial over the Beta(1, 2) prior:

$$p(x{=}5 \mid H_1) = \binom{6}{5}\frac{B(1{+}5,\ 2{+}1)}{B(1,2)} = 6 \cdot \frac{B(6,3)}{B(1,2)}.$$

Using $B(a,b) = \Gamma(a)\Gamma(b)/\Gamma(a{+}b)$:

$$B(6,3) = \frac{5! \cdot 2!}{8!}, \quad B(1,2) = \frac{0! \cdot 1!}{2!},$$

$$p(x{=}5 \mid H_1) = 6 \cdot \frac{1/168}{1/2} = \frac{1}{14}.$$

Prior centering: With $\theta \sim \text{Beta}(1,2)$, the mean $E[\theta] = \frac{1}{3}$ equals the null θ_0. Thus H_1 is centered at the null with no built-in bias, making the Bayes factor conservative—any evidence against θ_0 must come from the data.

3. Bayes Factor and Posterior

$$\text{BF}_{10} = \frac{p(x{=}5 \mid H_1)}{p(x{=}5 \mid H_0)} = \frac{1/14}{4/243} = \frac{243}{56} \approx 4.34.$$

Conjugate update under H_1: $\theta \mid \text{data}, H_1 \sim \text{Beta}(6,3)$, with posterior mean $6/9 = 2/3$ (pulled from prior mean $1/3$ toward the MLE $5/6$).
With equal prior model weights:

$$P(H_1 \mid \text{data}) = \frac{\text{BF}_{10}}{1 + \text{BF}_{10}} = \frac{243}{299} \approx 0.813.$$

4. Frequentist Comparison

$$p_{\text{exact}} = 2\,P(X \geq 5 \mid \theta_0{=}\tfrac{1}{3}) = 2 \cdot \frac{13}{729} \approx 0.036.$$

	Value	Interp.
BF_{10}	$243/56 \approx 4.34$	Positive for H_1
$P(H_1 \mid \text{data})$	$243/299 \approx 0.813$	(equal priors)
p_{exact}	$26/729 \approx 0.036$	Reject at 5%

Practitioner Insight: The Bayesian proportion test combines the binomial likelihood with an *informative* Beta prior centered at θ_0, ensuring the Bayes factor reflects evidence from the data rather than prior bias toward $\theta > \theta_0$. This centering yields a conservative test under H_1. The general form $\text{BF}_{10} = B(a{+}x,\ b{+}n{-}x)\ /\ \left[B(a,b)\,\theta_0^x(1{-}\theta_0)^{n-x}\right]$ extends the basic case to arbitrary priors and null values. In this example, $\text{BF}_{10} = 243/56 \approx 4.34$, indicating "positive" evidence, consistent with the frequentist rejection ($p \approx 0.036$). The posterior mean lies between the prior mean θ_0 and the MLE, with modest shrinkage reflecting the small sample size.

Common Pitfalls

- **Using a Beta(1,1) prior uncritically with small samples.** Although often called "non-informative," the Beta(1,1) prior still contributes information and can influence results when n is small. Consider informed or Jeffreys priors and report prior sensitivity.
- **Confusing credible intervals with confidence intervals.** A credible interval represents posterior probability for the parameter, whereas a confidence interval reflects long-run coverage of a procedure. Numerical values may be similar, but interpretations differ.
- **Ignoring stopping rules in sequential analyses.** Sequential monitoring is permissible, but ad-hoc stopping based on desired results can bias inference. Stopping criteria should be specified in advance and applied symmetrically to both hypotheses.
- **Assuming posterior estimates equal the MLE for large samples.** As n grows, posterior estimates often approach the MLE, but the prior still contributes information. The two coincide only approximately, not exactly.
- **Ignoring dependence among trials.** The binomial model assumes independent trials; clustering or repeated measures introduce overdispersion. Hierarchical or beta–binomial models are more appropriate when dependence is present.

HDI + ROPE Test (Combined Approach)

Highest Density Interval with Region of Practical Equivalence

BAYES.9 | **Category:** Bayesian Methods | **Data:** Any | **Null:** Posterior distribution

When to Use

- Test for practical equivalence (not just difference).
- Bayesian alternative to TOST equivalence testing.
- When small effects are negligible (define ROPE).
- Accept or reject null hypothesis region explicitly.

Assumptions

- Posterior distribution available for parameter of interest.
- Region of Practical Equivalence (ROPE) specified a priori.
- ROPE based on domain knowledge (not data-driven).
- HDI computed at specified credibility level (typically 95%).

Test Statistic

$$\mathrm{HDI}_\gamma = [L, U], \quad \mathrm{ROPE} = [R_L, R_U], \quad \text{Decision based on overlap/containment.}$$

Notation: HDI is highest posterior density interval, ROPE is region of practical equivalence

Plain English: The HDI+ROPE approach combines Bayesian estimation with practical decision-making by comparing a credible interval (highest density interval) to a region of practical equivalence. The ROPE defines parameter values so small they are practically equivalent to null (e.g., effect size $|\delta| < 0.1$). If the HDI falls entirely within ROPE, accept practical equivalence. If HDI falls entirely outside ROPE, reject equivalence (effect is practically significant). If HDI overlaps ROPE boundaries, the data are insufficient for a clear decision.

⚠ Practitioner Warning

ROPE boundaries must be specified based on domain expertise, not statistical convenience; inappropriate ROPE invalidates conclusions. No universal ROPE values; depends on context, measurement scale, and practical importance. Results depend on parameterization. HDI credibility level affects decisions; 95% conventional but not mandatory. Overlapping ROPE inconclusive, not evidence for null; requires more data or tighter priors. Does not provide Bayes factor; different decision framework focusing on practical significance. Very wide HDI (high uncertainty) often yields inconclusive results. ROPE typically symmetric around null but need not be.

Hypotheses

- H_0: θ lies within the ROPE (practical equivalence).
- H_1: θ lies outside the ROPE (meaningful effect).

Decision Rule: Accept H_0 if $\mathrm{HDI}_{95\%} \subset \mathrm{ROPE}$. Reject H_0 if $\mathrm{HDI}_{95\%} \cap \mathrm{ROPE} = \emptyset$. Otherwise, undecided.

Software References [Python, R, SAS]

Python	R	SAS
`arviz.hdi`	`bayestestR::hdi`	`Not Native (Manual)`

Additional Notes

Kruschke (2011, 2013) popularized the HDI+ROPE framework in *Doing Bayesian Data Analysis*. The highest density interval (HDI) is the narrowest interval containing a specified probability mass; for skewed posteriors, it is more informative than equal-tailed intervals. The ROPE defines a range of practically negligible effects, often based on standardized effect sizes or domain knowledge. Decision rule: (1) accept null if HDI $\subset$ ROPE, (2) reject null if HDI $\cap$ ROPE $= \emptyset$, (3) otherwise remain undecided. An alternative uses the posterior probability $P(\theta \in \mathrm{ROPE} \mid \mathrm{Data})$. The approach emphasizes practical significance and estimation, aligning with equivalence testing, and can be used sequentially until conclusions become clear.

Miscellanea: The HDI+ROPE framework, popularized by Kruschke (2010s), provides an intuitive Bayesian approach to decision-making by comparing credible intervals (HDI) with a region of practical equivalence (ROPE). A key distinction from Bayes factors is that it focuses on *practical significance*, answering "Is the effect meaningfully different from zero?" rather than "Which model is better supported?". This reflects a broader shift in statistical thinking—from testing whether an effect exists to assessing whether it is *large enough to matter* in practice. Emphasizing estimation and real-world relevance, the method gained traction through Kruschke's textbook and applied work in psychology and related fields.

Reference(s): Kruschke (2011, 2013, 2018)

HDI + ROPE Test (Combined Approach) — BAYES.9

Worked Example

Problem Statement A researcher tests whether a coin is fair by tossing it $n = 10$ times and observing $x = 7$ heads. The null value is $\theta_0 = 0.5$. Rather than testing a point null, the HDI+ROPE method defines a *region of practical equivalence* ROPE = $[0.45, 0.55]$: any bias within 5 percentage points of 0.5 is considered practically negligible. The prior $\theta \sim \text{Beta}(2, 2)$ encodes weak symmetrical belief (prior mean = 0.5, prior SD ≈ 0.22). Form the 95% HDI for the posterior and compare it with the ROPE.

1. Posterior Distribution
Conjugate update:

$$\theta \mid \text{data} \sim \text{Beta}(2{+}7,\ 2{+}3) = \text{Beta}(9,\ 5).$$

Exact posterior summaries:

$$E[\theta \mid \text{data}] = \frac{9}{14} \approx 0.643, \qquad \text{Mode} = \frac{8}{12} \approx 0.667,$$

$$\text{Var}[\theta \mid \text{data}] = \frac{9 \times 5}{14^2 \times 15} = \frac{3}{196}, \qquad \text{SD} = \frac{\sqrt{3}}{14} \approx 0.124.$$

2. Highest Density Interval (HDI)
The HDI is the *shortest* interval containing 95% of the posterior mass. For a unimodal Beta it is found by solving

$$f(\theta_L \mid \alpha, \beta) = f(\theta_H \mid \alpha, \beta), \qquad \int_{\theta_L}^{\theta_H} f(\theta)\, d\theta = 0.95,$$

where f is the Beta(9, 5) density. Numerical solution gives: 95% HDI = $[0.401,\ 0.874]$.

(Compare: equal-tailed 95% credible interval = $[0.370, 0.862]$, which is wider; the HDI is shorter because the posterior is skewed.)

3. Decision via HDI + ROPE

Outcome	Decision
HDI ⊂ ROPE	Accept H_0
HDI ∩ ROPE = ∅	Reject H_0
Partial overlap	Suspend judgment

HDI extends well below and above the ROPE, so

$$\text{HDI} \cap \text{ROPE} \neq \emptyset \quad \text{but} \quad \text{HDI} \not\subset \text{ROPE}.$$

$\Rightarrow$ **Suspend judgment.**

The posterior assigns only $P(\theta \in [0.45, 0.55] \mid \text{data}) \approx 0.158$ to the ROPE, yet the HDI still straddles it. Ten flips are insufficient to resolve the question.

4. Frequentist Comparison

$$p = 2P(X \geq 7 \mid \theta_0 = \tfrac{1}{2}) = 2 \cdot 176/1024 \approx 0.344.$$

Fail to reject H_0 at any standard level. Both methods agree: $n = 10$ provides insufficient evidence.

5. Contrast: $x = 9$ **heads** gives Beta(11, 3), 95% HDI = $[0.579,\ 0.968]$, entirely above the ROPE $\Rightarrow$ **Reject** H_0. Freq: $p = 11/512 \approx 0.021$.

Practitioner Insight: The HDI+ROPE approach replaces binary decisions with a three-way outcome: reject, accept, or suspend judgment. The ROPE formalizes "practical equivalence" and must be specified a priori, analogous to choosing α. The HDI is the shortest credible interval, adapting to posterior shape and typically narrower than equal-tailed intervals for skewed distributions. Unlike Bayes factor methods, HDI+ROPE requires no marginal likelihood— decisions follow directly from the posterior. The "undecided" outcome is a distinct advantage: here, despite the posterior shifting toward higher values, it remains too diffuse to exclude the ROPE, so suspending judgment is the appropriate conclusion.

Common Pitfalls

- **Choosing the ROPE width after inspecting the posterior.** The ROPE must be defined *a priori* based on substantive or practical significance. Adjusting the ROPE after viewing the HDI introduces confirmation bias and invalidates the decision rule.
- **Using a point null instead of an interval ROPE.** Equivalence testing requires a range of practically negligible effects. A point null ($\theta = 0$) cannot support meaningful equivalence decisions within the HDI+ROPE framework.
- **Treating partial HDI overlap with the ROPE as evidence of equivalence.** The decision rule requires the HDI to lie *entirely* inside the ROPE to accept practical equivalence. Any overlap outside the ROPE implies an inconclusive result.
- **Defining the ROPE narrower than measurement precision.** If the ROPE is smaller than the natural measurement uncertainty of the data, equivalence becomes practically unattainable. ROPE limits should reflect realistic measurement resolution and substantive relevance.
- **Applying HDI+ROPE without inspecting the posterior shape.** The interpretation assumes a reasonably unimodal posterior. Multimodal or highly skewed posteriors require careful inspection and possibly alternative summaries.

Posterior Predictive p-Value Test

Bayesian Model Checking via Posterior Predictive Replication

BAYES.10 | **Category:** Bayesian Methods | **Data:** Any | **Null:** Posterior predictive

When to Use

- Check model adequacy (Bayesian model checking).
- Detect model misspecification or violations.
- Test if observed data "extreme" under model.
- Complement parameter estimation with model diagnostics.

Assumptions

- Posterior distribution available for model parameters.
- Can generate data from posterior predictive distribution.
- Test statistic $T(\mathbf{y})$ chosen to detect departures of interest.
- Sufficient posterior samples for predictive distribution.

Test Statistic

$$p_{PPP} = P\Big(T(\mathbf{y}^{\text{rep}}) \geq T(\mathbf{y}^{\text{obs}}) \mid \mathbf{y}^{\text{obs}}\Big) = E\big[\mathbf{1}\{T(\mathbf{y}^{\text{rep}}) \geq T(\mathbf{y}^{\text{obs}})\} \mid \mathbf{y}^{\text{obs}}\big], \qquad \mathbf{y}^{\text{rep}} \sim P(\mathbf{y}^{\text{rep}}|\mathbf{y}^{\text{obs}}).$$

Notation: p_{PPP} is posterior predictive p-value, $\mathbf{y}^{\text{obs}}$ is observed data, $\mathbf{y}^{\text{rep}}$ is replicated data, T is test statistic

Plain English: Posterior predictive p-values assess model fit by generating replicated datasets from the fitted model and comparing them to observed data using a test statistic, $T(\mathbf{y})$. For each parameter value drawn from the posterior, simulate a new dataset, compute the test statistic, and compare to the observed test statistic. The p-value is the proportion of replications where the simulated statistic equals or exceeds the observed value. Values near 0 or 1 indicate the model fails to capture $T(\mathbf{y})$.

⚠ Practitioner Warning

Not classical p-values; interpretation differs—extreme values (< 0.05 or > 0.95) suggest model inadequacy. Choice of test statistic $T(\mathbf{y})$ critical; different statistics assess different model aspects (mean, variance, skewness, outliers). Using data-dependent test statistics can yield misleading p-values; prefer ancillary statistics. Values near 0.5 indicate good fit for that statistic but do not guarantee overall model adequacy. Multiple test statistics should be examined; single p-value insufficient. Not suitable for model selection; use for model checking and diagnostics. Sensitive to sample size; large samples detect trivial deviations.

Hypotheses

- H_0: Model adequately reproduces the observed feature $T(\mathbf{y})$
- H_1: Model fails to reproduce the observed feature.

Decision Rule: Extremes ($p_{PPP} < 0.05$ or > 0.95) suggest poor model fit for chosen $T(\mathbf{y})$. Near 0.5 indicate good fit.

Software References [Python, R, SAS]

`arviz.plot_ppc`	`bayesplot::ppc_stat`	`Not Native (Manual)`

Additional Notes

Posterior predictive checking (Rubin, 1984; Gelman et al., 1996) assesses model fit by comparing observed data to replicated data generated from the posterior. Algorithm: (1) draw $\theta^{(s)} \sim p(\theta|\mathbf{y}^{\text{obs}})$, (2) generate $\mathbf{y}^{\text{rep}(s)} \sim p(\mathbf{y}|\theta^{(s)})$, (3) compute test statistics T^{obs} and $T^{\text{rep}(s)}$, and (4) estimate $p_{PPP} = \frac{1}{S}\sum \mathbf{1}(T^{\text{rep}(s)} \geq T^{\text{obs}})$. Test statistics can target specific features such as location, spread, extremes, dependence, or distributional shape. Graphical checks (e.g., comparing T^{obs} to the distribution of T^{rep}) are often more informative than a single scalar summary. Posterior predictive p-values are not uniformly distributed under the null because the data are used twice—first to fit the model and then to assess it. As a result, they tend to be conservative and biased toward 0.5, so lack of extreme values does not guarantee good model fit, while extreme values indicate model misfit. The method evaluates absolute model adequacy and complements model comparison approaches such as Bayes factors.

Miscellanea: Donald Rubin (1980s) introduced the idea of posterior predictive checking as a Bayesian approach to model assessment, comparing observed data to replicated data generated from the fitted model. Posterior predictive checking reflects a shift from asking "Is the model true?" to "Does the model generate data that look like what we observed?", emphasizing model adequacy over hypothesis testing. The framework was later developed and popularized by Andrew Gelman, Xiao-Li Meng, and Hal Stern, and is now a central component of modern Bayesian model diagnostics.

Reference(s): Rubin (1984); Gelman et al. (1996); Gabry et al. (2019)

Worked Example

Problem Statement A service analyst observes $Y_{\text{obs}} = 3$ customer arrivals at a counter during a one-hour pilot. She fits a Poisson model, $Y \sim \text{Poisson}(\lambda)$, with prior $\lambda \sim \text{Gamma}(1, 1) = \text{Exp}(1)$ (prior mean = 1 arrival/hour). The following day she observes only $T_{\text{new}} = 1$ arrival during a two-hour window. Compute the posterior predictive p-value (PPP) to assess whether the new observation $T_{\text{new}} = 1$ is consistent with the fitted model.

1. Posterior for λ

Conjugate update after observing $Y_{\text{obs}} = 3$ in one hour:

$$\lambda \mid Y_{\text{obs}} \sim \text{Gamma}(1{+}3,\ 1{+}1) = \text{Gamma}(4,\ 2).$$

$$E[\lambda \mid Y_{\text{obs}}] = \tfrac{4}{2} = 2 \text{ arrivals/hour.}$$

2. Posterior Predictive for T_{rep}

The replicated total T_{rep} counts arrivals during a two-hour window ($t = 2$). Integrating Poisson(λt) over the Gamma(4, 2) posterior gives the Negative Binomial:

$$T_{\text{rep}} \mid Y_{\text{obs}} \sim \text{NegBin}\left(r = 4,\ p = \frac{b}{b+t} = \frac{2}{2+2} = \frac{1}{2}\right).$$

$$P(T_{\text{rep}} = k) = \binom{3+k}{k}\frac{1}{2^{4+k}}, \quad E[T_{\text{rep}}] = \frac{r(1-p)}{p} = 4$$

$$P(T_{\text{rep}} = 0) = \frac{1}{16}, \quad P(T_{\text{rep}} = 1) = \frac{4}{32}.$$

3. Posterior Predictive p-Value

$$\text{PPP} = P(T_{\text{rep}} \le 1 \mid Y_{\text{obs}}) = P(T_{\text{rep}} = 0) + P(T_{\text{rep}} = 1)$$

$$= \frac{1}{16} + \frac{4}{32} = \frac{3}{16} \approx 0.188.$$

	Value	Interp.
$E[T_{\text{rep}}]$	4	model expects 4
PPP	$3/16 \approx 0.188$	mild misfit
Frequentist p	$5e^{-4} \approx 0.092$	(Poisson(4))

4. Decision

Values near 0.5 indicate good fit; values near 0 or 1 indicate misfit. PPP ≈ 0.188 lies in the "mild concern" zone: the model expects four arrivals over two hours but only one was observed. The model may be overfitting the pilot observation, or the arrival rate may have changed. Conventional thresholds: PPP < 0.05 or > 0.95 indicates clear misfit.

Note: The frequentist p-value fixes $\lambda = E[\lambda \mid Y_{\text{obs}}] = 2$, giving $P(\text{Poisson}(4) < 1) = 5e^{-4} \approx 0.092$, which is smaller than PPP= 0.188 because it ignores posterior uncertainty about λ.

Practitioner Insight: The posterior predictive p-value (PPP) is a *model-checking* tool, not a hypothesis test or model comparison: it asks whether the fitted model can generate data resembling the observed data along a chosen dimension T. The choice of T targets specific features (e.g., totals, variance, or extremes). PPP is self-calibrating when T is a sufficient statistic, yielding values near 0.5; informative checks arise when T probes new data or features beyond sufficiency. Here, differing observation windows break this symmetry. The predictive distribution follows from Poisson–Gamma conjugacy, giving exact negative binomial probabilities. PPP exceeds the frequentist p-value because it averages tail probabilities over posterior uncertainty in λ, rather than conditioning on a fixed value.

Common Pitfalls

- **Treating posterior predictive p-values like classical p-values.** PPP are diagnostic measures rather than calibrated hypothesis tests. Values near 0 or 1 suggest model misfit, but there is no strict threshold such as 0.05 for rejection.
- **Relying on a single test statistic without graphical checks.** A single PPP rarely reveals the nature of model failure. Posterior predictive plots and multiple discrepancy statistics provide far more informative diagnostics.
- **Ignoring the double use of the data.** Posterior predictive checks use the observed data both to fit the model and to evaluate it, which can mask some forms of overfitting. Complement them with cross-validation or out-of-sample predictive checks when possible.
- **Using test statistics unrelated to the suspected model deficiency.** Effective checks require statistics sensitive to specific problems (e.g., outliers, skewness, heteroscedasticity). Generic summaries such as the mean may fail to detect important structural misfit.
- **Ignoring extreme posterior predictive p-values.** Values close to 0 or 1 indicate systematic discrepancies between the model and the data. Such results should prompt model revision, alternative distributions, or additional diagnostics.

XXV. BAYES — Bayesian Workflow: An Instructive Example

The Setting Suppose we observe n independent binary trials — for example, x successes out of n attempts. The data distribution (likelihood) is Binomial:

$$p(\mathbf{y} \mid p) = \binom{n}{x} p^x (1-p)^{n-x}, \quad p \in (0,1).$$

We wish to make inference about the unknown success probability p using two different priors, to illustrate how prior choice affects the posterior.

Conjugate Priors and Why They Matter A prior $\pi(\theta)$ is **conjugate** to a likelihood $p(\mathbf{y} \mid \theta)$ if the posterior $p(\theta \mid \mathbf{y})$ belongs to the same distributional family as the prior. Conjugacy is desirable for three reasons:

- The posterior has a **closed form**—no numerical integration or MCMC required.
- **Interpretability:** the hyperparameters of the posterior update additively with the data, making the influence of the prior transparent.
- **Sequential updating:** today's posterior becomes tomorrow's prior with no extra computation.

For the Binomial likelihood, the conjugate prior is the **Beta distribution**: $p \sim \text{Beta}(\alpha, \beta)$, with density $\pi(p) \propto p^{\alpha-1}(1-p)^{\beta-1}$. The hyperparameters α and β can be interpreted as *pseudo-counts*: $\alpha - 1$ prior successes and $\beta - 1$ prior failures.

Case 1: Weakly Informative Beta Prior Suppose we have little prior knowledge and choose a near-uniform prior:

$$\pi_1(p) = \text{Beta}(\alpha_0, \beta_0) \quad \text{with } \alpha_0 = \beta_0 = 1.$$

This is the uniform distribution on $(0,1)$, reflecting genuine prior ignorance about p. The posterior follows by Bayes' theorem:

$$\begin{aligned} p(p \mid \mathbf{y}) &\propto p(\mathbf{y} \mid p)\, \pi_1(p) \\ &\propto p^x (1-p)^{n-x} \cdot p^{\alpha_0 - 1}(1-p)^{\beta_0 - 1} \\ &= p^{x+\alpha_0-1}(1-p)^{n-x+\beta_0-1}. \end{aligned}$$

This is the kernel of a Beta distribution, so:

$$\boxed{p \mid \mathbf{y} \sim \text{Beta}(x+\alpha_0,\ n-x+\beta_0).}$$

With $\alpha_0 = \beta_0 = 1$, the posterior is simply $\text{Beta}(x+1,\ n-x+1)$. The posterior **mean** is:

$$E[p \mid \mathbf{y}] = \frac{x+\alpha_0}{n+\alpha_0+\beta_0} = \frac{x+1}{n+2},$$

a slight shrinkage of $\hat{p} = x/n$ toward 0.5 that vanishes as $n \longrightarrow \infty$.

Case 2: Informative Beta Prior Now suppose prior evidence suggests $p \approx 0.7$ with moderate confidence. We encode this as:

$$\pi_2(p) = \text{Beta}(\alpha_0, \beta_0) \quad \text{with } \alpha_0 = 7,\ \beta_0 = 3.$$

This prior has mean $7/10 = 0.7$ and places most mass above 0.5. The posterior is again Beta by conjugacy:

$$\boxed{p \mid \mathbf{y} \sim \text{Beta}(x+7,\ n-x+3).}$$

The posterior mean $E[p \mid \mathbf{y}] = (x+7)/(n+10)$ is a weighted average of the prior mean (0.7) and the data mean (x/n), with the prior contributing the equivalent of 10 additional observations. As n grows, the data dominate and the two posteriors converge regardless of prior choice.

General Update Rule For any $\text{Beta}(\alpha_0, \beta_0)$ prior, the posterior after observing x successes in n trials is:

$$p \mid \mathbf{y} \sim \text{Beta}(\underbrace{\alpha_0 + x}_{\text{prior+successes}},\ \underbrace{\beta_0 + n - x}_{\text{prior+failures}}).$$

The prior pseudocounts (α_0, β_0) simply add to the observed counts $(x, n-x)$. This additive update is the hallmark of conjugacy and makes the prior's influence explicit and quantifiable.

Posterior Summaries Given posterior $\text{Beta}(\alpha^*, \beta^*)$ where $\alpha^* = \alpha_0 + x$ and $\beta^* = \beta_0 + n - x$:

Summary	Formula
Mean	$\alpha^*/(\alpha^*+\beta^*)$
Mode (MAP)	$(\alpha^*-1)/(\alpha^*+\beta^*-2),\ \alpha^*, \beta^* > 1$
Variance	$\alpha^*\beta^*/[(\alpha^*+\beta^*)^2(\alpha^*+\beta^*+1)]$

Highest Density Interval (HDI). The HDI is the shortest interval $[p_L, p_U]$ such that $\int_{p_L}^{p_U} p(\theta \mid \mathbf{y})\, d\theta = 1-\alpha$. For a unimodal symmetric Beta posterior, the HDI coincides with the equal-tailed credible interval: $[F^{-1}(\alpha/2),\ F^{-1}(1-\alpha/2)]$ where F^{-1} is the Beta quantile function. For asymmetric posteriors (extreme p, small n), the HDI is shorter than the equal-tailed interval and should be computed numerically (e.g., `HDInterval::hdi()` in R).

Non-Conjugate Prior: Normal on Logit Scale For completeness, consider a non-conjugate prior placed on $\phi = \text{logit}(p) = \log[p/(1-p)]$:

$$\phi \sim N(0, \sigma_0^2), \quad p = \frac{e^\phi}{1+e^\phi}.$$

The posterior $p(\phi \mid \mathbf{y}) \propto p(\mathbf{y} \mid \phi)\, \pi(\phi)$ has no closed form. The Beta conjugate posterior and this logit-normal posterior will generally agree closely when n is moderate and neither is highly informative, but the logit-normal requires numerical integration or MCMC. This contrast illustrates why conjugate priors are preferred when available: identical inferential content at a fraction of the computational cost.

Prior Influence and Sensitivity The table below summarizes the posterior mean under both Beta priors for three hypothetical data outcomes, illustrating how prior influence fades with increasing n:

n	x	$\hat{p}$	**Case 1** Beta(1, 1)	**Case 2** Beta(7, 3)
10	3	0.30	0.333	0.500
50	15	0.30	0.311	0.367
200	60	0.30	0.302	0.319

With $n = 10$, the informative prior (Case 2) pulls the posterior mean substantially toward 0.7; by $n = 200$, both posteriors are close to the data mean $\hat{p} = 0.30$. This is **Bayesian consistency**: the posterior concentrates on the true value as data accumulate, regardless of the prior (provided $\pi(p_{\text{true}}) > 0$).

Bayesian Workflow Summary The complete workflow for this example is:

1. **Specify the likelihood:** Binomial(n, p).
2. **Choose a prior:** Beta(α_0, β_0) reflecting prior knowledge or ignorance.
3. **Derive the posterior:** Beta$(\alpha_0+x, \beta_0+n-x)$ by conjugacy.
4. **Summarize:** Posterior mean, MAP, HDI.
5. **Check sensitivity:** Compare posteriors under alternative priors; if conclusions are robust, the prior choice is not driving results.
6. **Use the posterior:** For testing, compute BF_{01} via Savage–Dickey at $p = p_0$; for decisions, apply HDI+ROPE; for prediction, compute the posterior predictive distribution $p(\tilde{y} \mid \mathbf{y}) = \int p(\tilde{y} \mid p)\, p(p \mid \mathbf{y})\, dp$, which for the Beta-Binomial model has the closed-form Beta-Binomial distribution.

Posterior vs. Posterior Predictive Distribution The **posterior** $p(\theta \mid \mathbf{y})$ describes uncertainty about the *parameter* after observing data. The **posterior predictive distribution** describes uncertainty about a *future observation* $\tilde{y}$, averaging over that parameter uncertainty:

$$p(\tilde{y} \mid \mathbf{y}) = \int p(\tilde{y} \mid \theta)\, p(\theta \mid \mathbf{y})\, d\theta.$$

The key distinction: the posterior is a distribution over θ; the posterior predictive is a distribution over observable data $\tilde{y}$. The integral propagates all remaining uncertainty about θ into the prediction — this is what separates Bayesian prediction from plug-in prediction (which simply substitutes $\hat{\theta}$ and ignores parameter uncertainty).

Beta-Binomial derivation. Continuing the example, predict $\tilde{x}$ successes in m future trials. With posterior $p \mid \mathbf{y} \sim \text{Beta}(\alpha^*, \beta^*)$, where $\alpha^* = \alpha_0 + x$ and $\beta^* = \beta_0 + n - x$:

$$\begin{aligned} p(\tilde{x} \mid \mathbf{y}) &= \int_0^1 p(\tilde{x} \mid p)\, p(p \mid \mathbf{y})\, dp \\ &= \int_0^1 \binom{m}{\tilde{x}} p^{\tilde{x}} (1-p)^{m-\tilde{x}} \\ &\qquad \times \frac{p^{\alpha^*-1}(1-p)^{\beta^*-1}}{B(\alpha^*, \beta^*)}\, dp \\ &= \binom{m}{\tilde{x}} \frac{1}{B(\alpha^*, \beta^*)} \int_0^1 p^{\tilde{x}+\alpha^*-1} \\ &\qquad \times (1-p)^{m-\tilde{x}+\beta^*-1}\, dp \\ &= \binom{m}{\tilde{x}} \frac{B(\alpha^* + \tilde{x},\ \beta^* + m - \tilde{x})}{B(\alpha^*,\ \beta^*)}. \end{aligned}$$

The integral in the third step is recognized as $B(\alpha^* + \tilde{x}, \beta^* + m - \tilde{x})$ by the definition of the beta function, giving the **Beta-Binomial** PMF in closed form. The result has heavier tails than Binomial$(m, \hat{p})$ because it propagates uncertainty about p into the prediction rather than treating $\hat{p}$ as exact. As $n \to \infty$ the posterior concentrates on p_{true} and the predictive converges to Binomial(m, p_{true}), recovering the plug-in result.

Three uses in this chapter.

- **Model checking:** the posterior predictive p-value (PPP test) compares observed $T(\mathbf{y})$ to the distribution of $T(\mathbf{y}^{\text{rep}})$ under $p(\mathbf{y}^{\text{rep}} \mid \mathbf{y})$.
- **Forecasting:** direct probability statements about future data without frequentist repeated-sampling arguments.
- **Model comparison:** LOO-CV (leave one out - crossvalidation) and WAIC (Watanabe-Akaike Information Criterion) both score models by their log posterior predictive density at held-out observations.

Common Conjugate Prior Pairs The table below lists the most frequently used conjugate pairs. In each case the posterior hyperparameters follow by the same additive logic as the Beta-Binomial example above. The *effective prior sample size* column indicates how many observations the prior is worth, making the prior's influence directly comparable to n.

Likelihood	**Prior**	**Posterior**	**Eff. prior** n
Binomial(n, p)	Beta(α, β)	Beta$(\alpha+x,\ \beta+n-x)$	$\alpha+\beta$
Poisson(λ)	Gamma(α, β)	Gamma$(\alpha+\sum x_i,\ \beta+n)$	β
Normal(μ, σ^2) σ^2 known	$N(\mu_0, \tau_0^2)$	$N(\mu_n, \tau_n^2)^\dagger$	σ^2/τ_0^2
Normal(μ, σ^2) μ known	Inv-$\chi^2(\nu_0, \sigma_0^2)$	Inv-$\chi^2(\nu_n, \sigma_n^2)^\ddagger$	ν_0
Exponential(λ)	Gamma(α, β)	Gamma$(\alpha+n,\ \beta+\sum x_i)$	β
Negative Bin.(r, p) r known	Beta(α, β)	Beta$(\alpha+nr,\ \beta+\sum x_i)$	$\alpha+\beta$
Multinomial$(\mathbf{p})$	Dirichlet$(\boldsymbol{\alpha})$	Dirichlet$(\boldsymbol{\alpha}+\mathbf{x})$	$\sum \alpha_k$
MVN$(\boldsymbol{\mu}, \boldsymbol{\Sigma})$ $\boldsymbol{\Sigma}$ known	MVN$(\boldsymbol{\mu}_0, \boldsymbol{\Lambda}_0^{-1})$	MVN$(\boldsymbol{\mu}_n, \boldsymbol{\Lambda}_n^{-1})^\S$	tr$(\boldsymbol{\Lambda}_0)/p$

† $\tau_n^{-2} = \tau_0^{-2} + n/\sigma^2;\ \mu_n = \tau_n^2(\mu_0/\tau_0^2 + n\bar{x}/\sigma^2)$.
‡ $\nu_n = \nu_0 + n;\ \nu_n\sigma_n^2 = \nu_0\sigma_0^2 + \sum(x_i - \mu)^2$.
§ $\boldsymbol{\Lambda}_n = \boldsymbol{\Lambda}_0 + n\boldsymbol{\Sigma}^{-1};\ \boldsymbol{\mu}_n = \boldsymbol{\Lambda}_n^{-1}(\boldsymbol{\Lambda}_0\boldsymbol{\mu}_0 + n\boldsymbol{\Sigma}^{-1}\bar{\mathbf{x}})$.

When no conjugate prior exists (e.g. logistic regression, hierarchical models), weakly informative priors combined with Markov Chain Monte Carlo (MCMC) are the standard approach. The conjugate pairs above cover the majority of closed-form Bayesian analyses encountered in this compendium.

XXV. BAYES — Companion Guide to Worked Examples

This companion guide provides the core formulas, distributions, and steps needed to work through the examples in this chapter. Each subsection corresponds directly to a worked example and presents the essential model setup, hypotheses, and computational steps—from likelihood to posterior and predictive quantities—so that readers can follow and reproduce the results without getting lost in technical detail.

BAYES.1: Bayes Factor for Binomial with Uniform Prior

1. Data Model

$$X \sim \text{Binomial}(n, \theta).$$

2. Hypotheses

$$H_0 : \theta = \theta_0 \qquad H_1 : \theta \sim \text{Uniform}(0, 1).$$

3. Marginal Likelihoods

Under H_0:

$$p(x \mid H_0) = \binom{n}{x} \theta_0^x (1 - \theta_0)^{n-x}.$$

Under H_1:

$$p(x \mid H_1) = \int_0^1 \binom{n}{x} \theta^x (1 - \theta)^{n-x} d\theta.$$

With Uniform(0, 1):

$$p(x \mid H_1) = \frac{1}{n+1}.$$

4. Bayes Factor

$$\text{BF}_{10} = \frac{p(x \mid H_1)}{p(x \mid H_0)} = \frac{1/(n+1)}{\binom{n}{x} \theta_0^x (1 - \theta_0)^{n-x}}.$$

5. Posterior Under H_1

$$\theta \mid \text{data} \sim \text{Beta}(x + 1,\ n - x + 1).$$

Key Insight

- Uniform prior implies equal marginal probability for all outcomes: $p(x) = 1/(n + 1)$
- Bayes factor compares a point null vs a fully flexible alternative
- Complexity of H_1 is penalized automatically (Occam's razor)

BAYES.2: Conjugate Normal Model

1. Data Model (Likelihood)

Let $\bar{x}$ denote the sample mean of n i.i.d. observations from $\mathcal{N}(\mu, \sigma^2)$. Then:

$$\bar{x} \mid \mu \sim \mathcal{N}\left(\mu, \frac{\sigma^2}{n}\right).$$

In this example: $n = 4$, $\sigma^2 = 1$, so

$$\bar{x} \mid \mu \sim \mathcal{N}(\mu, \tfrac{1}{4}).$$

2. Prior Distribution (under H_1)

The parameter μ is assigned a normal prior:

$$\mu \sim \mathcal{N}(\mu_0, \tau^2),$$

with $\mu_0 = 0$ and $\tau^2 = 1$.

3. Posterior Distribution

By conjugacy, the posterior is also normal:

$$\mu \mid \text{data} \sim \mathcal{N}(\mu_n, \sigma_n^2),$$

where μ_n and σ_n^2 are given below.

4. Posterior Update Formulas

$$\sigma_n^2 = \frac{1}{\frac{1}{\tau^2} + \frac{n}{\sigma^2}}, \qquad \mu_n = \sigma_n^2 \left(\frac{\mu_0}{\tau^2} + \frac{n\bar{x}}{\sigma^2} \right).$$

For this example:

$$\sigma_n^2 = \frac{1}{1+4} = \frac{1}{5}, \qquad \mu_n = \frac{1}{5}(0 + 4 \cdot 1) = \frac{4}{5}.$$

Remark: If no data are observed ($n = 0$), then

$$\sigma_n^2 = \frac{1}{1/\tau^2} = \tau^2, \qquad \mu_n = \sigma_n^2 \left(\frac{\mu_0}{\tau^2} \right) = \mu_0,$$

so the posterior reduces to the prior, $\mathcal{N}(\mu_0, \tau^2)$.

5. Notation Summary

- $\bar{x}$: sample mean
- n: sample size
- σ^2: known data variance
- μ: population mean (parameter of interest)
- μ_0, τ^2: prior mean and variance
- μ_n, σ_n^2: posterior mean and variance
- $\text{N}(x;\, \mu, \sigma^2)$ denotes the *normal density evaluated at* x for a normal distribution with mean μ and variance σ^2

Key Insight: The posterior mean μ_n is a precision-weighted average of the prior mean μ_0 and the data mean $\bar{x}$, with weights proportional to $1/\tau^2$ and n/σ^2.

BAYES.3: Poisson–Gamma Model Comparison

1. Data Model

Counts x_i follow independent Poisson distributions:

$$x_i \mid \lambda \sim \text{Poisson}(\lambda), \quad p(x_i \mid \lambda) = \frac{\lambda^{x_i} e^{-\lambda}}{x_i!}.$$

2. Prior

Rates follow Gamma priors:

$$\lambda \sim \text{Gamma}(\alpha, \beta), \quad p(\lambda) \propto \lambda^{\alpha-1} e^{-\beta\lambda}.$$

In this example: $\alpha = \beta = 1$ (Exponential prior).

3. Marginal Likelihood (Key Step)

For n observations with sum $S = \sum x_i$:

$$p(\mathbf{x} \mid M) = \frac{\beta^\alpha \Gamma(S+\alpha)}{\Gamma(\alpha)(n+\beta)^{S+\alpha} \prod x_i!}.$$

4. Model Comparison

$$\text{BF}_{10} = \frac{p(\mathbf{x} \mid M_1)}{p(\mathbf{x} \mid M_0)}.$$

Posterior model probabilities follow from Bayes' rule:

$$P(M_1 \mid \mathbf{x}) = \frac{p(\mathbf{x} \mid M_1)\, P(M_1)}{p(\mathbf{x} \mid M_0)\, P(M_0) + p(\mathbf{x} \mid M_1)\, P(M_1)}.$$

With equal priors: $P(M_0) = P(M_1) = 1/2$

$$P(M_1 \mid \mathbf{x}) = \frac{p(\mathbf{x} \mid M_1)}{p(\mathbf{x} \mid M_0) + p(\mathbf{x} \mid M_1)} = \frac{\text{BF}_{10}}{1 + \text{BF}_{10}}.$$

5. Posterior Updates (Conjugacy)

$$\lambda \mid \mathbf{x} \sim \text{Gamma}(S+\alpha,\ n+\beta), \quad E[\lambda \mid \mathbf{x}] = \frac{S+\alpha}{n+\beta}.$$

Key Insight: The marginal likelihood balances model fit and complexity; models with better fit to structured data (e.g. changepoints) can overcome the penalty for additional parameters.

BAYES.4: Bayesian t-Test (JZS)

1. Data Model

$$x_i \mid \mu, \sigma^2 \sim \mathcal{N}(\mu, \sigma^2), \quad i = 1, \ldots, n.$$

Sufficient statistics:

$$\bar{x}, \quad s^2, \quad t = \frac{\bar{x} - \mu_0}{s/\sqrt{n}}, \quad \nu = n - 1.$$

2. Parameterization

Define the standardized effect size:

$$\delta = \frac{\mu}{\sigma}.$$

This removes scale dependence and makes the test invariant to units.

3. Prior (JZS)

$$\delta \sim \text{Cauchy}(0, 1),$$

equivalently:

$$\delta \mid g \sim \mathcal{N}(0, g), \quad g \sim \text{Inv-}\chi^2(1).$$

Jeffreys prior for nuisance parameters:

$$\pi(\mu, \sigma) \propto \frac{1}{\sigma}.$$

4. Bayes Factor

From Rouder et al. 2009

$$\text{BF}_{10} = \left(1 + \frac{t^2}{\nu}\right)^{n/2} \int_0^\infty h(g)\, dg,$$

where

$$h(g) = \frac{(1+ng)^{-1/2} \left(1 + \frac{t^2}{(1+ng)\nu}\right)^{-n/2}}{(2\pi g^3)^{1/2} e^{1/(2g)}}.$$

5. Key Features

- Depends only on (t, n), not raw data.
- Integrates out both μ and σ^2.
- Prior on δ ensures scale invariance.
- Requires 1D numerical integration over g.

Key Insight: The Bayesian t-test balances evidence from the t-statistic against the flexibility of the alternative via the prior on effect size, producing a calibrated measure of evidence rather than a rejection decision.

BAYES.5: Bayesian ANOVA (g-Prior)

1. Data Model

$$x_{ij} = \mu + \alpha_j + \varepsilon_{ij}, \quad \varepsilon_{ij} \sim \mathcal{N}(0, \sigma^2),$$

where α_j are group effects with $\sum \alpha_j = 0$.

2. Sufficient Statistics

$$\text{SS}_b, \quad \text{SS}_w, \quad R^2 = \frac{\text{SS}_b}{\text{SS}_T}.$$

3. Prior (Unit-Information g-Prior)

$$\alpha \mid g, \sigma \sim \mathcal{N}(0,\ g\sigma^2 \mathbf{H}), \quad g = n.$$

Jeffreys prior for nuisance parameters:

$$\pi(\mu, \sigma) \propto \frac{1}{\sigma}.$$

4. Bayes Factor

From Rouder et al. 2012

$$\mathrm{BF}_{10} = (1+g)^{-(k-1)/2}\left(1 - \frac{gR^2}{1+g}\right)^{-(n-1)/2}.$$

5. Key Quantities

- k: number of groups
- n: total sample size
- R^2: proportion of variance explained
- g: prior scale (typically $g = n$)

Key Insight: The Bayes factor balances goodness-of-fit (R^2) against model complexity ($k - 1$ parameters), providing an automatic Occam's razor.

BAYES.6: Bayesian Test for Correlation

1. Data Model

$$(x_i, y_i) \sim \mathcal{N}_2(\boldsymbol{\mu}, \boldsymbol{\Sigma}), \quad \Sigma = \begin{pmatrix} 1 & \rho \\ \rho & 1 \end{pmatrix}.$$

Sufficient statistic:

$$r = \text{sample correlation}, \quad n = \text{sample size}.$$

2. Prior

$$\rho \sim \mathrm{Uniform}(-1, 1).$$

3. Likelihood

$$f(r \mid \rho, n) \propto \frac{(1-\rho^2)^{(n-1)/2}}{(1-r\rho)^{n-3/2}} \cdot {}_2F_1(\cdot).$$

4. Bayes Factor

$$\mathrm{BF}_{10} = \frac{\frac{1}{2}\int_{-1}^{1} f(r \mid \rho, n)\, d\rho}{f(r \mid 0, n)}.$$

5. Key Features

- Parameter space is bounded: $\rho \in (-1, 1)$.
- Prior is proper and uniform.
- ${}_2F_1(\cdot)$ is hypergeometric function.
- Requires numerical integration.
- Evidence depends strongly on sample size.

Key Insight: The Bayes factor averages evidence over all possible correlations under H_1, so with small n even a large sample correlation may yield only modest evidence.

BAYES.7: Bayesian Linear Regression (g-Prior)

1. Data Model

$$y = \beta_0 + \beta_1 x + \varepsilon, \quad \varepsilon \sim \mathcal{N}(0, \sigma^2).$$

2. Sufficient Statistics

$$R^2 = \frac{\mathrm{SS}_{\mathrm{reg}}}{\mathrm{SS}_T}, \quad n = \text{sample size}.$$

3. Prior

$$\beta_1 \mid \sigma^2 \sim \mathcal{N}\left(0,\ g\sigma^2 (X'X)^{-1}\right), \quad g = n.$$

Jeffreys prior:

$$\pi(\beta_0, \sigma) \propto \frac{1}{\sigma}.$$

4. Bayes Factor

$$\mathrm{BF}_{10} = (1+g)^{(n-2)/2}\left(1 + g(1 - R^2)\right)^{-(n-1)/2}.$$

5. Key Quantities

- R^2: proportion of variance explained
- g: prior scale (unit-information: $g = n$)
- n: sample size

Key Insight: The Bayes factor depends only on R^2 and balances explained variance against a built-in penalty for adding regression parameters (Occam's razor).

BAYES.8: Bayesian Test for a Proportion

1. Data Model

$$X \sim \mathrm{Binomial}(n, \theta).$$

2. Hypotheses

$$H_0 : \theta = \theta_0, \quad H_1 : \theta \sim \mathrm{Beta}(a, b).$$

3. Marginal Likelihoods

Under H_0:

$$p(x \mid H_0) = \binom{n}{x}\theta_0^x (1-\theta_0)^{n-x}.$$

Under H_1:

$$p(x \mid H_1) = \binom{n}{x}\frac{B(a+x,\ b+n-x)}{B(a,b)}.$$

4. Bayes Factor

$$\mathrm{BF}_{10} = \frac{B(a+x,\ b+n-x)}{B(a,b)\,\theta_0^x(1-\theta_0)^{n-x}}.$$

5. Posterior

$$\theta \mid \text{data} \sim \mathrm{Beta}(a+x,\ b+n-x).$$

6. Key Features

- Prior mean $a/(a+b)$ controls prior pull.
- Centering prior at θ_0 yields a conservative test.
- Evidence is averaged over all θ under H_1.

Key Insight: The Bayes factor compares a point null to a distributional alternative, with prior choice directly influencing the strength of evidence.

BAYES.9: HDI + ROPE Decision Method

1. Data Model

$$X \sim \text{Binomial}(n, \theta), \quad \theta \sim \text{Beta}(a, b).$$

Posterior:

$$\theta \mid \text{data} \sim \text{Beta}(a + x,\ b + n - x).$$

2. Highest Density Interval (HDI)

The $100(1-\alpha)\%$ HDI is the shortest interval $[\theta_L, \theta_H]$ such that:

$$\int_{\theta_L}^{\theta_H} p(\theta \mid \text{data})\, d\theta = 1 - \alpha.$$

3. Region of Practical Equivalence (ROPE)

$$\text{ROPE} = [\theta_0 - \epsilon,\ \theta_0 + \epsilon].$$

Defines values practically indistinguishable from H_0.

4. Decision Rule

- HDI $\subset$ ROPE $\Rightarrow$ Accept H_0
- HDI $\cap$ ROPE $= \emptyset \Rightarrow$ Reject H_0
- Otherwise $\Rightarrow$ Suspend judgment

5. Key Features

- Combines uncertainty (HDI) with practical relevance (ROPE)
- Produces a three-way decision
- Requires no Bayes factor computation

Key Insight: HDI + ROPE distinguishes statistical uncertainty from practical significance, allowing explicit "inconclusive" outcomes when data are insufficient.

BAYES.10: Posterior Predictive Check (Poisson–Gamma)

1. Model and Prior

$$Y \mid \lambda \sim \text{Poisson}(\lambda t), \qquad \lambda \sim \text{Gamma}(a, b).$$

2. Posterior Update

$$\lambda \mid Y \sim \text{Gamma}(a + Y,\ b + t).$$

3. Posterior Predictive Distribution

For a new exposure t_{new}:

$$T_{\text{rep}} \mid Y \sim \text{NegBin}\left(r = a + Y,\ p = \frac{b + t}{b + t + t_{\text{new}}}\right).$$

4. Posterior Predictive p-value (PPP)

$$\text{PPP} = P(T_{\text{rep}} \leq T_{\text{obs}} \mid Y).$$

5. Interpretation

- PPP $\approx 0.5 \Rightarrow$ good fit
- PPP ≈ 0 or $1 \Rightarrow$ model misfit
- PPP incorporates uncertainty in parameters

Key Insight

- PPP checks model *predictive adequacy*, not truth
- Based on simulated (replicated) data
- More conservative than frequentist p-values

Appendix: Reference Materials

Essential theory, formulas, and distributions

APPENDIX

Probability Theory: A Practical Reference

1. Why We Needed Axiomatic Probability

For most of human history, probability was intuitive and informal. The first attempt at a rigorous foundation — **classical probability** (Laplace, 1812) — defined the probability of an event as the ratio of favorable outcomes to total outcomes, assuming all outcomes are *equally likely*:

$$P(A) = \frac{\text{number of favorable outcomes}}{\text{total number of outcomes}}.$$

This works elegantly for dice, coins, and card games. It fails the moment outcomes are not symmetric.

Bertrand's paradox (1889). What is the probability that a chord drawn at random in a circle is longer than the side of an inscribed equilateral triangle? Bertrand showed the answer is $1/2$, $1/3$, or $1/4$ depending on how you define "random chord" — three equally reasonable methods give three different answers. The phrase "equally likely" is not self-defining; it depends on the sample space and how probability is distributed over it.

Frequentist probability (von Mises, 1919) offered a different foundation: the probability of an event is the limiting relative frequency of its occurrence in an infinitely long sequence of identical independent experiments. This is operationally meaningful — it is what we estimate when we repeat an experiment — but it has two fatal limitations. First, it cannot assign probabilities to single, unrepeatable events ("the probability that it rains tomorrow" has no frequency interpretation). Second, the limit is assumed to exist but never proved; it is circular to use limits of relative frequencies to define probability when the theory of limits itself uses probability.

Kolmogorov's axiomatic framework (1933) resolved the problem by separating two questions that had been conflated: *what probability means* and *what properties it must have.* Kolmogorov made no claim about what probability means — it could be a degree of belief, a limiting frequency, or a physical propensity. He simply required it to obey three axioms.

2. Kolmogorov's Three Axioms

Let Ω be the sample space (the set of all possible outcomes) and $\mathcal{F}$ a collection of subsets of Ω called **events**. A **probability measure** P assigns a number to each event in $\mathcal{F}$ satisfying:

A1. Non-negativity: $P(A) \geq 0$ for all events A.

A2. Normalization: $P(\Omega) = 1$.

A3. Countable additivity: if $A_1, A_2, \ldots$ are mutually exclusive events, then $P\big(\bigcup_{i=1}^{\infty} A_i\big) = \sum_{i=1}^{\infty} P(A_i)$.

Everything in probability theory — conditional probability, independence, expectation, distributions — follows from these three axioms. The collection $\mathcal{F}$ must be a **σ-algebra**: closed under complements and countable unions. This technical requirement ensures that probabilities are well-defined for all the events we might wish to consider. It also turns out to be the key to understanding why different types of convergence are genuinely different things.

Why this matters practically. The axioms license the rules every practitioner uses without thinking: $P(A^c) = 1 - P(A)$; $P(A \cup B) = P(A) + P(B) - P(A \cap B)$; $P(\varnothing) = 0$. None of these are definitions — they are *theorems* derived from three axioms. The axiomatic framework also makes clear what a probability model is: a triple $(\Omega, \mathcal{F}, P)$ specifying the sample space, the admissible events, and the measure. When two people argue about a probability, they are often arguing about Ω or $\mathcal{F}$, not about P itself.

3. Random Variables and Expectation

A **random variable** X is a function $X : \Omega \to \mathbb{R}$ mapping outcomes to numbers. It is not a variable in the algebraic sense — it is a function defined on the sample space. When we write $P(X \leq x)$, we mean $P(\{\omega \in \Omega : X(\omega) \leq x\})$, a probability assigned to a set of outcomes.

The **expectation** $E[X]$ is the probability-weighted average of the values X takes. For a discrete X, $E[X] = \sum_x x\, P(X = x)$. For continuous X, $E[X] = \int_{-\infty}^{\infty} x\, f(x)\, dx$.

What "on average" really means. When a statistician says "the sample mean equals μ on average," they mean $E[\bar{X}] = \mu$ — an expectation statement about the estimator as a random variable, averaged over all possible samples. This is a mathematical property of the estimator, not an empirical claim about any particular dataset. The *empirical* counterpart — that sample averages actually approach μ as we collect more data — is the content of the Law of Large Numbers, a *convergence theorem*, not an expectation statement. These are different claims and it is worth being precise about which one is being made.

4. Four Types of Convergence

Let $X_1, X_2, \ldots$ be a sequence of random variables and X a limiting random variable (or constant). There are four distinct ways the sequence can "converge to X," ordered from strongest to weakest.

Almost Sure Convergence ($X_n \xrightarrow{\text{a.s.}} X$). The sequence converges to X with probability one:

$$P\Big(\lim_{n\to\infty} X_n = X\Big) = 1.$$

This is the strongest notion. It says the sequence *actually converges* — like a sequence of real numbers — for almost every outcome in Ω. The qualifier "almost" allows a set of outcomes of probability zero to behave badly.

Example. Let $X_n = \bar{X}_n$ be the sample mean of n iid draws from a distribution with mean μ. The **Strong Law of Large Numbers** (SLLN) states $\bar{X}_n \xrightarrow{\text{a.s.}} \mu$. For every conceivable infinite sequence of draws (except a set of probability zero), the running average converges to μ. This is the rigorous foundation for the intuition that "more data gives a better estimate."

Convergence in Probability ($X_n \xrightarrow{p} X$)**.** For any $\varepsilon > 0$,

$$P(|X_n - X| > \varepsilon) \to 0 \quad \text{as } n \to \infty.$$

This is weaker than almost sure convergence. It says that the probability of being far from X vanishes, but occasional large deviations are permitted as long as they become increasingly rare. Almost sure convergence implies convergence in probability, but not vice versa.

Example. The **Weak Law of Large Numbers** (WLLN) states $\bar{X}_n \xrightarrow{p} \mu$. For any tolerance $\varepsilon > 0$, the probability that $|\bar{X}_n - \mu| > \varepsilon$ can be made arbitrarily small by taking n large enough. The WLLN requires only finite mean; the SLLN additionally requires finite variance (in the classical version) but holds under weaker conditions. Convergence in probability is what consistency of an estimator means (see the Statistical Foundations appendix).

Convergence in Mean Square ($X_n \xrightarrow{L^2} X$)**.**

$$E\left[(X_n - X)^2\right] \to 0 \quad \text{as } n \to \infty.$$

Also called L^2 convergence or convergence in quadratic mean. It implies convergence in probability (by Markov's inequality), but its relationship with almost sure convergence is subtle: neither implies the other in general. L^2 convergence is particularly natural in estimation because $E[(X_n - X)^2]$ is the mean squared error. Thus L^2 convergence is equivalent to MSE tending to zero.

Convergence in Distribution ($X_n \xrightarrow{d} X$)**.** For all continuity points x of F_X,

$$F_{X_n}(x) \to F_X(x) \quad \text{as } n \to \infty.$$

This is the *weakest* form of convergence. It says only that the *distribution* (CDF) of X_n converges to the distribution of X — not that the random variables themselves are close. Crucially, X_n and X need not even be defined on the same probability space (the same σ-algebra). The limiting object X is just a distribution, not necessarily a random variable in the original experiment.

This is where the CLT lives. The **Central Limit Theorem** states that for iid X_i with mean μ and variance σ^2:

$$\sqrt{n}(\bar{X}_n - \mu)/\sigma \xrightarrow{d} N(0, 1).$$

The standardized sample mean converges in distribution to a standard normal — but the sample mean itself, for any finite n, retains whatever distribution the X_i have. The normal is a limit, not the actual distribution. This explains why the normal approximation is better for some statistics than others, and why it can fail badly in small samples or when the X_i distribution is heavily skewed or heavy-tailed.

The Hierarchy.

$$\text{a.s.} \Rightarrow \xrightarrow{p} \Rightarrow \xrightarrow{d}, \qquad L^2 \Rightarrow \xrightarrow{p} \Rightarrow \xrightarrow{d}.$$

No other arrows hold in general. Convergence in distribution does not imply convergence in probability (the X_n may live on entirely different probability spaces). Almost sure convergence and L^2 convergence are incomparable.

5. The Laws of Large Numbers

Both laws formalize the intuition that sample averages stabilize around the population mean. They differ in the strength of the convergence and the conditions required.

Weak Law (Khinchin, 1929): If $X_1, X_2, \ldots$ are iid with $E[X_i] = \mu < \infty$, then $\bar{X}_n \xrightarrow{p} \mu$. Requires only a finite mean — no variance assumption.

Strong Law (Kolmogorov, 1933): Under the same finite-mean condition, $\bar{X}_n \xrightarrow{\text{a.s.}} \mu$. This is the theorem behind the intuition that a casino always wins in the long run, that insurance companies can price policies reliably, and that Monte Carlo integration converges.

When "on average" is really LLN. In everyday usage, statements like "the drug reduces blood pressure on average by 5 mmHg" or "the fund returns 8% per year on average" are empirical claims about long-run behavior. They are appeals to LLN — not to $E[X]$ as a mathematical expectation, but to the convergence of observed averages. Understanding the distinction clarifies what evidence these claims require: a single study (even large) establishes $E[X] \approx \mu$ at most approximately; repeated independent replications converge to μ almost surely.

6. The Central Limit Theorem

The CLT is arguably the most important result in applied statistics, yet it is frequently misunderstood.

Classical CLT (Lindeberg-Lévy): If $X_1, \ldots, X_n$ are iid with mean μ and finite variance σ^2, then as $n \to \infty$:

$$\frac{\bar{X}_n - \mu}{\sigma/\sqrt{n}} \xrightarrow{d} N(0, 1).$$

What the CLT says. The distribution of the standardized sample mean becomes approximately normal. The CLT makes no claim about the distribution of individual observations, about the sample mean for a fixed n, or about the tails of the distribution. It is a statement about a limiting distribution, not about any finite sample.

What the CLT does not say.

- It does not say the data are normally distributed. The X_i can have any distribution with finite variance.
- It does not say $n = 30$ is always sufficient. The rate of convergence depends on the skewness and kurtosis of the X_i distribution; heavy-tailed or highly skewed distributions require much larger n.
- It does not apply to statistics other than the sample mean without additional justification (though the delta method extends it to smooth functions of the mean).
- It does not apply when $\sigma^2 = \infty$ (e.g. Cauchy distribution) or when the X_i are strongly dependent.

Why CLT is the weakest convergence. The CLT asserts convergence in distribution — the weakest mode. The $N(0, 1)$ limit need not live on the same probability space as the X_i. The sample mean itself does not converge to anything (it is a consistent estimator of μ via LLN, but the standardized version $\sqrt{n}(\bar{X}_n - \mu)$ diverges — its scale grows without bound while its distribution stabilizes). The CLT captures the shape of the distribution while saying nothing about the actual values of the random variables.

Extensions. The **Lindeberg CLT** relaxes the iid assumption: independent but non-identically distributed X_i satisfy a CLT provided no single term dominates the sum (Lindeberg's condition). The **Lyapunov CLT** is a simpler sufficient condition requiring a finite moment of order $2 + \delta$ for some $\delta > 0$. **Functional CLTs** (Donsker's theorem) extend the result to stochastic processes: the rescaled partial-sum process converges in distribution to Brownian motion.

7. Two Useful Supporting Results

Borel-Cantelli Lemma. Let $A_1, A_2, \ldots$ be any sequence of events. If $\sum_{n=1}^{\infty} P(A_n) < \infty$, then $P(A_n \text{ occurs infinitely often}) = 0$. That is, if the probabilities sum to a finite number, only finitely many events occur almost surely. Conversely, if the A_n are independent and $\sum P(A_n) = \infty$, then $P(A_n \text{ i.o.}) = 1$.

Intuition. If you flip a fair coin infinitely often, you will see heads infinitely often (the series $\sum 1/2$ diverges). But if you play a game where the probability of winning round n is $1/n^2$, you will win only finitely many rounds (the series $\sum 1/n^2 = \pi^2/6$ converges). Borel-Cantelli is the workhorse behind almost sure convergence proofs.

Continuous Mapping Theorem. If $X_n \xrightarrow{d} X$ (or $\xrightarrow{p}$ or $\xrightarrow{\text{a.s.}}$) and g is a continuous function, then $g(X_n) \xrightarrow{d} g(X)$ (or $\xrightarrow{p}$ or $\xrightarrow{\text{a.s.}}$, respectively). This is why, for example, if $\hat{\sigma}^2 \xrightarrow{p} \sigma^2$ then $\hat{\sigma} \xrightarrow{p} \sigma$ (since the square root is continuous on $(0, \infty)$). It is invoked constantly in asymptotic theory, often without being named.

8. What Everyday Statistical Language Really Means

The table below translates common informal statements into their precise mathematical content:

Informal statement	Precise meaning
"On average, the estimator equals μ"	$E[\hat{\theta}] = \theta$ (unbiasedness; an expectation, not convergence)
"With enough data it gets close to μ"	$\hat{\theta}_n \xrightarrow{p} \mu$ (consistency; WLLN or direct argument)
"Eventually it equals μ almost surely"	$\hat{\theta}_n \xrightarrow{\text{a.s.}} \mu$ (SLLN; stronger than consistency)
"Approximately normally distributed for large n"	$\sqrt{n}(\hat{\theta}_n - \theta) \xrightarrow{d} N(0, V)$ (CLT or delta method)
"The drug works on average"	An appeal to LLN — requires repeated evidence, not one study
"The p-value would be < 0.05 if we repeated this many times"	Frequentist probability as long-run relative frequency (von Mises)

Summary

Classical probability failed because "equally likely" is undefined without a probability measure to define it. Kolmogorov's three axioms provide the foundation without specifying what probability means. Convergence in distribution (CLT) is the weakest notion of convergence — only the shape of the distribution stabilizes, not the random variables themselves. Convergence in probability (WLLN, consistency) is stronger; almost sure convergence (SLLN) is stronger still. When you say "on average," you are making an expectation claim; when you say "with more data," you are making a convergence claim. These are different mathematical statements requiring different justifications.

Statistical Foundations: A Practical Reference

1. Estimands, Estimators, and Estimates

Three words that sound similar but mean very different things:

Estimand: the quantity you actually want to know — a fixed, unknown feature of the population. Examples: the population mean μ, the proportion p who favor a policy, the regression coefficient β linking education to income. The estimand exists in the world independently of any data.

Estimator: a rule or formula applied to sample data to approximate the estimand. It is a *function* of the data — a random variable before the data are collected. The sample mean $\bar{X} = n^{-1} \sum_{i=1}^{n} X_i$ is an estimator of μ; the sample proportion $\hat{p} = x/n$ is an estimator of p.

Estimate: the specific number produced when the estimator is applied to a realized sample. Once you compute $\bar{x} = 47.3$ from your data, that is the estimate — a fixed number, no longer random.

Why the distinction matters. We evaluate estimators (as random variables) using properties like bias and variance. We report estimates (as numbers) with uncertainty measures like standard errors and confidence intervals. Confusing the two leads to statements like "the mean is unbiased" when what is meant is "the sample mean is an unbiased estimator of the population mean."

2. Properties of Estimators

Running example. Throughout this section, let $X_1, \dots, X_n \overset{\text{iid}}{\sim} (\mu, \sigma^2)$ and consider estimating μ.

Unbiasedness. An estimator $\hat{\theta}$ is **unbiased** for θ if $E[\hat{\theta}] = \theta$ for all θ. The bias is $\text{Bias}(\hat{\theta}) = E[\hat{\theta}] - \theta$.

$\bar{X}$ is unbiased for μ: $E[\bar{X}] = \mu$. The sample variance $S^2 = \frac{1}{n-1} \sum (X_i - \bar{X})^2$ is unbiased for σ^2; dividing by n instead of $n-1$ introduces a downward bias of $-\sigma^2/n$.

Unbiasedness means the estimator is correct *on average* over hypothetical repeated samples — not that it hits θ for any particular sample. An unbiased estimator can still have high variance and be practically useless.

Consistency. An estimator $\hat{\theta}_n$ is **consistent** for θ if $\hat{\theta}_n \xrightarrow{p} \theta$ as $n \to \infty$ — it converges in probability to the true value. Equivalently, $P(|\hat{\theta}_n - \theta| > \varepsilon) \to 0$ for any $\varepsilon > 0$.

$\bar{X}$ is consistent for μ by the Law of Large Numbers. Consistency is the *minimum* reasonable requirement for an estimator: an inconsistent estimator never gets better no matter how much data you collect.

Unbiasedness $\not\Rightarrow$ consistency and vice versa. A fixed constant $c \neq \mu$ is unbiased for nothing but is trivially "consistent" for c. Conversely, $\hat{\mu} = X_1$ (ignoring all other data) is unbiased but not consistent. In practice, unbiasedness matters for small samples; consistency matters asymptotically.

Efficiency. Among all unbiased estimators of θ, the most **efficient** one has the smallest variance. The **Cramér-Rao lower bound** (CRLB) gives the minimum achievable variance for any unbiased estimator. Let $f(x \mid \theta)$ denote the probability density (or mass) function of a single observation x given parameter θ, and let $\hat{\theta}$ denote an unbiased estimator based on a random sample of size n. Then:

$$\text{Var}(\hat{\theta}) \geq \frac{1}{\mathcal{I}_n(\theta)},$$

where $\mathcal{I}_n(\theta) = -nE\left[\frac{\partial^2}{\partial \theta^2} \log f(X \mid \theta)\right]$ is the **Fisher information** contained in n observations — a measure of how much information the data carry about θ. An estimator achieving the CRLB is called **UMVUE** (Uniformly Minimum Variance Unbiased Estimator). For normal data, $\bar{X}$ achieves the CRLB for μ: no unbiased estimator can have smaller variance.

Data and Information. Data and information are often treated as interchangeable, but they are not. Data are what we collect; information is what we learn. The distinction is subtle but fundamental.

A dataset may contain thousands of observations, yet only a fraction of what is recorded is relevant to the question being asked. Conversely, a small and carefully chosen summary can contain all the information needed for inference. Information, therefore, is not measured by volume — it is measured by relevance to the parameter of interest.

This raises a natural question: given a dataset, is there a function of the data — a statistic — that retains *everything* the data have to say about an unknown parameter θ, with nothing lost? The answer is yes, and the concept that formalises this is **sufficiency**.

Sufficiency. A statistic $T(\mathbf{X})$ is **sufficient** for θ if the conditional distribution of the data given T does not depend on θ. In other words, T captures all the information the data contain about θ. Once you know T, the individual observations add nothing for inference about θ.

For $X_i \overset{\text{iid}}{\sim} N(\mu, \sigma^2)$ with σ^2 known, $\bar{X}$ is sufficient for μ: knowing the sample mean, the remaining data scatter carries no additional information about μ. The **Factorization Theorem** provides a practical check: T is sufficient iff the likelihood factors as

$$L(\theta; \mathbf{x}) = g(T(\mathbf{x}), \theta) \cdot h(\mathbf{x}).$$

Minimal sufficiency is the coarsest sufficient statistic — it achieves maximum data reduction without losing information. **Completeness** rules out nuisance statistics: a sufficient statistic T is complete if no non-trivial function of T has expectation zero for all θ. A complete sufficient statistic is minimal sufficient (Bahadur's theorem), and by the Lehmann-Scheffé theorem, any unbiased estimator that is a function of a complete sufficient statistic is the UMVUE.

Ancillarity. A statistic $A(\mathbf{X})$ is **ancillary** for θ if its distribution does not depend on θ — it carries no information about θ. The sample range of a uniform distribution is ancillary for the location parameter. Ancillary statistics matter in conditional inference: Fisher argued that inference should be conducted conditional on the observed value of any ancillary statistic (the *conditionality principle*).

3. Maximum Likelihood Estimation

The **maximum likelihood estimator** (MLE) $\hat{\theta}_{\text{MLE}}$ maximizes the likelihood function $L(\theta;\mathbf{x}) = \prod_{i=1}^{n} f(x_i \mid \theta)$ — equivalently, the log-likelihood $\ell(\theta) = \sum_i \log f(x_i \mid \theta)$ — over the parameter space. Intuitively, the MLE picks the parameter value that makes the observed data most probable.

Example. For $X_i \overset{\text{iid}}{\sim} N(\mu, \sigma^2)$, the log-likelihood is $\ell(\mu, \sigma^2) = -\frac{n}{2}\log\sigma^2 - \frac{1}{2\sigma^2}\sum(x_i - \mu)^2 + \text{const}$. Setting $\partial\ell/\partial\mu = 0$ gives $\hat{\mu}_{\text{MLE}} = \bar{x}$. Setting $\partial\ell/\partial\sigma^2 = 0$ gives $\hat{\sigma}^2_{\text{MLE}} = n^{-1}\sum(x_i - \bar{x})^2$ — biased by a factor $(n-1)/n$ but consistent.

Key properties of MLEs. Under regularity conditions, MLEs are:

- **Consistent:** $\hat{\theta}_{\text{MLE}} \xrightarrow{p} \theta_0$.
- **Asymptotically normal:** $\sqrt{n}(\hat{\theta}_{\text{MLE}} - \theta_0) \xrightarrow{d} N(0, \mathcal{I}(\theta_0)^{-1})$.
- **Asymptotically efficient:** they achieve the CRLB in large samples — no consistent estimator has smaller asymptotic variance.
- **Invariant:** if $\hat{\theta}$ is the MLE of θ, then $g(\hat{\theta})$ is the MLE of $g(\theta)$ for any function g.

MLEs may be biased in small samples (as the normal variance example shows) but their large-sample properties make them the default estimation method across virtually every chapter of this compendium.

4. The Bias-Variance Tradeoff

The mean squared error (MSE) of an estimator decomposes as:

$$\text{MSE}(\hat{\theta}) = \text{Var}(\hat{\theta}) + \left[\text{Bias}(\hat{\theta})\right]^2.$$

This identity makes precise the tradeoff between bias and variance. An unbiased estimator with high variance may have larger MSE than a biased estimator with lower variance — and therefore perform worse in practice.

Example. Consider estimating μ from $n = 5$ observations. The sample mean $\bar{X}$ is unbiased with $\text{Var}(\bar{X}) = \sigma^2/5$. The estimator $\hat{\mu}_c = c\bar{X}$ for $c < 1$ introduces bias $(c-1)\mu$ but reduces variance to $c^2\sigma^2/5$. For small μ relative to σ, the shrunken estimator can have smaller MSE. This is the logic behind shrinkage estimators (James-Stein, ridge regression, Bayesian posterior means): a deliberate dose of bias can substantially reduce variance, lowering overall MSE.

The bias-variance tradeoff is most visible in prediction settings. An overfitted model has low bias (it fits the training data almost perfectly) but high variance (it changes drastically across different samples). An underfitted model has high bias but low variance. The optimal model complexity minimizes the sum, not either term alone.

5. Regularization: Ridge, Lasso, Elastic Net

In linear regression $\mathbf{y} = \mathbf{X}\boldsymbol{\beta} + \boldsymbol{\varepsilon}$, the OLS estimator $\hat{\boldsymbol{\beta}}_{\text{OLS}} = (\mathbf{X}'\mathbf{X})^{-1}\mathbf{X}'\mathbf{y}$ is BLUE (Best Linear Unbiased Estimator) by the Gauss-Markov theorem. However, when predictors are highly correlated (multicollinearity) or $p \approx n$, the OLS estimator has inflated variance — the matrix $\mathbf{X}'\mathbf{X}$ is nearly singular. **Regularization** deliberately introduces bias to reduce variance and improve prediction. All three methods below add a penalty term to the OLS objective.

Ridge Regression (Hoerl & Kennard, 1970).

$$\hat{\boldsymbol{\beta}}_{\text{ridge}} = \arg\min_{\boldsymbol{\beta}} \left\{ \|\mathbf{y} - \mathbf{X}\boldsymbol{\beta}\|^2 + \lambda\|\boldsymbol{\beta}\|_2^2 \right\}.$$

The closed-form solution is $\hat{\boldsymbol{\beta}}_{\text{ridge}} = (\mathbf{X}'\mathbf{X} + \lambda\mathbf{I})^{-1}\mathbf{X}'\mathbf{y}$. Adding $\lambda\mathbf{I}$ to the Gram matrix stabilises the inversion. Ridge shrinks all coefficients toward zero but never sets any to exactly zero — it performs no variable selection. The tuning parameter $\lambda \geq 0$ controls the degree of shrinkage: $\lambda = 0$ recovers OLS; $\lambda \to \infty$ shrinks all $\hat{\beta}_j \to 0$.

Bayesian interpretation: Ridge is equivalent to the posterior mode under a normal prior $\beta_j \overset{\text{iid}}{\sim} N(0, \sigma^2/\lambda)$.

Lasso (Tibshirani, 1996).

$$\hat{\boldsymbol{\beta}}_{\text{lasso}} = \arg\min_{\boldsymbol{\beta}} \left\{ \|\mathbf{y} - \mathbf{X}\boldsymbol{\beta}\|^2 + \lambda\|\boldsymbol{\beta}\|_1 \right\}.$$

The ℓ_1 penalty $\|\boldsymbol{\beta}\|_1 = \sum_j |\beta_j|$ is the key difference. Unlike the ℓ_2 penalty, the ℓ_1 penalty has corners at zero in the coefficient space, which causes some $\hat{\beta}_j$ to be set *exactly* to zero. Lasso therefore performs automatic variable selection: irrelevant predictors are excluded from the model. No closed-form solution exists; coordinate descent is the standard algorithm. Lasso can struggle when predictors are highly correlated (it tends to pick one and ignore the rest arbitrarily) and when $p \gg n$ (at most n predictors enter the model).

Bayesian interpretation: Lasso is the posterior mode under a Laplace (double exponential) prior $\beta_j \overset{\text{iid}}{\sim} \text{Laplace}(0, \sigma^2/\lambda)$.

Elastic Net (Zou & Hastie, 2005).

$$\hat{\boldsymbol{\beta}}_{\text{en}} = \arg\min_{\boldsymbol{\beta}} \left\{ \|\mathbf{y} - \mathbf{X}\boldsymbol{\beta}\|^2 + \lambda_1\|\boldsymbol{\beta}\|_1 + \lambda_2\|\boldsymbol{\beta}\|_2^2 \right\}.$$

A convex combination of Lasso and Ridge penalties. It inherits Lasso's sparsity (some coefficients set to zero) while

Ridge's ℓ_2 component handles correlated predictors by shrinking them together rather than arbitrarily selecting one. Elastic Net is preferred when groups of correlated predictors should enter or leave the model together (the "grouping effect"). The mixing parameter $\alpha = \lambda_1/(\lambda_1 + \lambda_2) \in [0, 1]$ interpolates between pure Ridge ($\alpha = 0$) and pure Lasso ($\alpha = 1$).

	Ridge	**Lasso**	**Elastic Net**
Penalty	ℓ_2	ℓ_1	$\ell_1 + \ell_2$
Sparsity	No	Yes	Yes
Correlated predictors	Stable	Unstable	Stable
$p \gg n$	Yes	Limited	Yes
Closed form	Yes	No	No

Choosing λ: All three methods require tuning. Cross-validation (typically 10-fold) is the standard approach: fit the model across a grid of λ values and select the λ minimizing held-out prediction error. The "one standard error rule" selects the largest λ within one standard error of the minimum, favoring parsimony.

6. Variable Selection: A Brief Survey

Variable selection — choosing which predictors to include in a model — is one of the most consequential decisions in applied statistics. The goal is to balance goodness of fit against model complexity, avoiding both underfitting (missing important predictors) and overfitting (including noise as signal).

Subset Selection. *Best subset selection* evaluates all 2^p possible models and selects the one minimizing a criterion (AIC, BIC, adjusted R^2). Computationally infeasible for $p > 30$–40. *Forward stepwise* starts from the null model and adds the predictor that most improves fit at each step. *Backward stepwise* starts from the full model and removes the weakest predictor. Both are greedy and do not guarantee finding the global optimum; they can be unstable — small data changes can produce very different models.

Information Criteria. Rather than testing, select the model minimizing:

$$\text{AIC} = -2\ell(\hat{\theta}) + 2k, \qquad \text{BIC} = -2\ell(\hat{\theta}) + k \log n,$$

where k is the number of parameters. AIC targets predictive accuracy; BIC is consistent (selects the true model as $n \to \infty$ if it is in the candidate set) but tends to favor sparser models than AIC. Neither requires a significance threshold; both reward fit and penalize complexity.

Regularization-Based Selection. Lasso and Elastic Net perform variable selection as a by-product of estimation (see Section 5). The advantage over stepwise methods is that the solution path varies continuously with λ rather than making hard include/exclude decisions at each step. LARS (Least Angle Regression, Efron et al. 2004) computes the entire Lasso solution path efficiently.

Post-Selection Inference. A critical warning: standard p-values and confidence intervals are **invalid** after variable selection. If you use the data to select variables and then test those same variables with the same data, Type I error rates are severely inflated. Remedies include:

- **Sample splitting:** use half the data to select variables, the other half to test them.
- **Cross-validation:** honest assessment of predictive performance without reusing the same observations for selection and evaluation.
- **Selective inference** (Lee et al. 2016): exact p-values conditional on the model selected by Lasso.
- **Bayesian variable selection:** posterior inclusion probabilities (PIPs) provide an alternative that does not rely on post-selection p-values.

Practical Guidance.

- With $p \ll n$ and a well-specified model: AIC/BIC or adjusted R^2 are adequate.
- With $p \approx n$ or correlated predictors: Lasso or Elastic Net with cross-validated λ.
- With $p \gg n$: Elastic Net or sparse Bayesian methods (spike-and-slab priors).
- Regardless of method: validate on held-out data and report uncertainty honestly. A selected model is a hypothesis, not a fact.

Summary

Estimands are what we want; estimators are how we get there; estimates are what we report. Good estimators are unbiased and consistent at minimum; efficient if possible. MLE provides a unified estimation framework with strong large-sample guarantees. The bias-variance tradeoff explains why regularization — deliberately biasing estimators — improves prediction when p is large relative to n. Variable selection requires care: the method that selects a model should not be the same analysis that claims to confirm it.

Statistical Distributions: Quick Reference

A Unifying Framework: The Pearson Differential Equation

Origin In 1895, Karl Pearson sought to generalize the Normal distribution after encountering markedly skewed data — specifically, the ratio of forehead breadth to body length of crabs collected at Naples by zoologist W.F.R. Weldon. The data clearly could not be described by the Normal curve. His question — *what differential equation must any "reasonable" frequency curve satisfy?* — led him to a single first-order equation that, depending on four parameters, generates a broad family of continuous distributions used in classical statistics.

The Differential Equation Pearson (1895) showed that the logarithmic derivative of a frequency function $f(x)$ satisfying natural regularity conditions can be written in the form:

$$\frac{d}{dx} \ln f(x) = \frac{-(x-a)}{b_0 + b_1 x + b_2 x^2},$$

or equivalently $f'(x)/f(x) = (a-x)/q(x)$, where $q(x) = b_0 + b_1 x + b_2 x^2$ is a quadratic denominator and a is the mode of the distribution. The four parameters $(a,\ b_0,\ b_1,\ b_2)$ are determined by the first four moments of the distribution, controlling location, scale, skewness, and tail weight respectively.

Why the Quadratic Denominator? The key insight is that $q(x)$ pins the tails of the distribution. When $q(x)$ has real roots of opposite sign, the density vanishes at those bounds, producing *bounded* support. When $q(x)$ has complex roots, the density has *unbounded* support with tail behavior governed by the leading coefficient b_2. The seven main Pearson types arise from these distinct geometric regimes of the roots.

Classification Pearson classified solutions by the nature of the roots of $q(x)$. In practice the classification is captured by the **κ-criterion**, expressed in terms of the standardized moment ratios $\beta_1 = \mu_3^2/\mu_2^3$ (squared skewness) and $\beta_2 = \mu_4/\mu_2^2$ (kurtosis):

$$\kappa = \frac{\beta_1(\beta_2+3)^2}{4(4\beta_2 - 3\beta_1)(2\beta_2 - 3\beta_1 - 6)}.$$

The three main types are determined as follows:

- $\kappa < 0$: roots of $q(x)$ are real and of opposite sign, confining the density to a bounded interval — **Type I** (Beta distribution).
- $0 < \kappa < 1$: roots are complex conjugates, producing unbounded asymmetric distributions — **Type IV** (no standard named form; arises in financial and extreme-value applications).
- $\kappa > 1$: roots are real and same-signed, giving a semi-infinite distribution — **Type VI** (F-distribution, Beta prime).

The symmetric cases ($b_1 = 0$) yield Types II and VII (symmetric Beta and Student's t); the degenerate case $b_2 = 0$ (one infinite root) yields Type III (Gamma, χ^2); the boundary case $\kappa = 1$ yields Type V (inverse Gamma).

The Normal Distribution as a Limiting Case When $b_1 = b_2 = 0$, the equation reduces to $f'(x)/f(x) = (a-x)/b_0$, integrating immediately to $f(x) \propto \exp\big(-(x-a)^2/2b_0\big)$ — the Normal distribution with mean $a = \mu$ and variance $b_0 = \sigma^2$. The Normal is therefore not a distinct Pearson *type* but the *degenerate limit* of the entire system as the denominator collapses to a constant. It is the limiting form to which every other member of the family converges as skewness and excess kurtosis simultaneously vanish.

Fitting by Method of Moments Pearson's original motivation was also computational: the four parameters (a, b_0, b_1, b_2) can be estimated directly from the first four sample moments $(\bar{x},\ s^2,\ \hat{\gamma}_1,\ \hat{\gamma}_2)$. Setting the theoretical moments equal to their sample counterparts gives a closed-form system for the parameters, making distribution fitting tractable long before maximum likelihood methods were widespread. The approach remains useful today in actuarial, financial, and reliability applications where parametric flexibility is needed without heavy computational overhead.

Practical Significance for Statistical Testing Every t-, F-, and χ^2-test statistic used in this compendium follows a distribution that is a Pearson family member. The family clarifies the structural cascade of relationships that practitioners rely on: χ^2_ν is Gamma(shape $= \nu/2$, scale $= 2$) (Type III); $F_{m,n}$ is a scaled Beta prime (Type VI); Student's t_ν is Type VII; and the χ^2, F, and t distributions all converge to the Normal as their degrees of freedom grow. These are not coincidences — they are structural consequences of sharing a single underlying generating differential equation.

The Pearson Family: Types, Criteria, and Named Distributions

Type	Criterion	Support	Shape	Named distributions
Limit	$b_1 = b_2 = 0$	$(-\infty, \infty)$	Symmetric, bell	Normal $N(\mu, \sigma^2)$
I	$\kappa < 0$; two real roots, opposite signs	$[x_1, x_2]$	Skewed, bounded	Beta(α, β); Uniform≡Beta$(1, 1)$
II	$b_1 = 0,\ b_2 < 0$	$[-a, a]$	Symmetric, bounded	Symmetric Beta; Uniform (special case)
III	$b_2 = 0$ (linear denominator)	$[c, \infty)$	Right-skewed, semi-infinite	Gamma(α, β); χ^2_ν ≡Gamma$(\nu/2, 2)$; Exponential
IV	$0 < \kappa < 1$ (complex roots)	$(-\infty, \infty)$	Skewed, unbounded	(No standard name; asymmetric heavy-tailed family)
V	$\kappa = 1$ (repeated root)	$(0, \infty)$	Right-skewed	Inverse Gamma; Lévy
VI	$\kappa > 1$ (two real roots, same sign)	$(0, \infty)$	Right-skewed, semi-infinite	Beta prime; $F_{m,n}$ (scaled Beta prime)
VII	$b_1 = 0,\ b_2 > 0$	$(-\infty, \infty)$	Symmetric, heavy-tailed	Student t_ν; Cauchy≡ t_1; Normal (limit $\nu \to \infty$)

Family Tree

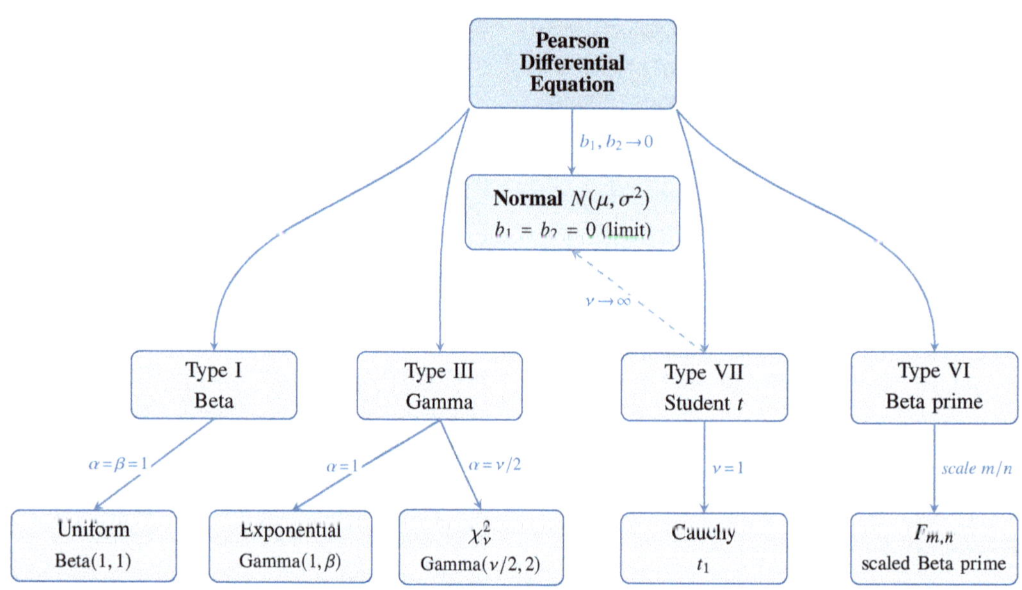

Types II, IV, and V are omitted from the diagram for clarity. Type II is the symmetric special case of Type I (Beta with equal shape parameters). Type IV has no standard named distribution. Type V corresponds to the Inverse Gamma. Reference: Pearson, K. (1895). *Contributions to the Mathematical Theory of Evolution II: Skew Variation in Homogeneous Material.* Philosophical Transactions of the Royal Society A, **186**, 343–414.

The following tables summarize the key properties of the individual distributions referenced throughout this compendium. Notation: $\Gamma(\cdot)$ is the gamma function, $B(a, b) = \Gamma(a)\Gamma(b)/\Gamma(a + b)$ is the beta function, $\psi(\cdot)$ is the digamma function, and log denotes the natural logarithm throughout.

1. Discrete Distributions

Property	**Bernoulli** Bern(p)	**Binomial** Bin(n, p)	**Poisson** Pois(λ)
Type	Discrete	Discrete	Discrete
PMF	$p^k(1-p)^{1-k}$, $k \in \{0, 1\}$	$\binom{n}{k}p^k(1-p)^{n-k}$, $k = 0, 1, \ldots, n$	$\frac{\lambda^k e^{-\lambda}}{k!}$, $k = 0, 1, 2, \ldots$
Support	$\{0, 1\}$	$\{0, 1, \ldots, n\}$	$\{0, 1, 2, \ldots\}$
Mean	p	np	λ
Variance	$p(1-p)$	$np(1-p)$	λ
MGF	$1 - p + pe^t$	$(1 - p + pe^t)^n$	$\exp(\lambda(e^t - 1))$
Char. fn.	$1 - p + pe^{it}$	$(1 - p + pe^{it})^n$	$\exp(\lambda(e^{it} - 1))$
Skewness	$\frac{1-2p}{\sqrt{p(1-p)}}$	$\frac{1-2p}{\sqrt{np(1-p)}}$	$\lambda^{-1/2}$
Ex. Kurtosis	$\frac{1-6p(1-p)}{p(1-p)}$	$\frac{1-6p(1-p)}{np(1-p)}$	λ^{-1}
Practitioner Note	Single binary trial; special case of Binomial with $n = 1$.	Counts successes in n independent trials; backbone of proportion testing.	Models rare-event counts; equidispersion assumption (mean = variance) often tested.

Property	**Negative Binomial** NB(r, p)	**Geometric** Geom(p)
Type	Discrete	Discrete
PMF	$\binom{k+r-1}{k}(1-p)^r p^k, \quad k = 0, 1, 2, \ldots$	$(1-p)^{k-1}p, \quad k = 1, 2, 3, \ldots$
Support	$\{0, 1, 2, \ldots\}$	$\{1, 2, 3, \ldots\}$
Mean	$\frac{r(1-p)}{p}$	$\frac{1}{p}$
Variance	$\frac{r(1-p)}{p^2}$	$\frac{1-p}{p^2}$
MGF	$\left(\frac{p}{1-(1-p)e^t}\right)^r, \quad t < -\log(1-p)$	$\frac{pe^t}{1-(1-p)e^t}, \quad t < -\log(1-p)$
Char. fn.	$\left(\frac{p}{1-(1-p)e^{it}}\right)^r$	$\frac{pe^{it}}{1-(1-p)e^{it}}$
Skewness	$\frac{2-p}{\sqrt{r(1-p)}}$	$\frac{2-p}{\sqrt{1-p}}$
Ex. Kurtosis	$\frac{p^2+6(1-p)}{r(1-p)}$	$\frac{p^2-6p+6}{1-p}$

Property	Negative Binomial $\text{NB}(r, p)$	Geometric $\text{Geom}(p)$
Practitioner Note	Overdispersed count data; variance exceeds mean. Common alternative to Poisson in GLMs when equidispersion fails. Dispersion parameter r (or θ) controls overdispersion.	Counts trials until first success; memoryless property. Used in reliability and sequential testing.

2. Continuous Distributions: Support $(0, \infty)$

Property	Exponential $\text{Exp}(\lambda)$	Gamma $\text{Gamma}(\alpha, \beta)$	Lognormal $\text{LN}(\mu, \sigma^2)$
Type	Continuous	Continuous	Continuous
PDF	$\lambda e^{-\lambda x}$	$\dfrac{\beta^\alpha x^{\alpha-1} e^{-\beta x}}{\Gamma(\alpha)}$	$\dfrac{1}{x\sigma\sqrt{2\pi}} \exp\left(-\dfrac{(\log x - \mu)^2}{2\sigma^2}\right)$
Support	$(0, \infty)$	$(0, \infty)$	$(0, \infty)$
Mean	λ^{-1}	α/β	$e^{\mu+\sigma^2/2}$
Variance	λ^{-2}	α/β^2	$(e^{\sigma^2} - 1)e^{2\mu+\sigma^2}$
MGF	$\dfrac{\lambda}{\lambda - t}, t < \lambda$	$\left(\dfrac{\beta}{\beta - t}\right)^\alpha, t < \beta$	Does not exist in closed form
Skewness	2	$2/\sqrt{\alpha}$	$(e^{\sigma^2} + 2)\sqrt{e^{\sigma^2} - 1}$
Ex. Kurtosis	6	$6/\alpha$	$e^{4\sigma^2} + 2e^{3\sigma^2} + 3e^{2\sigma^2} - 6$
Practitioner Note	Special case of Gamma ($\alpha = 1$); memoryless. Models time between events in a Poisson process.	Flexible right-skewed family; includes Exponential ($\alpha = 1$) and Chi-square ($\alpha = \nu/2$, $\beta = 1/2$) as special cases.	Log of a normal variable; right-skewed. Common in finance and insurance for loss severity. MGF does not exist; use log-transformed analysis.

Property	Pareto $\text{Pareto}(\alpha, x_m)$	Weibull $\text{Weibull}(k, \lambda)$	Tweedie $\text{Tw}_p(\mu, \phi)$
Type	Continuous	Continuous	Mixed/Continuous
PDF	$\dfrac{\alpha x_m^\alpha}{x^{\alpha+1}}, x \geq x_m$	$\dfrac{k}{\lambda}\left(\dfrac{x}{\lambda}\right)^{k-1} e^{-(x/\lambda)^k}$	No closed form in general
Support	$[x_m, \infty)$	$(0, \infty)$	$(0, \infty)$; mass at 0 if $1 < p < 2$
Mean	$\dfrac{\alpha x_m}{\alpha - 1}, \alpha > 1$	$\lambda\Gamma(1 + 1/k)$	μ
Variance	$\dfrac{x_m^2 \alpha}{(\alpha - 1)^2(\alpha - 2)}, \alpha > 2$	$\lambda^2\left[\Gamma(1 + 2/k) - \Gamma(1 + 1/k)^2\right]$	$\phi\mu^p$
MGF	Does not exist	Does not exist	Does not exist in general
Skewness	$\dfrac{2(1 + \alpha)}{\alpha - 3}\sqrt{\dfrac{\alpha - 2}{\alpha}}, \alpha > 3$	Depends on k	Depends on p
Ex. Kurtosis	$\dfrac{6(\alpha^3 + \alpha^2 - 6\alpha - 2)}{\alpha(\alpha - 3)(\alpha - 4)}$	Depends on k	Depends on p

Property	Pareto $\text{Pareto}(\alpha, x_m)$	Weibull $\text{Weibull}(k, \lambda)$	Tweedie $\text{Tw}_p(\mu, \phi)$
Practitioner Note	Heavy-tailed; power-law decay. MGF does not exist. Used in extreme value modeling and operational risk.	Generalizes Exponential ($k = 1$) and Rayleigh ($k = 2$). Widely used in reliability and survival analysis.	A family indexed by power p: Poisson ($p = 1$), Gamma ($p = 2$), compound Poisson-Gamma ($1 < p < 2$, default in GLM Tweedie). Variance proportional to μ^p.

3. Continuous Distributions: Support $(0, 1)$

Property	Beta $\text{Beta}(\alpha, \beta)$
Type	Continuous
PDF	$\dfrac{x^{\alpha-1}(1-x)^{\beta-1}}{B(\alpha,\beta)}, \quad x \in (0,1)$
Support	$(0, 1)$
Mean	$\dfrac{\alpha}{\alpha+\beta}$
Variance	$\dfrac{\alpha\beta}{(\alpha+\beta)^2(\alpha+\beta+1)}$
MGF	$1 + \sum_{k=1}^{\infty} \left(\prod_{r=0}^{k-1} \frac{\alpha+r}{\alpha+\beta+r} \right) \frac{t^k}{k!}$
Skewness	$\dfrac{2(\beta-\alpha)\sqrt{\alpha+\beta+1}}{(\alpha+\beta+2)\sqrt{\alpha\beta}}$
Ex. Kurtosis	$\dfrac{6[(\alpha-\beta)^2(\alpha+\beta+1) - \alpha\beta(\alpha+\beta+2)]}{\alpha\beta(\alpha+\beta+2)(\alpha+\beta+3)}$
Practitioner Note	Flexible distribution for proportions and probabilities. Special cases: Uniform(0, 1) when $\alpha = \beta = 1$; arc-sine distribution when $\alpha = \beta = 1/2$. Conjugate prior for the Binomial likelihood in Bayesian analysis.

4. Continuous Distributions: Support $(-\infty, \infty)$

Property	Normal $N(\mu, \sigma^2)$	Student's t t_ν	Logistic $\text{Logistic}(\mu, s)$
Type	Continuous	Continuous	Continuous
PDF	$\dfrac{1}{\sigma\sqrt{2\pi}} e^{-\frac{(x-\mu)^2}{2\sigma^2}}$	$\dfrac{\Gamma\left(\frac{\nu+1}{2}\right)}{\sqrt{\nu\pi}\,\Gamma\left(\frac{\nu}{2}\right)} \left(1 + \dfrac{x^2}{\nu}\right)^{-\frac{\nu+1}{2}}$	$\dfrac{e^{-(x-\mu)/s}}{s\left(1+e^{-(x-\mu)/s}\right)^2}$
Support	$(-\infty, \infty)$	$(-\infty, \infty)$	$(-\infty, \infty)$
Mean	μ	0 ($\nu > 1$)	μ
Variance	σ^2	$\dfrac{\nu}{\nu-2}$ ($\nu > 2$)	$\dfrac{\pi^2 s^2}{3}$
MGF	$e^{\mu t + \sigma^2 t^2/2}$	Does not exist ($\nu \leq 1$); $1 + \sum_{k=1}^{\infty} \frac{t^{2k} \prod_{j=1}^{k}(2j-1)\nu^{-k}}{(2k)!}$ for $\lvert t \rvert < \sqrt{\nu}$ (rarely used in practice)	$e^{\mu t} B(1 - st,\ 1 + st)$, $\lvert t \rvert < 1/s$
Skewness	0	0 ($\nu > 3$)	0

Property	Normal $N(\mu, \sigma^2)$	Student's t t_ν	Logistic $\text{Logistic}(\mu, s)$
Ex. Kurtosis	0	$\frac{6}{\nu - 4}$ $(\nu > 4)$	$6/5$
Practitioner Note	Foundation of classical statistics; central limit theorem guarantees approximate normality for large n.	Heavier tails than Normal; approaches $N(0, 1)$ as $\nu \to \infty$. Used when σ^2 is unknown.	Symmetric, heavier-tailed than Normal; CDF is the standard logistic function used in logistic regression.

Property	Laplace $\text{Lap}(\mu, b)$	Skew-Normal $\text{SN}(\xi, \omega, \alpha)$	Cauchy $\text{Cauchy}(\mu, \gamma)$
Type	Continuous	Continuous	Continuous
PDF	$\frac{1}{2b} e^{-\lvert x-\mu\rvert/b}$	$\frac{2}{\omega}\phi\left(\frac{x-\xi}{\omega}\right)\Phi\left(\alpha\frac{x-\xi}{\omega}\right)$	$\frac{1}{\pi\gamma\left[1+\left(\frac{x-\mu}{\gamma}\right)^2\right]}$
Support	$(-\infty, \infty)$	$(-\infty, \infty)$	$(-\infty, \infty)$
Mean	μ	$\xi + \omega\delta\sqrt{2/\pi}$, $\delta = \alpha/\sqrt{1+\alpha^2}$	Does not exist
Variance	$2b^2$	$\omega^2(1 - 2\delta^2/\pi)$	Does not exist
MGF	$\frac{e^{\mu t}}{1 - b^2t^2}$, $\lvert t\rvert < 1/b$	$2e^{\xi t + \omega^2 t^2/2}\Phi(\alpha\omega t)$	Does not exist
Skewness	0	$\frac{4-\pi}{2} \cdot \frac{(\delta\sqrt{2/\pi})^3}{(1 - 2\delta^2/\pi)^{3/2}}$	Does not exist
Ex. Kurtosis	3	Depends on α	Does not exist
Practitioner Note	Double exponential; heavier tails than Normal, sharper peak. Arises in Lasso regression as the prior on coefficients.	Generalizes Normal by adding skewness parameter α; reduces to $N(\xi, \omega^2)$ when $\alpha = 0$. Useful for modestly asymmetric data.	Pathological: no mean, variance, or MGF. Ratio of two independent standard normals. Illustrates failure of LLN and CLT.

5. Extreme Value Distributions

Property	GEV $\text{GEV}(\mu, \sigma, \xi)$	Gumbel $\text{Gumbel}(\mu, \beta)$
Type	Continuous	Continuous
PDF	$\frac{1}{\sigma}\left(1+\xi\frac{x-\mu}{\sigma}\right)^{-1/\xi-1}\exp\left[-\left(1+\xi\frac{x-\mu}{\sigma}\right)^{-1/\xi}\right]$	$\frac{1}{\beta}\exp\left[-\frac{x-\mu}{\beta} - e^{-(x-\mu)/\beta}\right]$
Support	$1 + \xi(x-\mu)/\sigma > 0$	$(-\infty, \infty)$
Mean	$\mu + \sigma[\Gamma(1-\xi) - 1]/\xi$, $\xi \neq 0$, $\xi < 1$	$\mu + \beta\gamma_E$, $\gamma_E \approx 0.5772$
Variance	$\sigma^2[\Gamma(1-2\xi) - \Gamma^2(1-\xi)]/\xi^2$, $\xi < 1/2$	$\pi^2\beta^2/6$
MGF	Does not exist in general	$\Gamma(1-\beta t)e^{\mu t}$, $t < 1/\beta$
Skewness	Depends on ξ via Γ functions	$12\sqrt{6}\,\zeta(3)/\pi^3 \approx 1.14$
Ex. Kurtosis	Depends on ξ	$12/5$

Property	GEV $\text{GEV}(\mu, \sigma, \xi)$	Gumbel $\text{Gumbel}(\mu, \beta)$
Practitioner Note	Unifies Fréchet ($\xi > 0$), Gumbel ($\xi = 0$), and Weibull ($\xi < 0$) families. Used in extreme value analysis for block maxima. Shape parameter ξ governs tail heaviness.	Special case of GEV with $\xi = 0$; light-tailed. Models maxima of many distributions including Normal and Exponential. Widely used in hydrology and climate extremes.

6. Multivariate Distributions

Formulas below use the following notation: $\mathbf{x} \in \mathbb{R}^p$, $\boldsymbol{\mu} \in \mathbb{R}^p$, $\boldsymbol{\Sigma}$ is a $p \times p$ positive definite covariance matrix, $|\cdot|$ denotes determinant, $\text{tr}(\cdot)$ denotes trace, $\Gamma_p(\cdot)$ is the multivariate gamma function, and $\mathbf{W}$ denotes a $p \times p$ positive definite matrix.

Multivariate Normal: $\text{MVN}(\boldsymbol{\mu}, \boldsymbol{\Sigma})$

Property	Formula / Value
Type	Continuous, $\mathbb{R}^p$
PDF	$\frac{1}{(2\pi)^{p/2}\lvert\boldsymbol{\Sigma}\rvert^{1/2}} \exp\left(-\frac{1}{2}(\mathbf{x}-\boldsymbol{\mu})^\top \boldsymbol{\Sigma}^{-1}(\mathbf{x}-\boldsymbol{\mu})\right)$
Mean	$\boldsymbol{\mu}$
Covariance	$\boldsymbol{\Sigma}$
MGF	$\exp\left(\mathbf{t}^\top \boldsymbol{\mu} + \frac{1}{2}\mathbf{t}^\top \boldsymbol{\Sigma}\mathbf{t}\right)$
Marginals	All marginals are normal; $X_i \sim N(\mu_i, \Sigma_{ii})$
Key property	Zero correlation $\Leftrightarrow$ independence (unique to the normal family)
Practitioner Note	Foundation of multivariate analysis. Quadratic form $(\mathbf{x}-\boldsymbol{\mu})^\top \boldsymbol{\Sigma}^{-1}(\mathbf{x}-\boldsymbol{\mu}) \sim \chi^2_p$. Closed under linear transformations, conditioning, and marginalization.

Multivariate *t*: $t_\nu(\boldsymbol{\mu}, \boldsymbol{\Sigma})$

Property	Formula / Value
Type	Continuous, $\mathbb{R}^p$
PDF	$\frac{\Gamma(\frac{\nu+p}{2})}{\Gamma(\frac{\nu}{2})\,(\nu\pi)^{p/2}\lvert\boldsymbol{\Sigma}\rvert^{1/2}} \left(1 + \frac{(\mathbf{x}-\boldsymbol{\mu})^\top \boldsymbol{\Sigma}^{-1}(\mathbf{x}-\boldsymbol{\mu})}{\nu}\right)^{-(\nu+p)/2}$
Mean	$\boldsymbol{\mu} \quad (\nu > 1)$
Covariance	$\frac{\nu}{\nu-2}\boldsymbol{\Sigma} \quad (\nu > 2)$
MGF	Does not exist
Key property	Heavier tails than MVN; approaches MVN as $\nu \to \infty$
Practitioner Note	Used when multivariate normality is suspected but tails are heavier. Arises as a scale mixture of normals. Applied in robust multivariate analysis and Bayesian models.

Dirichlet: $\mathrm{Dir}(\alpha)$, $\alpha = (\alpha_1, \ldots, \alpha_K)^\top$, $\alpha_i > 0$

Property	Formula / Value
Type	Continuous, simplex $\sum_{i=1}^{K} x_i = 1$, $x_i \in (0, 1)$
PDF	$\dfrac{\Gamma(\alpha_0)}{\prod_{i=1}^{K} \Gamma(\alpha_i)} \prod_{i=1}^{K} x_i^{\alpha_i - 1}, \quad \alpha_0 = \sum_{i=1}^{K} \alpha_i$
Mean	$E[X_i] = \alpha_i / \alpha_0$
Variance	$\mathrm{Var}(X_i) = \dfrac{\alpha_i(\alpha_0 - \alpha_i)}{\alpha_0^2(\alpha_0 + 1)}$
MGF	Does not exist in standard form
Key property	Multivariate generalization of the Beta distribution; marginals are $\mathrm{Beta}(\alpha_i, \alpha_0 - \alpha_i)$
Practitioner Note	Conjugate prior for the Multinomial likelihood. Standard prior for topic models (LDA) and Bayesian compositional data analysis. Concentration parameter α_0 controls dispersion of the distribution over the simplex.

Wishart: $\mathcal{W}_p(\mathbf{V}, n)$, $n \geq p$, $\mathbf{V}$ positive definite scale matrix

Property	Formula / Value
Type	Continuous, $p \times p$ positive definite matrices
PDF	$\dfrac{\lvert\mathbf{W}\rvert^{(n-p-1)/2} \exp\left(-\frac{1}{2}\mathrm{tr}(\mathbf{V}^{-1}\mathbf{W})\right)}{2^{np/2}\lvert\mathbf{V}\rvert^{n/2}\,\Gamma_p(n/2)}$
Mean	$n\mathbf{V}$
Variance	$\mathrm{Var}(W_{ij}) = n(V_{ij}^2 + V_{ii}V_{jj})$
MGF	$\lvert\mathbf{I} - 2\mathbf{V}\mathbf{T}\rvert^{-n/2}$
Key property	If $\mathbf{X}_i \overset{\text{iid}}{\sim} \mathrm{MVN}(\mathbf{0}, \mathbf{V})$, then $\sum_{i=1}^{n} \mathbf{X}_i\mathbf{X}_i^\top \sim \mathcal{W}_p(\mathbf{V}, n)$
Practitioner Note	Distribution of sample covariance matrices under multivariate normality. Central to MANOVA likelihood theory. Inverse-Wishart is the conjugate prior for $\mathbf{\Sigma}$ in Bayesian multivariate models.

7. Zero-Modified Distributions

Property	**ZIP** Zero-Inflated Poisson	**ZINB** Zero-Inflated Neg. Binomial
Type	Discrete mixture	Discrete mixture
PMF	$P(Y = 0) = \pi + (1 - \pi)e^{-\lambda}$ $P(Y = k) = (1 - \pi)\dfrac{\lambda^k e^{-\lambda}}{k!}, k \geq 1$	$P(Y = 0) = \pi + (1 - \pi)\left(\dfrac{r}{r + \mu}\right)^r$ $P(Y = k) = (1 - \pi)\binom{k + r - 1}{k}\left(\dfrac{r}{r + \mu}\right)^r\left(\dfrac{\mu}{r + \mu}\right)^k, k \geq 1$
Support	$\{0, 1, 2, \ldots\}$	$\{0, 1, 2, \ldots\}$
Mean	$(1 - \pi)\lambda$	$(1 - \pi)\mu$
Variance	$(1 - \pi)\lambda(1 + \pi\lambda)$	$(1 - \pi)\mu(1 + \mu/r + \pi\mu)$

Property	**ZIP** Zero-Inflated Poisson	**ZINB** Zero-Inflated Neg. Binomial
Parameters	$\pi \in [0, 1)$: structural-zero prob. $\lambda > 0$: Poisson rate	$\pi \in [0, 1)$: structural-zero prob. $\mu > 0$: mean count $r > 0$: dispersion
Practitioner Note	Use when count data has excess zeros beyond what Poisson predicts. Two-component mixture: structural zeros (always zero) and sampling zeros (from Poisson process). Test $H_0: \pi = 0$ using van den Broek score test (GLMS.11).	Combines zero-inflation with overdispersion. Preferred over ZIP when both excess zeros and overdispersion are present. More parameters require larger samples for stable estimation.

References

[1] Abdi, H., & Molin, P. (2007). Lilliefors/van soest's test of normality. *Encyclopedia of Measurement and Statistics*, 1–5.

[2] Agresti, A. (1992). A survey of exact inference for contingency tables. *Statistical Science*, *7*(1), 131–153.

[3] Agresti, A. (2013). *Categorical data analysis* (3rd ed.). Wiley.

[4] Agresti, A., & Caffo, B. (2000). Simple and effective confidence intervals for proportions and differences of proportions result from adding two successes and two failures. *The American Statistician*, *54*(4), 280–288.

[5] Agresti, A., & Coull, B. A. (1998). Approximate is better than "exact" for interval estimation of binomial proportions. *The American Statistician*, *52*(2), 119–126.

[6] Anderson, M. J. (2001). A new method for non-parametric multivariate analysis of variance. *Austral Ecology*, *26*(1), 32–46.

[7] Anderson, M. J., & ter Braak, C. J. F. (2003). Permutation tests for multifactorial analysis of variance. *Journal of Statistical Computation and Simulation*, *73*(2), 85–113.

[8] Anderson, S., & Hauck, W. W. (1983). A new procedure for testing equivalence in comparative bioavailability and other clinical trials. *Communications in Statistics - Theory and Methods*, *12*(23), 2663–2692.

[9] Anderson, T. W. (1962). On the distribution of the two-sample cramér–von mises criterion. *The Annals of Mathematical Statistics*, *33*(3), 1148–1159.

[10] Anderson, T. W., & Darling, D. A. (1952). Asymptotic theory of certain "goodness of fit" criteria based on stochastic processes. *The Annals of Mathematical Statistics*, *23*(2), 193–212.

[11] Anderson, T. W. (2003). *An introduction to multivariate statistical analysis* (3rd ed.). Wiley.

[12] Ansari, A. R., & Bradley, R. A. (1960). Rank-sum tests for dispersions. *Annals of Mathematical Statistics*, *31*(4), 1174–1189.

[13] Anscombe, F. J., & Glynn, W. J. (1983). Distribution of the kurtosis statistic b_2 for normal samples. *Biometrika*, *70*(1), 227–234.

[14] Anselin, L. (1988). *Spatial econometrics: Methods and models*. Kluwer Academic Publishers.

[15] Anselin, L. (1995). Local indicators of spatial association—lisa. *Geographical Analysis*, *27*(2), 93–115.

[16] Anselin, L., & Bera, A. K. (1998). Spatial dependence in linear regression models with an introduction to spatial econometrics. In A. Ullah & D. E. A. Giles (Eds.), *Handbook of applied economic statistics* (pp. 237–289). Marcel Dekker.

[17] Anselin, L., Bera, A. K., Florax, R., & Yoon, M. J. (1996). Simple diagnostic tests for spatial dependence. *Regional Science and Urban Economics*, *26*(1), 77–104.

[18] Arbuthnot, J. (1710). An argument for divine providence, taken from the constant regularity observed in the births of both sexes. *Philosophical Transactions of the Royal Society of London*, *27*, 186–190.

[19] Armitage, P. (1955). Tests for linear trends in proportions and frequencies. *Biometrics*, *11*(3), 375–386.

[20] ASTM International. (2016). Standard practice for dealing with outlying observations (astm e178-16a).

[21] Banerjee, A., & Carrion-i-Silvestre, J. L. (2017). Testing for panel cointegration using common correlated effects estimators. *Journal of Time Series Analysis*, *38*(4), 610–636.

[22] Barnard, G. A. (1945). A new test for 2×2 tables. *Nature*, *156*, 177.

[23] Barnard, G. A. (1959). Control charts and stochastic processes. *Journal of the Royal Statistical Society. Series B (Methodological)*, *21*(2), 239–271.

[24] Barnett, V., & Lewis, T. (1994). *Outliers in statistical data* (3rd ed.). Wiley.

[25] Bartels, R. (1982). The rank version of von neumann's ratio test for randomness. *Journal of the American Statistical Association*, *77*(377), 40–46.

[26] Bartlett, M. S. (1937). Properties of sufficiency and statistical tests. *Proceedings of the Royal Society of London. Series A*, *160*(901), 268–282.

[27] Bartlett, M. S. (1950). Tests of significance in factor analysis. *British Journal of Psychology, Statistical Section*, *3*(2), 77–85.

[28] Bartlett, M. S. (1951). The effect of standardization on a χ^2 approximation in factor analysis. *Biometrika*, *38*(3/4), 337–344.

[29] Barton, D. E., & David, F. N. (1957). Multiple rank tests and their applications. *Biometrika*, *44*(3/4), 295–308.

[30] Bayes, T. (1763). An essay towards solving a problem in the doctrine of chances. *Philosophical Transactions of the Royal Society of London*, *53*, 370–418.

[31] Beaulieu, J. J., & Miron, J. A. (1993). Seasonal unit roots in aggregate u.s. data. *Journal of Econometrics*, *55*(1–2), 305–328.

[32] Belsley, D. A., Kuh, E., & Welsch, R. E. (1980). Regression diagnostics: Identifying influential data and sources of collinearity. *Wiley*.

[33] Benjamini, Y., & Hochberg, Y. (1995). Controlling the false discovery rate: A practical and powerful approach to multiple testing. *Journal of the Royal Statistical Society: Series B (Methodological)*, *57*(1), 289–300.

[34] Benjamini, Y., & Yekutieli, D. (2001). The control of the false discovery rate in multiple testing under dependency. *The Annals of Statistics*, *29*(4), 1165–1188.

[35] Berger, R. L., & Hsu, J. C. (1996). Bioequivalence trials, intersection-union tests and equivalence confidence sets. *Statistical Science*, *11*(4), 283–319.

[36] Bergsma, W. (2013). A bias-correction for cramér's v and tschuprow's t. *Journal of the Korean Statistical Society*, *42*(3), 323–328.

[37] Bernoulli, J. (1713). *Ars conjectandi*. Thurneysen Brothers.

[38] Bickel, P. J., & Breiman, L. (1983). Sums of functions of nearest neighbor distances, moment bounds, limit theorems and a goodness-of-fit test. *The Annals of Probability*, *11*(1), 185–214.

[39] Bingham, C., & Nelson, L. (1981). An approximation for the distribution of the von neumann ratio. *Technometrics*, *23*(1), 85–89.

[40] Bishop, Y. M. M., Fienberg, S. E., & Holland, P. W. (1975). *Discrete multivariate analysis: Theory and practice*. MIT Press.

[41] Black, F. (1976). Studies of stock price volatility changes. In *Proceedings of the 1976 meetings of the american statistical association, business and economic statistics section* (pp. 177–181). American Statistical Association.

[42] Blum, J. R., Kiefer, J., & Rosenblatt, M. (1961). Distribution free tests of independence based on the sample distribution function. *Annals of Mathematical Statistics*, *32*(2), 485–498.

[43] Bollerslev, T. (1986). Generalized autoregressive conditional heteroskedasticity. *Journal of Econometrics*, *31*(3), 307–327.

[44] Boschloo, R. D. (1970). Raised conditional level of significance for the 2 × 2-table when testing the equality of two probabilities. *Statistica Neerlandica*, *24*(1), 1–9.

[45] Bowker, A. H. (1948). A test for symmetry in contingency tables. *Journal of the American Statistical Association*, *43*(244), 572–574.

[46] Bowman, K. O., & Shenton, L. R. (1975). Omnibus tests for univariate and multivariate normality. *Biometrika*, *62*(2), 445–448.

[47] Box, G. E. P. (1949). A general distribution theory for a class of likelihood criteria. *Biometrika*, *36*(3/4), 317–346.

[48] Box, G. E. P. (1953). Non-normality and tests on variances. *Biometrika*, *40*(3/4), 318–335.

[49] Box, G. E. P. (1954). Some theorems on quadratic forms applied in the study of analysis of variance problems, i. effect of inequality of variance in the one-way classification. *The Annals of Mathematical Statistics*, *25*(2), 290–302.

[50] Box, G. E. P., & Cox, D. R. (1964). An analysis of transformations. *Journal of the Royal Statistical Society: Series B*, *26*(2), 211–252.

[51] Box, G. E. P., & Pierce, D. A. (1970). Distribution of residual autocorrelations in autoregressive-integrated moving average time series models. *Journal of the American Statistical Association*, *65*(332), 1509–1526.

[52] Box, G. E. P., & Tidwell, P. W. (1962). Transformation of the independent variables. *Technometrics*, *4*(4), 531–550.

[53] Bradley, J. V. (1968). *Distribution-free statistical tests*. Prentice-Hall.

[54] Breitung, J. (2000). The local power of some unit root tests for panel data. *Advances in Econometrics*, *15*, 161–177.

[55] Breslow, N. E. (1970). A generalized kruskal-wallis test for comparing K samples subject to unequal patterns of censorship. *Biometrika*, *57*(3), 579–594.

[56] Breslow, N. E. (1974). Covariance analysis of censored survival data. *Biometrics*, *30*(1), 89–99.

[57] Breslow, N. E. (1984). Extra-poisson variation in log-linear models. *Applied Statistics*, *33*(1), 38–44.

[58] Breslow, N. E., & Day, N. E. (1980). *Statistical methods in cancer research, volume 1: The analysis of case-control studies*. IARC Scientific Publications.

[59] Breusch, T. S. (1978). Testing for autocorrelation in dynamic linear models. *Australian Economic Papers, 17*(31), 334–355.

[60] Breusch, T. S., & Pagan, A. R. (1979). A simple test for heteroscedasticity and random coefficient variation. *Econometrica, 47*(5), 1287–1294.

[61] Breusch, T. S., & Pagan, A. R. (1980). The lagrange multiplier test and its applications to model specification in econometrics. *The Review of Economic Studies, 47*(1), 239–253.

[62] Brock, W. A., Dechert, W. D., & Scheinkman, J. A. (1987). *A test for independence based on the correlation dimension* (tech. rep.). Department of Economics, University of Wisconsin at Madison.

[63] Brock, W. A., Dechert, W. D., Scheinkman, J. A., & LeBaron, B. (1996). A test for independence based on the correlation dimension. *Econometric Reviews, 15*(3), 197–235.

[64] Brown, G. W., & Mood, A. (1951). On median tests for linear hypotheses. *Proceedings of the Second Berkeley Symposium on Mathematical Statistics and Probability*, 159–166.

[65] Brown, L. D., Cai, T. T., & DasGupta, A. (2001). Interval estimation for a binomial proportion. *Statistical Science, 16*(2), 101–133.

[66] Brown, L. D., & Li, X. (2005). Confidence intervals for two-sample binomial distribution. *Journal of Statistical Planning and Inference, 130*(1–2), 359–375.

[67] Brown, M. B., & Forsythe, A. B. (1974). Robust tests for the equality of variances. *Journal of the American Statistical Association, 69*(346), 364–367.

[68] Brunner, E., Domhof, S., & Langer, F. (2002). *Nonparametric analysis of longitudinal data in factorial experiments*. John Wiley & Sons.

[69] Brunner, E., & Munzel, U. (2000). The nonparametric behrens-fisher problem: Asymptotic theory and a small-sample approximation. *Biometrical Journal, 42*(1), 17–25.

[70] Cameron, A. C., & Trivedi, P. K. (1990). Regression-based tests for overdispersion in the poisson model. *Journal of Business & Economic Statistics, 8*(4), 411–418.

[71] Canova, F., & Hansen, B. E. (1995). Are seasonal patterns constant over time? a test for seasonal stability. *Journal of Business & Economic Statistics, 13*(3), 237–252.

[72] Champ, C. W., & Woodall, W. H. (1987). Exact results for shewhart control charts with supplementary runs rules. *Technometrics, 29*(4), 393–399.

[73] Chatterjee, S., & Hadi, A. S. (1988). *Sensitivity analysis in linear regression*. John Wiley & Sons.

[74] Chauvenet, W. (1863). *A manual of spherical and practical astronomy* (Vol. 2). Lippincott.

[75] Cheng, S., & Long, J. S. (2007). Testing for iia in the multinomial logit model. *Sociological Methods & Research, 35*(4), 583–600.

[76] Chernoff, H. (1954). On the distribution of the likelihood ratio. *The Annals of Mathematical Statistics, 25*(3), 573–578.

[77] Cheung, Y.-W., & Lai, K. S. (1993). Finite-sample sizes of johansen's likelihood ratio tests for cointegration. *Oxford Bulletin of Economics and Statistics, 55*(3), 313–328.

[78] Chiang, C. L. (1968). *Introduction to stochastic processes in biostatistics*. John Wiley & Sons.

[79] Chib, S. (1995). Marginal likelihood from the gibbs output. *Journal of the American Statistical Association, 90*(432), 1313–1321.

[80] Choi, I. (2001). Unit root tests for panel data. *Journal of International Money and Finance, 20*(2), 249–272.

[81] Chow, G. C. (1960). Tests of equality between sets of coefficients in two linear regressions. *Econometrica, 28*(3), 591–605.

[82] Clark, T. E., & West, K. D. (2007). Approximately normal tests for equal predictive accuracy in nested models. *Journal of Econometrics, 138*(1), 291–311.

[83] Clarke, K. A. (2007). A simple distribution-free test for nonnested model selection. *Political Analysis, 15*(3), 347–363.

[84] Clemente, J., Montañés, A., & Reyes, M. (1998). Testing for a unit root in variables with a double change in the mean. *Economics Letters, 59*(2), 175–182.

[85] Cliff, A. D., & Ord, J. K. (1981). *Spatial processes: Models and applications*. Pion.

[86] Clopper, C. J., & Pearson, E. S. (1934). The use of confidence or fiducial limits illustrated in the case of the binomial. *Biometrika, 26*(4), 404–413.

[87] Cochran, W. G. (1952). The χ^2 test of goodness of fit. *The Annals of Mathematical Statistics, 23*(3), 315–345.

[88] Cochran, W. G. (1957). Analysis of covariance: Its nature and uses. *Biometrics, 13*(3), 261–281.

[89] Cochran, W. G. (1950). The comparison of percentages in matched samples. *Biometrika, 37*(3/4), 256–266.

[90] Cochran, W. G. (1954). Some methods for strengthening the common χ^2 tests. *Biometrics, 10*(4), 417–451.

[91] Cohen, J. (1960). A coefficient of agreement for nominal scales. *Educational and Psychological Measurement, 20*(1), 37–46.

[92] Cohen, J. (1968). Weighted kappa: Nominal scale agreement with provision for scaled disagreement or partial credit. *Psychological Bulletin, 70*(4), 213–220.

[93] Cohen, J. (1988). *Statistical power analysis for the behavioral sciences* (2nd ed.). Erlbaum.

[94] Collett, D. (2015). *Modelling survival data in medical research* (3rd ed.). CRC Press.

[95] Conover, W. J. (1999). *Practical nonparametric statistics* (3rd ed.). Wiley.

[96] Conover, W. J., Johnson, M. E., & Johnson, M. M. (1981). A comparative study of tests for homogeneity of variances, with applications to the outer continental shelf bidding data. *Technometrics, 23*(4), 351–361.

[97] Cook, J. D. (2008). Adaptive designs: A review of the literature. *Journal of Biopharmaceutical Statistics, 18*(6), 1047–1084.

[98] Cook, R. D. (1977). Detection of influential observation in linear regression. *Technometrics, 19*(1), 15–18.

[99] Cook, R. D. (1979). Influential observations in linear regression. *Journal of the American Statistical Association, 74*(365), 169–174.

[100] Copas, J. B., & Hosmer, D. W. (1989). Influence and sensitivity analysis in logistic regression. *Journal of the American Statistical Association, 84*(406), 654–664.

[101] Cox, D. R. (1972). Regression models and life-tables. *Journal of the Royal Statistical Society, Series B, 34*(2), 187–220.

[102] Cox, D. R. (1975). Partial likelihood. *Biometrika, 62*(2), 269–276.

[103] Cox, D. R., & Snell, E. J. (1968). A general definition of residuals. *Journal of the Royal Statistical Society, Series B, 30*(2), 248–275.

[104] Cox, D. R., & Stuart, A. (1955). Some quick sign tests for trend in location and dispersion. *Biometrika, 42*(1/2), 80–95.

[105] Cramér, H. (1928). On the composition of elementary errors. *Scandinavian Actuarial Journal, 1928*(1), 13–74.

[106] Cramér, H. (1946). *Mathematical methods of statistics*. Princeton University Press.

[107] D'Agostino, R. B. (1970). Transformation to normality of the null distribution of g_1. *Biometrika, 57*(3), 679–681.

[108] D'Agostino, R. B. (1971). An omnibus test of normality for moderate and large sample sizes. *Biometrika, 58*(2), 341–348.

[109] D'Agostino, R. B., Belanger, A., & D'Agostino, J., Ralph B. (1990). A suggestion for using powerful and informative tests of normality. *The American Statistician, 44*(4), 316–321.

[110] D'Agostino, R. B., & Pearson, E. S. (1973). Tests for departure from normality: Empirical results for the distributions of b_2 and $\sqrt{b_1}$. *Biometrika, 60*(3), 613–622.

[111] D'Agostino, R. B., & Stephens, M. A. (Eds.). (1986). *Goodness-of-fit techniques*. Marcel Dekker.

[112] Dallal, G. E., & Wilkinson, L. (1986). An analytic approximation to the distribution of lilliefors's test statistic for normality. *The American Statistician, 40*(4), 294–296.

[113] David, H. A. (1952). Upper 5 and 1% points of the maximum f-ratio. *Biometrika, 39*(3/4), 422–424.

[114] Davidson, R., & MacKinnon, J. G. (2004). *Econometric theory and methods*. Oxford University Press.

[115] Davison, A. C., & Hinkley, D. V. (1997). *Bootstrap methods and their application*. Cambridge University Press.

[116] de Moivre, A. (1738). *The doctrine of chances: Or, a method of calculating the probability of events in play* (2nd ed.). W. Pearson.

[117] Dean, C. B. (1992). Testing for overdispersion in poisson and binomial regression models. *Journal of the American Statistical Association, 87*(418), 451–457.

[118] Dean, R. B., & Dixon, W. J. (1951). Simplified statistics for small numbers of observations. *Analytical Chemistry, 23*(4), 636–638.

[119] DiCiccio, T. J., & Efron, B. (1996). Bootstrap confidence intervals. *Statistical Science, 11*(3), 189–228.

[120] Dickey, D. A., & Fuller, W. A. (1979). Distribution of the estimators for autoregressive time series with a unit root. *Journal of the American Statistical Association, 74*(366), 427–431.

[121] Dickey, J. M. (1971). The weighted likelihood ratio, linear hypotheses on normal location parameters. *Annals of Mathematical Statistics, 42*(1), 204–223.

[122] Diebold, F. X., & Mariano, R. S. (1995). Comparing predictive accuracy. *Journal of Business & Economic Statistics, 13*(3), 253–263.

[123] Dixon, W. J. (1950). Analysis of extreme values. *Annals of Mathematical Statistics, 21*(4), 488–506.

[124] Dixon, W. J. (1951). Ratios involving extreme values. *Annals of Mathematical Statistics, 22*(1), 68–78.

[125] Dixon, W. J., & Mood, A. M. (1946). The statistical sign test. *Journal of the American Statistical Association, 41*(236), 557–566.

[126] Duncan, A. J. (1986). *Quality control and industrial statistics* (5th). Irwin.

[127] Dunn, O. J., & Clark, V. A. (1969). Correlation coefficients measured on the same individuals. *Journal of the American Statistical Association, 64*(325), 366–377.

[128] Dunnett, C. W. (1955). A multiple comparison procedure for comparing several treatments with a control. *Journal of the American Statistical Association, 50*(272), 1096–1121.

[129] Dunnett, C. W. (1964). New tables for multiple comparisons with a control. *Biometrics, 20*(3), 482–491.

[130] Durbin, J., & Watson, G. S. (1950). Testing for serial correlation in least squares regression i. *Biometrika, 37*(3/4), 409–428.

[131] Durbin, J., & Watson, G. S. (1951). Testing for serial correlation in least squares regression. ii. *Biometrika, 38*(1/2), 159–178.

[132] Dwass, M. (1957). Modified randomization tests for nonparametric hypotheses. *Annals of Mathematical Statistics, 28*(1), 181–187.

[133] Edgington, E. S. (1995). *Randomization tests* (3rd). Marcel Dekker.

[134] Edwards, A. L. (1948). Note on the "correction for continuity" in testing the significance of the difference between correlated proportions. *Psychometrika, 13*(3), 185–187.

[135] Efron, B. (1979). Bootstrap methods: Another look at the jackknife. *Annals of Statistics, 7*(1), 1–26.

[136] Efron, B. (1982). *The jackknife, the bootstrap and other resampling plans* (Vol. 38). Society for Industrial; Applied Mathematics.

[137] Efron, B. (1987). Better bootstrap confidence intervals. *Journal of the American Statistical Association, 82*(397), 171–185.

[138] Efron, B., Hastie, T., Johnstone, I., & Tibshirani, R. (2004a). Least angle regression. *The Annals of Statistics, 32*(2), 407–499.

[139] Efron, B., Hastie, T., Johnstone, I., & Tibshirani, R. (2004b). Least angle regression. *The Annals of Statistics, 32*(2), 407–499.

[140] Efron, B., & Tibshirani, R. J. (1993). *An introduction to the bootstrap*. Chapman & Hall.

[141] Elkin, L. A., Kay, M., Higgins, J. J., & Wobbrock, J. O. (2021). An aligned rank transform procedure for multifactor contrast tests. *Proceedings of the 34th Annual ACM Symposium on User Interface Software and Technology*, 754–768.

[142] Elliott, G., Rothenberg, T. J., & Stock, J. H. (1996). Efficient tests for an autoregressive unit root. *Econometrica, 64*(4), 813–836.

[143] Enders, W. (2015). *Applied econometric time series* (4th ed.). John Wiley & Sons.

[144] Engle, R. F. (1982). Autoregressive conditional heteroscedasticity with estimates of the variance of united kingdom inflation. *Econometrica, 50*(4), 987–1007.

[145] Engle, R. F., & Granger, C. W. J. (1987). Co-integration and error correction: Representation, estimation, and testing. *Econometrica, 55*(2), 251–276.

[146] Engle, R. F., & Ng, V. K. (1993). Measuring and testing the impact of news on volatility. *Journal of Finance, 48*(5), 1749–1778.

[147] Ernst, M. D. (2004). Permutation methods: A basis for exact inference. *Statistical Science, 19*(4), 676–685.

[148] Farrington, C. P., & Manning, G. (1990). Test statistics and sample size formulae for comparative binomial trials with null hypothesis of non-zero risk difference or non-unity relative risk. *Statistics in Medicine, 9*(12), 1447–1454.

[149] Fine, J. P., & Gray, R. J. (1999). A proportional hazards model for the subdistribution of a competing risk. *Journal of the American Statistical Association, 94*(446), 496–509.

[150] Fisher, R. A. (1915). Frequency distribution of the values of the correlation coefficient in samples from an indefinitely large population. *Biometrika, 10*(4), 507–521.

[151] Fisher, R. A. (1918). The correlation between relatives on the supposition of mendelian inheritance. *Transactions of the Royal Society of Edinburgh, 52*(2), 399–433.

[152] Fisher, R. A. (1921). On the "probable error" of a coefficient of correlation deduced from a small sample. *Metron, 1*, 3–32.

[153] Fisher, R. A. (1924a). The distribution of the partial correlation coefficient. *Metron, 3*(3–4), 329–332.

[154] Fisher, R. A. (1924b). On a distribution yielding the error functions of several well known statistics. *Proceedings of the International Congress of Mathematics, 2*, 805–813.

[155] Fisher, R. A. (1932). *Statistical methods for research workers* (4th ed.). Oliver & Boyd.

[156] Fisher, R. A. (1935). *The design of experiments*. Oliver & Boyd.

[157] Fleiss, J. L. (1971). Measuring nominal scale agreement among many raters. *Psychological Bulletin, 76*(5), 378–382.

[158] Fleiss, J. L., Levin, B., & Paik, M. C. (2003). *Statistical methods for rates and proportions* (3rd ed.). Wiley.

[159] Fleming, T. R., & Harrington, D. P. (1981). A class of hypothesis tests for one and two sample censored survival data. *Communications in Statistics - Theory and Methods, 10*(8), 763–794.

[160] Fleming, T. R., & Harrington, D. P. (1991). *Counting processes and survival analysis*. Wiley.

[161] Fligner, M. A., & Killeen, T. J. (1976). Distribution-free two-sample tests for scale. *Journal of the American Statistical Association, 71*(353), 210–213.

[162] Florax, R. J. G. M., Folmer, H., & Rey, S. J. (2003). Specification searches in spatial econometrics: The relevance of hendry's methodology. *Regional Science and Urban Economics, 33*(5), 557–579.

[163] Franses, P. H. (1996). *Periodicity and stochastic trends in economic time series*. Oxford University Press.

[164] Franses, P. H., & Hobijn, B. (1997). Critical values for unit root tests in seasonal time series. *Journal of Applied Statistics, 24*(1), 25–47.

[165] Friedman, M. (1937). The use of ranks to avoid the assumption of normality implicit in the analysis of variance. *Journal of the American Statistical Association, 32*(200), 675–701.

[166] Fry, T. R. L., & Harris, M. N. (1996). A monte carlo study of tests for the independence of irrelevant alternatives property. *Transportation Research Part B: Methodological, 30*(1), 19–30.

[167] Fry, T. R. L., & Harris, M. N. (1998). Testing for independence of irrelevant alternatives: Some empirical results. *Sociological Methods & Research, 26*(3), 401–423.

[168] Gabry, J., Simpson, D., Vehtari, A., Betancourt, M., & Gelman, A. (2019). Visualization in bayesian workflow. *Journal of the Royal Statistical Society: Series A, 182*(2), 389–402.

[169] Gail, M. H. (1975). A review and critique of some models used in competing risk analysis. *Biometrics, 31*(1), 209–222.

[170] Games, P. A., & Howell, J. F. (1976). Pairwise multiple comparison procedures with unequal n's and/or variances: A monte carlo study. *Journal of Educational Statistics, 1*(2), 113–125.

[171] Gavaghan, D. J., Moore, R. A., & McQuay, H. J. (1996). An evaluation of homogeneity tests in meta-analyses in pain using simulations of individual patient data. *Pain, 66*(2–3), 273–281.

[172] Geary, R. C. (1954). The contiguity ratio and statistical mapping. *The Incorporated Statistician, 5*(3), 115–145.

[173] Gehan, E. A. (1965). A generalized wilcoxon test for comparing arbitrarily singly-censored samples. *Biometrika, 52*(1–2), 203–223.

[174] Gelman, A., & Loken, E. (2013). The garden of forking paths: Why multiple comparisons can be a problem, even when there is no "fishing expedition" or "p-hacking" and the research hypothesis was posited ahead of time [Technical Report]. *Department of Statistics, Columbia University*.

[175] Gelman, A., Meng, X.-L., & Stern, H. (1996). Posterior predictive assessment of model fitness via realized discrepancies. *Statistica Sinica, 6*(4), 733–760.

[176] Getis, A., & Ord, J. K. (1992). The analysis of spatial association by use of distance statistics. *Geographical Analysis, 24*(3), 189–206.

[177] Getis, A., & Ord, J. K. (2010). The analysis of spatial association. In M. M. Fischer & A. Getis (Eds.), *Handbook of applied spatial analysis* (pp. 255–278). Springer.

[178] Gibbons, J. D., & Chakraborti, S. (2011). *Nonparametric statistical inference* (5th ed.). CRC Press.

[179] Giesbrecht, F. G., & Burns, J. C. (1985). Two-stage analysis based on a mixed model: Large-sample asymptotic theory and small-sample simulation results. *Biometrics, 41*(2), 477–486.

[180] Glejser, H. (1969). A new test for heteroskedasticity. *Journal of the American Statistical Association, 64*(325), 316–323.

[181] Glosten, L. R., Jagannathan, R., & Runkle, D. E. (1993). On the relation between the expected value and the volatility of the nominal excess return on stocks. *Journal of Finance, 48*(5), 1779–1801.

[182] Godfrey, L. G. (1978a). Testing against general autoregressive and moving average error models when regressors include lagged dependent variables. *Econometrica*, *46*(6), 1293–1301.

[183] Godfrey, L. G. (1978b). Testing for multiplicative heteroskedasticity. *Journal of Econometrics*, *8*(2), 227–236.

[184] Goldfeld, S. M., & Quandt, R. E. (1965). Some tests for homoscedasticity. *Journal of the American Statistical Association*, *60*(310), 539–547.

[185] Good, P. I. (2005). *Permutation, parametric and bootstrap tests of hypotheses* (3rd ed.). Springer.

[186] Gosset, W. S. (1942). Comparison between balanced and random arrangements of field plots. *Biometrika*, *29*(3/4), 363–378.

[187] Gould, B. A. (1855). On peirce's criterion for the rejection of doubtful observations, with tables for facilitating its application. *Astronomical Journal*, *4*, 81–87.

[188] Grambsch, P. M., & Therneau, T. M. (1994). Proportional hazards tests and diagnostics based on weighted residuals. *Biometrika*, *81*(3), 515–526.

[189] Grant, E. L., & Leavenworth, R. S. (1996). *Statistical quality control* (7th ed.). McGraw-Hill.

[190] Gray, R. J. (1988). A class of K-sample tests for comparing the cumulative incidence of a competing risk. *Annals of Statistics*, *16*(3), 1141–1154.

[191] Greenhouse, S. W., & Geisser, S. (1959). On methods in the analysis of profile data. *Psychometrika*, *24*(2), 95–112.

[192] Greenwood, M. (1946). The statistical study of infectious diseases. *Journal of the Royal Statistical Society*, *109*(2), 85–110.

[193] Gregory, A. W., & Hansen, B. E. (1996). Residual-based tests for cointegration in models with regime shifts. *Journal of Econometrics*, *70*(1), 99–126.

[194] Gronau, Q. F., Sarafoglou, A., Matzke, D., Ly, A., Boehm, U., Marsman, M., et al. (2017). A tutorial on bridge sampling. *Journal of Mathematical Psychology*, *81*, 80–97.

[195] Gronau, Q. F., Singmann, H., & Wagenmakers, E.-J. (2020). Bridgesampling: An r package for estimating normalizing constants. *Journal of Statistical Software*, *92*(10), 1–29.

[196] Grubbs, F. E. (1950). Sample criteria for testing outlying observations. *Annals of Mathematical Statistics*, *21*(1), 27–58.

[197] Grubbs, F. E. (1969). Procedures for detecting outlying observations in samples. *Technometrics*, *11*(1), 1–21.

[198] Gu, K., Ng, H. K. T., Tang, M.-L., & Schucany, W. R. (2008). Testing the ratio of two poisson rates. *Biometrical Journal*, *50*(2), 283–298.

[199] Guilford, J. P. (1954). *Psychometric methods* (2nd ed.). McGraw-Hill.

[200] Gwet, K. L. (2002). Inter-rater reliability: Dependency on trait prevalence and marginal homogeneity. *Statistical Methods for Inter-Rater Reliability Assessment*, *2*, 1–9.

[201] Gwet, K. L. (2008). Computing inter-rater reliability and its variance in the presence of high agreement. *British Journal of Mathematical and Statistical Psychology*, *61*(1), 29–48.

[202] Gwet, K. L. (2014). *Handbook of inter-rater reliability: The definitive guide to measuring the extent of agreement among raters* (4th ed.). Advanced Analytics, LLC.

[203] Hall, P. (1988). Theoretical comparison of bootstrap confidence intervals. *The Annals of Statistics*, *16*(3), 927–953.

[204] Hall, P., & Wilson, S. R. (1991). Two guidelines for bootstrap hypothesis testing. *Biometrics*, *47*(2), 757–762.

[205] Hamed, K. H., & Rao, A. R. (1998). A modified mann–kendall trend test for autocorrelated data. *Journal of Hydrology*, *204*(1–4), 182–196.

[206] Harris, R. D. F., & Tzavalis, E. (1999). Inference for unit roots in dynamic panels where the time dimension is fixed. *Journal of Econometrics*, *91*(2), 201–226.

[207] Harrison, M. J., & McCabe, B. P. M. (1979). A test for heteroscedasticity based on ordinary least squares residuals. *Journal of the American Statistical Association*, *74*(366a), 494–499.

[208] Hart, B. I. (1942). The significance level of the ratio of the mean square successive difference to the variance. *The Annals of Mathematical Statistics*, *13*(4), 445–447.

[209] Hartley, H. O. (1950). The maximum f-ratio as a short-cut test for heterogeneity of variance. *Biometrika*, *37*(3/4), 308–312.

[210] Harvey, A. C., & Collier, P. (1977). Testing for functional form and autocorrelation in least squares regression. *Journal of the Royal Statistical Society: Series B (Methodological)*, *39*(1), 79–85.

[211] Harvey, A. C. (1976). Estimating regression models with multiplicative heteroscedasticity. *Econometrica*, *44*(3), 461–465.

[212] Harvey, D. I., Leybourne, S. J., & Newbold, P. (1997). Testing the equality of prediction mean squared errors. *International Journal of Forecasting*, *13*(2), 281–291.

[213] Hatemi-J, A. (2008). Tests for cointegration with two unknown regime shifts with an application to financial market integration. *Empirical Economics*, *35*(3), 497–505.

[214] Hauck, W. W., & Anderson, S. (1984). A new statistical procedure for testing equivalence in two-group comparative bioavailability trials. *Journal of Pharmacokinetics and Biopharmaceutics*, *12*(1), 83–91.

[215] Hauck, W. W., & Donner, A. (1977). Wald's test as applied to hypotheses in logit analysis. *Journal of the American Statistical Association*, *72*(360), 851–853.

[216] Hausman, J. A. (1978). Specification tests in econometrics. *Econometrica*, *46*(6), 1251–1271.

[217] Hawkins, D. M., & Olwell, D. H. (1998). *Cumulative sum charts and charting for quality improvement*. Springer.

[218] Hayes, A. F., & Krippendorff, K. (2007). Answering the call for a standard reliability measure for coding data. *Communication Methods and Measures*, *1*(1), 77–89.

[219] Heck, D. L. (1960). Charts of some upper percentage points of the distribution of the largest characteristic root. *Annals of Mathematical Statistics*, *31*(3), 625–642.

[220] Henze, N. (1994). A new consistent test for multivariate normality based on the empirical characteristic function. *Communications in Statistics - Theory and Methods*, *23*(4), 1035–1049.

[221] Henze, N., & Zirkler, B. (1990). A class of invariant consistent tests for multivariate normality. *Communications in Statistics - Theory and Methods*, *19*(10), 3595–3617.

[222] Higgins, J. J., Blair, R. C., & Tashtoush, S. (1990). The aligned rank transform procedure. *Proceedings of the Kansas State University Conference on Applied Statistics in Agriculture*, 185–195.

[223] Hirji, K. F. (2006). *Exact analysis of discrete data*. Chapman; Hall/CRC.

[224] Hirsch, R. M., & Slack, J. R. (1984). A nonparametric trend test for seasonal data with serial dependence. *Water Resources Research*, *20*(6), 727–732.

[225] Hirsch, R. M., Slack, J. R., & Smith, R. A. (1982). Techniques of trend analysis for monthly water quality data. *Water Resources Research*, *18*(1), 107–121.

[226] Hoaglin, D. C., & Welsch, R. E. (1978). The hat matrix in regression and anova. *The American Statistician*, *32*(1), 17–22.

[227] Hobijn, B., Franses, P. H., & Ooms, M. (2004). Generalizations of the kpss-test for stationarity. *Statistica Neerlandica*, *58*(4), 483–502.

[228] Hodges, J. L., & Lehmann, E. L. (1963). Estimates of location based on rank tests. *Annals of Mathematical Statistics*, *34*(2), 598–611.

[229] Hodges, J., J. L. (1957). The significance probability of the smirnov two-sample test. *Arkiv för Matematik*, *3*(5), 469–486.

[230] Hoeffding, W. (1948). A non-parametric test of independence. *Annals of Mathematical Statistics*, *19*(4), 546–557.

[231] Hoerl, A. E., & Kennard, R. W. (1970). Ridge regression: Biased estimation for nonorthogonal problems. *Technometrics*, *12*(1), 55–67.

[232] Hoeting, J. A., Madigan, D., Raftery, A. E., & Volinsky, C. T. (1999). Bayesian model averaging: A tutorial. *Statistical Science*, *14*(4), 382–401.

[233] Holgado-Tello, F. P., Chacón-Moscoso, S., Barbero-García, I., & Vila-Abad, E. (2010). Polychoric versus pearson correlations in exploratory and confirmatory factor analysis of ordinal variables. *Quality & Quantity*, *44*(1), 153–166.

[234] Hollander, M., Wolfe, D. A., & Chicken, E. (2014). *Nonparametric statistical methods* (3rd ed.). Wiley.

[235] Honda, Y. (1985). Testing the error components model with non-normal disturbances. *The Review of Economic Studies*, *52*(4), 681–690.

[236] Hope, A. C. A. (1968). A simplified monte carlo significance test procedure. *Journal of the Royal Statistical Society: Series B*, *30*(3), 582–598.

[237] Hosking, J. R. M. (1980). The multivariate portmanteau statistic. *Journal of the American Statistical Association*, *75*(371), 602–608.

[238] Hosmer, D. W., Hosmer, T., le Cessie, S., & Lemeshow, S. (1997). A comparison of goodness-of-fit tests for the logistic regression model. *Statistics in Medicine*, *16*(9), 965–980.

[239] Hosmer, D. W., & Lemeshow, S. (1980). Goodness of fit tests for the multiple logistic regression model. *Communications in Statistics - Theory and Methods*, *9*(10), 1043–1069.

[240] Hotelling, H. (1931). The generalization of student's ratio. *Annals of Mathematical Statistics*, *2*(3), 360–378.

[241] Hotelling, H. (1953). New light on the correlation coefficient and its transforms. *Journal of the Royal Statistical Society: Series B*, *15*(2), 193–232.

[242] Huber, P. J. (1964). Robust estimation of a location parameter. *Annals of Mathematical Statistics*, *35*(1), 73–101.

[243] Huber, P. J. (1981). *Robust statistics*. Wiley.

[244] Hunter, J. S. (1986). The exponentially weighted moving average. *Journal of Quality Technology*, *18*(4), 203–210.

[245] Huynh, H., & Feldt, L. S. (1976). Estimation of the box correction for degrees of freedom from sample data in randomized block and split-plot designs. *Journal of Educational Statistics*, *1*(1), 69–82.

[246] Hylleberg, S., Engle, R. F., Granger, C. W. J., & Yoo, B. S. (1990). Seasonal integration and cointegration. *Journal of Econometrics*, *44*(1–2), 215–238.

[247] Iglewicz, B., & Hoaglin, D. C. (1993). *Volume 16: How to detect and handle outliers*. ASQC Quality Press.

[248] Im, K. S., Pesaran, M. H., & Shin, Y. (2003). Testing for unit roots in heterogeneous panels. *Journal of Econometrics*, *115*(1), 53–74.

[249] Imbens, G. W., & Rubin, D. B. (2015). *Causal inference for statistics, social, and biomedical sciences*. Cambridge University Press.

[250] Inclán, C., & Tiao, G. C. (1994). Use of cumulative sums of squares for retrospective detection of changes of variance. *Journal of the American Statistical Association*, *89*(427), 913–923.

[251] International Conference on Harmonisation. (2000). *Choice of control group and related issues in clinical trials (ich e10)* (tech. rep.). International Conference on Harmonisation of Technical Requirements for Registration of Pharmaceuticals for Human Use. Geneva, Switzerland.

[252] Jaeschke, R., Singer, J., & Guyatt, G. H. (1989). Measurement of health status: Ascertaining the minimal clinically important difference. *Controlled Clinical Trials*, *10*(4), 407–415.

[253] James, G. S. (1951). The comparison of several groups of observations when the ratios of the population variances are unknown. *Biometrika*, *38*(3/4), 324–329.

[254] Jarque, C. M., & Bera, A. K. (1980). Efficient tests for normality, homoscedasticity and serial independence of regression residuals. *Economics Letters*, *6*(3), 255–259.

[255] Jarque, C. M., & Bera, A. K. (1987). A test for normality of observations and regression residuals. *International Statistical Review*, *55*(2), 163–172.

[256] Jeffreys, H. (1961). *Theory of probability* (3rd ed.). Oxford University Press.

[257] Johansen, S. (1980a). The welch–james approximation to the distribution of hotelling's T^2. *Biometrika*, *67*(1), 85–92.

[258] Johansen, S. (1980b). The welch–james approximation to the distribution of the residual sum of squares in a weighted linear regression. *Biometrika*, *67*(1), 85–92.

[259] Johansen, S. (1988). Statistical analysis of cointegration vectors. *Journal of Economic Dynamics and Control*, *12*(2–3), 231–254.

[260] Johansen, S. (1991). Estimation and hypothesis testing of cointegration vectors in gaussian vector autoregressive models. *Econometrica*, *59*(6), 1551–1580.

[261] John, S. (1971). Some optimal multivariate tests. *Biometrika*, *58*(1), 123–127.

[262] John, S. (1972). The distribution of a statistic used for testing sphericity of normal distributions. *Biometrika*, *59*(1), 169–173.

[263] Johnson, N. L., Kotz, S., & Kemp, A. W. (2005). *Univariate discrete distributions* (3rd ed.). Wiley.

[264] Johnson, R. A., & Wichern, D. W. (2015). *Applied multivariate statistical analysis* (6th ed.). Pearson.

[265] Jonckheere, A. R. (1954). A distribution-free k-sample test against ordered alternatives. *Biometrika*, *41*(1/2), 133–145.

[266] Kaiser, H. F. (1970). A second generation little jiffy. *Psychometrika*, *35*(4), 401–415.

[267] Kalbfleisch, J. D., & Prentice, R. L. (2002). *The statistical analysis of failure time data* (2nd ed.). Wiley.

[268] Kane, V. E. (1986). Process capability indices. *Journal of Quality Technology*, *18*(1), 41–52.

[269] Kao, C. (1999). Spurious regression and residual-based tests for cointegration in panel data. *Journal of Econometrics*, *90*(1), 1–44.

[270] Kapetanios, G., Shin, Y., & Snell, A. (2003). Testing for a unit root in the nonlinear star framework. *Journal of Econometrics*, *112*(2), 359–379.

[271] Kass, R. E., & Raftery, A. E. (1995). Bayes factors. *Journal of the American Statistical Association*, *90*(430), 773–795.

[272] Keenan, D. M. (1985). A tukey nonadditivity-type test for time series nonlinearity. *Biometrika*, *72*(1), 39–44.

[273] Kendall, M. G. (1938). A new measure of rank correlation. *Biometrika*, *30*(1/2), 81–93.

[274] Kendall, M. G. (1975). *Rank correlation methods* (4th ed.). Charles Griffin.

[275] Kendall, M. G., & Gibbons, J. D. (1990). *Rank correlation methods* (5th). Oxford University Press.

[276] Kendall, M. G., & Stuart, A. (1976). *The advanced theory of statistics, volume 3: Design and analysis, and time-series* (3rd ed.). Griffin.

[277] Kenward, M. G., & Roger, J. H. (1997). Small sample inference for fixed effects from restricted maximum likelihood. *Biometrics*, *53*(3), 983–997.

[278] King, M. L., & Wu, P. X. (1997). Locally optimal one-sided tests for multiparameter hypotheses. *Econometric Reviews*, *16*(2), 131–156.

[279] Klotz, J. (1962). Nonparametric tests for scale. *The Annals of Mathematical Statistics*, *33*(2), 498–512.

[280] Koenker, R. (1981). A note on studentizing a test for heteroscedasticity. *Journal of Econometrics*, *17*(1), 107–112.

[281] Kolmogorov, A. N. (1933). Sulla determinazione empirica di una legge di distribuzione [on the empirical determination of a distribution law]. *Giornale dell'Istituto Italiano degli Attuari*, *4*, 83–91.

[282] Kotz, S., & Johnson, N. L. (2002). Process capability indices—a review, 1992–2000. *Journal of Quality Technology*, *34*(1), 2–19.

[283] Kraemer, H. C. (2006). Correlation coefficients in medical research: From product moment correlation to the odds ratio. *Statistical Methods in Medical Research*, *15*(6), 525–545.

[284] Kramer, C. Y. (1956). Extension of multiple range tests to group means with unequal numbers of replications. *Biometrics*, *12*(3), 307–310.

[285] Krippendorff, K. (1970). Estimating the reliability, systematic error and random error of interval data. In *Educational and psychological measurement* (pp. 61–70).

[286] Krippendorff, K. (2018). *Content analysis: An introduction to its methodology* (4th ed.). SAGE Publications.

[287] Krishna Iyer, P. V. (1949). The distribution of the largest root of a determinantal equation in multivariate analysis. *Biometrika*, *36*(1/2), 107–121.

[288] Kruschke, J. K. (2011). Bayesian assessment of null values via parameter estimation and model comparison. *Perspectives on Psychological Science*, *6*(3), 299–312.

[289] Kruschke, J. K. (2013). Bayesian estimation supersedes the t test. *Journal of Experimental Psychology: General*, *142*(2), 573–603.

[290] Kruschke, J. K. (2018). Rejecting or accepting parameter values in bayesian estimation. *Advances in Methods and Practices in Psychological Science*, *1*(2), 270–280.

[291] Kruskal, W. H., & Wallis, W. A. (1952). Use of ranks in one-criterion variance analysis. *Journal of the American Statistical Association*, *47*(260), 583–621.

[292] Kuhfeld, W. F. (1986). A note on roy's largest root. *Psychometrika*, *51*(3), 479–481.

[293] Kuiper, N. H. (1960). Tests concerning random points on a circle. *Proceedings of the Koninklijke Nederlandse Akademie van Wetenschappen, Series A*, *63*, 38–47.

[294] Kundzewicz, Z. W., & Robson, A. J. (2004). Change detection in hydrological records—a review of the methodology. *Hydrological Sciences Journal*, *49*(1), 7–19.

[295] Kwiatkowski, D., Phillips, P. C. B., Schmidt, P., & Shin, Y. (1992). Testing the null hypothesis of stationarity against the alternative of a unit root: How sure are we that economic time series have a unit root? *Journal of Econometrics*, *54*(1–3), 159–178.

[296] Lambert, D. (1992). Zero-inflated poisson regression, with an application to defects in manufacturing. *Technometrics*, *34*(1), 1–14.

[297] Landis, J. R., & Koch, G. G. (1977). The measurement of observer agreement for categorical data. *Biometrics*, *33*(1), 159–174.

[298] Lawley, D. N. (1938). A generalization of fisher's z test. *Biometrika*, *30*(1/2), 180–187.

[299] le Cessie, S., & van Houwelingen, H. C. (1991). A goodness-of-fit test for binary regression models, based on smoothing methods. *Biometrics*, *47*(4), 1267–1282.

[300] le Cessie, S., & van Houwelingen, J. C. (1995). Testing the fit of a regression model via score tests in random effects models. *Biometrics*, *51*(2), 600–614.

[301] LeBaron, B. (1994). Nonlinear diagnostics and simple trading rules for high-frequency foreign exchange rates. *Journal of Business*, *67*(3), 331–356.

[302] Lee, J. D., Sun, D. L., Sun, Y., & Taylor, J. E. (2016a). Exact post-selection inference, with application to the lasso. *The Annals of Statistics*, *44*(3), 907–927.

[303] Lee, J. D., Sun, D. L., Sun, Y., & Taylor, J. E. (2016b). Exact post-selection inference, with application to the lasso. *The Annals of Statistics*, *44*(3), 907–927.

[304] Lee, J., & Strazicich, M. C. (2003). Minimum lagrange multiplier unit root test with two structural breaks. *Review of Economics and Statistics*, *85*(4), 1082–1089.

[305] Lee, J., & Strazicich, M. C. (2004). Minimum lm unit root test with one structural break. *Economics Bulletin*, *3*(33), 1–10.

[306] Lee, J., & Strazicich, M. C. (2013). Minimum lm unit root test with one structural break. *Economics Bulletin*, *33*(4), 2483–2492.

[307] Legendre, P., Oksanen, J., & ter Braak, C. F. J. (2015). Testing the significance of canonical axes in redundancy analysis. *Methods in Ecology and Evolution*, *6*(9), 1060–1069.

[308] Lehmann, E. L. (1986). *Testing statistical hypotheses* (2nd ed.). Wiley.

[309] Lehmann, E. L. (1975). *Nonparametrics: Statistical methods based on ranks*. McGraw-Hill.

[310] LeSage, J. P., & Pace, R. K. (2009). *Introduction to spatial econometrics*. CRC Press.

[311] Levene, H. (1960). Robust tests for equality of variances. In I. Olkin (Ed.), *Contributions to probability and statistics* (pp. 278–292). Stanford University Press.

[312] Levene, H., & Wolfowitz, J. (1944). The covariance matrix of runs up and down. *Annals of Mathematical Statistics*, *15*(1), 58–69.

[313] Levin, A., Lin, C.-F., & Chu, C.-S. J. (2002). Unit root tests in panel data: Asymptotic and finite-sample properties. *Journal of Econometrics*, *108*(1), 1–24.

[314] Li, W. K., & Mak, T.-K. (1994). On the squared residual autocorrelations in non-linear time series with conditional heteroskedasticity. *Journal of Time Series Analysis*, *15*(6), 627–636.

[315] Liang, F., Paulo, R., Molina, G., Clyde, M. A., & Berger, J. O. (2008). Mixtures of g priors for bayesian variable selection. *Journal of the American Statistical Association*, *103*(481), 410–423.

[316] Lilliefors, H. W. (1967). On the kolmogorov–smirnov test for normality with mean and variance unknown. *Journal of the American Statistical Association*, *62*(318), 399–402.

[317] Lilliefors, H. W. (1969). On the kolmogorov–smirnov test for the exponential distribution with mean unknown. *Journal of the American Statistical Association*, *64*(325), 387–389.

[318] Lin, R. S., Lin, J., Roychoudhury, S., Anderson, K. M., Hu, T., Huang, B., Leon, L. F., Liao, J. J. Z., Liu, R., Luo, X., Mukhopadhyay, P., Qin, R., Tatsuoka, K., Wang, X., Wang, Y., Zhu, J., Chen, T.-T., & Iacona, R. (2021). Alternative analysis methods for time-to-event endpoints under nonproportional hazards: A comparative analysis. *Statistics in Biopharmaceutical Research*, *13*(4), 613–622.

[319] Ljung, G. M., & Box, G. E. P. (1978). On a measure of lack of fit in time series models. *Biometrika*, *65*(2), 297–303.

[320] Lucas, J. M. (1976). The design and use of v-mask control schemes. *Journal of Quality Technology*, *8*(1), 1–12.

[321] Lucas, J. M., & Crosier, R. B. (1982). Fast initial response for cusum quality-control schemes: Give your cusum a head start. *Technometrics*, *24*(3), 199–205.

[322] Lucas, J. M., & Saccucci, M. S. (1990). Exponentially weighted moving average control schemes: Properties and enhancements. *Technometrics*, *32*(1), 1–12.

[323] Ly, A., Verhagen, J., & Wagenmakers, E.-J. (2016). Harold jeffreys's default bayes factor hypotheses: Explanation, extension, and application in psychology. *Journal of Mathematical Psychology*, *72*, 19–32.

[324] Machado, J. A. F., & Santos Silva, J. M. C. (2000). Glewwe's kernel estimator for censored data: An application to household expenditure data. *Journal of Applied Econometrics*, *15*(3), 253–274.

[325] MacKinnon, J. G., & White, H. (1985). Some heteroskedasticity-consistent covariance matrix estimators with improved finite sample properties. *Journal of Econometrics*, *29*(3), 305–325.

[326] MacKinnon, J. G. (1991). Critical values for cointegration tests. In R. F. Engle & C. W. J. Granger (Eds.), *Long-run economic relationships: Readings in cointegration* (pp. 267–276). Oxford University Press.

[327] MacKinnon, J. G. (1996). Numerical distribution functions for unit root and cointegration tests. *Journal of Applied Econometrics*, *11*(6), 601–618.

[328] MacKinnon, J. G. (2010). Critical values for cointegration tests. *Queen's Economics Department Working Paper*, (1227).

[329] Maddala, G. S., & Wu, S. (1999). A comparative study of unit root tests with panel data and a new simple test. *Oxford Bulletin of Economics and Statistics*, *61*(S1), 631–652.

[330] Magirr, D. (2023). The maxcombo test severely violates the type i error rate. *JAMA Oncology*, *9*(4), 589–590.

[331] Mahalanobis, P. C. (1936). On the generalized distance in statistics. *Proceedings of the National Institute of Sciences of India*, *2*(1), 49–55.

[332] Maki, D. (2012). Tests for cointegration allowing for an unknown number of breaks. *Economic Modelling*, *29*(5), 2011–2015.

[333] Mann, H. B. (1945). Nonparametric tests against trend. *Econometrica*, *13*(3), 245–259.

[334] Mann, H. B., & Whitney, D. R. (1947). On a test of whether one of two random variables is stochastically larger than the other. *Annals of Mathematical Statistics*, *18*(1), 50–60.

[335] Mantel, N. (1963). Chi-square tests with one degree of freedom: Extensions of the mantel-haenszel procedure. *Journal of the American Statistical Association*, *58*(303), 690–700.

[336] Mantel, N. (1966). Evaluation of survival data and two new rank order statistics arising in its consideration. *Cancer Chemotherapy Reports*, *50*(3), 163–170.

[337] Mantel, N. (1967). The detection of disease clustering and a generalized regression approach. *Cancer Research*, *27*(2), 209–220.

[338] Mantel, N., & Haenszel, W. (1959). Statistical aspects of the analysis of data from retrospective studies of disease. *Journal of the National Cancer Institute*, *22*(4), 719–748.

[339] Mardia, K. V. (1970). Measures of multivariate skewness and kurtosis with applications. *Biometrika*, *57*(3), 519–530.

[340] Mardia, K. V. (1974). Applications of some measures of multivariate skewness and kurtosis in testing normality and robustness studies. *Sankhyā, Series B*, *36*(2), 115–128.

[341] Mardia, K. V., Kent, J. T., & Bibby, J. M. (1979). *Multivariate analysis*. Academic Press.

[342] Maronna, R. A., Martin, D. R., & Yohai, V. J. (2006). *Robust statistics: Theory and methods*. Wiley.

[343] Marquardt, D. W. (1970). Generalized inverses, ridge regression, biased linear estimation, and nonlinear estimation. *Technometrics*, *12*(3), 591–612.

[344] Marsaglia, G., & Marsaglia, J. (2004). Evaluating the anderson-darling distribution. *Journal of Statistical Software*, *9*(2), 1–5.

[345] Mauchly, J. W. (1940). Significance test for sphericity of a normal n-variate distribution. *The Annals of Mathematical Statistics*, *11*(2), 204–209.

[346] Maxwell, A. E. (1970). Comparing the classification of subjects by two independent judges. *British Journal of Psychiatry*, *116*(535), 651–655.

[347] Maxwell, S. E., Delaney, H. D., & Kelley, K. (2017a). *Designing experiments and analyzing data: A model comparison perspective* (3rd ed.). Routledge.

[348] Maxwell, S. E., Delaney, H. D., & Kelley, K. (2017b). *Designing experiments and analyzing data: A model comparison perspective* (3rd ed.). Routledge.

[349] McCullagh, P., & Nelder, J. A. (1989). *Generalized linear models* (2nd ed.). Chapman & Hall.

[350] McFadden, D. (1974). Conditional logit analysis of qualitative choice behavior. In P. Zarembka (Ed.), *Frontiers in econometrics* (pp. 105–142). Academic Press.

[351] McKeon, J. J. (1974). F approximations to the distribution of hotelling's T^2. *Biometrika*, *61*(2), 381–383.

[352] McLeod, A. I., & Li, W. K. (1983). Diagnostic checking arma time series models using squared-residual autocorrelations. *Journal of Time Series Analysis*, *4*(4), 269–273.

[353] McNemar, Q. (1947). Note on the sampling error of the difference between correlated proportions or percentages. *Psychometrika*, *12*(2), 153–157.

[354] Mehta, C. R., & Hilton, J. F. (1993). Exact power of conditional and unconditional tests: Going beyond the 2×2 contingency table. *The American Statistician*, *47*(2), 91–98.

[355] Mehta, C. R., & Senchaudhuri, P. (2003). Conditional inference for categorical data. *Statistical Science*, *18*(1), 58–71.

[356] Meng, X.-L., Rosenthal, R., & Rubin, D. B. (1992). Comparing correlated correlation coefficients. *Psychological Bulletin*, *111*(1), 172–175.

[357] Miettinen, O., & Nurminen, M. (1985). Comparative analysis of two rates. *Statistics in Medicine*, *4*(2), 213–226.

[358] Miller, L. H. (1951). Table of percentage points of kolmogorov statistics. *Journal of the American Statistical Association*, *46*(253), 168–181.

[359] Montgomery, D. C. (2019). *Introduction to statistical quality control* (8th ed.). John Wiley & Sons.

[360] Monti, A. C. (1994). A proposal for a lagrange multiplier test for autocorrelation in arma models. *Biometrika*, *81*(4), 776–780.

[361] Mood, A. M. (1954). On the asymptotic efficiency of certain nonparametric two-sample tests. *Annals of Mathematical Statistics*, *25*(3), 514–522.

[362] Moran, P. A. P. (1947). Random associations on a line in two and three dimensions. *Journal of the Royal Statistical Society. Series B (Methodological)*, *9*(1), 92–98.

[363] Moran, P. A. P. (1950). Notes on continuous stochastic phenomena. *Biometrika*, *37*(1/2), 17–23.

[364] Morey, R. D., & Rouder, J. N. (2015). Bayesfactor: Computation of bayes factors for common designs. *Journal of Statistical Software*, *65*(1), 1–15.

[365] Moser, B. K., & Stevens, G. R. (1992). Homogeneity of variance in the two-sample means test. *The American Statistician*, *46*(1), 19–21.

[366] Mukhopadhyay, P., Ye, Z., Rubin, D., Halabi, S., & Chappell, R. J. (2022). Log-rank test vs MaxCombo and difference in restricted mean survival time tests for comparing survival under nonproportional hazards in immuno-oncology trials: A systematic review and meta-analysis. *JAMA Oncology*, *8*(8), 1160–1168.

[367] Mundlak, Y. (1978). On the pooling of time series and cross section data. *Econometrica*, *46*(1), 69–85.

[368] Murdoch, D. J., Tsai, Y.-L., & Adcock, J. (2008). P-values are random variables. *The American Statistician*, *62*(3), 242–245.

[369] Muthén, B. (1984). A general structural equation model with dichotomous, ordered categorical, and continuous latent variable indicators. *Psychometrika*, *49*(1), 115–132.

[370] Nagao, H. (1973). On some test criteria for covariance matrix. *The Annals of Statistics*, *1*(4), 700–709.

[371] Narayan, P. K. (2005). The saving and investment nexus for china: Evidence from cointegration tests. *Applied Economics*, *37*(17), 1979–1990.

[372] Nelder, J. A., & Wedderburn, R. W. M. (1972). *Generalized linear models* (Vol. 135).

[373] Nelson, D. B. (1991). Conditional heteroskedasticity in asset returns: A new approach. *Econometrica*, *59*(2), 347–370.

[374] Nelson, L. S. (1984). The shewhart control chart—tests for special causes. *Journal of Quality Technology*, *16*(4), 237–239.

[375] Nelson, L. S. (1985). Interpreting shewhart $\bar{X}$ control charts. *Journal of Quality Technology*, *17*(2), 114–116.

[376] Neubert, K., & Brunner, E. (2007). A studentized permutation test for the nonparametric behrens–fisher problem. *Computational Statistics & Data Analysis*, *51*(10), 5192–5204.

[377] Newcombe, R. G. (1998a). Interval estimation for the difference between independent proportions: Comparison of eleven methods. *Statistics in Medicine*, *17*(8), 873–890.

[378] Newcombe, R. G. (1998b). Two-sided confidence intervals for the single proportion: Comparison of seven methods. *Statistics in Medicine*, *17*(8), 857–872.

[379] Newey, W. K., & West, K. D. (1987). A simple, positive semi-definite, heteroskedasticity and autocorrelation consistent covariance matrix. *Econometrica*, *55*(3), 703–708.

[380] Neyman, J., & Pearson, E. S. (1928). On the use and interpretation of certain test criteria for purposes of statistical inference. *Biometrika*, *20A*(1/2), 175–240.

[381] Neyman, J., & Pearson, E. S. (1933). On the problem of the most efficient tests of statistical hypotheses. *Philosophical Transactions of the Royal Society of London. Series A, Containing Papers of a Mathematical or Physical Character*, *231*, 289–337.

[382] Ng, S., & Perron, P. (2001). Lag length selection and the construction of unit root tests with good size and power. *Econometrica*, *69*(6), 1519–1554.

[383] Nunes, L., Newbold, P., & Kuan, C.-M. (1997). Testing for unit roots with breaks: Evidence on the great crash and the unit root hypothesis reconsidered. *Oxford Bulletin of Economics and Statistics*, *59*(4), 435–448.

[384] Olson, C. L. (1976). On choosing a test statistic in multivariate analysis of variance. *Psychological Bulletin*, *83*(4), 579–586.

[385] Olson, C. L. (1979). Practical considerations in choosing a manova test statistic: A rejoinder to stevens. *Psychological Bulletin*, *86*(6), 1350–1352.

[386] Olsson, U. (1979). Maximum likelihood estimation of the polychoric correlation coefficient. *Psychometrika*, *44*(4), 443–460.

[387] Ord, J. K., & Getis, A. (1995). Local spatial autocorrelation statistics: Distributional issues and an application. *Geographical Analysis*, *27*(4), 286–306.

[388] Osterwald-Lenum, M. (1992). A note with quantiles of the asymptotic distribution of the maximum likelihood cointegration rank test statistics. *Oxford Bulletin of Economics and Statistics*, *54*(3), 461–472.

[389] Page, E. B. (1963). Ordered hypotheses for multiple treatments: A significance test for linear ranks. *Journal of the American Statistical Association*, *58*(301), 216–230.

[390] Page, E. S. (1954). Continuous inspection schemes. *Biometrika*, *41*(1/2), 100–115.

[391] Park, J. Y. (1990). Testing for unit roots and cointegration by variable addition. *Advances in Econometrics*, *8*, 107–133.

[392] Park, J. Y. (1992). Canonical cointegrating regressions. *Econometrica*, *60*(1), 119–143.

[393] Park, J. Y., Ouliaris, S., & Choi, B.-S. (1988). *Spurious regressions and tests for cointegration* (Cowles Foundation Discussion Paper No. 887). Cowles Foundation for Research in Economics, Yale University.

[394] Patterson, H. D., & Thompson, R. (1971). Recovery of inter-block information when block sizes are unequal. *Biometrika*, *58*(3), 545–554.

[395] Pearson, K. (1896). Mathematical contributions to the theory of evolution. iii. regression, heredity, and panmixia. *Philosophical Transactions of the Royal Society of London A*, *187*, 253–318.

[396] Pearson, K. (1900). Mathematical contributions to the theory of evolution. vii. on the correlation of characters not quantitatively measurable. *Philosophical Transactions of the Royal Society of London A*, *195*, 1–47.

[397] Pearson, K. (1909). On a new method of determining correlation between a measured character a, and a character b. *Biometrika*, *7*(1/2), 96–105.

[398] Pearson, K. (1922). *Tables for statisticians and biometricians*. Cambridge University Press.

[399] Pedroni, P. (1999). Critical values for cointegration tests in heterogeneous panels with multiple regressors. *Oxford Bulletin of Economics and Statistics*, *61*(S1), 653–670.

[400] Pedroni, P. (2004). Panel cointegration: Asymptotic and finite sample properties of pooled time series tests with an application to the ppp hypothesis. *Econometric Theory*, *20*(3), 597–625.

[401] Peirce, B. (1852). Criterion for the rejection of doubtful observations. *Astronomical Journal*, *2*, 161–163.

[402] Perron, P. (1989). The great crash, the oil price shock, and the unit root hypothesis. *Econometrica*, *57*(6), 1361–1401.

[403] Perron, P. (1997). Further evidence on breaking trend functions in macroeconomic variables. *Journal of Econometrics*, *80*(2), 355–385.

[404] Perron, P., & Ng, S. (1996). Useful modifications to some unit root tests with dependent errors and their local asymptotic properties. *Review of Economic Studies*, *63*(3), 435–463.

[405] Perron, P., & Vogelsang, T. J. (1992). Nonstationarity and level shifts with an application to purchasing power parity. *Journal of Business & Economic Statistics*, *10*(3), 301–320.

[406] Persyn, D., & Westerlund, J. (2008). Error-correction-based cointegration tests for panel data. *Stata Journal*, *8*(2), 232–241.

[407] Pesaran, M. H. (2004). *General diagnostic tests for cross section dependence in panels* (Cambridge Working Papers in Economics No. 0435). University of Cambridge.

[408] Pesaran, M. H. (2015). Testing weak cross-sectional dependence in large panels. *Econometric Reviews*, *34*(6-10), 1089–1117.

[409] Pesaran, M. H. (2021). General diagnostic tests for cross-section dependence in panels. *Empirical Economics*, *60*, 13–50.

[410] Pesaran, M. H., Shin, Y., & Smith, R. J. (2001). Bounds testing approaches to the analysis of level relationships. *Journal of Applied Econometrics*, *16*(3), 289–326.

[411] Peto, R., & Peto, J. (1972). Asymptotically efficient rank invariant test procedures. *Journal of the Royal Statistical Society, Series A*, *135*(2), 185–207.

[412] Pettitt, A. N. (1979). A non-parametric approach to the change-point problem. *Applied Statistics*, *28*(2), 126–135.

[413] Phillips, P. C. B., & Hansen, B. E. (1990). Statistical inference in instrumental variables regression with i(1) processes. *Review of Economic Studies*, *57*(1), 99–125.

[414] Phillips, P. C. B., & Ouliaris, S. (1990). Asymptotic properties of residual based tests for cointegration. *Econometrica*, *58*(1), 165–193.

[415] Phillips, P. C. B., & Perron, P. (1988). Testing for a unit root in time series regression. *Biometrika*, *75*(2), 335–346.

[416] Phipson, B., & Smyth, G. K. (2010). Permutation p-values should never be zero: Calculating exact p-values when permutations are randomly drawn. *Statistical Applications in Genetics and Molecular Biology*, *9*(1), Article 39.

[417] Pigeot, I., Schäfer, J., Röhmel, J., & Hauschke, D. (2003). Assessing non-inferiority of a new treatment in a three-arm clinical trial including a placebo. *Statistics in Medicine*, *22*(6), 883–899.

[418] Pillai, K. C. S. (1955). Some new test criteria in multivariate analysis. *The Annals of Mathematical Statistics*, *26*(1), 117–121.

[419] Pillai, K. C. S. (1960). On the distribution of the largest root of a matrix in multivariate analysis. *Biometrika*, *47*(3/4), 463–471.

[420] Poisson, S. D. (1837). *Recherches sur la probabilité des jugements en matière criminelle et en matière civile*. Bachelier.

[421] Pollak, M. (1985). Optimal detection of a change in distribution. *The Annals of Statistics*, *13*(1), 206–227.

[422] Pregibon, D. (1980). Goodness of link tests for generalized linear models. *Applied Statistics*, *29*(1), 15–24.

[423] Pregibon, D. (1981). Logistic regression diagnostics. *The Annals of Statistics*, *9*(4), 705–724.

[424] Przyborowski, J., & Wilenski, H. (1940). Homogeneity of results in testing samples from poisson series: With an application to testing clover seed for dodder. *Biometrika*, *31*(3/4), 313–323.

[425] Pyke, R. (1965). Spacings. *Journal of the Royal Statistical Society. Series B (Methodological)*, *27*(3), 395–449.

[426] Quade, D. (1979). Using weighted rankings in the analysis of complete blocks with additive block effects. *Journal of the American Statistical Association*, *74*(367), 680–683.

[427] Quandt, R. E. (1960). Tests of the hypothesis that a linear regression system obeys two separate regimes. *Journal of the American Statistical Association*, *55*(290), 324–330.

[428] Ramsey, J. B. (1969). Tests for specification errors in classical linear least-squares regression analysis. *Journal of the Royal Statistical Society: Series B*, *31*(2), 350–371.

[429] Rao, C. R. (1948). Large sample tests of statistical hypotheses concerning several parameters with applications to problems of estimation. *Proceedings of the Cambridge Philosophical Society*, *44*(1), 50–57.

[430] Rao, J. S. (1969). *Some contributions to the analysis of circular data* [Doctoral dissertation, Indian Statistical Institute].

[431] Rao, J. S. (1976). Some tests based on arc-lengths for the circle. *Sankhyā: The Indian Journal of Statistics, Series B*, *38*(4), 329–338.

[432] Roberts, S. W. (1959). Control chart tests based on geometric moving averages. *Technometrics*, *1*(3), 239–250.

[433] Roberts, S. W. (1966). A comparison of some control chart procedures. *Technometrics*, *8*(3), 411–430.

[434] Rorabacher, D. B. (1991). Statistical treatment for rejection of deviant values: Critical values of dixon's q parameter and related subrange ratios at the 95% confidence level. *Analytical Chemistry*, *63*(2), 139–146.

[435] Rosenbaum, P. R. (2002). *Observational studies* (2nd ed.). Springer.

[436] Rosner, B. (1983). Percentage points for a generalized esd many-outlier procedure. *Technometrics*, *25*(2), 165–172.

[437] Ross, S. M. (2003). Peirce's criterion for the elimination of suspect experimental data. *Journal of Engineering Technology*, *20*(2), 38–41.

[438] Rouder, J. N., & Morey, R. D. (2012). Default bayes factors for model selection in regression. *Multivariate Behavioral Research*, *47*(6), 877–903.

[439] Rouder, J. N., Morey, R. D., Speckman, P. L., & Province, J. M. (2012). Default bayes factors for anova designs. *Journal of Mathematical Psychology*, *56*(5), 356–374.

[440] Rouder, J. N., Speckman, P. L., Sun, D., Morey, R. D., & Iverson, G. (2009). Bayesian t tests for accepting and rejecting the null hypothesis. *Psychonomic Bulletin & Review*, *16*(2), 225–237.

[441] Rousseeuw, P. J. (1984). Least median of squares regression. In J. Franke, W. Härdle, & D. Martin (Eds.), *Robust and nonlinear time series analysis* (pp. 333–355). Springer.

[442] Rousseeuw, P. J. (1985). Multivariate estimation with high breakdown point. *Mathematical Statistics and Applications*, *8*, 283–297.

[443] Rousseeuw, P. J., & Croux, C. (1993). Alternatives to the median absolute deviation. *Journal of the American Statistical Association*, *88*(424), 1273–1283.

[444] Rousseeuw, P. J., & Driessen, K. V. (1999). A fast algorithm for the minimum covariance determinant estimator. *Technometrics*, *41*(3), 212–223.

[445] Rousseeuw, P. J., & Leroy, A. M. (1987). *Robust regression and outlier detection*. John Wiley & Sons.

[446] Rousseeuw, P. J., & van Zomeren, B. C. (1990). Unmasking multivariate outliers and leverage points. *Journal of the American Statistical Association*, *85*(411), 633–639.

[447] Roy, S. N. (1953). On a heuristic method of test construction and its use in multivariate analysis. *The Annals of Mathematical Statistics*, *24*(2), 220–238.

[448] Roychoudhury, S., et al. (2021). Beyond the hazard ratio: On the reporting of survival analyses in immuno-oncology trials. *Clinical Cancer Research*.

[449] Royston, P. (1982). An extension of shapiro and wilk's w test for normality to large samples. *Applied Statistics*, *31*(2), 115–124.

[450] Royston, P. (1983). Some techniques for assessing multivariate normality based on the shapiro–wilk w. *Applied Statistics*, *32*(2), 121–133.

[451] Royston, P. (1992). Approximating the shapiro–wilk w-test for non-normality. *Statistics and Computing*, *2*(3), 117–119.

[452] Royston, P. (1993). A toolkit for testing for non-normality in complete and censored samples. *The Statistician*, *42*(1), 37–43.

[453] Rubin, D. B. (1984). Bayesianly justifiable and relevant frequency calculations for the applied statistician. *Annals of Statistics*, *12*(4), 1151–1172.

[454] Russell, G. S., & Levitin, D. J. (1995). An expanded table of probability values for rao's spacing test. *Communications in Statistics – Simulation and Computation*, *24*(4), 879–888.

[455] Said, S. E., & Dickey, D. A. (1984). Testing for unit roots in autoregressive-moving average models of unknown order. *Biometrika*, *71*(3), 599–607.

[456] Sansó, A., Aragó, V., & Carrion-i-Silvestre, J. L. (2004). Testing for changes in the unconditional variance of financial time series. *Revista de Economía Financiera*, *4*, 32–53.

[457] Satterthwaite, F. E. (1946). An approximate distribution of estimates of variance components. *Biometrics Bulletin*, *2*(6), 110–114.

[458] Savin, N. E., & White, K. J. (1977). The durbin-watson test for serial correlation with extreme sample sizes or many regressors. *Econometrica*, *45*(8), 1989–1996.

[459] Schaalje, G. B., McBride, J. B., & Fellingham, G. W. (2002). Adequacy of approximations to distributions of test statistics in complex mixed linear models. *Journal of Agricultural, Biological, and Environmental Statistics*, *7*(4), 512–524.

[460] Scheffé, H. (1953). A method for judging all contrasts in the analysis of variance. *Biometrika*, *40*(1/2), 87–104.

[461] Scheffé, H. (1959). *The analysis of variance*. Wiley.

[462] Schmidt, P., & Phillips, P. C. B. (1992). Lm tests for a unit root in the presence of deterministic trends. *Oxford Bulletin of Economics and Statistics*, *54*(3), 257–287.

[463] Schoenfeld, D. (1982). Partial residuals for the proportional hazards regression model. *Biometrika*, *69*(1), 239–241.

[464] Scholz, F. W., & Stephens, M. A. (1987). K-sample anderson–darling tests. *Journal of the American Statistical Association*, *82*(399), 918–924.

[465] Schuirmann, D. J. (1987). A comparison of the two one-sided tests procedure and the power approach for assessing the equivalence of average bioavailability. *Journal of Pharmacokinetics and Biopharmaceutics*, *15*(6), 657–680.

[466] Schwert, G. W. (1989). Tests for unit roots: A monte carlo investigation. *Journal of Business & Economic Statistics*, *7*(2), 147–159.

[467] Self, S. G., & Liang, K.-Y. (1987). Asymptotic properties of maximum likelihood estimators and likelihood ratio tests under nonstandard conditions. *Journal of the American Statistical Association*, *82*(398), 605–610.

[468] Sen, P. K. (1968). Estimates of the regression coefficient based on kendall's tau. *Journal of the American Statistical Association*, *63*(324), 1379–1389.

[469] Shapiro, S. S., & Francia, R. S. (1972). An approximate analysis of variance test for normality. *Journal of the American Statistical Association*, *67*(337), 215–216.

[470] Shapiro, S. S., & Wilk, M. B. (1965). An analysis of variance test for normality (complete samples). *Biometrika*, *52*(3/4), 591–611.

[471] Sheskin, D. J. (2011). Handbook of parametric and nonparametric statistical procedures (5th ed.).

[472] Shewhart, W. A. (1931). *Economic control of quality of manufactured product*. Van Nostrand.

[473] Shin, Y. (1994). A residual-based test of the null of cointegration against the alternative of no cointegration. *Econometric Theory*, *10*(1), 91–115.

[474] Shiryaev, A. N. (1963). On optimum methods in quickest detection problems. *Theory of Probability and Its Applications*, *8*(1), 22–46.

[475] Shrout, P. E., & Fleiss, J. L. (1979). Intraclass correlations: Uses in assessing rater reliability. *Psychological Bulletin, 86*(2), 420–428.

[476] Siegel, S., & Tukey, J. W. (1960). A nonparametric sum of ranks procedure for relative spread in unpaired samples. *Journal of the American Statistical Association, 55*(291), 429–445.

[477] Silvapulle, M. J., & Silvapulle, P. K. (1995). *Constrained statistical inference: Inequality, order, and shape restrictions.* John Wiley & Sons.

[478] Silvey, S. D. (1959). The lagrangian multiplier test. *Annals of Mathematical Statistics, 30*(2), 389–407.

[479] Simonsohn, U., Nelson, L. D., & Simmons, J. P. (2014). P-curve: A key to the file-drawer. *Journal of Experimental Psychology: General, 143*(2), 534–547.

[480] Small, K. A., & Hsiao, C. (1985). Multinomial logit specification tests. *International Economic Review, 26*(3), 619–627.

[481] Smirnov, N. V. (1939). On the estimation of the discrepancy between empirical curves of distribution for two independent samples. *Bulletin of Moscow University, 2*(2), 3–16.

[482] Smirnov, N. V. (1948). Table for estimating the goodness of fit of empirical distributions. *The Annals of Mathematical Statistics, 19*(2), 279–281.

[483] Snedecor, G. W. (1934). *Calculation and interpretation of analysis of variance and covariance*. Collegiate Press.

[484] Snedecor, G. W., & Cochran, W. G. (1989). *Statistical methods* (8th ed.). Iowa State University Press.

[485] Sokal, R. R., & Rohlf, F. J. (1995). *Biometry: The principles and practice of statistics in biological research* (3rd). W. H. Freeman.

[486] Sollis, R. (2009). A simple unit root test against asymmetric star nonlinearity with an application to real exchange rates in nordic countries. *Economic Modelling, 26*(1), 118–125.

[487] Spearman, C. (1904). The proof and measurement of association between two things. *The American Journal of Psychology, 15*(1), 72–101.

[488] Steiger, J. H. (1980). Tests for comparing elements of a correlation matrix. *Psychological Bulletin, 87*(2), 245–251.

[489] Stephens, M. A. (1965). The goodness-of-fit statistic V_n: Distribution and significance points. *Biometrika, 52*(3/4), 309–321.

[490] Stephens, M. A. (1974). Edf statistics for goodness of fit and some comparisons. *Journal of the American Statistical Association, 69*(347), 730–737.

[491] Stephens, M. A. (1981). Edf statistics for goodness of fit and some comparisons. In G. Kallianpur, P. R. Krishnaiah, & J. K. Ghosh (Eds.), *Statistics and probability: Essays in honor of c. r. rao* (pp. 97–193). North-Holland.

[492] Stephens, M. A. (1986). Tests based on edf statistics. In R. B. D'Agostino & M. A. Stephens (Eds.), *Goodness-of-fit techniques* (pp. 97–193). Marcel Dekker.

[493] Stock, J. H. (1999). A class of tests for integration and cointegration. In R. F. Engle & H. White (Eds.), *Cointegration, causality, and forecasting* (pp. 137–167). Oxford University Press.

[494] Stock, J. H., & Watson, M. W. (1993). A simple estimator of cointegrating vectors in higher order integrated systems. *Econometrica, 61*(4), 783–820.

[495] Storey, J. D. (2002). A direct approach to false discovery rates. *Journal of the Royal Statistical Society: Series B (Statistical Methodology), 64*(3), 479–498.

[496] Stouffer, R. J., Steinsaltz, D., & Evans, S. N. (2019). The P-value uniformity test for model verification. *arXiv preprint arXiv:1903.07868.*

[497] Stram, D. O., & Lee, J. W. (1994). Variance components testing in the longitudinal mixed effects model. *Biometrics, 50*(4), 1171–1177.

[498] Stuart, A. (1955). A test for homogeneity of the marginal distributions in a two-way classification. *Biometrika, 42*(3/4), 412–416.

[499] Student. (1908). The probable error of a mean. *Biometrika, 6*(1), 1–25.

[500] Stukel, T. A. (1988). Generalized logistic models. *Journal of the American Statistical Association, 83*(402), 426–431.

[501] Sugiura, N. (1973). Asymptotic non-null distributions of the likelihood ratio criteria for covariance matrix tests. *The Annals of Statistics, 1*(4), 718–728.

[502] Suissa, S., & Shuster, J. J. (1985). Exact unconditional sample sizes for the 2×2 binomial trial. *Journal of the Royal Statistical Society A, 148*(4), 317–327.

[503] Székely, G. J., & Rizzo, M. L. (2009). Brownian distance covariance. *The Annals of Applied Statistics, 3*(4), 1236–1265.

[504] Székely, G. J., Rizzo, M. L., & Bakirov, N. K. (2007). Measuring and testing dependence by correlation of distances. *The Annals of Statistics*, *35*(6), 2769–2794.

[505] Tarone, R. E. (1985). On heterogeneity tests based on efficient scores. *Biometrika*, *72*(1), 91–95.

[506] Tarone, R. E., & Ware, J. (1977). On distribution-free tests for equality of survival distributions. *Biometrika*, *64*(1), 156–160.

[507] Tate, R. F. (1954). Correlation between a discrete and a continuous variable: Point-biserial correlation. *The Annals of Mathematical Statistics*, *25*(3), 603–607.

[508] Tate, R. F. (1955). The theory of correlation between two continuous variables when one is dichotomized. *Biometrika*, *42*(1–2), 205–216.

[509] Teräsvirta, T. (1994). Specification, estimation, and evaluation of smooth transition autoregressive models. *Journal of the American Statistical Association*, *89*(425), 208–218.

[510] Terman, L. M., & Merrill, M. A. (1937). *Measuring intelligence*. Houghton Mifflin.

[511] Terpstra, T. J. (1952). The asymptotic normality and consistency of kendall's test against trend, when ties are present in one ranking. *Indagationes Mathematicae*, *14*, 327–333.

[512] Theil, H. (1950). A rank-invariant method of linear and polynomial regression analysis. *Indagationes Mathematicae*, *12*, 85–91.

[513] Thompson, W. R. (1935). On a criterion for the rejection of observations and the distribution of the ratio of the deviation to the sample standard deviation. *Annals of Mathematical Statistics*, *6*, 214–219.

[514] Thursby, J. G., & Schmidt, P. (1977). Some properties of tests for specification error in a linear regression model. *Journal of the American Statistical Association*, *72*(359), 635–641.

[515] Tibshirani, R. (1996). Regression shrinkage and selection via the lasso. *Journal of the Royal Statistical Society: Series B (Methodological)*, *58*(1), 267–288.

[516] Tietjen, G. L., & Moore, R. H. (1972). Some grubbs-type statistics for the detection of several outliers. *Technometrics*, *14*(3), 583–597.

[517] Tong, H. (1990). *Non-linear time series: A dynamical system approach*. Oxford University Press.

[518] Tsay, R. S. (1986). Nonlinearity tests for time series. *Biometrika*, *73*(2), 461–466.

[519] Tsay, R. S. (1989). Testing and modeling threshold autoregressive processes. *Journal of the American Statistical Association*, *84*(405), 231–240.

[520] Tsay, R. S. (2010). *Analysis of financial time series* (3rd ed.). John Wiley & Sons.

[521] Tukey, J. W. (1949a). Comparing individual means in the analysis of variance. *Biometrics*, *5*(2), 99–114.

[522] Tukey, J. W. (1949b). One degree of freedom for non-additivity. *Biometrics*, *5*(3), 232–242.

[523] Urzúa, C. M. (1996). On the correct use of omnibus tests for normality. *Economics Letters*, *53*(3), 247–251.

[524] U.S. Food and Drug Administration. (2001). *Guidance for industry: Statistical approaches to establishing bioequivalence* (tech. rep.). Center for Drug Evaluation and Research (CDER). Rockville, MD.

[525] U.S. Food and Drug Administration. (2003a). Bioavailability and bioequivalence studies for orally administered drug products — general considerations.

[526] U.S. Food and Drug Administration. (2003b). *Guidance for industry: Exposure-response relationships — study design, data analysis, and regulatory applications* (tech. rep.). Center for Drug Evaluation and Research (CDER). Rockville, MD.

[527] U.S. Food and Drug Administration. (2016). *Non-inferiority clinical trials to establish effectiveness: Guidance for industry* (tech. rep.). Center for Drug Evaluation, Research (CDER), Center for Biologics Evaluation, and Research (CBER). Rockville, MD.

[528] U.S. Food and Drug Administration. (2019). *Adaptive designs for clinical trials of drugs and biologics: Guidance for industry* (tech. rep.). Center for Drug Evaluation, Research (CDER), Center for Biologics Evaluation, and Research (CBER). Silver Spring, MD.

[529] Utts, J. M. (1982). The effect of specification errors on forecasts. *Journal of the American Statistical Association*, *77*(377), 609–615.

[530] van den Broek, J. (1995). A score test for zero inflation in a poisson distribution. *Biometrics*, *51*(2), 738–743.

[531] Verbeke, G., & Molenberghs, G. (2000). *Linear mixed models for longitudinal data*. Springer.

[532] Verdinelli, I., & Wasserman, L. (1995). Computing bayes factors using a generalization of the savage–dickey density ratio. *Journal of the American Statistical Association*, *90*(430), 614–618.

[533] Vogelsang, T. J., & Perron, P. (1998). Additional tests for a unit root allowing for a break in the trend function at an unknown time. *International Economic Review*, *39*(4), 1073–1100.

[534] von Mises, R. (1931). *Wahrscheinlichkeit, statistik und wahrheit* (2nd ed.). Springer.

[535] von Neumann, J. (1941). Distribution of the ratio of the mean square successive difference to the variance. *Annals of Mathematical Statistics*, *12*(4), 367–395.

[536] Vuong, Q. H. (1989). Likelihood ratio tests for model selection and non-nested hypotheses. *Econometrica*, *57*(2), 307–333.

[537] Wagenmakers, E.-J., Lodewyckx, T., Kuriyal, H., & Grasman, R. (2010). Bayesian hypothesis testing for psychologists: A tutorial on the savage–dickey method. *Cognitive Psychology*, *60*(3), 158–189.

[538] Wald, A. (1943). Tests of statistical hypotheses concerning several parameters when the number of observations is large. *Transactions of the American Mathematical Society*, *54*(3), 426–482.

[539] Wald, A. (1947). *Sequential analysis*. John Wiley & Sons.

[540] Wald, A., & Wolfowitz, J. (1940). On a test whether two samples are from the same population. *Annals of Mathematical Statistics*, *11*(2), 147–162.

[541] Wallis, W. A., & Moore, G. H. (1941). *A significance test for time series and other ordered observations* (Technical Paper No. 1). National Bureau of Economic Research.

[542] Wasserstein, R. L., & Lazar, N. A. (2016). The asa statement on p-values: Context, process, and purpose. *The American Statistician*, *70*(2), 129–133.

[543] Welch, B. L. (1947). The generalization of "student"'s problem when several different population variances are involved. *Biometrika*, *34*(1/2), 28–35.

[544] Welch, B. L. (1951). On the comparison of several mean values: An alternative approach. *Biometrika*, *38*(3/4), 330–336.

[545] Wellek, S. (2010). *Testing statistical hypotheses of equivalence and noninferiority* (2nd ed.). Chapman & Hall/CRC.

[546] Welsch, R. E., & Kuh, E. (1977). Linear regression diagnostics. *Working Paper 923-77*.

[547] Westerlund, J. (2007). Testing for error correction in panel data. *Oxford Bulletin of Economics and Statistics*, *69*(6), 709–748.

[548] Westerlund, J., & Edgerton, D. L. (2007). A new panel bootstrap cointegration test. *Economics Letters*, *97*(3), 185–190.

[549] Western Electric Company. (1956). *Statistical quality control handbook*. AT&T.

[550] Wetzels, R., Matzke, D., Lee, M. D., Rouder, J. N., Iverson, G. J., & Wagenmakers, E.-J. (2011). Statistical evidence in experimental psychology: An empirical comparison using 855 t tests. *Perspectives on Psychological Science*, *6*(3), 291–298.

[551] Wetzels, R., & Wagenmakers, E.-J. (2012). A default bayesian hypothesis test for correlations and partial correlations. *Psychonomic Bulletin & Review*, *19*(6), 1057–1064.

[552] White, H. (1980). A heteroskedasticity-consistent covariance matrix estimator and a direct test for heteroskedasticity. *Econometrica*, *48*(4), 817–838.

[553] Wilcox, R. R. (1995). Comparing two independent groups via multiple quantiles. *The Statistician*, *44*(1), 91–99.

[554] Wilcox, R. R. (2017). *Introduction to robust estimation and hypothesis testing* (4th ed.). Academic Press.

[555] Wilcoxon, F. (1945). Individual comparisons by ranking methods. *Biometrics Bulletin*, *1*(6), 80–83.

[556] Wilks, S. S. (1932). Certain generalizations in the analysis of variance. *Biometrika*, *24*(3/4), 471–494.

[557] Wilks, S. S. (1938). The large-sample distribution of the likelihood ratio for testing composite hypotheses. *Annals of Mathematical Statistics*, *9*(1), 60–62.

[558] Wilks, S. S. (1946). Sample criteria for testing equality of means, equality of variances, and equality of covariances in a normal multivariate distribution. *The Annals of Mathematical Statistics*, *17*(3), 257–281.

[559] Williams, D. A. (1976a). Improved likelihood ratio tests for complete contingency tables. *Biometrika*, *63*(1), 33–37.

[560] Williams, D. A. (1976b). Improved likelihood ratio tests for complete contingency tables. *Biometrika*, *63*(1), 33–37.

[561] Williams, E. J. (1959). The comparison of regression variables. *Journal of the Royal Statistical Society. Series B (Methodological)*, *21*(2), 396–399.

[562] Wilson, E. B. (1927). Probable inference, the law of succession, and statistical inference. *Journal of the American Statistical Association*, *22*(158), 209–212.

[563] Wilson, P. (2015). The misuse of the vuong test for non-nested models to test for zero-inflation. *Economics Letters*, *127*, 51–53.

[564] Winsor, C. P. (1946). The percentile method of estimating certain population parameters. *Annals of Mathematical Statistics*, *17*(3), 299–310.

[565] Wobbrock, J. O., Findlater, L., Gergle, D., & Higgins, J. J. (2011). The aligned rank transform for nonparametric factorial analyses using only anova procedures. *Proceedings of the SIGCHI Conference on Human Factors in Computing Systems*, 143–146.

[566] Woolf, B. (1955). On estimating the relation between blood group and disease. *Annals of Human Genetics*, *19*(4), 251–253.

[567] Yang, S., & Prentice, R. L. (2005). Semiparametric analysis of short-term and long-term hazard ratios with two-sample survival data. *Biometrika*, *92*(1), 1–17.

[568] Yao, Y. (1965). An approximate degrees of freedom solution to the multivariate behrens–fisher problem. *Biometrika*, *52*(1–2), 139–147.

[569] Yates, F. (1934a). The analysis of multiple classifications with unequal numbers in the different classes. *Journal of the American Statistical Association*, *29*(185), 51–66.

[570] Yates, F. (1934b). Contingency tables involving small numbers and the χ^2 test. *Journal of the Royal Statistical Society Supplement*, *1*(2), 217–235.

[571] Yuen, K. K. (1974). The two-sample trimmed t for unequal population variances. *Biometrika*, *61*(1), 165–170.

[572] Yuen, K. K., & Dixon, W. J. (1973). The approximate behaviour and performance of the two-sample trimmed t. *Biometrika*, *60*(2), 369–374.

[573] Yule, G. U. (1907). On the theory of correlation for any number of variables, treated by a new system of notation. *Proceedings of the Royal Society of London A*, *79*(529), 182–193.

[574] Zivot, E., & Andrews, D. W. K. (1992). Further evidence on the great crash, the oil-price shock, and the unit-root hypothesis. *Journal of Business & Economic Statistics*, *10*(3), 251–270.

[575] Zou, H., & Hastie, T. (2005). Regularization and variable selection via the elastic net. *Journal of the Royal Statistical Society: Series B (Statistical Methodology)*, *67*(2), 301–320.

Index

About the Author

Maruthy Pannala holds a Ph.D. in Statistics from the University of Louisiana at Lafayette and an M.S. in Statistics from the Indian Institute of Technology (IIT) Kanpur—one of India's most selective technical universities.

He has spent more than twenty-five years building quantitative models in the banking and insurance industries, with work spanning risk-based pricing of insurance premiums, loss reserving, credit risk, stress testing, loss forecasting, and regulatory compliance. Throughout that career, his primary interest has been in making statistics useful: translating difficult technical ideas into language that practitioners and decision-makers can act on, and applying rigorous methods to real business problems without losing sight of the question that actually needs answering. He has mentored many analysts and statisticians over the years, and finds that side of the work as rewarding as the modeling itself.

He is also a Sudoku and chess enthusiast, and suspects there is more overlap between puzzles and statistical thinking than is generally acknowledged.

This is his first book. He plans to write several more.

Maruthy lives in Greensboro, North Carolina, with his family and Lucky, a Cockapoo who provided companionship through many late nights of writing.

Errata and Corrections

Despite careful verification of formulas, worked examples, and software implementations throughout this compendium, errors may remain. Corrections to known errata are maintained at the following address:

`https://sites.google.com/view/pannala-series/errata`

If you believe you have identified an error—whether in a test statistic formula, a worked example, a software implementation, or a bibliographic citation—the author welcomes correspondence at:

`pannala.series@gmail.com`

When reporting a potential error, please include the test code (e.g., `NORM.3`, `CTAB.7`), the section in which it appears, and a brief description of the discrepancy. Where applicable, references to primary sources or reproducible software output are especially helpful.

The author is particularly grateful to readers who verify results against original publications and report any errors, omissions, or discrepancies that may have escaped notice.

www.ingramcontent.com/pod-product-compliance
Ingram Content Group UK Ltd.
Pitfield, Milton Keynes, MK11 3LW, UK
UKHW051022091026
15624UKWH00012B/636

9 798996 151608

Steinar Løve Ellefmo and Fredrik Søreide
with contributions from
Georgy Cherkashov, Cyril Juliani,
Krishna Kanta Panthi, Sergey Petukhov,
Irina Poroshina, Richard Sinding-Larsen
and Ben Snook

Quantifying the Unknown

Marine Mineral Resource Potential on the Norwegian Extended Continental Shelf

This book is made possible with support from the Norwegian University of Science and Technology (NTNU).

A note about the images reproduced in this book: many of the detailed maps and illustrations are not legible in the small format of the print and PDF editions. All of the images are accessible in larger, fully legible form via links in the image captions in the digital editions.

This book is a peer-reviewed monograph.

Contents

Preface

The prospect of deep-sea mining was initially suggested by J. L. Mero in his book *Mineral Resources of the Sea* in the mid-1960s (Mero, 1965). Mero claimed that nearly limitless supplies of cobalt, nickel and other metals could be found throughout the oceans in the form of manganese nodules, which appear as lumps of compressed sediment on the sea floor at depths of about 5000 m. Nodules were originally discovered by the HMS *Challenger* expedition in the 1870s. Several nations started to investigate these claims, especially in the Clarion-Clipperton Zone in the Pacific Ocean in the 1960s. However, these initial resource estimates of deep-sea minerals turned out to be exaggerated. This overestimate, coupled with depressed metal prices, led to the near abandonment of nodule mining by 1982.

Norway was involved in several new research initiatives, primarily through the Fridtjof Nansen Institute in the 1980s and 1990s, to reinvigorate the case for seabed mining. This culminated in the establishment of the Norwegian Deep Seabed Mining Group, a consortium of five Norwegian companies, the Fridtjof Nansen Institute and Norwegian University of Science and Technology (NTNU), to develop new business cases from 1999 to 2001 (Søreide et al., 2001). However, due to the still-depressed metal prices, no profitable business cases could be identified.

Meanwhile, in 1977, a group of marine geologists discovered the first hydrothermal vents along the Galápagos Rift, part of the global Mid-Ocean Ridge (MOR) system where new oceanic crust is generated, while diving in the DSV *ALVIN*, a research submersible operated by Woods Hole Oceanographic Institution. It was slowly understood that these vent systems constitute mineral deposits in the making, delivering copper, zinc, gold and silver to the seafloor. Interest, therefore, shifted towards hydrothermal vents as another source of metals instead of scattered nodules.

Over the past decade, a new era of interest in deep-sea minerals has begun. Rising demand and diminishing supplies on land has pushed the search for new sources in many areas. Norway has a significant section of the Mid-Atlantic Ridge within its extended continental shelf. In its exploration of this part of the ridge since the early 2000s, the University of Bergen has discovered a handful of hydrothermal vent systems.

Consequently, in 2011 NTNU together with Statoil ASA (now Equinor ASA) and Nordic Ocean Resources AS established a new research initiative to investigate the deep marine minerals and mining potential in the Norwegian sector. This book presents the results of that project.

Due to the high resource potential indicated by these studies, the Norwegian government included marine minerals in its mineral strategy for the first time in 2013 and issued draft legislation for the management of marine minerals within the Norwegian extended continental shelf in 2017 (OED, 2017). This legislation (LOV-2019-03-22-7) was announced on 22 March 2019 and came into effect on 1 July 2019. Only time will tell, however, if marine mining will develop into a new industry in the future.

Trondheim, September 2019.
Steinar Løve Ellefmo and Fredrik Søreide

CHAPTER 1

Introduction

A mineral is a naturally occurring, inorganic substance with a definite chemical composition and crystal structure. There are about 3500 minerals in the Earth's crust. The minerals contain elements that humankind needs daily. We are big consumers of metals like copper, zinc, gold and iron and minerals like quartz and feldspar, to name just a few. It is likely that the demand for minerals will continue to increase, with, for example, an increasing demand for renewable energy production. Without exploitation of mineral resources, we will no longer have products like white paper, toothpaste, computers, batteries, cars, windmills, buildings, glass and the infrastructure that underpins modern civilisation.

About 71% of the Earth's surface is covered by water, and it should therefore come as no surprise that the sediments occurring on the ocean floor contain mineral deposits of potential economic value (Cronan, 1992, 1999). These include:

- Aggregates, which are nearshore deposits of detrital silicate minerals and calcium carbonate, principally used in the construction industry.
- Placers, which are detrital metallic deposits generally found on beaches and in nearshore areas.
- Phosphorites, which are mineral precipitates formed in situ on the sea floor, in generally deeper water than placers.
- Manganese nodules, which are concretionary deposits principally of iron, manganese, nickel, copper and cobalt, which occur typically in the deeper parts of the oceans.
- Ferromanganese oxide crusts, or the so-called cobalt rich crusts, which occupy seamount areas in and around island chains.
- Hydrothermal deposits, which are precipitates of principally iron, copper and zinc sulphides, formed as a result of submarine volcanic activity and hydrothermal venting, which can also contain enrichments of lead, silver and gold.

The processes of supply of elements to the sea floor will determine the occurrence of these deposits. Chemical elements enter seawater in a number of different ways. Sometimes their ultimate source is the dissolved material in rivers draining the continents. In other cases, they are derived from hydrothermal solutions entering seawater from submarine volcanoes. Because ocean mixing processes are slow, elements introduced into seawater from a specific source can sometimes be identified at considerable distances from that source. Thus, the oceans are by no means uniform in composition (Cronan, 1992).

Only about 10% of the ocean floor has been mapped at a resolution greater than 100 metres. Most mineral deposits, and many ocean floor features that can indicate such deposits, cannot be seen at this resolution. We need better, smarter and more efficient technologies and data-management methodologies and systems if we want to know more about the ocean floor and its resource potential.

Marine mineral deposits are present in many areas of the ocean, some in international waters and some in exclusive economic zones (EEZ). General recognition of exclusive economic zones as potential locations for marine mineral resources other than oil and gas dates from the early 1980s. Norway sent a proposal for the extension of its continental shelf to the Commission on the Limits of the Continental Shelf (CLCS) in 2006, which was found largely acceptable in 2009. However, Norway has not issued a final proclamation of the extended boundary because of potential overlaps with other countries. The current Norwegian jurisdiction is shown in Figure 1–1, including the waters surrounding the islands of Jan Mayen and Spitsbergen.

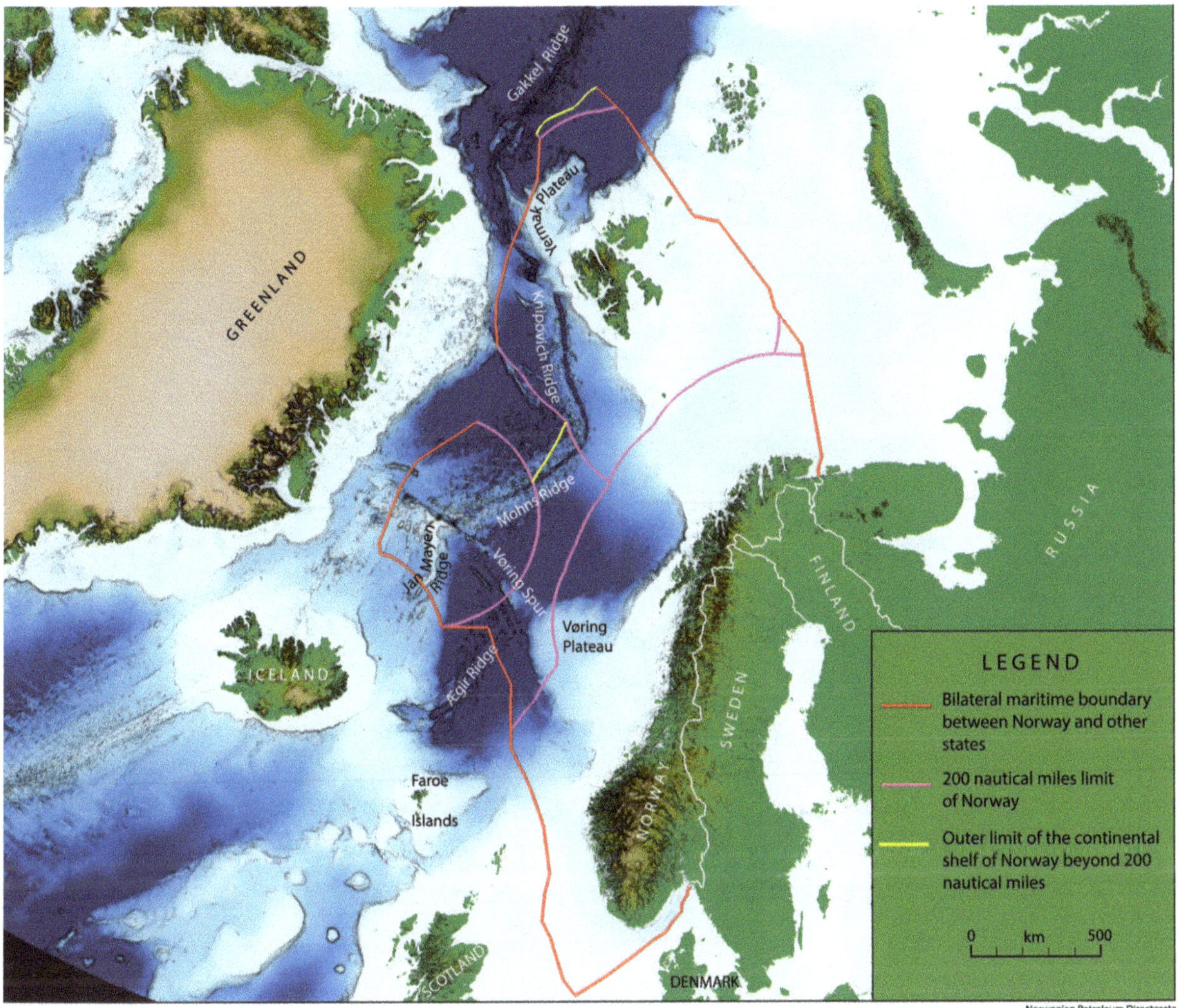

Figure 1-1 Norway manages one of the largest maritime zones in the world. The current Norwegian jurisdiction is defined by the red, yellow and purple lines on the map. Map source: Norwegian Petroleum Directorate / Harald Brekke and Becker et al. (2009). Click here for larger image.

It is possible that all of the marine mineral resources mentioned above are present to a certain extent within the Norwegian jurisdiction. However, to what extent needs to be quantified. Aggregates, hydrothermal deposits, crust and nodules have been positively identified. Figure 1–2 shows areas with potential for hydrothermal (sulphide) and crust deposits.

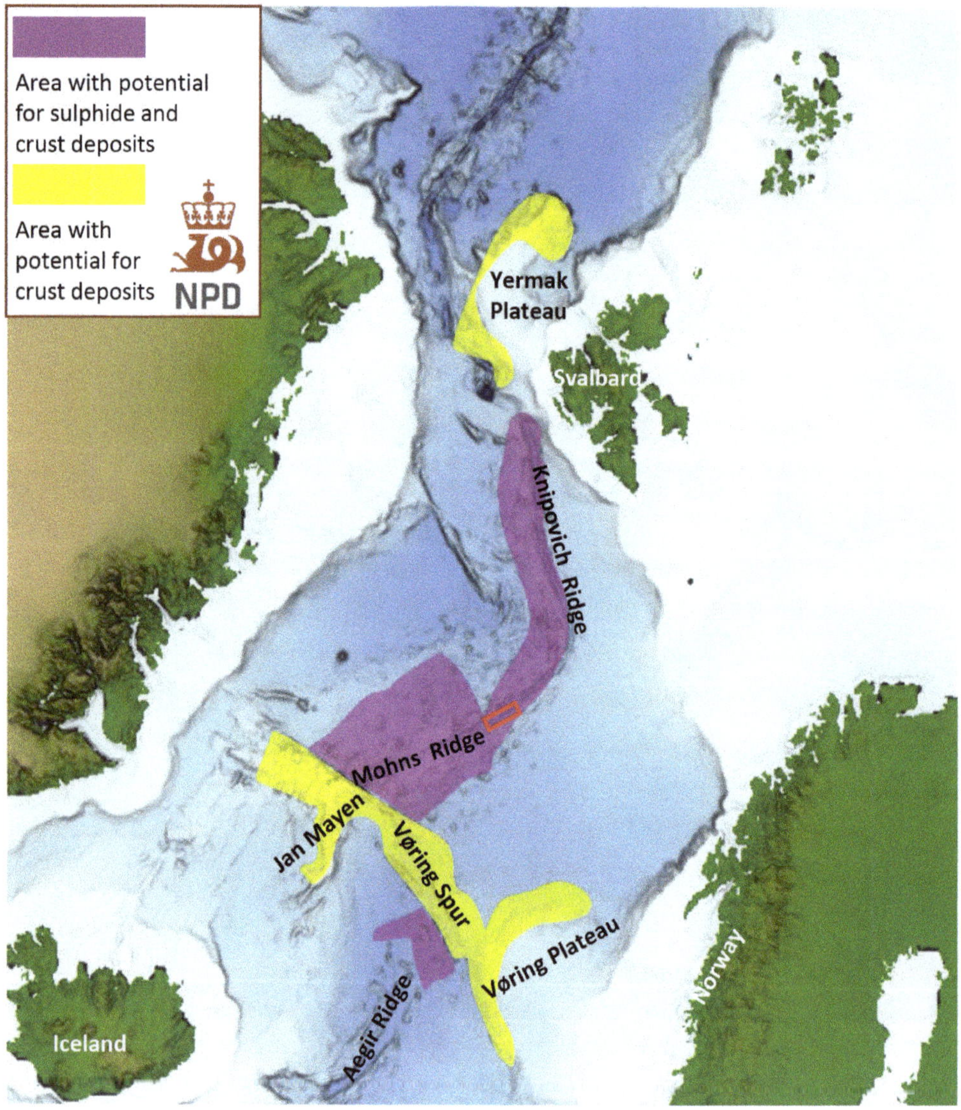

Figure 1-2 Map showing areas with a potential for sulphide and crust deposits inside Norwegian jurisdiction. The exploration area that was thoroughly investigated by the NPD in 2018 (NPD, 2018c) is indicated by the red rectangle. Map source: NPD, 2018c. Click here for larger image.

1.1 Nodules and crust

Nodules and crusts are concretions formed as a result of precipitation from the surrounding waters. Removal of elements from seawater and from the interstitial waters of sediments takes place through a number of reactions. These include direct precipitation, catalytic precipitation and absorption. The processes of direct removal of elements from seawater into sediments are termed hydrogenous phases, and generally tend to accumulate slowly. Manganese nodules and iron and manganese-rich (FeMn) crusts are primarily hydrogenous in origin and often cobalt rich (Cronan, 1992).

Exploration for nodules has been ongoing since the 1960s. Since nodules are primarily located in international waters, this is mainly handled by the International Seabed Authority (ISA). Interest in nodule mining has seen a revival in recent years with several new exploration licenses issued. The ISA has, to date, issued 16 exploration licenses as can be seen in Figure 1-3, 15 of which are located within the so-called Clarion-Clipperton Zone in the Pacific Ocean (ISA, 2018).

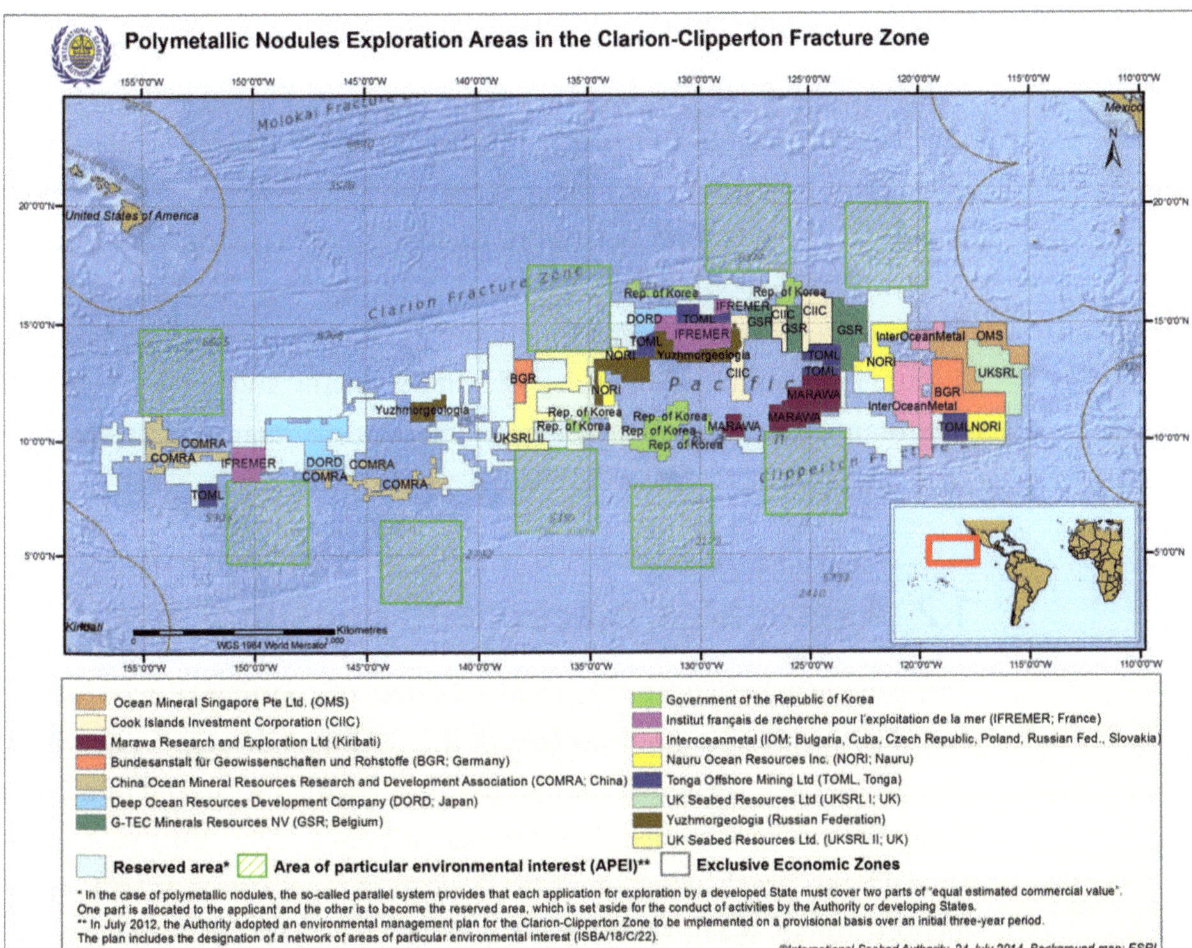

Figure 1-3 15 out of 16 exploration licenses for nodules issued by ISA are in the Clarion-Clipperton Zone in the Pacific (Image: © International Seabed Authority). Click here for larger image.

The Cook Islands also established a legal framework to issue nodule mining licenses in their EEZ in 2016. It is expected that they will begin to issue the first exploration license in 2019 (Cook Islands Seabed Minerals Authority, 2018). Four exploration licenses for FeMn crusts in the Western Pacific Ocean have currently been awarded by the International Seabed Authority (ISA, 2018).

Little is known about nodules and crusts in Norwegian jurisdiction, but findings from a 1980 cruise to ten locations in the Barents Sea indicate that ferromanganese concretions exist around Svalbard (Ingri, 1985). The deposits occur as discoidal and flat concretions and as coatings, in the latter case on lithified or detrital material or as extensive pavements on the Svalbard shelf. FeMn-rich coatings showed enrichment of Mo, Zn and Co in a Mn-oxide/oxyhydroxide mineral phase. The Fe-oxide/oxyhydroxide mineral phase holds high concretions of P and As. Two iron-rich concretions with high contents of P, Ca, St, Y, Yb and La were found east and northeast of Spitsbergen Banken, probably indicating upwelling of nutrient-rich, cold polar water along the Svalbard shelf. NPD (NPD, 2018d; Brekke, 2017) has done work on manganese crust inside the Norwegian jurisdiction. Initial chemical investigations indicate elevated values of rare earth elements (lanthanides, scandium and yttrium) and lithium compared to similar deposits in the Pacific and further south in the Atlantic Ocean. However, much more work is required to assess the economic value and density of these deposits.

1.2 Aggregates, placers and phosphorites

Terrigenous materials are the solid products of continental weathering and their composition reflects that of the continental rocks from which they are derived. They enter the oceans in a particulate form and remain in that form throughout their depositional processes. Clay minerals, quartz and feldspars are the most important silicate components in marine sediments (Cronan, 1992). These form a number of different deposit types.

Aggregates are non-metallic deposits consisting of sand, gravel, shell or coral debris and have been concentrated in their present occurrences by sea floor hydrodynamic processes, although their deposition may

The book initially gives a short geological background with focus on sea floor massive mineral resources. This is followed by an explanation of the play analysis methodology. One of the main challenges in play analysis, as it is used here, is to define the so-called permissive areas or tracts. How these are defined and their characteristics are explained in Chapters 4 and 5. Assessment results are presented in Chapter 6 and validated in Chapter 7. Chapter 8 describes the role of projects and activities like the MarMine Project (a research project funded by the Norwegian Research Council to study and sample one of the known active SMS systems) in validating the potential estimation results obtained in this and future studies. This chapter also includes a reference to the publicly-available results from two exploration cruises conducted by the Norwegian Petroleum Directorate, in 2018 and 2019. Some concluding remarks regarding resource potential are made in Chapter 9, along with a comparison and discussion of the results presented herein and the results from similar studies performed post 2014.

The book is the outcome of a collaborative project with participants from the Norwegian University of Science and Technology (NTNU), Nordic Mining ASA and Statoil ASA (now Equinor ASA) carried out between 2012 and 2019, and funded by Nordic Mining ASA and Statoil ASA. Steinar Løve Ellefmo and Fredrik Søreide have written most of the text and edited those portions drafted by the other researchers involved: Cyril Juliani, Krishna Panthi, Richard Sinding-Larsen and Ben Snook, all from NTNU, and Georgy Cherkashov, Sergey Petukhov and Irina Poroshina from VNIIOkeangeologia (Russian Institute for Geology and Mineral Resources of the Ocean).

Cyril Juliani and Ben Snook contributed with descriptions of the geology along the mid-Atlantic ridge, the geological processes and the analysis and characterization of sample material collected during the MarMine-cruise.

Richard Sinding-Larsen, along with Steinar Ellefmo and Fredrik Søreide, carried out the fundamental work in converting the analysis of the bathymetric data performed by researchers from VNIIOkeangeologia into the resource estimate. The results from this work were delivered to the project as a report referred to throughout the book as 'Ellefmo & Søreide, 2014'.

Krishna Kanta Panthi contributed the section 'Introduction to rock mechanics', which connects geodynamic analysis with rock mechanic principles, and he rewrote the 'Geodynamic analysis' section in chapter 4, 'Definition and description of permissive tracts in the absence of data'. He also contributed to the section on 'Morphostructural analysis'.

Georgy Cherkashov, Sergey Petukhov and Irina Poroshina carried out the analysis of the bathymetric /morphostructural data and thereby did the fundamental work necessary to define the permissive/favorable areas that were the input into the resource estimate. The results from their work were delivered to the project as a report referred to throughout the book as 'Cherkashov et al., 2013'.

CHAPTER 2

Geological Background

2.1 Regional setting and ridge morphology

2.1.1 Introduction

The Earth is dynamic. Landmasses and ocean floors have been in motion since the dawn of time and will continue to be into the (un) foreseeable future.

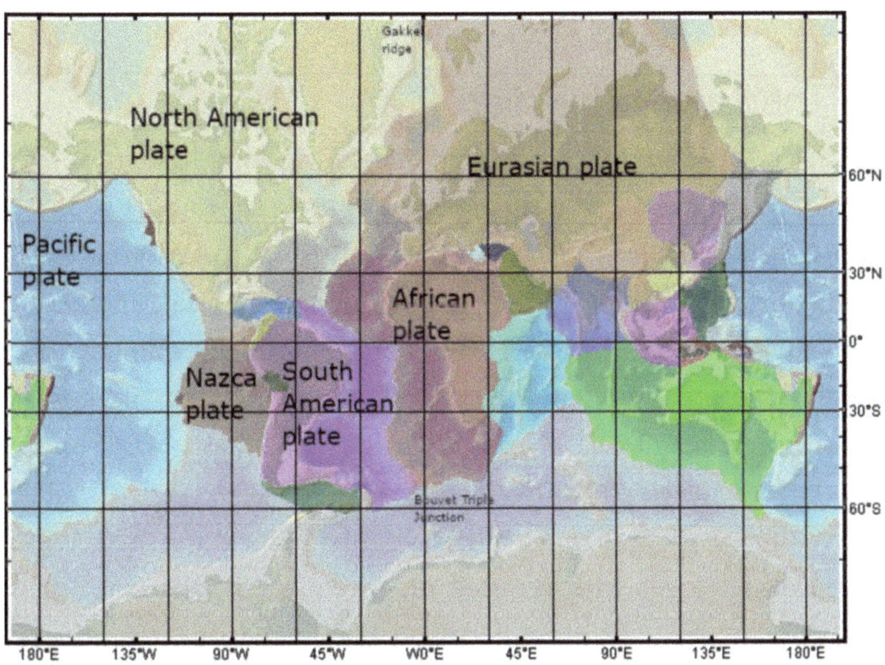

Figure 2-1 World map showing the landmasses, plate boundaries and mid-ocean ridges that run through the Atlantic, the Indian and the Pacific Oceans. Source: GeomapApp (http://www.geomapapp.org) version 3.6.8 and Global Multi-Resolution Topography Data Synthesis (GMRT Image Version 3.5). Click here for larger image.

Continents and ocean floors are destroyed at destructive plate boundaries where two plates meet, and new crust is made at divergent or constructive boundaries. We find destructive plate boundaries, for example, off the coast of Japan, where the Pacific Plate collides with and is pushed under the Eurasian Plate, and off the coast of South America, where the Nazca

Plate is pushed under the South American Plate and forms the Andes Mountains. The mid-ocean ridges belong to the longest mountain range in the world, constituting worldwide 64,000 kilometre long, volcanically-active and constructive plate boundaries. Water depths average about 2500 metres to the top of the ridge, reaching their deepest point of about 5500 metres in the Cayman Trough in the Western Caribbean and reaching above sea level in Iceland. In Iceland, the Mid-Atlantic Ridge (MAR) that splits the Atlantic Ocean floor into two parts, from the Gakkel Ridge northeast of Greenland in the north to the Bouvet Triple Junction in the south, crosses the Þingvellir National Park. The MAR separates the Eurasian and North American Plates in the north and the African and South American Plates in the south. See Figure 2–1.

2.1.2 The Knipovich Ridge

The Knipovich Ridge is an independent 550 km long system of spreading ridges in the North Atlantic, located in the eastern part of the Norwegian-Greenland Basin. See Figure 2–2. It joins the Mohns Ridge in the south, and is limited by the Molloy Transform Fault in the north. Its geological strike, practically in a north–south direction, is skewed (35°–65°) compared to that theoretically predicted for the spreading direction in this region (DeMets et al., 1990). All along the Knipovich Ridge, the well-developed rift valley has an average axial depth of about 3.5 km (Okino et al., 2002); its valley is rather typical for the sections of slow-spreading ridges with reduced magmatic feed. One can distinctly identify the 7–15 km wide inner floor and usually two or three ridge walls, often controlled by detachment faults, separated by rather wide terraces.

> A **detachment fault** is a large-scale fault structure associated with extensional tectonism. They can show tens of kilometres of displacements and play an important role in the formation of hydrothermal activity.

The ridge flanks that form the meridionally-extended plateau consist of volcanos and large massifs of complex origin, which are possibly oceanic detachments (Zonenshain et al., 1989). The eastern flank of the Knipovich Ridge is more subsided compared to the western side, a characteristic usually attributed to the accumulated sedimentary load (Kandilarov et al., 2010). On the other hand, no strong asymmetry is seen in the flange structure apart from the area north of 78 degrees north, where the eastern portion of the ridge valley joins directly with the continental margin of Svalbard. These characteristics of the Knipovich Ridge add a significant contrast to most other mid-ocean ridges situated far off continental margins. The system is situated close to the Svalbard and Barents Continental Margins, which were lifted up in the Late Cenozoic, some 60 million years ago. The area connecting the Mohns and Knipovich Ridges is close to the Bear Island Fan, which has a maximum thickness of more than 3 km of quaternary glacial sediments (Fiedler & Faleide, 1996; Faleide et al., 1996). The distal part of the fan reaches the spreading ridge. At the Mohns-Knipovich Ridge Bend, six multichannel seismic profiles have been collected and analysed in 2001 (Bruvoll et al., 2009). To the north of the bend, the Knipovich Ridge is situated close to the Bellsund and Isfjorden Fans, which have been in front of troughs on the continental shelf. About five million years old (late Plio-Pleistocene) fan complexes (Vorren et al., 1998) input sediments into the rift valley (Amundsen et al., 2011).

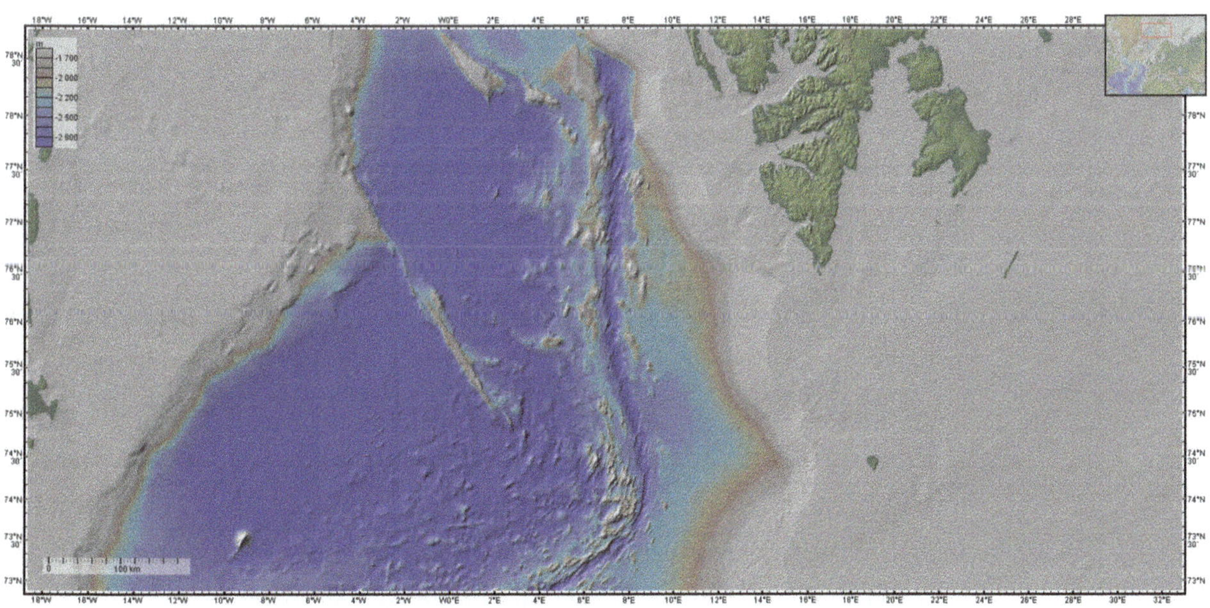

Figure 2-2 Knipovich Ridge characterised by a deep axial floor, high relief, highly oblique spreading and development of axial volcanic ridges. See inset for location. Source: GeoMapApp / GMRT Grid (Ryan et al., 2009). Click here for larger image.

Along the ridge, individual spreading segments are several tens of kilometres long and consist of two types: segments sustained by magmatic inputs and spaced at intervals of 85–100 km (Okino et al., 2002), and segments with lower relief that lack magmatism. Robust magmatic segments are considered to be loci of recent volcanic activity supported by mantle upwelling cells. The larger segment spacing observed on the Knipovich Ridge may reflect the large-scale thermal structure of the mantle that controls the focused magmatism. However, shallower geometric and kinematic factors, such as rift obliquity and plate separation rate, may also influence these processes. In particular, the southern Knipovich Ridge has lower relief spreading segment centres associated with high obliquity, slow spreading rates and reduced mantle melting, which results in a lower magma supply (Okino et al., 2002).

For the majority of the valley sections, one can hardly use the term 'volcanic segment' in its classical meaning, even though this issue is treated differently by different scientists, such as Crane et al. (2001) and Okino et al. (2002). It can be supposed that the morphostructural expressions for the Knipovich ridge develop in two stages, which is elaborated in the following. The topographic signature of breaks between segments, known as non-transform discontinuities, are clearly manifested on the valley flanks, along the zones running north-west and along the trend of the Paleo-Spitsbergen zone of displacement off-sets, which existed before the supposed beginning of spreading in this region (Crane et al., 2001). These signatures are expressed either by steep scarps with strike orientation having opposite inclination at the counter slopes of the valley or by sharp wedging of the axial structures directed north-east. In the first case, two systems of oppositely-inclining scarps intersect in the zones of non-transform displacement at the angle of 50° to 60°. Under these conditions, the elevation amplitude and volcanic activity are much higher. Thus, the tectonic processes at the segment boundaries control the arrangement of the areas of maximal magmatic activity, which clearly tends towards the zones of increased crustal permeability.

A **geomorphological (morphostructural) analysis** is an analysis of the morphology of the ocean floor that enables the identification of structures like flat-topped volcanoes, scarps, faults and axial volcanic ridges (AVRs), indicative of the probable presence of hydrothermal manifestations.

The Knipovich Ridge drops out of the formation scheme of an ordinary mid-ocean ridge. It lacks linearly-ordered magnetic anomalies (see Figure 2–4) and the major morphostructural elements are rather vaguely expressed. All this does not exclude a possible formation of sulphide ores on the Knipovich Ridge. Tectonism, as well as magmatism, are actively developed. Hence, hydrothermal venting followed by sulphide mineralisation is probable. Special studies are, however, necessary to reveal the criteria for predicting sites of possible SMS accumulation. These challenges are in this book addressed using geomorphological (morphostructural) analysis of the relief of the Knipovich Ridge and geodynamic modelling.

Basaltic lava flows of two age generations are visible on side-scan sonar images (Crane et al., 2001). Side-scan sonar is a method whereby acoustic images are acquired of the seafloor, often in great details, and show where young volcanic features and lavas are, or where the seafloor is covered in sediment. The youngest flows, scarcely covered with sediment, are less than several thousand years old (the sedimentation rate is 1 cm per 1000 years). In the northern part of the ridge, the young, sheet and pillow lava basalts are nearly equal and cover about the half of the valley floor. The largest volcanic structures form several linear ridges, 25–30 km long, with heights between 500 and 800 metres. These ridges cross the valley floor at a small angle and merge with the opposite scarps forming the valley walls. Moreover, many small irregular volcanic ridges have been recorded, including vast fields of predominantly developed sheet lavas, narrow basalt strips associated with small fracture outflows, and central type volcanoes. In the majority of cases, the volcanic manifestations are associated with linear tectonic structures (small faults and fractures) on the valley floor, and rarely with the large marginal faults. This is typical in spreading areas, where tangential stresses are close to zero. The maximal recent volcanism is recorded in the two segments south of 77 degrees north. Two regions, with somewhat more restricted development of the young volcanic rocks, are located in the intervals 77°15'–77°35' and 77°45'–77°58 north. There are also several volcanic centres located south of 76° N (Tamaki & Cherkashev, 2000).

The recent tectonic activity on the inner floor of the rift valleys is typically found at the areas of young volcanism and at the areas with thick sediment covers. It exhibits itself in the formation of an *en echelon* (oblique) system of small (the first tens of metres) fault scarps interfaced with linear fractures. The general strike of these structures varies between 25° and 35°, which approximately corresponds to the regional stress field developed due to sea floor spreading. Their end sections are often bent in the north-south direction, which suggests the existence of overall shear stresses. On the slopes of the rift valley, the modern tectonic activity is concentrated mostly at the zones of intersection of the longitudinal and transversal dislocations, corresponding to the non-transform discontinuities forming the spreading segment boundaries.

2.1.3 The Mohns Ridge

The Mohns Ridge, by the features of its structure, morphology and development, may be regarded as a mid-oceanic structure formed in the regime very close to a common MOR. See Figure 2–3.

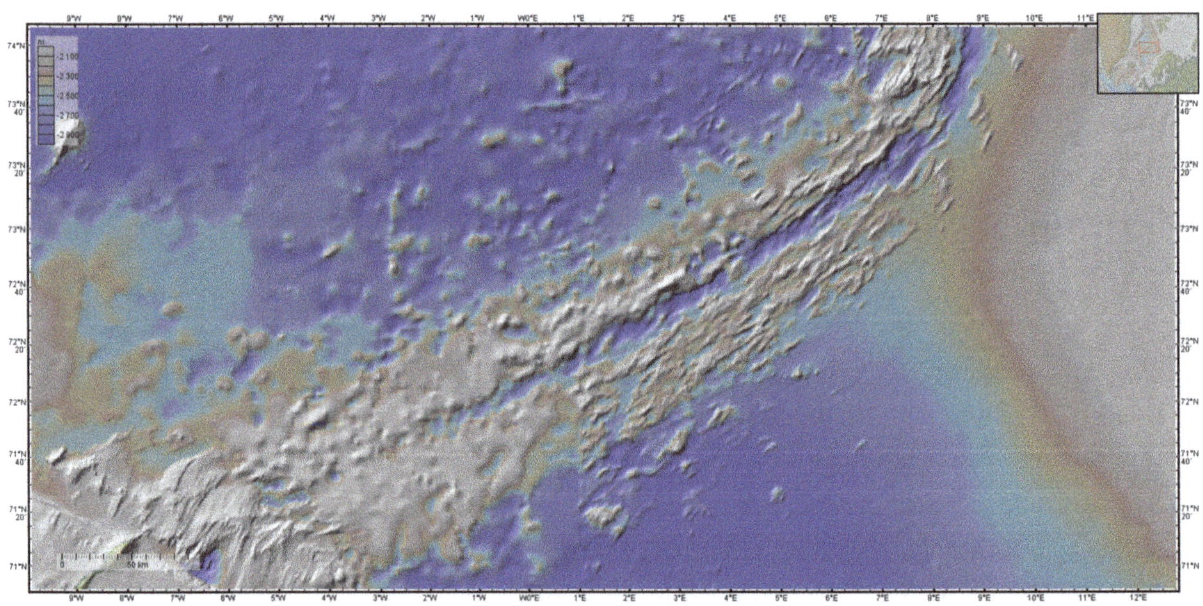

Figure 2–3 Mohns Ridge characterised by a deep axial floor, high relief, asymmetric development, oblique spreading and development of axial volcanic ridges. See inset for location. Source: GeoMapApp / GMRT Grid (Ryan et al., 2009). Click here for larger image.

The system of clear, linear magnetic anomalies indicates its spreading nature. See Figure 2–4. The absence of transform displacements and the skewed spreading are its abnormal features. Morphologically, it is well expressed. One may expect SMS type deposits in the valley among neo-volcanic elevations with rift structures, in the apical part on central-type volcanoes, at the edges with rotated blocks and flattened sub-horizontal terrace-like steps, outside the edges, between their upper rim and the toes of rift mounts. Several venting sites have been discovered in the region (Pedersen et al., 2010b). See Chapters 3.4 and 3.5.

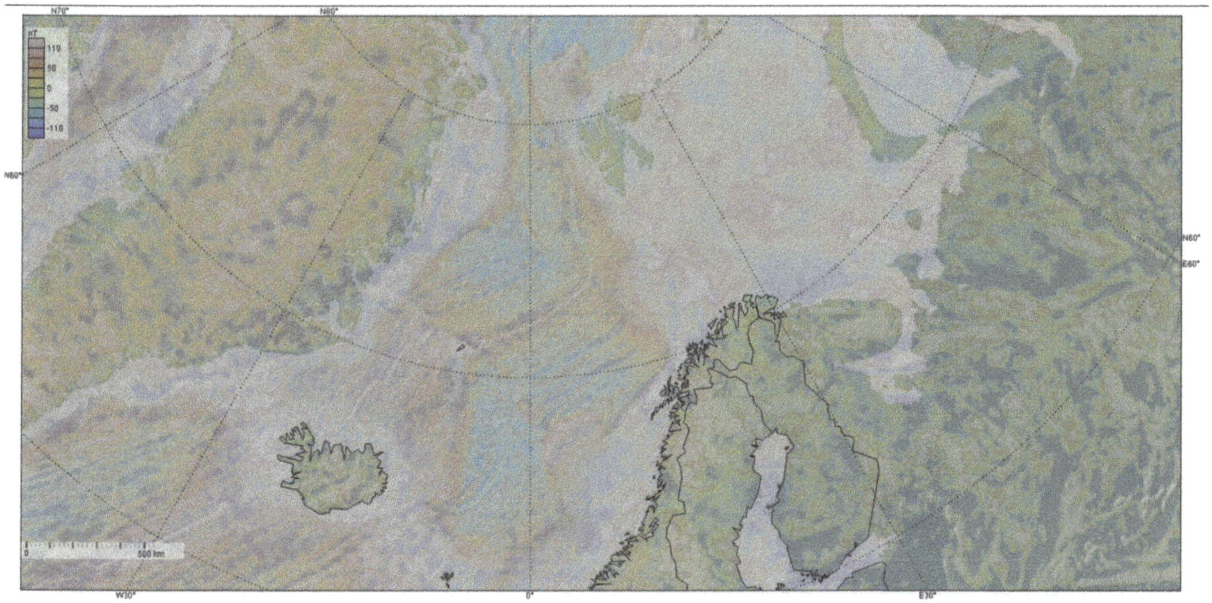

Figure 2-4 The magnetic anomalies on both sides of Mohns Ridge indicated by the blue and green repetitions indicating a varying direction of the magnetic field are not seen around Knipovich Ridge. Source: GeoMapApp (http://www.geomapapp.org) version 3.6.8 and the embedded 2-arc minute EMAG2-data from 2013 (http://www.geomag.us/models/emag2.html). Click here for larger image.

Mohns Ridge is an ultra-slow-spreading centre (<20 mm a^{-1}) situated within the Norwegian-Greenland Sea, that started opening approximately 53–55 Ma ago during the extension of the Mid-Atlantic Ridge (MAR) north of Iceland (Talwani & Eldholm, 1977; Torsvik et al., 2001; Vogt, 1986). It extends WSW-ENE for about 550 km from its southernmost part, delimited by the Jan Mayen Fracture Zone, to its northern junction with the Knipovich Ridge. The ridge presently spreads at a rate of about 15 to 16 mm/yr (Mosar et al., 2012) with an oblique relative displacement of 30° (Dauteuil and Brun, 1993). Its mean crustal thickness is of about 4.0 ± 0.5 km, a thickness considered lower than average for normal crust formed along the MAR (Klingelhöfer, 2000), but typical for ultra-slow-spreading ridges that generally show thinner and variable crustal thicknesses (Chen, 1992).

Along the ridge, water depth of the value floor ranges from 2.5 to 3.5 km below seawater level (Géli et al., 1994), and decreases progressively towards the south-west due to its proximity to the Icelandic hotspot (Talwani & Eldholm, 1977; Vogt, 1986). A characteristic aspect of the rift bathymetry is the strong asymmetric configuration of the flanks delimiting globally higher summits on the north-western sides (Vogt et al., 1982, Johansen et al., 2019). Moreover, the axial valley and the bounding flanks are marked by a series of en-echelon ridges interconnected by non-transform offsets (NTOs, see Figure 2–3) that commonly result from varying obliquity (Géli et al., 1994). This segmentation pattern occurs with intermittent volcanism characterised by basement highs elongated at the scale of approximately 20–30 km and oriented perpendicularly to the spreading direction. Previous studies distinguished these ridge segments to be volcanic edifices (Géli et al., 1994), while some involved tectonic origins solely with transtensional bounding faults (Dauteuil & Brun, 1993, 1996).

Non-transform offsets (NTOs) are areas with a gradual transition between spreading segments without any significant transform faulting. They are discontinuities (second-order) dividing the spreading centre into segments at intervals of 10 to 100 km. They represent significant descriptors of the tectonic processes along the MAR (Grindlay et al. 1991 and 1992; Sempéré et al. 1993; Gràcia et al. 2000).

Investigation of the bathymetry shows that the width of the axial rift valley, i.e. the distance between the rift walls, ranges in average from 5 to 15 km. The axial valley often deviates at NTOs close to deep basins (≥ 3000-m depth). These basins usually lack recognisable structures or volcanic edifices, but are commonly bounded by large, dextral east-west trending transverse faults possibly accommodating the depth variation of the brittle/ductile boundary (Dauteuil & Brun, 1996). NTOs have limited lateral structural expression, but their off-axis trace remains visible from alternating sets of continuous ridge-parallel terrains and from structures that trend slightly obliquely to the spreading direction. The volcanic activity at these transitions appears to be very restricted, and the importance of such zones for hydrothermal venting remains to be shown.

In the south-western part of the ridge, segments and transitions are difficult to identify because of irregular and complex bathymetric patterns associated with abnormally abundant volcanic activity, as well as to curved tectonic structures. The melting anomaly associates with more-shallow and smooth terrains as a result of the southward, hotspot interaction (Talwani & Eldholm, 1977; Vogt et al., 1981), whereas the curved structures respond to oblique extensions. In contrast, the northernmost Mohns-Knipovich Ridge Bend consists of a more symmetric ridge segments expressing a rough topography at the higher western side compared to the sediment-covered eastern flank, where only a few basement ridges are exposed. Along the overall ridge, the western flanks are, on average, 200 m higher than the opposite side, due to higher tectonic activity; their edges may sometimes correspond to rotated and uplifted fault blocks forming downward curved ridge flanks that are usually attributed to low-angle detachment surfaces (Smith et al., 2008).

Apart from the anomalous segment adjacent to Jan Mayen, most of Mohns Ridge has a well-developed rift valley with average depths of 2000–3200 m. In the north-east direction, a general drawdown of the inner rift-valley floor and the flange highs are noticed. The western flank is about 500 m higher than the eastern one. Hence, at the western flank, the tectonic deformations should be much higher compared to the eastern one.

Grabens are suppressions in the terrain or the ocean floor, caused by (normal) faulting in an extensional tectonic environment where the central part of the suppression sinks in compared to the flanks.

The morphology of the axial volcanic rises varies considerably, which may clearly be expressed by 400–800 m high axial rises. These rises are sometimes strongly deformed by grabens. There is an unusual abundance of large, flat-topped central-type volcanoes associated with different morphostructural settings in the valley floor and at its eastern flank.

The strike of the valley reaches 60 degrees, though the majority of its segments are separated by the right-slip offset and oriented at 40–50 degrees. The amplitude of its lateral offsets reaches 5 to 7 km in the northern part of the ridge, 8 to 15 km in the central part, and is rather irregular at the south where individual segments are often difficult to identify. In total, 15 spreading segments, with an average length of about 30 km, are recognised along Mohns Ridge.

2.2 Formation of seabed massive sulphide (SMS) deposits

Globally, there are some 64,000 km of mid-ocean ridges (MORs) where SMS deposits occur and, most of the time, these are outside the jurisdiction of national exclusive economic zones (EEZs). However, within the Norwegian extended continental shelf, the SMS deposits are formed at mid-ocean on the Mohns and Knipovitch Ridges, where Norway regulates exploration and potential mining of SMS deposits. Because of the extensional nature of MORs, melting due to pressure decreases produces mafic magmas that form new oceanic crust. The process involves crack propagation, where cold seawater flows in and hot, mineral-rich seawater flows out, creating chimney-like structures and plumes known as black smokers. Some of these cracks may extend thousands of metres into the lower crust. See Figure 2–5.

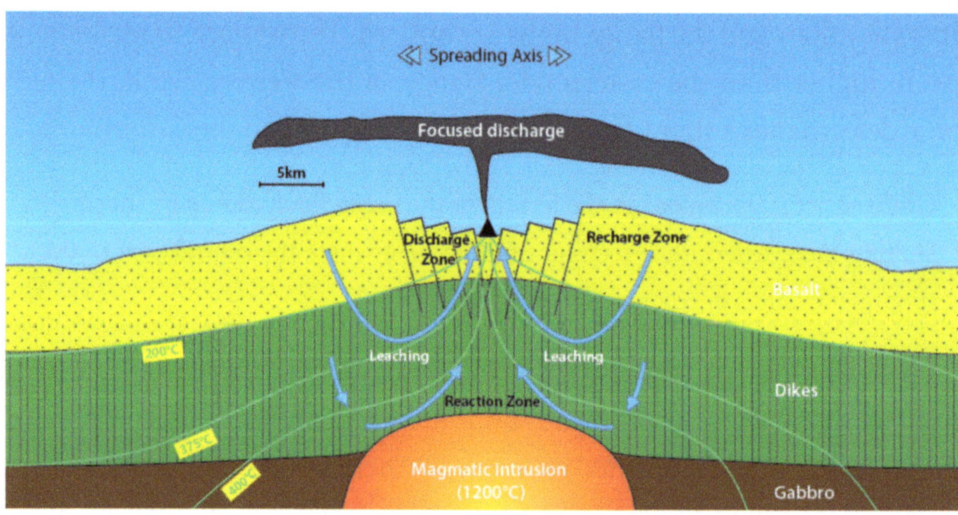

Figure 2–5 Formation of seabed massive sulphide deposits. A diagram depicting a mid-ocean ridge and hydrothermal circulation. Click here for larger image.

Conventional models of hydrothermal systems comprise sub-sea floor seawater circulations driven by a heat source, where seawater migrates downward through cracks and reacts with crustal or upper mantle rocks at several kilometres depth. During this process, seawater can be heated to temperatures of 400–500C, high enough to modify its physical properties and will eventually find its way upwards through features facilitating hydrothermal flow. This downward-upward convection scheme involves leaching of metals from the surrounding rock via chemical reactions. Correspondingly, the heated seawater evolves to a fluid more acidic, reduced and enriched in dissolved metals and sulphur. When these hydrothermal fluids rise to the sea floor, it expels most of its metal content in the water column through chimney-like structures (black smokers). A part of these components remains in the upper part of the sub-sea floor where metals precipitate into metal sulphides from conductive cooling and mixing of hot hydrothermal fluids with cold seawater. From these processes, metals of economic interest accumulate on the sea floor and underneath by mineralisation of the rock basement, which results in forming SMS deposits that contain significant amounts of sulphide minerals (e.g. Robb, 2004). These deposits may be comparable to or larger than some deposits on land (e.g. Scott, 2004).

SMS occurrences in the deep sea may occur at different tectonic settings like the mid-ocean ridges (MORs). Presently, all the known vents on slow or ultra-slow spreading centres, like the Mohns and Knipovitch Ridges, are distributed in various geologic settings; both magmatic and tectonic-dominated settings. The former are directly controlled by axial volcanic structures controlled by the presence of a high-temperature magma under the seabed. The latter are usually located at the more-or-less remote flankes of the rift valley, mainly in the areas of the development of large offset detachment faults forming structures called core complexes. In this setting, deep penetrating faults represent the dominant geological controlling process on hydrothermal circulations, and can, under certain circumstances, associate with underlying magmatic processes.

2.3 Factors controlling the location and the size of SMS deposits

Both volcanic and non-volcanic tectonic activity is usually long-lived at slow to ultra-slow mid-ocean ridges. This provides stability for hydrothermal systems (Hannington et al., 2005; Rona, 2008). The estimated frequency of venting sites along the mid-ocean ridges varies from 1 to every 400 km to 3 per 100 km (e.g. Baker & German, 2004; Hannington et al., 2010, 2011b). Faults play a vital role, both in charging and discharging hydrothermal systems, by acting as fluid conduits. Hydrothermal manifestations may, therefore, occur on structural highs many kilometres off-axis. The Snake Pit hydrothermal field occurs at a neo-volcanic ridge (Fouquet et al., 1993), and the TAG field is situated approximately 3 kilometres from the axis (Rona et al., 1993; Kleinrock et al., 1996). Hydrothermal systems can be driven by both serpentinisation and magmatic processes. These could act in combination, as in the Rainbow (Fouquet et al., 1997) and Logatchev fields (Mozgova et al., 2005). Slow-spreading ridges are structurally complex, showing evidence of tectonic extensions, large amounts of faulting (both detachment faulting and normal faulting) and exposure of gabbros and partly serpentinised ultramafic rocks in oceanic core complexes (Smith et al., 2008).

2.4 Loki's Castle and Mohns' Treasure venting sites

Loki's Castle is a hydrothermal vent field at the ultra-slow-spreading Arctic Mid-Ocean Ridge (Baumberger et al., 2010; Pedersen et al., 2010a, 2010b). The hydrothermal fluids show indications of having leached both sediments and mafic rocks in the ocean crust for metals. See below and Baumberger et al. (2010).

The ore-bearing vent, named Loki's Castle, is located at 73°30' N., 8° E. It lies on an axial volcanic ridge formed by young pillow lavas. The chimneys are up to 11 m high and show a maximum temperature of the fluids of 317° C. The content of CH_4 and H_2 is interpreted to indicate the presence of ultramafic rocks at depth (Pedersen et al., 2010b). The geochemical properties are similar to those of the Middle Valley and the Endeavour Ridge in the Pacific Ocean.

Loki's Castle is partly overlapped by a sediment cover. The region features high tectonic stress – a combination of fault disturbance and domical rises providing potential channels for the fluids at the formation of Loki's Castle.

Samples collected show a maximum metal content for Fe at 31wt%, for Zn at 5.4wt% and for Cu >1wt%, and the sediments contain non-insignificant amounts of metals and average 4.6 wt% Fe, 100 ppm Zn and 33 ppm Cu (Pedersen et al., 2010b).

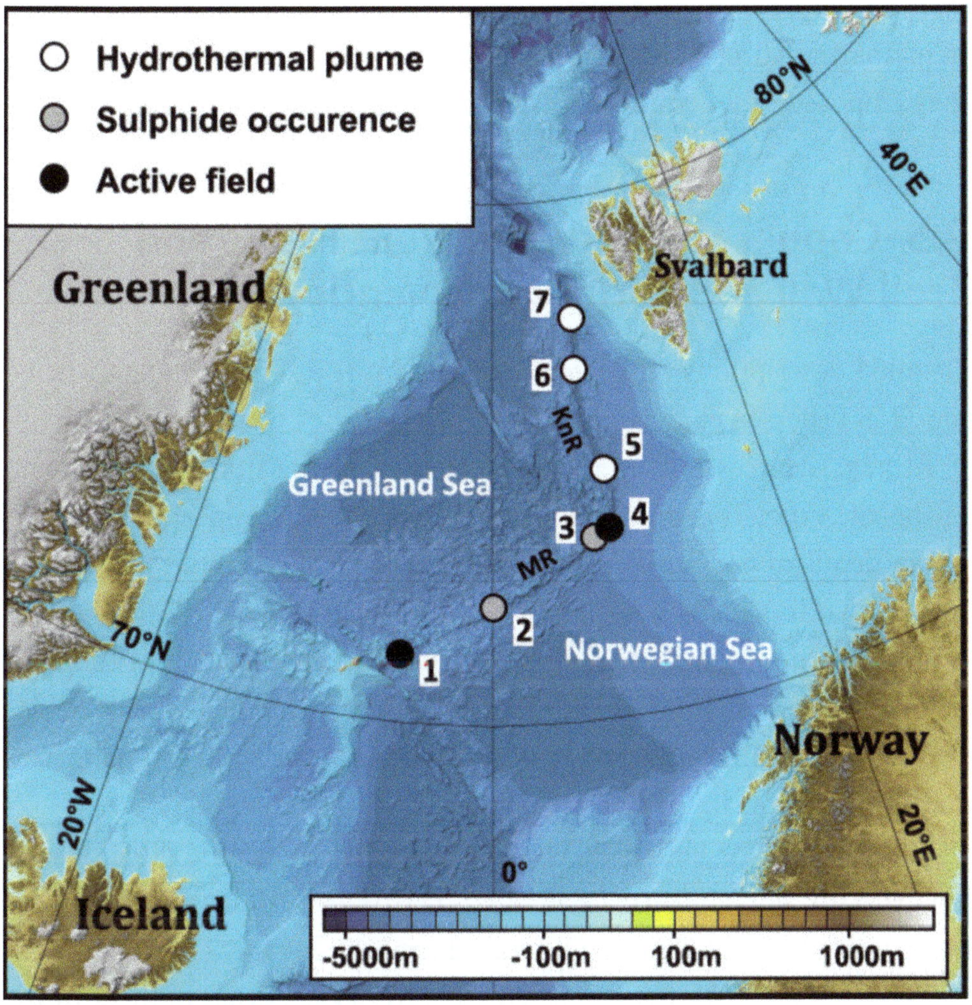

Figure 2-6 Active fields, sulphide occurrences and plumes along Mohns and Knipovich Ridges. MR = Mohns Ridge, KnR = Knipovich Ridge. Source: NTNU / Cyril Juliani. Click here for larger image.

1. Jan Mayen (Troll Wall, Soria Moria)
2. Copper Hill
3. Mohns' Treasure
4. Loki's Castle
5. 75°N. Plume with anomalies of LS, CH4, pH and ATP (Tamaki et al., 2001).
6. 76°48'N. Water-column temperature anomaly near north flank of submarine volcano (Sundvor, 1997).
7. 77°40'N. High CH4 and He-3 ratios observed. Potential occurrence of serpentinisation reactions (Connelly et al., 2002).

Baumberger et al. (2010) have investigated the chemical composition and the stable and radiogenic isotopes of the vent fluids exhaleded from the Loki's Castle vents and quantified the associated signatures. These signatures are compared to both sediment-influenced and sediment-starved systems. A pH of approximately 5.5 and relatively high concentrations of methane, hydrogen and ammonia are reported. Based on their study, they emphasise the sedimentary contribution.

2.5 Copper Hill and Ægir Vent Fields and Troll Wall and Soria Moria venting sites

Detailed descriptions of all vent sites except the Ægir vent site can be found in Pedersen et al. (2010b).

At 72.32 degrees north, 2.10 degrees east, in the north-western part of the rift valley is a mineralised vent zone called Copper Hill, located at the depth of 900 m (Pedersen et al., 2010b). The samples are shown by highly-mineralised breccias. The main mineral is quartz (more than 50%). Sulphides (mainly chalcopyrite) make up around 30% of the total bulk volume. The minerals are altered. It is accompanied by iron oxides and malachite. The temperature of formation ranges from 270° C to 330° C (Cherkashov et al., 2013).

In 2015, the Ægir Vent Field was discovered on a volcanic ridge close to Copper Hill, at 2200 metres water depth. This vent field is reported to be comparable to Loki's Castle (Olsen et al., 2016).

Two vents at 71° north are located 5 km from each other in the utmost-south section of Mohns Ridge.. In terms of depth, it is the shallowest section and its most elevated points reach 500 m. Seismic surveys have shown that the crust is 10 km thick. The major volcanic structure associated with the Troll Wall is a volcano with a 5 km caldera built on glacioclastites. A rift 150 m deep and 2 km wide transects the volcanic structure. The zone of venting is associated with the fracture of the eastern board of the rift. Deposits enriched in iron have been recorded here. The active part of the fracture is 1 km long. The vent consists of ten chimney structures 5–10 m high. The chimneys are distributed on the slopes of mounds of hydrothermal origin, and are emitting white smoke and CO2. The chimney material is composed of anhydrite, barite, sphalerite and pyrite. Around the focused sources, through the piles of blocks, diffused outflow of hydrothermal fluids can be seen. They are associated with bacterial mats, and the temperature of underlying sediments is 80°C (Cherkashov et al., 2013).

Soria Moria lies south of the Troll Wall, on the top of a volcanic edifice at a depth of 700 m below sea-level. The venting area is about 7 km^2. Two venting sites 100–200 metres in diameter each can be seen. The hydrothermal deposits overlap lava flows. The smokers are of two types: white smokers are represented by multiple sulphurised chimneys 8–9 metres high. The others are transparent smokers of lower temperature. They are built of barite and silica with a small part of pyrite, sphalerite and galena. The chimneys are 10 m high and 15–20 m wide. The walls of the white smoker chimneys are covered with bacterial mats.

Both fields at 71° north feature high-temperature venting (270° C), a pH ranging from 4.1 to 5.2, and emissions of methane and CO2.

CHAPTER 3

Play Analysis and 3-part Assessment

3.1 How undiscovered resources are estimated

The SMS resource assessment project on the Norwegian Continental Shelf estimates undiscovered resources with the aid of a method known as play analysis. This is a recognised approach, mostly used by the petroleum sector (NPD, 2013), but equally applicable to the mining sector (Juliani & Ellefmo, 2018b). Similar methodologies have been applied to assess mineral resource potential worldwide and in the USA for different commodities (Singer et al., 2010) as well as for oceanic seabed resources (Singer, 2014). The assessment process applied herein involves systematising and describing the SMS metallogenesis on the Mid-Atlantic Ridge, with additional information from appropriate ancient massive sulphide analogues. Play analysis shares many similarities with 3-part assessment (Singer et al., 2010). Five basic questions are addressed:

- Where and what represents the permissive tracts (favourable areas) for potential hydrothermal SMS resources?
  - The answer to this question includes a georeferenced region attributed an area in square kilometres.
- What are the chances that hydrothermal SMS resources exist inside the permissive tracts?
  - The answer to this question is a probability from 0 to 1 or 0 to 100%.
- If hydrothermal SMS resources exist and extensive exploration is performed, how many future accumulations will be found?
  - The answer to this question is basically a number. The number presupposes that the tract actually contains accumulations.

- What is the expected size distribution of future accumulations?
  - The answer to this question is the quantification of the min., max. and expected accumulation tonnage, including an assumption on the shape of the distribution.
- What types of metals and what grades will the future accumulations have?
  - The answer to this question is the naming of what elements and metals that are expected to be found and their grade distributions.

The SMS resources from both Mohns and Knipovitch Ridges are in this study regarded as coming from the same super-play. Each favourability area outlined in Chapter 4 will accordingly be modelled as a sub-play of the ridge super-play. Estimated resources are more uncertain the less is known about the SMS mineralisation process, and this project specifies accordingly resources with an uncertainty range. A probabilistic framework is chosen (GeoX, 2014) that can be used to estimate the amount of metals which could be proven and potentially produced from each sub-play given a deposit-modelling approach (Sinding-Larsen & Vokes, 1978). In this early assessment, the focus has been on aggregate contained metals rather than on grades and tonnages of individual SMS accumulations.

A **play** is a geographically and stratigraphically delimited area where a specific set of geological factors, such as recipient volumes, source rocks, leaching and hydrothermal flow paths, are present so that SMS deposits may be provable.

The project's estimate for undiscovered resources on the 1500 km MAR inside the extended continental shelf for Norway includes only the 1030 km of the ridge that has been interpreted by VNII Okeangeologia (Cherkashov et al., 2013). A small part of international sea floor intersects the assessment area on central Mohns Ridge, but without consequences for the assessment results in this study.

A permissive tract (Singer, 2010) is used in this work as a play.

Seventeen separate SMS sub-plays are defined along the ridges, based on morphostructural and geodynamical indicators. The plays encompass both the favourability areas that are outlined by the four morpho-favourable areas (MF-areas) on Knipovitch Ridge and the twelve morpho-favourable areas on Mohns Ridge, as well as the eight hydrothermal favourability areas (HF-areas) defined by discharged and stressed areas along the ridges (Cherkashov et al., 2013).

A play probability factor is a factor that needs to be present and effective in order for an accumulation to form. The two play probability factors (permissive tract probability factors) considered here are factors that combine the probability of having a metal and an energy source and effective migration of metal-laden hot fluids, and an effective precipitation and preservation of minerals and metals in or on the ocean floor:

1. The presence of an adequate source for metals, derived from the oceanic basement where a seawater hydrothermal convection system (migration) above a sub-axial magma chamber is capable of leaching elements such as Cu, Zn, together with Au, Ag from the crust.
2. The presence of adequate recipient volumes that could accommodate the precipitation and accumulation of sulphide concentrations, either on
   - neo-volcanic elevations within rift structures, associated with high-amplitude flank faults, e.g. axial volcanic ridges (AVRs)
   - in the central part of the central-type volcanoes
   - at the edges of the axial valley where angular relief blocks can be seen
   - on flattened sub-horizontal terrace-like steps, or
   - outside the spreading ridge flanks or between the upper rim and the toes of rift mountains with a large elevation contrast.

A confirmed SMS sub-play is characterised by the presence of at least one black smoker. Such a sub-play would have a play probability of 1, both for the adequacy of factor 1 as well as for factor 2. If the presence of black smokers has not been confirmed, the SMS sub-play remains unconfirmed. Since the sub-plays have been defined based on morphostructural elements referred to in adequacy factor 2, the confirmation of the SMS sub-play is associated with ensuring that the play probability for adequacy factor 1 is fulfilled.

The individual SMS probabilities at the local scale, given that the play is confirmed, are controlled by the following three principal conditions for the local presence of a potentially-producible massive sulphide accumulation within the SMS sub-play:

1. ...that a potentially black smoker with a given minimum tonnage will be found if the sub-play is thoroughly explored, i.e. the existence of a local manifestation of hydrothermal activity
2. ...that the hydrothermal activity indicates the presence of polymetallic sulphides (i.e. a black smoker)
3. ...that the potential SMS accumulation at least has a tonnage and grade that corresponds to the defined minimum values.

3.2 SMS accumulation; deposit density and size (tonnage) and grades

The local existence of vent fields that may host a SMS accumulation is a fundamental component of a SMS play analysis. The number of vent fields and the amount of metals each of them could produce determines the estimated resources for each SMS sub-play. The estimation of the number of vent fields with hydrothermal manifestations in a sub-play therefore requires an unambiguous definition of what constitutes a 'SMS accumulation'. Unfortunately, as stated by Hannington et al. (2011a), descriptions of SMS accumulations in literature have included everything from a single vent or chimney to a large mound or cluster of vent complexes. It is, therefore, important to ensure consistency between the distribution of sizes and the density of accumulations (Singer, 2014). The well-mapped Troodos massive in Cyprus has, for this reason, been chosen as an analogue for the mineralisation on the slow-spreading Mohns and Knipovich Ridges. Compelling arguments that the Troodos ophiolite is a valid analogue for MAR have been advanced (Granot et al., 2011), where it has been demonstrated how the Troodos ophiolite can unveil the process of gabbroic intrusions along a slow-spreading ridge.

In this study, information from the Cyprus analogue (Singer & Mosier, 1986), from MAR deposits (e.g. Fouquet et al., 2010 and Cherkashev et al., 2013) and global grades for MOR (Hannington, 2013a, 2013b) have been used to model vent SMS field sizes and metal contents. All sub-plays have been modelled with the same tonnage and grade distribution, due to a lack of information that could be used to differentiate between the sub-plays.

CHAPTER 4

Definition and Description of Permissive Tracts in the Absence of Data

4.1 Definition of permissive tracts

4.1.1 Introduction

This sub-chapter introduces rock mechanics and describes briefly how this is used in this study in geodynamic analyses, to highlight de-stressed zones prone to show hydrothermal manifestations. It continues with the basis for the morphostructural analysis and its link to deposit probabilities. Sub-chapters 4.1.3 and 4.1.4 describe the different permissive tracts that have been defined based on both geodynamical and morphostructural analysis. Details of the 17 sub-plays are provided in Sub-chapter 4.2. The identification of zones favourable for hydrothermal vent activity is based on the following indicators:

a. tensional structures (purple dotted line in Figure 4–12 and Figure 5–13) in the stress distribution maps associated with the highest gradient (black line in Figure 4–12 and Figure 4–13)
b. locations where the tangential stress acting on any fracture, fissure or discontinuity in the upper layer (see Sub-chapter 4.1.3) is close to zero (yellow line in Figure 4–12 and Figure 4–13. Area where a) and b) coincide is marked with a red line / red area in the figures.)
c. overlap between points described in a) and b) above and morpho-structural hydrothermal indicators, like axial volcanic ridges with graben structures, structures associated to detachment faults and ocean core complexes, longitudinal faults, transversal dislocations and large flat-top volcanoes.

Delineation of the most prospective areas for finding SMS, based on the geodynamical analysis, are outlined as white boxes in Figure 4–12 and Figure 4–13 . Prospective areas outlined based on the morphostructural analysis are shown in Figure 4–16 and Figure 4–17 with yellow dotted lines. These yellow dotted lines have formed the basis for the definitions of the permissive tracts.

4.1.2 Introduction to rock mechanics

The term mechanics means the study of the equilibrium and motion of bodies, which includes static and dynamic forces. Modern rock mechanics deals mainly with engineering application of a rock mass to civil, mining, petroleum and environmental engineering (Harrison & Hudson, 2000). However, the Earth's dynamic body can also be related to rock mechanic principles, because there is a considerable influence on rock mechanic principles by natural processes that occur in the Earth's crust. This comes from plate tectonic movement, resulting in folding, faulting, spreading and mountain-building processes as indicated in Figure 4–1.

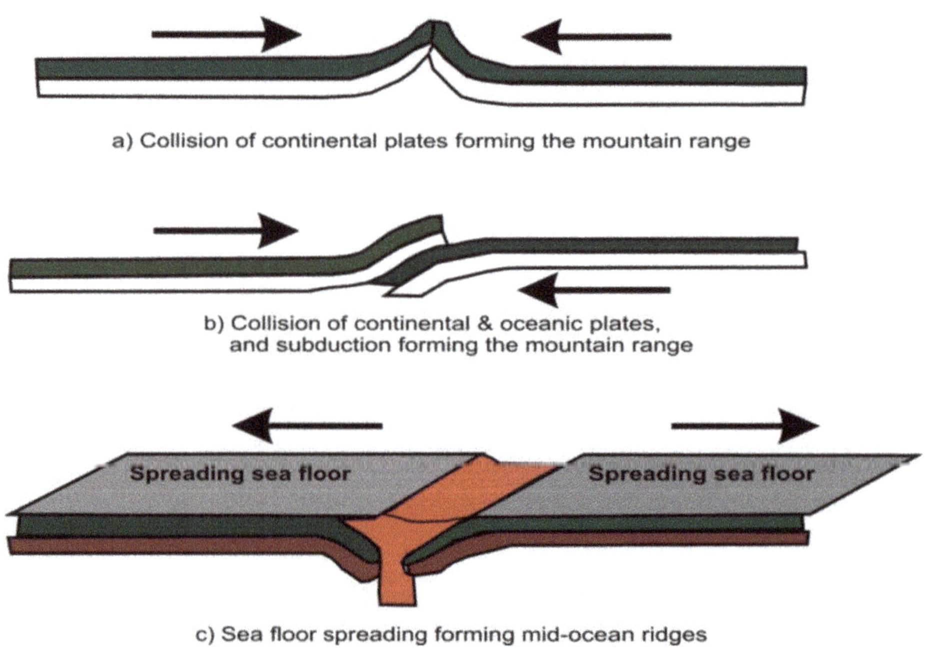

Figure 4–1 Conceptual mechanism of plate movement in the Earth's crust (Panthi, 2016). Click here for larger image.

Therefore, rock mechanics is a discipline that deals with forces that influence or change stress and strain environments in the rock mass. As seen in Figure 4–1, there are three main dynamic processes that are active in the Earth's crust:

a. mountain ranges are formed due to the collision of two continental plates of similar density, such as the Caledonides Mountain Range in Scandinavia
b. mountain ranges are formed due to the collision of continental and oceanic plates and subduction of the oceanic plate beneath the continental plate
c. formation of mid-ocean ridges due to sea floor spreading, a tensional environment where tangential stresses close to zero can be found

These dynamic processes influence the rock mass of the surrounding areas of collision, subduction and spreading considerably. These influences may be explained by the orientation of in-situ stresses in the vicinity of concern (Figure 4–2). These are mainly related to the dynamic tectonic movement of the Earth's plates. According to Panthi (2012), the magnitudes of total tectonic horizontal stress vary considerably and depend upon geographical location, geological environment (homogeneity in the rock mass), and distance from the main tectonic fault systems.

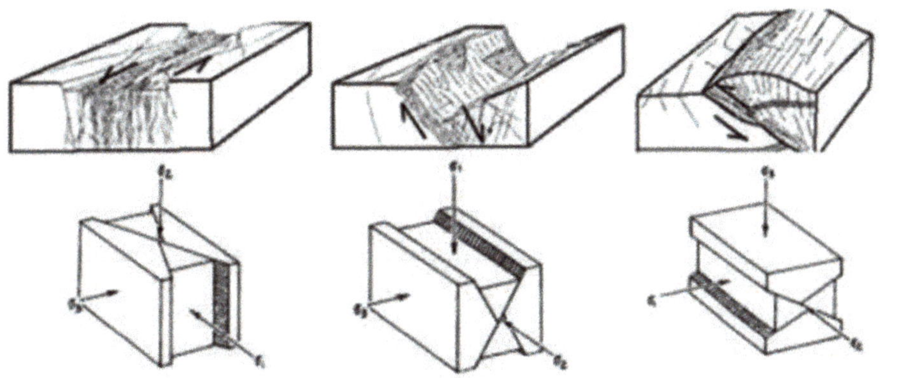

Figure 4–2 Anderson's classification of faults (in Panthi, 2006; modified from Braathen & Gabrielsen, 2000; and Rowland & Duebendorfer, 1994 . Illustration © K. K. Panthi; no reuse without permission). Click here for larger image.

In the course of the formation of mid-ocean ridges, many splays of long discontinuities (fracture systems) are formed, following the principle of the strike-slip faulting system (Figure 4–2a) caused by sea floor spreading, where major principle stress (σ1) and minor principle stress (σ3) are where the horizontal and the tangential stresses are close to zero. Figure 4–3 shows tectonic stress orientations worldwide.

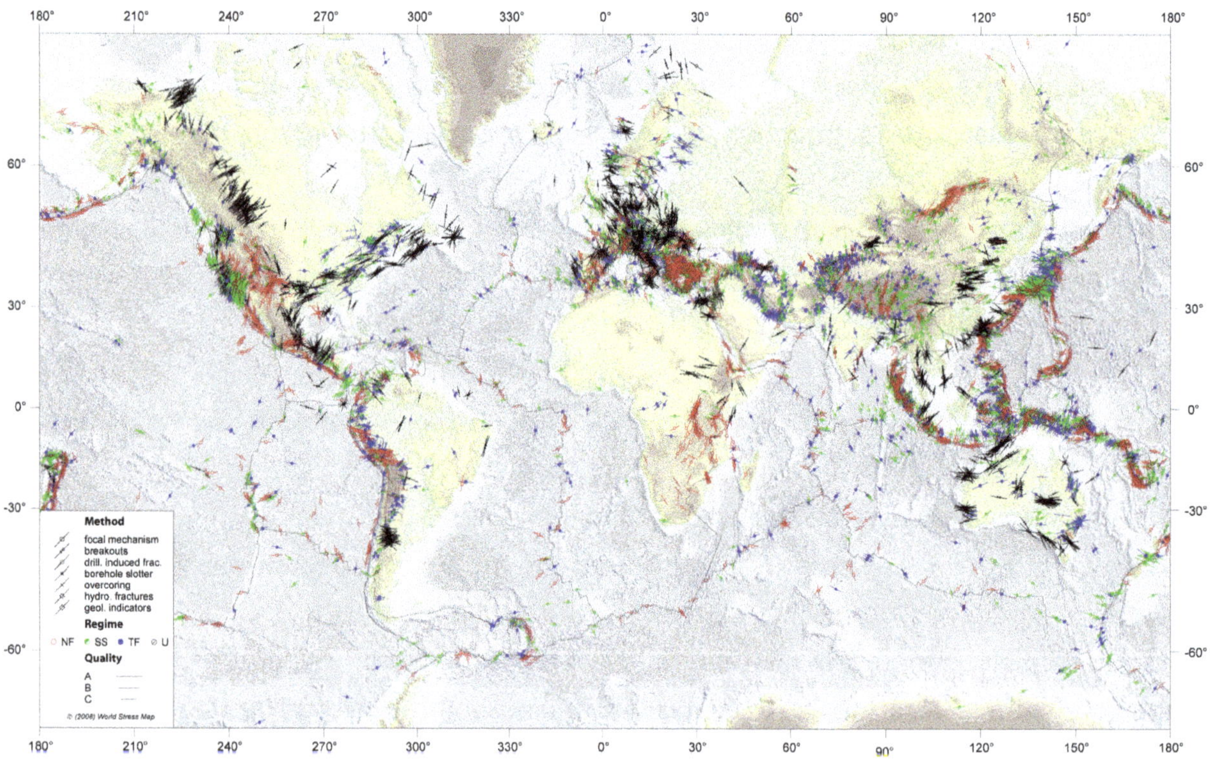

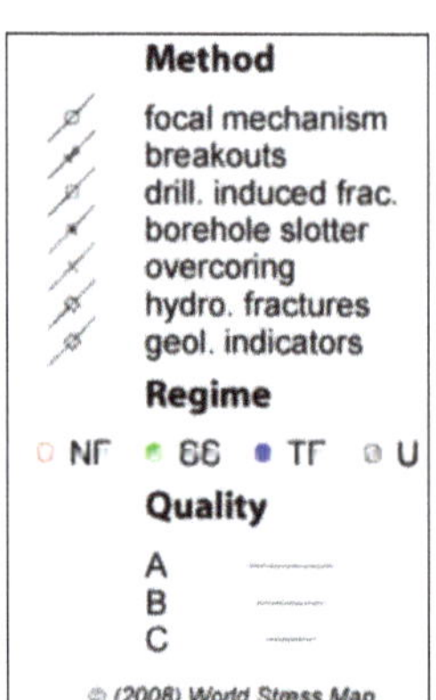

Figure 4-3 World Stress Map showing tectonic stress orientation and indicating plate division (Source: Heidbach, O., et al. (2018). The World Stress Map database release 2016: Crustal stress pattern across scales. *Tectonophysics, 744,* 484-498. http://doi.org/10.1016/j.tecto.2018.07.007). Click here for larger image.

This principle can be used to detect the areas that are prone to contain valuable minerals connected to sea floor massive sulphide deposits in the mid-oceanic ridge region. This chapter deals with assessing the possibilities for such anomalies along the Arctic Mid-Ocean Ridge (AMOR).

The tectonic stresses may be related to the current tectonic activities caused by the movement of tectonic plates. On one occasion, the rock mass may be strong enough to sustain these movements without any failure in the rock mass, leading to the storage of a considerable amount of in-situ tectonic stresses. On other occasions, the rock mass may become completely fractured, crumbled and folded, leading to a de-stressing situation. This is the reason why many regions in the world, where tectonic plates are in a state of limiting equilibrium due to strong rock mass, have high horizontal stress magnitudes, like in Scandinavia.

All three processes, shown in Figure 4–1, contribute to the overall stress state at a given point in a rock mass. In addition, other factors can alter the stress state, such as erosion, which will reduce the vertical stress component more than the horizontal components. Furthermore, in a condition where tangential stresses have magnitudes close to zero, a tensional environment will be created causing fractures at all scales and a typical de-stressed field.

Similarly, the concept of strain needs to be understood, so that rock mass deformation can be quantified. There are many applications of strain analysis. For example, some in-situ stress determination techniques measure strain and then compute the stresses. Therefore, it is necessary to be able to analyze the strain values so that the understanding of the anticipated deformation in rock masses can be understood. Strain is a measure of the deformation of a body and should be taken as being the same type of mathematical quantity as the stress. These strains, therefore, may be either normal strains or shear strains, which are directly analogous to normal stresses and shear stresses. Mathematically, they are linked through Hook's Law (Hudson, 2000).

4.1.3 Geodynamic analysis

Deformation modelling can be conducted to identify stressed areas of the crust where hydrothermal fluids may be discharged. The magnitude of stresses is calculated by geo-hydro-mechanical modelling. Such modelling helps to establish criteria correlating stressed zones of hydrothermal activity in the crust, which represent areas of tension, compression and discharge. The final aim of such modelling is to predict areas which potentially contain SMS deposits.

The mathematical framework follows Ludwig Prandtl's theoretical formulation of interface hydrodynamics (introduced by Prandtl, 1904), which states that the optimal channel configuration for fluid flow is circular and occurs at areas where tangential stress is zero ($\tau_t = 0$), see Figure 4–4.

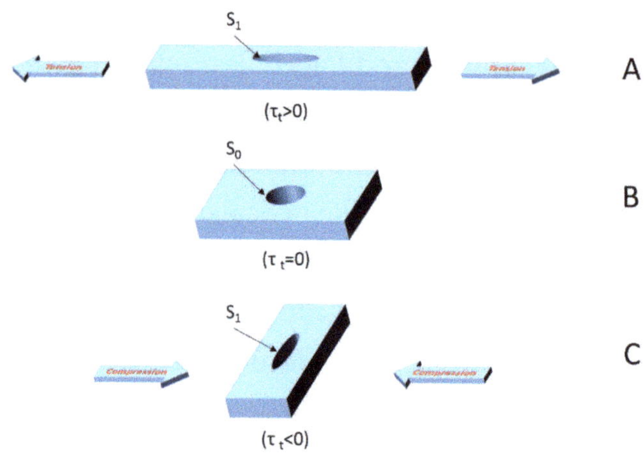

$$S_0(\tau_t=0) > S_1(\tau_t \neq 0)$$

Figure 4-4 The influence of tangential stress τ_t on the cross section of a fluid flowing channel. Top: tension; Middle: Discharge; Bottom: Compression . Click here for larger image.

Thus, the search for the hydrothermal areas with the most vigorous discharge of multi-phase fluid involves defining areas of the sea floor (Figure 4–1) where tangential stresses in the upper layer are near zero (Figure 4–5). In the model, the real sea floor relief is approximated by a double-layer structure, consisting of an elastic (upper) and a plastically deformable (lower) layer. Under the weight of the upper layer, the lower layer suffers plastic deformation where stress magnitude is directly related to the accumulated overlaying mass (Figure 4–5).

Figure 4-5 Double-layer structure, consisting of an elastic (upper, 1) and a deformable (lower, 2) layer. Click here for larger image.

The mechanical properties, consisting of modulus of elasticity, Poisson's ratio, specific weight of the model, are estimated based on the properties of a 5 kilometer (Klingenhöfer et al., 2000) thick oceanic crust, composed of tholeiitic basalts similar to those that Iceland is built up of. These are considered typical for the areas of the Arctic Mid-Ocean Ridge (AMOR). The assumption made is that the properties of this 5 kilometer stretch of basalt represent the average properties of the AMOR crust. Considering the complex physical structure of the oceanic crust, with basalts, sheeted dykes, gabbros and ultramafic rocks, this represents a simplification.

The real depth of the elastic layer is dependent on the thermal state of the crust. Young and warm crust is thinner than old and cold crust (e.g. Klingenhöfer et al., 2000). This has not been taken into account in the model developed here.

The AMOR contains a major fault system and has considerable influence on the surrounding area, where numerous faults have been and are being developed. These faults represent de-stressed areas, where compressional horizontal stresses caused by far-field stress are close to zero. Hence, only the horizontal compressional stress caused by gravity alone is relevant in the analysis.

In the model, the elastic layer is represented as an elastic membrane deflected on rods, with lengths and positions defined by the coordinates (x, y and z) at the point of contact between the rod and the membrane (Figure 4–6). The length is given by the z-coordinate. The coordinates are taken from the bathymetric data with a grid size of 100 x 100metres.

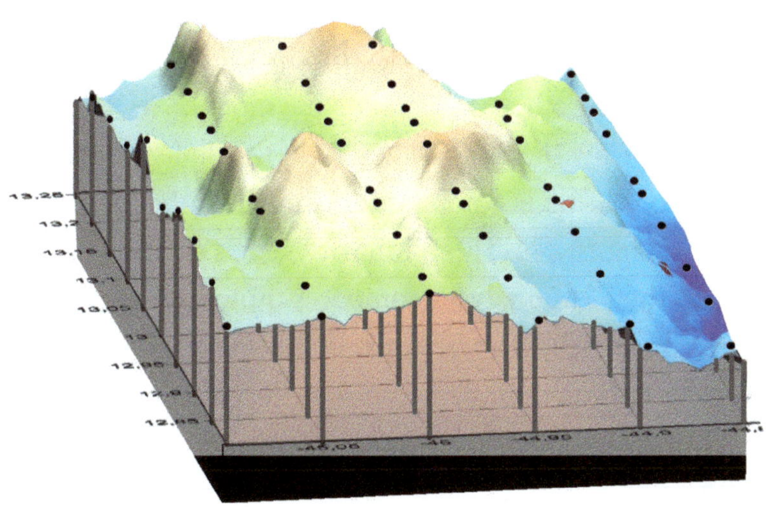

Figure 4–6 The simplified two-layered model of the Earth's crust used as a basis for the tangential-stress estimations. Click here for larger image.

The stress distribution in this model is calculated using the Finite Element Method (FEM). The idea of FEM is to model a continuum by discretizing into finite elements or small areas where the solutions can be reduced to a system of algebraic equations. This is indicated in Figure 4–7.

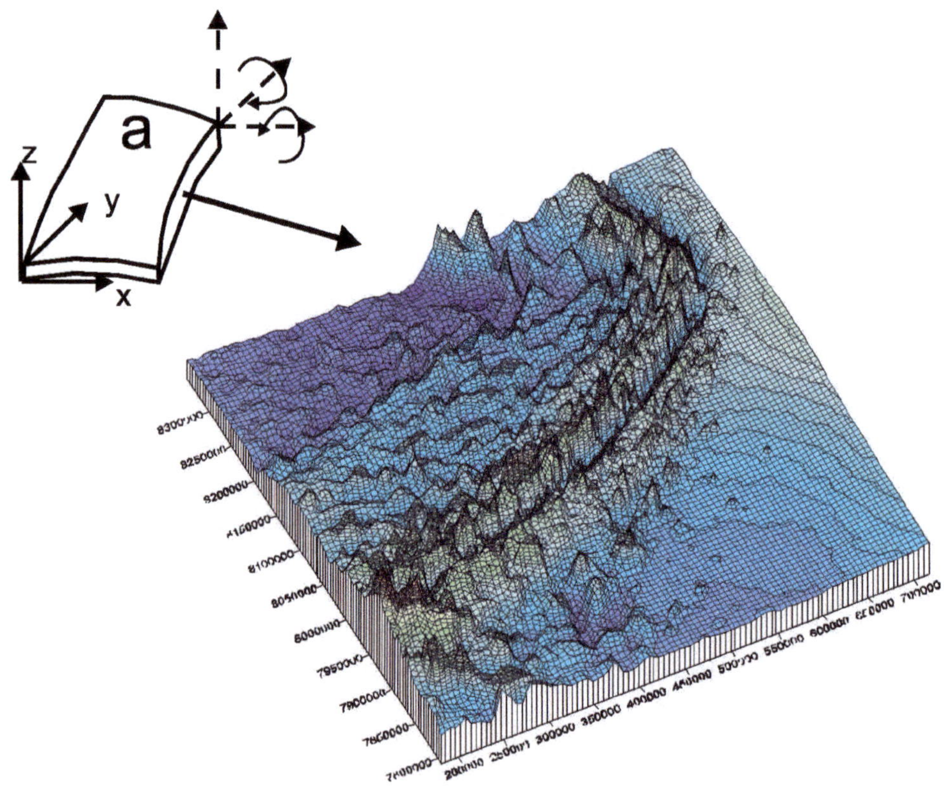

Figure 4–7 Discretisation of a finite element network; 'a' is the finite element, one of the small areas that build the model of Mohns Ridge. Click here for larger image.

In the modelling phase, the stress and displacement fields are modelled inside every finite element that builds the model.

Figure 4–8 below shows a 3D display of the deflection function used in the calculations. This is compared to the 3D bathymetric display of the same area in Figure 4 9. This similarity shows that the deflection matrix achieved by the FEM analyses compares quite well with the actual relief of the area under study, and confirms the strong link between the relief and the calculated crustal stresses using the chosen model of a 5 kilometer thick load.

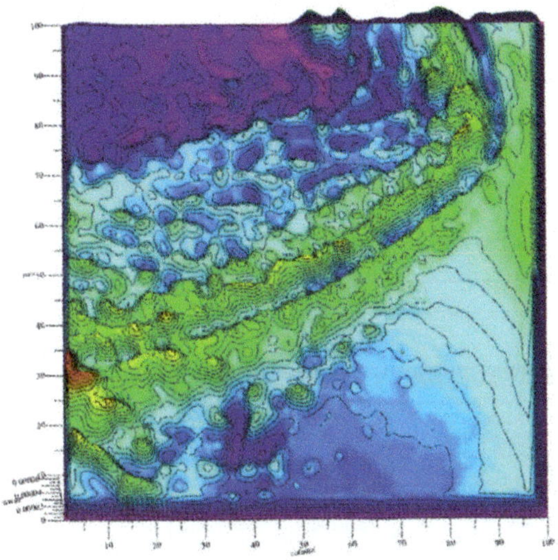

Figure 4-8 Relief of the deflection function.
Click here for larger image.

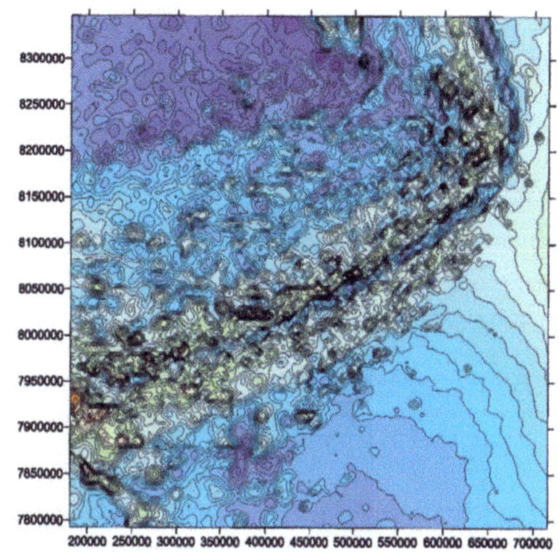

Figure 4-9 Multibeam relief.
Click here for larger image.

Knowing the deflection matrix, it is possible to define the inner forces acting within the plate and estimate the stresses.

Tests have demonstrated that the zero tangential stress isoline ($\tau_t = 0$) corresponds well with known active hydrothermal areas. Further tests have shown that relict and low-active hydrothermal manifestations are bound to the tensional areas where tangential stresses are larger than zero ($\tau_t > 0$) (Petukhov et al., 2004, 2010 and 2012, and Aleksandrov et al., 2009, in Cherkashov et al., 2013). See Figure 4–10.

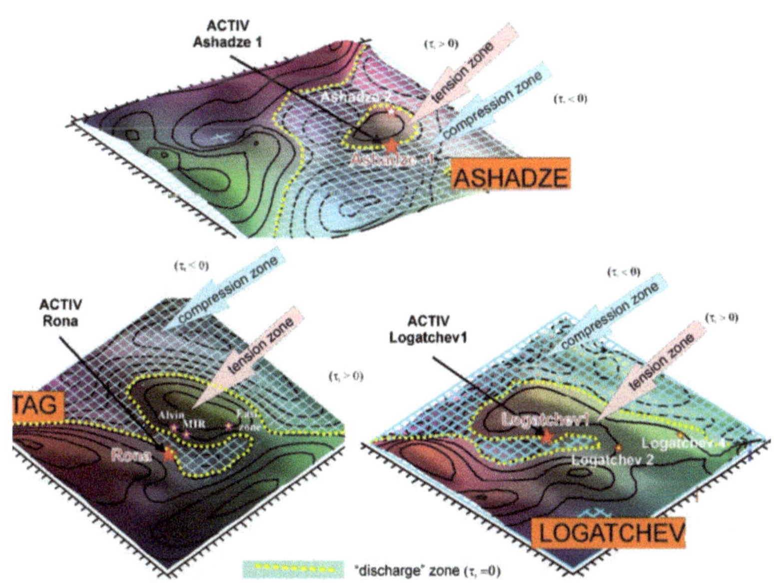

Figure 4-10 Models of tangential stress distribution MAR sections (Petukhov et al., 2004, 2010 and 2012, and Aleksandrov et al., 2009, in Cherkashov et al., 2013 . Illustration reproduced by permission; no reuse without rightsholder permission.).
a) Ashadze, b) TAG, c) Logatchev. The active fields are marked by a red star.
Click here for larger image.

The results presented for Ashadze, TAG and Logatchev in Figure 4–10 are crucial for the analysis of the deflection matrix and its interpretation. The results of calculating the tangential stresses in Mohns Ridge are shown in Figure 4–11.

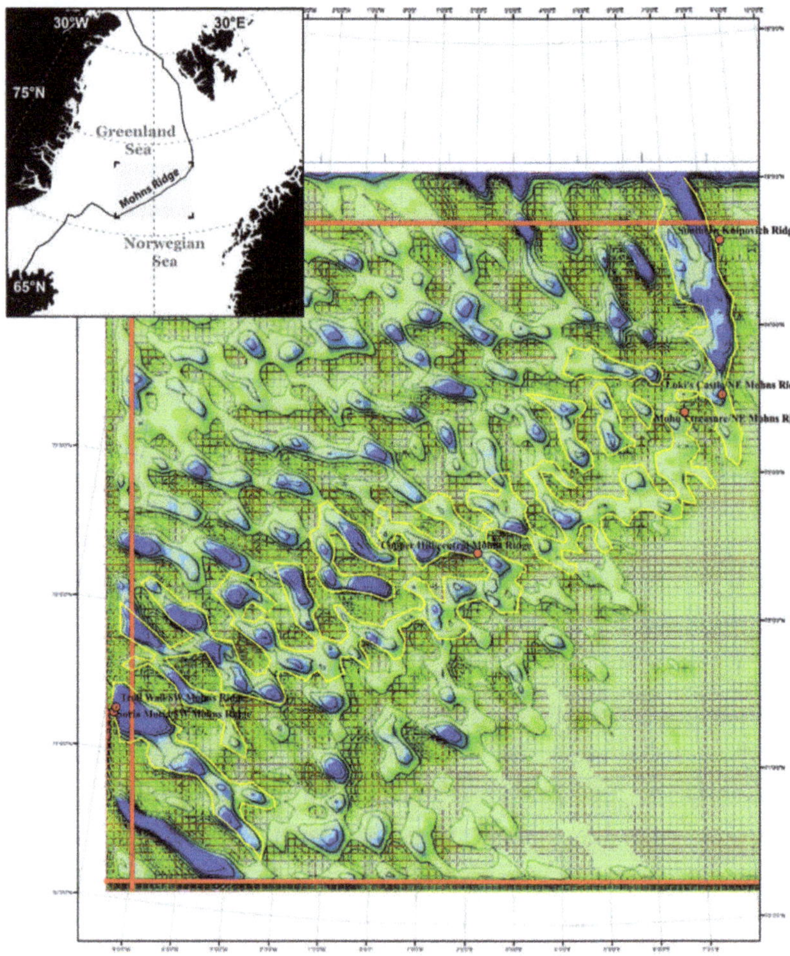

Figure 4–11 The distribution of tangential stress at the plate surface along Mohns Ridge. Yellow line is the zero stress isoline. Red dots are known hydrothermal manifestations (inactive and active) along Mohns Ridge (see Pedersen et al., 2010b). Click here for larger image.

A line with zero tangential stress can be plotted on the bathymetric map (yellow line in Figure 4–12 and Figure 4–13) that marks zones of possible discharge where stress changes from tension to compression. It is hypothesised that these zones are favourable to the presence of hydrothermal vents and match with the known hydrothermal sites along the Knipovich and Mohns Ridges shown in Figure 4–11, Figure 4–12, Figure 4–13, Figure 4–16 and Figure 4–17. Red lines in these figures indicate zones with a high concentration of favourable indicators for hydrothermal manifestation from the geodynamical analysis.

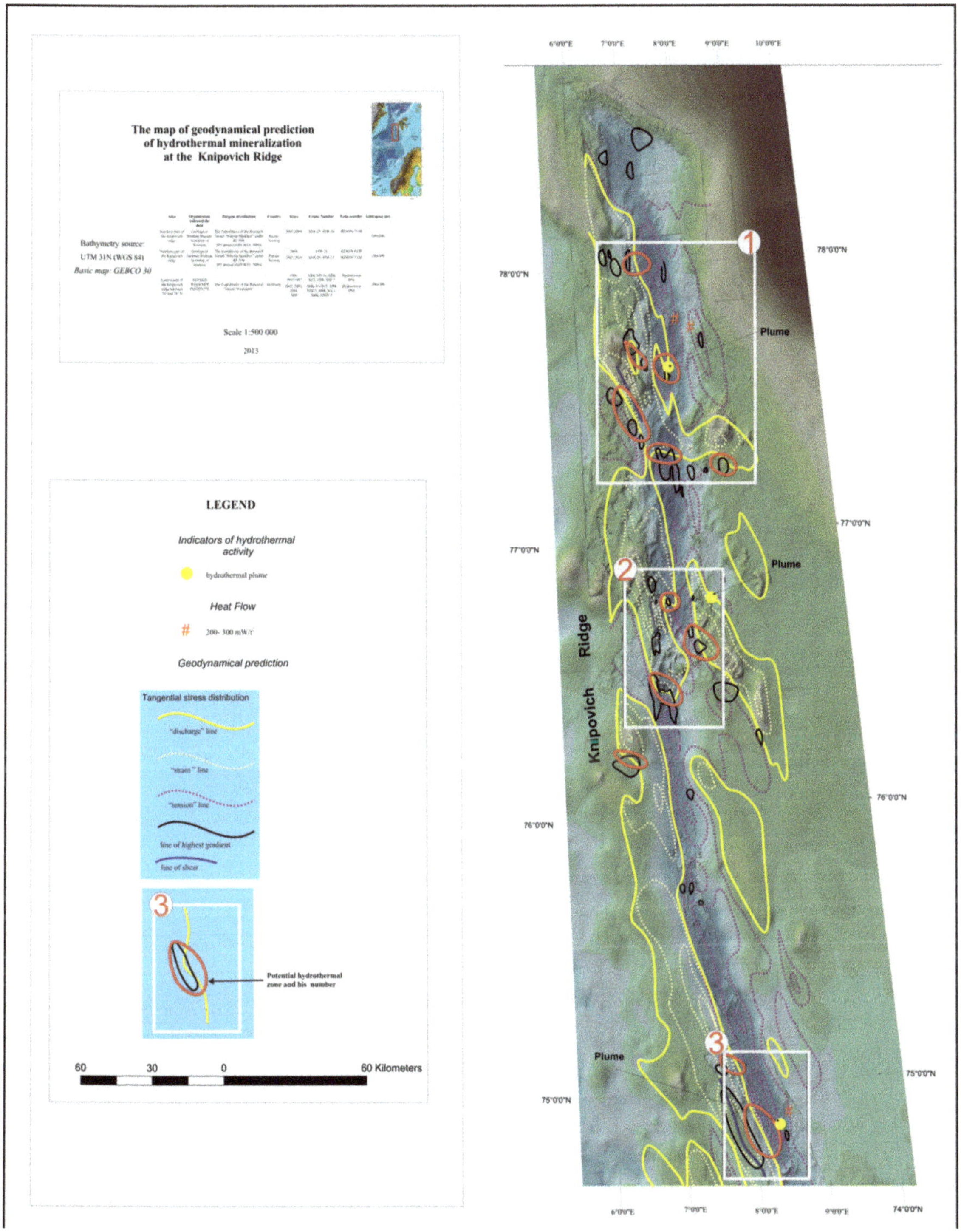

Figure 4-12 Areas most promising for finding SMS along the Knipovich Ridge, based on the geodynamic analysis. Areas having a concentration of prospective zones are shown as white boxes. These three areas (white box outline) and sub-areas have been modelled and named as separate plays in the play-database. Red lines indicate particularly promising areas, where the tensional structures can be found in close proximity to discharge zones and high gradients. Click here for larger image.

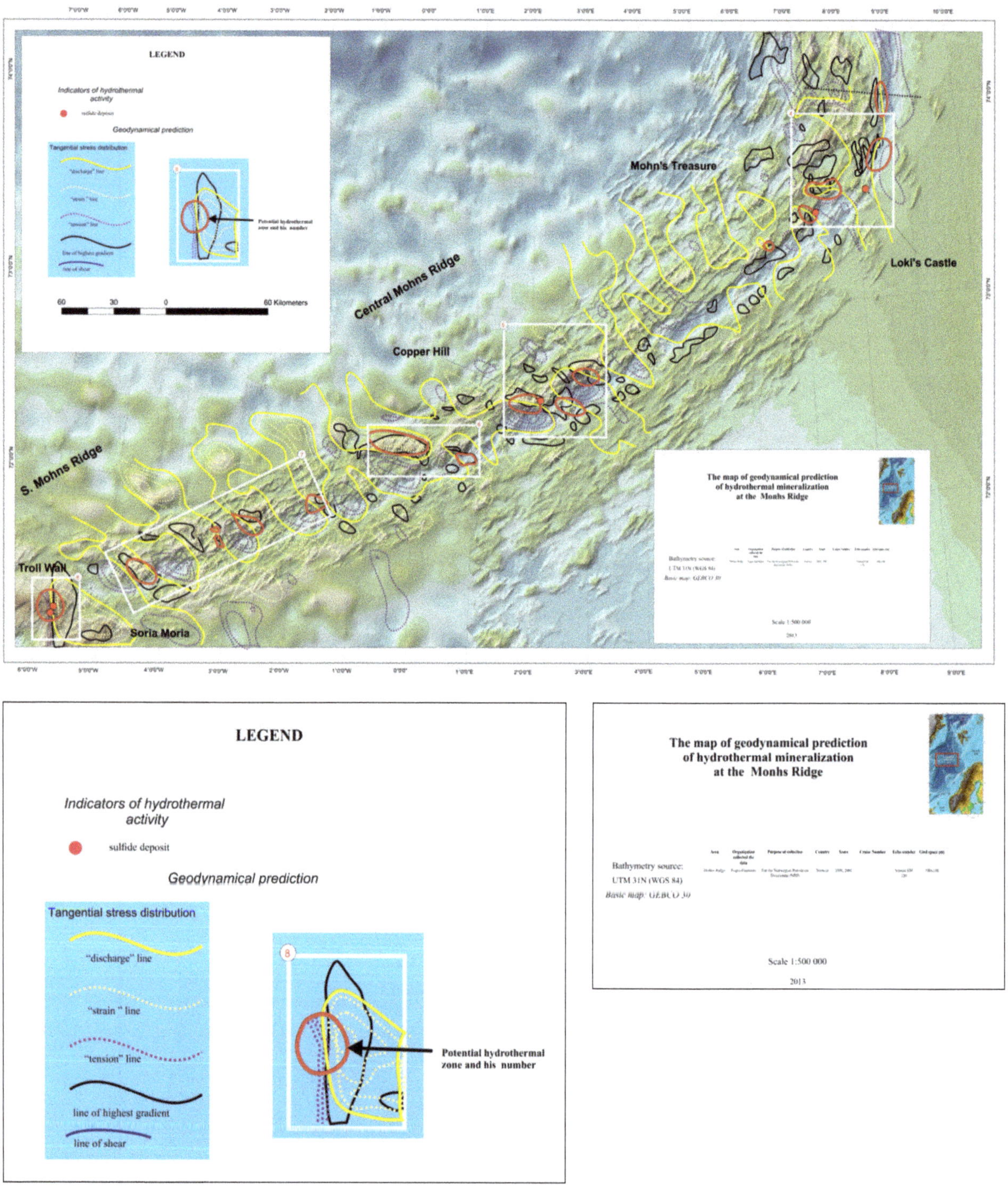

Figure 4-13 Areas most promising for finding SMS along Mohns Ridge, based on the geodynamical analysis. Areas having a concentration of prospective zones are shown as white boxes. These five areas (white box outline) and sub-areas have been modelled and named as separate plays in the play-database. Red lines indicate particularly promising areas, where the tensional structures can be found in close proximity to discharge zones and high gradients. Click here for larger image.

4.1.4 Morphostructural analysis

4.1.4.1 Data and methodology

Mohns Ridge

Morphostructural analysis of Mohns Ridge was based on the bathymetric surveys carried out by Fugro-Geoteam for the Norwegian Petroleum Directorate (NPD) in 2000 and 2001.

An original grid (100 x 100 m) was built using mean values of the primary data, which allowed locating details of the relief. Several bathymetric charts at different scales were drawn. However, the lack of a sonar survey has limited the accuracy of interpretations.

The morphostructural scheme of the axial zone of Mohns Ridge was drawn with the aim of determining tectonic and volcanic elements of the seabed structure, comprising both large and small tectonic and volcanic structures. Unfortunately, due to the absence of sonar images, it was not possible to differentiate between the seabed volcanic formations at the bottom of the valley by their age. However, it was possible to isolate structures associated with the detachment faults and oceanic core complexes (OCC).

Regarding the selection of promising areas for sulphide ore formation, general criteria were followed. All known vents on the ridges with slow or ultra-slow spreading are generally attributed to the axial or flank features (Krasnov et al., 1995). The former is directly controlled by the processes of axial volcanism. This could either be a thermal one, caused by the presence of a high-temperature magmatic heat source at shallow depths under the seabed or by fluids accompanying volcanic outflows from a deep source. The latter, on the other hand, do not demonstrate any noticeable association with the volcanic processes, although an indirect association is implied and is located at remote flanks or edges of the rift valley, mainly in the areas of the development of serpentinised peridotites.

In the majority of cases, the axial venting is concentrated at small grabens, which complicates apical surfaces of the axial volcanic ridges at their most elevated parts. The SMS deposits normally tend towards the areas of very young basalts in the zones of intersection of the longitudinal faults and the small transversal dislocations with the indications of tectonic tension at areas with tangential stresses close to zero. Large, central-type volcanoes may also serve as ore-controlling structures. This is valid in cases where they are located in the central parts of the inner floor of the rift valley and areas where they lie at the toes of the magma-conducting marginal faults having tangential stresses close to zero.

The flank SMS deposits are structurally controlled by zones of transection of deep-seated longitudinal strike-slip faults, mainly the detachment faults, with crossing tectonic dislocations. These provide an additional possibility for lateral subsurface migration of magmatic and hydrothermal fluids and, on some occasions, even for the melts far beyond the boundaries of the axial zone. In general, the ore-localising structures are represented by the flattened slopes, inter-ridge depressions and saddles. Recent research has proved that there exist quite high promises of ore-bearing OCC. This is especially the case when these are located immediately close to the axial zone (Cherkashov et al., 2013). One may notice that Mohns Ridge, from a morphostructural perspective, more or less meets the above-mentioned criteria.

The areas are outlined in such a way that the most promising structures are grouped together on the basis of the following:

- the complex tectonic areas with volcanic ridges having the axial grabens. In particular, with the volcanic steps and links/bridges connecting the volcanic rises and slopes.
- the structures spatially or generically associated with the detachment faults and OCC at the valley walls/slopes, especially areas where tensional environment exists due to close to zero tangential stresses.
- the large central-type volcanoes with the flat tops and craters and the chains of such volcanoes along the crossing faults with close-to-zero tangential stress environment.

Basically, the areas include axial and adjacent flank structures. However, in some cases, several remote flanks have also been included since their structures suggest the existence of extinct/relict vents.

Figure 4–14 shows the morphostructural elements along Mohns Ridge and Figure 4–16 show areas most promising for finding SMS deposits based on the morphostructural analysis.

Knipovich Ridge

Morphostructural analysis of the Knipovich Ridge axial zone has been made using the data from several bathymetric surveys. A summary bathymetric map has been drawn based on the original (R/V Haakon Mosby and R/V Polarstern data) and available (R/V Nikolai Strakhov data) grids. The general quality of the primary materials appeared to be poorer than that of Mohns Ridge, which limited the extent of details for interpretation. This drawback was partly compensated for by using side-scan sonar images (SEAMARC-II and ORETECH systems; (Crane et al., 2001.), which allowed for the outlining of areas of volcanic activity over time, as well as the zones of recent tectonic activity in the northern part of the ridge.

Echo-sounder survey data confirmed by video (Crane et al., 1995, 2001; Tamaki & Cherkashev, 2000) manifests that there have been recent volcanic activities on the major part of Knipovich Ridge, excluding its two sections in the interval of 74 to 75 degrees north and north of 78 degrees. One may notice that both sections are more skewed to the theoretical axis of spreading, where an increase in sedimentary cover of up to 2 km is recorded. The ORETECH materials (Crane et al., 2001) demonstrate somewhat lower volcanism in the southern half of the ridge. As the survey covered a relatively narrow strip (only 2 km) of the rift valley, this conclusion is uncertain.

Generally, hydrothermal manifestations and associated sulphide accumulations at Knipovich Ridge may be associated with several areas of modern tectonic and volcanic activities in the zones of intersection of multi-directional faults and structural features where tangential stresses are close to zero.

The morphostructural scheme of the Knipovich Ridge is based on the same principles and the same legend as that for the Mohns Ridge scheme. In addition, the northern half consists of fresh and older sheet and pillow lavas, as well as large elevated massifs at the flanks of the valley.

Figure 4–15 shows the morphostructural elements of the Knipovich Ridge and Figure 4–17 show areas most promising for finding SMS deposits, based on the morphostructural analysis along Knipovich Ridge.

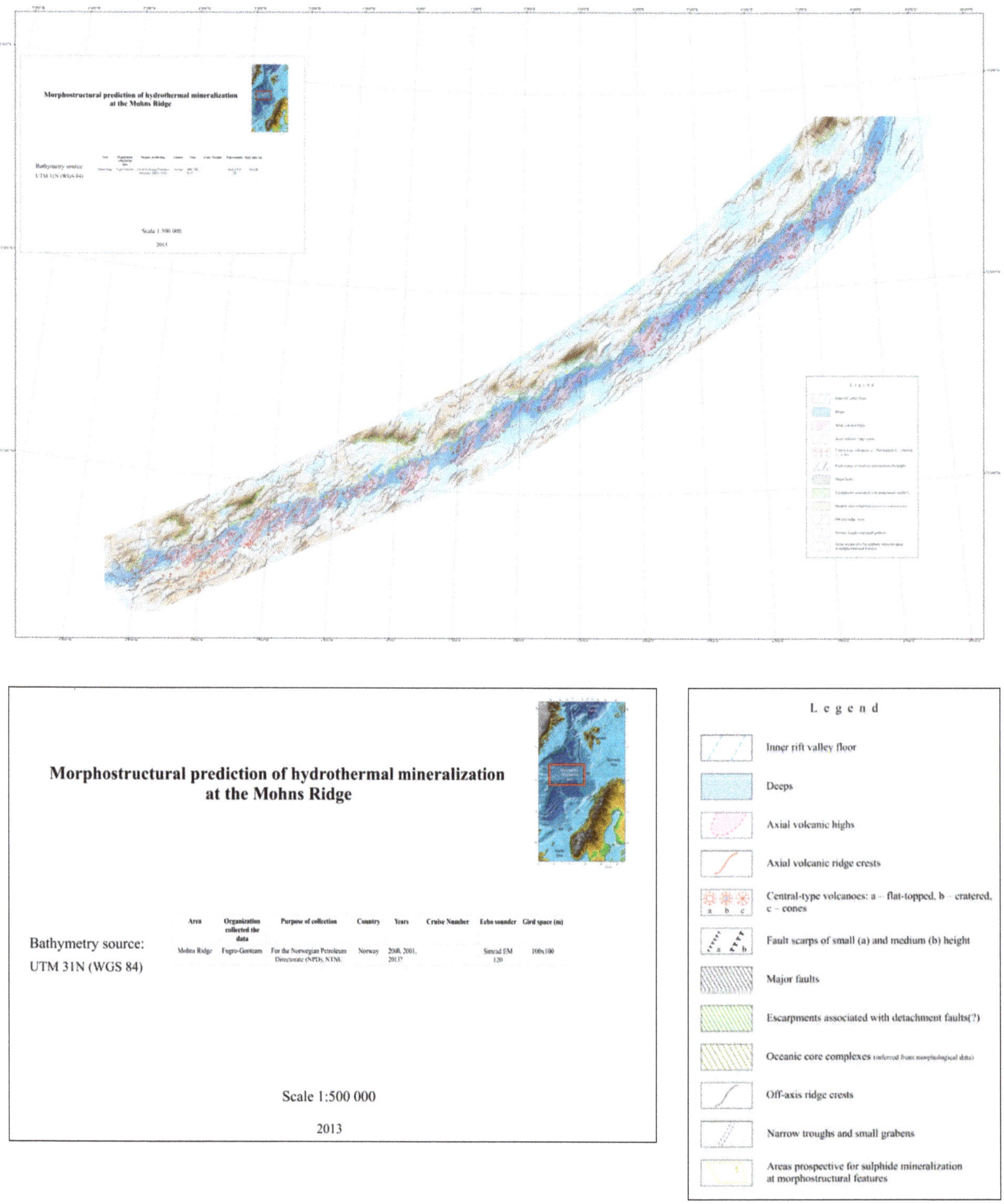

Figure 4–14 Morphostructural elements along Mohns Ridge. Click here for larger image.

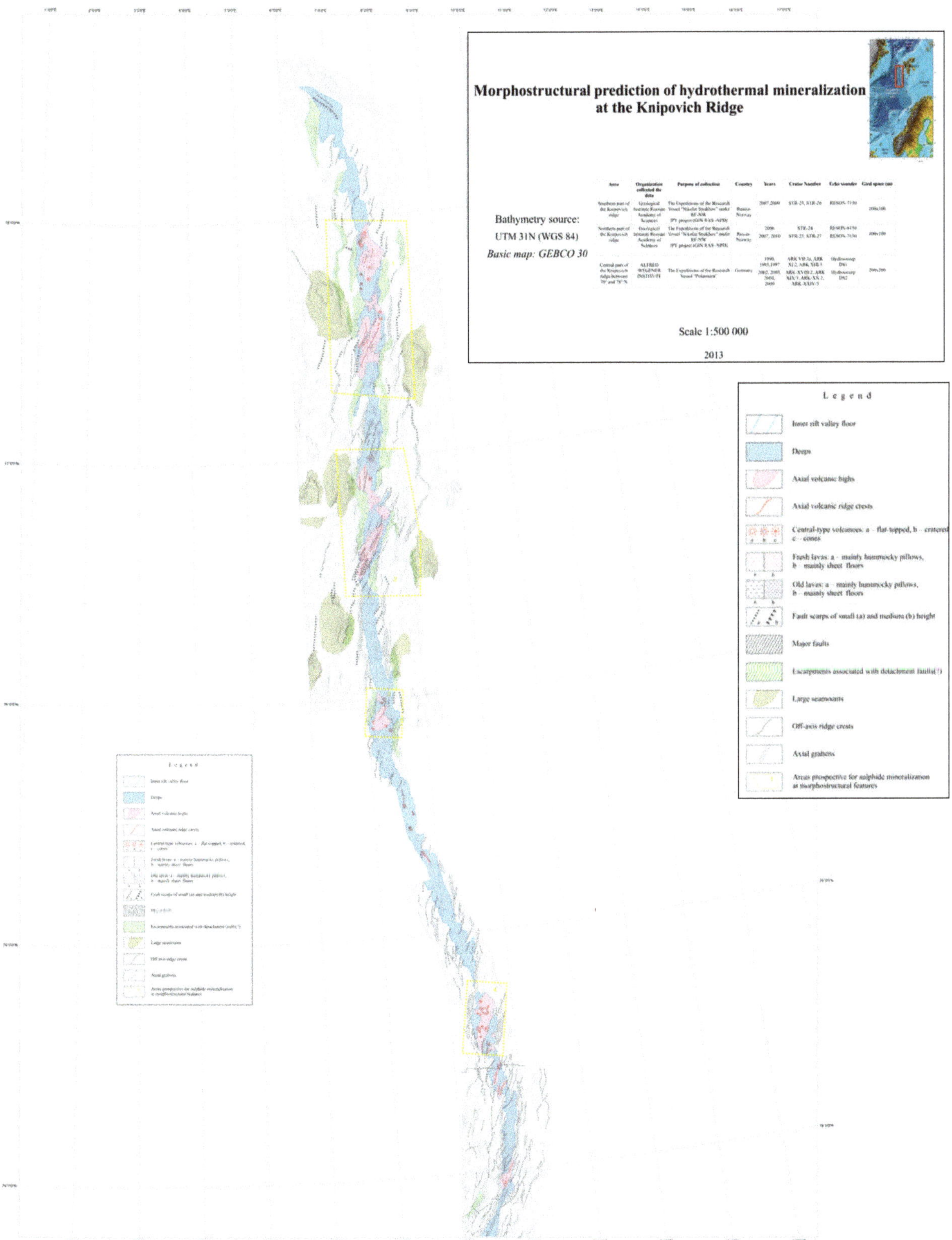

Figure 4-15 Morphostructural elements along Knipovich Ridge. Click here for larger image.

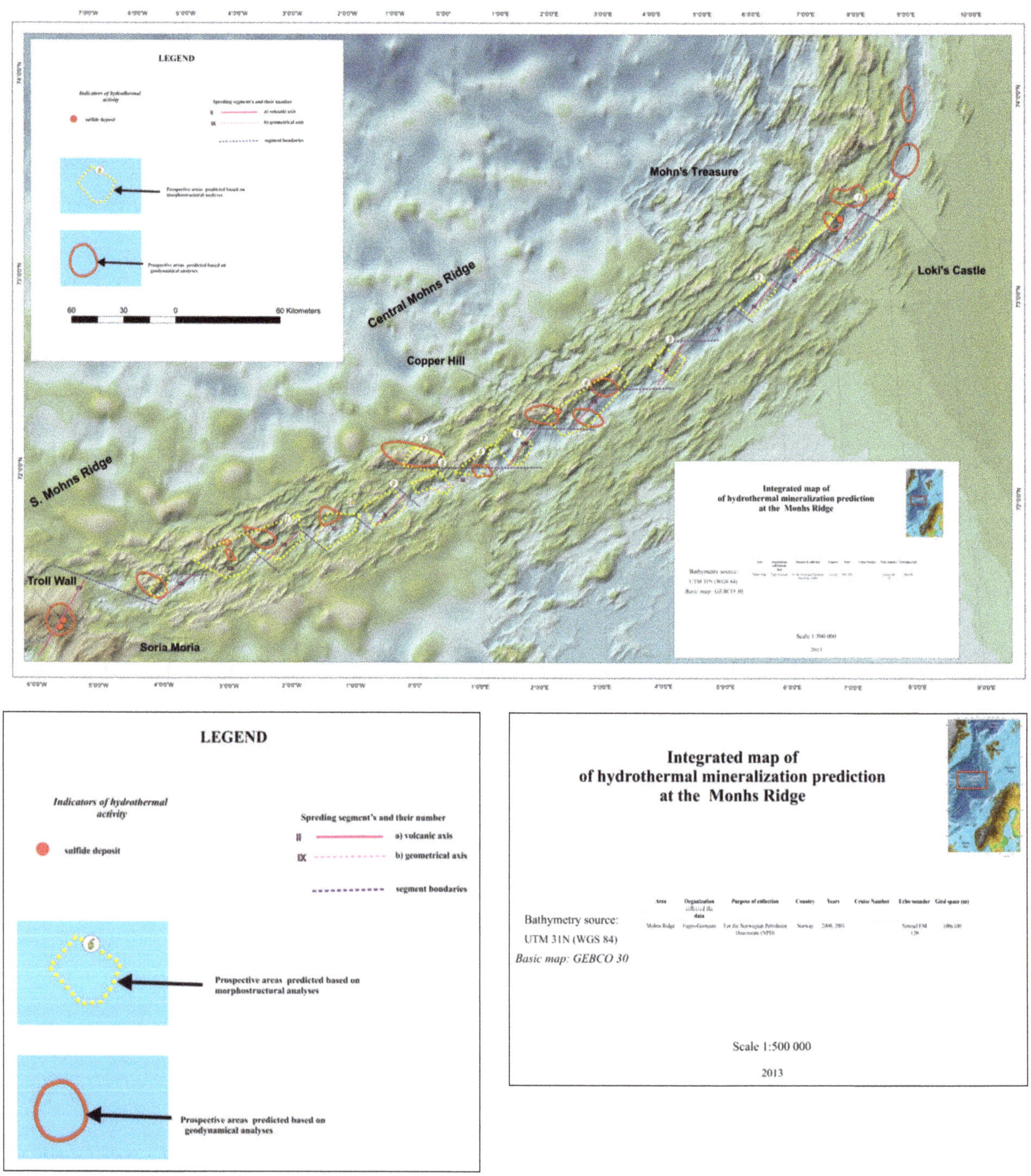

Figure 4-16 Areas most promising for finding SMS along Mohns Ridge, based on the morphostructural analysis. Segments also indicated. Areas having a concentration of prospective zones are shown as yellow dotted shapes. Click here for larger image.

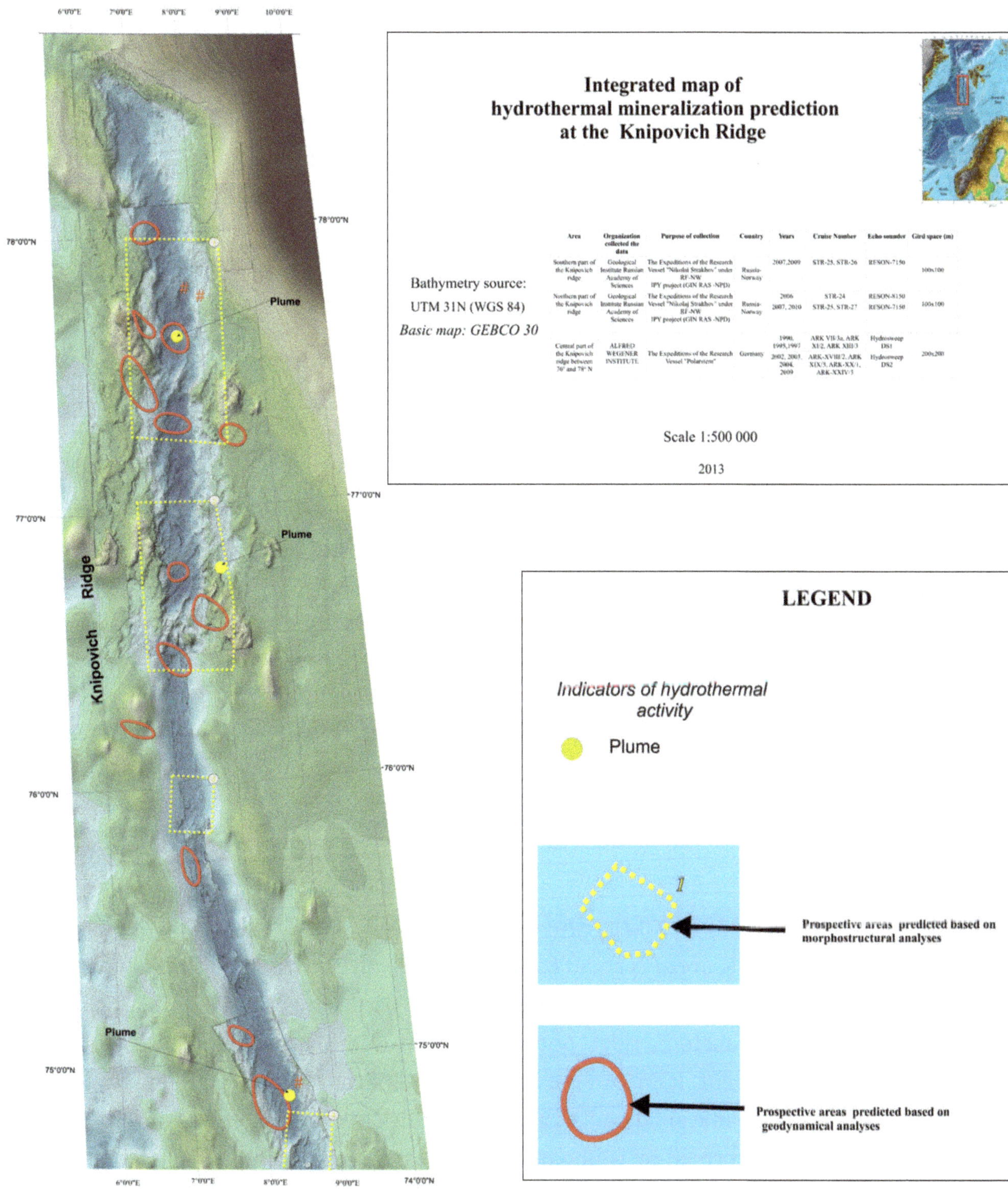

Figure 4-17 Areas most promising for finding SMS along the Knipovich Ridge, based on the morphostructural analysis. Areas having a concentration of prospective zones are shown as yellow dotted shapes. Click here for larger image.

4.2 Descriptions of single permissive tracts

4.2.1 Introduction

This sub-chapter contains a detailed description of the different permissive tracts that have been defined based on morphostructural and geodynamical analysis. Mohns Ridge has been segmented as shown in the previous sub-chapters. The descriptions are based on Cherkashov et al. (2013). The descriptions form the formal basis for the permissive tract characterisation in Chapter 6.

In total, eight areas have been defined and delineated based on the geodynamic analysis along Knipovich (3) and Mohns Ridges (5). These are numbered and outlined in white rectangles in Figure 4–12 and Figure 4–13 in the previous sub-chapter. From the morphostructural analysis, 16 favourable areas (MS-areas) have been identified, and these are indicated by yellow dotted lines in Figures 4–14 and 4–15. The permissive tracts are geographically centred on these 16 areas and around the southernmost part of Mohns Ridge, containing the Soria Moria and Troll Wall vent fields. The size of respective tracts in square kilometres has been calculated based on these yellow MS-areas. Favourable areas on Knipovich Ridge have been given these IDs: MS-K-area 1, MS-K-area 2 etc.; whereas favourable areas on Mohns Ridge have been given these IDs: MS-M-area 1, MS-M-area 2 etc.

4.2.2 Knipovich – permissive tract K1 and MS-K-area 1

The MS-K-area 1 is about 80 km long and covers two spreading segments of the rift valley, including its adjacent flanks. See Figure 4–18. A large volcanic-tectonic ridge, 600 metres high, dominates the southern part of the segment. It crosses the southern end of the area at a small angle and joins the marginal faults forming the ridge axis flanks. The surface of the ridge is formed by young pillow lavas, and a central graben is seen at its most elevated part. The slopes are controlled by tectonic scarps, spatially and perhaps genetically associated with the high-amplitude faults (possible detachment faults) of the valley walls. West of the axial ridge, a much smaller extinct volcanic ridge can be observed.

In its central part, the axial valley floor is largely covered with sediments and the lavas are seen only on a very small volcanic rises and in a strip to the north.

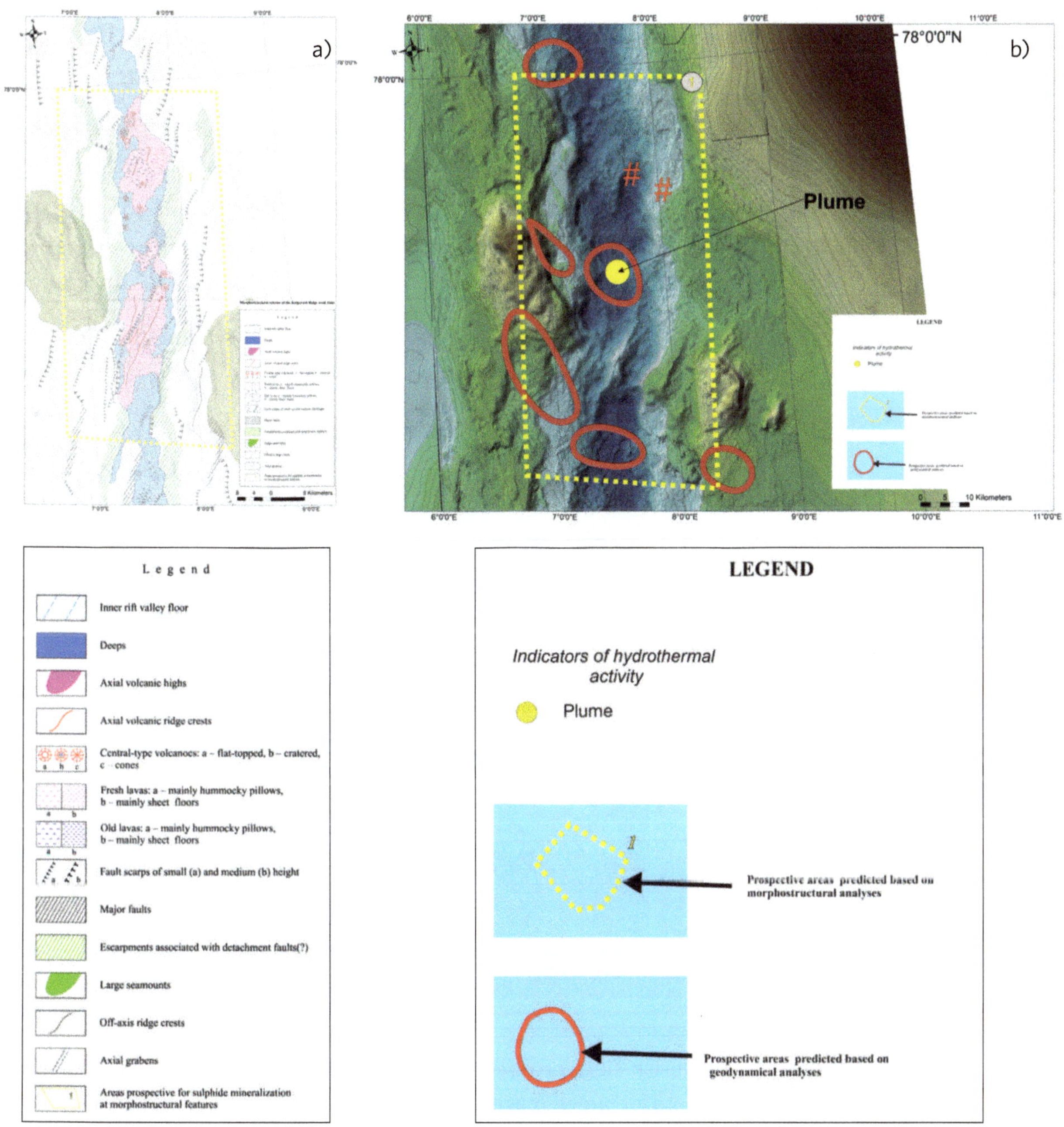

Figure 4-18 Knipovich Ridge interest area 1. Yellow dotted line is the area defined in the morphostructural analysis. Figure a shows the morphostructural elements, whereas figure b shows the prospective areas identified by the geodynamic analysis outlined in red. Red hashtags in b indicate the location of heat-flow measurements between 200 and 300 mW/m². Click here and here for larger image.

In the northern part of the area, there is also a large irregular volcanic rise with a relative height of 500 metres, with basalts of different ages. To the north of this rise, a cluster of central-type volcanoes, the largest being more than 3 kilometres in diameter, can be observed. An increased heat flow of up to 300 mW/m^2 is reported at two locations around 77048'N (Zayonchek et al., 2011 in Cherkashov et al., 2013).

The walls of the rift valley are mainly controlled by the large faults that could be interpreted as detachment faults. Of particular interest is a large zone of transverse tectonic deformations of north-west strike, crossing the rift valley at latitudes corresponding to the central parts of the southern axial ridge. On the flanks of the valley, several very large, elevated massifs of complex and uncertain genesis are located. Two of these are partly included in the area under study.

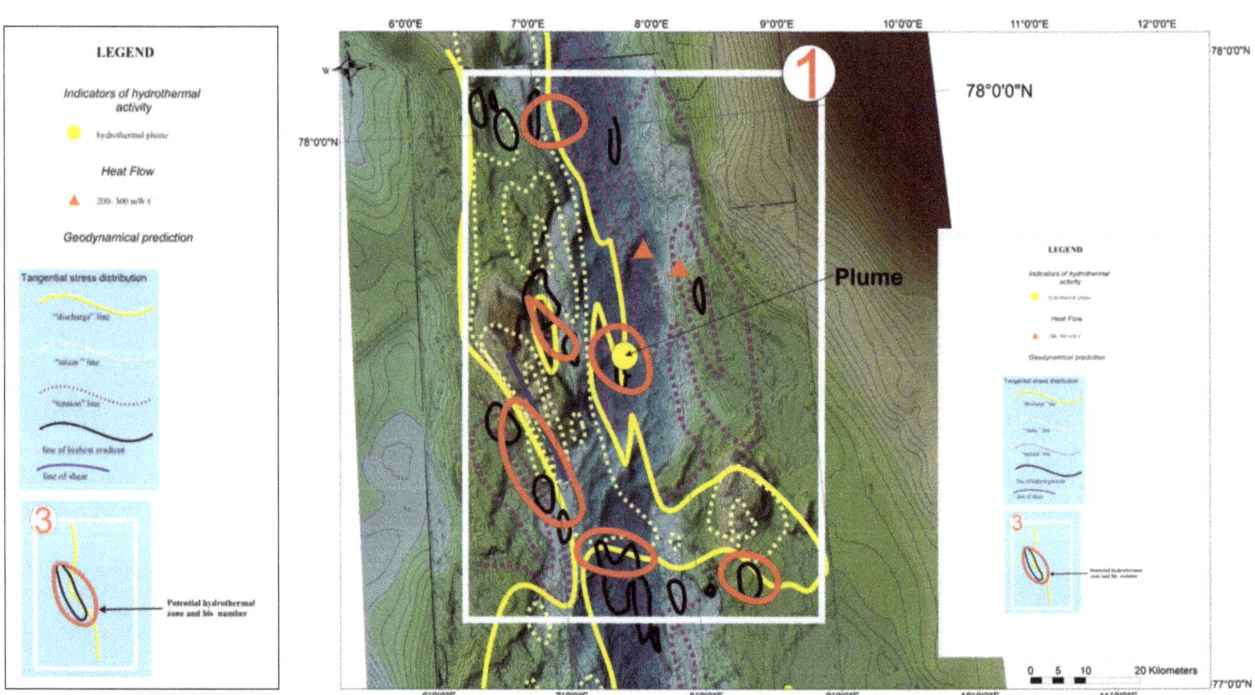

Figure 4-19 Knipovich Ridge prospectivity of interest from the geodynamical analysis in Area 1. Click here for larger image.

Area 1 includes several promising sites (outlined in red in Figure 4–19). In one of them, a hydrothermal plume has been recorded at 77°40′ N (Beaulieu, 2015). In area 1, an increased seafloor heat flow, of up to 300 mW/m², at 77°48′ N has been recorded and reported (Zayonchek et al., 2011 in Cherkashov et al., 2013).

By morphostructural analysis, the potential for finding SMS of an axial type is connected mainly with the central graben of the southern axial ridge and the volcanic cluster in the north. The location of the off-axial hydrothermal manifestations that one may expect will primarily be near the junction between volcanic structures with marginal faults, as well as along the tectonic dislocation zones, including the slopes of large flange massifs.

4.2.3 Knipovich - permissive tract K2 and MS-K-area 2

The MS-K-area 2 includes a section of the rift valley about 65 km long inside two adjacent spreading segments. See Figure 4–20 and Figure 4–21. The structure plane of its southern half is defined by the 500 m linear volcanic ridge, in many respects similar to the one described earlier, which at a small angle transects the inner bottom of the rift valley and merges with the marginal scarp of the western valley wall northwards of the zone of non-transform offset. According to the side-scan sonar data, the ridge has a multi-aged structure. The youngest part is practically uncovered by sediment. Basalts are found in the lower part of the eastern slope of the ridge and in the adjacent linear depression.

In the central part, the most interesting aspect is a bathymetric rise located in the zone of the non-transform discontinuity; it is a 120 m transverse terrace of volcanic origin, and its surface consists of very young pillow and sheet basalts without any noticeable sedimentary overburden. The south-east slope of the terrace is controlled by a tectonic scarp parallel to the spreading axis, and at this toe, young lava breccia have developed. Observations from MIR submersible dives here showed a very dense net of fractures in three main directions (N = North, E=East) N0E – N10E degrees, N30E – N40E degrees and N120E – N130E degrees. The other volcanic rises are relict structures.

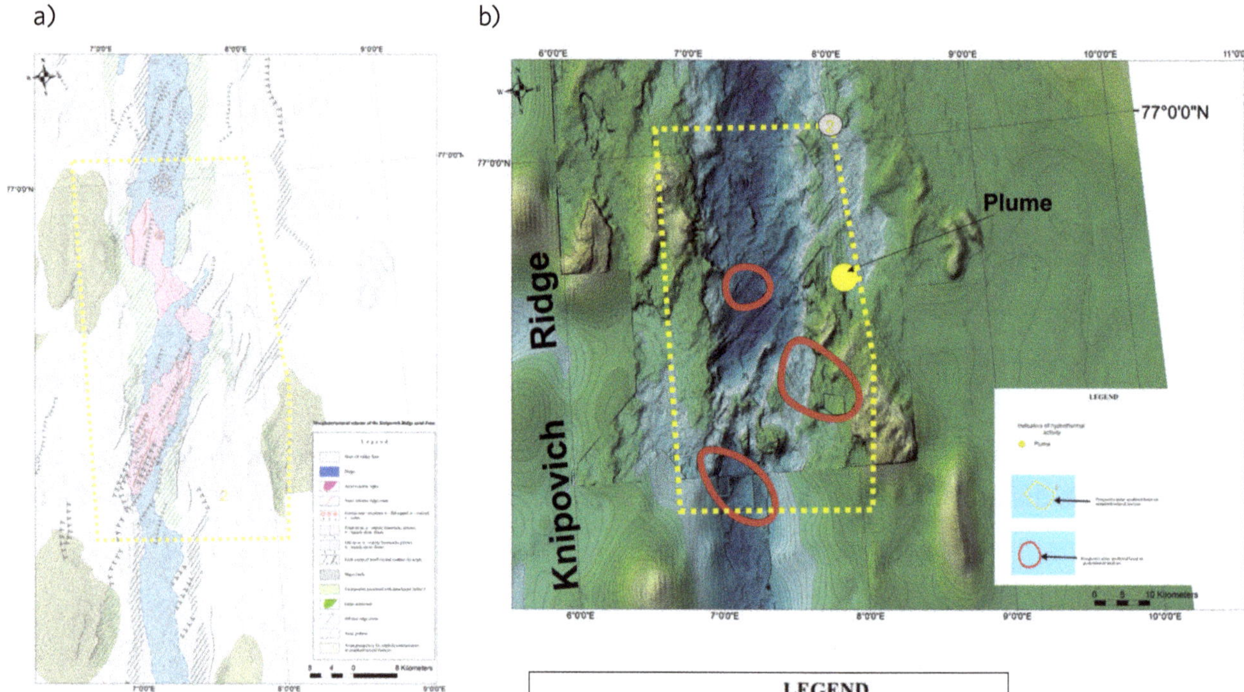

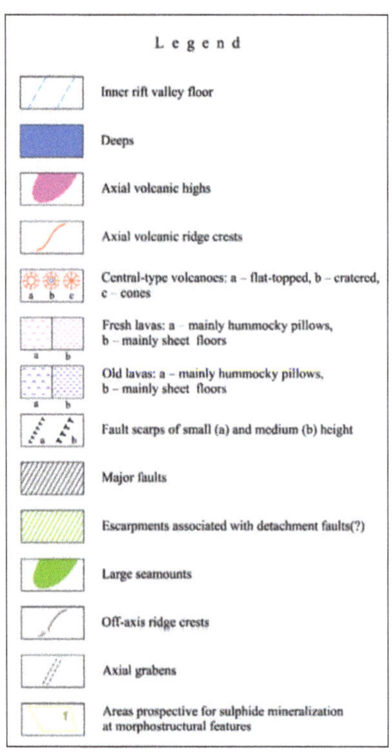

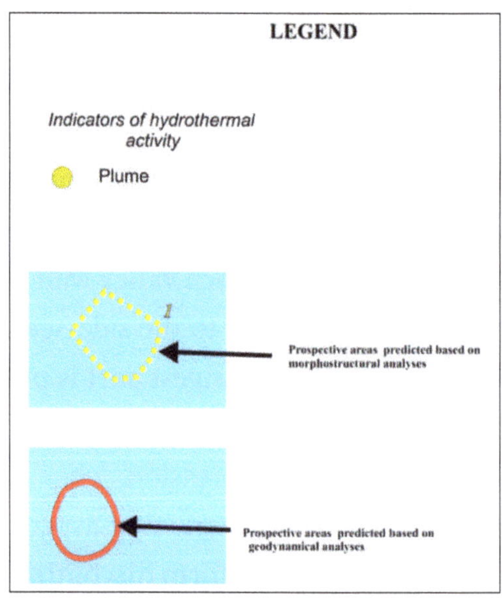

Figure 4-20 Knipovich Ridge interest area 2. Yellow dotted line is the area defined in the morphostructural analysis. Figure a shows the morphostructural elements, whereas figure b shows the prospective areas identified by the geodynamic analysis outlined in red. Click here and here for larger image.

Two 200 m central-type cratered volcanoes, 2 and 4 km in diameter, are located near the northern boundary. They are connected with the development of young sheet basalt lavas.

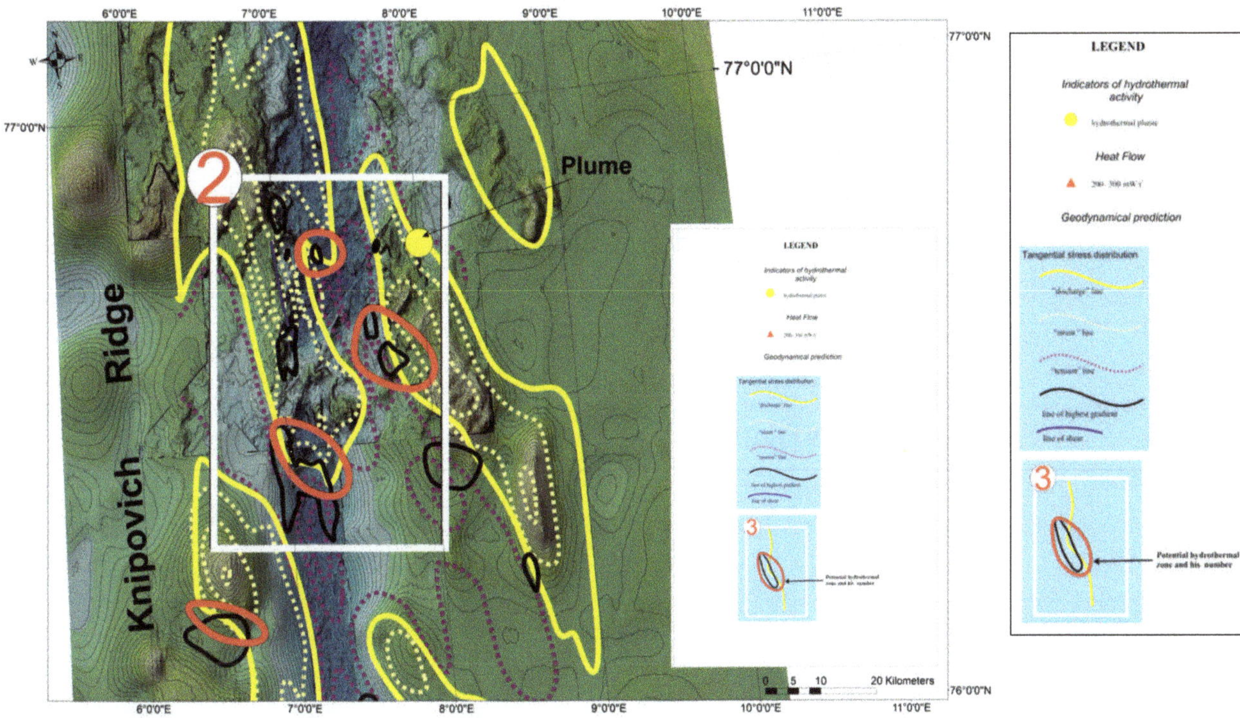

Figure 4-21 Knipovich Ridge prospectivity of interest from the geodynamical analysis in Area 2. Click here for larger image.

The prospect of finding sulphide mineralisations of an axial type here is connected mainly with the central graben of the southern axial ridge and the volcanic cluster in the north, similar to that described for Area 1. The inner rift-valley walls are mostly associated with high-amplitude faults (detachment faults).

Several promising areas have been found in Area 2 by the geodynamic analysis (indicated by red lines). All of them are discharged (indicated by the yellow line) and are situated near the high-gradient tension structures (indicated by black lines). See Figure 4–21. A plume has been detected in Area 2 at 76°48′ N (Beaulieu, 2015). This is located in the zone defined as a discharge zone (outlined in yellow).

The most promising areas are the axial volcanic ridge, the volcanic terrace, the valley slopes close to transverse tectonic disturbances, and in the zones of the junction between these transverse disturbances and the axial volcanic structures. See Figure 4–20.

4.2.4 Knipovich - permissive tract K3 and MS-K-area 3

MS-K-area 3 includes a 20 km section of the rift valley, located in an area of minor non-transform offset, marked by a 400 m axial volcanic rise and several flat-topped volcanoes of the central type. See Figure 4–22.

The tract is defined solely through the morphostructural analysis.

Young basalts are widely spread in this area (see also Crane et al., 2001). One may suppose, with a certain level of probability, that modern or extinct vents may exist in the axial zone and neighbouring flanges associated with the detachment faults interpreted in the area (see green shaded areas in Figure 4–22a).

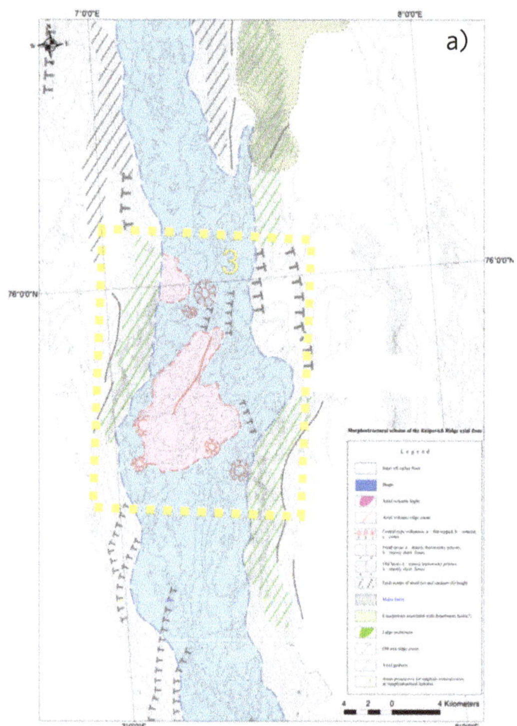

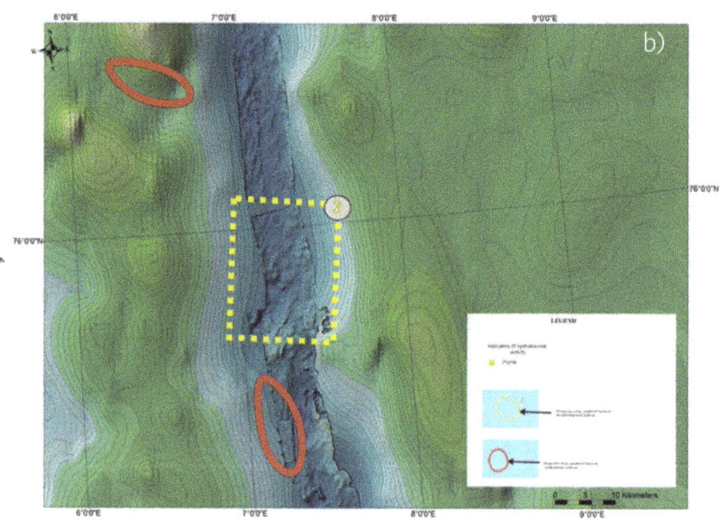

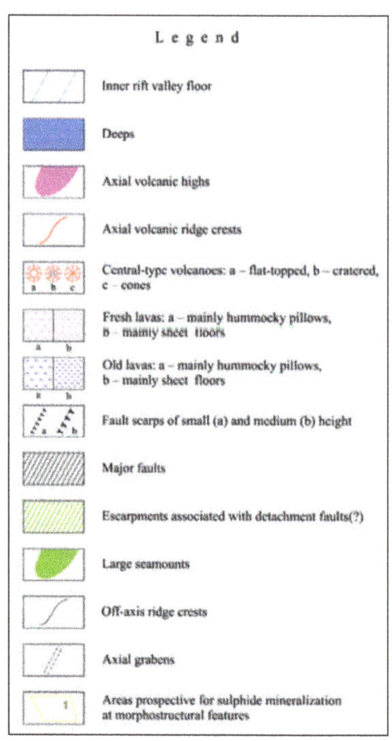

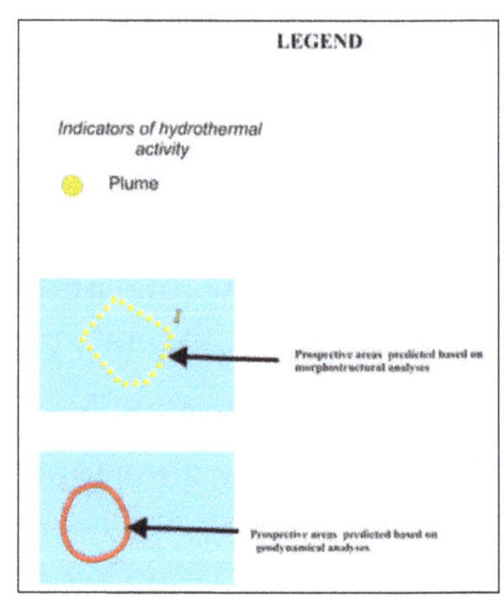

Figure 4–22 Knipovich Ridge permissive tract 3 from the morphostructural analysis. Figure a shows the morphostructural elements, whereas figure b shows the prospective areas identified by the geodynamic analysis, outlined in red. Click here and here for larger image.

4.2.5 Knipovich - permissive tract K4 and MS-K-area 4

MS-K-area 4 is connected to Area 3 from the geodynamic analysis (see Figure 4–23). It contains promising sites. There are hydrothermal plumes at 74°48′ N and 74°53′ N (Beaulieu, 2015). Although the plume at 74°53′ N is not confirmed, the nearby station (8°30′ E, 74°53′ N) has picked up a heat-flow measurement of 233 mW/m^2 (Zayonchek et al., 2010). Only the southernmost plume is shown in Figure 4–23.

Both hydrothermal plumes are in the estimated zone of discharge (indicated by the yellow line). However, due to uncertainties both with regard to the source of the plume and the position of the estimated zone of discharge, nothing conclusive can be stated. In Area 3, promising specific sites in addition to the plumes have been identified. They are located in the discharge zone close to the tension structure (western flank).

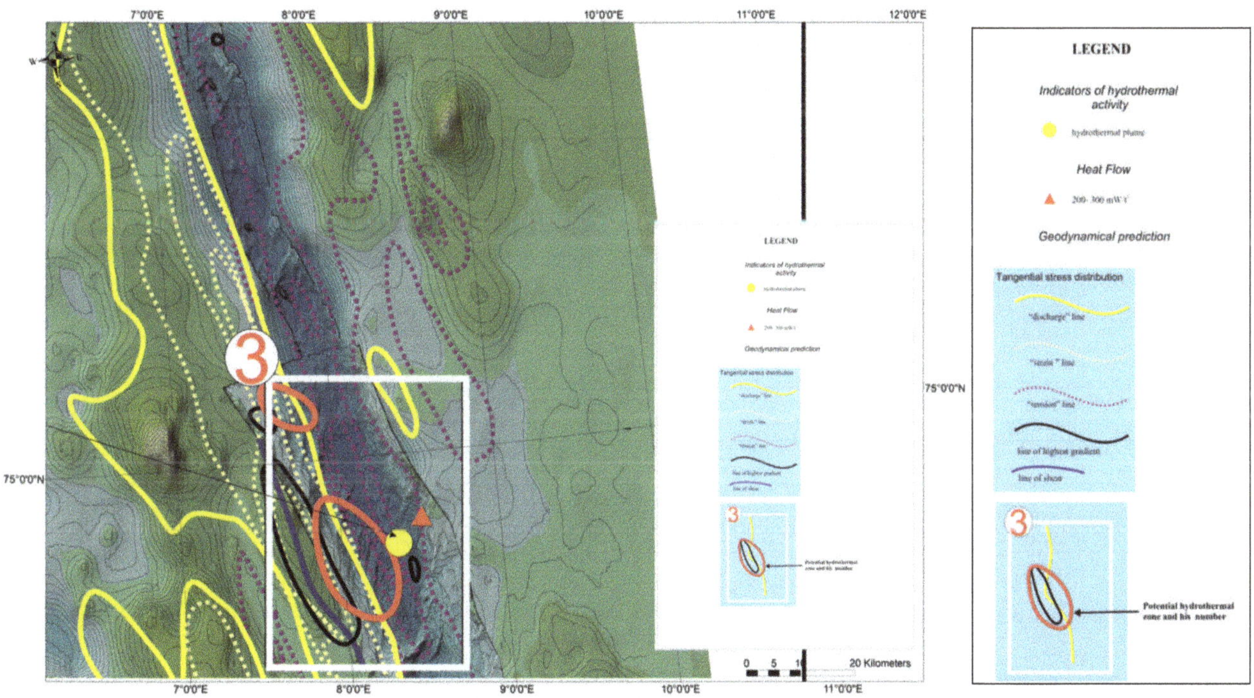

Figure 4–23 Knipovich Ridge prospectivity of interest area 3, defined by the geodynamical analysis. Orange triangle indicates the heat-flow anomaly. Click here for larger image.

a) b)

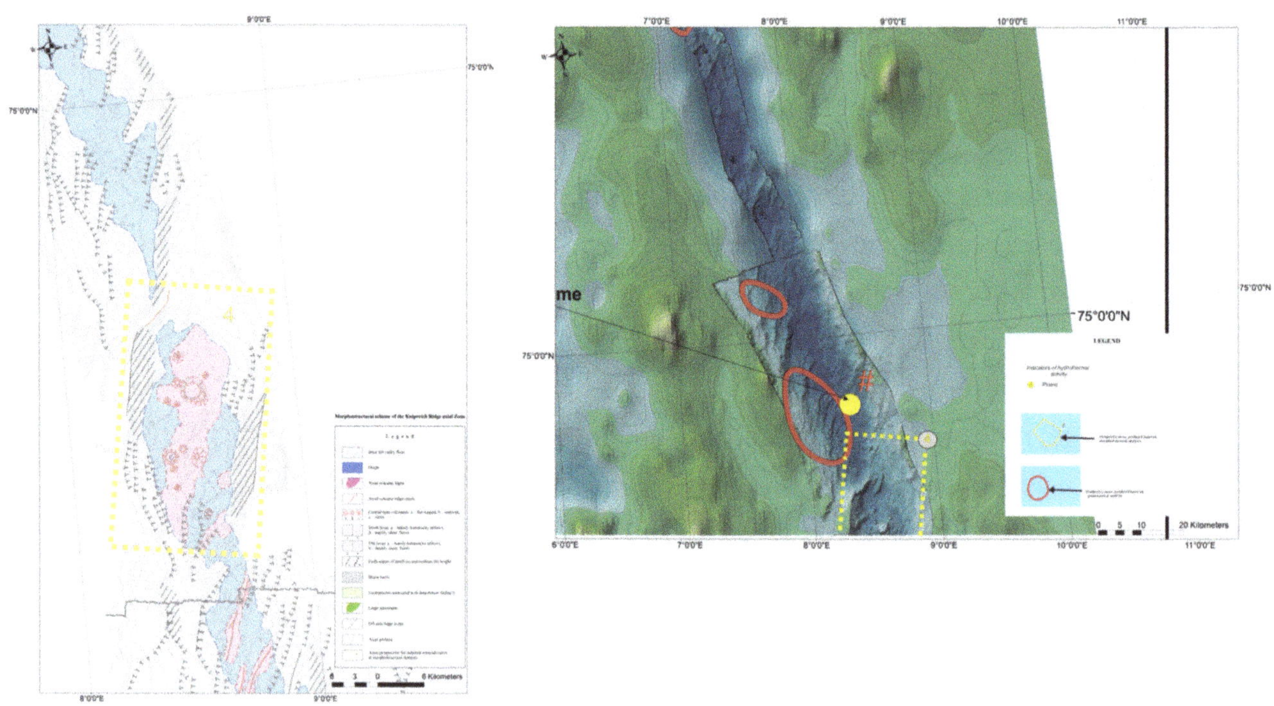

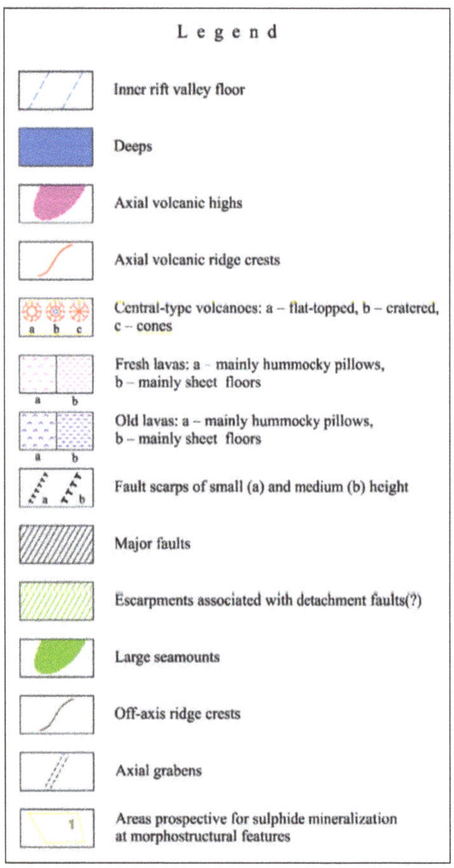

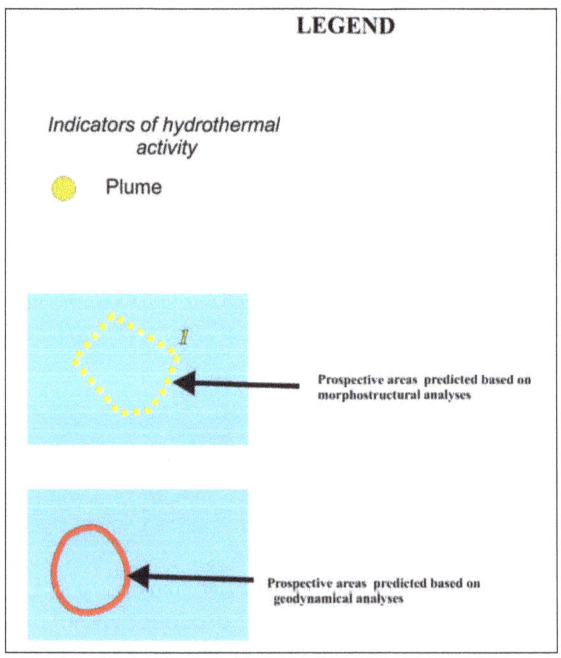

Figure 4-24 Knipovich Ridge permissive tract 4 from the morphostructural analysis. Figure a shows the morphostructural elements, whereas figure b shows the prospective areas identified by the geodynamic analysis, outlined in red. Click here and here for larger image.

In the region at 75.45° N. along the Knipovich Ridge, an abnormal turbidity was recorded at depths of 2600–2650 metres, along a 2 km zone (Cherkashov et al., 2013). Increased contents of Mn and dissolved CH_4 were observed. At the northern extremity, the anomaly is at a depth of 2850 metres, and at 2650–2700 metres at the southern extremity. One may suggest that the source of the plume anomalies lies further to the north, where outcrops of ultrabasic lithologies / rock types are likely, including serpentinites.

Area 3 in Figure 4–23 overlaps partly with segment approximately 30 km long, with distinct rhomboid outlines. See Figure 4–24. These are bound by rather specific non-transform offsets, where an abrupt contraction of the inner floor occurs due to spatial convergence of the opposite scarps of the valley walls of the adjacent segments. A low (200 m) sub-meridional linear volcanic rise without a clear crest runs along the whole section, and contains multiple central-type volcanoes, the largest reaching 3 km in diameter. The axial peak seems to be a very young structure and its surface is partly overlapped with young sheet lavas (Crane et al., 2001).

In the southern part of the section, the western rift valley wall represents a system of small steps. The eastern wall contains a monolithich structure and a steep scarp.

The promise of the area is associated with the high-amplitude flank faults. Of special interest is a narrow, high, probably tectonic ridge, which crosses the rift valley in the zone of non-transform offset in the north of the area. Also, the axial ridge volcanoes are of interest.

4.2.6 Mohns – permissive tract M1 and MS-M-area 1

Several promising areas have been identified by both geodynamical and morphostructural analysis, in and around interest area 4 from the geodynamical analysis. They are in the discharged condition and lie near the tension structures. These areas are promising for hydrothermal activity.

Hydrothermal manifestations found earlier confirm the prospectivity of permissive tract M1 / area 4 (Pedersen et al., 2010b). Loki's Castle is located at 73°33′ N in the discharge zone (indicated by yellow lines in Figure 4–25).

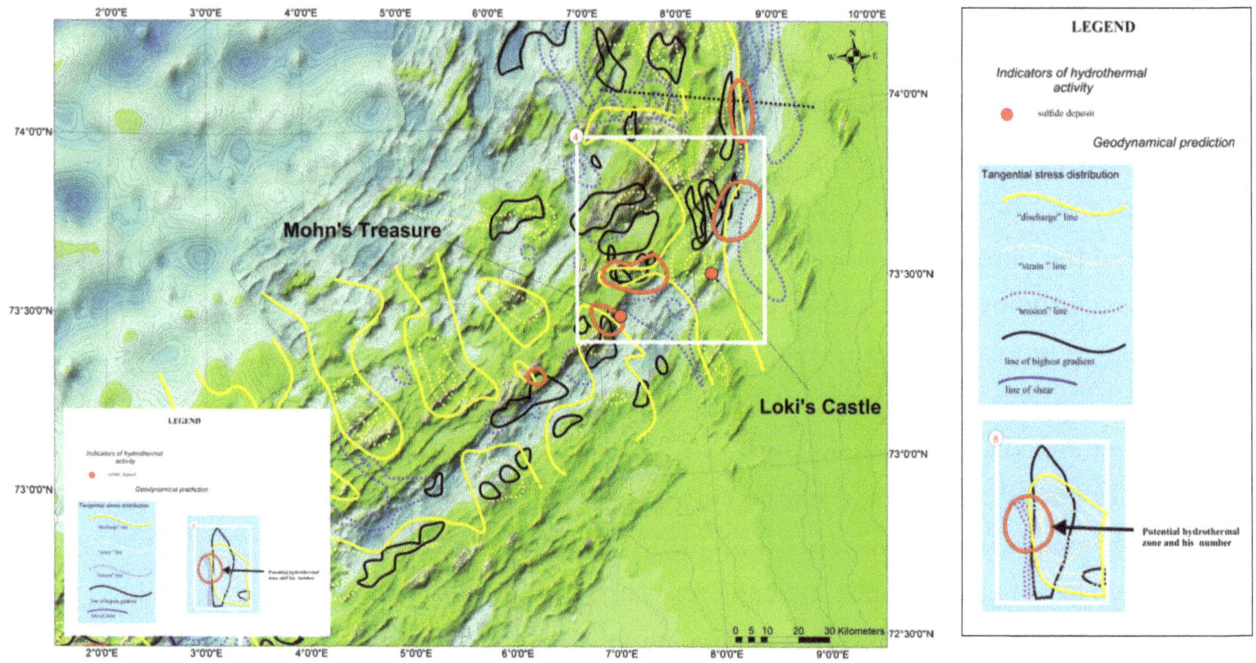

Figure 4–25 Mohns Ridge prospectivity of interest area 4, defined by the geodynamical analysis. Loki's Castle and Mohns' Treasure are both close to the yellow discharge line. Click here for larger image.

The second, Mohns' Treasure, at 73°27′ N is also in the discharged zone, and lies near the high-gradient tension structure. See again Figure 4–25. Their locations coincide with the areas identified as 'promising' by the geodynamical analysis, which is an indication of the appropriateness of the approach to revealing areas of hydrothermal activity.

MS-M-area M1 includes a section of the rift valley over 70 km long, having an almost complete set of morphostructural indicators for hydrothermal mineralisation (faults, central volcanos, scarps, axial volcanic ridges, graben structures etc.). This area has been divided into three segments. See Figure 4–26b. The most promising are the axial volcanic ridges with the central grabens and the multiple flat-topped and cratered central-type volcanoes, the rift-valley wall structures defined by high-amplitude faults (in particular, detachment faults) and the oceanic core complexes defined based on their morphology characterised by break-away ridges and corrugated surfaces. Thus, in the north-east, beside the known vent site on the axial graben (Pedersen et al., 2010b), one may assume a high probability for recent venting in the area, with transverse volcanic terraces at the toe of the eastern marginal scarp of the rift valley. Of special interest, is the western rift-valley wall in the middle and southern parts of the area, as well as a large central-type volcano at the foot of the eastern wall.

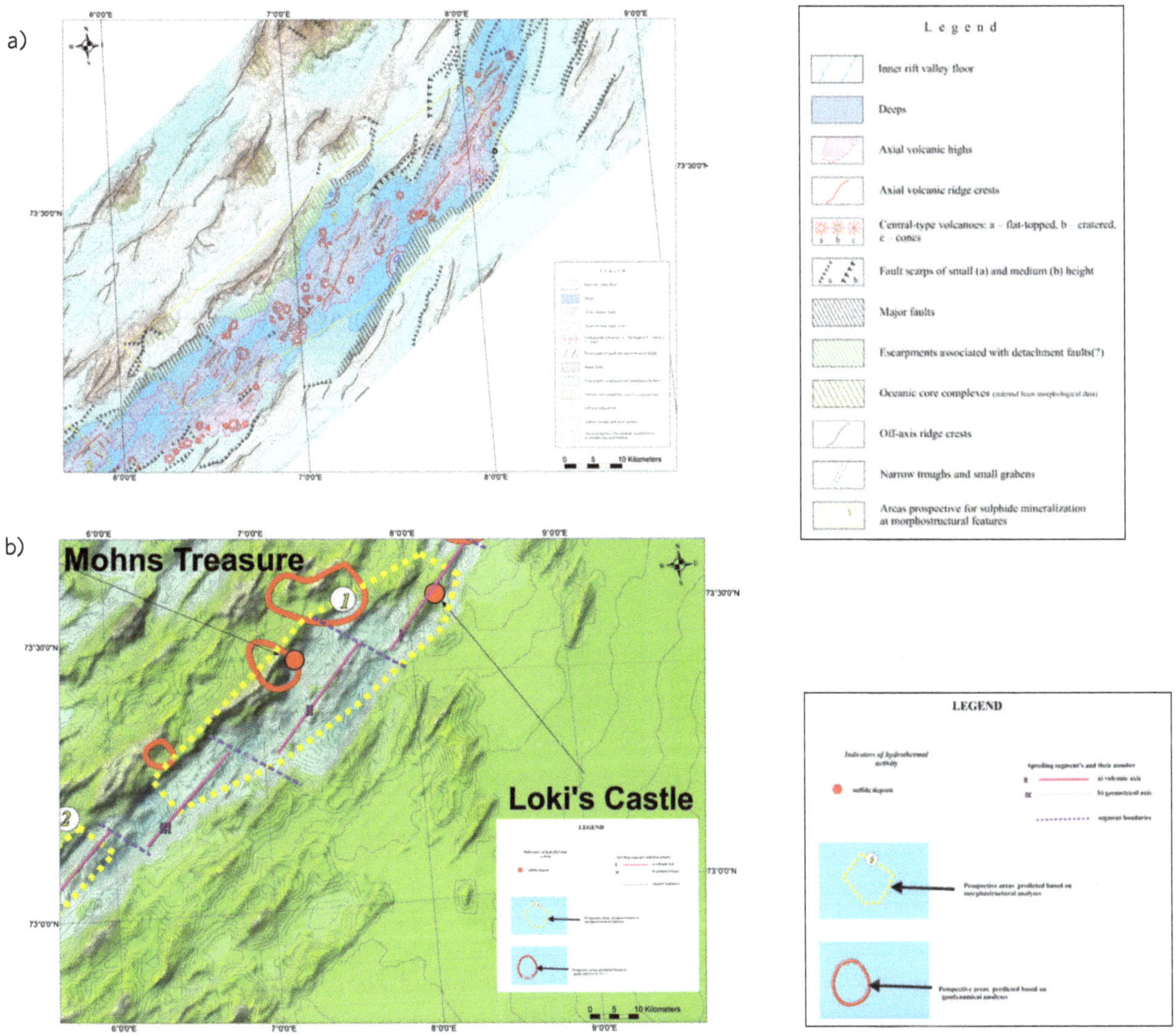

Figure 4-26 Mohns Ridge MS-M-area 1, close to interest area 4 defined by the geodynamical analysis. Figure a shows the morphostructural elements, whereas figure b shows the prospective areas identified by the geodynamic analysis, outlined in red. Click here and here for larger image.

4.2.7 Mohns - permissive tract M2 and MS-M-area 2

The MS-M-area 2 is in principal similar to the MS-M-area 1 described in the previous section. It includes a considerable part of segment IV, a rather high (700 m) axial volcanic ridge with the central graben in its most elevated part, and the western flank of the rift valley partly controlled by a detachment fault. See Figure 4–27a.

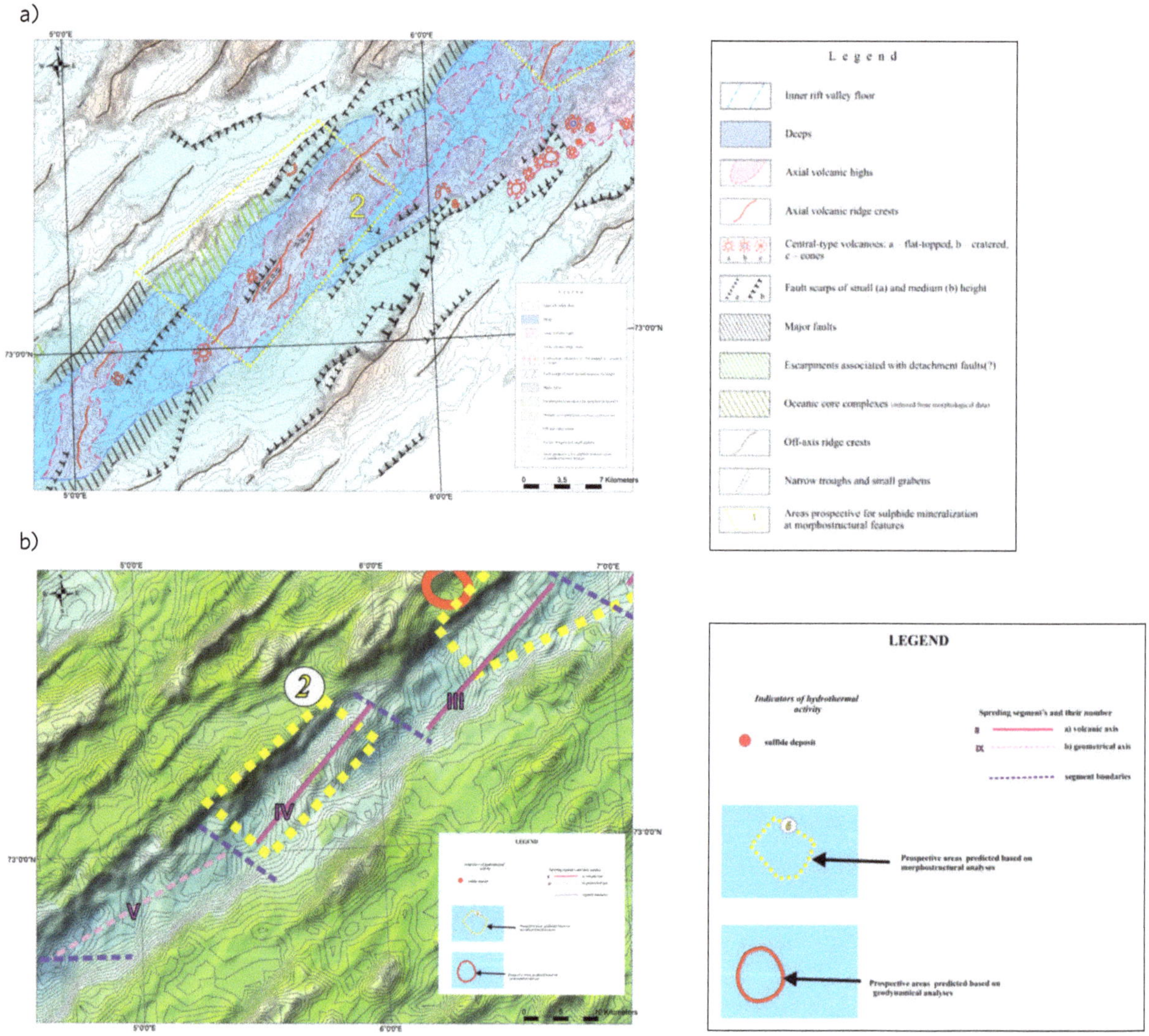

Figure 4-27 Mohns Ridge MS-M-area 2, southwest of interest area 4 defined by the geodynamical analysis. Figure a shows the morphostructural elements, whereas figure b shows the prospective areas identified by the geodynamic analysis, outlined in red. Click here and here for larger image.

This tract is solely defined from the morphostructural indicators. No significant geodynamic indicators are apparent.

4.2.8 Mohns - permissive tract M3 and MS-M-area 3

MS-M-area 3 lies in the central part of segment VI. See Figure 4–28. Its special feature is a very large and apparently rather old, massive volcano-tectonic ridge overlapping the eastern flank of the rift valley, and linking with the structures of the western edge by a well-expressed transverse terrace. The promises for venting in this area are associated with the slopes of the western flank which, however, do not show many of the morphostructural indicators deemed necessary for hydrothermal manifestations. In the axial ridge, according to the morphostructural indicators and the lack of geodynamical indications of active venting, a relict ore occurrence may be expected.

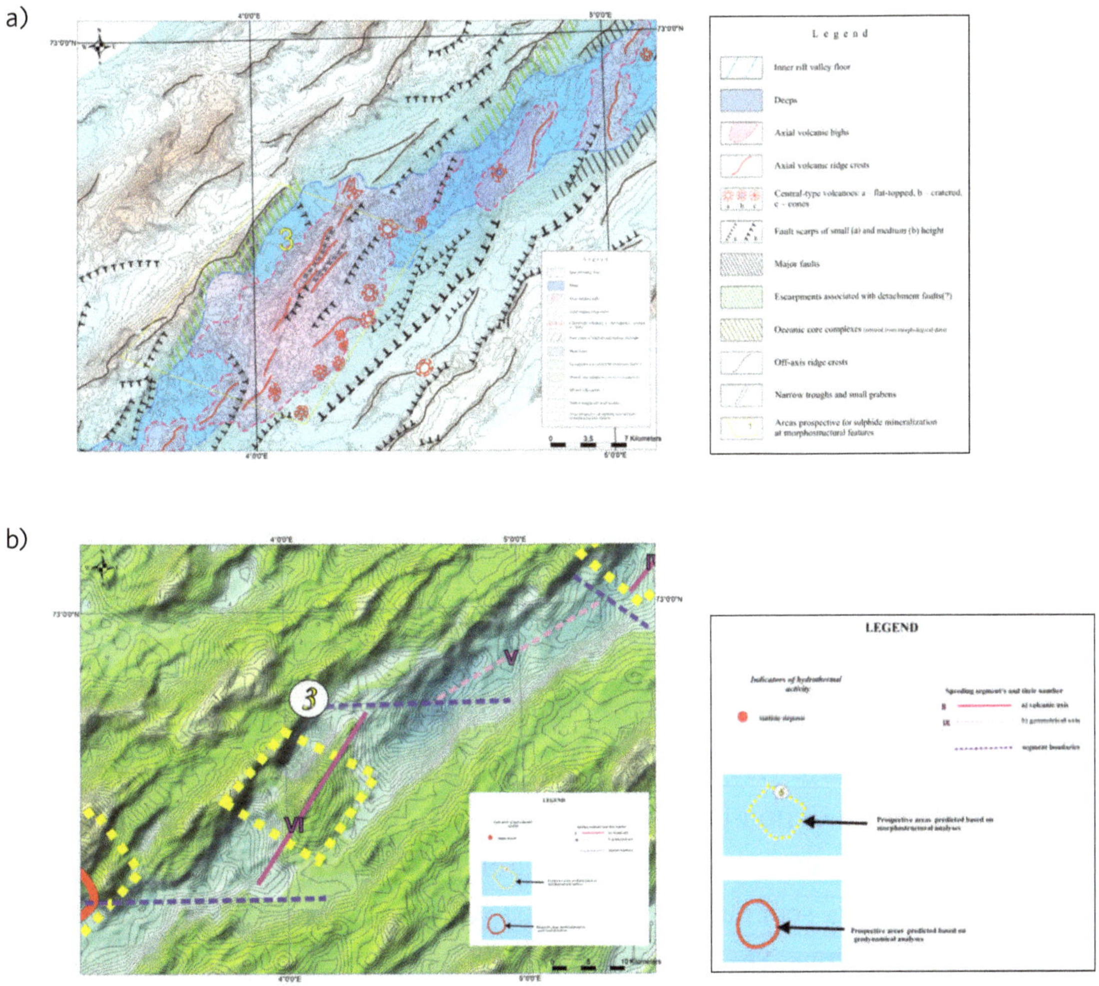

Figure 4-28 Mohns Ridge MS-M-area 3, northeast of interest area 5 defined by the geodynamical analysis. Figure a shows the morphostructural elements whereas figure b shows MS-M-area 3. No significant geodynamical indications are present. Click here and here for larger image.

As with tract M2, this tract is solely defined from the morphostructural indicators. No significant geodynamic indicators are present.

4.2.9 Mohns - permissive tract M4 and MS-M-area 4

The hydrothermal manifestation Copper Hill is in this area at 72°32' N (Pedersen et al., 2010b). Copper Hill is located in the discharged zone, and lies near the highest gradient tension structures (black outline in Figure 4–29). See also Figure 4–30. Apart from the area connected to Copper Hill, there are two other areas with the same conditions that could be regarded as promising; indicated in Figure 4–29 by red outlines.

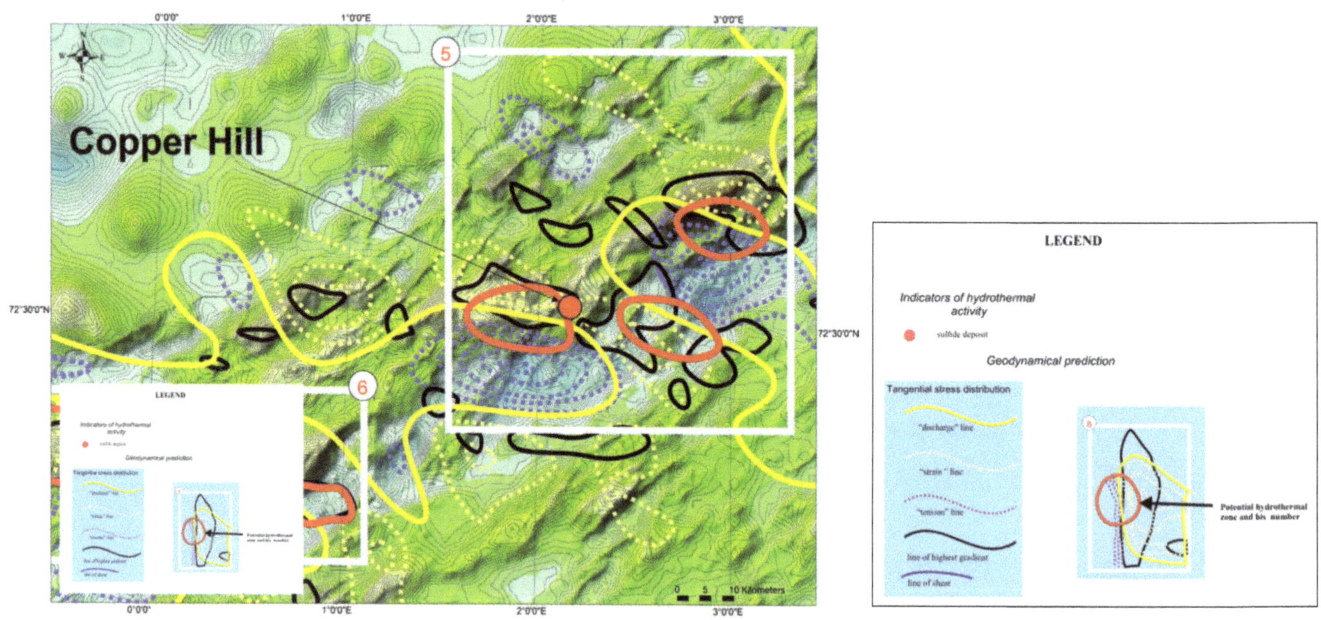

Figure 4–29 Mohns Ridge prospectivity of interest area 5, defined by the geodynamical analysis. Click here for larger image.

MS-M-area 4 (see Figure 4–30) fully covers segment VII and the adjacent zones of non-transform offsets of the rift valley, including the rather remote structures of its western flank. A funnel-like hole (crater?) in the depression, separating two (relict?) volcanic ridges is very noticeable in the axial zone. See Figure 4–30a. In the absence of an expressed volcanic rise, the presence of a crater may be indicative of a very shallow magma chamber (Vogt, 1974). The major prospects of this area are connected with the multiple, morphologically-attractive structures of the western flank, controlled by large faults.

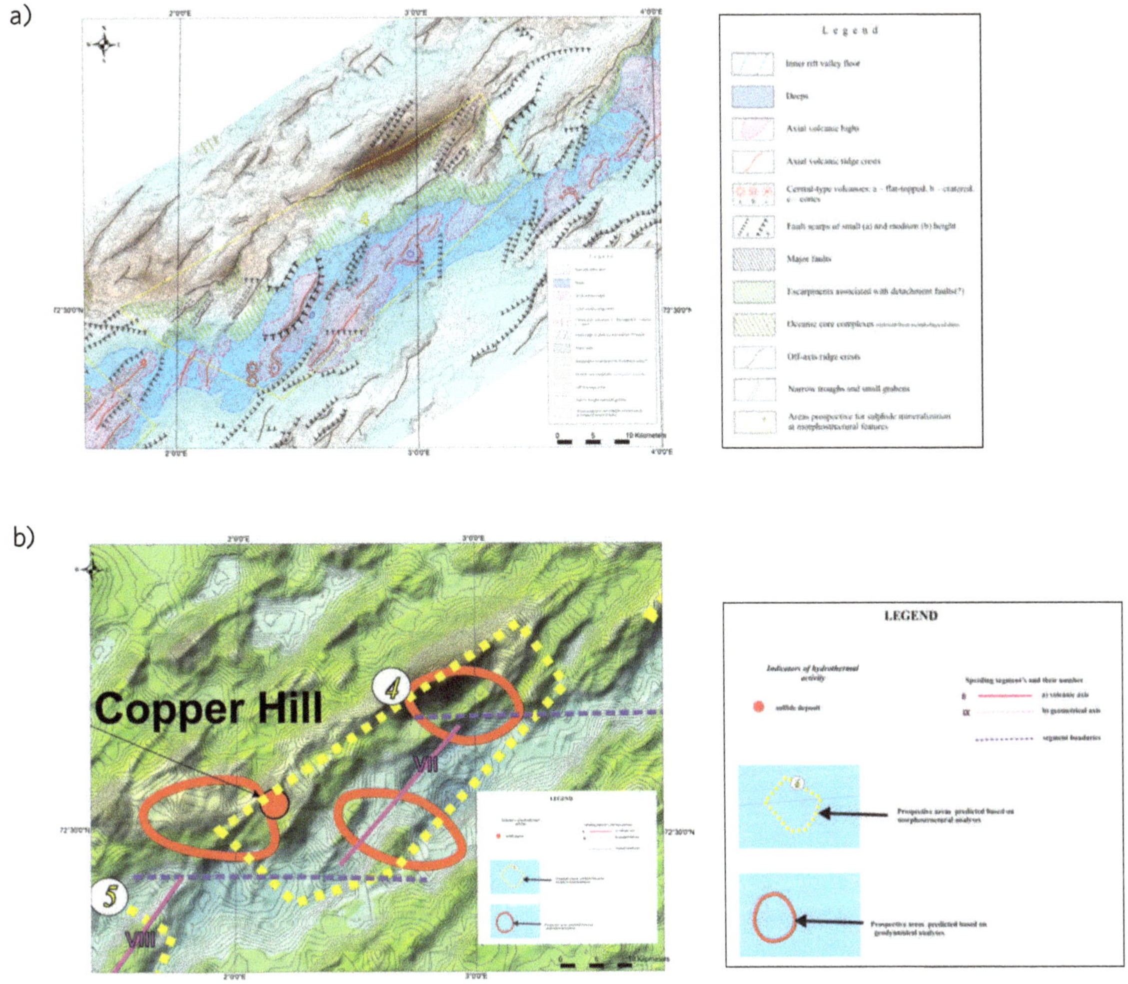

Figure 4-30 Mohns Ridge MS-M-area 4 (permissive tract 4), in area 5 defined by the geodynamical analysis. Figure a shows the morphostructural elements, whereas figure b shows the prospective areas identified by the geodynamic analysis, outlined in red. Click here and here for larger image.

4.2.10 Mohns - permissive tract M5 and MS-M-area 5

MS-M-area 5 is located in the central part of segment VIII, with many features similar to the MS-M-area 3 (segment VI). It is characterised by a well-defined structure with corrugations trending parallel to the spreading direction and occurring on a broad and smooth surface (see Figure 4-31a and b) of the western wall of the rift valley, making it comparable to a typical oceanic core complex.

a) b)

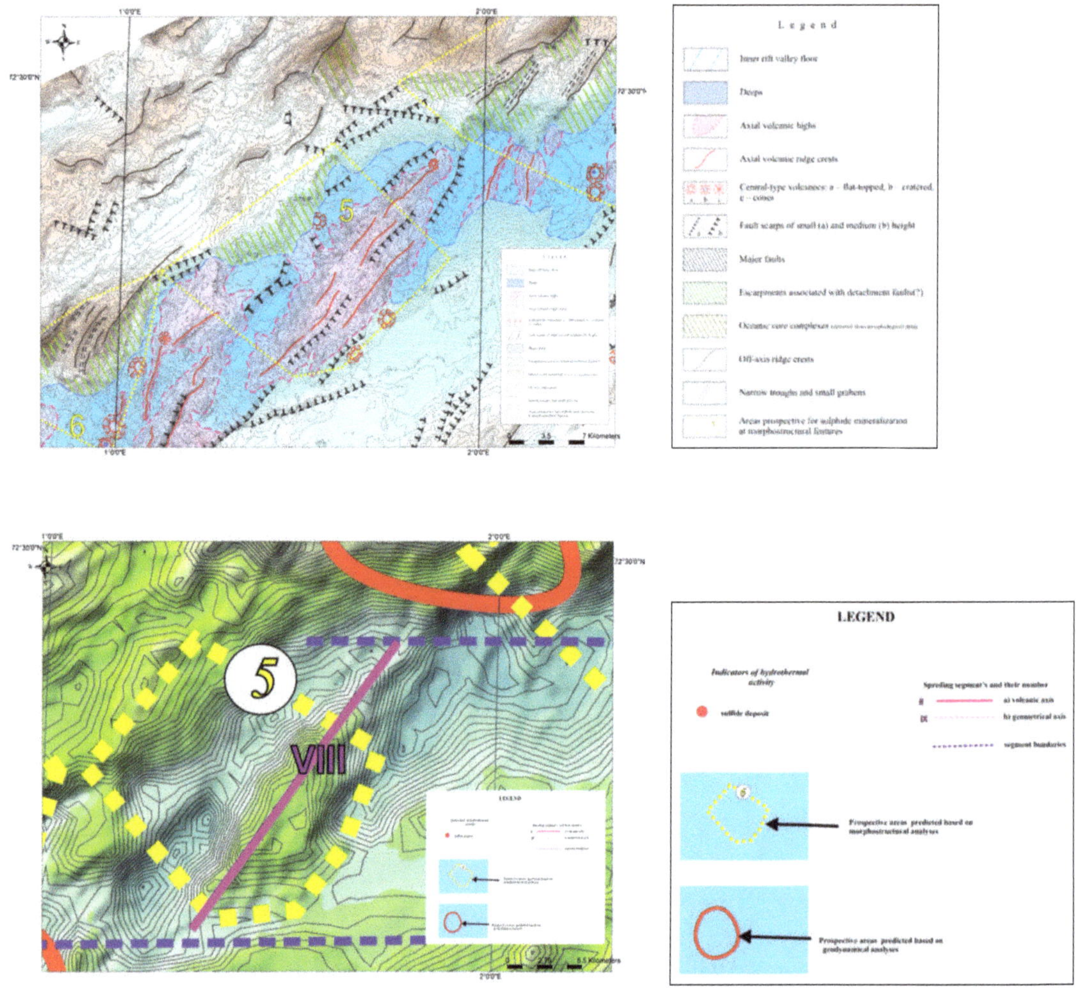

Figure 4-31 Mohns Ridge permissive tract M5, between interest area 5 and 6 defined by the geodynamical analysis. Figure a shows the morphostructural elements, whereas figure b shows the prospective areas identified by the geodynamic analysis, outlined in red. Corrugated structure interpreted in the north-western part of the tract. Click here and here or larger image.

No significant geodynamical anomalies are apparent in this tract, and it is solely defined based on the morphological analysis.

4.2.11 Mohns - permissive tract M6 to M9 and MS-M-area 6-9

Although no hydrothermal activity has been detected yet in this region, the results of the geodynamical analysis suggest that the area at the western flank of the ridge may be regarded as promising (the larger area within the red outline in Figure 4–32).

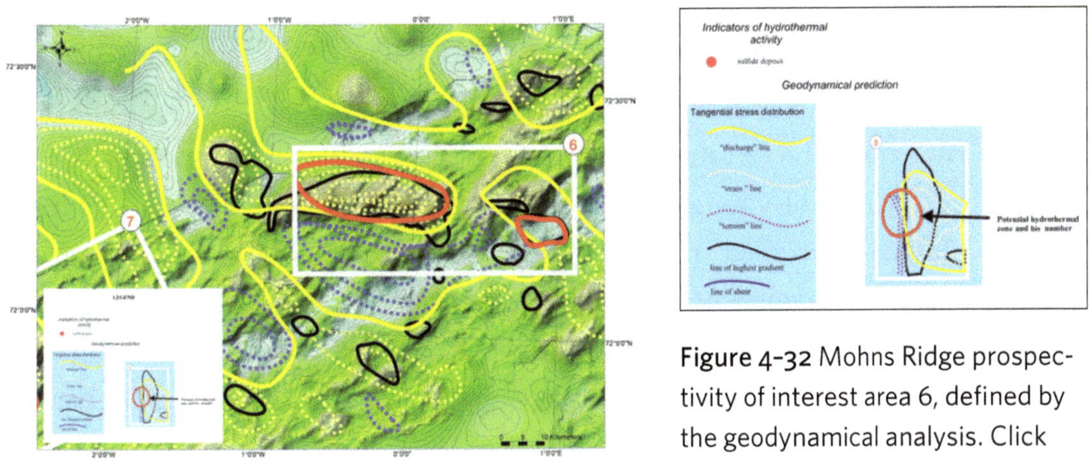

Figure 4-32 Mohns Ridge prospectivity of interest area 6, defined by the geodynamical analysis. Click here for larger image.

Connected to interest area 6 there are four promising permissive tracts; permissive tracts 6, 7, 8 and 9. See Figure 4–33. These tracts are, in the following, described separately.

a)

b)

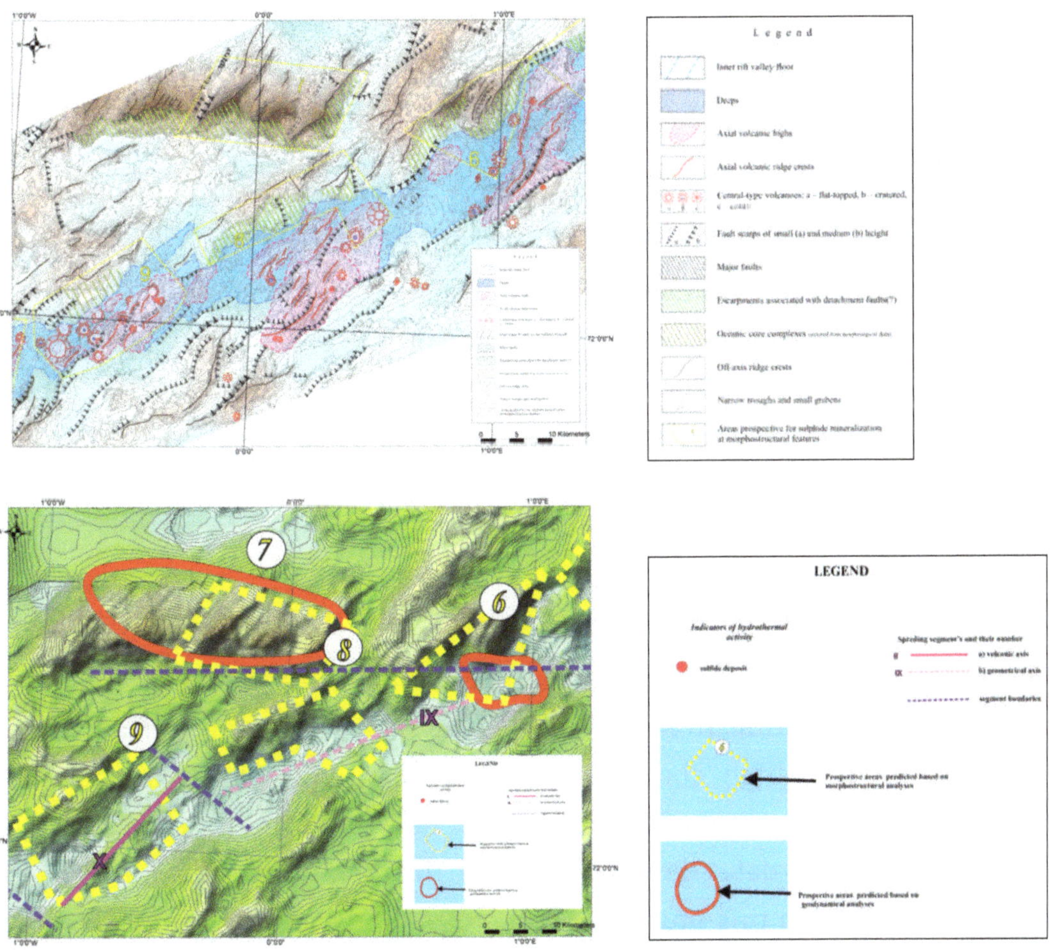

Figure 4-33 Mohns Ridge permissive tract M6 to M9, connected to interest area 6 by the geodynamical analysis. Figure a shows the morphostructural elements, whereas figure b shows the prospective areas identified by the geodynamic analysis, outlined in red. Click here and here for larger image.

4.2.12 Mohns - permissive tract M6 and MS-M-area 6

MS-M-area 6 is associated with a zone of the large lateral offset of the rift valley. The tectonic fault at the toe of a height is traceable southwards by a chain of young central-type volcanoes, which suggests its recent magma-controlled activity. Often, under such circumstances, certain extremely favourable conditions for high-temperature hydrothermal circulation appear in the area of the flank structures. In this case, sulphide deposits may occur in the graben-like depressions of the upper part of the elevation.

4.2.13 Mohns - permissive tract M7 and MS-M-area 7

MS-M-area 7 includes several relict highs of volcanic origin, similar to the above, formed at the earlier stages of the rift-valley development. The large transverse fault at their base seems to be still active. It is postulated that it is highly probable that relict sulphide deposits will be found here.

4.2.14 Mohns - permissive tract M8 and MS-M-area 8

MS-M-area 8 lies mainly on the western edge of segment IX of the rift valley. In the axial part of this segment, volcanic activity is rather low, apart from several large central-type volcanoes. This does not exclude any venting in the areas of detachment faults at the proximal flanks of the valley.

4.2.15 Mohns - permissive tract M9 and MS-M-area 9

MS-M-area 9 is located within the boundaries of segment X, which is very interesting from the morphological point of view, featuring an abnormally-high abundance of large flat-top central-type volcanoes on the valley floor and at the base of its western wall. According to the bathymetry data, the north part of this area represents an OCC-like structure. In this situation, it is hard to predict what type of hydrothermal circulation will prevail – the 'axial' or the 'flank' type. In general though, the area seems promising.

4.2.16 Mohns - permissive tract M10 and MS-M-area 10

No hydrothermal activity has been detected yet in this region. However, connected to area of interest 7, from the geodynamic analysis, there are three promising permissive tracts; numbers 10, 11 and 12. The line of discharge here coincides with pronounced tension structures (black outlines in Figure 4–34). The different permissive tracts are shown in Figure 4–36.

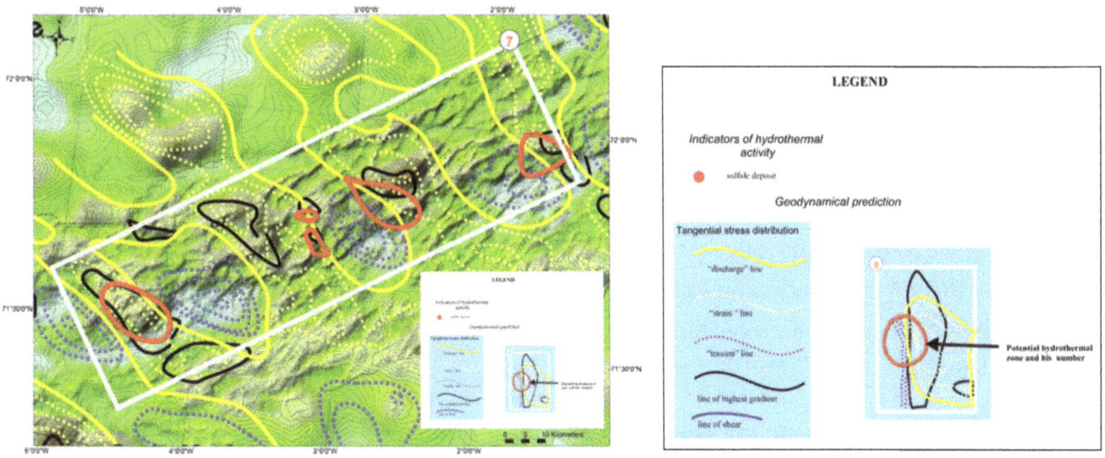

Figure 4–34 Mohns Ridge prospectivity of interest area 7, defined by the geodynamical analysis. Click here for larger image.

Permissive tract 10, like the preceding one, has many favourable morphostructural features, both in the axial zone and the western flank. Morphologically, its northers part looks like a large oceanic core complex and the area contains several volcanic rises. See Figure 4–35.

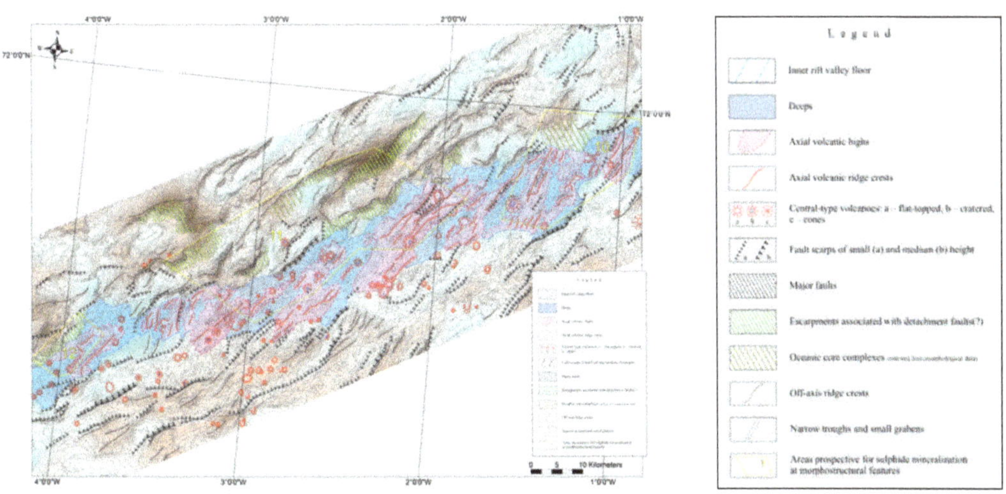

Figure 4–35 Mohns Ridge permissive tract M10 to M12, connected to interest area 7 by the geodynamical analysis. The figure shows the morphostructural elements. Click here for larger image.

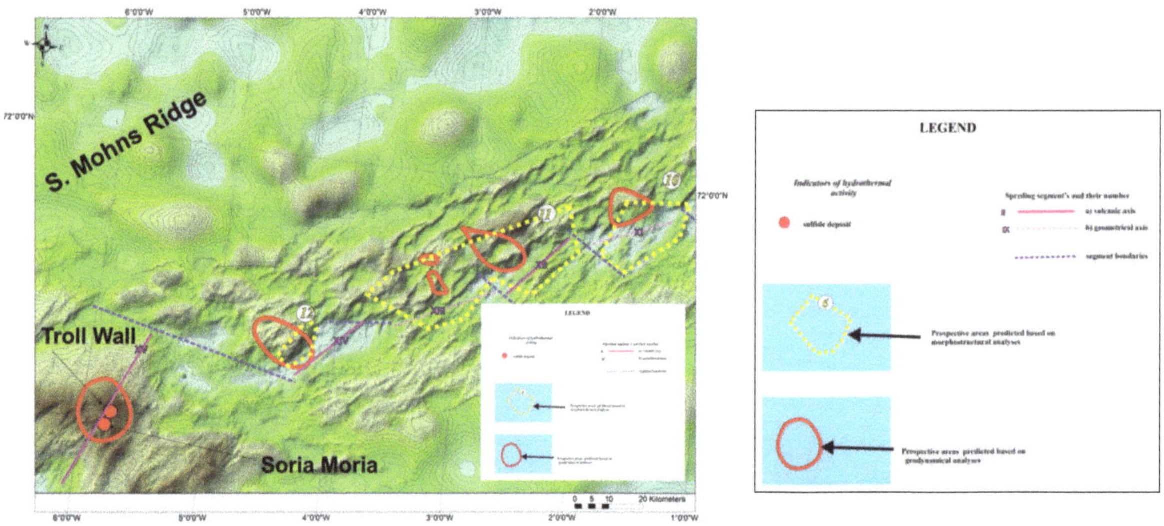

Figure 4-36 Mohns Ridge permissive tract M10 to M12, connected to interest area 7 by the geodynamical analysis. The figure shows the prospective areas identified by the geodynamic analysis, outlined in red. Click here for larger image.

4.2.17 Mohns - permissive tract M11 and MS-M-area 11

MS-M-area 11 combines multiple promising axial and flank structures of the two segments (XII and XIII) of the valley, including rather remote ones. See Figure 4–35 and Figure 4–36. It seems that its flank structures deserve special attention, since they are controlled by detachment faults demonstrating a variety of potentially ore-bearing morphostructural settings.

4.2.18 Mohns - permissive tract M12 and MS-M-area 12

MS-M-area 12 encompasses the western wall of the rift valley, defined by a detachment fault with steep inclination and a significant fault throw. See Figure 4–35 and Figure 4–36. At its base, a young volcanic zone is located and this fault may be interpreted as a magma-conducting channel. This situation suggests the high probability of hydrothermal fluid circulation in the zone of the flank tectonic dislocations, accompanied by the formation of sulphide deposits in the apical part of the structure.

4.2.19 Mohns - permissive tract M13 and MS-M-area 13

One may see on the map that the location of the hydrothermal manifestations found earlier, Troll Wall (71°17' N) and Soria Moria (71°15' N) (Pedersen et al., 2010b), coincides with the intersection points of the discharge line (yellow) and a pronounced tension structure. This correlation confirms the results of the geodynamical analysis indicating possible locations of hydrothermal activity. See Figure 4–37. Inside interest area 8 there is a site, north of the Troll Wall site, where the different isolines cross or are close to each other. This site should be regarded as promising.

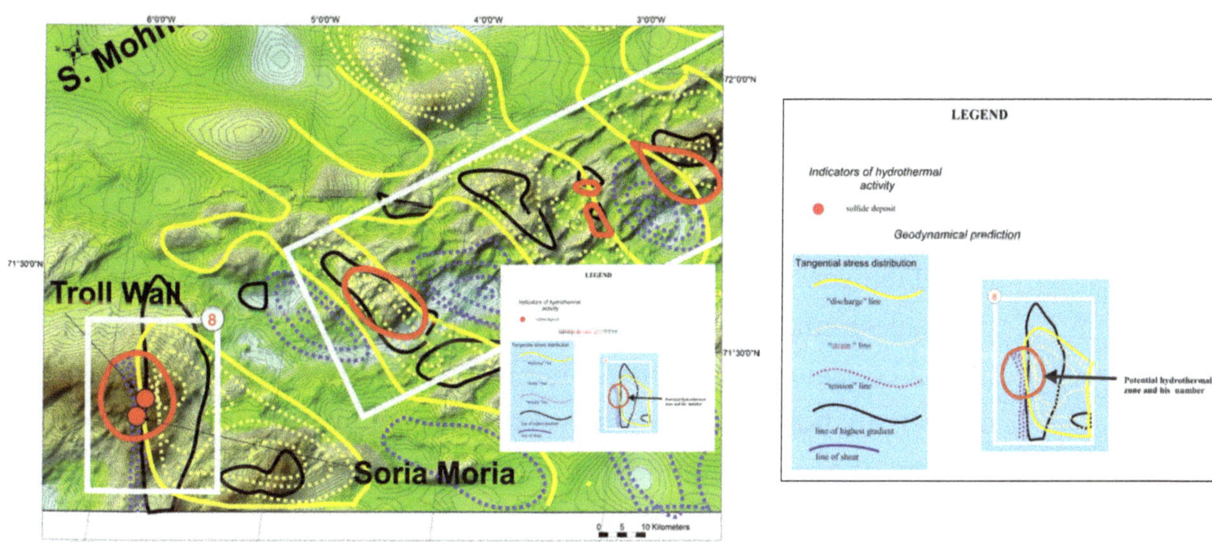

Figure 4–37 Mohns Ridge prospectivity of interest area 8, defined by the geodynamical analysis. Promising area north of Troll Wall, where the red, yellow and black lines cross. Click here for larger image.

CHAPTER 5

Permissive Tract Characterisation

5.1 Deposit density

The well-mapped Troodos massif in Cyprus has been chosen as an analogue for the slow-spreading Mohns and Knipovich Ridges.

A difficulty in applying a density from ancient deposits to those currently at the MAR is to mimic as closely as possible how the permissive tracts of the MAR translate into the permissive area of the analogue. Singer et al. (2001) estimated the massive sulphide deposit density within the 1010 km^2 exposed permissive extrusive rocks from Troodos to be 20 deposits per 1000 km^2, with an ore tonnage larger than 20000 tons that all fit the Cyprus volcanogenic massive sulphide grade and tonnage model (Singer, 1986). This represents the necessary consistency between the grade–tonnage model and the density model that Singer (2014) underlines is critical for the estimated numbers to be meaningful. All of the 20 deposits are hosted within pillow lavas that can be assumed to represent the depositional environment of the slow-spreading MAR that forms the permissible assessment areas along the Mohns and Knipovich Ridges. Although Singer et al. (2001) state that it is unlikely that any undiscovered Cyprus volcanogenic massive sulphide deposits are exposed in extrusive rocks, since the exposures are very well-explored, continued drilling programs may locate buried deposits. Up to eight areas with high mineralisation potential have been outlined for future sub-surface exploration (Jowitt et al., 2012). Additional bias may also come from any pre-erosional amounts of sulphides that have been lost to erosion before reaching the present landform which, for the Iberian pyrite belt, is estimated to be 10–15% (Barriga et al., 1997). Mosier et al. (2007) state that effects of erosion cannot be demonstrated by lower deposit densities over time, but he does not discuss the effects of changes during the evolution towards the present landform. The volcanogenic massive sulphide database (Mosier et al., 2009) quotes that both the Agrokipia and Limni deposits have had some their massive zones of sulphides eroded away.

To accommodate possible biases, the deposit density of 0.02 deposits/km^2 for the extrusives of Troodos has, in this study, been modelled to correspond to the 20 percentile of the lognormal number of accumulation distribution (Table 1). The green dots in Figure 5–1 display these percentiles of the distribution on the plotted density of deposits per 100000 km^2 versus the permissive area in km^2 (based on data in Singer et al., 2010). The Troodos density is also included in the plot. Our choice of accumulation density distribution (Table 1) for the Mohns and Knipovich permissive areas spans the whole population of volcanogenic massive sulphide deposit densities for permissive areas of approximately the same size as that of the Cyprus extrusives.

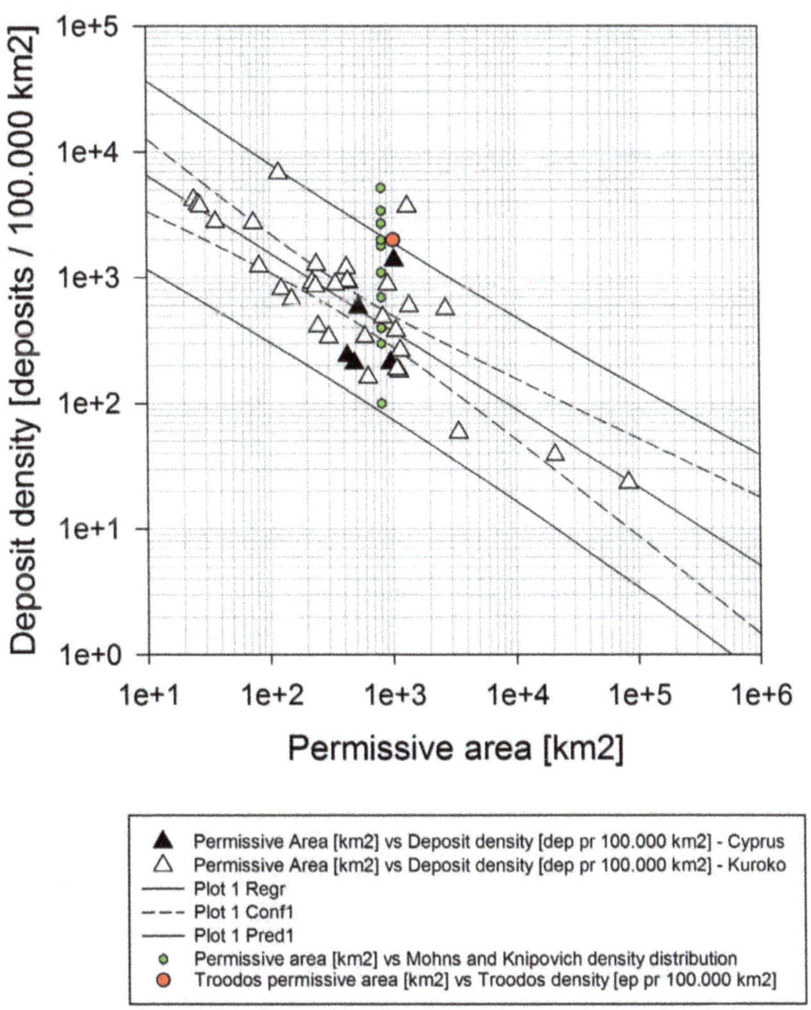

Figure 5–1 Plot of the density of deposits per 100.000 km^2 versus the permissive area in km^2. Percentiles of the SMS number of accumulations distribution in Table 1 is added to the plot with green dots. Figure based on data published in Singer et al. (2010). Troodos permissive area from Singer et al. (2001). Click here for larger image.

Assuming a lognormal distribution, the P20 and minimum and maximum values that span the population of volcanogenic massive sulphide deposit densities, the percentiles from the lognormal distribution can be derived. See Table 1.

Table 1 SMS density and aggregate number of accumulations inside the permissive tracts.

	Mode	Mean	Stdv	F100	F95	F90	F75	F50	F25	F20	F10	F5	F0
SMS density # acc per km²	0.007	0.014	10	0.001	0.003	0.004	0.007	0.011	0.018	0.020	0.027	0.034	0.052
# SMS acc	80	155	112	1	32	45	75	125	205	228	308	382	594

Given the density distribution, there is a 95% chance within the total 11389 km^2 of permissive area on the Mohns and Knipovich Ridges that there are 32 mineralised accumulations or more (percentiles can be seen in Table 1). The most likely number of accumulations, the mode, is 80 and the median is 125. The mean is as high as 155 due to the skew distribution, which has a coefficient of variation (CV) equal to 72%. This variability is of the same order of magnitude as the number of deposits distribution for mafic deposits, CV = 83%, in the recent assessment of the totality of SMS resources at mid-ocean ridges (MOR) by Singer (2014) and reflects the great uncertainty that accompanies assessments in frontier exploration areas.

The play methodology used for the assessment takes into account that only some of the hydrothermal activity indicators that will be investigated have the potential to lead to an SMS accumulation equal to, or larger than, the minimum size of 1700 metric tons of ore (50 metric tons of Cu metal) and a footprint equal to, or larger than, 0.04 km^2.

The number of hydrothermal activity indicators within each sub-play that could identify potential vent fields with surface expressions equal to, or larger than, 0.04 km^2, that could lead to SMS accumulations equal to, or larger than, 1700 metric tons, has been estimated to be proportional to the estimated number of accumulations and the size of the favourability zones, outlined by the morphological and geodynamical analysis. Due to the possibility of finding deposits well off-axis, like the TAG deposit connected to apparently long-lived boundary faults and small pillow volcanoes, some of the deposits could fall outside the stippled high favourability areas (outlined in Chapter 4).

The number of accumulations shown in Table 2 has been computed from the inferred, as well as the postulated, number of sites that could manifest hydrothermal activity within each of the 17 sub-plays, and subsequently aggregated to a total for the whole of the Knipovich and Mohns Ridges. The mean number has been estimated to be proportional to the deposit density and the size of the favourable areas outlined in Chapter 4. The expected total number of inferred and postulated hydrothermal manifestations in Table 2 has been calculated by dividing the number of accumulations by the probability 0.428 for having sulphides, given the presence of hydrothermal manifestations. The probability is calculated from the ratio between the number hydrothermal manifestations that show signs of sulphides and the total of hydrothermal manifestations along mid ocean ridges (Hannington et al., 2004 and 2013a) and the inferred and postulated manifestations is therefore the sum of manifestations with and without sulphides.

Table 2 Estimated number of accumulations and hydrothermal manifestations related to the permissive tracts (sub-plays).

Area	Permissive area [km^2]	Area fraction of total area	Expected number of accumulations	Expected number of inferred and postulated hydrothermal manifestations
MS-K-area-1	3100	0.272	42.2	98.6
MS-K-area-2	2375	0.209	32.3	75.5
MS-K-area-3	375	0.033	5.1	11.9
MS-K-area-4	500	0.044	6.8	15.9
MS-M-area-1	1150	0.101	15.7	36.6
MS-M-area-2	200	0.018	2.7	6.4
MS-M-area-3	225	0.020	3.1	7.2
MS-M-area-4	713	0.063	9.7	22.7
MS-M-area-5	288	0.025	3.9	9.2
MS-M-area-6	150	0.013	2.0	4.8
MS-M-area-7	200	0.018	2.7	6.4
MS-M-area-8	138	0.012	1.9	4.4
MS-M-area-9	275	0.024	3.7	8.7
MS-M-area-10	325	0.029	4.4	10.3
MS-M-area-11	975	0.086	13.3	31.0
MS-M-area-12	100	0.009	1.4	3.2
MS-M-area-13	300	0.026	4.1	9.5
Sum	11389		155	362.1

Table 3 Play areas and distribution of number of accumulations. Total area 11,389 km².

		Inverse cumulative distribution of the potential number of accumulations											
ID	**Area permissive tract [km²]**	**Mode**	**Mean**	**Std. dev.**	**F100**	**F95**	**F90**	**F75**	**F50**	**F25**	**F10**	**F5**
MS-K-area-1	3100	13.5	42.5	43.5	0.0	5.0	7.9	14.9	29.0	54.0	94.0	129.0
MS-K-area-2	2375	7.0	29.2	25.3	0.0	3.8	5.8	11.0	21.0	39.9	65.0	84.0
MS-K-area-3	375	2.0	5.2	4.7	0.0	0.0	1.0	2.0	4.0	7.0	11.1	14.5
MS-K-area-4	500	2.0	6.6	6.5	0.0	0.0	1.0	2.0	5.0	9.0	15.0	19.6
MS-M-area-1	1150	4.0	16.6	14.4	0.0	1.9	3.0	6.4	13.0	22.0	35.6	45.7
MS-M-area-2	200	0.0	2.7	2.8	0.0	0.0	0.0	1.0	2.0	4.0	6.3	8.0
MS-M-area-3	225	0.0	3.1	3.2	0.0	0.0	0.0	1.0	2.0	4.0	7.0	9.8
MS-M-area-4	713	3.0	10.4	10.6	0.0	1.0	2.0	4.0	7.0	13.0	22.8	31.8
MS-M-area-5	288	1.0	4.0	4.2	0.0	0.0	0.0	1.0	3.0	6.0	10.0	12.0
MS-M-area-6	150	0.0	2.4	2.4	0.0	0.0	0.0	1.0	2.0	3.0	6.0	7.0
MS-M-area-7	200	0.0	2.7	2.8	0.0	0.0	0.0	1.0	2.0	4.0	6.3	8.0
MS-M-area-8	138	0.0	2.4	2.4	0.0	0.0	0.0	1.0	2.0	3.0	6.0	7.0
MS-M-area-9	275	1.0	4.0	4.2	0.0	0.0	0.0	1.0	3.0	6.0	10.0	12.0
MS-M-area-10	325	1.0	4.3	4.7	0.0	0.0	0.0	1.0	3.0	6.0	10.1	14.1
MS-M-area-11	975	4.0	13.6	12.1	0.0	1.0	2.0	5.0	10.0	18.2	28.7	36.7
MS-M-area-12	100	0.0	1.3	1.5	0.0	0.0	0.0	0.0	1.0	2.0	3.0	4.1
MS-M-area-13	300	1.0	4.0	4.2	0.0	0.0	0.0	1.0	3.0	6.0	10.0	12.0
Sum	11389		155									

A reliability check on these numbers can be made by looking at the growth in the number of discovered active vents over the last 45 years. The discovered number of active vent fields on the MOR (64000 km) grows each year (Figure 5–2), and can be predicted to reach more than 1000 over the next 100 years.

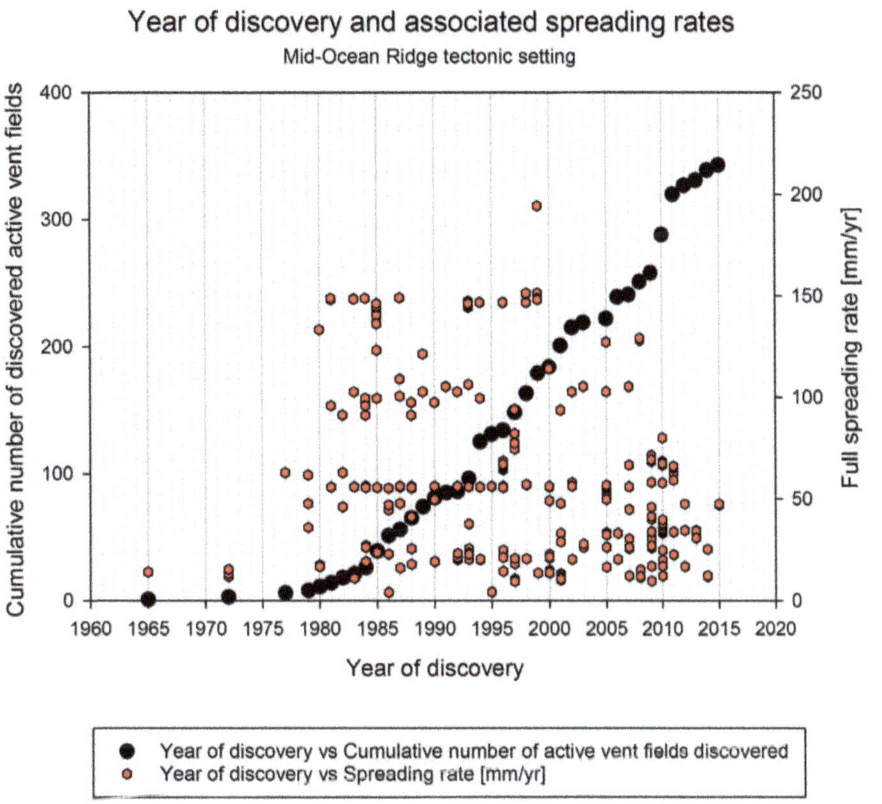

Figure 5-2 Evolution in time of the number of active fields on the MOR, based on data from Beaulieu et al. (2015). Click here for larger image.

If the length of the Knipovich Ridge is 460 km and Mohns Ridge is 570 km, then the 1030 km of ridge could, accordingly, be expected to ultimately have 16 active vent fields (being larger than the minimum used here). Inactive SMS accumulations, like 13030'N Semyenov, without hydrothermal manifestations are difficult to find. Postulating that, in addition to the surveyed and inferred vent fields, about an equal number of undiscovered vent fields might exist as concealed vent fields, with less bathymetric expression than the ones inferred from the morphostructural and geodynamic analysis of the bathymetric data. The 95% probability of having 32 or more SMS accumulations, as shown in Table 1, should therefore be regarded as a conservative projection for the ultimate number of both active and inactive undiscovered SMS accumulations on the Mohns and Knipovich Ridges, in view of the short SMS exploration period and low intensity of exploration until now.

Hannington et al. (2010) refer to estimations that all conclude that there are roughly 1000 active vent sites along the mid-ocean ridges. Furthermore, Hannington et al. (2010) conclude that the expected total number of what they term 'significant SMS occurrences along the neovolcanic zones of the world's oceans' is around 900, with a P90 of 500 and a P10 of 5000. Using the number of active vents at MOR settings in Beaulieu (2015) as an indicator of the distribution between the number of vents in different tectonic settings, around 500 out of these 900 could be found in MOR settings. This corresponds well with the minimum estimated value of deposits in neovolcanic zones presented in Hannington (2013a). Given 64000 km of mid-ocean ridge, this correlates to 8 active significant SMS occurrences per 1000 km of neovolcanic zones along the MOR.

5.2 Probability

Quantifying the probability on a regional level involves estimating the probability of favourable regional conditions that render it possible for local accumulations of metals. On the basis of the information from the known vent fields (Pedersen et al., 2010b) and the favourable and confirmed regional settings, this is set to 0.9 for unconfirmed plays. Confirmed plays have a play probability of 1.

Given the three SMS probabilities in Chapter 3, the first of these at the local scale is estimated to be 1, based on the criteria used for the seventeen permissive tracts outlined by the morphological and geodynamical analyses. This means that it is regarded as certain that there are local manifestations of hydrothermal activity, representing accumulations that potentially are within the required tonnage and grades inside the subplays, and furthermore, that these will be found if the plays are thoroughly explored.

The second of the individual SMS probabilities is estimated to be 0.428, given that from 327 sites with manifestations of hydrothermal activity, 140 sites showed active venting or polymetallic sulphides (Hannington et al., 2013a).

The third of the individual SMS probabilities is set to 1, based on our grade-tonnage and deposit-density requirement (Singer et al., 2008) that all accumulations must satisfy our minimum tonnage of 1700 metric tons used as the 100th percentile in the number of accumulations distribution shown in Table 6.

The SMS's technical discovery probability, given a favourable play, is the product of the probabilities for the adequacy of the three conditions above. This represents the probability that exploration activities can prove the presence of an accumulation with potentially exploitable amounts of metals, given the presence of hydrothermal activity related to the accumulation of 1700 metric tons or more of polymetallic sulphides with equal or larger metal content corresponding to minimum grades.

5.3 Tonnage and grade distributions

5.3.1 Introduction

The size and grade distributions that were used for the assessment of SMS on the Mohns and Knipovich Ridges are comparable to those compiled from Global SMS deposits at mid-ocean ridges and deposits in arc and back-arc settings, but have been modelled with less variability due to the expected lower frequency of extreme grades on the Norwegian ridges. The expected coefficient of variability is more than 3 times smaller than that calculated from the global data (Hannington et al., 2011b). The 25th percentiles for all these distributions are nearly half of those of the global SMS deposits, except for Cu where they are of similar magnitude. Copper has been modelled assuming a lognormal distribution with a 25% probability of having a grade of 7 wt % or more. It is important to underline that the grades represent average values over the whole tonnage of the SMS accumulation and are, therefore, lower than spurious grab samples from the surface or from individual samples.

The choice of grade and tonnage models should also take into account the geochemical redistribution related to how the metals are remobilised and re-deposited between pristine formations on the current oceanic crust and known VMS deposits in ancient terrains (e.g. Melekestseva et al., 2013).

5.3.2 Tonnage distribution

Hannington et al. (2010) present tonnages for 62 of the best-mapped SMS deposits. Only deposits that have surface areas larger than 100 m² (~3,000 tonnes, n = 42) have been included. These 42 deposits represent 67% of the deposits included in Hannington et al. (2010). The intercept for the 50th percentile indicates a median deposit size of 70,000 metric tons. Hannington et al. (2010) also present cumulative size frequency distribution of SMS deposits at MOR. This includes the 10% largest deposits, like Middle Valley, Zenith-Victory, Puys des Folles (~10 x 10^6 t), Semyenov (~9 x 10^6 t), Krasnov (~3 x 10^6 t), Sunrise (~3 x 10^6 t), the TAG Mound (2.7 x 10^6 t), and Solwara 1 (2.1 x 10^6 t). The Atlantis II Deposit in the Red Sea is excluded.

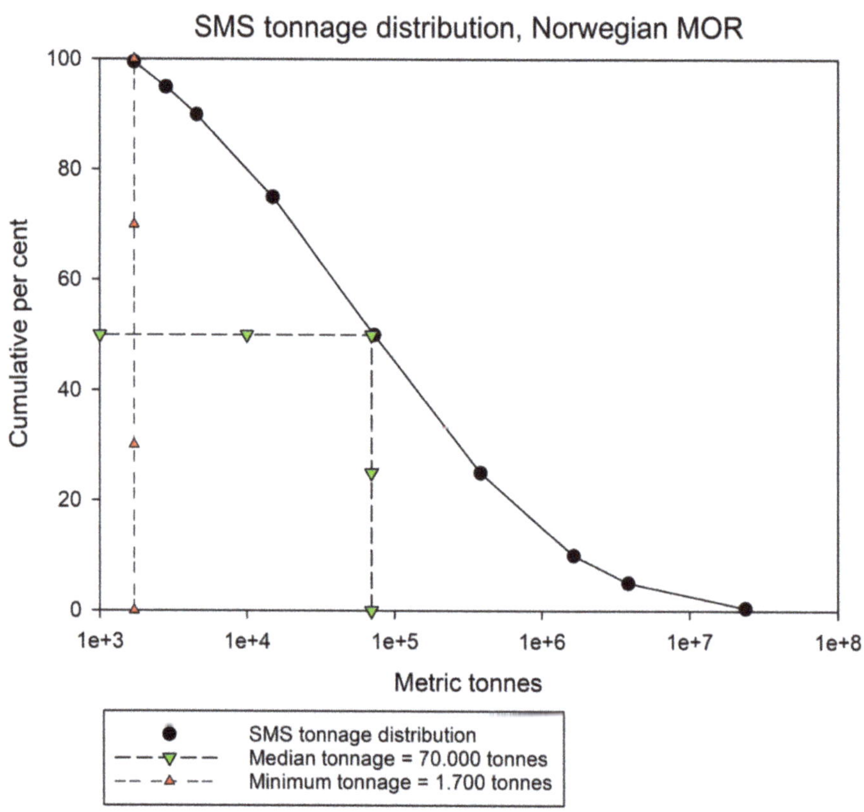

Figure 5-3 Cumulative size frequency distribution of sea floor massive sulphide (SMS) deposits used in this study. Click here for larger image.

Figure 5–3 and Table 4 show the size distribution used in this study.

Table 4 Ore-tonnage distribution

	Mode	Mean	Std. Dev.	F100	F95	F90	F75	F50	F25	F10	F5	F0
Ore tonnage [10³t]	2.0	855.4	3053.2	1.7	2.8	4.5	15	73	383	1643	3824	23856

It was decided to keep the same median size, of 70000 metric tonnes, as the distributions in Hannington et al. (2010). The minimum size is set to about 1700 metric tonnes, corresponding to a surface area of 40 m² of exposed sulphides. This is similar to the 90% percentile of the curve in Hannington et al. (2010). This choice is motivated by the requirement for having accumulations with a similar minimum contained metal in a SMS deposit on the Mohns and Knipovich Ridges as in the grade-tonnage models (Figure 5–4) in the recent assessment of the total MOR (Singer, 2014).

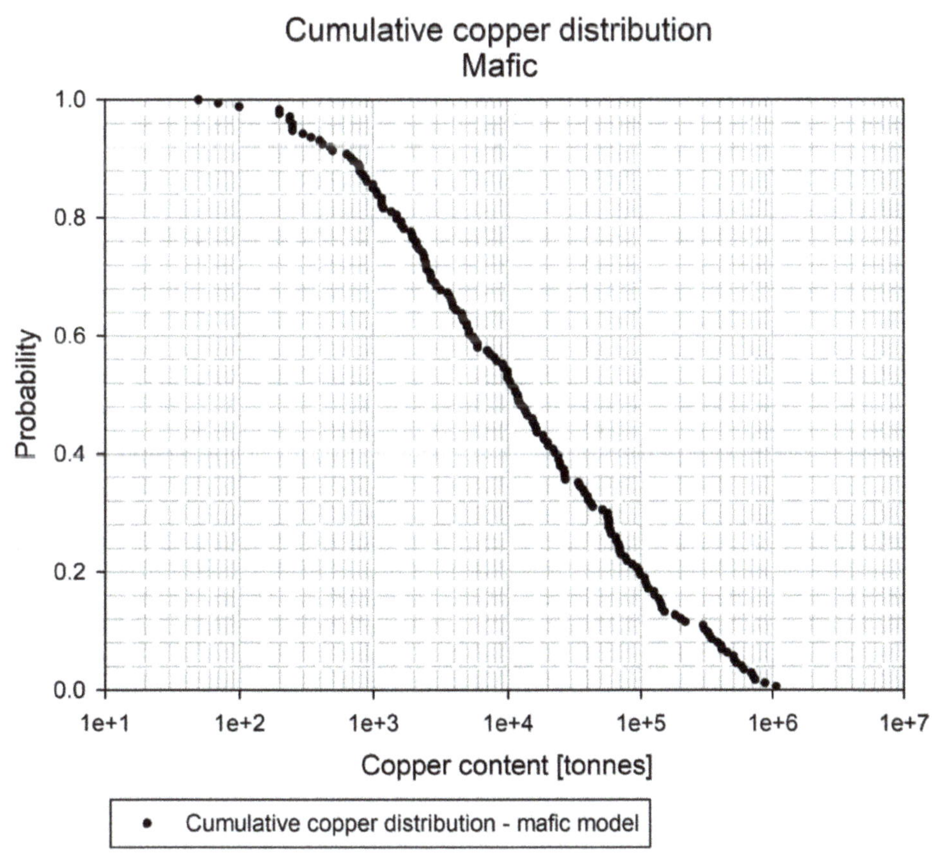

Figure 5–4 Metal distribution in mafic massive sulphide deposits. Data from Mosier et al. (2009). Click here for larger image.

5.4 Grades

The choice of grade distributions (Table 7) for SMS on the Mohns and Knipovich Ridges for Cu, Zn, Au and Ag deviates from the Global MOR distributions in Table 6 (Hannington et al., 2011a). The 18 deposits in our analogue area from the Troodos ophiolite have coefficients of variability (CV%) for Cu of 60%. The Mohns and Knipovich grades have coefficients of variability of less than 50%, except for gold with 121%. The grades in Table 6 all have coefficients of variability (CV%) larger than 80%, and a Cu/Zn ratio that is too much influenced by samples from intra-oceanic back-arc and arcs, as well as transitional arcs and continental margin arc environments, to represent valid distributions for our assessment.

Table 5 Average bulk compositions of 95 SMS deposits, 3,869 surface samples (from Hannington et al., 2013a).

Geological Setting	N	Cu [wt%]	Zn [wt%]	Pb [wt%]	Au [ppm]	Ag [ppm]
Mid-ocean ridges	2071	5.9	6.1	<0.1	1.6	89
Sedimented ridges	173	1.1	3.6	0.5	0.5	84
Intra-oceanic back-arc	668	3.9	16.4	0.9	6.6	210
Intra-oceanic arc	169	5.3	17.7	2.4	9.6	407
Transitional arc	728	6.4	14.8	2.0	12.2	692
Continental margin arc	60	3.1	20.3	10.0	2.3	953

Table 6 Percentiles for SMS grade distributions for SMS deposits, all tectonic settings (from Hannington et al., 2011a).

SMS (N=62)	75th Percentile	50th Percentile	25th Percentile	Mean	Std. dev	Std. dev/ Mean %
Cu metal [wt %]	1.70	4.30	7.00	5.58	4.61	83 %
Zn metal [wt %]	4.30	10.60	18.60	15	14.9	99 %
Au metal [ppm]	0.30	1.70	5.30	7.05	28.38	403 %
Ag metal [ppm]	37.00	107.00	234.00	209.8	353.7	169 %

Table 7 Percentiles for SMS grade Norwegian MAR

Metal grades	75th Percentile	50th Percentile	25th Percentile	Mean	Std. dev	Std. dev/ Mean %
Cu metal [wt %]	5.04	5.78	6.62	5.90	1.35	23 %
Zn metal [wt %]	4.88	5.84	6.94	6.08	1.82	30 %
Au metal [ppm]	0.61	1.06	1.82	1.58	1.91	121 %
Ag metal [ppm]	62.3	82.9	109	90.9	44.1	49 %

The grade distribution that was used for the assessment of SMS on the Mohns and Knipovich Ridges was modelled as shown in Table 8.

Table 8 Grade distributions for individual SMS accumulations.

Metal grades	Mode	Mean	Std. dev	F100	F95	F90	F75	F50	F25	F10	F5	F0
Cu metal [wt %]	5.45	5.90	1.35	3.00	3.93	4.33	5.04	5.78	6.62	7.56	8.25	10.90
Zn metal [wt %]	5.74	6.08	1.82	2.66	3.62	4.10	4.88	5.84	6.94	8.33	9.38	13.34
Au metal [ppm]	0.67	1.58	1.91	0.07	0.23	0.34	0.61	1.06	1.82	3.18	4.58	12.8
Ag metal [ppm]	76.7	90.9	44.1	19.2	37.1	46.5	62.3	82.9	109	143	171	284

The individual distributions are shown in Figures 5–5 and 5–6.

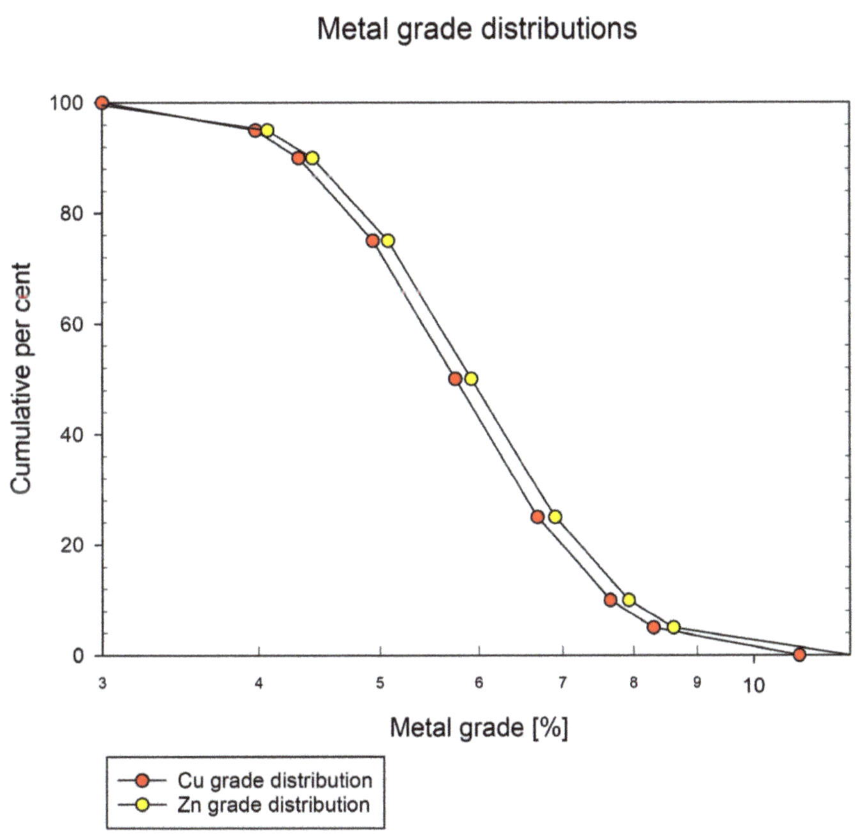

Figure 5-5 Copper and zinc grades. Mean grades given in Table 8. Click here for larger image.

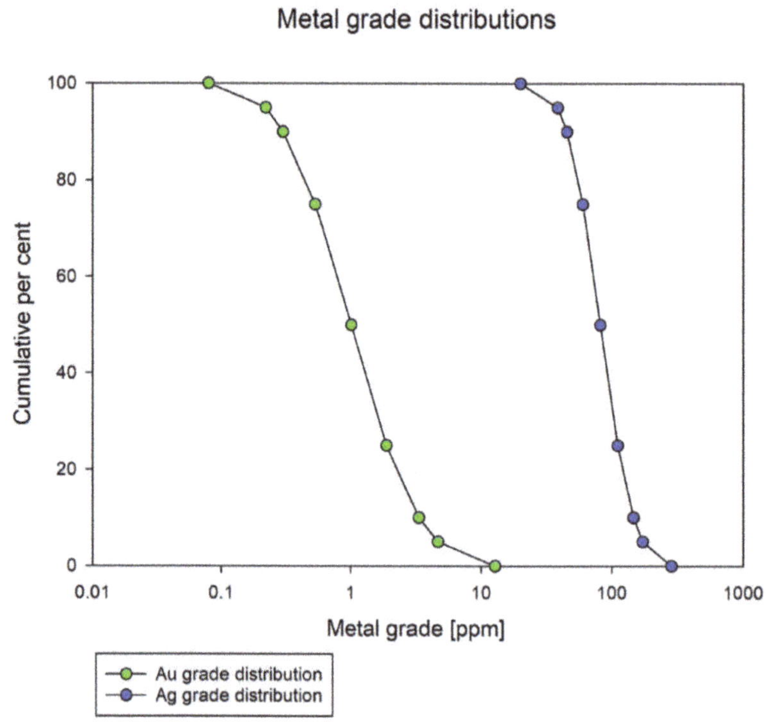

Figure 5-6 Ag and Au grades. Mean grades given in Table 8. Click here for larger image.

Given the grade and tonnage distributions, one can calculate the distribution of contained metric tons of metal in a single undiscovered SMS accumulation. See Table 9.

The metal distributions in the SMS deposit model from the Mohns and Knipovich Ridges (Table 9) are similar to the mafic models used by Singer (2014), especially for high-end percentiles, but deviates below the upper quartile (See Figure 5-4). If we compare the medians in Table 9 with Figure 5-4, which are based on Singer (2014), we note that the models used in this study represent a compound model dominated by the mafic model for the higher tonnages, and the felsic-bimodal model for the lower tonnages.

Table 9 Distribution of contained metric tonnes of metal in a single undiscovered SMS accumulation

Resource Type [Units]	Mode	Mean	Std. dev.	F95	F75	F50	F25	F10	F5	F1	F0
Cu metal [tonnes]	146	50684	187408	159	859	4209	22318	95677	221013	956006	1445386
Accumulation size											
Au metal [tonnes]	0,0	1,4	8,0	0,0	0,0	0,1	0,4	2,0	5,1	24,5	39,6
Accumulation size											
Zn metal [tonnes]	152	52122	191568	164	871	4303	22843	98204	229640	979139	1401190
Accumulation size											
Ag metal [tonnes]	0,2	76,7	300,1	0,2	1,2	5,9	32	143	335	1421	2099
Accumulation size											

In view of the great uncertainties regarding pristine grade and tonnages of VMS accumulations, this study chose deposit densities from the Cyprus VMS deposits. For grades, the project chose the mean values according to the average values from mid-ocean ridges, and the maximum values according to the observed tonnages and grades in the Russian MAR exploration areas (Cherkashov et al., 2010, 2013). It must, however, be noted that there are only a few drilled deposits. Sample material comes primarily from surface samples. The appropriateness of the grade distributions used here has been assessed by comparing the total amount of metal potential with the total metal potential calculated by Hannington (2013b), Cathles (2013) and Singer (2014). See Chapter 7.

CHAPTER 6

Results

6.1 Assumptions

The probabilistic methodology was applied to each of the 17 sub-play assessments to produce metal estimates in the form of probability distributions. The sub-plays are assumed to belong to the same geological play, i.e. the MAR, even if local ridge segments are individually different. To take into account the limited dependency that can be inferred between sub-plays, aggregation (summation of the potential in the 17 sub-plays) was conducted with dependency factors.

The Knipovich Ridge shows similar architecture to the north-eastern Mohns Ridge but spreading along the Knipovich is up to 60° oblique, and the axial valley ridges (AVR) are more widely spaced, typically 100 km, compared with 30–50 km at the central Mohns Ridge (Rona et al., 2010). Due to the narrower spacing of the AVRs, more diversity can be expected along the Mohns Ridge than along the Knipovich Ridge. A dependency factor of 0.45 is therefore used for Mohns play aggregation, compared to a higher dependency factor of 0.6 for the aggregation of the Knipovich plays. These dependency factors regulate the dependencies between the sub-plays within each of the two ridges.

If one or more of the Mohns Ridge sub-plays will produce economically-recoverable amounts of Cu, Zn, Au and Ag, then the likelihood is high for having favourable mineralisation conditions also in one or more of the Knipovich sub-plays. A dependency factor of 0.9 is therefore used when the resources of the Mohns and Knipovich Ridges are combined to represent the total expected metal resources from the Norwegian part of the MAR. This dependency factor regulates the dependency between the two ridges.

In the assessment of in-situ gross monetary value, a gram to ounce (oz) ratio of 28.35 has been used to accommodate the gold price quotes (USD / oz).

6.2 Metal resource potential

Each of the 17 sub-play probability distributions for Cu, Zn, Au and Ag were aggregated using a Monte Carlo simulation (GeoX, 2014), to produce resource estimates for the entire Mohns and Knipovich Ridges (Table 10). For each commodity, the following distributions are calculated:

- Aggregated number of accumulations
- Accumulation size distribution
- Conditional aggregate potential, given that the assessment area has at least one sub-play with a confirmed SMS
- The unconditional aggregate potential.

The conditional aggregated potential estimate presupposes that at least one of the 17 sub-plays are confirmed, whereas the unconditional aggregated potential does not make this assumption. The unconditional potential is therefore lower than the conditional potential because the latter is multiplied by the probability that the plays are confirmed. The metal distributions are all skewed to the left, generally with a standard deviation equal to or substantially larger than the mean (for aggregate potential).

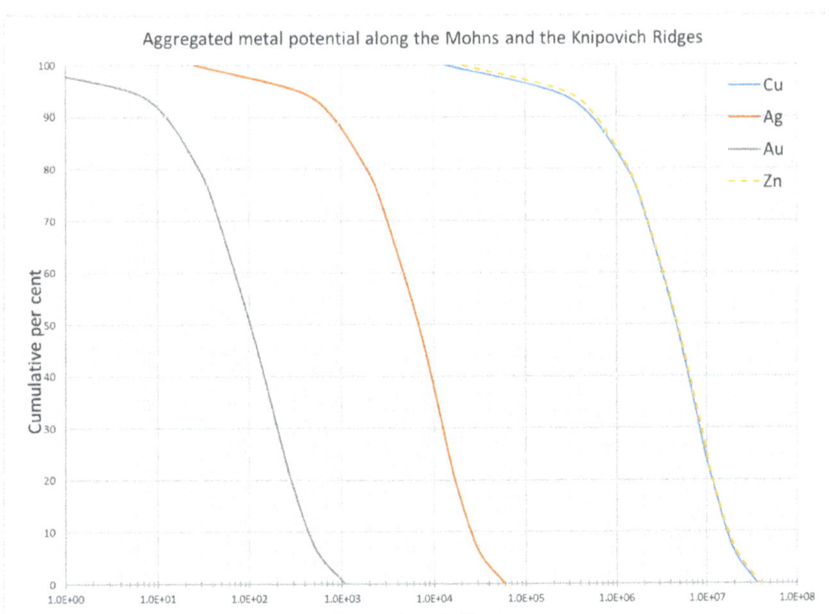

Figure 6-1 Aggregated metal potential along the Mohns and Knipovich Ridges. Click here for larger image.

Table 10 shows the conditional and unconditional aggregated potential. The mean of the skewed distributions is approximately 7 million metric tons of copper and zinc, 175 metric tons of gold and about 10,000 metric tons of silver. Figure 6–1 shows the metal distributions, illustrating that there is a 5% probability that the metal potential along the 1030 kilometres of studied ridge is about 21 million metric tons of copper, 22 million metric tons of zinc, just under 600 metric tons of gold and 34000 metric tons of silver or more. The uncertainties and the skewness are large, illustrated by the large standard deviations and the large difference between the minimum and maximum values (min. / max. values illustrated only in Figure 6–1). For silver, copper and zinc, the ratio between the expected value, the mean, and the standard deviation is slightly above 1 or 100%. For gold, this ratio is, as expected, significantly higher than one. The copper and zinc distributions in Figure 6–1 are overlapping. The zinc distribution is illustrated with a dotted yellow line. Table 10 also show the percentiles P95, P75, P50 (the median), P25 and P5. For copper, these indicate that there is a 95% chance that the aggregated copper tonnage along both the Mohns and Knipovich Ridges is larger than ~200 thousand metric tons. For zinc, there is a 5% chance that the tonnage is ~23 million metric tons or more. For gold there is a 95% chance that he aggregated tonnage is larger than 5 tonnes and a 5% chance that it is larger than ~600 tonnes. For silver the corresponding tonnages are 340 and 34000 tonnes for 95% and 5% respectively. Given the skewness of the distributions, which is also apparent in Figure 6–1 (remembering that these distributions are plotted on a logarithmic scale) the 'mode' is the most likely value and significantly smaller than the mean.

We observe in Table 10 that the conditional and unconditional aggregate potentials are nearly equal, indicating that there is a negligible risk that none of the 17 sub-plays will confirm at least one SMS accumulation. The assessment results will therefore only show the distributions for the conditional aggregate commodity potential that was used as a basis for calculating the total gross metal value for each commodity.

Table 10 Aggregated accumulation number, size and in-situ metal resources on the Mohns and Knipovich Ridges

Resource Type [Units]	Mode	Mean	Std. dev.	F95	F75	F50	F25	F5
Cu metal [metric tons]								
Number of Cu accumulations	92	161	124	30	74	126	210	416
Accumulation size Cu	0	45834	137948	0	1680	4585	21791	224533
Conditional aggregate potential Cu	14558	6884938	7033961	195906	1752433	4613919	9614568	21678550
Uncond, aggregate potential Cu	14558	6882185	7033901	194216	1749347	4611112	9611228	21675930
Au metal [metric tons]								
Number of Au accumulations	72	157	117	28	74	126	211	385
Accumulation size Au	0	1	3	0	0	0	0	5
Conditional aggregate potential Au	0	175	201	5	38	103	238	597
Uncond, aggregate potential Au	0	175	201	5	38	103	238	596
Zn metal [metric tons]								
Number of Zn accumulations	68	161	121	29	74	129	213	402
Accumulation size Zn	0	51913	160575	6	1667	4446	21340	257048
Conditional aggregate potential Zn	36855	7095203	7338270	258783	1776559	4703893	9945984	22746144
Uncond, aggregate potential Zn	36855	7092365	7338174	256571	1773835	4700878	9943090	22743314
Ag metal [metric tons]								
Number of Ag accumulations	73	161	122	29	74	127	210	399
Accumulation size Ag	0	68	207	0	2	6	33	338
Conditional aggregate potential Ag	25	10524	11237	339	2528	6821	14527	34074
Uncond, aggregate potential Ag	25	10514	11236	332	2519	6813	14517	34064

6.3 Metal prices

Metals are traded on stock exchanges worldwide, and their prices are controlled by varying supply and demand mechanisms and are highly volatile on local and regional scales. It is, however, outside the scope of this book to do a robust and in-depth forecast and sensitivity analysis of future metal prices. The basis for the metal price assumptions are prices reported by Thomson Reuters / SNL Metals and Mining. See Figure 6–2.

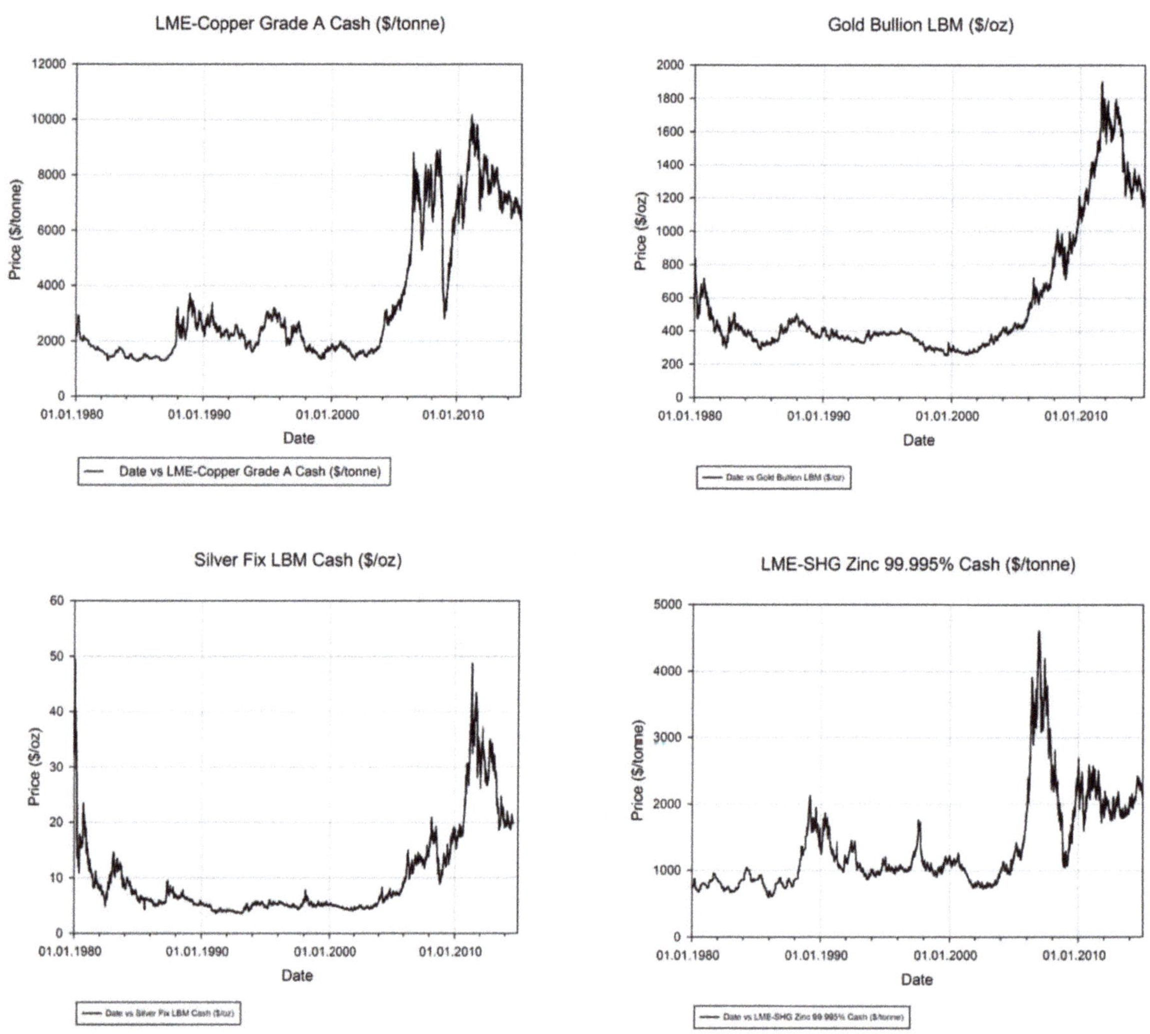

Figure 6–2 Metal prices from January 1980 until recently. Used as a basis for deciding the commodity prices used in the valuation. Source: Thomson Reuters, SNL Metals & Mining, an offering of S&P Global Market Intelligence. Click here, here, here and here for larger image.

The following rounded values close to the prices at the time of the assessment (15/12-2016) have been assumed in the calculation of the gross value:

- Au=1200$/oz
- Ag=20$/oz
- Cu=7000$/metric ton
- Zn=2000$/metric ton

6.4 Gross value of yet-to-find metal resources along the Knipovich and Mohns Ridges

The aggregated gross values (in-situ metal value) for copper, zinc, gold and silver are given in Table 11.

The metal gross values are given by multiplying the 'Conditional aggregate potential' given in Table 10 with the commodity prices in Table 11.

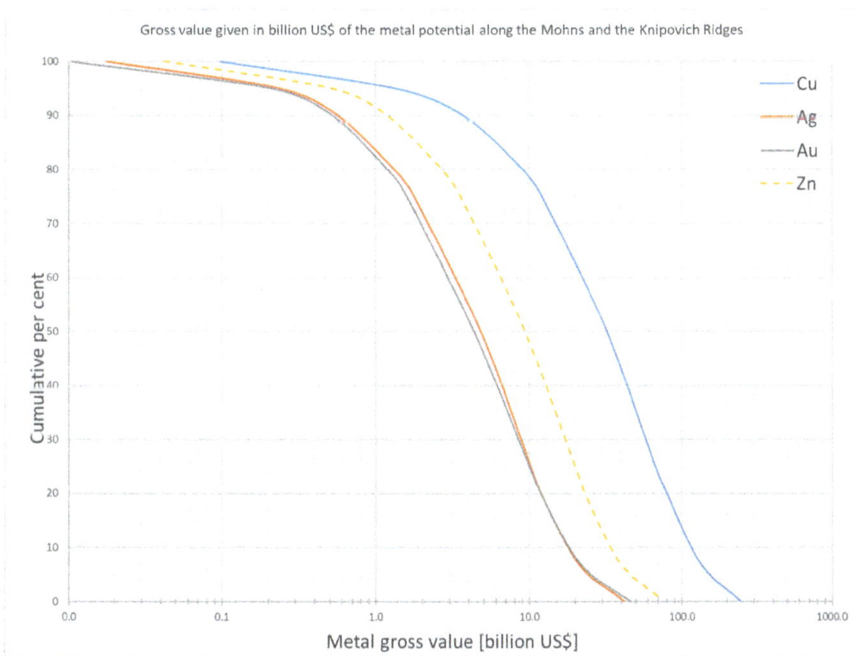

Figure 6-3 Aggregated gross value of copper, zinc, gold and silver along the Mohns and Knipovich Ridges. Click here for larger image.

Figure 6–3 and Table 11 show the gross value distributions. The figure illustrates that given the commodity prices above, the *in-situ* value potential along the 1030 kilometres of ridge is between US$ 0.1 billion and US$ 246 billion for copper (blue graph in Figure 6–3), between US$ 0.01 billion and US$ 46 billion for gold (grey graph in Figure 6–3), between US$ 0.04 billion and US$ 75 billion for zinc (dashed yellow graph in Figure 6–3) and between US$ 0.02 billion and US$ 41 billion for silver (orange graph in Figure 6–3). This gives a total upside potential of about 400 billion US$. Clearly, these numbers are associated with a high degree of uncertainty, given both the uncertainty in commodity prices, the deposit density, the deposit tonnage and the deposit grades. It is also apparent in Figure 6–3 that copper is the major financial driver.

Table 11 show the percentiles P95, P75, P50 (the median), P25 and P5 for the Aggregated gross values of the commodities. For copper, these quantify a 95% chance (P95) that the gross copper value (*in-situ* value) along both the Mohns and Knipovich Ridges is larger than 1.4 billion US$. For zinc, there is a 5% chance (P5) that the gross value is 45.5 billion US$ or larger. For gold and silver there is a 5% chance that the *in-situ* value is larger than 25 billion US$ and 24 billion US$ respectively. The distributions are highly skewed, with consistently slightly higher standard deviation than mean and a much lower mode than mean, emphasising the uncertainty.

The gross values are compared to the Norwegian Oil Fund. This fund had a market value of 8,300 billion Norwegian kroner (NOK) by November 5, 2018. The mean gross value of the total metal resources along the Mohns and the Knipovich Ridges is thereby about 8% of the Norwegian Oil Fund.

Table 11 Aggregated gross value Mohns and Knipovich Ridges

Resource Type and gross value [Units]	US$ / metric ton	Mode	Mean	Std. dev.	F95	F75	F50	F25	F5
Cu metal gross value [Billion US$]									
Conditional aggregate potential	7000	0.10	48.2	49.2	1.4	12.3	32.3	67.3	151.7
Au metal gross value [Billion US$]									
Conditional aggregate potential	42328754	0.009	7.4	8.5	0.2	1.6	4.4	10.1	25.2
Zn metal gross value [Billion US$]									
Conditional aggregate potential	2000	0.07	14.2	14.7	0.5	3.6	9.4	19.9	45.5
Ag metal gross value [Billion US$]									
Conditional aggregate potential	705479	0.018	7.4	7.9	0.2	1.8	4.8	10.2	24.0
Total gross metal value [10^9 US$]		0.2	77.2	80.3	2.3	19.2	50.8	107.5	246.5
Norwegian Oil Fund 5/11-18 [NOK and US$]. Exchange rate 8.37	8.3E+12	9.9E+11							
% of Norwegian Oil Fund 4/1-17		0.02 %	7.80 %	8.12 %	0.02 %	1.94 %	5.14 %	10.86 %	24.91 %

CHAPTER 7

Validation of Aggregated Play Resources

Despite excellent reviews (e.g. Pedersen et al., 2010b; Hannington, 2005–2013), it remains difficult to distinguish the fundamentals from the interesting details regarding the ultimate potential for sea-floor massive sulphide (SMS) deposits. Several influential publications have addressed the issue of estimating the resource base of sea-floor massive sulphide (SMS) deposits (Cathles, 2011a, 2011b; Hannington et al., 2011b; Hannington 2011, 2013b; Singer, 2014; Juliani & Ellefmo, 2018a). From a cursory evaluation of these and related articles, it may appear that some authors are optimists, whereas others are pessimists. Thus Cathles (op. cit.) estimates that with an accumulation efficiency of 3% (3% of the metals in the hydrothermal fluids is precipitated as sulphides), the amount of copper in sea-floor mineral deposits, 106 billion metric tonnes that would last for more than 6000 years at the current rate of extraction, whereas Hannington and co-authors found the amount of copper and zinc in the more easily-accessible (i.e. surface deposits) of the neovolcanic zones of the global oceans to be just 30 million metric tonnes. The large discrepancy in the two estimates is not only a matter of optimism versus pessimism, but rather reflects the fact that two different reservoirs have been estimated. Cathles evaluates the total metal retained in mineral deposits in the oceanic crust (exposed plus buried), whereas Hannington focuses his attention on surface deposits in the more easily-accessible neovolcanic zones of the global oceans. The estimate by Cathles differs in two main aspects:

- the whole area of the global oceans is included; and
- the implicated mineralisation is not only on, but also under the sea floor.

From an ore deposit perspective, questions are mainly reduced to:

- What fraction of the metal carried by the venting fluid is actually accumulated on the sea floor (what is the accumulation efficiency)?
- To what extent are the accumulated metal sulphides concentrated spatially (how are the sulphides swept into piles)?

Cathles' starting point is the fact that the heat introduced into the crust at the ridge equals that lost by 350°C venting. This means that the thickness of the crust alone determines the mass of 350°C seawater that can circulate, on average, through each square metre of sea floor. He then combines this with the observed chemistry (from MORs in the Pacific) of the high-temperature axial vent fluids, and calculates that 0.67 kg/m^2 of copper metal could be on each square metre of sea floor.

The crustal thickness along Mohns Ridge is well below the global average for typical oceanic crust, and shows a high variability with a mean thickness of around 4 km (Klingelhöfer et al., 2000). With 2D seismic imaging of the sub-seafloor and a given density model, Krysiński et al. (2013) documented that the oceanic crust formed at the Knipovich Ridge is between 6 and 8 km thick.

The total deposition of Cu in the neovolcanic zone (including known sub-sea-floor stockwork mineralisation and near field sediments) is, according to Hannington (2013b), ~1.74 kg/m^2, which comprises: 0.44 kg in near field sediments, 0.34 kg in SMS deposits and 0.96 kg in stockworks. Using the tonnage factor of 3 indicated above, this corresponds to 5.19 kg/m^2, which comprises: 1.31 kg in near field sediments, 1.01 kg in SMS deposits and 2.87 kg in stockworks, for the total Cu deposited in the neovolcanic zone along an ultra-slow ridge. This corresponds to about 9% of the total high-temperature axial flow, assuming a minimum flux to the ridges of ~58.2kg Cu/m^2. Following Hannington's (2013b) statements about the balance and the adjustments to MAR conditions, the difference between the minimum flux and the total deposition of 5.19 kg/m^2 in the neovolcanic zone of 53 kg/m^2 Cu must be deposited in the crust as sub-sea-floor mineralisation associated with diffuse axial flow.

The discrepancy in the two estimation approaches of Cu metal in SMS deposits/m^2 applied to the Mohns–Knipovich Ridges may, according to Barriga et al. (2013), not be a matter of optimism versus pessimism, but rather reflects the fact that Cathles' approach, leading to 1.88 kg Cu/m^2, evaluates the total metal retained in mineral deposits in the oceanic crust (exposed plus concealed). On the other hand, Hannington's approach leads to 3.88 kg Cu/m^2 for SMS and sub-sea-floor alteration and mineralisation (1.01 kg Cu/m^2 + 2.87 kg Cu/m^2). Computed exclusively for the 'significant massive sulphide accumulation' *on* the sea floor, in the easily-accessible neovolcanic zone, this value of 1.01 kg Cu/m^2 gives 10.4 million metric tons of copper, which corresponds to the 22% percentiles for Cu in the aggregated resources in Table 10, and a smaller upside on the assessed resources than that calculated using Cathles' approach.

Singer (2014) makes a probabilistic estimate of the Cu, Zn, Pb, Au and Ag tonnages inside a 4 km narrow neovolcanic zone around the ridge, with a total area of 256,000 km^2. He uses the three-part assessment framework and onshore analogues to model the contained metal and deposit density (number of deposits per km^2). Estimations are thereby computed with much smaller deposit densities and only pertain to the 4 km wide zone, with resulting Cu resources that represent a clear downside which corresponds closely to the 95% percentiles for Cu in the aggregated resources in Table 10.

Figure 7–1 shows a comparison between the potential resource of copper in percentiles, obtained by this study, and the mean values obtained by Cathles (2013), Hannington (2013b) and Singer (2014). One may notice that the percentiles defined in this study span the sample space defined by the three authors, yielding a degree of further confidence in our estimates.

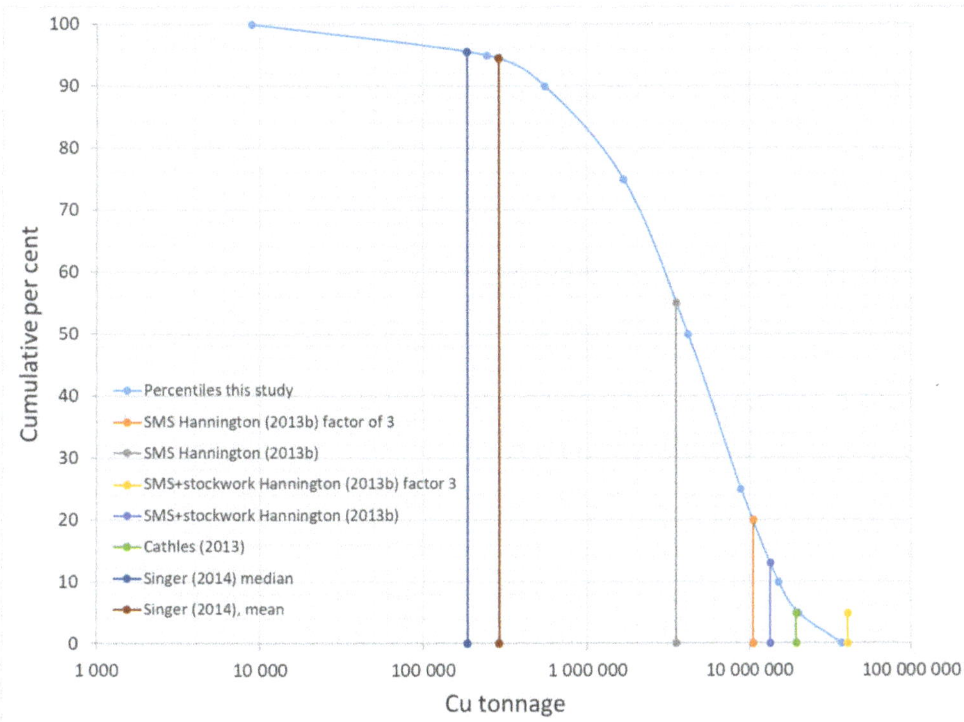

Figure 7-1 Validation of Cu tonnage obtained in the mineral resource copper potential estimates inside the Norwegian Extended continental shelf. Click here for larger image.

CHAPTER 8

Testing the Hypothesis

8.1 The MarMine Project

To reduce the uncertainty of the presented estimates of the mineral resource potential along the extended Norwegian continental shelf, more exploration and sampling is needed.

NTNU therefore applied for a grant from the Norwegian Research Council, together with 13 Norwegian companies, in 2015. The main goals of the MarMine Project were to test exploration technologies and to collect mineralised material to seek to improve the accuracy of the estimate of undiscovered mineral resources on the extended Norwegian continental shelf, focusing on seafloor massive sulphide (SMS) deposits along the Arctic Mid-Ocean Ridge (AMOR), and to evaluate exploitation technologies for potential future mining activities. An integrated part of this is to test to see whether an exploration strategy based on a morphostructural analysis of the ocean floor bathymetric data can be used to find undiscovered accumulations of sulphide mineralisation. In the summer of 2016, the MarMine Project carried out a three-week research cruise to the AMOR (Ludvigsen et al., 2016) with the multipurpose subsea vessel, Polar King (see Figure 8–1).

The main purpose of the research cruise was to collect samples for further analysis and in the course of its duration, several hundreds of kilos of collapsed chimney fragments were collected from the Loki's Castle mounds. Most of these were collected from the surface, but the project also used a Remotely Operated Vehicle (ROV) mounted drilling system to obtain sub-surface samples. One of the interesting research questions is whether the mineralisation differs at different depths. However, core drilling on the Loki's Castle mounds was difficult due to the mixture of unconsolidated sediments and chimney fragments (Ludvigsen et al., 2017; Søreide et al., 2017). One of the Loki's Castle chimneys can be seen in Figure 8–2.

Figure 8-1 M/V Polar King multipurpose subsea vessel with the Hugin AUV surfacing with new data from the seafloor. Photo: NTNU MarMine; no reuse without rightsholder permission. Click here for larger image.

Figure 8-2 One of the chimneys on the Loki's Castle site. Hydrothermal fluid can be seen flowing out of the chimney. Photo: NTNU MarMine; no reuse without rightsholder permission. Click here for larger image.

To investigate the appropriateness of Autonomous Underwater Vehicles (AUV) as a platform to find hydrothermal manifestations, multivariate datasets were collected from the Loki's Castle site, the Mohns' Treasure area, and two exploration areas south-south-east of Mohns' Treasure. The benthic fauna at the presumed inactive site Mohns' Treasure was naturally distinctively different (low biomass, high biodiversity) from the active site at Loki's Castle (high biomass and low biodiversity). Sponges and crinoids on rocky substratum and stalked crinoids on sediments dominated. Echinoderms (small asteroids and holothurians) were also observed on sediment. An Axial Volcanic Ridge (AVR) was defined as exploration area 1 and 2. This exploration area definition was primarily done based on the morphostructural work presented herein identifying faults, graben structures and volcanic features. Based on high-resolution bathymetric maps, geological features were selected for ground-truthing using the ROV. A dramatic bathymetry, similar to alpine terrain, was identified but no new hydrothermal manifestations were discovered during these surveys.

8.2 Post-cruise results

In order to better understand the composition of SMS, it is necessary to directly investigate the material of which they are comprised. Physical samples are needed to make microscopically-thin sections to understand the mineralogy, and to make powders suitable for chemical analysis using X-ray and electron beam techniques. On terrestrial sites of geological interest, this is a relatively simple process, but at a depth of 2500 metres under water in the middle of the Arctic Ocean, physical sampling is more complicated. During the MarMine research cruise, AUVs and ROVs were sent to the ocean floor to investigate the sites, with the aim of collecting samples and data to return to the laboratory.

The metal-laden fluids escape back into the ocean from the crust via holes and cracks referred to as hydrothermal vents, much like when we see geysers or gas escaping from vents in hot water spring areas. As the hot, buoyant metal-enriched water is expelled from the hydrothermal vents, it forms tall sediment-laden plumes. Vent plumes dominated by Zn and Pb are known as 'white smokers', and those dominated by Cu and Fe as 'black smokers'. The minerals which precipitate out of the solution from these vents build up over time (like inside a kettle), literally forming chimneys for the vents. These chimneys are composed of the elements carried by the hydrothermal fluids, and therefore the minerals which form in these chimneys are likely to be representative of those which form in potentially-significant quantities at lower depths. This is clearly visible at Loki's Castle, where several chimneys are present in the two main mounds, and some of these chimneys are active black smokers.

The hydrothermal vent chimneys are prone to collapse periodically, due to internal fluid-pressure building up and subsequently breaking off fragments, earthquakes that destroy the structures completely, or simply because the smokers grow too large to support their own weight. These fragments of the vent material fall down the sides of the mounds, forming scree slopes and debris fields.

It is this material that was successfully collected by the ROV and eventually investigated by the project scientists, along with samples of the muddy sediments from the sea floor, and drill cores of basalt which represent the oceanic crust.

In total, more than 200 kg of loose boulders from the flanks of the mound and collapsed fragments from the chimneys were recovered. Due to the often elevated levels of Fe in the SMS material, the rocks have a strong rusty coating masking the internal features. Upon first cutting and coring the boulders, initial impressions were all related to the variability of the material. The rock is often highly porous, with the cavities lined with a variety of crystals. The texture is chaotic, with billowing laminations of rusty and white carbonate, sulphate and silicate minerals, grey clay-dominated areas, and fine-grained black zones. It is the black zones which typically contain the elements and sulphide minerals of interest: copper in chalcopyrite and isocubanite, zinc in sphalerite, and, occasionally, silver and gold. In some material, these phases (including pyrite) amalgamate into interlocking lenses, hence the term 'massive sulphide'. These can be seen in Figure 8-3, where coarser sulfide layers are hosted within a porous, black fine-grained groundmass (scale bar is 4 cm).

Figure 8-3 Metallic layers hosted within a porous, black fine-grained groundmass (scale bar is 4 cm). Click here for larger image.

Initial petrographical and geochemical investigations, based on the surface samples from Loki's Castle, indicate the potential for grades and tonnages comparable to those of economically-significant volcanogenic massive sulphide (VMS) deposits; ancient terrestrial versions of SMS deposits which have long been utilised as a major source of Cu, Zn and Pb from onshore mines.

The majority of the samples which display visible sulphide mineralisation typically contain 0.5wt% Cu and 1wt% Zn, but samples with as much as 2wt% Cu, and 7.4wt% Zn have been recorded. These grades are considerably lower than the average grades used in the resource potential estimation presented herein, but the amount of data these samples represent is not assessed to be great enough to change the defined grade distributions. In order to really lower the uncertainty, even more data must be collected and, in particular, along the 3rd dimension (i.e. sub-seafloor). Both Ag, Au and, surprisingly, some Pb are present in the collected material. Understanding the form of the Cu and Zn-bearing minerals is important in order to provide information necessary for subsequent mineral-processing considerations, ensuring efficient extraction. More information on mineralisation, content and potential exploitation technologies will become available as the MarMine Project continues.

8.3 Updating and improving the play definitions

During the MarMine research cruise, data from Loki's Castle, the greater Mohns' Treasure area, including the scarp edges, and two exploration areas on an axial volcanic ridge were collected. Collected data included high resolution sea floor bathymetry, sediment acoustic reflectivity (acoustic backscatter), salinity, temperature, pressure, turbidity, magnetism and methane content, using the Multi-Beam EchoSounder (MBES), CTD, turbidity sensor, Self-Compensating Magnetometer (SCM) and CH4-sensor. This data will undergo detailed studies to improve the understanding, descriptions and definitions of the plays included in the play analysis. In the future, this will reduce uncertainty because it will be feasible to develop more specific play analyses, e.g. an AVR, a sediment hosted, and an oceanic core complex play analysis.

8.4 Governmental exploration cruises to Mohns Ridge

Early in 2018, the Norwegian Petroleum Directorate (NPD) asked for qualifying documents to participate in a request for tenders in relation to the '*Acquisition of Seabed Massive Sulfide Data in the Norwegian Sea*' (NPD, 2018a). After the tendering process was concluded, NPD engaged Swire Seabed AS and Ocean Floor Geophysics Inc. to map for potential sulphide mineralisations inside a predefined area, along central parts of Mohns Ridge (NPD, 2018b). The defined area overlaps significantly with permissive tract M3 and M4 defined in this study. See Chapter 4.2. NPD required spontaneous potential data (SP), multi-beam bathymetry (MBES), backscatter data, and grav/mag with high frequency seismic (HFS) and sub-bottom profiler (SBP) data as optional. Electromagnetic (EM) and magnetotelluric (MT) data were also defined as relevant. In addition, relevant geological samples from areas of special interest, sampled with gravity cores or grabs, were required (NPD, 2018a, and NPD, 2018b).

The area defined by NPD is interesting from a deposit perspective. It would be expected that the collection of high resolution bathymetric data would give local and pertinent details of the sea floor morphologies rich in textural information. The northern and southern AVRs may reveal fissured hummocky terrains with intermittent growth of lava domes and cones (several to hundreds of metres in diameter), or smooth lava flows where eruption rates are higher. In addition, larger quasi-circular volcanic structures (1 km in diameter, 100–200 m height) may disclose local fissuring and terrains that go from relatively smooth upper surface (flat-topped type) to steep-sided topographic depressions (crater type). Vertically-elongated shapes rising up from the sea floor may also dissect the hummocky terrains, i.e. eruptive fissures which consist of melt flows that are often in conjunction with other features, such as faults. The sediment thickness, which is expected to be low in these young and volcanically-active areas, increases with distance away from the active spreading centre as the crust gets older and less volcanically active. Combined with acoustic backscatter images, important details can be tied to sedimentary processes along all the multibeam bathymetry lines, which provide information on local mass movements or sediment-covered flat sea floor that is often found in arch-shaped depressions. If used, the sub-bottom geo-acoustic profiler may help decipher the underlying rugged topography, and differentiate areas of little sediment accumulation to those where sediment transport is higher, notably in structurally-controlled off-axis provinces. Off-axis normal faulting may expose valuable outcrops that preserve distinguishable mineralised horizons. Gravitational collapses and landslides may, however, participate greatly in covering smaller features, especially within tectonic terrains of large fault scarps.

The focus of the exploration cruise was not to find active systems, but rather inactive systems assumed to still contain mineralised material. The project commenced on 1st August 2018, involving the ROV support vessel Seabed Worker from Swire Seabed AS and a Kongsberg Hugin AUV with various sensors from Ocean Floor Geophysics Inc. After 25 days, the project was successfully completed finding new deposits, both associated with active and inactive systems (NPD, 2018c). It is stated that '*The NPD has identified a large area of sulphide minerals, that was previously unknown. The deposits could include important industrial metals such as copper, zinc, cobalt, nickel, vanadium, wolfram and silver.'* Furthermore, '*The new area of sulphide minerals that was discovered contains many such piles of gravel and collapsed black smokers, including a 26 metre high, non-active tower, in addition to some active systems.'* This confirms the interesting potential along the Mohns Ridge. Figure 8.4 shows the approximate location of the newly discovered sulphide occurrence in the central parts of the Mohns ridge.

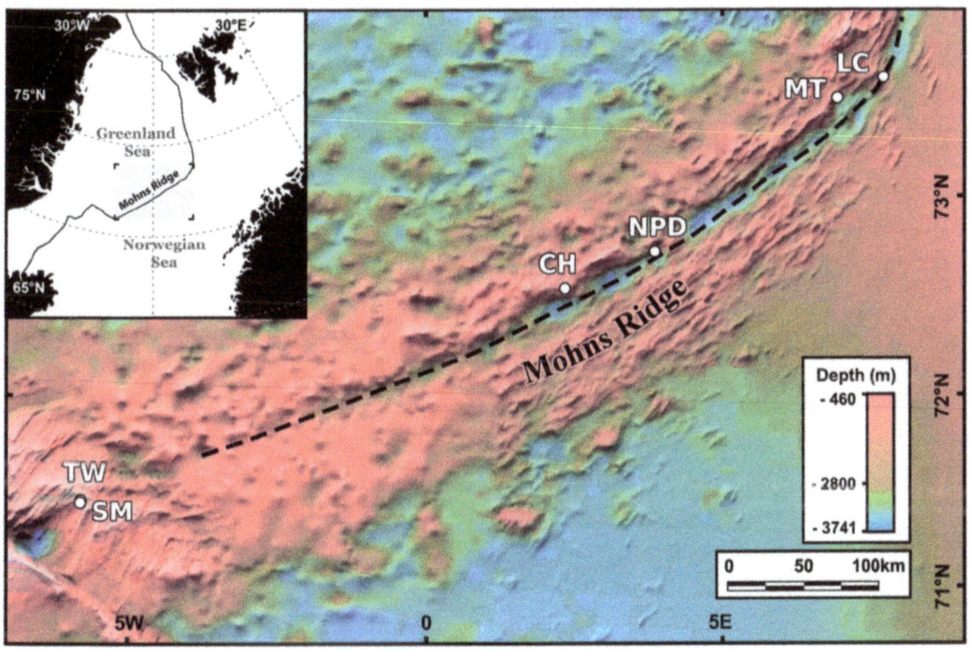

Figure 8-4 Location of the newly discovered sulphide occurrence ('NPD' in map) in the central parts of the Mohns Ridge. The newly discovered occurrence is plotted with the occurrences Loki's Castle (LC), the Mohns Treasure (MT), Copper Hill (CH), the Troll Wall (TW) and the Soria Moria (SM). Ægir vent field is not included. Source map: NTNU / Cyril Juliani. Source occurrence location: Norwegian Petroleum Directorate. Click here for larger image.

The location confirms play M3 (see section 5.2) which enables an update of the resource potential estimates already build, using the play analysis framework. In a new analysis, the sub-play M3 is then confirmed and will therefore be associated with a higher play probability, i.e. a lower regional scale risk. No grade data have been made available and collected sample material will in the coming years undergo detailed mineralogical, textural and chemical analysis. A reliable tonnage estimate requires drilling in combination with geophysics to understand the occurrence geometry and grade distribution. The necessary data for this is not yet available.

During another exploration cruise to Mohns Ridge in August-September 2019, NPD discovered additional 'new sulphide deposits' (NPD, 2019). Exact location and characteristics have not yet been published but the exploration area overlaps permissive tracts identified and described herein. Simultaneous deployment of more than one AUV enabled timely, cost-efficient data collection, and large amounts of high resolution data were collected. This is promising, an when more information from the cruise is made available, the present resource potential estimate can be updated.

CHAPTER 9

Discussion and Concluding Remarks

9.1 Classification of the marine minerals discovered on the AMOR

Deposits can be defined both according to geological knowledge and according to the so-called modifying factors that include the mining method, processing route and environmental, social, economic, governmental, market and legal aspects. US Geological Survey (USGS, 1980) classified mineral resources into identified and undiscovered resources (Figure 9.1). Mineral resources that are identified are further separated into inferred, indicated and measured resources, dependent on the knowledge base. Undiscovered resources are postulated and can be, like identified resources, sub-divided dependent on the level of geological knowledge into hypothetical and speculative resources. Hypothetical resources are postulated resources in confirmed regions, whereas speculative resources are inside unconfirmed regions.

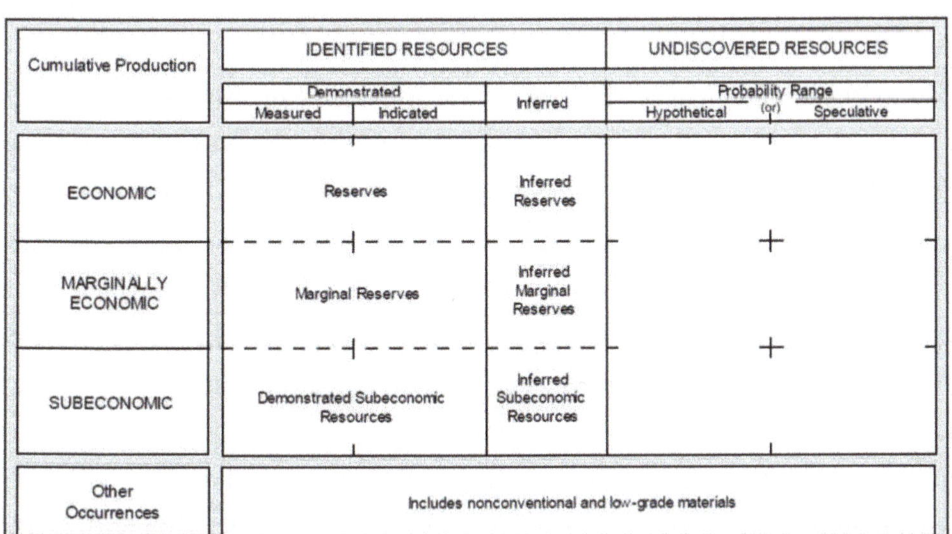

Figure 9-1 USGS classification scheme for mineral resources and reserves. Figure from USGS (2009). Click here for larger image.

Worldwide, there are a number of classification systems for mineral resources and reserves in use. The National Instrument 43-101 (NI, 2018) and the Australasian Joint Ore Reserves Committee Code (JORC Code; JORC, 2012) are two commonly-used standards. The JORC Code is aligned with the classification scheme of the USGS (1980), and classifies resources and reserves into the following sub-groups:

- An inferred mineral *resource* is a deposit which has been sampled sufficiently to make an estimate of its grade and tonnage at low confidence.
- An indicated mineral *resource* is a deposit where the continuity is not yet confirmed but in which tonnage and other deposit characteristics can be estimated with a reasonable level of confidence.
- A measured mineral *resource* is a sampled deposit with a high level of geological confidence and confirmed geological continuity.
- A probable ore *reserve* is the economically-mineable part of an indicated or a measured mineral resource, dependent on the uncertainty in the modifying factors.
- A proven ore *reserve* is the economically-mineable part of a measured mineral resource. The deposit has been proved mineable also in terms of the modifying factors.

JORC (2012) continues with a definition of an exploration target. This is defined as an '*estimate of the exploration potential of a mineral deposit in a defined geological setting where the statement or estimate, quoted as a range of tonnes and a range of grade (or quality), relates to mineralization for which there has been insufficient exploration to estimate a Mineral Resource*'.

The mineralisation that has been located on the AMOR could be termed as exploration targets (JORC, 2012), whereas the mineral resource assessment results, presented herein, are quantifying hypothetical resources, according to the classification scheme of USGS (1980). The sampled material and the data collected during the MarMine research cruise will initiate the definition of inferred resources along the AMOR. This will, in the future, also be used to update the models presented in this book. It will, however, take many more research cruises before the AMOR deposits can be quantified as indicated or measured resources, and considerable work on all the modifying factors must be performed to render it possible to define any reserves.

Worldwide, the Solwara 1 Project inside the national jurisdiction of Papua New Guinea is the project that has come farthest in defining and classifying any SMS mineral resources on the deep ocean floor. The project owner, Nautilus Minerals Ltd, have published a report documenting extensive mapping and mineral classification according to NI43-101 (Lipton et al., 2012).

9.2 Resource potential

The results presented in this book were developed to review the knowledge about seabed mineral resources in Norway, with a special focus on possible massive sulphide mineralisation along the Mid-Atlantic Ridge. The work was carried out by NTNU and supported by Nordic Ocean Resources AS and Statoil ASA (now Equinor ASA).

The main area of interest was the extended Norwegian Continental Shelf, on the northern part of the Mid-Atlantic Ridge, an area with volcanic activity that has created the islands of Jan Mayen and Iceland. In the volcanic areas, boiling seawater extracts minerals and metals from the oceanic crust. When this comes into contact with cold water, gold, silver, copper and zinc are precipitated to a varying degree, dependent on the geological settings.

Analyses and interpretations of seabed topography, structures and geology have been carried out in order to disclose relevant areas for the formation of massive sulphides deposits on the ocean floor. Based on these data, a probabilistic mineral resource assessment was done using the same methodological framework as is used for oil and gas resources, by the Norwegian Petroleum Directorate and oil and gas companies worldwide. In the summer of 2016, NTNU together with 13 industry partners and funded by the Norwegian Research Council through the MarMine Project, carried out fieldwork in the area to collect samples from known sea floor massive sulphide deposits. The collected samples will, in the coming years, be used to update the input into the probabilistic analysis and thereby update the estimates. Work along these lines has been initiated (Juliani & Ellefmo, 2018a, 2018b and 2018c) and will continue in the years to come.

Information on the general locations and amounts of undiscovered seabed massive sulphide resources within the Norwegian Economic Exclusive Zone will be increasingly important to exploration and resource managers, environmental planners, economists and policymakers, as we possibly progress towards a sustainable extraction of these resources.

This book contains the results of probabilistic estimates of the amounts of metal of copper (Cu), zinc (Zn), silver (Ag), and gold (Au) in known and undiscovered seabed massive sulphide vent fields on the Mid-Atlantic Ridge, inside the Norwegian jurisdiction on the extended Norwegian Continental Shelf. The play analysis methodology that was used to make these estimates allows for an explicit expression of the estimates of undiscovered resources and their associated uncertainty in a form that is useful to decision makers. A combination of the probability distributions of the estimated number of undiscovered vent fields, the grades and the tonnages, was used to obtain the probability distributions for undiscovered metals. However, populating the input distributions consistently with robust data is not trivial. One has to think outside the box. It is essential to connect and combine the dots and pieces of data and information available. The grade data used as a basis in this study is based on global datasets, but all from mid-ocean ridge settings. Some data presented by Pedersen et al. (2010b) and data obtained from samples collected during the MarMine research cruise, show grade values lower than the grade data used in this play analysis. This data is scanty and based on surface samples only, and its significance relative to the much larger datasets presented by Cherkashev et al. (2013) and Hannington (2013a and 2013b) should not be emphasised.

A framework for the analysis has been established in this project. While we believe this framework is solid, the input parameters and data can always be improved, and the effects of improvements can be quantified immediately, especially in terms of reducing uncertainty. One issue that could be discussed is, of course, the use of a dependency factor quantifying the correlation between and inside plays in the aggregation of the plays. It goes almost without saying that it is wrong not to use one, but the real question lies in what the dependency factor should be. New ways of collecting and analysing data must be developed in order to quantify this parameter more precisely. Another issue not worked on here is to differentiate between the characteristics of the different permissive tracts. Given more local variability and coincidence of both geodynamical and morphostructural indicators, it would be possible to assign different properties to different permissive tracts. Further, a third issue is the possible correlation between, for example, the deposit grade and the deposit tonnage. Based on ancient deposits found and studied onshore, no clear correlations can be identified between these two parameters. However, large high-grade deposits are very rare. The effects of incorporating such trends or correlations are beyond the scope of this publication, but it would be interesting to look into the effects and how they could be incorporated into the play analysis.

There are currently five surveyed and a number of inferred seabed massive sulphide vent fields on the Norwegian Mid-Atlantic Ridge. To define favourable zones for vent fields, morphostructural and geodynamic analysis of bathymetric data was used according to the criteria established by the Russian Institute for Geology and Mineral Resources of the Ocean (VNIIOkeangeologia). In addition to the inferred vent fields, concealed vent fields may exist with less bathymetric expression than the ones inferred. This has been shown by internationally-acknowledged researchers who have published results that the assessment results obtained in this study have been validated against (Cathles, 2011a, 2011b; Hannington et al., 2011b; Hannington 2011, 2013b; Singer, 2014)

The total of inferred and postulated undiscovered metal (hypothetical resource) amounts to an expected endowment of 6.9 million metric tons of copper metal, in addition to zinc (7.1 million metric tons), gold (175 metric tons), and silver (10524 metric tons). Due to the lack of detailed data along this frontier exploration area, a great uncertainty is attached to these results with a huge upside potential. There is accordingly a 5% probability of having metal resources of more than or equal to 21 million metric tons of copper metal, 22 million metric tons of zinc, 597 metric tons of gold, and 34000 metric tons of silver. This corresponds to a gross rock value for the seabed minerals within Norwegian jurisdiction of NOK 600 billion. The estimation further indicates a possible potential of more than NOK 2000 billion. These values are, of course, very much dependent on exchange rates, because the commodities assessed here are traded on industrial metals markets like the London Metal Exchange and thereby traded in US dollars. The commodity prices are also volatile and highly dependent on trading conditions and states of the market. Further, the gross rock values are the 'value of the mineral resource in the ground' and the uncertainty is significant. The values do not take unknown cost of extraction, trans-shipment and processing and their uncertainties into account. Much must be done and investigated before the uncertainty is reduced and a decision can be made regarding a potential exploitation of these resources.

Nevertheless, due to the high potential indicated by this study, the Norwegian Government included marine minerals in their official minerals strategy in 2013, and in 2017, the Norwegian Government presented a proposal for a new Act on Offshore Mineral Exploration (Prop. 106 L (2017–2018) *Lov om mineralvirksomhet på kontinentalsokkelen (havbunnsmineralloven)*). It states that the Norwegian State is the owner of minerals on the Norwegian Continental Shelf and the bill introduces a new regime for private mineral exploration and extraction inspired by the Petroleum Act. The new law came into effect on 1 July 2019.

Further exploration in the years to come will hopefully confirm the existence of the inferred resources, as well as the costs and viability of their development, and thereby reduce the great uncertainty that at the present time is related to their tonnages and gross value. It is our strong recommendation that the Norwegian authorities provide the necessary funding to map the entire ridge, so that the uncertainty can be reduced and the viability of marine minerals as a future mining resource can be properly evaluated. This includes finalising a predictable and efficient administrative system, and encouraging and rendering it possible for universities and research institutions to do the necessary research.

About the Authors

Steinar L. Ellefmo is Associate Professor at the Department of Geoscience and Petroleum, Norwegian University of Science and Technology (NTNU). His research and teaching interests are within the broad field of mineral resource management and mining engineering, specifically geostatistics, 3D ore body modelling, mine planning and mineral resource potential assessment. Ellefmo is a member of the Australasian Institute of Mining and Metallurgy (AusIMM) and the Society of Economic Geologists (SEG).

Fredrik Søreide holds a PhD in Marine Engineering from the Norwegian University of Science and Technology (NTNU). He is currently Adjunct Professor at the Department of Marine Technology at NTNU focusing on underwater technology and marine science. He has been involved in marine minerals research and ocean mining activities for the past 20 years and is currently mainly engaged in the JPIO project, Ecological Aspects of Deep-sea Mining. Søreide is the author of three other books.

Georgy Cherkashov is Deputy Director of the Institute for Geology and Mineral Resources of the Ocean (VNIIOkeangeologia, St. Petersburg, Russia). He holds a Dr. Sci. for research of seafloor massive sulphide deposits of the Mid-Atlantic Ridge (2004). Cherkashov is a member of the Legal and Technical Commission of the International Seabed Authority (since 2012) and Professor of Marine Geology at St. Petersburg State University (since 2005).

Cyril Juliani received his PhD from the Norwegian University of Science and Technology (NTNU), Department of Geoscience and Petroleum. His dissertation focused on quantifying undiscovered seabed mineral resources and evaluating the exploration risk for deep-sea mineral prospects. Juliani is currently involved in the NTNU oceans pilot program on deep-sea mining, a research project focusing on the different value chains of seabed mineral extraction and exploration.

Krishna Kanta Panthi is Professor of Geological Engineering at the Department of Geoscience and Petroleum, Norwegian University of Science and Technology (NTNU). He holds a PhD (Dr.Ing.) in rock engineering, MSc in hydropower development and MSc in tunnelling. He has over 27 years of experience in research, design and implementation of rock engineering, tunnelling, hydropower, mining and slope engineering projects. Panthi is the author of over 70 scientific articles and book chapters published in internationally recognized journals, publication canals and conference proceedings.

Sergey Petukhov heads the analytical group Department of Geology and Mineral Resources of the Ocean, VNIIOkeangeologia. Petukhov holds a PhD in mining engineering and is an expert in data management and modelling of geodynamic processes.

Irina Poroshina is a senior scientist at VNIIOkeangeologia. She is an expert in geomorphology and structural analysis of mid-ocean ridges.

Richard Sinding-Larsen has been Professor of Resource Geology at the Norwegian University of Science and Technology (NTNU) since 1980, Emeritus since 2012. His research focuses on the estimation of undiscovered mineral resources, with particular attention to the process of discovering metals and petroleum. He has a long-standing interest in deposit modelling and pioneered probabilistic assessment of Caledonian massive sulphide resources and initiated the development of methodology and tools (GeoX, now owned by Schlumberger) used as decision support for risk, resource and economic evaluation of exploration. Sinding-Larsen's current research interest is in how to appraise and quantify resource uncertainties within mineral and petroleum systems whose components are logically related in complex ways – such as sulphides linked to volcanic spreading ridges beneath the oceans and petroleum systems.

Ben Snook achieved a PhD in granitic petrology after working on a variety of exploration geology projects in South America. He has continued to pursue research in Norwegian mineral deposits, specialising in the characterisation of seafloor massive sulphide ores and in automated mineralogy analytical systems.

References

Aleksandrov, P.A., Anikeeva, L.I., Andreev, S.I., & Petukhov, S.I. (2009). Thalassochemistry of the ocean ore genesis (in Russian). St. Petersburg: Federal State Unitary Enterprise 'VNIIOkeangeologia'.

Amundsen, I.M.H., Blinova, M., Hjelstuen, B.O., Mjelde, R., & Haflidason, H. (2011). The Cenozoic western Svalbard margin: Sediment geometry and sedimentary processes in an area of ultraslow oceanic spreading. *Marine Geophysical Research.* https://doi.org/10.1007/s11001-011-9127-z

Barriga, F.J.A.S., Carvalho, D. & Ribeiro, A. (1997). Introduction to the Iberian Pyrite Belt. In F.J.A.S. Barriga & D. Carvalho (Eds.), *Geology and VMS deposits of the Iberian Pyrite Belt guidebook series.* Volume 27. Society of Economic Geologists. https://doi.org/10.5382/GB.27

Barriga, F.J.A.S, Relvas, J.M.R.S, Santos, R., & Pascoal, A. (2013). Estimating and finding seafloor and sub-seafloor sulfide mineralization: Optimists versus pessimists? Conference report, Recent Developments in Atlantic Seabed Minerals Exploration and Other Topics. 21-28 October 2013. Rio de Janeiro and Porto de Galinhas, Brazil: Underwater Mining Institute.

Baker, E.T., & German, C.R. (2004). On the global distribution of mid-ocean ridge hydrothermal vent-fields. *Am Geophys Union Geophys Monogr, 148*, 245–66.

Baker, E.T. (2017). Exploring the ocean for hydrothermal venting: New techniques, new discoveries, new insights. *Ore Geology Reviews.* http://dx.doi.org/10.1016/j.oregeorev.2017.02.006

Baumberger, T., Frueh-Green, G.L., Pedersen, R., Thorseth, I.H., Lilley, M.D., & Moeller, K. (2010). Loki's Castle: A sediment-influenced hydrothermal vent field at the ultra-slow spreading Arctic Mid-Ocean Ridge. American Geophysical Union, Fall Meeting 2010, abstract #OS21A-1481.

Beaulieu, S.E. (2015). InterRidge global database of active submarine hydrothermal vent fields. Prepared for InterRidge, Version 3.3. Version 3.4 retrieved 03.01.2017 from http://vents-data.interridge.org

Becker, J.J., Sandwell, D.T., Smith, W.H.F., Braud, J., Binder, B., Depner, J., Fabre, D., Factor, J., Ingalls, S., Kim, S-H., Ladner, R., Marks, K., Nelson, S., Pharaoh, A., Trimmer, R., Von Rosenberg, J., Wallace, G. & Weatherall, P. (2009). Global Bathymetry and Elevation Data at 30 Arc Seconds Resolution: SRTM30_PLUS. *Marine Geodesy*, 32(4), 355–371. http://dx.doi.org/10.1080/01490410903297766. Data retrieved from http://www.pacioos.hawaii.edu/metadata/srtm30plus_v11_bathy.html

Braathen, A., & Gabrielsen, R.H. (2000). Bruddsoner i fjell – oppbygning og definisjoner. *Gråsteinen* 7. NGU.

Brekke, H. (2017). Framtidens mineraler. *Geo365*. Retrieved 03.06.2017 from https://www.geo365.no/bergindustri/framtidens-mineraler/

Bruvoll, V., Breivik, A.J., Mjelde, R., & Pedersen, R.B. (2009). Burial of the Mohn-Knipovich seafloor spreading ridge by the Bear Island Fan: Time constraints on tectonic evolution from seismic stratigraphy. *Tectonics*, 28, TC4001. https://doi.org/10.1029/2008TC002396

Cathles, L. M. (2011a). What processes at mid-ocean ridges tell us about volcanogenic massive sulfide deposits. *Mineralium Deposita,* 46(5), 639-657. https://doi.org/10.1007/s00126-010-0292-9

Cathles, L. M. (2011b). Reply to comment by Hannington: 'Are black smokers copper plating the ocean floor?'. *Mineralium Deposita, 46*(5), 665–669.

Cathles, L. M. (2013). Future Rx: Optimism, preparation, acceptance of risk. In G.R.T. Jenkin, P.A.J. Lusty, I. Mcdonald, M.P. Smith, A.J. Boyce & J.J. Wilkinson (Eds.), *Ore deposits in an evolving earth.* London: Geological Society, London. Retrieved from https://doi.org/10.1144/SP393.6

Charlou, J.L, Donval, J.P, Fouquet, Y, Jean-Baptiste P, & Holm, N. (2002). Geochemistry of high H2 and CH4 vent fluids issuing from ultramafic rocks at the Rainbow hydrothermal field (36j14VN, MAR). *Chemical Geology, 191,* 345–359. http://dx.doi.org/10.1016/S0009-2541(02)00134-1

Chen, Y.J. (1992). Oceanic crustal thickness versus spreading rate. *Geophysical Research Letters, 19,* 753–756. https://doi.org/10.1029/92GL00161

Cherkashov, G.A., Ivanov, V.N., Bel'tenev, V.I., Lazareva, L.I., Rozhdestvenskayab, I.I., Samovarov, M.L., Poroshina, I.M., Sergeev, M.B., Stepanova, T.V., Dobretsova, I.G., & Kuznetsov, V.Y. (2013). Massive sulfide ores of the Northern Equatorial Mid Atlantic Ridge. *Oceanology, 53*(5) 607–619.

Cherkashov, G., Poroshina, I., Stepanova, T., Ivanov, V., Bel'tenev, V., Lazareva, L., Rozhdestvenskaya, I., Samovarov, M., Shilov, V., Glasby, G. P., Fouquet, Y, & Kuznetsov, V. (2010). Seafloor massive sulfides from the northern Equatorial Mid-Atlantic Ridge: New discoveries and perspectives. *Marine Georesources and Geotechnology 28,* 222–239.

Cherkashov, G.A., Andreev, S., Petukhov, S., Poroshina, I., & Zayonchek, A. (2013). Identification of SMS prospective areas at the mid-ocean ridges within the Norwegian EEZ based on analysis of multibeam bathymetry data (Unpublished report). Saint-Petersburg: VNII Okeangeologia.

Connelly, D.P., German, C.R., Naganuma, T., Pimenov, N., Egorov, A., & Dohsik, H. (2002). Hydrothermal plume signals along the Knipovich Ridge. American Geophysical Union, Fall Meeting 2002, abstract #T11A-1229.

Cook Islands Seabed Minerals Authority. (2018). Retrieved 05.11.2018 from http://www.seabedmineralsauthority.gov.ck

Crane, K., & Solheim, A. (1995). Seafloor atlas of the Norwegian-Greenland Sea. Tromsø: Norsk Polarinstitutt.

Crane, K., Doss, H., Vogt, P., Sunsvor, E., Cherkashov, G., Poroshina, I., & Joseph, D. (2001). The role of the Spitsbergen shear zone in determining morphology, segmentation and evolution of the Knipovich ridge. *Marine Geophysical Research, 22,* 153–205.

Cronan, D.S. (1992). *Marine minerals in exclusive economic zones.* London: Chapman & Hall.

Cronan, D.S. (Ed.) (1999). *Handbook of marine mineral deposits.* Boca Raton, FL: CRC Press.

Dauteuil, O., & Brun, J.-P. (1993). Oblique rifting in a slow-spreading ridge. *Nature, 361,* 145–148. https://doi.org/10.1038/361145a0

Dauteuil, O., & Brun, J.P. (1996). Deformation partitioning in a slow spreading ridge undergoing oblique extension: Mohns Ridge, Norwegian Sea. *Tectonics, 15,* 870–884.

DeMets, C., Gordon, R., Argus, D., & Stein, S. (1990). Current plate motion. *Geophysical Journal International, 101,* 425–478.

Douville, E., et al. (2002). The rainbow vent fluids (36°14 N, MAR): The influence of ultramafic rocks and phase separation on trace metal content in Mid-Atlantic Ridge hydrothermal fluids. *Chemical Geology, 184,* 37–48.

Ellefmo, S.L., & Søreide, F. (2014). *Marine minerals in Norwegian waters, a survey of knowledge and research needs.* Trondheim: NTNU.

Faleide, J.I., Solheim, A., Fiedler, A., Hjelstuen, B.O., Andersen, E.S., & and Vanneste, K. (1996). Late Cenozoic evolution of the western Barents Sea-Svalbard continental margin. *Global Planet Change, 12,* 53–74. https://doi.org/10.1016/092 -8181(95)00012-7

Fiedler, A., & Faleide, J.I. (1996). Cenozoic sedimentation along the southwestern Barents Sea margin in relation to uplift and erosion of the shelf. *Global and Planetary Change, 12,* 75–93. https://doi.org/10.1016/0921-8181(95)00013-5

Fouquet, Y., Wafik, A., Cambon, P., Mevel, C., Meyer, G., & Gente, P. (1993). Tectonic setting and mineralogical and geochemical zonation in the Snake Pit sulfide deposit (Mid-Atlantic Ridge at 23°N). *Economic Geology, 88,* 2018–2036.

Fouquet, Y., Charlou, J.L., Ondréas, H., Radford-Knoery, J., Donval, J.P., Douville, E., Apprioual, R., Cambon, P., Pellé, H., Landuré, J.Y., Normand, A., Ponsevera, E., German, C., Parson, L., Barriga, F., Costa, I., Relvas, J., & Ribeiro, A. (1997). Discovery and first submersible investigations on the Rainbow hydrothermal field on the MAR (36°14′ N). American Geophysical Union Fall Meeting, San Francisco. *Eos Transactions American Geophysical Union, 78*(46), 832.

Fouquet, Y., Cambon, P., Etoubleau, J., Charlou, J. L., Ondréas, H., Barriga, F. J. A. S. Cherkashov, G., Semkova, T., Poroshina, I. Bohn, M. Donval, J. P., Henry, K., Murphy, P., & Rouxel, O. (2010). Geodiversity of hydrothermal processes along the Mid-Atlantic Ridge and ultramafic-hosted mineralization: A new type of oceanic Cu-Zn-Co-Au volcanogenic massive sulfide deposit. In P.A. Rona, et al., *Diversity of hydrothermal systems on slow spreading ocean ridges* (pp. 321–368). Washington, DC: American Geophysical Union.

Fournier, M., & Petit, C. (2007). Oblique rifting at oceanic ridges: Relationship between spreading and stretching directions from earthquake focal mechanisms. *Journal of Structural Geology, 29*, 201–208. https://doi.org/10.1016/j.jsg.2006.07.017

Géli, L., Renard, V., & Rommevaux, C. (1994). Ocean crust formation processes at very slow spreading centers: A model for the Mohns Ridge, near 72°N, based on magnetic, gravity,and seismic data. *Journal of Geophysical Research, 99*, 2995–3013. https://doi.org/0148-0227/94/9 3JB-02966505.00

GeoX (2014). Schlumberger. Retrieved January 2014 from http://www.software.slb.com/lists/salesandmarketingdocuments/attachments/313/geox_prospect_assessment_solution.pdf

German, C.R., Thurnherr, A.M., Knoery, J., Charloub, J.L., Jean-Baptiste, P., & Edmonds, H.N. (2010). Heat, volume and chemical fluxes from submarine venting: A synthesis of results from the Rainbow hydrothermal field, 36 degrees N MAR. *Deep-sea Research Part I, Oceanographic Research Papers, 57*(4) 518–527. Retrieved from http://archimer.ifremer.fr/doc/00003/11423/

Gràcia, E., Charloub, J.L,. Radford-Knoery, J., & Parson, L.M. (2000). Non-transform offsets along the Mid-Atlantic Ridge south of the Azores (38°N–34°N): Ultramafic exposures and hosting of hydrothermal vents. *Earth and Planetary Science Letters 177*(1–2), 89–103.

Granot, R., Abelson, M., Ron, H., Lusk, M.W., & Agnon, A. (2011). Direct evidence for dynamic magma supply fossilized in the lower oceanic crust of the Troodos ophiolite. *Geophysical. Research Letters, 38*, L16311. https://doi.org/10.1029/2011GL048220

Grindlay, N.R., Fox, P.J., & MacDonald, K.C. (1991). Second-order ridge axis discontinuities in the south Atlantic: Morphology, structure, and evolution. *Marine Geophysical Researches, 13*(1), 21–49.

Grindlay, N.R., Fox, P.J., & Vogt, P.R. (1992). Morphology and tectonics of the Mid-Atlantic Ridge (25°- 27°30′S) from Sea Beam and magnetic data. *Journal of Geophysical Research, 97*(B5), 6983–7010.

Hannington, M.D, Petersen, S., Herzig, P.M., & Jonasson, I.R. (2004). A global database of seafloor hydrothermal systems, including a digital database of geochemical analyses of seafloor polymetallic sulfides. Geological Survey of Canada, Open File, 4598, 2004. Retrieved from https://oceanrep.geomar.de/33243/1/Hannington.pdf

Hannington, M.D., de Ronde, C.E.J., & Petersen, S. (2005). Sea-floor tectonics and submarine hydrothermal systems. In J. Hedenquist et al., (Eds.), *100th anniversary volume of economic geology* (pp. 111–141). Littleton, CO: Society of Economic Geologists, Inc.

Hannington, M.D. & Monecke, T. (2009). Global exploration models for polymetallic sulphides in The Area. Assessment of the draft regulations on prospecting and exploration for polymetallic sulphides. *Marine Georesources & Geotechnology, 27*, 132–159.

Hannington, M., Jamieson, J., Monecke, T., & Petersen, S. (2010). Modern sea-floor massive sulfides and base metal resources: Toward an Estimate of Global SMS Potential Seafloor Massive Sulfide Deposits. *Society of Economic Geologists Special Publication, 15*, 317–338.

Hannington, M.D., Jamieson, J.W., Monecke, T., & Petersen, S. (2011a). Estimating the metal content of SMS deposits. (conference report, pp. 1–4). *OCEANS, IEEE.*

Hannington, M., Jamieson, J., Monecke, T., Petersen, S., & Beaulieu, S. (2011b). The abundance of seafloor massive sulfide deposits. *Geology. 39*(12), 1155–1158.

Hannington, M. D., Jamieson, J. W, Monecke, T. and Petersen, S. (2013a). Towards an estimate of global SMS potential. Kiel, Germany: Futureocean Society-Integrated School of Ocean Sciences (ISOS), Christian-Albrechts-Universität. Retrieved from http://fileserver.futureocean.org/forschung/r3/Hannington_SMS_resource_potential.pdf

Hannington, M.D. (2013b). The role of black smokers in the Cu mass balance of the oceanic crust. *Earth and Planetary Science Letters, 374*, pp. 215–226. http://doi.org/10.1016/j.epsl.2013.06.004

Harrison, J.P. & Hudson, J.A. (2000). *Engineering rock mechanics: Part 2. Illustrative worked examples*. Oxford: Pergamon.

Havbunnsmineralloven (2019). Lov om mineralvirksomhet på kontinentalsokkelen (LOV-2019-03-22-7). Retrieved from https://lovdata.no/dokument/NL/lov/2019-03-22-7

Ingri, J. (1985). Geochemistry of Ferromanganese Concretions in the Barents Sea. *Marine Geology, 67*, 101–119.

ISA (2018). Website of the International Seabed Authority. Retrieved 18.03.2018 from https://www.isa.org.jm/deep-seabed-minerals-contractors/overview

Johansen, S.E., Panzner, M., Mittet, R., Amundsen, H.E.F., Lim, A., Vik, E., Landrø, M., & Arntsen, B. (2019). Deep electrical imaging of the ultraslow-spreading Mohns Ridge. *Nature, 567*, 379–383. https://doi.org/10.1038/s41586-019-1010-0

JORC (2012). The JORC Code. Retrieved 18.03.2018 from http://www.jorc.org/docs/JORC_code_2012.pdf

Jowitt, S.M., Jenkin, G.R.T., Coogan, L.A., & Naden, J. (2012). Quantifying the release of base metals from source rocks for volcanogenic massive. *Journal of Geochemical Exploration, 1*(18), 47–59.

Juliani, C.J., & Ellefmo, S.L. (2018a). Resource assessment of undiscovered seafloor massive sulfide deposits on an Arctic mid-ocean ridge: Application of grade and tonnage models. *Ore Geology Reviews, 102*.

Juliani, C.J., & Ellefmo, S.L. (2018b). Multi-scale quantitative risk analysis of seabed minerals: Principles and application to seafloor massive sulfide prospects. *Natural Resources Research*. https://doi.org/10.1007/s11053-018-9427-y

Juliani, C.J., & Ellefmo, S.L. (2018c). Probabilistic estimates of permissive areas for undiscovered seafloor massive sulfide deposits on an Arctic Mid-Ocean Ridge. *Ore Geology Reviews, 95.*

Kandilarov, A., Landa, H., Mjelde, R., Pedersen, R.B., Okino, K., & Murai, Y. (2010). Crustal structure of the ultra-slow spreading Knipovich Ridge, North Atlantic, along a presumed ridge segment center. *Marine Geophysical Researches, 31*(3), 173–195. https://doi.org/10.1007/s11001-010-9095-8

Kleinrock, M.C., & Humphris, S.E. (1996). Structural control on sea-floor hydrothermal activity at the TAG active mound. *Nature, 382*,149–53.

Klingelhöfer, F., Géli, L., Matias, L., Steinsland, N., & Mohr, J. (2000). Crustal structure of a super-slow spreading centre: A seismic refraction study of Mohns Ridge. *Geophysical Journal International, 14*(2), 509–526.

Krasnov, S.G., Cherkashev, G.A., Stepanova, T.V., Batuyev, B.N., Krotov, A.G., Malin, B.V., Maslov, M.N., Markov, V.F., Poroshina, I.M., Samovarov, M.S., Ashadze, A.M., Lazareva, L.I., & Ermo-layev, I.K. (1995). Detailed geological studies of hydrothermal fields in the North Atlantic. In L.M. Parson, C.L. Walker & D.R. Dixon (Eds.), *Hydrothermal vents and processes* (pp. 43–64). Geological Society Special Publication, vol. 87.

Krysiński, L., Grad, M., & Mjelde, R. (2013). Seismic and density structure of the lithosphere–asthenosphere system along transect Knipovich Ridge–Spitsbergen–Barents Sea– geological and petrophysical implications. *Polar Polish Research, 34*(2), 111–138.

Lipton, I. (2012). Mineral Resource Estimate. Solwara Project, Bismarck Sea, PNG (report written for Nautilus Minerals).

Ludvigsen, M., Aasly, K., Ellefmo, S.L., Hilário, A., Ramirez-Llodra, E., Søreide, F. X., Falcon-Suarez, I., Juliani, C.J., Kieswetter, A., Lim, A., Malmquist, C., Nornes, S.M., Reimers, H., Paulsen, E., & Sture, Ø. (2016). NTNU Cruise reports 2016 no 1 - MarMine cruise report Arctic Mid-Ocean Ridge 15.08.2016–05.09.2016. Trondheim: NTNU.

Ludvigsen, M., Søreide, F., Aasly, K., Ellefmo, S.L., Zylstra, M., & Pardey, M. (2017). ROV based drilling for deep sea mining exploration. In conference proceedings, IEEE OCEANS 2017. Aberdeen: IEEE.

Mero, J.L. (1965). *Mineral resources of the sea.* First edition. Amersterdam/London/ New York: Elsevier Publishing Co.

Melekestseva, I.Y., Zaykov, V.V., Nimis, P., Tret'yakov, G.A. & Tessalina, S.G. (2013). Cu–(Ni–Co–Au) -bearing massive sulfide deposits associated with mafic–ultramafic rocks of the Main Urals Fault, South Urals: Geological structures, ore textural and mineralogical features, comparison with modern analogs. *Ore Geology Reviews, 52*, 18–36.

Mosar, J., Lewis, G., & Torsvik, T.H. (2002). North Atlantic sea-floor spreading rates: Implications for the Tertiary development of inversion structures of the Norwegian–Greenland Sea. *Journal of the Geological Society*, *159*, 503–515.

Mosier, D.L., Singer, D.A., & Berger, V.I. (2007). Volcanogenic massive sulfide deposit density. s.l. USGS Scientific Investigations Report 2007–5082. Retrieved from http://pubs.usgs.gov/sir/2007/5082/sir2007-5082.pdf

Mosier, D.L., Berger, V.I. & Singer, D.A. (2009). Volcanogenic massive sulfide deposits of the world—database and grade and tonnage models. Retrieved from http://pubs.usgs.gov/of/2009/1034/

Mozgova, N.N., Bordaev, Y.S., Gablina, I.F., Cherkashev, G.A., & Stepanova, T.V. (2005). Mineral assemblages as indicators of the maturity of oceanic hydrothermal sulfide mounds. *Lithology and Mineral Resources, 40*, 293–319.

NI (2018). National Instrument 43-101. Retrieved 18.03.2018 from http://web.cim.org/standards/

NPD (2013). Norwegian Petroleum Directorate resource report 2013. Retrieved from https://www.npd.no/en/facts/publications/reports2/resource-report/archive-for-resource-reports/

NPD (2018a). Acquisition of seabed massive sulfide data in the Norwegian Sea. Qualifying documentation SMS data acquisition 2018. Retrieved 03.11.2018 from https://www.mercell.com/m/file/GetFile.ashx?id=80348247&version=0

NPD (2018b). Norwegian Petroleum Directorate maps deep sea mineral deposits. Retrieved 03.11.2018 from http://www.npd.no/en/news/News/2018/Norwegian-Petroleum-Directorate-maps-deep-sea-mineral-deposits/

NPD (2018c). The Norwegian Petroleum Directorate has found new deep sea mineral deposits. Retrieved 03.11.2018 from http://www.npd.no/en/news/News/2018/New-deep-sea-mineral-deposits/

NPD (2018d). Ministry of Petroleum and Energy responsible for mineral deposits on the shelf. Retrieved 29.12.2018 from http://www.npd.no/en/news/News/2017/Ministry-of-Petroleum-and-Energy-responsible-for-mineral-deposits-on-the-shelf-/

NPD (2019). Successful exploration for seabed minerals. Retrieved 10.09.2019 from https://www.npd.no/en/facts/news/general-news/2019/successful-exploration-for-seabed-minerals/

OED (2017). Lov om mineralvirksomhet på kontinentalsokkelen. Prop. 106 L (2017–2018). Draft legislation for the management of marine minerals on the Norwegian Continental Shelf. Retrieved from https://www.regjeringen.no/no/dokumenter/prop.-106-l-20172018/id2605252/

Okino, K., Curewitz, D., Asada, M., Tamak, i K., Vogt, P., & Crane, K. (2002). Preliminary analysis of the Knipovich Ridge segmentation: Influence of focused magmatism and ridge obliquity on an ultraslow spreading system. *Earth and Planetary Science Letters*, *202*, 375–288.

Olsen, B.R, Økland, I.E, Thorseth, I.H., Pedersen, R.B., & Rapp, H.T. (2016). Environmental challenges related to offshore mining and gas hydrate extraction. Report M-532 2016. Trondheim: Miljødirektoratet.

Petukhov, S.I., & Aleksandrov P.A. (2004). Application of a method of geodynamic zoning for identification of fields of hydrothermal activity in the region of mid-Atlantic ridge (for MAR – 15 0 30′ – 17030′ N.). Minerals of the ocean – integrated strategies-2. Conference abstract, SPb, 2004, pp. 41–42.

Petukhov, S.I, Aleksandrov, P.A., & Andreev, S.I. (2010). Deformation model of hydrothermal sulfide ore fields for prediction of hydrothermal activity locations (for different areas of the Atlantic and Indian oceans). Minerals of the ocean – 5 and deep-sea minerals and mining-2. Conference abstract, SPb, 2010, pp. 82–83.

Petukhov, S.I, Aleksandrov, P.A., & Tao, C. (2012). Modeling of the environments favourable for hydrothermal activity using relief analysis (for South Atlantic Ridge). Minerals of the ocean – 6 and deep-sea minerals and mining-3. Conference abstract, SPb, 2012, pp. 136–137.

Pedersen, R.B, Rapp, H.P., Thorseth, I.H., Dilley, M.D., Barriga, F.J.A.S., Baumberger, T., Flesland, K., Fonseca, R., Früh-Green, G.L., & Jørgensen, S.L. (2010a). Discovery of a black smoker vent field and vent fauna at the Arctic Mid- Ocean Ridge. *Nature communication.* https://doi.org/10.1038/ncomms1124

Pedersen, R.B, Thorseth, I.H., Nygård, T.E., Dilley, M.D., & Kelley, D.S. (2010b). Hydrothermal activity at the arctic mid-ocean ridges. In P.A. Rona, et al., *Diversity of hydrothermal systems on slow spreading ocean ridges* (pp. 67–89). Washington, DC: American Geophysical Union.

Panthi, K.K. (2006). Analysis of engineering geological uncertainties related to tunneling in Himalayan rock mass conditions. (PhD thesis.) Trondheim: Norwegian University of Science and Technology.

Panthi, K.K. (2012). Evaluation of rock bursting phenomena in a tunnel in the Himalayas. *Bulletin of Engineering Geology and the Environment*, *71*(4), 761–769.

Panthi, K.K. (2016). Introduction to engineering geology. Lecture notes on TGB5110 Engineering Geology and Tunneling BC. Norwegian University of Science and Technology (NTNU).

Prandtl, L. (1904). Über Flussigkeitsbewegungen bei sehr kleiner Reibung (Motion of fluids with very little viscosity). In A. Krauzer (Ed.), Proceeding of the Third International Mathematics Congress, Heidelberg, Germany (pp. 484–491). Publisher unknown. See English translation from 2001 in J. A. K. Ackroyd, B. P. Axcell & A. I. Ruban (Eds.), *Early developments of modern aerodynamics* (p. 77). Oxford: Butterworth-Heinemann.

Robb, L. (2004). *Introduction to ore-forming processes.* Oxford: Wiley-Blackwell.

Rona, P.A. & Scott, S.D. (1993). A special issue on sea-floor hydrothermal mineralization: New perspectives. *Economic Geology*, *88*, 1935–75.

Rona, P.A. (2008). The changing vision of marine minerals. *Ore Geology Review*, *33*, 618–66.

Rona, P.A., Devey, C.W., Dyment, J., & Murton, B.J. (2010). Diversity of hydrothermal systems on slow spreading ocean ridges. Washington, DC: American Geophysical Union.

Rowland, S.M., & Duebendorfer, E.M. (1994). *Structural analysis and synthesis. A laboratory course in structural geology,* 2nd Edition. Malden, MA/Oxford/ Melbourne/Berlin: Blackwell Scientific Publications.

Ryan, W.B.F., Carbotte, S.M., Coplan, J.O., O'Hara, S., Melkonian, A., Arko, R., Weissel, R.A., Ferrini, V., Goodwillie, A., Nitsche, F., Bonczkowski, J., & Zemsky, R. (2009). Global multi-resolution topography synthesis, *Geochemistry. Geophysics. Geosystems*, *10*. https://doi.org/10.1029/2008GC002332

Schmidt, K., Koschinsky, A., Garbe-Schönberg, D., de Carvalho, L. M., Seifert, R. (2007). Geochemistry of hydrothermal fluids from the ultramafic-hosted Logatchev hydrothermal field, 15°N on the Mid-Atlantic Ridge: Temporal and spatial investigation. *Chemical Geology, 242*(1–2), 1–21. Retrieved from http://dx.doi.org/10.1016/j.chemgeo.2007.01.023

Scott, S. (2004). Proposed exploration and mining technologies for polymetallic sulphides. In proceedings from workshop for the establishment of environmental baselines at deep seafloor cobalt-rich crusts and deep seabed polymetallic sulphide mine sites in the area for the purpose of evaluating the likely effects of exploration and exploitation on the environment. International Seabed Authority, Kingston, Jamaica, 6-10 September 2004. Retrieved from https://ran-s3.s3.amazonaws.com/isa.org.jm/s3fs-public/documents/EN/Workshops/2004/Proceedings-ae.pdf

Sempéré, J.C., Lin, J., Brown, H.S. Schouten, H., & Purdy, G.M. (1993). Segmentation and morphotectonic variations along a slow-spreading center: The Mid-Atlantic Ridge (24°00′ N- 30°40′ N). *Marine Geophysical Researches, 15*(3), 153–200.

Seyfried Jr., W.E., et al. (2011). Vent fluid chemistry of the Rainbow hydrothermal system (36 N, MAR): Phase equilibria and in situ pH controls on subseafloor alteration processes. *Geochimica et Cosmochimica Acta*, *75*, 1574–1593.

Sinding-Larsen, R. & Ellefmo, S. (2014). Seabed massive sulphide resource assessment of undiscovered potentially recoverable copper, zinc, silver, and gold related to hydrothermal vent fields on the Mid-Atlantic ridge within the Norwegian Economic Exclusive Zone. In S.L. Ellefmo & F. Søreide, *Marine minerals in Norwegian waters, a survey of knowledge and research needs.* Trondheim: NTNU.

Sinding-Larsen, R. & Vokes, F. (1978). The use of deposit modeling in the assessment of potential resources as exemplified by Caledonian stratabound sulfide deposits. *Mathematical Geology*, *10*(5), 565–579. https://doi.org/10.1007/BF02461986